# RHODODENDRON HYBRIDS

# RHODODENDRON HYBRIDS
## A Guide To Their Origins

*(Includes selected, named forms of rhododendron species)*

## Homer E. Salley
### *and*
## Harold E. Greer

**TIMBER PRESS**
*Portland, Oregon*

Published by Timber Press, 1986

All rights reserved.
1986
Homer E. Salley
Harold E. Greer
Printed in Hong Kong

ISBN 0-88192-061-4

TIMBER PRESS
9999 SW Wilshire
Portland, Oregon 97225

# TABLE OF CONTENTS

# PREFACE

How did it all begin? That is the question:

Nearly 30 years ago, in my early years of interest in rhododendrons, my parents and I would discuss the parentage of hybrid crosses I was considering making, or ones that had already been made. Frequently, we would list the species parents of the plants we were considering. Suddenly, the idea came to one of us, why not begin a file of hybrids that had been traced to their species parents?

So, in the days before computer printers and copy machines, we began to make hand written notebook paper pages with carbon copies of the lineage of many hybrids. This was gradually done to include all of the rhododendrons listed in the *International Rhododendron Register*, tracing them to their species parentage, if their parentage was known. To this was added the constant flow of newly registered hybrids being made throughout the world.

Soon this became a sizeable manuscript. Some efforts were made to get it printed, as it was thought to be of greater use than the *International Rhododendron Register* because it contained additional information about hardiness and ratings. Further, tracing the parentage of each hybrid required difficult and expensive typesetting, not easily done in those days, which would cause the book to sell for more than $150.00! The feeling was who would pay that much? So, with my increasing responsibilities and my father's failing health, the book was set aside for many years.

In the mid '70's after the death of my father, Richard Abel of Timber Press recognized the need for this book and entered into a contract with me to publish the book. I again went back to work on the book, and spent many, many hours on it, but the duties of a growing business, plus the number of new hybrids that were continually being named, made it an unending task which I could not finish.

But, a savior came on the scene, Homer Salley. So in 1979, I turned the work over to his capable hands. After years of hard labor, he completed the work! My thanks go to him for finishing the book and for adding so much more to it. It's a much more usable book with the many additions he has provided.

I must dedicate and give my special thanks for this book to my parents who provided me with the opportunity to pursue my interest in rhododendrons and to my father especially, who spent countless hours hand writing some of the early parts of the book.

Time has also made it possible to add hundreds of color pictures which improve the book, and my thanks go to those who provided pictures I did not have.

So, it has taken many years for this book to finally reach your hands. I hope you feel it was worth the wait!

Harold E. Greer

# ACKNOWLEDGMENTS

All who gave assistance are too numerous to list, but I especially want to thank those who come to mind and later I will regret having left out names that should also be here.

My wife, Sarah Salley, should be considered one of the authors and not just "a word processor" even though she was chained to the computer terminal for many months. She was determined even more than I "to get it all right."

Edwin K. Parker, as the Registrar for the American Rhododendron Society was the original author of hundreds of the published descriptions which were greatly condensed. Any lack of clarity is not his error but ours. Some of the color illustrations are from his slide collection of recently registered hybrids.

Dr. Alan C. Leslie, International Registrar, Royal Horticultural Society, always answered promptly the many inquiries concerning hybrids, their valid names, parentage, etc. I look forward to the publication of his newly revised Register.

Harold E. Greer, a very busy rhododendron nurseryman, allowed us to complete what he and his father started many years ago. Most of the color illustrations are from his photographic talents. And much parentage information is from the catalog of Greer Gardens.

Dr. David G. Leach contributed color photographs of his own hybrids for cold climates which are now being made available by the thousands through the modern miracle of tissue propagation. He was always able to find time to offer helpful information and advice.

Mrs. Mary Hartsfield did many hours of proof-reading and made helpful recommendations as to arrangements of some material.

Members of the RHS Rhododendron Group—Maj. Walter Magor and Steve Thompson edited the names list for the British Isles—Lady Cynthia Postan lent her collection of old nursery catalogues (and James Russell authored one of the best)—"Mac" Speed so pleasantly toured us on the "wrong side" of the road—and the many others who may have been great lovers of the species rhododendrons, but were tolerant of this American with an equally great obsession for hybrids.

Walter Schmalscheidt translated descriptions from his book on the rhododendron hybrids of Germany before 1939.

Members of the Great Lakes Chapter:
Dr. Louis B. Martin who cheerfully lent us his complete collection of RHS Yearbooks and ARS Journals. Wells Kneirim contributed slides for some of the color. Some initial encouragement with lasting meaning came from Donald Lewis and past presidents Dr. D. L. Hinerman and Felix A. Robinson. And a moving inspiration always came with the memory of such original creators who have recently passed on: Lanny Pride, Tony Shammarello and Peter Girard.

A. F. George of Hydon Nurseries granted permission to reproduce from his rhododendron catalog those beautiful line drawings of flower and leaf shapes.

Graham F. Smith of New Zealand supplied parentage information from his side of the world.

Peter A. Cox read the species list in Appendix A.

And thanks to all others who answered my inquiries and wondered if it would all ever get done.

Now I must go "scythe the bracken and the brambles" to help save my own rhododendrons that have survived despite three years' neglect.

October, 1985                          H. E. S.

# INTRODUCTION

The book was written as a reference for the amateur grower and hybridizer as well as the professional specialist. The enthusiastic neophyte will find information he needs, with an incentive to further study of a fascinating-to-engrossing area of horticulture.

The book describes over 4,800 named hybrids and selected forms of species. It traces the parents, (if known) and if a parent was a hybrid its parents are shown also. The species contribution to parentage is shown in fractions. The contributions are the inherited determinants of flower color, resistance to heat and cold, disease and insects, plant size and habit, shape and texture of foliage and its persistence.

The book could not include "everything by everybody." We did cover a broad spectrum, chronologically, from the early 1800s to the summer of 1985. Although most of the entries came from Great Britain and the United States, others are included from Australia, New Zealand, Canada, and western Europe.

The information came from many sources; the bibliography is necessarily selective. If some data vary from that published earlier, they were changed only after verification with hybridizer, or family, grower, distributor or the Registrar.

Hundreds of hybrids of uncertain or unknown parentage are shown in Appendix D. Little information is supplied on "unknowns," unless they are known to have received an award or have been a parent of another hybrid. This is a parentage book; it may even become known as "The Stud Book" of rhododendrons. Many hybridizers were too excited about new creations to keep very good records. We do not record here speculations about parentage without the use of (?) question marks in the diagrams. We have not reduced the quantity of the unknown, but have compiled more into this book than you will find in any other. What C. O. Dexter did on Cape Cod and what William Whitney did on Hood Canal we leave to others. In no way would or could we detract from the beauty of their creations.

We give the names of the people who created these hybrids and selected the unusual forms of species. We give the names of the plant explorers who brought back the seed. We tell who grew them on and who introduced them in the nursery trade; who gave them a name and registered a valid name so that anywhere in the world a rhododendron called "Naomi" would be the same. We also indicate by an asterisk (*) unregistered names.

Registration is done internationally by the Royal Horticultural Society, Vincent Sq. London. Many countries have their own registrars who act as coordinators with the RHS. Mrs. Jay Murray is the Registrar for members of the American Rhododendron Society at this time. New registrations are released quarterly by the ARS and annually by the RHS, beginning with the publication of *The International Rhododendron Register*, 1958. So it is incorrect to say a rhododendron was "Registered in 1898." There were introductions even in the early 1800s and we give many dates after the name of hybridizers, but formal registration began only in 1958. That date appears often and may only mean the hybrid was known at the time of the first published registrations. Another edition of the Register is expected soon.

To read descriptive entries given on the 350 pages that follow some explanations are needed:

Abbreviations: ARS American Rhododendron Society
RBG Royal Botanic Garden(s)
RHS Royal Horticultural Society
cl. Clone, an asexual reproduction true to its parent in every way.
g. Grex, a group of sister seedlings all with the same parents, but with some variations in traits.
- - - We don't know which it was.

The abbreviations used in citing published color illustrations are shown with the Bibliography.

Color names from horticultural color charts are listed in Appendix F, p. 389. At times the only color designation available was a color number from the color chart used for reference at the time of registration. These color descriptions by number have not been used in the text but rather the color names as shown in the cross reference chart. In case of any error in the printing of numbers in published records our colors would seem odd indeed! A bright blue flower on an elepidote rhododendron would be most unusual! We found other hazards in making color translations. One is the eye of one beholder does not always perceive the same color as the eye of another. We agree the color chart number is the most accurate means of recording colors for others who have the same color chart. At present the RHS Color Chart of 1966 is out of print but may be reprinted.

GIVEN BELOW IS AN EXPLANATION OF THE PARENTAGE DIAGRAMS SHOWN ON PAGES 1-350:

```
        (Female or seed parent)
                    /                    - griffithianum (1/8)      > Species contribution to the
                   /        - unnamed hybrid -                        parentage is always shown in
   cl.       - Mrs. Furnival -            - unknown (1/4)            fractions; species plus un-
             -              -             - caucasicum (1/8)         knowns must total one.
   Party Pink -             - unnamed hybrid -
   (Valid     -                          - unknown
     name)  - catawbiense   var. album (1/2)                       > Species names underscored.
            (Pollen or male parent)
```

5ft (1.5m)        -20°F (-29°C)        M          5/5
(At 10 yrs)       (Bud hardiness)   (Blooming time) (Ratings)

> Est. of height at 10 yrs. if raised from seed. From cutting, size may be as much as 50% greater.

Flowers 3in (7.6cm) wide, openly funnel-shaped, of purplish pink shading lighter in the center, with conspicuous dorsal spotting of strong yellow. Ball-shaped truss 6in (15cm) across, 18-flowered. Plant broad, branching well, with leaves widely elliptic, held 3 years. See also Cyprus, Persia. (David G. Leach, 1973) C. A. (ARS) 1981; A. E. (ARS) 1982; S. P. A. (ARS) 1983 See color illus. ARS J 38.2 (Spring 1984): cover.

> Descriptions of flowers and foliage.

> "See also" refers to sisters.
> Hybridizer; registration date.
> Awards received.
> Published illustrations cited.

Species names follow the revised classifications of Chamberlain and Cullen, RBG, Edinburgh. When hybridizers of the past used names that have no clearly known present-day synonyms those names were retained. Vireyas were not revised by Edinburgh so our authority for vireya names was *The Rhododendron Handbook*, 1980 or the names as they appeared in the registrations. Lepidotes and elepidotes of this volume are compiled in Appendix A, but not vireyas. For azaleas see *Azaleas* by Galle.

Awards of the ARS:  S.P.A.    Superior Plant Award
                    A.E.      Award of Excellence
                    P.A.      Preliminary Award
Awards of the RHS:  A.G.M.    Award of Garden Merit
                    F.C.C.    First Class Certificate
                    A.M.      Award of Merit

We regret diacritical marks could not be used since they were not available on our Leading Edge Personal Computer keyboard.

The authors welcome comments, corrections, additions for future revisions.

cl.

A. Bedford or -mauve seedling--unknown (1/2)
Arthur Bedford-
                    -ponticum (1/2)

6ft(1.8m)        -5°F(-21°C)        ML        4/3

Habit large, upright and vigorous. Tolerates full sun. Leaves
large, up to 6.5in(16.5cm) long, dark, glossy green, held only
one year. Open funnel-shaped flowers, pale mauve in tube, darker on lobes with dark rose madder to nearly black markings within; up to 16 flowers in dome-shaped trusses. Named for the Head
Gardener to Lionel de Rothschild. (Lowinsky) A. M. (RHS) 1936;
F. C. C. (RHS) 1958   See color illus. VV p. 2.

cl.
              -campylocarpum ssp. campylocarpum (1/2)
A. Gilbert-
              -fortunei ssp. discolor (1/2)

Flowers pale creamy buff with saffron rose pink. (Lowinsky)
A. M. (RHS) 1925

cl.

A. M. S.     See Spring Delight

g.
        -campylocarpum ssp. campylocarpum (1/2)
Abalone-
        -callimorphum ssp. callimorphum (1/2)

Flowers cream to pink. Top buds cut by frost at Exbury in se-
vere winters. (Rothschild, 1933)


---

Abbeyi    Parentage unknown

Flowers delicate pink, frosted and veined with rose on the reverse. (G. Abbey)  A. M. (RHS) 1900

g.
      -thomsonii ssp. thomsonii (1/2)
Abbot-
      -arboreum ssp. delavayi var. delavayi (1/2)

Deep red flowers, blooming very early. Just hardy at Exbury.
(Rothschild, 1933)

cl.                                    -maximum (1/4)
              -Marchioness of Lansdowne-
Abe Arnott-                            -unknown (1/2)
          -                    -catawbiense (1/4)
              -Lee's Dark Purple-
                              -unknown

5ft(1.5m)        -5°F(-21°C)        ML

Orchid purple flowers with a heavy blotch on upper lobe spreading into the throat and adjacent lobes. Ball-shaped trusses of
16. Leaves narrowly elliptic. (E. O. Weber, 1974)

cl.                            -catawbiense (1/8)
              -Album Elegans-
          -Loder's White-            -unknown (1/8)
Abegail-              -griffithianum (1/4)
          -calophytum var. calophytum (pink form) (1/2)

6ft(1.8m)        0°F(-18°C)        E

(For other possible parentage information see Loder's White.)
Leaves 5.5in(14cm) by 2in(5cm). Flowers 3in(7.6cm) wide, pink
with small dark dorsal blotch, up to 14 per ball-shaped truss.
(Carl Phetteplace, 1979) See color illus. ARS Q 79:3 p. 170.

        -Laura    -griersonianum (1/2)
cl.  -Aberconway-          -thomsonii ssp. thomsonii (1/8)
     -        -Barclayi-    -arboreum ssp. arboreum
Abessa-                -Glory of -            (1/16)
     -              Penjerrick-griffithianum (1/16)
     -    -forrestii ssp. forrestii Repens Group (1/4)
        -Elizabeth-
                  -griersonianum

Scarlet flowers. (Lord Aberconway, intro. 1946)

cl.
          yakushimanum ssp. yakushimanum (1/2)
Abigail Jury-    -dichroanthum ssp. dichroanthum (1/4)
          -Dido-
                -decorum (1/4)

Trusses of 9-10 flowers, 6-lobed, white, flushed carmine rose,
upper lobe with green and red spots. Young leaves have fawn,
suede-like indumentum. (F. M. Jury, New Zealand, 1982)

cl.
              -catawbiense (1/2)
Abraham Lincoln-
              -unknown (1/2)

Red flowers. (S. B. Parsons, before 1875)

cl.                    -griffithianum (1/8)
              -Mars-
          -Vulcan-    -unknown (1/8)
Abundant Life*    -griersonianum (1/4)
          -Gunrei--Satsuki azalea (1/2)

Unregistered azaleodendron. (A. Martin, 1961)

cl.

Acadia      See Cornwallis

cl.
      -Pygmalion--unknown
Acclaim-              -haematodes ssp. haematodes
      -unnamed hybrid-    -fortunei ssp fortunei
                  -Wellfleet-
                  (Dexter#8)-decorum

5ft(1.5m)        -5°(-21°C)        M

The exact combination of the parentage is uncertain. Flowers of
good substance, 4in(10cm) wide, of strong purplish red with a
small dark eye; 16 held in ball-shaped trusses 6in(12.5cm)
across. Plant rounded, well-branched. See also Accomac, Aronimink, Avondale. (C. O. Dexter before 1943; named by Wister;
reg. 1978)

cl.

Accomac      For parentage see Acclaim

4ft(1.2m)        -5°F(-21°C)        M

Plant rounded, moderately branched; leaves held 2 years, yellow
green. Flowers openly funnel-shaped  2.5in(6.4cm) across by
2in(5cm) long, strong purplish red with spotting on dorsal lobes
to very dark in throat. Truss high, ball-shaped, 14 flowers.
See also Aronimink, Avondale. (Dexter before 1943; named by
John Wister & reg. 1978) See color illus. LW pl. 11.

              -thomsonii ssp. thomsonii (1/4)
cl.  -Barclayi-        -arboreum ssp. arboreum (3/8)
     -        -Glory of -
Achilles-      Penjerrick-griffithianum (1/8)
     -    -arboreum ssp. arboreum
        -Werei-
              -barbatum (1/4)

Cherry red flowers, 5-lobed, 3in(7.6cm) across by 2.25in(5.7cm)
long; trusses hold 16 flowers. (Gen. Harrison, 1969)

```
                    -javanicum (5/16)
---       -Amabile-              -jasminiflorum (3/16)
          -Princess Alexandra-          -jasminiflorum
Acidalia-                     -Princess Royal-
        -                             -javanicum
        -robinsonii (teysmanni) (1/2)
```

Vireya hybrid.  Flowers of pale primrose yellow.  (Veitch, 1891)

```
                              -wardii var. wardii (?) (1/8)
cl.              -Chlorops-
      -unnamed hybrid-         -vernicosum ? (1/8)
Acierto-        -diaprepes (1/4)
      -       -fortunei ssp. discolor (1/4)
      -Bonito-              -fortunei ssp. fortunei
           -Pride of Leonardslee-              (1/8)
           (Luscombei)        -thomsonii ssp. thomsonii
                                              (1/8)
6ft(1.8m)        0°F(-18°C)        M
```

Fragrant flowers with 7 lobes, light neyron rose with flares of geranium lake, held in 8in(20cm) trusses of 16-18; upright plant half as wide as tall; yellow green leaves held 3 years.  (Cross by Dr. David W. Goheen, 1963; Dr. L. Bulgini, intro. 1982; reg. 1984 by Goheen)

```
cl.                 -brookeanum var. gracile (1/4)
    -Duchess of Edinburgh-
Acis-                  -longiflorum (lobbii) (1/4)
    -              -jasminiflorum (3/8)
    -Princess Alexandra-         -jasminiflorum
              -Princess Royal-
                       -javanicum (1/8)
```

Vireya hybrid; salmon-colored.  (Veitch, 1891)

```
cl.
    -Lady Alice Fitzwilliam--unknown (1/2)
Actress-
    -edgeworthii (bullatum) (1/2)
```

Flowers white inside, stained red outside, and fragrant.  Blooms early.  (Gen. Harrison, 1965)

cl.

Acubaefolium     See Acubifolium

cl.

Acubifolium*--form of ponticum

Synonym Acubaefolium.

```
cl.                 -diaprepes (1/4)
           -Polar Bear-
Ada Agnes Archer-       -auriculatum (1/4)
    -                -griersonianum (1/4)
       -Vulcan's Flame-    -griffithianum (1/8)
                  -Mars-
                       -unknown(1/8)
3ft(.9m)        -5°F(-21°C)        VL
```

Flowers of good substance, 5in(12.7cm) by 3in(7.6cm), fragrant, red purple, in trusses 13in(33cm) wide by 6in(15.2cm) high of 10 flowers.  Plant upright, rounded.  (Holden, 1981)

```
              -thomsonii ssp. thomsonii (1/4)
cl. -Barclayi-    -arboreum ssp. arboreum (3/8)
    -       -Glory of -
Adamant-    Penjerrick-griffithianum (1/8)
    -    -arboreum ssp. arboreum
       -Werei-
          -barbatum (1/4)
```

Flowers deep red, no spots, 3in(7.6cm) across, 2in(5cm) long, 15 per tight truss.  See also Achilles.  (Gen. Harrison, 1962)

```
cl.          -fortunei ssp. fortunei (1/4)
    -unnamed hybrid-
Adana-       -diaprepes (1/4)
    -Noyo -arboreum ssp. nilagiricum (1/4)
      Chief-
          -unknown (1/4)
5ft(1.5m)        5°F(-15°C)        E
```

Ball-shaped trusses 6in(15cm) across, of 14-16 flowers, vermillion red  with jasper red throat, 7-lobed.  Plant broader than high with leaves held 4 years; new growth bronze.  (Dr. David W. Goheen, intro. 1980; reg. 1985)

```
g.
    -thomsonii ssp. thomsonii (1/2)
Adder-
    -diphrocalyx (1/2)
```

Flowers of deep rose.  (Rothschild, 1933)

```
                    -griffithianum (1/8)
g. & cl.     -Kewense-
     -Aurora-      -fortunei ssp. fortunei (1/8)
Adelaide-    -thomsonii ssp. thomsonii (3/4)
     -thomsonii ssp. thomsonii
5ft(1.5m)        0°F(-18°C)        M
```

Scarlet with faint primrose, five dark nectar pouches at base and lightly spotted inside the tube at the base.  (Rothschild, 1930)  A. M. (RHS) 1935

```
cl.
     -maximum (1/2)
Adele's Pink-              -griersonianum (1/8)
     -              -Azor-
       -unnamed hybrid-    -fortunei ssp. discolor (1/8)
                  -unknown (1/4)
3ft(.9m)        -15°F(-26°C)        L
```

Buds of deep rose pink.  Flowers openly funnel-shaped 2.5in(6.4cm) across by 1.5in(4cm) long, medium rose shading to creamy white in lobe centers and throat.  Dome-shaped truss of 14-16 flowers.  Leaves retained 2 years.  Plant upright, as broad as tall.  (William Fetterhoff cross, 1961; reg. 1981)

```
cl.
     -maximum (1/2)
Adele's Yellow-
     -wardii var. wardii (1/2)
3ft(.9m)        -10°F(-23°C)        L
```

Buds are cream and pink; flowers primrose yellow, widely funnel-campanulate.  Truss dome-shaped, somewhat lax with 13-14 flowers.  Plant is broad, semi-dwarf, branching well; leaves held two years.  (William Fetterhoff, 1973)

```
g.
     -neriiflorum ssp. neriiflorum (1/2)
Adjutant-
     -sperabile var. sperabile (1/2)
```

Scarlet color and very free flowering, "but not as good as either parent"--L. de R.  (Rothschild, 1933)

```
                    -souliei (1/16)
              -Soulbut-
       -Vanessa-        -Sir Chas. Butler--fortunei ssp.
                   -griersonianum (1/4)      fortunei (5/16)
g. -Adonis-       -griffithianum (3/8)
   -      -Sunrise-
Adlo-         -griersonianum
   -      -griffithianum
    -Loderi-
       -fortunei ssp. fortunei
```

Flowers are pale pink.  (Lord Aberconway, intro. 1950)

cl.
```
              -Sir Charles Butler--fortunei ssp. fortunei (1/2)
Admiral Piet-        -maximum (1/4)
  Hein       -Halopeanum-
                         -griffithianum (1/4)
```

Flowers are light rose lilac, and fragrant.    (C. B. van Nes)
A. M. (RHS) 1957

```
              -dichroanthum ssp. dichroanthum (1/4)
cl.    -Jasper-
              -Lady      -fortunei ssp. discolor (1/8)
Adolph  -  -Bessborough-
Heineman-                -campylocarpum ssp. campylocarpum
        -                            Elatum Group (1/8)
        -        -griersonianum (1/4)
        -Tally Ho-
                  -facetum (eriogynum) (1/4)
```

Straw yellow, pink tips; semi-dwarf. Late. (Seabrook, 1964)

```
                  -souliei (1/8)
g.         -Soulbut-
    -Vanessa-      -Sir Chas. Butler--fortunei ssp. fortunei
Adonis-  -griersonianum (1/2)                         (1/8)
      -  -griffithianum (1/4)
      -Sunrise-
                -griersonianum
```

Pale pink flowers. (Lord Aberconway, intro. 1942)

```
                      -souliei (1/16)
              -Soulbut-
        -Vanessa-      -Sir Chas. Butler--fortunei ssp.
    -Adonis-  -griersonianum (1/4)        fortunei (3/16)
g.  -     -griffithianum (1/4)
    -    -Sunrise-
Adota-          -griersonianum
    -          -griffithianum
    -    -Loderi-
    -Coreta-   -fortunei ssp. fortunei
          -arboreum ssp. zeylanicum (1/4)
```

Red-flowered trusses. (Lord Aberconway, intro. 1946)

```
g.
    -williamsianum (1/2)
Adrastia-
    -neriiflorum ssp. neriiflorum (1/2)
```

3ft(.9m)        0°(-18°C)        EM        3/2

Selected from a group, but apparently only one form is in the U.S.A. Flowers deep pink like most williamsianum hybrids, bell-shaped, appearing in clusters of 1-4. (Lord Aberconway, 1948)

```
g.      -williamsianum (1/4)
    -Adrastia-
Adrean-    -neriiflorum ssp. neriiflorum (1/4)
    -beanianum (1/2)
```

Parentage of Adrastia as in Leach. Red flowers. (Lord Aberconway, intro. 1946)

```
              -unnamed-campylocarpum ssp. campylocarpum (1/4)
cl.        -hybrid -
          -        -unknown (1/4)
Adriaan Koster-            -griffithianum (1/8)
          -        -George-
          -        - Hardy-catawbiense (1/8)
          -Mrs. Lindsay-
            Smith   -Duchess of Edinburgh--unknown (1/4)
```

4ft(1.2m)        5°F(-15°C)        M

Creamy white with a red spot in the throat. See also Diane, Harvest Moon, Mrs. Betty Robertson, Zuiderzee.    (M. Koster, 1920)
A. M. (RHS) 1935

```
g.          -arboreum ssp. arboreum (1/4)
      -Cornubia-      -thomsonii ssp. thomsonii (1/8)
Advie-      -Shilsonii-
    -              -barbatum (1/8)
    -diphrocalyx (1/2)
```

Compact plant of average size. Deep rose flowers. (Rothschild, 1933)

cl.

Aestivale*--form of cinnabarinum ssp. cinnabarinum

Flowers yellow, tipped orange; leaves narrower than other forms. (Hutchinson)

```
cl.
      -ambiguum (1/2)
Affectation-
      -cinnabarinum ssp. cinnabarinum Blandfordiiflorum
                              Group (1/2)
```

Pale yellow flowers with rose tips. (Sunningdale Nurs., 1951)

---

Afghan      Parentage unknown

Flowers deep blood red.  A parent of an Exbury hybrid, Michele.

```
cl.
      -fortunei ssp. discolor (1/2)
Afterglow-
      -hardy hybrid--unknown (1/2)
```

Flowers are pale pink and mauve.  (W. C. Slocock, 1935)

```
cl.
      -dichroanthum ssp. dichroanthum (1/2)
Agate Pass Jewel-
      -unknown (1/2)
```

3ft(.9m)        5°F(-15°C)        L

Flowers 1.5in(3.8cm) wide, pale orange with faint green stripes inside and out; up to 5 per truss.  Leaves 3.5in(7.6cm) by 1.5in (3.8cm) and held 3 years.  (Ben Nelson cross; intro. by Florence Putney, 1975)

```
cl.
    -yakushimanum ssp. yakushimanum (1/2)
Agateen-      -catawbiense (1/4)
    -Henriette Sargent-
              -unknown (1/4)
```

3ft(.9m)        -10°F(-23°C)        ML

Semi-dwarf plant spreads broader than tall with leaves having plastered brown indumentum, retained 3 years.  Buds neyron rose, of good substance, opening to flowers spinel red, held in trusses of 15.  (Frank Arsen cross, 1969; reg. 1983)

```
cl.          -sanguineum ssp. didymum (1/4)
    -Arthur Osborn-
Aglare-      -griersonianum (1/2)
    -        -griersonianum
    -Rapture-
          -arboreum ssp. zeylanicum (1/4)
```

Flowers are deep red, 12 per truss.  (Gen. Harrison, reg. 1955)

```
* * * * * * * * * * * * * * * * * * * * * *
* Hybrids of unknown parentage are described briefly in App. D *
* * * * * * * * * * * * * * * * * * * * * *
```

cl.                        -sanguineum ssp. didymum (1/4)
      -Arthur Osborn-
Agleam-                   -griersonianum (1/4)
      -                   -griffithianum (1/4)
        -Isabella-
                       -auriculatum (1/4)

Bright red flowers.  (Gen. Harrison, reg. 1954)

cl.
    -diaprepes (1/2)
Aglow-                -sanguineum ssp. didymum (1/4)
      -Arthur Osborn-
                      -griersonianum (1/4)

Flowers pink with darker center.  (Gen. Harrison, 1954)

cl.

Agnes     Parentage unknown

A parent of Homer, a 1906 cross by Seidel,  and probably not the
Agnes (A. M. 1943) by Lord Swathling.  Might Seidel have used
Agnes Beaufort or Agnes Mangles?

cl.
    -griersonianum (1/2)
Agnes-                          -griffithianum (3/8)
              -Beauty of-
   -Norman Gill- Tremough-arboreum ssp. arboreum (1/8)
              -griffithianum

Deep pink in bud, opening lighter pink with darker pink on edges
and base of tube. (Lord Swaythling) A. M. (RHS) 1943

g.
            -griffithianum (1/2)
Agnes Beaufort-
              -unknown (1/2)

Color, etc. unknown.  (Possibly intro. by Methven or Mangles)

g.

        -Loder's White, q.v. for 2 possible parentage diagrams
Agnes -
Lamont-
       -thomsonii ssp. thomsonii

Flowers of various shades of pink.  (Royal Botanic Garden, Edinburgh)

g.
            -griffithianum (1/4)
   -George Hardy-
Aida-              -catawbiense (1/4)
   -auriculatum (1/2)

Flowers white with dark red markings.  (Discontinued at Exbury)
Rothschild, 1933)

cl.
            -griffithianum (1/2)
Aileen Henderson-
              -unknown (1/2)

Flowers  yellow to creamy white,  with a brownish yellow blotch,
held in large conical trusses.  (M. Koster & Sons, before 1958)

g. & cl.
        -tephropeplum (1/2)
Ailsa-Jean-
        -moupinense(1/2)

Pale pink flowers. (Adams-Acton, 1942)   A. M. (RHS) 1946

    * * * * * * * * * * * * * * * * * *
    * Hybridizers, registrants, raisers, etc., with their *
    *      locations are listed in App. B              *
    * * * * * * * * * * * * * * * * * *

cl.
    -lutescens (1/2)
Airy Fairy-
    -Cornell Pink--mucronulatum (1/2)

4ft(1.2m)          0°F(-18°C)          E

Flowers pink with deeper red spotting, widely funnel-shaped,
1.75in(4.4cm) across, in trusses of 3; floriferous. Leaves narrowly elliptic and almost deciduous. (Francis Maloney cross,
1969; intro. by Mae Granston, 1976) A. M. (Wisley Trials, 1984)

cl.

Ajax      Parentage unknown

Flowers of bright rose.  (G. Waterer, before 1860)   A. M. (RHS)
1890

cl.                       -brookeanum var. gracile (1/4)
      -Crown Princess of Germany-     -jasminiflorum (1/4)
Ajax-                        -Princess-
    -javanicum              Royal -javanicum (5/8)

Vireya hybrid.  Yellow to red orange flowers.   (Veitch)   A. M.
(RHS) 1890

g. & cl.                -griffithianum (1/4)
      -Loderi King George-
Akbar-                -fortunei ssp. fortunei (1/4)
    -fortunei ssp. discolor (1/2)

Flowers are rose pink  with  a small streaked crimson stain in
the throat; sweetly scented. (Rothschild, 1933) A. M. (RHS) 1952

cl.          -griffithianum (1/4)
        -Mars-
Al Jolson-   -unknown (1/4)
        -sanguineum ssp. sanguineum var. haemaleum (1/2)

Compact plant with small leaves. Flowers cardinal red, campanulate, 2.25in(3.2cm) long by 2in(5cm)wide, 4-8 in a loose truss.
(C. S. Seabrook, 1969)

g. & cl.
    -griersonianum (1/2)
Aladdin-
    -auriculatum (1/2)

6ft(1.8m)        0°F(-18°C)           VL        3/3

A large plant  with leaves 7in(17.8cm) by 2in(5cm). Improved by
pruning.  Appearance much like its parent auriculatum. Flowers
medium pink , deeper in the throat; trusses of about 14.   Heat-
tolerant. (Crosfield, 1930) A. M. (RHS) 1935

cl.              -griersonianum (1/4)
        -Aladdin-
Aladdin's-   -auriculatum (1/4)
  Light -          -neriiflorum ssp. neriiflorum (1/8)
        -   -Nereid-
        -Peach-     -dichroanthum ssp. dichroanthum (1/8)
        Lady-
          -fortunei ssp. discolor (1/4)

A blend of shell pink and Delft rose.  Late.  (Lancaster, 1965)

g.                    -neriiflorum ssp. neriiflorum (3/8)
      -F. C. Puddle-
   -Ethel-       -griersonianum (3/8)
Alan-   -forrestii ssp. forrestii Repens Group (1/4)
   -          -neriiflorum ssp. neriiflorum
    -F. C. Puddle-
              -griersonianum

Blood red flowers.  (Lord Aberconway, 1946)

'Accomplishment' by Dexter
Photo by Greer

'All East' by Delp
Photo by Delp

'Alpine Glow' by Rothschild
Photo by Greer

'Altaclarense' by Gowen, 1831
Photo by Salley

'Angelo' by Rothschild
Photo by Greer

'Ann Carey' by Lem/Mrs. S. Anderson
Photo by Greer

'Anna' by Lem
Photo by Greer

'Anna Delp' by Delp
Photo by Delp

'Anna H. Hall' by Leach
Photo by Leach

'Applause' by Leach
Photo by Leach

'Apricot Nectar' by Lyons
Photo by Greer

'Apricot Sherbet' by Greer
Photo by Greer

cl.

Alaska     See Finlandia


g. & cl.         -griffithianum (1/4)
         -Loderi-
Albatross-     -fortunei ssp. fortunei (1/4)
          -fortunei ssp. discolor (1/2)

6ft(1.8m)          0°F(-18°C)          L          4/3

A large, beautiful hybrid. Open habit, leaves 6in(15.2cm) by
2.5in(6.4cm). Deep pink in bud, opening slightly bluish pink;
scented; flowers white with reverse of petals slightly tinged
pink. See also Exbury Albatross, Townhill Albatross.
(Rothschild, 1930) A. M. (RHS) 1934 A. M. (Wisley Trials) 1953
A. G. M. (RHS) 1968 See color illus. PB pl. 58.


cl.        -caucasicum (1/4)
        -Viola-
Albert-     -unknown (1/2)
        -          -catawbiense (1/4)
        -Everestianum-
                  -unknown

Flowers delicate lilac. (Seidel, 1899)


cl.
        -maximum (1/2)
Albert Close-
          -macrophyllum (1/2)

5ft(1.5m)        -5°F(-21°C)          ML          3/2

Flowers bright rose pink, with heavy chocolate red spotting in
the throat; compact conical truss. Straggly open growth of blue
green foliage. Tolerates heat. (Fraser and Gable, 1951) See
color illus. VV p. 22; also LW pl 42.


cl.

Albert Schweitzer     Parentage unknown

Flowers of rose Bengal with blotch of currant red, held in large
pyramidal trusses of 13-14. (Adr. van Nes) S. M. (Rotterdam)
1960


---
        -falconeri ssp. falconeri (1/2)
Alberti-
        -ponticum ? (1/2)

Synonym Albertus. Purple magenta, bell-shaped corolla. (Origin
unknown)


---

Albertus     See Alberti


cl.

Albescens     Parentage unknown

Large white flowers marked sulphur yellow at base; fragrant. A
parent of Harry Tagg, Basileos. (Origin unknown)


cl.

Albiflorum Laciniatum*--form of hirsutum

Flowers in terminal clusters of white; leaves at the base quite
deeply incised. Native of central and eastern Alps. This form
grown by Froebel of Zurich.


g.
        -campylocarpum ssp. campylocarpum
Albino-
        -Loder's White, q.v. for 2 possible parentage diagrams

Flowers are near-white. See also Serin. (Stephenson Clarke;
exhibited by Whitaker, 1935)


cl.

Albion Ridge--form of macrophyllum

6ft(1.8m)          15°F(-10°C)          M

White buds open to widely funnel-shaped flowers, 2.5in(6.4cm)
across, frilled, with small green blotch; up to 15 per truss.
Plant habit upright; leaves elliptic 6.5in(16.5cm) by 3in(7.6cm)
and held 2 years. (Layered in the wild by J. Drewry; E. German
named and reg., 1976)


cl.
        -arboreum ssp. cinnamomeum var. album (1/2)
Album*
        -caucasicum (1/2)

Flowers in compact trusses, pink in bud, opening to white or
blush, with a faint ray of yellowish green. Early. Habit compact. (Cunningham & Fraser of Edinburgh) A. G. M. (RHS) 1969


cl.
        -catawbiense (1/2)
Album Elegans-
        -unknown (1/2)

6ft(1.8m)          -20°F(-29°C)          ML          2/3

Flowers open pale lilac, fading white, with yellow green blotch;
about 16 in rounded compact trusses. Vigorous plant of open habit, with good foliage. (H. Waterer, before 1876) See color
illus. VV p. 14.


cl.
        -catawbiense (1/2)
Album Grandiflorum-
        -unknown (1/2)

5ft(1.5m)          -15°F(-25°C)          ML          3/3

Pale mauve flowers with a brownish green flare. (J. Waterer,
before 1851)


cl.
        -catawbiense (1/2)
Album Novum-
        -unknown (1/2)

Flowers white, tinged rose lilac, with greenish yellow spots.
(L. van Houtte)


g. & cl.
        -lutescens (1/2)
Alcesta-
        -burmanicum (1/2)

Pale creamy yellow with a darker yellow blotch. (Lord
Aberconway) A. M. (RHS) 1935


g.              -haematodes ssp. haematodes (1/4)
        -Hiraethlyn-
Alcibiades-        -griffithianum (1/4)
        -          -neriiflorum ssp. neriiflorum (1/4)
        -F. C. Puddle-
                  -griersonianum (1/4)

Turkey red flowers. (Lord Aberconway, intro. 1941)

```
cl.    -yakushimanum ssp. yakushimanum (1/2)
       -                          -catawbiense (1/16)
Aldham-            -Atrosanguineum-
         -Atrier-              -unknown (1/16)
      -Gable's -     -griersonianum (1/8)
       Flamingo-     -decorum (1/8)
             -Dechaem-
                  -haematodes ssp. haematodes (1/8)

1.5ft(.45m)          -5°F(-21°C)            M
```

Flowers 3in(7.6cm) across, 6-lobed, pink with rose red throat,
and held in trusses of about 12.  Plant as broad as tall; leaves
4.75in(12.1cm) by 1.5in(3.8cm), held 2 years.  (L. Bagoly cross,
1969; raiser C. Herbert, reg. 1980)

```
g.          -neriiflorum ssp. neriiflorum Euchaites Group (1/4)
       -Portia-
Alesia-     -strigillosum (1/4)
       -meddianum var. meddianum (1/2)
```

Flowers of crimson scarlet.  (Lord Aberconway, intro. 1941)

```
cl.

Alexander Adie      Parentage unknown
```

Synonym Jay Gould.  A parent of Eidam, Holbein.  Flowers dark
pink.  (J. Waterer, before 1871)

```
cl.
             -catawbiense (1/2)
Alexander Dancer-
             -unknown (1/2)
```

Flowers reddish magenta with a lighter center.  (A. Waterer, before 1865)

```
cl.
             -dalhousiae var. dalhousiae (1/2)
Alf Bramley-
            -Kallistos--nuttallii var. stellatum (1/2)
```

Yellow in bud, opening to white flowers flushed pink and faintly
scented.  (Alfred Bramley, 1970)

```
cl.
                    -catawbiense (1/2)
Alfred--Everestianum (selfed)-
                    -unknown (1/2)
```

Flowers lilac, with faint green yellow markings on lighter background.  See also Allah, Anton, August, Bertha, Etzel.  (T. J.
R. Seidel, cross 1899)

```
cl.
           -fortunei ssp. discolor (1/2)
Alfred Coates-
           -facetum (eriogynum) (1/2)
```

Rose madder flowers; trusses hold up to 13.  (Cross about 1938;
raised by Coates; intro. by RBG, Kew, 1977)

```
             -fortunei ssp. discolor (1/4)
cl.    -Margaret-      -dichroanthum ssp. dichroanthum (1/8)
       - Dunn   -Fabia-
Algeriene-             -griersonianum (3/8)
              -haematodes ssp. haematodes (1/4)
        -May Day-
              -griersonianum
```

Ivory, flushed pale rose at tips, slight olive stain in the
throat.  (Sunningdale Nurseries, 1955)

```
cl.
       -griffithianum (1/2)
Alice-
       -unknown (1/2)

6ft(1.8m)          -5°F(-21°C)            M          4/3
```

Flowers frosty pink, lighter in center, held in large upright
trusses.  Vigorous upright plant, improved by some shade.  Named
after Mrs. Gomer Waterer.  Some records suggest it is a seedling
of Pink Pearl.  (J. Waterer)  A. M.  (RHS) 1910   See color illus.
F p. 31.

```
cl.       -campylocarpum ssp. campylocarpum Elatum Group
       -Ole -                            (1/4)
Alice  -Olson-fortunei ssp. discolor (1/4)
Franklin-      -griffithianum (1/4)
         -Loderi King-
          George   -fortunei ssp. fortunei (1/4)
```

Flowers medium yellow, with a green throat.  Foliage  resembles
Loderi.  (Cross by Halfdan Lem; intro. by  H. L. Larson,  1959)
P. A. (ARS) 1960

```
cl.
             -fortunei ssp. fortunei (1/2)
Alice in Wonderland-
             -unknown (1/2)

6ft(1.8m)          -10°F(-23°C)            M
```

Flowers openly funnel-shaped, 2in(5cm) across by 1.5in(3.8cm)
long, with 7 wavy lobes of light purplish pink; 16 held in a
ball-shaped truss 4in(10.2cm) across by 5in(12.7cm) high.  Free-
flowering.  Plant upright, rounded, well-branched; leaves held 2
years.  (Dexter cross, before 1943; Tyler Arboretum, reg. 1979)

```
cl.
          -griffithianum (1/2)
Alice Mangles-
          -ponticum (1/2)
```

The first hybrid of griffithianum to receive an award.  (James
Mangles) F.C.C. (RHS) 1882

```
cl.
          -fortunei ssp. discolor (1/2)
Alice Martineau-
          -unknown (1/2)
```

Rosy crimson, dark blotch and throat; very late.  (Slocock)

```
                              -griffithianum (1/16)
                  -George Hardy-
cl.    -Mrs. Lindsay-       -catawbiense (1/16)
    -Diane- Smith    -Duchess of Edinburgh--unknown (1/4)
Alice -    -           -campylocarpum ssp. campylocarpum
Street-    -unnamed hybrid-                (1/8)
       -              -unknown
       -wardii var. wardii (1/2)

4ft(1.2m)          -5°F(-21°C)            M          3/4
```

Synonym  Miss Street.  Flowers soft yellow; foliage grass green.
Requires shade.  (M. Koster, 1953)

```
cl.           -racemosum (1/4)
          -unnamed hybrid-
Alice Swift*          -mucronulatum (pink form) (1/4)
             -minus var. minus Carolinianum Group (1/2)
```

Grows to about 3ft(.9m).  Flowers bright clear pink in terminal
clusters.  Floriferous; blooms later than P.J.M.  Winter foliage
a crisp bright green.  No frost damage in New Jersey.  Deadhead-
ing not necessary.  (Leon Yavorsky)

cl.
      -Sir John Waterer--unknown (1/2)
Alinta-      -catawbiense (1/4)
    -Vesuvius-
                -arboreum ssp. arboreum (1/4)

Solferino purple flowers, 14-16 in a spherical truss.  (V. Boulter, Australia, 1962)

g. & cl.

Alison   -yunnanense (1/2)
Johnstone-
      -cinnabarinum ssp. xanthocodon Concatenans Group (1/2)

5ft(1.5m)        5°F(-15°C)        EM        4/3

The plant has the appearance of the Concatenans Group with blue
green waxy foliage.  Buds best in full sun; likes cool roots.
Trusses have 9-10 flowers, delicate amber with pink frosting.
(G. H. Johnstone)  A. M. (RHS) 1945

g. & cl.
     -barbatum (1/2)
Alix-
     -hookeri (1/2)

Brilliant clear crimson flowers.  Tall, gaunt habit.  Barely
hardy at Exbury.  (Rothschild, 1930)  A. M. (RHS) 1935

cl.
                 -catawbiense (1/2)
Allah--Everestianum, selfed-
                 -unknown (1/2)

Light purplish pink flowers with yellow markings and pale center.  (T. J. R. Seidel, intro. 1913)

cl.      -griersonianum (1/4)
     -Azor-
Allegro-    -fortunei ssp. discolor (1/4)
     -            -griffithianum (1/4)
       -Loderi King George-
                     -fortunei ssp. fortunei (1/4)

This parentage produced a plant with pink flowers.  (Lancaster)

cl.

Alley Cat    Parentage unknown

Coral pink, darker blotch.  (Endre Ostbo selection; intro. by
Owen Ostbo)  P. A. (ARS) 1960

cl.      -dichroanthum ssp. dichroanthum (1/2)
     -Fabia-
Allure-   -griersonianum (1/4)
     -    -dichroanthum ssp. dichroanthum
     -Dante-
          -facetum (eriogynum) (1/4)

Scarlet flowers.  (Lord Aberconway, 1942)

g.           -ciliatum (1/4)
          -Rosy Bell-
Almond Thacker-     -glaucophyllum var. glaucophyllum (1/4)
          -racemosum (1/2)

Flowers deep almond pink.  Parentage as in Leach, reversed in
the Register.  (Thacker, 1946)

g.          -Blood Red--arboreum ssp. arboreum (1/4)
       -Cornubia-          -thomsonii ssp. thomsonii (1/8)
Almondtime-       -Shilsonii-
       -            -barbatum (1/8)
       -sutchuenense (1/2)

Flowers bright cerise pink with a few dark crimson dots.  See also Cornsutch.  (E. J. P. Magor)  A. M. (RHS) 1925

g.           -arboreum ssp. arboreum (1/4)
        -Sir Charles Lemon-
Alpaca-        -unknown (1/4)
     -neriiflorum ssp. neriiflorum (1/2)

Pink.  Cultivation discontinued at Exbury.  (Rothschild, 1933)

g.
        -brachyanthum ssp. brachyanthum (1/2)
Alpine Gem-
        -ferrugineum (1/2)

Flowers are yellow, shaded pink.  (Thacker, 1942)

cl.      -griffithianum (1/4)
     -Loderi-
Alpine-     -fortunei ssp. fortunei (1/4)
 Glow -calophytum var. calophytum (1/2)

6ft(1.8m)        0°F(-18°C)        E        4/4

Avalanche g.  An  attractive  foliage plant all year around.
Vigorous new stem growth often is larger than a man's finger.
Very large trusses of pale pink appear in early spring.  Rather
difficult to propagate.  See also Alpine Rose.  (Rothschild,
1933)  A. M. (RHS) 1938

cl.

Alpine Rose    For parentage see Alpine Glow.

Avalanche g.  Delicate pink flowers, slightly deeper than Alpine
Glow.  (Rothschild, 1933)

          -campylocarpum ssp. campylocarpum (1/8)
       -Ethel-
g. -Alan-   -orbiculare ssp. orbiculare (1/8)
 -     -        -neriiflorum ssp. neriiflorum (1/8)
Alrc   -F. C. Puddle-
 -        -griersonianum (1/8)
    -forrestii ssp. forrestii Repens Group (1/2)

Blood red flowers.  (Lord Aberconway, intro. 1946)

        -wardii var. wardii (1/4)
cl.  -Hawk-
 -     -Lady     -fortunei ssp. discolor (1/8)
Alston-   -Bessborough-
 -            -campylocarpum ssp. campylocarpum
    -griffithianum (1/2)          Elatum Group (1/8)

Leaves ovate-lanceolate, large, dark green. Flowers cream white
with base stain of cream yellow, open funnel-shaped,  in trusses
of 5-7.  (L. S. Fortescue, 1971)

g.            -catawbiense (1/4)
        -unnamed hybrid-
Altaclarense-        -ponticum (1/4)
        -arboreum ssp. arboreum (1/2)

One of the first man-made rhododendron hybrids.  Grows to become
a large tree.  Deep red.  (J. R. Gowen, 1831)  F. C. C. 1865

g.
     -Ivery's Scarlet--unknown (1/2)
Alvinda-     -griffithianum (1/4)
      -Loderi-
           -fortunei ssp. fortunei (1/4)

Pale rose pink flowers.  (Lord Aberconway cross, 1926;  intro.
1933)

g.
          -javanicum (5/8)
Amabile-                      -jasminiflorum (3/8)
      -Princess Alexandra-              -jasminiflorum
                          -Princess Royal-
                                      -javanicum

Vireya hybrid.   Flowers pinkish yellow to salmon pink.  (Origin
unknown) F. C. C. (RHS) 1886

g. & cl.    -Blood Red--arboreum ssp. arboreum (1/4)
     -Cornubia-          -thomsonii ssp. thomsonii (1/8)
Amalfi-      -Shilsonii-
     -              -barbatum (1/8)
     -calophytum var. calophytum (1/2)

Carmine rose, shaded white on tube, spotted dark red with crimson splash at base.   (Rothschild, 1933)   A. M. (RHS) 1939

cl.
         -catawbiense (1/2)
Amaranthora-
            -unknown (1/2)

Description and dates unknown.  (Parsons)

cl.
         -hemsleyanum (1/2)
Amarillo-         -fortunei ssp. discolor (1/4)
     -Autumn Gold-    -dichroanthum ssp. dichroanthum (1/8)
               -Fabia-
                     -griersonianum (1/8)

6ft(1.8m)         0°F(-18°C)         ML

Fragrant flowers of 7 lobes, bright lemon yellow, reverse light
mars orange, held in 8in(20.3cm) trusses of 12-14.  Tall plant
2/3 as wide as tall with leaves 6.5in(16.5cm) long, deep dark
green, retained 3-4 years.  (Dr. David W. Goheen cross, 1963;
reg. 1984)

g.               -griffithianum (1/4)
         -Halopeanum-
Amaryllis-          -maximum (1/4)
         -haematodes ssp. haematodes (1/2)

Color unknown.  (Robert Wallace, 1934)

g.                     -campylocarpum ssp. campylocarpum Elatum
         -Penjerrick-                            Group (1/8)
     -Amaura-             -griffithianum (3/8)
Amasun-   -griersonianum (1/2)
     -         -griffithianum
     -Sunrise-
               -griersonianum

Pale rose flowers.  (Lord Aberconway cross, 1939; intro. 1946)

g.          -williamsianum (1/4)
     -Adrastia-
Amata-          -neriiflorum ssp. neriiflorum (1/2)
     -          -neriiflorum ssp. neriiflorum
     -F. C. Puddle-
               -griersonianum (1/4)

Red flowers.  (Lord Aberconway cross, 1935; intro. 1941)

g.          -campylocarpum ssp. campylocarpum Elatum Group
     -Penjerrick-                                  (1/4)
Amaura-          -griffithianum (1/4)
     -griersonianum (1/2)

Pale pink flowers.  (Lord Aberconway, 1933)

cl.
         -fortunei ssp. fortunei (1/2)
Amazement-
         -wardii var. wardii (1/2)

6ft(1.8m)         -5°F(-21°C)         ML

Flowers 3in(7.6cm) wide, 7 lobes, very fragrant, yellow, up to
13 in ball-shaped trusses.  Plant broad, upright, rounded, well-
branched with stiff branches; leaves held 2 years.  (Cross by D.
Hardgrove; Doris Royce, intro. 1978)

cl.         -catawbiense (5/16)
       -Mrs. C. S.-
Amazing- Sargent   -unknown (5/16)
  Grace -               -catawbiense var. album (3/8)
       -      -Belle Heller-          -catawbiene
         -Swansdown-          -Madame Carvalho-
               -catawbiense var. album     -unknown

3ft(.9m)         -30°F(-35°C)         L

Buds orchid lavender; flowers phlox pink with yellow green spotting, 12 per truss, of heavy substance.  Floriferous.  Plant upright, broad, branching well.  (Orlando Pride, 1979)   See color
illus.  ARS Q 34:3 (Summer 1980): p.138.

g.               -campylocarpum ssp. campylocarpum
       -Penjerrick-            Elatum Group (1/8)
     -Amaura-         -griffithianum (1/8)
Amazor-     -griersonianum (1/2)
     -     -griersonianum
     -Azor-
         -fortunei ssp. discolor (1/4)

Pink flowers.  (Lord Aberconway cross, 1934; intro. 1946)

g.
     -racemosum (white form) (1/2)
Amba-
     -burmanicum (1/2)

Pale yellow flowers.  (Lord Aberconway cross, 1926; intro. 1934)

cl.       -Jan Dekens--unknown (13/16)
     -                         -ponticum (1/16)
Ambassadeur-               -Michael Waterer-
     -            -Prometheus-          -unknown
     -Madame   -          -Monitor--unknown
      de Bruin-      -arboreum ssp. arboreum (1/8)
               -Doncaster-
                     -unknown

Deep pink flowers 3in(7.6cm) across, 20 per truss.  Upper petal
lighter pink.  (P. van Nes, reg. 1962) S. M. (Rotterdam) 1960

cl.  -Fawn- -fortunei ssp. fortunei (1/4)
     -       -dichroanthum ssp. dichroanthum (1/4)
     -     -Fabia-
Amber-     -griersonianum (1/8)
  Gem -     -Lady     -fortunei ssp. discolor (1/8)
     -       -Bessborough-
     -Jalisco-          -campylocarpum ssp. campylocarpum
     -                       Elatum Group (1/8)
     -     -dichroanthum ssp. dichroanthum
       -Dido-
         -decorum (1/8)

3ft(.9m)         -5°F(-21°C)         ML

A superior plant with fragrant 7-lobed flowers of heavy substance, lobes evenly rounded and slightly wavy.  Flowers funnel-
shaped, 4.5in(11.4cm) wide, Venetian pink, shading to nasturtium
orange with strong orange red spotting, in loose ball trusses
of 10.   Leaves medium green; plant broad.  (Del James cross,
1959; Arthur Childers intro., 1977)

cl.
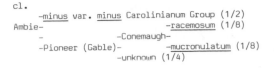
```
        -minus var. minus Carolinianum Group (1/2)
Ambie-                    -racemosum (1/8)
              -Conemaugh-
     -Pioneer (Gable)-          -mucronulatum (1/8)
                          -unknown (1/4)
```

3ft(.9m)          -20°F(-29°C)           M

Buds dark purplish red; flowers widely funnel-shaped, creamy
white, irregularly variegated, with one lobe of each flower all
pink, clusters of 5-18.  Plant semi-dwarf, rounded, compact,
tightly branched to the ground.  Leaves held 1 year only.  See
also Billy Bear. (William Fetterhoff, 1976)

g.
```
         -javanicum (5/8)
Ambient-                    -jasminiflorum (3/8)
       -Princess Alexandra-          -jasminiflorum
                           -Princess Royal-
                                      -javanicum
```

Vireya hybrid.  Flowers pinkish yellow to salmon. (Veitch, 1891)

cl.
```
       -ambiguum (1/2)
Ambition-
       -lutescens (1/2)
```

Pale primrose yellow with a yellowish flare.  (Cross by Raffil;
named by Collingwood Ingram, 1961)

g.
```
                 -griffithianum (1/4)
       -Queen Wilhelmina-
Ambrose-               -unknown (1/4)
       -haematodes ssp. chaetomallum (1/2)
```

Flowers are rose pink. (Rothschild, 1933)

cl.
```
                  -catawbiense (1/4)
       -Parsons Grandiflorum-
America-                    -unknown (3/4)
       -dark red hybrid--unknown
```

5ft(1.5m)          -20°F(-29°C)           ML           3/3

Medium-sized, sprawling plant with matte green foliage. Compact
truss, ball-shaped of very dark, blue-toned red flowers.  One of
true ironclad hardiness for severely cold climates. (M. Koster
cross, 1904; intro. about 1920)  See color illus. VV p. viii.

cl.
```
                          -griffithianum
                -Queen Wilhelmina-     (1/4)
Americana--sport of Britannia-      -unknown (3/4)
                          -Stanley Davies--unknown
```

5ft(1.5m)          -10°F(-23°C)           ML           3/3

Sport of Britannia occuring at the graft union.  The understock
was a hybrid of fortunei ssp. discolor.  Crimson red flowers.
(Benjamin F. Lancaster, before 1958)

---

Amethyst      Parentage unknown

Bright rose purple.  (Noble, before 1850)  A. M. (RHS) 1931

cl.
```
       -fortunei ssp. fortunei ? (1/2)
Amethyst-
       -unknown (1/2)
```

Lavender flowers, frilled. (C. O. Dexter cross; intro. by Westbury Rose Co., 1959)

cl. -
```
        -aberconwayi (1/2)         -arboreum ssp. arboreum (1/16)
            -Doncaster-
     -     -unnamed-          -unknown (1/8)
Amigo-  -hybrid -     -neriiflorum ssp. neriiflorum
     -     -     -Nereid-                    (1/16)
     -Witch -     -dichroanthum ssp. dichroanthum
     Doctor-     -griffithianum (1/16)        (1/16)
          -Mars-
          -Vulcan-  -unknown
                 -griersonianum (1/8)
```

4ft(1.2m)          5°F(-15°C)           EM           3/3

Buds deep pink; flowers flat saucer-shaped, of heavy substance,
reddish edges shading to yellow center, heavily spotted red;  up
to 15 per truss.  Plant upright, branching well; new growth is
bronze green.  (Dr. David Goheen, 1972)

---

Amilcar      Parentage unknown

Deep violet purple, spotted black.  (Origin unknown) F. C. C.
(RHS) 1860

cl.
```
                           -haematodes ssp. haematodes (1/8)
                -Grosclaude-
    -unnamed hybrid-     -facetum (eriogynum) (1/8)
Amity-                       -griffithianum (1/16)
    -               -Queen
    -     -Britannia-Wilhelmina-unknown (3/16)
    -               -
    -               -Stanley Davies--unknown
    -Koichiro Wada--yakushimanum ssp. yakushimanum (1/2)
```

3ft(.9m)          10°F(-12°C)           M

Flowers widely campanulate, 2.5in(6cm) across, rose opal fading
to jasper red with spotting on dorsal lobe.  Trusses 5in(13cm)
across, 4.5in(11cm) high, ball-shaped with 15 flowers.  Leaves
held 3 years with heavy indumentum.  New foliage, silvery tomentose.  Plant rounded, semi-dwarf. (James A. Elliott, reg. 1983)

cl.
```
       -ambiguum (1/2)
Amkeys-
       -keysii (1/2)
```

Flowers orange outside fading to yellow at opening of tube. (E.
J. P. Magor cross, 1914; intro. 1926)

g.
```
                           -griffithianum (1/8)
                -Queen Wilhelmina-
       -Britannia-          -unknown (3/8)
Ammerlandense*      -Stanley Davies--unknown
       -williamsianum (1/2)
```

4ft(1.2m)          0°F(-18°C)           M           3/3

Flowers of coral rose. (Dietrich Hobbie, 1946)

---
```
       -ciliatum (1/2)
Amoenum-
       -dauricum (1/2)
```

Flowers pink.  Plant similar to Praecox, but larger flowers.

g. & cl.
```
       -griersonianum (1/2)
Amor-
       -thayerianum (1/2)
```

Flowers white, flushed with irregular pink staining, tinged pink
on the outside; trusses of about 10 flowers.  (J. B. Stevenson,
1927)  A. M. (RHS) 1951

cl.
        -wardii var. wardii (1/2)
Amour-
      -Lady          -fortunei ssp. discolor (1/4)
        Bessborough-
                    -campylocarpum ssp. campylocarpum Elatum Group
                                                              (1/4)
6ft(1.8m)          5°F(-15°C)          M          3/2

Hawk g. Pale yellow flowers. One of the 7 named selections in
this family, the most famous being Crest, which is a deeper
yellow. (Rothschild, 1940)


cl.
        -catawbiense (1/2)
Amphion-
        -unknown (1/2)

Synonym F. L. Ames.  Red or rose pink flowers.  (A. Waterer)


---
        -catacosmum (1/2)
Amrie-
        -haematodes ssp. chaetomallum (1/2)

Blackish crimson flowers.  (J. B. Stevenson, 1951)


g.                    -maximum (1/4)
        -Burgemeester Aarts-
Amsel-                    -L. L. Liebig--unknown (1/4)
        -williamsianum (1/2)

Flowers of clear rose color.  (Dietrich Hobbie, 1947)


cl.
        -griffithianum (1/2)
Amy-
        -unknown (1/2)

5ft(1.5m)          -5°F(-21°C)          ML          3/3

Sturdy plant with medium green leaves,  6in(15.2cm) by 2.5in
(6.4cm). Deep pink flowers in compact, upright trusses. Tolerant of heat and sun. (J. Waterer, Sons & Crisp)


cl.
        -yakushimanum ssp. yakushimanum (1/2)
Amy -                              -griffithianum (1/8)
Jane-               -Queen Wilhelmina-
        -Earl of Athlone-               -unknown (3/8)
                    -Stanley Davies--unknown

3ft(.9m)          10°F(-12°C)          M

Flowers light neyron rose with narrow edging of rose red. Dome-
shaped trusses hold 11 flowers.   Plant rounded, almost as broad
as tall. (Cross by Donald Stanton; raiser Edw. Anderson; intro.
1980; reg. 1983)


cl.                              -ponticum (3/4)
                    -Purple Splendour-
Anah Kruschke-                    -unknown (1/4)
                    -ponticum

5ft(1.5m)          -15°F(-26°C)          ML          3/4

Parentage given as in Van Veen, though registered as a seedling
of ponticum. Dense plant with glossy foliage; leaves
4.5in(11.4cm) by 1.5in(3.8cm), retained 3 years. Flowers 3in
(7.6cm) wide, up to 12 per tight, ball-shaped truss, deep purple. Tolerates heat and sun. (Cross by Kruschke; intro. by
Arthur Wright, Sr., 1955; reg. by Van Veen, 1973) See color
illus. VV p. 56.


* * * * * * * * * * * * * * * * * * * * *
* Species used in hybrids of this volume are listed in App. A. *
* * * * * * * * * * * * * * * * * * * * *


               -Queen      -griffithianum (1/8)
cl.              -Wilhelmina-
        -Britannia-               -unknown (3/8)
Ananouri-          -Stanley Davies--unknown
        -fortunei ssp. discolor (1/2)

3ft(.9m)          -5°F(-21°C)          ML

Leaves 7in(17.8cm) by 2.5in(6.4cm). Flowers medium red; trusses
hold up to 12. (Howard Phipps intro., 1955; reg. 1971)   C. A.
(ARS) 1973


               -wardii var. wardii (1/4)
               -               -griffithianum (1/32)
cl.    -Idealist-               -Kewense-
       -          -Aurora-          -fortunei ssp.
Anchorage-    `-Naomia-          -          fortunei (21/32)
       -               -thomsonii ssp. thomsonii (1/16)
       -          -fortunei ssp. fortunei
       -fortunei ssp. fortunei

Trusses of 9 flowers, very light yellow green deepening
to chartreuse green in eye, fading to greenish white. (Edmund
de Rothschild, 1981)


                         -griffithianum (1/8)
cl.          -Queen Wilhelmina-
       -Britannia-               -unknown (3/8)
Andre-          -Stanley Davies--unknown
       -yakushimanum ssp. yakushimnaum (1/2)

Truss of 11-14 flowers, deep rose in bud, opening pink fading to
pale pink. (Cross by de Belder, Belgium; intro. by the Research
Station for Woody Nursery Crops, Boskoop)

g. & cl.
        -arboreum ssp. arboreum (1/2)
Androcles-
        -calophytum var. calophytum (1/2)

Large trusses of up to 30 flowers each, rhodamine pink with 4
lines of darker spots from the base of the tubes to the ovary.
(Rothschild, 1933) A. M. (RHS) 1948


cl.
        -stewartianum (scarlet form) (1/2)
Andromeda-
        -neriiflorum ssp. neriiflorum (1/2)

Crimson scarlet flowers.  (Reuthe, 1941)


               -Sir Charles Butler--fortunei ssp. fortunei
cl.  -Van Nes  -                              (1/4)
     -Sensation-          -griffithianum (3/8)
Angel-          -White Pearl-
     -               -maximum (3/8)
     -          -griffithianum
     -White Pearl-
               -maximum

White flowers, pale mauve on the outside, with red speckling on
the upper lobe. (R. L. Rowarth, Australia, 1972)


g. & cl.
        -griffithianum (1/2)
Angelo-
        -fortunei ssp. discolor (1/2)

6ft(1.8m)          -5°F(-21°C)          L          5/3

Blush pink outside, white with pale green spots within. Tall
truss of about 12 flowers.  See also Exbury Angelo, Sheffield
Park Angelo, Solent Queen, Solent Snow, Solent Swan.
(Rothschild, 1930) A. M. (RHS) 1935; F. C. C. (RHS) 1948  See
color illus. PB pl. 10.

cl.

Angelo Diadem    Parentage as above

Angelo g. White flowers; cross first made by Rothschild. (Haworth-Booth)

cl.

Angelo Solent Queen    Parentage as above

Angelo g. Flowers very pale rose fading to almost white, spotted green within. A small, late-blooming shrub. (Rothschild) A. M. (RHS) 1939

g.
    -campylocarpum ssp. campylocarpum (1/2)
Anita-
    -griersonianum (1/2)

Flowers are yellow, flushed rose and bell-shaped. (Lord Aberconway, 1941)

cl.
    -keiskei (1/2)
Ann Carey-
    -spinuliferum (1/2)

5ft(1.5m)        5°F(-15°C)        E        3/3

Campanulate flowers 1in(2.5cm) wide, opening chartreuse, changing to coral pink. Plant as broad as tall; leaf margins rolled. (Lem cross; Mrs. S. Anderson intro., reg. 1977) P.A. (ARS) 1966

cl.
    -yakushimanum ssp. yakushimanum (1/2)
Ann Elizabeth-
    -unknown (1/2)

2.5ft(.75m)        -10°F(-23°C)        ML

Strong purplish red buds open to flowers of good substance, deep reddish purple, throat pale pink, held in trusses of 11-15. Plant upright, semi-dwarf, as broad as tall, with leaves medium olive green, held 2 years. (John Ford, Secrest Arboretum, reg. 1984)

cl.
    Advanced generation hybrid including catawbiense
Ann Luettgen    and maximum, and possibly unknown hybrids

5ft(1.5m)        -15°F(-26°C)        M

Buds deep purplish red; flowers 4in(10.2cm) wide, deep pink with dark purplish red dorsal spotting. Trusses hold about 17. (W. David Smith, 1979)

cl.  -unnamed-dichroanthum ssp. apodectum (1/4)
    - hybrid-
Ann   -          -campylocarpum ssp. campylocarpum (Hooker's form)
Pascoe-      -wardii var. wardii (1/4)                    (1/4)
    -            -                -griffithianum (1/32)
    -Idealist-        -Kewense-
              -Aurora-    -fortunei ssp. fortunei
          -Naomi-    -thomsonii ssp.thomsonii  (5/32)
          -                    (1/16)
          -fortunei ssp. fortunei

Plant habit open, low, leaves 4.5in(11.5cm) by 1.5in(3.8cm). Flowers yellow, slightly flushed yellow orange, 12 to 14 per truss. (Lester E. Brandt, 1970)

cl.         -griffithianum (5/8)
    -Beauty of Tremough-
  -Norman Gill-        -arboreum ssp. arboreum (1/8)
Anna-        -griffithianum
    -                    -griffithianum
    -Jean Marie de Montague-
                    -unknown (1/4)

6ft(1.8m)        0°F(-18°C)        ML        4/4

A plant of unusual foliage, heavily textured and deeply veined. Flower buds deep reddish pink, opening to deep pink with reddish flare in the throat, and fading to pale pink, creating a two-toned effect. Trusses hold about 12 flowers. (Halfdan Lem, before 1958) P. A. (ARS) 1952

cl.
                        -racemosum (1/4)
                  -Conemaugh-
Anna Baldsiefen--Pioneer (selfed)-    -mucronulatum (1/4)
                  -unknown (1/2)

3ft(.9m)        -20°F(-29°C)        VE        4/3

Heavily clothed in small foliage, more evergreen than its parent Pioneer. Flowers clear phlox pink, with deeper margins; trusses hold about 17. Very free-flowering. (Warren Baldsiefen, 1964) P. C. (RHS) 1978; H. C. (RHS) 1979  See color illus. K p. 48.

cl.
    -adenopodum (1/2)
Anna Caroline Gable*
    -metternichii var. tsukusianum ? (1/2)

Pale pink flowers appear early. Plant has good tight form; hardy to -10°F(-23°C). (Joseph Gable)

                                  -catawbiense (1/8)
                  -Parsons Grandiflorum-
          -America-        -unknown (9/16)
    -unnamed-        -dark red hybrid--unknown
cl. - hybrid-        -griffithianum (3/16)
    -      -      -Mars-
Anna-        -Blaze-    -unknown
Delp*        -catawbiense var. rubrum (1/8)
    -            -griffithianum
    -        -Mars-
  -Red Brave-    -unknown        -catawbiense
    -        -Parsons Grandiflorum-
    -America-            -unknown
          -dark red hybrid--unknown

6ft(1.8m)        -15°F(-26°C)        June (zone 4)        4/3

Red flowers in trusses rounded to pyramidal. Habit and foliage as catawbiense. (Weldon Delp, 1968)

cl.
    -catawbiense var. album (1/2)
Anna H. Hall-
    -Koichiro Wada--yakushimanum ssp. yakushimanum (1/2)

5ft(1.5m)        -25°F(-32°C)        M        4/4

Plant compact, densely foliaged, with thin tannish brown indumentum. New growth and buds conspicuously indumented. White flowers open from fresh pink buds, resembling the inflorescence of yakushimanum. Bloomed after -32°F(-36°C), in 1961. See also Great Lakes, May Time, Pink Frosting, Spring Frolic. (David G. Leach, 1962)

cl.
    -catawbiense (1/2)
Anna Parsons-
    -unknown (1/2)

Purplish red flowers. (S. B. Parsons, before 1875)

cl.    -griersonianum (1/2)
    -                -griffithianum (3/16)
Anna Rose-
  Whitney -        -George-
    -        -Pink - Hardy-catawbiense (3/16)
    -        -Pearl-        -arboreum ssp. arboreum
    -Countess-        -Broughtonii-            (1/16)
    of Derby-            -unknown (1/16)
    -            -catawbiense
    -Cynthia-
                -griffithianum

6ft(1.8m)        -5°F(-21°C)        ML        4/3

Anna Rose Whitney in flower is an impressive sight with huge 4in (10cm) flowers of deep rose pink and very light brown spotting on upper lobes. Handsome foliage with matte fern green leaves 8in(20.3cm) by 3in(7.6cm). Grows well in full sun, but better flowers produced under some shade. (Cross by Van Veen, Sr.; intro. by Whitney) P. A. (ARS) 1954 See color illus. VV p. 60.

cl.

Anna Strelow--form of annae Laxiflorum Group (F 27706)

5ft(1.5m)          10°F(-12°C)          EM          3/3

White flowers  shading near base to yellow white,  lobes faintly
flushed reddish purple, in loose trusses of 14-16.  Leaves
narrowly oblong, dull green above, glaucous beneath. (Raised by
Col. S. R. Clarke, intro. by R. N. Stephenson Clarke, reg. 1979)
A. M. (RHS) 1979

g.
           -campanulatum ssp. campanulatum (1/2)
Annabella-      -griffithianum (1/4)
         -Loderi-
                 -fortunei ssp. fortunei (1/4)

White flowers are flushed light mauve.  (Rothschild, 1933)


cl.
              -falconeri ssp. falconeri (1/2)
Annapolis Royal-
              -sinogrande (1/2)

Trusses of 25 flowers, 11-lobed, light chartreuse green  faintly
striped in magenta.  Large leaves  with woolly fawn indumentum.
See also Charles.  (Cross by Lionel de Rothschild, 1938; reg. by
Edmund de Rothschild, 1983)


cl.

Annapurna--form of anthopogon ssp. hypenanthum (S S & W 9090)

18in(46cm)          0°F(-18°C)          EM          4/3

This form  was collected in Nepal  by Stainton, Sykes, and Williams.  Light yellow, narrowly tubular flowers held in trusses of
8 or 9.  Small leaves glossy deep green above, densely covered
below with dark brown, scaly indumentum. (Peter A. Cox,  Glendoick Gardens, Perth, Scotland) A. M. (RHS) 1974

cl.
    -thomsonii ssp. thomsonii (1/2)
Anne-
    -unknown (1/2)

A pink-flowered shrub.  (Messel)  A. M. (RHS) 1928


cl.
           -arboreum ssp. cinnamomeum var. roseum (1/2)
Anne Clarke-
           -sutchuenense (1/2)

Pink,  fading to white; 20-22 flowers per truss.  Leaves narrowly elliptic, dark green above, silvery grey felted below. (R. N.
S. Clarke, Borde Hill, 1977)  A. M. (RHS) 1977


                         -fortunei ssp. discolor (3/8)
            -Lady     -
            -Bessborough-
cl.      -Day  -          -campylocarpum ssp. campylocarpum
         -Dream-                Elatum Group (1/8)
Anne George-    -griersonianum (1/4)
         -          -dichroanthum ssp. dichroanthum (1/8)
         -     -Dido-
         -Ice Cream-  -decorum (1/8)
                     -fortunei ssp. discolor

Flowers salmon pink in bud, fading paler; held in loose trusses.
(A. F. George, Hydon Nurseries, 1965)

cl.
          -Catalgla--catawbiense var. album Glass (1/2)
Anne Glass*
          -decorum (1/2)

Synonyms Mrs. Powell Glass, Mrs. Carter Glass.  Pure white flowers, in large lax trusses.  (Joseph Gable, exhibited 1949)


                                        -griffithianum
                       -Queen Wilhelmina-      (1/16)
cl.         -Brittania-            -unknown
         -C. P. Rafill-     -Stanley Davies--unknown (11/16)
Anne -           -griersonianum (1/4)
Hardgrove-
         -Moser's Maroon--unknown

6ft(1.8m)          0°F(-18°C)          L

Currant red flowers, 5-lobed, 3.5in(8.9cm) wide, about 11 to a
truss.  Leaves 6in(15.2cm) by 2in(5cm), held 1 year.  (Crossed
1949, raised by D. L. Hardgrove; intro. by S. Burns, reg. 1978)

cl.                -griffithianum (1/8)
              -Loderi-
Anne -Avalanche-    -fortunei ssp. fortunei (1/8)
Henny-        -calophytum var. calophytum (3/4)
     -calophytum var. calophytum

Buds deep pink.  Flowers campanulate, light pink, fading to
white with red blotch.  Truss dome-shaped, of 15 flowers.  Plant
upright, branching moderately.  Leaves held 3 years.  (Cross by
Anne Henny, 1951; intro. by Carl Heller; reg. 1980)

cl.
      -ciliicalyx (1/2)
Anne Teese-
      -formosum var. formosum (1/2)

White flowers, flushed solferino purple, and striped with purple
on the outside.  (A. J. Teese, Australia, reg. 1968)

cl.
      -catawbiense (1/2)
Annedore-
      -unknown (1/2)

Shown in the Register of 1958 as a synonym of Lady Armstrong,
credited to A. Waterer, before 1870.  Flowers bright, light carmine red with a white throat, ochre markings. (Offered by T. J.
Hermann Seidel, 1926)

cl.
      -Marion--unknown (3/4)
Annelies  -
Van de Ven-              -arboreum ssp. arboreum (1/4)
          -Mrs. Henry Shilson-
                          -unknown

Flowers red purple with a lighter center.  (Van de Ven, 1974)

cl.
              -fortunei ssp. discolor (1/2)
Annette Weyerhaeuser-   -griersonianum (1/4)
              -unnamed hybrid-
                     -unknown (1/4)

Trusses hold 11-12 light magenta flowers, 5-lobed.  (H. L. Larson, 1982)

                   -decorum (1/4)
cl.       -unnamed hybrid-
          -          -griersonianum (1/4)
Annie Dalton-                  -catawbiense (1/8)
          -     -Parsons Grandiflorum -
          -America-           -unknown (3/8)
                -dark red hybrid--unknown

5ft(1.5m)          -15°F(-26°C)          ML          4/2

Synonym Degram.  Large, lax trusses of apricot pink flowers,  of
good substance,  4in(10.2cm) wide.  Leaves 8.25in(21cm) long  by
2.25in(5.7cm) wide,  with a slight twist.  (Joseph Gable, 1960)
A. E. (ARS) 1960   See color illus. VV p. 16.

cl.
      -Marion--unknown
Annie Dosser-
      -unknown

Bright pink flowers shading to white in center, 13 per truss.
(D. Dosser, Australia, reg. 1982)

```
cl.            -griffithianum (1/4)
         -Loderi-
Annie Dring-      -fortunei ssp. fortunei (1/4)
         -Corona--unknown (1/2)
```

Flowers reddish purple with a basal blotch of purple, in trusses
of 14.  (W. V. Joslin cross; intro. by H. Dring, British Columbia; Mrs. Lillian Hodgson, reg. 1977)

```
                         -griffithianum (1/8)
cl.            -George Hardy-
     -Pink Pearl-      -catawbiense (1/8)
Annie E.-     -      -arboreum ssp. arboreum (1/8)
 Endtz -      -Broughtonii-
                     -unknown (5/8)
         -unknown
```

5ft(1.5m)        0°F(-18°C)        ML        3/3

Flowers medium pink  with frilled margins, in rounded trusses of
7-9.  Plant upright, spreading, vigorous; deep green leaves, 6in
(15.2cm) long.  Reasonably heat tolerant.  (L. J. Endtz, 1939)

```
cl.
     -griffithianum (1/2)
Anthaea-
     -unknown (1/2)
```

Also called Anthea.  (Mangles)

```
cl.
         -Scintillation--unknown (5/8)
Anthony Wayne*         -catawbiense (1/8)
         -Atrosanguineum-
     -Atrier-         -unknown
              -griersonianum (1/4)
```

(Cross by Charles Herbert)

```
cl.                -campylocarpum ssp. campylocarpum (3/4)
         -Lady Primrose-
Anthony Webb-         -unknown (1/4)
     -campylocarpum ssp. campylocarpum
```

Soft yellow.  (Thacker, 1938)

```
                         -catawbiense (1/16)
              -Atrosanguineum-
         -Atrier-         -unknown (9/16)
cl.    -Mary Belle-    -griersonianum (1/8)
              -decorum (1/8)
Antigua       Dcchaem-
     -              -haematodes ssp. haematodes (1/8)
     -Dexter's Apricot--unknown
```

5ft(1.5m)        -5°F(-21°C)        M

Flowers azalea pink shading to Naples yellow, 5-lobed, openly
funnel-shaped; base of throat blood red, with dorsal blotch and
some spotting on lobes.  Lax trusses hold 10 flowers 3.25in(8cm)
wide.  Plant as broad as tall. (J. Becales cross, 1970; C. Herbert intro.; reg. 1977)

```
cl.
              -catawbiense (1/2)
Anton--Everestianum (selfed)-
              -unknown (1/2)
```

Frilled flowers, lilac with ochre markings.  (Seidel, 1906)

```
*  *  *  *  *  *  *  *  *  *  *  *  *  *  *  *  *  *  *
*  Plant names followed by asterisks were not registered at *
*   press time, according to latest available information  *
*  *  *  *  *  *  *  *  *  *  *  *  *  *  *  *  *  *  *
```

```
              -fortunei ssp. discolor (3/8)
cl.   -Sir Frederick-
    - Moore  -         -arboreum ssp. zeylanicum (1/8)
Anton -         -St. Keverne-
Rupert-              -griffithianum (1/8)
         -elliottii (1/4)
     -Kilimanjaro-
              -Moser's Maroon--unknown (1/8)
         -Dusky-
         Maid-
         -fortunei ssp. discolor
```

Pink flowers, color deepening in the throat.  (Edmund de Rothschild, 1971)

```
g. & cl.         -arboreum ssp. arboreum (1/4)
     -Gill's Triumph-
Antonio-         -griffithianum (1/4)
     -fortunei ssp. discolor (1/2)
```

Synonym Exbury Antonio.  Suffused rose pink, becoming lighter
with age, crimson blotch and spotted within.  (Rothschild, 1933)
A. M. (RHS) 1939

```
cl.
```

Antonio Omega    Parentage as above  (Antonio g.)

Pale pink flowers.  (Rothschild, 1933)

```
                     -griffithianum (1/8)
cl.            -George-
         -Pink Pearl- Hardy-catawbiense (1/8)
Antoon van Welie-    -      -arboreum ssp. arboreum
         -              -Broughtonii-         (1/8)
         -unknown         -unknown (5/8)
```

6ft(1.8m)        -5°F(-21°C)        ML        4/3

Huge flowers, deep pink, in trusses exceptionally nice.  Large,
attractive, waxy leaves. (L. J. Endtz) See color illus. VV p. x.

```
                     -griffithianum (1/8)
              -Mars-
cl.            -      -unknown (5/16)    -catawbiense
         -Sammetglut-    -Parsons Grandiflorum-    (1/16)
Anuschka-         -Nova -         -unknown
         -         Zembla-red hybrid--unknown
         -Koichiro Wada--yakushimanum ssp. yakushimanum (1/2)
```

Trusses of 11-14 flowers, 5-lobed, spinel red, shading inwards
to neyron rose; calyx greenish pink.  Leaves  densely hairy.
(Cross by H. Hachmann; reg. by G. Stuck, 1983)

```
cl.
```

Aola--form of valentinianum

2.5ft(.75m)        15°F(-10°C)        E        3/2

Flowers 3in(7.6cm) wide, double and fimbriated, canary yellow to
Chinese yellow, in trusses of 3-5.  Plant bushy, upright; leaves
typical of the species.  (Benjamin Lancaster, 1962)

```
g.            -griffithianum (1/4)
     -Gill's Triumph-
Apache-         -arboreum ssp. arboreum (1/4)
     -thomsonii ssp. thomsonii (1/2)
```

Bright rose flowers.  Not in cultivation at Exbury.  (Rothschild,
1933)

```
g.
     -aperantum (1/2)
Apar-
     -arboreum ssp. arboreum (1/2)
```

Deep red flowers.  (Lord Aberconway cross, 1938; intro. 1946)

```
g.               -haematodes ssp. haematodes (1/4)
          -Choremia-
Aperemia-        -arboreum ssp. arboreum (1/4)
          -aperantum (1/2)
```

Deep red flowers.  (Lord Aberconway cross, 1937; intro. 1946)

```
g.
          -aperantum (1/2)
Aperme-
     -meddianum var. meddianum (1/2)
```

Red flowers.  (Lord Aberconway cross, 1937; intro. 1946)

```
cl.
   -edgeworthii (bullatum) (1/2)
Apis-
   -nuttallii (1/2)
```

Pure white with a canary yellow base.  (S. R. Clarke, 1936)

```
g.             -sperabile var. sperabile (1/4)
        -Eupheno-
Apodeno-      -griersonianum (1/4)
       -dichroanthum ssp. apodectum (1/2)
```

Vermilion flowers.  (Lord Aberconway cross, 1935; intro. 1946)

```
g.
        -dichroanthum ssp. apodectum (1/2)
Apodorum-
        -decorum (1/2)
```

A parent of Fire Flame by Scrase-Dickins.  Description and intro-
ducer unknown.  (1942)

```
cl.
        -dichroanthum ssp. apodectum (1/2)
Apotrophia-
        -facetum (eriogynum) (1/2)
```

A small shrub, leaves oblong-elliptical 2.75in(7cm) to 4.75in
(12cm) long, mucronate with thin fawn indumentum underneath.
Flowers fleshy, shrimp pink, tubular campanulate, marked within
on the lobes with darker tinted areas. (E. J. P. Magor cross;
reg. by Maj. E. W. M. Magor, 1963)

```
cl.   -Catalgla--cataubiense var. album Glass (1/2)
      -                    -unnamed-campylocarpum ssp. cam-
Applause-                  - hybrid-  pylocarpum (1/16)
      -   -Adriaan Koster-      -unknown (1/8)
      -          -           -            -griffithi-
   -unnamed-     -           -George- anum (1/32)
      hybrid-          -Mrs. Lindsay- Hardy-
      -              Smith       -       -cataubiense
   -williamsianum             -             (1/32)
         (1/4)                 -Duchess of
                                 Edinburgh--unknown
```

5ft(1.5m)        -20°F(-29°C)        EM        4/3

Flowers white, faint ivory shading, openly campanulate, 5-lobed,
about 2.5in((6.4cm) wide; globular trusses of 11.  Plant dense,
broader than tall; medium green leaves 2.25in(5.7cm) long.  See
also  Flair, Finlandia, Robin Leach.  (David G. Leach, 1972)

```
cl.
        -fortunei ssp. fortunei (1/2)
Apple Blossom-
              -campylocarpum ssp. campylocarpum
   -unnamed hybrid-                    (1/4)
              -unknown (1/4)
```

Parentage as in Slocock catalog.  Low-growing shrub with flowers
pink, cream and white.  (Slocock)

```
   *  *  *  *  *  *  *  *  *  *  *  *  *  *  *  *  *
   * Hybrids of unknown parentage that have won awards, or are  *
   * a parent of another hybrid, are in the text; for all oth-  *
   * ers of unknown parentage, see App. D.                      *
   *  *  *  *  *  *  *  *  *  *  *  *  *  *  *  *  *
```

```
cl.            -Catalgla--cataubiense var. album Glass (1/4)
          -unnamed-
Apple    - hybrid-Koichiro Wada--yakushimanum ssp. yakushimanum
Dumpling-                                              (1/4)
          -unnamed-Round Wood--lanigerum (1/4)
          -hybrid -
                   -cataubiense (red clone) (1/4)
```

2ft(.6m)          -10°F(-23°C)          ML

Pale lemon yellow flowers with greenish yellow throat, 7-lobed,
fragrant, in trusses of 7. Plant upright, well-branched. (Henry Yates cross, 1967; Mrs. Yates intro. 1976)

```
cl.            -neriiflorum ssp. neriiflorum Euchaites Group
          -Portia-                                (1/4)
Appleford-       -strigillosum (1/4)
          -barbatum (1/2)
```

Flowers of bright cardinal red open in a compact, dome-shaped
truss of 20.  Leaves are 3 times as long as wide, narrowly
oblong and slightly revolute.  (Crown Estate, Windsor)  A. M.
(RHS) 1966

```
cl.
            -cinnabarinum ssp. cinnabarinum Roylei Group (1/2)
Apricot Lady-        -cinnabarinum ssp. cinnabarinum (1/4)
 Chamberlain-Royal Flush  -
             (orange form)-maddenii ssp. maddenii (1/4)
```

5ft(1.5m)          10°F(-12°C)          ML          4/3

Lady Chamberlain g.  Flowers colored apricot. (Rothschild, 1930)

```
                          -dichroanthum ssp. dirchroanthum (5/16)
                  -unnamed-
          -unnamed- hybrid-
          - hybrid-       -neriiflorum ssp. neriiflorum (1/8)
cl.  -         -          -fortunei ssp. discolor (1/4)
     -        -unnamed-
Apricot-     hybrid-    -dichroanthum ssp. dichroanthum
 Nectar-           -Fabia-
     -                 -griersonianum (1/16)
     -        -Lady   -fortunei ssp. discolor
     -        -Bessborough-
     -Jalisco-       -campylocarpum ssp. campylocarpum
     -                           Elatum Group (1/8)
     -      -dichroanthum ssp. dichroanthum
     -Dido-
          -decorum (1/8)
```

4ft(1.2m)          -5°F(-21°C)          M

A ball-shaped truss of 14 flowers, cadmium orange with a scarlet
edge.  Compact plant with large leaves.  (Marshall Lyons, 1971)

```
                          -George-griffithianum (3/32)
                  -Mrs. Lindsay- Hardy-
                  - Smith      -cataubiense (1/32)
          -Diane-          -Duchess of Edinburgh--unknown
          -          -campylocarpum ssp. campylocarpum
   -unnamed-   -unnamed-                    (3/16)
cl.  - hybrid-   hybrid-unknown (1/8)
     -       -        -campylocarpum ssp. campylocarpum
Apricot-   -Citronella-   -griffithianum
Spinner-           -Kewense-
     -                  -fortunei ssp. fortunei (1/16)
     -      -wardii var. wardii (1/4)
   -unnamed-
      hybrid-   -dichroanthum ssp. dichroanthum (1/8)
          -Dido-
             -decorum  (1/8)
```

Yellow flowers tinged with orange.  (John Waterer, Sons & Crisp,
1975)

APRICOT #3 (see p 244)

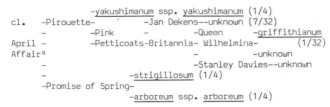

```
                    -yakushimanum ssp. yakushimanum (1/4)
cl.   -Pirouette-        -Jan Dekens--unknown (7/32)
  -       -Pink   -       -Queen   -griffithianum
April   -Petticoats-Britannia- Wilhelmina-          (1/32)
Affair*  -              -          -unknown
                              -Stanley Davies--unknown
  -               -strigillosum (1/4)
   -Promise of Spring-
                    -arboreum ssp. arboreum (1/4)
```

Flowers of hot pink; wavy margins. Early. (John G. Lofthouse, 1982)

```
cl.
          -minus var. minus Carolinianum Group (white form)
April Blush-                                       (1/2)
          -mucronulatum (pink form) (1/2)
```

Flowers are very abundant and early, blush pink fading to white. Plant is semi-deciduous. (G. Guy Nearing, 1968)

```
cl.
          -hippophaeoides var. hippophaeoides (1/2)
April Chimes-
          -mollicomum (1/2)
```

3ft(.9m)          0°F(-18°C)          EM          4/3

Leaves are sharply pointed and attractive. Flowers are rosy mauve. Excellent as a cut bloom for indoor decoration. (Intro. by Hillier & Sons, 1969)

```
cl.
        -Wilgen's Ruby--unknown (1/2)
April Glow-
        -williamsianum (1/2)
```

3ft(.9m)          -10°F(-23°C)          EM          4/3

A round, dense-growing plant with rosy pink flowers 3.5in(9cm) wide and 2.25in(6.4cm) long. Reddish new foliage. (Wilgen's Nursery, Holland, 1975) A. M. (Boskoop) 1965; Gold Medal (Boskoop) 1966; H. C. (RHS) 1975

cl.

April Showers      See April Glow

```
---
        -ponticum (1/2)
Aprilis-
        -dauricum (1/2)
```

One of the few hybrids resulting from a cross between an elepidote (non-scaly) and a lepidote (scaly). See also Crierdal, John Marchand. (Herbert, 1843)

```
g.
   -williamsianum (1/2)
Arab-
   -sperabile var. sperabile (1/2)
```

Low, compact plant with early-blooming pink flowers. (Rothschild, 1933)

```
cl.
   -konorii (1/2)
Aravir-               -Pink Delight--unknown (1/4)
   -Nancy Adler Miller-
                  -jasminiflorum (1/4)
```

Low          32°F(0°C)          Nov.-Mar.

Vireya hybrid. Flowers white, 2.25in(5.7cm) wide, tubular funnel-shaped, very fragrant, with 7-10 in a dome-shaped truss. At 8 years, rounded plant was 2ft(.6m) broad and tall. Leaves held 3-4 years. (Peter Sullivan cross, 1971; Wm. A. Moynier intro., reg. 1980)

```
g.
        -arboreum ssp. cinnamomeum var. album (1/2)
Arbad-
        -adenogynum (1/2)
```

Flowers white, boldly spotted crimson, pink outside. (E. J. P. Magor cross, 1917; intro. 1926)

```
g.
        -arboreum ssp. cinnamomeum var. album (1/2)
Arbcalo-
        -calophytum var. calophytum (1/2)
```

White, pink suffused bell flowers with a conspicuous claret blotch. (E. J. P. Magor, 1940)

```
g.
        -campylocarpum ssp. campylocarpum (1/2)
Arbcamp-
        -arboreum ssp. cinnamomeum var. album (1/2)
```

Flowers creamy white with faint crimson spots. (E. J. P. Magor cross, 1915; intro. 1928)

```
               -wardii var. wardii (1/4)
cl.   -Crest-
  -       -Lady     -fortunei ssp. discolor (1/8)
Arborfield-   Bessborough-
  -               -campylocarpum ssp. campylocarpum
  -     -griffithianum (1/4)       Elatum Group (1/8)
   -Loderi-
       Julie-fortunei ssp. fortunei (1/4)
```

Twelve mimosa yellow flowers on long pedicels form a lax truss. Corolla campanulate, 4.5in(11.5cm) across, darkening toward the throat. (Crown Estate, Windsor) A. M. (RHS) 1963

```
---
        -arboreum ssp. arboreum (pink form) (1/2)
Arbsutch-
        -sutchuenense (1/2)
```

Pink flowers. (E. J. P. Magor, 1915)

```
g.
        -minus var. minus Carolinianum Group (1/2)
Arbutifolium-
        -ferrugineum (1/2)
```

Clusters of about 12 small, bell-shaped flowers, crimson pink, slightly spotted. Very dense, dwarf plant flowering late. Foliage small, leathery, dark dull green, purplish in winter. Very hardy. (Origin unknown, before 1917)

cl.

Arctic Pearl--dauricum (white form)

2ft(.6m)          -15°F(-26°C)          E

Flowers pure white, 2in(5cm) wide, up to 5 per truss. Upright plant with small elliptic leaves. May be difficult in warm climates; needs excellent drainage. (Warren B. Baldsiefen, raised 1960; intro. 1970)

```
cl.
        -maximum (white selection) (1/2)
Arctic Snow-
        -brachycarpum ssp. brachycarpum (1/2)
```

4ft(1.2m)          -35°F(-37°C)          L

Light pink buds opening to white, with considerable lemon yellow spotting on dorsal lobe. Flowers of good substance, widely funnel-shaped, 2in(10cm) wide, fragrant, in dome-shaped trusses of 19. Plant upright, well-branched; leaves held 2 years. (Rudy Behring cross, 1973; reg. 1984)

cl.
```
              -glaucophyllum var. glaucophyllum (3/4)
Arden Belle-           -ciliatum (1/4)
          -Rosy Bell-
                   -glaucophyllum var. glaucophyllum
```

Rose pink flowers; smooth, mahogany-colored bark.  Parentage as in Leach; reversed in the Register.  (Thacker, 1942)


cl.
```
              -racemosum (1/2)
Arden Fairy-
              -lutescens (1/2)
```

Creamy pink flowers.  (Thacker, 1946)


cl.
```
              -wardii var. wardii
Arden Primrose-                      -griffithianum (1/8)
          -                -George Hardy-
              -Pink Pearl-           -catawbiense (1/8)
                   -                 -arboreum ssp. arboreum
                      -Broughtonii-                    (1/8)
                              --unknown (1/8)
```

Pale yellow flowers.  (Thacker, 1940)


g.
```
         -forrestii ssp. forrestii Repens Group (1/2)
Ardis-
         -arboreum var. wearii ? (1/2)
```

Bright red flowers.  (Lord Aberconway cross, 1936; intro. 1946)


cl.
```
                             -griffithianum (1/8)
         -Queen Wilhelmina-
      -Britannia-            -unknown (3/8)
Ardy-        -Stanley Davies--unknown
       -williamsianum (1/2)
```

4ft(1.2m)          -10°F(-23°C)          EM

Compact, globular-growing plant with broadly ovate leaves. Flowers rose pink and bell-shaped; 7-8 per truss.  (Experiment Station, Boskoop, 1968)


cl.
```
         -haematodes ssp. haematodes (1/2)
Arena-           -griersonianum (1/4)
       -Matador-
                 -strigillosum (1/4)
```

Red flowers.  (Cross by Francis Hanger, RHS Garden, Wisley; intro. 1953)


---
```
         -griffithianum (1/2)
Arethusa-
         -unknown (1/2)
```

Delicate pink flowers.  (Waterer, Sons & Crisp)


cl.
```
         -augustinii ssp. augustinii (1/2)
Argiolus-
         -concinnum (1/2)
```

Pale mauve flowers.  See also Danube.  (Lord Aberconway cross 1926; intro. 1946)


g.
```
       -fortunei ssp. discolor (1/2)
Argosy-
       -auriculatum (1/2)
```

6ft(1.8m)          -5°F(-21°C)          VL          3/3

Sweet scented white flowers  on an upright plant. Leaves long, narrow.  (Rothschild, 1933)


cl.
```
              -fortunei ssp. discolor (1/2)
Argosy Snow White-
              -auriculatum (1/2)
```

Argosy g.   White flowers.  (Rothschild cross; intro. by Waterer, Sons and Crisp) A. M. (RHS) 1938


cl.
```
         -griffithianum (1/2)
Ariadne-
         -Grand Duke of Wurtemberg--unknown (1/2)
```

Rose flowers, heavily spotted.  Plant of poor habit.  (G. B. van Nes)


cl.

Ariel--form of javanicum

Vireya.  Clear yellow flowers.   A. M. (RHS) 1893


g.
```
       -fortunei ssp. discolor (1/2)
Ariel-
       -Memoir--unknown (1/2)
```

White or pale pink flowers; tall, well-shaped plant. (Rothschild, 1933)


g. & cl.
```
       -thomsonii ssp. thomsonii 1/2)
Aries-
       -neriiflorum ssp. neriiflorum (1/2)
```

A plant of medium height with flowers of blood red, bell-shaped, in flat trusses.  (Sir John Ramsden, 1922)    A. M. (RHS) 1932; F. C. C. (RHS) 1938


cl.
```
                             -griffithianum (1/8)
          -George Hardy-
      -Pink Pearl-           -catawbiense (1/8)
Aristide-        -           -arboreum ssp. arboreum (1/8)
 Briand -     -Broughtonii-
          -                 --unknown (5/8)
          -unnamed hardy hybrid--unknown
```

Deep rose flowers in large trusses, on a plant of medium height. (L. J. Endtz & Co.)


cl.

Aristocrat     See Buttermint


```
              -sanguineum ssp. didymum (1/4)
         -Carmen-
         -         -forrestii ssp. forrestii Repens Group (1/4)
         -                      -campylocarpum ssp. campylo-
Arkle-              -unnamed-                   carpum (1/16)
      -          -Adriaan- hybrid--unknown (1/8)
      -          - Koster-                 -griffithianum (1/32)
      -          -        -          -George-
    -Moonshine-       -Mrs. Lindsay- Hardy-catawbiense (1/32)
      Supreme -         Smith       -
                   -          -Duchess of Edinburgh--
                   -                         unknown
              -wardii var. wardii Litiense Group (1/4)
```

Carmine red flowers.  (Hydon Nurseries, 1972)


cl.

Arlene Utz--Advanced generation hybrid including catawbiense, maximum and possibly unknown hybrids

6ft(1.8m)          -15°F(-26°C)          ML

Flowers 3in(7.6cm) wide, 7-lobed,  of deep pink with darker edging and throat, in trusses of 11.   (W. David Smith cross, 1968; reg. 1978)

g.
       -dichroanthum ssp. dichroanthum (1/2)
Arma-
    -forrestii ssp. forrestii Repens Group (1/2)

Flowers are orange scarlet. (Lord Aberconway, 1933)


g.
     -cinnabarinum ssp. xanthocodon Concatenans Group (1/2)
Armia-
    -lutescens (1/2)

Yellow flowers. (J. B. Stevenson, intro. 1947)


cl.
               -griffithianum (1/4)
     -unnamed hybrid-
Armistice Day-         -unknown (3/4)
     -Maxwell T. Masters--unknown

Scarlet red flowers in trusses of 10-14. (C. B. van Nes & Sons, 1930)


cl.
        -catawbiense (1/4)
   -Everestianum-
Arno*        -unknown (3/4)
   -unknown

Pale lilac flowers with red markings. (T. J. R. Seidel, 1906)


                       -griffithianum (9/16)
           -Beauty of-
         -Norman- Tremough-arboreum ssp. arboreum (1/16)
cl.     -Anna- Gill -griffithianum
    -      -         -griffithianum
Arnold Piper-   -Jean Marie -
    -     de Montague-unknown (3/8)
    -         -griffithianum
    -Marinus Koster-
           -unknown

2.5ft(.8m)       0°F(-18°C)       M

Synonym Pied Piper. Currant red buds; flowers rose red with a greyed purple blotch, widely funnel-shaped, 5in(12.7cm) across, in trusses of 15. Elliptic leaves held 3 years; brownish green new growth. (Halfdan Lem cross; raiser Arnold Piper; intro. by J. G. Lofthouse, 1971; reg. 1981)


---
             -auriculatum (1/2)
Aromana--Flameheart (selfed)-   -griersonianum (1/4)
              -Azor-
                   -fortunei ssp. discolor (1/4)

White flowers with intense, far-reaching fragrance; blooms late. (Michael Haworth-Booth, 1968)


cl.
             -Pygmalion--unknown
      -unnamed hybrid-
Aronimink-        -haematodes ssp. haematodes
   -            -fortunei ssp. fortunei
    -Wellfleet (Dexter #8)-
           -decorum

6ft(1.8m)      -5°F(-21°C)     M

The exact combination of the parentage is uncertain. Flowers of strong purplish red, openly funnel-shaped, 4in(10.2cm) wide, in spherical trusses of 20. Plant as broad as tall; leaves olive green, to 6in(15.2cm) long, held 2 years. (Dexter cross c.1940; Scott Horticultural Foundation, Swarthmore College, reg. 1980)

    * * * * * * * * * * * * * * * * * * * * *
    * Hybridizers, registrants, raisers, etc., with their *
    *      locations are listed in App. B      *
    * * * * * * * * * * * * * * * * * * * * *


cl.
        -Queen    -griffithianum (1/8)
        -Wilhelmina-
   -Britannia-      -unknown (3/8)
Art      -Stanley Davies--unknown
Wickens-   -dichroanthum ssp. dichroanthum (1/4)
    -Jasper-
        -Lady    -fortunei ssp. discolor (1/8)
       -Bessborough-
          -campylocarpum ssp. campylocarpum
                     Elatum Group (1/8)

Plant habit is flat and spreading. Turkey red flowers held in loose trusses. Blooms late. (C. S. Seabrook, reg. 1968)


---           -javanicum (7/16)
     -Ophelia-  -javanicum
Artemis-    -Princess -       -jasminiflorum (1/16)
  -      Alexandra-Princess Royal-
   -teysmanii (1/2)        -javanicum

Vireya hybrid with primrose yellow flowers. (Veitch, 1891)


---
        -arboreum ssp. cinnamomeum var. album (1/2)
Artemis-
    -griffithianum (1/2)

A hybrid with white flowers. (E. J. P. Magor, 1921)


cl.
      -williamsianum (1/2)
Arthur J. Ivens-
    -fortunei ssp. discolor Houlstonii Group (1/2)

4ft(1.2m)      -5°F(-21°C)     M     3/3

A dome-shaped plant with ovate, medium-sized leaves. Flowers are bell-shaped, a colorful Persian rose. Easy to propagate; slow to start flowering. Arthur J. Ivens was nursery manager to the Messrs. Hillier. (Hillier) A. M. (RHS) 1944 See color illus. F p. 32.


cl.
        -unknown (3/4)
Arthur John Holden-    -ponticum (1/4)
       -Purple Splendour-
          -unknown

2.5ft(.75m)      0°F(-18°C)     M

Flowers of good substance, red purple, held in ball-shaped trusses of 13-15. Upright plant, branching well, wider than tall, leaves held 3 years. (Paul Holden cross, 1968; raised and intro. by Mrs. A. J. Holden; reg. 1981)


cl.
      -sanguineum ssp. didymum (1/2)
Arthur Osborn-
    -griersonianum (1/2)

3ft(.9m)      5°F(-15°C)     VL     3/3

Dark red flowers and deep green foliage, indumented. Small shrub; requires some shade. (RBG, Kew, 1929) A. M. (RHS) 1933


cl.
    -maximum (1/2)
Arthur Pride-
    -catawbiense (1/2)

5ft(1.5m)      -30°F(-35°C)     L

Flowers 2in(5cm) across, white center with orchid edges and blotch of chartreuse spots in ball-shaped trusses of 29. A natural hybrid collected wild in North Carolina, 1934. (Orlando S. Pride, reg. 1979) See color illus. ARS Q 34:3 (Summer 1980): p. 138.

g.
                -wardii var. wardii (1/2)
Arthur Smith-
                -decorum (1/2)

Yellow  flowers.  See also Ightham Yellow.  (Lord Digby, 1951)


cl.
                -souliei (1/2)         -griffithianum (1/4)
Arthur Stevens-          -Pink Shell-
                -Coronation-              -unknown (1/8)
                  Day    -      -griffithianum
                          -Loderi-
                                  -fortunei ssp. fortunei (1/8)

Trusses full but not tight, of 7-10 flowers.   Pink buds open to
white with blotch of oxblood red at base of flower.  Dark green
leaves narrowly oblong, paler beneath.  (Crossed, raised, intro.
by Hillier; exhibited by Crown Estate, Windsor) A. M. (RHS) 1976


cl.
        -griersonianum (3/4)
Arthuria-          -griersonianum
        -Master Dick-           -arboreum ssp. arboreum
        -            -Doncaster-                    (1/16)
             -The Don-         -unknown (1/16)
                     -griffithianum (1/8)

Flowers rose with a darker throat, brown spotting on the 3 upper
lobes.  Trusses hold about 13 flowers.  A. M. (RHS) 1952


cl.                      -lacteum (1/8)
                  -Lacs-
     -unnamed hybrid-   -sinogrande (1/8)
Artist-               -unknown (1/4)
     -
     -irroratum ssp. irroratum (1/2)

Flowers white, at first touched with pink, faintly spotted, 3.25
in(8.3cm) across and 2.5in(6.4cm) long, in trusses of 11.  (Gen.
Harrison, 1964)


                      -campylocarpum ssp. campylocarpum
g.          -Penjerrick-            Elatum Group (1/8)
     -Amaura-          -griffithianum (1/8)
Artus-      -griersonianum (1/2)
     -      -dichroanthum ssp. dichroanthum (1/4)
     -Fabia-
           -griersonianum

Red flowers.  (Lord Aberconway cross, 1934; intro. 1941)


g.           -campylocarpum ssp. campylocarpum
     -Penjerrick-            Elatum Group (1/4)
Aruna-          -griffithianum (1/4)
     -wightii (1/2)

Yellow flowers.  (Lord Aberconway cross, 1926; intro. 1933)


cl.
                -thomsonii ssp. thomsonii (1/2)
Ascot Brilliant-
                -unknown (1/2)

Medium-sized plant  of lax habit.   Red or deep crimson flowers,
trumpet-shaped.  Very free flowering.  (Standish, 1861)


cl.

Ascreavie--form of maddenii ssp. maddenii (L & S 1141)

4ft(1.2m)         15°F(-10°C)          L          3/3

Flowers white with reverse flushed red purple in trusses of 3-5.
(A. E. Hardy, 1978)  A. M. (RHS) 1978


cl.

Ashcombe--form of veitchianum Cubittii Group

4ft(1.2m)          20°F(-7°C)          M-L          4/3

White flowers with large yellow orange blotch in throat, up to
4.5in(11.5cm) across, in loose flat trusses of about 5; scented.
May show pale pink tinge.  (Crown Estate Commissioners, Windsor,
named 1962)  F. C. C. (RHS) 1962


cl.
        -fortunei ? (1/2)
Ashes of Roses-
            -unknown (1/2)

Red flowers.  At times confused with Anna Rose Whitney.   (C. O.
Dexter cross; named by H. E. DuPont; intro. by Gladsgay Gardens;
reg. 1958)


Ashleyi--form of maximum

Considered an extreme mutant; found in North Carolina.


g. & cl.         -dichroanthum ssp. dichroanthum (1/4)
          -Astarte-             -campylocarpum ssp. campylocarpum
Aspansia-       -Penjerrick-           Elatum Group (1/8)
          -              -griffithianum (1/8)
          -haematodes ssp. haematodes (1/2)

Brilliant red flowers; dwarf plant with spreading habit. (Lord
Aberconway cross, 1932; intro. 1945)  A. M. (RHS) 1945


cl.

Aspansia Ruby       Parentage as above

Aspansia g.  Blood red flowers.  (Lord Aberconway, 1932)


                      -brookeanum var. gracile (1/4)
---    -Maiden's Blush-          -jasminiflorum (3/16)
       -            -Princess -            -jasminiflorum
Aspasia-          -Alexandra-Princess Royal-
       -                              -javanicum (1/16)
       -javanicum var. teysmannii (1/2)

Vireya hybrid.  Pure yellow flowers.  (Origin unknown)   A. M.
(RHS) 1889


cl.
        -calophytum var. calophytum (1/2)
Assaye-
        -sutchuenense (1/2)

Robin Hood g.  White flowers flushed with pale magenta, blotched
with ruby red; about 22 per truss, large and heavy. Leaves 10.5
in(26cm) long.  (J. C. Williams, 1963)  A. M. (RHS) 1963


g.          -dichroanthum ssp. dichroanthum (1/2)
     -Fabia-
Astabia-     -griersonianum (1/4)
     -          -dichroanthum ssp. dichroanthum
     -Astarte-             -campylocarpum ssp. campylocarpum
          -Penjerrick-           Elatum Group (1/8)
                  -griffithianum (1/8)

Red flowers.   Parentage as in Leach;  reversed in the Register.
(Lord Aberconway, 1942)


g.
     -forrestii ssp. forrestii Repens Group (1/2)
Asta-
     -haematodes ssp. chaetomallum (1/2)

Deep red flowers.  (Lord Aberconway, 1941)

cl.
       -dichroanthum ssp. dichroanthum (1/2)
Astarte-        -campylocarpum ssp. campylocarpum Elatum Group
   -Penjerrick-                     (1/4)
         -griffithianum (1/4)

Flowers are apricot or salmon pink, and pendulous.  (Lord Aber
conway cross, 1924; intro. 1931)  A. M. (RHS) 1931

cl.


Astarte    See Naomi Astarte


          -dichroanthum ssp. dichroanthum (1/4)
g.  -Astarte-      -campylocarpum ssp. campylocarpum
  -     -Penjerrick-       Elatum Group (1/8)
Astel-       -griffithianum (1/8)
  -    -forrestii ssp. forrestii Repens Group (1/4)
   -Elizabeth-
       -griersonianum (1/4)

Red flowers.  (Lord Aberconway cross, 1937; intro. 1946)

         -sperabile var. sperabile (1/4)
g.  -Eupheno-
  -    -griersonianum (1/4)
Asterno-  -dichroanthum ssp. dichroanthum (1/4)
    -Astarte-
  -     -campylocarpum ssp. campylocarpum
    -Penjerrick-      Elatum Group (1/8)
       -griffithianum (1/8)

Deep red flowers.  (Lord Aberconway cross, 1937; intro. 1946)

g.      -caucasicum (1/4)
   -Dr. Stocker-
Asteroid-    -griffithianum (1/4)
  -thomsonii ssp. thomsonii (1/2)

Flowers of rich, rosy pink.  See also Crimson Banner, Rosy Queen.
(Rothschild, 1933)

g.       -sanguineum ssp. didymum (1/4)
  -Arthur Osborn-
Astos-     -griersonianum (1/4)
  -    -dichroanthum ssp. dichroanthum (1/4)
   -Astarte-      -campylocarpum ssp. campylocarpum
     -Penjerrick-      Elatum Group (1/8)
       -griffithianum (1/8)

Flowers of brilliant scarlet.  (Lord Aberconway cross,1930;
intro. 1946)

g.         -catawbiense (1/4)
  -Catawbiense Grandiflorum-
Astraa-       -unknown (1/4)
  -fortunei ssp. discolor (1/2)

Large flowers, colored mallow to violet. (Dietrich Hobbie, 1948)

cl.
     -Koichiro Wada--yakushimanum ssp. yakushimanum (1/2)
Astroglow-
    -aureum var. aureum (1/2)

Dwarf     -25°F(-32°C)      E

Flowers openly funnel-shaped, Egyptian buff (very light yellow
orange) with greenish spotting; reverse is pink to pale pink.
Truss 4in(10cm) across, ball to dome-shaped with 10 flowers.
Leaves held 2-3 years with medium amount of grey brown indumentum.  Dwarf plant broad and rounded.  (B. C. Potter, reg. 1981)

g.      -dichroanthum ssp. dichroanthum (1/4)
  -Astarte-     -campylocarpum ssp. campylocarpum
Astrow-  -Penjerrick-      Elatum Group (1/8)
  -hookeri (1/2)  -griffithianum (1/8)

Dark red flowers.  (Lord Aberconway cross, 1936; intro. 1946)

---
      -catawbiense (1/2)
Athenia-
    -unknown (1/2)

Flowers blush white with yellow blotch.  (Origin unknown)  Com-
mendation, 1860.

cl.


Athens    See Last Hurrah

g.         -catawbiense (1/4)
  -Atrosanguineum-
Atkar*       -unknown (1/2)
  -       -brachycarpum ssp. brachycarpum (1/4)
  -Kentucky Cardinal-
      -Essex Scarlet--unknown

Clones variable; a few very good: AS-1, Best Atkar; AS-2, Atkar
Star.  Flowers probably red.  (Joseph Gable)

g.         -catawbiense (1/4)
  -Atrosanguineum-
Atrier-       -unknown (1/4)
  -griersonianum (1/2)

4ft(1.2m)      -10°F(-23°C)       M

The five named clones of this group are: Atrier Hardy, Atrier #7,
Atrier Oak, Red Head, William Montgomery.  All have red flowers,
and a rather open habit. (Joseph Gable, exhibited 1945)

g.         -catawbiense (1/4)
  -Atrosanguineum-
Atroflo-      -unknown (1/4)
  -floccigerum ssp. floccigerum (1/2)

5ft(1.5m)      -5°F(-21°C)     ML     4/3

More than one  form exists.  One form possibly has smirnowii as
the seed parent, and has long thin foliage with thick fawn
indumentum.  Rose-colored flowers of good substance.  (Joseph
Gable, 1940)  A. E. (ARS) 1960   See color illus. ARS J 36:1
(Winter 1982): p 11.

cl.         -catawbiense (1/4)
  -Atrosanguineum-
Atror-       -unknown (1/4)
  -     -orbiculare ssp. orbiculare (1/4)
  -unnamed hybrid-
      -williamsianum (1/4)

Parentage as in Leach; reversed in American Rhododendron Hybrids.  Flowers pinkish crimson fading to blush pink. (Joseph
B. Gable exhibited, 1945)

cl.
      -catawbiense (1/2)
Atrosanguineum-
    -unknown (1/2)

5ft(1.5m)      -20°F(-29°C)     ML     2/3

Dark red flowers, good foliage and habit.  (H. Waterer, 1851)

---         -catawbiense (1/4)
  -Atrosanguineum-
Atsonii-      -unknown (1/4)
  -thomsonii ssp. thomsonii (1/2)

Red flowers.  (Joseph Gable, exhibited 1937)


\* \* \* \* \* \* \* \* \* \* \* \* \* \* \* \* \* \* \* \* \*
\* Species used in hybrids of this volume are listed in App. A. \*
\* \* \* \* \* \* \* \* \* \* \* \* \* \* \* \* \* \* \* \* \*

cl.

Attar--form of <u>decorum</u> from the Hu Yu expedition

6ft(1.8m)              5°F(-15°C)              M          3/4

Neyron rose flowers shaded sap green in the throat, funnel-campanulate, 6 to 8 lobes, 4.5in(11.5cm) across.  (Raised by James
Barto; Rudolph Henny intro., 1962)


cl.
                -<u>ponticum</u> (1/2)
Attraction-
                -<u>unknown</u> (1/2)

Plant has spreading habit.  Mauve flowers.  Late.  (M. Koster &
Sons)


g.
         -<u>augustinii</u> ssp. <u>augustinii</u> (1/2)
Augfast-
         -<u>fastigiatum</u> (1/2)

2.5ft(.75m)           0°F(-18°C)           EM          4/4

Many  dark lavender blue  star-shaped flowers  in small trusses.
Plant dome-shaped, very dense;  creamy yellow new growth appears
soon after flowering.  See also Ightham.  (E. J. P. Magor, 1921)
See color illus. F p. 33.


cl.
                                -<u>catawbiense</u> (1/2)
August--Everestianum (selfed)-
                                -<u>unknown</u> (1/2)

Pink flowers, strong yellowish green markings.  (Seidel, 1906)


cl.
         -Marion--unknown (3/4)
August Moon*     -<u>campylocarpum</u> ssp. <u>campylocarpum</u> (1/4)
         -Unique-
                -unknown

Hybridized in Australia, where it blooms in August.  Large flowers of bright yellow; very early.  Medium-sized shrub with good
foliage.


cl.
                -<u>ponticum</u> (1/2)
Auguste van Geert-
                -<u>unknown</u> (1/2)

Purplish red flowers in large trusses.  Blooms early.  (Charles
van Geert, before 1867)


cl.
                -<u>ponticum</u> (1/2)
Aunt Martha-
                -<u>unknown</u> (1/2)

5ft(1.5m)            -10°F(-23°C)           ML          4/3

Flowers bright red purple, speckled gold in the center, held in
large trusses.   Foliage dense, leathery.  Vigorous growth when
young.  (Roy W. Clark, reg. 1958)


cl.                      -<u>griffithianum</u> (3/8)
             -Marinus Koster-
Aunty Thora-                -unknown (1/4)
     -             -<u>fortunei</u> ssp. <u>fortunei</u> (1/4)
         -Pilgrim-        -<u>arboreum</u> ssp. <u>arboreum</u> (1/8)
                -Gill's Triumph-
                             -<u>griffithianum</u>

Buds Tyrian rose; flowers glowing pink with red blotch, 5-lobed,
about 12 in tall trusses.  Plant broad, rounded; elliptic leaves
held 2 years; new stems red.  (Sigrid Laxdall, reg. 1979)     See
color illus. ARS Q 33:4 (Fall 1979): p. 239.


g.
    -<u>xanthostephanum</u> (1/2)
Auredge-
    -<u>edgeworthii</u> (1/2)

A hybrid by E. J. P. Magor, intro. 1938.


g.
      -<u>griersonianum</u> (1/2)
Auriel-           -<u>diaprepes</u> (1/4)
      -Polar Bear-
                  -<u>auriculatum</u> (1/4)

Pink flowers.  (Sir James Horlick, 1942)


___                     -<u>brookeanum</u> var. <u>gracile</u> (1/4)
         -Crown Princess-        -<u>jasminiflorum</u> (1/8)
Aurora- of Germany   -Princess Royal-
     -                        -<u>javanicum</u> (5/8)
         -<u>javanicum</u>

Vireya hybrid.  Yellow to red and orange flowers.  (Veitch, 1891)


g. & cl.     -<u>griffithianum</u> (1/4)
         -Kewense-
Aurora-      -<u>fortunei</u> ssp. <u>fortunei</u> (1/4)
         -<u>thomsonii</u> ssp. <u>thomsonii</u> (1/2)

6ft(1.8m)            0°F(-18°C)            M          4/4

Flowers of 8-10 in a broad lax truss, soft pink, sweetly
scented.  A parent of Naomi and many other hybrids.  (Cross by
Richard Gill of Tremough; named by Lionel de Rothschild)  A. M.
(RHS) 1922


                      -<u>dichroanthum</u> ssp. <u>dichroanthum</u> (3/8)
g.   -Fabia-
     -     -<u>griersonianum</u> (1/4)
Autumn-          -<u>dichroanthum</u> ssp. <u>dichroanthum</u>
     -     -Neda-
     -Clotted-   -Cunningham's Sulphur--<u>caucasicum</u> (1/8)
      Cream -<u>auriculatum</u> (1/4)

Orange flowers.  (Lord Aberconway cross, 1942; intro. 1950)


cl.
         -Searchlight--unknown
Autumn Beauty-
         -unknown

Rose-colored flowers.  (F. Lovegrove, Australia, 1972)


cl.
         -<u>fortunei</u> ssp. <u>discolor</u> (1/2)
Autumn Gold-    -<u>dichroanthum</u> ssp. <u>dichroanthum</u> (1/4)
         -Fabia-
                -<u>griersonianum</u> (1/4)

5ft(1.5m)            0°F(-18°C)            ML          4/3

Flowers apricot salmon, shaded pink,  with orange eye at center,
up to 10 in a truss.  Slightly twisted light green leaves.  Tolerates heat.  (Van Veen, Sr., 1956)


g. & cl.      -<u>griffithianum</u> (1/4)
         -Loderi-
Avalanche-    -<u>fortunei</u> ssp. <u>fortunei</u> (1/4)
         -<u>calophytum</u> var. <u>calophytum</u> (1/2)

6ft(1.8m)            0°F(-18°C)            E          5/4

Huge trusses of pure white flowers  with a small rosy blotch  at
the base.  Fragrant.  Leaves 8.5in(21.6cm) by 2.5in(6.4cm), held
one year.  See also Alpine Glow, Alpine Rose.  (Rothschild, 1933)
A. M. (RHS) 1934; F.C.C. (RHS) 1938  See color illus. PB pl. 18.

'April Glow' by Wilgen's Nursery
Photo by Greer

'Arctic Tern' by Cox
Photo by Greer

'Arnold Piper' by Lem/Piper/Lofthouse
Photo by Greer

'Aronimink' by Dexter/Wister
Photo by West

'Arthur Osborn' by RBG, Kew Gardens
Photo by Greer

'Ascot Brilliant' by Standish, 1861
Photo by Salley

'Augfast' by Magor
Photo by Greer

'Aunt Martha' by R. W. Clark
Photo by Greer

'Aurora' by Gill & Rothschild
Photo by Greer

'Avocet' by Rothschild
Photo by Greer

'Award' by James/Mossman/Ward
Photo by Greer

'Axel Olsen' by Hobbie
Photo by Greer

'Babylon' by Reuthe
Photo by Greer

'Bali' by Leach
Photo by Greer

'Ballad' by Leach
Photo by Leach

'Bambino' by Brockenbrough
Photo by Greer

'Bangkok' by Leach
Photo by Leach

'Barto Alpine' by Barto/Greer
Photo by Greer

'Barto Ivory' by Barto/Greer
Photo by Greer

'Barto Lavender' by Barto/Greer
Photo by Greer

'Beau Brummel' by Rothschild
Photo by Greer

'Beautiful Day' by Whitney
Photo by Greer

'Beefeater' by RHS, Wisley
Photo by Greer

'Belle Heller' by Shammarello
Photo by Greer

```
cl.         -griersonianum (1/2)
     -Tally Ho-
Avania-        -facetum (eriogynum) (1/4)
     -          -griersonianum
     -Rapture-
                -arboreum ssp. zeylanicum (1/4)
```

Geranium red flowers, 18 per truss.  (Gen. Harrison, 1955)

```
cl.
     -thomsonii ssp. thomsonii (1/2)
Avis-                -arboreum ssp. arboreum (1/4)
     -Glory of Penjerrick-
                          -griffithianum (1/4)
```

Barclayi g.  (Barclay Fox)

```
cl.
     -azalea occidentale (1/2)
Avita-
     -           -fortunei ssp. discolor (1/4)
     -Margaret-      -dichroanthum ssp. dichroathum (1/8)
      Dunn  -Fabia-
                     -griersonianum (1/8)
```

Azaleodendron.  Tangerine orange flowers.  (Lester Brandt, 1951)

```
g.
     -fortunei ssp. discolor (1/2)
Avocet-
     -fortunei ssp. fortunei (1/2)
```

Large, sweet-scented, white flowers in June.  (Rothschild, 1933)

```
cl.              -Pygmalion--unknown
     -unnamed hybrid-
Avondale-           -haematodes ssp. haematodes
     -                   -fortunei ssp. fortunei
     -Wellfleet (Dexter #8)-
                            -decorum
```

6ft(1.8m)        -5°F(-21°C)          M

Exact combination of above parentage is not known.  Flowers of
heavy substance, fragrant, widely funnel-campanulate, 3in(7.6cm)
across by 2in(5cm) long,  strong red with black spotting.  Ball-
shaped trusses, 5in(12.7cm) across, hold 14 flowers.  Plant upright, well-branched; leaves held 2 years.  (C. O. Dexter cross,
c.1940; John J. Tyler Arboretum, reg. 1980)

```
cl.
     -ciliatum (1/2)
Avril-
     -uniflorum var. imperator (1/2)
```

dwarf        5°F(-15°C)          E

Amaranth rose, deeper in bud; flowers in trusses of 2 or 3,
funnel-campanulate, up to 2in(5cm) across  by 1.5in(3.8cm) long.
(Brian O. Mulligan, 1965)

```
                                    -griffithianum (5/16)
                   -Beauty of Tremough-
          -Norman Gill-          -arboreum ssp. arboreum
cl. -Anna-        -griffithianum              (1/16)
    -    -              -griffithianum
Award- -Jean Marie de Montague-
    -                 -unknown (1/8)
    -            -fortunei ssp. discolor (1/4)
    -Margaret Dunn-   -dichroanthum ssp. dichroanthum (1/8)
                 -Fabia-
                        -griersonianum (1/8)
```

5ft(1.5m)        0°F(-18°C)          ML          4/3

Flowers white with light yellow flare in throat, margins shading
pink, 4.5in(11.5cm) wide, fragrant; in ball-shaped trusses of 12
to 15.  Plant upright, about as broad as high; new growth greenish copper or bronze.  (Cross by Del James; grown by F. Mossman;
intro. by C. Ward, 1973)    C. A. (ARS) 1974

```
            -Essex Scarlet--unknown (1/2)
Axel Olsen*
       -forrestii ssp. forrestii  Repens Group (1/2)
```

2ft(.6m)          -10°F(-23°C)          M          4/4

Possibly a sister seedling of Elizabeth Hobbie (parentage as
above).  Leaves very bullate, much like Scarlet Wonder.  Red
flowers.  (Dietrich Hobbie)

```
g.
     -fortunei ssp. discolor (1/2)
Ayah-
     -facetum (eriogynum) (1/2)
```

5ft(1.5m)          -10°F(-23°C)          L          3/3

Several forms are known; some are hardier and more upright than
others and all are late to flower.  Flower colors are pastel
pinks with light yellow in the throat.  (Rothschild, 1933)

```
g.
     -fortunei ssp. discolor (1/2)
Ayesha-
     -arboreum ssp. arboreum (1/2)
```

Tall, compact plant; flowers bright pink.  (Rothschild, 1933)

```
              -wardii var. wardii (1/4)
        -Crest-          -fortunei ssp. discolor (1/8)
        -    -Lady  -
cl. -      Bessborough-campylocarpum ssp. campylocarpum
    -                              Elatum Group (1/8)
Ayton-                  -griffithianum (1/32)
    -              -Kewense-
    -       -Aurora-     -fortunei ssp. fortunei (5/32)
    -    -Naomi-    -thomsonii ssp. thomsonii (1/16)
    -Golden-   -fortunei ssp. fortunei
    Dream-campylocarpum ssp. campylocarpum (1/4)
```

Trusses hold 12 flowers of light chartreuse yellow.  (Edmund de
Rothschild, 1982)

```
cl.
     -ponticum (1/2)
Azaleoides-
     -azalea periclymenoides (nudiflorum) (1/2)
```

Azaleodendron.  Flowers pale lilac with yellow center; fragrant.
(Thompson, c.1820, London)

```
g.
     -russatum (cantabile) (1/2)
Azamia-
     -augustinii ssp. augustinii (1/2)
```

Blue mauve flowers opening early.  (J. B. Stevenson cross, 1934;
intro. 1947)

```
g.
     -griersonianum (1/2)
Azma-
     -fortunei ssp. fortunei (1/2)
```

5ft(1.5m)          10°F(-12°C)          ML          3/2

Flowers soft salmon pink, up to 4in(10.2cm) across,  in rounded
lax trusses.  Plant upright, rounded; long, narrow leaves.  (J.
B. Stevenson cross, 1927; intro. 1933)

```
cl.       -griersonianum (1/4)
     -Azor-
Azonea-   -fortunei ssp. discolor (1/4)
     -Catanea--catawbiense var. album (1/2)
```

6ft(1.8m)          -15°F(-26°C)          L

Flowers neyron rose with a few pale spots, 3in(7.6cm) across, 12
in a ball-shaped truss 6in(15.2cm) wide.  Plant upright, branching moderately; elliptic leaves held 2 years.   (G. Guy Nearing,
1973)

g. & cl.
```
    -griersonianum (1/2)
Azor-
    -fortunei ssp. discolor (1/2)
```

6ft(1.8m)          5°F(-15°C)          L          3/3

Several forms exist, mostly salmon pink with ruddy brown
fleckings toward the base; loose trusses. Plant grows into a
large shrub or small tree. Some forms hold foliage less than one
year. See also C. F. Wood, Lily Dache. (J. B. Stevenson, 1927)
A. M. (RHS) 1933    See color illus. VV p. 74.

cl.

Azor Sister     Parentage same as Azor but a later cross

Flowers rhodamine pink, shading to red at the base, currant red
speckling on upper petal at throat. Trusses 7in(17.8cm) wide,
compact, dome-shaped, holding 9 flowers. Floriferous. (J. B.
Stevenson, reg. 1962)   F.C.C. (RHS) 1960

g.
```
    -griersonianum (1/2)
Azrie-
    -diaprepes (1/2)
```

Flowers soft salmon pink. (J. B. Stevenson cross, 1927; intro.
1933)

g.
```
    -arboreum ssp. arboreum (1/2)
Aztec-
    -irroratum ssp. irroratum (1/2)
```

Flowers pink with maroon spotting. (Rothschild cross, 1926; intro. 1933)

cl.
```
           -dichroanthum ssp. scyphocalyx (1/4)
     -Indiana-
Aztec-        -kyawii (1/4)
 Gold-              -wardii var. wardii ? (1/8)
    -    -Chlorops-
    -Inca-          -vernicosum ? (1/8)
     Gold-
         -unknown (1/4)
```

4ft(1.2m)                        EM

Plant has glossy green leaves and coppery new growth. Rounded
trusses of about 18 bell-shaped flowers, of deep, clear primrose
yellow. (Benjamin F. Lancaster, 1965)

cl.
```
    -russatum (1/2)
Azurika-
    -impeditum (1/2)
```

Flowers held in trusses of 7-8, 5-lobed, violet color. (Cross
by H. Hachmann; reg. by G. Stuck, 1983)

cl.
```
        -russatum (1/2)
Azurwolke-                    -intricatum (1/8)
     -          -Intrifast-
     -Blue Diamond-          -fastigiatum (1/8)
                  -augustinii ssp. augustinii (1/4)
```

Flowers held in trusses of 6-7, 5-lobed, medium violet. Leaves
scaly. See also Gletschernacht. (Cross by H. Hachmann; reg. by
G. Stuck, 1983)

```
    * * * * * * * * * * * * * * * * * * * *
    * Text includes some selected named forms of species.  See *
    *                   others in App. C.                       *
    * * * * * * * * * * * * * * * * * * * *
```

g.
```
                        -arboreum ssp. arboreum (1/8)
              -Doncaster-
        -The Don-        -unknown (1/8)
B. B. C.-    -griffithianum (1/4)
        -neriiflorum ssp. neriiflorum Euchaites Group (1/2)
```

Red flowers. Foliage similar to the species parent. No longer
in cultivation at Exbury. (Rothschild, 1934)

cl.
```
           -cataubiense (1/2)
B. de Bruin or-
Bas de Bruin  -unknown (1/2)
```

Flowers of dark rich scarlet, spotted black; about 20 flowers in
conical trusses. Midseason. Straggly habit. (A. Waterer)

---
```
        -griffithianum (1/2)
B. Griffith-
        -unknown (1/2)
```

(Mangles)

cl.
```
    -haematodes ssp. haematodes (1/2)
Baba-          -griersonianum (1/4)
   -Sarita Loder-    -griffithianum (1/8)
           -Loderi-
                -fortunei ssp. fortunei (1/8)
```

Deep red, waxy flowers, funnel-shaped, 3in(7.6cm) across, held
in loose trusses. (Sir John Ramsden cross; intro. by William
Pennington-Ramsden, 1964)

cl.
```
            -souliei (1/4)
     -Rosy Morn-    -griffithianum (1/8)
Baby Bonnet-   -Loderi-
    -              -fortunei ssp. fortunei (1/8)
    -    -dichroanthum ssp. dichroanthum (1/4)
     -Dido-
         -decorum (1/4)
```

Flowers peach color, lighter at center, with dark fuchsia pink
margins, openly campanulate, 8-lobed, 12 per truss. Plant habit medium to upright; leaves to 4in(10.2cm) long. Blooms late
midseason. (Rudolph & Leona Henny, 1965)

cl.
```
    -yakushimanum ssp. yakushimanum (1/2)
Baby Doll-                -caucasicum (1/8)
    -      -Cunningham's White-
    -Holden-                -ponticum (white form) (1/8)
          -cataubiense (1/4)
```

2ft(.6m)          -5°F(-21°C)          EM

Cherry pink buds opening to near-white flowers; about 15 held in
a ball-shaped truss. Dark olive green leaves with thin, tan indumentum beneath. (Raised and intro. by Elmer Morris from a Benjamin Lancaster seedling, 1973)

cl.

Baby Mouse--form of campylocarpum ssp. campylocarpum

Flowers deep plum purple. (C. Ingram, 1973) A. M. (RHS) 1973

cl.
```
    -calophytum var. calophytum (1/2)
Babylon-
    -praevernum (1/2)
```

6ft(1.8m)          -5°F(-21°C)          E          5/5

Plant habit is a dense, rounded mound of strong, rigid red stems
and large lustrous leaves. Huge satiny white flowers with chocolate blotch, in early spring. One of the best hybrids of calophytum. (Reuthe, 1955)

cl.
          -unnamed hybrid--unknown (7/8)
Baccarat-                                     -caucasicum (1/8)
        -               -Chevalier F. de Savage-
          -Max Sye-                          -unnamed hybrid--unknown
                  -unknown

Flowers crimson with large black blotch, in round trusses of 14-
15. See also Boskoop. (Adr. van Nes; reg. 1962) A. M. (after
Boskoop trial) 1960

cl.

Bacchus     Parentage unknown

A parent of Emeline Buckley. Crimson flowers in a large truss.
(A. Waterer, before 1915)

                                    -griffithianum (1/8)
cl.          -Unknown-Queen Wilhelmina-
        -Warrior-                   -unknown (3/8)
Bacher's Gold-      -Stanley Davies--unknown
        -        -dichroanthum ssp. dichroanthum (1/4)
          -Fabia-
                -griersonianum (1/4)

4ft(1.2m)        5°F(-15°C)          ML          4/4

Edges of lobes are pink, shading to yellow in the center;
upright truss. (John Bacher) P. A. (ARS) 1955

cl.
          -Essex Scarlet--unknown (1/2)
Bad Eilsen-
          -forrestii ssp. forrestii Repens Group (1/2)

2ft(.6m)        -10°F(-23°C)          M          4/4

Vigorous, free-flowering plant of upright and compact habit.
Small leaves twice as long as broad, heavily veined. Cardinal
red flowers with faint spotting, in lax spherical trusses. Dis-
likes heat. (Hobbie, 1969)   A. M. (Wisley Trials) 1969; H. C.
(Wisley Trials) 1969

cl.
          -Essex Scarlet--unknown (1/2)
Baden Baden*
          -forrestii ssp. forrestii Repens Group (1/2)

May be from the same cross as Elisabeth Hobbie, Bad Eilsen. Vigorous, compact, upright habit, over twice as wide as high. Foliage dark glossy green.  Flowers of deep cherry red, in trusses
6in(15.2cm) across. Very floriferous.  (Hobbie)   H. C. (RHS)
1972

cl.              -griffithianum (1/4)
          -Mars-
Bagoly's Beauty-   -unknown (1/4)
          -Koichiro Wada--yakushimanum ssp. yakushimanum
                                                    (1/2)

3ft(.9m)        -10°F(-23°C)          M

Buds rose red open to funnel-shaped flowers, roseine purple,
reverse darker. Bell-shaped trusses, 5in(12.7cm) across, hold
16 flowers. Plant broader than high, well-branched with leaves
held 3 years, bright yellow green. (Lewis Bagoly cross 1966;
raised by Marie Tietjens; reg. 1984)

cl.
          -thomsonii ssp. thomsonii (1/2)
Bagshot Ruby-
          -unknown (1/2)

The pollen parent may be John Waterer, (catawbiense x). Large,
conical trusses of ruby red flowers. Foliage dark, dull green.
(J. Waterer, c.1900) A. M. (RHS) 1916

cl.

Bagshot Sands--form of lutescens

6ft(1.8m)        5°F(-15°C)          EM          3/3

A slender-growing shrub with lanceolate leaves. Funnel-shaped
trusses of primrose yellow flowers. (Mrs. R. M. Stevenson)
A. M. (RHS) 1953

cl.
          -catawbiense var. album (1/2)
Bali-              -neriiflorum ssp. neriiflorum (1/8)
        -unnamed-
     -unnamed- hybrid-dichroanthum ssp. dichroanthum (1/8)
        hybrid-
              -fortunei ssp. discolor (1/4)

6ft(1.8m)        -15°F(-26°C)          ML          4/4

Wide trusses hold 14-17 flowers colored neyron rose with yellow
throat. Plant rounded, spreading 1.5 times wider than tall
with yellow green leaves held 2 years. See also Nuance. (David
G. Leach, intro. 1974; reg. 1983)  See color illus. HG p. 78.

cl.
          -Dexter hybrid--unknown (7/8)
Ballad-                        -catawbiense (1/8)
        -        -Parsons Grandiflorum-
     -America-                -unknown
              -dark red hybrid--unknown

3ft(.9m)        -15°F(-26°C)          ML          3/3

Flowers very pale purple with edging of purplish pink, bold red
blotch, darker on reverse; full pyramidal trusses hold 15. Foliage density average; narrowly obovate leaves, twice as long as
wide. (David G. Leach, 1972)

cl.          -griffithianum (1/4)
        -Alice-
Ballerina-    -unknown (1/4)
        -fortunei ssp. discolor (1/2)

White flowers with yellow flare on upper petals, and 14 flowers
per truss. (Waterer, Sons, & Crisp)

cl.
          -Koichiro Wada--yakushimanum ssp.yakushimanum (1/2)
Bally Cotton-
          -aureum var. aureum (1/2)

Dwarf          -5°F(-21°C)          E

Dwarf plant, branching moderately; leaves held 3 years with
light tan indumentum underneath. Flowers 2in(5cm) wide, white
with yellowish green dorsal blotch and spots, up to 7 per ball-
shaped truss. (W. A. Reese, reg. 1979)

cl.
          -minus var. minus Carolinianum Group (3/4)
Balta*          -minus var. minus Carolinianum Group
        -P. J. M. -
              -dauricum (1/4)

Light pink to white flowers, one week later than P. J. M.  Very
floriferous; slow-growing--3in to 5in(7.6cm to 12.7cm) per year.
Broad, upright plant; small, lustrous, convex, dark green foliage. (Edmund V. Mezitt, Weston Nurseries)

cl.
          -yakushimanum ssp. yakushimanum (1/2)
Bambi*          -dichroanthum ssp. dichroanthum (1/4)
        -Fabia Tangerine-
              -griersonianum (1/4)

3ft(.9m)        5°F(-15°C)          M

Plant of compact habit; flowers red in bud, opening to soft pink
with a yellow tinge. (Waterer, Sons & Crisp, 1964?)

cl.
          -keiskei (1/2)
Banana Boat*
          -Arctic Pearl--dauricum (white form) (1/2)

Plant 2ft(.6m) high by 2.5ft(.75m) wide in 10 years.  Blooms
early like keiskei, but flowers more open with much yellow.  Aromatic foliage stays a good dark green in winter.  (Dr. G. David
Lewis)

cl.
          -catawbiense var. album (1/2)
Bangkok-            -dichroanthum ssp. dichroanthum (1/4)
     -unnamed hybrid-            -griffithianum (1/8)
                    -unnamed hybrid-
                              -auriculatum 1/8)

4ft(1.2m)        -20°F(-29°C)         M         4/4

Flowers open funnel-shaped, of heavy substance, 5 rounded lobes,
light to strong mauve pink  shading in the center to pale orange
yellow; blotch and spotting of dark reddish orange.  Lax trusses
hold about 13 flowers.  Rounded elliptic leaves, dark green, 4in
(10.2cm) long, held 2 years.  (David G. Leach, 1972)   See color
illus. ARS J 36:1 (Winter 1982): p. 6.

g.
          -auriculatum (1/2)
Banshee-            -arboreum ssp. arboreum (1/4)
     -John Tremayne-
                    -griffithianum (1/4)

A tall, well-branched shrub with flowers in late season, blush
pink.  (Rothschild, 1934)

g.
          -campylocarpum ssp. campylocarpum Elatum Group (1/2)
Barbara-       -griffithianum (1/4)
     -Loderi-
          -fortunei ssp. fortunei (1/4)

Loose trusses of deep cream flowers with a soft pink flush.  See
also Pinafore.  (Lionel de Rothschild cross, 1934; intro. by Edmund de Rothschild, 1948)

                    -souliei ? (1/4)
cl.           -Virginia-
          - Scott  -unknown (3/8)
Barbara Houston-            -arboreum ssp. arboreum (1/8)
          -      -Doncaster-
          -Belvedere-            -unknown
                    -dichroanthum ssp. dichroanthum (1/4)

4ft(1.2m)        10°F(-12°C)         L

Flowers 3in(7.6cm) wide, orange buff, large red spots on dorsal
lobes, margins pink; up to 9 per truss.  (H. L. Larson, 1979)

cl.
          -maddenii ssp. maddenii (calophyllum) (1/2)
Barbara Jury-       -maddenii ssp. crassum (1/4)
          -Sirius-
               -cinnabarinum ssp. cinnabarinum Roylei Group
                                                    (1/4)
Trusses of 5-6 flowers, 5-lobed, Naples yellow deepening on
outside of tube and flushed pink, orange  yellow within.  Leaves
have brown indumentum.  See also Christine Denz, Felicity Fair.
(F. M. Jury, 1982)

cl.
               Complex parentage including catawbiense
Barbara Tanger   and maximum and possibly unknown hybrids

10ft(3m)        -15°F(-26°C)         M

Leaves twice as long as wide.  Flowers 3.5in(8.8cm) wide, 17 per
truss; buds strong reddish purple, flowers medium purplish red
with deep purplish red dorsal spotting.  (W. David Smith, 1979)

cl.                    -griffithianum (1/4)
          -Queen Wilhelmina-
Barbara Wallace-            -unknown (3/4)
          -Helen Waterer--unknown

Flowers are reddish pink with a white center.  (C. B. van Nes &
Son, before 1958)

cl.
          -barbatum (1/2)
Barbarossa-
     -unknown (1/2)

Early-blooming, bright red flowers, similar to barbatum.  (Sir
James Horlick, before 1958)

g.
          -barbatum (1/2)
Barbsutch-
          -sutchuenense (1/2)

Deep rose pink flowers with a crimson blotch.  (E. J. P. Magor,
intro. 1930)

g.
          -thomsonii ssp. thomsonii (1/2)
Barclayi-
     -                    -arboreum ssp. arboreum (1/4)
          -Glory of Penjerrick-
                         -griffithianum (1/4)

Flowers are intense crimson.  Very early.  See also Romala.
(Barclay Fox, early this century.)

cl.

Barclayi Robert Fox      Parentage as above

5ft(1.5m)        15°F(-10°C)         E         4/3

Rounded trusses of deep blood red flowers, waxy texture.  Habit
upright, open; leaves flat, oval, round at ends with red streak
in upper side of midrib.  (Barclay Fox) A. M. (RHS) 1921

* * * * * * * * * * * * * * * * * * * * * * * * * * * *
*                                                     *
* Descriptive color names compiled from horticultural color *
* charts are listed in Appendix F, p. 389.             *
*                                                     *
* * * * * * * * * * * * * * * * * * * * * * * * * * * *

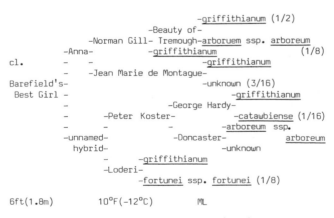

cl.

6ft(1.8m)        10°F(-12°C)         ML

Fragrant flowers of heavy substance, 4.5in(11.cm) wide, red with
purple stain at base of throat, 15 in trusses 7.5in(19cm) across
by 8.5in(21.5cm) high. Leaves held 4-5 years; plant upright and
branching moderately.  (Grady Barefield cross; Mary Barefield,
intro. 1978)

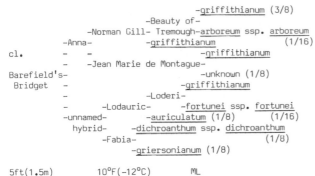

```
                            -griffithianum (3/8)
                    -Beauty of-
            -Norman Gill- Tremough-arboreum ssp. arboreum
    -Anna-            -griffithianum                (1/16)
cl.    -     -                   -griffithianum
    -   -Jean Marie de Montague-
Barefield's-                       -unknown (1/8)
  Bridget -                  -griffithianum
           -            -Loderi-
            -Lodauric-      -fortunei ssp. fortunei
  -unnamed-      -auriculatum (1/8)      (1/16)
    hybrid-      -dichroanthum ssp. dichroanthum
          -Fabia-                        (1/8)
              -griersonianum (1/8)
```

5ft(1.5m)          10°F(-12°C)        ML

Buds cardinal red; flowers Venetian pink, fragrant, tubular funnel shaped, of heavy substance, 4in(10.2cm) across; large, dome-
shaped trusses hold 14-16 flowers.  Floriferous.  Plant broad,
well-branched;  leaves 7in(17.8cm) by 2in(5cm),  held 2-3 years.
(Mary W. Barefield, intro. 1978)

```
                    -griffithianum (3/16)
                -Mars-
cl.       -Vulcan-    -unknown (1/8)
          -       -griersonianum (1/4)
Barefield's -                      -griffithianum
Granddaughter-          -George Hardy-
     -   -Mrs. Lindsay-          -catawbiense (1/16)
    -Dot- Smith   -Duchess of Edinburgh--unknown
      -fortunei ssp. fortunei (1/4)        (1/8)
```

5ft(1.5m)          10°F(-12°C)        M

Buds Tyrian purple, flowers rhodamine pink  shading to rose red
margins, openly funnel-shaped, 4in(10.2cm) across;  14-21 flowers
in trusses 8in(20.3cm) tall.  Plant broad, well-branched; leaves
held 3 years.  (Mary W. Barefield, intro. 1975)

```
                        -griffithianum (3/8)
                    -Beauty of-
            -Norman Gill- Tremough-arboreum ssp. arboreum
cl.    -Anna-           -griffithianum                (1/16)
    -       -                       -griffithianum
Barefield's-   -Jean Marie de Montague-
  Liza   -                       -unknown (3/8)
          -        -Tudelu (Lem)--unknown
    -unnamed-           -lacteum (1/8)
      hybrid-unnamed hybrid-       -griffithianum
                      -Loderi-
                       -fortunei ssp.
                           fortunei (1/16)
```

6ft(1.8m)          10°F(-12°C)        M

Buds of neyron rose opening to flowers of paler neyron rose, red
at margins with ruby red stain in throat, of good substance, and
5.5in(14cm) across.  Trusses 8.5in(21.5cm) wide hold 13-16.  Upright, well-branched plant with bronze new growth; leaves held 3
years.  (Mary W. Barefield, intro. 1978)

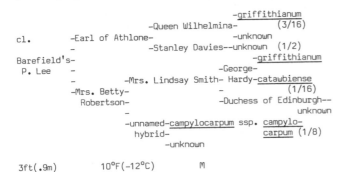

```
                        -griffithianum
                -Queen Wilhelmina-    (3/16)
cl.      -Earl of Athlone-       -unknown
            -Stanley Davies--unknown (1/2)
Barefield's-                      -griffithianum
  P. Lee  -               -George-
    -        -Mrs. Lindsay Smith- Hardy-catawbiense
  -Mrs. Betty-                       (1/16)
    Robertson-              -Duchess of Edinburgh--
    -                             unknown
        -unnamed-campylocarpum ssp. campylo-
          hybrid-                 carpum (1/8)
              -unknown
```

3ft(.9m)          10°F(-12°C)        M

Rose red buds opening to flowers of neyron rose, tubular funnel-
shaped, 3in(7.6cm) across, 14 held in high, ball-shaped trusses.
Floriferous.   Broad, well-branched plant;  leaves held 2 years.
(Mary W. Barefield, 1978)

```
                        -griffithianum (9/32)
                -Mars-
        -Vulcan-   -unknown (3/16)
cl.     -      -griersonianum (1/4)
        -                      -griffithianum
Barefield's -          -Beauty of-
Virginia Anne-    -Norman- Tremough-arboreum ssp.
            -Anna- Gill -        arboreum (1/32)
        -       -        -griffithianum
   -unnamed-    -                      -griffithianum
    - hybrid-   -Jean Marie de Montague-
                -                -unknown
                -   -campylocarpum ssp. campylo-
                -Ole Olson-      carpum Elatum Group (1/8)
                -fortunei ssp. discolor (1/8)
```

4ft(1.2m)          5°F(-15°C)         M

Flowers 3in(7.6cm) across, of dark pink deepening in throat, exterior with darker stripes, up to 12 per truss.  Free-flowering.
Plant broad, well-branched;  leaves 8in(20.3cm) by 2.5in(6.4cm)
held 3 years.  (Mary W. Barefield, intro. and reg. 1972)

```
                    -griffithianum (1/8)
                -Mars-
cl.     -Sammetglut-    -unknown (5/16)
        -                         -catawbiense
Barmstedt-      -Nova -Parsons Grandiflorum-       (1/16)
        -      Zembla-            -unknown
        -            -dark red hybrid--unknown
    -Koichiro Wada--yakushimanum ssp. yakushimanum (1/2)
```

Trusses hold 18-22 flowers, 5-lobed, bright neyron rose shading
lighter inwards, marked with bright neyron rose.  Leaves strongly revolute with white indumentum.  (Cross by H. Hachmann; reg.
G. Stuck, 1983)  See col. ill. ARS J 39:2 (Summer 1885): p. 63.

```
cl.
        -yakushimanum ssp. yakushimanum (Exbury form) (1/2)
Baron Lionel-
        -yakushimanum ssp. yakushimanum (1/2)
```

Not a hybrid, but a selected form of the species.  Trusses of
14-17 flowers colored light orchid pink opening to clear white
with speckles on upper lobe of absinthe green.  (Cross by Lionel
de Rothschild; reg. by Edmund de Rothschild, 1982)

```
                        -griffithianum (1/16)
                    -Kewense-
g. & cl.       -Aurora-   -fortunei ssp. fortunei
        -Exbury-     -            (5/16)
Baron Phillipe- Naomi-    -thomsonii ssp. thomsonii (1/8)
  de Rothschild-  -fortunei ssp. fortunei
        -      -wardii var. wardii (1/4)
        -Crest-            -fortunei ssp. discolor
            -Lady Bessborough-            (1/8)
                    -campylocarpum ssp. campy-
                      locarpum Elatum Gp. (1/8)
```

Upright, compact habit like Naomi.  Bears well-formed trusses of
large, pale yellow flowers in profusion.  At Exbury, withstands
cold in an exposed location, blossoms in May.  (Edmund de Rothschild, 1962)

```
cl.
            -javanicum (3/4)
Baroness Henry Schroder-     -jasminiflorum (1/4)
            -Princess Royal-
                    -javanicum
```

Vireya hybrid.   White, finely spotted.   (J. Waterer)  F. C. C.
(RHS) 1883   According to Bean, another javanicum hybrid of the
same name, by Veitch, was the plant that received the award.


\* \* \* \* \* \* \* \* \* \* \* \* \* \* \* \* \* \* \* \* \*
\* Hybrids of unknown parentage are described briefly in App. D \*
\* \* \* \* \* \* \* \* \* \* \* \* \* \* \* \* \* \* \* \* \*

cl.
    -catawbiense (1/2)
Barryb-                          and perhaps other hybrids
    -maximum (1/2)

6ft(1.8m)          -10°F(-23°C)          ML

Flowers 2in(5cm) wide, reddish purple with yellow green dorsal
blotch, up to 14 in a truss. Floriferous. Plant upright, well-
branched; leaves held 2 years.  (W. David Smith, 1978)

g.                -thomsonii ssp. thomsonii (1/4)
& cl. -Barclayi-
-           -Glory of  -arboreum ssp. arboreum (1/8)
Bartia-            Penjerrick-
-                      -griffithianum (1/8)
-        -neriiflorum ssp. neriiflorum (1/4)
    -Portia-
       -strigillosum (1/4)

Turkey red flowers in trusses of 12.  (Lord Aberconway, 1936)
A. M. (RHS) 1948

cl.

Barto Blue--form of augustinii ssp. augustinii

6ft(1.8m)          -5°F(-21°C)          EM          4/4

An excellent selection of this species.  It is a clean, clear,
beautiful blue. (Raised by Barto; intro. by Phetteplace, 1958)

cl.
    -fortunei ssp. fortunei (1/2)
Barto Ivory-
    -unknown (1/2)

6ft(1.8m)          -10°F(-23°C)          EM          4/3

Much like fortunei, with rich ivory flowers; fragrant. Thick-
textured, heavy foliage. (James Barto cross; Joseph Steinmetz
raiser; intro. by Harold E.Greer, 1962)

cl.

Barto Rose--form of oreodoxa var. fargesii

5ft(1.5m)          -10°F(-23°C)          E          4/3

Flowers rose madder with deep purple spotting in upper corolla,
campanulate, 7-lobed, 2.25in(5.7cm) wide; trusses hold about 14.
Pyramidal plant, 3/4 as wide as tall; elliptic leaves. (Raiser
J. Barto; named by Dr. C. H. Phetteplace; reg. 1963)

cl.
    -yakushimanum ssp. yakushimanum (1/2)
Bashful-       -arboreum ssp. arboreum (1/4)
    -Doncaster-
       -unknown (1/4)

Flowers camellia rose with deeper shading, reddish brown blotch.
Compact growth habit. (John Waterer, Sons & Crisp, 1971)

cl.
    -Albescens--unknown (1/2)
Basileos-
    -ciliicalyx (1/2)

4ft(1.2m)          25°F(-4°C)          M

Flowers white with pink flush, deep pink stripe on back of petals.  Richly fragrant. (Intro. by Strybing Arboretum, 1970)

cl.          -Moser's Maroon--unknown (1/4)
    -Bibiani-
Bastion-       -arboreum ssp. arboreum (1/4)
    -elliottii (1/2)

Gibralter g.  Flowers dark red with the black spotting of Bibi-
ani, in tight, upright trusses.  Growth habit like Gibralter.
(Rothschild, 1961) P. C. (RHS) 1961  See color illus. PB pl. 50.

g.                      -griffithianum (1/4)
    -Gill's Goliath-
Battle Axe-          -unknown (1/4)
    -fortunei ssp. discolor (1/2)

Rich pink flowers.  (Rothschild, 1934)

g.                      -griffithianum (1/4)
    -Dawn's Delight-
Bauble-             -unknown (1/4)
    -campylocarpum ssp. campylocarpum (1/2)

Flowers are straw yellow, spotted, campanulate.  A dense plant
with small, rounded, glossy leaves. (Rothschild, 1934)

g.          -dichroanthum ssp. dichroanthum (1/4)
    -Fabia-
Beacon-     -griersonianum (1/2)
-                 -sanguinem ssp. didymum (1/4)
    -Arthur Osborn-
       -griersonianum

See also Fabos.  Brilliant scarlet flowers.  (Lord Aberconway
cross, 1935; intro. 1942)

                   -dichroanthum ssp. dichroanthum (1/8)
g.          -Astarte-          -campylocarpum ssp. campylo-
-Cuida-       -Penjerrick-    carpum Elatum Group (1/16)
Beada-       -                   -griffithianum (1/16)
-        -griersonianum (1/4)
    -beanianum (1/2)

Pink flowers.  (Lord Aberconway cross, 1939; intro. 1942)

g.          -haematodes ssp. haematodes (1/4)
    -Choremia-
Beatrice-       -arboreum ssp. arboreum (1/4)
    -meddianum var. meddianum (1/2)

Pink flowers.  (Lord Aberconway cross, 1937; intro. 1950)

cl.
    -lacteum (1/2)
Beatrice-                -caucasicum (1/8)
  Keir   -      -Dr. Stocker-
    -Logan  -          -griffithianum (1/8)
    Damaris-campylocarpum ssp. campylocarpum (1/4)

Flowers light chartreuse green, openly campanulate, in firm
trusses of 18-20. (Crown Estate, Windsor) A. M. (RHS) 1974

cl.                      -catawbiense (1/4)
    -Charles Dickens-
Beatrice Pierce-          -unknown (1/4)
-                      -decorum (1/4)
    -unnamed hybrid-
       -griffithianum (1/4)

Rose flowers with maroon center.  (G. Guy Nearing, 1944)

cl.
    -burmanicum (1/2)
Beatrix Anderson-
    -unknown (1/2)

A plant for the cool greenhouse.  Flowers up to 4in(10.2cm) wide
in loose trusses of 4 or 5; buds yellowish green, opening paler.
(Cross by E. B. Anderson; raised and exhibited by Crown Estate,
Windsor, 1968) A. M. (RHS) 1968

g. & cl.
    -Essex Scarlet--unknown (1/2)
Beau Brummel-
    -facetum (eriogynum) (1/2)

5ft(1.5m)          -5°F(-21°C)          L          3/3

A good late variety of dark blood red flowers speckled within;
trusses form compact balls. Foliage dull dark green. (Rothschild, 1934) A. M. (RHS) 1936

```
cl.                      -caucasicum (1/4)
           -Boule de Neige-             -catawbiense (1/8)
Beaufort-                -hardy hybrid-
           -                           -unknown (1/8)
           -fortunei ssp. fortunei (1/2)
```

5ft(1.5m)          -15°F(-26°C)          M          3/3

Insect-resistant, compact plant.  Fragrant white flowers tinted mauve; trusses of 14.  (J. Gable)    See color illus. LW pl. 50.

---

```
           -griffithianum (1/2)
Beaulieu-
           -unknown (1/2)
```

Flowers of peach pink with whitish pink edges. (Waterer, Sons & Crisp)

```
cl.        - wardii var. wardii (1/2)
Beaulieu -
Hawk     - Lady      - fortunei ssp. discolor (1/4)
           Bessborough - campylocarpum ssp. campylocarpum
                                  Elatum Group (1/4)
```

6ft(1.8m)          5°F(-15°C)          M          3/2

Hawk g.  Another member of a family of seven made famous, mainly by Crest, for the rich yellow flowers.  This one is a pale yellow.  (Rothschild, 1940)

```
                        -dichroanthum ssp. dichroanthum (1/8)
             -Goldsworth-
cl.      -Hotei- Orange  -fortunei ssp. discolor (1/4)
         -        -            -souliei (1/8)
Beautiful-   -unnamed hybrid-
Day      -                    -wardii var. wardii (3/8)
         -   -wardii var. wardii        -fortunei ssp. discolor
         -Crest-
             -Lady Bessborough-
                         -campylocarpum ssp. campylo-
                                  carpum Elatum Group (1/8)
```

3ft(.9m)          -5°F(-21°C)          L

Buds show brick red shading to orange buff; flowers open to empire yellow with narrow, diffused stripes of tangerine orange from throat to outer edge.  Trusses of 12 flowers, large, open funnel-shaped, five wavy lobes.  Leaves retained 3 years. (Cross by William E. Whitney; intro. by George & Anne Sather, 1975)

```
cl.                -catawbiense (1/2)
Beauty of Bagshot-
                  -unknown (1/2)
```

White flowers with a soft mauve tinge and dark blotch.  (J. Waterer)

cl.

Beauty of Berry Hill       Parentage unknown

Synonym Beauty of Surrey.  Flowers scarlet, finely spotted. (A. Waterer)  F. C. C. (RHS) 1872

```
cl.
              -unknown (possibly campanulatum) (1/2)
Beauty of Littleworth-
              -griffithianum (1/2)
```

6ft(1.8m)          -5°F(-21°C)          M          4/3

One of the largest trusses of any rhododendron, 16-19 flowers of pure white with reddish purple spotting; each flower about 5in(12.7cm) across.  New stems often as large as a man's finger and over 12in(30.4cm) long.  Upright habit with large, dark green leaves, slightly glossy.  (J. H. Mangles, before 1884) F. C. C. (RHS) 1904; F. C. C. (Wisley Trials) 1953  See color illus. F p. 33.

cl.

Beauty of Surrey     See Beauty of Berry Hill

g.

```
              -griffithianum (1/2)
Beauty of Tremough-
              -arboreum ssp. arboreum (1/2)
```

Flowers rose pink, fading to pale rose.  See also Bodnant Beauty of Tremough, Glory of Penjerrick, Gill's Triumph, John Tremayne, Gillii.  (R. Gill & Son)  F. C. C. (RHS) 1902

```
cl.
                    -fortunei ssp. discolor (1/2)
Beckyann--(unnamed hybrid selfed)-
                    -campylocarpum ssp. campylo-
                                  carpum (1/2)
```

3ft(.9m)          -10°F(-23°C)          ML

Flowers of heavy substance, openly funnel-shaped, 5-lobed, cream with greyed red dorsal spotting, in ball-shaped trusses of 12. Leaves glossy green, held 2 years; plant broadly upright. (H. R. Yates cross; intro. by Mrs. Yates, 1970; reg. 1977)

```
cl.                -catawbiense (1/4)
           -Atrosanguineum-
Beechwood Pink-        -unknown (1/4)
           -fortunei ssp. fortunei (1/2)
```

5ft(1.5m)          -5°F(-21°C)          M          4/3

Flowers bright pink in large trusses. (Joseph B. Gable cross; reg. by Charles Herbert, 1962)  A. E. (ARS) 1960

```
cl.                -elliottii (3/4)
           -Fusilier-
Beefeater-         -griersonianum (1/4)
           -elliottii
```

Excessively large, poorly formed, flat-topped trusses of up to 26 flowers, geranium lake red with some limited pale spotting. Leaves dull green above, bright green underneath with traces of indumentum.  (RHS Garden, Wisley)  A. M. (RHS) 1958; F. C. C. (RHS) 1959

```
cl.
              -macabeanum (1/2)
Beekman's Delight-
              -unknown (1/2)
```

Flowers of very pale yellow with a purple blotch in the throat; trusses hold about 20.  (J. Beekman, 1981)

cl.

Beer Sheba--form of cerasinum

Flowers white, stained rose madder, in trusses of 5.  Leaves 3.5in(9cm) by 1.5in(3.8cm).  (Lionel de Rothschild selection; reg. 1963)

```
cl.    -Jan Dekens--unknown (13/16)
       -                           -ponticum (1/16)
Bel Air-             -Michael-
       -        -Prometheus-Waterer--unknown
       -        -                       -Monitor--unknown
       -Madame de Bruin-     -arboreum ssp. arboreum (1/8)
       -             -Doncaster-
                           -unknown
```

Flowers dark rose with a pink blotch.  (P. van Nes)  S. M. (Rotterdam) 1960;  A. M. (Boskoop) 1960

```
cl.
              -macabeanum (1/2)
Belinda Beekman-
              -unknown (1/2)
```

White flowers, spotted purple, held in trusses of 17-20. (J. Beekman, Australia, reg. 1982)

cl.                  -laetum (1/4)
     -unnamed hybrid-
Belisar-                 -zoelleri (1/2)
     -                    -macgregoriae (1/4)
     -unnamed hybrid-
                              -zoelleri


6ft(1.8m)         32°F(0°C)         Oct.-Dec.

Synonym Ioane.  Vireya hybrid.  Flowers tubular funnel-shaped,
2.5in(5.7cm) across, persimmon orange with yellow orange throat.
Conical trusses 5.5in((14cm) wide hold 5-7 flowers.  (Peter Sullivan cross; intro. and named by William A. Moynier, 1981)

g.
            -Essex Scarlet--unknown (1/2)
Belisha Beacon-
            -arboreum ssp. arboreum (1/2)

Medium height; dusky scarlet flowers in tight trusses. (Rothschild, 1934)


cl.          -Corona--unknown (1/4)
     -Bow Bells-
Bell-          -williamsianum (1/4)
Song*                      -fortunei ssp. discolor
     -                -Lady    -              (1/8)
     -Day Dream (red form)-Bessborough-campylocarpum ssp. campylo-
     -                              carpum Elatum Group (1/8)
                    -griersonianum (1/4)

3ft(.9m)         0°F(-18°C)         M

Ruffled flowers in 2 shades of pink.  Plant habit spreading and
bushy.  (The Bovees Nursery)

g.           -thomsonii ssp. thomsonii (1/4)
     -Shilsonii-
Bella-          -barbatum (1/4)
     -griffithianum (1/2)

Pale rose flowers.  (Lord Aberconway cross, 1926; intro. 1936)


cl.            -griffithianum (1/4)
        -Isabella-
Bellbird-          -auriculatum (1/4)
        -souliei (1/2)

Pure white flowers in loose trusses.  (Collingwood Ingram, 1964)

cl.
     -fortunei ssp. discolor (1/2)
Belle-
     -campylocarpum ssp. campylocarpum Elatum group (1/2)

Lady Bessborough g.  Flowers of very pale lemon yellow.  (Rothschild, 1933)

cl.                    -catawbiense (1/2)
         -Catawbiense Album-
Belle Heller-              -unknown (1/2)
     -                        -catawbiense
        -white catawbiense hybrid-
                             -unknown

5ft(1.5m)         -10°F(-23°C)         M         4/3

Good plant habit, a vigorous grower with large, dark green
leaves.  Flowers large, white with a conspicuous golden blotch,
held in large, globular trusses.  (A. M. Shammarello, before
1958)  See color illus. VV p. 111.

cl.
            -rigidum var. album (1/2)
Belle of Tremeer-
            -augustinii ssp. augustinii (1/2)

A slender-growing natural hybrid with pale mauve flowers.  (Gen.
Harrison, 1957)  P. C. (RHS) 1969


cl.
        -fortunei ssp. fortunei (1/2)
Bellefontaine-
        -smirnowii (1/2)

6ft(1.8m)         -15°F(-26°C)         L

Buds rose opal opening to fragrant flowers, neyron rose with
upper lobes flecked olive brown, in large trusses of 10.  Plant
upright, well-branched, wider than tall with leaves glossy, dark
green.  (R. B. Pike cross, before 1958; intro. by D. L. Craig,
Canada Agriculture Dept, 1977)

g.                -fortunei ssp. discolor (1/4)
        -Norman Shaw-        -catawbiense (1/8)
Bellerophon-         -B. de Bruin-
     -                       -unknown (1/8)
        -facetum (eriogynum) (1/2)

Late-blooming flowers, bright crimson red in tight trusses of 10
or more.  Tall plant of tidy habit.  (Rothschild, 1934)

cl.
     -minus var. minus Carolinianum Group (1/2)
Bellvale-
     -dauricum var. album (1/2)

3ft(.9m)         -25°F(-32°C)         EM

Floriferous plant, with broadly elliptic small leaves, turning a
rich coppery purple in autumn, and retained 2 years.  Broad, upright, well-branched shrub.  Flowers light pink with a mauve
flare, widely funnel-shaped, of heavy substance.  (Warren Baldsiefen, 1973)

cl.            -arboreum ssp. arboreum (1/4)
     -Doncaster-
Belvedere-      -unknown (1/4)

     -dichroanthum ssp. dichroanthum (1/2)

Soft pink.  (Not now cultivated at Exbury.) (Rothschild, 1927)


cl.                            -wardii var. wardii ? (1/8)
                      -Chlorops-
            -Lackamas Spice-      -vernicosum ? (1/8)
Ben Lancaster-          -diaprepes (1/4)
     -               -fortunei ssp. discolor (1/4)
        -Evening Glow-    -dichroanthum ssp. dichroanthum
                -Fabia-                           (1/8)
                    -griersonianum (1/8)

Flowers of mimosa yellow, deepening with age, 3.5in(8.9cm) wide,
campanulate, fragrant; ball-shaped, 12-flowered trusses.  Plant
3ft(.9m) in 7 years, bushy, with persistent leaves 5in(12.7cm).
Late-flowering.  (Benjamin Lancaster, 1963; reg. 1971)

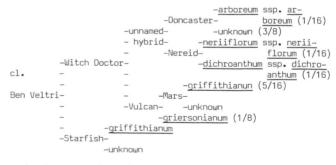

2ft(.6m)         0°F(-18°C)         M

Buds rose red opening to flowers of reddish purple with greyed
orange spotting, widely funnel-campanulate 2.25in(5.7cm) across;
rather lax trusses, 6in(15.2cm) wide, hold 15-20 flowers.  Floriferous.  Plant broad, well-branched; elliptic leaves held three
years; very light indumentum on new foliage. (Paul Holden, 1980)

cl.

Bengal (Leach)     See Madras


cl.
        -Essex Scarlet--unknown (1/2)
Bengal*
        -forrestii ssp. forrestii Repens Group (1/2)

Scarlet.  Leaves dark green; branches arched.  (D. Hobbie, 1960)


cl.
            -fortunei ssp. discolor (1/2)
Bengal Rose-       -griersonianum (1/4)
              -Tally Ho-
                   -facetum (eriogynum) (1/4)

Flowers Bengal rose, open bell-shaped, 4in(10cm) across, 12-15
held in upright trusses.  Vigorous, upright plant.  (Ben Lancaster, 1963)

g.
                -Sir Charles Butler--fortunei ssp. fortunei (1/2)
Bengt M. Schalin-
               -wardii var. wardii (L S & T 5679) (1/2)

Bright yellow with red patches in the throat.  (D. Hobbie, 1952)

g.
        -fortunei ssp. discolor (1/2)
Benito-      -fortunei ssp. fortunei (1/4)
        -Luscombei-
              -thomsonii ssp. thomsonii (1/4)

White with pink buds; brownish rose markings on the upper lobe.
(Origin unknown)  A. M. (RHS) 1934


cl.

Benmore--form of montroseanum (mollyanum)

5ft(1.5m)         10°F(-12°C)          E-M          3/3

Pink flowers with crimson blotch at base.  Compact growth habit;
large foliage.  Known only in cultivation.  (Younger Botanic
Garden, Benmore)  A. M. (RHS) 1957; F. C. C. (RHS) 1957


cl.

Bennan--form of meddianum var. atrokermesinum (K-W 19452)

Cardinal red flowers with darker markings held in trusses of 17.
(Collected by F. Kingdon-Ward; raised and intro. by Brodick Castle, 1977)  A. M. (RHS) 1977


cl.

Bergie*--form of ciliatum

White flowers.  (H. L. Larson)

cl.
              -wardii var. wardii (1/2)
Bergie Larson-      -dichroanthum ssp. dichroanthum (1/4)
               -Jasper-
                   -Lady       -fortunei ssp. discolor (1/8)
                    Bessborough-
                            -campylocarpum ssp. campylo-
                                            carpum (1/8)
3.5ft(1m)         10°F(-12°C)          L

Buds red. Flowers of heavy substance, widely funnel-campanulate,
3in(7.6cm) across, cadmium orange with slight red spotting;
trusses 5in(12.7cm) wide, ball-shaped, 12-flowered.  Floriferous.  Plant broad, well-branched; leaves retained 3 years.  (Mr.
& Mrs. H. L. Larson, 1978)

cl.
    -williamsianum (1/2)
Berlin*
    -unknown (1/2)

Large pink flowers, bell-shaped, pendant.  Plant rounded; very
hardy.  (J. Burns, 1964)

                    -catawbiense var. album (1/4)
cl.    -Anna H.-
     - Hall  -Koichiro Wada--yakushimanum ssp. yakushimanum
Bermuda-                                               (1/4)
     -      -fortunei ssp. fortunei (1/4)
     -Nestucca-
                -yakushimanum ssp. yakushimanum (1/4)

4ft(1.2m)        -20°F(-29°C)          ML          4/4

Plant rounded, semi-dwarf, spreading much wider than tall.
Dome-shaped trusses of 14 flowers colored phlox pink with dorsal
spots of yellow green.  (David G. Leach, intro. 1975; reg. 1983)


cl.
    -decorum (1/2)
Bern-
    -hybrid unknown (1/2)

Flowers pastel mauve with lighter center; prominent rose blotch
on upper lobes.  (John Bacher) P. A. (ARS) 1955


                             -griffithianum (1/8)
                   -George Hardy-
cl.      -Pink Pearl-       -catawbiense (1/8)
       -          -           -arboreum ssp.arboreum
Bernard Crisp-       -Broughtonii-              (1/8)
       -                      -unknown (5/8)
            -hardy hybrid--unknown

Flowers of pale rose pink.  (Waterer, Sons & Crisp, 1920)  A. M.
(RHS) 1921


cl.
            -barbatum (1/2)
Bernard Gill-
            -unknown (1/2)

Flowers  soft rose red  with a few dots of  deeper carmine pink.
(R. Gill & Son)  A. M. (RHS) 1925


cl.
            -calophytum var. calophytum (1/2)
Bernard Shaw-
            -           -griffithianum (1/8)
            -George Hardy-
         - Pink -       -catawbiense (1/8)
          Pearl-       -arboreum ssp. arboreum (1/8)
            -Broughtonii-
                    -unknown (1/8)

Clear pink flowers.  (Reuthe, 1955)


cl.
        -maddenii ssp. maddenii (calophyllum) (3/4)
Bernice-      -cinnabarinum ssp. cinnabarinum
      -Royal Flush-        Blandiiflorum Group (1/4)
            -maddenii ssp. maddenii (calophyllum)

Trusses hold 7 flowers, 5-lobed, crimson becoming blush white on
margins.  Leaves elliptic with brown scaly indumentum.  (F. M.
Jury, reg. 1982)

cl.                     -dichroanthum ssp. dichroanthum
        -Goldsworth Orange-                       (1/4)
Bernstein-        -fortunei ssp. discolor (1/4)
      -Mrs. J. G. Millais--unknown(1/2)

Trusses hold 12-15 flowers, 5-lobed, deep  yellow orange with
conspicuous blotch of bright orange red.  Leaves are elliptic
and hairy.  (Cross by H. Hachmann; reg. by G. Stuck, 1983)

g. & cl.          -arboreum ssp. arboreum (1/4)
          -Doncaster-
Berryrose-          -unknown (1/4)
          -
          -dichroanthum ssp. dichroanthum (1/2)

3ft(.9m)          0°F(-18°C)          ML          4/2

Apricot orange with yellow bicolor.  Plant has a good rounded
shape, but foliage is held only one year with frequent unattrac-
tive spots in winter.  See also Minterne Berryrose.(Rothschild)
A. M. (RHS) 1934

cl.          -griersonianum (1/4)
          -Diva-          -fortunei ssp. discolor (1/8)
Bert Larson-   -Ladybird-
          -          -Corona--unknown (1/8)
          -strigillosum (1/2)

5ft(1.5m)          10°F(-12°C)          E

Flowers blood red with darker brownish red spotting in trusses
6in(15cm) across with 12-14 per truss.  Leaves held 4-5 years
with woolly, greyed orange indumentum. (H. L. Larson, reg. 1979)

cl.
          -catawbiense (1/2)
Bertha--Everestianum (selfed)-
          -unknown (1/2)

Flowers carmine pink with orange red markings. (T. J. R. Seidel,
cross 1892)

cl.

Beryl     Parentage unknown.

Dull purple flowers.  (Origin unknown)  A. M. (RHS) 1931

cl.

Beryl Taylor--form of campylocarpum ssp. campylocarpum
          (LS & T 4738)
1ft(.3m)          -10°F(-23°C)          EM          3/3

Compact, upright dwarf with small leaves, dark green, glossy.
Flowers in clusters of 4-5, bell-shaped, red purple.  (Lord
Aberconway)  A. M. (RHS) 1975

g.
          -spinuliferum (1/2)
Berylline-
          -valentinianum (1/2)

Dense, low-growing shrub.  Very showy trusses of pale gold flowers flushed rose, in small, compact trusses.  (Rothschild, 1934)

          -campanulatum ssp. campanulatum (1/4)
cl. -Constant Nymph-          -ponticum (3/8)
     -          -Purple Splendour-
Bess-          -unknown (3/8)
     -          -ponticum
     -Purple Splendour-
          -unknown

Lavender.  (Donald Waterer cross, 1947; intro. Knap Hill, 1975)

          -caucasicum (1/4)
cl.     -Boule de Neige-          -catawbiense (3/8)
     -          -hardy hybrid-
Besse Howells-          -unknown (3/8)
     -          -catawbiense
     -unnamed red hybrid-
          -unknown

4ft(1.2m)          -20°F(-29°C)          EM          3/4

Compact habit; dark green leaves. Ruffled flowers, burgundy red
with dark red blotch; globular trusses. (A. M. Shammarello, intro. 1961)     See color illus. ARS Q 27:2 (Spring 1973): p. 102;
LW pl. 85;  VV p. 131.

          -griffithianum (1/8)
cl.     -Loderi King George-
     -Olympic-          -fortunei ssp. fortunei (3/8)
Bessie- Lady  -williamsianum (1/4)
Farmer-   -fortunei ssp. fortunei
     -Fawn-          -dichroanthum ssp. dichroanthum (1/8)
          -Fabia-
          -griersonianum (1/8)

3ft(.9m)          0°F(-18°C)          ML

Compact mound is the growth habit.  Fragrant, flat-faced flower,
pale pink in bud, opening pure white.  (A. A. Childers, 1970)

cl.

Beta--form of lanigerum

Flowers rose opal.  (Sir Edward Bolitho, 1963)

cl.
          -griersonianum (1/2)
Betelgeuse-
          -Madame Colijn--unknown (1/2)

Flowers of salmon pink.  (G. Reuthe, 1940)

          -thomsonii ssp. thomsonii (1/8)
cl.     -Bagshot Ruby-
     -Princess -          -unknown (3/8)
Betsie-Elizabeth-unknown
Balcom-          -forrestii ssp. forrestii Repens Group (1/4)
     -Elizabeth-
          -griersonianum (1/4)

4ft(1.2m)          0°F(-18°C)          M

Flowers openly funnel-shaped with 5 wavy lobes, currant red;
conical trusses 6in(15.2cm) across, 13-flowered.  Plant fairly
dense, rounded, branching well; leaves are bullate, narrowly elliptic, and retained 3 years; new growth tinged red. (Thomas J.
McGuire, 1974)

cl.          -Catalgla--catawbiense var. album Glass
     -Cataldi-          (1/4)
Betsy Kruson-     -wardii var. wardii (1/4)
     -     -griffithianum (1/4)
          -Mars-
          -unknown (1/4)

4ft(1.2m)          -10°F(-23°C)          ML

Flowers reddish purple shading to almost white in throat, with a
small flare of greyed yellow;  flowers openly funnel-shaped with
5 wavy lobes, 10 held in conical trusses 6in(15.2cm) wide.  Narrow elliptic leaves held 2 years on a broad well-branched plant.
(Mrs. Henry Yates, 1975)

cl.          -forrestii ssp. forrestii Repens Group
     -Elizabeth-          (1/4)
Better Half*          -griersonianum (1/4)
     -unknown (1/2)

3ft(.9m)          0°F(-18°C)          EM          3/4

The above parentage is uncertain.  Attractive red trusses.
Dense, compact foliage.  The "better half" of two seedlings once
growing together as one in Whitney Gardens.  (Selected by the
Sathers)

* * * * * * * * * * * * * * * * * * * *
* Hybrids of unknown parentage that have won awards, or are *
* a parent of another hybrid, are in the text; for all oth- *
* ers of unknown parentage, see App. D.                     *
* * * * * * * * * * * * * * * * * * * *

```
cl.
        -fortunei ssp.fortunei (1/2)
Betty-
        -thomsonii ssp. thomsonii (1/2)
```

Deep pink flowers. (Loder)  A. M. (RHS) 1927

```
                            -Sir Charles Butler--fortunei ssp.
        -Van Nes Sensation-                     fortunei (1/4)
cl.     -                 -          -griffithianum (1/4)
        -                 -Halopeanum-
Betty   -                             -maximum (1/8)
Anderson-        -griffithianum
        -         -Mars-
        -unnamed- -unknown (1/4)
         hybrid-             -ponticum (1/8)
                -Purple Splendour-
                              -unknown
```

5ft(1.5m)          10°F(-12°C)          M

Buds of reddish purple open to flowers of amethyst violet shading lighter to near white in centers of lobes and throat, reverse darker with stripes, held in ball-shaped trusses of 10.
Upright plant, 3/4 as wide as tall with yellow green leaves
7in(17.8cm) long by 3in(7.6cm) wide. See also Bonnie Snaza.
(Edwin (Ted) Anderson, 1983)

```
cl.                -caucasicum (1/4)
              -Nobleanum-
Betty Boulter-         -arboreum ssp. arboreum (1/4)
             -         -campylocarpum ssp. campylocarpum (1/4)
             -Unique-
                      -unknown (1/4)
```

Tyrian purple flowers, lightly spotted, in ball-shaped trusses.
(V. J. Boulter, 1962)

```
cl.
        -smirnowii (1/2)
Betty Breene-
            -Fluffy Ruffles F2, Dexter hybrid--unknown (1/2)
```

4ft(1.2m)          -25°F(-32°C)          ML          2/4

Plant compact, well-branched, wider than tall, with leaves to 6
in(15.2cm) long. Flowers of pale pink, flushed mauve. (David
G. Leach, 1960)

```
cl.
Betty Graham--form of anthopogon ssp. anthopogon (L & S 1091)
```

1ft(.3m)          -5°F(-21°C)          EM          3/3

Flowers narrowly tubular with 5 small joined lobes in trusses of
8-10, strongly fragrant, rose purple pink deeper toward base.
Aromatic leaves, dark green above, undersurface densely covered
with dark brown scales.  (E. H. M. Cox & P. A. Cox)  A. M. 1969

```
cl.
        -fortunei ssp. fortunei (1/2)
Betty Hume-
          -unknown (1/2)
```

5ft(1.5m)          -5°F(-21°C)          ML          4/4

Flowers are pink, ruffled, fragrant, up to 4in(10.2cm) across.
(Raised by Dexter; intro. by W. Baldsiefen and Efinger, 1962)

```
g.                  -fortunei ssp. fortunei (1/4)
            -Luscombei-
Betty King-         -thomsonii ssp. thomsonii (3/4)
          -thomsonii ssp. thomsonii
```

A named group that varies in color from red to rose pink. (Sir
James Horlick cross, 1930; intro. 1942)

```
cl.                 -fortunei ssp. fortunei (1/4)
            -Luscombei-
Betty Royal-        -thomsonii ssp. thomsonii (3/4)
          -thomsonii ssp. thomsonii
```

Betty King g.  Red or pink flowers. (Sir James Horlick, 1942)

```
cl.

Betty Stewart      Parentage unknown
```

Flowers cherry red, spotted and suffused white on upper lobes.
(C. B. van Nes & Sons)  A. M. (RHS) 1936

```
cl.                 -griffithianum (1/4)
            -George Hardy-
Betty Wormald-      -catawbiense (1/4)
            -red hybrid--unknown (1/2)
```

6ft(1.8m)          -5°F(-21°C)          M          4/3

Very large, almost flat pastel pink flowers with a pale center
and light purple spotting on the upper lobe, held in dome-shaped
trusses. Plant upright, spreading; large foliage.  (M. Koster,
before 1922)  A. M. (RHS) 1935;  F. C. C. (RHS) 1964;  A. G. M.
(RHS) 1969   See color illus. VV p. 58.

```
cl.

Betty's Purple Russatum      See Night Editor
```

```
                      -dichroanthum ssp. dichroanthum (3/16)
                 -Dido-
                 -    -decorum (3/16)
        -unnamed-              -fortunei ssp. discolor
        - hybrid-      -Lady -                  (1/16)
cl.     -             -Bessborough-campylocarpum ssp. campylo-
        -     -Jalisco-          carpum Elatum Group (1/16)
Beverley-     -       -dichroanthum ssp. dichroanthum
Tasker -      -Dido-
        -             -decorum
        -     -Sarita-griersonianum (1/8)
        -     - Loder-        -griffithianum (5/64)
        -Yellow-     -Loderi-
         Creek-             -fortunei ssp. fortunei (9/64)
        -             -wardii var. wardii (1/8)
        -             -                  -griffithianum
        -Idealist-            -Kewense-
        -             -Aurora-        -fortunei ssp.
                -Naomi-      -               fortunei
                -             -thomsonii ssp. thomsonii
                -fortunei ssp. fortunei        (1/32)
```

4ft(1.2m)          14°F(-10°C)          M

Flowers of ripe lemon yellow, green in base of throat, pink stamens, green pistil, frilled, 3in(7.6cm) across; loose trusses
hold 7-11. Floriferous. Leaves medium green, slightly glossy.
Plant grows as wide as high. (Raiser H. R. Tasker; intro. 1977)

```
cl.        -racemosum (1/4)
        -Racil-
Bewitched-      -ciliatum (1/4)
        -Cornell Pink--mucronulatum (1/2)
```

4ft(1.2m)          0°F(-18°C)          E          3/3

Upright plant, open-growing, leaves 1.5in(3.8cm) by .5in(1.3cm),
persistent. Flowers 2in(5cm) across, 2-5 in ball-shaped truss,
light purple pink. (W. L. Guttormson, 1967)

```
g.

Bianca      Parentage unknown
```

Different forms are known--salmon pink, delicate crimson, pure
white with yellow spots. (Origin unknown)  F. C. C. (RHS) 1866

cl.                -catawbiense (3/4)
        -Mrs. Milner-
Bibber-              -unknown (1/4)
        -catawbiense

Flowers carmine red with faint brown markings.  See also Granat.
(T. J. R. Seidel, cross 1900)


g. & cl.
        -Moser's Maroon--unknown (1/2)
Bibiani-
        -arboreum ssp. arboreum (1/2)

6ft(1.8m)          5°F(-15°C)          EM          4/4

Bright blood red color in the spring--early.  Truss of about 14,
dense and rounded.  Foliage is exceptionaly glossy, deep green.
Easily grown, above average in vigor, but needs wind protection.
Easy to propagate.  (Rothschild) A. M. (RHS) 1934


cl.
        -grande (1/2)
Big Ben-
        -unknown (1/2)

Red flowers, in trusses of 23-28.  (J. Beekman, 1980)


cl.
        -white Dexter hybrid--unknown (1/2)
Big Girl-        -fortunei ssp. discolor (1/4)
        -Avocet-
                -fortunei ssp. fortunei (1/4)

3ft(.9m)          -10°F(-23°C)          ML

Flowers are pale purplish pink with olive green blotch, tubular
funnel-shaped; loose trusses about 7.5in(19cm) across hold up to
11 flowers.  (Howard Phipps, 1973)


cl.
        -LaBar's White--catawbiense var. album (1/2)
Big Mac-
        -macabeanum (1/2)

Barium yellow with a dark red blotch, 11-12 flowers per truss.
(Mrs. John F. Knippenberg cross, 1955; reg. 1977)


cl.
        -Catalgla--catawbiense var. album Glass (1/2)
Big 'O'-         -catawbiense (1/4)
        -Pink Twins-
                -haematodes ssp. haematodes (1/4)

5ft(1.5m)          -15°F(-26°C)

Phlox purple buds open to lighter flowers, fading white, with a
beetroot purple blotch on upper 3 lobes.  The first clone given
this name is now extinct.  (Weldon E. Delp)


                        -Catalgla--catawbiense var. album Glass
cl.     -unnamed hybrid-                               (1/4)
        -              -fortunei ssp. fortunei (1/4)
Big Savage-           -decorum ? (1/8)
        -      -Caroline-
        -Cadis-        -brachycarpum ssp. brachycarpum ? (1/8)
                -fortunei ssp. discolor (1/4)

5.5ft(1.7m)          -10°F(-23°C)          ML

Flowers lavender pink with greenish throat, widely funnel-shaped
and 5- or 6-lobed; trusses hold 13.  Plant very dense, vigorous,
with large glossy foliage.  (Raised by H. R. Yates, 1962; intro.
1970)


cl.
                -griffithianum (1/2)
Big Savage Red--Mars (selfed)-
                -unknown (1/2)

3ft(.9m)          -20°F(-29°C)          ML

Flowers of Bengal rose  with white dorsal flare containing a few
green spots, of heavy substance, openly funnel-shaped, 5 frilled
lobes.  About 16 flowers in conical trusses.  Plant upright and
well-branched; glossy, elliptical leaves.  (Henry Yates cross,
1956; intro. by Mrs. Yates, 1977)


cl.
        -caucasicum (1/2)
Bijou-
        -unknown (1/2)

Introduced by Moser, 1914.


cl.

Bill Browning--form of mallotum (Farrer 815)

Trusses of 12 crimson flowers.  (Raiser Col. S. R. Clarke; reg.
by R. N. S. Clarke, 1978)  A. M. (RHS) 1933


cl.
        -Bergii--ciliatum (1/2)
Bill Massey-
        -nuttallii (1/2)

4ft(1.2m)          20°F(-7°C)          EM

Buds apricot pink opening to tubular funnel-shaped flowers, near
white with five wavy lobes.  Reverse is slight pink shading at
lobes' edges and a stripe.  Very fragrant.  Truss has 4-6
flowers.  Leaves retained 2-3 years, narrowly elliptical,
glossy, bullate.  A rounded plant, branching well.  (John S.
Drucker cross, 1973; intro. by Trillium Lane Nursery, 1978)


cl.
        -minus var. minus Carolinianum Group (1/2)
        -                       -racemosum (1/8)
Billy Bear-              -Conemaugh-
        -Pioneer (Gable)-       -mucronulatum (1/8)
                        -unknown (1/4)

2.5ft(.75m)          -20°F(-29°C)          M

Buds cyclamen purple; flowers widely funnel-shaped, amaranth
rose, shaded at lobe edges with red spotting.  Ball-shaped terminal flower clusters of 1-4 trusses, each 2-4 flowers.  Glossy
leaves held 1 year.  Semi-dwarf.  (William M. Fetterhoff, 1979)


cl.
        -haematodes ssp. haematodes (1/4)
        -May Day-
Billy Budd-     -griersonianum (1/4)
        -elliottii (1/2)

Flat-topped trusses of Turkey red flowers--10-12 make a dense
truss in May.  Leaves elliptic, slightly wavy with traces of
tomentum beneath. (Cross by Francis Hanger, RHS Garden, Wisley;
intro. 1954)  A. M. (RHS) 1957


cl.        -wightii (1/4)
        -China-
Binfield-   -fortunei ssp. fortunei (1/4)
        -     -wardii var. wardii (1/4)
        -Crest-        -fortunei ssp. discolor (1/8)
        -Lady     -
        Bessborough-campylocarpum ssp. campylocarpum
                                     Elatum Group (1/8)

Flowers primrose yellow with slight red stain in throat, openly
campanulate, up to 17 per rounded truss.  Leaves 6in(15cm) by
2.5in(6.4cm).  (Crown Estate, Windsor)  A. M. (RHS) 1964

```
                       -fortunei ssp. fortunei (13/16)
                 -Kewense-
cl.      -Aurora-          -griffithianum (1/16)
      -Naomi-      -thomsonii ssp. thomsonii (1/8)
Birthday- -fortunei ssp. fortunei
Greeting-

         -fortunei ssp. fortunei
```

Flowers yellow orange diffusing towards edges to red purple,
reverse diffused shades with some deep red markings; trusses of
6-7. (Edmund de Rothschild) A. M. (RHS) 1979

```
cl.
         -fortunei ssp. discolor (1/2)
Biscuit Box-
         -elliottii (1/2)
```

Flowers barium yellow, fading outward to empire yellow with
throat faintly spotted greenish yellow, outside tinged pink.
Trusses hold ten. Corolla is open funnel-shaped, 2.5in(6.4cm)
long by 3in(6.6cm) wide. (Cross by Francis Hanger, RHS Garden, Wisley) A. M. (RHS) 1964

```
                          -fortunei ssp. discolor
             -Lady Bessborough-              (1/4)
        -Jalisco-          -campylocarpum ssp. campylo-
cl.      - Elect-               carpum Elatum Group (1/4)
      -          -dichroanthum ssp. dichroanthum (1/8)
Bishopsgate-  -Dido-
      -          -decorum (1/8)
      -    -wardii var. wardii (1/4)
      -Crest-          -fortunei ssp. discolor
         -Lady Bessborough-

                          -campylocarpum ssp. campylo-
                               carpum Elatum Group
```

Flowers openly campanulate, of yellow variably flushed red, with
blotched upper throat. Full rounded trusses hold 10-12 flowers.
(Crown Estate Commisioners) P. C. (RHS) 1974

```
g. & cl.
      -cinnabarinum ssp. cinnabarinum Roylei Group (1/2)
Biskra-
      -ambiguum (1/2)
```

5ft(1.5m)        10°F(-12°C)          M          4/3

An erect and slender bush up to 15ft(4.6m) bearing distinctive
trumpets of vermilion blended with biscuit, in tight clusters of
8. Smooth, dark green, elliptic leaves 2.5in(6.4cm) long,
covered beneath with minute, crimson, resin-bearing cells.
(Rothschild, 1934) A. M. (RHS) 1940 See color illus. ARS Q 27:
2 (Spring 1973): p. 102.

```
                -campanulatum ssp. campanulatum (1/8)
cl.        -Madame Naun-
      -Viola-          -unknown (1/8)
Bismarck-   -          -catawbiense (5/8)
      -    -unnamed hybrid-
      -catawbiense          -ponticum (1/8)
```

Flowers open pale purplish pink with reddish brown markings.
(T. J. R. Seidel cross, 1900)

```
                    -griffithianum (1/16)
             -George Hardy-
      -Mrs. Lindsay-          -catawbiense (1/16)
cl.  -Diane- Smith     -Duchess of Edinburgh--unknown (1/4)
      -          -campylocarpum ssp. campylocarpum (3/8)
Bitter-    -unnamed-
  Lemon-      hybrid-unknown
      -          -campylocarpoum ssp. campylocarpum
      -Gladys Rose-
             -fortunei ssp. fortunei (1/4)
```

Flowers pale primrose with a red blotch. (John Waterer, Sons &
Crisp, 1975)

```
            -griffithianum (5/16)
cl.      -Jean Marie -
      -de Montague-unknown (7/16)
Black Magic-                -griffithianum
            -          -Queen Wilhelmina-
      -    -Britannia-          -unknown
      -Leo-      -Stanley Davies--unknown
            -elliottii (1/4)
```

4ft(1.2m)        -5°F(-21°C)          ML          4/4

Look down with pleasure at these unusual blackish red flowers in
late spring; abundant foliage, matte forest green. Floriferous.
See also Heat Wave. (Harold E. Greer, intro. 1982)

```
cl.
            -thomsonii ssp. thomsonii (1/2)
Black Prince-                -Moser's
      -                -Maroon--unknown (1/4)
      -Romany Chal (F. C. C. form)-
                          -facetum (1/4)
```

5ft(1.5m)        0°F(-18°C)          ML          4/2

Oxblood red. Flowers in trusses of 15-16. (Lester Brandt,
1962)

```
cl.
            -sanguineum ssp. sanguineum var. sanguineum
Black Prince's Ruby-                          (1/2)
            -thomsonii ssp. thomsonii (1/2)
```

3ft(.9m)        10°F(-12°C)          EM

Buds black red opening to currant red flowers of heavy waxy
substance; openly campanulate with 5 wavy veins; trusses of 5.
Broad plant with leaves elliptic, held 3 years. (Ben Nelson
cross, 1965; intro. by Howard Short, 1977)

```
cl.
            -dichroanthum ssp. dichroanthum (1/2)
Black Strap-
            -sanguineum ssp. didymum (1/2)
```

Flowers of deep blood red. (Sunningdale Nurseries, 1951)

```
cl.

Blackhills--form of lacteum
```

5ft(1.5m)        -5°F(-21°C)          EM          4/4

Flowers primrose yellow without blotch or spots. (S. V.
Christie) F. C. C. (RHS) 1965

```
cl.
      -Essex Scarlet--unknown (1/2)
Blackie*
      -brachycarpum ssp. brachycarpum ? (1/2)
```

-5°F(-21°C)          ML          3/3

Sibling of Kentucky Cardinal. Very dark red flowers having a
fruity fragrance. (Joseph Gable)

```
cl.
      -Moser's Maroon--unknown (1/2)
Blackie-          -sanguineum ssp. didymum (1/4)
      -Arthur Osborn -
            -griersonianum (1/4)
```

Flowers cardinal red. (Rudolph Henny, 1960) See color illus.
ARS Q 31:2 (Spring 1977): p. 103.

```
cl.

Blackwater-form of spinuliferum
```

4ft(1.2m)        5°F(-15°C)          EM          3/3

Superior Turkey red tubular flowers, greenish white at the base.
Stamens held free, more or less equal;  filaments white, anthers
black.  Style yellowish, held free of stamens.  (Brodick Castle)
A. M. (RHS) 1977


g.                        -griffithianum (1/2)
              -Gauntlettii-
Blanc de Chine-           -unknown (1/4)
          -        -griffithianum
           -Loderi-
                   -fortunei ssp. fortunei (1/4)


Flowers light pink to white.  There may be two different hybrids
named Gauntlettii; the other one may have the same parentage as
Loder's White, q.v.  (Collingwood Ingram, 1939)


g. & cl.              -griffithianum (1/4)
          -Godesberg-
Blanc-mange-          -unknown (1/4)
          -auriculatum (1/2)


See also Swan Lake.  Flowers pure white, openly funnel-shaped,
frilled at margins, in a shapely, well-poised truss of up to 18
blossoms.  Shrub 15-20ft(4.5-6m), as broad as high; matte green
leaves 9in(22.8cm) by 4in(10.2cm).  (Rothschild, 1934)    A. M.
(RHS) 1947


cl.
          -catawbiense (5/8)
Blandyanum-                        -catawbiense
          -           -unnamed hybrid-
          -Altaclarense-           -ponticum (1/8)
                   -arboreum ssp. arboreum (1/4)


Rosy crimson flowers, about 25 held in dense trusses, blooming
midseason.  Elliptic leaves with thin brown indumentum, grey
felted petioles.  Shrub dense, rounded.  See also Towardii.
(Standish and Noble)  Silver Knightian Medal, 1848


cl.
          -augustinii ssp. augustinii (Tower Court)
Blaney's Blue-            -intricatum (1/8)
                 -Intrifast-
          -Blue Diamond-      -fastigiatum (1/8)
                    -augustinii ssp. augustinii (3/4)


4.5ft(1.4m)          -5°F(-21°C)          EM          4/4


Vigorous plant, producing masses of misty blue flowers.  Foliage
is dense and attractive, forest green in summer with a bronze
tone in winter.  Propagates easily and flowers young.  (L. T.
Blaney cross;  intro. after testing by Dr. Robert Ticknor, 1978)


cl.       -griffithianum (1/4)
      -Mars-
Blaze-    -unknown (1/4)
      -catawbiense (red form) (1/2)


4ft(1.2m)          -25°F(-32°C)          M          2/3


Flowers of strong red, edged in deeper red, with blotch of light
pink, funnel-shaped; firm pyramidal trusses of 16-18.  Leaves of
medium size, elliptic, with dull red pubescent petioles.  (David
G. Leach, 1980)


                        -neriiflorum ssp. neriiflorum
              -Nereid-                       (1/8)
cl.       -unnamed hybrid-   -dichroanthum ssp. dichroanthum
          -           -                            (1/8)
Blazen Sun*          -fortunei ssp. discolor (1/4)
          -          -maximum (1/4)
          -Russell Harmon-
                   -catawbiense (1/4)


Synonym Sunburst.  Large trusses of brilliant orange red flowers
in midseason.  Good plant habit; hardy to -15°F(-26°C).  (David
G. Leach cross; raiser Orlando S. Pride; named 1976)  See color
illus. ARS Q 34:3 (Summer 1980): p. 142.


                        -arboreum ssp. ar-
              -Grand Arab-     boreum (1/16)
cl.       -unnamed hybrid-        -catawbiense (1/16)
      -John Coutts-            -griffithianum (1/4)
Blessed-      -griersonianum (1/2)
 Event -          -griersonianum
      -Sarita Loder-      -griffithianum
               -Loderi-
                   -fortunei ssp. fortunei (1/8)


Flowers Tyrian rose with cherry spots, held in upright trusses
of 9-10.  (Marshall Lyons, 1961)


cl.
      -roxieanum var. roxieanum (1/2)
Blewbury-
      -maculiferum ssp. anhweiense (1/2)


Flowers white, upper throat spotted reddish purple, openly campanulate; rather tight, full trusses of 18-20.  Leaves convex as
roxieanum, lightly covered with loose, pale brown, woolly indumentum.  (Crown Estate, Windsor, 1968) A. M. (RHS) 1968


                        -catawbiense (1/8)
cl.       -Parsons Grandiflorum-
      -Nova Zembla-            -unknown (5/8)
Blinklicht-      -dark red hybrid--unknown
      -      -griffithianum (1/4)
      -Mars-
          -unknown


Flowers held in trusses of 15-21, rich cardinal red with slight
marks of deep oxblood red.  Leaves are elliptic, hairy.  See
also Sammetglut, Hachmann's Feuerschein.  (H. Hachmann cross;
reg. by G. Stuck, 1983)


---
      -haematodes ssp. haematodes (1/2)
Blitz-
      -G. S. Sims--unknown (1/2)


3ft(.9cm)          5°F(-15°C)          M          2/4


Dark red flowers.  Plant very floriferous, dense-growing, with
dark green leaves.  "An admirable little hybrid."  (Clark, 1945)


                        -neriiflorum ssp. neriiflorum (1/8)
cl.       -Nereid-
      -unnamed hybrid-   -dichroanthum ssp. dichroanthum
Blondie*      -fortunei ssp. discolor (1/4)          (1/8)
      -          -maximum (1/4)
      -Russell Harmon-
                   -catawbiense (1/4)


5ft(1.5m)          -12°F(-24°C)          ML


See also Blazen Sun.  Pale yellow flowers; very compact shrub.
Parent plant for yellows.  (Cross by David G. Leach; intro. by
Orlando S. Pride, 1969)


cl.
      -dichroanthum ssp. dichroanthum (1/2)
Blood Orange*      -griersonianum (1/4)
      -May Day-
               -haematodes ssp. haematodes (1/4)


(Sunningdale, 1951)


---
      -forrestii ssp. forrestii Repens Group (1/2)
Blood Ruby-      -haematodes ssp. haematodes (1/4)
      -Mandalay-
          -venator (1/4)


2ft(.6m)          5°F(-15°C)          EM          3/3


Flowers blood red; small rounded foliage.  (L. E. Brandt, 1954)

cl.

Blue Admiral*     Parentage unknown

A parent of Florence Mann.  May be an Australian hybrid.


cl.
          -Blue Ensign--unknown (3/4)
Blue Boy-          -ponticum (1/4)
          -Purple Splendour-
                              -unknown

5ft(1.5m)         0°F(-18°C)          L

Violet flowers of good substance 2.5in(6.4cm) across with promi-
inent, almost black, triangular blotch in throat; trusses ball-
shaped 6in(15.2cm) wide, of 18-20 flowers.  Floriferous.  Leaves
held 3 years.  (Elsie M. Watson, 1981)

cl.
          -russatum (1/2)          -intricatum (1/8)
Blue Chip-          -Intrifast-
          -Blue Diamond-          -fastigiatum (1/8)
                      -augustinii ssp. augustinii (1/4)

Purple violet flowers, widely funnel-shaped, 3.5in(9cm) across;
5-7 flowered axillary clusters form rounded terminal trusses.
Dark green, narrowly elliptic leaves, 2.25in(5.5cm) long.  (A.
F. George, Hydon Nurseries)  A. M. (RHS) 1978

cl.

Blue Cloud--form of augustinii ssp. chasmanth*hum*

6ft(1.8m)         0°F(-18°C)          EM          4/3

Small leaves typical of species.  Powder blue flowers, 2.5in(6.4
cm) across, very profuse.  (Hansen, 1958)

cl.                          -ponticum (1/4)
          -Purple Splendour-
Blue Crown-          -unknown (3/4)
          -Blue Peter--unknown

Trusses hold 20 flowers of spectrum violet with lighter center,
blotched with medium greyed purple.  (K. Van de Ven, 1981)


---
          -fortunei ssp. discolor (1/2)
Blue Danube-          -ponticum (1/4)
          -Purple Splendour-
                    -unknown (1/4)

Large trusses of rich lavender blue flowers with an almost green
blotch.  Foliage dark green.  Blooms same time as Blue Peter.
(Knap Hill)  H. C. (RHS) 1959

g. & cl.          -intricatum (1/4)
          -Intrifast-
Blue Diamond-          -fastigiatum (1/4)
          -augustinii ssp. augustinii (1/2)

3ft(.9m)          -5°F(-21°C)          EM          5/4

One of the best "blues," it flowers with a profusion of small,
bright blue flowers.  Plant habit upright, well-branched; leaves
small, bronze green in winter.  Propagates well.  Two clones are
grown in the U. S. (Crossfield, 1935)  A. M. (RHS) 1935; F. C.
C. (RHS) 1939  See color illus. F p. 34

cl.

Blue Ensign--unknown (may be ponticum hybrid)

4ft(1.2m)          -15°F(-26°C)          M          4/3

Flowers pale lavender blue  with prominent dark blotch, held in
rounded trusses of 6-9.  Large leaves, dark matte green, tend to
spot like Blue Peter, but provide a better growth habit.   Plant
well-clothed, attractive, propagates easily, blooms earlier than
Blue Peter.  (W. C. Slocock,1934)  A. M. (RHS) 1959    See color
illus. HG p. 142;  VV p. 126.

cl.
          -Blue Ensign--unknown (3/4)
Blue Hawaii-          -ponticum (1/4)
          -Purple Splendour-
                      -unknown

5ft(1.5m)         0°F(-18°C)          L

Purple flowers of heavy substance, with yellow green blotch, 3.5
in(9cm) wide, openly funnel-shaped.  Trusses 7in(17.8cm) across
hold 16 flowers.  Plant upright, broad; narrowly elliptic glossy
leaves held 2 years. See also Blue Boy.  (Elsie M. Watson, 1981)

cl.
          -hippophaeoides var. hippophaeoides Fimbriatum Group
Blue Haze-          -russatum (1/4)                    (1/2)
          -Russantinii-
                      -augustinii ssp. augustinii (1/4)

Dwarf plant with small blue flowers, funnel-shaped, in midseason.  (Sir John Ramsden cross; intro. by William Pennington-
Ramsden, 1965)


cl.
          -ponticum (1/2)
Blue Jay-
          -Blue Ensign ?--unknown (1/2)

5ft(1.5m)         -10°F(-23°C)          ML          3/3

Reg. as a ponticum seedling.  Flowers essentially blue, 3in
(7.6cm) across, with pansy violet edging and a blotch of dahlia
purple in rounded, upright trusses of 15. Dense, glossy green
foliage; tolerates heat and sun.  (H. L. Larson, 1964)

cl.                          -ponticum (1/2)
          -Purple Splendour-
Blue Lagoon*          -unknown (1/2)
          -          -mauve seedling--unknown
          -A. Bedford-
                      -ponticum

5ft(1.5m)         -10°F(-23°C)          ML          3/4

Smoky blue purple flower with a large deep purple eye.  Deep forest green recurved leaves. (Harold Greer cross; raised by G. Bax-
ter, W. Thompson)

cl.

Blue Light--form of brachyanthum ssp. hypolepidotum

Flowers pale aureolin yellow.  Leaves covered beneath with
dense, silvery scales. (Crown Estate, Windsor)  A. M. (RHS) 1951

                              -intricatum (1/8)
---                -Intrifast-
          -Blue Diamond-          -fastigiatum (1/8)
Blue Moon-          -augustinii ssp. augustinii (3/4)
          -augustinii ssp. augustinii

(Lord Digby, 1955)

cl.                          -ponticum (1/4)
          -Purple Splendour-
Blue Pacific-          -unknown (1/4)
          -          -campanulatum ssp. campanulatum (1/4)
          -Susan-
                    -fortunei ssp. fortunei (1/4)

Bluish purple flowers with a distinct deep blotch, of good substance, openly funnel-shaped with 5 wavy lobes.  Ball-shaped
trusses hold 10-16 flowers.  Plant broad, branching moderately;
glossy, narrowly elliptic leaves.  Prefers some shade.  (William
E. Whitney cross, 1965; reg. by the Sathers, 1976)  See color
illus. HG p. 142;  K p. 109.


cl.

Blue Peter      Parentage unknown.  (May be ponticum hybrid)

4.5ft(1.4m)         -15°F(-26°C)          M          4/3

Flowers light lavender blue with a striking purple flare,  very
frilly edges, held in tight conical trusses.  Glossy dark green
leaves tending to roll under, on a plant wider than tall.  Grows
in sun or shade;  propagates easily. (Waterer & Crisp)    A. M.
(RHS) 1933; F. C. C. (RHS) 1958: A. G. M. (RHS) 1969  See color
illus. F p. 34;  VV p. 17, 68.


                            -impeditum (3/8)
cl.            -Blue Tit-
        -Sapphire-         -augustinii ssp. augustinii (5/8)
Blue Pool-      -impeditum
        -augustinii ssp. augustinii

A compact shrub of slow growth, 3-4ft(.9m-1.2m) in height and in
breadth.  Flowers lavender blue.  (Knap Hill, 1966)


cl.               -ponticum (1/2)
        -A. Bedford-
Blue Rhapsody-         -unknown (1/2)
        -                      -ponticum
        -Purple Splendour-
                           -unknown

5ft(1.5m)          -5°F(-21°C)          ML          4/3

Flowers 3.5in(8.9cm) wide, medium purple with darker spotting in
throat, up to 12 in ball-shaped trusses.  Upright plant, stiffly
branched; leaves yellowish green, held 2 years.  (William Whitney cross, 1962; intro. by the Sathers, 1970)


g.               -impeditum (1/4)
        -Blue Tit-
Blue Ribbon-      -augustinii ssp. augustinii (3/4)
        -augustinii ssp. augustinii

Blue flowers.  See also Greeneye, Plaineye, Purple Eye.  (G. H.
Johnstone, 1954)


cl.               -russatum (1/4)
        -Russautinii-
Blue Ridge-         -augustinii ssp. augustinii (3/4)
        -
-augustinii ssp. augustinii

2ft(.6m)          -12°F(-24°C)          M

Flowers campanula violet, paler in throat, slightly fragrant, in
dome-shaped trusses of 4-5.  Floriferous.  Plant rounded, well-
branched, about as wide as high, with fragrant foliage.  (Velma
& Russell Haag cross, 1966; T. Richardson, intro. 1975; reg. by
the Haags, 1981)


                        -Sir Charles Butler--fortunei ssp.
cl.      -Van Nes Sensation-                fortunei (1/4)
        -                      -griffithianum (1/8)
Blue River-            -White Pearl-
        -                      -maximum (1/8)
        -Emperor de Maroc--unknown (1/2)

5ft(1.5m)          5°F(-15°C)          M          3/3

Flowers campanula violet, openly bell-shaped,  4in(10.2cm) wide;
upright trusses hold about 17.  Compact bush; glossy, dark green
foliage.  (Ruth Lyons, reg. 1962)  A. E. (ARS) 1961


---
            -augustinii ssp. augustinii (1/2)
Blue Sky-
        -rubiginosum Desquamatum Group (1/2)

Blue flowers.  (Lord Digby cross, 1940; intro. 1955)


cl.               -impeditum (3/4)
        -Saint Tudy-
Blue Star-         -augustinii ssp. augustinii (1/4)
        -impeditum

Mauve blue flowers, 1.25in(3.2cm) across.  (Gen. Harrison, 1961)


cl.

Blue Steel--form of impeditum  (selfed)

1ft(.3m)          -15°F(-26°C)          EM          4/4

Densely twiggy  dwarf shrub, with silver grey leaves .5in(1.3cm)
long.  Flowers light violet blue in clusters of 3.  (Raiser G. H.
White; reg. by J. P. C. Russell, 1983)


g. & cl.
        -intricatum (1/2)
Bluebird-
        -augustinii ssp. augustinii (1/2)

3ft(.9m)          0°F(-18°C)          EM          4/3

Dense, small leaves create a finely textured plant, wider than
high, having numerous electric blue flowers.  Best in open sun.
(Lord Aberconway)  P. C. (RHS) 1937; A. M. (RHS) 1943;  A. G. M.
(RHS) 1968


* * * * * * * * * * * * * * * * * * * * * * * * * * * * * * * *
*                                                             *
*                                                             *
*  The named clones of a prolific hybridizer, Weldon E.       *
*  Delp, are listed in App. E.                                *
*                                                             *
*                                                             *
*                                                             *
* * * * * * * * * * * * * * * * * * * * * * * * * * * * * * * *


g.               -intricatum (1/4)
        -Bluebird-
Bluestone-         -augustinii ssp. augustinii (3/4)
        -augustinii ssp. augustinii

Terminal clusters of deep lilac blue flowers; early.  (Lord
Aberconway, 1950)


cl.
        -impeditum (1/2)
Bluette-
        -Lackamas Blue--augustinii ssp. augustinii (1/2)

3ft(.9m)          -5°F(-21°C)          M          3/4

Abundant flowers hyacinth blue, in trusses of 8.  Twiggy, globe-
shaped plant; scaly leaves  1.5in(3.8cm) by .5in(1.3cm).  Heat-
tolerant; best in full sun.  (Benjamin A. Lancaster, reg. 1958)


cl.
        -Blue Peter--unknown (1/2)
Blurettia-
        -Koichiro Wada--yakushimanum ssp. yakushimanum (1/2)

Flower trusses hold 11-14, color medium mallow purple on wavy
margins, shading inwards much lighter.  Leaves ovate, hairy.
(H. Hachmann cross; reg. by G. Stuck, 1983)

'Ben Moseley' by Dexter/Moseley
Photo by Greer

'Berryrose' by Rothschild
Photo by Greer

'Betsie Balcom' by McGuire
Photo by Greer

'Betty Arrington' by Dexter/Arrington
Photo by Greer

'Betty Hume' by Dexter/Baldsiefen
Photo by Leach

'Biskra' by Rothschild
Photo by Greer

'Black Magic' by Greer
Photo by Greer

'Black Prince' by Brandt
Photo by Greer

'Blaney's Blue' by Blaney/Ticknor
Photo by Greer

'Blaze' by Leach
Photo by Leach

'Blazen Sun' by Leach/Pride
Photo by Leach

'Blue Boy' by Watson
Photo by Greer

'Blue Crown' by K. Van de Ven
Photo by Greer

'Blue Jay' by Larson
Photo by Greer

'Blue Lagoon' by Greer/Baxter/Thompson
Photo by Greer

'Blue Rhapsody' by Whitney/Sather
Photo by Greer

'Blue Ridge' by Haag
Photo by Greer

'Blue River' by R. Lyons
Photo by Greer

'Bob's Blue' by Rhodes (Canada)
Photo by Greer

'Bob Peters Special' by Peters
Photo by Peters

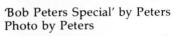
'Bonito' by Rothschild
Photo by Greer

'Bonnie Babe' by J. A. Elliott
Photo by Greer

'Bo-peep' by Rothschild
Photo by Greer

'Boule de Rose' by Shammarello/Leach
Photo by Leach

```
                    -dichroanthum ssp. dichroanthum (1/8)
g.       -Astarte-        -campylocarpum ssp. campylocarpum
    -Ouida-    -Penjerrick-               Elatum Group (1/16)
Blush-   -                -griffithianum (1/16)
     -   -griersonianum (1/4)
     -aperantum (1/2)
```

Pale pink flowers. (Lord Aberconway cross, 1933; intro. 1944)


cl.


Blush     See Carex Blush


cl.

              -C. O. D. (Dexter hybrid)--unknown
Blush Button-
              -Honeydew (Dexter hybrid)--unknown

5ft(1.5m)          -10°F(-23°C)          M

Coral red buds open to cream flowers 3in(7.6cm) across, held in
flat trusses of 8. Plant spreads 1.5 times its height. (Mrs. J.
Knippenberg, reg. 1976)

---
                 -arboreum ssp. arboreum (1/2)
Blushing Beauty-
                 -unknown (1/2)

Blush pink flowers. (R. Gill & Son)


```
g.                    -griffithianum (1/4)
          -Dawn's Delight-
Blushing Bride-       -unknown (1/4)
          -fortunei ssp. discolor (1/2)
```

Flowers of rosy carmine. Not in cultivation at Exbury. (Rothschild, 1934)


```
g.
       -thomsonii ssp. thomsonii (1/2)
Boadicea-
       -fortunei ssp. discolor (1/2)
```

A tall, upright plant with waxy crimson flowers like thomsonii.
Free-flowering; blooms early. (Rothschild, 1934)


cl.
      -Koichiro Wada--yakushimanum ssp. yakushimanum (1/2)
Bob Bovee-
      -wardi var. wardii (1/2)

2.5ft(.75m)          0°F(-18°C)          M

Flowers primrose yellow with red spot changing to green in the
throat, 3in(7.6cm) wide; trusses hold up to 12. Plant dense and
well-branched; glossy dark green leaves retained 3 years. (Robert M. Bovee, raiser; intro. by Sorensen and Watson; reg. 1976)


cl.         -williamsianum (1/4)
      -Adrastia-
Bob Cherry-      -neriiflorum ssp. neriiflorum (1/4)
      -forrestii ssp. forrestii Repens Group (1/2)

Flowers of deep crimson red. (Sunningdale Nurseries, 1951)


Bob Peters Special--seedling of fortunei ssp. fortunei

4ft(1.2m)          -15°F(-26°C)          M

Very fragrant flowers of light purplish pink with pale orange
yellow throat, openly funnel-shaped, 3.5in(8.9cm) across with 7
wavy, crinkled lobes. Flat trusses 7in(17.8cm) wide, hold 8-10
blooms. Floriferous. Medium olive green leaves held 2 years;
new twigs bright yellow green. Plant upright, 1.5 times wider
than tall. (Raised and reg. 1983 by Robert K. Peters; intro. by
Ray Carter)


cl.
      -campylogynum (1/2)
Bobbet-
      -campylogynum Cremastum Group (1/2)

2ft(.6m)          -5°F(-21°C)          M          3/3

Flowers of greenish golden yellow on long pedicels; leafy green,
prominent calyces. Small obovate leaves tend to have yellowish
edges. (Mr. & Mrs. James Caperci, 1966)


g.
      -fortunei ssp. discolor (1/2)
Bobolink-
      -neriiflorum ssp. neriiflorum (1/2)

5ft(1.5m)          0°F(-18°C)          L          3/3

A free-flowering shrub with deep yellow or apricot flowers.
(Rothschild, 1934)


```
                       -augustinii ssp. chasmanthum (1/8)
                   -Electra-
cl.    -Ilam Violet-    -augustinii ssp. augustinii (3/8)
       -          -russatum (1/4)
Bob's Blue-            -intricatum (1/8)
          -Intrifast-
       -Blue Diamond-   -fastigiatum (1/8)
              -augustinii ssp. augustinii
```

Violet-colored flowers, in trusses of 3-5. (Robert C. Rhodes,
1979)


```
cl.  -wardii var. wardii (1/2)
     -                      -caucasicum (3/32)
Bob's -           -Jacksonii-    -caucasicum
Yellow-    -Goldsworth-    -Nobleanum-
     -    - Yellow -            -campylocarpum
     -Goldfort-                 ssp. campylocarpum
          -campylocarpum
     -    ssp. campylocarpum (5/32)
     -fortunei ssp. fortunei (1/4)
```

Trusses of 10 yellow flowers with yellow green flare in throat.
(Robert C. Rhodes, 1979)


cl.
        -campanulatum ssp. campanulatum (1/2)
Boddaertianum-
(Bodartianum)-
        -arboreum ssp. cinnamomeum var. album (1/2)

6ft(1.8m)          0°F(-18°C)          E

Compact, rounded trusses of widely funnel-shaped flowers, lavender pink in bud, opening white, with a ray of crimson purple
markings. A striking rhododendron with a long blooming season.
(Van Houtte, 1863)


cl.             -ciliatum (1/4)
         -Cilpinense-
Bodega Crystal Pink-   -moupinense (1/4)
         -Cornell Pink--mucronulatum (1/2)

3.5ft(1m)          5°F(-15°C)          E

Flowers openly funnel-shaped, flax pink with red purple spotting
in trusses of 3-7. Leaves small, elliptic, light green, semi-
deciduous. Plant upright, branching moderately. (Carl Heller,
1975)


```
* * * * * * * * * * * * * * * * * * * *
* Hybrids of unknown parentage that have won awards, or are *
* a parent of another hybrid, are in the text; for all oth- *
* ers of unknown parentage, see App. D.                     *
* * * * * * * * * * * * * * * * * * * *
```

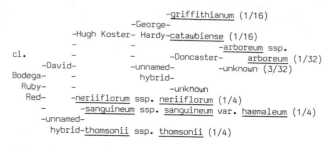

```
                           -griffithianum (1/16)
                     -George-
          -Hugh Koster- Hardy-catawbiense (1/16)
cl.        -            -              -arboreum ssp.
           -            -    -Doncaster-    arboreum (1/32)
     -David-            -unnamed-        -unknown (3/32)
Bodega-    -            hybrid-
  Ruby-    -                        -unknown
  Red-     -neriiflorum ssp. neriiflorum (1/4)
      -    -sanguineum ssp. sanguineum var. haemaleum (1/4)
    -unnamed-
       hybrid-thomsonii ssp. thomsonii (1/4)
```

5ft(1.5m)          5°F(-15°C)          EM

Flowers currant red with black spotting on upper lobe,  of heavy
substance, 5-lobed.  Flattened truss holds 7 flowers.  Plant upright; narrowly elliptic leaves retained 3-4 years.  (Carl Heller, 1975)

```
cl.                    -venator (1/4)
              -Vanguard-
Bodega Toreador-       -griersonianum (1/2)
     -                 -griersonianum
              -Matador-
                       -strigillosum (1/4)
```

3ft(.9m)          5°F(-15°C)          M

Flowers signal red with red spotting on upper lobes, widely funnel-campanulate.  Lax truses hold 8 flowers.  Floriferous.  Foliage narrowly elliptic  to lanceolate with tan indumentum, held
3-4 years on a rounded, well-branched plant  (Carl Heller, 1975)

```
                         -griffithianum (9/16)
                  -Kewense-
             -Aurora-       -fortunei ssp. for-
cl.      -Yvonne -    -              tunei (1/16)
         -Opaline-   -thomsonii ssp. thomsonii (1/8)
Bodega y Quadra-    -griffithianum
             -          -arboreum ssp. cinnamomeum var.
          -Loder's White-            album (1/4)
                      -griffithianum
```

5.5ft(1.65m)          12°F(-11°C)          M

(See Loder's White for other possible parentage.)  Flowers
bright pink blending to lighter throat, red spot, 4.25in(11cm)
across, widely funnel-campanulate, of heavy substance.  Conical
trusses 10in(25.4cm) wide, hold 9 flowers.  Plant nearly as wide
as high; leaves retained 1-2 years.  (Carl G. Heller, reg. 1977)

```
                        thomsonii ssp. thomsonii
cl.      -Ascot Brilliant-                (1/8)
    -unnamed hybrid-         -unknown (1/8)
Bodil-         -neriiflorum ssp. neriiflorum (1/4)
     -haematodes ssp. haematodes (1/2)
```

Red flowers.  (Exhibited by Acton, 1950)

```
cl.
                   -griffithianum (1/2)
Bodnant Beauty of Tremough-
                   -arboreum ssp. arboreum (1/2)
```

Beauty of Tremough g.  Pale pink flowers; scented.  (Lord Aberconway, 1948)

cl.

Bodnant Day    See Budget Farthing

```
cl.
            -haematodes ssp. haematodes (1/2)
Bodnant May Day-
            -griersonianum (1/2)
```

May Day g.  Deep scarlet flowers, scented.  (Lord Aberconway,
1939)

cl.

Bodnant Pink--form of arboreum ssp. cinnamomeum var. album

introduced by Lord Aberconway.

cl.

Bodnant Red--campylogynum Cremastum Group

2.5ft(.75m)          -10°F(-23°C)          EM          4/4

Flowers purplish red, 1in(2.5cm) across, 5-lobed, on long pedicels singly or in small clusters.  Leaves small, elliptic, dark
green.  (Crown Estate Commissioners)  A. M. (RHS) 1971

```
cl.
            -thomsonii ssp. thomsonii (1/2)
Bodnant Thomwilliams-
            -williamsianum (1/2)
```

Thomwilliams g.  Waxy rose magenta.  (Lord Aberconway, 1935)
A. M. (RHS) 1935

```
cl.
          -cinnabarinum ssp. cinnabarinum (yellow form)
Bodnant Yellow-                          (1/2)
     -            -cinnabarinum ssp. cinnabarinum (1/4)
          -Royal Flush-
          (orange form)-maddenii ssp. maddenii (1/4)
```

5ft(1.5m)          10°F(-12°C)          ML          4/3

Lady Chamberlain g.  Flowers butter yellow to orange buff with a
deeper reddish flush.  (Lord Aberconway)  F. C. C. (RHS) 1944

```
                           -griffithianum (1/8)
cl.            -Queen Wilhelmina-
       -Unknown Warrior-        -unknown (3/8)
Bold Lad-          -Stanley Davies--unknown
       -arboreum ssp. arboreum (1/2)
```

Trusses of 22 red flowers, slightly spotted.  (T. Lelliot cross;
intro. by G. Langdon, 1976)

```
cl.            -elliottii (1/4)
         -Fusilier-
Bombardier-       -griersonianum (1/4)
         -griffithianum (1/2)
```

Bright pink flowers 4.5in(10.8cm) across, 3in(7.6cm) high; about
16 to a truss.  (Gen. Harrison, 1961)

```
                 -unnamed-dichroanthum ssp. scyphocalyx (1/8)
cl.    -unnamed- hybrid-
     - hybrid-        -kyawii (1/8)
Bombay-      -Catalgla--catawbiense var. album Glass (1/2)
         -Catalgla--catawbiense var. album Glass
       -unnamed-
         hybrid-wardii var. wardii (1/4)
```

3ft(.9m)          -10°F(-23°C)          EM          3/2

Flowers of greenish yellow, darker on the outside, with a blotch
of vivid greenish yellow; mound-shaped trusses hold 12-14.  Foliage dark green, elliptic, on a plant of superior density.  (David G. Leach, 1972)

```
g.
     -souliei (1/2)
Bonbon-
     -maximum (1/2)
```

Trusses  of  cream-colored flowers  on a  medium-sized,  compact
shrub.  (Rothschild cross, 1927; intro. 1934)

cl.                      -fortunei ssp. discolor (1/4)
          -unnamed hybrid-
Bonfire-          -Mrs. R. G. Shaw--unknown (1/4)
          -griersonianum (1/2)

5ft(1.5m)        -5°F(-21°C)        ML        3/3

Fiery red flowers in large, loose, conical trusses. Leaves dull green, 6in(15.2cm) by 2in(5cm). Plant habit better in open sun. (Waterer, Sons & Crisp, 1928) A. M. (RHS) 1933

g. & cl.
        -fortunei ssp. discolor (1/2)
Bonito-          -fortunei ssp. fortunei (1/4)
        -Luscombei-
                  -thomsonii ssp. thomsonii (1/4)

5ft(1.5m)        -5°F(-21°C)        L        4/4

Flowers pale blush pink, fading to white, with heavy brown spotting on the upper corolla, up to 5in(12.7cm) wide; large rounded trusses. Dark green leaves 7in(17.9cm) by 2.75in(7cm) on an upright, vigorous plant. (Rothschild, 1934) A. M. (RHS) 1934

cl.        -griffithianum (1/4)
        -Mars-
Bonnie*    -unknown (1/4)
        -          -catawbiense (1/4)
        -unnamed hybrid-
                  -haematodes ssp. haematodes (1/4)

Plant hardy to at least -10°F(-23°C). Pale buff flowers. Midseason. (Joseph Gable)

                          -wardii var. wardii (?) (1/8)
cl.              -Chlorops-
        -Inca Gold-          -vernicosum ? (1/8)
Bonnie Babe-          -unknown (3/4)
        -orange hybrid--unknown

3ft(.9m)        10°F(-12°C)        ML

Synonym Cecil's Choice. Flowers barium yellow, more intense in throat, spotted cardinal red, held in dome-shaped trusses of 10-14. Plant semi-dwarf, broader than high with leaves yellow green. (James A. Elliott, 1983)

cl.        -Scintillation--unknown (5/8)
        -                          -catawbiense (1/8)
Bonnie Brae-          -Atrosanguineum-
        -Red Head (Gable)-          -unknown
                          -griersonianum (1/4)

6ft(1.8m)        -5°F(-21°C)        M

Flowers 4in(10.2cm) wide, 7-lobed, fragrant, orchid pink with edges darker, yellowish green blotch; trusses hold about 17. Leaves 7in(17.8cm) by 3in(7.6cm). (Charles Herbert, reg. 1977)

cl.
        -fortunei ssp. fortunei (1/2)
Bonnie Maid-
        -unknown (1/2)

5ft(1.5m)        -10°F(-23°C)        M

Saucer-shaped flowers 2.75in(7cm) wide, 7-lobed, pink with red dorsal rays; up to 13 in a ball-shaped truss. Leaves 5.5in(14cm) by 2in(5cm). (Cross by Maurice E. Hall; intro. by Edward J. Brown; reg. 1979)

                  -Sir Charles Butler--fortunei ssp.
cl.    -Van Nes Sensation-                      fortunei (1/4)
        -          -          -maximum (1/8)
Bonnie-          -Halopeanum-
Snaza -          -griffithianum (1/4)
        -          -griffithianum
        -          -Mars-
        -unnamed-    -unknown (1/4)
          hybrid-          -ponticum (1/8)
                  -Purple Splendour-
                          -unknown

5ft(1.5m)        10°F(-12°C)        M

Flowers 4in(10cm) wide, lilac purple shading lighter with heavy spotting of ruby red in ball-shaped trusses 7in(17.8cm) tall. Well branched plant 2/3 as wide as tall with leaves yellow green, held 2 years. See also Betty Anderson. (Edwin (Ted) Anderson, 1983)

g. & cl.
        -lutescens (1/2)
Bo-peep-
        -moupinense (1/2)

3ft(.9m)        5°F(-15°C)        E        3/3

Flat, funnel-shaped flowers of pale greenish yellow, with darker yellow spotting on the back of the tube. Truss of 2-3 flowers; often multiple buds on each stem gives the appearance of many more flowers per truss. Leaves small, somewhat glossy, light green. Worth cultivating for bright, early flowers; a good companion plant for P.J.M. (Rothschild, 1934) A. M. (RHS) 1937 See color illus. F p. 36.

cl.              -arboreum ssp. arboreum (1/4)
        -Doncaster-
Borde Hill-          -unknown (1/2)
        -          -griffithianum (1/4)
        -Mrs. A. M. Williams-
                  -unknown

3ft(.9m)        5°F(-15°C)        ML        4/3

Waxy flowers, dark scarlet, medium-sized, held in rounded trusses. Large leaves, dark green, pointed. Plant of upright habit, medium density; not vigorous. (C. B. van Nes) A. M. (Wisley Trials) 1948

cl.
        -sanguineum ssp. didymum (1/2)
Borde Hill Red Cap-
        -facetum (eriogynum) (1/2)

Red Cap g. Flowers are red. (Col. S. R. Clarke, before 1958)

cl.
        -microgynum Gymnocarpum Group (1/2)
Bordeaux-
        -beanianum (1/2)

5ft(1.5m)        5°F(-15°C)        EM        2/2

Synonym Burgundian. Flowers dark red, waxen. Compact plant; small, dark green, glossy foliage with brown indumentum beneath. (E. de Rothschild, 1963)

                  -facetum (eriogynum) (1/8)
        -unnamed-
        - hybrid-unknown (5/16)
cl.    -unnamed-    -dichroanthum ssp. dichroanthum (1/8)
        - hybrid-Fabia-
Borderer-          -griersonianum (1/8)
        -          -yakushimanum ssp. yakushimanum (1/4)
        -unnamed-          -griffithianum (1/16)
          hybrid-    -Queen Wilhelmina-
                  -Britannia-          -unknown
                          -Stanley Davies--unknown

Flowers of China rose, deepening at margins. (Waterer, Sons & Crisp, cross 1958; intro. 1975)

                          -catawbiense (1/8)
g.              -Parsons Grandiflorum-
        -America-          -unknown (3/8)
Borkum-        -dark red hybrid--unknown
        -williamsianum (1/2)

Clear rose flowers. (Dietrich Hobbie, 1947)

cl.
```
        -garden hybrid--unknown (7/8)   -caucasicum (1/8)
Boskoop-        -Chevalier F. de Sauvage-
        -Max Sye-                    -unknown
                -unknown
```

These rosy red flowers have a dark brown blotch and are held in
round trusses of about 13.  (Adr. van Nes) A. M. (after Boskoop
trial) 1960


cl.
```
                -caucasicum (1/4)
        -Boule de Neige-        -catawbiense (1/8)
Bosutch*            -unnamed-
        -            hybrid-unknown (1/8)
        -sutchuenense (1/2)
```

5ft(1.5m)          -5°F(-21°C)          E          3/4

Good plant habit with dark green foliage.  Flowers white with
lavender spotting; campanulate.  (Joseph B. Gable)


cl.

Bosutch Pink*      Parentage as above

5ft(1.5m)          -5°F(-21°C)          E          3/4

Sibling of Bosutch, blooming a week later.  Pink flowers.  (Joseph Gable)


cl.
```
                        -catawbiense (1/2)
Botha--Everestianum (selfed)-
                        -unknown (1/2)
```

Flowers pale lilac pink with ochre markings on a lighter ground;
frilled.  (T. J. R. Seidel, cross 1892)


cl.
```
            -caucasicum (1/2)
Boule de Neige-        -catawbiense (1/4)
            -unnamed hybrid-
                        -unknown (1/4)
```

4ft(1.2m)          -25°F(-32°C)          M          4/4

"Ball of Snow" is considered one of the ironclads.  Super hardy.
Flowers appear white in medium-sized, compact, round trusses.
Leaves are heavily textured, medium large, light green.  Plant
forms a dense, round bush; vigorous and fairly easy to grow with
good drainage.  An old hybrid, still popular.  (Oudieu, 1878)
See color illus. VV p. 31.


cl.
```
                        -catawbiense (3/8)
            -unnamed hybrid-
Boule de Rose-                -unknown (3/8)
        -                -caucasicum (1/4)
            -Boule de Neige-        -catawbiense
                    -unnamed hybrid-
                                -unknown
```

3ft(1.2m)          -25°F(-32°C)          EM          2/4

Flowers bright rose pink, slight brownish flare on upper lobe.
Low, compact plant.  (Cross by A. M. Shammarello; raised and intro. by David G. Leach, 1957)


```
---               -Duchess of Teck--unknown (1/4)
        -Lord Wolseley-
Boule D'Or-        -javanicum (1/4)
        -javanicum var. teysmannii ? (1/2)
```

Vireya hybrid exhibited by J. Veitch, 1891.


cl.
```
                -caucasicum (1/4)
        -Boule de Neige-            -catawbiense (1/8)
        -            -unnamed hybrid-
Boulodes-                    -unknown (3/8)
        -                -griffithanum (1/8)
        -            -Loderi-
        -unnamed hybrid-    -fortunei ssp. fortunei (1/8)
                -unknown
```

5ft(1.5m)          -15°F(-26°C)          ML

Flowers fragrant, light pink with faint yellow spotting, turning
white, openly funnel-shaped, 3.5in(8.9cm) across, frilled edges.
Trusses ball-shaped, 10-flowered.  Plant rounded, well-branched;
elliptic deep green leaves, very flat, retained 2 years.  (G. Guy
Nearing, 1963; reg. 1973)


g.
```
        -davidsonianum (1/2)
Bouquet-    -rigidum (caeruleum var. album) (1/4)
        -Peace-
            -cinnabarinum ssp. xanthocodon Concatenans
                                Group (1/4)
```

White flowers.  (Lord Aberconway, 1950)


cl.

Boursault      See Catawbiense Boursault


g. & cl.
```
        -Corona--unknown (1/2)
Bow Bells-
        -williamsianum (1/2)
```

Light pink flowers deeper on reverse; trusses of 4 to 7.  Floriferous.  Rounded, medium-sized leaves; new growth bronze.  Requires some shade; propagates easily.  (Rothschild, 1934)  A. M.
(RHS) 1935    See color illus. F p. 36;  HG p. 131;  VV p. 10.


cl.
```
                -Corona--unknown (1/4)
        -Bow Bells-
        -            -williamsianum (1/4)
Bow Street-                -griffithianum (1/8)
        -            -Socrianum-
        -unnamed-            -dichroanthum ssp. scyphocalyx (1/8)
            hybrid-    -wardii var. wardii (1/8)
                -Rima-
                    -decorum (1/8)
```

Pale yellow flowers with slight pink flushing.  Trusses of 8-10.
(A. F. George, Hydon Nurseries; intro. 1975)


cl.
```
        -minus var. chapmanii (1/2)
Bowie-
        -minus var. minus Carolinianum Group (1/2)
```

5ft(1.5m)          0°F(-18°C)          L

Many pink flowers 1.5in(3.8cm) wide, pale brownish green blotch;
terminal ball-shaped clusters from 2 or 3 buds, each 10-14 flow-
ed.  Plant vigorous, compact; leaves held 1-2years.  (Henry T.
Skinner cross, 1953; intro. by U. S. National Arboretum, 1971)


g.
```
        -brachyanthum ssp. brachyanthum (1/2)
Brachbooth-
        -boothii (1/2)
```

Flowers butter yellow with a green tinge, darker spots.  (E. J.
P. Magor cross, 1920; intro. 1926)


g.
```
        -brachycarpum ssp. brachycarpum (1/2)
Brachdis-
        -fortunei ssp. discolor (1/2)
```

Flowers blush yellow, with green spotting.  (E. J. P. Magor, intro. 1925)

g.
        -brachyanthum ssp. brachyanthum (1/2)
Brachlep-
        -lepidotum (1/2)

Flowers yellow to pink. (E. J. P. Magor cross, 1920; intro. 1924)

g.
        -brachycarpum ssp. brachycarpum (1/2)
Brachsoul-
        -souliei (1/2)

Flowers spotted crimson, light rose shading to white, darker outside. (E. J. P. Magor cross, 1917; intro. 1927)

g.
        -brachyanthum ssp. brachyanthum (1/2)
Brachydum-
        -flavidum var. flavidum (1/2)

Yellow flowers. (J. Waterer, 1921)

cl.
        -brachyanthum ssp. brachyanthum (Wilson 6771) (1/2)
Brachydum-
 Primum  -flavidum var. flavidum (Wilson 1773) (1/2)

Pale yellow flowers, tubular-shaped. (Collector, Wilson; hybrid's origin unknown) A. M. (RHS) 1924

cl.
        -souliei (1/2)
Bradfield*
        -yakushimanum ssp. yakushimanum (1/2)

White flowers, edges flushed reddish purple, upper throat spotted red purple, reverse more heavily flushed with same. (Crown Estate, Windsor) P. C. (RHS) 1979

cl.
        -brachycarpum ssp. brachycarpum (1/2)
Bramax*
        -maximum (1/2)

White flowers. Late-blooming; hardy to -20°F(-29°C). (J. Gable)

cl.
        -pubescens (1/2)
Brandywine-
        -keiskei (1/2)

Low        -10°F(-23°C)        EM        4/2

Small cream-colored flowers edged in rose, in 2in(5cm) spherical trusses. Leaves 3in(7.6cm) by .5in(1.3cm) wide. (G. Guy Nearing, 1950)

cl.
        -auriculatum (1/2)
Brass Rubber-
        -griersonianum (1/2)

Flowers white, irregularly suffused red, with coloring more concentrated in throat and at edges; reverse is red, densely covered with glutinous hairs. Leaves narrowly elliptic, lightly covered with reddsh brown indumentum beneath. (Crossed at Bodnant 1938; Lord Aberconway intro. and reg., 1978) A. M. (RHS) 1978

cl.
        -catawbiense var. album (1/2)
Bravo!-              -fortunei ssp. fortunei (1/4)
    -unnamed hybrid-              -arboreum ssp arboreum (1/8)
              -unnamed hybrid-
                      -griffithianum (1/8)

6ft(1.8m)        -25°F(-32°C)        E        4/4

Trusses of 11-12 flowers, light purplish pink shading to lighter in center with sparse dorsal brown spotting. See also Dolly Madison. (David G. Leach, 1970)   See color illus. HG p. 79.

cl.
            -griffithianum (1/4)
  -unnamed hybrid-
Bray-        -unknown (1/4)
    -    -wardii var. wardii (1/4)
    -Hawk-        -fortunei ssp. discolor (1/8)
        -Lady    -
        Bessborough-campylocarpum ssp. campylocarpum
                              Elatum Group (1/8)

Deep pink buds opening to soft mimosa yellow, upper lobes paler; reverse shaded pink. (Crown Estate, Windsor) A. M. (RHS) 1960

g. & cl.                -griffithianum (1/4)
        -Dawn's Delight-
Break of Day-        -unknown (1/4)
        -dichroanthum ssp. dichroanthum (1/2)

4ft(1.2)        -5°F(-21°C)        3/1

Open, uncrowded trusses of 14 flowers, funnel-shaped, deep orange at the base and edged with orange pink--unusual colors for a broad-leafed rhododendron. (Rothschild, 1934) A. M. 1936

cl.
        -haematodes ssp. haematodes (1/2)
Bremen-
        -red hybrid--unknown (1/2)

Dwarf plant with leaves elliptic to ovate, 2-3.5in(5-9cm) long, 1.25in(3cm) broad, silver green. Loose trusses of up to 12 flowers, campanulate, scarlet with small red calyx; early. (Cross by Georg Arends; intro. by G. D. Bohlje, 1963)

cl.
        -J. H. Agnew--unknown (1/2)
Brenda-
        -griersonianum (1/2)

A plant of medium size, good habit and compact, pink trusses. This shrub no longer in cultivation at Exbury. (Rothschild, 1927) P. C. (RHS) 1935

cl.
        -catawbiense var. album (1/2)
Brenda Lee-              -ponticum (1/4)
        -Purple Splendour-
                      -unknown (1/4)

3ft(.9m)        -20°F(-29°C)        ML

Flowers openly funnel-shaped with 5 frilled lobes, bishop's violet with a blotch of lighter color. Truss is 7in(17.8cm) across, ball-shaped, of 11 flowers. Plant wider than tall with leaves held 2 years. (Henry Yates cross; intro. by Mrs. Yates, 1977)

cl.
        -griffithianum (1/2)
Brentor-    -wardii var. wardii (1/4)
    -Hawk-        -fortunei ssp. discolor (1/8)
        -Lady    -
        Bessborough-campylocarpum ssp. campylocarpum
                              Elatum Group (1/8)

Flowers red purple fading to cream with a basal red blotch. (L. S. Fortescue, 1972)

g. & cl.
        -leucaspis (1/2)
Bric-a-brac-
        -moupinense (1/2)

3ft(.9m)        5°F(-15°C)        VE        4/3

Flowers snowy white with faint pink markings on upper lobes; anthers chocolate, making an interesting contrast. Small, round, pubescent leaves; new growth bronze. Bark on older plants shiny and papery, peeling. Needs frequent pruning. (Rothschild, 1934) A. M. (R.H.S.) 1945

cl.
        -williamsianum (1/2)
Brickdust-    -dichroanthum ssp. dichroanthum (1/4)
        -Dido-
            -decorum (1/4)

3ft(.9m)       -5°F(-21°C)      M    4/4

Flowers 3in(7.6m) wide, rose madder shaded to rhodonite red, 6-8
in lax trusses. Rounded leaves, 2in(5cm) by 1in(2.5cm).  Plant
grows vigorously, propagates well.  (R. Henny, 1960)

cl.

Bride    See The Bride

cl.

Bridge    See The Bridge

cl.           -wardii var. wardii Litiense Group (1/4)
      -Henry R. Yates-
Bridge-         -unknown (3/8)
 North-         -caucasicum (1/4)
    -Boule de Neige-       -catawbiense (1/8)
              -unnamed hybrid-
                    -unknown

3.5ft(1m)      -5°F(-21°C)     ML

Flowers openly funnel-shaped, 5 or 6-lobed, neyron rose,  fading
lighter with age, throat spotted gold, in ball-shaped trusses of
14.  Plant upright, branching moderately; leaves held 2 years.
(Charles Herbert cross, 1968; intro. by W. A. Reese, 1977)


cl.             -griffithianum (1/4)
      -Loderi Helen-
Bridgeport*      -fortunei ssp. fortunei (1/4)
     -strigillosum (1/2)

(H. L. Larson)


g. & cl.        -griffithianum (1/4)
      -Dawn's Delight-
Brigadier-        -unknown (1/4)
     -arboreum ssp. arboreum (1/2)

A tall plant with loose habit and  large, pale pink trusses.
(Rothschild, 1934)


             -unnamed-fortunei ssp. discolor (5/32)
        -unnamed- hybrid-
       - hybrid-    -unknown (1/8)
    -unnamed-    -facetum (eriogynum) (1/8)
   - hybrid-
   -     -unnamed-fortunei ssp. discolor
   -   -unnamed- hybrid-
cl.  -    - hybrid-    -unknown
Bright-            -griersonianum (5/16)
Future-              -fortunei ssp. dis-
 -         -Lady   -      color
 -        -Bessborough-campylocarpum ssp.
 -     -Jalisco-    campylocarpum Elatum Group
 -     -Eclipse-            (1/32)
 -    -unnamed-    -dichroanthum ssp. dichro-
 -    - hybrid-    -Dido-      anthum (5/32)
 -    -     -   -decorum (1/32)
 -unnamed-     -   -ellottii (1/16)
  hybrid-    -Fusilier-
     -        -griersonianum
     -        -dichroanthum ssp. dichroanthum
    -Fabia (Waterer)-
            -griersonianum

Orient pink flowers, upper lobes heavily marked in citron green.
(Cross by John Waterer, Sons & Crisp, 1958; intro. 1975)

cl.
      -griffithianum (1/2)
Bright Eyes-
      -diphrocalyx (1/2)

White flowers flushed pink.  No longer grown at Exbury.
(Rothschild, 1934)

cl.
      -forrestii ssp. forrestii Repens Group (K-W 6832)
Brightwell-                (1/2)
      -barbatum (1/2)

Leaves 2.5in(6.4cm) by 1in(2.5cm), narrowly elliptic.  Loose
trusses of 6 flowers, widely funnel-campanulate, 2.25in(5.7cm)
across, currant red. (Crown Estate, Windsor)  A. M. (RHS) 1966

cl.
      -insigne (1/2)
Brigitte-
     -Mrs. J. G. Millais--unknown (1/2)

Trusses hold 19-23 flowers, medium roseine purple shading
toward the center to rhodamine purple and white, with an olive
green blotch.  Lanceolate, hairy leaves.  (Cross by H. Hachmann;
reg. by G. Stuck, 1983)

---                  -brookeanum var. gracile (1/4)
      -Duchess of Edinburgh-
Brilliant-           -longiflorum (lobbii) (1/4)
     -javanicum (1/2)

Vireya hybrid.  Flowers of crimson. (Origin unknown)   F. C. C.
(RHS) 1883

cl.
      -thomsonii ssp. thomsonii (1/2)
Brilliant-
     -unknown (1/2)

Bright red flowers.  (J. Waterer)

cl.
     -Labrador Tea--ledum (1/2)
Brilliant*      -forrestii ssp. forrestii Repens Group (1/4)
     -Elizabeth-
            -griersonianum (1/4)

Not a true rhododendron.  Small bright red flowers; new growth
also bright red.  Low, and slow-growing.  (Halfdan Lem)

cl.        -griffithianum (1/4)
    -Loderi-
Brinco-    -fortunei ssp. fortunei (1/4)
    -thomsonii ssp. thomsonii (1/2)

A tall, erect shrub  with big trusses of large, bright rose pink
flowers.  (Rothschild, 1928; reg. 1955)

               -fortunei ssp. discolor (3/16)
         -Lady   -
      -Day -Bessborough-campylocarpum ssp. campylo-
cl.  -unnamed-Dream-        carpum Elatum Group (1/16)
  - hybrid-    -griersonianum (3/16)
Brinny-    -     -fortunei ssp. discolor
  -    -Margaret Dunn-    -dichroanthum ssp. dichro-
  -           -Fabia-     anthum (1/16)
  -unknown (1/2)        -griersonianum

4ft(1.2m)      -5°F(-21°C)     M    3/3

Flower buds bronze, opening to straw yellow flowers  with bronze
red center, campanulate, 4.5in(11.5cm) wide, in  compact trusses
of 8-10.  Leaves light green, lanceolate, curling down at edges,
on a rather compact plant.  (Wilbur Graves cross; intro. by Mr.
& Mrs. K. Janeck, 1964; reg. 1964)

cl.                         -griffithianum (1/4)
        -Queen Wilhelmina-
Britannia-                  -unknown (3/4)
        -Stanley Davies--unknown

4ft(1.2m)         -5°F(-21°C)        ML        4/4

A well-known rhododendron, but difficult to propagate! Flowers
bright scarlet, bell-shaped in a rounded truss. Large leaves,
deeply veined, light dull green, held 3 years, making a dense
plant. Performs well in sun but with even lighter foliage.
(C. B. van Nes & Sons, 1921) A. M. (RHS) 1921; F. C. C. (RHS)
1937; A. G. M. (RHS) 1968 See color illus F p. 37.

                                -griffithianum (1/8)
cl.                  -Queen  -
        -Britannia-Wilhelmina-unknown (3/8)
Britannia's Bells-   -
        -          -Stanley Davies--unknown
        -williamsianum (1/2)

2ft(.6m)         0°F(-18°C)        EM

Widely bell-shaped flowers, 5-lobed, of good substance and rose
red; trusses of 7-8. A sturdy, bushy mound with small to medium
leaves, oval-shaped. (Benjamin F. Lancaster, 1965)

g.
        -Vervaeniana (Indian azalea) (1/2)
Brocade-
        -williamsianum (1/2)

3ft(.9m)         0°F(-18°C)        M        3/3

Azaleodendron. Flowers in clusters of vivid carmine in bud, o-
pening to peach pink bells on slightly drooping branches. Plant
attractive, rounded, of open habit. (Rothschild, 1934)

cl.

Brodick--form of rex ssp. arizelum

5ft(1.5m)         5°F(-15°C)        EM        3/4

Purple flowers with almost-black shade of crimson in the throat,
about 20 in globular trusses. Leaves thickly indumented beneath.
(Brodick Castle Gardens) A. M. (RHS) 1963

                                -fortunei ssp. discolor (1/8)
                -Lady Bessborough-
cl.     -Jalisco-               -campylocarpum ssp. campylo-
        -Goshawk-               carpum Elatum Group (1/8)
Brookside-      -  -dichroanthum ssp. dichroanthum (1/8)
        -       -Dido-
        -               -decorum (1/8)
        -griersonianum (1/2)

Flowers of yellow ochre shaded darker with tinges of Delft rose,
from blood red buds; lax, flat-topped trusses hold 11. Foliage
broadly lanceolate, 7in(18cm) by 2in(5cm). (Crown Estate, Windsor) A. M. (RHS) 1962

cl.
        -arboreum ssp. arboreum (1/2)
Broughtonii-
        -unknown (1/2)

Rosy crimson flowers with dark spotting; 20-flowered pyramidal
trusses. A large, dense shrub, growing to 20ft(6m). (Broughton, before 1853) See color illus. F p. 7.

cl.                         -maximum (1/4)
        -unnamed hybrid-
Broughtonii Aureum-         -ponticum (1/4)
        -molle (1/2)

4ft(1.2m)         -5°F(-21°C)        ML        4/2

Synonym Norbitonense Broughtonianum. Azaleodendron. Flowers of
soft yellow with orange yellow spots, about 2.5in(6.4cm) across,
in small round trusses. Foliage rough-textured, rather sparse.
(Origin unknown; raised by Smith of Norbiton c.1830) F. C. C.
(RHS) 1935 See color illus. VV p. 129.

cl.
                        -campylocarpum ssp. campy-
Bruce Brechtbill--bud sport of Unique-       locarpum (1/2)
                        -unknown (1/2)

4ft(1.2m)         0°F(-18°C)        EM        4/5

Identical to Unique, except for flower color, which is pale pink
with some yellow in the throat. (Brechtbill Nursery, 1970)

cl.             -griffithianum (1/4)
        -Sincerity-
Brumas-         -unknown (1/2)
        -               -adenogynum (1/4)
        -Xenosporum (detonsum)-
                        -unknown

The "species" detonsum is now considered a hybrid of adenogynum.
Flowers pure white with purple honey pouches, very lightly spot-
ted, 4.5in(11.5cm) across; conical trusses of 13 flowers. (Gen.
Harrison, 1961)

---
        -javanicum (5/8)
Brunette-               -brookeanum var. gracile (1/4)
        -Princess Frederica-       -jasminiflorum (1/8)
                        -Princess Royal-
                                -javanicum

Vireya hybrid. Yellow to yellow orange flowers. (Exhibited by
Veitch, 1891)

g.
        -wardii var. wardii (1/2)
Brunhilde-      -griffithianum (1/4)
        -Loderi-
                -fortunei ssp. fortunei (1/4)

Pale cream flowers. (J. B. Stevenson cross, 1930; intro. 1951)

cl.
        -fortunei ssp fortunei ? (1/2)
Bryantville-
        -unknown (1/2)

3ft(.9m)         -5°F(-21°C)        M

Flowers medium purplish pink with faint yellow green spotting,
openly funnel-shaped, of good substance, 3.5in(9cm) across; 18-
flowered ball-shaped trusses 6in(15cm) wide. Plant broader than
high and rounded; glossy olive green leaves held 2 years. (C. O.
Dexter before 1943; named by J. Wister, H. Howard; intro. by Tyler Arboretum; reg. 1981)

cl.    -King of-fortunei ssp. discolor (1/4)
       - Shrubs-       -dichroanthum ssp. dichroanthum (1/8)
Bryce -        -Fabia-
Canyon-                -griersonianum (1/8)
       -               -wardii var. wardii (1/4)
       -                       -griffithianum (1/32)
       -Idealist-              -Kewense-
       -       -Aurora-        -fortunei ssp. for-
       -Naomi-         -               tunei (5/32)
       -               -thomsonii ssp. thomsonii (1/16)
       -fortunei ssp. fortunei

3ft(.9m)         0°F(-18°C)        M

Flowers carrot red with red blotch deep in throat, of good substance, openly funnel-shaped, 3.5in(8.3cm) across; trusses dome-
shaped, 7in(18cm) across, of 10-14 flowers. Floriferous. Plant
upright, well-branched; leaves held 2 years; new foliage bronze.
(Arthur Childers cross, 1961; intro. by Bovees Nursery, and reg.
1982)

cl.
                -oreodoxa var. fargesii (erubescens) (1/2)
Buchanan Simpson-
                    -unknown (1/2)

5ft(1.5m)          -5°F(-21°C)        EM        3/4

Flowers phlox pink, 3.5in(8.9cm) wide, throat speckled reddish
olive, base of corolla yellowish; trusses hold 8-10. Plant compact with sturdy stems and bright green leaves, bullate. (Mr. &
Mrs. E. J. Greig, reg. 1963)

                        -souliei (1/8)
cl.              -Soulbut-
        -Vanessa-            -Sir Charles Butler--fortunei ssp. for-
Buckland-       -griersonianum (1/4)                   tunei (1/8)
        -yakushimanum ssp. yakushimanum (1/2)

Loose, open truss of 8 flowers, openly funnel-shaped, 3in(7.6cm)
in diameter, red purple with darker markings. Leaves 3.5in(9cm)
by 1.25in(3.2cm), ovate to narrowly lanceolate, dark green. (L.
S. Fortescue) P. C. (RHS) 1970

cl.
                -griffithianum (1/2)
Buckland Beauty-            -campylocarpum ssp. campylocarpum
        -Letty Edwards-            Elatum Group (1/4)
                    -fortunei ssp. fortunei (1/4)

Flowers yellow with slight basal red spotting on upper lobes, in
trusses of 11. (L. S. Fortescue, 1977)

cl.                 -wardii var. wardii (1/4)
        -Windsor Hawk-          -campylocarpum ssp. campylo-
Buckland-       -Lady     -   carpum Elatum Group (1/8)
  Cream -       Bessborough-fortunei ssp. discolor (1/8)
        -              -            -griffithianum (1/4)
        -Loderi Julie (Loderi selfed)-
                                -fortunei ssp. fortunei
                                            (1/4)

Yellow flowers with central inner stains of red, in trusses of 7
or 8. (L. S. Fortescue cross, 1960; intro. by Fortescue Garden
Trust, 1980; reg. by Keith Wiley)

cl.
                -unknown (1/2)
Bud Flanagan-
                -ponticum (1/2)

Flowers of sparkling mauve with a large flash of deep chestnut;
enormous conical trusses of 18-20. (Edmund de Rothschild, 1965)
See color illus. PB pl. 20.

cl.

Budget Farthing--oreodoxa var. fargesii

5ft(1.5m)          -10°F(-23°C)         E        3/3

White flowers suffused with reddish purple, 7-lobed, openly campanulate, 2.5in(6.4cm) wide; trusses of 8-9. Dark green leaves,
broadly elliptic, 3in(7.6cm) long. (Lord Aberconway)  A. M.
(RHS) 1969

cl.            -neriiflorum ssp. neriiflorum (1/4)
        -Nereid-
Buff Lady-      -dichroanthum ssp. dichroanthum (1/4)
        -fortunei ssp. discolor (1/2)

3ft(.9m)          -5°F(-21°C)          L        3/3

Flowers of Egyptian buff with coral pink shading,  widely bell-
shaped, 6-lobed, held in spherical trusses of 12. Large leaves,
medium green, elliptic.  Plant rounded; heat-tolerant. (Seed
from Rose of England; intro. by Benjamin Lancaster in the 1950s)

cl.           -griffithianum (1/4)
        -Spitfire-
Buketta-        -unknown (1/2)
        -            -Essex Scarlet--unknown
        -Fruhlingszauber-
                    -forrestii ssp. forrestii Repens Group
                                            (1/4)

Trusses of 10-12 flowers cardinal red marked ruby red.  Elliptic
leaves, hairy. (H. Hachmann cross; reg. by G. Stuck, 1983)

g. & cl.
        -edgeworthii (bullatum) (1/2)
Bulbul-
        -moupinense (1/2)

Rather straggly, somewhat tender; flowers white with faint yellow spotting. (Rothschild, 1934)  A. M. (RHS) 1949

g.
        -elliottii (1/2)                 -griffithianum (1/8)
Bulldog-            -Queen Wilhelmina-
        -Earl of Athlone-            -unknown (3/8)
                    -Stanley Davies--unknown

Deep red flowers. (Bolitho, 1937)

cl.
                        -auriculatum (1/2)
Bullseye--Flameheart (selfed)-   -griersonianum (1/4)
                        -Azor-
                            -fortunei ssp. fortunei (1/4)

Flowers 3in(7.6cm) across, 7-lobed, crimson with red eye.
(Michael Haworth-Booth, 1968)

cl.
        -macgregoriae (1/2)
Bulolo Gold-
        -aurigeranum (1/2)

Vireya hybrid with trusses of 16 flowers, soft nasturtium orange
with buttercup yellow throat. (Cross by T. Lelliot; intro. by
R. Cutten, 1981)

cl.                     -griffithianum (1/2)
        -Loderi King George-
Bulstrode Beauty-        -fortunei ssp. fortunei (1/4)
        -              -arboreum ssp. cinnamomeum
        -Loder's White-            var. album (1/4)
                    -griffithianum

(See Loder's White for other possible parentage.)  White flowers. (Sir John Ramsden, 1934)

cl.

Bulstrode Belle      Parentage as above

White flowers. (Sir John Ramsden, 1934)

cl.                     -griffithianum (1/4)
        -unnamed hybrid-
Bulstrode Park-            -unknown (1/2)
        -      -catawbiense (1/4)
        -Sefton-
            -unknown

5ft(1.5m)          -5°F(-21°C)          M        3/2

Flowers bright scarlet crimson, waxy; trusses rather large and
loose. Pointed, dark green leaves. (C. B. van Nes & Sons)

cl.           -dichroanthum ssp. dichroanthum (1/4)
        -Fabia-
Bunting-    -griersonianum (1/4)
        -          -caucasicum (1/4)
        -Dr. Stocker-
                -griffithianum (1/4)

Trusses of 9-11 flowers, tubular-campanulate, 3in(7.6cm) across,
Dresden yellow and spotted green. (Rudolph Henny,  intro. 1960)

cl.
                    -maximum (1/2)
Burgemeester Aarts-
                    -L. L. Liebig--unknown (1/2)

A plant of tall habit and dark red flowers. (M. Koster, 1915)

cl.


Burgundian      See Bordeaux


                              -griffithianum (1/8)
cl.          -Queen Wilhelmina-
      -Britannia-                -unknown (5/8)
Burgundy-     -Stanley Davies--unknown
                         -ponticum (1/4)
      -Purple Splendour-
                         -unknown


5ft(1.5m)        -15°F(-26°C)        ML        4/3


Burgundy red flowers in trusses of 15. (Cross by Rose, of England; intro. by Halfdan Lem, before 1958)


          -griffithianum (1/4)
cl. -Mars-
    -     -unknown (9/16)                 -catawbiense (3/16)
Burma-              -Parsons Grandiflorum-
    -       -America-              -unknown
    -Fanfare-      -dark red hybrid--unknown
    -             -catawbiense
         -Kettledrum-
                    -uknown


5ft(1.5m)        -20°F(-29°C)        ML


A plant 1.5 times as broad as tall and well-branched; yellowish
green leaves, retained 2-3 years. Conical trusses of about 17
flowers, brighter than cardinal red, with heavy black spotting.
(David G. Leach cross, 1958; intro. 1982; reg. 1984)


cl.          -dichroanthum sp. dichroanthum (1/4)
      -Fabia   -
Burma Road-Tangerine-griersonianum (1/4)
      -           -Moser's Maroon--unknown (1/4)
      -Romany Chal-
                  -facetum (eriogynum) (1/4)


Compact, flat-topped trusses with about 11 flowers. Each flower
3.5in(9cm) across, open bell-shaped, rich yellowish apricot,
suffused pale rose pink, deeper pink on reverse. (Cross by
Francis Hanger, RHS Gardens, Wisley, reg. 1964) A. M. (RHS) 1958


g.
          -haematodes ssp. haematodes (1/2)
Burning Bush-
          -dichroanthum ssp. dichroanthum (1/2)


A low, spreading shrub, with narrow bell-shaped flowers of a
bright tangerine red. (Rothschild, 1934)


g.
      -auriculatum (1/2)
Bustard-          -campylocarpum ssp. campylocarpum Elatum
      -Penjerrick-                        Group (1/4)
                 -griffithianum (1/4)


6ft(1.8m)        5°F(-15°C)        L        3/3


Very large flowers, white with a crimson spot in the throat, of
good substance; large upright trusses. (Rothschild, 1934)


cl.


Butcher Wood--form of tephropeplum (K-W 20844)

2.5ft(.75m)        5°F(-15°C)        EM        3/2

Pink flowers. (Maj. A. E. Hardy) A. M. (RHS) 1975


cl.


Butheana--form of mucronatum

Selection of the "species" mucronatum, now consdidered a hybrid
of unknown origin. The name is now properly written Mucronatum.

g.
    -Sir Charles Butler--fortunei ssp. fortunei (3/4)
Butkew-     -griffithianum (1/4)
    -Kewense-
            -fortunei ssp. fortunei


Pink flowers. (E. J. P. Magor cross 1918; intro. 1929)


                         -dichroanthum ssp. dichroanthum
            -Goldsworth Orange-                    (1/4)
      -Hotei-              -fortunei ssp. discolor (1/8)
cl.   -    -unnamed-souliei (1/8)
      -     hybrid-
Butter -          -wardii var. wardii (1/8)
Brickle-     -dichroanthum ssp. dichroanthum
    -     -Dido-
    -Lem's-     -decorum (1/8)
      Cameo-     -Beauty of-griffithianum (5/32)
         -Norman- Tremough-
         -   - Gill -     -arboreum ssp. arboreum
      -Anna-     -griffithianum          (1/32)
         -                -griffithianum
         -Jean Marie de Montague-
                              -unknown (1/16)


3ft(.9m)        0°F(-18°C)        M


Long lasting flowers of heavy substance, 6-7 lobes, chrome
yellow, edges buttercup yellow, reverse strong gold, cardinal
red blotch in throat. Trusses hold 12. (Cross 1976; intro. 1982,
John G. Lofthouse)


cl.
      -xanthostephanum (1/2)
Butterball-
          -triflorum (1/2)


Trusses hold 3-4 open-campanulate flowers, canary yellow lightly speckled with deeper brownish yellow. Sturdy upright plant,
4ft(1.2m) in 10 years. (E. J. Greig, reg. 1968)


cl.
      -campylocarpum ssp. campylocarpum (1/2)
Buttercup-
      -A. W. Hardy Hybrid--unknown (1/2)


Yellow flowers shaded apricot. (Walter C. Slocock, 1924)

cl.
      -campylocarpum ssp. campylocarpum (1/2)
Butterfly-     -catawbiense (1/4)
      -Mrs. Milner-
                 -unknown (1/4)


5ft(1.5m)        0°F(-18°C)        M        3/3


Rounded but not tight trusses of pale yellow flowers with red
spotting on upper petal. Plant of medium compact habit, but
often appears leggy. Easily propagated. Does well in sun or
shade and is heat-tolerant. (W. C. Slocock) A. M. (RHS) 1940


                    -campylocarpum ssp. campylocarpum (1/4)
cl.   -Unique-
      -     -unknown (1/4)
Buttermint-              -dichroanthum ssp. dichroanthum
      -             -Fabia-                    (1/8)
      -unnamed hybrid-     -griersonianum (1/8)
                     -dichroanthum ssp. apodectum (1/4)


3ft(.9m)        -5°F(-21°C)        M        4/4


Synonym Compact Yellow. Orange buds open to yellow flowers
with red dorsal spotting and streaks on reverse, up to 15 per
lax truss that is 6in(15cm) across. Plant rounded, as wide as
high. (Richard Mauritsen cross; reg. by Harold E. Greer, 1979)

```
           -Lady        -fortunei ssp. discolor (1/4)
           -Bessborough-
       -Jalisco-          -campylocarpum ssp. campylo-
cl.        -        -             carpum Elatum Group (1/4)
           -        -     -dichroanthum ssp. dichroanthum (1/8)
Buttersteep-   -Dido-
       -          -decorum (1/8)
       -     -wardii var. wardii (1/4)
       -Crest-              -fortunei ssp. discolor
           -Lady Bessborough-
                        -campylocarpum ssp. campylo-
                           carpum Elatum Group
```

Flowers barium yellow with a small blotch of currant red, openly
funnel-shaped, 7-lobed, in loose trusses of 12-14.   Pale green,
elliptic leaves 4in(10.2cm) long.   (Crown Estate Commissioners)
A. M. (RHS) 1971

```
cl.        -griffithianum (1/4)
     -Loderi-
Buxom-     -fortunei ssp. fortunei (1/4)
     -     -caucasicum (1/4)
     -Sophia Gray-
              -unknown (1/4)
```

Deep pink flowers with a dark throat. (Sir James Horlick, 1962)

```
cl.                     -griffithianum (1/4)
          -Queen Wilhelmina-
C. B. van Nes-          -unknown (3/4)
          -Stanley Davies--unknown
```

5ft(1.5m)         5°F(-15°C)         EM         3/2

Symmetrical, compact, upright plant  with ample foliage.  Built-
up trusses of glowing scarlet flowers, 3.5in(8.9cm) across.  See
also Britannia, Earl of Athlone, Langley Park, London, Trilby,
Unknown Warrior.  (C. B. van Nes, 1921)

```
cl.
          -griersonianum (1/2)
C. F. Wood-
          -fortunei ssp. discolor (1/2)
```

Azor g.  Exhibited by Preston, 1947.

```
cl.                -arboreum ssp. cinnamomeum var. album (1/4)
          -Loder's White-
C. I. S.-          -griffithianum (1/4)
     -          -dichroanthum ssp. dichroanthum (1/4)
          -Fabia-
               -griersonianum (1/4)
```

4ft(1.2m)         5°F(-15°C)         M         4/3

See Loder's White for other possible parentage.  Flowers orange
yellow changing to creamy apricot, bright orange red throat, to
4in(10.2cm) across; loose trusses hold up to 15.  Very floriferous.  Plant upright, almost as wide as high; leaf tips twist
distinctively.  Named for Claude I. Sersanous, Past President,
ARS.  (R. Henny, before 1958)   P. A. (ARS) 1952; A. M. (RHS)
1975   See color illus. K p. 160; VV p. 57.

```
cl.
     -fortunei ssp. fortunei ? (1/2)
C. O. D.-
     -unknown (1/2)
```

4ft(1.2m)         -5°F(-21°C)         ML

Named in honor of C. O. Dexter. Fragrant flowers of pale yellow,
darker in throat, deep yellow blotch, with pink flush on the reverse.  A parent of Blush Button.  (C. O. Dexter, 1925-42;
intro. by Everitt, reg. 1958)

```
                           -griffithianum (1/8)
             -Queen Wilhelmina-
g.     -Britannia-            -unknown (3/8)
C. P. Rafill-     -Stanley Davies--unknown
     -griersonianum (1/2)
```

5ft(1.5m)         0°F(-18°C)         L         4/3

Deep orange red flowers in large rounded trusses.  Plant rather
dense, spreading; light green leaves 6in(15.2cm) long, with reddish brown petioles.  Other forms exist; above ratings for Kew
form.  (Exhibited by RBG, Kew, 1949)

```
cl.
     -catawbiense (1/2)
C. S. Sargent-
     -unknown (1/2)
```

Red flowers.  A parent of Sardis.  (A. Waterer, 1888)

```
cl.
```

C. Sabine Seabrook     See Cecil S. Seabrook

```
cl.        -griersonianum (1/4)
     -Tally Ho-
Cable Car-     -facetum (eriogynum) (1/4)
     -fortunei ssp. fortunei (1/2)
```

4ft(1.2m)                    ML

Neyron rose flowers of heavy substance, openly funnel-shaped, 5-
to 7-lobed, 4.5in(10.8cm) across; 9-13 held in ball-shaped truss
9in(23cm) across.  Plant upright, well-branched; leaves medium
green with pale green indumentum, held 2 years.  (Albert Golden
cross, 1962; intro. 1974)

```
cl.        -decorum ? (1/4)
     -Caroline-
Cadis-     -brachycarpum ssp. brachycarpum ? (1/4)
     -fortunei ssp. discolor (1/2)
```

5ft(1.5m)         -15°(-26°C)         ML         4/4

Flowers of light pink, large, very fragrant, in flat trusses.
Plant of good habit, with dense, pretty foliage of long narrow
leaves.  (Joseph Gable, 1958)   A. E. (ARS) 1959   See color
illus. LW pl. 39;  VV p. 69.

```
g.
          -cinnabarinum ssp. cinnabarinum (1/2)
Caerhays John*
          -cinnabarinum ssp. xanthocodon Concatenans Group
                                                    (1/2)
```

Medium size, bushy habit and erect.  Funnnel-shaped flowers,
waxy, deep apricot, later than Caerhays Lawrence.  Not a hybrid,
but a form of the species cinnabarinum.  (J. C. Williams, 1967)
See color illus. Hillier Colour Dictionary, p. 197.

```
cl.
          -cinnabarinum ssp. cinnabarinum (1/2)
Caerhays Lawrence*
          -cinnabarinum ssp. xanthocodon Concatenans
                                         Group (1/2)
```

Rich yellow, waxy flowers, otherwise similiar to Caerhays John.
J. C. Williams, 1967)   See color illus. ARS J 37:4 (Fall 1983):
p. 182.

```
cl.
     -cinnabarinum ssp. xanthocodon Concatenans Group (1/2)
Caerhays-
 Philip -cinnabarinum ssp. cinnabarinum Blandfordiiflorum
                                         Group (1/2)
```

Yellow flowers 2 1/2in(6.4cm) across, funnel-shaped, with 7 per
loose truss.  Leaves 4 1/2in(11.5cm) long, elliptic, with very
scanty brown indumentum.  Blooms early.  (Raised by Charles
Williams and Charles Michael.)  A. M. (RHS) 1966

cl.

Caerhays Pink*--form of <u>davidsonianum</u>

Flowers are clear pink, spotted red.  (J. C. Williams)


cl.
```
                    -edgeworthii (1/2)
Caerhays Princess Alice*
                    -ciliatum (1/2)
```

This plant has a dwarf and compact growth habit and does not re-
quire routine pruning.  White flowers tinged pink outside; fragrant.  See also Princess Alice.  (Caerhays)


cl.
```
          -fortunei ssp. fortunei (1/2)
Caernarvon-
          -unknown (1/2)
```

3.3ft(1m)          -5°F(-21°C)          M-ML

Flowers dark pink with deeper spotting, fragrant, openly funnel-
shaped, 3.25in(8.3cm) across; ball-shaped trusses hold up to 14.
Plant upright, stiffly branched; leaves retained 2 years. (Cross
by Anthony Consolini; raised by R. de Longchamp, W. A. Reese and
intro. by Reese, 1977)


cl.

Caeruleum*--form of <u>rigidum</u>

6ft(1.8m)          -5°F(-21°C)          EM          4/3

A plant of neat, glaucous foliage and a mass of pure white flowers with olive markings.  A lovely Triflora which draws much attention.  (Origin unknown; sold by Cox of Glendoick)

```
                              -brookeanum var.
                    -Duchess of-   gracile (1/16)
                    - Edinburgh-
cl.        -Triumphans-        -longiflorum (lobbii)
        -unnamed-          -javanicum           (1/16)
Cair Paravel- hybrid-javanicum (3/8)
           -
           -leucogigas (1/2)
```

3ft(.9m)          32°F(0°C)          VE & L

Vireya hybrid.  Claret rose flowers of heavy substance, tubular
funnel-shaped, 6 wavy lobes, 3.25in(8.3cm) wide,  very fragrant;
dome-shaped trusses hold about 13 flowers.  Plant broad, rounded
and branching moderately; smooth, glossy leaves  held 3-4 years.
(Peter Sullivan cross, 1972; intro. by William A. Moynier, 1979)

```
           -Catalgla--catawbiense var. album Glass (1/4)
cl. -unnamed-
    - hybrid-fortunei ssp. fortunei (1/4)
Cairo-          -japonicum var. japonicum (metternichii)
    -unnamed-Eidam-                          (1/8)
      hybrid-    -Alexander Adie--unknown (1/8)
             -williamsianum (1/4)
```

6ft(1.8m)          -20°F(-29°C)          EM          3/3

White flowers  with small blotch and spotting of bright greenish
yellow, 3.25in(8.3cm) across; ball-shaped trusses hold up to 16.
Plant rounded, branching moderately; dark green elliptic foliage
held 2 years.  (David G. Leach, reg. 1973)


cl.
```
          -konorii (1/2)
Calavar-
          -zoelleri (1/2)
```

2ft(.6m)          32°F(0°C)          VE


Vireya hybrid.  Very fragrant flowers of carmine red with barium
yellow throat, of good substance, 3.25in(8.3cm) across, 5 wavy
lobes; flat trusses hold 5 flowers.  Plant broad, well-branched;
glossy leaves held 3 years.  (Cross by Peter Sullivan, Strybing
Arboretum; W. A. Moynier intro., 1978)


cl.
```
                    -dichroanthum ssp. dichroanthum (1/4)
          -unnamed hybrid-
Calcutta-           -kyawii (1/4)

          -Catalgla--catawbiense var. album Glass (1/2)
```

3ft(.9m)          -15°F(-26°C)          L          2/3

Orange buds open to brilliant yellow flowers  with orange edges,
5-lobed, tubular-campanulate, 1.5in(3.8cm) wide; 13-flowered lax
and pendulous trusses.  Plant twice as broad as tall, with dense
foliage.  (David G. Leach, reg. 1972)


g. & cl.
```
          -calophytum var. calophytum (1/2)
Calfort-
          -fortunei ssp. fortunei (1/2)
```

Pink in bud, and pink flowers externally marked with rosy lilac.
(Collingwood Ingram)    A. M. (RHS) 1932


cl.
```
                    -ciliicalyx (1/4)
          -Else Frye-
California Gold-     -unknown (1/4)
          -          -valentinianum (1/4)
          -Eldorado-
                    -johnstoneanum (1/4)
```

6ft(1.8m)          15°F(-10°C)          VE

Flowers primrose yellow, fragrant, tubular funnel-shaped, five
wavy lobes, 3.25in(8.3cm) across, in trusses of 6.  Floriferous.
Plant broad, well-branched; spinach green elliptic leaves with
silvery scales below, held 2 years.  (Paul and Ruby Bowman cross,
1962; reg. 1976)


cl.
```
          -dichroanthum ssp. dichroanthum (1/2)
Callagold-
          -wardii var. wardii (1/2)
```

Flowers reddish-colored capped with amber gold, 1.5in(3.8cm), on
a plant of dwarf habit, blooming late.  Leaves up to 3in(7.6cm).
(Rudolph Henny; Leona Henny intro. and reg., 1964)


cl.
```
          -calophytum var. calophytum (1/2)
Callight-
          -Neon Light--unknown (1/2)
```

Flowers of light spirea red paling with age, maroon eye; trusses
hold 22 flowers.  (K. Van de Ven, Australia, 1973)


---
```
          -caucasicum (1/2)
Calliope*
          -unknown (1/2)
```

Flowers open pink and fade white; reddish brown markings.  (T.
J. R. Seidel, before 1894)


g.
```
                    -caucasicum (1/4)
          -Dr. Stocker-
Callirhoe-          -griffithianum (1/4)
          -arboreum ssp. arboreum (blood red form) (1/2)
```

Flowers rose neyron red with crimson spots.  (E. J. P. Magor,
intro. 1928)

g.                              -griffithianum (1/4)
        -Queen Wilhelmina-
Calomina-                       -unknown (1/4)
        -calophytum var. calophytum (1/2)

Synonym Armantine.  Medium to tall plant bearing trusses of pink
flowers with a darker blotch.  Blooms early.  (Lady Loder, 1934)

cl.

Calostrotum Pink     See Cutie

cl.

Calostrotum Rose     See Cutie


g.
        -calophytum var. calophytum (1/2)
Calrose-
        -griersonianum (1/2)

Funnel-shaped flowers, deep rose in bud, opening pink with deep
rose stain.  A large plant with good foliage.  (Lord Aberconway,
1939)

cl.
        -Catalgla--catawbiense var. album Glass (1/2)
Calsap*
        -Sappho--unknown (1/2)

5ft(1.5m)        -25°F(-32°C)          L

A very hardy plant with white flowers, boldly blotched dark purple.  (M. W. Michener, 1970)     Best of Show award, Great Lakes
Ch., ARS, 1985

g.
            -calophytum var. calophytum (1/2)
Calstocker-               -caucasicum (1/4)
            -Dr. Stocker-
                          -griffithianum (1/4)

6ft(1.8m)        0°F(-18°C)           E          5/4

See Exbury Calstocker, a better-known form.  (Cross by Whitaker
of Pylewell Park; intro. 1935)

g.              -griffithianum (1/4)
          -Gilian-
Calypso-        -thomsonii ssp. thomsonii (1/4)
        -smithii (1/2)

Intro. by E. J. P. Magor, 1934.  Parentage of Gilian as shown
differs from that in the published Register of 1958.  Information obtained from Maj. E. W. M. Magor.


g.
        -campylocarpum ssp. campylocarpum (1/2)
Caman-
        -beanianum (1/2)

Red flowers.  (Lord Aberconway, intro. 1946)


g. & cl.         -campylocarpum ssp. campylocarpum
        -Penjerrick-             Elatum Group (1/4)
Camilla-        -griffithianum (1/2)
        -               -griffithianum
        -Loderi King George-
                        -fortunei ssp. fortunei (1/4)

Large white flowers.  (Lord Aberconway)  A. M. (RHS) 1944

g.
        -campylocarpum ssp. campylocarpum (1/2)
Campbut-
        -Sir Charles Butler--fortunei ssp. fortunei (1/2)

A very pale yellow flower of 5-6 lobes with dense spotting of
crimson.  (E. J. P. Magor, 1926)

cl.
        -campylocarpum ssp. campylocarpum (1/2)
Campdis*
        -fortunei ssp. discolor (1/2)

A parent of E. J. P. Magor, by D. Hohhie, c.1954.

g.
        -campylocarpum ssp. campylocarpum (1/2)
Campirr-
        -irroratum ssp. irroratum (1/2)

A fine old hybrid from Lamellen with crimson flowers, spotted in
pale yellow.  (E. J. P. Magor, 1926)

g.
            -campylocarpum ssp. campylocarpum (1/2)
Campkew-        -griffithianum (1/4)
            -Kewense-
                    -fortunei ssp. fortunei (1/4)

Creamy white flowers with crimson spots.  (E. J. P. Magor, 1925)

g.
            -campylocarpum ssp. campylocarpum (1/2)
Campxen-                    -adenogynum (1/4)
            -Xenosporum (detonsum)-
                            -unknown (1/4)

Flower white or pale yellow, red spots.  (Former name for detonsum was xenosporum now considered a hybrid of adenogynum, hence
the new hybrid name, Xenosporum.)  (E. J. P. Magor, 1940)

cl.
        -campylogynum ? (1/2)
Canada-
        -unknown (1/2)

1ft(.3m)         -5°F(-21°C)          M          3/3

Very small, tubular, deep rose pink flowers.  Terminal inflores-
cence of 3-5 small trusses, each with 3-5 flowers.  Floriferous.
New foliage has reddish edge; retained 2 years.  (Cross by E. J.
Greig; intro. and reg. by J. F. Caperci, 1977)


                            -griersonianum (1/4)
            -Mrs. Horace Fogg-        -griffithianum (15/32)
cl.          -              -Loderi-
            -              Venus-fortunei ssp. fortunei (1/8)
Canadian-                         -griffithianum
 Beauty -            -Beauty of-
        -              -Norman- Tremough-arboreum ssp. arboreum
        -              - Gill -                        (1/32)
        -       -Anna-     -griffithianum
       -Point -  -griffithianum
        Defiance-          -griffithianum
            -Marinus Koster-
                          -unknown (1/8)

6ft(1.8m)        0°F(-18°C)          ML

Flowers with light pink centers shading darker on edges,  openly
funnel-shaped,  5-lobed, 4.5in(10.8cm) across;  trusses hold 16.
Leaves to 7in(17.8cm).  (John Lofthouse, 1971)

                            -dichroanthum ssp. dichroanthum
cl.        -Goldsworth Orange-                       (1/8)
        -Hotei-              -fortunei ssp. discolor (1/8)
        -    -                   -souliei (1/8)
Canadian-    -unnamed hybrid-
  Gold  -              -wardii var. wardii (3/8)
        -              -wardii var. wardii
        -unnamed hybrid-
                    -unknown (1/4)

Golden yellow buds opening to pale sulphur yellow flowers with a
reddish maroon blotch; trusses of 9-11. Midseason. (John Lofthouse cross; intro. 1979)

```
                                  -caucasicum (1/8)
cl.              -Cunningham's White-
          -Rocket-              -ponticum (white form) (1/8)
Canadian-      -catawbiense (red form) (1/4)
  Lilac -        -Catalgla--catawbiense var. album Glass (1/4)
        -unnamed-              -fortunei ssp. discolor (1/8)
         hybrid-Lady -
              Bessborough-campylocarpum ssp. campylocarpum
                                            Elatum Group (1/8)
```

15in(37cm)          -25°F(-32°C)          ML

Flowers with triangular petals, light purple with light tan
spots in trusses of 13. Dwarf plant with creeping and upright
branches; leaves dull yellow green. See also Canadian Magenta.
(Rudy Behring, 1983)

```
                                  -caucasicum (1/8)
cl.              -Cunningham's White-
          -Rocket-              -ponticum (white form) (1/8)
Canadian-      -catawbiense (red form) (1/4)
  Magenta-       -Catalgla--catawbiense var. album Glass (1/4)
        -unnamed-              -fortunei ssp. discolor (1/8)
         hybrid-Lady -
              Bessborough-campylocarpum ssp. campylocarpum
                                            Elatum Group (1/8)
```

2ft(.6m)           -25°F(-32°C)           ML

Flowers with pie-shaped petals, very light cyclamen purple with
darker spotting, in trusses of 13. Dwarf plant equally as wide
as tall with leaves holding 2 years. See also Canadian Lilac.
(Rudy Behring, 1983)

```
                              -caucasicum (1/16)
                    -Cunningham's-
              -Rocket- White     -ponticum (white form)
              -       -                             (1/16)
cl.      -unnamed-     -catawbiense (red form) (1/8)
         - hybrid-Catalgla--catawbiense var. album Glass
Canadian Pink-                                     (1/4)
                     -fortunei ssp. discolor (1/4)
         -Lady Bessborough-
                     -campylocarpum ssp. campylocar-
                                 pum Elatum Group (1/4)
```

3ft(.9m)           -25°F(-32°C)          L

Trusses of 15 flowers, light mallow purple with cyclamen purple
spotting. Plant upright, open, over half as wide as high; foliage glossy, held 2 years. (Rudy Behring cross, 1974; reg. 1985)

```
cl.
              -yakushimanum ssp. yakushimanum (1/2)
Canadian Sunset-              -griffithianum (1/8)
         -        -King George-
          -Gipsy King-        -unknown (1/8)
              -haematodes ssp. haematodes (1/4)
```

Compact plant blooming in midseason. Red buds open to red flowers fading in center to orange with yellow cast. (Cross by John
Lofthouse; reg. 1974)

```
cl.           -campylocarpum ssp. campylocarpum (1/4)
        -unnamed hybrid-
Canary-              -fortunei ssp. discolor (1/4)
         -griffithianum (1/4)
         -Loderi-
              -fortunei ssp. fortunei (1/4)
```

Pale yellow flowers. (W. T. Stead)

```
cl.
       -campylocarpum ssp. campylocarpum Elatum Group (1/2)
Canary-
       -caucasicum (1/2)
```

3ft(.9m)           -10°F(-23°C)          EM          3/3

One of the most hardy, deep yellow rhododendrons. Compact-grow-
ing plant with deeply veined, rounded leaves. As yellow as Crest
but not quite as bright in color. (M. Koster & Sons, 1920)

```
cl.
       -campylogynum Cremastum Group (1/2)
Candi-
       -racemosum (1/2)
```

3ft(.9m)           -10°F(-23°C)          M          3/3

Small leaves, 1in(2.5cm) long; flowers bright rose, growing in
clusters of up to 6 along the branches. (J. F. Caperci, 1963)

```
cl.          -catawbiense ? (1/2)
Candidissimum-
             -maximum ? (1/2)
```

5ft(1.5m)          -15°F(-26°C)          L

Dark green foliage with nice trusses of blush white flowers.
Trusses 6in(15cm) wide, holding 12. (Reg. 1966 without origin
stated, but possibly S. B. Parsons, before 1871)

```
              -Winter Favourite--unknown (1/4)
cl.     -Denise-              -chrysodoron (3/8)
        -        -Chrysomanicum-
Candle Gleam-              -burmanicum (3/8)
        -        -chrysodoron
        -Chrysomanicum-
                    -burmanicum
```

Trusses of 4-6 flowers, tubular funnel-shaped, primrose yellow
with pink edges. Shrub up to 3ft(.9m) tall. (F. Boulter, reg.
1984)

cl.

Candy     See Sham's Candy

```
              -wardii var. wardii (1/4)
cl.     -Hawk-       -fortunei ssp. discolor (1/8)
        -     -Lady -
Candy Floss-  Bessborough-campylocarpum ssp. campylocarpum
        -                                  Elatum Group (1/8)
        -           -griffithianum (1/4)
        -Mrs. Randall-
         Davidson  -campylocarpum ssp. campylocarpum (1/4)
```

Campanulate flowers 3.5in(8.9cm) wide, cream suffused with pink,
darker at margins, in lax trusses of 10. Dark green leaves 4in
(10.2cm) long; petiole stained purple. (Crown Estate, Windsor,
reg. 1963)

cl.

Canton Consul--form of hanceanum Nanum Group

Dwarf habit. Flowers creamish green in bud, and opening cream,
campanulate, 1in(2.5cm) wide. (Crown Estate, Windsor, reg. 1962)
A. M. (RHS) 1957

```
cl.
       -williamsianum (1/2)
Caper-       -haematodes ssp. haematodes (1/4)
       -Hiraethylin-
              -griffithianum (1/4)
```

Flowers light azalea pink, funnel-campanulate, 4in(10.2cm) wide,
in trusses of 5-7. Compact habit; leaves 3in(7.6cm) by 2in(5cm)
Blooms early. (Rudolph Henny, reg. 1964)

cl.

Capri     See Cyprus

cl.

Caprice--form of augustinii ssp. augustinii

Once considered a superior form of augustinii selected by Hansen
but shown in the Register of 1958 as a hybrid.

```
                         -souliei (1/16)
                   -Soulbut-
           -Vanessa-           -Sir Charles Butler--fortunei ssp.
           -           -griersonianum             fortunei (1/16)
g.      -Etna-      -dichroanthum ssp. dichroanthum (1/8)
        -      -Fabia-
Capriole-           -griersonianum (1/2)
        -           -facetum (eriogynum) (1/4)
        -Jacquetta-
                   -griersonianum
```

Dark red flowers.  (Lord Aberconway, intro. 1942)

```
cl.
                -griersonianum (1/2)
Captain Blood-           -griffithianum (1/4)
              -Queen Wilhelmina-
                         -unknown (1/4)
```

Juliana g.  Flowers of claret rose with slight spotting on upper
lobes.  (Collingwood Ingram)   A. M. (RHS) 1947

```
cl.           -griffithianum (1/4)
            -Mars-
Captain Jack-   -unknown (1/4)
            -facetum (eriogynum) (1/2)
```

6ft(1.8m)          5°F(-15°C)          ML          4/3

Flowers blood red, 3.5in(8.9cm) across, with a waxy shine, about
15 in a truss.  Foliage large, with an unusual roll to the margins. (Rudolph Henny, 1956)  P. A.(ARS) 1956   See color illus.
ARS Q 29:1 (Jan. 1975): p. 35.

```
                                      -thomsonii ssp.
cl.                      -Bagshot Ruby-   thomsonii (1/8)
            -Princess Elizabeth-         -unknown (3/8)
Captain Kidd-                   -unknown-
            -                -haematodes ssp. haematodes (1/4)
            -May Day-
                   -griersonianum (1/4)
```

4ft(1.2m)          0°F(-18°C)          ML

Bright red, waxy flowers, 3.5in(8.9cm) across, in trusses of 13-
15.  Leaves up to 7in(17.8cm) long.  Growth habit spreading and
contorted.  Blooms midseason. (Rudolph Henny, reg. 1961)  P. A.
(ARS) 1960

```
g.
            -maximum (1/2)
Captivation-                   -catawbiense (1/8)
           -             -unnamed hybrid-
           -Altaclarense-           -ponticum (1/8)
                       -arboreum ssp. arboreum (1/4)
```

Flowers rosy crimson with black spots. (Standard & Noble, before
1850)

```
                   -griffithianum (1/8)
cl.         -Kewense-
          -Aurora-           -fortunei ssp. fortunei (1/8)
Cara Mia-   -thomsonii ssp. thomsonii (1/4)
        -    -wardii var. wardii (1/4)
        -Crest-                -fortunei ssp. discolor (1/8)
              -Lady Bessborough-
                            -campylocarpum ssp. campylo-
                             carpum Elatum Group (1/8)
```

Flowers pale rose in bud opening to cream with yellow throat and
basal crimson stain, 4.5in(11.5cm) across, with 12-15 per truss.
Plant tall, compact. (Edmund de Rothschild, 1966)   A. M. (RHS)
1966

```
cl.
            -catawbiense (1/2)
Caractacus-
          -unknown (1/2)
```

6ft(1.8m)          -25°F(-32°C)          L          2/3

Flowers purplish red; medium-sized leaves, deeply veined.  Compact growth habit; foliage tends to yellow in full sun. (A. Waterer)  F. C. C. (RHS) 1865

```
g. & cl.
            -arboreum ssp. arboreum (5/8)
Cardinal-           -thomsonii ssp. thomsonii (1/4)
        -Barclayi-           -arboreum ssp. arboreum
                   -Glory of  -
                   Penjerrick-griffithianum (1/8)
```

Bright scarlet flowers.  (Lord Aberconway)  F. C. C. (RHS) 1937

```
cl.                            -brookeanum var. gracile (1/4)
            -Duchess of Edinburgh-
Cardinale-                      -longiflorum (lobbii) (1/4)
        -javanicum (1/2)
```

Vireya hybrid with scarlet crimson flowers.  F. C. C. (RHS) 1885

```
                            -neriiflorum ssp. neriiflorum (1/8)
g.          -F. C. Puddle-
        -Ethel-           -griersonianum (1/8)
Careth-    -forrestii ssp. forrestii Repens Group (1/4)
       -          -arboreum ssp. arboreum (5/16)
       -Cardinal-           -thomsonii ssp. thomsonii (1/8)
                -Barclayi-
                   -Glory of  -arboreum ssp. arboreum
                   Penjerrick-
                        -griffithianum (1/16)
```

Red flowers.  (Lord Aberconway, intro. 1946)

```
g. & cl.

     -irroratum ssp. irroratum (1/2)
Carex-
     -oreodoxa var. fargesii (1/2)
```

Flowers of rich pink within, darker without,  about 8 in medium-
sized neat trusses.  Tall, pyramidal, early-blooming bush, with
greyish green foliage. Floriferous. (Rothschild, 1932)   A. M.
(RHS) 1932

```
cl.

Carex Blush     Parentage as above
```

Very early flowering and tall, forming a pyramidal bush.  Leaves
are soft greyish green resembling fargesii.  Pink buds; campanulate, pink-tinged flowers with dark spots,  in trusses of about
8. (Rothschild, 1932)

```
cl.

Carex White     Parentage as above
```

Carex g.  Many moderate-sized trusses of bell-shaped flowers,
pink in bud, opening to almost pure white with maroon spots.
See Carex Blush for foliage and growth habit. (Rothschild, 1932)

```
cl.                   -griffithianum (1/4)
     -Loderi King George-
Carioca-                -fortunei ssp. fortunei (1/4)
     -Ostbo's Y3--unknown (1/2)
```

4ft(1.2m)          0°F(-18°C)          ML

Flowers to 4in(10.2cm) wide,  center of cream blending into wide
border of purplish pink with spotted upper petal of dark reddish
orange. (Bovee, 1963)

```
                 -griffithianum (1/16)
             -Kewense-
g.       -Aurora-      -fortunei ssp. fortunei (5/16)
    -Naomi-      -thomsonii ssp. thomsonii (1/8)
Carita-    -fortunei ssp. fortunei
      -campylocarpum ssp. campylocarpum (1/2)
```

5ft(1.5m)          5°F(-15°C)          M          4/4

Pink buds, opening primrose yellow. Leaves rich green, held one
year only. See also Golden Dream and two other clones below.
(Rothschild, 1935) A. M. (RHS) 1945 See color illus. PB pl. 7;
VV p. 79.


cl.


Carita Charm     Parentage as above

5ft(1.5m)          5°F(-15°C)          M          4/4

Similar to above with deep pink buds opening to cream, flushed
peach pink. (Rothschild, 1935) See color illus. PB pl. 6 & 7.


cl.


Carita Golden Dream     See Golden Dream


cl.


Carita Inchmery     Parentage as above

5ft(1.5m)          5°F(-15°C)          M          4/4

Deep rose red buds open to a blend of biscuit yellow and rose or
salmon, fading to opal tints. Bush well-proportioned, rather
open; rich green leaves. (Rothschild, 1935) See color illus.
F p. 37; PB pl. 7.


cl.
         -catawbiense (1/2)
Carl Mette-
         -unknown (1/2)

A parent of Hassan and others by Seidel, Mrs. W. R. Dykes by van
Ness. Red flowers. (Origin unknown)


```
                          -griffithianum (1/16)
                   -Kewense-
cl.            -Aurora-      -fortunei ssp. fortunei
        -Naomi Pink-      -                   (13/16)
Carl    - Beauty   -      -thomsonii ssp. thomsonii (1/8)
Phetteplace-      -fortunei ssp. fortunei
          -fortunei ssp. fortunei (pink form)
```

6ft(1.8m)          -10°F(-23°C)          M          4/3

Flowers of intense pink with random chocolate brown spotting, of
heavy substance, fragrant, 4in(10.2cm) wide; ball-shaped trusses
of 13. Floriferous. Plant upright; greyish olive green foliage
retained 2 years. (Dr. Carl Phetteplace cross; intro. by Robert
Guitteau, 1975)


```
                    -campylocarpum ssp. campylocarpum
             -Ole Olson-                    Elatum Group (1/8)
cl.   -Lem's Goal-      -fortunei ssp. discolor (1/4)
      -          -griersonianum (1/8)
Carlene-      -Azor-
      -          -fortunei ssp. discolor
     -williamsianum (1/2)
```

3ft(.9m)          5°F(-15°C)          M          3/3

Loose bells of warm golden cream flushed with glowing pink at
centers. Plant a mound of bright green with rounded leaves held
on yellow stems. (Halfdan Lem; intro. by Carl Fawcett, 1962)
See color illus. HG p. 130.


g.
       -Souvenir of Anthony Waterer--unknown (1/2)
Carmania-
       -facetum (eriogynum) (1/2)

The Hillier Manual calls it a first-class Exbury hybrid, with
bright pink flowers, blooming midseason; it is also shown in the
International Rhododendron Registry of 1958. Not listed in the
Rothschild Rhododendrons by Phillips and Barber. (Rothschild,
1935)


cl.


Carmelita*--form of chamaethomsonii var. chamaethauma

Collected and named by Kingdon-Ward (5847) from Doshong La, Him-
alaya. Flowers of luminous carmine in threes. Has larger foliage than Scarlet Runner, Scarlet Pimpernel (App. C).


g.
       -sanguineum ssp. didymum (1/2)
Carmen-
       -forrestii ssp. forrestii Repens Group (1/2)

1ft(.3m)          -5°F(-21°C)          EM          4/5

Several forms of this hybrid are distributed. One of the best
dwarfs. Blooms when quite young; flowers deep bright red, campanulate, in clusters of 2-5. Shrub sturdy, spreading; foliage
small, dense, glossy. (Rothschild, 1935) See color illus. ARS
Q 28:4 (Fall 1974): p. 238; F p. 37; VV p. 41.


```
g.              -haematodes ssp. haematodes (1/4)
        -May Day-
Carnival-      -griersonianum (1/4)
        -      -griffithianum (1/4)
        -Loderi-
              -fortunei ssp. fortunei (1/4)
```

Pink flowers. (Lord Aberconway, intro. 1950)


```
cl.              -diaprepes (1/4)
        -Polar Bear-
Carol Amelia-      -auriculatum (1/4)
        -          -fortunei ssp. discolor (1/4)
        -Evening Glow-      -dichroanthum ssp. dichro-
              -Fabia-                    anthum (1/8)
                    -griersonianum (1/8)
```

4ft(1.2m)          -5°F(-21°C)          VL

Very fragrant white flowers 5in(12.7cm) across, 7 wavy lobes, in
somewhat open and lax trusses of 8-10. Plant upright and well-
branched. (A. John Holden, 1979)


```
cl.              -forrestii ssp. forrestii Repens Group (1/4)
        -Elizabeth-
Carol High-      -griersonianum (1/4)
        -          -fortunei ssp. fortunei ? (1/4)
        -French Creek-
              -unknown (1/4)
```

5ft(1.5m)          -10°F(-23°C)          EM

Very fragrant white flowers with chartreuse spotting; trusses
6in(15cm) wide hold about 14. Plant upright, broad, with stiff
branches; new growth bronze green. (Charles Herbert, reg. 1979)


```
                 -griffithianum (1/8)
cl.       -Mars-
      -Vulcan-   -unknown (1/8)
Carol Jean-      -griersonianum (1/4)
      -          -calophytum var. calophytum (1/4)
      -Robin Hood-
              -sutchuenense (1/4)
```

Flowers carmine with dark blotch on dorsal lobe, deeper red in
throat; trusses hold 14. (J. H. Klupenger, reg. 1977) A. M.
(RHS) 1957

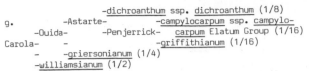

```
                    -dichroanthum ssp. dichroanthum (1/8)
g.         -Astarte-          -campylocarpum ssp. campylo-
      -Ouida-    -Penjerrick-  carpum Elatum Group (1/16)
Carola-    -                 -griffithianum (1/16)
     -      -griersonianum (1/4)
      -williamsianum (1/2)
```

Pink flowers.  (Lord Aberconway, intro. 1941)


```
cl.
           -minus var. minus Carolinianum Group (1/2)
Carolina Rose*
           -azalea prinophyllum (roseum) (1/2)
```

3ft(.9m)          -10°F(-23°C)            M

A well-branched azaleodendron  with a profusion of small, dark
pink flowers. (Mrs. J. F. Knippenberg)


```
cl.
      -decorum ? (1/2)
Caroline-
      -brachycarpum ssp. brachycarpum ? (1/2)
```

6ft(1.8m)          -15°F(-26°C)          ML          3/3

Pale orchid pink flowers; light fragrance  enhanced by warm sun.
Highly resistant to root rot; not easy to propagate.  Gable's
first named hybrid. (Joseph Gable, 1956)    See color illus. LW
pl. 43; VV p. 44.


```
cl.
              -yakushimanum ssp. yakushimanum (1/2)
Caroline Allbrook-            -ponticum (1/4)
              -Purple Splendour-
                     -unknown (1/4)
```

4ft(1.2m)          -5°F(-21°C)            M

Rose  purple flowers, about 20 in a truss,  with cream blotching
in the upper center.  Plant vigorous, spreading as wide as high.
See also Ernest Inman.     (A. F. George,  Hydon Nurseries, 1975)
A. M. (RHS) 1977


```
cl.        -Corona--unknown (1/4)
      -Bow Bells-
Caroline-      -williamsianum (1/4)
de Zoete-              -dichroanthum ssp. scyphocalyx (1/8)
      -          -Socrianum-
      -unnamed-        -griffithianum (1/8)
       hybrid-   -wardii var. wardii (1/8)
              -Rima-
                  -decorum (1/8)
```

Pure white, well-shaped flowers, in trusses of about 10.  Plant
of medium growth; glossy foliage. (A. F. George, Hydon Nurseries, 1974)


```
cl.           -forrestii ssp. forrestii Repens Group
       -Elizabeth-                      (1/4)
Caroline Gem-      -griersonianum (1/4)
      -            -decorum ? (1/4)
       -Caroline-
              -brachycarpum ssp. brachycarpum ? (1/4)
```

2.5ft(.75m)          -5°F(-21°C)            M

Empire rose buds opening to carmine rose flowers with a flush of
coral, bell-shaped, up to 4in(10.2cm) across; conical trusses of
10-12.  Plant compact, rounded, as broad as tall. (Mrs. J. Knippenberg, reg. 1966)


```
g.
           -fortunei ssp. fortunei (1/2)
Caroline Spencer-
           -williamsianum (1/2)
```

Scarlet flowers.  See also Hubert Robert. (Adams-Acton, 1950)


```
                    -maximum (1/8)
         -Halopeanum-
g.  -Snow Queen-     -griffithianum (1/2)
      -        -griffithianum
Caroline-    -Loderi-
Whitner-        -fortunei ssp. fortunei (3/8)
              -griffithianum
      -Loderi Sir Edmund-
                  -fortunei ssp. fortunei
```

Flowers of delicate pink. (Lady Loder)   A. M. (RHS) 1935


```
cl.
      -wardii var. wardii (1/2)
Carolyn Grace-
      -unknown (1/2)
```

4ft(1.2m)          0°F(-18°C)          EM          3/3

Flowers of chartreuse yellow with green tinge, 3in((7.8cm) wide,
in rounded trusses of 7-10.  (George Grace)    A. E. (ARS) 1960
See color illus. ARS Q 29:2 (Spring 1975): p. 103.


```
cl.
                        -griffithianum
              -George Hardy-     (1/8)
      -Mrs. Lindsay Smith-     -catawbiense (1/8)
Carolyn Hardy-        -Duchess of Edinburgh--unknown
      -wardii var. wardii (1/2)          (1/4)
```

Flowers light sap green, open funnel-shaped, 3.5in(8.9cm) wide,
5-lobed, in loose trusses.  Leaves oblong, 4.5in(11.5cm) long.
(Collingwood Ingram, 1967)  A. M. (RHS) 1967


```
cl.
      -minus var. minus Carolinianum Group (1/2)
Carousel-
      -saluense var. saluense (1/2)
```

Upright spreading plant, 29in(73cm) high by 41in(102cm) wide in
20 years.  Dark green leaves 1.5in(3.8cm) long.  Flowers lavender pink with spotted throat, 1in(2.5cm) across; trusses hold up
to 10. (Mr. & Mrs. James Caperci, 1965)


```
cl.
```

Carse--form of irroratum ssp. irroratum

White-flowered,  blooming early.  Reg. as seedling of irroratum.
(P. Cox, 1965)


```
cl.
      -Corona--unknown (5/8)
Cary Ann-        -griffithianum (1/8)
      -      -Mars-
      -Vulcan-    -unknown
              -griersonianum (1/4)
```

3ft(.9m)          5°F(-15°C)            M          3/4

Flowers coral red, trumpet-shaped, 2in(5cm) across; conical
trusses hold about 17.  Very floriferous.  Attractive dark green
foliage.  Tolerant of full sun. (Wright Sr. & Jr., 1962)  P. A.
(ARS) 1961   See color illus. VV p. 25.


```
                    -caucasicum (3/16)
cl.      -Jacksonii-      -caucasicum
      -Goldsworth-      -Nobleanum-
Cary's- Yellow   -              -arboreum ssp. arboreum
Cream-      -                          (1/8)
      -          -campylocarpum ssp. campylocarpum (1/4)
      -unnamed-brachycarpum ssp. brachycarpum (1/4)
       hybrid-              -catawbiense (1/16)
              -Mrs. P.  -Atrosanguinem-
               den Duden-      -unknown (1/8)
              -              -arboreum ssp. arboreum
              -Doncaster-
                     -unknown
```

'Bravo!' by Leach
Photo by Leach

'Bremen' by Arends/Bohlje
Photo by Greer

'Britannia' by C. B. van Nes
Photo by Greer

'Bronze Wing' by Teese (Aus)
Photo by Greer

'Broughtonii' by Broughton, 1853
Photo by Greer

'Broughtonii Aureum' by Smith of Norbiton, 1830
Photo by Greer

'Brown Eyes' (BD 1046) by Dexter/Bosley
Photo by Leach

'Bruce Brechtbill' by Brechtbill
Photo by Greer

'Burgundy' by Rose/Lem
Photo by Greer

'Burma' by Leach
Photo by Leach

'C. B. van Nes' by van Nes
Photo by Greer

'C. I. S.' by R. Henny
Photo by Greer

'Cairo' by Leach
Photo by Knierim

'California Gold' by Bowman
Photo by Greer

'Calsap' by Michener
Photo by Michener

'Canada' by Greig/Caperci
Photo by Greer

'Canary' (Koster) by M. Koster, 1920
Photo by Greer

'Carl Phetteplace' by Phetteplace/Guitteau
Photo by Greer

'Catawbiense Boursault' by Boursault
Photo by Greer

'Cathy Jo' by Disney
Photo by Greer

'Celeste' by Waterer
Photo by Greer

'Ceylon' by Gable/Leach
Photo by Leach

'Chapeau' by Broxson/Disney
Photo by Greer

'Chapmanii Wonder' by Wada
Photo by Greer

2ft(.6m)        -15°F(-26°C)        ML

Light yellow flowers with spotting of darker yellow, in trusses
of 16.  Plant 3/4 as wide as tall, well-branched; leaves held 2
years.  Other named selections from this cross: Cary's Yellow,
Josephine V. Cary, Strawberries and Cream, William P. Cary.  (E.
A. Cary, reg. 1980)

```
                       -caucasicum (3/16)
cl.           -Jacksonii-       -caucasicum
     -Goldsworth-       -Nobleanum-
Cary's- Yellow  -                -arboreum ssp. arboreum
Yellow-       -campylocarpum ssp. campylocarpum (1/4)   (1/8)
        -brachycarpum ssp. brachycarpum (1/4)
      -unnamed-                     -catawbiense (1/16)
       hybrid-          -Atrosanguineum-
        -Mrs. P. den-              -unknown (1/8)
           Ouden    -       -arboreum ssp. arboreum
                     -Doncaster-
                              -unknown
```

2ft(.6m)        -15°F(-26°C)        ML

Chartreuse flowers 1.75in(4.5cm) across, 5-lobed, funnel-shaped;
12 held in ball-shaped, open trusses.  Plant rounded, as wide as
tall; narrowly elliptic dull green leaves held 2 years.  See also
Cary's Cream, Josephine P. Cary, Strawberries and Cream, William
P. Carey.  (Edward A. Cary, cross 1969; reg. 1980)

```
      -Jan Dekens--unknown (11/16)   -griffithianum (9/32)
cl.     -         -Norman-Beauty of-
        -         - Gill - Tremough-arboreum ssp. arboreum
Castanets-    -Anna-     -                        (1/32)
        -     -    -      -griffithianum
     -Point  -       -      -griffithianum
     -Defiance-  -Jean Marie -
        -          de Montague-unknown
        -                  -griffithianum
        -Marinus Koster-
                        -unknown
```

5ft(1.5m)       0°F(-18°C)        M

Red buds open to very light pink with large spotting of light
Indian lake and fragrant.  Reverse roseine purple with radial
ribs of fuchsia purple.  Plant as wide as high with large,
glossy leaves, yellow green.  (John G. Lofthouse, intro. 1974;
reg. 1984)

cl.
```
      -barbatum (pink form) (1/2)
Castle of Mey-
      -haematodes ssp. chaetomallum (1/2)
```

Named after the Scottish home of the Queen Mother.  Plant bushy,
medium-sized, compact.  Flowers deep red, waxy, campanulate, in
flat trusses.  Pale green leaves with slight indumentum beneath.
Blooms midseason.  (Rothschild, 1963)

cl.
```
     -Catalgla--catawbiense var. album Glass (1/2)
Cataldi*
     -wardii var. wardii (1/2)
```

A parent of Betsy Kruson.

cl.

Catalgla--catawbiense var. album Glass

6ft(1.8m)       -25°F(-32°C)        ML        3/3

Light pink buds opening to pure white, held in high-domed, many-
flowered trusses.  A parent of many very hardy hybrids.  (Joseph
Gable, 1959)  See color illus. LW pl. 53.

cl.
```
             -catawbiense (1/4)
        -Catawbiense Album-
Catalode-            -unknown (1/4)
        -              -griffithianum (1/4)
        -Loderi King George-
                     -fortunei ssp. fortunei (1/4)
```

6ft(1.8m)       -15°F(-26°C)        M        3/3

Synonym County of York.  Flowers of good substance, pale
chartruese in bud, opening white with olive throat in tall
truss.  Beautiful long leaves, convex, dark green.  (Gable,
1936)  A. E. (ARS) 1960  See color illus. LW pl. 46; VV p. 64.

cl.

Catanea--form of catawbiense var. album

This is a white, shapely form of catawbiense.  (Selected by Joseph Gable; intro. by G. Guy Nearing; reg. 1959)

cl.
```
             -catawbiense (1/2)
Catawbiense Album-
             -unknown (1/2)
```

6ft(1.8m)       -25°F(-32°C)        ML        3/3

A plant of great hardiness and vigor.  Buds flushed lilac, opening to pure white flowers with greenish yellow spotting, held in
compact, rounded trusses.  Sometimes listed as a form of catawbiense, rather than a hybrid.  (A. Waterer, before 1900)  See
color illus. VV p. 55.

cl.
```
             -catawbiense (1/2)
Catawbiense Boursault-
             -unknown (1/2)
```

5ft(1.5m)       -25°F(-32°C)        ML        3/3

Purple flowers in rounded trusses; a plant of good sturdy habit.
Sometimes called a select form of catawbiense. (Boursault)

cl.
```
             -catawbiense (1/2)
Catawbiense Grandiflorum-
             -unknown (1/2)
```

6ft(1.8m)       -25°F(-32°C)        ML        2/3

Long-lasting, distinctive lilac flowers; full rounded trusses on
a very hardy, attractive plant.  (A. Waterer)  See color illus.
F p. 73; VV p. 4.

---
```
                 -catawbiense var. album (1/2)
Catawbiense Grandiflorum Album-
                 -unknown (1/2)
```

Possibly a selected form of Catawbiense  Album or Grandiflorum
(both, A. Waterer).  Full rounded trusses of white flowers.  At
maturity plant is 15ft(4.5m) or taller.  Very hardy, vigorous,
floriferous.  See color illus. F p. 72.

g.
```
                 -catawbiense (1/4)
       -unnamed hybrid-
Catfortcampy*      -fortunei ssp. fortunei (1/4)
       -campylocarpum ssp. campylocarpum (1/2)
```

5ft(1.5m)       -5°F(-21°C)        M        4/4-4/2

Flowers of apricot pink to coral.  Named clones are Catfortcampy
#2, Dupont's Apricot, Salmon Bamboo; none registered.  (Joseph
B. Gable)  See color illus. LW pl. 51

cl.

Catfortcampy Two (#2)*      Parentage as above

5ft(1.5m)          -10°F(-23°C)          M          4/4

Catfortcampy g.  Hardiest and most vigorous clone.  Flowers of
soft apricot pink.  (Joseph Gable)


cl.

Cathaem #1      See Conoco-cheague


cl.
       -Corona--unknown (5/8)
Cathy-       -fortunei ssp. discolor (1/4)
     -Dondis-        -arboreum ssp. arboreum (1/8)
              -Doncaaster-
                      -unknown

Flowers of bright pink to white, with a yellow throat.  (Rudolph
Henny, 1956)


cl.

Cathy Carter (fortunei ssp. discolor x Nereid x Tally Ho) X
           Autumn Gold  (Exact combination unknown)

3ft(.9m)          -10°F(-23°C)          ML

Flowers of heavy substance funnel-shaped 4in(10cm) across with
7-8 lobes, bicolor of carmine rose and lemon yellow in trusses
of 8-11.  Plant broad, semi-dwarf.  (Bernice I. Jordan, 1976)


                        -ponticum (3/8)
             -Purple   -
     -unnamed-Splendour-unknown (9/16)
cl.     - hybrid-                    -griffithianum (1/16)
        -       -          -Queen Wilhelmina-
Cathy Jo*    -Britannia-               -unknown
        -              -Stanley Davies--unknown
        -         -ponticum
     -unnamed hybrid-
                   -unknown

Deep wine purple flowers; each lobe folds back, making the flower seem flat-faced.  (Disney)


g.
      -caucasicum (1/2)
Cauapo-
      -dichroanthum ssp. apodectum (1/2)

A hybrid with apricot colored flowers.  (E. J. P. Magor, 1927)


g.
      -caucasicum (1/2)
Caubut-
      -Sir Charles Butler--fortunei ssp. fortunei (1/2)

Flowers are 7-lobed, white shaded yellow inside, and spotted
green, held in trusses of 8.  (E. J. P. Magor, 1926)


g.
      -caucasicum var. stramineum ? (1/2)
Caucamp-
      -campylocarpum ssp. campylocarpum (1/2)

Flowers light yellow with red spotting.  (E. J. P. Magor)


cl.
             -caucasicum (1/2)
Caucasicum Pictum-
             -unknown (1/2)

Pink, frilled flowers with a dark blotch.  Of medium height.


g.
      -caucasicum var. stramineum ? (1/2)
Cauking-       -griffithianum (1/4)
     -Mrs. Randall-
        Davidson  -campylocarpum ssp. campylocarpum
                      (Hooker's form) (1/4)

Cream colored flowers, crimson spotting.  (E. J. P. Magor, 1928)


cl.
       -Essex Scarlet--unknown (1/2)
Cavalcade-
       -griersonianum (1/2)

Red flowers.  A parent of Dusky Wood.  (Waterer, Sons & Crisp)


                          -souliei (1/16)
                    -Soulbut-
g.          -Vanessa-     -Sir Charles Butler--fortunei
       -Radiance-    -              ssp. fortunei (1/16)
Cavalier-     -         -griersonianum
     -         -griersonianum (3/8)
     -facetum (eriogynum) (1/2)

Brilliant scarlet flowers.  (Lord Aberconway, intro. 1950)


cl.
       -Pygmalion--unknown (1/2)
Cavalier-    -griersonianum (1/4)
       -Tally Ho-
              -facetum (eriogynum) (1/4)

5ft(1.5m)          5°F(-15°C)          ML          3/3

Flowers bright red, heavily spotted; leaves very dark green, of
medium size.  Trusses small but abundant.  (R. Henny, reg. 1958)

cl.

Cecil Nice--form of pocophorum var. pocophorum

4ft(1.2m)          5°F(-15°C)          EM          3/3

Flowers dark cardinal red with deeper markings in upper throat.
Foliage heavily indumented.  (Anne, Countess of Rosse, Nymans
Garden, The National Trust)  A. M. (RHS) 1971


cl.                   -ellottii (1/4)
              -Fusilier-
Cecil S. Seabrook-    -griersonianum (1/4)
         -    -dichroanthum ssp. dichroanthum (1/4)
         -Jasper-
               -Lady    -fortunei ssp. discolor
               -Bessborough-          (1/8)
                      -campylocarpum ssp. campylo-
                          carpum Elatum Group (1/8)

Flowers campanulate, 2.5in(6.4cm) wide, buttercup yellow  fading
to orange buff, up to 22 in flat trusses.  Plant resembles Fusilier; of medium height; dark green leaves with beige indumentum.
Reg. also as C. Sabine Seabrook.  (C. S. Seabrook, 1967)


                   -fortunei ssp. discolor (3/8)
cl.      -King of Shrubs-    -dichroanthum ssp. dichroanthum
         -         -Fabia-               (1/8)
Cecil Smith-             -griersonianum (1/8)
         -    -wardii var. wardii (1/4)
         -Crest-          -fortunei ssp. discolor
              -Lady Bessborough-
                         -campylocarpum ssp. campylo-
                             carpum Elatum Group (1/8)

5ft(1.5m)          0°F(-18°C)          M

Dresden yellow flowers, 4in(10.2cm) wide, of heavy substance, 5-
or 6-lobed; open, ball-shaped trusses hold 10-11.  Plant almost
as broad as tall; ivy green leaves held 1-2 years.  (Cecil Smith
cross, 1958; named by C. Phetteplace; reg. 1976)

cl.
```
                    -yakushimanum ssp. yakushimanum (1/2)
Cecily Fortescue-      -wardii var. wardii (1/4)
            -Hawk-          -fortunei ssp. discolor (1/8)
            -Lady -
                  -Bessborough-campylocarpum ssp. campylocar-
                                    pum Elatum Group (1/8)
```

White flowers with pale pink spotting, held in trusses of 11-14.
(L. S. Fortescue, 1981)

cl.
```
            -griffithianum (1/4)
      -Alice-
Celeste-      -unknown (1/4)
      -fortunei ssp. discolor (1/2)
```

6ft(1.8m)          -10°F(-23°C)          ML

Beautiful trusses of fuchsia pink flowers on a vigorous plant.
(Waterer & Crisp) A. M. (RHS) 1944

cl.
```
                  -dichroanthum ssp. dichroanthum (1/4)
            -Fabia-
Celestial Bells-      -griersonianum (1/4)
          -              -wardii var. wardii (1/4)
          -unnamed hybrid-
                        -wardii var. wardii (1/4)
```

2ft(.6m)          5°F(-15°C)          M

Flowers of apricot yellow fading to cream yellow; drooping
trusses of 7. Compact, bushy plant; leaves elliptic, 4in(10.cm)
long with reddish petioles. (Dr. Rollin G. Wyrens, 1964)

cl.
```
            -fortunei ssp. discolor (1/2)
Century Twentyone-
            -Cunningham's Sulphur--caucasicum (1/2)
```

Pale sulphur yellow buds, opening to white flowers 3in(7.6cm)
wide; rounded trusses of about 15. Plant 6ft(1.8m) tall; foliage very shiny. (Halfdan Lem, 1963)

cl.

Ceramic--form of wardii var. wardii (selfed)

4ft(1.2m)          0°F(-18°C)          M          4/3

Upright trusses of 6 flowers, white with sap green throat, bell-
shaped, 2.25in(5.7cm) wide. Midseason. (Rudolph Henny, 1960)

cl.
```
      -elliottii (1/2)
Cerisette-
      -decorum (1/2)
```

Sheila Moore g. Flowers fuchsine pink, with scarlet band around
the throat. (Lord Digby) A. M. (RHS) 1954

cl.

Cetewayo     Parentage unknown

A plant of medium height with flowers very dark, almost black
purple. Foliage is dark and glossy. (A. Waterer) A. M. (RHS)
1958

cl.
```
      -Catalgla--catawbiense var. album Glass (1/2)
Ceylon-
      -fortunei ssp. fortunei (1/2)
```

6ft(1.8m)          -25°F(-32°C)          M          3/4

Flowers openly funnel-shaped 3.25in(8.3cm) across, rose purple,
in trusses of 12. Floriferous. Glossy leaves 5in(12.7cm) long,
held 2 years. (Joseph Gable cross, before 1950; intro. by David
G. Leach, 1973; reg. 1982)

g.
```
                        -ciliatum (3/4)
            -Countess of Haddington-
Chaffinch-                  -dalhousiae var. dalhousiae
            -ciliatum                              (1/4)
```

Plant of low-growing, open habit with hairy foliage. Pale pink,
fragrant flowers, in numerous small trusses. Blooms early midseason. (Rothschild, 1935)

g.
```
      -cinnabarinum ssp. cinnabarinum (1/2)
Chalice-
      -maddenii ssp. maddenii (1/2)
```

Tubular flowers pale pink within, outside creamy white suffused
with rose. See also Rose Mangles. A. M. (RHS) 1932

g.
```
                  -thomsonii ssp. thomsonii (1/4)
      -Chanticleer-
Challenger-        -facetum (eriogynum) (1/4)
      -              -griffithianum (1/4)
      -Sunrise-
            -griersonianum (1/4)
```

Rose-colored flowers. (Lord Aberconway, intro. 1950)

cl.
```
                  -dichroanthum ssp. dichroanthum (1/4)
      -Goldsworth Orange-
Champagne-              -fortunei ssp. discolor (1/4)
      -griersonianum (1/2)
```

Tortoiseshell g. Medium to tall growing, wider than high;
unusual creamy apricot flowers and 14 per truss. (W. C. Slocock,
1946) H. C. (RHS) 1962; A. M. (Wisley Trials) 1967

g.
```
                        -cinnabarinum ssp. cinnabarinum
  -Lady Chamberlain-          Roylei Group (1/4)
  -                  -cinnabarinum ssp.
Chan-      -Royal Flush-      cinnabarinum (1/8)
  -                  -maddenii ssp. maddenii (1/8)
  -cinnabarinum ssp. xanthocodon (1/2)
```

Orange yellow flowers. See also Comely. (Lord Aberconway, 1946)

g.
```
                        -souliei (1/16)
                  -Soulbut-
            -Vanessa-      -Sir Charles Butler--fortunei ssp.
            -      -                        fortunei (1/16)
g.  -Adonis-      -griersonianum (1/4)
    -      -              -griffithianum (1/8)
Chandon-      -Sunrise-
    -              -griersonianum
    -          -thomsonii ssp. thomsonii (1/4)
    -Chanticleer-
            -facetum (eriogynum) (1/4)
```

Pink flowers. (Lord Aberconway, 1946)

g.
```
            -thomsonii ssp. thomsonii (1/4)
    -Chanticleer-
    -      -facetum (eriogynum) (1/4)
Chaneta-      -griffithianum (1/8)
    -      -Loderi-
    -Coreta-      -fortunei ssp. fortunei (1/8)
      -arboreum ssp. zeylanicum (1/4)
```

Red flowers. (Lord Aberconway, 1946)

```
cl.                       -dichroanthum ssp. apodectum (1/4)
            -unnamed hybrid-
Chang   -         -campylocarpum ssp. campylocarpum
Tso Lin-                          (Hooker form) (1/4)
    -          -wardii var. wardii (1/4)
    -Idealist   -                    -griffithianum (1/32)
    (A. M. form)-                 -Kewense-
                 -Aurora-        -fortunei ssp. fortunei
              -Naomi-     -                      (5/32)
                 -      -thomsonii ssp. thomsonii (1/16)
                    -fortunei ssp. fortunei
```

Flowers cadmium orange in bud, opening to a medium lemon yellow,
darker in throat.  Trusses hold about 9.  Dark green leaves;  no
indumentum.  (Lester E. Brandt, 1968)


```
g.
           -thomsonii ssp. thomsonii (1/2)
Chanticleer-
           -facetum (eriogynum) (1/2)
```

6ft(1.8m)`         15°F(-9°C)         ML          3/2

Waxy, bright red flowers, 3in(7.6cm) across, in trusses of 6-9.
Plant upright with medium green foliage; vigorous, easy to grow.
(Rothschild, 1935)


```
                                    -griffithianum (1/16)
                      -Queen Wilhelmina-
cl.        -Britannia-              -unknown (11/16)
       -unnamed-        -Stanley Davies--unknown
Chapeau* hybrid-unknown
     -                   -ponticum (1/4)
       -Purple Splendour-
                      -unknown
```

5ft(1.5m)          -5°F(-21°C)         ML          4/3

Flowers have a purple edge and white center on all lobes with a
large, deep purple blotch on the upper lobe.  Foliage matte
green.  (Broxson-Disney)


cl.

Chapel Wood--form of lanigerum (R 03913)

4ft(1.2m)          5°F(-15°C)         EM          3/3

A handsome shrub with very tight trusses of 40-50 flowers, medium neyron rose color.  Smooth dark green leaves, with indumentum
beneath changing through white, grey and brown.  (Crown Estate,
1962)  A. M. (RHS) 1961; F. C. C. (RHS) 1967


```
cl.
           -minus var. chapmanii (1/2)
Chapmanii Wonder*
           -dauricum var. album (1/2)
```

5ft(1.5m)          -15°F(-26°C)         E          3/3

Vigorous plant with foliage similar to P.J.M.  Flowers bright
lavender pink; tolerant of heat and cold.  (Raiser K. Wada)


cl.

Charisma--form of ciliicalyx

4ft(1.2m)          15°F(-9°C)         EM          3/3

Truss holds 2-4 flowers, rose madder in bud, opening rose pink
with darker lines on each lobe, yellow orange blotch on the upper lobe.  Leaves scaly, elliptic.  (Raised by the New Zealand
Rhododendron Association; reg. by Graham Smith, Pukeiti Rhododendron Trust, 1982)


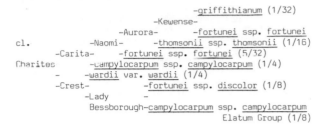

```
                      -griffithianum (1/32)
                 -Kewense-
            -Aurora-      -fortunei ssp. fortunei
cl.      -Naomi-      -thomsonii ssp. thomsonii (1/16)
      -Carita-    -fortunei ssp. fortunei (5/32)
Charites     -campylocarpum ssp. campylocarpum (1/4)
     -      -wardii var. wardii (1/4)
      -Crest-       -fortunei ssp. discolor (1/8)
         -Lady       -
      Bessborough-campylocarpum ssp. campylocarpum
                          Elatum Group (1/8)
```

Pale yellow flowers held in large, well-built trusses.  Plant
habit upright and compact.  (Edmund de Rothschild)  P. C. (RHS)
1965


```
cl.
       -falconeri ssp. falconeri (1/2)
Charles-
       -sinogrande (1/2)
```

Trusses of 25 flowers, 8-lobed, very light chartruese green with
blotch of reddish purple deep in the center. Leaves 18.4in(46cm)
by 7.2in(18cm), oblanceolate, with woolly fawn indumentum.  See
also Annapolis Royal. (Lionel de Rothschild cross, 1938; reg. by
Edmund de Rothschild, 1983)


```
cl.
           -catawbiense (1/2)
Charles Bagley-
           -unknown (1/2)
```

6ft(1.8m)          -20°F(-29°C)         ML          2/1

Rather open and spreading habit.  Flowers cherry red, held in
upright trusses.  Survives very cold winters. (A. Waterer, 1865)


```
cl.                -campylocarpum ssp. campylocarpum
            -Letty Edwards-            Elatum Group (1/4)
Charles Begg-        -fortunei ssp. fortunei (1/4)
           -unknown (1/2)
```

Flowers of Tyrian rose, paler in center. (A. Bramley, 1966)


```
cl.
           -catawbiense (1/2)
Charles Dickens-
           -unknown (1/2)
```

6ft(1.8m)          -25°F(-32°C)         ML          2/2

Flowers purplish crimson red, spotted, in conical trusses.  Dark
green foliage.  Slow-growing; difficult to propagate. (A. Waterer, 1865)  F. C. C. (RHS) 1865


```
cl.
           -haematodes ssp. haematodes (1/2)
Charles Michael-
           -thomsonii ssp. thomsonii (1/2)
```

Major g.  This selected clone exhibited by C. Williams, 1948.


```
cl.
           -Catalgla--catawbiense var. album Glass (1/2)
Charles Robinson*        -griffithianum (1/8)
        -       -Mars-
         -Vulcan-    -unknown (1/8)
              -griersonianum (1/4)
```

The deep pink flower buds have survived temperatures below -25°F
(-32°C).  (Orlando S. Pride, 1977)


cl.

Charles Waterer     Parentage unknown

A parent of Souvenir de D. A. Koster.  Red flowers with lighter
center. (J. Waterer, before 1922)

cl.
            -fortunei ssp. fortunei ? (1/2)
Charlestown-
            -unknown (1/2)


6ft(1.8m)          -5°F(-21°C)          M


Flowers of shades of pink with chartreuse throat, 7-lobed, fragrant, 4in(10.2cm) across; about 16 in trusses 8in(20.3cm) wide.
Upright plant,  as broad as tall; leaves retained 2 years.  (C.
O. Dexter cross; raiser S. Everitt; intro. by C. Herbert; reg.
1976)


cl.
            -wardii var. wardii (croceum) (1/2)
Charley-    -dichroanthum ssp. dichroanthum (1/4)
      -Fabia-
                  -griersonianum (1/4)


A parent of Lonny. White flowers tinted amaranth rose,  turning
creamy yellow with yellow blotch, widely funnel-shaped, 5in(12.7
cm) across, sweetly fragrant; full, rounded trusses. Leaves to
8in(20.3cm) long. (Del W. James, 1962)


cl.                        -thomsonii ssp. thomsonii (1/4)
      -Barclayi Robert Fox-       -arboreum ssp. arboreum
Charlotte-            -Glory of  -                  (1/8)
  Currie -              -Penjerrick-griffithianum (1/8)
      -williamsianum (1/2)


Flowers deep red. (Gen. Harrison, 1975)


cl.
                  -fortunei ssp. fortunei (1/2)
Charlotte de Rothschild-       -arboreum ssp. zeylanicum
                  -St. Keverne-                  (1/4)
                        -griffithianum (1/4)


6ft(1.8m)          -5°F(-21°C)          ML          2/1


Sir Frederick Moore g.  Flowers of clear pink, spotted, campanulate, in trusses of 14. (Rothschild, 1962) A. M. (RHS) 1958


g.
      -forrestii ssp. forresii Repens Group (1/2)
Charm-      -thomsonii ssp. thomsonii (1/4)
      -Shilsonii-
            -barbatum (1/4)


Blood red flowers. A parent of Ruby Gem. (Lord Aberconway, 1941)


g. & cl.      -haematodes ssp. haematodes (1/4)
         -May Day-
Charmaine-      -griersonianum (1/4)
      -      -forrestii ssp. forrestii Repens Group (1/4)
      -Charm-      -thomsonii ssp. thomsonii (1/8)
            -Shilsonii-
                  -barbatum (1/8)


Blood red flowers. (Lord Aberconway)  A. M. (RHS) 1946


cl.
         -flavidum var. flavidum (1/2)
Chartreuse-      -cinnabarinum ssp. cinnabarinum
      -Lady           Roylei Group (1/4)
      -Rosebery-      -cinnabarinum ssp. cinnabarinum
            -Royal Flush-                  (1/8)
            (pink form)-maddenii ssp. maddenii (1/8)


Yellow flowers held in loose trusses of 10 or more. (Lester E.
Brandt, 1960)


g.
      -campylocarpum ssp. campylocarpum (1/2)
Chaste-            -Pink Bell--unknown (1/4)
      -Queen o' the May-
                  -griersonianum (1/4)


Loose clusters of bell-shaped flowers, light primrose yellow. A
rose form was exhibited by Lord Aberconway in 1939. (Crosfield,
1930)


            -griffithianum (1/8)
cl.      -Loderi-
   -Albatross-      -fortunei ssp. fortunei (1/8)
Chat-      -fortunei ssp. discolor (1/4)
   -wardii var. wardii (1/2)


Flowers are marigold orange changing to sap green, 3.5in(9cm)
across. Midseason. (Rudolph Henny, 1960)


cl.      -Moser's Maroon--unknown (9/32)
      -Impi-
Chateau-      -sanguineum ssp. didymum (1/4)
  Lafite-      -elliottii (1/4)
      -Jutland-            -fortunei ssp. discolor (1/16)
                  -Norman-      -catawbiense(1/32)
            -Bellerophon- Shaw -B. de Bruin-
            -                  -unknown
            -facetum (eriogynum) (1/8)


Small shrub as broad as high with deep green foliage; dark red
flowers in medium-sized trusses. (Rothschild, 1963)


cl.            -sanguineum ssp. didymum (3/4)
      -Arthur Osborn-
Chatterbox-      -griersonianum (1/4)
      -sanguineum ssp. didymum


Flowers currant red, funnel-campanulate, 2in(5cm) wide;  6-8 per
truss. (Rudolph Henny, 1960)


cl.            -ponticum (1/4)
      -Purple Splendour-
Chauncy-            -unknown (1/4)
  Alcott-      -dichroanthum ssp. dichroanthum (1/4)
      -Jasper-      -fortunei ssp. discolor (1/8)
      (Exbury -Lady Bessborough-
      Special)-      -campylocarpum ssp. campylo-
                  carpum Elatum Group (1/0)


5ft(1.5m)          -5°F(-21°C)          M


A spreading plant with rounded light green leaves.  Flowers are
porcelain rose, lighter at the base, 2.5in(6.4cm) wide,  held in
trusses of 8-10.   (C. S. Seabrook, 1967)


                  -Corona--unknown (1/4)
cl.      -Bow Bells-
      -      -williamsianum (1/4)
Cheapside-      -dichroanthum ssp. scyphocalyx (1/8)
      -      -Socrianum-
      -unnamed-      -griffithianum (1/8)
      hybrid-      -wardii var. wardii (1/8)
            -Rima-
                  -decorum (1/8)


3ft(.9m)          0°F(-18°C)          M


Pale yellow flowers with reverse flushed apricot. (A. F. George,
Hydon Nurseries, 1975)


cl.            -caucasicum (1/4)
      -Cunningham's White-
Cheer-            -ponticum (white form) (1/4)
      -            -catawbiense (1/4)
      -red catawbiense hybrid-
                  -unknown (1/4)


5ft(1.5m)          -15°F(-26°C)          EM          3/3


Shell pink flowers 2.5in(6.4cm) wide, conspicuous red blotch, in
conical trusses.  Plant rounded, compact; glossy leaves 4in
(10cm) long. (Anthony M. Shammarello, 1958)  See color illus.
VV p. 4.

cl.
       -wardii var. wardii (1/2)
Cheerio-        -souliei (1/4)
  -Rosy Morn-     -griffithianum (1/8)
       -Loderi-
            -fortunei ssp. fortunei (1/8)

Flowers light pink to white, with red markings; imbricate lobes.
(R. Henny, 1944)

cl.
      -williamsianum (1/2)
Cheetans-
      -souliei ? (1/2)

White flowers held in lax trusses. (Gibson Bros.)  P. C. (RHS)
1960

cl.

Cheiranthifolium*--form of ponticum

4ft(1.2m)      -10°F(-23°C)      L     2/4

A form of ponticum with long, narrow leaves having irregular
wavy margins.  The flower is the typical light purple of the
species.  Floriferous.  (Origin unknown)

cl.
     -cinnabarinum var. cinnabarinum Roylei Group (1/2)
Chelsea-      -cinnabarinum ssp. cinnabarinum (1/4)
  -Royal Flush  -
   (orange form)-maddenii ssp. maddenii (1/4)

5ft(1.5m)     10°F(-12°C)     ML    4/3

Lady Chamberlain g.  One of several clones having quite similar
traits; this form has orange and pink flowers. (Rothschild,1930)
See color illus. HG p. 67.

cl.

Chelsea Chimes--form of coryanum (K-W 6311)

6ft(1.8m)     0°F(-18°C)     E-M    3/3

Clear white flowers  with sparse spotting in throat, in trusses
of 8-9.  (Collector Kingdon-Ward; raised by Col. S. R. Clarke;
intro. by R. N. Stephenson Clarke)  A. M. (RHS) 1979

cl.  -yakushimanum ssp. yakushimanum (1/2)
   -          -Lady     -fortunei ssp. discolor (1/16)
Chelsea-         -Bessborough-
Seventy-   -Jalisco-      -campylocarpum ssp. campylo-
  -     -Eclipse-       carpum Elatum Group (1/16)
  -unnamed-    -    -dichroanthum ssp. dichroanthum
    hybrid-    -Dido-         (1/16)
    -       -decorum (1/16)
   -     -elliottii (1/8)
    -Fusilier-
        -griersonianum (1/8)

3ft(.9m)      0°F(-18°C)     M

Salmon pink flowers shaded to rose pink.  See also Vintage Rose.
(John Waterer, Sons and Crisp, 1972)

g.
     -Rose Perfection--unknown (1/2)
Cheronia-
     -orbiculare ssp. orbiculare (1/2)

Rose pink flowers. (Lord Aberconway, 1933)

cl.

Cherry Brandy*--cerasinum (K-W 6923)

3ft(.9m)     0°F(-18°C)     EM    4/4

Crisp foliage and bell-shaped flowers of cherry red, in trusses
of 5-7.  Named in the field by Kingdon-Ward. (Countess of Rosse
and National Trust, Nymans Garden)  A. M. (RHS) 1973    Another
form, from KW 6923, creamy white flowers with cherry red edges,
had received an A. M. in 1930. (Lt. Col. Messel, Nymans Garden)

cl.
      -thomsonii ssp. thomsonii (1/2)
Cherry Bright-
      -williamsianum (1/2)

1.5ft(.45m)     0°F(-18°C)     M    3/3

Flowers very bright cherry red, openly campanulate; trusses of 7
or 8.  Small cordate leaves. (Benjamin Lancaster, 1964)

cl.          -campylocarpum ssp. campylocarpum
    -Elsie Straver-         (1/4)
Cherry Custard-     -unknown (1/4)
      -       -dichroanthum ssp. dichroanthum
     -Roman Pottery-         (1/4)
        -griersonianum (1/4)

20in(.51m)     0°F(-18°C)     M

Flower buds mandarin red, opening to medium saffron yellow, with
coral red stripes from throat to margins; on reverse also; flowers 2in(5cm) wide, in flat trusses of 10-12.  Plant 50% broader
than tall, decumbent branches; leaves held 3 years. (John Lofthouse cross, 1970; reg. 1981)

cl.        -fortunei ssp. fortunei (1/4)
cl.  -Fawn-    -dichroanthum ssp. dichroanthum (1/8)
  -    -Fabia-
Cherry -      -griersonianum (1/4)
Jubilee-       -fortunei ssp. discolor (1/4)
  -   -Lady    -
  -Lem's-Bessborough-campylocarpum ssp. campylocarpum
    Goal-            Elatum Group (1/8)
    -   -griersonianum
    -Azor-
      -fortunei ssp. discolor

3ft(.9m)     0°F(-18°C)     M

Flowers up to 5in(13cm) across, yellowish pink suffused tangerine, banded a deeper shade; trusses of 9. Open habit; leaves up
to 8in(20.3cm) long, medium dark green. (A. A. Childers, 1971)

cl.

Cherry Tip--form of rex ssp. fictolacteum (R 11395)

5ft(1.5m)     -5°F(-21°C)     EM    3/4

Bright cherry pink buds opening to white flowers, flushed pink,
deep crimson blotch  on upper segments. (Collector, Rock; Lord
Digby)  A. M. (RHS) 1953

cl.
     -pubescens (1/2)
Chesapeake-
     -keiskei (1/2)

3ft(.9m)     -10°F(-23°C)     EM    3/2

Small leaves, up to 2in(5cm) long.  Apricot-colored flowers fading white, on every terminal. (G. Guy Nearing, 1950)

cl.

Chesterland     Parentage unknown

7ft(2.1m)     -25°F(-32°C)     EM    4/4

A parent of Hillsdale. Flowers pale pink with bold yellow spotting on back of upper lobe, funnel rotate, 6-lobed; about 12 per
truss.  Very large, glossy leaves, elliptic oblanceolate; plant
slightly taller than wide, very dense.  Difficult to propagate.
(Raiser Julian Pot; intro. by David G. Leach, 1964)

cl.
                  -caucasicum (1/2)
Chevalier Felix de Sauvage-
                  -hardy hybrid--unknown (1/2)

4.5ft(1.35m)          -5°F(-21°C)          EM          3/3

Flowers coral rose accentuated by a rich dark blotch in center.
Trusses tight, upright, cone-shaped, of 9-15 flowers.
(Sauvage, c.1870)    See color illus. F p. 38; VV p. 19.

```
                           -fortunei ssp. discolor (1/8)
                  -Lady    -
                  -Bessborough-campylocarpum ssp. campylocarpum
cl.      -Jalisco-                             Elatum Group (1/8)
         -        -       -dichroanthum ssp. dichroanthum (1/8)
Cheyenne-        -Dido-
         -        -decorum (1/8)
         -       -griffithianum (1/4)
         -Loderi-
                 -fortunei ssp. fortunei (1/4)
```

5ft(1.5m)          -5°F(-21°C)          M          4/3

Flowers of very deep yellow with faint brown throat markings, in
trusses of 9. Leaves medium to large. (Harold E. Greer, 1964)
See color illus. HG p. 129.

```
cl.      -yakushimanum ssp. yakushimanum (1/2)
         -                      -fortunei ssp. discolor
Chickamauga*        -Lady     -                  (3/16)
         -Day  -Bessborough-campylocarpum ssp. campylo-
    -Shah -Dream-                   carpum Elatum Group (1/16)
    -Jehan-       -griersonianum (3/16)
         -                -fortunei ssp. discolor
    -Margaret Dunn-      -dichroanthum ssp. dichro-
              -Fabia-                        anthum (1/16)
                        -griersonianum
```

3ft(.9m)          -15°F(-26°C)          M

Rose pink buds open to flowers light pink on outside, white with
gold eye inside. Compact plant habit; narrow dark green leaves.
(Carl P. Fawcett cross; intro. by Sweetbriar Nursery, 1985)

```
cl.      -dichroanthum ssp. scyphocalyx (1/2)
Chief -                  -sanguineum ssp. didymum (1/8)
Joseph-      -Rubina-         -griersonianum (3/16)
     -unnamed-        -Tally Ho-
        hybrid-              -facetum (eriogynum) (1/16)
        -        -dichroanthum (1/8) sp dichroanthum
              -Fabia-
                   -griersonianum
```

1.5ft(.45m)                              M

Low, spreading shrub, twice as wide as tall; elliptic leaves  to
4in(10.2cm) long, dark green. Flowers open funnel-shaped, 2.5in
(5.7cm) across, bright spinel red. (J. A. Witt, University of
Washington Arboretum, 1968)

cl.

Chief Paulina--form of concinnum var. pseudoyanthinum

5ft(1.5m)     5°F(-15°C)          EM          4/3

Synonym James' Purple. A form with royal purple flowers spotted
dark brown; the Exbury form has more red. Small clusters of 3-6
flowers. Foliage typical of species: a dark bluish green, scaly
on both sides. (Del James, before 1950) P. A. (ARS) 1954    See
color illus. ARS Q 27:3 (Summer 1973): p. 170.

```
cl.      -hanceanum Nanum Group (1/2)
Chiffchaff-
         -fletcheranum (1/2)
```

2ft(.6m)          5°F(-15°C)          EM          3/3

Small, erect, loose trusses of 5-6 flowers, lemon yellow, widely
funnel-campanulate, 5-lobed; red winter foliage. (Peter A. Cox,
1976) A. M. (RHS) 1976

```
cl.      -rupicola var. chryseum (1/2)
Chikor-
       -ludlowii (1/2)
```

1.5ft(.45m)          0°F(-18°C)          EM          4/3

The first hybrid of ludlowii shown to the RHS.    A true yellow
dwarf with twiggy stems as in a miniature tree. Foliage bronze
red in winter. (Peter A. Cox, 1962) A. M. (RHS) 1962; F. C. C.
(RHS) 1968    See color illus. ARS Q 27:2 (Spring 1973): p. 102;
HG p. 129.

```
g. & cl.
     -wightii (1/2)
China-
     -fortunei ssp. fortunei (1/2)
```

5ft(1.5m)          0°F(-18°C)          M          4/3

A fine truss of creamy flowers with red centers. Large leaves
are deeply veined. (Slocock, 1936) A. M. (RHS) 1940; F. C. C.
(Wisley Trials) 1982

```
cl.            -wightii (1/4)
           -China-
China Moon*    -fortunei ssp. fortunei (1/4)
         -        -campylocarpum ssp. campylocarpum (1/4)
           -Moonstone-
                    -williamsianum (1/4)
```

3ft(.9m)          -5°F(-21°C)          M          4/4

Bright, clear lemon yellow flowers, without markings; floriferous. Plant habit neat, compact, with rounded moss green leaves.
(Eichelser)    See color illus. HG p. 98.


China Sea*--form of longiflorum (lobbii)

Vireya. Collected from the shore of the S. China Sea, Sarawak,
Malaysia (Borneo). Bright pink flowers, long tubes. (Swisher,
1980)

```
cl.
     -yungningense (1/2)
Chinese Blue-       -intricatum (1/4)
         -Bluebird-
                 -augustinii ssp. augustinii (1/4)
```

2ft(.6m)          0°F(-18°C)          E

Compact, slow-growing plant with leaves like Bluebird but more
convex; flowers a brighter blue, with long red stamens. (Del W.
James, cross 1953; intro. by Sorensen & Watson, Bovees Nursery,
1982)

cl.


Chinese Silver--form of argyrophyllum ssp. nankingense

5ft(1.5m)          0°F(-18°C)          EM          3/5

Attractive foliage, glossy dark green above and silvery beneath.
Delicate deep pink flowers. (Crown Estate, Windsor) A. M. (RHS)
1957

```
cl.      -keiskei (1/2)
Chink-
      -trichocladum (1/2)
```

3ft(1.5m)          -5°F(-21°C)          EM

Pale chartreuse yellow blossoms, with some darker spotting; lax
trusses of 5 drooping, flat campanulate flowers. Tendency to be
semi-deciduous; red winter foliage. (Crown Estate, Windsor)
A. M. (RHS) 1961

cl.          -wightii (1/4)
      -China-
Chinmar-      -fortunei ssp. fortunei (1/4)
      -          -wardii var. wardii (1/4)
        -Margaret Findlay-
                      -griersonianum (1/4)

5ft.(1.5m)        0°Γ(-18°C)            EM

Light yellow, openly campanulate flowers with small red rays and
spotting, 3in(7.6cm) wide, in flat trusses of 13.  Bush rounded,
well-branched; olive green leaves with maroon petioles, held two
years. (Karl Sifferman cross, c.1960; intro. by Sifferman & Ben
Nelson; reg. 1982 by D. K. McClure)


cl.
          -ponticum (1/2)
Chionoides-
          -unknown (1/2)

4ft(1.2m)          -10°F(-23°C)        ML        3/4

Flowers white with yellow centers; numerous dome-shaped trusses.
Broad, dense plant  with attractive narrow foliage.  Tolerant of
sun and cold.  (J. Waterer, before 1883)   See color illus. VV
p. 51.


g.              -thomsonii ssp. thomsonii (1/4)
      -Barclayi-          -arboreum ssp. arboreum (1/8)
Chiron-      -Glory of  -
      -          Penjerrick-griffithianum (1/8)
      -haematodes ssp. haematodes (1/2)

Deep red flowers.  (Lord Aberconway, 1946)


cl.
      -dichroanthum ssp. apodectum (1/2)
Chivalry-
      -griersonianum (1/2)

Orange salmon flowers, hose-in-hose, 3in(7.6cm) across;  trusses
hold 8.  (Gen. Harrison, 1962)


cl.

Chlorops

4ft(1.2m)        -5°F(-21°C)          M        3/3

Once considered a species but known only in cultivation;  may be
a chance hybrid between wardii and vernicosum.  See Chamberlain,
p. 430 [Notes RBG Edinb. 39:3 (1982)]   Saucer-shaped flowers in
trusses of 6-8, cream to pale yellow with light purple blotch.
Dense foliage of bluish olive green.


cl.
      -haematodes ssp. haematodes (1/2)
Choremia-
      -arboreum ssp. arboreum (1/2)

Waxy, bell-shaped flowers of crimson scarlet appear early; plant
medium-sized, much like haematodes in habit.   (Lord Aberconway,
1933)   A. M. (RHS) 1933; F. C. C. (RHS) 1948


cl.
      -wardii var. wardii (3/4)
Chough-                          -griffithianum (1/16)
      -              -George Hardy-
      -Carolyn-Mrs. Lindsay-          -catawbiense (1/16)
      Hardy - Smith      -Duchess of Edinburgh--unknown (1/8)
              -wardii var. wardii

Truss of 10-12 flowers, light greenish yellow darker in throat,
dark purplish red blotch in upper throat. (Crossed, raised, exhibited by Collingwood Ingram; reg. by G. A. Hardy)  A. M. (RHS)
1981


cl.          -catawbiense (1/4)
      -Kettledrum-
Chris-      -unknown (1/4)
      -wardii var. wardii (1/2)

3ft(.9m)          -10°F(-23°C)            ML

Flowers phlox pink in throat, shading to primrose yellow at margins, openly funnel-shaped, 3in(7.6cm) across; lax, dome-shaped
trusses hold 13-15.  Upright, well-branched plant; very glossy
leaves held 2 years.  (W. M. Fetterhoff cross, 1962; reg. 1981)


cl.
        -Tacoma--unknown (3/4)
Christabel-                    -griffithianum (1/8)
  Tomes    -unnamed-Loderi (pink form)-
          hybrid-              -fortunei ssp.
              -unknown                fortunei (1/8)

Pale cream flowers, pink markings. (W. E. Glennie, reg. 1984)


cl.
      -Catalgla--catawbiense var. album Glass (1/2)
Christina-          -campylocarpum ssp. campylocarpum
Delp*  -Mrs. W. C. Slocock-                    (1/4)
              -unknown (1/4)

8ft(2.4m)        -15°F(-26°C)      Early June (Z4)        3/3

Synonym Christina.  White flowers, orchid-tipped, with a yellow
eye and long stamens recurving toward the pistil; neat pyramidal
trusses.  Compact plant; large leaves. (Weldon Delp)  See color
illus. LW pl. 94.


cl.
          -maddenii ssp. maddenii (1/2)
Christine Denz-      -maddenii ssp. crassum (1/4)
            -Sirius-
              -cinnabarinum ssp. cinnabarinum
                      Roylei Group (1/4)

Trusses hold 5-6 flowers, carmine rose in bud, opening Naples
yellow flushed pink, with bright yellow throat.  Brown scaly indumentum beneath elliptic leaves.  See also Felicity Fair, Barbara Jury.  (F. M. Jury, 1982)


cl.
          -caucasicum (1/2)
Christmas Cheer-
          -unknown (1/2)

4ft(1.2m)          -10°F(-23°C)            VE        3/4

An old hybrid of rather dense, compact habit.  Flowers pink in
bud, opening blush pink  and fading to very pale pink, 2in(5cm)
across.  Sun-tolerant.  Can be forced for Christmas.  (Methven)
See color illus. F p. 39 & front cover; VV p. 81.


g.
      -chrysodoron (1/2)
Chrycil-      -ciliatum (1/4)
      -Cilpinense-
              -moupinense (1/4)

A compact dwarf plant with flowers lemon to yellow. (Lord Aberconway, 1947)


g.
          -moupinense (1/2)
Chrypinense-
          -chrysodoron (1/2)

Pale yellow flowers.  (Lord Aberconway, 1946)


g. & cl.
          -chrysodoron (1/2)
Chrysaspis-
          -leucaspis (1/2)

Yellow flowers.  (Lord Aberconway)  A. M. (RHS) 1942

g. & cl.
```
            -chrysodoron (1/2)
Chrysomanicum-
            -burmanicum (1/2)
```

Primrose yellow flowers.  A parent of Denise.  (Lord Aberconway, 1947)  A. M. (RHS) 1947

cl.

Church Lane--yakushimanum ssp. yakushimanum, selfed

2ft(.6m)          -15°F(-26°C)          EM          5/5

Pink with yellow spotting on upper lobe.  (P. G. Valder, 1972)

cl.
```
            -falconeri ssp. falconeri (1/2)
Churchill-
            -sinogrande (1/2)
```

7ft(2.1m)          5°F(-15°C)          M          4/4

Fortune g.  Flowers of very light primrose yellow, 10-lobed; 32-35 in trusses 9.5in(23cm) across.  Leaves with brown indumentum like falconeri.  (Edmund de Rothschild; reg. 1972)  A. M. (RHS) 1971

---
```
            -ciliatum (1/2)
Cilaspis-
            -leucaspis (1/2)
```

White flowers.  (E. J. P. Magor, 1931)

g.
```
            -ciliatum (1/2)
Cilbooth-
            -boothii (1/2)
```

An introduction of E. J. P. Magor, 1928.

cl.
```
            -ciliatum (1/2)
Cilkeisk-
            -keiskei (1/2)
```

Flowers white, tinged pink.  (E. J. P. Magor, 1926)

g. & cl.
```
            -ciliatum (1/2)
Cilpinense-
            -moupinense (1/2)
```

3ft(1.5m)          5°F(-15°C)          E          3/3

Compact plant;  deep forest green, shiny leaves.  Flowers  blush pink touched with deeper pink.  Floriferous.  Can be forced for Christmas.  (Lord Aberconway, 1927)  A. M. (RHS) 1927; F. C. C. (RHS) 1968    See color illus. ARS Q 31:3 (Summer 1977): p. 170; F p. 39; VV p. 18.

g.
```
            -cinnabarinum ssp. cinnabarinum (1/2)
Cincrass-
            -maddenii ssp. crassum (1/2)
```

Flowers white, changing to cream at base.  (E. J. P. Magor) A. M. (RHS) 1935

cl.
```
        -calostrotum ssp. calostrotum (1/2)
Cindy-
        -Bergie--ciliatum (1/2)
```

Blush pink flowers.  (H. L. Larson, 1953)

```
                            -catawbiense (1/4)
                -Parsons Grandiflorum-
            -America-              -unknown (9/16)
      -unnamed-     -dark red hybrid--unknown
cl.   - hybrid-      -griffithianum (3/16)
                  -Mars-
Cindy Lou*    -Blaze-   -unknown
                -catawbiense (red form)
         -          -griffithianum
         -       -Mars-
      -Red Brave-   -unknown          -catawbiense
         -          -Parsons Grandiflorum-
            -America-              -unknown
                  -dark red hybrid--unknown
```

10ft(3m)          -25°F(-32°C)          L          4/3

Red flowers in rounded trusses.  Foliage and  plant habit as in catawbiense.  See also Anna Delp.  (Weldon Delp, 1968)

g.
```
            -cinnabarium ssp. cinnabarium Roylei Group (1/2)
Cinnandrum-
            -maddenii ssp. maddenii (polyandrum) (1/2)
```

Flowers of pinkish peach.  (Lord Aberconway, cross 1931; intro. 1937)

cl.

Cinnandrum Tangerine     Parentage as above

Outside of flower deep rose, pale apricot within.  (Lord Aberconway, 1937)  A. M. (RHS) 1937

g. & cl.
```
            -cinnabarinum ssp. cinnabarinum (1/2)
Cinnkeys-
            -keysii (1/2)
```

Clusters of tubular flowers, light clear red in tube with lobes soft yellow.  Shrub has oval,  glossy green leaves.  Midseason.  (E. J. P. Magor, 1926)  A. M. (RHS) 1935    Three different descriptions exist, the above from I.R.R., others from Hillier, and Hydon.  Which form received the A. M.?    See color illus. ARS Q 32:3 (Summer 1978): p. 170.

g.
```
            -cinnabarinum ssp. cinnabarinum (1/2)
Cinnmadd-
            -maddenii ssp. maddenii (1/2)
```

An introduction of E. J. P. Magor.

g.
```
        -cinnabarinum ssp. cinnabarinum Blandfordiiflorum Group
Cinzan-                                                (1/2)
        -cinnabarinum ssp. xanthocodon (1/2)
```

A medium-sized, slender shrub with pendant apricot flowers.  The Registry of 1958 shows a coral pink form--others may exist.  (J. B. Stevenson, 1951)

g.
```
            -haematodes ssp. haematodes (1/4)
      -Humming Bird-
Circe-            -williamsianum (1/4)
      -            -neriiflorum ssp. neriiflorum (1/4)
      -F. C. Puddle-
                  -griersonianum (1/4)
```

Bright red flowers.  (Lord Aberconway, intro. 1942)  A. M. (RHS) 1942

cl.
```
            -dichroanthum ssp. dichroanthum (1/4)
      -Fabia-
Circus-    -griersonianum (1/4)
      -            -ponticum (1/4)
      -Purple Splendour-
                  -unknown (1/4)
```

3ft(.9m)          0°F(-18°C)         ML

Flowers of azalea pink aging to cadmium orange, spotting of deep
Tuscan yellow, margins rhodonite red; 12-14 held in ball-shaped
trusses.  Plant upright, moderately branched; narrowly elliptic
leaves with scant golden brown indumentum, held 2 years.  (Cross
by George Grace, 1961; Louis Grothaus, intro. 1973; reg. 1981)

```
cl.      -williamsianum (1/2)
         -                                -griffithianum (1/16)
Citation-                    -George Hardy-
         -      -Mrs. Lindsay-             -catawbiense (1/16)
         -Diane- Smith       -Duchess of Edinburgh--unknown (1/4)
              -unnamed-campylocarpum ssp. campylocarpum (1/8)
                   hybrid-
                          -unknown
```

Clear red flowers and dark green foliage; plant 1ft(.3m) high,
spreading.  (Leona & Rudolph Henny, 1965)

```
cl.
        -Catanea--catawbiense var. album (1/2)
Citrine-    -wardii var. wardii (1/4)
        -Crest-          -fortunei ssp. discolor (1/8)
            -Lady
            Bessborough-campylocarpum ssp. campylocarpum
                                        Elatum Group (1/8)
```

2.5ft(.75)          -5°F(-21°C)         M

Semi-dwarf, wider than tall at maturity;  leaves of medium size,
glossy, yellow green, held 3 years. Fragrant flowers, funnel-
shaped, 6-lobed, of light greenish yellow,  held in dome-shaped
trusses of 13. (Alfred A. Raustein, 1983)

```
cl.
        -yakushimanum ssp. yakushimanum (1/2)
Claire-             -griffithianum (1/4)
      -      -Pink Shell-          -fortunei ssp.
      -Coronation-         -H. M. Arderne- fortunei (3/16)
       Day    -      -griffithianum -
              -Loderi-              -unknown (1/16)
                  -fortunei ssp. fortunei
```

1.5ft(.45m)          5°F(-15°C)         M

Dwarf plant, wider than high, with leathery, medium-sized leaves
held 2 years. Flowers 4in(10.2cm) wide, 5-lobed, of very light
neyron rose fading white, held in large, conical trusses of 8-
10.  (H. L. Larson cross; intro. J. A. Davies, 1980; reg. 1983)

```
                     -neriiflorum ssp. neriiflorum (1/8)
g.           -Nereid-
      -Euryalus-     -dichroanthum ssp. dichroanthum (1/8)
Clansman-     -griersonianum (1/4)
      -arboreum ssp. zeylanicum (1/2)
```

Red flowers.  (Lord Aberconway, 1950)

```
cl.
    -decorum (1/2)
Clara-         -griffithianum (1/4)
    -Loderi King George-
                   -fortunei ssp. fortunei (1/4)
```

4ft(1.2m)          5°F(-15°C)         M

Pure white flowers, openly funnel-shaped, 5in(12.7cm) wide, in
trusses of 10-12. Bushy plant with light green foliage. (Slonecker, 1962)

```
                     -griffithianum (1/8)
              -Mars-
cl.       -Vulcan-   -unknown (1/2)
      -         -griersonianum (1/4)
Clara Curry-              -catawbiense (1/8)
      -      -Parsons Grandiflorum-
      -America-              -unknown
              -dark red hybrid--unknown
```

3ft(.9m)          0°F(-18°C)          M

Ruby red flowers, 3in(7.6cm) wide, in round trusses of 12-14.  A
plant of upright and broad habit, leaves medium to large, held 2
years.  New growth reddish.  (J. H. Hughes, 1977)

```
                -decorum (1/4)
cl.     -unnamed-
        - hybrid-fortunei ssp. discolor (1/4)
Clara Raustein-          -wardii var. wardii (1/8)
        -      -unnamed-
        -unnamed- hybrid-dichroanthum ssp.  dichroanthum
             hybrid-                                (1/8)
                -fortunei ssp. fortunei (1/4)
```

6ft(1.8m)          -5°F(-21°C)          M

Flowers apricot and warm yellow fading paler yellow, widely funnel-campanulate, 3in(7.6cm) wide; conical trusses of 13.  Plant
broad, spreading; leaves held 2 years.  (Raustein, raised 1962;
reg. 1972)  See color illus. ARS Q 27:3 (Summer 1973): p. 170;
LW pl. 96.

```
cl.                     -haematodes ssp. haematodes (3/4)
             -Choremia-
Clara Weyerhauser*    -arboreum ssp. arboreum (1/4)
             -haematodes ssp. haematodes
```

An unregistered selection by Lester E. Brandt.

```
cl.
        -christianae (1/2)
Clare Rouse-
        -laetum (1/2)
```

Vireya hybrid.  Trusses of 5-7 flowers, tubular campanulate, of
deep Indian yellow, with some lobes nasturtium red.  Shrub up to
6ft(1.8m)  (D. B. Stanton cross; J. Rouse reg., 1984)

```
            -yakushimanum ssp. yakushimanum (1/4)
     -unnamed-                     -griffithianum (1/16)
cl.  - hybrid-      -Queen Wilhelmina-
     -      -Britannia-          -unknown (3/16)
Claret-     -Stanley Davies--unknown
Bumble-          -fortunei ssp. discolor (1/8)
     -     -Lady Bessborough-
     -          -campylocarpum ssp. campylocarpum
     -Jalisco-                   Elatum Group (1/8)
     -     -dichroanthum ssp. dichroanthum (1/8)
         -Dido-
            -decorum (1/8)
```

4ft(1.2m)          0°F(-18°C)          ML

Plum-colored edges with a lighter plum center. (John Waterer,
Sons & Crisp, intro. 1975)

```
                     -griffithianum (1/8)
cl.          -Queen Wilhelmina-
      -Earl of Athlone-          -unknown (3/8)
Claribel-      -Stanley Davies--unknown
      -haematodes ssp. haematodes (1/2)
```

An erect dwarf 12in(30.4cm) by 12in. in 10 years. Leaves small,
dark green, insect-resistant. Flowers currant red, 3-6 in loose
trusses. Midseason. (C. S. Seabrook, 1965)

```
cl.
        -sanguineum ssp. didymum (1/2)
Clarice of Langau-
        -haematodes ssp. haematodes (1/2)
```

Dark red flowers in loose trusses. Low, compact bush, dense and
rounded like haematodes. Light brown indumentum on leaves. For
the small garden. (Edmund de Rothschild, 1966)

```
                      -Corona--unknown (5/8)
cl.         -Bow Bells-
            -           -williamsianum (1/4)
Clatsop Belle-                          -griffithianum
            -                  -Queen Wilhelmina-      (1/8)
            -Earl of Athlone-            -unknown
                        -Stanley Davies--unknown
```

4ft(1.2m)          0°F(-18°C)          EM

Bell-shaped flowers of soft rose to deep pink, slightly ruffled,
2in(5cm) wide, about 15 in loose trusses. Plant compact; olive
green foliage. (George Baker, 1966)

```
---
            -arboreum ssp. arboreum (blood red form) (1/2)
Cleopatra-                          -thomsonii ssp. thomsonii
            -              -unnamed hybrid-              (1/8)
            -unnamed hybrid-            -unknown (1/8)
                        -sutchuenense (1/4)
```

Flowers of deep satiny pink. (E. J. P. Magor, 1931)

```
cl.
            -wardii var. wardii Litiense Group (1/2)
Clewer-     -wardii var. wardii (1/4)
    -Crest-                  -fortunei ssp. discolor (1/8)
        -Lady Bessborough-
                        -campylocarpum ssp. campylocarpum
                                Clatum Group (1/8)
```

Primrose yellow, 6- or 7-lobed, 2.25in(5.7cm) wide, in a compact
truss. (Crown Estate, Windsor, 1967)   P. C. (RHS) 1966

```
cl.                 -leucaspis (1/4)
            -Bric-a-brac-
Cliff Garland-      -moupinense (1/4)
            -mucronulatum (pure pink form) (1/2)
```

1ft(.3m)          -20°F(-29°C)          E          4/4

Leaves are small and nearly circular. Flowers shell pink, up to
2in(5cm) wide; about 4 flowers per truss. See also Cliff Spangle. (G. Guy Nearing, 1972)

cl.

Cliff Spangle    Same parentage as Cliff Garland

1ft(.3m)          -20°F(-29°C)          E          4/4

Leaves quite small, green all year. Flowers in clusters of 5,
deep pink and about 2.5in(6.3cm) wide. (G. Guy Nearing, 1972)

```
g.                      -adenogynum (1/4)
        -Xenosporum (detonsum)-
Clio-                       -unknown (1/4)
    -       -griffithianum (1/4)
    -Gilian-
            -thomsonii ssp. thomsonii (1/4)
```

Flowers colored carmine lake. (E. J. P. Magor, 1931)

```
cl.                         -phaeopeplum (1/4)
        -Dr. Herman Sleumer-
Clipsie-                        -zoelleri (1/4)
        -Nancy Adler-Pink Delight--unknown (1/4) (Veitch hybrid)
            Miller       -
                -jasminiflorum (1/4)
```

1ft(.3m)          32°F(0°C)          Sept.-Dec.

Vireya hybrid. Flowers 2in(5cm) wide, tubular funnel-shaped, 5
to 6 lobes, red with dark pink edges, fragrant; lax, dome-shaped
trusses of 6-8 flowers. Plant twice as broad as tall. (P. Sullivan cross; intro. by W. Moynier and reg. 1980)

```
cl.         -ciliatum (1/4)
        -Praecox-
Clodion-    -dauricum (1/4)
        -
            -cinnabarinum ssp. cinnabarinum Roylei Group (1/2)
```

Flowers a dull claret red. (Adams-Acton, 1934)

```
g. & cl.            -dichroanthum ssp. dichroanthum (1/4)
                -Neda-
Clotted Cream-      -Cunningham's Sulphur--caucasicum (1/4)
                -auriculatum (1/2)
```

Flowers of deep cream. (Lord Aberconway) A. M. (RHS) 1942

```
            -fortunei ssp. fortunei (1/4)
        -unnamed-       -decorum (1/16)
cl.  - hybrid-    -unnamed-
            -Madonna- hybrid-griersonianum (1/16)
Cloud-      (Gable)-                       -catawbiense
Nine-           -           -Parsons Grandiflorum-      (5/32)
    -               -America-               -unknown
    -                       -dark red hybrid--unknown   (7/32)
    -               -caucasicum (1/4)
    -Boule de Neige-            -catawbiense
                -unnamed hybrid-
                        -unknown
```

2ft(.6m)          -15°F(-26°C)          M

White flowers with a fern green blotch, fragrant, 3.5in(8.9cm)
across, in 12-flowered trusses. Plant as broad as wide; leaves
dull spinach green, held 3 years. (Russell & Velma Haag cross,
1962; reg. 1981)

```
cl.                 -catawbiense var. album (7/16)
        -Great Lakes-
Cloud-          -Koichiro Wada--yakushimanum ssp. yakushimanum
Twelve*                 -catawbiense (5/32)      (1/4)
    -       -Mrs. C. S. Sargent-
    -unnamed-               -unknown (5/32)
    hybrid-      -catawbiense var. album
        -Lodestar-              -catawbiense var. album
            -Belle Heller-          -catawbiense
                        -Madame Carvalho-
                                -unknown
```

Pure white flowers held in neat, tight trusses. Excellent plant
habit and foliage. Hardy to at least -20°F(-29°C). (Orlando S.
Pride, 1975)

```
g.
        -sperabile var. sperabile (1/2)
Clove-
        -sanguineum ssp. sanguineum var. haemaleum (1/2)
```

Campanulate crimson flowers. A semi-dwarf plant of dense habit.
(Rothschild, 1935)

```
cl.
        -grande (1/2)
Clyne Blush-
        -hodgsonii (1/2)
```

Elsae g. Three other named clones from this grex by Heneage-
Vivian in 1933: Clyne Cerise, Clyne Elsae, Clyne Pearl.

```
cl.
        -grande (1/2)
Clyne Cerise-
        -hodgsonii (1/2)
```

Elsae g. Three other named clones from this grex by Heneage-
Vivian in 1933: Clyne Blush, Clyne Elsae, Clyne Pearl.

cl.
            -grande (1/2)
Clyne Elsae-
            -hodgsonii (1/2)

Elsae g. Trusses of 30 flowers, fuchsia purple, edges stained a
deeper shade. (Heneage-Vivian, 1933)  A. M. (RHS) 1940


cl.
            -grande (1/2)
Clyne Pearl-
            -hodgsonii (1/2)

Elsae g.   Three other named clones from this grex by Heneage-
Vivian in 1933:  Clyne Blush, Clyne Cerise, Clyne Elsae.


g.

Coalition     Parentage unknown

Bright salmon red flowers; some forms paler.  (R. Gill & Son)
A. M. (RHS) 1922


cl.

Coals of Fire--form of cerasinum (K-W 5830)

"...trusses of five, and of an intense burning scarlet; at the
base of the corolla are five circular jet-black honey-glands,
each about the size of a shirt-button." Described and named by
Frank Kingdon-Ward, 1924; quoted by W. J. Bean (Trees and Shrubs
... v. III, 8th ed. rev., p. 626)


                          -intricatum (1/8)
cl.               -Intrifast-
      -Blue Diamond-        -fastigiatum (1/8)
Cobalt-           -augustinii ssp. augustinii 1/4)
      -russatum (1/2)

Deep blue flowers on a small plant. (Collingwood Ingram, 1970)


cl.

Coccinea*--form of ferrugineum

Red form of dwarf plant Swiss Alpine Rose.  Easily propagated.


                    -yakushimanum ssp. yakushimanum (1/4)
cl.         -unnamed-        -dichroanthum ssp. dichroanthum
      - hybrid-Fabia -                             (1/8)
Coch-y-bondu-     Tangerine-griersonianum (3/8)
                    -griersonianum
           -Tally Ho-
                    -facetum (eriogynum) (1/4)

Flowers are very light sap green. (Waterer, Sons & Crisp, 1975)


cl.
                -cinnabarinum ssp. cinnabarinum (3/4)
Cock of the Rock-            -cinnabarinum ssp. cinnabarinum
                -Rose Mangles-
                          -maddenii ssp. maddenii (1/4)

Flowers clear orange, outside suffused ruby red on orange. (Stephenson Clarke)  A. M. (RHS) 1932


g.
      -racemosum (1/2)
Codorus-
      -minus var. minus Carolinianum Group (1/2)

Pink flowers resembling trailing arbutus.  Hardy to -28°F(-34°C)
(Joseph Gable cross; intro. about 1934)


                    -griffithianum (1/4)
            -Halopeanum-
cl.     -Snow Queen-        -maximum (1/8)
      -          -          -griffithianum
Coldstream-       -Loderi-
      -                    -fortunei ssp. fortunei (1/8)
      -unknown (1/2)

Flowers white, suffused yellow in throat, openly funnel-shaped;
rounded trusses 7.5in(19cm) across, hold 10-12 flowers. (Major
Hardy, 1972)  A. M. (RHS) 1972


cl.
      -fortunei ssp. discolor Houlstonii Group (1/2)
Colehurst-       -Sir Charles Butler--fortunei ssp. fortunei
      -Van Nes  -          -griffithianum (1/8)      (1/4)
      Sensation-Halopeanum-
                          -maximum (1/8)

An Australian hybrid with flowers of light neyron rose with center of creamy white; 14-flowered trusses. (V. J. Boulter, 1978)


cl.

College Pink     Parentage unknown

A parent of Posy, a hybrid by R. C. Gordon.


cl.

Collingwood Ingram--form of trichostomum Ledoides Group

2.5ft(.75m)        -5°F(-21°C)           M          4/3

Truss of 16-20 flowers, light fuchsia purple, 5-lobed, tubular,
less than .5in wide. Leaves quite small, about 1in(2.5cm) long,
narrow, aromatic, non-scaly. (Collector Forrest; raised by Cox
of Glendoick; exhibited by Lady Anne Palmer) F. C. C. (RHS) 1976


cl.        -wardii var. wardii (1/4)
      -Hawk-            -fortunei ssp. discolor (1/8)
Colonel-   -Lady      -
 Remy  -   Bessborough-campylocarpum ssp. campylocarpum Elatum
      -campylocarpum ssp. campylocarpum (1/2)      Group (1/8)

Compact trusses of up to 13 flowers, 5-lobed, bell-shaped, very
light greenish yellow, reverse stronger color. (Edmund de
Rothschild)  A. M. (RHS) 1978


g.
            -falconeri ssp. falconeri (1/2)
Colonel Rogers-
            -niveum (1/2)

Tall plant with fine foliage; rose purple flowers. Early. (Rogers, cross 1926; exhibited by Heneage-Vivian, 1933)


cl.
            -campylocarpum ssp. campylocarpum (1/2)
Colonel Thorneycroft-
            -unknown (1/2)

A small, early hybrid with rich pink buds opening to pale yellow
flowers. Medium-sized trusses. (Harry White, Sunningdale Nurseries, 1955)


cl.
            -minus var. minus Carolinianum Group (1/2)
Colts Neck Rose*
            -dauricum (1/2)

Pinkish lavender flowers. Small leaves with good winter color;
growth habit good, compact; easily propagated. Attractive land-
scape plant. (Dr. G. David Lewis)    See color illus. ARS J 37:2
(Spring 1983): p. 70.

cl.
                -griersonianum (3/4)
Columbia Sunset-      -griersonianum
                  -Azor-
                        -fortunei ssp. discolor (1/4)

5ft(1.5m)          5°F(-15°C)          L

Flowers red in bud opening deep pink with darker center; trusses
hold about 14. Dark green leaves with light fawn indumentum beneath. (George L. Baker, 1968)

cl.

Colville--form of minus var. minus Carolinianum Group

4ft(1.2m)          -25°F(-32°C)          M          3/3

Flowers in clusters of 9-10, 5-lobed, light red purple, shading
deeper. Leaves narrowly ovate with pitted reddish brown scales
beneath. (Mrs. N. R. Colville, reg. 1983) A. M. (RHS) 1968

                -cinnabarinum ssp. cinnabarinum
g.    -Lady          Roylei Group (1/4)
   - Chamberlain-    -cinnabarinum ssp. cinnabarinum (1/8)
Comely-          -Royal-
   -              Flush-maddenii ssp. maddenii (1/8)
   -cinnabarinum ssp. xanthocodon Concatenans Group (1/2)

Orange yellow flowers. See also Chan. (Lord Aberconway, 1946)

                        -fortunei ssp. discolor (1/4)
              -Lady Bessborough-
   -                    -campylocarpum ssp. campylocar-
cl.   -Jalisco-                   pum Elatum Group (1/4)
   -        -        -dichroanthum ssp. dichroanthum (3/8)
Comstock-    -Dido-
   -              -decorum (1/8)
   -        -dichroanthum ssp. dichroanthum
   -Jasper-              -fortunei ssp. discolor
              -Lady Bessborough-
                        -campylocarpum ssp. campylocar-
                              pum Elatum Group

3ft(.9m)          -5°F(-21°C)          M          4/4

Large flowers of warm orange streaked apricot and yellow; showy
calyx. Dense, rounded foliage of cool forest green. Difficult
to propagate. (Del James & Harold Greer, 1979)

cl.
      -augustinii ssp. augustinii (1/2)
Concerto-
      -searsiae (1/2)

Purple flowers with green spotting, in umbels of 9-16. Medium-
sized leaves, dark dull green. (Collector, Wilson; raiser E. de
Rothschild) A. M. (RHS) 1979

cl.
      -ciliicalyx (1/2)
Conchita-
      -moupinense (1/2)

4.5ft(1.35m)          20°F(-7°C)          E          3/3

Flowers phlox pink, spotted crimson, 4in(10cm) wide. Leaves are
medium-sized. (Raiser John S. Drucker; R. German, reg. 1976)

                  -wightii (1/8)
              -China-
cl.   -The Master-    -fortunei ssp. fortunei (1/4)
   -        -        -campylocarpum ssp. campylocarpum
Concord-          -Letty -          Elatum Group (1/8)
   -              Edwards-fortunei ssp. fortunei
   -yakushimanum ssp. yakushimanum (1/2)

3ft(.9m)          -10°F(-23°C)          M

White flowers with a green throat; nice-looking, compact plant.
(A. F. George, 1978)

                -griffithianum (1/8)
cl.      -Isabella-
   -Muy Lindo-    -auriculatum (1/8)
Condor-      -      -decorum (1/8)
   -          -unnamed hybrid-
   -Zella--unknown (1/2)    -souliei (1/8)

Flowers open from pink buds then fade to creamy white. Leaves
8.5in(22cm) long, dark green. (Collingwood Ingram) A. M. (RHS)
1976

g. & cl.
      -racemosum (1/2)
Conemaugh-
      -mucronulatum (1/2)

5ft(1.5m)          -15°F(-26°C)          VE          3/3

A rather open, twiggy plant, small-leafed, semideciduous. Star-
shaped pink lavender flowers; small spherical trusses. Various
clones exist. (Joseph Gable, 1934)

g.
      -minus var. minus Carolinianum Group (1/2)
Conestoga-
      -racemosum (1/2)

Hardy to -20°F(-29°C). Pink flowers. Early. (Gable; cross of
1954-5 replaced 1929's extinct Conestoga)

g.
      -minus var. minus Carolinianum Group (1/2)
Conewago-
      -mucronulatum (1/2)

5ft(1.5m)          -25°F(-32°C)          E          3/2

Flowers of lilac rose on a plant of rather open habit, with many
branches; leaves 2.5in (6.4cm) long. A very floriferous form,
Conewago Improved, may have higher rating. (Joseph Gable, 1934)
See color illus. LW pl. 35.

g.
      -haematodes ssp. haematodes (1/2)
Conewingo-
      -diphrocalyx (1/2)

(Joseph Gable, intro. 1934)

cl.
      -Corona--unknown (5/8)
Confection-          -arboreum ssp. arboreum (1/8)
   -        -Doncaster-
   -Dondis-      -unknown
      -fortunei ssp. discolor (1/4)

5ft(1.5m)          0°F(-18°C)          ML          3/3

Rather compact plant; large dark green leaves held over 2 years.
Flowers of rose madder, 3.5in(8.9cm) across, funnel-campanulate;
upright trusses hold 16. (R. Henny) P. A. (ARS) 1956

cl.   -catawbiense var. album (1/2)
   -                    -caucasicum (3/32)
Congo-          -Jacksonii-    -caucasicum
   -        -Goldsworth-    -Nobleanum-
   -        - Yellow -          -arboreum ssp.
   -Goldfort-                  arboreum (1/32)
   -              -campylocarpum ssp. campylocarpum (1/8)
      -fortunei ssp. fortunei (1/4)

5ft(1.5m)          -20°F(-29°C)          M

White flowers from buds flushed spirea red in ball-shaped
trusses of 16. Plant much broader than high with leaves
5.5in(14cm) long, held 2 years. See also Luxor. (David G.
Leach, intro. 1972; reg. 1984)

cl.                    -caucasicum (1/4)
          -Christmas Cheer-
Connetquot-                   -unknown (3/4)
          -Pygmalion--unknown


1.5ft(.45m)         -5°F(-21°C)            M


Cardinal red buds and Tyrian purple flowers, in trusses of 15.
Plant broader than tall; leaves held 3 years.  (A. A. Raustein,
1983)


cl.
               -kyawii (1/2)
Connie Hatton-
               -fortunei ssp. discolor (1/2)


6ft(1.8m)          0°F(-18°C)            VL


Fragrant dark pink flowers, paler in throat; trusses 9in(23cm)
wide hold about 12.  Plant upright, broad; large leaves retained
up to 3 years.  (H. A. Short, 1974)


                              -thomsonii ssp. thomsonii (1/8)
cl.            -Bagshot Ruby-
      -Princess -              -unknown (1/2)
Connie -Elizabeth-unknown
Stanton-                -caucasicum (1/4)
      -Boule de Neige-              -catawbiense (1/8)
                        -unnamed hybrid-
                                  -unknown


3ft(.9m)          -10°F(-23°C)           M


Flowers watermelon pink to 3in(7.6cm) across, in shape and truss
similiar to Boule de Neige.  Plant compact, as broad as tall;
dark green leaves 4in(10.2cm) long.  (Ernest N. Stanton, 1969)


cl.             -griffithianum (1/4)
           -Mars-
Connie Yates-    -unknown (1/4)
           -Koichiro Wada--yakushimanum ssp. yakushimanum (1/2)


3ft(.9m)          -10°F(-23°C)           ML


Flowers 2.5in(6.3cm) across, dark pink with white blaze, golden
spots; 14-flowered trusses.  Medium-sized leaves.  (H. R. Yates
cross; intro. by Mrs. Yates, 1977)  See color illus. LW pl. 103.


g. & cl.
              -catawbiense (red form) (1/2)
Conoco-cheague -
(Conoco Cheague)-haematodes ssp. haematodes (1/2)


                  -10°F(-23°C)        EM         3/3


Synonym Cathaem #1.  Rose red flowers in lax trusses; corolloid
calyx, reflexed.  Gable thought this haematodes (from Arnold
Arboretum) was a hybrid; in other crosses he used pollen from E.
J. P. Magor in England.  (Joseph Gable, 1934)


cl.
      -cinnabarinum ssp. cinnabarinum Roylei Group (1/2)
Conroy-
      -cinnabarinum ssp. cinnabarinum Concatenans Group (1/2)


Bushy habit.  Trumpet-shaped, waxy flowers, light orange tinged
with rose, in loose trusses; midseason.  (Lord Aberconway, 1937)
A. M. (RHS) 1950


                  -wardii var. wardii (1/4)
cl.       -Jervis-
      - Bay  -Lady       -fortunei ssp. discolor (1/8)
Constable-      -Bessborough-
          -                -campylocarpum ssp. campylocarpum
                                      Elatum Group (1/8)
          -wardii var. wardii Litiense Group (1/2)


Open truss with up to 20 flowers, bell-shaped, greenish yellow
with some reddish spotting.  (Cross by Francis Hanger, RHS
Garden, Wisley) A. M. (RHS) 1961


g.                 -arboreum ssp. cinnamomeum var. album
          -unnamed hybrid-                            (1/4)
Constance-                   -griffithianum (1/4)
          -auriculatum (1/2)


A tall plant with very large cream to ivory flowers.  It is not
shown in Rothschild Rhododendrons.  (Rothschild, 1936)


cl.
               -campanulatum ssp. campanulatum (1/2)
Constant Nymph-              -ponticum (1/4)
               -Purple Splendour-
                             -unknown (1/4)


White flowers, faintly flushed, held in large, dome-shaped
trusses of 12-14.  Plant upright, broader than tall; dull, dark
green foliage. (Knap Hill Nursery, 1955)   H. C. (Wisley Trials)
1969;  A. M. (RHS) 1971


g.
      -cinnabarinum ssp. xanthocodon Concatenans Group (1/2)
Conyan-
      -concinnum var. pseudoyanthinum (1/2)


5ft(1.5m)          5°F(-15°C)           EM


Unusual hybrid that shows the traits of both parents--a yellow
green and a purple.  Slender-growing but compact with dainty
tubular flowers.  Conyan Apricot is creamy apricot, Conyan Pink
a soft shell pink, Conyan Salmon of salmon orange, flushed pink.
(J. B. Stevenson, 1953)


                   -Pink Bell--unknown (1/4)
cl.   -Queen o' the May-
      -                -griersonianum (3/8)
Cookie-    -fortunei ssp. fortunei (1/4)
      -Fawn-      -dichroanthum (1/8) *ssp dichroanthum*
           -Fabia-
                -griersonianum


4ft(1.2m)          -10°F(-23°C)           EM


Large flowers in very large trusses, neyron rose with spotting
of maroon.  Very fragrant.  Upright plant, spreading; large ivy
green leaves held 2 years. (Del W. James cross; reg. 1980 by
Hendricks Park)


cl.
               -campylocarpum ssp. campylocarpum (1/4)
          -Chaste-          -Pink Bell--unknown (1/8)
Cool Haven-      -Queen o'-
                the May -griersonianum (1/8)
          -wardii var. wardii Litiense Group (1/2)


Medium-sized shrub with medium leaves.  Flowers in trusses of 18
-20, Dresden yellow with crimson markings  and slight fragrance.
(Raised at Embley Park; intro. by Hillier, 1969)


g.
          -griffithianum (1/2)
Coombe Royal-
          -unknown (1/2)


Flowers a delicate pink, spotted brown on the upper lobe.  (Origin unknown)  A. M. (RHS) 1900


cl.
          -catawbiense (1/2)
Coplen's White-
          -unknown (1/2)


White flowers.  (Cross by Coplen; intro. by J. Gable; reg. 1958)


cl.
Copper--form of cinnabarinum ssp. xanthocodon Concatenans Group
                                            (L&S 6560)


3ft(.9m)          5°F(-15°C)           EM         3/4

Coral colored flowers, suffused with orange and red in trusses of 5-8. (Selected by Collingwood Ingram) A. M. (Wisley Trials) 1954

```
                    -wardii var. wardii (1/4)
cl.    -Crest-              -fortunei ssp. discolor (1/8)
       -     -Lady     -
Copper -     Bessborough-campylocarpum Elatum Group (1/8)
Kettles-                         -campylocarpum ssp.
       -     -Souvenir of W. C. Slocock- campylocarpum (1/8))
       -unnamed-                  -unknown (1/8)
          hybrid-dichroanthum ssp. apodectum (1/4)
```

Buds open bright copper, changing to yellow. Trusses hold 9-11 flowers. (John Lofthouse, 1979)

```
                       -souliei (1/16)
                   -Soulbut-
            -Vanessa-        -Sir Charles Butler--fortunei ssp.
g. -Eudora-    -griersonianum (1/8)          fortunei (1/16)
   -       -facetum (eriogynum) (1/4)
Cora-          -griffithianum (1/8)
   -      -Loderi-
  -Coreta-    -fortunei ssp. fortunei (1/8)

           -arboreum ssp. zeylanicum (1/4)
```

Red flowers. (Lord Aberconway, 1946)

```
cl.             -neriiflorum ssp. neriiflorum (1/4)
      -unnamed hybrid-
Coral-          -dichroanthum ssp. dichroanthum (1/4)
     -fortunei ssp. discolor (1/2)
```

Flowers of orange red. See also Phyllis Ballard. (Endre Ostbo) P. A. (ARS) 1956

```
cl.
         -elliottii (1/2)
Coral Island-    -dichroanthum ssp. dichroanthum (1/4)
         -Fabia-
               -griersonianum (1/4)
```

F. Hanger, cross 1947; reg. 1954 by RHS Garden, Wisley, but no longer known at Wisley except by the records.

```
                -thomsonii ssp. thomsonii (1/4)
cl.     -Barclayi-              -arboreum ssp. arboreum
        -       -Glory of Penjerrick-            (1/8)
Coral Pink-                    -griffithianum (1/8)
        -    -lacteum (1/4)
        -Lacs-
             -sinogrande (1/4)
```

Flowers of deep coral pink, 4in(10.2cm) across, 7-lobed; trusses hold about 16. (Gen. Harrison, 1971)

```
              -fortunei ssp. discolor (1/8)
          -King of-    -dichroanthum ssp. dichroanthum (3/8)
          - Shrubs-Fabia-
cl. -unnamed-         -griersonianum (1/8)
    - hybrid-  -fortunei ssp. fortunei (1/8)
Coral-      -Fawn-   -dichroanthum ssp. dichroanthum
Queen-         -Fabia-
    -                -griersonianum
    -    -dichroanthum ssp. dichroanthum
    -Dido-
        -decorum (1/4)
```

A New Zealand hybrid with trusses of 9-12 flowers, a medium rhodonite red. (Mrs. R. J. Coker, 1979)

```
cl.       -dichroanthum ssp. dichroanthum (1/4)
     -Fabia-
Coral-  -griersonianum(1/4)
 Reef-      -caucasicum (1/4)
    -Goldsworth-
       Orange  -campylocarpum ssp. campylocarpum (1/4)
```

Gloriana g. A medium-sized shrub, with flowers of salmon pink, pink margins, slightly flushed in the throat, and held in open, lax trusses. (Francis Hanger, RHS Garden, Wisley) A. M. (RHS) 1954

```
              -dichroanthum ssp. dichroanthum (3/16)
         -Dido-
         -   -decorum (1/8)
      -Lem's-      -Beauty of-griffithianum (5/32)
      -Cameo-  -Norman- Tremough-
cl.   - Gill -        -arboreum ssp. arboreum (1/32)
      -   -Anna-   -griffithianum
Coral-    -         -griffithianum
Skies-    -Jean Marie -
      -         de Montague-unknown (1/16)
      -              -dichroanthum ssp. dichroanthum
      -         -Fabia-
      -   -unnamed-   -griersonianum (1/16)
      -unnamed- hybrid-bureavii (1/8)
        hybrid-   -wardii var. wardii (1/8)
           -Crest-        -fortunei ssp. discolor (1/16)
              -Lady     -
              Bessborough-campylocarpum ssp. campylocarpum
                              Elatum Group (1/16)
```

3ft(.9m)          5°F(-15°C)          EM

The exact combination of above parentage is unknown, hence the species fractions may not be correct. Buds open claret rose; flowers empire rose with radial stripes; the colors do not fade. Trusses large and long-lasting. Plant broad as tall with yellow green foliage, held 2 years. (John Lofthouse, reg. 1983)

```
cl.
         -yakushimanum ssp. yakushimanum (1/2)
Coral Velvet-
         -unknown (1/2)
```

2.5ft(.75m)          -15°F(-26°C)          EM          3/3

Once considered a selected form of yakushimanum. Small leaves and stem are thick with velvety indumentum. Long-lasting flowers open coral pink, and fade to light salmon, 2in(5cm) across. (Swanson & Greer, 1970)

```
cl.           -thomsonii ssp. thomsonii (1/4)
      -Cornish Cross-
Coralia-      -griffithianum (1/2)
     -             -griffithianum
     -Loderi Pink Diamond-
                -fortunei ssp. fortunei (1/4)
```

Ruthelma g. Salmon-colored flowers. See also H. Witner. (Sir E. Loder)

```
           -dichroanthum ssp. dichroanthum (1/4)
g.   -Dante-
     -    -facetum (eriogynum) (1/4)
Cordan-      -griffithianum (1/8)
     -    -Loderi-
    -Coreta-    -fortunei ssp. fortunei ((1/8)
            -arboreum ssp. zeylanicum (1/4)
```

Flowers of deep red. (Lord Aberconway, intro. 1946)

```
g.            -sanguineum ssp. didymum (1/4)
      -Arthur Osborn -
Cordelia-     -griersonianum (1/4)
     -           -griffithianum (1/4)
      -Loderi King George-
                -fortunei ssp. fortunei (1/4)
```

Red flowers. (Lord Aberconway, 1950)

```
cl.              -dichroanthum ssp. dichroanthum
      -Goldsworth Orange-                 (1/4)
Cordy Wagner-        -fortunei ssp. discolor (1/4)
      -              -griffithianum (1/4)
       -Loderi King George-
                -fortunei ssp. fortunei (1/4)
```

Trusses of 8-9 flowers, coral pink at edges with rays of burnt orange flushed amber.   Leaves 10in(25.4cm) long by 3in(7.6cm). (Lester E. Brandt, 1966)

```
                  -sperabile var. sperabile (1/4)
g.    -Eupheno-
-               -griersonianum (1/4)
                       -griffithianum (1/8)
Coreno-        -Loderi-
-    -Coreta-         -fortunei ssp. fortunei (1/8)
              -arboreum ssp. zeylanicum (1/4)
```

Crimson flowers.  (Lord Aberconway, 1946)

```
g.            -campylocarpum ssp. campylocarpum (1/4)
     -Penjerrick-            Elatum Group (1/4)
Coresia-      -griffithianum (1/2)
-                   -thomsonii ssp. thomsonii (1/4)
     -Cornish Cross-
                   -griffithianum
```

Flowers of pale pink.  (Lord Aberconway, 1933)

```
g. & cl.    -griffithianum (1/4)
        -Loderi-
Coreta-      -fortunei ssp. fortunei (1/4)
        -arboreum ssp. zeylanicum (1/2)
```

Deep crimson scarlet.  (Lord Aberconway)  F. C. C. (RHS) 1935

```
g.               -thomsonii ssp. thomsonii (1/4)
       -Cornish Cross-
Coreum-          -griffithianum (1/4)
       -arboreum ssp. arboreum (1/2)
```

Deep red flowers.  (Lord Aberconway, intro. 1950)

```
cl.                  -griffithianum (1/4)
                 -Mars-
Corinne--Vulcan (selfed)-    -unknown (1/4)
                 -griersonianum (1/2)
```

The trumpet-shaped flowers are pink with petaloid stamens. (Halfdan Lem cross; reg. by McClure, 1958)

```
g.           -haematodes ssp. haematodes (1/4)
      -Choremia-
Corma-       -arboreum ssp. arboreum (1/4)
      -haematodes ssp. chaetomallum (1/2)
```

Dark red flowers.  (Lord Aberconway, 1950)

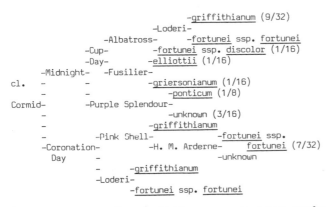

```
                           -griffithianum (9/32)
                       -Loderi-
              -Albatross-      -fortunei ssp. fortunei
         -Cup-        -fortunei ssp. discolor (1/16)
         -Day-   -Fusilier-   -elliottii (1/16)
  -Midnight-       -griersonianum (1/16)
cl.  -      -               -ponticum (1/8)
     -      -Purple Splendour-
Cormid-       -       -unknown (3/16)
     -              -griffithianum
     -      -Pink Shell-       -fortunei ssp.
     -Coronation-     -H. M. Arderne-  fortunei (7/32)
     Day   -            -unknown
           -     -griffithianum
           -Loderi-
                -fortunei ssp. fortunei
```

Trusses of about 14 flowers, light mauve, spotted deep purple, with a green center.  (K. van de Ven, 1980)

cl.

Cornell Pink--form of mucronulatum

5ft(1.5m)          -15°F(-26°C)          VE          3/3

Deciduous, open, upright shrub.  Before the leaves appear, the bright, clear pink flowers are showing in clusters of 2-3, sometimes when the ground is snow-covered.  Flowers are 1.5in(3.8cm) across, widely funnel-shaped.  (Selected by Skinner; reg. 1958) A. M. (RHS) 1965

```
cl.
           -campylocarpum ssp. campylocarpum (1/2)
Cornish Cracker-       -caucasicum (1/4)
           -Dr. Stocker-
                      -griffithianum (1/4)
```

Damaris g.  (Gen. Harrison, 1956)

```
g.
           -thomsonii ssp. thomsonii (1/2)
Cornish Cross-
           -griffithianum (1/2)
```

6ft(1.8m)          10°F(-12°C)          EM          5/3

Large glossy leaves, medium green.  Flowers 5in(13cm) across, of heavy substance, deep pink to red fading through shades of pink to nearly white inside.  Large, loose truss, up to 9 flowers. Upright habit having smooth maroon brown to copper brown colored trunk.  Some forms superior to others.  See also Exbury Cornish Cross, Pengaer.  (S. Smith, Penjerrick; reg. 1958)  See color illus. F  p. 29 and 40.

```
---
           -fortunei ssp. discolor (1/2)
Cornish Loderi-
           -griffithianum (1/2)
```

A parent of Nimbus.  Flowers off-white.  (J. C. Williams, reg. 1958)

```
g.          -Blood Red--arboreum ssp. arboreum (1/4)
      -Cornubia-        -thomsonii ssp. thomsonii (1/8)
Cornsutch-      -Shilsonii-
      -              -barbatum (1/8)
      -sutchuenense (1/2)
```

Flowers are carmine purple with a dark blotch. See also Almondtime.  (E. J. P. Magor, 1926)  A. M. (RHS) 1925

```
g.
        -Blood Red--arboreum ssp. arboreum (1/2)
Cornubia-      -thomsonii ssp. thomsonii (1/4)
        -Shilsonii-
              -barbatum (1/4)
```

6ft(1.8m)          15°F(-9°C)          VE          4/3

A very striking early red.  Large conical truss of blood red. Easily propagated.  In mild climates grows to be a small tree in sheltered gardens.  Large leaves, medium green, slightly bullate.  (Barclay Fox)  A. M. (RHS) 1912  See color illus. V V p. 105.

```
cl.
        -fortunei ssp. fortunei (1/2)
Cornwallis-
        -unknown (1/2)
```

1.5ft(.45m)          -15°F(-26°C)          L

Synonym Acadia.  Pink flowers, heavily spotted, very fragrant; up to 11 held in trusses 7in(17.8cm) across.  Plant much broader than tall; medium green foliage retained 2 years.  (G. D. Swain & D. L. Craig, Agriculture Research Station, N. S., Canada, 1977)

'Cherry Bright' by Lancaster
Photo by Greer

'Chesterland' by Pot/Leach
Photo by Knierim

'Chevalier Felix de Sauvage' by Sauvage
Photo by Greer

'Chiffchaff' by Cox
Photo by Greer

'China' by Slocock
Photo by Greer

'Choremia' by Aberconway
Photo by Greer

'Cinnkeys' by Magor
Photo by Greer

'Circus' by Grace/Grothaus
Photo by Greer

'Cloud Nine' by C. R. Haag
Photo by Haag

'Comstock' by James/Greer
Photo by Greer

'Conchita' by Drucker/German
Photo by Greer

'Congo' by Leach
Photo by Leach

'Connie Stanton' by Stanton
Photo by Knierim

'Coral Velvet' by Swanson/Greer
Photo by Greer

'Corinne' by Lem/McClure
Photo by Greer

'Corry Koster' by M. Koster
Photo by Greer

'Countess of Athlone' by C. B. van Nes
Photo by Greer

'Countess of Haddington' by unknown
Photo by Greer

'Crater Lake' by Phetteplace
Photo by Greer

'Cream Glory' by Greer
Photo by Greer

'Creamy Chiffon' (double-flowered) by Whitney/Briggs
Photo by Greer

'Creeping Jenny' by Aberconway
Photo by Greer

'Crete' by Lancaster/Leach
Photo by Leach

'Crossroads' by Larson
Photo by Greer

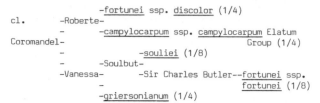

```
                     -fortunei ssp. discolor (1/4)
cl.       -Roberte-
          -               -campylocarpum ssp. campylocarpum Elatum
Coromandel-                                        Group (1/4)
          -                    -souliei (1/8)
          -         -Soulbut-
          -Vanessa-        -Sir Charles Butler--fortunei ssp.
          -                                     fortunei (1/8)
                    -griersonianum (1/4)
```

A small plant with the vigor of Roberte (Lady Bessborough g.),
foliage like Vanessa. Flowers are bell-shaped, tangerine pink
and yellow buds  opening to pale creamy yellow with faint red
spots; 15-flowered trusses.  Neat, compact, small plant; blooms
very late. (Sunningdale Nursery, 1957)

cl.

Corona     Parentage unknown

4ft(1.2m)         -5°F(-21°C)          M          4/3

Conical trusses of about 15 rose pink flowers.  Plant slow-growing, forms a compact mound.  Insect-resistant; tolerates sun and
heat. (J. Waterer)  A. M. (RHS) 1911    See color illus. ARS Q
29:3 (Summer, 1975): p. 170.

```
                    -griffithianum (1/2)
cl.         -Pink Shell-      -fortunei ssp. fortunei
          -           -H. M. Arderne-              (3/8)
Coronation Day-              -unknown (1/8)
          -       -griffithianum
          -Loderi-
                  -fortunei ssp. fortunei
```

Very large flowers up to 5in(12.7cm) wide, fragrant, colored a
delicate China rose with a crimson basal blotch, in large loose
trusses. (Crosfield, 1937) A. M. (RHS) 1949   See color illus.
ARS Q 26:3 (Summer 1972): p. 180.

```
g.
       -Corona--unknown (1/2)
Coronet-
       -wardii var. wardii (1/2)
```

Two clones Freckle Face, Panther. (Exhibited by Wallace, 1937)

```
g.
       -Corona--unknown (1/2)
Coronis-    -griffithianum (1/4)
       -Loderi-
           -fortunei ssp. fortunei (1/4)
```

Rose-colored flowers. (Lord Aberconway, 1926)

```
                      -campylocarpum ssp. campylocarpum
           -Penjerrick-              Elatum Group (1/16)
g.      -Amaura-       -griffithianum (5/16)
     -Eros-   -griersonianum (3/8)
Corros- -griersonianum
     -            -thomsonii ssp. thomsonii (1/4)
     -Cornish Cross-
                -griffithianum
```

Red flowers. (Lord Aberconway, 1946)

```
cl.              -arboreum ssp. arboreum (1/4)
          -Doncaster-
Corry Koster-     -unknown (1/4)
          -              -griffithianum (1/4)
          -George Hardy-
                   -catawbiense (1/4)
```

5ft(1.5m)         -5°F(-21°C)          EM          2/2

Parentage uncertain but given by Millais as above.  Flowers of
pale pink, heavily spotted red in throat, with frilled margins;
large, rounded trusses.  Upright habit; medium green leaves that
stand upright. (M. Koster & Sons, 1909)

```
g.          -haematodes ssp. haematodes (1/4)
     -Choremia-
Corsa-    -arboreum ssp. arboreum (1/4)
     -sanguineum ssp. sanguineum var. sanguineum (1/2)
```

Blood red flowers. (Lord Aberconway, 1946)

cl.

Corsock--form of phaeochrysum var. phaeochrysum

5ft(1.5m)         5°F(-15°C)          EM          3/3

Leaves 6.5in(16.5cm) long, with brown indumentum beneath.  Flow-
ers of China rose, lightly spotted, in tight trusses of 18-20.
It is no longer in cultivation at Corsock House.  (Collected by
Forrest;  raised by Gen. D. MacEwen;  exhibited by F. L. Ingall)
P. C. (RHS) 1966

```
cl.                 -catawbiense (1/2)
Cosima--Everestianum (selfed)-
                    -unknown (1/2)
```

Frilled, rose pink flowers with yellow markings. (T. J. R. Seidel, cross 1893)

```
cl.                  -caucasicum (1/4)
          -Cunningham's White-
Cosmopolitan-            -ponticum var. album (1/4)
          -Vesuvius--unknown (1/2)
```

5ft(1.5m)         -10°F(-23°C)          ML          3/4

Pink flowers fading to pale pink, brownish red blotch.  Attractive glossy foliage. (Hagen, Boskoop)  A. M. (RHS) 1957   See
color illus. HG p. 79.

```
cl.        -campylocarpum ssp. caloxanthum (1/2)
Costa del Sol-
          -lacteum (1/2)
```

Flowers of yellow orange flushed red in throat, openly campanulate, 2.5in(6.4cm) wide; loose trusses of 14-16.  A low, rounded
plant; medium-sized, dark green leaves with light brown indumentum below.  The reverse cross of Joanita.  See Rothschild Rhododendrons, p. 112. (Raised by L. de Rothschild; exhibited by E.
de Rothschild)  A. M. (RHS) 1969

```
                  -dichroanthum ssp. dichroanthum (1/4)
cl.    -Fabia-
          -griersonianum (1/4)
Cotillon-              -griffithianum (1/16)
     -            -Kewense-
     -       -Aurora-      -fortunei ssp. fortunei (5/16)
     -Naomi-    -thomsonii ssp. thomsonii (1/8)
          -fortunei ssp. fortunei
```

When registered this plant was 4ft(1.2m) high, 6ft(1.8m) broad;
the leaves 8in(20.3cm) long.  Flowers Mars orange inside, jasper
red outside, 7-lobed, 4in(10.2cm) across; trusses of 9-10.  (R.
Henny, reg. 1960)

```
cl.              -griffithianum (1/2)
          -Marinus Koster-
Cotton Candy-         -unknown (1/4)
          -              -griffithianum
          -Loderi Venus-
                   -fortunei ssp. fortunei (1/4)
```

6ft(1.8m)         0°F(-18°C)          M          4/4

Tall conical trusses of very large pastel pink flowers.  Foliage
dark green, 6in(15.2cm) long.  Grows well in full sun. (J. Henny & Wennekamp, 1958)   See color illus. VV p. 104.

```
cl.                        -griffithianum (1/4)
        -Loderi King George-
Cougar-                    -fortunei ssp. fortunei (1/4)
        -unknown (1/2)


5ft(1.5m)          20°F(-7°C)          EM
```

Synonym Loderi Cougar.  Buds cardinal red; flowers bright neyron
rose, tubular funnel-shaped, fragrant, in 9in(23cm) trusses of
15.  Flowers held upright, not "floppy" as other Loderis.  Plant
well branched, as broad as tall.  (Jim Drewry-Gene German-Cath-
erine Weeks, reg. 1983)

```
cl.                        -griffithianum (1/4)
        -Geoffrey Millais-
Countess of Athlone-       -unknown (1/2)
        -                       -catawbiense (1/4)
        -Catawbiense Grandiflorum-
                                -unknown

6ft(1.8m)          -10°F(-23°C)          ML          2/3
```

Purple buds open to large mauve flowers with greenish yellow
basal markings, widely funnel-shaped, in tight conical trusses.
Plant of upright habit; glossy dark green foliage.  (C. B. van
Nes, 1923)

```
                                -griffithianum (3/8)
                        -George Hardy-
        -Pink Pearl-            -catawbiense (3/8)
cl.     -               -       -arboreum ssp. arboreum
        -               -Broughtonii-                (1/8)
Countess of Derby-              -unknown (1/8)

        -               -catawbiense
        -Cynthia-
                -griffithianum
```

Synonym Eureka Maid.  Deep pink buds opening to very large widely funnel-shaped flowers of rich rose pink, fading to pale pink,
with reddish brown spots and streaks inside; trusses large, conical. Plant open, spreading;  abundant foliage of medium green.
(Harry White, Sunningdale, c.1913) A. M. (RHS) 1930   See color
illus. VV p. 30.

```
cl.
        -ciliatum (1/2)
Countess of Haddington-
                -dalhousiae var. dalhousiae (1/2)

5ft(1.5m)          20°F(-7°C)          E          4/4
```

Flowers pink in bud, opening white flushed rose, funnel-shaped,
3in(7.6cm) wide, in very lax trusses of 3 or 4. Floriferous and
fragrant. Large, bright green leaves with hairy edges. (Origin
unknown) F. C. C. (RHS) 1962

```
cl.
        -edgeworthii (1/2)
Countess of Sefton-         -ciliatum (1/4)
                -Multiflorum-
                        -virgatum ssp. virgatum (1/4)

5ft(1.5m)          20°F(-7°C)          E          3/3
```

Flowers pink in bud, opening white, funnel-shaped, 2.5in(6.4cm)
wide; 2 or 3 in lax trusses. Corollas usually split lengthwise.
Bushy plant. (Isaac Davies, 1877)

```
cl.


County of York     See Catalode
```

```
g.
        -ciliatum (1/2)
Cowbell-
        -edgeworthii (bullatum) (1/2)

3ft(.9m)          15°F(-10°C)          EM          4/4
```

One of the better tender hybrids, with large pure white flowers
and fuzzy dark green foliage.  Other forms exist. (Rothschild,
1935)

```
g.
        -williamsianum (1/2)
Cowslip-
        -wardii var. wardii (1/2)

2.5ft(.75m)          0°F(-18°C)          EM          3/3
```

Flowers light primrose yellow to cream, pale pink blush, saucer-
to bell-shaped; several in loose trusses.  Plant a compact mound
with small oval leaves. (Lord Aberconway)  A. M. (RHS) 1937

```
cl.

Cowtye--form of charitopes ssp. tsangpoense (K-W 5844)

3ft(.9m)          0°F(-18°C)          M
```

Purple flowers with dark spotting, only .5in(1.3cm) long,
tubular bell-shaped, in trusses of 3-5.  Leaves 2in(5cm) by 1in
(2.5cm), dull dark green; plant upright, open. (Kingdon-Ward
collector; A. E. Hardy raiser)  A. M. (RHS) 1972

```
cl.
        -dichroanthum ssp. dichroanthum (1/2)
Crackerjack-
        -wardii var. wardii (1/2)
```

Campanulate flowers of empire yellow, 2.5in(6.7cm) wide, 8-10 in
a truss.  Plant dwarfish; blooms late midseason. (Rudolph & Leona Henny, 1965)

```
cl.
        -gracilentum (1/2)
Craig Faragher-
        -jasminiflorum (1/2)
```

Vireya hybrid.  Red flowers, in trusses of 6-8. (Cross by Craig
Faragher; intro. & reg. by Graham Snell, Australia, 1982)

```
cl.
        -niveum ? (1/2)
Crarae-
        -unknown (1/2)
```

Flowers smoky purple tinged with red, campanulate, 7-lobed, and
1.5in(3.8cm) wide; 20-23 per truss.  Dull green leaves, 7in(17.8
cm) long, mealy white indumentum beneath. (Sir George Campbell)
A. M. (RHS) 1965

```
cl.

Crarae Crimson--form of cinnabarinum ssp. cinnabarinum Roylei
                                                        Group
```

Received a P. C. award in 1975 as a hardy flowering plant. (Sir
Ilay Campbell, raiser; intro. by Collingwood Ingram)

```
cl.            -griffithianum (1/4)
        -Mars-
Craswell-    -unknown (1/4)
        -Koichiro Wada--yakushimanum ssp. yakushimanum (1/2)

3.5ft(1m)          -5°F(-21°C)          M
```

Carmine rose flowers, openly funnel-shaped, 2in(5cm) across, in
14-flowered ball-shaped trusses.  Plant upright, almost as wide
as tall, with stiff branches; leaves held 3 years. (Dr. Whildin
A. Reese, cross 1968; reg. 1977)

```
cl.
        -Barto Blue--augustinii ssp. augustinii (3/4)
Crater Lake-        -intricatum (1/4)
        -Bluebird-
                -augustinii ssp. augustinii
```

5ft(1.5m)          -5°F(-21°C)          EM          4/3

Brilliant electric, violet blue flowers, saucer-shaped, up to 2in(5cm) wide, ruffled edges, held in small clusters, many per branch.  Plant of upright habit, nicely branched; new growth bronze yellow, turning later to Irish green. (Dr. Carl Phette-place) C. A. (ARS) 1975

```
cl.
          -Marion--unknown (11/16)
Crazy Joe-     -griersonianum (1/4)
     -Radium-                    -griffithianum (1/16)
     -            -Queen Wilhelmina-
          -Earl of-          -unknown
          Athlone-Stanley Davies--unknown
```

Trusses hold 15 cardinal red flowers. See also Maybe. (G. Langdon, 1982)

```
cl.
          -rupicola var. chryseum (1/2)
Cream Crest-     -ciliatum (1/4)
     -Cilpinense-
               -moupinense (1/4)
```

3ft(.9m)          0°F(-18°C)          EM          3/3

Bright creamy yellow, cup-shaped flowers, 1in(2.5cm) across, 6-8 in tight trusses on a compact plant that has vigor, good foliage and likes the sun. (A. Wright, reg. 1963) See color illus. VV p. 54.

```
          -Lady     -fortunei ssp. discolor (3/16)
          -Bessborough-
     -Jalisco-          -campylocarpum ssp.campylocar-
     -     -                    pum Elatum Group (3/16)
     -     -     -dichroanthum ssp. dichroanthum (1/4)
-Comstock-     -Dido-
-     -          -decorum (1/8)
-     -     -dichroanthum ssp. dichroanthum
cl. -     -Jasper-          -fortunei ssp. discolor
-     -          -Lady Bessborough-
Cream-                    -campylocarpum ssp. campy-
Glory-                         locarpum Elatum Group
-          -Lady     -fortunei ssp. discolor
-          -Bessborough-
-     -Jalisco-          -campylocarpum ssp. campylo-
-     -     -                    carpum Elatum Group
-     -     -     -dichroanthum ssp. dichroanthum
-Cheyenne-     -Dido-
-     -          -decorum
-     -griffithianum (1/8)
-Loderi-
          fortunei ssp. fortunei (1/8)
```

4ft(1.2m)          -5°F(-21°C)          M          3/3

Flowers are deep cream, of heavy substance, 4.5in(11.5cm) wide, openly funnel-campanulate, 7-lobed, fragrant; calyx of same color.  High dome-shaped trusses hold 8-9.  Plant rounded, almost as wide as tall. (Harold Greer, cross 1965; intro. 1978)    See color illus. HG p. 98.

```
cl.
          -dalhousiae var. dalhousiae (1/2)
Cream Trumpet-
          -nuttallii (1/2)
```

Victorianum g.  White flowers with large orange blotch, funnel-campanulate, in trusses of 3-4.  Before 1879. (RBG, Edinburgh, reg. 1962) F. C. C. (RHS) 1958

```
cl.
          -griersonianum (1/2)
Creeping Jenny-
          -forrestii ssp. forrestii Repens Group (1/2)
```

2ft(.6m)          5°F(-15°C)          EM          3/4

Synonym Jenny.  Flowers bright red, campanulate, in lax trusses of 5-6; compact, small-leaved plant. See also Elizabeth. (Lord Aberconway)

g. & cl.          -fortunei ssp. fortunei (1/4)
     -Luscombei-
```
Cremorne-          -thomsonii ssp. thomsonii (1/4)
          -campylocarpum ssp. campylocarpum (1/2)
```

A well-shaped plant of medium height.  Flowers are medium-sized, shading from rose to soft yellow and creamy white within.  See also Townhill Cremorne. (Rothschild, 1935) A. M. (RHS) 1947

```
          -griffithianum (3/16)
cl.     -Mars-
     -Vulcan-     -unknown (3/8)
Creole-     -griersonianum (1/4)     -griffithianum
Belle*          -George Hardy-
     -Mrs. Lindsay-          -catawbiense
-Harvest- Smith     -Duchess of Edinburgh--unknown
  Moon  -
          -unnamed-campylocarpum ssp. campylocarpum (1/8)
           hybrid-
                -unknown
```

4ft(1.2m)          -5°F(-21°C)          ML          3/4

Bright pink flowers.  Plant well-branched; forest green foliage. (Thompson, 1982)    See color illus. HG p. 100.

```
cl.
     -wardii var. wardii (1/2)
Crest-          -fortunei ssp. discolor (1/4)
     -Lady Bessborough-
               -campylocarpum ssp. campylocarpum Elatum
                                        Group (1/4)
```

6ft(1.8m)          -5°F(-21°C)          M          5/3

Hawk g.  Synonym Hawk Crest.  Primrose yellow flowers, slight darkening around throat, up to 4in(10cm) across, in large dome-shaped trusses.  Heavy budding after plant is well established.  Plant habit upright, open.  Leaves oval and glossy, held only 1 year. (Rothschild, 1940) F. C. C. (RHS) 1953  See color illus. ARS Q 28:4 (Fall 1974): p. 239; F p. 47; HG p. 68; PB pl. 23.

```
cl.
     -smirnowii (1/2)
Crete-
     -yakushimanum ssp. yakushimanum (1/2)
```

7ft(2.1m)          -15°F(-26°C)          M          4/5

Magenta rose buds open to flowers of light mallow purple with a few ochre spots, aging to white, of good substance, 2.75in(7cm) across, 5-lobed;  dome-shaped trusses hold 12; free-flowering.  Plant wider than tall; leaves heavily indumented, held 4 years. (B. Lancaster cross; David G. Leach, raised & intro.; reg. 1982)

```
          -thomsonii ssp. thomsonii (1/4)
cl.  -Barclayi-          -arboreum ssp. arboreum (1/8)
-          -Glory of -
Cretonne-          Penjerrick-griffithianum (3/8)
-          -griffithianum
     -Loderi-
          -fortunei ssp. fortunei (1/4)
```

White flowers stained by rose Bengal within and without. (Sir Giles Loder) A. M. (RHS) 1940

```
               -griffithianum (1/8)
cl.          -Queen Wilhelmina-
     -Britannia-          -unknown (3/8)
Cricket*     -Stanley Davies--unknown
     -Koichiro Wada--yakushimanum ssp. yakushimanum (1/2)
```

3ft(.9m)          5°F(-15°C)          EM

Rosy pink flowers fading lighter pink.  New growth silvery green turning to medium dark green. (James Elliott)

cl.                      -caucasicum (1/4)
           -Dr. Stocker-
Crimson Banner-           -griffithianum (1/4)
           -thomsonii ssp. thomsonii (1/2)

Asteroid g.  (Intro. by Sunningdale Nursery, 1934)

cl.                      -neriiflorum ssp. neriiflorum (1/4)
           -F. C. Puddle-
Crimson Bells-           -griersonianum (1/4)
           -williamsianum (1/2)

1.5ft(.45m)        0°F(-18°C)        EM

Crimson, bell-shaped flowers, 3in(7.6cm) across; trusses hold 7.
Plant twice as wide as tall; oval leaves 3in(7.6cm) long. (Benjamin Lancaster, 1965)

cl.
           -Marion--unknown (1/2)
Crimson Glow-           -thomsonii ssp. thomsonii (1/4)
           -Shilsonii-
                      -barbatum (1/4)

Red flowers with lighter center, slightly spotted on all lobes;
trusses of 20. (K. Van de Ven, Australia, 1973)

cl.
           -yakushimanum ssp. yakushimanum (1/2)
Crimson Pippin-
           -sanguineum ssp. sanguineum var. haemaleum (1/2)

Currant red flowers; trusses of 10-11. Leaves have satinwood or
greyed orange woolly indumentum. (H. L. Larson, 1983)

cl.
           -Moser's Maroon--unknown
Crimson Queen-           -fortunei ssp. discolor (1/4)
           -Azor (Harm's form)-
                      -griersonianum (1/4)

2.5ft(.75m)        -5°F(-21°C)        M

Crimson flowers, held in trusses of 12. Plant slow-growing and
dense with long, narrow leaves. Easily propagated. (B. Lancaster, 1962)

cl.         -griffithianum (1/4)
           -Mars-
Crimson Stain-   -unknown (1/4)
           -yakushimanum ssp. yakushimanum (1/2)

3ft(.9m)        -15°F(-26°C)        ML

Purple red buds open to flowers of diffused white, fading white;
early throat deep pink, margins purplish red. Flowers of good
substance, 3in(7.6cm) across, with 5 wavy lobes, in round trusses
of 12-16. Floriferous. Plant rounded, well-branched; foliage
olive green, lightly indumented, held 2-3 years. (A. F. Serbin
& R. G. Shanklin, 1983)

                      -griffithianum (1/4)
           -Queen Wilhelmina-
cl.         -Britannia-           -unknown (3/4)
           -         -Stanley Davies--unknown
Crimson Star-           -griffithianum
           -Queen Wilhelmina-
           -Unknown-           -unknown
           Warrior-Stanley Davies--unknown

Red flowers, from a cross of two sisters. (Donald Hardgrove,
exhibited 1951)

cl.
        -Blue Peter--unknown (5/8)
Crinkles-           -arboreum ssp. arboreum (1/8)
        -           -Doncaster-
        -Corry Koster-   -unknown
        -           -griffithianum (1/8)
           -George Hardy-
                      -catawbiense (1/8)

Flowers reddish purple with a purple blotch; trusses hold about
21. (K. Van de Ven, reg. 1981)

g.
           -spinuliferum (1/2)
Crossbill-
           -lutescens (1/2)

Slim, erect bush with striking red new growth; tubular flowers
of yellow, flushed apricot. Early. Other forms exist. (J. C.
Williams, 1933)

cl.
           -strigillosum (1/2)
Crossroads*
           -unknown (1/2)

5ft(1.5m)        0°F(-18°C)        EM        3/3

Cardinal red flowers of good substance. Plant stems, and midveins of narrow leaves are woolly with coarse bristles; slight
indumentum beneath. (H. L. Larson; intro. by H. E. Greer)

cl.

Crown Equerry--form of niveum

5ft(1.5m)        5°F(-15°C)        M

Purplish lilac flowers, darker at lip, tubular-campanulate, 2in.
(5cm) across. Leaves to 6.5in(16.5cm) long; felted brown indumentum beneath. (Collector unknown; raised by J. B. Stevenson;
intro. by Crown Estate, Windsor) F. C. C. (RHS) 1979

---
                      -brookeanum var. gracile (1/2)
Crown Princess of Germany-           -jasminiflorum (1/4)
           -Princess Royal-
                      -javanicum (1/4)

Vireya hybrid. Yellow flowers; parent of Ajax and other hybrids.
(Veitch, 1891)

cl.
           -aberconwayi (1/2)
Crowthorne-
           -souliei (1/2)

Called a dwarf in its registration, but award description gives
height as 5.5ft(1.65m). Upright, compact habit; glossy medium-
sized leaves. White flowers, tinged pink with deep pink edges;
tight trusses of about 11. (Crown Estate, Windsor, 1963)

cl.

Cruachan--form of temenium var. gilvum Chrysanthum Group
                                      (R 22272)
Flowers clear pure yellow, on a dwarf, spreading shrub up to 3ft
(1.5m) tall. (Collected by Rock; raised & exhibited by Mrs. K.
L. Kenneth; reg. 1965)  A. M. (RHS) 1958 [as R. chrysanthemum];
F. C. C. (RHS) 1964

cl.

Cuff-link      Parentage unknown

A plant for the cool greenhouse with leaves medium-sized, dark
glossy green. Trusses of 5 or 6 flowers, tubular bell-shaped,
very light yellow. (Origin unknown; reg. by Geoffrey Gorer)
A. M. (RHS) 1978

cl.
           -caucasicum (1/2)
Cunningham's Album Compactum*
           -unknown (1/2)

Flowers blush pink in bud, opening white with a ray of greenish
brown markings. Sturdy, slow-growing shrub; dense, leafy habit.
Slower growth, narrower leaves (indumentum beneath) than in Cunningham's White. (Cunningham)

cl.

Cunningham's Sulphur--form of caucasicum

3ft(.9m)          -5°F(-21°C)          VE          2/2

Synonym White's Cunningham's Sulphur.  Small pale yellow flowers
in rounded trusses of about 15.  Plant compact, bushy; elliptic,
medium green, glossy leaves.  Heat-tolerant, but best in partial
shade.  (Cunningham, in the 1800s)

cl.
                    -caucasicum (1/2)
Cunningham's White-
                    -ponticum var. album (1/2)

4ft(1.2m)          -15°F(-26°C)          ML          2/3

Buds flushed pink open to small white flowers with a pale yellow
eye; small, upright many-flowered trusses.  Plant spreading;
dark green foliage.  Tolerates slightly alkaline soil and industrial smoke.  (Cunningham, 1830)  See color illus. VV p. 22.

cl.                    -griffithianum (1/8)
          -Albatross-     -fortunei ssp. fortunei (1/8)
Cup Day-          -Loderi-
          -        -fortunei ssp. discolor (1/4)
          -Fusilier-     -elliottii (1/4)
                    -griersonianum (1/4)

Flowers of Tyrian rose, heavily spotted on all lobes.  See also
Success.  (K. Van de Ven, Australia, 1965)

cl.
          -ludlowii (1/2)
Curlew-
          -fletcherianum (1/2)

1.5ft(.45m)          -5°F(-21°C)          EM          4/4

Bright yellow flowers with upper petals spotted, 2in(5cm)
wide, openly campanulate, 5-lobed; compact trusses of 2 or 3.
Plant much wider than tall; small dark leaves with sparse, scaly
indumentum.  (Peter Cox)  F. C. C. (RHS) 1969; A. M. (RHS) 1981
See color illus. in  Cox: Dwarf Rhododendrons (1st ed.)  pl. I.

cl.
          -thomsonii ssp. thomsonii (blood red form) (1/2)
Currant Bells-
          -williamsianum (1/2)

2ft(.6m)          5°F(-15°C)          EM

Currant red flowers, bell-shaped, in trusses of 9.  Plant wider
than tall, dense, with dark green leaves.  (B. Lancaster, 1967)

cl.

Curtisii--form of citrinum var. discoloratum (multicolor)

Vireya.  Crimson flowers.  (Origin unknown)  F. C. C. (RHS) 1883

cl.          -wightii (1/4)
          -China-
Custard-          -fortunei sp. fortunei (1/4)
          -decorum (1/2)

Primrose yellow flowers, funnel-campanulate, 5- or 6-lobed, almost 4in(10.2cm) wide; 12-14 in racemose umbels.  Upright, tall
plant, to 10ft(3m); leaves 6in(15cm) long.  (R. & L. Henny, 1963)

cl.
          -calostrotum ssp. calostrotum (1/2)
Cutie-
          -unknown (1/2)

3ft(.9m)          -15°F(-26°C)          M          3/3

Synonyms Calostrotum Pink, Calostrotum Rose.  Heavy blooming of
small, lilac-shaded pink flowers.  Compact rounded plant; leaves
1in(2.5cm) long with slight tan indumentum.  Height also given
as 1.5ft(.45m) at 10 years.  (Cross by Greig; intro. by Larson)
P. A. (ARS) 1959; A. E. (ARS) 1962  See color illus. VV p. 24.

g.                    -neriiflorum ssp. neriiflorum (1/2)
          -F. C. Puddle-
Cyclops-          -griersonianum (1/4)
          -          -neriiflorum ssp. neriiflorum
          -Neriihaem-
                    -haematodes ssp. haematodes (1/4)

Red flowers.  (Lord Aberconway, 1941)

cl.
          -cataubiense (1/2)
Cynthia-
          -griffithianum (1/2)

6ft(1.8m)          -15°F(-26°C)          M          4/3

Synonym Lord Palmerston.  Fine, showy, conical trusses of rosy
crimson flowers with blackish crimson markings, 3in(7.6cm) wide.
Large, vigorous, dome-shaped bush grows in sun or shade, resists
pests; strong, compact foliage.  Raised at Sunningdale Nurseries
(Sussex) in 1865.  (Standish & Noble) A. G. M. (RHS) 1969  See
color illus. VV p. 71.

cl.
          -zoelleri (1/2)
Cyprian-
          -Chlorinda (Veitch hybrid)--unknown (1/2)

5ft(1.5m)          32°F(0°C)          July-Sept.

Vireya hybrid.  Flowers fleshy and long-lasting, tubular funnel-
shaped, 2.75in(7cm) across, maize yellow with wide orange edge,
not fragrant; dome-shaped trusses of 7-8.  Plant rounded, well-
branched; leaves 4in(10.2cm) long, held 3 years.  (Peter Sullivan
& William Moynier, 1983)

                              -griffithianum (1/8)
               -unnamed hybrid-
cl.   -Mrs. Furnival-          -unknown (1/4)
          -          -          -caucasicum (1/8)
Cyprus-          -unnamed hybrid-
          -          -unknown
          -cataubiense var. album (1/2)

3ft(1.2m)          -20°F(-29°C)          ML          4/4

Synonym Capri.  Flowers 3.25in(8.3cm) across, funnel-shaped, 5-
lobed, white with bold ochre blotch, in large ball-shaped trusses of up to 20.  Well-branched, broader than high with leaves retained 3 years.  See also Party Pink, Persia.  (David G. Leach,
intro. 1973; reg. 1983)  C. A. (ARS) 1982; A. E. (ARS) 1983

---
          -Sir Charles Lemon--arboreum ssp. arboreum (1/2)
Cyrene-
          -lanatum (1/2)

The white flowers have a tinge of lemon.  (E. J. P. Magor, 1921)

cl.

Cyril Berkeley--form of forrestii ssp. forrestii var. tumescens

Turkey red flowers, 2.5in(6.4cm) wide, in loose trusses of 3-4.
Slow-growing: a 3ft(.9m) compact bush in 10 years.  (Greig, Roy-
sten Nursery, 1965)

                              -caucasicum (3/16)
               -Jacksonii-          -caucasicum
cl.          -Goldsworth-          -Nobleanum-
          - Yellow     -          -arboreum ssp. arbor-
Czech Beauty-                              eum (1/16)
          -          -campylocarpum ssp. campylocarpum (1/4)
          -unknown (1/2)

4ft(1.2m)          -10°F(-23°C)          M

Large trusses of 11 flowers, primrose yellow  with greyed orange
spotting, 3.5in(8.9cm) wide.  Plant rounded, broader than tall;
glossy leaves held 3 years.  (Louis Hindla, 1979)

g.
        -dichroanthum ssp. scyphocalyx (1/2)
Dacia-          -neriiflorum ssp. neriiflorum (1/4)
    -F. C. Puddle-
                -griersonianum (1/4)

Orange red flowers.  (Lord Aberconway, intro. 1941)


g. & cl.        -haematodes ssp. haematodes (1/4)
    -May Day-
Dainty-         -griersonianum (1/2)
    -              -forrestii ssp. forrestii Repens Group (1/4)
    -Elizabeth-
                -griersonianum

Brilliant scarlet.  (Lord Aberconway) A. M. (RHS) 1942; F. C. C.
(RHS) 1944


cl.
        -williamsianum (1/2)
Dainty Jean-
        -Helene Schiffner--unknown (1/2)

1ft(.3m)        0°F(-18°C)        M

Compact dwarf plant.  White flowers flushed pink outside, quite
flat, slightly ruffled, to 3in(7.6cm) wide; 6 in loose trusses.
(Bovee, 1963)


cl.
        -campylocarpum ssp. campylocarpum (1/2)
Dairymaid-
        -unknown (1/2)

A compact shrub.  Flowers of cream with pink lines on petals and
red spotting in throat, in tight trusses. (Slocock, 1930)  A. M.
(RHS) 1934  (Exbury Trial)    See color illus. F p. 22, 41, 50;
PB pl. 55.


cl.
        -smirnowii (1/2)
Daisy-          -catawbiense (1/4)
    -Mrs. Milner-
                -unknown (1/4)

Bright carmine rose flowers, with yellowish green markings on a
lighter ground.  (T. J. R. Seidel, before 1902)


cl.
        -catawbiense (1/2)
Daisy Rand-
        -unknown (1/2)

Rose red flowers.  (Samuel Parsons, c.1870s)


g.
        -dalhousiae var. dalhousiae (1/2)
Dalbull-
        -edgeworthii (bullatum) (1/2)

White flowers.  (E. J. P. Magor, intro. 1936)


cl.

Dalkeith--dwarf chance seedling, possibly natural hybrid.  Has
        some characteristics of Triflora, but is Uniflora.

Purple violet flowers in trusses of 10-17.  (Intro. by C. A. Mc-
Laughlin; reg. by Mrs. P. J. Warren, 1977)


cl.

Dalriada--form of japonicum var. pentamerum (degronianum)

3ft(.9m)        -5°F(-21°C)        EM        3/3

A Japanese native, selected in Scotland, grown in cold climates
around the world.  Bright rose-colored flowers; attractive
foliage indumented less heavily than yakushimanum. Differs from
the species only in its spreading habit.  (Mrs. K. L. Kenneth,
1962)  P. C. (RHS) 1960


g.              -caucasicum (1/4)
    -Dr. Stocker-
Damaris-        -griffithianum (1/4)
    -campylocarpum ssp. campylocarpum (1/2)

4ft(1.2m)        5°F(-15°C)        M        4/3

A plant with pale yellow trusses and glossy new foliage.  See
also Cream Cracker, Cornish Cracker, Logan Damaris, Townhill Damaris. (E. J. P. Magor, 1926)

cl.

Damaris Logan      See Logan Damaris

cl.

Dame Edith Sitwell--form of lindleyi

5ft(1.5m)        25°F(-4°C)        E-ML        4/2

A cool greenhouse plant.  Flowers white, tinged with pale pink,
tubular funnel-shaped, 4.3in(11cm) across; held in trusses of 4-
5.  Leaves 5.5in(14cm) long. (Geoffrey Gorer)  A. M. (RHS) 1965


cl.              -maximum (1/4)        -catawbiense
        -Standishii-        -unnamed hybrid-        (1/16)
Dame Nellie-        -Altaclarense-        -ponticum
    Melba    -        -        (1/16)
        -        -arboreum ssp. arboreum
        -arboreum ssp. arboreum        (5/8)

6ft(1.8m)        0°F(-18°C)        M        3/4

Bright pink flowers with crimson spots.  Upright plant with dark
green, glossy leaves.  (Sir Edmund Loder)  A. M. (RHS) 1926


                -caucasicum (1/8)
            -Dr. Stocker-
cl.    -Damaris-        -griffithianum (1/8)
    -        -campylocarpum ssp. campylocarpum (1/4)
Damophyle-        -adenogynum (1/8)
    -        -Xenosporum (detonsum)-
    -unnamed-        -unknown (3/8)
    hybrid-unknown

A shapely bush, 15ft(4.5m) at 25 years, with bell-shaped, white
flowers held in racemose umbels of 10-12. (Cross by E. J. P.
Magor; named and registered by Maj. E. W. M. Magor, 1962)


g. & cl.
    -A. W. Bright Rose--unknown (1/2)
Damozel-
    -griersonianum (1/2)

6ft(1.8m)        0°F(-18°C)        4/4

Widely spreading, with too few narrow, dark green leaves, thinly
indumented.  Dome-shaped trusses of about 17 deep ruby red flowers, spotted darker. (Rothschild, 1936)  A. M. (RHS) 1948 [to
a deep rose pink form]  See color illus. F p. 42.


                -forrestii ssp. forrestii Repens Group
cl.        -Elizabeth-        (1/4)
    -        -griersonianum (1/4)
Dan Laxdall-        -griffithianum (1/8)
    -        -Coombe Royal-
    -Mrs. G. W. Leak-        -unknown (1/4)
    -        -caucasicum (1/8)
        -Chevalier Felix-
        de Sauvage  -unknown

3ft(.9m)        10°F(-12°C)        ML

Buds and flowers of neyron rose, deeper in throat and as narrow
stripes up centers of lobes, in trusses of 7-10. Flowers have a
"glow" (fluorescence). Dark green foliage held 2 years. (Sigrid Laxdall, 1976) See color illus. ARS Q 30:2 (Spring 1976):
p. 103.

```
---                              -brookeanum var. gracile (1/4)
      -Crown Princess of Germany-              -jasminiflorum
Dante-                     -Princess Royal-             (1/8)
      -                                 -javanicum (5/8)
      -javanicum
```

A vireya hybrid by Veitch. (1891)

```
g.
      -dichroanthum ssp. dichroanthum (1/2)
Dante-
      -facetum (eriogynum) (1/2)
```

Flowers described in the Register as being "yellow to red orange
or vermilion." (Lord Aberconway, 1936)

```
cl.
      -concinnum (1/2)
Danube-
      -augustinii ssp. augustinii (1/2)
```

Argiolus g. Previously known as Argiolus var. Danube. (Lord
Aberconway, 1950)

```
cl.
      -japonicum var. japonicum (metternichii) (1/2)
Daphne-
      -Alexander Adie--unknown (1/2)
```

White flowers with greenish yellow markings. (T. J. R. Seidel,
cross 1894; intro. 1902)

```
cl.              -arboreum ssp. arboreum (1/4)
      -Red Admiral-
Daphne-          -thomsonii ssp. thomsonii (1/4)
      -neriiflorum ssp. neriiflorum (1/2)
```

Flowers are hose-in-hose, bright crimson. See also Eithne. (E.
J. P. Magor, 1928) A. M. (RHS) 1933

```
cl.
      -griffithianum (1/2)
Daphne Daffarn-
      -unknown (1/2)
```

Rounded trusses of campanulate flowers, bright rose pink shading
to white on lobes, with light red speckles. (Originated before
1884)

```
cl.                    -arboreum ssp. arboreum (1/8)
      -Red Admiral-
      -Daphne-         -thomsonii ssp. thomsonii (1/8)
Daphne-    -neriiflorum ssp. neriiflorum (1/4)
Magor-         -adenogynum (1/4)
      -Xenosporum-
                -unknown (1/4)
```

A plant of 10ft(3.0m) at 25 years with leaves 3.25in(8.3cm)
long. Flowers are fleshy, blood red with large petaloid calyx,
held in groups of 6-8. (Cross by E. J. P. Magor; named & reg.
by Maj. E. W. M. Magor, 1962)

```
cl.
      -virgatum ssp. virgatum (1/2)
Daphnoides-
      -unknown (1/2)
```

4ft(1.2m)        -15°F(-26°C)        ML        2/4

Most unusual foliage of rolled, glossy leaves tightly spaced on
the stems of a dense mound. Prolific trusses of purple flowers.
(T. Methven, cultivated 1868) See color illus. VV p. 98.

```
cl.
      -smirnowii (1/2)
Darius-        -catawbiense (1/4)
      -Mrs. Milner-
                -unknown (1/4)
```

Purplish red flowers, ochre markings. (T. J. R. Seidel, cross
1894; intro. 1902)

```
cl.              -catawbiense (1/4)
      -Kettledrum-
Dark Eyes-       -unknown (3/8)
      -                    -adenogynum (1/8)
      -unnamed-Xenosporum (detonsum)-
       hybrid-           -unknown
             -griersonianum (1/4)
```

5ft(1.5m)        -5°F(-21°C)        M

Bright rose flowers with dark red blotch in trusses of about 6.
Rounded leaves, 6in(15cm) by 3in(7.6cm). (G. Guy Nearing, 1977)

```
cl.
      -sanguineum ssp. sanguineum var. haemaleum (1/2)
Dark Stranger-
      -williamsianum (1/2)
```

Trusses of 3-4 flowers, bell-shaped, ruby red. Leaves dull dark
green, 2in(5cm) long. (Probably crossed by Col. S. R. Clarke;
raised by R. N. Stephenson Clarke) A. M. (RHS) 1980

```
cl.
      -griersonianum (1/2)
Darlene-              -griffithianum (1/8)
      -              -unnamed hybrid-
      -Armistice Day-          - unknown (3/8)
                -Maxwell T. Masters--unknown
```

Bright red flowers. (Halfdan Lem, reg. 1958) P. A. (ARS) 1952

```
cl.              -Monsieur Thiers--unknown (1/4)
      -J. H. Van Nes-
Darlene's Pink-   -griffithianum (1/2)
      -              -griffithianum
      -Loderi King George-
                -fortunei ssp. fortunei (1/4)
```

4ft(1.2m)        15°F(-10°C)        EM

Flowers of strong purplish red, 2.75in(7cm) wide, openly funnel-
campanulate, 5-6 wavy lobes; dome-shaped trusses of 11-15.
Floriferous. Plant upright, taller than broad; long narrow foliage held 2-3 years. (Dr. Ben Briggs, cross 1963; intro. Edgar
L. Knight; reg. 1980)

```
cl.                    -griffithianum (1/8)
             -George Hardy-
cl.          -              -catawbiense (1/8)
      -Hugh Koster-              -arboreum ssp. arbor-
David-       -              -Doncaster-       eum (1/16)
      -       -unnamed hybrid-       -unknown (3/16)
      -                    -unknown
      -neriiflorum ssp. neriiflorum (1/2)
```

6ft(1.8m)        5°F(-15°C)        EM        4/3

Fine, brilliant, blood red flowers with white anthers, frilly-
margined, campanulate; loosely formed trusses. Plant of upright
habit; dark green foliage. (Lord Swaythling) F. C. C. (RHS)
1939; A. M. (Wisley Trials) 1957 See color illus. VV p. 31.

```
cl.
      -catawbiense var. compactum (1/2)
David Forsythe-   -griffithianum (1/4)
      -Mars-
             -unknown (1/4)
```

2.5ft(.75m)        -10°F(-23°C)        ML

Buds scarlet, opening to dark scarlet flowers, 3in(7.6cm) wide;
ball-shaped trusses of about 8. Plant broader than tall; leaves
dark green, 4in(10.2cm) long, held 2 years. (Raiser S. Baldanza;
reg. 1972)

```
cl.                           -catawbiense (1/4)
                -Atrosanguineum-
David Gable-                       -unknown (1/4)
                -fortunei ssp. fortunei (1/2)

5ft(1.5m)           -15°F(-26°C)            M           4/4
```

Flowers fuchsine pink, 3.5in(8.9cm) across, with basal blotch of
Indian red, campanulate; large dome-shaped trusses; floriferous.
Good plant habit; dark green leaves, 7in(17.8cm) long.  (Joseph
Gable)  A. E. (ARS) 1960   See color illus. LW pl. 44; VV p. 91.

```
cl.
               -minus var. minus Carolinianum Group (1/2)
               -                       -racemosum (1/8)
David John-                    -Conemaugh-
            -Pioneer (Gable)-          -mucronulatum (1/8)
                      -unknown (1/4)

3ft(.9m)           -20°F(-29°C)            M
```

Ball-shaped terminal flower clusters, 2-6 trusses, each 2-5
flowered or a total of 4-20 flowers, mallow purple with faint
red spotting. See also Gary. (William Fetterhoff, 1980)

```
cl.               -griersonianum (1/4)
                -Karkov-             -arboreum ssp. arboreum (1/8)
David       -      -Red Admiral-
Rockefeller-             -thomsonii ssp. thomsonii (1/8)
           -          -griffithianum (1/8)
              -Gipsy-King George-
                 King-           -unknown (1/8)
                   -haematodes ssp. haematodes (1/4)
```

Trusses of 10 flowers, medium rose speckled red.  (Edmund de
Rothschild, 1982)

```
cl.
         -griffithianum (1/2)
Dawn-
         -unknown (1/2)
```

White flowers blushed phlox pink; 8-flowered trusses. (Waterer,
Sons & Crisp)  A. M. (RHS) 1950

```
cl.
           -griffithianum (1/2)
Dawn's Delight-
           -unknown (1/2)
```

Good foliage; stems show pink tinge of griffithianum. Carmine
buds open to large, frilly soft pink flowers, marked in crimson,
shading white in tube; tall conical trusses. Midseason. (J. H.
Mangles, raised before 1884)  A. M. (RHS) 1911

```
g. & cl.                  -fortunei ssp. discolor (1/4)
             -Lady Bessborough-
Day Dream-             -campylocarpum ssp. campylocarpum
             -griersonianum (1/2)            Elatum Group (1/4)

5ft(1.5m)           5°F(-15°C)            L          3/2
```

Buds rich rose, flowers opening the same, then turning to creamy
pale yellow with a glowing rose eye and faint spotting, fading
to creamy white; large loose trusses. Another form, Bisciut, is
considered somewhat hardier than this one.    (Rothschild, 1936)
A. M. (RHS) 1940    See color illus. PB pl. 55, 65.

```
                -cinnabarinum ssp. cinnabarinum Roylei Group
cl. -Lady       -                                    (1/4)
    -Chamberlain-          -cinnabarinum ssp. cinnabarinum
Dayan-       -Royal Flush-                             (1/8)
     -          (orange form)-maddenii ssp. maddenii (1/8)
     -cinnabarinum ssp. xanthocodon Concatenans Group (1/2)
```

Comely g. Trusses of 8-9 flowers, 2in(5cm) wide by 2.25in
(5.7cm) long; waxy orange yellow. (Edmund de Rothschild) A. M.
(RHS) 1967

```
cl.        -haematodes ssp. haematodes (1/4)
        -May Day-
Debbie-    -griersonianum (1/4)
        -         -sanguineum ssp. didymum (1/4)
         -Carmen-
                -forrestii ssp. forrestii Repens Group (1/4)

2ft(.6m)           0°F(-18°C)            EM          3/4
```

Round and compact dwarf plant, with dense, dark green foliage.
Bright blood red flowers in great profusion. (J. Henny)

```
                 -griersonianum (1/4)
cl.    -Creeping-
       - Jenny -forrestii ssp. forrestii Repens Group (1/4)
Deborah-                 -griffithianum (1/8)
            -Queen Wilhelmina-
        -Unknown-           -unknown (3/8)
        -Warrior-Stanley Davies--unknown
```

Red flowers in trusses of 10-12. See also Florence. (Dr. E. E.
Smith, Surrey, reg. 1976)

```
               -dichroanthum ssp. dichroanthum
           -Fabia-
cl.    - C.I.S. -     -griersonianum
       -         -Loder's White, q.v. for 2 possible
Debra Tuomala-                parentage diagrams
       -         -arboreum ssp. nilagricum
          -Noyo Chief-
               -unknown

4ft(1.2m)           15°F(-10°C)            E
```

Buds very light cardinal red  open to flowers of 7-8 wavy lobes,
light chrome yellow mottled azalea pink, in trusses of 8.  Plant
nearly as wide as tall; emerald green leaves held 2 years. (Carl
Tuomala; reg. by E. R. German, 1984)

```
cl.
        -Catalgla--catawbiense var. album Glass (1/2)
Decalgla-
        -decorum (1/2)

4ft(1.2m)           -15°F(-26°C)            M
```

Round trusses, 6in(15cm) wide, hold 12 flowers, white with green
spots. Glossy leaves, with grayish brown new growth, on a well-
branched shrub. (Cross by J. Gable; reg. by G. Nearing, 1973)

```
g.
        -decorum (1/2)
Decatros*          -catawbiense (1/4)
          -Atrosanguineum-
                 -unknown (1/4)
```

A parent of Millard Kepner. Pink flowers. (Joseph Gable, cross
1932)

```
g.
        -decorum (1/2)
Decauck-
        -griffithianum (aucklandii) (1/2)
```

White flowers. (E. J. P. Magor, 1912)

cl. *Dechaem* (p 146) (261)(265) (201)

Decimus--form of zoelleri

A vireya species  for the cool  greenhouse. Flowers of saffron
yellow with a band of burnt orange, in trusses of 7. Each flower is less than .35in(.9cm) across; leaves are .7in(1.8cm) long,
by .3in(.75cm) wide. From W. New Guinea at 6000ft(1800m) (Collected by J. C. van Steenis; exhibited by RBG, Kew) A. M. (RHS)
1973

```
                      -souliei (1/16)
               -Soulbut-
        -Vanessa-        -Sir Charles Butler--fortunei
g.  -Radiance-                      ssp. fortunei (5/16)
-          -        -griersonianum (3/8)
Decora-        -griersonianum
-          -griffithianum (1/4)
    -Loderi-
             -fortunei ssp. fortunei
```

Pink flowers.  (Lord Aberconway, intro. 1946)


cl.
```
      -decorum (1/2)
Decsoul-
      -souliei (1/2)
```

Rosy buds opening to almost pure white flowers. (W. I. Whitaker)
A. M. (RHS) 1937


cl.

Deer Dell--form of glaucophyllum var. glaucophyllum

3ft(.9m)        0°F(-18°C)        EM        3/2

Trusses of 4-7 flowers, 5-lobed, ruby red. Leaves with the scaly
indumentum of a glaucophyllum.  (J. F. McQuire, reg. 1984)


cl.
```
    -elliottii (1/2)
Degas-
    -haematodes ssp. haematodes (1/2)
```

Compact trusses of 9 bell-shaped flowers, currant red with very
dark red spotting.  Upright habit, broader than tall, compact,
with medium-sized leaves, dark green. (F. Hanger cross, 1950,
RHS Garden, Wisley) H. C. (RHS) 1961


cl.

Degram      See Annie Dalton


cl.
```
  -wardii var. wardii (croceum) (1/2)
Del-    -dichroanthum ssp. dichroanthum (1/4)
  -Fabia-
        -griersonianum (1/4)
```

3ft(.9m)        5°F(-15°C)        M

Rose-colored flowers with greenish yellow blotch, 4.5in(11.5cm)
wide, held in trusses of about 12.  (Cross by Del W. James; Dr.
C. D. Thompson, intro. 1962)


cl.

Del James--form of taggianum

8ft(2.4m)        20°F(-7°C)        VE        5/2

Flowers are white with yellow blotch, rose pink in bud, held in
loose trusses of 2-5, and very fragrant.  Plant is straggly but
upright. (Dr. & Mrs. P. J. Bowman, 1965)


cl.
```
        -Morio--unknown (3/4)
Delamere Belle-                -griffithianum (1/4)
        -Mrs. E. C. Stirling-
                    -unknown
```

A hybrid with trusses of 4 flowers of very light rose, fading to
cream. (D. Dosser, Australia, 1980)


cl.
```
      -pubescens (1/2)
Delaware-
      -keiskei (1/2)
```

3ft(.9m)        -10F°(-23°C)        EM        3/2

Small leaves and small flowers, apricot fading to white. (G. Guy
Nearing, 1958)


cl.
```
              -catawbiense (1/4)
      -unnamed hybrid-
Delendich-        -unknown (1/4)
      -maximum (1/2)
```

7ft(2.1m)        -25°F(-32°C)        M

Truss is 6in(15.2cm) wide with up to 18 flowers, beetroot
purple.  Leaves are yellow green, held 3 years on an upright
plant with stiff branches.  (W. David Smith intro. 1979; reg.
1983)


```
                -griffithianum (1/8)
cl.        -Pink -        -fortunei ssp. fortunei
      -Rosabel-Shell-H. M. Arderne-        (1/16)
Delicious-    -        -unknown (1/16)
      -        -griersonianum (1/4)
      -arboreum ssp. arboreum (1/2)
```

An Australian hybrid with trusses of 20 flowers of light carmine
rose.  See also Hot Wonder.  (G. Langdon, 1982)


cl.
```
              -catawbiense (1/4)
      -Mrs. Milner-
Delila-        -unknown (1/4)
    -smirnowii (1/2)
```

Flowers carmine red with small dark markings; frilled edges.
See also Donar, Desiderius, Dietrich, Ella.  (T. J. R. Seidel,
1902)


```
                  -dichroanthum ssp. dichroanthum (1/8)
g. & cl.  -Astarte-        -campylocarpum ssp. campylocar-
      -Ouida-    -Penjerrick-    pum Elatum Group (1/16)
Delius-    -        -griffithianum (1/16)
-        -griersonianum (1/4)
    -elliottii (1/2)
```

Blood red flowers shaded carmine. (Lord Aberconway, 1936)  A. M.
(RHS) 1942


cl.

Demi-john--form of johnstoneanum

4ft(1.2m)        15°F(-9°C)        M

A plant for the cool greenhouse.  Fragrant white flowers, semi-
double, throat flushed medium yellowish green, 2.75in(7cm) wide;
trusses of 3-4.  Elliptic dark green leaves, 2.5in(6.4cm) long,
scaly beneath.  (Sir Giles Loder)  A. M. (RHS 1975


```
                  -souliei (1/8)
            -Soulbut-
    -Vanessa Pastel-        -Sir Charles Butler--fortunei ssp.
    -        -griersonianum (1/4)        fortunei (1/8)
cl.    -                -griffithianum (9/32)
-                -Beauty of-
Denali*        -Norman- Tremough-arboreum ssp. arbor-
-        - Gill -            eum (1/32)
-        -Anna-    -griffithianum
-    -Pink Walloper-    -Jean Marie de Montague-
-            -unknown (3/16)
-            -griffithianum
    -Marinus Koster-
            -unknown
```

6ft(1.8m)          -5°F(-21°C)          M

Many large rose pink flowers in huge trusses. Beautiful, dense,
deep green foliage. Denali--the Alaskan Indian name for Mt. Mc-
Kinley, meaning 'Big One'. (James Elliott, intro. 1984)


cl.
          -griersonianum (1/2)
Dencombe-           -griffithianum (1/4)
          -Loderi-
                   -fortunei ssp. fortunei (1/4)

Sarita Loder g. Flowers of shrimp pink. (Col. G. H. Loder)


cl.
          -Winter Favourite--unknown (1/2)
Denise-              -chrysodoron (1/4)
          -Chrysomanicum-
                        -burmanicum (1/4)

Flowers are greyed yellow, flushed Venetian pink. (V. J.
Boulter, Australia, 1972)


g.
          -dalhousiae var. dalhousiae (1/2)
Denisonii-            -edgeworthi (1/4)
          -unnamed hybrid-
                        -Gibsonii--unknown (1/4)

Pure white flowers with lemon stain. (Origin unknown) F. C. C.
(RHS) 1862


cl.   -fortunei ssp. discolor Houlstonii Group (1/2)
-                          -soulei (1/32)
Desert-                -Soulbut-
 Sun -       -Vanessa-      -fortunei ssp. fortunei (1/32)
   -   -Etna-       -griersonianum (1/8)
   -   -   -   -dichroanthum ssp. dichroanthum (1/16)
   -Veta-   -Fabia-
   -            -griersonianum
   -venator (1/4)

Plant has trusses of 10 red flowers with yellow orange centers.
(V. J. Boulter & Sons, Australia, 1978)


cl.                  -catawbiense (1/4)
          -Mrs. Milner-
Desiderius-          -unknown (1/4)
          -smirnowii (1/2)

Frilled flowers, pure dark carmine pink with greenish markings.
(T. J. R. Seidel, 1902)


g.          -dichroanthum ssp. scyphocalyx (1/4)
          -Medusa-
Desna-          -griersonianum (3/4)
          -griersonianum

Flowers orange red. (Lord Aberconway, intro. 1946)


cl.              -griffithianum (1/4)
          -Sincerity-
Destiny-          -unknown (1/4)
          -thomsonii ssp. thomsonii (1/2)

Deep coral pink flowers 3.5in(9cm) wide; trusses hold about 12.
(Gen. Harrison, 1962)


g.                          -adenogynum (1/4)
          -Xenosporum (detonsum)-
Detonhaem-                   -unknown (1/4)
          -haematodes ssp. haematodes (1/2)

Introduced in 1932 by E. J. P. Magor.


cl.

Detonsum     See Xenosporum

cl.
          -fortunei ssp. discolor (1/2)
Devagilla-      -arboreum ssp. arboreum (1/4)
          -Cornubia-      -thomsonii ssp. thomsonii (1/8)
                    -Shilsonii-
                            -barbatum (1/8)

Rosy pink flowers on a tall compact hybrid. Midseason. (Rothschild, 1936)

g.
          -auriculatum (1/2)
Devaluation-
          -arboreum ssp. arboreum (1/2)

Tall, compact plant. Large white flowers, flushed pink. Late.
(Rothschild, 1936)

cl.
          -A. W. Hardy Hybrid--unknown (1/2)
Devonshire Cream-
          -campylocarpum ssp. campylocarpum (1/2)

3ft(.9m)          0°F(-18°C)          M          3/3

Creamy yellow flowers with a deep red blotch at base, in compact
ball-shaped trusses. Very slow growing, compact plant; prefers
partial shade. Dark green leaves are 2in(5cm) long. (Slocock,
1924) A. M. (RHS) 1940

cl.
          -Catanea--catawbiense var. album (1/2)
Dextanea-
          -white Dexter hybrid--unknown (1/2)

7ft(2.1m)          -20°F(-29°C)          M

White fragrant flowers, 3in(7.6cm) wide, in ball-shaped trusses
of 10. Plant rounded, well-branched, taller than wide; yellowish green leaves, held 2 years. (Hess cross; intro. by G. Nearing; reg. 1973)

DEXTER HYBRIDS OF UNKNOWN PARENTS     SEE APPENDIX D

cl.                      -Pygmalion--unknown
          -unnamed hybrid-
Dexter's Agatha-          -haematodes ssp. haematodes
          -            -fortunei ssp. fortunei
          -Wellfleet -
          (Dexter#8)-decorum

4ft(1.2m)          -5°F(-21°C)          M

(The exact combination of the parentage is uncertain) Flowers
purplish red, with sparse flecks of dark red, 3in(7.8cm) across,
6-7 lobes, in ball-shaped trusses of 10. Floriferous. Plant as
broad as high, well-branched; olive green leaves held 2 years.
(C. O. Dexter, 1925-42; John J. Tyler Arboretum, reg. 1981)

cl.
          -Pygmalion--unknown
Dexter's -          -haematodes ssp. haematodes
Harlequin-unnamed hybrid-
          -          -fortunei ssp. fortunei
          -Wellfleet-
          (Dexter#8)-decorum

5ft(1.5m)          -5°F(-21°C)          M

(The exact combination of the parentage is uncertain) Synonym
Harlequin. Flowers deep purplish pink at edges, paling to almost white in centers, 3.25in(8.3cm) wide, 5-7 very wavy lobes;
ball-shaped trusses hold 10-15. Floriferous. Plant upright and
rounded, nearly as broad as tall, with glossy olive green leaves
retained 2 years. (C. O. Dexter, 1925-1942; reg. by John J. Tyler Arboretum, 1981)  See color illus. improperly captioned as
Halesite in ARS Q 29:2 (Spring 1975): p. 102.

```
---                         -brookeanum var. gracile (1/4)
        -Duchess of Edinburgh-
Diadem-                      -longiflorum (lobbii) (1/4)
        -javanicum (1/2)
```

Vireya hybrid. Flowers pink to scarlet crimson. F. C. C. (RHS) 1896

cl.

Diadema      Parentage unknown

A parent of Herme, by Seidel.

cl.
```
                -yunnanense (1/2)
Diamond Wedding-
                -davidsonianum (1/2)
```

White flowers, suffused purple with light orange red spotting in upper throat, 2.5in(6.4cm) wide; up to 24 flowers, axillary and terminal, in firm rounded trusses. Glossy dark green leaves. (Maj. A. E. Hardy; reg. 1978) A. M. (RHS) 1978

cl.
```
                -yunnanense (1/2)
Diana Colville-
                -unknown (1/2)
```

Rose purple flowers, fading almost white at margins, rich brown spotting in throat, 2in(5cm) wide; terminal and axillary clusters form loose rounded trusses of about 25. Vigorous plant 6ft (1.8m) tall and slightly wider; leaves 3.5in(8.9cm) long, with scaly brown indumentum below. (Col. N. R. Colville; reg. 1968) P.C. (RHS) 1966; A. M. (Wisley Trials) 1972; F. C. C. (RHS) 1972

cl.
```
               -yakushimanum ssp. yakushimanum (1/2)
Diana Pearson-        -Margaret--unknown (1/4)
              -Glamour-
                       -griersonianum (1/4)
```

Margaret may be a griffithianum hybrid. Plant 2ft(.6m) by 2.5ft (.75m) wide in 12 years, compact and vigorous. Very pale pink flowers, lightly flushed with tinge of magenta, about 2in(5cm) wide; trusses of 13. See also Luise Verey. (A. F. George, Hydon Nurseries) A. M. (RHS) 1980

cl.
```
                           -griffithianum (1/8)
             -George Hardy-
        -Mrs. Lindsay Smith-
Diane-                      -catawbiense (1/8)
      -                -Duchess of Edinburgh--unknown (1/2)
      -             -campylocarpum ssp. campylocarpum (1/4)
      -unnamed hybrid-
                      -unknown
```

6ft(1.8m)        -5°F(-21°C)        EM        3/2

Crowded trusses of large cream yellow flowers flushed with primrose yellow. Upright, vigorous plant that prefers shade; glossy green foliage. (M. Koster & Sons, 1920) A. M. (Wisley Trials) 1948  See color illus. F p. 42.

cl.
```
           -macabeanum (1/2)
Diane Beekman-
           -unknown (1/2)
```

Creamy white flowers in trusses of about 16. (J. Beekman, 1980)

cl.
```
             -forrestii ssp. forrestii Repens Group (1/4)
        -Elizabeth-
Diane Lux-        -griersonianum (1/4)
      -       -campylocarpum ssp. campylocarpum (1/4)
        -Unique-
                -unknown (1/4)
```

1.5ft(.45m)        10°F(-12°C)        E

Flowers claret rose fading to Venetian pink, widely funnel-campanulate, 2in(5cm) wide, in high ball-shaped trusses of 12; very floriferous. Plant twice as broad as high; yellow green leaves, about 3in(7.6cm) long, held 2 years. (Edward F. Drake, cross 1962; intro. and reg. by Mrs. Vincent A. Lux, Jr., 1981)

cl.
```
            -Marinus-griffithianum (1/2)
       -Koster -
       -         -unknown (1/4)
Diane Titcomb-         -griffithianum
       -       -Halopeanum-
       -Snow -         -maximum (1/8)
       Queen-         -griffithianum
              -Loderi-
                      -fortunei ssp. fortunei (1/8)
```

6ft(1.8m)        0°F(-18°C)        M        4/3

A shrub of large foliage and fine trusses of white flowers with pink edging. (H. L. Larson, 1942) P. A. (ARS) 1958

g.
```
        -dichroanthum ssp. dichroanthum (1/2)
Dicharb-
        -arboreum ssp. arboreum (1/2)
```

Flowers of apricot. (E. J. P. Magor, intro. 1936)

g.
```
        -diaprepes (1/2)
Dichdiap-
        -dichroanthum ssp. dichroanthum (1/2)
```

Introduced by E. J. P. Magor in 1938.

cl.
```
      -dichroanthum ssp. dichroanthum (1/2)
Dido-
      -decorum (1/2)
```

3ft(.9m)        5°F(-15°C)        ML        3/4

Orange pink flowers with yellow center, in lax trusses. Sturdy, compact plant; light green leaves, stiff and rounded. (Wilding, 1934)

cl.
```
              -catawbiense (1/4)
        -Mrs. Milner-
Dietrich-       -unknown (1/4)
        -smirnowii (1/2)
```

Light purplish pink flowers with faint yellow brown markings. See also Dellla, Desiderius, Donar. (Seidel, 1902)

g.
```
              -griffithianum (1/4)
     -Glory of Leonardslee-
Dignity-             -unknown (1/4)
     -thomsonii ssp. thomsonii (1/2)
```

Rose red flowers. (Lord Aberconway, 1950)

g.
```
            -neriiflorum ssp. neriiflorum (1/4)
     -F. C. Puddle-
Dimity-      -griersonianum (1/4)
     -        -campylocarpum ssp. campylocarpum Elatum Group
     -Penjerrick-                              (1/4)
              -griffithianum (1/4)
```

Pale rose flowers. (Lord Aberconway, 1946)

cl.
```
        -fortunei ssp. discolor (1/2)
Ding Dong-
        -lacteum (1/2)
```

Silky white campanulate flowers, pale yellow center. (Sir John Bolitho, reg. 1962)  See color illus. ARS J 37:4 (Fall 1983): p. 186.

cl.
```
            -yakushimanum ssp. yakushimanum (1/2)
Dinty Moores-            -ponticum (1/4)
            -Purple Splendour-
                         -unknown (1/4)
```

27in(.69m)        -15°F(-26°C)          M

Flowers white with ruby red spots, 2in(5cm) wide, 5 wavy lobes,
held in trusses of 15-17.  Plant as wide as high;  leaves held 3
years.  (Dr. R. Rhodes cross; raised & intro. by M. Wildfong, R.
Behring; reg. 1980)

cl.
```
                    -wardii var. wardii (1/4)
        -Mrs. Lammot Copeland-        -souliei (?) (1/8)
Diny Dee-                   -Virginia Scott-
        -deep yellow hybrid--unknown      -unknown (5/8)
```

3ft(.9m)        0°F(-18°C)          M

Lemon yellow flowers open from buds of orange red in trusses
6in(15cm) across, of 9-12.  Plant almost as broad as high;  deep
green leaves, held 2 years.  (Cross 1969 by H. L. Larson; Joseph
Davis, intro. 1982; reg. 1984)

cl.
```
            -arboreum ssp. arboreum (1/4)
        -Red Argenteum-
Diogenes-            -grande (1/4)
        -calophytum var. calophytum (1/2)
```

Very large shrub with creamy pink flowers, dark blotch.  Foliage
and flowers resemble the species parent calophytum.  (Rothschild, 1936)

g.
```
        -neriiflorum ssp. neriiflorum (1/2)
Dione-          -arboreum ssp. arboreum (1/8)
    -       -Cornubia-      -thomsonii ssp. thomsonii
    -Cornsutch-      -Shilsonii-            (1/16)
    -                   -barbatum (1/16)
        -sutchuenense (1/4)
```

Flower color is carmine lake.  (E. J. P. Magor, 1936)

cl.
```
        -griffithianum (1/2)
Diphole Pink-
        -unknown (1/2)
```

Flowers deep rose pink or cerise pink, faintly spotted in brown.
(J. Waterer, Sons & Crisp)   A. M. (RHS) 1916

cl.

Direcktor E. Hjelm

A second generation fortunei hybrid.  Flowers dark carmine rose,
with bronze blotch.  (D. A. Koster, reg. 1958)

cl.
```
        -fortunei ssp. discolor (1/2)
Disca-        -decorum ? (1/4)
        -Caroline-
            -brachycarpum ssp. brachycarpum ? (1/4)
```

5ft(1.5m)        -10°F(-23°C)          ML        3/3

Fragrant flowers, white tinged with pink, frilled edges, held in
large dome-shaped trusses.  Plant habit good; a vigorous grower;
prefers light shade.  (Joseph B. Gable, 1944)    See color illus.
ARS J 38:1 (Winter 1984): p. 6;  VV p. 88.

```
* * * * * * * * * * * * * * * * * * * * * * *
* Hybrids of unknown parentage are described briefly in App. D *
* * * * * * * * * * * * * * * * * * * * * * *
```

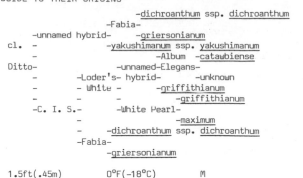

cl.  -
```
                -dichroanthum ssp. dichroanthum
            -Fabia-
-unnamed hybrid-      -griersonianum
            -yakushimanum ssp. yakushimanum
-       -            -Album  -catawbiense
Ditto-        -unnamed-Elegans-
-       -Loder's- hybrid-      -unknown
-       - White -      -griffithianum
-       -    -      -griffithianum
-C. I. S.-      -White Pearl-
-                   -maximum
-       -dichroanthum ssp. dichroanthum
        -Fabia-
        -griersonianum
```

1.5ft(.45m)        0°F(-18°C)          M

See Loder's White  for other possible parentage.  Flowers Egyptian buff with azalea pink overlay; lax, dome-shaped trusses of 5
to 7.  Free-flowering.  Rounded plant, almost as broad as tall;
glossy leaves retained only 1 year.  (Arthur A. Childers, 1982)

g. & cl.    -fortunei ssp. discolor (1/4)
```
    -Ladybird-
Diva-        -Corona--unknown (1/4)
    -griersonianum (1/2)
```

6ft(1.8m)        0°F(-18°C)          ML        3/2

Foliage with brown indumentum shows the griersonianum in the
parentage.  Medium-sized trusses of deep pink flowers with
specks of brown.  (Rothschild, 1936)  A. M. (RHS) 1937

cl.
```
                        -griffithianum (1/8)
                    -George-
        -Mrs. Lindsay Smith- Hardy-catawbiense (1/16)
    -Zuiderzee-            -Duchess of Edinburgh--
-   -       -                  unknown (1/4)
-   -       -      -campylocarpum ssp. campylo-
Dixie Lee-      -unnamed hybrid-            carpum (1/8)
  Ray  -                  -unknown
-                   -griffithianum
-               -Kewense-
-       -Aurora-      -fortunei ssp. fortunei (5/16)
-Naomi Pink-      -thomsonii ssp. thomsonii (1/8)
  Beauty  -fortunei ssp. fortunei
```

Flowers of orchid pink with slight red spotting, a white throat,
4.5in(11.5m) wide, 7 wavy lobes; high dome-shaped trusses of 14-
17.  Floriferous.  Plant upright, branching moderately, as wide
as tall; glossy leaves 5in(12.7cm) long, held 3 years.  (Larson,
1979)

cl.
```
    -yakushimanum ssp. yakushimanum (1/2)
Doc-
    -unknown (1/2)
```

3ft(.9m)        -15°F(-26°C)          M

Flowers rose pink  with deeper rose on margins,  spotting on upper lobe,  corolla 1.5in(3.5cm) across;  rounded trusses hold 9.
Plant vigorous, upright, with compact growth habit.   (John Waterer, Sons & Crisp, 1972)   H. C. (RHS) 1978

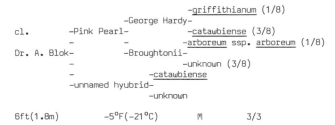

cl.
```
                        -griffithianum (1/8)
                    -George Hardy-
cl.     -Pink Pearl-      -catawbiense (3/8)
-       -            -arboreum ssp. arboreum (1/8)
Dr. A. Blok-      -Broughtonii-
-                   -unknown (3/8)
-               -catawbiense
-unnamed hyubrid-
                -unknown
```

6ft(1.8m)        -5°F(-21°C)          M        3/3

Synonym  Dr. O. Blok.  Large pink flowers with a lighter center,
pale yellow marking on upper lobe; in round trusses.  Plant has
good upright habit; medium green leaves  6in(15.2cm) long.   (L.
Endtz & Co.)   A. M. (RHS) 1937   See color illus. VV p. 32.

```
cl.                      -dichroanthum ssp. dichroanthum (1/4)
              -Goldsworth Orange-
Dr. Ans-                 -fortunei ssp. discolor (1/4)
Heyting-
         -Stanley Davies--unknown (1/2)
```

A Dutch hybrid with flowers of deep salmon pink and an orange
brown blotch. (Boskoop Research Sta. for Arboriculture, 1982)

```
                                   -griffithianum (1/8)
                  -George Hardy-
cl.      -Pink Pearl-              -catawbiense (3/8)
         -           -            -arboreum ssp. arboreum (1/8)
Dr. Arnold-         -Broughtonii-
W. Endtz -                        -unknown (3/8)
         -                -catawbiense
         -unnamed hybrid-
                  -unknown
```

Flowers fringed, carmine red, and tinted lilac.  Plant similiar
to Dr. A. Blok.  (L. Endtz & Co., 1937)

```
cl.            -oreotrephes (1/4)
         -Goldstrike-            -cinnabarinum ssp. cinnabarinum
Doctor -          -Royal Flush-                        (1/8)
Birdbath-                       -maddenii ssp. maddenii (1/8)
         -cinnabarinum ssp. xanthocodon (1/2)
```

5ft(1.5m)           10°F(-12°C)          EM

Flowers are Chinese yellow with orange shading, tubular bell-
shaped in lax trusses of 8.  Plant branches well with leaves
3in(7.6cm) long, very aromatic, held 3 years.  (Carl G. Heller,
1980)

cl.

Doctor Bowman--selected arboreum lying between ssp. arboreum
              and ssp. delavayi  (K-W 21976)

2.5ft(.75m)         10°F(-12°C)          Dec.-Feb.

Cardinal red flowers with plum purple nectaries at base; trusses
hold 12.  Spreading plant habit, over 3 times as broad as tall;
silvery indumentum on mature leaves.  (Collected by F. Kingdon-
Ward in Burma, 1956; raised by Dr. P. Bowman, et al.; reg. by E.
Gorman, 1982) See color illus. ARS J 38:2 (Spring 1984): p. 55.

```
                         -neriiflorum ssp. neriiflorum
              -unnamed-Nereid-                   (1/16)
              - hybrid-    -dichroanthum ssp. dichroan-
cl.      -Blondie-       -                        thum (1/16)
         -           -   -fortunei ssp. discolor (1/8)
Doctor Bruce-        -   -maximum (1/8)
   Bradley*  -     -Russell Harmon-
         -           -catawbiense (1/8)
         -Catalgla--catawbiense var. album Glass (1/2)
```

Pale yellow flowers.  (Orlando S. Pride, 1977)

```
                         -griffithianum (1/16)
              -George Hardy-
         -Pink Pearl-    -catawbiense (7/16)
         -          -    -arboreum ssp. arboreum
cl.      -Marion-    -Broughtonii-           (1/16)
         -     -          -unknown (7/16)
Dr. C. H.-     -          -catawbiense
   Felix   -  -Catawbiense Grandiflorum-
         -                        -unknown
         -                -catawbiense
         -Prof. C. S. Sargent-
                        -unknown
```

Flowers fuchsine pink.  (Felix & Dijkhuis, 1951)

```
cl.                      -catawbiense (1/4)
              -Atrosanguineum-
Dr. H. C. Dresselhuys-   -unknown (1/2)
         -          -arboreum ssp. arboreum (1/4)
         -Doncaster-
                  -unknown
```

6ft(1.8m)           -15°F(-26°C)          M          2/3

Flowers of aniline red with a lighter blotch, in tall trusses.
Plant upright; foliage fairly good but rather scant.  See also
Dr. H. J. Lovink.  (H. den Ouden & Son, 1920)

```
                         -Queen     -griffithianum
              -Britannia-Wilhelmina-         (1/16)
g.            -Linswegeanum-        -     -unknown (3/16)
Dr. H. C. Karl-          -    -Stanley Davies--unknown
   Forster    -          -forrestii ssp. forrestii
              -               Repens Group (1/4)
              -williamsianum (1/2)
```

Flowers deep rose to scarlet.  (Dietrich Hobbie, 1952)

```
cl.                      -catawbiense (1/4)
              -Atrosanguineum-
Dr. H. J. Lovink-        -unknown (1/2)
              -    -arboreum ssp. arboreum (1/4)
              -Doncaster-
                  -unknown
```

5ft(1.5m)           -15°F(-26°C)          M          1/3

Flowers of aniline red.  See also Dr. H. C. Dresselhuys. (H. den
Ouden & Son, 1929)

```
cl.
         -arboreum ssp. arboreum (1/2)
Doctor Henry Wade-
         -griersonianum (1/2)
```

Lilac pink flowers with darker edges.  (Mrs. Henry Wade, 1972)

```
cl.
         -phaeopeplum (1/2)
Dr. Herman Sleumer-
         -zoelleri (1/2)
```

A vireya hybrid from Australia,  named for the botanist who introduced many vireyas.  A parent of Clipsie, q.v.  Flowers light
currant red. (T. Lelliott, 1972)   See color illus. ARS Q 29:4
(Fall 1975): p. 238.

```
              -wardii var. wardii (1/4)
         -Hawk-
cl.      -  -Lady     -fortunei ssp. discolor (1/8)
         -  -Bessborough-
Dr. John-          -campylocarpum ssp. campylocarpum
   Yeates -                        Elatum Group (1/8)
         -          -griffithianum (1/8)
         -     -Loderi-
         -Ilam Cream-  -fortunei ssp. fortunei (1/8)
              -unknown (1/4)
```

A hybrid with trusses holding 15-16 flowers, magenta rose in
bud, opening to rhodamine pink.  (Cross of Dr. J. S. Yeates;
reg. by Mr. & Mrs. R. Perry, 1981)

```
                         -maximum (1/8)
              -unnamed hybrid-
         -Tahiti-        -catawbiense (1/8)
cl.      -  -unnamed-dichroanthum ssp. dichroanthum (1/8)
         -     hybrid-
Doctor Lewis-        -unnamed-fortunei ssp. discolor
   Santini*  -          hybrid-            (1/16)
         -               -campylocarpum ssp. campylo-
         -                        carpum (1/16)
         -Catalgla--catawbiense var. album Glass (1/2)
```

6ft.(1.8m)          -25°F(-32°C)          ML

Mr. Pride described this as a very hardy plant with pale yellow
flowers, in an informal truss.  (Orlando S. Pride, 1975)

cl.                    -Mira--unknown (1/4)
          -Leopold-
Dr. Masters-         -catawbiense (1/4)
          -azalea japonicum (1/2)


Azaleodendron.  One parent may be  Prince Camille de Rohan,  a
caucasicum hybrid.  Bright pink in bud, opening salmon with a
yellowish tinge.  (G. Vander Meulen, 1892)


cl.

Dr. O. Blok       See Dr. A. Blok


cl.                              -ciliicalyx (1/4)
                    -Else Frye-
Doctor Richard Anderson-         -unknown (1/4)
                    -johnstoneanum (1/2)

3ft(.9m)          15°F(-26°C)          E


Large white flowers, to 4.25in(10.8cm) wide, edges blushed pink,
an orange blotch; trusses of about 6.  Plant upright, with stiff
branches; medium-sized glossy leaves.  (R. Anderson & H. Braafladt, 1979)


cl.                              -caucasicum (1/4)
                    -Boule de Neige-         -catawbiense
Dr. Richard Meriam*         -unnamed hybrid-      (1/8)
                    -              -unknown (1/8)
                    -Mt. Siga--vernicosum (1/2)

4ft(1.2m)          -20°F(-29°C)          M


Clear pink flowers in loose trusses.  Plant very compact.  See
also William Fleming.  (Orlando S. Pride, 1977)


cl.
          -griersonianum (1/2)
Dr. Ross-                    -arboreum ssp. arboreum (1/8)
     -              -Doncaster-
     -Borde Hill-         -unknown (1/4)
          -              -griffithianum (1/8)
          -Mrs. A. M. Williams-
                    -unknown


Deep maroon in bud, with flowers bright red, funnel-shaped, in
rounded trusses.  Plant habit spreading; pointed, medium green
leaves 6in(15.2in) long.  (Rudolph Henny; reg. 1958)


cl.
          -caucasicum (1/2)
Dr. Stocker-
          -griffithianum (1/2)

5ft(1.5m)          5°F(-15°C)          M          3/3


Plant is broader than tall, with loose trusses of large campanulate flowers of ivory white, 3in(7.6cm) wide, with faint brownish red markings.  Roots easily; pest-resistant.  (Raised by G.
Abbey, gardener to Dr. Stocker of Avery Hill; intro. by Veitch)
A. M. (RHS) 1900


cl.

Doctor Tjebbes       Parentage unknown

Flowers are roseine purple, held in trusses of 14-16.  Leaves
are  dark green, 5.5in(14cm) long.  Blooms late and  considered
very hardy in Holland.  (C. A. van den Akker) A. M. (Boskoop)
1967


cl.                              -catawbiense (1/4)
                    -Charles Dickens-
Dr. V. H. Rutgers-               -unknown (3/4)
                    -Lord Roberts--unknown

5ft(1.5m)          -15°F(-26°C)          ML          3/3


Flowers bright crimson, fringed. Plant broad, dense; leaves dark
green. (H. den Ouden & Son, 1925)   See color illus. VV p. 66.


cl.                              -griffithianum (1/4)
                    -Queen Wilhelmina-
Dr. W. F. Wery-                  -unknown (3/4)
                    -Stanley Davies--unknown


Flowers of scarlet red.  See also Britannia. (C. B. van Nes &
Sons)


cl.                              -caucasicum (1/4)
                    -Boule de Neige-         -catawbiense (1/8)
Doctor William-               -unnamed hybrid-
   Fleming*        -                  -unknown (1/8)
                    -Mt. Siga--vernicosum (1/2)

4ft(1.2m)          -10°F(-23°C)          M


Flowers of a very pretty peach color; compact plant habit.   See
also Doctor Richard Meriam.  (Orlando S. Pride, 1975)


cl.

Doker-la--form of primuliflorum var. cephalanthoides
               (Not now recognized as a separate species by the RBG,
               Edinburgh.)

A compact truss of 10-12 flowers, spirea red, paling to near
white at edges.  Inside corolla tube  densely hairy.  Aromatic
leaves up to 1in(2.5cm) long, glossy dark green, with heavy indumentum.  (Raised by Sir James Horlick; reg. by  P. A. Cox)
A. M. (RHS) 1980


cl.
   -Corona--unknown (5/8)
Doll-     -fortunei ssp. discolor (1/4)
   -Dondis-     -arboreum ssp. arboreum (1/8)
          -Doncaster-
               -unknown


Clear bright pink flowers, in high trusses.  (R. Henny, intro.
1956; reg. 1958)


cl.                    -fortunei ssp. discolor (1/4)
     -Sir Frederick-          -arboreum ssp. zeylanicum
Dollar  -  Moore     -St. Keverne-               (1/8)
Princess-                  -griffithianum (1/8)
          -          -griersoniaum (1/4)
          -Tally Ho-
               -facetum (eriogynum) (1/4)


A tall upright plant with neat trusses of very deep carmine pink
flowers.  (Edmund de Rothschild, 1966)


g.                    -griffithianum (1/4)
     -Dawn's Delight-
Dolly-               -unknown (1/4)
     -griersonianum (1/2)


Pink flowers held in drooping trusses.  Plant habit tall, open.
(Rothschild, 1940)


cl.
               -catawbiense var. album (1/2)
Dolly Madison-          -fortunei ssp. fortunei (1/4)
          -unnamed hybrid-     -arboreum ssp. arboreum
                    -unnamed-               (1/8)
               hybrid-griffithianum (1/8)

7ft(2.1m)          -20°F(-29°C)          M          4/4


Flowers white  with reddish brown dorsal blotch, openly campanulate, 5-lobed, 3.25in(8.3cm) across, in full trusses of 12 or 13
blossoms.  Leaves flat, dark green, semi-glossy; average foliage
density.  (David G. Leach, 1971)

cl.
       -dichroanthum ssp. dichroanthum (1/2)
Dolphin-
       -crinigerum var. crinigerum (1/2)

Salmon pink flowers, hose-in-hose, 2.5in(6.4cm) across; 12 held
in flat trusses. (Gen. Harrison, 1962)

                            -arboreum ssp. arboreum (1/8)
            -Glory of Penjerrick-
g.  -Barclayi-             -griffithianum (1/4)
          -thomsonii ssp. thomsonii (1/2)
Domino-                -griffithianum
-      -Glory of Leonardslee-
  -Dignity-             -unknown (1/8)
         -thomsonii ssp. thomsonii

Red flowers. (Lord Aberconway, 1950)

cl.
       -griffithianum (1/2)
Don Ernesto-      -arboreum ssp. arboreum (1/4)
       -Doncaster-
             -unknown (1/4)

The Don g. Flowers of rich, rosy scarlet. (Lowinsky) A. M.
(RHS) 1920

cl.
       -laetum (1/2)
Don Stanton-
       -macgregoriae (1/2)

A vireya hybrid for the cool greenhouse. Flowers are Indian
yellow held in trusses of 10-11, tubular bell-shaped. Leaves
lightly scaly, glossy, dark green, up to 3in(7.6cm) long. (Seed
from Don Stanton, Australia; reg. E. F. Allen) A. M. (RHS) 1981

cl.
       -griffithianum (1/2)
Dona Tizia-      -arboreum ssp. arboreum (1/4)
       -Doncaster-
            -unknown (1/4)

The Don g. Flowers pale pink fading to white, deeper pink outside. Bright red pedicels. (Lowinsky) A. M. (RHS) 1921

cl.
       -lochiae (1/2)
Donald Stanton-
       -laetum (1/2)

A vireya hybrid for the cool greenhouse. Drooping flowers, Turkey red, 5-lobed, tubular funnel-shaped; trusses hold 9-10.
Glossy dark green leaves 3.5in(9cm) long, slightly scaly. (D.
Stanton, Australia; exhibited & reg. by RBG, Kew) A. M. (RHS)
1978

cl.
          -griffithianum (1/4)
       -Alice-
Donald Waterer-   -unknown (1/2)
-            -catawbiense (1/4)
      -Gomer Waterer-
              -unknown

Deep rose pink flowers, paler in center. (J. Waterer, Sons &
Crisp) A. M. (RHS) 1916

cl.
         -catawbiense (1/4)
  -Mrs. Milner-
Donar-       -unknown (1/4)
  -smirnowii (1/2)

Frilled flowers, light carmine red with white throat and very
dark red markings. (T. J. R. Seidel, 1902)

cl.
       -arboreum ssp. arboreum (1/2)
Doncaster-
       -unknown (1/2)

3ft(.9m)        -5°F(-21°C)       M     2/3

Brilliant crimson scarlet flowers, faint markings, 2.5in(6.4cm)
wide, in rounded trusses. Plant broadly dome-shaped; very dark
green, glossy leaves. Easy to grow and heat-tolerant. A standby since the late 1800s. (A. Waterer)

g.
       -fortunei ssp. discolor (1/2)
Dondis-      -arboreum ssp. arboreum (1/4)
    -Doncaster-
           -unknown (1/4)

A parent of Cathy, Confection. (RBG, Kew)

cl.
       -griffithianum (1/2)
Donna Anita-      -arboreum ssp. arboreum (1/4)
       -Doncaster-
            -unknown (1/4)

The Don g. Shell pink flowers. (Lowinsky) A. M. (RHS) 1920

cl.
       -griffithianum (1/2)
Donna Florenza-      -arboreum ssp. arboreum (1/4)
       -Doncaster-
            -unknown (1/4)

The Don g. Flowers of deep rich rose. (Lowinsky) A. M. (RHS)
1920

cl.
       -fortunei ssp. fortunei (1/2)
Donna Hardgrove-     -wardii var. wardii (1/4)
       -unnamed-
          hybrid-dichroanthum ssp. dichroanthum (1/4)

3ft(.9m)        -5°F(-21°C)       ML

Plant rounded, well-branched, broader than tall; leaves held 1
year. Flowers apricot pink, flushed yellow, 2.5in(6.4cm) wide;
lax trusses of 8. Floriferous. (D. Hardgrove & S. Burns, 1978)

cl.
       -racemosum ? (1/2)
Donna Tullen*
       -unknown (1/2)

4ft(1.2m)       -5°F(-21°C)       EM     3/3

A pretty, medium-sized plant with attractive foliage like that
of racemosum. Flowers pink, with a touch of cream. (Origin unknown)

cl.
       -campylocarpum ssp. campylocarpum (1/4)
   -Unique-
Donnie-    -unknown (1/4)
  -wightii (1/2)

Flowers cream-colored, spotted with port wine red on upper lobe;
15-flowered trusses. (Sir James Horlick, 1962)

cl.
       -haematodes ssp. chaetomallum (1/2)
Donnington-     -neriiflorum ssp. neriiflorum Euchaites Group
     -Portia-                  (1/4)
           -strigillosum (1/4)

Loose trusses of 9 cardinal red flowers, 2in(5cm) wide, funnel-
campanulate. Leaves 4.5in(11.5cm) long, with loose light brown
indumentum beneath. (Crown Estate, Windsor) P. C. (RHS) 1966

cl.
                    -yakushimanum ssp. yakushimanum (1/2)
Donvale Pearl-          -arboreum ssp. arboreum (1/4)
              -unnamed hybrid-
                               -unknown (1/4)

Trusses hold 28 flowers, bright spinel red, tubular bell-shaped.
(R. J. O'Shannosy, 1983)


cl.                 -scabrifolium var. spinuliferum (1/4)
              -Crossbill-
Donvale Pink-      -lutescens (1/4)
 Drift    -scabrifolium var. spiciferum (1/2)

Trusses of 5-10 flowers, tubular funnel-shaped, bright neyron
rose. Height about 3ft(.9m). (J. O'Shannassy, reg. 1984)


                                   -griffithianum
                         -Queen    -          (1/16)
                  -Britannia- Wilhelmina-unknown (5/16)
cl.               -
          -Lamplighter-        -Stanley Davies--unknown
Donvale Ruby-      -             -fortunei ssp.
          -       -Madame Fr. J. Chauvin-  fortunei (1/8)
            -arboreum ssp. arboreum (1/2)    -unknown

Trusses of 14-16 flowers, tubular bell-shaped, scarlet red.
Height about 3ft(.9m). (J. O'Shannassy, reg. 1984)


                              -facetum (eriogynum) (1/8)
                    -unnamed hybrid-
cl. -unnamed hybrid-        -unknown (1/8)
    -              -dichroanthum ssp. dichroanthum (1/4)
Dopey-       -Fabia-
    -              -griersonianum (1/4)
    -         -yakushimanum ssp. yakushimanum (1/4)
    -unnamed hybrid-
              -Fabia  -dichroanthum ssp. dichroanthum
              -Tangerine-
                    -griersonianum

3ft(.9m)        5°F(-15°C)         M

Flowers glossy red, paling toward margins with dark brown spots
on upper lobe, campanulate, to 2.5in(6.4cm) across; 16-flowered
spherical trusses. Floriferous. Plant upright, compact, vigorous; dull green leaves 4in(10.2cm) long. (Waterer, Sons & Crisp,
1971)   A. M. (RHS) 1977; F. C. C. (RHS) 1979


cl.
          -minus var. minus Carolinianum Group (1/2)
Dora Amateis-
          -ciliatum (1/2)

3ft(.9m)        -15°F(-26°C)         EM        4/4

White flowers lightly spotted green, about 2in(5cm) across, in
clusters of 3-6. Floriferous. Plant vigorous, twice as wide as
high; deep green, dense foliage with bronze highlights if grown
in full sun, and aromatic. (Edmond Amateis, 1955)   A. E. (ARS)
n.d.;   P. C. (RHS) 1972; A. M. (RHS) 1976; F. C. C. (RHS) 1981
See color illus. HG p. 99; VV p. 16.


g.              -thomsonii ssp. thomsonii (1/4)
      -Bagshot Ruby-
Dorcas-           -unknown (1/4)
      -fortunei ssp. discolor (1/2)

Rose pink flowers; medium-sized trusses on a tall plant. Midseason. (Rothschild, 1936)


cl.
      -Jay Gould--unknown (1/2)
Doria*
      -japonicum var. japonicum (metternichii) (1/2)

Flowers salmon pink with yellow markings. (T. J. R. Seidel)


g. & cl.
          -griersonianum (1/2)
Dorinthia-      -haematodes ssp. haematodes (1/4)
          -Hiraethlyn-
                    -griffithianum (1/4)

Flowers of clear shiny red  with deep wavy lobes. (Lord Aberconway, 1938)   F. C. C. (RHS) 1938


cl.

Doris Bigler  Complex parentage including catawbiense, maximum,
              and possibly unknown hybrids

6ft(1.8m)        -15°F(-26°C)         M

Small flowers 2in(5cm) across, of light purple with strong reddish purple edging and yellow blotch, 5 very frilled lobes; 17-
flowered trusses, ball-shaped. Free-flowering. Plant rounded,
well-branched; rather glossy, olive green leaves, held 2 years.
(W. D. Smith, 1979)


cl.              -griffithianum (1/2)
          -Loderi-
Doris Caroline-      -fortunei ssp. fortunei (1/4)
          -          -griffithianum
          -Lady Bligh-
                    -unknown (1/4)

Truss of 10-12 upright flowers, rose colored; 6-7 lobes.  Plant
habit upright and tall with leaves 6in(15cm) long. (Rudolph
Henny, reg. 1961) P. A. (ARS) 1960


cl.              -Moser's Maroon--unknown (1/2)
      -Romany Chai-
Doris Moss-      -griersonianum (1/4)
          -      -ponticum (1/4)
          -Purple Splendour-
                    -unknown

A hybrid from Wales with flowers ruby red.  See also Gwerfyl
Moss. (William Moss, 1975)


cl.              -griffithianum (1/4)
      -Dawn's Delight-
Dormouse-           -unkown (1/4)
      -williamsianum (1/2)

3ft(.9m)        0°F(-18°C)         EM        3/4

Forms a neat, compact, dome-shaped bush. Bell-shaped flowers of
delicate pink with deeper edges, in loose clusters. Oval leaves
about 3in(7.6cm) long, copper-colored when young.  (Rothschild,
1936)


cl.
      -griffithianum (1/2)
Dorothea-
      -decorum (1/2)

Flowers of white flushed pink, with green center.  Very sweetly
fragrant. (Lowinsky 1925) A. M. (RHS) 1925


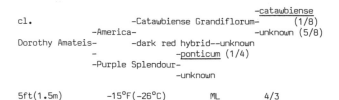
cl.                             -catawbiense
                -Catawbiense Grandiflorum-      (1/8)
          -America-                -unknown (5/8)
Dorothy Amateis-      -dark red hybrid--unknown
          -              -ponticum (1/4)
          -Purple Splendour-
                    -unknown

5ft(1.5m)        -15°F(-26°C)         ML        4/3

Large trusses of rosy purple flowers with a deeper purple eye, 3
in(7.6cm) across.  Rounded plant, wider than high; leaves matte
olive green 5in(12.7cm) long, held in upright position. (Raiser
Edmond Amateis; W. Baldsiefen, intro. 1971) See color illus. HG
p. 141.

'Cup Day' by K. Van de Ven
Photo by Greer

'Curlew' by Cox
Photo by Greer

'Cyprus' by Leach
Photo by Leach

'David Harris' by Delp
Photo by Delp

'Day Dream' (biscuit form) by Rothschild
Photo by Greer

'Day Dream' (red form) by Rothschild
Photo by Greer

'Denali' by J. A. Elliott
Photo by Greer

'Devonshire Cream' by Slocock
Photo by Greer

'Direcktor E. Hjelm' by D. A. Koster
Photo by Greer

'Dixy Lee Ray' by Larson
Photo by Greer

'Doc' by J. Waterer & Crisp
Photo by Greer

'Dr. Arnold W. Endtz' by Endtz
Photo by Greer

'Dr. Tjebbes' by van den Akker
Photo by Greer

'Dolly Madison' by Leach
Photo by Leach

'Donna Totten' by unknown
Photo by Greer

'Dopey' by J. Waterer & Crisp
Photo by Greer

'Dora Amateis' by Amateis
Photo by Greer

'Dorinthia' by Aberconway
Photo by Greer

'Dormouse' by Rothschild
Photo by Greer

'Dorothy Robbins' by Brandt
Photo by Greer

'Double Date' (double-flowered) by Whitney
Photo by Greer

'Douglas R. Stephens' by J. F. Stephens
Photo by Greer

'Drum Major' by Rothschild
Photo by Greer

cl.                          -griffithianum (1/4)
          -Jean Marie de Montague-
Dorothy Lee-                      -unknown (1/4)
          -sanguineum ssp. sanguineum var. haemaleum (1/2)

Semi-dwarf plant; dark green leaves 2.5in(6.4cm) long. Oxblood
red flowers, tubular campanulate; loose trusses of 10-12. (C.
S. Seabrook, 1968)


cl.              -forrestii ssp. forrestii Repens Group (1/4)
          -Elizabeth-
Dorothy -          -griersonianum (1/4)
Lonsdale-Barclayi -thomsonii ssp. thomsonii (1/4)
          Helen Fox-              -arboreum ssp. arboreum
                    -Glory of Penjerrick-            (1/8)
                              -griffithianum (1/8)

An Australian hybrid. Scarlet red flowers, with slight spotting
on the upper lobes. (H. D. Rose, 1973)


cl.
                    -Vincent van Gogh--unknown (1/2)
Dorothy Peste Andersen-
              -Loderi      -griffithianum (1/4)
                King George-
                          -fortunei ssp. fortunei (1/4)

4ft(1.2m)          5°F(-15°C)          ML

Tight trusses of 24 fragrant flowers, light solferine purple, 7-
lobed. Plant as wide as tall, with leaves up to 8in(20cm) long.
(Fred Peste cross, 1975; reg. by Anne Sather, 1985)


cl.
       -campylocarpum ssp. campylocarpum (Hooker's form) (1/2)
Dorothy-
Robbins-           -fortunei ssp. discolor (1/4)
          -Margaret Dunn-      -dichroanthum ssp. dichroanthum
                    -Fabia-                    (1/8)
                        -griersonianum (1/8)

A compact plant 3ft(.9m) tall and equally as broad at 10 years;
leaves 4in(10.2cm) long. Flowers 4in(10.2cm) wide, 5-lobed, of
empire yellow. (Lester E. Brandt, 1970)


cl.

Doshong La--form of mekongense var. mekongense Viridescens Group

3ft(.9m)          0°F(-18°C)          ML          3/3

Rounded trusses of 4-5 flowers of light yellow, flushed old rose
at the tips, flecked with olive green, 1.5in((3.8cm) wide. Aromatic small leaves tinged bronze when young; foliage blue green
until autumn; first year wood is green, later red brown. (Seed
by Kingdon-Ward, Doshong La Pass, Tibet; selected and raised by
E. H. M. & P. A. Cox) A. M. (RHS) 1972


cl.                          -griffithianum (1/8)
                    -George Hardy-
     -Mrs. Lindsay Smith-          -catawbiense (1/8)
Dot-                    -Duchess of Edinburgh--unknown (1/4)
       -fortunei ssp. fortunei (1/2)

5ft(1.5m)          -10°F(-23°C)          M          4/3

Large white flowers of unusual satiny texture, spotted, in big
trusses. Plant habit upright, open. (Lord Swaythling, 1945)
A. M. (RHS) 1945


cl.

Double Diamond--form of johnstoneanum (double form)

Unusual, fully double flowers like creamy yellow gardenias, with
trusses of 2 or 3 flowers, each corolla 4in(10.2cm) across, 3in.
(7.6cm) long. (Crown Estate, Windsor) A. M. (RHS) 1956


            -griersonianum (3/8)
          -                    -griffithianum (9/64)
     -Anna Rose-          -George-
     - Whitney -     -Pink - Hardy-catawbiense (9/64)
cl.  -          -          -Pearl-          -arboreum ssp.
     -     -Countess-     -Broughtonii-     arboreum (3/64)
     -     of Derby-          -unknown (19/64)
Double-     -          -catawbiense
Drake-          -Cynthia-
     -               -griffithianum
     -          -griersonianum
     -                         -griffithianum
     -     -Anna Rose-          -George-
     -     - Whitney -     -Pink - Hardy-catawbiense
     -unnamed-     -          -Pearl-          -arboreum ssp.
       hybrid-          -Countess-     -Broughtonii-
          -unknown-     of Derby-          -unknown
                    -          -catawbiense
                         -Cynthia-
                              -griffithianum

3ft(1.2m)          10°F(-12°C)          M

Red flowers with slight greyed orange spotting, tubular funnel-
shaped, 2in(5cm) wide, 5 frilled lobes. Floriferous. Plant as
broad as tall, rounded, with arching branches; narrow leaves are
held 2 years. (Edward Drake, cross 1965; Diane Lux, intro. and
reg. 1981)


cl.
       -unknown (3/4)
Double Dutch-          -wardi var. wardii (1/4)
          -Carolyn Grace-
                    -unknown

2.5ft(.75m)          15°F(-10°C)          EM

Flowers medium pink with light shell pink throat and stripe down
center of lobes, deep pink edging; 2.5in(6.4cm) wide, 7 frilled
lobes, hose-in-hose. Plant well-branched, about as wide as tall;
matte olive green leaves held 2 years. (Cross by Benj. Briggs;
intro. & reg. Edgar L. Knight, 1980)


cl.              -wardii var. wardii (1/4)
          -Carolyn Grace-
Doubloons-          -unknown (1/4)
       -          -campylocarpum ssp. campylocarpum (1/4)
       -Moonstone-
                    -williamsianum (1/4)

4ft(1.2m)          -5°F(-21°C)          EM          3/4

A rounded, compact plant. Creamy yellow flowers, to 5in(12.7cm)
across, widely funnel-shaped; loose trusses of 10. (A. Wright,
1962)


cl.                          -griffithianum (1/4)
              -unnamed hybrid-
Douglas McEwan-          -unknown (3/4)
          -Monsieur Thiers--unknown

Rosy red flowers. (C. B. van Nes & Sons, before 1958)


cl.                              -griffithianum (1/4)
              -Jean Marie de Montague-
Douglas R. Stephens-                    -unknown (3/4)
              -unnamed white hybrid--unknown

5ft(1.5m)          -10°F(-23°C)          EM

Rose red buds open to spinel red flowers with dark red blotch in
throat and dorsal spotting, 4.5in(11.5cm) wide, 5 wavy lobes; 14
in tall conical trusses. Upright plant, broader than tall; dark
green, glossy leaves, 7in(17.8cm) long. See also Freeman R. Stephens. (Mr. & Mrs. J. Freeman Stephens, 1973)

```
            -King of-fortunei ssp. discolor (1/4)
cl.         -Shrubs -    -dichroanthum ssp. dichroanthum
      -         -Fabia-                          (1/8)
Doug's Greeneyes-        -griersonianum (3/8)
      -                        -Corona-unknown (1/8)
      -       -unnamed hybrid-    -griffithianum
      -Flame-              -Loderi-        (1/16)
            -griersonianum        -fortunei ssp.
                                    fortunei (1/16)
```

8ft(2.4m)        0°F(-18°C)         M

Flowers in loose trusses of 13-15, 7 lobes, neyron rose with
green center.  Plant habit tall and open  with  leaves 5.5in
(13.3cm) long, dark green. (Halfdan Lem cross; reg. by Britt
Smith, 1972)

cl.

Down Under--form of lochiae

Vireya species.  Flowers tubular funnel-shaped, a light shade of
geranium lake, in clusters of 2-7. (Crown Estate, Windsor, 1962)
A. M. (RHS) 1957

```
cl.            -neriiflorum ssp. neriiflorum (1/4)
      -F. C. Puddle-
Dragon-          -griersonianum (1/4)
      -haematodes ssp. haematodes (1/2)
```

See also Phoebus, sibling seedling. (Lord Aberconway, 1941)

```
g.
            -auriculatum (1/2)
Dragonfly-
            -facetum (eriogynum) (1/2)
```

A late-flowering plant that grows to a large, dense shrub, with
long, narrow, hairy leaves.  Flowers resemble auriculatum, but
are carmine rather than pink or white. (Rothschild, 1936)

```
cl.
            -wardii var. wardii (1/2)
Drake's Orchid-
            -fortunei ssp. discolor (1/2)
```

4ft(1.2m)        10°F(-12°C)        ML

Lavender flowers with mimosa yellow throat, greyed orange spotting, openly funnel-shaped, 2.5in(6.4cm) across, 5 to 7 frilled
lobes.  Upright plant with stiff branches; glossy leaves held 2
years. (Ed. Drake, cross 1971; Diane A. Lux, intro., reg. 1981)

```
              -Lady    -fortunei ssp. discolor (3/8)
g. & cl.  -Day Dream-Bessborough-
      -                -campylocarpum ssp. campylocarpum
Dream Girl-    -griersonianum (3/8)      Elatum Group (1/8)
      -            -fortunei ssp. discolor
      -Margaret Dunn-    -dichroanthum ssp. dichroanthum
            -Fabia-                          (1/8)
                  -griersonianum
```

Flowers orange buff with a blood red throat.  See also Gold
Mohur, Mohur, Golden Pheasant, Shah Jehan. (L. E. Brandt, 1953)

```
cl.                -mauve seedling--unknown (1/2)
            -A. Bedford-
Dream of Kings-        -ponticum (1/2)
      -                    -ponticum
            -Purple Splendour-
                        -unknown
```

4ft(1.2m)        0°F(-18°C)         ML

Lavender flowers with large purple blotch, 3.5in(8.9cm) across,
openly funnel-shaped, 5 wavy lobes; trusses of 16-20.  Plant habit upright, as broad as tall; glossy leaves 6in(15.2cm) long,
retained 2 years.   (Mrs. Halsey A. Frederick, Jr., reg. 1977)

```
cl.
            -ciliatum (1/2)
Dresden China-        -edgeworthii (bullatum) (1/4)
      -Sesterianum-
                  -formosum var. formosum ? (1/4)
```

White flowers.  (Gen. Harrison)

```
cl.         -griffithianum (1/4)
      -Mars-
Dress Up-    -unknown (1/4)
      -          -griersonianum (1/4)
      -Tally Ho-
                  -facetum (eriogynum) (1/4)
```

When registered,  plant was 6ft(1.8m) tall with leaves 8in(20cm)
long.  Flowers double or semi-double, Delft rose, in trusses of
10.  Late midseason. (Rudolph Henny, 1964)

```
g.
            -arboreum ssp. arboreum (1/2)
Drum Major-
            -griersonianum (1/2)
```

Bright red flowers in compact trusses on a plant of open habit.
Early midseason. (Rothschild, 1936)

```
g.
            -forrestii ssp. forrestii Repens Group (1/2)
Dryad-
      -hookeri (1/2)
```

Deep red flowers. (Lord Aberconway, 1946)

```
cl.         -souliei (1/4)
      -Latona-
Dryope-    -dichroanthum ssp. dichroanthum (1/4)
      -          -haematodes ssp. haematodes (1/4)
      -May Day-
                  -griersonianum (1/4)
```

Rose flowers, flushed coral red. (Sunningdale Nurseries, 1955)

```
cl.
            -brookeanum var. gracile (1/2)
Duchess of Connaught-
            -longiflorum (lobbii) (1/2)
```

Vireya hybrid.  Bright scarlet flowers.  See also Duchess of Edinburgh (vireya). (J. Mason) F. C. C. (RHS) 1881

```
cl.
```

Duchess of Cornwall      Parentage unknown

A parent of Orient Express.  Round trusses of 18-20 flowers,
deep  neyron rose with dark red spots.  Leaves 5.5in(14cm) long,
covered underneath with silvery plastered indumentum. (R. Gill,
1910; shown by Hillier)   A. M. (RHS) 1974

```
cl.
            -brookeanum var. gracile (1/2)
Duchess of Edinburgh-
            -longiflorum (lobbii) (1/2)
```

Much confusion exists between 2 hybrids of the same name.  This
vireya hybrid has flowers of light crimson  with paler centers.
See also  Duchess of Connaught. (J. Waterer)  F. C. C. (RHS)
1874

cl.

Duchess of Edinburgh*      Parentage unknown

This originated before 1905, according to Walter Schmalscheidt,
Oldenburg, West Germany, who cares for a collection of old rhododendron hybrids including this one. Flowers are rosy red with
a lighter center and small brown markings, producing an attractive bi-color effect; a parent of many hybrids. There are two by
this name: the 1958 International Rhododendron Register had recorded the other, a vireya, and mistakenly listed it as a parent
of elepidote hybrids.

```
cl.                         -jasminiflorum (1/4)
                  -Princess Royal-
Duchess of Fife-            -javanicum (1/4)
                  -teysmannii (1/2)
```

Vireya hybrid with large cream-colored flowers, flushed pale red.
A. M. (RHS) 1889

```
cl.
                  -barbatum (1/2)
Duchess of Portland-          -caucasicum (1/4)
                  -Handsworth White-
                             -unknown (1/4)
```

Fine foliage with brown indumentum. Soft lavender buds opening
pure white with pale yellowish eye, in shapely trusses. (Fisher, Son & Sibray) A. M. (RHS) 1903

```
                             -dichroanthum ssp.
                  -Goldsworth-      dichroanthum (1/8)
                  - Orange  -fortunei ssp. discolor
cl.               -Hotei-                        (1/8)
                  -          -souliei (1/8)
Duchess of Rothesay-   -unnamed-
                  -       hybrid-wardii var. wardii (1/8)
                  -decorum (1/2)
```

Trusses of 17 flowers, 5-lobed, pale aureolin yellow shading
deeper to center of lobe. (Edmund de Rothschild) A. M. (RHS)
1983

cl.

Duchess of Teck     Parentage unknown

A parent of Lord Wolseley.   White flowers, deeply edged in rosy
mauve, with bronze blotch. (Waterer, Sons & Crisp, 1892) A. M.
(RHS) 1916

```
cl.
                  -fortunei ssp. fortunei (1/2)
Duchess of York-      -catawbiense (1/4)
                  -Scipio-
                       -unknown (1/4)
```

5ft(1.5m)          5°F(-15°C)          M          3/3

Large fragrant flowers of salmon pink, with brownish markings on
the throat. Foliage resembles that of the species fortunei. (G.
Paul)  A. M. (RHS) 1894

```
cl.
    -catawbiense var. album (1/2)
Duet-                    -dichroanthum ssp. dichroan-
    -          -unnamed hybrid-              thum (1/8)
    -unnamed hybrid-          -griffithianum (1/8)
                  -auriculatum (1/4)
```

5ft(1.5m)          -25°F(-32°C)          ML          4/2

A very hardy and beautiful hybrid for the cold country. Flowers
pale yellow, edged pink, with a green blotch and green spotting
in throat. Plant grows as broad as tall; leaves 4.25in(10.8cm).
See also Peach Parfait. (David G. Leach, 1960)

g. & cl.
```
                  -arboreum ssp. arboreum (1/2)
Duke of Cornwall-
                  -barbatum (1/2)
```

Crimson flowers. (R. Gill & Son)   A. M. (RHS) 1907

```
cl.
                  -brookeanum var. gracile (1/2)
Duke of Teck-            -jasminiflorum (1/4)
                  -Princess Royal-
                             -javanicum (1/4)
```

Vireya hybrid.  Rosy lilac flowers. (J. Waterer, before 1875)

```
cl.
                  -fortunei ssp. fortunei (1/2)
Duke of York-      -catawbiense (1/4)
                  -Scipio-
                       -unknown (1/4)
```

Flowers of rosy pink spotted with cream.  A. M. (RHS) 1894

```
            -elliottii (1/4)
cl.    -Kiev-          -thomsonii ssp. thomsonii (1/8)
    -    -Barclayi -          -arboreum ssp. arboreum
Dukeshill-    Robert Fox-Glory of  -            (1/16)
    -          Penjerrick-griffithianum (1/16)
    -          -facetum (1/4)
    -Lady Digby-
                  -strigillosum (1/4)
```

Flowers of scarlet red on a hardy plant. (Crown Estate, Windsor)
A. M. (RHS) 1973

```
            -diaprepes (1/4)
cl.    -Vibrant-
    -          -wardii var. wardii (1/4)
Dusky Dawn-                  -fortunei ssp. discolor (1/8)
    -          -Lady Bessborough-
    -          -          -campylocarpum ssp. campylo-
    -Jalisco-          carpum Elatum Group (1/8)
    -    -dichroanthum ssp. dichroanthum (1/8)
       -Dido-
            -decorum (1/8)
```

Flowers pale apricot, turning near-white. (Gen. Harrison, 1973)

```
g.
            -Moser's Maroon--unknown (1/2)
Dusky Maid-
            -fortunei ssp. discolor (1/2)
```

A tall and erect shrub with a compact truss of dark red flowers
similar to the seed parent, Moser's Maroon. One of the parents
of Kilimanjaro. (Rothschild, 1936)

```
                  -Essex Scarlet--unknown (1/8)
            -Cavalcade-
cl.    -unnamed-      -griersonianum (1/8)
    - hybrid-Mary      -campylocarpum ssp. campylocarpum
Dusky Wood-    -Swaythling-            (1/8)
    -          -fortunei ssp. fortunei (1/8)
    -    -wardii var. wardii (1/4)
    -Crest-          -fortunei ssp. discolor (1/8)
       -Lady Bessborough-
                  -campylocarpum ssp. campylocar-
                       pum Elatum Group (1/8)
```

5ft(1.5m)          5°F(-15°C)          M

Flowers of soft rose pink, flushed red. (John Waterer, Sons &
Crisp, intro. 1975)

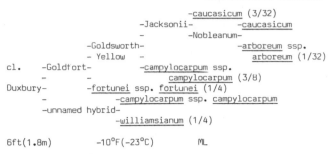

```
                              -caucasicum (3/32)
                 -Jacksonii-          -caucasicum
                 -          -Nobleanum-
           -Goldsworth-              -arboreum ssp.
           - Yellow -                  arboreum (1/32)
cl.   -Goldfort-      -campylocarpum ssp.
      -                   campylocarpum (3/8)
Duxbury-     -fortunei ssp. fortunei (1/4)
      -            -campylocarpum ssp. campylocarpum
      -unnamed hybrid-
                   -williamsianum (1/4)


6ft(1.8m)        -10°F(-23°C)        ML
```

Buds amber yellow and salmon open to flowers light primrose
yellow, held in 7in(17.8cm) trusses of 7-11.  Plant tall, as
wide as high with leaves retained 3 years, very dark green.
(Cross by Lewis Bagoly, 1970; raised by Mr. & Mrs. F. B. Lawson,
reg. 1984)

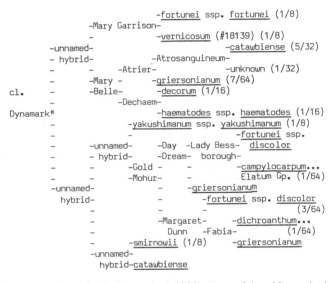

```
                         -fortunei ssp. fortunei (1/8)
              -Mary Garrison-
              -               -vernicosum (#18139) (1/8)
        -unnamed-                   -catawbiense (5/32)
        - hybrid-        -Atrosanguineum-
        -      -    -Atrier-          -unknown (1/32)
        -    -Mary -   -griersonianum (7/64)
cl.     -    -Belle-   -decorum (1/16)
        -          -Dechaem-
Dynamark*              -haematodes ssp. haematodes (1/16)
        -          -yakushimanum ssp. yakushimanum (1/8)
        -        -           -fortunei ssp.
        -    -unnamed-   -Day -Lady Bess- discolor
        -    - hybrid-   -Dream- borough-
        -    -    -Gold  -       -campylocarpum...
        -    -    -Mohur -       Elatum Gp. (1/64)
     -unnamed-        -griersonianum
       hybrid-           -fortunei ssp. discolor
        -       -        -              (3/64)
        -       -Margaret-    -dichroanthum...
        -        Dunn   -Fabia-       (1/64)
        -     -smirnowii (1/8)  -griersonianum
        -unnamed-
          hybrid-catawbiense
```

Deep purplish pink buds open to brilliant greenish yellow, edged
with deep purplish to strong pink, strong red throat.  (W. Delp)

```
                    -arboreum ssp. cinnamomeum roseum (1/4)
cl.        -Fair Lady-        -griffithianum (3/8)
      -             -Loderi Venus-
Earl J. Reed-              -fortunei ssp. fortunei (3/8)
      -          -griffithianum
      -Loderi Venus-
                  -fortunei ssp. fortunei

5ft(1.5m)         10°F(-12°C)        EM
```

White flowers, roseine purple at edges, 3.5in(8.9cm) across,  6-
lobed, held in ball-shaped trusses of 13.  Plant upright; leaves
medium-sized, yellow green, held 2 years.  (George Clarke, cross
1969; Edwin Anderson, intro. & reg. 1983)

```
                              -griffithianum (3/8)
cl.              -Queen Wilhelmina-
            -Earl of Athlone-          -unknown (5/8)
Earl Moore-          -Stanley Davies--unknown
      -                   -griffithianum
      -Jean Marie de Montague-
                      -unknown
```

Flowers cherry red, in tight trusses of 10-15.  Tall plant with
long, thin, vine-like branches, and glossy, medium green leaves.
(C. S. Seabrook, 1967)

```
cl.                        -griffithianum (1/4)
                 -Queen Wilhelmina-
Earl of Athlone-          -unknown (3/4)
                 -Stanley Davies--unknown


5ft(1.5m)        0°F(-18°C)        EM        5/2
```

One of the finest red trusses, dome-shaped, compact. Campanulate
flowers of bright blood red. Plant habit open, spreading; long,
narrow, dark green leaves.  (C. B. van Nes)  F. C. C. (RHS) 1933

```
cl.
          -griersonianum (1/2)      -griffithianum (1/16)
Earl of   -            -George Hardy-
Donoughmore-    -Mrs. L. A.-      -catawbiense (1/16)
          -unnamed-  Dunnett -unknown (3/8)
            hybrid-unknown

5ft(1.5m)         -5°F(-21°C)        M        3/3
```

Bright red flowers with an orange glow. (M. Koster & Sons, 1953)
See color illus. F p. 43.

```
cl.
           -campylocarpum ssp. campylocarpum (1/2)
Earl of Morley-
           -arboreum ssp. cinnamomeum var. album (1/2)
```

Flowers of glistening ivory white with crimson spots, in rounded
trusses. A very large, slow-growing shrub or small tree of pyramidal habit, first raised at Westonbirt c.1910.  Blooms early.
See also Leonardslee Primrose.  (Sir Edmund Loder, before 1958)

```
                          -catawbiense (red form) (1/8)
                 -Pink Twins-
cl.   -Shaazam-        -haematodes ssp. haematodes (1/8)
      -        -               -griffithianum (1/8)
Earlene-    -Leah Yates--Mars (selfed)-
      -            -unknown (1/8)
      -Koichiro Wada--yakushimanum ssp. yakushimanum (1/2)

2ft(.6m)         -15°F(-26°C)        ML
```

Plant rounded, semi-dwarf, as wide as tall, leaves medium-sized.
New growth with orange brown indumentum. Flowers 2in(5cm) wide,
5-lobed, salmon-colored, claret rose on reverse, in  ball-shaped
trusses of about 14. (Maletta Yates, 1977)

```
g.                       -thomsonii ssp. thomsonii (1/4)
              -Ascot Brilliant-
Early Brilliant-          -unknown (1/4)
             -barbatum (1/2)
```

Synonym Fireball.  Bright red flowers.  (W. C. Slocock)

```
cl.               -arboreum ssp. arboreum (1/4)
         -Red Admiral-
Early Days-          -thomsonii ssp. thomsonii (1/4)
         -unknown (1/2)
```

Trusses of 12-14 bell-shaped flowers, cardinal red.  Height 5ft
(1.5m).  (V. J. Boulter, reg. 1983)

```
cl.             -ciliatum (1/4)
         -Praecox-
Early Gem-      -dauricum (3/4)
         -dauricum
```

Flowers rosy lilac.  F. C. C. (RHS) 1874

```
g.        -strigillosum (1/2)
Early Stir-
          -irroratum ssp. irroratum (1/2)
```

Currant red flowers.  (Lord Digby, 1934)

cl.
          -williamsianum (1/2)
Earlybird-
          -oreodoxa var. fargesii (1/2)

Rose-colored flowers shading to almost white in the tube, funnel
campanulate, 3in(7.6cm) wide, in trusses of 6.  Rounded leaves,
1.5in(3.8cm) across.  Very early.  (Rudolph Henny, 1960)


g.
          -griersonianum (1/2)
East Knoyle-
        -       -griffithianum (1/4)
          -Loderi-
                  -fortunei ssp. fortunei (1/4)

Carmine red buds opening to large rose-colored flowers darker in
throat, widely funnel-shaped, in loose trusses.  (Sir James Horlick, 1930)


cl.
           -williamsianum (1/2)
Easter Bells-    -wightii (1/4)
          -China-
                 -fortunei ssp. fortunei (1/4)

2ft(.6m)          5°F(-15°C)          EM          3/4

Medium-sized, oval leaves on a plant wider than tall.  Flowers
opening cream and fading white; lax trusses of about 8.  (Benjamin Lancaster, 1967)


cl.

Easter Island--alutaceum var. russotinctum (F 20425)

Compact trusses of 10-12 flowers, white with ruby red spotting.
Matte green leaves 4in(10.2cm) long; fawn to rusty brown indu-
mentum beneath.  (Collector George Forrest; raiser Col. S. R.
Clarke; reg. by R. N. Stephenson Clarke, 1980)  A. M. (RHS) 1980


cl.
          -minus var. minus Carolinianum Group (3/4)
Ebony-      -minus var. minus Carolinianum Group
       -P.J.M.-
                -dauricum (1/4)

1ft(.3m)          -15°F(-26°C)          EM

Light purple flowers, 1in(2.5cm) across, 5 wavy lobes; 4-10 in
dome-shaped trusses.  Floriferous.  Plant rounded, well-branched
(decumbent); very small, shiny leaves are deep maroon in winter,
held 1 year.  (Edmund V. Mezitt, Weston Nurseries, 1979)


cl.
         -decorum (1/2)
Echo-                 -fortunei ssp. discolor (1/4)
       -Lady Bessborough-
                         -campylocarpum ssp. campylocarpum Elatum
                                            Group (1/4)

Creamy flowers, fading white, with a claret eye.  (Sunningdale
Nurseries, 1955)


cl.            -griffithianum (1/2)
        -Mars-
Ed Knight-   -unknown (1/4)
                         -griffithianum
        -Loderi King George-
                         -fortunei ssp. fortunei (1/4)

4ft(1.2m)          15°F(-10°C)          M

Very large trusses--19 flowers of medium magenta color, 5 frilly
lobes, openly funnel-shaped, 4in(10.2cm) across.  Upright plant,
taller than wide; dark green leaves held 2 years.  (Cross by Dr.
B. Briggs; intro. by E. Knight; reg. 1983)


cl.              -dichroanthum ssp. dichroanthum
          -Fabia-                              (1/8)
     -unnamed hybrid-   -griersonianum (1/8)
Ed Long*               -unknown (1/4)
     -fortunei ssp. fortunei (1/2)

A medium-sized, compact plant; buds chartreuse green opening to
cream, with green center and blotch, to 4in(10.2cm) wide; rounded trusses.  (Intro. by D. W. James, 1963)


          -Jervis-wardii var. wardii (1/4)
cl.     - Bay -
        -     -Lady       -fortunei ssp. discolor (1/8)
Eddystone-    -Bessborough-
        -                 -campylocarpum ssp. campylocarpum
        -                             Elatum Group (1/8)
        -wardii var. wardii Litiense Group (1/2)

Trusses of 14 flowers, primrose yellow with small crimson stain
in throat, openly campanulate, 2.5in(6.4cm) wide. (Francis Hanger, RHS Garden, Wisley)  A. M. (RHS) 1964


                 -arboreum ssp. zeylanicum (1/4)
cl.      -Ilam Alarm-          -griffithianum (1/8)
                     -unnamed hybrid-
Edgar Stead-                    -unknown (1/8)
        -            -thomsonii ssp. thomsonii (1/4)
          -Shilsonii-
                    -barbatum (1/4)

(Stead)


---
         -nuttallii (1/2)
Edinense-
         -dalhousiae var. dalhousiae (1/2)

White flowers stained yellow at base.  Reverse cross to Victorianum, q.v.


cl.
      -fortunei ssp. discolor (1/2)
Edith-
      -hardy hybrid--unknown (1/2)

Flowers deep rose pink with dark blotch.  Plant medium to tall;
late.  (Slocock, reg. 1958)   A. M. (RHS) 1931


cl.
               -sanguineum ssp. sanguineum
Edith Berkeley-   var. didymoides (consanguineum) (1/2)
                         -griffithianum (1/4)
               -Loderi King George-
                         -fortunei ssp. fortunei (1/4)

Flowers of rose madder with white stamens, no spotting, 6-lobed,
3in(7.6cm) across.  Late midseason.  (Mr. & Mrs. Greig, Canada,
1963)

cl.
          -Marion--unknown (3/4)
Edith Boulter-    -campylocarpum ssp. campylocarpum (1/4)
          -Unique-
                 -unknown

Lavender pink flowers with darker margins, frilled;  ball-shaped
trusses of 15-18.  (V. J. Boulter, Australia, 1962)


                              -griffithianum (1/8)
cl.               -Coombe Royal-
          -Mrs. G. W.-         -unknown (3/4)
        - Leak                  -caucasicum (1/8)
Edith Brown-     -Chevalier Felix-
        -            de Sauvage  -hardy hybrid--unknown
          -unknown

Trusses of 18-24 flowers, pink with brown blotch like Mrs. G. W.
Leak, but darker.  Height to 9.8ft(3m).  (W. E. Glennie, reg.
1984)

cl.                          -arboreum ssp. arboreum (1/4)
                  -Doncaster-
Edith Mackworth Praed-          -unknown (3/4)
                  -unknown

Flowers of bright crimson or cherry scarlet. (M. Koster & Sons)
A. M. (RHS) 1934


cl.                          -catawbiense (1/4)
                  -English Roseum-
Edith Pride-                 -unknown (1/4)
               -maximum (1/2)

5ft(1.5m)          -25°F(-32°C)          ML

Pink flowers with pale yellow spotting and a small white blotch,
2in(5cm) across, 5-lobed; large ball-shaped trusses hold up to
22. Plant rounded, well-branched, as wide as tall; narrowly elliptic leaves held 3 years. (Orlando S. Pride, 1979)


cl.
            -catawbiense var. album (1/2)
Edmond Amateis-
            -Dexter seedling--unknown (1/2)

6ft(1.8m)          -25°F(-32°C)          EM          4/4

White flowers with bold, twin-rayed dorsal blotch of red, 3.5in
(8.3cm) wide, rotate funnel-shaped; spherical trusses hold up to
13. Plant habit upright; good foliage density. (Edmond Amateis
cross; raised and intro. by David G. Leach, 1969)


cl.              -catawbiense (1/4)
            -unnamed hybrid-
Edmond Moulin-          -unknown (1/2)
       -               -maximum (1/4)
            -unnamed hybrid-
                  -unknown

5ft(1.5m)          -25°F(-32°C)          M

Rounded trusses 6in(15.2cm) wide,  holding 22 flowers of fuchsia
purple with prominent blotch of greyed yellow.  Plant upright,
nearly as broad as high; leaves medium-sized, dark green, held 3
years. (W. David Smith, 1983)


g.
            -arboreum ssp. arboreum (1/2)
Edmondi-
            -barbatum (1/2)

Orange scarlet flowers. (R. Gill, 1876)


cl.                      -elliottii (1/2)
            -Kilimanjaro-          -Moser's Maroon--unknown (1/8)
Edmund de -          -Dusky Maid-
Rothschild-               -fortunei ssp. discolor (1/8)
       -          -elliottii
            -Fusilier-
                  -griersonianum (1/4)

6ft(1.8m)          15°F(-9°C)          ML

Deep red flowers, heavily and uniformly spotted inside, widely
funnel-campanulate, 3.5in(6.9cm) wide; large trusses hold up to
22 flowers. Upright plant, vigorous growth habit; smooth leaves
6in(15.2cm) long. (Edmund de Rothschild, 1963)      A. M. (RHS)
1968      See color illus. PB pl. 21.


cl.              -griffithianum (1/4)
            -Alice-
Edna McCarty-     -unknown (1/4)
            -auriculatum (1/2)

6ft(1.8m)          -5°F(-21°C)          L          3/2

A sterile hybrid, without stamens. Large, fragrant white flowers; large foliage. (Endre Ostbo, 1962) P. A. (ARS) 1959


cl.            -griffithianum (1/4)
       -Mars-
       -     -unknown (5/8)
Ed's Red-                     -catawbiense (1/8)
       -          -Parsons Grandiflorum-
       -America-               -unknown
            -dark red hybrid--unknown

4ft(1.2m)          -15°F(-26°C)          M

Cardinal red flowers with a light brown blotch, in ball-shaped
trusses of 20. Plant spreads equally as broad as tall; leaves
dark green, retained 2 years. (Harold E. Reiley, reg. 1984)


g.              -campylocarpum ssp. campylocarpum Elatum Group
       -Penjerrick-                              (1/4)
Edusa-          -griffithianum (1/4)
       -campylocarpum ssp. campylocarpum (1/2)

Yellow flowers. (Lord Aberconway, intro. 1933)


cl.                      -neriiflorum ssp. neriiflorum (1/4)
            -unnamed hybrid-
Edward Dunn-               -dichroanthum ssp. dichroanthum (1/4)
            -fortunei ssp. discolor (1/2)

5ft(1.5m)          -5°F(-21°C)          VL          4/3

Apricot pink flowers, 3in(7.6cm) across, in dome-shaped trusses.
Plant habit spreading; moss green foliage.  Tolerates heat; best
in part shade. (Endre Ostbo, 1958) P. A. (ARS) 1958


cl.              -dichroanthum ssp. dichroanthum (1/4)
            -Fabia-
Edward Long-     -griersonianum (1/4)
            -fortunei ssp. fortunei (1/2)

Medium-sized, compact plant with leaves 5in(12.7cm) long.  Buds
chartreuse green, opening to creamy flowers 4in(10.2cm) across,
with faint green center blotch; rounded trusses of 8-10.  Midseason. (Del W. James, 1963)


cl.
            -catawbiense (1/2)
Edward S. Rand-
            -unknown (1/2)

6ft(1.8m)          -15°F(-26°C)          ML          2/3

Flowers of crimson red with bronze yellow eye.  Plant habit
compact; foliage slightly yellowish green.  Resistant to root
rot. (A. Waterer, 1870)


cl.                      -catawbiense (1/4)
            -Purpureum Elegans-
Edwin O. Weber-               -unknown (3/4)
            -Madame Albert Moser--unknown

6ft(1.8m)          -5°F(-21°C)          ML

Flowers of royal purple  with large uranium green dorsal blotch,
3.5in(8.9cm) wide; up to 28 in large trusses.  Floriferous.  Upright plant, moderately branched; dark green leaves, 6.5in(16.5
cm) long, retained 3 years. (E. O. Weber, 1974)


            -Lady      -fortunei ssp. discolor (3/8)
            -Bessborough-
cl.       -Day Dream-          -campylocarpum ssp. campylo-
       -          -                carpum Elatum Group (1/8)
Edwin Parker-     -griersonianum (1/4)
       -          -griffithianum (1/8)
       -          -Loderi-
       -Albatross-     -fortunei ssp. fortunei (1/8)
            -fortunei ssp. discolor

5ft(1.5m)          0°F(-18°C)          L

Flowers pink to peach, creamy pink center on each of 5 lobes, to 4in(10.2cm) wide, in 16-flowered trusses 9in(23cm) broad; leaves to 5in(12.7cm) long.  (Rudolph & Leona Henny, 1968)

```
cl.                        -cataubiense (1/2)
       -Alfred--Everestianum (selfed)-
Effner-                              -unknown (1/2)
       -                -cataubiense
       -Everestianum-
                       -unknown
```

Lilac-colored flowers, greenish yellow markings. (Seidel, 1903)

```
cl.                      -haematodes var. haematodes (3/4)
                -May Day-
Egbert van Alstyne-      -griersonianum (1/4)
                -haematodes ssp. haematodes
```

Orient red, bell-shaped flowers, slightly larger than haematodes 2in(5cm) long, in open trusses.  Plant of medium size, shapely, compact; yellowish green, medium-sized leaves.  Midseason.  See also Harry von Tilzer. (C. S. Seabrook, reg. 1968)

```
cl.
```

Eggbrechtii*        Parentage unknown

A parent of Gudrun, Holger both by Seidel.

```
cl.
       -campylogynum (white form) (1/2)
Egret-
       -White Lace--racemosum (1/2)
```

Plant of compact dwarf habit with small leaves, medium glossy green.  Trusses of 6 flowers, white slightly tinged pale green towards the calyx. (Peter A. Cox)  A. M. (RHS) 1982  H. C. (Wisley Trials) 1983

```
                 -soulei (1/4)
cl.   -Halcyone-          -fortunei ssp. discolor (3/8)
      -          -Lady Bessborough-
Egyptian-                 -campylocarpum ssp. campylo-
      -                           carpum Elatum Group (1/8)
      -              -fortunei ssp. discolor
      -Margaret Dunn-     -dichroanthum ssp. dichroanthum
                     -Fabia-                        (1/8)
                          -griersonianum (1/8)
```

Yellow flowers overlaid with salmon; brown blotch. (Sunningdale Nurseries, 1955)

```
cl.   -japonicum var. japonicum (metternichii) (1/2)
Eidam-
       -Alexander Adie--unknown (1/2)
```

White flowers, flushed rose. (T. J. R. Seidel cross, 1903)  See color illus. WS p. 27.

```
cl.   -minus var. minus Carolinianum Group (white form) (1/2)
Eider-
       -leucaspis (1/2)
```

Compact, spreading plant 1.5ft(.45m) high; matte green leaves 2.25in(5.7cm) long.  White flowers, 2.5in(6.4cm) wide, funnel-shaped; trusses of 5-8. (Raised & intro. by P. A. Cox)  H. C. (RHS) 1979; A. M. (RHS) 1981

```
cl.          -griffithianum (1/4)
            -Mars-
Eileen Hall-  -unknown (1/2)
      -         -Pygmalion--unknown
      -unnamed-
         hybrid-haematodes ssp. haematodes (1/4)
```

5ft(1.5m)          -15°F(-26°C)          M

Amaranth rose flowers with dorsal red spotting, 3in(7.6cm) wide, in ball-shaped trusses of 10; free-flowering. Plant broader than tall; 7in(17.8cm) leaves held 2 years. (Maurice Hall, reg. 1980)

```
cl.             -arboreum ssp. arboreum (1/4)
       -Red Admiral-
Eithne-          -thomsonii ssp. thomsonii (1/4)
       -neriiflorum ssp. neriiflorum (1/2)
```

Daphne g.  Deep crimson scarlet. (E. J. P. Magor)  A. M. (RHS) 1937

```
                    -Beauty of-griffithianum (9/16)
             -Norman- Tremough-
             - Gill -      -arboreum ssp. arboreum (1/16)
cl.   -Anna-  -griffithianum
      -                   -griffithianum
El Camino-  -Jean Marie de Montague-
      -                    -unknown (3/8)
      -           -griffithianum
      -Marinus Koster-
                  -unknown
```

7ft(2.1m)          -5°F(-21°C)          M          4/4

The above parentage is uncertain; it is thought to be the same as Halfdan Lem's Walloper. Flowers of deep pink, with a darker blotch and spotting, 5in(12.7cm) wide. Plant upright, stiffly branched; leaves to 5.5in(14cm), held 3 years. (Lem?, Whitney, and the Sathers, 1976)

```
             -Lady       -fortunei ssp. discolor (1/8)
             -Bessborough-
      -Jalisco-          -campylocarpum ssp. campylocarpum
cl.   -       -                  Elatum Group (1/8)
      -       -dichroanthum ssp. dichroanthum (3/8)
El Greco-  -Dido-
      -       -decorum (1/8)
      -       -Dawn's -griffithianum (1/8)
      -Break of-Delight-
         Day   -      -unknown (1/8)
             -dichroanthum ssp. dichroanthum
```

Flowers of saffron yellow changing to azalea pink at edges, carmine rose flush, 3.5in(8.9cm) across, 10 per truss.  Plant is compact, broader than tall [5ft(1.5m)]; dull green leaves 6.5in(16.5cm) long. (RHS, Wisley)  A. M. (Wisley Trials) 1961

```
cl.
       -macabeanum (1/2)
Elaine Rowe-
       -unknown (1/2)
```

Trusses of 15 flowers, 7- or 8-lobed, neyron rose in bud, lighter on outside when open, inside cream with maroon flare. Leaves 8in(20cm) long by 4in(10cm) wide, elliptic, with silvery grey woolly indumentum. (G. Huthnance, reg. 1984)

```
cl.             -pubescens (1/4)
      -Chesapeake-
Elam-          -keiskei (1/4)
      -unknown (1/2)
```

2ft(.6m)          -20°F(-29°C)          M

Deep rose flowers are 1.5in(3.8cm) wide, held in clusters of 3-4. Very small leaves.  See also Elsmere. (G. Guy Nearing, 1972)

```
g.
       -valentinianum (1/2)
Eldorado-
       -johnstoneanum (1/2)
```

4ft(1.2m)          15°F(-10°C)          E          3/3

Flowers primrose yellow, medium-sized, funnel-shaped, clusters of 3-4.  Open-growing plant; small, scaly, dark yellowish green leaves. (Rothschild, 1937)

cl.

Eleanor Black--form of <u>konorii</u>

Vireya species.  White flowers, flushed with rose pink, deeper
in the throat, 4in(10.2cm) wide, tubular funnel-shaped; strongly
fragrant at night.  Both surfaces of the broadly elliptic leaves
covered with very tiny knobs, creating a rough texture.  (Collector & raiser Michael Black)  A M. (RHS) 1969

g. & cl.
          -<u>rubiginosum</u> Desquamatum Group (1/2)
Eleanore-
          -<u>augustinii</u> ssp. <u>augustinii</u> (1/2)

A tall plant with flowers larger than either parent, in clusters
of 4-5.  One form is lavender blue, another blends violet and
amethyst.  Listed in some catalogs as Eleanor.  (Rothschild,
intro. 1937; reg. 1958)  A. M. (RHS) 1943

g. & cl.
          -<u>augustinii</u> ssp. <u>chasmanthum</u> (1/2)
Electra-
          -<u>augustinii</u> ssp. <u>augustinii</u> (1/2)

5ft(1.5m)         5°F(-15°C)         EM         5/4

Electra is not really a hybrid, since it is a cross of two forms
of the same species.  The flower is one of the bluest of the <u>au</u>-
<u>gustinii</u> but not the largest.  Plant is upright; leaves are 2.5
in(6.4cm) long, medium green.  (Rothschild, 1937)  A. M. (RHS)
1940

cl.               -<u>augustinii</u> ssp. <u>chasmanthum</u> (1/4)
          -Electra-
Electra's Son-    -<u>augustinii</u> ssp. <u>augustinii</u> (1/4)
          -
              -unknown (1/2)

6ft(1.8m)         10°F(-12°C)         ML

Flowers lavender, with yellow green blotch, reddish lavender on
the reverse, flat, 3in(7.6cm) wide; terminal inflorescence of 1-
4 trusses, each 2-4 flowered.  Plant upright, well-branched; elliptic leaves to 4in(10.2cm).  (Edwin K. Parker, 1977)

                            -<u>catawbiense</u> (5/8)
g.              -unnamed hybrid-
      -Altaclarense-         -<u>ponticum</u> (1/8)
Elegans-              -<u>arboreum</u> ssp. <u>arboreum</u> (1/4)
      -<u>catawbiense</u>

Flowers deep rose, some spotting, in fine trusses.  (Standish &
Noble, before 1850)

cl.               -<u>caucasicum</u> (1/4)
      -Dr. Stocker-
Elfin-             -<u>griffithianum</u> (1/4)
      -<u>orbiculare</u> ssp. <u>orbiculare</u> (1/2)

Flowers cream-colored, flushed pink.  (Sir James Horlick, 1933)

cl.               -<u>williamsianum</u> (1/4)
      -Jock-
Elfin Hill-       -<u>griersonianum</u> (1/4)
          -<u>haematodes</u> ssp. <u>haematodes</u> (1/2)

1ft(.3m)          0°F(-18°C)         EM

Flowers of China rose to 1.75in(3.8cm) wide, funnel-campanulate.
Compact plant, as broad as tall; dark green, oval leaves.  (Vernon Wyatt, 1968)

cl.               -<u>catawbiense</u> (pink form) (1/2)
      -unnamed hybrid-
Elie-             -unknown (1/2)
      -                 -<u>catawbiense</u> (pink form)
      -unnamed hybrid-
                      -unknown

6ft(1.8m)         -10°F(-23°C)         M

Flowers vibrant deep pink with red blotch, 2.5in(6.4cm) across,
in conical trusses.  Glossy dark green leaves.  (A. M. Shammarello, 1955)  See color illus. VV p. 75.

cl.
              -Essex Scarlet--unknown (1/2)
Elisabeth Hobbie-
              -<u>forrestii</u> ssp. <u>forrestii</u> Repens Group (1/2)

2.5ft(.75m)       -5°F(-21°C)         EM         4/4

Scarlet red flowers with faint dark red spotting, in lax trusses
of 5-7.  Mound-shaped, compact plant, with deep green elongated
leaves.  (Dietrich Hobbie, 1945)  H. C. (Wisley Trials) 1974   See
color illus. VV p. 130.

cl.
              -<u>griersonianum</u> (1/2)
Elisabeth Inglis-Jones-     -<u>haematodes</u> ssp. <u>haematodes</u> (1/4)
              -May Day-
                      -<u>griersonianum</u> (1/4)

Red flowers, held in loose trusses.  (Sir James Horlick, 1962)

                  -<u>griffithianum</u> (1/8)
cl.           -Mars-
          -Vulcan-    -unknown (1/4)
Elise Whipple-    -<u>griersonianum</u> (1/4)
                      -<u>caucasicum</u> (1/4)
          -Boule de Neige-         -<u>catawbiense</u> (1/8)
                  -hardy hybrid-
                          -unknown
1.5ft(.45m)       -5°F(-21°C)         ML

Red buds opening to white flowers suffused red, the reverse red,
2.5in(6.4cm) across, in flat trusses of 15.  Plant broader than
tall; leaves held 2 years.  (R. Kruse; M. Whipple, intro. 1973)

cl.
      -<u>minus</u> var. <u>minus</u> Carolinanum Group (1/2)
Elite*
      -<u>dauricum</u> (1/2)

One of three selected forms of P.J.M. propagated and distributed
by Weston Nurseries; flowers are deeper pink.  (Edmund V. Mezitt)

g.
          -<u>forrestii</u> ssp. <u>forrestii</u> Repens Group (1/2)
Elizabeth-
          -<u>griersonianum</u> (1/2)

3ft(.9m)          0°F(-18°C)         EM         4/4

Bright red flowers 3.5in(8.9cm) wide, funnel-campanulate, in lax
trusses of 6-8; may have a few blooms in autumn.  Compact plant,
wider than tall; dark green leaves held 3 or more years.  (Lord
Aberconway, 1929)   A. M. (RHS) 1939; F. C. C. (RHS) 1943   See
col. illus. ARS Q 34:3 (Summer 1980): p. 194; F p. 44; VV p. 72.

cl.

Elizabeth Bennet--form of <u>formosum</u> var. <u>inaequale</u> (C & H 301)

A flowering plant for the cool greenhouse.  White flowers with a
yellow green blotch, 4.75in(12cm) wide, 5-lobed; trusses of 5 or
6.  Dark green leaves 2.75in(7cm) long, scaly below.  (Raised &
reg. by Mrs. Elizabeth Mackenzie)    F. C. C. (RHS) 1981

cl.

Elizabeth David--form of <u>burmanicum</u>

4ft(1.2m)         15°F(-10°C)         EM         4/3

A cool greenhouse plant.  Flowers light chartreuse yellow with
the reverse slightly darker, campanulate, 2in(5cm) wide; trusses
of 4.  Leaves up to 2in(5cm) long, glossy dark green, both sides
covered with scaly brown indumentum. (Collector unknown; raised
& exhibited by Mrs. Elizabeth Mackenzie)  A. M. (RHS) 1980

```
                    -lacteum (1/4)        -griffithianum (1/16)
        -Lionel's-                 -Kewense-
       - Triumph-      -Aurora-         -fortunei ssp. fortunei
cl.    -           -Naomi-     -                      (5/16)
       Elizabeth de-          -thomsonii ssp.thomsonii (1/8)
       Rothschild -       -fortunei ssp. fortunei
                   -           -griffithianum
                   -               -Kewense-
                   -        -Aurora-     -fortunei ssp. fortunei
                   -   -Exbury-   -thomsonii ssp. thomsonii
                   -   -Naomi -fortunei ssp. fortunei
                   -unnamed-
                     hybrid-unknown (1/4)

5ft(1.5m)         -5°F(-21°C)          M          4/3
```

Very pale yellow flowers with maroon spotting in throat, openly
funnel-shaped, 4.5in(11.5cm) broad; rounded, well-packed trusses
of about 18. Plant upright; dark green elliptic leaves. (Edmund
de Rothschild) A. M. (RHS) 1965

```
cl.         -auriculatum (1/2)
      Elizabeth Lock-
            -unknown (1/2)
```

Leaves up to 9in(23cm) long, 3in(7.6cm) wide, dark green.
Flowers held in round trusses of 12-13, 5-lobed, 3in(7.6cm) wide
with slight fragrance, color carmine rose. (Origin unknown;
shown by J. A. N. Lock) A. M. (RHS) 1968

```
cl.                              -haematodes ssp. haema-
      Elizabeth Lockhart--sport of Humming Bird-      todes (1/2)
                                 -williamsianum (1/2)

2.5ft(.75m)       0°F(-18°C)        EM         3/3
```

Flowers very dark red--darker than haematodes. Free-flowering.
Leaves to 2in(5cm), dark green flushed with reddish brown, petioles very dark red; young foliage very deep, glossy red. (R. D.
Lockhart, 1965) P. C. (RHS) 1964; H. C. (RHS) 1972   See color
illus. F p. 44.

```
                                -griffithianum (3/8)
cl.           -Queen Wilhelmina-
       -Britannia-              -unknown (3/8)
Elizabeth-        -Stanley Davies--unknown
  Mount   -     -griffithianum
       -Loderi-
               -fortunei ssp. fortunei (1/4)
```

Trusses of 9 flowers, Tyrian red purple in bud, opening rose
Bengal, shading to nearly white in center of lobes, blotch at
base of upper lobe scarlet red, strongly suffused black brown.
(Cross by Mr. White, Embley Park; intro. by Hillier Nurseries;
reg. by G. Mount, 1982)

```
cl.          -maximum (1/2)
      Elizabeth Scholtz-          -catawbiense (1/4)
                   -unnamed hybrid-
                            -unknown (1/4)

4ft(1.2m)       -25°F(-32°C)        M
```

Dome-shaped trusses, 7in(17.8cm) wide with 22 flowers, each
3.25in(8.3cm) across, spirea red. Upright plant, well-branched,
leaves held 3 years. (W. David Smith, 1983)

```
                     -griffithianum (1/2)
cl.         -Marinus Koster-
       -             -unknown (1/4)
Elizabeth Titcomb-          -maximum (1/8)
       -         -Halopeanum-
          -Snow Queen-     -griffithianum
          -          -griffithianum
             -Loderi-
                     -fortunei ssp. fortunei (1/8)
```

```
6ft(1.8m)        0°F(-18°C)         EM          4/4
```

Pink buds opening to white flowers, 4.5in(11.5cm) wide, of heavy
substance, about 16 in tall conical trusses. Strong growth habit; dark green foliage. Sun-tolerant. See also Diane Titcomb,
Julie Titcomb. (H. L. Larson, 1958) P. A. (ARS) 1958

```
cl.       -Betsy Trotwood--unknown (5/8)
      -Mars Novus-   -griffithianum (3/8)
      -           -Mars-
Ella*           -unknown
      -       -griffithianum
      -Kohinoor-
             -unknown
```

Frilled flowers fiery dark carmine red; new foliage of red
shoots. (Seidel, 1894)

```
cl.
      -smirnowii (1/2)
Ella-        -catawbiense (1/4)
      -Mrs. Milner-
             -unknown (1/4)
```

Carmine red flowers. (T. J. R. Seidel, 1915)

```
cl.
      -dichroanthum ssp. dichroanthum (1/2)
Ella-
      -wardii var. wardii (1/2)

4ft(1.2m)        10°F(-12°)         ML          4/3
```

Apricot copper buds opening orange to yellow flowers, flat, 3in.
(7.6cm) across; trusses hold up to 7. Floriferous. Leaves held
3 years. Likes partial shade; roots easily. (R. Henny, 1958)

```
cl.
```

Ellestee--form of wardii var. wardii (L S & T 5679)

Bright lemon yellow flowers, a deep crimson blotch. (Collectors
Ludlow, Sherriff, Taylor; raiser C. Ingram) A. M. (RHS) 1959

```
                          -souliei (1/16)
g.          -Soulbut-
&       -Vanessa-     -Sir Charles Butler--fortunei ssp.
cl. -Etna-      -griersonianum (1/4)        fortunei (1/16)
      -     -     -dichroanthum ssp. dichroanthum (1/8)
Elna-    -Fabia-
      -         -griersonianum
      -elliottii (1/2)
```

Scarlet flowers with pronounced spotting. (Lord Aberconway)
A. M. (RHS) 1949

```
g.
      -elliottii (1/2)
Elrise-      -griffithianum (1/4)
      -Sunrise-
             -griersonianum (1/4)
```

Red flowers. (Lord Aberconway, 1950)

```
                    -campylocarpum ssp. campylocarpum
g. & cl.      -Penjerrick-        Elatum Group (1/16)
      -Amaura-      -griffithianum (1/16)
   -Eros-      -griersonianum (3/8)
Elros-   -griersonianum
   -elliottii (1/2)
```

Trusses of 15 flowers, salmon pink with darker spotting toward
base of petal. (Lord Aberconway, 1945) A. M. (RHS) 1948

cl.

Elsa Reid--form of <u>souliei</u>

Flowers cream-colored with crimson spot in throat, 2.5in(6.4cm)
wide; trusses of 5-8.  Plant upright, spreading, 5ft(1.5m) tall.
(Collector unknown; raised by Mrs. A. C. U. Berry; intro. by The
Bovees Nursery, 1968; reg. 1970)


g. & cl.
        -<u>grande</u> (1/2)
Elsae-
        -<u>hodgsonii</u> (1/2)

Ivory white flowers with crimson basal blotch.  (Reuthe)  A. M.
(RHS) 1925


cl.
          -<u>ciliicalyx</u> (1/2)
Else Frye-
          -unknown (1/2)

5-8ft(1.5-2.4m)          15°F(-10°C)          E          4/3

White flowers with chrome yellow throat, exterior flushed rose,
4in(10.2cm) wide, very fragrant; loose trusses of 3-6. Glossy,
deeply veined, leathery leaves.  (P. Bowman, reg. 1963)      See
color illus. VV p. 78.


cl.

Elsie Louisa--form of <u>macgregoriae</u>

Vireya.  A cool greenhouse plant with leaves up to 2.5in(6.4cm)
long.  Flowers tubular in loose trusses of up to 10, color
bright carrot red to cadmium orange.  (Collected and raised by
Michael Black)  A. M. (RHS) 1977


g.
              -<u>souliei</u> (1/2)
Elsie Phipps-                  -<u>campylocarpum</u> ssp. <u>campylocarpum</u> Elatum
              -Penjerrick-                                        Group (1/4)
                               -<u>griffithianum</u> (1/4)

Flowers of pale yellow, flushed pink.  (Lord Aberconway, 1941)


cl.
            -<u>campylocarpum</u> ssp. <u>campylocarpum</u> (1/2)
Elsie Straver-
              -unknown (1/2)

Flowers creamy yellow, dark red blotch in throat, campanulate;
trusses of 12-16.  Yellowish green, wrinkled leaves.  Late midseason. (Straver, Boskoop, reg. 1969) Gold Medal (Boskoop) 1966


cl.           -<u>pubescens</u> (1/4)
        -Chesapeake-
Elsmere-        -<u>keiskei</u> (1/4)
        -unknown (1/2)

2ft(.6m)          -25°F(-32°C)          M

Lemon yellow flowers 1in(2.5cm) across; clusters of 5-6 scattered over the plant.  Leaves to 2in(5cm).  See also Elam. (G. Guy
Nearing, reg. 1972)


cl.
        -<u>campylocarpum</u> ssp. <u>campylocarpum</u> (1/2)
Elspeth-
        -hardy hybrid--unknown (1/2)

4ft(1.2m)          0°F(-18°C)          M          3/4

Light green leaves on an upright, compact plant.  Bright scarlet
buds opening to deep pinkish apricot, fading to cream, in small
rounded trusses. (Slocock) A. M. (RHS) 1937


                          -<u>campylocarpum</u> ssp. <u>campylocarpum</u>
              -Penjerrick-                      Elatum Group (1/16)
g.      -Amaura-          -<u>griffithianum</u> (1/16)
      -Venus-    -<u>griersonianum</u> (1/8)
Elvus-    -<u>facetum</u> (<u>eriogynum</u>) (1/4)
      -<u>elliottii</u> (1/2)

Red flowers.  (Lord Aberconway, intro. 1946)


                     -<u>dichroanthum</u> ssp. <u>dichroanthum</u> (1/8)
cl.         -Fabia-
      -unnamed hybrid-    -<u>griersonianum</u> (1/8)
Elya-            -<u>bureavii</u> (1/4)
      -<u>yakushimanum</u> ssp. <u>yakushimanum</u> (1/2)

Truss of 7-8 flowers, 5-6 lobed, neyron rose.  Leaves have
orange tan, felt-like indumentum.  (H. L. Larson, 1982)


cl.
         -<u>ciliatum</u> (1/2)
Emasculum-
         -<u>dauricum</u> (1/2)

5ft(1.5m)          -10°F(-23°C)          VE          4/3

Pale rosy lilac flowers, funnel-shaped, 2in(5cm) wide, 5-lobed,
in trusses of 2 or 3; very free-flowering.  Dense, twiggy plant;
small, lightly scaly leaves.  (Waterer, 1958) A. M. (RHS) 1976


cl.
          -<u>campylocarpum</u> ssp. <u>campylocarpum</u> (1/2)
Ember Glow-
          -<u>forrestii</u> ssp. <u>forrestii</u> Repens Group (1/2)

A rounded, dwarf shrub.  Blood red flowers, campanulate.  Early.
(Cross by Sir John Ramsden; intro. by Sir William Pennington-
Ramsden, 1965)


cl.
            -<u>thomsonii</u> ssp. <u>thomsonii</u> (1/2)
Embley Blush-
            -<u>campylocarpum</u> ssp. <u>campylocarpum</u> (1/2)

Exminster g.  (J. J. Crosfield, 1935)


cl.

Embley Park     Parentage as above

Exminster g. Flowers of pale rose with yellowish tinge in tube.
(J. J. Crosfield, 1936)  A. M. (RHS) 1936


cl.

Embley Pink     Parentage as above

Exminster g.  Flowers of rose. (J. J. Crosfield, 1930)   A. M.
(RHS) 1930


cl.                      -<u>griffithianum</u> (1/4)
              -unnamed hybrid-
Emeline Buckley-              -unknown (3/4)
              -Bacchus--unknown

Flowers bright pink with dark blotch.  (C. B. van Nes & Sons)


              -<u>wardii</u> var. <u>wardii</u> (1/4)
              -          -<u>griffithianum</u> (3/32)
       -Idealist-        -Kewense-
g.     -      -Aurora-      -<u>fortunei</u> ssp. <u>fortunei</u>
       -      -Naomi-      -              (15/32)
Emerald-      -      -<u>thomsonii</u> ssp. <u>thomsonii</u> (3/16)
  Isle -      -<u>fortunei</u> ssp. <u>fortunei</u>
       -              -<u>griffithianum</u>
       -      -Kewense-
   -      -Aurora-      -<u>fortunei</u> ssp. <u>fortunei</u>
   -Exbury-      -<u>thomsonii</u> ssp. <u>thomsonii</u>
     Naomi-<u>fortunei</u> ssp. <u>fortunei</u>

Bell-shaped flowers of chartreuse green, darker in the throat.
See also New Comet. (RHS Garden, Wisley)   A. M. (RHS) 1956

```
cl.                              -haematodes ssp. haematodes (3/4)
            -May Day (dwarf form)-
Emett Adams-                     -griersonianum (1/4)
            -haematodes ssp. haematodes
```

Blood red flowers, tubular-campanulate, in loose trusses of 6-8.
Plant semi-dwarf, rather compact. Leaves heavily indumented below; new growth also hairy on top. (C. S. Seabrook, reg. 1969)

```
                  -caucasicum (1/4)
cl. -Boule de Neige-            -catawbiense (3/8)
    -              -hardy hybrid-
Emil-              -              -unknown (3/8)
    -              -catawbiense
    -Mrs. Milner-
                  -unknown
```

White flowers with a soft pink touch and greenish yellow
markings. (T. J. R. Seidel, intro. 1903)

```
                                    -griffithianum (3/8)
cl.                   -unnamed hybrid-
            -J. H. van Nes-          -unknown (3/8)
Emily Allison-        -Monsieur Thiers--unknown
            -                        -griffithianum
            -Loderi King George-
                                -fortunei ssp. fortunei (1/4)

4ft(1.2m)          10°F(-12°C)          ML
```

Flowers white with currant red ring in throat, very fragrant, 7-
lobed, of heavy substance, to 5in(12.7cm) across; trusses spherical, 8in(20.3cm) wide, hold 11. Floriferous. Plant upright,
well-branched, nearly as broad as tall; leaves held 1-2 years.
(Cross by Halfdan Lem; intro. & reg. by Joy & Jos. Bailey, 1983)

```
cl.
            -griffithianum (1/2)
Emily Mangles-
            -unknown (1/2)
```

Synonym Polly Peachum. Flowers pink with crimson blotch. (Mangles at Littleworth)

```
cl.          -catawbiense (1/2)
    -Annedore-
Emma-          -unknown (1/2)
    -          -catawbiense
    -Mrs. Milner-
                -unknown
```

Bright carmine red flowers with darker markings on a light background. (T. J. R. Seidel, intro. 1917)

cl.

Empereur de Maroc     Parentage unknown

A parent of Blue River. Brilliant purple flowers. (Origin unknown)

```
cl.
        -Moser's Maroon--unknown (1/2)
Empire Day-
        -griersonianum (1/2)

5ft(1.5m)          0°F(-18°C)          ML          2/2
```

Romany Chai g. of Rothschild. Blood red flowers, otherwise
similar to Romany Chai. (Shown by Knap Hill)   A. M. (RHS) 1932

```
---                        -brookeanum var. gracile (1/4)
        -Crown Princess of Germany-
Empress-                            -jasminiflorum
        -                  -Princess Royal-      (1/8)
        -javanicum                   -javanicum (5/8)
```

Vireya hybrid. Flowers yellow to orange.   F. C. C. (RHS) 1884

cl.

Empress Eugenie     Parentage unknown

A parent of Euthom. Flowers creamy white, finely spotted.

```
cl.
        -aberconwayi (1/2)
Enborne-
        -maculiferum ssp. anhweinse (1/2)
```

Compact trusses of 12-14 flowers, white flushed phlox pink, upper throat speckled pink, widely funnel-campanulate. A dwarf
plant with reflexed leaves 3in(7.6cm) long. (Crown Estate, Windsor) A. M. (RHS) 1966

```
g.
        -arboreum ssp. cinnamomeum var. album (1/2)
Endeavour-
        -lacteum (1/2)
```

A tall shrub, early-blooming; white flowers lightly flushed with
pale yellow, 5-lobed, tubular-campanulate; full trusses of 13-16
flowers. Leaves matte dark green above, silvery brown plastered
indumentum below. (L. de Rothschild; exhibited by E. de Rothschild) A. M. (RHS) 1983

```
cl.
        -souliei (1/2)
Endre Ostbo-
        -fortunei ssp. discolor (1/2)

6ft(1.8m)          0°F(-18°C)          ML          3/3
```

Flowers pale blush pink, fringed darker pink, spotted red in upper corolla, saucer-shaped, to 4in(10.2cm) wide; lax trusses of
about 9. (Endre Ostbo) P. A. (ARS) 1954

```
cl.
        -arboreum ssp. arboreum (1/2)
Endsleigh Pink-
        -unknown (1/2)
```

Small, pale rose flowers, lightly dotted, in neat trusses, on a
large bush. More cold-tolerant than arboreum. (Origin unknown)

```
                  -elliottii (1/2)
cl.    -Jutland-               -fortunei ssp. discolor
    -          -         -Norman-        (1/16)
Englemere-    -Bellerophon- Shaw -        -catawbiense
    -         -elliottii    -    -B. de Bruin-     (1/32)
    -Royal-             -         -unknown (9/32)
    Blood-          -facetum (eriogynum) (1/8)
             -Rubens--unknown
```

Tight rounded trusses 7in(17.8cm) wide, hold about 32 flowers of
medium cardinal red. Leaves 6in(15cm) long, dark green. (Crown
Estate) A. M. (RHS) 1971

```
cl.
                            -catawbiense (1/2)
English Roseum*--sport of Roseum Elegans ? -
                            -unknown (1/2)

6ft(1.8m)          -25°F(-32°C)          ML          3/3
```

Flowers are soft rosy pink, with smooth, glossy leaves of dark
green. A vigorous, fast-growing hybrid, tolerant of extremes in
cold, heat, and humidity. (A. Waterer)

cl.

Epoch--form of <u>minus</u> var. <u>minus</u> Carolinianum Group (white form)
     grown from colchicine-treated seed

3ft(.9m)          -10°F(-23°C)          M          3/2

A tetraploid rhododendron. Flowers white, shaded pink, frilled,
of good substance, saucer-shaped, up to 2.5in(6.4cm) wide; ball-
shaped trusses hold about 12. Plant rounded, wider than tall;
deep green foliage held 2-3 years. (Dr. A. E. Kehr, 1972)  See
color illus. K p. 56.

cl.
        -C. O. D.--unknown (3/4)
Erchless-                             -<u>griffithianum</u> (1/8)
       -                 -unnamed hybrid-
       -Mrs. Furnival-                -unknown
       -                               -<u>caucasicum</u> (1/8)
                         -unnamed hybrid-
                               -unknown

When registered this plant was 6ft(1.8m) tall, but its age was
not stated. Flowers of pale purplish pink with dark red eye, 5-
lobed, frilly-edged. Open, spreading habit. (Intro. by Howard
Phipps, 1955; reg. 1972)

g.          -<u>dichroanthum</u> ssp. <u>dichroanthum</u> (1/4)
        -Fabia-
Erebus-     -<u>griersonianum</u> (3/4)
        -<u>griersonianum</u>

Flowers brilliant deep scarlet, funnel-shaped, in loose trusses.
A compact plant. Midseason. (Lord Aberconway, 1936)

cl.

Eric Rudd--form of <u>facetum</u> (<u>eriogynum</u>) (F 13508)

Loose trusses of up to 13 flowers, 5-lobed, 2.75in(7cm) across,
color Delft rose. Leaves are 6.5in(16.5cm) long, narrowly
elliptic, dull dark green. (Collector George Forrest; raised by
Col. S. R. Clarke; reg. by R. N. Stephenson Clarke)  A. M. 1980

cl.                             -<u>dichroanthum</u> ssp. <u>dichroanthum</u>
            -Fabia High Beeches-                              (1/4)
Eric Stockton-                  -<u>griersonianum</u> (1/2)
          -                 -<u>griersonianum</u>
            -Tally Ho-
                      -<u>facetum</u> (<u>eriogynum</u>) (1/4)

Leaves are 6.75in(17cm) long with brown indumentum. Flowers of
jasper red with darker veins, widely funnel-campanulate; trusses
of 5-9. (Raised by G. H. Loder;  exhibited by  H. E. Boscawen)
A. M. (RHS) 1967

g.          -<u>dichroanthum</u> ssp. <u>scyphocalyx</u> (1/4)
        -Medusa-
Eridusa-    -<u>griersonianum</u> (1/4)
        -<u>facetum</u> (<u>eriogynum</u>) (1/2)

Deep red flowers. (Lord Aberconway, intro. 1946)

cl.
        -<u>japonicum</u> var. <u>japonicum</u> (<u>metternichii</u>) (1/2)
Erika-
        -Alexander Adie--unknown (1/2)

Light carmine pink flowers with brown markings. (T. J. R. Seidel, intro. 1903)

g.          -<u>griersonianum</u> (1/4)
        -Azor-
Eriozor-    -<u>fortunei</u> ssp. <u>discolor</u> (1/4)
        -<u>facetum</u> (<u>eriogynum</u>) (1/2)

Red flowers. (Lord Aberconway, 1946)

cl.                             -<u>griffithianum</u> (1/4)
              -Queen Wilhelmina-
       -Britannia-              -unknown (3/8)
Ermine-        -Stanley Davies--unknown
       -                    -Sir Charles Butler--<u>fortunei</u> ssp.
       -Mrs. A. T. de la Mare-                  <u>fortunei</u> (1/4)
                              -<u>maximum</u> (1/8)
                      -Halopeanum-
                              -<u>griffithianum</u>

3ft(.9m)          0°F(-18°C)          ML          3/3

Flowers 3.5in(9cm) across, white, unspotted, funnel-campanulate;
compact, conical trusses hold about 11. Plant wider than tall,
with a stiff growth habit; medium green leaves 6in(15.2cm) long.
Tolerates full sun. (Rudolph Henny, 1959)

cl.
        -<u>fortunei</u> ssp. <u>fortunei</u> (1/2)
Ernest Gill-
        -<u>arboreum</u> ssp. <u>arboreum</u> (1/2)

Flowers bright pink with crimson basal blotch. (R. Gill & Son)
A. M. (RHS) 1918

cl.
        -<u>yakushimanum</u> ssp. <u>yakushimanum</u> (1/2)
Ernest Inman-           -<u>ponticum</u> (1/4)
        -Purple Splendour-
                        -unknown (1/4)

4ft(1.2m)          -10°F(-23°C)          EM

Lavender fading to white at base. Very free-flowering. Dense,
dark foliage on a very compact plant. See also Caroline All-
brook. (A. F. George, Hydon Nurseries, 1976)  A. M. (RHS) 1979

cl.                             -<u>griffithianum</u> (1/8)
              -Queen Wilhelmina-
       -Britannia-              -unknown (3/8)
Ernest R.-     -Stanley Davies--unknown
  Ball  -              -<u>dichroanthum</u> ssp. <u>dichroanthum</u> (1/4)
        -Jasper    -          -<u>fortunei</u> ssp. <u>discolor</u> (1/8)
         (Exbury form)-Lady   -
                     -Bessborough-<u>campylocarpum</u> ssp. <u>campylo-</u>
                                  carpum Elatum Group (1/8)

Flowers of orange yellow with some darker unfading stripes, flat
funnel-campanulate, to 3.5in(8.9cm) wide; tight trusses hold up
to 14. Plant of open habit; foliage like that of Britannia. (C.
S. Seabrook, 1967)

g.
        -<u>yunnanense</u> (1/2)
Ernestine-
        -<u>cinnabarinum</u> ssp. <u>cinnabarinum</u> Roylei Group (1/2)

Lilac pink, trumpet-shaped flowers in great profusion. Growth
habit tall, compact; foliage has bluish cast. (Rothschild, 1937)

cl.
        -<u>dauricum</u> (1/2)
Ernie Dee-
        -<u>racemosum</u> (1/2)

1ft(.3m)          0°F(-18°C)          EM          3/4

Slightly frilled purple flowers with dorsal red spotting, flattened, to 1in(2.5cm) wide; terminal inflorescence of 4-8 trusses
each 2-flowered; long-lasting, abundant blooms. Plant more than
twice as broad as high, well-branched; small, dark green leaves
held 2 years. (J. F. Caperci, reg. 1976)

g.          -<u>campylocarpum</u> ssp. <u>campylocarpum</u> Elatum
        -Penjerrick-                           Group (1/8)
   -Amaura-          -<u>griffithianum</u> (1/8)
Eros-       -<u>griersonianum</u> (3/4)
   -<u>griersonianum</u>

Pale pink flowers. (Lord Aberconway, 1936)

```
g.            -dichroanthum ssp. dichroanthum (1/8)
     -Fabia-
  -Erebus-    -griersonianum (3/8)
Erso-     -griersonianum
    -souliei (1/2)
```

(Lord Aberconway, intro. 1946)

```
g.                    -griffithianum (1/4)
      -Loderi King George-
Esmeralda-            -fortunei ssp. fortunei (1/4)
        -neriiflorum ssp. neriiflorum (1/2)
```

Flowers of pale pink to deep rose. (Rothschild, 1937)

```
g.
      -barbatum (1/2)
Espernza-
      -strigillosum (1/2)
```

Deep red flowers, tubular campanulate, similiar to strigillosum.
Plant tall, not compact. Early. (Rothschild, 1937)

```
cl.
      -griersonianum (1/2)
Esquire-
      -unknown (1/2)
```

4ft(1.2m)        5°F(-15°C)        M        4/3

Buds long, pointed, opening to watermelon pink flowers about 4in
(10.2cm) across, in loose round trusses. Habit upright, spreading; dark green leaves with red brown petioles. (Raised by Barto; intro. by D. James before 1958)

```
cl.
      -falconeri ssp. eximium (1/2)
Essa-
      -unknown (1/2)
```

This plant first flowered over 20 years after the seeds germin-
ated. Large leaves, 10in(25.4cm) long by 4in(10.2cm) wide with
patchy thin tomentum beneath. White flowers with slight staining of pink, deep crimson blotch; rounded trusses of 23. (A. C.
& E. F. A. Gibson) A. M. (RHS) 1964

```
cl.

Essex Scarlet    Parentage unknown
```

Often a parent. Flowers deep crimson scarlet with almost black
blotch. (G. Paul, 1899) A. M. (RHS) 1899

```
cl.                -griffithianum (1/4)
      -Loderi Venus-
Estelle Gatke-      -fortunei ssp. fortunei (1/4)
      -              -griersonianum (1/4)
        -Tally Ho-
             -facetum (eriogynum) (1/4)
```

Introduced by R. M. Gatke, 1958.

```
                          -caucasicum (3/32)
                -Jacksonii-
            -Goldsworth-     -caucasicum
cl.         - Yellow   -Nobleanum-
   -Goldfort-     -          -arboreum ssp.
Ester -      -    -campylocarpum ssp. arboreum (1/32)
Dudley-      -fortunei ssp. fortunei (1/4)  campylocarpum
        -      -griffithianum (1/4)           (1/8)
        -King George-
             -unknown (1/4)
```

Yellow flowers, deeper in throat. (H. D. Rose, Australia, 1973)

```
g.            -arboreum ssp. cinnmomeum var. album (1/2)
Esterel-
      -meddianum var. meddianum (1/2)
```

Rosy pink, waxy flowers; medium-sized plant. (Intro. by Rothschild, 1937)

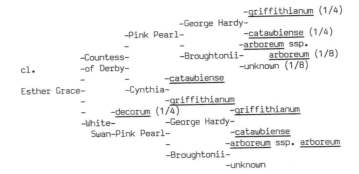

```
                              -griffithianum (1/4)
                    -George Hardy-
            -Pink Pearl-      -catawbiense (1/4)
            -              -arboreum ssp.
      -Countess-     -Broughtonii-  arboreum (1/8)
cl.      -of Derby-          -unknown (1/8)
Esther Grace-     -Cynthia-  -catawbiense
      -              -griffithianum
      -    -decorum (1/4)      -griffithianum
      -White-          -George Hardy-
        Swan-Pink Pearl-     -catawbiense
             -              -arboreum ssp. arboreum
             -Broughtonii-
                  -unknown
```

White flowers with few reddish brown spots, quite flat, ruffled,
to 4.5in(11.5cm) wide; trusses 14-flowered. Plant to 5ft(1.5m),
open; pointed leaves 5in(12.7cm) long. Late midseason. (The Bovees, 1972)

```
cl.            -griffithianum (3/8)
      -Mars-
Esther Packard-   -unknown (1/2)
      -              -griffithianum
      -        -Loderi-
      -unnamed hybrid-   -fortunei ssp. fortunei
        -unknown          (1/8)
```

2.5ft(.75m)        -5°F(-21°C)        M

Flowers fuchsine pink, campanulate, to 4in(10.2cm) wide; rounded
trusses. Plant low, spreading; glossy dark green leaves. (Melvin V. Love, 1964)

```
g. & cl.
      -neriiflorum ssp. neriiflorum (1/4)
   -F. C. Puddle-
Ethel-      -griersonianum (1/4)
   -forrestii ssp. forrestii Repens Group (1/2)
```

1.5ft(.45m)        5°F(-15°C)        EM        3/3

Blazing scarlet flowers, campanulate, double calyx; trusses hold
3-5. Creeping, spreading habit, very compact; leaves glossy and
purple-backed. (Lord Aberconway, 1934) F. C. C. (RHS) 1940

```
cl.
      -brachycarpum ssp. brachycarpum (1/2)
Ethel V. Cary-            -catawbiense (1/4)
      -Mrs. Charles S. Sargent-
             -unknown (1/4)
```

2ft(.6m)        -20°F(-29°C)        ML

Plant broader than tall; convex leaves have thin indumentum; new
growth silver green with red stems. Flowers with orchid edge,
fading white in center, reverse solid orchid; spherical trusses
of about 20. (Edward A. Cary, reg. 1973)

```
g. & cl.
   -campylocarpum ssp. campylocarpum (1/2)
Ethyl-
   -orbiculare ssp. orbiculare (1/2)
```

A medium-sized shrub with foliage like orbiculare. Flowers of
cream, tinged pink. Early. (Rothschild, 1937)

```
                  -souliei (1/8)
g.          -Soulbut-
    -Vanessa-         -Sir Charles Butler--fortunei ssp. fortunei
Etna-       -griersonianum (1/2)                      (1/8)
    -       -dichroanthum ssp. dichroanthum (1/4)
    -Fabia-
            -griersonianum
```

Magenta red flowers.  (Lord Aberconway, intro. 1936)


cl.


Etna      See Lady Chamberlain Etna (Rothschild)

```
                        -campylocarpum ssp. campylocarpum
              -Penjerrick-                Elatum Group (1/16)
g.    -Amaura-        -griffithianum (1/16)
    -Eros-    -griersonianum (3/8)
Etos-    -griersonianum
    -thomsonii ssp. thomsonii (1/2)
```

Deep red flowers.  (Lord Aberconway, intro. 1946)

```
cl.            -elliottii (1/4)
        -Fusilier-
Etta Burrows-    -griersonianum (1/4)
        -strigillosum (1/2)
```

6ft(1.8m)          5°F(-15°C)         E         4/4

Plant broader than tall; fir green leaves to 9in(22.9cm) long,
indumentumed like strigillosum.  Bright blood red flowers, 3in.
(7.6cm) wide, campanulate; rounded trusses of about 30.  (H. L.
Larson, 1965)  See color illus. HG p. 80.

```
cl.                -catawbiense (1/2)
Etzel--Everestianum (selfed)-
                   -unknown (1/2)
```

Pale carmine pink flowers with light green markings.  See also
Effner. (T. J. R. Seidel cross, 1895; intro. 1903)

```
cl.
    -hanceanum Namum Group (1/2)
Euan Cox-
    -ludlowii (1/2)
```

Plant spreads much wider than tall, with leaves 1in(2.5cm) long,
dull dark green. A dwarf of compact habit.  Flowers light yellow
spotted with brick dust red, 2in(5cm) wide, in trusses of 3.
(E. H. M. Cox cross, 1965; reg. by Glendoick Gardens, 1981)
A. M. (Wisley Trials) 1981

```
---
    -campylocarpum ssp. campylocarpum (1/2)
Eucamp-
    -Empress Eugenie--unknown (1/2)
```

Cream-colored flowers.  (E. J. P. Magor)

```
cl.
    -forrestii ssp. forrestii Repens Group (1/2)
Euchelia-
    -unknown (1/2)
```

Flowers bright deep crimson, of heavy substance.  (Lord Aberconway) A. M. (RHS) 1935

```
                -souliei (1/8)
g.          -Soulbut-
    -Vanessa-        -Sir Charles Butler--fortunei ssp.
Eudora-    -griersonianum (1/4)        fortunei (1/8)
    -facetum (eriogynum) (1/2)
```

Flowers reddish scarlet.  (Lord Aberconway, intro. 1936)

```
cl.            -campylocarpum ssp. campylocarpum (1/4)
        -Unique-
Eugene*    -unknown (1/2)
    -                    -griffithianum (1/4)
        -Jean Marie de Montague-
                        -unknown
```

4ft(1.2m)          -5°F(-21°C)          M          3/4

Unusually large "pine cone" flower buds with interesting bracts.
Flowers coral red, in tight rounded trusses.  Foliage deep green
and very dense.  (Harold E. Greer)

```
---
    -Empress Eugenie--unknown (1/2)
Euking-        -griffithianum (1/4)
    -Mrs. Kingsmill-
            -campylocarpum ssp. campylocarpum (1/4)
```

Flowers of clear lemon yellow with faint red spotting. (E. J. P.
Magor, 1935)

```
cl.            -Monsieur Thiers--unknown (1/4)
        -J. H. Van Nes-
Eulalie Wagner-        -griffithianum (1/2)
    -                    -griffithianum
            -Loderi King George-
                    -fortunei ssp. fortunei (1/4)
```

Flowers of fuchsine pink with darker veins, 4.5in(11.5cm) wide,
funnel-shaped,  in 12-flowered trusses, 8in(20.3cm) wide by 9in.
(22.9cm) high.  Elliptic dark green leaves, 5in(12.7cm) long.
(Raised by Halfdan Lem; intro. by C. Fawcett; reg. 1974)  P. A.
(ARS) 1963

```
                -souliei (1/16)
            -Soulbut-
g.      -Vanessa-    -Sir Charles Butler--fortunei ssp.
    -Eudora-    -griersonianum (1/8)        fortunei (5/16)
Eulo-    -facetum (eriogynum) (1/4)
    -        -griffithianum (1/4)
    -Loderi-
        -fortunei ssp. fortunei
```

Pale pink flowers.  (Lord Aberconway, 1941)

```
g.
    -sperabile var. sperabile (1/2)
Eupheno-
    -griersonianum (1/2)
```

Deep red flowers.  (Lord Aberconway, 1936)

```
g. & cl.
    -arboreum ssp. arboreum (1/2)
Euphrosyne-    -griffithianum (1/4)
        -Loderi-
            -fortunei ssp. fortunei (1/4)
```

Large flowers of bright carmine are speckled with crimson.
Clone Euphrosyne Ruby was exhibited by Sir Edmund Loder. (Rothschild)  A. M. (RHS) 1923

```
g. & cl.
    -arboreum ssp. arboreum (1/2)
Eureka-
    -hookeri (1/2)
```

A medium-sized, compact plant. Blood red flowers; trusses of 15
to 18. (Rothschild, 1937)  A. M. (RHS) 1939

```
g.
    -ungernii (1/2)
Europa-
    -kyawii (1/2)
```

6ft(1.8m)          5°F(-15°C)          L          3/3

Flowers rosy lilac, tubular, to 3in((7.6cm) across, in rounded open trusses. Attractive foliage: elliptic, glossy, dark green leaves with prominent veining and tan indumentum. (Not shown in Rothschild Rhododendrons) (Rothschild, 1937)  See color illus. ARS J 38:2 (Spring 1984): p. 54.

```
g.            -neriiflorum ssp. neriiflorum (1/4)
        -Nereid-
Euryalus-     -dichroanthum ssp. dichroanthum (1/4)
        -griersonianum (1/2)
```

Deep red flowers. (Lord Aberconway, 1936)

```
g. & cl.
        -arboreum ssp. cinnamomeum var. album (1/2)
Eurydice-
        -kyawii (1/2)
```

A medium-sized plant; attractive foliage.  Large white flowers tinged with rose and crimson. (Rothschild) A. M. (RHS) 1937

```
cl.
        -oreotrephes (exquisitum) (1/2)
Euterpe-
        -augustinii ssp. augustinii (1/2)
```

Rose-colored flowers. (Reuthe, 1941)

```
cl.
        -caucasicum (1/2)
Euterpe*
        -unknown (1/2)
```

Pale pink buds open to pure white flowers with greenish yellow markings. See also  Herman Seidel, Jewess. (T. J. R. Seidel cross, 1894)

```
g.
        -Empress Eugenie--unknown (1/2)
Euthom-
        -thomsonii ssp. thomsonii (1/2)
```

Flowers are light carmine shading into pink, spotted crimson. (E. J. P. Magor)

```
cl.            -diaprepes (1/4)
        -Polar Bear-
Eva Rebecca-     -auriculatum (1/4)
      -        -fortunei ssp. discolor (1/4)
      -Autumn Gold-  -dichroanthum ssp. dichroanthum
              -Fabia-                      (1/8)
                  -griersonianum (1/8)
```

3.5ft(1m)        -5°F(-21°C)        L

Flowers orange shading to yellow, 4in(10.2cm) wide, openly funnel-shaped, 7-lobed, fragrant; trusses 8in(20.3cm) across  hold 8-10. Upright plant, broader than high; leaves 7.5in(19cm) long by 2in(5cm) wide, retained 2 years. (John A. Holden, 1979)

```
cl.

Evangeline      See Fundy
```

```
cl.            -griffithianum (3/8)
       -Queen Wilhelmina-
   -Britannia-         -unknown (3/8)
Evelyn-      -Stanley Davies--unknown
   -            -griffithianum
   -Loderi Venus-
            -fortunei ssp. fortunei (1/4)
```

Flowers blush pink and white. (Rudolph Henny, intro. 1956)

```
g. & cl.
        -hodgsonii (1/2)
Evening-      -falconeri ssp. falconeri (1/4)
        -Muriel-
            -grande (1/4)
```

See also Mist. (Sir Giles Loder, 1950)

```
cl.
        -fortunei ssp. discolor (1/2)
Evening Glow-      -dichroanthum ssp. dichroanthum (1/4)
        -Fabia-
            -griersonianum (1/4)
```

5ft(1.5m)        -5°F(-21°C)        L        4/3

Flowers bright yellow with a prominent calyx; lax trusses. Compact plant habit; light green, narrow leaves, 5in(12.7cm) long. Prefers light shade. (Van Veen)  See color illus. VV p. 19.

```
g.            -griffithianum (1/4)
        -Sunrise-
Eventide-      -griersonianum (3/4)
        -griersonianum
```

Flowers of pale rose pink. (Lord Aberconway, intro. 1941)

```
cl.

Everchoice      Parentage unknown
```

A parent of Robert Huber, a hybrid reg. by Charles Herbert.  May be a Dexter hybrid.

```
cl.
        -catawbiense (1/2)
Everestianum-
        -unknown (1/2)
```

6ft(1.8m)        -15°F(-26°C)        ML        2/3

A vigorous plant that flowers heavily in sun or light shade, and is pest-resistant.  Flowers rosy lilac, spotted in throat, edges frilled, 2in(5cm) wide; rounded trusses of about 15.  Dark green oval leaves, 5in(12.7cm) long. (A. Waterer, before 1850)

```
cl.
        -fortunei ssp. fortunei ? (1/2)
Everitt's Hardy Mauve-
        -unknown (1/2)
```

A hybrid by C. O. Dexter; introduced by Everitt.

4ft(1.2m)        -10°F(-23°C)        EM        4/4

Flowers clear pink with lighter pink rays; the calyx rolls back its lobes as the flower ages, making a star-shaped appearance at the center of the truss.  Plant habit compact; dull green leaves of medium size. (Harold E. Greer, 1982)  See color illus. HG p. 78.

g.
```
          -sanguineum ssp. didymum (1/2)
Exburiense-
          -kyawii (1/2)
```

One of the latest-flowering red rhododendrons. Flowers campanulate, waxy, very dark red.  Medium-sized, spreading plant; dark
green, rounded leaves. (Rothschild, 1937)

cl.

Exbury--form of praestans (coryphaeum)

Large leaves, 11in(28cm) by 4in(10.2cm), on a very large shrub
or small tree.  Flowers white, a translucent shade of very pale
yellow, throat blotched with crimson, 2in(5cm) wide; large globular trusses of 32. (Edmund de Rothschild) A. M. (RHS) 1963

cl.
```
                    -griffithianum (1/4)
               -Loderi-
Exbury Albatross-       -fortunei ssp. fortunei (1/4)
               -fortunei ssp. discolor (1/2)
```

6ft(1.8m)          0°F(-18°C)          L          4/3

Albatross g.  Flowers of blush pink specked with mahogany in the
throat.  A fine, tall plant with large flowers and long, broad
leaves. (Rothschild, 1930) F. C. C. (RHS) 1935

cl.
```
               -griffithianum (1/2)
Exbury Angelo-
               -fortunei ssp. discolor (1/2)
```

Angelo g.  A tree-like plant with leaves 9in(23cm) long.  Flowers white  with the green spotting typical of the grex. (Rothschild, 1930) F. C. C. (RHS) 1947

cl.

Exbury Antonio      Not known at Exbury, but shown in the Register.
                    See Antonio.

cl.
```
               -calophytum var. calophytum (1/2)
Exbury Calstocker-       -caucasicum (1/4)
               -Dr. Stocker-
                         -griffithianum (1/4)
```

6ft(1.8m)          -5°F(-21°C)          E          5/4

Calstocker g.  Flowers creamy light pink with rich red flare in
the center, held in huge trusses of about 23.  A tree-like rhododendron; large, thick leaves. (Rothschild) A. M. (RHS) 1948

cl.
```
               -thomsonii ssp. thomsonii (1/2)
Exbury Cornish Cross-
               -griffithianum (F.C.C. form) (1/2)
                    (aucklandii)
```

See also the Cornish Cross grex made by Samuel Smith, Head Gardener at Penjerrick. The synonym aucklandii above, was the name
used by Lionel de Rothschild.  Exbury Cornish Cross is a large
plant of open habit, the flowers showing variations in crimson
instead of the rose pinks of the original Cornish Cross. (Rothschild)  A. M. (RHS) 1935   See color illus. PB pl. 31.

cl.
```
               -dichroanthum ssp. dichroanthum (1/2)
Exbury Fabia-
               -griersonianum (1/2)
```

4ft(1.2m)          5°F(-15°C)          ML          3/3

Original Fabia g. by Aberconway.  The Rothschild selection is
considered the largest-flowered of all the Fabias, colored apricot yellow, tinged with pink.  A tall, spreading plant; leaves
dull green, coated with indumentum below.  See also Fabia High
Beeches. (Rothschild)

cl.
```
          -wardii var. wardii (1/2)
Exbury Hawk-          -fortunei ssp. discolor (1/4)
          -Lady Bessborough-
                         -campylocarpum ssp. campylocarpum
                                        Elatum Group (1/4)
```

6ft(1.8m)          5°F(-15°C)          M          3/2

Hawk g.  One of the seven named clones from the search for pure
yellow.  This one is a clear yellow, but not as rich in color as
others in the group. (Rothschild, 1940) A. M. (RHS) 1949

cl.
```
               -griffithianum (1/2)
Exbury Isabella-
               -auriculatum (1/2)
```

6ft(1.8m)          5°F(-15°C)          L          3/3

Isabella g.  A dense plant with large trusses holding fragrant
white flowers. (Rothschild, 1948)

cl.
```
                         -fortunei ssp. discolor (1/4)
          -Lady Bessborough-
          -                -campylocarpum ssp. campylocarpum
Exbury Jalisco*                          Elatum Group (1/4)
          -     -dichroanthum ssp. dichroanthum (1/4)
          -Dido-
               -decorum (1/4)
```

4ft(1.2m)          5°F(-15°C)          ML          4/4

Jalisco g.  One of the best selected clones in the group.  Plant
is medium-sized, with richly green leaves and dusky yellow widely funnel-campanulate flowers. (Rothschild, 1942)

cl.
```
          -cinnabarinum ssp. cinnabarinum Roylei Group (1/2)
Exbury Lady-          -cinnabarinum ssp. cinnabarinum (1/4)
Chamberlain-Royal Flush -
               (orange form)-maddenii ssp. maddenii (1/4)
```

5ft(1.5m)          10°F(-12°C)          ML          4/3

Lady Chamberlain g.  Flowers of orange or yellow flushed with
salmon, pendulous and tubular as cinnabarinum, but larger.
Plant medium-sized, erect, with beautiful foliage. (Rothschild)
F. C. C. (RHS) 1931

cl.
```
               -griersonianum (1/2)
Exbury Matador-
               -strigillosum (1/2)
```

3ft(.9m)          15°F(-9°C)          M          3/3

Original Matador g. by Lord Aberconway.  The Exbury form has orange scarlet flowers.  Plant habit is open; dark green foliage
has buff indumentum beneath. (Rothschild)   See color illus. PB
pl. 65.

cl.
```
               -haematodes ssp. haematodes (1/2)
Exbury May Day-
               -griersonianum (1/2)
```

3ft(.9m)          5°F(-15°C)          EM          3/4

May Day g.  The Rothschild cross produced this clone, with dull
green foliage coated beneath with pale buff indumentum; bright
scarlet flowers of very high quality. See also Bodnant May Day,
and May Day g. by A. Williams. (Rothschild)   See color illus.
PB pl. 53.

cl.

Exbury Merlin      See Hawk Merlin

'Duchess of Edinburgh' (Parents unknown) by unknown
Photo by Schmalscheidt

'Duet' by Leach
Photo by Greer

'Earl of Athlone' by C. B. van Nes
Photo by Greer

'Edmond Amateis' by Amateis/Leach
Photo by Knierim

'Edna McCarty' by Ostbo
Photo by Greer

'Edward Dunn' by Ostbo
Photo by Greer

'Edwin O. Weber' by Weber
Photo by Greer

'Eider' by Cox
Photo by Greer

'El Camino' by Lem or Whitney ?
Photo by Greer

'Elizabeth de Rothschild' by Rothschild
Photo by Greer

'Elsae' by Reuthe
Photo by Greer

'Emasculum' by Waterer
Photo by Greer

'Endre Ostbo' by Ostbo
Photo by Greer

'Ernest Gill' by Gill
Photo by Greer

'Ernie Dee' by Caperci
Photo by Greer

'Etta Burrows' by Larson
Photo by Greer

'Euan Cox' by Cox
Photo by Greer

'Euclid' by Noble, 1850
Photo by Greer

'Eulalie Wagner' by Lem/Fawcett
Photo by Greer

'Eurydice' by Rothschild
Photo by Greer

'Everestianum' by Waterer
Photo by Greer

'Exbury Calstocker' by Rothschild
Photo by Greer

'Exbury Naomi' by Rothschild
Photo by Greer

'Exotic' by Bovee
Photo by Greer

```
                    -griffithianum (1/8)
cl.        -Kewense-
      -Aurora-        -fortunei ssp. fortunei (5/8)
Exbury-    -thomsonii ssp. thomsonii (1/4)
 Naomi-
        -fortunei ssp. fortunei
```

5ft(1.5m)          -5°F(-21°C)        M         4/4

Naomi g. Richly textured flowers of pink, with shadings of yellow and apricot. Medium compact shrub; leaves 7in(17.8cm) long, 3in(7.6cm) wide, both ends rounded. (Rothschild, 1926) H. C. (Wisley Trials) 1968; A. M. (RHS) 1933   See color illus. ARS J 36:4 (Fall 1982): p. 135; PB pl. 12, 64.


cl.

Exbury Pink*--form of souliei

5ft(1.5m)          -5°F(-21°C)        M         4/3

Exhibited by Lionel de Rothschild. F. C. C. (RHS) 1936


```
cl.
              -sanguineum ssp. didymum (1/2)
Exbury Red Cap-
              -facetum (eriogynum) (1/2)
```

3ft(.9m)          5°F(-15°C)        L         2/2

Orig. Red Cap g. by J. B. Stevenson. The Exbury form from this cross is deep crimson and blooms late. For small gardens. See also Borde Hill Red Cap, Townhill Red Cap. (Rothschild)


```
cl.
            -souliei (1/2)
Exbury Souldis-
            -fortunei ssp. discolor (1/2)
```

See also the original Souldis g. by E. J. P. Magor. The Exbury form is tall but compact; pink buds open to white flowers, with notched petals and a crimson blotch. (Rothschild, 1948)


```
cl.
               -spinuliferum (1/2)
Exbury Spinulosum-
               -racemosum (1/2)
```

Spinulosum g. Flowers are small, orange red tubes 1in.(2.5cm) long, held in clusters. Leaves are narrow, dark green and glossy, 3in(7.6cm) long. (Rothschild) A. M. (RHS) 1948


```
                    -dichroanthum ssp. dichroanthum (1/8)
            -Dido-
            -      -decorum (1/8)
      -Lem's-                    -griffithianum (7/32)
      -Cameo-           -Beauty of-
cl.   -      -Norman- Tremough-arboreum ssp. arboreum
      -      - Gill -                      (1/32)
Excalibur   -Anna-    -griffithianum
      -      -        -griffithianum
      -      -Jean Marie -
      -       de Montague-unknown (1/2)
      -
      -Pink    -Jan Dekens--unknown
      Petticoats-                    -griffithianum
             -      -Queen Wilhelmina-
             -Britannia-            -unknown
                    -Stanley Davies--unknown
```

5ft(1.5m)          0°F(-18°C)        M

Trusses 7.5in(19cm) tall, cone-shaped with 28-32 flowers rose Bengal, shading lighter in center. Upright plant with stiff branches; leaves 6.5in(16.5cm) long, retained 2 years. (Intro. by John G. Lofthouse, 1981; reg. 1983)


```
g. & cl.
            -thomsonii ssp. thomsonii (1/2)
Exminster-
            -campylocarpum ssp. campylocarpum (1/2)
```

Flowers pink over cream. (Barclay Fox, 1923)   A. M. (RHS) 1923


```
g.
        -ciliatum (1/2)
Exoniense-
        -veitchianum (1/2)
```

Creamy white flowers. (Veitch)   F. C. C. (RHS) 1881


```
cl.               -griffithianum (1/4)
    -Loderi King George-
Exotic-           -fortunei ssp. fortunei (1/4)
    -Ostbo Y3--unknown (1/2)
```

5ft(1.5m)          5°F(-15°C)        M         4/3

Red flowers blended with yellow and pink 4.5in(11.5cm) broad, in trusses of about 12. Dark green leaves 8.5in(21.6cm) long. (R. M. Bovee, 1962) P. A. (RHS) 1961


```
g.
        -javanicum (1/2)
Exquisite-
        -teysmanni (1/2)
```

Vireya hybrid. Rich canary yellow flowers, with prominent crimson anthers.   A. M. (RHS) 1899


```
cl.
        -macabeanum (1/2)
Eyestopper-
        -unknown (1/2)
```

A hybrid in New Zealand with trusses of 20 flowers primrose yellow, fading to near white. See also Pink Frills. (C. Huthnance, 1980)


```
g. & cl.
            -neriiflorum ssp. neriiflorum (1/2)
F. C. Puddle-
            -griersonianum (1/2)
```

5ft(1.5m)          5°F(-15°C)        EM         3/3

The habit is fairly upright with medium-sized, dull, dark green leaves. Flowers are a brilliant orange scarlet; trusses are lax and open. Named for the now-retired Curator/Garden Supervisor at Bodnant. (Lord Aberconway cross, 1926)   A. M. (RHS) 1932


```
cl.
        -catawbiense (1/2)
F. D. Godman-
        -unknown (1/2)
```

5ft(1.5m)          -10°F(-23°C)        M         2/3

An old hardy catawbiense hybrid; flowers dark magenta red with a black blotch. (A. Waterer, 1888)


cl.

F. L. Ames     See Amphion

g. & cl.
   -dichroanthum ssp. dichroanthum (1/2)
Fabia-
   -griersonianum (1/2)

3ft(.9m)          5°F(-15°C)        M         3/4

Several have made this cross, with various selected forms named.
Plant  usually neat and compact, with flowers carried in loose,
drooping trusses.  The Bodnant form is a wide, dome-shaped bush
with loose, flat trusses of funnel-shaped scarlet flowers shaded
orange in the tube, and speckled with light brown.  See also Exbury Fabia, Fabia High Beeches, Fabia Tangerine, Fabia Tower
Court, Minterne Apricot, Roman Pottery, Solveig.  (Lord Aberon-
way, 1934)  A.M. (RHS) 1934  See color illus. F p. 45; VV p. 72.

cl.
           -dichroanthum ssp. dichroanthum (1/2)
Fabia High Beeches-
           -griersonianum (1/2)

3ft(.9m)          5°F(-15°C)        M         3/4

Fabia g.  A variety with orange flowers.  See also  Exbury Fabia
and other clones of the Fabia group.  (Col. G. H. Loder, 1937)

cl.
         -dichroanthum ssp. dichroanthum (1/2)
Fabia Tangerine-
         -griersonianum (1/2)

3ft(.9m)          5°F(-15°C)        M         3/4

Fabia g.  Low and compact, with reddish brown indumentum on the
foliage.  Flowers slightly lighter orange color than Fabia, with
a crimson flush at the tips, and held in loose drooping trusses.
(Lord Aberconway, 1934)  A. M. (RHS) 1940

cl.
       -dichroanthum ssp. dichroanthum (1/2)
Fabia Tower Court-
     -griersonianum (1/2)

Fabia g. This form is semi-dwarf  with attractive dark foliage;
fragrant bell-shaped flowers of soft orange pink, with a deep
rose flush at the margins.  (Stevenson, 1937)

g.          -dichroanthum ssp. dichroanthum (1/4)
   -Fabia-
Fabiola-     -griersonianum (1/2)
   -         -neriiflorum ssp. neriiflorum (1/4)
   -F. C. Puddle-
          -griersonianum

Red flowers.  (Lord Aberconway, 1946)

            -souliei (1/16)
         -Soulbut-
    -Vanessa-      -Sir Charles Butler--fortunei
    -                        ssp. fortunei (1/16)
g.   -Adonis-     -griersonianum (1/2)
   -     -          -griffithianum (1/8)
Fabionis-    -Sunrise-
   -                   -griersonianum
   -     -dichroanthum ssp. dichroanthum (1/4)
   -Fabia-
       -griersonianum

Pale rose flowers.  (Lord Aberconway, 1946)

cl.          -sanguineum ssp. sanguineum (1/4)
   -Arthur Osborn-
Fabos-          -griersonianum (1/2)
  -       -dichroanthum ssp. dichroanthum (1/4)
   -Fabia-
     -griersonianum

Beacon g.  Crimson flowers.  (Reuthe, 1952)

cl.
      -fortunei ssp. fortunei (1/2)
Faggetter's Favourite-
        -unknown (1/2)

6ft(1.8m)         -5°F(-21°C)       M         5/4

Flowers are flushed pink and bronze speckled, in large trusses,
and fragrant. It is best grown with some shade. (Slocock) A. M.
(Wisley Trials) 1955   See color illus. VV p. 7.

cl.
   -arboreum ssp. cinnamomeum var. roseum (1/2)
Fair Lady-        -griffithianum (1/4)
   -Loderi Venus-
         -fortunei ssp. fortunei (1/4)

6ft(1.8m)         0°F(-18°C)        M         4/3

Upright trusses of about 12 flowers, rose pink with darker shading.  Leaves up  to 8in(20.3cm) long and 3in(7.6cm) wide.  (Rudolph Henny, 1959)  P. A. (ARS) 1959

           -thomsonii ssp. thomsonii (1/8)
g.     -Cornish Cross-
  -Coreum-         -griffithianum (1/8)
Fair -     -arboreum ssp. arboreum (1/4)
Maiden-
   -griersonianum (1/2)

Bright red flowers.  (Lord Aberconway, intro. 1942)

cl.

Fair Sky--form of augustinii ssp. augustinii

6ft(1.8m)         -5°F(-21°C)       M         3/3

Foliage typical of species--long, narrow, smooth, dark green,
the underside densely scaly. Flowers flat-faced and square, of
light purple blue, spotted yellowish green.  (James Barto & Dr.
Carl Phetteplace, 1976)

       -dichroanthum ssp. dichroanthum
        -Fabia-                    (1/4)
cl.     -unnamed hybrid-  -griersonianum (1/4)
   -          -yakushimanum ssp. yakushimanum (1/4)
Fairweather-          -dichroanthum ssp. dichroanthum
   -          -Fabia-
   -Hello Dolly-   -griersonianum
      -smirnowii (1/4)

Synonym Seafoam.  Plant vigorous and easy to grow; outstanding
foliage, heavily indumented  from the parentage of yakushimanum
and smirnowii; leaves retained 3 years.  Flowers of warm yellow,
in ball-shaped trusses of about 16.  (Brockenbrough, 1974)

g.          -dichroanthum ssp. dichroanthum (1/4)
   -Fabia-
Fairy-    -griersonianum (1/2)
   -         -facetum (1/4)
   -Jacquetta-
       -griersonianum

Flowers of bright red.  (Lord Aberconway, 1942)

g.
     -Lady Mar--unknown (1/2)
Fairy Light-
     -griersonianum (1/2)

Name shown as one word in the I.R.R., as two in Rothschild Rhododendrons.  A fast-growing plant of medium height. Two forms exist: bright pink and salmon pink flowers.  (Rothschild, 1948)

cl.                           -adenogynum (1/4)
              -Xenosporum (detonsum)-
Fairy Tale-                         -unknown (3/4)
              -unknown

Flowers of pale pink, 3.5in(8.9cm) across, 7-lobed, in compact
trusses of about 13. Floriferous. (Gen. Harrison, 1965)

cl.             -griffithianum (1/4)
           -Loderi-
Fairy Tale-     -fortunei ssp. fortunei (1/4)
           -arboreum ssp. cinnamomeum var. roseum (1/2)

Synonym Rudy's Fairy Tale. The plant has fuschine pink flowers,
4in(10.2cm) across; trusses of about 11. Leaves 7in(17.8cm)
long. Name as above from American Rhododendron Hybrids. Entry
in The Register under Rudy's Fairytale; reg. in 1963 and 1972.
(Rudolph Henny)

cl.              -forrestii ssp. forrestii Repens Group
           -Elizabeth-                          (1/4)
Faith Henty-      -griersonianum (1/4)
        -      -Queen     -griffithianum (1/8)
           -Earl of-Wilhelmina-
           Athlone-          -unknown (3/8)
                   -Stanley Davies--unknown

An Australian hybrid. Turkey red flowers in trusses of 5-7. (J.
V. Boulter, 1978)

cl.

Fake    See Purple Fake

cl.
       -falconeri ssp. falconeri (1/2)
Faltho-
       -thomsonii ssp. thomsonii (1/2)

Surprise g. Rose madder flowers, with faint spotting on upper
lobes, dark red basal blotches. (Loder)   A. M. (RHS) 1954

g.
       -wardii var. wardii (1/2)
Falvia-
       -campylocarpum ssp. campylocarpum (1/2)

Yellow flowers. (Lord Aberconway, shown 1933)

g.        -dichroanthum ssp. scyphocalyx Herpesticum Group (1/4)
      -Metis-
Fame-     -griersonianum (1/2)
      -   -dichroanthum ssp. dichroanthum (1/4)
      -Fabia-
          -griersonianum

Orange red flowers. (Lord Aberconway, intro, 1946)

cl.                       -griffithianum (1/8)
       -Mrs. Lindsay Smith-  -George Hardy-
Fancy-                    -catawbiense (1/4)
      -               -Duchess of Edinburgh--unknown (5/8)
                    -Mrs. J. J. Crosfield--unknown
      -Mrs. Helen Koster-                   -catawbiense
                    -Catawbiense Grandiflorum-
                              -unknown

Flowers violet to mauve, large red blotch.   (M. Koster & Sons)
A. M. (Boskoop) 1955

g. & cl.
          -unknown (1/2)
Fancy Free-
          -facetum (eriogynum) (1/2)

Medium-sized, compact plant. Clear bright pink flowers, stained
salmon with dark speckling. (Rothschild) A. M. (RHS) 1938

g.                   -griffithianum (1/8)
          -Queen Wilhelmina-
      -Britannia-          -unknown (3/8)
Fandango-       -Stanley Davies--unknown
        -haematodes ssp. haematodes (1/2)

Fine crimson scarlet flowers, waxy, campanulate. Plant of moderate height. (Rothschild, 1938)

cl.                        -catawbiense (3/8)
          -Parson's Grandiflorum-
       -America-           -unknown (5/8)
Fanfare-      -dark red hybrid--unknown
       -          -catawbiense
       -Kettledrum-
                  -unknown

5ft(1.5m)        -20°F(-29°C)         ML        3/2

Parentage according to Leach. Flowers a nonfading bright scarlet; dome-shaped trusses. Leaves 5.25in(1.5m) long. Shade- and
sun-tolerant. (A. Shammarello & D. Leach, c.1962)

g.
       -Lowinsky hybrid--unknown (1/2)
Fantasy-
       -griersonianum (1/2)

Plant has a loose, spreading habit, with trusses of deep pink in
May. (Rothschild, 1938)

g.
          -cinnabarinum ssp. cinnabarinum Roylei Group (1/2)
Fantin Latour-
          -oreotrephes (1/2)

Mauve flowers. (Adams-Acton, 1934)

cl.

Far Horizon--form of principis Vellereum Group

5ft(1.5m)         -5°F(-21°C)          E        3/4

Trusses hold up to 24 flowers, roseine purple fading to white in
throat, lightly spotted magenta. Leaves up to 4in(10.2cm) long,
dull dark green with heavy silvery brown indumentum beneath.
(Collector Kingdon-Ward; raiser Col. S. R. Clarke; exhibitor R.
N. Stephenson Clark) A. M. (RHS) 1979

cl.
       -Moser's Maroon--unknown (1/2)
Farall-
       -facetum (eriogynum) (1/2)

6ft(1.8m)        5°F(-15°C)           ML        3/3

Romany Chal g. A selection of Michael Haworth-Booth of Farall
Nurseries, intro. 1954.

cl.
                    -auriculatum (1/2)
Farall Target--Flameheart (selfed)-  -griersonianum (1/4)
                    -Azor-
                         -fortunei ssp. discolor
                                       (1/4)

Flameheart improved by selfing. Late-blooming. Flowers orange
red. (M. Haworth-Booth, Farall Nurseries, 1969)

cl.
       -diaprepes? (1/2)
Farewell Party-
       -unknown (1/2)

5ft(1.5m)        -5°F(-21°C)          L

Leaves medium to large, a glossy yellow green. Plant almost
twice as wide as high. Flowers up to 4.5in(11.5cm) across, white
with throat spotting of yellow green, held in flat trusses of
12. (Roy Kersey cross; reg. by Mrs. H. A. Frederick, Jr., 1976)

g.
          -oreodoxa var. fargesii (1/2)
Fargarb-
          -arboreum ssp. arboreum (blood red form) (1/2)

Flowers are violet rose, deepest shade.  (E. J. P. Magor, 1928)

g.
          -oreodoxa var. fargesii (1/2)
Fargcalo-
          -calophytum var. calophytum (1/2)

A hybrid introduced by E. J. P. Magor, 1940.

cl.
          -oreodoxa var. fargesii (1/2)
Fargsutch-
          -sutchuenense (1/2)

Flowers light pink, heavily spotted with crimson.  (E. J. P. Magor, 1933)

cl.                 -catawbiense (1/4)
          -Everestianum-
Farnese-             -unknown (3/4)
          -Eggebrechtii--unknown

White flowers flushed pale lilac, light red brown markings.  (T.
J. R. Seidel, 1914)

g.
          -oreodoxa var. fargesii (1/2)
Farola-   -griffithianum (1/4)
          -Loderi-
                   -fortunei ssp. fortunei (1/4)

Very pale pink flowers.  (Loder, 1940)

cl.
          -oreodoxa var. fargesii (1/2)
Farther-
          -morii (1/2)

Flowers white, lightly flushed orchid pink, openly campanulate,
2.5in(6.4cm) across; loose trusses of 10.  Leaves 5in(12.7cm)
long.  (Collingwood Ingram)  P. C. (RHS) 1969

g. & cl.
          -forrestii ssp. forrestii Repens Group (1/2)
Fascinator-         -haematodes ssp. haematodes (1/4)
          -Hiraethlyn-
                   -griffithianum (1/4)

1.5ft(.45m)      5°F(-15°C)         EM        3/4

A low mound covered with small leaves.  Cherry red, bell-shaped
flowers in profusion.  (Lord Aberconway)  A. M. (RHS) 1950

g.
          -fastigiatum (1/2)
Fasthip-
          -hippophaeoides var. hippophaeoides (1/2)

Flowers very pale lavender,  near white.  (E. J. P. Magor, 1926)

cl.
                   -catawbiense (1/2)
Fastuosum Flore Pleno-
                   -ponticum (1/2)

6ft(1.8m)        -15°F(-26°C)        ML        3/3

Plant habit is slightly open and rounded, with medium-sized dull
green leaves. Sun-tolerant.  Flowers lavender blue with darker
spotting, semi-double; rather loose trusses. (Francoisi, before
1846) A. G. M. (RHS) 1928   See color illus. F p. 7;  VV p. 28.

cl.

Fastuosum Plenum      See Fastuosum Flore Pleno

---
          -javanicum (5/8)
Favourite-               -jasminiflorum (3/8)
          -Princess Alexandra-     -jasminiflorum
                              -Princess Royal-
                                        -javanicum

Vireya hybrid.  Flower pinkish yellow to salmon.  F. C. C. (RHS)
1882

cl.
          -fortunei ssp. fortunei (1/2)
Fawn-     -dichroanthum ssp. dichroanthum (1/4)
          -Fabia-
          -griersonianum (1/4)

6ft(1.8m)        5°F(-15°C)        M        4/3

Plant upright, with medium green leaves, to 8in(20.3cm).  Flowers salmon pink shading orange yellow in center, very flat, up
to 5in(12.7cm) wide;  open-topped trusses of about 9.  (Del  W.
James) P. A. (ARS) 1955

cl.              -griersonianum (1/4)
          -Tally Ho-
Fayetta-         -facetum (eriogynum) (1/4)
          -              -dichroanthum ssp. dichroanthum (1/4)
          -Golden Horn-
                   -elliottii (1/4)

(William Whitney, 1956)

                              -souliei (1/16)
                         -Soulbut-
g. & cl.           -Vanessa-    -Sir Charles Butler--fortunei
          -Radiance-   -                    ssp. fortunei (1/16)
          -            -griersonianum (5/8)
Felicity-    -griersonianum
          -              -neriiflorum ssp. neriiflorum (1/4)
          -F. C. Puddle-
                   -griersonianum

Called a red amaranth rose in the Register.  (Lord Aberconway)
A. M. (RHS) 1942

cl.
          -maddenii ssp. maddenii (1/2)
Felicity- -maddenii ssp. crassum (1/4)
  Fair   -Sirius-
                   -cinnabarinum ssp. cinnabarinum Roylei Gp. (1/4)

Trusses hold 7-8 flowers, 5-lobed, amber yellow, orange throat.
Elliptic leaves with brown scaly indumentum.  See also Barbara
Jury, Christine Denz.  (F. M. Jury, 1982)

cl.
          -fulgens (1/2)
Felicity Magor-
          -barbatum (1/2)

Deep crimson flowers, with spots of darker crimson.  (E. J. P.
Magor cross, 1928; reg. by Mrs. Magor, 1956)

cl.                 -phaeopeplum (1/4)
          -unnamed hybrid-
Felinda-             -lochiae (1/4)
          -leucogigas (1/2)

Vireya hybrid.  Tubular funnel-shaped flower, very light reddish
purple in winter but lighter in summer.  Very fragrant.  Truss
8.5in(21.6cm) across, dome-shaped, holding 7-8 flowers.  Plant
equally as wide as tall, holding leaves 4 years. (Cross by Peter
Sullivan; reg. by William A. Moynier, 1981)

g.
          -dichronathum ssp. dichroanthum (1/2)
Felis-
          -facetum (1/2)

Synonym Felix.  Similiar to Fabia, but later and taller.  Low,
spreading hybrid with bell-shaped flowers of orange to yellow in
loose trusses.  Late.  (Rothschild, 1938)

cl.

Felix    See Felis

cl.

Fernhill*--form of arboreum ssp. cinnamomeum var. roseum

Flowers rose pink with darker markings; plant attains the height
of the species. First raised at Glasnevin before 1850, and distributed to the Darley family, who owned Fernhill in County Dublin. See color pl. in Irish Florilegium, p. 165.

```
                    -souliei (1/4)
cl.   -Peregrine-    -wardii var. wardii (1/8)
  -            -Hawk-
Fernhill-          -Lady        -fortunei ssp. discolor
  -            Bessborough-               (1/16)
  -yakushimanum ssp.          -campylocarpum ssp. campylo-
         yakushimanum (1/2)        carpum Elatum Group (1/16)
```

Large trusses 7in(17.8cm) across, of 6-9 flowers of creamy white
flushed pale pink. Plant of compact habit, broader than high;
glossy green leaves about 4in(10.2cm) long. (Crown Estate, Windsor) A. M. (Wisley Trials) 1973

```
cl.       -Moser's Maroon--unknown (1/4)
   -Romany Chal-
Festival-       -facetum (eriogynum) (1/4)
   -griersonianum (1/2)
```

Theresa g. Red flowers. (Stavordale, 1950)    See color illus.
F p. 45.

```
               -Koichiro Wada--yakushimanum ssp.
cl.   -Hachmann's Polaris-        yakushimanum (1/4)
  -          -    -catawbiense ? (1/8)
Festival*        -Omega-
  -                  -unknown (1/8)
  -wardii var. wardii (1/2)
```

A low, compact bush. Orange buds opening golden yellow with an
orange blotch; floriferous. Hardy to -8°F(-22°C). To be introduced in 1986. (H. Hachmann)

```
g.        -griffithianum (1/4)
   -King George-
Fez-       -unknown (1/4)
   -sanguineum ssp. sanguineum var. haemaleum (1/2)
```

Plant habit low and dense. Flowers waxy, crimson, and carried
in loose trusses. Midseason. (Rothschild, 1938)

```
                  -campylocarpum ssp. campylocarpum
         -Penjerrick-      Elatum Group (1/16)
g.   -Amaura-     -griffithianum (1/16)
  -Eros-    -griersonianum (5/8)
Fiesta-   -griersonianum
  -          -neriiflorum ssp. neriiflorum (1/4)
   -F. C. Puddle-
         -griersonianum
```

Red flowers. (Lord Aberconway, intro. 1950)

```
                    -griffithianum
           -George Hardy-      (1/16)
     -Mrs. Lindsay-       -catawbiense (1/16)
cl.   - Smith    -Duchess of Edinburgh--unknown
  -Diane-               (1/4)
Fifth Avenue-   -       -campylocarpum ssp.
  Red   - -unnamed hybrid-    campylocarpum (1/8)
  -              -unknown
  -haematodes ssp. haematodes (1/2)
```

Flowers neyron rose, campanulate, glossy, and held in trusses of
about 10; leaves 4in(10cm) long by 2in(5cm) wide. A dwarf plant
growing to 2ft(.6m) in sun, and 3ft(.9m) in shade. (Rudolph and
Leona Henny, 1964)

cl.

Fifty-Fine    See Golden Gala

```
               -dichroanthum ssp. dichroanthum (1/8)
       -Astarte-    -campylocarpum ssp. campylocarpum
  -        -Penjerrick-       Elatum Group (1/16)
  -Solon-        -griffithianum (5/16)
g.   -    -griffithianum
   -Sunrise-
Figaro-      -griersonianum (1/8)
  -         -griffithianum
  -   -Loderi-
  -Coreta-    -fortunei ssp. fortunei (1/8)
   -arboreum ssp. zeylanicum (1/4)
```

Red flowers. (Lord Aberconway, 1950)

```
cl.       -catawbiense (1/4)
  -Russell Harmon-
Fiji-     -maximum (1/4)
  -       -dichroanthum ssp. dichroanthum (1/4)
   -Goldsworth Orange-
         -fortunei ssp. discolor (1/4)
```

| | | | |
|---|---|---|---|
| 4.5ft(1.35m) | -20°F(-29°C) | L | 4/2 |

Buds of dark red and yellow open to flowers of claret rose, held
in trusses of 13. Plant much broader than tall, with matte yellow green leaves held 2 years. A hardy, late-flowering, handsome
hybrid. (David G. Leach, intro. 1976; reg. 1982)

cl.

Finch--form of rubiginosum Desquamatum Group

| | | | |
|---|---|---|---|
| 6ft(1.8m) | 0°F(-18°C) | EM | 3/2 |

Flowers of rosy lilac; compact trusses. Leaves fold in winter.
(Selected by James Barto; intro. by R. Henny, 1958)

cl.

Fine Bristles--form of pubescens

| | | | |
|---|---|---|---|
| 4ft(1.2m) | -5°F(-21°C) | EM | 3/3 |

Upright, compact plant; small leaves covered on both sides with
long hairs. Deep pink buds open to white flowers suffused rose,
appearing along the branches. (Crown Estate, Windsor, reg. 1962)
A. M. (RHS) 1955

```
g.         -ciliatum (1/4)
   -Cilpinense-
Fine Feathers-    -moupinense (1/4)
     -lutescens (1/2)
```

| | | | |
|---|---|---|---|
| 3ft(.9m) | 5°F(-15°C) | E | 3/3 |

Delicate flowers white with yellow and flushed pink. (Lord
Aberconway, 1946)

```
cl.        -ciliatum (1/4)
Fine   -Cilpinense-
Feathers-   -moupinense (1/4)
Primrose-lutescens (1/2)
```

Fine Feathers g. Flowers of pale primrose yellow, blooming
early. (Lord Aberconway, 1946)

```
cl.   -souliei (1/2)
Finesse-   -Corona--unknown (1/4)
   -Bow Bells-
      -williamsianum (1/4)
```

Flowers of Persian rose, 2.5in(6.25cm) wide, 7 in a truss. (Rudolph Henny, reg. 1958)

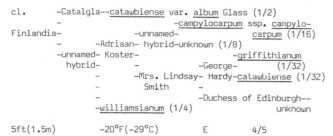

```
cl.    -Catalgla--catawbiense var. album Glass (1/2)
                        -campylocarpum ssp. campylo-
Finlandia-       -unnamed-              carpum (1/16)
      -           -Adriaan- hybrid-unknown (1/8)
      -unnamed- Koster-              -griffithianum
         hybrid-        -           -George-      (1/32)
              -      -Mrs. Lindsay- Hardy-catawbiense (1/32)
              -         Smith       -
              -                 -Duchess of Edinburgh--
              -williamsianum (1/4)            unknown

5ft(1.5m)       -20°F(-29°C)        E      4/5
```

Synonym Alaska. Many white flowers of heavy substance from pink
buds, in ball-shaped trusses of 12. Leaves glossy medium green,
retained 3 years. Plant rounded, branching well.  See also Applause, Flair, Robin Leach. (David G. Leach, 1972)   See color
illus. ARS J 36:1 (Winter 1982): p. 6.

```
cl.          -Corona--unknown (1/4)
      -Bow Bells-
Fiona-        -williamsianum (1/4)
      -        -griffithianum (1/4)
      -Loderi Pink-
        Diamond  -fortunei ssp. fortunei (1/4)
```

Flowers 4in(10.2cm) wide, 7-lobed, phlox pink. (L. Brandt, 1962)

```
cl.                    -griffithianum (3/8)
       -Loderi King George-
Fiona -           -fortunei ssp. fortunei (1/4)
Wilson-              -griffithianum
      -         -George Hardy-
      -Pink Pearl-       -catawbiense (1/8)
      -        -arboreum ssp. arboreum (1/8)
         -Broughtonii-
                    -unknown(1/8)
```

Light mallow purple flowers, in trusses of 9-10.  See also Irene
Hall. (Cross by Geoffrey Hall; reg. by Lord Harewood, 1979)

```
g.              -fortunei ssp. discolor (1/4)
       -Norman Shaw-         -catawbiense (1/8)
Fire Bird-           -B. de Bruin-
      -                 -unknown (1/8)
      -griersonianum (1/2)

6ft(1.8m)        5°F(-15°C)        ML      3/3
```

Tall hybrid with trusses of salmon red flowers. The green young
foliage is a striking contrast against crimson bracts.  (Rothschild, 1938)

```
cl.          -dichroanthum ssp. apodectum (1/4)
       -Apodorum-
Fire Flame-    -decorum (1/4)
      -griersonianum (1/2)
```

Bright red flowers. (Scrase-Dickens, 1942)   A. M. (RHS) 1942

```
g.
       -griersonianum (1/2)
Fire Glow-     -griffithianum (roseum superbum ?) (1/4)
      -unnamed-          -fortunei ssp. fortunei (1/8)
         hybrid-H. M. Arderne-
                      -unknown (1/8)
```

A large-growing shrub with Turkey red flowers, faintly fragrant.
Foliage wavy and twisted. (Crosfield, Embley Park, 1920)  A. M.
(RHS) 1935

```
cl.
       -venator (1/2)
Fire Music-
         -dichroanthum ssp. dichroanthum (1/2)
```

Fiery orange flowers. (Thacker, 1942)

```
cl.                    -griffithianum (1/8)
          -Queen Wilhelmina-
       -Britannia-        -unknown (3/8)
Fire Prince-    -Stanley Davies--unknown
       -arboreum ssp. delavayi var. delavayi (1/2)
```

An Australian hybrid with cardinal red flowers, lightly spotted
on upper lobe.  (K. van de Ven)

```
cl.

Fire Walk      For parentage see Fire Prince
```

Australian hybrid.  Turkey red flowers.  (K. van de Ven, 1972)

```
cl.    -Purple Splendour-     -ponticum (1/4)
       -          -unknown (5/16)
Fire Wine-          -fortunei ssp. discolor (1/8)
       -        -Norman Shaw-    -catawbiense (1/16)
       -Fire Bird-      -B. de Bruin-
                  -            -unknown
              -griersonianum (1/4)

3.5ft(1m)         0°F(-18°C)        ML      4/3
```

Flowers an unusual bright purplish red, darker dorsal spotting,
3in(7.6cm) across, ruffled and frilled; large flat trusses hold
about 16.  Plant wider than tall; narrow grass green leaves held
2 years. (Harold E. Greer, reg. 1979)

```
cl.
       -barbatum (1/2)
Fireball-      -thomsonii ssp. thomsonii (1/4)
      -Ascot Brilliant-
                 -unknown (1/4)
```

Synonym for Early Brilliant (by Slocock)? Glowing carmine scarlet flowers, bell-shaped, with frilled edges; rounded trusses.
Very early. (Richard Gill & Sons)  A. M. (RHS) 1925

```
g.           -fortunei ssp. discolor (1/4)
       -Sardis-          -catawbiense (1/8)
Firedrake-    -C. S. Sargent-
      -kyawii (1/2)        -unknown (1/8)
```

Large, bright red flowers appear late in the season, on the tall
but compact plant. (Rothschild, 1938)

```
                         -souliei (1/16)
                  -Soulbut-
             -Vanessa-     -Sir Charles Butler--fortunei
g.     -Radiance-     -           ssp. fortunei (1/16)
      -        -        -griersonianum (5/8)
Firefinch-     -griersonianum
      -        -facetum (1/4)
      -Jacquetta-
                -griersonianum
```

Scarlet flowers. (Lord Aberconway, 1942)

```
g.           -spinuliferum (3/4)
       -Crossbill-
Firefly-       -lutescens (1/4)
       -spinuliferum
```

Small flowers, shaded yellow to apricot.  Shrub medium-sized and
open. (Rothschild, 1938)

```
cl.                -griffithianum (1/4)
          -Jean Marie de-
Fireman Jeff- Montague    -unknown (1/4)
      -            -haematodes ssp. haematodes (1/4)
      -Grosclaude-
                 -facetum (eriogynum) (1/4)

3ft(.9m)        0°F(-18°C)        M      4/4
```

Bright blood red flowers with large bright red calyces, held in
trusses of about 10. Plant well-branched and wider than high;
medium green foliage retained 3 years. (Brandt-Eichelser, 1977)
See color illus. HG p. 144.

```
                           -griffithianum (1/8)
g. & cl.          -Queen Wilhelmina-
          -Britannia-              -unknown (3/8)
Firetail-         -Stanley Davies--unknown
          -
          -facetum (eriogynum) (1/2)
```

Tall plant with flowers of deep scarlet, spotted brown; compact
trusses. See also Trident. (J. J. Crosfield) A.M. (RHS) 1934;
F. C. C. (RHS) 1937

cl.

Firewine     See Fire Wine

cl.
```
          -oreotrephes (1/2)
First Love-          -cinnabarinum ssp. cinnabarinum (1/4)
          -Royal Flush-
                    -maddenii ssp. maddenii (1/4)
```

5ft1.5m)          5°F(-15°C)          M

Flowers pink with maroon eye; 8-flowered trusses. Shrub about 6
ft(1.8m) tall, bushy; leaves to 2.5in(6.4cm) long by 1in(2.5cm).
Midseason. (Rudolph & Leona Henny, 1966)

cl.
```
          -racemosum (1/2)
Fittianum*
          -unknown (1/2)
```

Once considered a separate species, then thought to be a form of
dauricum, and now called a natural hybrid of racemosum. Rose
purple flowers. Very early.

cl.
```
          -racemosum (3/4)
     -Fittianum-
Fittra-          -unknown (1/4)
     -racemosum
```

A compact dwarf plant with vivid deep pink flowers, in trusses
of about 30; plant may be completely covered. Early to midseason. (Hillier, 1938)  A. M. (RHS) 1949

```
cl. -Catalgla--catawbiense var. album Glass (1/2)
     -                    -campylocarpum ssp. campylo-
Flair-          -unnamed-          carpum (1/16)
     -          -Adriaan- hybrid-unknown (1/8)
     -          - Koster-          -griffithianum (1/32)
     -unnamed-          -          -George-
     hybrid-     -Mrs. Lindsay- Hardy-catawbiense (1/32)
     -          Smith     -
     -                    -Duchess of Edinburgh--unknown
     -williamsianum (1/4)
```

5ft(1.5m)          -15°F(-26°C)          EM          4/3

Buds light to strong purplish pink. Flowers of heavy substance,
white with faint ivory shading. Ball-shaped trusses hold 10-11
flowers. Free-flowering. Medium green leaves retained 3 years,
on a rounded, well-branched plant. See also Applause, Finlandia
and Robin Leach. (David G. Leach, 1970)

```
                    -griffithianum (1/8)
g.          -Loderi-
     -unnamed hybrid-          -fortunei ssp. fortunei (1/8)
Flame-          -Corona--unknown (1/4)
     -griersonianum (1/2)
```

A hybrid by Halfdan Lem, before 1958.

```
                    -fortunei ssp. discolor (3/8)
          -Lady          -
          -Bessborough-campylocarpum ssp. campylo-
cl.     -Day Dream-          carpum Elatum Group (1/8)
Flame Tips-     -griersonianum (3/8)
     -          -fortunei ssp. discolor
     -Margaret Dunn-     -dichroanthum ssp. dichroanthum
          -Fabia-                    (1/8)
          -griersonianum
```

Dark flame red buds opening flame red, suffused with yellow; the
flowers tubular funnel-shaped 2.25in(5.7cm) across, 5-lobed, 6-7
per truss. Upright plant 3.5in(9cm) tall. Late midseason. (H.
Graves cross; intro. by Mr. & Mrs. K. Janeck, 1965)

cl.
```
          -auriculatum (1/2)
Flameheart-     -griersonianum (1/4)
          -Azor-
                    -fortunei ssp. fortunei (1/4)
```

A parent of Bullseye. (Haworth-Booth, 1955)

```
                    -Parsons     -catawbiense (3/16)
               -America- Grandiflorum-
               -                    -unknown (1/2)
cl.     -Fanfare-          -dark red hybrid--unknown
     -          -          -catawbiense
Flamenco-     -Kettledrum-
     -               -unknown
     -          -forrestii ssp. forrestii Repens Group (1/4)
     -Gertrud-               -ponticum (1/16)
     Schale-          -Michael-
          -Prometheus-Waterer-unknown
               -
               -Monitor--unknown
```

1.5ft(.45m)          -15°F(-26°C)          M          3/4

Flowers 2.5in(6.4cm) wide, crimson with spotting of blood red,
in flat trusses of 6-9. A very dense dwarf plant, much broader
than high, with yellow green leaves held 2 years. See also Sing-
apore, Small Wonder. (David G. Leach, intro. 1968; reg. 1982)

cl.
```
          -aurigeranum (?) (1/2)
Flamenco Dancer-
          -macgregoriae (1/2)
```

Vireya hybrid. Trusses of 20-22 flowers, 5-lobed, 1.25in(3.2cm)
across, tubular funnel-shaped, with deeply cut lobes giving a
a star-like form; flowers chrome yellow suffused with Spanish
orange, darkening with age. Very small brown scales on elliptic
leaves. (Cross by I. Lelliot; G. F. Smith, reg. 1984)

cl.
```
          -laetum (1/2)
Flaming Ball-
          -javanicum (1/2)
```

A vireya hybrid with trusses of 8 flowers, poppy red shading to
nasturtium orange in throat. (Cross by Don Stanton; reg. by G.
Langdon, 1982)

```
g.
     -Loder's White          See Loder's White for 2 possible
Flamingo-                    parentage diagrams
     -griersonianum
```

Vivid crimson pink buds open rich bright rose; shapely trusses.
(Shown by Sir James Horlick, 1941)  See color illus. LW pl. 37.

cl.
```
          -catawbiense (1/2)
Flamme-
          -Mira--unknown (1/2)
```

(Not Stevenson's Mira, 1951) Light purplish violet flowers with
paler centers, yellow brown markings. (T. J. R. Seidel, 1904)

cl.
```
     -Mrs. R. S. Holford--unknown (1/2)
Flare-          -auriculatum (1/4)
     -unnamed hybrid-
                    -griersonianum (1/4)
```

A tall, late-flowering plant with large trusses of bright salmon
red flowers. (Slocock, reg. 1958)

```
g.        -dichroanthum ssp. dichroanthum (1/4)
       -Fabia-
Flash-      -griersonianum (1/4)
     -dichroanthum ssp. scyphocalyx (1/2)
```

Orange red flowers.  (Lord Aberconway, 1946)


```
cl.
           -arboreum ssp. arboreum (1/2)
Flashing Red-
              -unknown (1/2)
```

Flowers red  with black spots; trusses of 17.  (G. Langdon, Australia, 1980)


```
g.
           -callimorphum ssp. callimorphum (1/2)
Flashlight-
           -campylocarpum ssp. campylocarpum (1/2)
```

Trusses like campylocarpum; flower buds orange pink, opening to
lemon yellow, fading to pale yellow with a red blotch.  Slow-
growing plant of neat habit, reaches maturity at about 7ft(2.1m)
at Exbury.  (Rothschild, 1938)


```
cl.
         -Corona--unknown (1/2)
Flatterer-                    -fortunei ssp. discolor (1/8)
           -Lady       -
     -Day Dream- Bessborough-campylocarpum ssp. campylo-
          -                      carpum Elatum Group (1/8)
               -griersonianum (1/4)
```

An interesting rhododendron with watermelon red flowers and narrow, pointed leaves.  (Rudolph Henny, 1957)  P. A. (ARS) 1957


```
cl.
         -Seedling K. 30--unknown (1/2)
Flavour-    -dichroanthum ssp. dichroanthum (1/4)
       -Dido-
            -decorum (1/4)
```

Flowers bell-shaped, slightly darker than maize yellow with a
light chartreuse throat, held in round trusses of 11.  Leaves of
medium green, about 5in(12.7cm) long.  (Walter C. Slocock)  H. C.
(Wisley Trials) 1960


```
cl.         -griffithianum (1/4)
       -Loderi-
Fleece-      -fortunei ssp. fortunei (1/2)
     -        -fortunei ssp. fortunei
       -Luscombei-
                 -thomsonii ssp. thomsonii (1/4)
```

Loose trusses of 7 flowers, rhodamine pink with some deep pink
tinges and red spotting.  (Heneage-Vivian cross, 1927; named by
Mrs. R. M. Stevenson, 1959; reg. 1962)


```
cl.
         -thomsonii ssp. thomsonii (1/2)
Fleur de Roi-
             -campanulatum ssp. campanulatum (1/2)
```

Cream-colored flowers, suffused pink.  (Wright, 1903)


```
cl.
```

Fleurie--form of smithii

Turkey red flowers, tubular-campanulate, 5-lobed; tight rounded
trusses of about 25.  Leaves, margins reflexed, up to 6in(15cm)
long, covered beneath with woolly brown indumentum.  (Collector
T. J. Booth; raised and reg. by R. N. Stephenson Clark)  A. M.
(RHS) 1978


```
                      -dichroanthum ssp. dichroanthum
                   -Fabia-                            (3/16)
              -Umpqua-      -griersonianum (1/4)
cl.  -unnamed- Chief-     -griersonianum
     - hybrid-     -Azor-
Flicker-        -         -fortunei ssp. discolor (1/16)
     -          -War Paint--elliottii (1/4)
     -  -fortunei ssp. fortunei (1/4)
     -Fawn-     -dichroanthum ssp. dichroanthum
          -Fabia-
               -griersonianum
```

4ft(1.2m)          -10°F(-23°C)          M

Delft rose flowers held in large, rounded trusses 8in(20cm) wide
by 7in(17.8cm) high, with 10-12 per truss.  Plant well-branched,
nearly as wide as tall, with dark green leaves 5in(12.7cm) held
2 years. (Del James cross; Ray James intro. and reg., 1980)


```
                                  -catawbiense (1/8)
                      -Parsons    -
                -America-Grandiflorum-unknown (9/32)
         -unnamed-      -dark red hybrid--unknown
         - hybrid-        -griffithianum (3/32)
         -          -    -Mars-
         -        -Blaze-   -unknown
    -Cindy Lou-         -catawbiense (red form)
         -             -griffithianum
cl.  -        -    -Mars-
     -      -Red -   -unknown            -catawbiense
Flipper*  -Brave-      -Parsons Grandiflorum-
         -America-                      -unknown
         -              -dark red hybrid--unknown
         -              -dichroanthum ssp. dichroanthum (1/4)
     -Goldsworth Orange-
                       -fortunei ssp. discolor (1/4)
```

Buds of bright guardsman red opening to crimson flowers.  Medium-sized plant.  (Weldon Delp, 1985)


```
           -Queen   -griffithianum (1/8)
cl.      -Wilhelmina-
     -Britannia-    -unknown (3/8)
Flirt-    -Stanley Davies--unknown
     -Koichiro Wada--yakushimanum ssp. yakushimanum (1/2)
```

3.5ft(1m)          10°F(-12°C)          M

Flowers rose red with a white throat; as color fades, the white
throat extends to half of each lobe.  Lax, rounded trusses, 6in
(15.2cm) across, hold about 14.  Plant broader than tall; leaves
retained 3 years.  (James A. Elliott, reg. 1980)


```
cl.      -yakushimanum ssp. yakushimanum (1/4)
     -Bambi-       -dichroanthum ssp. dichroanthum (1/8)
Floor-   -Fabia -
Show-    -Tangerine-griersonianum (1/8)
     -arboreum ssp. arboreum (1/2)
```

An Australian hybrid with trusses of 17 flowers, Delft rose.
(R. Cutten, 1981)


```
cl.
     -fortunei ssp. discolor (1/2)
Flora -        -griffithianum (1/8)
Donald-Lodauric-Loderi-
     Iceberg -    -fortunei ssp. fortunei (1/8)
             -auriculatum (1/4)
```

Light orchid pink flowers, flushed phlox pink on reverse.  (Donald Smith, UK, 1972)


```
cl.
     -thomsonii ssp. thomsonii (5/8)
Flora -        -campylocarpum ssp. campylocarpum (1/8)
Markeeta-   -Unique-
     -unnamed-    -unknown (1/8)
        hybrid-      -fortunei ssp. fortunei (1/8)
            -Pride of  -
       Leonardslee-thomsonii ssp. thomsonii
```

4ft(1.2m)          -5°F(-21°C)          EM          3/4

A bush broader than high; round glossy leaves. Coral pink buds,
opening to ivory white flowers rimmed in bright pink. (Flora
Markeeta Nursery, 1967) P. A. (ARS) 1967

cl.
                 -nuttallii (1/2)
Floral Dance-
                 -edgeworthii (1/2)

Camellia rose flowers with a yellow blotch, 5 frilly lobes, funnel-shaped; trusses of 4. Ovate leaves with tan indumentum. (F.
M. Jury 1982)

cl.
                 -forrestii ssp. forrestii Repens Group (1/2)
Flora's Boy*                     -griffithianum (1/4)
                 -Jean Marie de Montague-
                                 -unknown (1/4)

2ft(.6m)          5°F(-15°C)          M          3/3

Flowers large, bright waxy red, held upright. Plant habit also
upright; foliage glossy green. (Flora Markeeta Nursery)

                         -griersonianum (1/4)
cl.     -Creeping Jenny-
        -                -forrestii ssp. forrestii Repens Group
Florence-                        -griffithianum (1/8)     (1/4)
        -                -Queen Wilhelmina-
        -Unknown-                        -unknown (3/8)
        -Warrior-Stanley Davies--unknown

Scarlet red flowers in trusses of 10. See also Deborah. (Dr. E.
E. Smith, Surrey, 1976)

cl.
                 -wardii var. wardii (1/2)
Florence Archer-          -campylocarpum ssp. campylocarpum (3/8)
                 -Marcia-
              GLOMMS -fortunei ssp. fortunei (1/8)
                 - CANPY
3ft(.9m)          10°F(-12°C)          L

Flowers of heavy substance, empire yellow flushed and edged with
jasper red, dorsal spotting of absinthe green; corolla of 5 wavy
lobes, 3.75in(9.5cm) across. High trusses of 8 flowers. Plant
as wide as tall, well-branched; leaves held 3 years. (H. L. Larson, reg. 1979)

g.
                 -arboreum ssp. arboreum (1/2)
Florence Gill-
                 -unknown (1/2)

White flowers with pink margins. (R. Gill & Son, reg. 1958)

cl.
                 -rigidum (1/2)
Florence Mann*
                 -Blue Admiral--unknown (1/2)

Flowers of sea lavender violet. (A. Bramley, Australia, 1963)

cl.
                 -maximum (1/2)
Florence Rinehimer-          -catawbiense (1/4)
                 -unnamed-
                     hybrid-unknown (1/4)

7ft(2.1m)          -25°F(-32°C)          ML

Upright, well-branched shrub; yellow green leaves to 5.5in(14cm)
across, held 3 years. Flowers roseine purple with dorsal spotting of dark olive yellow, 3in(7.6cm) across; rounded trusses of
16. (W. David Smith, reg. 1983)

Florence Smith*     Parentage unknown

A parent of Mrs. Harold Terry. May be Florence Sarah Smith, a
registered clone (Smith of Darley Dale). Pink flowers (both).

g.                 -oreotrephes ? (1/4)
                 -Oreoroyle-
Florescent-          -cinnabarinum ssp. cinnabarinum
        -                     Roylei Group ? (1/4)
                 -cinnabarinum ssp. xanthocoden Concatenans Group (1/2)

Plum-colored flowers. (Lord Aberconway, 1950)

cl.

Floriade     Parentage unknown

5ft(1.5m)          -5°F(-21°C)          ML          3/2

May be a hybrid of Britannia. Flowers are Turkey red with a
dark brown blotch, held in trusses of 16-20. (Adr. van Nes,
1962) Gold Medal (Rotterdam) 1960

cl.
                 -catawbiense (1/2)
Flushing-
                 -unknown (1/2)

Trusses of crimson flowers. (S. Parsons, intro. before 1875)

cl.
                 -edgeworthii (bullatum) (1/2)
Folies Bergere-                 -valentinianum (1/4)
                 -Parisienne-
                                 -burmanicum (1/4)

A flowering plant for the cool greenhouse. (Raised and exhibited by Sir Giles Loder, Leonardslee.) P. C. (RHS) 1967

cl.

Folius Purpureus*--form of ponticum

4ft(1.2m)          -10°F(-23°C)          L          2/4

Described in Cox, Larger Species... p. 251, as having green fol-
age turning copper color in winter.

cl.

Folk's Wood  form of annae Laxiflorum Group

5ft(1.5m)          10°F(-12°C)          E-M          2/3

Flowers bell-shaped, 5-lobed, pure white, in trusses of 14.
Leaves 6in(15cm) by 2in(5cm). (Maj. A. E. Hardy) A. M. (RHS)
1977

cl.
                 -haematodes ssp. haematodes (1/2)
Forerunner-          -venator (1/4)
                 -Vanguard-
                                 -griersonianum (1/4)

Glowing scarlet flowers. (Sunningdale Nurseries, 1951)

                         -forrestii ssp. forrestii Repens Group
cl.     -Elizabeth-                          (1/4)
        -                -griersonianum (1/4)
Forest Blaze-          -thomsonii ssp. thomsonii (1/4)
        -Barclayi -          -arboreum ssp. arboreum (1/8)
             Helen Fox-Glory of  -
                     Penjerrick-griffithianum (1/8)

Trusses of 8-10 flowers, campanulate, guardsman red. Height
about 7ft(2.1m). (K. van de Ven, reg. 1984)

```
                    -griersonianum (1/4)
cl.       -Tally Ho-
          -          -facetum (eriogynum) (1/4)
Forest Fire-                    -griffithianum (1/8)
          -         -Queen Wilhelmina-
          -Britannia-                  -unknown (3/8)
                    -Stanley Davies--unknown
```

Flowers red, waxy, funnel-campanulate, to 3in(7.6cm) wide, held
in loose trusses. Medium-sized, bushy plant. Late. (R. Henny,
cross 1944; reg. 1958)

```
                              -griffithianum (1/8)
cl.            -Queen Wilhelmina-
          -Britannia-              -unknown (3/8)
Forest Flame-    -Stanley Davies--unknown
          -arboreum ssp. delavayi var. delavayi (1/2)
```

Trusses of 16 flowers, campanulate, currant red.  Height about
7ft(2m). (K. van de Ven, reg. 1984)

```
                         -Essex Scarlet--unknown (1/8)
                 -Cavalcade-
cl.    -unnamed-          -griersonianum (1/8)
       - hybrid-Mary      -campylocarpum ssp. campylocarpum
Forever-      -Swaythling-                              (1/8)
 Amber -              -fortunei ssp. fortunei (1/8)
       -   -wardii var. wardii (1/4)
       -Crest-        -fortunei ssp. discolor (1/8)
          -Lady       -
          Bessborough-campylocarpum ssp. campylocarpum
                                  Elatum Group (1/8)
```

5ft(1.5m)          5°F(-15°C)          M

An open-growing bush with flowers amber, a smoky orange, edged
red. (John Waterer, Sons & Crisp cross, 1959; reg. 1975)

```
cl.
          -veitchianum (1/2)
Forsterianum-
          -edgeworthii (1/2)
```

5ft(1.5m)          20°F(-7°C)          M          4/4

Funnel-shaped  large white flowers with a yellow flare,  heavily
fragrant; lax trusses of 3-4.  Plant upright, compact; lanceolate, hairy-edged leaves of dark rich green. (Forster, 1917?)

```
cl.
          -griersonianum (1/2)
Fort Langley-
          -unknown (1/2)
```

Deep red flowers in profusion. Attractive plant habit with dull
citrus green foliage. (R. Clark)

```
g. & cl.
       -falconeri ssp. falconeri (1/2)
Fortune-
       -sinogrande (1/2)
```

A large, highly rated shrub, considered superior to its parents.
Not hardy in cold climates.  Flowers bell-shaped, primrose yellow, 25 held in trusses 9in(23cm) tall.  Mature leaf is 15in(41
cm) long, and holds a light coat of tan indumentum. (Rothschild)
F. C. C. (RHS) 1938   See color illus. PB pl. 15 & 16.

```
cl.
          -fortunei ssp. fortunei (1/2)
Fortwilliam-
          -williamsianum (1/2)
```

3ft(.9m)          -5°F(-21°C)          ML

Flowers 4in(10cm) wide, light pink with darker pink stripes and
yellow throat, in trusses of up to 12; fragrant. (Charles
Herbert, 1967)

```
                    -fortunei ssp. discolor (1/2)
cl.  -Lady Bessborough-
                    -campylocarpum ssp. campylocarpum Elatum
Fotis-                                        Group (1/4)
     -            -fortunei ssp. discolor
     -Margaret Dunn-    -dichroanthum ssp. dichroanthum (1/8)
                  -Fabia-
                    -griersonianum (1/8)
```

Flowers tangerine pink  fading to nankeen yellow,  crimson brown
spots in throat. (Sunningdale Nurseries, 1955)

```
                    -dichroanthum ssp. dichroanthum
cl.            -Fabia-                        (3/16)
     -unnamed hybrid-    -griersonianum (3/16)
Four  -            -bureavii (1/4)
Crosses-           -fortunei ssp. discolor (1/8)
     -       -King of-    -dichroanthum ssp. dichroanthum
     -unnamed- Shrubs-Fabia-
        hybrid-          -griersonianum
          -smirnowii (1/4)
```

Flower buds blood red, opening medium scarlet red, fading to
light carrot orange, center currant red. Leaves have champagne
colored indumentum. (Cross made at New Zealand Rhod. Assoc.;
grown by Mrs. R. Pinney; reg. 1982)

```
cl.            -griersonianum (1/4)
     -Matador-
Fox   -        -strigillosum (1/4)
Hunter-             -thomsonii ssp. thomsonii (1/8)
     -   -Shilsonii-
     -Gaul-      -barbatum (1/8)
        -elliottii (1/4)
```

Bright red flowers in trusses of 12; each flower 2.75in(7cm)
wide by 2.5in(6.4cm) long. (Gen. Harrison, 1967)

```
                         -arboreum ssp. arboreum (1/8)
cl.            -Doncaster-
     -Thunderstorm-       -unknown (3/8)
Foxhunter*           -unknown
     -           -griersonianum (1/4)
     -Tally Ho-
          -facetum (eriogynum) (1/4)
```

Tall-growing with flowers dark scarlet, opening late. (Slocock)

```
cl.
```

Foxy--a form of fortunei ssp. fortunei

A selected form of this species having trusses of 11-13 flowers,
color amaranth rose and lighter shades of same. (John Fox, 1982)

```
                         -fortunei ssp. fortunei (1/8)
cl.            -Luscombei-
     -Betty King-      -thomsonii ssp. thomsonii (5/8)
Fragonard-       -thomsonii ssp. thomsonii
     -           -thomsonii ssp. thomsonii
     -Cornish Cross-
          -griffithianum (1/4)
```

Flowers Indian red. (Adams-Acton, 1942)

```
g.
     -catawbiense (1/2)
Fragrans-
     -azalea viscosum (1/2)
```

A sweet-scented azaleodendron, fast-growing and compact.
Trusses of small flowers, pale mauve with centers lighter to
white. Very late season. (Paxton, of Chandler & Sons, 1843)

cl.
           -edgeworthii (1/2)
Fragrantissimum-
           -formosum var. formosum (1/2)

3ft(.9m)          15°F(-10°C)          EM          4/2

A rather leggy, tender plant that is one of the most fragrant of
all shrubs. Flowers blushed carmine in bud, opening white with
tinge of pink, a creamy yellow center, up to 4in(10.2cm) across,
in lax trusses of 4.     Narrow, downy leaves.     (Origin unknown)
F. C. C. (RHS) 1868

cl.                      -Dexter hybrid--unknown
        -Helen Everitt-
Fran Labera-             -Dexter hybrid--unknown
        -Honeydew (Dexter)--unknown

Buds greenish white opening white to cream, chartreuse throat,
up to 4in(10.2cm) wide, of heavy substance, 7 frilly lobes; 8-11
per truss. Very floriferous, with very strong nutmeg fragrance.
Plant upright, broader than tall; medium to light green leaves,
held 2 years. See also Sally Fuller. (H. & S. Fuller, 1978)

cl.                           -griffithianum (1/8)
        -Queen Wilhelmina-
   -Britannia-              -unknown (7/8)
Francesca-      -Stanley Davies--unknown
   -Dexter #202--unknown

8ft(2.4m)          -10°F(-23°C)          ML

Patented hybrid, no. 3709, sold by Bald Hill Nurseries, Exeter,
RI. Flower buds are black red, opening bright carmine red, tu-
bular-campanulate, 3.5in(8.9cm) wide; large trusses of about 25.
Plant habit broad, globular. (A. Consolini & A. Savella, 1971)

cl.
        -macabeaneanum (1/2)
Francesca Beekman-
        -unknown (1/2)

An Australian hybrid of a beautiful species. About 26 pale yellow flowers held in one truss. (J. Beekman, 1980)

g. & cl.
        -dichroanthum ssp. dichroanthum (1/2)
Francis Hanger-      -griffithianum (1/4)
        -Isabella-
        -auriculatum (1/4)

5ft(1.5m)          -5°F(-21°C)          L          2/2

Large flowers of deep yellow tinged pale rose at the edges, held
in loose trusses of about 7. Plant well-branched, with new foliage appearing about the time the flowers open, thus partially
concealing the effect. Named in memory of a former curator of
the RHS Garden, Wisley. (Rothschild, 1942) A. M. (RHS) 1950

cl.
   -Koichiro Wada--yakushimanum ssp. yakushimanum (1/2)
Frango-      -arboreum ssp. nilagiricum (1/4)
   -Noyo Chief-
        -unknown (1/4)

5ft(1.5m)          5°F(-15°C)          EM

Slightly fragrant flowers, empire rose with orange spotting held
in trusses 5.5in(14cm) wide, holding 14-18 flowers each. Leaves
retained 4 years, glossy, yellow green with beige indumentum.
Plant rounded, branching well. (Dr. David Goheen, reg. 1982)

cl.          -griffithianum (1/4)
    -Mars-
Frank Baum-      -unknown (1/4)
   -      -dichroanthum ssp. dichroanthum (1/4)
   -Jasper-          -fortunei ssp. discolor (1/8)
      -Lady      -
   Bessborough-campylocarpum ssp. campylocarpum
                Elatum Group (1/8)

5ft(1.5m)          -5°F(-21°C)          L          4/3

Trusses round and full; flowers colored coral watermelon. Fir
green leaves with red petioles produce an unusual effect. (C.
S. Seabrook, 1968)    See color illus. HG p. 80.

cl.
        -ponticum (1/2)
Frank Galsworthy-
        -unknown (1/2)

Flowers rich reddish purple with a bold flare, orange brown to
off-white with faint greenish spotting; round trusses of 15-
20. When plant was 5.5ft(1.65m) tall in 1960, it was 9ft(2.7m)
wide, fairly compact, with leaves about 6in(15cm) long. (Raiser
A. Waterer; shown by W. C. Slocock) A. M. (Wisley Trials) 1960

cl.
      -fortunei ssp. discolor (1/2)
   -Golden-      -dichroanthum ssp. dichroanthum (1/4)
 - Belle-Fabia-
Frank -      -griersonianum (1/4)
Heuston-
     -fortunei ssp. discolor
   -Autumn-      -dichroanthum ssp. dichroanthum
    Gold -Fabia-
      -griersonianum

4ft(1.2m)          0°F(-18°C)          M

Flowers 6-lobed, light brick red edged in orange buff, held in
7in(17.8cm) trusses of 10-14. Plant spreads as broad as high
with yellow green leaves held 2 years. (Walter Elliott cross,
1972; reg. 1983)

cl.

Frank Kingdon Ward--form of glischrum ssp. rude

Flowers of 7 lobes, white with outside stained purple,
3in(7.6cm) wide, held in loose trusses of 16. Leaves are
7in(17.8cm) long, covered with yellowish green hair. (A. C. &
J. F. A. Gibson) A. M. (RHS) 1969

cl.

Frank Ludlow--form of dalhousiae var. dalhousiae (L S & T 6694)

A plant for the cool greenhouse. White flowers heavily stained
yellow, tubular-campanulate, 3.75in(9.5cm) wide; loose trusses.
Leaves 5in(12.7cm) long, scaly. (Collectors Ludlow, Sherriff &
Taylor; raised by Maj. A. E. Hardy) A. M. (RHS) 1974; F. C. C.
(RHS) 1974

cl.     -Margaret--unknown (1/4)
   -Glamour-
Franz-      -griersonianum (1/4)
Lehar-      -dichroanthum ssp. dichroanthum (1/4)
   -      -      -griffithianum (1/32)
    -Jester-          -Kewense-
      -Aurora-      -fortunei ssp. fortunei (5/32)
     -Naomi-      -thomsonii ssp. thomsonii (1/16)
      -fortunei ssp. fortunei

Blood red flowers in loose, semi-erect trusses. Plant is only
18in(46cm) tall after 10 years. Blooms late midseason. (C. S.
Seabrook, 1965)

cl.
        -ponticum (1/2)
Frau Minna Hartl-
        -azalea--unknown (1/2)

Azaleodendron, in the Register. Double flowers of carmine rose.
(Hartl, 1891)

cl.
      -Corona--unknown (1/2)
Freckle Face-
        -wardii var. wardii (1/2)

Coronet g. (Collingwood Ingram, 1954)

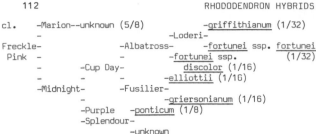

cl.    -Marion--unknown (5/8)          -griffithianum (1/32)
                                -Loderi-
Freckle-           -Albatross-    -fortunei ssp. fortunei
   Pink -          -            -fortunei ssp.        (1/32)
       -      -Cup Day-          discolor (1/16)
       -        -       -        -elliottii (1/16)
     -Midnight-     -Fusilier-
                   -             -griersonianum (1/16)
             -Purple  -ponticum (1/8)
             -Splendour-
                       -unknown

Trusses hold 12 flowers, bell-shaped, solferino purple.  Height
4ft(1.2m).  (K. Van de Ven, reg. 1984)

cl.
          -minus var. minus Carolinianum Group (1/2)
Fred Clark-
          -yungningense (glomerulatum) (1/2)

1.5ft(.45m)          -15°F(-26°C)          M

Mauve pink flowers, saucer-shaped, about 1.5in(3.5cm) wide, in
trusses of 5-7, at terminal.  Floriferous.  Plant broader than
tall, well-branched; glossy spinach green leaves held one year;
winter foliage greenish mahogany.  (William Fetterhoff, 1981)

cl.                      -neriiflorum ssp. neriiflorum (1/4)
          -unnamed hybrid-
Fred Hamilton-          -griersonianum (1/4)
          -dichroanthum ssp. dichroanthum (1/2)

3ft(.9m)          -5°F(-21°C)          ML          3/3

Trusses of yellow flowers with pink bands between lobes.  Dense,
low plant, well-branched, wider than tall.  (Halfdan Lem cross;
intro. by Theodore Van Veen, 1972)

cl.
          -Corona--unknown (1/2)
Fred Harris-
          -fortunei ssp. fortunei (1/2)

Clear pink flowers, 4in(10.2cm) wide, 5-7 lobes slightly reflexed; conical trusses of 14.  (Gen. Harrison)

cl.          -wardii var. wardii (1/4)
          -Prelude-
Fred Holden-          -fortunei ssp. fortunei (1/4)
          -Sappho--unknown (1/2)

3ft(.9m)          0°F(-18°C)          EM

Fragrant flowers of good substance, white with some lobes shaded
light yellow, in trusses of 16-19.  Well-branched plant, broader
than tall with yellow green leaves held 3 years.  (Cross by Paul
Holden; raised by Mrs. A. J. Holden, reg. 1981)

cl.          -sanguineum ssp. didymum (1/4)
          -Carmen-
Fred Robbins-    -forrestii ssp. forrestii Repens Group (1/4)
       -          -haematodes ssp. haematodes (1/4)
          -Choremia-
                   -arboreum ssp. arboreum (1/4)

After 9 years from seed, plant was 20in(51cm) high, and about as
wide; leaves 2.25in(5.7cm) long, with silvery indumentum below.
Flowers of chrysanthemum crimson, 2.25in(5.7in) wide, with calyx
same hue; trusses of 7.  See also Honore Hacanson, Little Nemo.
(Lester E. Brandt, 1965)

cl.
                   -campylocarpum ssp. campylocarpum (1/4)
          -Mary Swathling-
Fred Rose-          -fortunei ssp. fortunei (1/4)
          -lacteum (1/2)

Lemon yellow flowers with throat lightly spotted red, widely
campanulate, 2.5in(6.4cm) across; compact rounded trusses of 10.
Free-flowering.  Plant habit upright, as broad as tall; leaves
dull green.  (Raiser F. Rose; Sunningdale Nurseries, reg. 1962)
H. C. (PHS) 1973   See col. ill. ARS Q 29:1(Winter 1975): p. 35.

g.
          -fortunei ssp. fortunei (1/2)
Fred     -                    -fortunei ssp. discolor (1/8)
Wynniatt-     -Lady Bessborough-
                              -campylocarpum ssp. campylo-
          -Jalisco-            carpum Elatum Group (1/8)
                   -    -dichroanthum ssp. dichroanthum (1/8)
                   -Dido-
                       -decorum (1/8)

Flat-topped, lax trusses  hold up to 10 flowers  of maize yellow
with  rose-tinged edges,  4.5in(11.5cm) wide.  Leaves 6.5in(16.5
cm) long by 3in(7.6cm) broad.   Named after the Head Gardener at
Exbury.    See also  Jerez, Joyful, Simita, Trianon, and Stanway.
(Exhibited by L. de Rothschild)  A.M. (RHS) 1963;  F. C. C. (RHS)
1980

cl.
          -fortunei ssp. fortunei (1/2)
Freda-
          -unknown (1/2)

Light pink flowers.  (Paul)

cl.                    -souliei (1/4)
          -Virginia Scott-
Freda Rosage-          -unknown (1/4)
          -    -dichroanthum ssp. dichroanthum (1/4)
          -Jasper-          -fortunei ssp. discolor (1/8)
                   -Lady    -
                   -Bessborough-campylocarpum ssp. campylocarpum
                                      Elatum Group (1/8)

Buds of geranium lake opening pale barium yellow, the upper lobe
dark citron yellow; corolla split with overlapping lobes so that
flower seems double.  Trusses 13-flowered.  See also Golden Pip-
pin.  (H. L. Larson cross, 1964; reg. 1982)

cl.                    -griffithianum (1/4)
          -Jean Marie  -
Freeman R. Stephens- de Montague-unknown (3/4)
                   -
                   -unnamed white hybrid--unknown

5ft(1.5m)          -5°F(-21°C)          M

Plant grows much wider than high, with glossy dark green leaves.
Fragrant flowers are  4in(10cm) wide by 5in(12.7cm)long,  cherry
red with deep red spots; up to 15 are held in trusses  9in(23cm)
tall.  See also Douglas R. Stephens.  (Mr. & Mrs. J. Freeman Stephens, 1973)

cl.
          -fortunei ssp. fortunei ? (1/2)
French Creek-
          -unknown (1/2)

5ft(1.5m)          -15°F(-26°C)          ML

Upright plant, moderately branched with medium-sized leaves held
3 years.  Flowers up to 3.5in(9cm) across, pink fading to white
in large trusses of up to 20.  (Charles Herbert, 1977)

cl.                    -ponticum (1/4)
          -Purple Splendour-
Friday*                    -unknown (3/4)
          -unknown

5ft(1.5m)          -5°F(-21°C)          ML

Large flowers of satin purple with a black velvet blotch on the
upper lobe.  Leaves medium matte green, with leathery texture.
Better-looking than  Purple Splendour.  (Henry Landauer, Forest
Grove, Oregon)

```
                         -dichroanthum ssp. dichroanthum (1/8)
                 -Goldsworth-
cl.        -Hotei- Orange    -fortunei ssp. discolor (1/8)
  -          -               -souliei (1/8)
Frilled    -         -unnamed hybrid-
Petticoats-                    -wardii var. wardii (3/8)
  -             -Pink     -Jan Dekens--unknown (7/32)
  -          -Petticoats-        -Queen    -griffithianum
-unnamed-             -Britannia-Wilhelmina-          (1/32)
  hybrid-                   -          -unknown
                             -Stanley Davies--unknown
                 -wardii var. wardii
```

2.5ft(.75m)          0°F(-18°C)          M

A rounded plant, broader than tall; glossy, yellow green leaves 4.5in(11.5cm) long, held 3 years.  Flowers very pale chartreuse yellow, darker on reverse; 16 in round trusses 6in(15.2cm) wide. (John Lofthouse, 1981)

```
cl.                -campylocarpum ssp. campylocarpum Elatum
   -Letty Edwards-                              Group (3/8)
Frontier-          -fortunei ssp. fortunei (1/4)
   -     -wardii var. wardii (1/4)
      -Crest-
         -Lady     -fortunei ssp. discolor (1/8)
         -Bessborough-
              -campylocarpum ssp. campylocarpum
                                   Elatum Group
```

4.5ft(1.35m)         10°F(-12°C)          ML

Rose-colored flowers shading to barium yellow at center, openly funnel-shaped, to 4in(10.2cm) wide, 7-lobed; 14-flowered spherical trusses.  Plant upright, rounded, well-branched, as broad as tall; medium-sized leaves held 3 years.  (J. A. Elliott, 1978)

```
cl.
          -Mucronatum--a hybrid with parentage unknown (1/2)
Frosty Morn-
          -Blaauw's Pink--Kurume azalea ? (1/2)
```

An azaleodendron(?) from Australia with one or more parents uncertain.  Flower is very light purple with spotting.  (H. van de Ven, 1972)

```
cl.          -yakushimanum ssp. yakushimanum (1/4)
       -Bambi-              -dichroanthum ssp. dichroanthum
Frou Frou-   -Fabia Tangerine-                      (1/8)
    -                  -griersonianum (1/8)
       -arboreum ssp. arboreum (1/2)
```

Flowers of soft claret rose  with darker stripes; trusses of 22. (R. Cutten, Australia, reg. 1982)

```
cl.
     -caucasicum (1/2)
Fruehauf-
     -unknown (1/2)
```

A ruby red German hybrid which normally flowers in April but also used for forcing at Christmas.  (Albertzland, Germany, 1962)

```
cl.
          -Essex Scarlet--unknown (1/2)
Fruhlingszauber*
          -forrestii ssp. forrestii Repens Group (1/2)
```

An unregistered scarlet hybrid, by Dietrich Hobbie.  A parent of Buketta.

```
cl.
     -grande ? (1/2)
Fulbrook-
     -unknown (1/2)
```

Phlox pink flowers spotted maroon on upper lobe, bell-shaped, in large trusses.  Leaves 9.5in(24cm) long, 3.5in(9cm) wide. (Mrs. Douglas Gordon)  A. M. (RHS) 1961

```
cl.
     -fulgens (1/2)
Fulgarb-
     -Blood Red--arboreum ssp. arboreum (1/2)
```

Very large plant with small, compact trusses of 15 rich crimson flowers.  Very early.  (E. J. P. Magor)  A. M. (RHS) 1937

```
g.
     -cinnabarinum ssp. cinnabarinum Blandfordiiflorum Gp.
Full House-                                        (1/2)
     -maddenii ssp. maddenii (1/2)
```

Trumpet-shaped flowers of deep pink, flushed rose.  (Johnstone)

```
            -wardii var. wardii (1/4)
   -Crest-              -fortunei ssp. discolor (1/8)
cl. -    -Lady Bessborough-
Full-            -campylocarpum ssp. campylocarpum
Moon-                          Elatum Group (1/8)
   -                 -griffithianum (1/16)
   -            -George Hardy-
   -       -Mrs. Lindsay-        -catawbiense (1/16)
   -       - Smith   -Duchess of Edinburgh--unknown
   -Harvest Moon-                           (1/4)
   -          -campylocarpum ssp. campylocarpum (1/8)
        -unnamed-
         hybrid-unknown
```

4ft(1.2m)         -5°F(-21°C)          M          4/3

Flowers a deep yellow like Crest, but with a more compact growth habit.  Needs partial shade.  Foliage of bright shiny green, and prominently veined.  (John Henny, 1958)  P. A. (ARS) 1955

```
cl.
     -fortunei ssp. fortunei (1/2)
Fundy-
     -smirnowii (1/2)
```

6ft(1.8m)         -15°F(-26°C)          L

Synonym Evangeline.  Medium-sized leaves with light indumentum on a plant wider than tall.  Pleasantly scented flower, 7-lobed, rose-colored on edges, fading lighter to center with olive brown blotch, in trusses of up to 10.  (Cross by L. M. Hancock; raiser G. S. Swain; D. I. Craig, intro. and reg. 1977)

```
cl.
Furnivall's Daughter     Parentage unknown, but may be
                         Mrs. Furnival selfed
```

5ft(1.5m)         -5°F(-21°C)          M          5/4

Beautifully shaped tall trusses of pink flowers, each with cherry pink blotch.  Plant of much vigor, with lively green, large leaves.  (Knap Hill, 1957)  H. C. (RHS) 1957; A. M. (RHS) 1958; F. C. C. (RHS) 1961    See color illus. HG p. 77.

```
cl.
     -oreotrephes (1/2)
Furore-
     -cinnabarinum ssp. xanthocodon Concatenans Group (1/2)
```

Pale mauve pink flowers.  (Gen. Harrison, reg. 1958)

```
g. & cl.
     -elliottii (1/2)
Fusilier-
     -griersonianum (1/2)
```

5ft(1.5m)         10°F(-12°C)          ML          4/3

Upright growth habit, with large, dark green indumented leaves. Bright orange red flowers in open top, medium size trusses.  Several clones are known.  (Rothschild)  A. M. (RHS) 1938; F. C. C. (RHS) 1942

cl.

G. A. Sims     Parentage unknown

A parent of Griersims.  Flowers deep currant red,  2.5in(6.4cm)
wide, frilled edges; trusses of 16.  Plant upright and spreading
with leaves 6-7in(17.8cm) long, glossy dark green.  (A. Waterer,
Jr., Knap Hill Nursery)  A. M. (RHS) 1938; A. M. (RHS) 1972

```
                        -catawbiense (1/8)
             -Atrosanguineum-
cl.   -Dr. H. C. -             -unknown (1/4)
      -Dresselhuys-
Gabriel-             -arboreum ssp. arboreum (1/8)
    -       -Doncaster-
      -smirnowii (1/2)      -unknown
```

3ft(.9m)        -15°F(-26°C)        L

Plant as broad as high, rounded, semi-compact with medium-sized
leaves.  Flowers 2.75in(7cm) across, pink with olive brown
spotting, held in trusses of 18.  (G. Swain & D. Craig, Canada
Dept. of Agriculture Research Station, N. S., 1978)

```
cl.
          -arboreum ssp. arboreum (1/2)
Gabriele Liebig-
          -ponticum (1/2)
```

Frilled flowers white with pink touch and very bold red brown
markings.  (In Register as cross of Otto Schulz, 1860; in
Germany credited to Emil Liebig, before 1863)

```
             -griffithianum (1/8)
cl.       -Loderi-
     -Avalanche-    -fortunei ssp. fortunei (1/8)
Galactic-    -calophytum var. calophytum (1/4)
     -lacteum (1/2)
```

Early flowers, deep cream with two small crimson rays,  slightly
scented, 3.5in(8.9cm) wide; very large trusses of about 22.  A
tall, handsome plant bearing leaves 8in(20.3cm) long by 3in(7.6
cm) wide, with plastered indumentum below.  (Cross by Lionel de
Rothschild; shown by Edmund de Rothschild)  A. M. (RHS) 1964; F.
C. C. (RHS) 1970    See color illus. PB pl. 35.

```
cl.
          -azalea--unknown
Galloper Light-
          -rhododendron--unknown
```

5ft(1.5m)        -5°F(-21°C)        M        3/1

A deciduous azaleodendron as beautiful as Glory of Littleworth,
and more hardy.  Rose pink buds opening salmon pink, then creamy
yellow; good trusses of comparatively large flowers.  Strong
foliage. (A. Waterer, to L. de Rothschild?  Not in Exbury Register)  A. M. (RHS) 1927)

```
cl.             -griersonianum (1/4)
     -Gladys Rillstone-
Gardis-             -unknown (1/4)
     -fortunei ssp. discolor (1/2)
```

Flowers of light neyron rose with a maroon blotch and speckles;
trusses of about 14.  (E. de Rothschild, 1982)

cl.

Gargantua--form of diaprepes (F 11958)

6ft(1.8m)        0°F(-18°C)        L        4/4

A triploid clone  with thick stems.  Very large, fragrant, white
flowers with a green basal flush.  A broad, vigorous plant.  (G.
Forrest, collector; raised by J. B. Stevenson and Crown Estate;
exhibited by Crown Estate, Windsor)    A. M. (RHS) 1953; F. C. C.
(RHS) 1974

```
g.
     -griffithianum (1/2)
Garnet-          -arboreum ssp. arboreum (1/4)
     -Broughtonii-
              -unknown (1/4)
```

5ft(1.5m)        5°F(-15°C)        M        3/4

Flowers deep salmon rose, flushed red.  Upright plant, spreading
habit, rather compact; glossy dark green leaves, 5.25in(13.3cm).
P. D. Williams, 1942)

```
cl.             -arboreum ssp. arboreum (1/4)
          -Doncaster-
Gartendirektor-     -unknown (1/4)
 Glocker    -williamsianum (1/2)
```

3ft(.9m)        -5°F(-21°C)        M

Flowers deep reddish rose partly with darker edges, in umbels of
8 12.  Heavy, dark green foliage; tolerates sun.  (Dietrich Hobbie, 1952)

```
                   -campylocarpum ssp. campylocarpum
             -unnamed-                   (1/8)
cl.       -Adriaan- hybrid-unknown (1/4)
     - Koster-          -griffithianum (1/16)
Gartendirektor-   -          -George-
 Rieger    -    -Mrs. Lindsay- Hardy-catawbiense (1/16)
     -    - Smith    -
     -             -Duchess of Edinburgh--
     -williamsianum (1/2)          unknown
```

Cream-colored flowers with dark red spotting.  A vigorous plant,
growing to 6ft(1.8m); rounded leaves.  Floriferous; bud hardy to
-10°F(-23°C).  Early.  (Dietrich Hobbie)  See color illus. ARS Q
34:2 (Spring 1980): cover.

```
cl.
     -minus var. minus Carolinianum Group (1/2)
Gary-             -racemosum (1/8)
     -          -Conemaugh-
     -Pioneer (Gable)-          -mucronulatum (1/8)
             -unknown (1/4)
```

2.5ft(.75m)        -20°F(-29°C)        M

Flowers in terminal spherical clusters, composed of 1-4 trusses
each of 3-5 flowers (total 3-20), of rhodamine pink with light
spotting.  Semi-dwarf plant, rounded, wider than tall; yellowish
green leaves held only 1 year.  See also David John.  (William
Fetterhoff, 1980)

cl.

Gary Herbert--form of vernicosum (R 18139)

4ft(1.2m)        -5°F(-21°C)        M

Fragrant flowers of shaded salmon tones, openly funnel-shaped,
3in(7.6cm) wide, ruffled;  flat trusses of 7.  Plant rounded and
moderately branched; glossy leaves held 3 years.  New growth has
showy, burgundy red bracts.  (Collector Rock, 1929; J. Gable selection; raised & intro. by C. Herbert; reg. 1976)

```
g. & cl.     -thomsonii ssp. thomsonii (1/4)
     -Shilsonii-
Gaul-     -barbatum (1/4)
     -elliottii (1/2)
```

Waxy, ruby red flowers, widely funnel-shaped; compact, spherical
trusses.  Deep green leaves; top branches need pinching.  Early.
See also Gaul Mastadon.  (Rothschild, 1939)  A. M. (RHS) 1939

cl.

Gaul Mastodon     Parentage as above

A richer red than that of Gaul, otherwise the two are similar.
(Rothschild, 1939)

cl.
                  According to <u>Hillier's</u> <u>Manual</u> same
Gauntlettii      parentage as Loder's White, q.v. for 2
                  possible parentage diagrams.

Rich pink in bud, opening blush white, speckled crimson.
Foliage dark green, glossy.  (Origin uncertain, before 1934)


cl.
         -<u>davidsonianum</u> (1/2)
Gavotte-
         -<u>mollicum</u> (1/2)

Flowers pink to mauve.  (Adams-Acton, 1945)


cl.
          -<u>yakushimanum</u> ssp. <u>yakushimanum</u> (1/2)
Gay Arsen-                   -<u>catawbiense</u> (1/4)
          -Henriette Sargent-
                              -unknown (1/4)

3ft(.9m)         -10°F(-23°C)          ML

Flowers of neyron rose fading lighter,  sap green dorsal blotch,
openly funnel-shaped, 3in(7.6cm) wide; dome-shaped trusses hold
15.  Floriferous.  Plant rounded, broader than high; leaves held
3 years.  (Frank Arsen, intro. 1974; reg. 1983)


cl.
                      -Essex Scarlert--unknown (1/4)
          -Beau Brummell-
Gay Gordon-                  -<u>facetum</u> (<u>eriogynum</u>) (1/4)
          -<u>elliottii</u> (1/2)

A hybrid of medium size, loose habit, and waxy scarlet flowers--
which were described as "hunting coat pink."  (Rothschild, 1939)
See color illus. ARS J 36:1 (Winter 1982): p. 14.


cl.                           -<u>catawbiense</u> (1/8)
                -Atrosanguineum-
          -Atroflo-              -unknown (5/8)
Gay Princess-       -<u>floccigerum</u> ssp. <u>floccigerum</u> (1/4)
          -unknown

4ft(1.2m)        -5°F(-21°C)          M

Synonym Pink Princess.  Plant upright, rounded, well-branched;
long narrow leaves held 2 years.  Flowers medium fuchsia purple,
lighter on edges, throat and back, greyed yellow orange blotch,
and about 3in(7.6cm) wide; dome-shaped trusses hold 18.  (C. Herbert, 1979)


cl.
          -Marion--unknown (11/16)
Gay Song-   -<u>griersonianum</u> (1/4)      -<u>griffithianum</u> (1/16)
          -Radium-         -Queen Wilhelmina-
             -Earl of-                 -unknown
                Athlone-Stanley Davies--unknown

Trusses hold 14 flowers colored neyron rose, with darker spotting.  (G. Langdon, 1980)


cl.       -<u>griffithianum</u> (1/4)
          -Azma-
Gayblade-    -<u>fortunei</u> ssp. <u>fortunei</u> (1/4)
          -    -<u>griffithianum</u> (1/4)
          -Mars-
             -unknown (1/4)

3.5ft(1m)        5°F(-15°C)          ML

Plant of compact habit.  Flowers of gaudy pink, in trusses of
15-18.  (Raised by A. Wright; intro. by A. Lindsley; reg. 1971)


                            -<u>thomsonii</u> ssp. <u>thomsonii</u>
                 -Bagshot-                            (1/8)
cl.      -Princess Elizabeth- Ruby -unknown (5/8)
-                -
Gee Whiz*            -unknown
      -        -Catalgla--<u>catawbiense</u> var. <u>album</u> Glass (1/4)
         -Calsap-
            -Sappho--unknown

Medium-sized plant, hardy to -15°F(-26°C).  Flowers mallow purple with blotch of dark beetroot purple.  (Weldon E. Delp)

g.
      -Pineapple--unknown (1/2)
Geisha-
      -<u>dichroanthum</u> ssp. <u>dichroanthum</u> (1/2)

Bush of low, spreading habit, with cream-colored, campanulate
flowers in loose trusses.  (Rothschild cross, 1932; intro. 1939)

g.          -<u>maximum</u> (1/4)
   -Halopeanum-
Gem-          -<u>griffithianum</u> (1/4)
   -<u>thomsonii</u> ssp. <u>thomsonii</u> (1/2)

Flowers light blush or rose, deep pink edges.  (Loder, 1926)

                      -<u>griffithianum</u> (1/8)
cl.       -Dawn's Delight-
          -Bauble-         -unknown (1/8)
Gemstone-        -<u>campylocarpum</u> ssp. <u>campylocarpum</u> (1/4)
          -<u>wardii</u> var. <u>wardii</u> (1/2)

Medium-sized, light yellow flowers carried in trusses of about
7.  (Edmund de Rothschild, 1979)

cl.
      -<u>scabrifolium</u> var. <u>spiciferum</u> (1/2)
Gene-
      -<u>ciliatum</u> (1/2)

2ft(.6m)         5°F(-15°C)          E          4/2

Plant broader than tall; leaves about 2in(5cm) long.  Rosy purple flowers, funnel-shaped, 1.75in(4.5cm) long, in ball-shaped
trusses. (University of Washington Arboretum & B. Mulligan; reg.
1963)

cl.
          -Scintillation--unknown (5/8)
General Anthony-              -<u>catawbiense</u> (1/8)
   Wayne        -   -Atrosanguimeum-
             -Atrier-              -unknown
                -<u>griersonianum</u> (1/4)

6ft(1.8m)        -5°F(-21°C)          M

Plant broad, as wide as tall at maturity, well-branched; large,
glossy, rounded leaves, held 3 years.  Fragrant flowers, openly
funnel-shaped, 4in(10cm) wide,  neyron rose tinged with glowing
orange, light green throat with chartreuse green dorsal spots;
large conical trusses of 15-17.  (Charles Herbert, intro. 1975)

cl.
          -<u>griffithianum</u> (1/2)
General Eisenhower-
          -unknown (1/2)

6ft(1.8m)        10°F(-12°C)          M          3/3

Large carmine red flowers with ruffled edges, in large trusses.
Compact, sturdy, vigorous plant; big, dark green, waxy leaves.
(Anthony Kluis, 1946)   See color illus. F p. 45;  VV p. 122.

cl.
                 -<u>yakushimanum</u> ssp. <u>yakushimanum</u> (1/2)
General Eric Harrison-         -<u>thomsonii</u> ssp. <u>thomsonii</u> (1/4)
                 -Shilsonii-
                            -<u>barbatum</u> (1/4)

Bright crimson flowers with darker red spotting. (Hydon Nurseries, 1972)

cl.
                    -catawbiense (1/2)
General Grant-
                    -unknown (1/2)

A hardy hybrid with flowers of soft rose. (Intro. before 1875 by
Samuel B. Parsons (1819-1906) of Long Island, N.Y.)


cl.

General                              -yakushimanum ssp.
Practitioner--(unnamed hybrid, selfed)-    yakushimanum (1/2)
                                     -unknown (1/2)

A yakushimanum hybrid with flowers of very light sap green. (J.
Waterer, Sons & Crisp, intro. 1975)


cl.
                    -thomsonii ssp. thomsonii (1/2)
General Sir John du Cane-
                        -fortunei ssp. discolor (1/2)

Named to honor the officer of World War I.  A tall plant, with
large lax trusses of fragrant, funnel-shaped flowers, rose madder fading lighter, a dark crimson basal blotch. (Rothschild,
1933)


                                      -griffithianum (1/8)
cl.                    -Queen Wilhelmina-
          -Unknown Warrior-           -unknown (3/8)
Geneva-              -Stanley Davies--unknown
     -          -dichroanthum ssp. dichroanthum (1/4)
          -Fabia-
                  -griersonianum (1/4)

Campanulate flowers of camellia rose, lighter in center, faint
Majolica yellow spotting on upper lobe, 3in(7.6cm) across, and
frilled. (John Bacher, reg. 1958)   P. A. (ARS) 1955


                                      -griffithianum (1/8)
cl.                    -Queen Wilhelmina-
          -Britannia-                 -unknown (3/8)
Genghis Khan-          -Stanley Davies--unknown
     -          -dichroanthum ssp. dichroanthum (1/4)
          -Felis-
                -facetum (eriogynum) (1/4)

5ft(1.5m)        -5°F(-21°C)        M        4/4

Flowers guardsman red, 3in(7.6cm) across, with enlarged calyx of
the same color; rounded trusses of about 15.  Compact plant;
dense dark green foliage with silver plastered indumentum below.
(Lester A. Brandt, 1969)    See color illus. HG p. 111.


cl.
        -catawbiense (1/2)
Genoveva-
        -unknown (1/2)


Synonym Lucy Neal.  Flowers of pale lilac white with a yellow
green blotch.  A parent of Moerheim's Pink.  (T. J. R. Seidel)


cl.
                -edgeworthii (bullatum) (1/2)
Geoffrey Judson-
                -johnstoneanum (1/2)

Creamy white flowers with yellow blotch in throat. (G. A. Judson, 1973)


cl.
                -griffithianum (1/2)
Geoffrey Millais-
                -unknown (1/2)

Pink buds opening to gleaming white flowers flushed pink, fragrant and frilled; conical trusses.  Large, open shrub; foliage
dark green, glossy. (Otto Schulz cross, 1892; intro. C. B. van
Nes) A. M. (RHS) 1922


cl.

Geordie Sherriff--form of lindleyi

5ft(1.5m)        25°F(-4°C)        E-ML        4/2

Trusses hold 5-7 very fragrant white flowers stained spirea red
on upper lobes.  Leaves are dark green above with thin brown indumentum beneath. (A. C. & J. F. A. Gibson)  A. M. (RHS) 1969


cl.
        -lateum (1/2)
George Budgen-
        -zoelleri (1/2)

4ft(1.2m)        25°F(-4°C)        Sept.-Dec.

Vireya hybrid.  Flowers of golden yellow, tubular funnel-
shaped, in trusses 7.5in(19cm) wide, holding 5 flowers.  Upright
plant, moderately branched;  dark green leaves 4in(10.2cm) long,
held 3-4 years. (T. Lelliott cross; raiser P. Sullivan; reg. by
William Pollard, 1977)


cl.
            -arboreum ssp. arboreum (1/2)
George Cunningham-
            -campanulatum ssp. campanulatum ? (1/2)

A plant with unusual flowers: the outside of the flower is white
but the inside is heavily mottled with black; in neat trusses.
Foliage resembles cinnamomeum. See also Boddaertianum. (Geo.
Cunningham, before 1875)


                    -griffithianum (3/8)
cl.        -Loderi-
              -fortunei ssp. fortunei (1/4)
George Grace-              -arboreum ssp. arboreum (1/8)
         -          -Doncaster-
          -Borde Hill-          -unknown (1/4)
         -                  -griffithianum
               -Mrs. A. M. Williams-
                              -unknown

Synonym Grace.  Flowers are rhodamine pink, with lobes slightly
recurved, 4.5in(11.5cm) across; trusses hold about 15. (R. Henny
cross, 1943)  P. A. (ARS) 1952


cl.
        -griffithianum (1/2)
George Hardy-
        -catawbiense (1/2)

Parentage as recorded in the Register,  but some have suggested
ponticum x griffithianum.  A parent of Mrs. Lindsay Smith, and
others.  Flowers of blush pink, fading white. (Mangles, before
1922)


cl.
            -cinnabarinum ssp. xanthocodon Concatenans Group (1/2)
George    -          -cinnabarinum ssp. cinnabarinum
Johnstone-Full House-          Blandfordiiflorum Group (1/4)
                    -maddenii ssp. maddenii (1/4)

From the Trewithen Orange g. of George Johnstone.  Flowers
bright nasturtium orange held in loose trusses of 7-9.  Leaves
are 2.25in(5.7cm) long, indumented beneath and aromatic. (Shown
by Collingwood Ingram)  A. M. (RHS) 1967


cl.          -haematodes ssp. haematodes (1/4)
       -May Day-
George M.-       -griersonianum (1/4)
  Cohan  -       -dichroanthum ssp. dichroanthum (1/4)
       -Jasper-              -fortunei ssp. discolor (1/8)
       (Exbury-Lady Bessborough-
        form)              -campylocarpum ssp. campylocar-
                              pum Elatum Group (1/8)

Medium-sized, open plant.  Flowers azalea pink with base of carrot red; to 3in(7.6cm) across; open trusses.  Midseason. (C. S.
Seabrook, 1967)

'Fair Lady' by R. Henny
Photo by Greer

'Fairweather' by Brockenbrough
Photo by Greer

'Fanfare' by Leach
Photo by Leach

'Fascinator' by Aberconway
Photo by Greer

'Fastuosum Flore Pleno' (double-flowered) by Francoisi
Photo by Greer

'Fawn' by James
Photo by Greer

'Fiji' by Leach
Photo by Leach

'Finlandia' by Leach
Photo by Leach

'Fire Bird' by Rothschild
Photo by Greer

'Fire Wine' by Greer
Photo by Greer

'Fireman Jeff' by Brandt/Eichelser
Photo by Greer

'First Love' by R. Henny
Photo by Greer

'Flair' by Leach
Photo by Leach

'Flamenco' by Leach
Photo by Leach

'Flirt' by J. A. Elliott
Photo by Greer

'Flora Markeeta' by Flora Markeeta Nursery
Photo by Greer

'Flora's Boy' by Flora Markeeta Nursery
Photo by Greer

'Florence Sarah Smith' by Smith of Darley Dale
Photo by Greer

'Fragrantissimum' by unknown
Photo by Greer

'Floriade' by Adr. van Nes
Photo by Greer

'Frank Galsworthy' by A. Waterer
Photo by Greer

'Frank Baum' by Seabrook
Photo by Greer

'Fred Rose' by Rose (Sunningdale Nurs.)
Photo by Greer

'Freeman R. Stephens' by J. F. Stephens
Photo by Greer

```
         -yakushimanum ssp. yakushimanum (1/4)
         -                    -Parsons    -catawbiense
         -                    -Grandiflorum-      (1/16)
   -unnamed-        -America-        -unknown (9/64)
   - hybrid-   -unnamed-       -dark red hybrid--unknown
   -        - hybrid-        -griffithianum (3/64)
   -        -        -      -Mars-
cl. -    -Cindy-   -Blaze-  -unknown
   -     Lou-     -         -catawbiense (red form)
George-   -              -griffithianum
Ring*-    -          -Mars-
   -     -Red Brave-  -unknown
   -              -Parsons   -catawbiense
   -              -America-Grandiflorum-
   -              -          -unknown
   -              -dark red hybrid--unknown
   -        -meddianum var. meddianum (1/4)
   -unnamed hybrid-
              -strigillosum (1/4)
```

Dark red flowers on a medium-sized plant.  Hardy to -5°F(-21°C).
(Weldon E. Delp)


cl.
          -griersonianum (1/2)
George Ritter-
          -unknown (1/2)

4ft(1.2m)        15°F(-9°C)         L

Fuchsine pink flowers with crimson throat, widely funnel-shaped,
to 4.25in(10.8cm), 5-7 lobes; lax trusses of about 6.  Plant
wider than tall, well-branched; spinach green leaves, narrowly
elliptic, about 7.25in(18.5cm) long, held 3 years.  (Else Frye
cross; raiser J. S. Drucker; reg. by E. German, 1976)


cl.                 -fortunei ssp. fortunei (1/4)
          -Luscombei-
George Sansom-        -thomsonii ssp. thomsonii (1/4)
          -lacteum (1/2)

Leaves 5in(12.7cm) long by 2in(5cm) wide.  Flowers pale yellow,
flushed pink, from rosy pink buds.  (Geoffrey Gorer, 1968)
A. M. (RHS) 1965


```
                    -wardii var. wardii (1/16)
              -unnamed-        -neriiflorum ssp.
              - hybrid-F.C.Puddle-  neriiflorum (1/32)
              -        -
        -Virginia-        -griersonianum (1/32)
        -Richards-        -griffith.
        -        -      -George-
        -     -Mrs. Lindsay- Hardy-cataw.
        -   -Mrs. Betty- Smith  -
   -Vera       Robertson -        -Duchess of
   -Elliott-        -          Edinburgh--unk.
cl. -    -        -campylocarpum ssp.
   -    -     -unnamed- campylocarpum (1/32)
George-   -      hybrid-
Sweesy-   -        -unknown (1/4)
   -    -fortunei ssp. fortunei (1/4)
   -        -griffithianum (5/64)
   -     -George-
   -    -Pink - Hardy-catawbiense (13/64)
   -Dr. A. Blok-Pearl-    -arboreum ssp. arboreum
   -    -Broughtonii-        (1/16)
   -        -unknown
   -        -catawbiense
        -unnamed hybrid-
              - unknown
```

4ft(1.2m)        5°F(-15°C)         EM

Growth habit is rounded and upright, branching wider than tall
into maturity.  Leaves medium-sized, dark green.  Flowers up to
3.25in(8.3cm) wide, open funnel-shaped, wavy margins, soft pink
spotted golden brown, in rounded trusses of 12.  (Walter Elliott,
1981)


cl.

George Taylor--form of araiophyllum

5ft(1.5m)        10°F(-12°C)         EM

Flowers are slightly fragrant, white with red purple markings
and carried in loose trusses of 7-8.  Leaves are dark green and
have brown indumentum only along the midrib.  (Raised by RBG,
Kew)  A. M. (RHS) 1971


cl.             -New Yellow #6002 (Whitney) (1/2)
George's Delight*  -wardii var. wardii (1/4)
          -Crest-        -fortunei ssp. discolor (1/8)
              -Lady    -
          Bessborough-campylocarpum ssp. campylo-
                        carpum Elatum Group (1/8)

4ft(1.2m)        0°F(-18°C)         M

Flowers have dark pink margins blending into soft yellow, darker
yellow throat.  Shrub well-branched, shapely; matte green foliage.  Plant blooms young, likes some shade;  easy to propagate.
(Cross by W. Whitney; raisers A. & E. Sather)  See color illus.
ARS J 38:3 (Summer 1984): p. 115.


cl.          -yakushimanum ssp. yakushimanum (1/2)
Georgette-        -thomsonii ssp. thomsonii (1/4)
          -Exbury Cornish Cross-
                        -griffithianum (1/4)

3ft(.9m)        0°F(-18°C)         EM

A compact plant with slightly glossy foliage, medium dark green.
Globular trusses are 6in(15.2cm) high, 12-flowered.  White flowers, outside flushed pink when young, campanulate, 2.5in(6.4cm)
across.  (A. F. George, Hydon Nurseries, 1972)  H. C. (RHS) 1977


cl.

Gerald Loder--form of japonicum var. pentamerum (degronianum)

3ft(.9m)        -5°F(-21°C)         E         3/3

Plant compact with dense foliage, dark green above and heavy
coat of brown indumentum below.  Flowers erect, 10-12 per truss,
white with shades of roseine purple, darker lines down center of
petals.  (Collector unknown; shown by RBG, Kew,  from Wakehurst
Pl.)  A. M. (RHS) 1974  See color illus. ARS J 37:4 (Fall 1983)
p. 186.


g.          -praevernum (1/2)
Geraldii-
          -sutchuenense (1/2)

A large shrub bearing white flowers with a deep purple blotch;
handsome foliage.  Early.  (May be a natural hybrid; origin unknown)  A. M. (RHS) 1945


g.          -forrestii ssp. forrestii Repens Group (1/2)
Gertrud Schale-        -ponticum (1/8)
          -        -Michael Waterer-
          -Prometheus-        -unknown (3/8)
              -Monitor--unknown

Translucent scarlet red flowers.  (Dietrich Hobbie, 1951)


cl.                    -griffithianum (1/4)
          -Loderi King George-
Gertrude Bovee-        -fortunei ssp. fortunei (1/4)
          -Ostbo's Y No. 3--unknown (1/2)

5ft(1.5m)        -5°F(-21°C)         ML

Very dark green leaves, to 8in(20.3cm) long by 3in(7.6cm) wide, on a compact plant.  Flowers large and ruffled, soft cream color, blushed pink to margins, upper petal spotted dark red; about 12 per truss.  (The Bovees, 1972)

```
g.               -Moser's Maroon--unknown (1/4)
          -Bibiani-
Gibraltar-     -arboreum ssp. arboreum (1/4)
          -elliottii (1/2)
```

Not to be confused with the Exbury azalea of the same name.  Rhododendron Gibralter  is tall, with a loose habit; it has very attractive foliage of bright chestnut color when young.  Large deep red flowers held in dome-shaped trusses. (Rothschild, 1939) See color illus. ARS J 36:1 (Winter 1982): p. 14.

cl.

Gibsonii    Parentage unknown

A parent of Denisonii.  Flowers pure white.  (T. Methven, 1868)

```
cl.
          -catawbiense (1/2)
Giganteum-
          -unknown (1/2)
```

6ft(1.8m)         -15°F(-26°C)         EM         3/3

Should not be confused with the species of the same name.  Very hardy plant, with dark green medium-sized leaves.  Light crimson flowers held in nicely shaped trusses. (H. Waterer, before 1851) See color illus. VV p. 99.

cl.

Gigha--form of calostrotum ssp. calostrotum

2ft(.6m)         -5°(-21°C)         M         4/4

Grey green leaves less than an inch long.  Flowers of rose crimson instead of the usual purple, about 2in(5cm) across,  saucer-shaped, 5-lobed, 2-3 per cluster. (Sir James Horlick; Hillier reg., 1970) A. G. M. (RHS) 1969; F. C. C. (RHS) 1971

cl.

Gigi    Parentage unknown

5ft(1.5m)         -5°F(-21°C)         ML         4/3

Rose red flowers with deeper spotting; about 18 in ball-shaped trusses.  PLant wider than tall, well-branched; yellowish green leaves held 3 years.    (C. Dexter, 1925-42; S. Burns, reg. 1973) A. E. (ARS) 1973        See color illus. ARS Q 27:4 (Fall 1973): p. 238; LW pl. 17.

```
cl.      -wardii var. wardii (1/4)
     -Crest-          -fortunei ssp. discolor (1/8)
Gilded-   -Lady      -
 Gown -    Bessborough-campylocarpum ssp. campylocarpum
     -unknown (1/2)            Elatum Group (1/8)
```

Orange red buds opening light primrose yellow; trusses of 9-10 flowers. (Mrs. R. J. Coker, New Zealand, 1979)

```
cl.
          -aurigeranum (?) (1/2)
Gilded Sunrise-
          -laetum (1/2)
```

Vireya hybrid.    Trusses of 7-8 flowers, 5-lobed, 2in(5cm) wide, tubular funnel-shaped, lemon yellow suffused with apricot, darkening with age.  Minute brown scales on elliptic leaves. (E. B. Perrot cross; reg. by G. F. Smith, New Zealand registrar, 1984)

```
cl.
     -thomsonii ssp. thomsonii (1/2)
Gilian-
     -griffithianum (1/2)
```

Incorrect parentage listed in the Register.   Large cardinal red flowers, lightly spotted. (E. J. P. Magor)   A. M. (RHS) 1923

```
g.
     -griffithianum (1/2)
Gillii-
     -arboreum ssp. arboreum (blood red form) (1/2)
```

Flowers of soft rose. (R. Gill & Son) A. M. (RHS) 1919

```
cl.
          -griffithianum (1/2)
Gill's Crimson-
          -unknown (1/2)
```

6ft(1.8m)         5°F(-15°C)         E         4/3

Upright, sturdy plant.  Bright blood red, long-lasting flowers, in rounded trusses.  Slow to start blooming. (R. Gill & Son)

```
cl.
          -griffithianum (5/8)
          -                    -griffithianum
Gill's Gloriosa-        -George Hardy-
          -Pink Pearl-          -catawbiense (1/8)
          -               -arboreum ssp. arboreum
          -Broughtonii-                (1/8)
               -unknown (1/8)
```

Bright cerise flowers. (R. Gill & Son)   A. M. (RHS) 1925

```
cl.
          -griffithianum (1/2)
Gill's Goliath-
          -unknown (1/2)
```

Carmine pink. (R. Gill & Son)  A. M. (RHS) 1914

cl.

Gill's Nobleanum Album    Parentage unknown

Pure white flowers, green tinge at base. (R. Gill & Son)  A. M. (RHS) 1926

```
cl.
          -arboreum ssp. arboreum (1/2)
Gill's Triumph-
          -griffithianum (1/2)
```

Beauty of Tremough g.  Flowers of strawberry red fading to pink. (R. Gill & Son)  A. M. (RHS) 1906

cl.

Gina    Parentage unknown

A tall plant with dark green rounded leaves; large, deep ruby red flowers in erect trusses. (T. Lowinsky)   A. M. (RHS) 1931

```
cl.
          -Marion--unknown (3/4)
Gina Lollobrigida-          -catawbiense (1/4)
          -Mrs. C. S. Sargent-
               -unknown
```

Two hybrids are named  Marion, q.v., and which was used here is not known.  Flowers of orchid purple, spotted brown, are held in compact trusses of about 15. (Felix & Dijkhuis, Boskoop, 1965)

```
cl.
          -japonicum var. japonicum (metternichii) (1/2)
Ginny Beale-
          -adenopodum (1/2)
```

2.5ft(.75m)        -10°F(-23°C)        M

Flowers of rose, marked only by pale pink stripes on reverse, 2
in(5cm) wide; trusses of about 16. Plant rounded, much broader
than high; narrow, indumented leaves held 3 years. (J. B. Gable
cross, c.1964; R.B. Davis, Jr., raised & intro.; reg. 1979)

```
                        -griffithianum (1/8)
cl.             -Mars-
        -unnamed hybrid-       -unknown (5/8)
Ginny Delp*           -yakushimanum ssp. yakushimanum (1/4)
        -Dexter's Apple Blossom--unknown
```

Medium-sized plant; hardy to -15°F(-26°C). Buds roseine purple
opening white; flowers speckled with yellowish green. (Weldon
E. Delp)

```
cl.
        -keiskei (prostrate form) (1/2)
Ginny Gee-
        -racemosum (F 19404) (1/2)
```

1.5ft(.45m)        0°F(-18°C)        EM        5/5

Plant broader than tall, with both creeping and stiff branches;
very small leaves, some red in winter. Flowers 1in(2.5cm) wide,
white with light pink blush, held in groups of about 11 buds, of
4-5 flowers each. (W. E. Berg, 1979) C. A. (ARS) 1983; A. E.
(ARS) 1984  See color illus. ARS J 38:2 (Spring 1984): p. 103.

```
                    -catawbiense (1/4)
cl. -unnamed hybrid-       -dichroanthum ssp. dichroanthum (5/32)
    -            -Fabia-
Ginny-                    -griersonianum (7/32)
Mae* -    -yakushimanum ssp. yakushimanum (1/4)
    -    -                -fortunei ssp. discolor
    -    -         -Lady   -              (3/32)
 -Si Si-    -Day -Bessborough-campylocarpum ssp. campylo-
    -      -Dream-                carpum Elatum Gr. (1/32)
    -Gold-    -griersonianum
    -Mohur-        -fortunei ssp. discolor
        -Margaret-    -dichroanthum ssp. dichroanthum
        Dunn  -Fabia-
                    -griersonianum
```

Flowers very light cadmium orange with red blotch. Medium-sized
plant, hardy to -15°F(-26°C) (Weldon E. Delp)

```
                -Catalgla--catawbiense var. album Glass (1/4)
cl.   -Gosh Darn-                -decorum ? (1/16)
   -        -         -Caroline-
Ginny's-       -Mrs. H. R.-       -brachycarpum ssp. brachy-
Delight*    Yates   -unknown (1/8)       carpum ? (1/16)
   -      -fortunei ssp. fortunei (1/4)
   -Donna   -       -wardii var. wardii (1/8)
   -Hardgrove-unnamed-
        hybrid-dichroanthum ssp. dichroanthum (1/8)
```

Strong yellowish pink buds open medium to light yellowish pink,
edged in spinel red, with throat of light greenish yellow. Tall
plant; hardy to -5°F(-21°C). (Weldon E. Delp)

```
g.            -griffithianum (1/4)
        -King George-
Gipsy King-       -unknown (1/4)
        -haematodes ssp. haematodes (1/2)
```

3ft(.9m)        5°F(-15°C)        M        4/3

Grows with a loose, spreading habit. Flowers of heavy waxy substance, a rich deep red, in rather lax trusses. See also Memory. (Rothschild, 1939)

GIPSY MAID  (p 302)
```
cl.
        -fortunei ssp. discolor (1/2)
Gipsy Moth-                -griffithianum (1/8)
    -                -Loderi-
    -Lodauric Iceberg-    -fortunei ssp. fortunei (1/8)
            -auriculatum (1/4)
```

Fragrant flowers, bright spirea red with outside deeply marked
medium magenta rose, held in large trusses. See also Southern
Cross, Starcross. (A. F. George, Hydon Nurseries) P. C. (RHS)
1967

```
cl.        -Corona--unknown (1/4)
        -Bow Bells-
Gipsy Queen*    -williamsianum (1/4)
        -thomsonii ssp. thomsonii (1/4)
    -Barclayi -        -arboreum ssp. arboreum (1/8)
    Robert Fox-Glory of -
            Penjerrick-griffithianum (1/8)
```

3ft(.9m)        0°F(-18°C)        EM        3/3

Deep red flower bud scales contrast nicely with deep green
foliage. (Origin unknown)

```
cl.            -catawbiense (3/8)
        -Everestianum-
Gisela-            -unknown (3/8)
    -            -caucasicum (1/4)
    -Boule de Neige-        -catawbiense
            -hardy hybrid-
                -unknown
```

White flowers tinted pale lilac with yellow green markings. (T.
J. R. Seidel cross, 1905)

```
cl.            -wightii (1/4)
        -China-
Glad Tidings-    -fortunei ssp. fortunei (1/4)
        -williamsianum (1/2)
```

4ft(1.2m)        0°F(-18°C)        EM        3/3

Upright, rounded plant; dark green leaves to 6in(15.2cm) long,
cordate. Flowers 4in(10.2cm) wide, cream tinged pink, yellow
throat with red basal flare; upright trusses of about 12. (Benjamin Lancaster, 1965)

```
g.            -neriiflorum ssp. neriiflorum (1/4)
        -F. C. Puddle-
Gladiator-        -griersonianum (3/4)
    -griersonianum
```

The trusses hold deep scarlet flowers. (Lord Aberconway, 1941)

```
g. & cl.
        -campylocarpum ssp. campylocarpum (1/2)
Gladys-
        -fortunei ssp. fortunei (1/2)
```

6ft(1.8m)        0°F(-18°C)        EM        3/3

Cream-colored to pale yellow flowers with crimson markings, 2in
(5cm) across; small rounded trusses of 9-11. Tall, open plant;
medium green leaves 5in(12.7cm) long. See also Gladys Rose, Mary
Swathling. (Col. S. R. Clarke) One form received the A. M.
(RHS) 1950; another form: A. M. (RHS) 1934

```
cl.            -fortunei ssp. discolor (1/8)
        -Ladybird-
    -Diva-        -Corona--unknown (1/8)
Gladys Johnson-    -griersonianum (1/4)
        -fortunei ssp. fortunei (1/2)
```

5ft(1.5m)        0°F(-18°C)        ML

Large leaves, and large flowers, up to 4in(10.2cm) across, rose
pink fading lighter; rounded trusses of 15. Fragrant. (Johnson,
1958) P. A. (ARS) 1958

```
cl.            -griersonianum (1/2)
Gladys Rillstone*
        -unknown (1/2)
```

A parent of Exbury hybrids.

cl.
        -campylocarpum ssp. campylocarpum (1/2)
Gladys Rose-
        -fortunei ssp. fortunei (1/2)

Gladys g.  Creamy white flowers with crimson blotch in throat.
(Col. S. R. Clarke)  A. M. (RHS) 1950


g. & cl.
        -Margaret--unknown (1/2)
Glamour-
        -griersonianum (1/2)

4ft(1.2m)         5°F(-15°C)         M         2/2

Margaret may be a griffithianum hybrid. Plant low and spreading
with deep cherry red flowers in trusses of 10.  (Rothschild,
1939) A. M. (RHS) 1946    See color illus. PB pl. 3.


cl.
        -griersonianum (1/2)
Glasgow Glow-                    -fortunei ssp. fortunei (1/8)
        -            -Luscombei-
        -Scarlet Lady-          -thomsonii ssp. thomsonii
        -                                           (1/8)
                -haematodes ssp. haematodes (1/4)


Stirling Maxwell, 1953.


g.
        -glaucophyllum var. glaucophyllum (1/2)
Glaucoboothii-
        -boothii (1/2)

Flowers are pinkish with brownish red spotting.  (E. J. P.
Magor, 1922)


cl.
        -glaucophyllum var. glaucophyllum (1/2)
Glautescens-
        -lutescens (1/2)

Yellow flowers suffused with pink.  (Reuthe, 1939)


cl.
        -cinnabarinum ssp. cinnabarinum Roylei Group (1/2)
Gleam-            -cinnabarinum ssp. cinnabarinum (1/4)
        -Royal Flush-
                -maddenii ssp. maddenii (1/4)

6ft(1.8m)         5°F(-15°)         ML         4/3

Lady Chamberlain g.  Long, tubular flowers of orange and yellow
tones with crimson-tipped petals.  Upright small branches with
shiny round leaves form a narrow, columnar shrub typical of the
species cinnabarinum.  (Rothschild, 1930)


cl.
        -Marion--unknown (1/2)
Glen Bruce-              -griersonianum (1/4)
        -unnamed hybrid-
                -fortunei ssp. fortunei (1/4)

Trusses of 22 flowers, medium neyron rose with  deeper rose center.  (G. Langdon, 1975)


cl.
        -sinogrande (1/2)
Glen Catocal-
        -macabeanum (1/2)

Rounded trusses of 28 white flowers,  flushed cream with blotch
of greyed red. Leaves are 20.5in(52cm) long by 10in(25cm) wide,
dark green above with grey indumentum beneath.  (Raised by J. F.
A. Gibson; intro. by Brodick Castle Gardens)  P. C. (RHS) 1976


cl.
Glen Cloy--form of luteiflorum (glaucophyllum var. luteiflorum)

2ft(.6m)         0°F(-18°C)         E-M         4/3

Small shrub that may grow to 4ft(1.2m),  with aromatic foliage.
Flowers Dresden yellow,  7-11 per truss,  openly bell-shaped.
(Brodick Castle Gardens)   F. C. C. (RHS) 1966

cl.
        -President Roosevelt--unknown (7/8)
Glen Crimson-                    -griffithianum (1/8)
        -Earl of -Queen Wilhelmina-
         Athlone-             -unknown
                -Stanley Davies-unknown

An Australian hybrid with crimson flowers, in trusses of about
22. (A. Raper, Victoria, 1976)


cl.        -haematodes ssp. haematodes (1/4)
        -May Day-
Glen Glow-      -griersonianum (1/4)
        -unknown (1/2)

An Australian hybrid blooming there in October.  Flowers colored
geranium lake. (Cross by A. Brandy; intro. by A. Raper, reg.
1964)   See color illus. ARS J 37:2 (Spring 1983): p. 75.


cl.

Glen Rosa--form of sidereum

6ft(1.8m)         5°F(-15°C)         EM         3/3

The award description called it a  "half-hardy flowering plant";
a prospective grower might wonder which half is hardy.  Flowers
of primrose yellow, crimson ring at base of throat, ventricose-
campanulate, 2.5in(5.6cm) wide, with as many as 8 lobes; about
20 per truss. Leaves 9in(23cm) long, dull silver below.  (Brodick Castle, Isle of Arran)  A. M. (RHS) 1965


cl.
        -sanguineum ssp. didymum (1/2)
Glenarn-
        -chamaethomsonii ssp. chamaethomsonii (1/2)

Small flowers held in trusses of 4, deep cardinal red, 5-lobed.
Smooth ovate leaves, about 2.5in(6cm) long.  (Cross by A. & F.
Gibson; reg. by P. Urlwin-Smith, 1983)


cl.

Glendoick--form of racemosum

4ft(1.2m)         -5°F(-21°C)         EM         4/3

An unusually deep-colored form of this very floriferous species.
Rich, dark pink flowers along the young shoots in April.  (P. A.
Cox, Glendoick Gardens, 1971)


cl.        -diaprepes (1/4)
        -Polar Bear-
Glenrose Beauty-      -auriculatum (1/4)
        -unknown (1/2)

White flowers with a pale pink stripe.  (W. E. Glennie, NZ, reg.
1980)


                        -griffithianum (5/16)
                -Kewense-
cl.      -Aurora-      -fortunei ssp. fortunei (1/16)
        -Naomi-      -thomsonii ssp. thomsonii (1/8)
Glenrose Star-    -fortunei ssp. fortunei
        -red seedling--unknown (1/2)

Trusses resemble Naomi, with flowers cherry red in bud, opening
white with red margins.  Plant habit like Naomi.  (W. E. Glennie, reg. 1984)

cl.
        -grande (1/2)
Glenshant-
        -macabeanum (1/2)

Leaves are 10in(25.4cm) long by 4in(10.2cm) wide, with silvery
indumentum beneath.  Flowers creamy white, flushed pale yellow,
in flat trusses of about 26.  (Brodick Castle Gardens)    A. M.
(RHS) 1964

cl.
             -russatum (1/2)        -fastigiatum (1/8)
Gletschernacht -              -Intrifast-
(Glacier Night)-Blue Diamond-        -intricatum (1/8)
               -augustinii ssp. augustinii (1/4)

Flowers an unusual deep violet blue color, 5-lobed; trusses of
8.  Small, scaly leaves.  Hardy to -5°F(-21°C)  Sold in the U.S.
as Starry Night.  See also Azurwolke.  (H. Hachmann cross; reg.
by G. Stuck, 1983)

cl.

Gloire de Gandavensis      Parentage unknown

Synonym Gloria Gandavensis. A parent of a few old hybrids; still
in cultivation.  White flowers, beautifully spotted.  (Methven,
c.1868)

cl.

Gloria Gandavensis      See Gloire de Gandavensis

cl.            -dichroanthum ssp. dichroanthum (1/2)
    -Fabia-
Gloriana-    -griersonianum (1/4)
    -                    -dichroanthum ssp. dichroanthum
   -Goldsworth Orange-
               -fortunei ssp. discolor (1/4)

Trusses of 12 flowers brightly colored a variable shade between
claret rose and empire rose, suffused in the throat with orange.
See also Coral Reef.  (RHS Garden, Wisley)  A. M. (RHS) 1953

cl.

Glory of Keston      Parentage unknown

A parent of Pele, q.v.

cl.
            -griffithianum (1/2)
Glory of Leonardslee-
            -unknown (1/2)

Flowers of light strawberry red.  A parent of Dignity.  (R. Gill
& Son)

cl.

Glory of Littleworth      Parentage unknown

5ft(1.5m)         -5°F(-21°C)          M          5/3

Azaleodendron.  Snowy white with intense orange blotch.  Takes
full sun.  Grey blue foliage in summer.  (Mangles)  A. M. (RHS)
1911 See color illus. HG p. 99.

cl.
          -arboreum ssp. arboreum (1/2)
Glory of Penjerrick-
           -griffithianum (1/2)

Beauty of Tremough g.  Flowers of deep strawberry red fading to
pink.  (R. Gill & Son)  A. M. (RHS) 1904

cl.
     -griersonianum (1/2)        -griffithianum (1/8)
Glow-            -unnamed hybrid-
   -Armistice Day-              -unknown (3/8)
          -Maxwell T. Masters--unknown

Introduced by Robert M. Bovee, before 1958.

cl.
    -dichroanthum ssp. dichroanthum (1/2)
Glow Worm-        -souliei (1/4)
    -Halcyone-            -fortunei ssp. discolor (1/8)
         -Lady -
     Bessborough-campylocarpum ssp. campylocarpum
                    Elatum Group (1/8)

Flowers of butter yellow.  (Sunningdale Nurseries, 1951)

cl.                    -Moser's Maroon--unknown (1/4)
       -Romany Chal-
Glowing Embers-        -facetum (eriogynum) (1/4)
     -griersonianum (1/2)

4ft(1.2m)          5°F(-15°C)          ML          4/3

Plant fairly compact; long, narrow, dark green leaves.  Flowers
of waxy texture, without stamens, glowing orange red.  Easy to
grow; sun-tolerant. (John Henny, before 1958)  See color illus.
ARS Q 27:2 (Spring 1973): p. 103.

cl.
   -fortunei ssp. fortunei (1/2)
Glowing-                          -griffithianum (1/16)
  Star -              -Queen -
    -              -Britannia-Wilhelmina-unknown (3/16)
   -C. P. Raffill-        -
        -              -Stanley Davies--unknown
        -griersonianum (1/4)

Flowers coral rose with red throat.  (Donald Hardgrove, 1954)

cl.
    -fortunei ssp. fortunei ? (1/2)
Gloxineum-
     -unknown (1/2)

5ft(1.5m)          -10°F(-23°C)          M

A plant of attractive habit and vigorous growth.  Flowers pink,
slightly fragrant. (C. Dexter, 1925-42; reg. by de Wilde, 1958)

cl.
   -haematodes ssp. haematodes (1/2)
Gnom-
    -red hybrid--unknown (1/2)

Plant slow-growing; leaves about 2.5in(6.4cm) long, scanty brown
felted indumentum beneath.  Flowers scarlet, 2.5in(6.4cm) wide,
broadly funnel-shaped, with a large scarlet calyx; loose trusses
of about 15. (Raiser Georg Arends; intro. by G. D. Bohlje, 1959)

cl.

Goat Fell--form of arboreum ssp. arboreum

6ft(1.8m)          5°F(-15°C)          E          3/4

Flowers cherry red with throat lightly spotted, tubular-campanulate, 2.25in(5.7cm) across; globular trusses of 15. Leaves about
7in(17.8cm) long by 2in(5cm) wide, silvery underneath.  (Brodick
Castle Gardens)  A. M. (RHS) 1964

```
                                -griffithianum (1/8)
g. & cl.          -Dawn's Delight-
      -Break of Day-                 -unknown (1/8)
Goblin-              -dichroanthum ssp. dichroanthum (1/4)
      -griersonianum (1/2)

4ft(1.2m)        5°F(-15°C)         M        2/1
```

Reaches maturity at about 5ft(1.5m); dark sage green foliage on
a neat bush.  Flowers bright salmon pink, funnel-shaped, 3in(7.6
cm) across, calyx large and flower-colored; loose trusses of 7.
See also Goblin Pink.  (Rothschild) A. M. (RHS) 1939

```
                                -griffithianum (1/8)
cl.               -Dawn's Delight-
      -Break of Day-                 -unknown (1/8)
Goblin Pink-         -dichroanthum ssp. dichroanthum (1/4)
      -griersonianum (1/2)

4ft(1.2m)        5°F(-15°C)         M        2/1
```

Goblin g.   Similar to Goblin, but the flower color is soft rose
with a gold throat. (Rothschild, 1939)

```
cl.
      -griffithianum (1/2)
Godesberg-
      -unknown (1/2)
```

White flowers with brown spotting.  (Otto Schulz cross, 1892;
intro. C. B. van Nes & Sons)

```
cl.                           -catawbiense (1/2)
      -Catawbiense Grandiflorum-
Goethe-                        -unknown (1/2)
      -              -catawbiense
      -Mrs. Milner-
                    -unknown
```

Light pink flowers with pale brown markings.    (T. J. R. Seidel
cross, 1897; intro. 1905)

```
g.                    -Essex Scarlet--unknown (1/4)
      -Beau Brummell-
Golconda-             -facetum (eriogynum) (1/4)
      -dichroanthum ssp. dichroanthum (1/2)
```

A medium-sized, compact Exbury hybrid, with bright pink flowers.
(Rothschild, cross 1932; intro. 1939)

```
cl.           -dichroanthum ssp. dichroanthum (1/4)
      -Fabia-
Gold Braid-   -griersonianum (1/4)
              -fortunei ssp. discolor (1/4)
      -Dondis-        -arboreum ssp. arboreum (1/8)
              -Doncaster-
                       -unknown (1/8)
```

A semi-dwarf with flowers of pale pink and bright yellow.  Large
calyx. (Rudolph Henny cross, 1942)

```
cl.
      -yakushimanum ssp. yakushimanum (Exbury form) (1/2)
Gold Spread-
      -aureum var. aureum (1/2)

7in(17.8cm)        -25°F(-32°C)          EM
```

A very dwarf plant, 5 times as wide as high, well-branched; yellow green leaves, slightly glossy and indumented.  Flowers Dresden yellow, 5-lobed, 12 in trusses 3.5in(9cm) across.  (Basil C.
Potter, cross 1967; reg. 1981)

```
              -wardii.var. wardii (1/2)
      -Crest-              -fortunei ssp. discolor (3/16)
      -     -Lady
      -     Bessborough-campylocarpum ssp. campylo-
      -                  carpum Elatum Group (3/16)
      -Lemon -                          -griffith.
      -Custard-                  George-      (1/64)
cl.   -      -         -Mrs. Lindsay- Hardy-cataw.
      -      -              - Smith        (1/64)
Gold  -      -Mrs. Betty-      -Duchess of
Incense-     - Robertson-       Edinburgh--unk.
      -      -unnamed-        -campylocarpum ssp.
      -       hybrid-      -unnamed- campylocarpum (1/32)
      -       -              hybrid-
      -       -                     -unknown (1/16)
      -       -     -wardii var. wardii
      -     -wardii var. wardii
      -Crest-        -fortunei ssp. discolor
      -     -Lady
            Bessborough-campylocarpum ssp. campylocarpum
                                          Elatum Group

6ft(1.8m)        5°F(-15°C)          M
```

Very fragrant flowers buttercup yellow with edges light neyron
rose, held in large trusses of 14-18.  Upright plant spreading
nearly as wide as tall with leaves held 2 years. (Walter
Elliott cross, 1970; reg. 1983; raiser Fred Peste)

```
                     -fortunei ssp. discolor (3/8)
           -Lady
cl.   -Day Dream-Bessborough-campylocarpum ssp. campylo-
      -         -                carpum Elatum Group (1/8)
Gold Mohur-    -griersonianum (3/8)
              -fortunei ssp. discolor
      -Margaret Dunn-    -dichroanthum ssp. dichroanthum
              -Fabia-                          (1/8)
                    -griersonianum

5ft(1.5m)        0°F(-18°C)          ML        4/3
```

Dream Girl g.  Flowers 3in(7.6cm) wide, funnel-shaped, golden
yellow with rainbow pink marking the division of each petal and
held in trusses of 10-12.  Large leaves, myrtle green with a
downward roll.  See also Golden Pheasant, Mohur, Shah Jehan. (L.
E. Brandt, 1953) P. A. (ARS) 1955

```
                                -caucasicum (3/32)
                    -Jacksonii-
                    -         -caucasicum
         -Goldsworth-      -Nobleanum-
         - Yellow  -               -arboreum ssp.
cl. -Goldfort-       -campylocarpum    arboreum (1/32)
      -            -campylocarpum (3/16)
Gold- -fortunei ssp. fortunei (1/4)
Moon- -wardii var. wardii (1/8)
      - -Hawk-        -fortunei ssp. discolor (1/16)
      - -     -Lady
      -Full-   Bessborough-campylocarpum ssp. campylocarpum
       Moon-                       Elatum Group (1/16)
      -               -campylocarpum ssp. campylocarpum
      -       -unnamed-
      -Adriaan- hybrid-unknown (1/8)
       Koster-            -griffithianum (1/32)
              -         -George-
              -Mrs. Lindsay- Hardy-catawbiense (1/32)
               Smith    -
                        -Duchess of Edinburgh--unknown

3ft(.9m)        -10°F(-23°C)          EM
```

Upright plant, broader than tall, branching well, large glossy
leaves.  Flowers of heavy substance, 4in(10cm) wide, 7-lobed,
white with throats tinged yellowish green, held in trusses of up
to 11.  (Lewis Bagoly cross; reg. by Marie Tietjens, 1976)

```
cl.

Gold Spread    See column to the left
```

cl.

Goldbuckett-Scintillation--unknown (1/2)
(Golden    -
 Bouquet)  -wardii var. wardii (1/2)

Creamy yellow flowers, a red brown to ruby blotch, 5 wavy lobes;
trusses of 10-14. Foliage resembles Scintillation's.   Hardy to
-15°F(-26°C).   (H. Hachmann cross; G. Stuck, reg. 1983)   Gold
Medal, IV International Garden Exhibition (Munich), 1983    See
color illus. ARS J 39:2 (Spring 1985): cover.


cl.
        -wardii var. wardii (1/2)
Goldbug-    -dichroanthum ssp. dichroanthum (1/4)
        -Fabia-
              -griersonianum (1/4)

3ft(.9m)         5°F(-15°C)           M          3/3

Compact, very broad plant;  deep green leaves, slightly twisted.
Flowers open brick red with maroon speckling, then fade to yellow. (Rudolph Henny, reg. 1958)

                  -wardii var. wardii ? (1/4)
cl.      -Chlorops-
         -            -vernicosum ? (1/4)
Golden   -               -fortunei ssp. fortunei (1/8)
Anniversary-      -Golden-
         -        - West -campylocarpum ssp. campylocarpum
         -unnamed-                                    (1/8)
         hybrid-                     -caucasicum (1/16)
         -          -Dr. Stocker-
              -Mariloo-            -griffithianum (1/16)
                        -lacteum (1/8)

Flowers primrose yellow with a reddish flare, bell-shaped, held
in round trusses.  When reg. it was 32in(81cm) high and
15in(38cm) wide with leaves 5in(12.7cm) long.  Blooms midseason.
(Benjamin Lancaster, 1970)


cl.
        -Yaku Fairy--keiskei (1/2)
Golden Bee-
        -mekongense var. melinanthum (1/2)

2ft(.6m)         0°F(-18°C)           EM

A semi-dwarf, broader than tall, with small leaves, held 2
years. Lax trusses of 5-6 flowers, bright mimosa yellow. (W. E.
Berg, 1982) C. A. (ARS) 1983; A. E. (ARS) 1984   See color illus. ARS J 38:2 (Spring 1984): p. 103.


cl.
            -fortunei ssp. discolor (1/2)
Golden Belle-   -dichroanthum ssp. dichroanthum (1/4)
          -Fabia-
                -griersonianum (1/4)

4ft(1.2m)        0°F(-18°C)           ML         3/3

Margaret Dunn g. Large flowers of orange and yellow tones with
deep pink edges, blooming late mid-season on a low, spreading,
compact plant. (John Henny, reg. 1958)


                  -haematodes ssp. haematodes (1/4)
cl.        -May Day-
           -            -griersonianum (3/8)
Golden Cockerel-            -dichroanthum ssp. dichroanthum
           -        -Fabia-                          (1/8)
               -Lascaux-    -griersonianum
                        -wardii var. wardii Litiense Group (1/4)

3ft(.9cm)        5°F(-15°C)            M

Scarlet buds opening soft yellow with deeper yellow spotting, a
large double calyx; lax trusses of 10-12 flowers. (A. F. George,
Hydon Nurseries, 1969)


cl.
        -macabeanum (1/2)
Golden Dawn-
        -unknown (1/2)

Flowers of primrose yellow with faint spots at base, 7-lobed, in
trusses of about 12.  Leaves oblong-elliptic. (H. G. Huthnance,
1982)


cl.        -fortunei ssp. discolor (1/4)
       -Dondis-          -arboreum ssp. arboreum (1/8)
Golden Days-   -Doncaster-
                        -unknown (1/8)
           -dichroanthum ssp. dichroanthum (1/2)

4ft(1.2m)        0°F(-18°C)           EM         3/3

Flowers of maize yellow, fading to straw yellow and edged carrot
red, funnel-campanulate, 3in(7.6cm) wide; trusses of 8-10.   Upright plant; dark green foliage. (Rudolph & Leona Henny, 1959)


                      -griffithianum (1/16)
                  -Kewense-
cl.        -Aurora-      -fortunei ssp. fortunei (5/16)
       -Naomi-    -thomsonii ssp. thomsonii (1/8)
Golden Dream-   -fortunei ssp. fortunei
           -campylocarpum ssp. campylocarpum (1/2)

5ft(1.5m)        5°F(-15°C)           EM         4/3

Carita g. Taller and more open than Carita (A.M.)  Many trusses
of creamy yellow flowers from dusky pink buds.  Medium-sized,
oval leaves. (Rothschild, 1935)   See color. illus. PB pl. 7.


                  -dichroanthum ssp. scyphocalyx (1/8)
             -Socrianum-
cl.  -unnamed-          -griffithianum (1/8)
     - hybrid-  -wardii var. wardii (3/8)
Golden-       -Rima-
Falcon-       -decorum (1/8)
     -   -wardii var. wardii
      -Crest-            -fortunei ssp. discolor (1/8)
          -Lady Bessborough-
                        -campylocarpum ssp. campylocarpum
                                 Elatum Group (1/8)

5ft(1.5m)        5°F(-15°C)            M

Very pale yellow flowers, in trusses of 10.  Plant of open habit. (A. F. George, Hydon Nurseries; reg. 1976)


cl.

Golden Fleece (Reuthe)      See Princess Anne


cl.        -dichroanthum ssp. dichroanthum (1/4)
     -Goldsworth Orange-
     -              -fortunei ssp. discolor (1/4)
Golden-            -griffithianum (5/16)
Fleece-       -Kewense-
     -     -Aurora-    -fortunei ssp. fortunei (1/16)
      -Yvonne-    -thomsonii ssp. thomsonii (1/8)

          -griffithianum

Large golden yellow flowers with frilled lobes. (Slocock, 1967)


cl.        -dichroanthum ssp. dichroanthum (1/4)
       -Fabia-
Golden Folly-    -griersonianum (1/4)
       -            -campylocarpum ssp. campylocarpum (1/4)
           -Moonstone-
                  -williamsianum (1/4)

Grows only 3ft(.9m) in 10 years, wider than high with medium-
sized leaves. Flowers bell-shaped, deep yellow, held in round
trusses. (H. L. Larson, 1965)

cl.
                    -catawbiense var. album (1/4)
      -Great Lakes-
Golden-              -yakushimanum ssp. yakushimanum (1/4)
  Gala -            -catawbiense var. rubrum (1/4)
      -Good Hope-
                    -wardii var. wardii (1/4)

2.5ft(.75m)         -20°F(-29°C)          ML

Synonym Fifty-Fine. Flowers of soft ivory, with a sparse dorsal
spotting of yellowish green, broadly funnel-shaped, 2.5in(6.4cm)
wide, 5-lobed, in trusses of 14. Plant well-branched, broader
than tall; leaves glossy olive green, 4.5in(11.5cm) long, held 2
years. (David G. Leach, reg. 1983)

cl.
              -dichroanthum ssp. scyphocalyx (1/2)
Golden Gate*
              -unknown (1/2)

3ft(.9m)            -5°F(-21°C)          M          3/4

Flowers colored an attractive orange. Plant of compact habit.
Easily propagated. (Origin unknown)

cl.

Golden Gate--form of zoelleri

5ft(1.5m)           30°F(-1°C)                season continuous

Vireya. Plant habit loose, and grows slightly wider than tall;
leaves broadly elliptic, 3in(7.6cm) long. Flowers open a yellow
chartreuse changing to orange yellow, funnel-shaped; trusses of
5-6. (Strybing Arboretum, 1970)

cl.
              -rupicola var. chryseum (1/2)
Golden Gift-
              -leucaspis (1/2)

Early flowers, yellow and saucer-shaped, 1in(2.5cm) wide, 4 or 5
per truss. Plant low and spreading, 1.5ft(.45m) across when described in 1962. (A. Wright, reg. 1963)

cl.
                    -cinnabarinum ssp. cinnabarinum (1/4)
        -unnamed hybrid-
Golden Gift-         -maddenii ssp. maddenii (1/4)
        -cinnabarinum ssp. xanthocodon Concatenans Group (1/2)

Flowers of bright orange yellow, with pale olive green mottling
in upper throat, campanulate, 2in(5cm) wide, 5-lobed; trusses of
8. Leaves 5in(12.7cm) long. (Gen. Harrison)  A. M. (RHS) 1970

cl.
                    -dichroanthum ssp. dichroanthum (1/4)
        -Fabia-
Golden Glow-        -griersonianum (1/2)
        -           -griersonianum
        -Azor-
                    -fortunei ssp. discolor (1/4)

A hybrid by Benjamin F. Lancaster, reg. 1958.

g. & cl.
              -dichroanthum ssp. dichroanthum (1/2)
Golden Horn-
              -elliottii (1/2)

4ft(1.2m)           10°F(-12°C)          ML         3/3

Flowers of orange and deep salmon, tubular trumpet shaped, with
a large double calyx; loose, flat-topped trusses of 9-11. Plant
compact; dark sage green foliage with brown indumentum. (Rothschild, 1939)  A. M. (RHS) 1945

cl.

Golden Horn Persimmon      See Persimmon

cl.
              -decorum (1/2)
Golden Jubilee-
              -unknown (1/2)

2.5ft(.75m)         0°F(-18°C)           M

Synonym Pearce's Golden Jubilee. Flowers 3.5in(9cm) wide, very
fragrant, 7-lobed, light neyron rose, held in round trusses of
about 9. Plant upright, nearly as broad as high; leaves yellow
green, 7in(17.8cm) long, held 2 years. (Cross by R. A. Pearce;
reg. by Elsie M. Watson, 1983)

                         -dichroanthum ssp. dichroanthum (1/8)
              -Goldsworth-
cl.    -Hotei- Orange   -fortunei ssp. discolor (1/8)
  -        -            -souliei (1/8)
Golden -   -unnamed hybrid-
Moments-             -wardii var. wardii (3/8)
  -          -wardii var. wardii
      -unnamed hybrid-
              -unknown (1/4)

4ft(1.2m)           0°F(-18°C)           EM

Flowers of mimosa yellow, widely funnel-campanulate, 2.5in(6.4
cm) across, 6-lobed; tall, open trusses hold 15. Plant rounded,
as wide as high; leaves about 4.5in(11.5cm) long, glossy yellow
green, held 3 years. (John G. Lofthouse, 1981)

cl.
         -cinnabarinum ssp. xanthocodon Concatenans Group (1/2)
Golden-              -cinnabarinum ssp. cinnabarinum
  Orfe -Lady -            Roylei Group (1/4)
     -Chamberlain-   -cinnabarinum ssp. cinnabarinum (1/8)
              -Royal-
              -Flush-maddenii ssp. maddenii (1/8)

4ft(1.2m)           5°F(-15°C)           M

Flowers of drooping habit in trusses of 7, yellow shades of nasturtium orange. Leaves medium-sized, scaly beneath. Originally
from Tower Court. (Mrs. Roza Harrison)  A. M. (RHS) 1964

cl.
              -moupinense (pink form) (1/2)
Golden Oriole Busaco-
              -sulfurem (1/2)

Synonym Golden Oriole Venetia. Flowers of primrose yellow suffused pale reddish pink, heavily spotted on upper lobes, 1.75in
(4.5cm) wide, with prominent brown anthers. (J. C. Williams)
A. M. (RHS) 1963

cl.
              -moupinense (white form) (1/2)
Golden Oriole Talavera-
              -sulfureum (1/2)

Dresden yellow flowers, spotted deep yellow on upper lobes, up
to 2in(5cm) wide, with prominent brown anthers; trusses of 3-4.
Leaves 2in(5cm) long, with deep red petioles. See also Golden
Oriole Busaco. (J. C. Williams, Caerhays Castle)  A. M. (RHS)
1947; F. C. C. (RHS) 1963

cl.

Golden Oriole Venetia      See Golden Oriole Busaco

```
cl.                            -fortunei ssp. discolor (3/8)
              -Lady Bessborough-
      -Day Dream-                   -campylocarpum ssp. campylo-
Golden  -       -griersonianum       carpum Elatum Group (1/8)
Pheasant-              -fortunei ssp. discolor
      -Margaret Dunn-     -dichroanthum ssp. dichroanthum
                    -Fabia-                          (1/8)
                         -griersonianum (3/8)

4ft(1.2m)         5°F(-15°C)          ML          3/3
```

Dream Girl g. Flowers 4.5in(11.5cm) across, flat, 8 per truss, golden yellow with orange blotch in throat. See also Gold Mohur, Mohur, Shah Jehan. (L. E. Brandt, 1966)

```
cl.                     -souliei (1/4)
              -Virginia Scott-
Golden Pippin-          -unknown (1/4)
      -           -dichroanthum ssp. dichroanthum (1/4)
      -Jasper-
            -Lady      -fortunei ssp. discolor (1/8)
            Bessborough-
                    -campylocarpum ssp. campylo-
                     carpum Elatum Group (1/8)
```

Flowers held in trusses of 7-9, 5-6-lobed, lemon yellow, upper lobes spotted red. See also Freda Rosage. (H. L. Larson, 1983)

```
cl.
          -cinnabarinum ssp. cinnabarinum Roylei Group (1/2)
Golden Queen-      -cinnabarinum ssp. cinnabarinum (1/4)
      -Royal Flush -
      (orange form)-maddenii ssp. maddenii (1/4)

5ft(1.5m)        10°F(-12°C)         ML          4/3
```

Lady Chamberlain g. Drooping flowers, soft yellow, flushed orange, in trusses of 6-7. Erect plant, medium-sized with nice foliage. (Rothschild) F. C. C. (RHS) 1931

```
cl.                            -catawbiense (1/4)
              -Atrosanguineum-
      -unnamed hybrid-          -unknown (1/4)
Golden Salmon-    -griersonianum (1/2)
      -as above (Probably a sibling cross)

5ft(1.5m)        -10°F(-23°C)        ML
```

Flowers of salmon with darker blotch of radiating stripes, open funnel-campanulate, 3.5in(8.9cm) across, 7-lobed, of heavy substance, fragrant; lax trusses hold 10. Plant wider than high, with stiff upright branches; dark green leaves retained 2 years. (J. B. Gable cross; raised & intro. by G. G. Nearing; reg. 1973)

```
              -haematodes ssp. haematodes (1/4)
cl.      -May Day-
      -          -griersonianum (3/8)
Golden Spur-          -dichroanthum ssp. dichroanthum (1/8)
      -          - Fabia-
      -Lascaux-          -griersonianum
              -wardii var. wardii Litiense Group (1/4)
```

Red flowers, shading to golden apricot. See also Golden Cockerel. (A. F. George, Hydon Nurseries, 1975)

```
cl.
      -fortunei ssp. fortunei (1/2)
Golden Star-
      -wardii var. wardii (1/2)

5ft(1.5m)        0°F(-18°C)          ML          4/3
```

Prelude g. Flowers mimosa yellow, 3in(12.6cm) wide, with 7 wavy lobes, in ball-shaped trusses of 7. Free-flowering. Plant almost as broad as tall; elliptic leaves held 2 years. (Donald L. Hardgrove cross; Sidney Burns, intro. 1966; reg. 1978) See color illus. ARS Q 26:2 (Spring 1972) p. 106.

```
cl.
      -wardii var. wardii (1/2)
Golden Tear-
      -brachycarpum ssp. tigerstedtii ? (1/2)

2ft(.6m)         -15°F(-26°C)         ML
```

Shrimp red buds open to flowers of light chartreuse green, spotted currant red, in trusses of 13-15. Plant 2/3 as wide as high; dark green leaves held 2 years. (Cross by Dietrich Hobbie, 1960; reg. by Rudy Behring, 1985)

```
          -yakushimanum ssp. yakushimanum (1/4)
cl.  -Bambi-          -dichroanthum ssp. dichroanthum (1/8)
  -    -Fabia Tangerine-
Golden-          -griersonianum (3/8)
  Torch-          -haematodes ssp. haematodes (1/8)
  -    -Grosclaude-
  -unnamed-          -facetum (eriogynum) (1/8)
  hybrid-
      -griersonianum

4ft(1.2m)         5°F(-15°C)          ML
```

Compact truss 6in(15cm) across, globe-shaped, 13-15 flowers, soft yellow. Upright habit, fairly compact, leaves medium-sized, dull green. (John Waterer, Sons & Crisp, 1972) H. C. (RHS) 1977; A. M. (Wisley Trials) 1984

```
cl.    -yakushimanum ssp. yakushimanum (1/2)
Golden -      -wardii var. wardii (1/4)
Wedding-Mrs. Lammot-      -souliei (1/8)
      Copeland -Virginia Scott-
                    -unknown (1/8)

2ft(.6m)         0°F(-18°C)           M
```

Brick red buds open to chrome yellow flowers of 7 lobes, held in 6in(15cm) trusses of 10-14. Plant semi-dwarf, as wide as high, with dark green leaves 4.75in(12cm) long. (Cross by H. L. Larson, 1969; intro. by Joe A. Davis 1980; reg. 1984)

```
cl.
      -fortunei ssp. fortunei (1/2)
Golden West-
      -campylocarpum ssp. campylocarpum (1/2)
```

Flowers of Dresden yellow with a small red blotch, funnel campanulate, 3in(7.6cm) across; rounded trusses of 14. (Intro. by Del W. James, 1956)

```
cl.
      -dichroanthum ssp. scyphocalyx (1/2)
Golden Witt-      -campylocarpum ssp. campylocarpum
  -    -Moonstone-                          (1/8)
  -unnamed-      -williamsianum (1/4)
  hybrid-      -williamsianum
      -Adrastia-
              -neriiflorum ssp. neriiflorum (1/8)

3ft(.9m)         0°F(-18°C)           M          4/4
```

Low plant with medium-sized foliage, dark green. Flowers primrose yellow spotted red, in trusses of 7-9. (J. A. Witt cross; reg. by L. J. Michaud, 1977) See color illus. HG p. 98.

```
cl.

Goldendale     See p. 126 for this entry.
```

```
g.
      -wardii var. wardii (1/2)
Goldfinch-
      -Mrs. P. D. Williams--unknown (1/2)
```

A hardy hybrid of medium height. Pink flowers with a golden blotch. Late. (R. Collyer cross; exhibited by Collingwood Ingram, 1945)

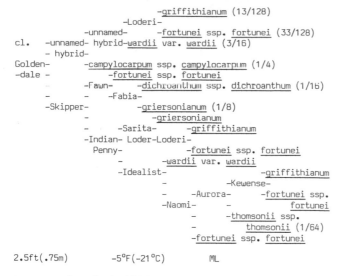

```
                        -griffithianum (13/128)
                -Loderi-
        -unnamed-       -fortunei ssp. fortunei (33/128)
cl.   -unnamed- hybrid-wardii var. wardii (3/16)
      - hybrid-
Golden-        -campylocarpum ssp. campylocarpum (1/4)
-dale -            -fortunei ssp. fortunei
      -Fawn-   -dichroanthum ssp. dichroanthum (1/16)
      -     -Fabia-
      -Skipper-        -griersonianum (1/8)
      -                -griersonianum
      -     -Sarita-        -griffithianum
      -Indian- Loder-Loderi-
          Penny-        -fortunei ssp. fortunei
            -        -wardii var. wardii
              -Idealist-        -griffithianum
              -                -Kewense-
              -     -Aurora-   -fortunei ssp.
                -Naomi-    -          fortunei
                -        -thomsonii ssp.
                -        -        thomsonii (1/64)
                -fortunei ssp. fortunei
```

2.5ft(.75m)        -5°F(-21°C)        ML

Leaves medium-sized, light green, held 3 years.  Flowers of heavy substance, butter yellow (darker than Crest) with dark red center, 3in(7.6cm) across; trusses hold 8-14, up to 10in(25.4cm) broad.  (Charles McNew, 1974)


cl.
        -burmanicum (1/2)
Goldfinger-
        -valentinianum (1/2)

3ft(.9m)        15°F(-9°C)        EM        2/2

Parisienne g.  Lax trusses of 4 flowers, primrose yellow, each flower 3in(7.6cm) wide.  Described as a cool greenhouse plant with leaves about 3in(7.6cm) long, dark olive green, densely scaly.  (Sir Giles Loder, Leonardslee)  A. M. (RHS) 1965


```
                        -caucasicum (3/16)
                -Jacksonii-        -caucasicum
        -Goldsworth Yellow-    -Nobleanum-
Goldfort-        -                -arboreum ssp.
        -                        arboreum (1/16)
        -            -campylocarpum ssp. campylocarpum
        -fortunei ssp. fortunei (1/2)        (1/4)
```

5ft(1.5m)        -15°F(-26°C)        M        3/3

A  "yellow" that is more hardy than most, but it's a light ivory yellow flower, pale greenish yellow in center, held in rounded medium-sized trusses.  Upright, open habit; leaves 6in(15cm) long, yellowish green, held 4 years.  (Slocock, 1937)  See color illus. VV p. 127.


```
                -wardii var. wardii (5/8)        -griffithianum
                -                -George-        (1/64)
        -unnamed-        -Mrs. Lindsay- Hardy-catawbiense
        - hybrid-    -Diane- Smith    -
        -    -Alice -    -        -Duchess of
        -    Street-    -        Edinburgh--unknown
cl.    -        -        -        (6/32)
        -        -        -campylocarpum ssp.
Goldkrone-    -        -unnamed- campylocarpum (1/32)
(Gold    -        -    hybrid-
  Crown) -    -        -unknown
        -        -wardii var. wardii
        -        -catawbiense (9/64)
        -    -Omega-
        -Hachmann's-    -unknown
          Marina    -wardii var. wardii
```

Reverse cross of Sandra.  Truss of 16-18 flowers, 5-lobed, golden yellow, marked ruby red with darker spots; compact trusses. Bud hardy to -5°F(-21°C).  See also Graf Lennart. (H. Hachmann cross, 1969; intro. 1981; reg. by G. Stuck, 1983)


```
                -catawbiense (9/64)
cl.    -Hachmann's-Omega-
      - Marina  -    -unknown (6/32)
Goldrausch*    -wardii var. wardii (5/8)
(Golden  -wardii var. wardii        -griffithianum
 Ecstasy) -        -George-        (1/64)
        -unnamed-        -Mrs.Lindsay- Hardy-catawbiense
        hybrid-    -Diane- Smith    -Duchess of Edinburgh
        -            -    -unknown
        -Alice -    -unnamed-campylocarpum ssp. cam-
        -Street-    hybrid-        pylocarpum (1/32)
        -            -unknown
        -wardii var. wardii
```

Reverse cross of Goldkrone.  Orange buds opening golden yellow. Bud hardy to -8°F(-22°C).  See also Hachmann's Brasilia, Sandra. (H. Hachmann cross; intro. 1983)


cl.
        -oreotrephes (1/2)
Goldstrike-        -cinnabarinum ssp. cinnabarinum (1/4)
        -Royal Flush-
                -maddenii ssp. maddenii (1/4)

4ft(1.2m)        0°F(-18°C)        M        4/4

Unusual plant with up to 8 tubular-shaped flowers per truss, golden yellow, the upper lobes buttercup yellow.  Leaves medium-sized, glossy olive green.  (Rudolph Henny. 1962)


cl.
            -griffithianum (1/2)
Goldsworth Crimson-        -arboreum ssp. arboreum (1/4)
            -Doncaster-
                -unknown (1/4)

5ft(1.5m)        0°F(-18°C)        EM        2/3

Above parentage from Slocock's catalogue.  Bright crimson flowers in rounded trusses.  Plant spreading but compact; dark green leaves, long and glossy.  (Slocock, 1926)  A. M. (Wisley Trials) 1960;  F. C. C. (RHS) 1971   See color illus. VV p. 24.


cl.
            -dichroanthum ssp. dichroanthum (1/2)
Goldsworth Orange-
        -fortunei ssp. discolor (1/2)

5ft(1.5m)        -5°F(-21°C)        L        3/3

Lax trusses of many flowers, funnel-shaped, pale orange tinted pink.  Leaves held on red petioles.  Plant open and spreading, heat-tolerant but flowers best in shade.  (Slocock, 1938)


cl.
        -griffithianum (1/2)
Goldsworth Pink-
        -unknown (1/2)

A tall shrub with large flowers, deep rose in bud, opening mottled rose pink fading to white; lax conical trusses.  Midseason. (Slocock)  A. M. (Wisley Trials) 1958


cl.
                -caucasicum (3/8)
        -Jacksonii-        -caucasicum
Goldsworth Yellow-    -Nobleanum-
        -            -arboreum ssp. arboreum
            -campylocarpum ssp. campylocarpum        (1/8)
                        (1/2)

5ft(1.5m)        -15°F(-26°C)        ML

Parentage as in Leach.  Dense plant with yellowish green leaves, medium-sized.  Apricot buds open to buff yellow flowers of thin texture, held in compact trusses.  (Slocock)  A. M. (RHS) 1925 See color illus. F p. 46; VV p. 125.


cl.
        -yakushimanum ssp. yakushimanum (1/2)
Golfer-
        -pseudochrysanthum (Exbury form, A.M.) (1/2)

15in(38cm)          -15°F(-26°C)          M

Neyron rose buds opening to clear pink flowers 2in(5cm) across,
widely funnel-campanulate; domed trusses of 13. Plant twice as
wide as tall; leaves 3in(7.6cm) long, margins recurved; held 2-
3 years; silvery tomentum on new growth, much retained; heavy
pale grey indumentum beneath.  Named for Pat Berg whose hobby is
golfing. (W. Berg cross, 1966; J. Caperci, intro. & reg., 1980)
See color illus. ARS Q 34:4 (Fall 1980): p. 206.

cl.
            -Antoon van Welie, q.v.
Goliath* Professor J. H. Zaayer, q.v.
            -Annie E. Endtz, q.v.

The exact combination of the above parents is unknown, but all
are Pink Pearl hybrids.  Truss of 15 flowers, deep pink.  (P.
van Nes) F. C. C. (Boskoop) 1959

cl.
            -catawbiense (1/2)
Gomer Waterer-
            -unknown (1/2)

6ft(1.8m)          -15°F(-26°C)          ML          3/4

An old standby, and one of the best-looking plants year round.
Plant is broadly upright, with large, glossy, dark green leaves.
Buds have slight rose tinge, opening white. (J. Waterer, before
1900) A. M. (RHS) 1906    See color illus. VV p. 20.

g.
            -Lady Harcourt--unknown (1/2)
Gondolier-
            -griersonianum (1/2)

Large, bright red flowers  in midseason.  Plant of medium size,
and loose habit. (Rothschild, 1947)

g.
            -Lord Milner--unknown (1/2)
Good Cheer-
            -sutchuenense (1/2)

A tall plant bearing compact trusses of white flowers, marked
with pink.  Early. (Rothschild, 1939)

cl.
            -fortunei ssp. fortunei (1/2)
Good Fortune-
            -unknown (1/2)

4ft(1.2m)          -15°F(-26°C)          M

Rose-colored buds open to fragrant flowers,  white with a tinge
of purplish rose.  Trusses 5.5in(14cm) wide by 4.5in(11.5cm)
high, hold up to 10 flowers.  Plant is rounded, broader than
tall, holding matte dark green leaves to 3 years.   (Russell and
Velma Haag, 1981)

cl.
            -catawbiense var. rubrum (1/2)
Good Hope-
            -wardii var. wardii (1/2)

6ft(1.8m)          -20°F(-29°C)          EM          4/4

Buds are jasper red opening to canary yellow flowers, 3in(7.6cm)
wide, about 13 held in trusses 6in(15cm) across.  Plant is wider
than tall, holding glossy yellow green leaves 2 years.   (David
G. Leach, intro. 1976; reg. 1982)  See color illus. ARS J 36:1
(Winter 1982): p. 3.

                                    -griffithianum (1/8)
cl.              -Queen Wilhelmina-
        -Britannia-              -unknown (5/8)
Good News-      -Stanley Davies--unknown
        -             -Moser's Maroon--unknown
        -Romany Chal-
                    -griersonianum (1/4)

4ft(1.2m)          -5°F(-21°C)          VL          3/3

A plant favored for its later blooming season.  Red buds open to
crimson scarlet flowers (similiar to Britannia), widely funnel-
campanulate, 2.5in(6.4cm) across; lax trusses hold 12-15.  Shrub
is compact, wider than tall.  Leaves with slight tan indumentum,
retained 2 years; new growth pale green. (John Henny, 1974)

cl.
            -ludlowii (1/2)
Goosander-
            -lutescens (1/2)

When shown in 1981 plant was 8.25in(21cm) high and 17in(43cm)
wide with small, dark green leaves.  Flowers are yellow, held in
trusses of 3-5.  Each flower is about 1.6in(4cm) across and 1.25
in(3.2cm) long. (Raiser P. A. Cox) A. M. (Wisley Trials) 1981

                        -catawbiense (1/8)
cl.              -Mrs. Milner-
        -Rinaldo-        -unknown (1/8)
Gorlitz-      -smirnowii (1/4)
        -williamsianum (1/2)

Pink flowers. (Raiser Joh. Bruns, W. Germany; reg. 1972)

cl.

Goshawk      See Jalisco Goshawk

g.
        -azalea--unknown (1/2)
Govenianum-              -catawbiense (1/4)
        -unnamed hybrid-
                    -ponticum (1/4)

A fragrant azaleodendron resembling Fragrans, q.v.  This has a
richer color of soft rose. (Methven, 1868)

cl.

Grace      See George Grace

cl.                          -griffithianum (1/4)
            -Jean Marie de Montague-
Grace Seabrook-              -unknown (1/4)
            -strigillosum (1/2)

5ft(1.5m)          -5°F(-21°C)          EM

Flowers currant red at margins, shading at center to blood red,
funnel-shaped, 3in(7.6cm) across, in tight trusses.  Plant habit
similiar to female parent; very large dark green leaves. (C. S.
Seabrook, 1965)

cl.
            -lutescens (1/2)
Graduation-
            -mekongense var. mekongense Viridescens Group (1/2)

3ft(.9m)          10°F(-12°C)          M

Flowers widely funnel-campanulate, 5-lobed.  Color varies even
within individual truss, Naples yellow with edging light brick
red and dorsal spotting; trusses of 5.   Small scaly leaves held
2-3 years. (Cross by Dr. C. Heller; reg. by Peggy Moore, 1984)

cl.

Graf Lennart      See p. 128

cl.

Graf Zeppelin      See p. 128

cl.                          -catawbiense (1/4)
            -Scharnhorst-
Grafin Kirchbach-              -unknown (1/4)
            -forrestii ssp. forrestii Repens Group (1/2)

Scarlet red flowers. (Joh. Bruns, W. Germany, 1972)

```
                    -wardii var. wardii (5/8)    -griffithianum
                                  -George-                 (1/64)
         -unnamed-        -Mrs.Lindsay- Hardy-catawbiense
         - hybrid-        - Smith -                        (9/64)
cl.      -        -Diane-            -Duchess of Edinburgh
         -        -      -           --unknown (3/16)
Graf     -      -Alice -      -    -campylocarpum ssp. campy-
Lennart* -      -Street-  -unnamed-          locarpum (1/32)
(Count   -      -           hybrid-unknown
Lennart)-                -wardii var. wardii
         -                    -catawbiense
         -          -Omega-
         -Hachmann's-      -unknown
           Marina    -wardii var. wardii
```

Beautiful orange yellow buds opening pure light yellow, in lax
trusses. Floriferous. Bluish green foliage typical of wardii.
Hardy to -8°F(-22°C). Midseason.  See also Goldkrone. (H. Hachmann, intro. 1983)

```
                          -griffithianum (1/8)
cl.          -George Hardy-
       -Pink -           -catawbiense (3/8)
Graf   -Pearl-     -arboreum ssp. arboreum (1/8)
Zeppelin- -Broughtonii-
       -              -unknown (3/8)
       -                    -catawbiense
       -Mrs. C. S. Sargent-
                        -unknown
```

5ft(1.5m)        -10°F(-23°C)        ML        3/5

Bright clear pink flowers, blushing darker to edges, held in
rounded trusses of up to 10.  A very good-looking plant, vigorous, with outstanding foliage, glossy and dark green. (C. B.
van Nes & Sons, reg. 1958)

```
cl.              -catawbiense (3/4)
       -Mrs. Milner-
Granat-            -unknown (1/4)
       -catawbiense
```

Ruby red flowers with pale markings on a lighter background, and
somewhat frilled. (T. J. R. Seidel cross, 1897; intro. 1905)

cl.

Grand Arab     See Vesuvius (Waterer's)

```
cl.                -arboreum ssp. arboreum (1/4)
              -Doncaster-
Grand Finale-          -unknown (1/4)
              -facetum (eriogynum) (1/2)
```

Vivid red flowers in very late season. (Haworth-Booth, intro.
1932; reg. 1962)

cl.

Grand Marquis*     Parentage unknown

An unregistered parent of Longbeach Dream. (G. E. Grigg, NZ)

```
cl.
           -grande (1/2)
Grand Prix-
           -falconeri ssp. eximium (1/2)
```

Ivory-colored flowers shaded pale carmine on reverse. (Heneage-
Vivian)  A. M. (RHS) 1940

```
                                     -griffithianum (7/32)
                        -George-
            -Pink Pearl- Hardy-catawbiense (7/32)
            -          -            -arboreum ssp.
         -Countess -     -Broughtonii- arboreum (3/32)
         - of Derby-              -unknown (7/32)
   -Trude -          -      -catawbiense
   -Webster-      -Cynthia-
cl. -        -          -griffithianum
         -Countess of Derby--(as above)
Grand-                    -griffithianum
  Slam*               -George-
       -           -Pink Pearl- Hardy-catawbiense
       -  -Antoon van-    -            -arboreum ssp.
       -   - Welie -      -Broughtonii-     arboreum
       -        -unknown           -unknown
       -Lydia-                -fortunei ssp. discolor (1/16)
       -     -Lady Bessborough-
         -Day -                 -campylocarpum ssp. campylo-
         -Dream-                     carpum Elatum Group (1/16)
              -griersonianum (1/8)
```

6ft(1.8m)       -5°F(-21°C)        ML        4/3

Brilliant pink flowers, in a cone-shaped truss up to 1ft(.3m)
high. Large glossy green leaves, and sturdy stems for the enormous flowers. Proper care makes it vigorous and tidy. (Greer,
intro. 1982) See color illus. HG p. 142.

```
g.
       -falconeri ssp. eximium (1/2)
Grandex-
       -sinogrande (1/2)
```

Yellow flowers, flushed pink. (Lord Aberconway, 1950)

```
                                    -griffithianum (5/8)
cl.                   -Queen Wilhelmina-
              -Britannia-            -unknown (3/8)
Great Britain-        -Stanley Davies--unknown
              -griffithianum
```

A hybrid by Lord Digby, 1953.

```
cl.
          -Catalgla--catawbiense var. album Glass (1/2)
Great Day*
          -calophytum var. calophytum (1/2)
```

5ft(1.5m)       -25°F(-32°C)        M

A very hardy hybrid, bearing white flowers with a deep burgundy
blotch. Very compact growth habit, with leaves 10in(25.4cm) or
longer. (Sam Baldanza cross; raiser Orlando S. Pride, 1977)

```
cl.
          -catawbiense var. album (1/2)
Great Lakes-
          -Koichiro Wada--yakushimanum ssp. yakushimanum (1/2)
```

4ft(1.2m)       -25°F(-32°C)        M        4/4

Clear pink buds, opening light pink and fading to chalk white;
flowers 2.5in(6.4cm) wide; up to 15 per truss. Leaves 3.5in(8.9
cm) by 1.5in(3.8cm). See also Anna H. Hall, Pink Frosting, and
Spring Frolic. (David G. Leach, 1960)  P. A. (ARS) 1960

cl.

Great Laurel--synonym maximum  (Not a selected form)

```
cl.
     -Mrs. J. G. Millais--unknown (1/2)
     -                               -fortunei ssp. discolor
Great-                   -Lady      -                    (1/16)
Scott-         -Jalisco- Bessborough-campylocarpum ssp. campylo-
     -      -         -                  carpum Elatum Group (1/16)
     -      -         -       -dichroanthum ssp. dichroanthum
     -Cheyenne-      -Dido-                            (1/16)
     -                    -decorum (1/16)
     -                 -griffithianum (1/8)
                 -Loderi-
                       -fortunei ssp. fortunei (1/8)

5ft(1.5m)         -5°F(-21°C)          M          4/2
```

Large flowers, pure white, are held in very round tight trusses.
Flowers 5in(12.7cm) across, with a dark purple maroon blotch on
the upper lobes.  Smooth foliage is deep green.  See also Razzle
Dazzle and White Gold.  (Greer, 1979)

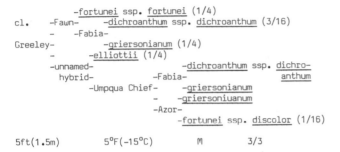

```
             -fortunei ssp. fortunei (1/4)
cl.   -Fawn-    -dichroanthum ssp. dichroanthum (3/16)
     -    -Fabia-
Greeley-        -griersonianum (1/4)
     -      -elliottii (1/4)
     -unnamed-              -dichroanthum ssp. dichro-
      hybrid-      -Fabia-                       anthum
           -Umpqua Chief-     -griersonianum
                       -       -griersoniuanum
                        -Azor-
                             -fortunei ssp. discolor (1/16)

5ft(1.5m)          5°F(-15°C)          M          3/3
```

Watermelon pink flowers 4in(10cm) across, upper 3 lobes densely
spotted; trusses of about 13.  Long, narrow leaves.  (Greer,
1961)

```
cl.
         -augustinii ssp. augustinii (3/4)
Green Eye-
     -             -impeditum (1/4)
         -Saint Tudy-
                  -augustinii ssp. augustinii
```

An older registration record showed Blue Tit as pollen parent of
Greeneye.  Correction was made in the Rhododendron Yearbook,
1965.  The plant has mauve blue flowers with green spots.  (Gen.
Harrison, intro. 1955)

```
                                 -griffithianum (1/8)
cl.                    -George Hardy-
         -Mrs. Lindsay-            -catawbiense (1/8)
Green Goddess- Smith    -Duchess of Edinburgh--unknown (1/4)
     -              -fortunei ssp. fortunei (1/4)
     -unnamed-        -dichroanthum ssp. dichroanthum
         hybrid-Fabia-                          (1/8)
                    -griersonianum (1/8)
```

Chartreuse flowers with a green blotch, flattened, 3.75in(9.5cm)
across; spherical trusses of 12.  Oval leaves 7.5in(19cm) long.
Midseason.  (Marshall Lyons, 1971)

```
cl.

Green Highlander    See upper right column
```

```
cl.

Greeneye    See Green Eye
```

```
cl.                          -griffithianum (1/8)
         -Mrs. Lindsay-George Hardy-
Greenfinch- Smith              -catawbiense (1/8)
     -           -Duchess of Edinburgh--unknown (1/4)
     -campylocarpum ssp. campylocarpum (1/2)

Zuiderzee g.  (Collingwood Ingram, 1945)
```

```
             -wardii var. wardii (1/4)
                             -griffithianum
          -Idealist-        -Kewense-         (1/64)
     -              -Aurora-      -fortunei ssp.
     -                                 fortunei
     -unnamed-       -Naomi-   -thomsonii ssp.
     - hybrid-                     thomsonii (1/32)
     -        -             -fortunei ssp. fortunei (5/64)
cl.  -       -wardii var. wardii
     -    -Hawk-          -fortunei ssp. discolor
Green -    -Lady         -                (3/16)
Highlander-    -Bessborough-
     -                   -campylocarpum ssp. campylo-
     -                        carpum Elatum Group (3/16)
     -                 -fortunei ssp. discolor
     -       -Lady     -
     -       -Bessborough-campylocarpum ssp. campylocarpum
     -Jalisco-                    Elatum Group
     -          -dichroanthum ssp. dichroanthum (1/8)
          -Dido-
                -decorum (1/8)

5ft(1.5m)        5°F(-15°C)          M          3/3
```

Flowers of very soft yellow, deepening at center, on an upright,
open-growing plant.  (John Waterer, Sons & Crisp, 1975)

```
cl.

Greenmantle--form of phaeochrysum var. phaeochrysum (R 59229)

5ft(1.5m)        5°F(-15°C)          EM         3/3
```

Round trusses of 11-15 flowers, very light greenish white with
small red blotch.  Leaves are 5.25in(13.3cm) long with light
coverage below of dark brown indumentum.  (Collector Joseph
Rock; raised by Col. S. R. Clarke; exhibited by R. N. Stephenson
Clarke)  A. M. (RHS) 1977

```
cl.              -decorum (1/4)
         -Golden Jubilee-
Greensprite-          -unknown (1/4)
     -       -wardii var. wardii (1/4)
         -Crest-      -fortunei ssp. discolor (1/8)
              -Lady  -
              Bessborough-campylocarpum ssp. campylocarpum
                         (Elatum Group) (1/8)

4ft(1.2m)        10°F(-12°C)          L
```

Flowers of very pale greenish white with a faint yellow green
blotch, 8 wavy lobes, about 3.5in(8.9cm) wide; spherical trusses
of 10.  Gardenia-like fragrance.  Plant upright and as broad as
tall; leaves held 2 years; new growth may hide some blooms.  (Elsie Watson, 1982)

```
                             -griffithianum (5/32)
                     -George-
             -Mrs. Lindsay- Hardy-catawbiense (1/32)
             - Smith
          -Diane-        -Duchess of
     -                       Edinburgh--unknown (1/8)
     -unnamed-    -             -campylocarpum ssp.
cl.  - hybrid- -unnamed hybrid-    campylocarpum (7/16)
     -        -             -unknown
Greenwell-         -campylocarpum ssp. campylocarpum
 Glory  -    -Gladys Rose-
     -              -fortunei ssp. fortunei (1/4)
     -         -campylocarpum ssp. campylocarpum
     -unnamed hybrid-    -griffithianum
                  -Kewense-
                       -fortunei ssp. fortunei

4ft(1.2m)        5°F(-15°C)          M
```

A very pale yellow flower on a deep green plant.  Requires a
little shade.  (John Waterer, Sons & Crisp, 1975)

g.                    -thomsonii ssp. thomsonii (1/4)
          -Chanticleer-
Greeting-              -facetum (eriogynum) (1/4)
      -         -griffithianum (1/4)
          -Loderi-
                    -fortunei ssp. fortunei (1/4)

A red-flowered  hybrid introduced in 1950.  (Lord Aberconway)


g.
          -Lady Rumbold--unknown (1/2)
Grenada-
          -griersonianum (1/2)

A medium-sized plant with trusses of red flowers, in midseason.
(Rothschild, 1939)


g. & cl.
          -Moser's Maroon--unknown (1/2)
Grenadier-
          -elliottii (1/2)

6ft(1.8m)          5°F(-15°C)          L          4/3

Large ball-shaped trusses of waxy flowers, deep rich red, black
in throat.  Plant very strong, fast-growing; foliage large, dark
green.  (Rothschild, 1939)  F.C.C. (RHS) 1943  See color illus.
PB pl. 54.


g. & cl.
          -Pauline--unknown (1/2)
Grenadine-
          -griersonianum (1/2)

A compact plant of average size, that needs some protection.
Flowers cherry-colored, stained with orange and brown.  (Rothschild, 1939)  A. M. (RHS) 1956; F. C. C. (Wisley Trials) 1982


g.                    -decorum (1/4)
          -unnamed hybrid-
Gretchen-              -grffithianum (1/4)
      -          -catawbiense (1/4)
          -Kettledrum-
                    -unknown (1/4)

5ft(1.5m)          -15°F(-26°C)          ML          4/4

Plant habit and foliage both good, on a very floriferous hybrid.
Pink flowers, red-throated; large dome-shaped trusses.  Gretchen
No. One is considered the best of the grex.  (G. G. Nearing & J.
B. Gable, 1958)


                    -wardii var. wardii (1/2)
          -                    -griffithianum (1/32)
cl.    -Idealist-          -Kewense-
    -          -Aurora-          -fortunei ssp. fortunei
Gretchen-          -Naomi-    -thomsonii ssp. thomsonii (1/16)
 Gossler-              -fortunei ssp. fortunei (5/32)
    -    -wardii var. wardii
      -Crest-              -fortunei ssp. discolor (1/8)
          -Lady Bessborough-
                    -campylocarpum ssp. campylocarpum
                              Elatum Group (1/8)
5ft(1.5m)          -5°F(-21°C)          ML          4/3

Plant rounded, branching moderately, nearly as wide as high,
leaves olive green, held 2 years.  Flowers large, 4.5in(11.5cm)
wide, yellowish green with red blotch, in trusses of 12.  (Carl
H. Phetteplace, 1977)


                    -caucasicum (1/4)
cl.          -Boule de Neige-          -catawbiense (3/8)
    -                    -unnamed hybrid-
Gretchen Medlar-                    -unknown (3/8)
    -                    -catawbiense
      -Henriette Sargent-
                    -unknown

Hybrid by Dr. Henry T. Skinner, before 1958.


g. & cl.      -neriiflorum ssp. neriiflorum Euchaites Group
      -Portia-                              (1/4)
Gretia-    -strigillosum (1/4)
      -griersonianum (1/2)

Blood red flowers, 3in(7.6cm) across, in trusses of 14.  (Lord
Aberconway) A. M. (RHS) 1946


cl.
      -bureavii (1/2)
Gretsel*    -dichroanthum ssp. dichroanthum (1/4)
      -Fabia-
          -griersonianum (1/4)

3ft(.9m)          -5°F(-21°C)          M          3/4

Superb dark green foliage with indumentum.  Salmon orange edging on pastel pink to shell pink, large flowers.  A compact and
shapely plant.  See also Hansel.  (Halfdan Lem)  See col. illus.
ARS Q 33:3 (Summer 1979): cover.


cl.
          -griersonianum (1/2)
Griercalyx-
          -megacalyx (1/2)

A parent of Muriel Holman.  (Exhibited 1941)


cl.
          -griesonianum (1/2)
Grierdal-
          -dalhousiae var. dalhousiae (1/2)

Scarlet flowers, tinged magenta.  According to Davidian (Vol. 1,
p.15) the "only one authentic hybrid between" a lepidote (scaly)
and an elepidote (non-scaly).  See also Aprilis and John
Marchand.  (Walker Heneage-Vivian)

g.
                    -griersonianum (1/2)
Grierocaster    -          -arboreum ssp. arboreum (1/4)
(High Beeches form)-Doncaster-
                    -unknown (1/4)

Cross by Col. G. H. Loder at High Beeches; intro. in 1937.  The
same cross was made by Sir Gerald Loder, and intro. in 1939 from
Wakehurst.

g.
                    -griersonianum (1/2)
Grierosplendour-          -ponticum (1/4)
          -Purple Splendour-
                    -unknown (1/4)

4ft(1.2m)          -5°F(-21°C)          ML          3/3

Red purple flowers with darker blotch are produced in profusion
even on young shrubs.  Upright plant spreads when mature.  Long,
narrow, light green leaves.  (Col. G. H. Loder, 1937)


g.
      -G. A. Sims--unknown (1/2)
Griersims-
          -griersonianum (1/2)

Late flowers of intense scarlet.  (Sir John Ramsden, 1938)


cl.          -elliottii (1/4)
    -Fusilier-
    -          -griersonianum (1/4)
Grilse-              -fortunei ssp. discolor (1/8)
      -Lady
    -Jalisco-Bessborough-campylocarpum ssp. campylocarpum
    -Eclipse-                    Elatum Group (1/8)
    -          -dichroanthum ssp. dichroanthum (1/8)
      -Dido-
          -decorum (1/8)

Flowers of reddish rose, with dark crimson spots in the throat.
(Crown Estate, 1948)  A. M. (RHS) 1957

```
cl.          -dichroanthum ssp. dichroanthum (3/8)
        -Fabia-
Griselda-    -griersonianum (3/8)
        -              -fortunei ssp. discolor (1/4)
        -Margaret Dunn-    -dichroanthum ssp. dichroanthum
                    -Fabia-
                        -griersonianum
```

Plant of medium dwarf habit, and rather slow-growing; leaves up
to 6in(15cm) long.  Flowers orange, flushed pink and scarlet,
heavily spotted, flat, 3in(7.6cm) wide.  (R. Henny, before 1958)

```
g.
        -arboreum ssp. cinnamomeum var. album (1/2)
Grisette-    -caucasicum (1/4)
        -Dr. Stocker-
                -griffithianum (1/4)
```

Tall-growing shrub, dark foliage; pure white flowers with dark
markings in throat.  (Rothschild, 1939)

```
g. & cl.
        -haematodes ssp. haematodes (1/2)
Grosclaude-
        -facetum (eriogynum) (1/2)
```

3ft(.9m)          5°F(-15°C)          M          3/4

Plant of neat, compact habit; foliage rich dark green with rusty
orange indumentum.  Buds of sealing wax red open brilliant scarlet; flowers waxy, campanulate, heavy-textured, in flat trusses
of 9-11.  (Rothschild, 1941)  A. M. (RHS) 1945

```
cl.
        -Bodnant Red--campylogynum Cremastum Group (1/2)
Grouse-
        -Gigha--calostrotum ssp. calostrotum (1/2)
```

1.5ft(.45m)          -5°F(-21°C)          E

Deep rose flowers, campanulate, to 1.5in(3.8cm) wide, 5-lobed,
in clusters of 2-4; elliptic leaves about 1in(2.5cm) long.  (P.
A. Cox, Glendoick Gardens)  A. M. (RHS) 1977

```
cl.
        -yakushimanum ssp. yakushimanum (1/2)
Grumpy-
        -hybrid--unknown (1/2)
```

2ft(.6m)          -5°F(-21°C)          M

Plant of compact habit, wider than tall; medium-sized leaves of
dull dark green.  Flowers cream, tinged pink; globular trusses
of about 11.  (John Waterer, Sons & Crisp)  A. M. (RHS) 1979

```
                    -griffithianum (1/8)
cl.          -Loderi-
        -Albatross-    -fortunei ssp. fortunei (1/8)
Guardian Fir-        -fortunei ssp. discolor (1/2)
            -          -fortunei ssp. discolor
        -unnamed-          -griersonianum (1/8)
            hybrid-Tally Ho-
                    -facetum (eriogynum) (1/8)
```

6ft(1.8m)          0°F(-18°C)          L

Plant broad as high, branching well; leaves dark green, retained
one year.  Flowers 5in(12.7cm) wide, of strong substance, medium
light pink shading to light pink on upper lobes with small red
blotch, in trusses of 10.  (Halfdan Lem)  C. A. (ARS) 1975  See
color illus. ARS Q 28:4 (Fall 1974): cover.

```
g.
        -Ivery's Scarlet--unknown (1/2)
Guardsman-
        -arboreum ssp. arboreum (1/2)
```

Dark red flowers.  (Loder)

```
cl.
        -Eggebrechtii--unknown
Gudrun-
        -Madame Linden--unknown
```

White flowers tinted soft purple, strongly marked with dark
brownish red.  (T. J. R. Seidel, 1905)

```
cl.
        -Cunningham's Sulphur--caucasicum (1/2)
Guepier-
        -lacteum (1/2)
```

Small, round truss of 17 flowers chartreuse green with slightly
darker shading on upper lobe.  Each flower is 2in(5cm) wide by
1.5in(3.8cm) long.  (Collingwood Ingram)  A. M. (RHS) 1961

```
cl.
        -catawbiense (1/2)
Guido-
        -unknown (1/2)
```

Crimson-colored flowers.  (A. Waterer, before 1850)

```
cl.                    -catawbiense (1/4)
        -Mrs. C. S. Sargent-
Guy Bradour-              -unknown (1/2)
        -              -ponticum (1/4)
        -Purple Splendour-
                -unknown
```

2.5ft(.75m)          -15°F(-26°C)          ML

Semi-dwarf, broader than tall with leaves medium size.  Widely
funnel-shaped flowers, 3.5in(9cm) across, fragrant, violet with
a very dark blotch, in trusses of up to 10.  (Henry R. Yates
cross; reg. by Mrs. Yates, 1977)

```
cl.                    -adenogynum (1/4)
        -Xenosporum (detonsum)-
Guy Nearing-              -unknown (1/4)
        -          -thomsonii ssp. thomsonii (1/4)
        -Gilian-
                -griffithianum (1/4)
```

2ft(.6m)          -15°F(-26°C)          ML

Rounded plant, as broad as high; leaves medium to large, held 2
years.  Flowers red purple, with wine-colored blotch, 2in(5cm)
wide, in trusses of 25.  (Nearing cross; reg. by Raustein, 1973)
See color illus. ARS Q 27:3 (Summer 1973): p. 171.

```
                        -griffithianum (9/32)
                -Beauty of-
            -Norman- Tremough-arboreum ssp. arboreum
            - Gill -                (1/32)
        -Anna-    -griffithianum
        -    -              -griffithianum
cl. -Walloper-    -Jean Marie de Montague-
Gwen-    -    -              -unknown (11/16)
Bell*    -Marinus Koster-    -griffithianum
    -              -unknown
    -unknown
```

5ft(1.5m)          -5°F(-21°C)          ML          4/4

Synonym Walloper #6.  Gigantic trusses of  flowers, medium pink
with small dark eye.   Large leaves have a pleasing bronze green
color.  (Cross by Halfdan Lem; raised by Fisher)

```
cl.                    -griffithianum (1/4)
        -Loderi King George-
Gwen's Pink-              -fortunei ssp. fortunei (1/4)
        -    -griersonianum (1/4)
        -Azor-
                -fortunei ssp.discolor (1/4)
```

6ft(1.8m)          5°F(-15°C)          L

Very fragrant flowers of 7 lobes, 4.5in(11.5cm) wide, light
neyron rose shading to claret rose, in trusses of 9. Plant upright with stiff branches. Leaves 6in(15.2cm) long, held 1-2
years. (Cross by Grady Barefield; reg. by Mary Barefield, 1983)


cl.                          -ponticum (1/4)
                 -Purple Splendour-
Gwerfyl Moss-               -unknown (1/2)
                          -Moser's Maroon--unknown
                 -Romany Chai-
                          -griersonianum (1/4)


Flowers of cyclamen purple, with blotch of beetroot purple; 15-
to 17-flowered trusses. See also Doris Moss. (William Moss,
Flint, Wales, 1974)


cl.
                 -griersonianum (1/2)
Gwillt-King -
(Gwyllt King)-arboreum ssp. zeylanicum (1/2)


Large Turkey red flowers, with spotting on upper lobe. A vigorous plant. (Caton Haig, Portmeirion, Wales, 1938)  A. M. (RHS)
(1952)


cl.
                 -griffithianum (1/2)
Gylla MacGregor-
                 -unknown (1/2)


Light red flowers. (M. Koster & Sons)


cl.
                 -catawbiense (1/2)
H. H. Hunnewell-
                 -unknown (1/2)


Purple red flowers.  There is also a red azalea of this name by
A. Waterer, c.1920. (A. Waterer)


---
                 -fortunei ssp. fortunei (1/2)
H. M. Ardene-
                 -unknown (1/2)


Flowers pink with a dark blotch. (G. Paul)  A. M. (RHS) 1896


---
                 -fortunei ssp. fortunei (1/2)
H. T. Gill-
                 -thomsonii ssp. thomsonii (1/2)


Rose-colored flowers. (R. Gill & Son)  A. M. (RHS) 1921


cl.
                 -catawbiense (1/2)
H. W. Sargent-
                 -unknown (1/2)

5ft(1.5m)         -25°F(-32°C)         ML          2/2

Very hardy and very old hybrid with magenta flowers. (A. Waterer) F. C. C. (RHS) 1865


cl.                      -thomsonii ssp. thomsonii (1/4)
            -Cornish Cross-
H. Whitner-                 -griffithianum (1/2)
          -                 -griffithianum
            -Loderi Pink Diamond-
                          -fortunei ssp. fortunei (1/4)


Ruthelma g.  Clear pink flowers. (Sir E. Loder)  A. M. (RHS)
1935


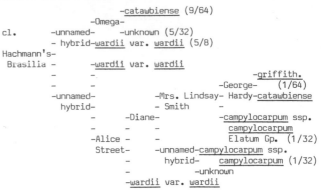

                         -catawbiense (9/64)
               -Omega-
cl.     -unnamed-    -unknown (5/32)
        - hybrid-wardii var. wardii (5/8)
Hachmann's-
  Brasilia -   -wardii var. wardii
           -                              -griffith.
           -                    -George-    (1/64)
           -unnamed-        -Mrs. Lindsay- Hardy-catawbiense
             hybrid-        - Smith
             -     -Diane-          -campylocarpum ssp.
             -     -     -            campylocarpum
             -Alice-     -            Elatum Gp. (1/32)
              Street-   -unnamed-campylocarpum ssp.
             -         hybrid-     campylocarpum (1/32)
             -             -unknown
             -wardii var. wardii


1. The reverse cross of Goldkrone, q.v. Trusses of 16-18 flowers,
5-lobed, spinel red with darker veins and markings, and slight
markings of bright yellow. Leaves ovate, slightly hairy.  The
first unnamed hybrid in diagram is an orange sister seedling of
Hachmann's Marina (Omega x wardii).  See also Goldrausch, and
Sandra. (Hans Hachmann cross; G. Stuck, reg. 1983)


                 -catawbiense (1/4)
cl.      -Humboldt-
          -            -unknown (11/16)
Hachmann's-                                    -griffithianum
  Constanze-                 -Queen Wilhelmina-     (1/16)
          -           -Britannia-           -unknown
          -Kluis    -       -Stanley Davies--unknown
          Sensation-unnamed seedling--unknown


Trusses of 18-20 flowers, 5-lobed, magenta rose with blotch of
ruby red; calyx reddish green; leaves elliptic, hairy. (H. Hachmann cross, 1959; G. Stuck, reg. 1983)


                               -catawbiense (1/8)
                 -Parsons Grandiflorum-
cl.     -Nova Zembla-                 -unknown (5/8)
Hachmann's -         -dark red hybrid--unknown
Feuerschein-     -griffithianum (1/4)
(Firelight)-Mars-
                 -unknown


Large scarlet flowers with brownish spots, white-tipped anthers,
5-lobed, in trusses of 14.  Plant 4.5ft-5ft(1.35m-1.5m) at maturity, well-branched; very dense foliage; leaves hairy. Unusually floriferous and vigorous. Bud hardy to -17°F(-27°C).  See
also Blinklicht, Sammetglut. (Cross by H. Hachmann; reg. by G.
Stuck, 1983)  See color illus. ARS J 39:1 (Winter 1985): p. 19.


cl.              -catawbiense (1/4)
            -Omega-
Hachmann's Marina*   -unknown (1/4)
            -wardii var. wardii (1/2)


A parent of other Hachmann hybrids.  Flowers of pure light yellow.  The Omega is Seidel's, a carmine rose; the wardii came
from Dietrich Hobbie and has no interior red markings. (H. Hachmann cross, 1963; reg. 1978)


cl.              -catawbiense (1/4)
            -Humboldt-
Hachmann's-      -unknown (1/2)
  Ornament -                 -fortunei ssp. fortunei (1/4)
            -Direcktor E. Hjelm-
                          -unknown


Trusses of 12 flowers, 5-lobed, mallow purple spotted ruby red,
blotched a dark greyed purple; leaves elliptic, slightly hairy.
Blooms late midseason. Hardy to -8°F(-22°C).  See also Kokar-
dia. (H. Hachmann cross, 1962; G. Stuck, reg. 1983)


cl.
            -Koichiro Wada--yakushimanum ssp. yakushimanum (1/2)
Hachmann's-      -catawbiense (1/4)
  Polaris -Omega-
                 -unknown (1/4)

'Friday' by Landauer
Photo by Greer

'Frontier' by J. A. Elliott
Photo by Greer

'Full House' by Johnstone
Photo by Greer

'Full Moon' by John Henny
Photo by Greer

'Furnivall's Daughter' by Knap Hill
Photo by Greer

'Fusilier' by Rothschild
Photo by Greer

'Gary Herbert' by Rock/Gable
Photo by Greer

'Gee Whiz' by Delp
Photo by Delp

'Gemmiferum' by Methven
Photo by Greer

'General Eisenhower' by Kluis
Photo by Greer

'George Sweesy' by W. Elliott
Photo by W. Elliott

'Gertrud Schale' by Hobbie
Photo by Greer

'Gertrude Bovee' by Bovee
Photo by Greer

'Gill's Crimson' by Gill
Photo by Greer

'Ginny's Delight' by Delp
Photo by Delp

'Gina' by Lowinsky
Photo by Greer

'Gleam' by Rothschild
Photo by Greer

'Glenda Farrell' by Dexter/Wister
Photo by Greer

'Gletschernacht' ('Glacier Night') by Hachmann
('Starry Night,' in USA)
Photo by Greer

'Glory of Leonardslee' by Gill
Photo by Greer

'Glory of Penjerrick' by Gill
Photo by Greer

'Goblin Pink' by Rothschild
Photo by Greer

'Gold Mohur' by Brandt
Photo by Greer

'Goldbug' by R. Henny
Photo by Greer

Carmine red buds, flowers light fuschia purple at edges, blending into light rhodamine purple. Unusually floriferous, even as
a young plant. A low, compact mound with dense foliage; leaves
elliptic-oval and strongly hairy. Hardy to -17°F(-27°C); wind-
and sun-tolerant. (H. Hachmann cross, 1963; G. Stuck, reg. 1983)
Gold Medal (Basel) 1980

```
cl.
                 -insigne (1/2)
Hachmann's Rosarka*      -griffithianum (1/4)
                 -Spitfire-
                        -unknown (1/4)
```

Light carmine red flowers on a compact bush. Has an unusually
long blooming period. Bud hardy to -12°F(-24°C). (H. Hachmann
cross; intro. 1983)

```
                        -catawbiense (1/8)
                 -Humboldt-
cl.       -Kokardia-      -unknown (3/8)
          -      -Direktor E.-fortunei ssp. fortunei (1/8)
Hachmann's-       Hjelm    -
 Rosita   -              -unknown
          -       -griffithianum (1/8)
          -unnamed-Mars-
          - hybrid-   -unknown
                 -Koichiro Wada--yakushimanum ssp. yakushimanum
                                                            (1/4)
```

Trusses of 8-9 flowers, 5-lobed, roseine purple on edges blending into lighter roseine purple, to very light purple; outside a
yellow green. Leaves elliptic, hairy. (H. Hachmann cross, 1969;
G. Stuck, reg. 1983)

```
cl.
                 -russatum (1/2)
Hachmann's Violetta-
                 -impeditum (1/2)
```

Many trusses of 13-22 flowers, 1in(2.5cm) across, 5-lobed, of a
light violet bluer than Azurika, q.v. Hardy to -8°F(-22°C). (H.
Hachmann cross, 1963; G. Stuck, reg. 1983)

```
g.
          -barbatum (1/2)
Haembarb-
          -haematocheilum (1/2)
```

Flowers of bright rose pink with brown blotches. (E. J. P. Magor, 1915)

```
g.
          -souliei (1/2)
Halcyone-            -fortunei ssp. discolor (1/4)
          -Lady Bessborough-
                 -campylocarpum ssp. campylocarpum
                                       Elatum Group (1/4)
```

5ft(1.5m)         -5°F(-21°C)         M          3/3

Much like the parent souliei with cup-shaped pink flowers held
in lax trusses. Another form has pale yellow flowers. See also Perdita and Sandling. (Rothschild, 1940)  See color illus.
PB pl. 51.

```
                        -griffithianum (5/16)
                 -Beauty of-
                 - Tremough-arboreum ssp. arboreum (1/16)
cl.       -Norman Gill-
    -Anna-       -griffithianum
Half -    -              -griffithianum
Penny-    -Jean Marie de Montague-
          -              -unknown (1/8)
          -       -fortunei ssp. discolor (1/4)
          -Margaret Dunn-   -dichroanthum ssp. dichroanthum (1/8)
                 -Fabia-
                        -griersonianum (1/8)
```

Pink buds opening primrose yellow, with large red blotches sur-
rounding corolla center; flowers 5-6in.(12.7-15.2cm) across, and
widely campanulate, 12-14 per truss. Medium-sized, compact bush
with leaves 8in(20.3cm) long. Late. (Del W. James, 1960)

```
                        -griffithianum (1/2)
cl.       -Jean Marie de Montague-
          -              -unknown (3/8)
Halfdan Lem-            -griffithianum
          -       -Loderi King George-
          -       -              -fortunei (1/8)
          -Red Loderi-            -griffithianum
                 -Earl of-Queen Wilhelmina-
                  Athlone-            -unknown
                        -Stanley Davies--unknown
```

5ft(1.5m)         -5°F(-21°C)         M          4/4

Bright red flowers 3.5in(8.3cm) wide, with darker spots on dorsal lobe; about 13 in large, tight trusses. Deep green leaves,
8in(20.3cm) long. Vigorous plant. (Cross by Lem; intro. by the
Seattle Chapter of the ARS; reg. 1974)

```
cl.
          -venator (1/2)
Hallali-
          -sanguineum ssp. sanguineum var. haemaleum (1/2)
```

Trusses hold 4-7 flowers carried terminally, forming many umbels, cardinal red with darker spotting in throat. Leaves dull
green above, lightly felted with grey indumentum beneath. (Collingwood Ingram)  Published registration information shows the
F.C.C. - 1979; published award information shows a P.C. in 1979.
(Both in the RHS yearbook, Rhododendrons 1981-82)

```
cl.               -williamsianum (1/4)
          -Kimberly-
Hallelujah-       -fortunei ssp. fortunei (1/4)
          -              -griffithianum (1/4)
          -Jean Marie de Montague-
                        -unknown (1/4)
```

4ft(1.2m)         -15°F(-26°C)         M          4/5

Large, dark green leaves are heavy-textured with a downward bend
midway in the leaf. Flowers are rose red, of good substance and
held in a large, tight truss. May be a tetroploid. (Harold E.
Greer, 1976)  A. E. (ARS) 1983  See color illus. HG p. 97.

```
cl.
          -maximum (1/2)
Halopeanum-
          -griffithianum (1/2)
```

6ft(1.8m)  {      5°F(-15°C)         M          3/3

Parentage as in Leach. Synonym White Pearl. Flowers in tall
conical trusses have pastel pink tint, fading to pristine white.
A very vigorous plant; dark green, slightly rough-textured foliage. (Halope, Belgium, 1896) A. M. (RHS) 1908

```
cl.
    -lacteum (1/2)       -griffithianum (1/16)
Halton-       -Kewense-
    -    -Aurora-       -fortunei ssp. fortunei (5/16)
    -Naomi-    -thomsonii ssp. thomsonii (1/8)
          -fortunei ssp. fortunei
```

Lionel's Triumph g. Round trusses 6in(15.2cm) across, holding
about 14 flowers of very light sap green. Leaves to 6in(15.2cm)
long by 2.75in(7cm) wide, and without indumentum. (Rothschild)
A. M. (RHS) 1967

```
cl.               -dichroanthum ssp. dichroanthum (1/4)
          -Fabia-
Hamma Hamma-     -griersonianum (1/4)
          -unknown (1/2)
```

5ft(1.5m)         15°F(-9°C)         ML          3/2

Leaves medium-to-large, rather narrow. Red flowers, heavily
spotted black, 2.5in(6.4cm) across, in trusses of about 18.
(Roy W. Clark, 1979)

```
                          -griffithianum (1/8)
cl.             -George Hardy-
     -Mrs. Lindsay Smith-         -catawbiense (1/8)
Handel-             -Duchess of Edinburgh--unknown (1/2)
     -                -campanulatum ssp. campanulatum (1/4)
     -unnamed hybrid-
                   -unknown
```

Flowers of yellow flushed green, with green spots.  Another form
has creamy white flowers. (M. Koster & Sons, 1920)  A. M. (RHS)
1937

```
cl.
                  -caucasicum (1/2)
Handsworth White-
                  -unknown (1/2)
```

White flowers with pink blush.  A parent of Duchess of Portland.
(Fisher, Son & Sibray, before 1928)

```
cl.
       -bureavii (1/2)
Hansel*   -dichroanthum ssp. dichroanthum (1/4)
       -Fabia-
             -griersonianum (1/4)
```

3ft(.9m)          -5°F(-21°C)         M          3/4

Foliage with buff-colored indumentum.  Flowers of salmon orange
in soft shaded tones, held in lax trusses.  See also Gretsel.
(Halfdan Lem)

```
cl.
            -fortunei ssp. fortunei (1/2)
Happiness-
            -unknown (1/2)
```

Maitland Dougall, intro. 1939; reg. as fortunei seedling, 1958.

```
g.
     -Pauline--unknown (1/2)
Happy-
     -griffithianum (1/2)
```

A medium-sized plant having pink flowers with a darker blotch.
(Rothschild, 1940)

```
cl.                      -maximum (1/4)
          -Lady Clementine Mitford-
Happy Day-                -unknown (1/2)
     -        -fortunei ssp. discolor (1/4)
          -Ladybird-
                 -Corona--unknown
```

Ball-shaped trusses of phlox pink flowers, partially tubular.
Plant habit compact with long leaves.  (Marshall Lyons, 1971)

```
                        -maximum (1/8)
                -Halopeanum-
cl.   -Snow Queen-         -griffithianum (1/4)
    -         -        -griffithianum
Happy  -        -Loderi-
Occasion-              -fortunei ssp. fortunei (1/8)
    -    -wardii var. wardii (1/4)
      -Crest-              -fortunei ssp. discolor (1/8)
            -Lady Bessborough-
                 -campylocarpum ssp. campylocarpum
                              Elatum Group (1/8)
```

An Exbury hybrid.  Trusses of 9-12 white flowers with centers of
yellow green.  (Edmund de Rothschild, 1980)

```
cl.
                  -racemosum (1/2)
Hardijzer's Beauty-
                  -Kurume azalea--unknown (1/2)
```

3ft(.9m)          -5°F(-21°C)         EM          4/3

A semi-dwarf evergreen azaleodendron from Holland.  Vigorous and
sun-tolerant.  Clear pink flowers open all along the stems.  The
small glossy foliage has a purplish tinge.  See also Ria Hardijzer.  (W. H. Hardijzer, 1965)  A. M. (RHS) 1970   See color
illus. F p. 46.

```
cl.
         -fortunei ssp. fortunei (1/2)
Hardy Giant-
         -rex ssp. fictolacteum (1/2)
```

6ft(1.8m)          -5°F(-21°C)          ML

Leaves are quite large,  9in(23cm) by 3in(7.6cm).  Flowers 3.5in
(8.9cm) across, are creamy white with raspberry blotch  and held
in conical trusses of about 14.   (Mrs. J. F. Knippenberg, 1967)
See color illus. LW pl. 107.

```
cl.
        -racemosum (1/2)
Hariet-
        -ciliatum (1/2)
```

Pale pink flowers 1in(2.5cm) long, about 20 per cluster. (Noble)
A. M. (RHS) 1957

```
cl.

Harlequin       See Dexter's Harlequin
```

```
                -dichroanthum ssp. dichroanthum (1/8)
g.        -Fabia-
     -Vega-     -griersonianum (1/8)
Harmony-    -haematodes ssp. haematodes (1/4)
     -        -griffithianum (1/4)
     -Loderi-
            -fortunei ssp. fortunei (1/4)
```

A pink-flowered hybrid by Lord Aberconway, 1950.

```
                        -griffithianum (1/8)
cl.              -Loderi-
     -Albatross-       -fortunei ssp. fortunei (1/8)
Harnden's-       -fortunei ssp. discolor (3/8)
  White   -    -wardii var. wardii (1/4)
      -Hawk-              -fortunei ssp. discolor
            -Lady Bessborough-
                 -campylocarpum ssp. campylocarpum
                              Elatum Group (1/8)
```

5ft(1.5m)          0°F(-18°C)          ML          4/3

Flowers off-white, 4.5in(1.5cm) across, in full rounded trusses
of 10.  Growth habit upright; leaves glossy, heavy-textured, and
citrus green.  (Harold E. Greer, 1979)

```
cl.
         -maximum (1/2)
Harold Amateis-
         -strigillosum (1/2)
```

3ft(1.2m)          -10°F(-23°C)          M

Cardinal red flowers with dark maroon throat, campanulate, 2.25
in(5.7in) across; trusses of about 20.  Plant 3ft at 14 years,
wider than tall; heavy rugose leaves 6in(15.2cm) long, light indumentum.  (E. Amateis cross; W. & S. Baldsiefen, intro. 1967)

```
                        -thomsonii ssp. thomsonii
---             -Cornish Cross-              (1/8)
       -unnamed hybrid-       -griffithianum (3/8)
Harold-             -wardii var. wardii (1/4)
Heal -              -griffithianum
     -Loderi King George-
                 -fortunei ssp. fortunei (1/4)
```

A parent of Willy-Nilly.  (Collinwood Ingram, 1955)

cl.

Harp Wood--form of hodgsonii

5ft(1.5m)            10°F(-12C)            EM           3/4

Flowers light cyclamen purple with darker veined markings and
darker reddish purple blotch. (Maj. A. E. Hardy)    P. C. (RHS)
1971


cl.
                         -catawbiense (1/2)
Harrisville*--Newport (F4)-
                         -unknown (1/2)

A tall shrub, hardy to -25°F(-32°C).  Flowers bright purple violet with white upper lobe, green spotting.  (Weldon E. Delp)


cl.
            -strigillosum (1/2)
Harry Carter*
            -sutchuenense (1/2)

5ft(1.5m)            0°F(-18°C)            E            3/3

Olive green leaves have the appearance of strigillosum.  Plant
covered with medium pink trusses.  Branching upright and sturdy,
combining the best of both parents.    (Mrs. E. G. Greig, Royston
Nursery; intro. by A. R. Cook)


cl.
            -Albescens--unknown (1/2)
Harry Tagg-
            -ciliicalyx (1/2)

White flowers with faint greenish yellow stain, held in  trusses
of 3-4, each flower 4.5in(11.5cm) wide. Named for the Assist-
ant at RBG, Edinburgh, later Keeper of the Museum. (RBG, Edinburgh; reg. 1962) A. M. (RHS) 1958


cl.                            -haematodes ssp. haematodes (3/4)
            -May Day (dwarf form)-
Harry von-                     -griersonianum (1/4)
 Tilzer   -haematodes ssp. haematodes

Plant very small, compact with small leaves, dark green, glossy,
with heavy indumentum.  Flowers blood red, a little larger than
haematodes, held in loose trusses.  See also Egbert van Al-
styne. (Cecil S. Seabrook, reg. 1967)


g.
            -arboreum var. kermesinum ? (1/2)
Harry White-                -thomsonii ssp. thomsonii (1/4)
            -Ascot Brilliant-
                            -unknown (1/4)

Scarlet or blood red flowers. (Loder and Rothschild, 1922)


cl.                            -griffithianum (1/8)
            -Mrs. Lindsay-
Harvest Moon- Smith    -Duchess of Edinburgh--unknown (1/2)         -catawbiense (1/8)
            -                  -campylocarpum ssp. campylocarpum
            -unnamed hybrid-                              (1/4)
                            -unknown

4ft(1.2m)            -5°F(-21°C)            M            3/3

Upright plant; leaves 4in(10.2cm) long.  Interesting foliage of
glossy yellowish green, heavily textured.  Flowers pale lemon or
creamy yellow  with reddish flare, widely campanulate; rounded,
compact trusses of 10-12. (M. Koster & Sons)  A. M. (RHS) 1948
See color illus. F p. 47; VV p. 102.


cl.              -fortunei ssp. discolor (1/2)
            -Lady Bessborough-
Harvest-                  -campylocarpum ssp. campylocarpum
 Queen -                                    Elatum Group (1/4)
       -                  -fortunei ssp. discolor
            -Margaret Dunn-    -dichroanthum ssp. dichroanthum (1/8)
                          -Fabia-
                                -griersonianum (1/8)

Flowers canary yellow, darkening in throat. (J. Russell, 1972)


cl.              -catawbiense (3/4)
            -Carl Mette-
Hassan-            -unknown (1/4)
            -catawbiense

Flowers a fiery carmine red with reddish brown markings.  (T. J.
R. Seidel cross, 1898; intro. 1906)


g. & cl.
            -wardii var. wardii (1/2)
Hawk-                  -fortunei ssp. discolor (1/4)
            -Lady Bessborough-
                          -campylocarpum ssp. campylocarpum
                                            Elatum Group (1/4)

6ft(1.8m)            5°F(-15°C)            M            3/2

This cross was made twice in an effort to find a superior yellow.  The first cross used the Exbury A. M. form of wardii (K-W
4170) Apricot buds opening daffodil yellow with slight red eye,
funnel-shaped, in neat trusses.  Habit tall, compact.  See also
Amour, Beaulieu Hawk, Crest, Hawk Kestrel, Hawk Merlin,  Jervis
Bay. (Rothschild, 1940) A. M. (RHS) 1949


cl.

Hawk Crest      See Crest


cl.

Hawk Kestrel      Parentage as above

6ft(1.8m)            5°F(-15°C)            M            3/2

Hawk g.  A selection from the Hawk family; called a rich yellow.
(Rothschild, 1940)


cl.

Hawk Merlin      Parentage as above

6ft(1.8m)            5°F(-15°C)            M            3/2

Synonym Exbury Merlin.  "A different shade of yellow" from the
other clones in the Hawk grex. (Rothschild)    P. C. (RHS) 1950


cl.
            -hodgsonii (1/2)
Haze-      -falconeri ssp. falconeri (1/4)
            -Muriel-
                  -grande (1/4)

Evening g.  Light pinkish mauve flowers. (Sir G. Loder, 1954)


cl.
            -bureavii (1/2)
Hazel-
            -unknown (1/2)

5ft(1.5m)            -15°F(-26°C)            EM           3/4

Light pink flowers edged and striped with deep pink, slight
brown spotting, 3in(7.6cm) wide; compact trusses of 12-15. Upright plant; deep green foliage covered with moderate amount of
woolly tan indumentum. (Origin unknown; intro. & reg. by H. E.
Greer, 1979)

cl.
           -azalea <u>occidentale</u> (1/2)
Hazle Smith-
           -Corona--unknown (1/2)

An azaleodendron; white flowers with chrysanthemum crimson
blotch held in trusses of 16-18.  Made a bush of 5ft(1.5m) by
5ft(1.5m) in 14 years.  Late midseason. (Vernon Wyatt, 1965)


cl.

Heane Wood--form of <u>argyrophyllum</u> ssp. <u>hypoglaucum</u>

5ft(1.5m)          5°F(-15°C)          M          3/4

Loose trusses of 8 flowers, pink in bud, opening white  suffused
with red purple, some lighter spotting in throat.  Leaves to 4.5
in(11.5cm) long, undersurface grey plastered indumentum. (Major
A. E. Hardy)  A. M. (RHS) 1972


cl.          -Sir Charles Butler--<u>fortunei</u> ssp. <u>fortunei</u>
     -Mrs. A. T.-           -<u>griffithianum</u> (1/4)          (1/4)
           -de la Mare-Halopeanum-
Heart's-                    -<u>maximum</u> (1/8)
Delight-                           -<u>griffithianum</u>
     -              -Queen Wilhelmina-
     -Britannia-                    -unknown (3/8)
           -Stanley Davies--unknown

5ft(1.5m)          10°F(-12°C)          ML          4/4

Synonym Heart's Desire.  Light vibrant red with deeper marking.
A well-shaped plant with dark green leaves and dense leaf cover.
(Alma Manenica, 1983)


cl.

Heart's Desire     See Heart's Delight


                         -<u>griffithianum</u> (5/16)
cl.      -Jean Marie de Montague-
     -                    -unknown (7/16)
Heat Wave*                      -<u>griffithianum</u>
     -              -Queen Wilhelmina-
     -  -Britannia-                    -unknown
     -Leo-           -Stanley Davies--unknown
           -<u>elliottii</u> (1/4)

5ft(1.5m)          0°(-18°C)          ML          4/3

Deep blood red flowers in large, upright trusses.  Growth habit
is strong to support the big trusses.  Leaves medium large and
deep forest green, contrasting with the unusually red flowers.
See also Black Magic. (Harold E. Greer, intro. 1982)

                              -<u>cataubiense</u>
                -Parsons Grandiflorum-     (1/8)
           -America-                    -unknown (1/2)
cl.      -unnamed-      -dark red hybrid--unknown
     - hybrid-<u>yakushimanum</u> ssp. <u>yakushimanum</u> (1/4)
Heatarama*      -<u>griffithianum</u> (1/8)
     -     -Mars-
     -Moniz-   -unknown
           -America--see above
Dark cardinal red buds  opening lighter red,  speckled with dark
red. (Weldon E. Delp)


cl.

           -unknown (1/2)
Heather Boulter-      -<u>chrysodoron</u> (1/4)
           -Chrysomanicum-
                    -<u>burmanicum</u> (1/4)

Trusses of  6 flowers, 2.5in(6.4cm) across, funnel-shaped, light
greyed yellow, darker spotting. Height 2ft(.6m).  (V. Boulter,
reg. 1984)


                    -<u>griffithianum</u> (1/8)
cl.      -Queen Wilhelmina-
     -Britannia-          -unknown (5/8)
Heather-      -Stanley Davies--unknown
 Moth -      -<u>yakushimanum</u> ssp. <u>yakushimanum</u> (1/4)
     -unnamed-
        hybrid-hybrid unknown

Flowers very light purple, flushed a darker shade. (John Waterer, Sons & Crisp, 1975)


cl.
           -<u>caucasicum</u> (1/2)
Heatherside Beauty-
           -unknown (1/2)

5ft(1.5m)          -10°F(-23°C)          E          3/3

Shrub upright and bushy.  White flowers in attractive trusses.
Early. (Frederick Street, before 1958)


cl.
           -<u>fortunei</u> ssp. <u>fortunei</u> (1/2)
Heavenly Scent*
           -unknown (1/2)

5ft(1.5m)          0°F(-18°C)          EM

Large, medium pink flowers, upper lobe faintly spotted red; high
lax trusses. Delightfully fragrant.  Plant habit rounded, well-
branched; matte green foliage held 2 years.  Prefers some shade.
Propagation rather difficult. (William Whitney; A. & E. Sather)


g.          -<u>neriiflorum</u> ssp. <u>neriiflorum</u> (1/4)
     -Neriihaem-
Hebe-      -<u>haematodes</u> ssp. <u>haematodes</u> (1/4)
     -<u>williamsianum</u> (1/2)

A small bush bearing campanulate flowers of the deepest shade of
rose pink. (E. J. P. Magor, 1927)


g.
     -<u>dichroanthum</u> ssp. <u>scyphocalyx</u> Herpesticum Group (1/2)
Heca-
     -<u>campylocarpum</u> ssp. <u>campylocarpum</u> (1/2)

A hybrid with yellow flowers. (Lord Aberconway, intro. 1941)


g.
     -<u>thomsonii</u> ssp. <u>thomsonii</u> (1/2)
Hecla-
     -<u>griersonianum</u> (1/2)

Called a good red.  See also Red Dragon. (Lord Aberconway, 1941)


cl.
     -<u>decorum</u> (1/2)
Helen-      -<u>souliei</u> (1/4)
     -Souldis-
           -<u>fortunei</u> ssp. <u>discolor</u> (1/4)

White flowers with a yellow throat. (Lester E. Brandt, 1952)


cl.          -<u>fortunei</u> ssp. <u>fortunei</u> (1/4)
     -unnamed hybrid-
Helen Child-      -unknown (1/4)
     -<u>williamsianum</u> (1/2)

2.5ft(.75m)          10°F(-12°C)          EM

Dark red buds open to rose pink flowers of heavy substance, 3in
(7.6cm) across, spotted pink in throat, in 9-flowered trusses.
Floriferous. Elliptic dark green leaves retained 3 years.  (H.
L. Larson, 1977)

```
                                    -griffithianum (1/8)
cl.                -Queen Wilhelmina-
        -Unknown Warrior-            -unknown (1/2)
Helen-            -Stanley Davies--unknown
Deehr-                    -arboreum ssp. nilagricum (1/8)
-               -Noyo Chief-
        -Noyo Brave-            -unknown
                    -Koichiro Wada--yakushimanum ssp. yakushimanum
                                                        (1/4)

3ft(.9m)        (Untested in cold)        E
```

Flowers openly bell-shaped, 5-lobed, colored crimson, held in ball-shaped trusses of 15. New foliage has medium amount of tan indumentum. Leaves held 3 years. (William A. Moynier, 1984)

```
cl.
            -elliottii (1/2)
Helen Druecker-                    -griffithianum (1/8)
                -George Hardy-
            -Betty Wormald-            -catawbiense (1/8)
                        -red garden hybrid--unknown (1/4)

4.5ft(1.35m)        5°F(-15°C)        M        4/3
```

Large leaves, 8in(20.3cm) by 4in(10.2cm). Trusses hold 16 flowers, to 4.5in(11.5cm) wide, rose madder shading darker on edges. (John S. Druecker, 1964)

```
cl.

Helen Everitt        Parentage two unknown Dexter hybrids

6ft(1.8m)        -15°F(-26°C)        M
```

Leaves are 4in(10.2cm) by 2in(5cm). Very large flowers, 5in(12.7 cm) across, with overlapping lobes, pure white, vestigial stamens; globular trusses hold 7-9. Very fragrant. (Cross by Sam Everitt; intro. by H. & S. Fuller, 1958; reg. 1975)

```
cl.
        -williamsianum (1/2)            -griffithianum (1/16)
Helen-                -Queen Wilhelmina-
Fosen-        -Britannia-            -unknown (5/16)
    -Burgundy-        -Stanley Davies--unknown
        -            -ponticum (1/8)
            -Purple Splendour-
                -unknown

Truss holds 6-7 flowers, 5-lobed, fuchsia purple. Leaves glabrous. (H. L. Larson, 1983)
```

```
                        -griffithianum (1/4)
cl.                -Loderi-
        -Irene Stead (=I.M.S.)-        -fortunei ssp. fortunei
Helen Holmes-        -Loderi (as above)        (1/4)
        -unknown (1/2)
```

Truss of 12 flowers, 6-lobed, white with a prominent yellow stigma. See also Holmeslee Triumph. (A. G. Holmes, 1982)

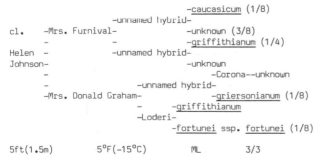

```
                    -caucasicum (1/8)
                -unnamed hybrid-
cl.    -Mrs. Furnival-        -unknown (3/8)
    -        -            -griffithianum (1/4)
Helen-        -unnamed hybrid-
Johnson-                -unknown
    -                -Corona--unknown
    -            -unnamed hybrid-
    -Mrs. Donald Graham-        -griersonianum (1/8)
                -griffithianum
            -Loderi-
                -fortunei ssp. fortunei (1/8)

5ft(1.5m)        5°F(-15°C)        ML        3/3
```

Flowers carmine rose with a striking blotch. Flower trumpet-shaped, with a narrow throat opening to a wide flare. Plant is covered with lush, narrow, fir green leaves. (Endre Ostbo) P. A. (ARS) 1956

```
                -fortunei ssp. fortunei (3/8)
cl.    -Fawn-    -dichroanthum ssp. dichroanthum (1/8)
    -    -Fabia-
Helen -        -griersonianum (1/8)
Louise*            -griffithianum (1/8)
            -Loderi-
    -Albatross-    -fortunei ssp. fortunei
            -fortunei ssp. discolor (1/4)

(P. Saunders) C. A. (ARS) 1972
```

```
cl.                    -racemosum (1/4)
            -unnamed hybrid-
Helen Scott Richey-            -moupinense (1/4)
            -Cornell Pink--mucronulatum (1/2)

2.5ft(.75m)        -5°F(-21°C)
```

Fuchsine pink flowers lightly spotted in rose, 5-lobed, widely funnel-shaped, 1.25in(3.2cm) wide, blooming at branch terminals holding up to 5 clusters of 2 flowers each. Upright, rounded plant; leaves 1in(2.5cm) long, scaly below; new growth bronze. (Robert W. Scott, 1977)

```
cl.

Helen Waterer        Parentage unknown
```

Red-edged flowers with white center. A parent of Barbara Wallace. (J. Waterer, before 1890)

```
cl.                -fortunei ssp. discolor (1/4)
        -unnamed hybrid-
Helen Webster-            -unknown (1/4)
    -            -fortunei ssp. fortunei (1/4)
    -Richard Gill-
            -thomsonii ssp. thomsonii (1/4)
```

Large trusses of widely bell-shaped flowers, phlox pink spotted with orange brown. (Crown Lands) (A. M. (RHS) 1954)

```
cl.
        -fortunei ssp. fortunei ? (1/2)
Helene Huber-
    -unknown (1/2)

6ft(1.8m)        -5°F(-21°C)        M
```

Dark purplish pink buds opening medium mauve pink, spotted with almond shell brown, openly funnel-shaped, 3in(7.6cm) across, and very fragrant; spherical trusses of 14-16. Floriferous. Plant upright, rounded, half as broad as tall. (C. O. Dexter cross; raised by Swarthmore College & C. Herbert; reg. by Herbert 1978)

---

```
Helene Schiffner        Parentage unknown

4ft(1.2m)        -5°F(-21°C)        M        4/4
```

Unusually dark buds open to pale lilac-tinted flowers, becoming pure white with very faint yellowish brown markings, held in upright, dome-shaped trusses. Narrow leaves of deep mistletoe green with reddish stems. (Seidel) F. C. C. (RHS) 1893 See color illus. VV p. 83.

```
cl.        -decorum (1/4)
    -unnamed hybrid-
Helios-            -fortunei ssp. discolor (1/4)
    -            -fortunei ssp. fortunei (1/4)
    -unnamed hybrid-    -wardii var. wardii (1/8)
                -unnamed-
            hybrid-dichroanthum ssp. dichroanthum
                                (1/8)
3ft(.9m)        -5°F(-21°C)        ML
```

Neyron rose buds opening Venetian pink, and shading to soft yellow. Flowers openly funnel-shaped, 3.5in(8.9cm) across; rounded trusses of 10. Plant habit rounded, well-branched; glossy medium green leaves held 3 years. (Alfred A. Raustein, 1979)

```
cl.           -dichroanthum ssp. dichroanthum (1/4)
         -Fabia-
Hello Dolly-      -griersonianum (1/4)
            -smirnowii (1/2)
```

3ft(.9m)        -10°F(-23°C)        EM        3/3

Flowers yellow with warm tones of orange and rose, 2.75in(7cm)
wide; large calyx gives the effect of a double.  Plant rounded,
well-branched; medium green leaves with beige indumentum, held 2
years.   (Halfdan Lem cross; reg. by James Elliott, 1974)    See
color illus. HG p. 129.

```
cl.
         -fortunei ssp. fortunei (1/2)
Helloween-
         -unknown (1/2)
```

Flowers pale cream, flushed pink, like fortunei.   (W. E. Glennie, reg. 1984)

```
cl.    -            -catawbiense (3/8)
         -Mrs. Milner-
Helmholtz-            -unknown (3/8)
       -                 -caucasicum (1/4)
         -Boule de Neige-           -catawbiense
                       -hardy hybrid-
                                 -unknown
```

Dark pink flowers with few markings.   (T. J. R. Seidel,  before
1920)

```
                        -fortunei ssp. discolor (1/8)
                 -Lady         -
cl.      -Jalisco-Bessborough-campylocarpum ssp. campylocarpum
       - Elect -                       Elatum Group (1/8)
Hendrick's-      -      -dichroanthum ssp. dichroanthum (1/4)
  Park    -      -Dido-
       -           -decorum (1/8)
       -    -fortunei ssp. fortunei (1/4)
       -Fawn-    -dichroanthum ssp. dichroanthum
               -Fabia-
                   -griersonianum (1/8)
```

6ft(1.8m)        -10°F(-23°C)          M

Very large trusses of 10-12 fragrant flowers, neyron rose with
currant red throat. Plant upright, well-branched; leaves to 4in
(10.2cm) by 2in(5cm), ivy green. (Cross by Del W. James; raised
by Hendrick's Park; reg. by Ray James, 1980)

```
cl.
         -catawbiense (1/2)
Henriette Sargent-
         -unknown (1/2)
```

5ft(1.5m)        -25°F(-32°C)        ML        2/3

Synonym Henrietta Sargent. Large rounded trusses of dark rose
pink; similar to Mrs. C. S. Sargent,  but smaller, more compact,
slower growing. (A. Waterer, 1891) See color illus. VV p. 108.

```
cl.
         -dichroanthum ssp. dichroanthum (1/2)
Henry E. Burbridge-      -campyocarpum ssp. campylocar-
            -Letty Edwards-   carpum Elatum Group (1/4)
                -fortunei ssp. fortunei (1/4)
```

Apricot yellow flowers. (Thacker, 1944)

```
cl.
         -wardii var. wardii Litiense Group (1/2)
Henry R. Yates-
            -unknown (1/2)
```

4ft(1.2m)        -5°F(-21°C)          E

Flowers 2.5in(6.4cm) across, creamy ivory with a bold yellow
flare, funnel-campanulate. Plant compact and dense; leaves to 6
in(15.2cm) long. (Jos. Gable, 1971)  See color illus. K p. 136;
LW pl. 57.

```
cl.
```

Henry Shilson--form of arboreum ssp. arboreum

A named, selected form of this species with red flowers.   (R.
Gill & Son, reg. 1958)

```
cl.
         -dalhousiae var. dalhousiae (1/2)
Henryanum-
         -formosum var. formosum (1/2)
```

White flowers, tinged blush pink. (Shown by Henry, 1862; perhaps related to Prof. Augustine Henry of Dublin?)  F. C. C. 1865

```
g.
         -arboreum ssp. arboreum (1/2)
Her Majesty-      -veitchianum (1/4)
         -Fosterianum-
                  -edgeworthii (1/4)
```

Soft crimson flowers.  F. C. C. (RHS) 1889

```
g.
   -haematodes ssp. haematodes (1/2)
Hera-    -williamsianum (1/4)
   -Jock-
          -griersonianum (1/4)
```

Flowers of scarlet red. (Lord Aberconway, 1941)

```
cl.
```

Herbert Mitchell--form of cerasinum (K-W 6923)

3ft(.9m)        -5°F(-21°C)        EM        4/4

Raised by the Countess of Rosse, Nymans Garden.  A.M. (RHS) 1973

```
cl.
```

Herbert Parsons     See President Lincoln

```
g.
     -javanicum (1/2)
Hercules-
         -unknown (1/2)
```

Vireya. Flowers of apricot yellow suffused rose pink. (Davies,
before 1890)  A. M. (RHS) 1899

```
                        -griffithianum (1/8)
g.             -Dawn's Delight-
   -Break of Day-        -unknown (1/8)
Herga-        -dichroanthum ssp. dichroanthum (1/4)
   -            -fortunei ssp. discolor (1/4)
      -Lady Bessborough-
              -campylocarpum ssp. campylocarpum
                       Elatum Group (1/4)
```

Pale yellow flowers with a dark throat. May. (Rothschild, 1940)

```
g.
     -Kaiser Wilhelm--unknown
Hermann*
     -Agnes--unknown
```

Bright, light carmine red flowers. See also Homer. (Cross by
Seidel, 1898; intro. c.1916)

g.
                    -caucasicum (1/2)
Hermann Seidel*
                    -unknown (1/2)

Frilled flowers, light carmine red. (Seidel, before 1926)


                    -japonicum var. japonicum (metternichii) (1/2)
Hermann Nitzschner*
                    -catawbiense (1/2)

Salmon pink flowers and greenish yellow markings.  Named for the
Head Gardner at Grungrabschen.  (T. J. R. Seidel)


g.                  -japonicum var. japonicum (metternichii)
                    -Eidam-                                (1/4)
Hermann Seidel-     -Alexander Adie--unknown (1/4)
                    -williamsianum (1/2)

Bright rose flowers in umbels of 10-12. (Dietrich Hobbie, 1952)


g.        -dichroanthum ssp. apodectum (1/2)
Hermes-                 -fortunei ssp. discolor (1/4)
          -Lady Bessborough-
                        -campylocarpum ssp. campylocarpum Elatum
                                                      Group (1/4)

Bell-shaped flowers of chrome yellow with pink markings, held in
loose trusses.  Plant below  average height,  but spreading and
compact.  Midseason. (Rothschild, 1940)


cl.             -thomsonii ssp. thomsonii (1/4)
          -Gilian-
Hermione-       -griffithianum (1/4)
          -arboreum ssp. arboreum (1/2)

An award-winning hybrid  with flowers of blood red.  (E. J. P.
Magor, 1935)  A. M. (RHS) 1941


cl.       -catawbiense (1/2)
Hero-
          -Gloria Gandavensis (white)--unknown (1/2)

Pure white with olive green markings.  (T. J. R. Seidel cross,
1906; intro. 1915-16)


cl.                   -adenogynum (1/4)
            Xenosporum (dotoneum)-
Hesperia-                   -unknown (3/4)
            -Empress Eugenie--unknown

E. J. P. Magor cross, 1921.


g.          -fortunei ssp. discolor (1/4)
          -Ayah-
Hesperides-     -facetum (eriogynum) (1/4)
                -griersonianum (1/2)

A tall, compact plant; rose pink flowers held in full trusses in
June. (Rothschild, 1940)


cl.               -campanulatum ssp. campanulatum (1/4)
          -Constant Nymph-
Hess*                       -ponticum (3/8)
          -              -Purple Splendour-
          -Purple Splendour-    -ponticum--unknown (3/8)
                          -unknown

(Crossed by Donald Waterer, 1947; raised and introduced by Knap
Hill Nurseries, 1975)


                              -dichroanthum ssp. dichroanthum
cl.             -Goldsworth-                            (1/8)
          -Tortoiseshell- Orange   -fortunei ssp. discolor (1/8)
Hestia-  Wonder            -
                    -griersonianum (1/4)
          -Goliath--unknown (1/2)

Trusses hold 11-16 flowers 3.2in(8cm) wide, cardinal red in bud,
opening neyron rose with bronze yellow blotch  and bronze dorsal
spotting.  (Research Station for Woody Nursery Crops,  Boskoop,
reg. 1983)


cl.
          -cinnabarinum ssp. cinnabarinum (1/2)
Hethersett-
          -maddenii ssp. maddenii (1/2)

Royal Flush  g.  Trusses of 7 tubular-shaped  flowers,  sweetly
scented, primrose yellow fading to buff, the reverse tinged nasturtium orange. (Cross by H. Mangles; shown by Mrs. Douglas Gordon)  A. M. (RHS) 1962


g.              -williamsianum (1/4)
          -Adrastia-
Hiawatha-       -neriiflorum ssp. neriiflorum (1/4)
          -griersonianum (1/2)

Flowers of carmine red.   (Lord Aberconway, 1941)


cl.             -dichroanthum ssp. dichroanthum (1/4)
          -Fabia-
High Curley-    -griersonianum (1/2)
          -            -griersonianum
          -unnamed hybrid-    -fortunei ssp. discolor (1/8)
                          -unnamed-
                          hybrid-unknown (1/8)

Flowers light primrose yellow. (Waterer, Sons & Crisp, 1975)


cl.

High Flier--form of vesiculiferum (K-W 10952)

5ft(1.5m)        5°F(-15°C)          EM         3/3

Loose trusses of 10-12 flowers, 5-lobed, ruby red, outside
strongly flushed magenta.  Leaves 7in(17.8cm) long, covered with
reddish-tipped glandular hairs. (Collector F. Kingdon-Ward;
shown by Crown Estate, Windsor) A. M. (RHS) 1968


                          -caucasicum (1/8)
cl.       -Dr. Stocker-
          -Damaris-       -griffithianum (1/8)
High Gold*      -campylocarpum ssp. campylocarpum (1/4)
          -    -wardii var. wardii (1/4)
          -Crest-             -fortunei ssp. discolor (1/8)
                -Lady Bessborough-
                          -campylocarpum ssp. campylo-
                          carpum Elatum Group (1/8)

5ft(1.5m)        -5°F(-21°C)          M          4/4

A deep, bright, true lemon yellow.  Flowers large, open-faced,
and held in big trusses.  Vigorous plant with excellent branching habit; leaves of bright apple green. (John Eichelser, intro.
1982)


cl.       -Mrs. J. G. Millais--unknown (1/2)
High Summer-    -wardii var. wardii (1/4)
          -Inamorata-
                -fortunei ssp. discolor (1/4)

Parentage is a reverse of Romy.  Trusses of 8-10 flowers, very
light yellow shading to citrus yellow in center. (E. G. Millais
Nurseries, reg. 1981)

cl.
                    -yakushimanum ssp. yakushimanum (1/2)
Highfield Cream-
                    -unknown (1/2)

Flowers in trusses of 9, mimosa yellow fading to pale cream with
a slightly green throat. (Cross by Mrs. R. J. Coker; intro. by
Mrs. J. S. Clyne, reg. 1979)

                                    -souliei (1/16)
                            -Soulbut-
                    -Vanessa-         -Sir Charles Butler--fortunei
g.        -Adonis-      -griersonianum ssp. fortunei (1/16)
          -          -          -griffithianum (1/8)
Highlander-     -Sunrise-
          -               -griersonianum (1/4)
          -arboreum ssp. zeylanicum (1/2)

Rose red flowers.  (Lord Aberconway, 1950)

                                    -griffithianum (1/8)
cl.                 -Queen Wilhelmina-
          -Earl of Athlone-            -unknown (3/8)
Highnoon-               -Stanley Davies--unknown
          -     -dichroanthum ssp. dichroanthum (1/4)
          -Fabia-
                    -griersonianum (1/4)

Flowers are bright waxy red in loose trusses. Plant is very slow
growing and has sprawling habit. (Rudolph Henny, reg. 1958)


cl.
          -fortunei ssp. discolor (1/2)
Hill Ayah-
          -facetum (eriogynum) (1/2)

Leaves 6.5in(15.5cm) by 2.5in(6.4cm).  Flowers rose madder
opening from cardinal red buds, funnel-shaped, 7-lobed with 8-11
in compact trusses. (E. J. P. Magor; reg. by Maj. E. W. M. Mag-
or, 1966)


cl.
          -Chesterland--unknown (1/2)
Hillsdale-
          -fortunei ssp. fortunei (1/2)

4ft(1.2m)          -25°F(-32°C)          M

Tall conical trusses of about 18 flowers, light neyron rose with
prominent spotting of chartreuse green. Plant upright, rounded,
well-branched; leaves 5.75in(14.5cm) long, a glossy dark green,
held 3 years. Plant resistant to disease and insects, and heat-
tolerant. The only hybrid registered by this prolific hybridizer before his tragic death. (Cross by H. Roland Schroeder; Stephen Schroeder, intro. 1976; reg. 1983)

cl.
          -hodgsonii (1/2)
Himalayan Child-
          -falconeri ssp. falconeri (1/2)

Loose trusses hold 28-30 flowers, rhodamine purple with darker
staining and small blotch of ruby red in throat. Leaves to 11in
28cm) long, with silvery brown indumentum. (From K-W 13681, as
sinogrande, collector Kingdon-Ward; raised and reg. by the
Crown Estate, Windsor) A. M. (RHS) 1981

cl.

Hindustan     See upper right column

---
                    -citrinum var. discoloratum (1/2)
Hippolyta-        -javanicum (5/16)
          -Queen of the-          -brookeanum var. gracile (1/8)
          Yellows   -Princess -          -jasminiflorum (1/16)
                    -Frederica-Princess-
                              Royal  -javanicum

Vireya hybrid, scarlet-flowered.  Seed parent formerly called
multicolor var. curtisii.     F. C. C. (RHS) 1888


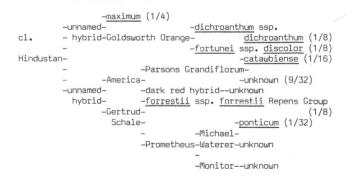
                    -maximum (1/4)
          -unnamed-          -dichroanthum ssp.
cl.       - hybrid-Goldsworth Orange-     dichroanthum (1/8)
          -               -fortunei ssp. discolor (1/8)
Hindustan-               -catawbiense (1/16)
          -     -Parsons Grandiflorum-
          -     -America-          -unknown (9/32)
          -unnamed-     -dark red hybrid--unknown
          hybrid-     -forrestii ssp. forrestii Repens Group
          -Gertrud-               (1/8)
          Schale-          -ponticum (1/32)
          -          -Michael-
          -Prometheus-Waterer-unknown
          -
                    -Monitor--unknown

6ft(1.8m)          -20°F(-29°C)          M

Flowers of good substance, orange buff flushed soft pink (garden
effect is orange), 3in(7.6cm) across, 5 wavy lobes; dome-shaped
trusses of 17.  Leaves 5.25in(13.3cm) long by 2.5in(6.4cm) wide;
plant as wide as tall. (David G. Leach cross, 1967; reg. 1983)


g.       -hippophaeoides var. hippophaeoides (1/2)
Hipsal-
          -saluense var. saluense (1/2)

Flowers purplish mauve. (E. J. P. Magor, 1926)


g.
          -haematodes ssp. haematodes (1/2)
Hiraethlyn-
          -griffithianum (1/2)

Flowers are rose to deep red. (Lord Aberconway, intro. 1933)


cl.

His Lordship--form of aberconwayi

3ft(.9m)          5°F(-15°C)          EM          4/3

Saucer-shaped flowers, white, spotted red on upper lobes. (Crown
Estate, Windsor, reg. 1962) A. M. (RHS) 1945


cl.                 -griffithianum (1/4)
          -Isabella-
Hispaniola-     -auriculatum (1/4)
          -griersonianum (1/2)

Infanta g.  (Collingwood Ingram, before 1958)


cl.                         -wardii var. wardii (1/4)
          -Mrs. Lammot-          -souliei (1/8)
Hjalmar L. Larson- Copeland  -Virginia Scott-
          -               -unknown (5/8)
          -yellow hybrid--unknown

Edges and reverse of flowers pale brick red, inside lemon
yellow, upper lobe spotted scarlet red. Leaves elliptic,
glabrous. (Cross by H. L. Larson; reg. by Mrs. L. Hodgson,
1982)


cl.

Ho Emma--form of japonicum var. japonicum (metternichii)

3ft(.9m)          -15°F(-26°C)          EM          4/4

Trusses hold up to 15 flowers, 7-lobed, white flushed light
fuchsia purple, with stronger veinal color of deep fuchsia purple; upper throat spotted ruby red.  Leaves glossy, dark green
with tawny felted indumentum beneath. (Collector unknown; shown
by R. N. S. Clarke, Borde Hill) A. M. (RHS) 1976

cl.
          -pubescens (1/2)
Hockessin-
          -keiskei (1/2)

3ft(.9m)          -25°F(-32°C)          E          3/2

Flowers apricot, fading white, with clusters on all terminals.
Leaves rich green and hairy, 2in(5cm) long by .5in(1.3cm) wide.
Plant similiar to Chesapeake, but larger and more open.  (G. Guy
Nearing, intro. c.1950)


cl.


Hokkaido--form of dauricum

4ft(1.2m)          -25°F(-32°C)          EM          4/3

Very hardy flowering species with trusses of 1-2, this form is
white, faint green markings  on dorsal lobe. Leaves are scaly.
Blooms a month later than other strains. (Collector unknown; intro. in the US by Warren Berg, into the UK by P. A. Cox)  A. M.
(RHS) 1979


cl.
          -Alexander Adie--unknown (3/4)
Holbein-          -catawbiense (1/4)
          -Carl Mette-
                    -unknown

Ruby pink with faint reddish brown markings on light background;
wavy margins. (T. J. R. Seidel cross, 1906; intro. 1918)


cl.                    -caucasicum (1/4)
          -Cunningham's White-
Holden-          -ponticum (white form) (1/4)
     -          -catawbiense (1/4)
          -red hybrid-
                    -unknown (1/4)

4ft(1.2m)          -15°F(-26°C)          EM          3/4

A compact hybrid with lustrous dark green foliage and rose red
flowers marked by small red spots.  See also Rocket, Spring
Glory, Spring Parade and Vernus. (A. M. Shammerello, reg. 1958)
See color illus. LW pl. 86; VV p. 124.


cl.
          -Madame Linden--unknown
Holger-
          -Eggbrechtii--unknown

Light purplish violet; markings may be yellowish green or
reddish brown.  See also Gudrun. (T. J. R. Seidel, intro. 1918)


cl.


Holker Hall--form of arboreum ssp. arboreum

Still growing at Holker Hall in 1983; reg. in 1958.


                              -griffithianum (1/8)
                    -George Hardy-
cl.     -Pink Pearl-          -catawbiense (3/8)
     -          -          -arboreum ssp. arboreum (1/8)
Hollandia-          -Broughtonii-
     -                    -unknown (3/8)
     -          -catawbiense
          -Charles Dickens-
                    -unknown

6ft(1.8m)          0°F(-18°C)          ML          3/3

Synonym G. Streseman.  Attractive foliage with vibrant red
flowers. (L. J. Endtz, before 1958) See color illus. F p. 48.


cl.                              -griffithianum (1/4)
          -Irene Stead--Loderi (selfed)-
Holmeslee-          -fortunei ssp. fortunei
Barndance-                              (1/4)
          -M30 seedling--unknown (1/2)

Trusses of 9-10 flowers, 7- or 8-lobed, chartreuse green in bud,
opening white with dawn pink flush; fragrant. (A. Graham Holmes,
NZ, reg. 1984)


                              -griffithianum (1/8)
                    -Coombe Royal-
cl.          -Mrs. G. W. Leak-          -unknown (3/4)
Holmeslee Flair-     -          -caucasicum (1/8)
          -unknown          -Chevalier Felix-
                    de Sauvage     -unknown

Truss of 16 flowers, 5-lobed, phlox pink in bud, opening lighter
with red flare and spots on upper lobe. (A. G. Holmes, 1982)


cl.
          -unknown
Holmeslee Missie-          -dichroanthum ssp. dichroanthum
          -          -Fabia-
          -C. I. S.-          -griersonianum
                    -
                    -Loder's White, q.v. for 2 possible
                              parentage diagrams

See Loder's White for possible alternate parentage.  Flower buds
are cream overlaid dawn pink, opening to Naples yellow flushed
pink, red in throat. (A. G. Holmes, 1982)


                              -griffithianum (1/4)
                    -Loderi-
cl.          -Waxeye-          -fortunei ssp. fortunei (3/8)
          -          -unknown (1/4)
Holmeslee Opal-          -Sir Charles Butler--fortunei ssp.
          -Van Nes -                    fortunei
          -Sensation-          -maximum (1/8)
                    -Halopeanum-
                              -griffithianum

Trusses of 12-18 flowers, white with tinge of very light neyron
rose. (A. G. Holmes, 1982)


cl.
                    -unknown (1/2)
Holmeslee Sunrise-          -decorum (1/4)
                    -Little -     -dichroanthum ssp. dichroanthum
                    -Pudding-Fabia-                    (1/8)
                              -griersonianum (1/8)

Trusses of 7 flowers of medium spinel red, fading to cream in
the upper lobe. (A. G. Holmes, 1982)


cl.                              -griffithianum (1/4)
          -Irene Stead-Loderi (selfed)-
Holmeslee- (I.M.S.)          -fortunei ssp. fortunei
 Triangle -                              (1/4)
          -unknown (1/2)

Truss of 10 flowers, 7-lobed, rhodamine pink, upper lobe flecked
red.  See also Helen Holmes. (A. G. Holmes, 1982)


cl.          -catawbiense (1/2)
          -Humboldt-
Holstein-          -unknown (1/2)
     -          -catawbiense
          -Catawbiense Grandiflorum-
                    -unknown

Flowers in trusses of 15-16, 5-lobed, lilac purple, shading paler in center, dark red blotch.  Leaves oval, sparsely hairy.
(Cross by H. H. Hachmann; reg. by G. Stuck, 1983)

cl.                      -fortunei ssp. discolor (1/4)
         -King of Shrubs-      -dichroanthum ssp. dichroanthum
Holy Moses*        -Fabia-                                (1/8)
         -                    -griersonianum (1/8)
         -Souvenir of Anthony Waterer--unknown (1/2)

4ft(1.2m)          -5°F(-21°C)          ML          4/4

Above parentage is uncertain;  another parent may be smirnowii.
Flowers are an orange and yellow bicolor.  Exceptional foliage,
similar to smirnowii.  (Halfdan Lem)


cl.
       -Kaiser Wilhelm--unknown
Homer-
       -Agnes--unknown

(The Agnes by Swaything was too recent to be used by Seidel)  A
parent of Royal Pink.  Intense pure pink with faint red brown
markings.  Two forms exist; one may be more hardy.  (T. J. R.
Seidel cross, 1906; intro. 1916)


cl.
       -wardii var. wardii (1/2)
Honey-          -Corona--unknown (1/4)
       -Bow Bells-
               -williamsianum (1/4)

Flowers of Egyptian buff fading paler, 3.5in(8.9cm) wide.  Plant
1ft(.3m) high, 3ft(.9m) broad; leaves 2.5in(6.4cm) across, orbicular.  (Rudolph Henny, 1960; reg. 1962)


         -wardii var. wardii (1/4)
cl. -Hawk-
    -      -Lady        -fortunei ssp. discolor (1/8)
Honey*    Bessborough-
    -                  -campylocarpum ssp. campylocarpum Elatum
    -      -wightii (1/4)                              Group (1/8)
    -China-
         -fortunei ssp. fortunei (1/4)

A vigorous plant, growing to 5ft(1.5m) tall and 5ft(1.8m) wide.
Flowers of clear yellow, opening with a pink flush, held in
large trusses.  (Slocock Nurseries)  H. C. (RHS) 1975


cl.


Honey Dew*     Parentage unknown

Synonyms Dexter's Honeydew, Honeydew.  A parent of Blush Button,
Fran Labera.  (C. O. Dexter, before 1943)


cl.          -wardii var. wardii (1/4)
       -Crest-        -fortunei ssp. discolor (1/8)
Honey Glow-    -Lady
       -       Bessborough-campylocarpum ssp. campylocarpum
       -unknown (1/2)                        Elatum Group (1/8)

Trusses hold 12-14 flowers, light primrose yellow flushed very
light neyron rose.  (Cross by Mrs. R. J. Coker; intro. & reg. by
Mr. & Mrs. C. A. Grant, 1979)


cl.


Honey Wood--form of trichanthum

6ft(1.8m)          -5°F(-21°C)          L          2/3

Flowers in clusters of 3, widely funnel-shaped, 5-lobed,
spectrum violet, outside light cyclamen purple.  Leaves elliptic
and medium-sized, both sides covered with scattered brown
indumentum.  (Raised at Tower Court; intro. by Maj. A. E. Hardy)
P. C. (RHS) 1969; A. M. (RHS) 1971


cl.          -wardii var. wardii (1/4)
       -Carolyn Grace-
Honeydew-      -unknown (1/4)
       -        -campylocarpum ssp. campyloarpum (1/4)
       -Moonstone-
               -williamsianum (1/4)

4ft(1.2m)          5°F(-15°C)          M          3/4

A compact, rounded plant.  Flowers sunny yellow like Moonstone,
to 4in(10.2cm) wide; trusses hold 5-7.  (Art Wright, 1962)


               -wardii var. wardii (Barto form)
cl.    -unnamed hybrid-      -A. W. Hardy Hybrid--unknown
       -              -Devonshire-                (1/8)
Honeymoon-        Cream  -campylocarpum ssp. campy-
       -                              locarpum (1/8)
       -wardii var. wardii (3/4)

4ft(1.2m)          -5°F(-21°C)          EM          4/4

A compact plant, heavy, glossy, dark green foliage held 3 years.
Chartreuse green buds opening pale chartreuse yellow with dull
orange blotch, 2in(5cm) across; about 14 in domed trusses.  Plant
rounded, as wide as tall; blooms very young, propagates easily.
(Cross by W. E. Whitney; reg. by the Sathers, 1976)


cl.
       -Catalgla--catawbiense var. album Glass (1/2)
Hong Kong-     -wardii var. wardii (1/4)
         -Crest-
               -Lady        -fortunei ssp. discolor (1/8)
               -Bessborough-
                       -campylocarpum ssp. campylocarpum
                                      Elatum Group (1/8)
7ft(2.1m)          -20°F(-29°C)          M          4/4

Flat trusses of 13 flowers, primrose yellow with greenish yellow
blotch.  Plant well-branched, broader than tall; leaves glossy
yellow green, held 2 years.   (David G. Leach, intro. 1974; reg.
1983)


                       -fortunei ssp. discolor (1/8)
               -Lady    -
               -Bessborough-campylocarpum ssp. campylocarpum
cl.    -Jalisco-                      Elatum Group (1/8)
       -Eclipse-    -dichroanthum ssp. dichroanthum (1/8)
Honiton-       -Dido-
       -               -decorum (1/8)
       -       -       -catawbiense (1/4)
       -Album Elegans-
               -unknown (1/4)

Strong open-growing plant.  Flowers cream-colored, flushed pink,
with a red eye.  (Knap Hill Nurseries, reg. 1968)


cl.          -sanguineum ssp. didymum (1/4)
       -Carmen-
Honore-    -forrestii ssp. forrestii Repens Group (1/4)
Hacanson-      -haematodes ssp. haematodes (1/4)
       -Choremia-
               -arboreum ssp. arboreum (1/4)

Plant grew to 1ft(.3m) by 1.5ft(.45m) in 9 years, from seed.
Flowers cardinal red, 2in(5cm) across, with a red calyx; trusses
hold 8.  Leaves 3.5in(8.9cm) long, silvery beneath.  See also
Fred Robbins, Little Nemo.  (Lester E. Brandt, 1965)


               -souliei (1/4)
cl.    -Rosy Morn-    -griffithianum (1/8)
       -       -Loderi-
Hoopskirt-     -fortunei ssp. fortunei (1/8)
       -   -dichroanthum ssp. dichroanthum (1/4)
       -Dido-
         -decorum (1/4)

Orange to peach to yellow, with a yellow basal blotch.  (Rudolph
Henny cross, 1944; reg. 1958)

cl.
                    -lindleyi (1/2)
Hope Braafladt-                -ciliatum (1/4)
                -Countess of-
                    Haddington-dalhousiae var. dalhousiae (1/4)

3ft(.9m)           20°F(-7°C)          E          4/3

White flowers 3.25in(8.3cm) wide, with dull lavender blotch and
faint stripes, held in lax trusses of about 15.  Plant upright,
with willowy branches;  yellow green leaves 3.75in(9.5cm) long,
golden brown scales beneath, held 2-3 years.  (Dr. J. H. Braafladt, 1979)


                        -griffithianum (3/16)
                -Loderi-
cl.   -unnamed-    -fortunei ssp. fortunei (1/8)
      - hybrid-                    -griffithianum
Hope    -      -      -Queen Wilhelmina-
Findlay-   -Earl of-            -unknown (3/16)
      -    Athlone-Stanley Davies--unknown
      -        -griersonianum (1/4)
      -Creeping Jenny-
              -forrestii ssp. forrestii Repens Group
                                (1/4)

Plant vigorous, spreading, compact; dark green, glossy leaves.
Flowers currant red, funnel-shaped, about 2.5in(6.4cm) wide, 8-
10 in globular, lax trusses.  (Windsor Great Park)    P.C. (RHS)
1974;  A. M. 1979 1979


                    -neriiflorum ssp. neriiflorum (1/4)
        -F. C. Puddle-
Hopeful-            -griersonianum (1/4)
        -hookeri (1/2)

A red-flowered hybrid.  (Lord Aberconway, intro. 1946)


cl.
      -yakushimanum ssp. yakushimanum (1/2)
Hoppy-        -arboreum ssp. arboreum (1/4)
        -Doncaster-
                -unknown (1/4)

3ft(.9m)         -10°F(-23°C)          M

White flowers with greenish speckling on upper segment, 2in(5cm)
across; ball-shaped trusses of 18; free-flowering.  Plant vigorous, upright, compact; dark, dull green foliage.  See also Bashful, Pink Cherub, Sleepy, and Sneezy.  (Waterer, Sons & Crisp,
1972)  A. M. (RHS) 1977


cl.                  -griffithianum (1/4)
        -unnamed hybrid-
Horsham-            -unknown (3/4)
        -Monsieur Thiers--unknown

Deep red flowers.  (C. B. van Nes & Sons, before 1922)


cl.

Hot Pants    See Ooh-la-la


                -sanguineum ssp. didymum (1/4)
cl.    -Carmen-
       -     -forrestii ssp. forrestii Repens Group (1/4)
Hot Stuff*                    -griffithianum (1/16)
              -Queen Wilhelmina-
       -   -Britannia-            -unknown (3/16)
       -Leo-        -Stanley Davies--unknown
        -elliottii (1/4)

2ft(.6m)         5°F(-15°C)          M          3/4

Waxy, deep red flowers on a small plant.  Leaves very dark
green, highly polished.  Nice plant habit with strong branches.
(Harold E. Greer, intro. 1982)


                    -griffithianum (1/8)
cl.       -Pink Shell-        -fortunei ssp.
        -Rosabel-        -H. M. Arderne-    fortunei (1/16)
Hot Wonder-    -griersonianum (1/4)    -unknown (1/16)
        -arboreum ssp. arboreum (1/2)

Flowers are carmine with darker throat.  See also Delicious.
(G. Langdon, 1982)


cl.          -dichroanthum ssp. dichroanthum (1/4)
    -Goldsworth Orange-
Hotei-          -fortunei ssp. discolor (1/4)
    -          -souliei (1/4)
      -unnamed hybrid-
              -wardii var. wardii (1/4)

3ft(.9m)          5°F(-15°C)          M          5/4

Flowers canary yellow with darker throat, open-capanulate,
2.5in(6.4cm) wide, 6-lobed; spherical trusses hold 12.  Compact
plant; narrowly elliptic dark green leaves to 4.75in(12cm) long.
(Cross by K. Sifferman; raised by Glendoick Gardens;  B. Nelson,
intro. 1968)    P. A. (RHS) 1964;  A. M. (RHS) 1974    See color
illus. ARS Q 31:1 (Winter 1977): p. 35; HG p. 129; K p. 120.


cl.
      -facetum (eriogynum) (1/2)
Hotshot-    -griffithianum (1/4)
      -Mars-
          -unknown (1/4)

A dwarfish plant with red flowers.  (Rudolph Henny cross, 1944;
reg. 1958)


cl.
        -fortunei ssp. fortunei (1/2)
Hubert Robert-
        -williamsianum (1/2)

Caroline Spencer g.  Shell pink flowers.  (Adams-Acton, 1950)


cl.
        -minus var. minus Carolinianum Group (white) (1/2)
Hudson Bay-
        -dauricum (white) (1/2)

3ft(.9m)         -20°F(-29°C)          E

Terminal flower clusters of 3-4, each 3-flowered; white flowers
1.75in(4.5cm) wide, 5 wavy lobes; plant broad, rounded and well-
branched; glossy yellow green leaves with rusty brown scales beneath, held 2 years.  (David G. Leach, intro. 1979; reg. 1983)


                    -griffithianum (1/4)
cl.     -George Hardy-
        -          -catawbiense (1/4)
Hugh Koster-        -arboreum ssp. arboreum (1/8)
            -Doncaster-
        -unnamed-        -unknown (3/8)
          hybrid-unknown

6ft(1.8m)          5°F(-15°C)          ML          2/3

Medium-sized, fine crimson flowers with lighter center; trusses
rounded.  Upright and spreading growth; grooved leaves.  (M. Koster & Sons, 1915)  A. M. (RHS) 1933


cl.
        -unknown (3/4)
Hugo Casciola-        -ponticum (1/4)
        -Purple Splendour-
                -unknown

2.5ft(.75m)          0°F(-18°C)          M

Ball-shaped trusses of 10-15 flowers, medium reddish purple with
lighter margins, a large flare of Indian orange on dorsal lobe,
openly funnel-shaped, 3.25in(8.3cm) across, 5-6 wavy lobes.
Plant upright, well-branched, wider than tall; narrow leaves 5in
(12.7cm) long, held 3 years.  (Paul Holden, 1981)

cl.

```
         -cinnabarinum ssp. xanthocodon (A. M. form) (1/2)
Hulagu-      -cinnabarinum ssp. cinnabarinum Roylei Group (1/4)
  Khan -Lady  -          -cinnabarinum ssp. cinnabarinum (1/8)
     Rosebery-Royal Flush-
                         -maddenii ssp. maddenii (1/8)
```

Plant grew to 6ft(1.8m) in 9 years.  The truss has 10 pendulous flowers, bright Indian yellow, reverse deeper shade.  Leaves dark green, scaly below.  (Lester Brandt, 1969)


cl.

```
         -fortunei ssp. fortunei (1/2)
Hullaballoo-
         -thomsonii ssp. thomsonii (1/2)
```

Flowers rose Bengal with paler edges, darker staining on the reverse, small throat blotch of cardinal red, openly bell-shaped, 3.25in(8.4cm) wide; loose trusses of 10-12.  Elliptic dark green leaves.  (Cross by Sir Edmund Loder; raiser H. E. Boscawen, High Beeches)  A. M. (RHS) 1974


---

```
         -catawbiense (1/2)
Humboldt-
         -unknown (1/2)
```

Rose-colored flowers, dark markings.  (T. J. R. Seidel, 1926)


cl.

```
                    -ciliicalyx (1/4)
            -Else Frye-
Humboldt Sunrise-      -unknown (1/4)
            -johnstoneanum (1/2)
```

3ft(.9m)           15°F(-10°C)              EM

Flowers Naples yellow with darker blotch, openly funnel-shaped about 3.25in(8.3cm) across, in trusses of 6.  Free-flowering, with heavy carnation fragrance. Plant upright, wider than tall; leaves with hairy margins, held 2 years.  (Cross by Dr. Richard Anderson; intro. by Dr. J.H. Braaflad, 1972)  See color illus. ARS Q 33:1 (Winter 1979): cover.


cl.


Hummel--form of rufum

4ft(1.2m)          -5°F(-21°C)         EM         3/3

Flowers of this species are white or pinkish purple, spotted crimson.  Leaves have thick brown indumentum.  (Origin of this selection unknown)  A. M. (RHS) n.d.


g.

```
         -haematodes ssp. haematodes (1/2)
Humming Bird-
         -williamsianum (1/2)
```

2.5ft(.75m)        0°F(-18°C)         EM         3/4

Flowers a deeper rich rose red than the species williamsianum, of very heavy substance, bell-shaped; lax trusses of 4-5.  Plant compact, dense; leaves round, dark green, leathery.  Slow-growing; needs some shade.  (J. C. Williams, Caerhays Castle, 1933)  See color illus. F p. 48.


cl.

```
         -augustinii var. chasmanthum (1/2)
Hunter's Moon-
         -cinnabarinum ssp. xanthocodon Concatenans Group
                                                    (1/2)
```

Flowers dull white with greenish brown spotting, tubular, 1in (2.5cm) wide and 1.5in(3.8cm) long, held in trusses of about 8. (Gen. Harrison, 1963)


g.

```
         -barbatum (1/2)
Huntsman-
         -campylocarpum ssp. campylocarpum Elatum Group (1/2)
```

A parent of the Exbury hybrid, Nehru.  (Loder)

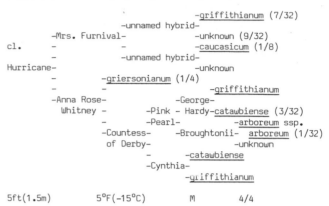

```
                             -griffithianum (7/32)
                -unnamed hybrid-
      -Mrs. Furnival-       -unknown (9/32)
cl.   -          -          -caucasicum (1/8)
      -           -unnamed hybrid-
Hurricane-       -           -unknown
      -       -griersonianum (1/4)
      -       -              -griffithianum
      -Anna Rose-       -George-
       Whitney -      -Pink - Hardy-catawbiense (3/32)
              -       -Pearl-      -arboreum ssp.
              -Countess-    -Broughtonii- arboreum (1/32)
               of Derby-           -unknown
              -        -catawbiense
              -Cynthia-
                       -griffithianum
```

5ft(1.5m)          5°F(-15°C)          M          4/4

Flowers a beautiful shade of pink, with markings of deeper pink, of good substance, 3in(7.6cm) across; ball-shaped trusses of 15.  Compact, vigorous plant; glossy leaves held 3 years.  (William Whitney cross; intro. & reg. by Anne & George Sather, 1976)  See color illus. VV p. 118.


```
                    -facetum (eriogynum) (1/4)
cl.  -unnamed hybrid-     -dichroanthum ssp. dichroanthum (1/8)
     -              -Fabia-
Hussar-                 -griersonianum (1/8)
     -        -haematodes ssp. haematodes (1/4)
     -May Day-
            -griersonianum (1/4)
```

4ft(1.2m)          5°F(-15°C)              M

Blood red flowers, upper lobe speckled brown; plant of compact habit.  Free-flowering.  (John Waterer, Sons & Crisp, reg. 1971)


g.

```
         -azalea viscosum (1/2)
Hybridum-
         -maximum (1/2)
```

Azaleodendron.  Fragrant yellow flowers, spotted and edged with pink.  (Herbert, 1817; offered by Methven, 1868-69)


cl.

```
         -yakushimanum ssp. yakushimanum (1/2)
Hydon Ball-       -griersonianum (1/4)
      -Springbok-       -ponticum (1/8)
              -unnamed hybrid-
                       -unknown (1/8)
```

3ft(.9m)          0°F(-18°C)              M

Low, compact habit, wider than tall; leaves 3.5in(9cm) long, dull dark green.  Trusses of 16-18 flowers, pale cream, spotted brownish maize yellow.  See also Hydon Glow, Hydon Hunter, Morning Cloud, Morning Magic.  (A. George, 1969)  A. M. (RHS) 1977


cl.

```
         -yakushimanum ssp. yakushimanum (1/2)
Hydon Glow-       -griersonianum (1/4)
      -Springbok-       -ponticum (1/8)
              -unnamed hybrid-
                       -unknown (1/8)
```

3ft(.9m)          0°F(-18°C)              M

Large, compact rounded truss of about 14 flowers, rosy pink with slight spotting. See also Hydon Ball. (A. F. George, Hydon Nurseries, 1969)

cl.                      -haematodes ssp. haematodes (1/4)
                   -May Day-
Hydon Harrier-          -griersonianum (1/4)
           -                -wardii var. wardii (1/4)
              -Jervis Bay-               -fortunei ssp. discolor
                       -Lady   -                      (1/8)
                        -Bessborough-campylocarpum ssp. campylo-
                                            carpum Elatum Group (1/8)

6ft(1.8m)          5°F(-15°C)           M

Orange scarlet flowers on an open-growing plant. (A. F. George,
Hydon Nurseries, 1975)


cl.
                 -yakushimanum ssp. yakushimanum (1/2)
Hydon Hunter-          -griersonianum (1/4)
           -Springbok-                -ponticum (1/8)
                       -unnamed hybrid-
                                 -unknown (1/8)

3ft(.9m)          5°F(-15°C)           M

Vigorous, upright, compact habit with leaves 4in(10cm) long by
1.5in(3.7cm) wide, dark green. Dome-shaped truss of 14 flowers,
white flushed with light orchid pink. (A. F. George, Hydon Nurseries, 1972) A. M. (RHS) 1976; F. C. C. (RHS) 1979


cl.
              -russatum (1/2)         -intricatum (1/8)
Hydon Mist-          -Intrifast-
        -Blue Diamond-          -fastigiatum (1/8)
                        -augustinii ssp. augustinii (1/4)

2ft(.6m)          -5°F(-21°C)          EM

Very deep violet flowers held in trusses of 6-7 on a dwarf,
compact plant. See also Blue Chip. (A. F. George, Hydon
Nurseries, 1976)


cl.
              -Moser's Maroon--unknown (1/2)
Hydon Pink-                -dichroanthum ssp. dichroanthum (1/8)
        -             -Dido-
        -Ice Cream-          -decorum (1/8)
                         -fortunei ssp. discolor (1/4)

4ft(1.2m)          0°F(-18°C)          ML

Flowers of pinkish lavender with a crimson throat; large compact
trusses of about 20. (A. F. George, Hydon Nurseries, 1969)


cl.
              -augustinii ssp. augustinii (3/4)
Hydon Rodney-          -russatum (1/4)
          -Azamia-
                  -augustinii ssp. augustinii

4ft(1.2m)          -5°F(-21°C)          EM

Flowers of vibrant violet with darker spotting. (A. F. George,
Hydon Nurseries, 1972)


cl.                          -fortunei ssp. discolor (1/4)
                   -unnamed hybrid-
Hydon Salmon-                -unknown (1/4)
             -griersonianum (1/2)

5ft(1.5m)          0°F(-18°C)           M

Flowers a glowing salmon red, unfading, about 2.5in(6.4cm) wide,
open funnel-shaped; full rounded trusses of 12-14. Matte green
narrow leaves to 6in(15.2cm) long. (A. F. George, Hydon Nurseries, 1970) A. M. (RHS) 1976


cl.
                   -orthocladum var. microleucum (1/2)
Hydon Snowflake-          -rupicola var. chryseum (1/4)
                -Chikor-
                        -ludlowii (1/4)

2ft(.6m)          0°F(-18°C)           EM

White flowers in trusses of 3. Dwarf plant of open habit. (A.
F. George, Hydon Nurseries, 1978)


                 -caucasicum (1/4)
cl. -Boule de Neige-          -catawbiense (3/8)
   -               -hardy hybrid-
Hymen-                          -unknown (3/8)
    -                 -catawbiense
     -Everestianum-
                   -unknown

Flowers light purple violet with yellow brown or greenish brown
markings. (T. J. R. Seidel cross, 1898; intro. 1906)


g.
       -Mrs. R. S. Holford--unknown (1/2)
Hypatia-
       -kyawii (1/2)

Red flowers on a tall plant, in late season. (Rothschild, 1940)


                      -arboreum ssp. arboreum (5/16)
g.      -Cardinal-                -arboreum ssp. arboreum
    -        -               -Glory of -
Hyperion-        -Barclayi-Penjerrick-griffithianum (1/16)
    -        -
    -               -thomsonii ssp. thomsonii (1/8)
     -forrestii ssp. forrestii Repens Group (1/2)

Blood red flowers. (Lord Aberconway, 1941)


cl.
       -hyperythrum (1/2)
Hypermax*
       -maximum (1/2)

Large flowers of pale lavender with purple freckles, held in big
trusses of 14. Plant medium-sized; hardy to -15°F(-26°C) (Weldon E. Delp)


cl.

I. M. S.      See Irene Stead


               -Moser's Maroon--unknown (1/4)
g. -Romany Chai-
   -               -griersonianum (1/4)
Iago-
   -               -fortunei ssp. discolor (1/4)
    -Lady-
     -Bessborough-campylocarpum ssp. campylocarpum Elatum Group
                                                   (1/4)
Rosy crimson flowers with darker spots; shrub about 5ft(1.5m),
upright. Midseason. (Rothschild, 1941)


cl.
         -Marion--unknown (1/2)
Ian Wallace-          -griersonianum (1/4)
          -Tally Ho-
                   -facetum (eriogynum) (1/4)

Light red flowers slightly spotted in throat. (A. Howells, 1972)

```
cl.             -griffithianum (1/4)
         -Isabella-
Iberia-         -auriculatum (1/4)
       -wardii var. wardii (1/2)
```

Well-filled trusses of 10-12 flowers, funnel-shaped and creamy
white, with conspicuous dark ruby crimson blotch in the throat,
widely funnel-campanulate, 2.75in(7cm) across.  Elliptic leaves
5in(12.7cm) long.  (Collinwood Ingram)   A. M. (RHS) 1971

```
g. & cl.
     -griersonianum (1/2)
Ibex-
     -pocophorum var. pocophorum (1/2)
```

5ft(1.5m)        5°F(-15°C)         EM         4/3

Flowers bright crimson scarlet with darker spots, funnel-shaped,
in dome-shaped trusses.  Plant of medium height; dark sage green
leaves with brown felted indumentum beneath.  Other forms have
flowers of rose carmine, or orange scarlet.  (Rothschild, 1941)
A. M. (RHS) 1948

```
                       -griffithianum (1/16)
                 -Kewense-
g.        -Aurora-      -fortunei ssp. fortunei (1/16)
    -Adelaide-    -thomsonii ssp. thomsonii (3/8)
Ibis-       -thomsonii ssp. thomsonii
    -griersonianum (1/2)
```

A rather compact plant about 5ft(1.5m) tall, bearing shapely
trusses of cherry pink flowers.  Early midseason.  (Rothschild,
1941)

```
g.          -campylocarpum ssp. campylocarpum (1/4)
     -A. Gilbert-
Icarus-         -fortunei ssp. discolor (1/4)
     -dichroanthum ssp. scyphocalyx Herpesticum Group (1/2)
```

Unusual coloring, with deep rose pink buds opening to biscuit-
colored, campanulate flowers shaded rose, described in some catalogs as "orange"; flattened trusses.  Plant small, low-growing,
compact.  (Rothschild)  A. M. (RHS) 1947

```
cl.        -dichroanthum ssp. dichroanthum (1/4)
      -Dido-
Ice Cream-    -decorum (1/4)
      -fortunei ssp. discolor (1/2)
```

Vigorous  plant, spreading  wider than tall, leaves 7.5in(19cm)
long, medium dull green.  Flowers of camellia rose, with white
throat and pale olive spots on upper petal, funnel-shaped, about
3.5in(8.9cm) wide; dome-shaped trusses of 12-14.  (W. C. Slocock,
1962)  A. M. (Wisley Trials) 1960

```
cl.
      -Catalgla--catawbiense var. album Glass (1/2)
Ice Cube-
      -              -catawbiense var. album
      -      -Catawbiense Album-        (3/8)
      -Belle Heller-        -unknown (1/8)
              -catawbiense var. album
```

4ft(1.2m)       -20°F(-29°C)        ML        4/3

Flowers a delightful ivory white with a blotch of lemon yellow,
funnel-shaped, 2.5in(6.4cm) wide, in conical trusses.  Olive
green foliage in profusion, on a vigorous, bushy plant.  (A. M.
Shammarello, 1973)  See color illus. LW pl. 84; VV p. 35.

```
cl.
      -auriculatum (1/2)
Iceberg*    -griffithianum (1/4)
      -Loderi-
              -fortunei ssp. fortunei (1/4)
```

A tall plant with long leaves, and large trusses of sweetly fragrant flowers, white with green center.  (Slocock)  A. M. (RHS)
1950

```
g.
     -Moser's Maroon--unknown (1/2)
Icenia-         -fortunei ssp. discolor (1/4)
       -Lady Bessborough-
                -campylocarpum ssp. campylocarpum Elatum
                                            Group (1/4)
```

Compact trusses of pink flowers on a medium-sized plant. (Rothschild, 1941)

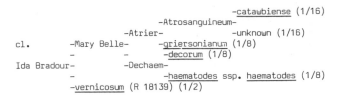

```
                            -catawbiense (1/16)
                      -Atrosanguineum-
               -Atrier-         -unknown (1/16)
cl.     -Mary Belle-    -griersonianum (1/8)
        -       -decorum (1/8)
Ida Bradour-    -Dechaem-
      -              -haematodes ssp. haematodes (1/8)
      -vernicosum (R 18139) (1/2)
```

Flowers in shades of pink  with a yellow green throat, saucer-
shaped, 3in(7.6cm) across; ball-shaped trusses hold 12-14.  Plant
upright with arching branches, broader than high; narrow leaves
held 2 years. (Joseph Gable cross; raiser Henry Yates; reg. by
Mrs. Yates, 1977)

```
                       -griffithianum (1/8)
g.        -Dawn's Delight-
   -Dolly-         -unknown (1/8)
Idaho-    -griersonianum (1/4)
   -elliottii (1/2)
```

Plant medium-sized and erect, with large trusses of brick red
flowers.  (Rothschild, 1941)

```
g. & cl.
      -wardii var. wardii (1/2)
      -              -griffithianum (1/16)
Idealist-       -Kewense-
      -Aurora-      -fortunei ssp. fortunei (5/16)
   -Naomi-    -thomsonii ssp. thomsonii (1/8)
      -fortunei ssp. fortunei
```

5ft(1.5m)        5°F(-15°C)         M         4/3

Upright, vigorous plant with leaves of medium green, 4.5in(11.5
cm) long.  Flowers pale greenish yellow, with a dark red throat,
widely campanulate, 3in(7.6cm) across.  (Rothschild, 1941)  A.M.
(RHS) 1945    See color illus. PB pl. 4 & 5; VV p. 97.

```
                   -griffithianum (3/8)
cl. -Loderi King George-
   -              -fortunei ssp. fortunei (1/4)
Idol-               -griffithianum
   -       -Queen Wilhelmina-
   -Britannia-         -unknown (3/8)
             -Stanley Davies--unknown
```

Rose flowers with a lighter center. (R. Henny)  P. A. (ARS) 1957

```
cl.
      -augustinii ssp. augustinii (Bodnant form) (1/2)
Ightham-
      -fastigiatum (1/2)
```

Augfast g.  Exhibited by G. Reuthe, 1952.

```
---
              -wardii var. wardii (1/2)
Ightham Yellow-
              -decorum (1/2)
```

Arthur Smith g.  Yellow flowers.  (Reuthe, 1952)

```
cl.
         -catawbiense (1/2)
Ignatius Sargent-
         -unknown (1/2)
```

5ft(1.5m)          -25°F(-32°C)          ML          2/2

A very hardy hybrid of open habit, with large leaves. Rose-colored flowers, large, slightly fragrant.  (Waterer, before 1900)
See color illus. VV p. 99.

```
cl.                   -griffithianum (1/4)
              -unnamed hybrid-
Ilam Alarm-           -unknown (1/4)
              -arboreum ssp. zeylanicum (1/2)
```

Bright red flowers, lighter in center.  (Stead, NZ)

```
                        -griffithianum (1/16)
                  -George-
              -Pink - Hardy-catawbiense (1/16)
cl.     -unnamed-Pearl-        -arboreum ssp. arboreum
        - hybrid-    -Broughtonii-              (1/16)
Ilam Apricot-                  -unknown (5/16)
        -        -arboreum ssp. zeylanicum (1/4)
        -              -dichroanthum ssp. dichroanthum
        -unnamed hybrid-                    (1/4)
                      -unknown
```

A hybrid by Stead in the 1940s?

```
cl.              -griffithianum (1/4)
          -Loderi-
Ilam Cream-      -fortunei ssp. fortunei (1/4)
          -unknown (1/2)
```

Trusses of 12-14 flowers, pink in bud, opening pale greenish
white, suffused pink around lobes. (Cross by Edgar Stead c.1950;
reg. by L. Roland Stead, 1981)

```
cl.
          -formosum var. formosum (1/2)
Ilam Pearl-
          -unknown (1/2)
```

Truss of 2-3 flowers, pure white, Tuscan yellow base inside, reverse tinted cardinal red. (Raised by Edgar Stead; reg. by Dunedin Rhododendron Group, 1981)

```
cl.
          -facetum (1/2)
Ilam Pink-      -griffithianum (1/4)
Splendour-Loderi-
                -fortunei ssp. fortunei (1/4)
```

Truss of 10-12 flowers, neyron rose shading lighter, with throat
of spinel red.  (Cross by E. Stead; reg. Mrs. R. J. Coker, 1982)

```
cl.              -augustinii ssp. chasmanthum (1/4)
          -Electra-
Ilam Violet-     -augustinii ssp. augustinii (1/4)
          -russatum (1/2)
```

4ft(1.2m)          5°F(-15°C)          M          4/3

Plant vigorous, upright; dark green leaves, slightly glossy, and
bronze in winter; foliage aromatic. Compact trusses of 11 flowers of deep violet blue. (Raiser W. T. Stead; sent for trial by
Slocock Nurseries)  A. M. (Wisley Trials) 1983   See color illus. VV p. 36.

```
g.      -dichroanthum ssp. dichroanthum (1/4)
    -Nereid-
Iliad-      -neriiflorum ssp. neriiflorum (1/4)
    -kyawii  (1/2)
```

A small, compact plant with deep green foliage, and blood red
flower clusters in late season. (Rothschild)  A. M. (RHS) 1949

```
                      -caucasicum (1/8)
              -unnamed hybrid-
cl.   -Mrs. Furnival-          -unknown (1/4)
      -              -griffithianum (1/8)
      Illahee-       -unnamed hybrid-
      -                    -unknown
      -      -fortunei ssp. discolor (1/4)
      -Evening-    -dichroanthum ssp. dichroanthum (1/8)
      Glow  -Fabia-
              -griersonianum (1/8)
```

3ft(.9m)          -5°F(-21°C)          ML

Flowers are 3in(7.6cm) across, lavender pink with yellow tinged
throat, orange yellow flare, held in flat trusses of 12.  Leaves
are 6in(15cm) by 2in(5cm). (Allen Van Veen, 1977)

```
g.              -Moser's Maroon--unknown (1/4)
      -Romany Chal-
Illyria-      -facetum (eriogynum) (1/4)
      -kyawii (1/2)
```

A tall plant with large leaves.  Crimson flowers appear late in
the season. (Rothschild, 1941)

```
g.
      -valentinianum (1/2)
Ilona-
      -auritum (1/2)
```

Many small trusses of pale golden yellow flowers.  Shrub blooms
early, and is not hardy. (Rothschild, 1941)

```
g. & cl.
      -impeditum (1/2)
Impeanum-
      -hanceanum (1/2)
```

One for the rock garden: almost prostrate--seldom as much as 1ft
(.3m) high.  Small flowers, cobalt blue. (RBG, Kew before 1932)
F. C. C. (RHS) 1934

```
cl.              -ponticum (1/2)
      -Purple Splendour-
Imperial*              -unknown (1/2)
      -              -mauve seedling--unknown
      -A. Bedford-
              -ponticum
```

4ft(1.2m)          -5°F(-21°C)          ML          4/2

Flower buds often appear on very young plants.  The blooms are
blue purple with a prominent black flare in the  throat.  Leaves
are glossy green with red petioles.  See also Plum Beautiful.
(Harold Greer, intro. 1982)

```
g. & cl.
      -Moser's Maroon--unknown (1/2)
Impi-
      -sanguineum ssp. didymum (1/2)
```

4.5ft(1.35m)          5°F(-15°C)          L          3/3

Medium-sized plant; deep ruddy green leaves on red petioles.
The black red flowers in loose trusses gleam brightly when back-
lighted by the afternoon sun. (Rothschild)  A. M. (RHS) 1945

```
cl.
      -macabeanum (1/2)
Ina Hair-
      -unknown (1/2)
```

Flowers rose madder in bud, fading to cream, with spotting of
rose madder; trusses of 20 flowers. (Raised at Pukeiti Rhododendron Trust; reg. by G. F. Smith, NZ, 1979)

g.
              -wardii var. wardii (1/2)
Inamorata-
              -fortunei ssp. discolor (1/2)

The foliage resembles discolor, with long, narrow leaves in fine
rosettes, but darker and more glossy; leaf tips and stalks are
purple. The flowers are shaped like wardii, of soft yellow or
sulphur yellow, and large. A tall vigorous plant. (Rothschild)
A. M. (RHS) 1950

                    -griffithianum (3/8)
cl.       -Mars-
          -     -unknown (3/8)
Inca Chief-                   -griffithianum
                        -Mars-
          -unnamed hybrid-  -unknown
                        -catawbiense var. rubrum (1/4)

5ft(1.5m)        -20°F(-29°C)           M

Flowers of strong purplish red, lighter at center, with a deeper
blotch, rotate-campanulate, to 3in(7.6cm) across; firm spherical
trusses of 17-18. Plant upright, almost as broad as tall; elliptical leaves of medium green, 5in(12.7cm) long.  See also
Scarlet Blast. (David G. Leach, 1972)

cl.              -wardii var. wardii ? (1/4)
          -Chlorops-
Inca Gold-       -vernicosum ? (1/4)
          -unknown (1/2)

3ft(.9cm)        -5°F(-21°C)          EM        4/4

Leaves of medium size. Flowers 3in(7.6cm) wide, barium yellow,
somewhat rayed with mahogany; trusses of 12. (Benjamin Lancaster, 1962)   P. A. (ARS) 1961

---                   -brookeanum var. gracile
          -Maiden's-                 (1/2)
          - Blush  -       -jasminiflorum (3/16)
Incarnatum -        -Princess-
Floribundum-        Royal  -javanicum (1/16)
          -                -brookeanum var. gracile
          -Prince Leopold-
                    -longiflorum (lobbii) (1/4)

Vireya hybrid. Rosy salmon flowers. F. C. C. (RHS) 1885

                    -arboreum ssp. arboreum (1/8)
g.            -Doncaster-
      -The Don-      -unknown (1/8)
Inchmery-   -griffithianum (1/4)
      -facetum (eriogynum) (1/2)

Waxy, deep pink flowers on a tall plant blooming late midseason.
(Rothschild, 1941)

cl.
          -catawbiense (1/2)
Indian Chief-
          -unknown (1/2)

Red flowers. (Warren Stokes, before 1958)

              -griersonianum (1/4)
cl.  -Sarita Loder-    -griffithianum (5/32)
     -          -Loderi-
Indian-              -fortunei ssp. fortunei (9/32)
Penny-    -wardii var. wardii (1/4)
     -     -                  -griffithianum
     -Idealist-       -Kewense-
     -        -Aurora-      -fortunei ssp. fortunei
          -Naomi-    -thomsonii ssp. thomsonii (1/16)
               -fortunei ssp. fortunei

Synonym Penny. Trusses of 12-14 flowers, light coppery aureolin
yellow. (Del W. James, reg. 1958)

g.
          -dichroanthum ssp. scyphocalyx (1/2)
Indiana-
          -kyawii (1/2)

4ft(1.2m)        10°F(-12°C)           L         3/3

Orange red flowers in medium-sized, lax trusses. A strong- and
slow-growing plant, spreading habit; glossy, dark green foliage
with edges sharply turned down. (Rothschild, 1941)

                    -wardii var. wardii (1/4)
          -Crest-          -fortunei ssp. discolor (1/8)
cl.       -      -Lady    -
          -      -Bessborough-campylocarpum ssp. campylocarpum
Indiridiva-                    Elatum Group (1/8)
          -              -maximum (1/8)
          -        -Halopeanum-
          -Snow Queen-      -griffithianum (1/4)
          -        -      -griffithianum
                 -Loderi-
                    -fortunei ssp. fortunei (1/8)

Flowers pale chartreuse yellow deeper in throat, with dark yellow green markings, widely funnel-campanulate, to 4in(10.2cm)
across, 7-lobed; firm rounded trusses of about 10. Dull green,
oblong leaves 6.5in(15.8m) long. (Edmund de Rothschild)  P. C.
(RHS) 1978

g.
          -souliei (1/2)
Indomitable-      -thomsonii ssp. thomsonii (1/4)
          -General Sir -
          John du Cane-fortunei ssp. discolor (1/4)

Flower similar to parent souliei: large, white, flushed pink, in
medium-sized trusses; a tall, compact plant. (Rothschild, 1941)

g. & cl.       -griffithianum (1/4)
          -Isabella-
Infanta-       -auriculatum (1/4)
          -griersonianum (1/2)

Flowers of light crimson deepening to rose madder at base of the
tube. Another form has huge white flowers flushed in throat with
salmon pink. (Collingwood Ingram, before 1956)

cl.
          -catawbiense (1/2)
Ingeborg-
          -unknown (1/2)

Dark crimson flowers, lighter in throat, with umber spotting.
(T. J. R. Seidel)

                    -griffithianum (1/8)
          -Isabella-
cl.   -Infanta-      -auriculatum (1/8)
      -        -griersonianum (1/4)
Ingos -               -fortunei ssp. discolor (1/8)
Rubellum-    -Lady
      -Jalisco-Bessborough-campylocarpum ssp. campylocarpum
      -Goshawk-                  Elatum Group (1/8)
          -      -dichroanthum ssp. dichroanthum (1/8)
          -Dido-
               -decorum (1/8)

Flowers of medium neyron rose, fading lighter, with a dark red
blotch. (Collingwood Ingram, 1973)

g.
      -insigne (1/2)
Ingre-
      -griersonianum (1/2)

Deep pink flowers. (Lord Aberconway, intro. 1936)

'Golden Anniversary' by Lancaster
Photo by Greer

'Golden Gala' by Leach
Photo by Leach

'Golden Gate' by unknown
(*dichroanthum* ssp. *scyphocalyx* ×)
Photo by Greer

'Golden Star'
Photo by Greer

'Golden Horn' by Rothschild
Photo by Greer

'Golden Torch' by J. Waterer & Crisp
Photo by Greer

'Golden Wit' by Witt/Michaud
Photo by Greer

'Goldendale' by McNew
Photo by Greer

'Goldfort' by Slocock
Photo by Greer

'Goldilocks' by Kerrigan
Photo by Greer

'Goldstrike' by R. Henny
Photo by Greer

'Goldsworth Orange' by Slocock
Photo by Greer

'Goldsworth Pink' by Slocock
Photo by Greer

'Goldsworth Yellow' by Slocock
Photo by Leach

'Gomer Waterer' by J. Waterer
Photo by Greer

'Good Hope' by Leach
Photo by Leach

'Good News' by J. Henny
Photo by Greer

'Goosander' by Cox
Photo by Greer

'Grace Seabrook' by Seabrook
Photo by Greer

'Graf Zeppelin' by C. B. van Nes
Photo by Greer

'Great Lakes' by Leach
Photo by Leach

'Great Scott' by Greer
Photo by Greer

'Greeley' by Greer
Photo by Greer

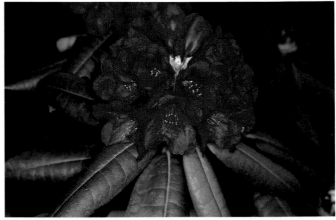

'Grenadier' by Rothschild
Photo by Greer

```
g.                -griersonianum (1/4)
      -Tally Ho-
Ingrid-        -facetum (eriogynum) (1/4)
      -griffithianum (1/2)
```

Large pink flowers held in open trusses on a tall shrub. Late-flowering. (Rothschild, intro. 1941)

```
g.         -insigne (1/4)
      -Ingre-
Insun-        -griersonianum (1/2)
      -            -griffithianum (1/4)
      -Sunrise-
                  -griersonianum
```

Pink-flowered. (Lord Aberconway, intro. 1946)

```
---
           -ferrugineum (1/2)
Intermedium-
           -hirsutum (1/2)
```

A natural hybrid, 1891.

```
g.                -Essex Scarlet--unknown (1/4)
      -Beau Brummell-
Intrepid-         -facetum (eriogynum) (1/4)
      -kyawii (1/2)
```

A tall plant with rosy red flowers in compact trusses. Blooms late midseason. (Rothschild, 1941)

```
g.
      -intricatum (1/2)
Intrifast-
      -fastigiatum (1/2)
```

A dwarf with clusters of small, violet blue, early-blooming flowers; new foliage is bright metallic blue. (Lowinsky, 1958)

```
            -yakushimanum ssp. yakushimanum (1/4)
      -unnamed-                     -griffithianum (1/4)
cl.   - hybrid-        -Queen Wilhelmina-
      -      -Britannia-            -unknown (3/8)
Invicta-              -Stanley Davies--unknown
      -              -griffithianum
      -      -Loderi-
      -unnamed-      -fortunei ssp. fortunei (1/8)
       hybrid-            -griffithianum
                    -Queen Wilhelmina-
             -Britannia-            -unknown
                    -Stanley Davies--unknown
```

4ft(1.2m)        0°F(-18°C)        ML

An upright plant with pale purple flowers. (John Waterer, Sons & Crisp, 1975)

```
g.
      -valentinianum (1/2)
Iola-
      -edgeworthii (1/2)
```

Clear primrose yellow flowers with a darker eye; trusses of 2-3. Grey green leaves, very hairy. A small, tender plant. (Rothschild, 1941)

```
                        -griffithianum (1/8)
g.              -Godesburg-
      -Blanc-mange-          -unknown (1/8)
Iolanthe-         -auriculatum (1/4)
      -kyawii (1/2)
```

Large cerise-colored flowers. Tall, tender, and very late flowering. (Rothschild, 1941)

```
g.               -ciliatum (1/4)
      -Countess of Haddington-
Ione-            -dalhousiae var. dalhousiae (1/4)
      -edgeworthii (bullatum) (1/2)
```

Flowers open pale primrose yellow and fade to white. (E. J. P Magor, 1926)

```
g.
           -haematodes ssp. haematodes (1/2)
Iphigenia-            -arboreum ssp. arboreum (1/4)
      -Red Admiral-
                  -thomsonii ssp. thomsonii (1/4)
```

Red flowers. (E. J. P. Magor, intro. 1934)

```
cl.
           -yakushimanum ssp. yakushimanum (1/2)
Irene Bain-        -haematodes ssp. haematodes (1/4)
      -May Day-
                  -griersonianum (1/4)
```

Pale pink flowers fading to cream, with pink frilled margins. (Dr. Yeates cross; intro. & reg. by Mrs. Irene Bain, NZ, 1979)

```
                        -griffithianum (3/8)
cl.   -Loderi King George-
      -                 -fortunei ssp. fortunei (1/4)
Irene Hall-            -griffithianum
      -            -George Hardy-
      -Pink Pearl-            -catawbiense (1/8)
      -            -arboreum ssp. arboreum (1/8)
                  -Broughtonii-
                         -unknown (1/8)
```

Trusses of 11-12 flowers, white flushed pink outside, upper petals slightly spotted, with a small deep red mark at the base. See also Fiona Wilson. (Geoffrey Hall; intro. by Lord Harewood; reg. 1979)

```
cl.
                  -griffithianum (1/2)
Irene Stead--Loderi (selfed)-
                  -fortunei ssp. fortunei (1/2)
```

Synonym I.M.S. A shrub similar to the Loderi group, with pink flowers. (W. T. Stead, NZ, reg. 1958)

```
cl.
      -yakushimanum ssp. yakushimanum (Exbury form) (1/2)
Irish Yaku-   -griffithianum (1/4)
      -Mars-
            -unknown (1/4)
```

3ft(.9m)        -10°F(-23°C)        M

Buds of bright neyron rose opening to very light neyron rose. Truss of 22 flowers, 6-lobed, openly funnel-shaped. Plant semi-dwarf, broader than tall; leaves 6in(15cm) long with light brown indumentum beneath. (Raiser Comerford Nursery; intro. by P. J. McGuiness; reg. 1983)

```
g.               -maximum (1/4)
      -Midsummer-
Ironside-        -unknown (1/4)
      -kyawii (1/2)
```

A tall plant with large crimson flowers, late-blooming. (Rothschild, 1941)

```
cl.                  -griffithianum (1/4)
                  -Mars-
Irresistible Impulse*   -unknown (1/2)
      -                  -ponticum (1/4)
                  -Purple Splendour-
                        -unknown
```

5ft(1.5m)        -15°F(-26°C)        ML        4/4

Large trusses of deep red purple cover the plant; flowers have a
large dark purple blotch on upper lobe.  Like many other Purple
Splendour hybrids the plant flowers very young. Leathery leaves
of deep green. (Olin Dobbs cross; intro. by Greer, 1982)   See
color illus. HG p. 100.

---

```
              -irroratum ssp. irroratum (1/2)
Irrifarg-
              -oreodoxa var. fargesii (1/2)
```

Light pink flowers, heavily spotted with crimson.  (E. J. P. Magor, intro. 1926)

```
g.                        -thomsonii ssp. thomsonii (1/4)
              -unnamed hybrid-
Isaac Newton-             -catawbiense (1/4)
              -forrestii ssp. forrestii Repens Group (1/2)
```

Flowers of carmine red.  (Dietrich Hobbie, 1952)

```
g.
        -maddenii ssp. maddenii (calophyllum) (1/2)
Isabel-
        -maddenii ssp. maddenii (1/2)
```

Flowers of light rose.  Early.  (Lord Aberconway, 1947)

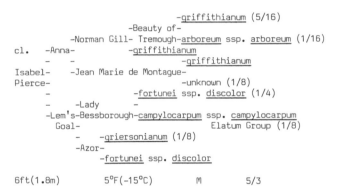

```
                            -griffithianum (5/16)
                 -Beauty of-
            -Norman Gill- Tremough-arboreum ssp. arboreum (1/16)
cl.   -Anna-        -griffithianum
        -      -           -griffithianum
Isabel-    -Jean Marie de Montague-
Pierce-                     -unknown (1/8)
        -                  -fortunei ssp. discolor (1/4)
        -     -Lady
        -Lem's-Bessborough-campylocarpum ssp. campylocarpum
          Goal-                     Elatum Group (1/8)
              -        -griersonianum (1/8)
              -Azor-
                   -fortunei ssp. discolor
```

6ft(1.8m)         5°F(-15°C)          M          5/3

Flowers of rich pink, lighter in center,  prominent brown blotch
and spotting in throat, narrow stripes of deep pink from base to
edges, openly campanulate, 4in(10.2cm) wide, wavy margins; about
10 in a truss. Glossy deep green leaves, with a yellow midvein.
(Cross by Halfdan Lem; Lawrence Pierce, reg. 1975)      See color
illus. ARS Q 29:4 (Fall 1975): p. 238;  K p. 36.

```
g. & cl.
        -griffithianum (1/2)
Isabella-
        -auriculatum (1/2)
```

Clear pink flowers.  See also Exbury Isabella, Isabella Nevada.
A Leonardslee form has very large, white, fragrant flowers on a
tall bush. (Col. G. H. Loder, High Beeches, 1934)

```
cl.
              -griffithianum (1/2)
Isabella Mangles-
              -unknown (1/2)
```

A tall plant with large leaves; big trusses of light pink, trumpet-shaped flowers.  (J. H. Mangles, before 1880)

```
cl.
              -griffithianum (1/2)
Isabella Nevada-
              -auriculatum (1/2)
```

Isabella g.  Pink flowers. (Collingwood Ingram, 1936)

cl.

Island Gem--form of oreotrephes

5ft(1.5m)            10°F(-12°C)           EM          3/4

Orchid flowers with a reddish blotch, in terminal inflorescence
of 3 buds, each 4-flowered.  Leaves typical of the species.  The
flowers are similiar in color to the award-winning Exbury form
(A.M. 1937). (Seed from RHS; raiser H. A. Short; reg. 1978)

cl.

Island Sunset--form of zoelleri

A vireya with flowers medium Indian yellow in tube, shading into
flame red on lobes.  (Don B. Stanton, New Guinea, reg. 1973)

```
g.
      -wardii var. wardii (1/2)
Isme-
      -venator (1/2)
```

Flowers yellow with rose-colored spots, in small trusses.  Medium-sized, attractive plant.  (Rothschild, 1941)

```
cl.
              -fortunei ssp. discolor (1/2)
Isobel Baillie-                     -griffithianum (1/8)
      -                -George Hardy-
      -Betty Wormald-          -catawbiense (1/8)
                      -red hybrid--unknown (1/4)
```

Flowers ruby red with flare of red spots.  (Haworth-Booth, 1972)

```
g.         -dichroanthum ssp. dichroanthum (1/4)
      -Fabia-
Ispahan-      -griersonianum (1/4)
      -wardii var. wardii (1/2)
```

Plant low and broad.  Waxy flowers, orange to yellow in loose
trusses.  Blooms midseason.  (Rothschild, 1941)

```
g.
      -Mrs. H. Stocker--unknown (1/2)
Istanbul-
      -elliottii (1/2)
```

A tall hybrid, with full red trusses.  Late.  (Rothschild, 1941)

```
g.  -dichroanthum ssp. dichroanthum (1/2)
    -                -griffithianum (1/16)
Istar-        -Kewense-
    -   -Aurora-        -fortunei ssp. fortunei (5/16)
    -Naomi-    -thomsonii sp. thomsonii (1/8)
          -fortunei ssp. fortunei
```

Low, compact plant.  Yellow flowers, in loose trusses.  (Rothschild, 1941)

```
g.         -dichroanthum ssp. dichroanthum (1/4)
    -Astarte-        -campylocarpum ssp. campylocarpum
Istria-    -Penjerrick-        Elatum Group (1/8)
    -                -griffithianum (5/8)
    -griffithianum
```

Pale pink flowers.   (Lord Aberconway, 1946)

cl.

Ita--form of hirtipes (L S & T  3624)

5ft(1.5m)         5°F(-15°C)           EM

Loose trusses of about 5 flowers, 5-lobed, funnel-shaped, phlox pink, stained and striped deeper shades. (Collectors Ludlow, Sherriff, Taylor; shown by A. & J. F. Gibson) A. M. (RHS) 1965

```
g.              -fortunei ssp. fortunei (1/4)
     -Duke of York-    -catawbiense (1/8)
Italia-         -Scipio-
     -                  -unknown (1/8)
     -griersonianum (1/2)
```

Flowers of salmon red. (Dietrich Hobbie, 1954)

```
g.              -catawbiense (1/4)
     -B. de Bruin-
Ivan-              -unknown (1/4)
     -kyawii (1/2)
```

Plant tall, upright. Large bright red flowers. Late. (Rothschild, 1941)

```
                    -fortunei ssp. discolor (1/8)
             -King of-     -dichroanthum ssp. dichroanthum
     -unnamed- Shrubs-Fabia-                        (3/8)
cl.  - hybrid-          -griersonianum (1/8)
     -         -          -fortunei ssp. fortunei (1/8)
Ivan D.-   -Fawn-    -dichroanthum ssp. dichroanthum
Wood  -         -Fabia-
     -                  -griersonianum
     -    -dichroanthum ssp. dichroanthum
     -Dido-
         -decorum (1/4)
```

3ft(.9m)          5°F(-15°C)          ML

Flowers light neyron rose at edges, orange buff in center fading to Naples yellow. Plant of vigorous habit. (Mrs. R. J. Coker, New Zealand, 1979)

```
g. & cl.        -thomsonii ssp. thomsonii (1/4)
     -Chanticleer-
Ivanhoe-         -facetum (eriogynum) (1/4)
     -griersonianum (1/2)
```

6ft(1.8m)          5°F(-15°C)          M          2/2

Glistening, brilliant scarlet flowers, spotted with deeper red, funnel-shaped; flat-topped trusses. Plant tall, of loose habit. Blooms early to midseason. (Rothschild, 1941) A. M. (RHS) 1945 See color illus. PB pl. 57.

cl.

Ivery's Scarlet      Parentage unknown

6ft(1.8m)          10°F(-12°C)          EM          3/3

Synonym Ivorianum. Brilliant red flowers in small, dome-shaped trusses. Growth habit upright, inclined to be willowy. Narrow leaves of light green, 5.5in(14cm) long. A parent of Alvinda, and others. (Origin unknown; reg. 1958) See color illus. F p. 50; PB pl. 55.

```
g.              -dichroanthum ssp. dichroanthum (1/4)
     -Fabia-
Iviza-     -griersonianum (1/4)
     -          -auriculatum (1/4)
     -Bustard-     -campylocarpum ssp. campylocarpum
          -Penjerrick-                Elatum Group (1/8)
               -griffithianum (1/8)
```

Plant about 5ft(1.5m) tall. Leaves spoon-shaped, glossy green. Coral yellow buds opening to orange salmon tubes, like Fabia, in lax trusses. A clone, Philomene, has straw-color flowers tinted pink with green markings. (Rothschild, intro. 1941)

cl.

Ivorianum      See Ivery's Scarlet

```
cl.              -wardii var. wardii ? (1/4)
          -Chlorops-
Ivory Bells-      -vernicosum ? (1/4)
     -williamsianum (1/2)
```

3ft(.9m)          -5°F(-21°C)          4/4

Chinese yellow buds opening straw yellow; flowers flattened, to 3in(7.6cm) wide, 7-10 in graceful trusses. Plant sturdy, compact; stiff oval leaves. (Benjamin Lancaster, 1966)

```
                         -griffithianum (3/16)
               -George Hardy-
          -Mrs. Lindsay-     -catawbiense (1/16)
cl.  -Diane- Smith   -Duchess of Edinburgh--unknown (1/4)
     -      -          -campylocarpum ssp. campylocarpum
Ivory-    -unnamed hybrid-                    (3/8)
Queen-         -unknown
     -    -campylocarpum ssp. campylocarpum
     -Phyrne-     -griffithianum
          -Loderi-
               -fortunei ssp. fortunei (1/8)
```

4ft(1.2m)          0°F(-18°C)          EM

Large, creamy white flowers, with red blotch. Attractive glossy green foliage. (Waterer, Sons & Crisp, 1965)

```
cl.
          -catawbiense var. album (1/2)
Ivory Tower-          -wardii var. wardii (1/4)
     -unnamed hybrid-
               -fortunei ssp. fortunei (1/4)
```

5ft(1.5m)          -25°F(-32°C)          EM          4/3

Ivory flowers, with dorsal blushing and greenish yellow stripes, 3in(7.6cm) wide, 5- to 7-lobed; ball-shaped trusses hold about 13. Plant as broad as tall with leaves 4.5in(11.5cm) long. See also Limelight. (David G. Leach, 1964)

```
cl.
     -cinnabarinum ssp. cinnabarinum Roylei Group (1/2)
Ivy-          -cinnabarinum ssp. cinnabarinum (1/4)
     -Royal Flush-
     (orange form)-maddenii ssp. maddenii (1/4)
```

5ft(1.5m)          10°F(-12°C)          ML          4/3

Lady Chamberlain g. Salmon-colored flowers. One of ten named clones in this family. (Rothschild, 1930)

```
cl.
          -falconeri ssp. falconeri (1/2)
J. E. Harris-
          -thomsonii ssp. thomsonii (1/2)
```

Surprise g. See also Faltho. (J. G. E. Harris, 1940)

```
cl.
          -Ivery's Scarlet--unknown (1/2)
J. Edgar Hoover-     -dichroanthum ssp. dichroanthum (1/4)
          -Francis-     -griffithianum (1/8)
          Hanger-Isabella-
                    -auriculatum (1/8)
```

Plant of medium size; large, light green leaves. White flowers edged pink, outside rose madder to Turkey red, campanulate, to 4 in(10.2cm) across; loose trusses, but flowers upstanding. Late. (C. S. Seabrook, 1965)

```
                              -thomsonii ssp. thomsonii (1/4)
cl.          -Ascot Brilliant-
             -                 -unknown (3/8)
J. G. Millais-                 -griffithianum (1/8)
                  -George Hardy-
             -Pink Pearl-        -catawbiense (1/8)
             -                   -arboreum ssp. arboreum (1/8)
                  -Broughtonii-
                              -unknown

6ft(1.8m)          5°F(-15°C)          E          3/2
```

Flowers deep blood red with heavy dark spotting, held in rounded
trusses.  Plant tall, upright, vigorous, of open habit; medium
green leaves.  (J. Waterer)


cl.


J. H. Agnew      Parentage unknown

A parent of the Exbury hybrid, Brenda.  (Origin unknown)


cl.
            -Monsieur Thiers--unknown (1/2)
J. H. van Nes-
            -griffithianum (1/2)

5ft(1.5m)          -5°F(-21°C)          M          3/3

Plant habit is rather compact; pale green leaves.  Flowers of
glowing soft red, lighter in center, about 2.5in(2.6cm) across,
with pointed lobes; conical, compact trusses.  Needs some shade.
(C. B. van Nes, before 1958)    See color illus. VV p. 82.


cl.
            -nivale ssp. nivale (F 16450) (1/2)
J. Hutton Edgar-
            -unknown (1/2)

Truss holds 3 bluish purple flowers.  (Collector George Forrest;
raised by Col. G. H. Loder; intro. 1919; reg. by The Hon. H. E.
Boscawen, High Beeches, 1982)


cl.
                         -arboreum ssp. arboreum (1/8)
            -Doncaster-
            -unnamed-       -unknown
J. J. de Vink- hybrid-unknown (5/8)
            -        -griffithianum (1/4)
            -unnamed-
             hybrid-unknown

Trusses of 11-13 flowers, rose red with brown blotch.  (M. Koster & Sons) A. M. (Wisley Trials) 1946


cl.


J. W. Fleming      See John Willis Fleming


cl.
            -veitchianum Cubittii Group (1/2)
Jabberwocky-
            -moupinense (1/2)

2ft(.6m)          15°F(-10°C)          E

A plant for the cool greenhouse.  Flowers of light violet pink
with dark reddish pink spotting in upper throat, funnel-shaped,
to 1.5in(3.8cm) wide; trusses of 3.  (A. F. George, Hydon Nurs-
ries; reg. 1976)  P. C. (RHS) 1975


cl.


Jack Hext--form of spinuliferum

4ft(1.2m)          5°F(-15°C)          EM          3/3

Currant red axillary flowers in threes and fours held in loose
terminal clusters.  Elliptic leaves, sparsely scaly.  New growth
has soft hairs.  (Collector unknown; raiser N. T. Holman; reg.
1974)  A. M. (RHS) 1971


```
                         -arboreum ssp. arboreum (1/8)
cl.          -Doncaster-
            -Borde Hill-       -unknown (3/4)
Jack Lyons-  -                 -griffithianum (1/8)
            -        -Mrs. A. M. Williams-
            -Rose Red--unknown          -unknown
```

Plant of compact habit; large, oval, pointed leaves.  Flowers of
crimson to cardinal red, campanulate; ball-shaped trusses of 13.
Midseason.  (Marshall Lyons, 1972)


```
cl.          -unnamed-Catalgla--catawbiense var. album Glass
            - hybrid-                            (1/4)
Jack Owen Yates-    -wardii var. wardii (1/4)
            -    -griffithianum (1/4)
                 -Mars-
                    -unknown (1/4)

3ft(.9m)          -10°F(-23°C)          ML
```

Upright plant, with arching branches; medium green leaves to 7in
(17.8cm) long, held 2 years.  Flowers light burgundy  shading to
white center, with 2 yellow flares, openly funnel-shaped, about
2.5in(6.4cm) across; conical trusses of 12-13.  (Raiser Henry
Yates; reg. by Mrs. Yates, 1977)


```
                         -dichroanthum ssp. dichroanthum
g. & cl.        -Fabia-                            (1/8)
            -Erebus-       -griersonianum (5/8)
Jack the Ripper-    -griersonianum
            -           -facetum (1/4)
                 -Jacquetta-
                    -griersonianum
```

Blood red flowers  with faint spots on upper lobes.  (Lord Aberconway)  A. M. (RHS) 1949


```
cl.
            -caucasicum (3/4)
Jacksonii-       -caucasicum
            -Nobleanum-
                 -arboreum ssp. arboreum (1/4)
```

Medium-sized bush, rounded,  slow growing.  Flowers bright rose
pink with maroon markings and paler spots, widely funnel-shaped,
in large, well-formed trusses.  Floriferous; blooms early.  Tolerates industrial pollution.  (Herbert, 1835)


```
cl.
            -williamsianum (1/2)
Jackwill*
            -unknown (1/2)

4ft(1.2m)          -10°F(-23°C)          EM
```

A recent import with rounded leaves, flowers of light rose pink.
(Dietrich Hobbie)


```
g.              -Moser's Maroon--unknown (1/4)
            -Bibiani-
Jacobean-       -arboreum ssp. arboreum (1/4)
            -sanguineum ssp. sanguineum var. haemaleum (1/2)
```

Plant about 5ft(1.5m) tall, bearing medium-sized flowers of deep
maroon.  (Rothschild, 1942)


```
cl.
            -griersonianum (1/2)
Jacob's Red-
            -unknown (1/2)
```

A hybrid with bright scarlet flowers.  (Knap Hill, reg. 1962)

g.
```
        -facetum (1/2)
Jacqueline-                  -griffithianum (1/8)
            -Loderi-
       -Albatross-      -fortunei ssp. fortunei (1/8)
              -fortunei ssp. discolor (1/4)
```

Tall, compact plant. Large pink flowers, rather late. (Rothschild, 1942)

g.
```
        -dichroanthum ssp. dichroanthum (1/2)
Jacques-                  -fortunei ssp. discolor (1/8)
            -Lady
       -Day Dream-Bessborough-campylocarpum ssp. campylocarpum
                    -                            Elatum Group (1/8)
              -griersonianum (1/4)
```

Flowers pink to orange; medium-sized bush. (Rothschild, 1942)

g.
```
        -facetum (1/2)
Jacquetta-
        -griersonianum (1/2)
```

Funnel-shaped red flowers, in loose trusses. A large plant with
dull, dark green foliage. Midseason. (Lord Digby, 1953)

cl.
```
        -dichroanthum ssp. dichroanthum (1/4)
    -Fabia-
Jade-       -griersonianum (1/4)
    -Corona--unknown (1/2)
```

| | | | |
|---|---|---|---|
| 3ft(.9m) | 5°F(-15°C) | EM | 3/3 |

Very colorful flowers, a combination of orange, pink, and later
greenish yellow; in tight trusses. Leaves 3.5in(8.9cm) long by
2in(5cm) wide. (Rudolph Henny, 1958)

cl.
```
        -forrestii ssp. forrestii Repens Group
    -Jaipur-                           (1/4)
J'aime-     -meddianum var. meddianum (1/4)
    -            -haematodes ssp. haematodes (1/4)
    -May Day (Reuthe's)-
                    -griersonianum (1/4)
```

Compact, spreading habit; leaves 3in(7.6cm) long by 1.5in(3.8cm)
with thin indumentum. Lax truss of 8-12 tubular bell-shaped
flowers, orient red with faint crimson speckling. (Lester
Brandt raiser; reg. 1964 by C. E. Simons)

g.
```
        -forrestii ssp. forrestii Repens Group (1/2)
Jaipur-
        -meddianum var. meddianum (1/2)
```

| | | | |
|---|---|---|---|
| 3ft(.9m) | 5°F(-15°C) | E | 3/2 |

Flowers of heavy substance, of deep rich crimson, medium-sized,
drooping. Blooms young; difficult to grow. (Rothschild, 1942)

cl.
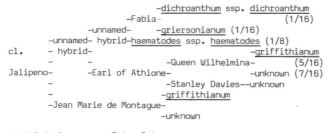
```
                          -dichroanthum ssp. dichroanthum
                 -Fabia-                          (1/16)
            -unnamed-      -griersonianum (1/16)
        -unnamed- hybrid-haematodes ssp. haematodes (1/8)
cl.    - hybrid-                       -griffithianum
        -          -           -Queen Wilhelmina-     (5/16)
Jalipeno-    -Earl of Athlone-            -unknown (7/16)
        -                       -Stanley Davies--unknown
        -                       -griffithianum
        -Jean Marie de Montague-
                              -unknown
```

| | | | |
|---|---|---|---|
| 2.5ft(.75m) | 5°F(-15°C) | EM | |

Flowers bright red  with dark brown nectaries and spots, widely
funnel-campanulate, 2in(5cm) across; conical trusses of 16-18.
Floriferous.  Plant rounded, well-branched, almost as broad as
tall; leaves bullate, dark green and held 2-3 years. (Dr. David
Goheen, 1977)

g. & cl.
```
            -fortunei ssp. discolor (1/4)
    -Lady -
    -Bessborough-campylocarpum ssp. campylocarpum
Jalisco-                    Elatum Group (1/4)
    -      -dichroanthum ssp. dichroanthum (1/4)
    -Dido-
        -decorum (1/4)
```

| | | | |
|---|---|---|---|
| 4ft(1.2m) | 5°F(-15°C) | ML | 4/4 |

A plant often  used in hybridizing.  Straw yellow flowers, with
some dark red spotting, 3.5in(8.9cm) across; trusses loose and
drooping, of 8-10 flowers.  Broad, rich green leaves.  (Rothschild, 1942)

cl.

Jalisco Eclipse      Parentage as above

| | | | |
|---|---|---|---|
| 4ft(1.2m) | 5°F(-15°C) | ML | 4/3 |

Jalisco g.  Very like the others of the family.  Flowers primrose yellow, blotched and spotted crimson at base inside, crimson streaks outside. (Rothschild, 1942)  A. M. (RHS) 1948

cl.

Jalisco Elect      Parentage as above

| | | | |
|---|---|---|---|
| 4ft(1.2m) | 5°F(-15°C) | ML | 4/3 |

Jalisco g.  Flowers primrose yellow with paler lobes, brownish
red spotting within.  Loose trusses of about 10. (Rothschild,
1942)  A. M. (RHS) 1948

cl.

Jalisco Emblem      Parentage as above

| | | | |
|---|---|---|---|
| 4ft(1.2m) | 5°F(-15°C) | ML | 4/3 |

Jalisco g.  This clone has pale yellow flowers, with a dark basal blotch. (Rothschild, 1948)

cl.

Jalisco Goshawk      Parentage as above

| | | | |
|---|---|---|---|
| 4ft(1.2m) | 5°F(-15°C) | ML | 4/3 |

Jalisco g.  Considered the best of this grex.  Flowers of mimosa
yellow, spotted crimson. (Rothschild)  F. C. C. (RHS) 1954

cl.

Jalisco Janet      Parentage as above

| | | | |
|---|---|---|---|
| 4ft(1.2m) | 5°F(-15°C) | ML | 4/3 |

Jalisco g.  Apricot yellow flowers. (Rothschild, 1948)

cl.

Jalisco Jubilant      Parentage as above

| | | | |
|---|---|---|---|
| 4ft(1.2m) | 5°F(-15°C) | ML | 4/3 |

Jalisco g.  A fine late addition to this grex.  Up to 14 buttercup yellow flowers with faint green spots in the throat, each
held in a yellow  calyx,  make attractive trusses. (Edmund de
Rothschild) A. M. (RHS) 1966      See color illus. ARS J 36:4
(Fall 1982): p. 139.

g.
```
                              -griffithianum (1/8)
             -Dawn's Delight-
    -Break of Day-            -unknown (1/8)
Jamaica-          -dichroanthum ssp. dichroanthum (1/4)
    -facetum (eriogynum) (1/2)
```

Low to medium-sized spreading plant, with trusses of deep orange
red campanulate flowers.  Early to midseason. (Rothschild, 1942)

cl.
```
              -orbiculare ssp. orbiculare ? (1/2)
James Barto-
              -williamsianum ? (1/2)
```

5ft(1.5m)        -5°F(-21°C)        EM        3/4

Rather compact habit; tough-textured elliptic leaves, 3in(7.6cm)
long.  Flowers funnel-shaped, fuchsine pink, slightly fragrant,
in loose trusses of 3-5.  (James Barto cross; shown by  Clarence
Prentice)  P. A. (ARS) 1953


cl.
```
              -fortunei ssp. discolor (1/2)
James Burchett-
              -unknown (1/2)
```

Vigorous plant, spreading 1.5 times as wide as tall with leaves
up to 8in(20.3cm) long, dark green.  Flower truss is 6in(15.2cm)
across, compact, 15-17 flowers, funnel-shaped, white with slight
touch of pink at edges.  (Slocock, reg. 1962) A. M. (Wisley
Trials) 1960


cl.
```
                            -griffithianum (1/4)
           -Jean Marie de Montague-
James C. Stephens-              -unknown (3/4)
           -unknown white hybrid
```

4ft(1.2m)        -10°F(-23°C)        EM

Flowers of heavy substnce, spinel red with red blotch and spots,
openly funnel-shaped, 4.5in(11.5cm) wide, 5-lobed.  Plant erect,
well-branched, broader than tall; glossy, dark green leaves 6in
(15.2cm) long.  (Mr. & Mrs. J. Freeman Stephens, reg. 1973)


cl.

James Comber--form of sanguineum ssp. sanguineum var. haemaleum

Collector unknown; raised and intro. by the Countess of Rosse,
Nymans Garden.  A. M. (RHS) 1973


```
                         -griffithianum (3/16)
                 -Mars-
           -Vulcan-      -unknown (1/16)
cl. -Old  -       -griersonianum (1/4)
     -Copper-      -dichroanthum ssp. dichroanthum (1/8)
James-      -Fabia-
Drive-           -griersonianum
      -            -Sir Charles Butler-fortunei ssp. fortunei
     -Van Nes Sensation-      -maximum (1/8)        (1/4)
                 -Halopeanum-
                         -griffithianum
```

5ft(1.5m)        5°F(-15°C)        L

Trusses of 7 flowers, light greyed orange with darker throat and
light red margins; the reverse a very pale orange white.  Floriferous.  Plant upright, broader than tall; dull, dark green foliage held 2 years.  (Walter Elliott, 1981)


cl.

James' Purple    See Chief Paulina


g.
```
              -facetum (eriogynum) (1/2)
Jan Baptiste-
              -griffithianum (1/2)
```

Coral-colored flowers.  (Adams-Acton, 1934)


cl.
```
                    -griffithianum (1/8)
              -Mars-
      -Captain Jack-    -unknown (3/8)
Jan Bee-          -facetum (1/4)
      -            -arboreum ssp. arboreum ? (1/4)
      -Kingianum-
              -unknown
```

4ft(1.2m)        5°F(-15°C)        E

Tubular funnel-shaped flowers of heavy substance, guardsman red
with spotting of dark cardinal red, held in trusses of 18-20.
Plant upright, broad, leaves held 3-4 years.  (W. E. Berg, 1981)
C. A. (ARS) 1982


cl.

Jan Dekens     Parentage unknown

5ft(1.5m)        -5°F(-21°C)        EM        3/4

A parent of Pink Petticoats.  Strong, vigorous plant with large,
boldly curled leaves.  Large fringed flowers of vibrant bright
pink, deeper on edges, rapidly fading to pale pink.  Trusses compact, upstanding.  (J. Blaauw & Co., Boskoop, 1940)    See color
illus. VV p. 41.


g. & cl.
```
           -dichroanthum ssp. dichroanthum (1/4)
      -Fabia-
Jan Steen-    -griersonianum (1/4)
      -            -fortunei ssp. discolor (1/4)
      -Lady Bessborough-
                 -campylocarpum ssp. campylocarpum
                              Elatum Group (1/4)
```

Flowers have cream margins, shading to pink, then to orange, and
a maroon throat blotch; upright trusses.  Shrub is tall, erect,
rather sparsely branched.  (Rothschild)   P. C. (RHS) 1950   See
color illus. PB pl. 8.


cl.
```
     -fortunei ssp. fortunei (1/2)
Jancio-    -wardii var. wardii (1/4)
    -Crest-        -fortunei ssp. discolor (1/8)
         -Lady     -
      Bessborough-campylocarpum ssp. campylocarpum
                         Elatum Group (1/8)
```

Exbury hybrid.  Trusses of 10-14 flowers, light yellow green
deepening in center, with a small eye of currant red.    (Edmund
de Rothschild, 1981)


cl.
```
           -lacteum (1/4)
     -unnamed-                -campylocarpum ssp. campylo-
     - hybrid-Mary Swaythling-              carpum (1/8)
Jan-Di-Lyn-              -fortunei ssp. fortunei (1/8)
     -          -campylocarpum ssp. campylocarpum
     -Ole Olson-           Elatum Group (1/4)
           -fortunei ssp. discolor (1/4)
```

5ft(1.5m)        -5°F(-21°C)        M

Flowers light cream, shading deeper in throat, blush pink in
throat,  short trumpet-shaped, up to 4in(10cm) broad, in compact
trusses of 9.  Leaves dark green, 6in(15.2cm) long.  (Rollin G.
Wyrens, reg. 1963)


cl.

Jane Banks--form of ambiguum

5ft(1.5m)        -5°F(-21°C)        EM        3/3

Flowers in clusters of 3-7, tubular funnel-shaped, 5-lobed,
chartreuse green, upper throat with greenish spotting.  Aromatic
foliage, dark green.  (W. L. & R. A. Banks) A. M. (RHS) 1976

cl.
        -nuttallii (L & S 12117) (1/2)
Jane Hardy-
        -lindleyii (L & S 2744) (1/2)

Flowers Chinese yellow in throat, fading to creamy white towards
edges, the latter flushed with magnolia purple. Trusses of 3-6.
(G. A. Hardy) P. C. (RHS) 1981

cl.
            -griffithianum (1/2)
      -Lady Bligh-
Jane Henny-      -unknown (1/4)
-           -griffithianum
      -Loderi Venus-
            -fortunei ssp. fortunei (1/4)

6ft(1.8m)      -10°F(-23°C)      ML

Synonym Rosebud. White flowers with soft pink shading and mottling, pale brick red rays in throat, openly funnel-shaped, 4in
(10.2cm) across, held in trusses of 13. Plant upright, rounded;
leaves held 3 years. (Rudolph and Leona Henny, reg, 1978)

cl.
           -diaprepes (1/4)
      -Polar Bear-
Jane Holden-      -auriculatum (1/4)
-           -fortunei ssp. discolor (1/4)
      -Autumn Gold-   -dichroanthum ssp. dichroanthum
           -Fabia-            (1/8)
             -griersonianum (1/8)

4ft(1.2m)      -5°F(-21°C)      L

Lax trusses hold 8-10 fragrant flowers, empire rose with light
brick red flare; center of lobes and throat darker. Plant wider than tall; leaves 7in(17.8cm) long, held 2 years. (A. John
Holden, 1979)

cl.
        -yakushimanum ssp. yakushimanum (1/2)
Jane Redford-
        -decorum (1/2)

Truss of 9-13 flowers, 7-lobed, neyron rose in bud, opening
white, flushed lighter red externally (especially on mid-ribs)
with spots of citron green in throat. Leaves oblong to
oblanceolate, sparsely hairy. (Hybridizer, Knap Hill Nurseries; reg. by R. Redford, 1983)

                    -Corona--unknown (5/8)
              -unnamed-
cl.      -Mrs. Donald- hybrid-griersonianum (1/8)
-   Graham    -     -griffithianum (1/8)
Jane Rogers-      -Loderi-
              -fortunei ssp. fortunei (1/8)
      -Mrs. R. S. Holford--unknown

5ft(1.5m)      -5°F(-21°C)    ML     4/3

A clear pink flower with a vibrant deep rose flare, the color
contrast resembling Mrs. Furnival's. (Endre Ostbo, 1957)

              -griffithianum (3/8)
g.      -Loderi-
  -Avalanche-   -fortunei ssp. fortunei (1/8)
Janet-      -calophytum var. calophytum (1/4)
-        -caucasicum (1/4)
  -Dr. Stocker-
      -griffithianum

6ft(1.8m)     5°(15°C)     E    4/3

A large plant with extra large, pure white flowers, stained deep
crimson in the throat; big flat-topped trusses. Plant semi-compact; leaves 8in(20.3cm) long. Ranked among the best of the
tall hybrids at Exbury. (Rothschild, 1942) A. M. (RHS) 1950

cl.
      -Dexter hybrid--unknown
Janet Blair-
      -unknown

6ft(1.8m)     -15°F(-26°C)    M    4/4

Synonym John Wister. A parent of many hybrids. Frilled, light
pink flowers, paler centers, golden bronze rays on upper lobes;
large trusses. Plant vigorous, compact, wider than tall; dense,
dark foliage. (D. G. Leach, 1962) See color illus. VV p. 82.

cl.
      Complex parentage includes catawbiense,
Janet Rice    maximum, and possibly unknown hybrids.

7ft(2.1m)     -15°F(-26°C)    M

Buds deep purplish pink, opening lighter purplish pink with yellowish green spotting, openly funnel-shaped, 3.5in(8.9cm) wide,
5 frilled, pointed lobes; rather loose, ball-shaped trusses of
10. Free-flowering. Plant habit upright, broad, well-branched;
glossy leaves held 2 years. (W. David Smith, reg. 1979)

cl.
             -souliei ? (1/4)
      -Virginia Scott-
Janet Scroggs-      -unknown (1/4)
-      -dichroanthum ssp. dichroanthum (1/4)
      -Jasper-
          -Lady    -fortunei ssp. fortunei (1/8)
        -Bessborough-
             -campylocarpum ssp. campylocar-
              pum Elatum Group (1/8)

6ft(1.8m)     10°F(-12°C)    ML

Yellow flowers with dorsal spotting of dull orange, openly funnel-shaped, 2.75in(7cm) wide, 5 wavy lobes; domed-shaped trusses
of 7-9. Floriferous. Plant rounded, well-branched, as broad as
tall; leaves held 3 years. (H. L. Larson, cross 1958; reg. 1979)

cl.
           -catawbiense (1/2)
Janet Ward*--sport of Cynthia-
           -griffithianum (1/2)

A hardy, flowering plant exhibited by Slocock Nurseries. P. C.
(RHS) 1974

cl.
          -campanulatum ssp. campanulatum (1/2)
Janet Warrilow-
          -unknown (1/2)

It is uncertain if this is a hybrid or a form of the species
campanulatum. Flowers lavender blue with a pale center. (Cross
by Donald Waterer; reg. 1975 by Knap Hill Nursery)

cl.
    -yellow hybrid--unknown (1/2)
Janielle-    -campylocarpum ssp. campylocarpum (1/4)
      -Marcia-   -campylocarpum ssp. campylocarpum (1/8)
          -Gladys-
            -fortunei ssp. fortunei (1/8)

Pink flowers with red blotch. (Harold Greer, 1961)

               -caucasicum (3/16)
cl.       -Jacksonii-    -caucasicum
    -Goldsworth-     -Nobleanum-
Janine  - Yellow  -       -arboreum ssp. arboreum
Alexandre-     -            (1/16)
Debray  -      -campylocarpum ssp. campylocarpum (1/4)
    -yakushimanum ssp. yakushimanum (1/2)

Compact plant with indumented foliage. Creamy pink flowers with
faint yellow center. (Raised by F. J. Street; reg. 1965)

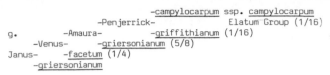

```
                          -campylocarpum ssp. campylocarpum
              -Penjerrick-             Elatum Group (1/16)
g.      -Amaura-          -griffithianum (1/16)
   -Venus-        -griersonianum (5/8)
Janus-    -facetum (1/4)
      -griersonianum
```

A scarlet-flowered hybrid.  (Lord Aberconway, intro. 1946)


```
g.
      -lacteum (1/2)
Jason-           -campylocarpum ssp. campylocarpum Elatum
     -Penjerrick-                     Group (1/4)
            -griffithianum (1/4)
```

Habit and appearance show much of lacteum.  Color between chartreuse green and primrose yellow, fading to ivory at base; flowers campanulate, 3.75in(9.5cm) across, 15-17 per truss.  Leaves
7.5in(19cm) long. (Rothschild, 1942) A. M. (RHS) 1966


cl.
```
            -maximum (1/2)
Jason's Maxim-
                -unknown (1/2)
```

3ft(.9m)          -25°F(-32°C)            L

Flowers are funnel-shaped, 5-lobed, white with yellow spotting,
held in ball-shaped trusses of 16.  Plant very similar to
maximum except more compact with leaves held 3 years. (Cross by
Clement Bowers; raised at Planting Fields Arboretum; reg. by R.
F. Miller, 1977)


g. & cl.
```
       -dichroanthum ssp. dichroanthum (1/2)
Jasper-          -fortunei ssp. discolor (1/4)
     -Lady Bessborough-
                  -campylocarpum ssp. campylocarpum
                         Elatum Group (1/4)
```

4ft(1.2m)        -5°F(-21°C)          M          2/3

A low spreading plant with pale orange flowers.  Considered very
hardy at Exbury. (Rothschild, 1942)


cl.

Jasper Pimento     Parentage as above

Jasper g.  Plant similiar to above, but flowers are deep orange
red. (Rothschild, 1942)


cl.

Jaune--form of brachyanthum ssp. brachyanthum

3ft(.9m)          -5°F(-21°C)          2/3

Flowers held in trusses of 3-4, bell-shaped, primrose yellow.
Aromatic leaves 1.5in(3.8cm) long with scaly indumentum on both
sides. (Shown by Collingwood Ingram) A. M. (RHS) 1966


```
       -griffithianum (3/8)
cl. -Mars-
     -      -unknown (3/8)
Java-                 -griffithianum
     -            -Mars-
  -unnamed hybrid-     -unknown
                  -catawbiense (red form) (1/4)
```

5ft(1.5m)         -10°F(-23°C)          ML          4/2

Cardinal red flowers with neyron rose throat, widely funnel-campanulate, 2.75in(7cm) wide, 5 wavy lobes; dome-shaped trusses of
16-17. Plant slightly wider than tall; leaves held 2 years. See
also Inca Chief, Scarlet Blast. (Cross by David G. Leach, 1958;
reg. 1983)


cl.
```
      -laetum (1/2)
Java Light-
      -javanicum (1/2)
```

Vireya hybrid.  Trusses of 12 flowers, 5-lobed, funnel-shaped,
of capsicum red, shading to orange in throat and tube.  Leaves
ovate with sparse golden brown scales. (D. B. Stanton cross; E.
B. Perrott, raiser; G. F. Smith, reg. 1984)


cl.

Jay Gould     See Alexander Adie


```
                                -griffithianum
                 -Queen Wilhelmina-       (1/16)
cl.        -Britannia-          -unknown (11/16)
   -C. P. Raffill-     -Stanley Davies--unknown
Jay   -          -griersonianum (1/4)
McMartin-
     -Moser's Maroon--unknown
```

6ft(1.8m)        10°F(-12°C)          ML          4/3

Dense trusses hold 18 flowers, chrysanthemum crimson with black
spots on upper lobe. Plant straggly with large leaves, 9in(30cm)
by 1.5in(3.8cm). (Halfdan Lem cross; reg. by Paul Bowman, 1962)


g.
```
   -griersonianum (1/2)
Jean-
   -decorum (1/2)
```

Pink flowers.  A parent of Mrs. Helen Weyerhauser. (Stirling-
Maxwell, 1936)


```
         -thomsonii ssp. thomsonii (1/4)
g.   -Barclayi-            -arboreum ssp. arboreum
     -      -Glory of Penjerrick-           (1/8)
Jean -              -griffithianum (1/4)
Goujon-                   -griffithianum
     -              -Queen Wilhelmina-
     -Unknown Warrior-          -unknown (3/8)
                 -Stanley Davies--unknown
```

Flowers vivid scarlet. (Adams-Acton, 1936)


cl.
```
           Parentage includes catawbiense, maximium,
Jean Leppo     and perhaps unknown hybrids.
```

5ft(1.5m)        15°F(-26°C)          M

Purplish red buds opening medium pink, slight yellow blotch, and
reverse strong purplish pink; flowers 2.5in(6.4cm) across, widely funnel-shaped, in trusses of 17.  Plant upright, rounded, as
broad as tall. (W. David Smith, reg. 1978)


---

Jean Marie de Montague     See The Honorable Jean Marie de Mon-
                               tague


```
                  -griffithianum (3/16)
            -Kewense-
cl.    -Aurora-    -fortunei ssp. fortunei (7/16)
   -Naomi-  -thomsonii ssp. thomsonii (1/8)
Jean  -fortunei ssp. fortunei
Rhodes-       -griersonianum (1/4)
  -Mrs. Horace Fogg-    -griffithianum
              -Loderi Venus-
                  -fortunei ssp. fortunei
```

Trusses of 14 flowers, neyron rose with dark throat chrysanthemum crimson. (Dr. R. C. Rhodes cross; raiser Mrs. Lillian Hodgson; reg. 1979)

'Gretchen' by Nearing/Gable
Photo by Knierim

'Gretchen Gossler' by Phetteplace
Photo by Greer

'Grosclaude' by Rothschild
Photo by Greer

'Grumpy' by J. Waterer, Sons & Crisp
Photo by Greer

'Gwen Bell' by Lem/Fisher
Photo by Greer

'Hallelujah' by Greer
Photo by Greer

'Happy Day' by Lyons
Photo by Greer

'Harnden's White' by Greer
Photo by Greer

'Harold Amateis' by Amateis/Baldsiefen
Photo by Leach

'Harry Carter' by Greig/Cook
Photo by Greer

'Hazel' by Greer
Photo by Greer

'Heart's Delight' by Manenica
Photo by Greer

'Heat Wave' by Greer
Photo by Greer

'Heatherside Beauty' by F. Street
Photo by Greer

'Helen Johnson' by Ostbo
Photo by Greer

'Hello Dolly' by Lem/Elliott
Photo by Greer

'Henry R. Yates' by Gable
Photo by Knierim

'High Gold' by Eichelser
Photo by Greer

'Hindustan' by Leach
Photo by Leach

'Hiraethlyn' by Aberconway
Photo by Greer

'Holmeslee Sunrise' by Holmes
Photo by Greer

'Holy Moses' by Lem
Photo by Greer

'Homestead' by Edgar Stead
Photo by Greer

'Honeydew' by Wright
Photo by Greer

```
---                 -caucasicum (1/2)
             -
Jean Verschaffelt*
                 -unknown (1/2)
```

Flowers dark crimson, with a few black markings, frilled edges.
(T. J. R. Seidel, before 1862)

```
cl.
               -macabeanum (1/2)
Jeanne Church-     -campylocarpum ssp. campylocarpum (1/4)
             -Unique-
                   -unknown (1/4)
```

Truss holds 21 flowers, primrose yellow  with pink flush,  and 3
red flares inside the base.  (Raised by New Zealand Rhododendron
Association; reg. by Mrs. Wynne Rayner, 1982)

```
cl.                 -griffithianum (3/8)
         -Marinus Koster-
Jenice-             -unknown (1/4)
Coffey-    -fortunei ssp. fortunei (1/4)
     -Pilgrim-          -arboreum ssp. arboreum (1/8)
             -Gill's Triumph-
                         -griffithianum
```

3ft(.9m)          10°F(-12°C)          L

Trusses hold 15 flowers, rose pink with minor deep pink spotting
and deep pink stripes down reverse. Plant semi-dwarf, upright;
dark green leaves held 2 years. (Sigrid Laxdall, 1979)

```
                           -griffithianum (1/4)
             -Queen Wilhelmina-
cl.      -Britannia-           -unknown (3/4)
     -         -Stanley Davies--unknown
Jennie Lewis-            -griffithianum
     -        -Queen Wilhelmina-
     -Trilby-           -unknown
             -Stanley Davies--unknown
```

5ft(1.5m)         -5°F(-21°F)           E

Both parents have a common ancestry.  White flowers tinged
green, saucer-shaped, 7-lobed, 6in(15cm) wide, held in rounded
trusses of 10.  Plant is round with  leaves 7in(17.8cm) long,
dark green.  See also Jennie Dosser. (Lillie Dosser, 1964)

```
                           -griffithianum (1/4)
             -Queen Wilhelmina-
cl.      -Britannia-           -unknown (3/4)
     -         -Stanley Davies--unknown
Jennie Dosser-            -griffithianum
     -        -Queen Wilhelmina-
     -Trilby-           -unknown
             -Stanley Davies--unknown
```

10ft(3m)          -5°F(-21°C)           L

Light purple flowers with a deep purplish red blotch, funnel-
shaped, to 4in(10.2cm) across, in trusses of about 14.  Upright
plant; leaves 7in(17.8cm) long, medium dark green.  See also
Jennie Lewis. (Lillie Dosser, 1964)

```
---
      -campylocarpum ssp. campylocarpum (1/2)
Jennifer-
      -griffithianum (1/2)
```

Flowers of lemon yellow.  (J. H. Johnstone)

```
                      -Bagshot-thomsonii ssp. thomsonii (1/8)
cl.          -Princess - Ruby  -
             -Elizabeth-    -unknown (3/8)
Jennifer Harris*     -unknown
     -        -smirnowii (1/4)
     -unnamed-
          hybrid-yakushimanum ssp. yakushimanum (1/4)
```

Medium-sized plant, with flowers cream and pink. Hardy to -15°F
(-26°C).  (Weldon E. Delp)

```
                 -dichroanthum ssp. dichroanthum (1/4)
        -Jasper-       -fortunei ssp. discolor (1/4)
cl.    -     -Lady  -
       -    Bessborough-campylocarpum ssp. campylocarpum
Jennifer-                       Elatum Group (1/4)
 Marshal-       -campylocarpum ssp. campylocarpum
       -    -Ole Olsen-         Elatum Group
     -Alice  -    -fortunei ssp. discolor
     -Franklin-      -griffithianum (1/8)
         -Loderi King-
         George    -fortunei ssp. fortunei (1/8)
```

Flowers held in trusses of 9-12, 5-lobed, deep aureolin yellow,
flecked poppy orange red.  (H. L. Larson, 1983)

```
cl.

Jenny      See Creeping Jenny
```

```
cl.
   -minus var. minus Carolinianum Group (pink form) (1/2)
Jenny*
   -calostrotum ssp. calostrotum (1/2)
```

4ft(1.2m)         -15°F(-26°C)          ML          3/3

Clear pink flowers on a compact plant spreading wider than tall.
(Weldon E. Delp)

```
cl.
        -cinnabarinum ssp. cinnabarinum Roylei Group (1/2)
Jenny Gordon-
        -trichanthum (1/2)
```

Trusses of 8 flowers, 5-lobed, color garnet lake.  Leaves
elliptic, glabrous.  (Lady Adam Gordon, 1983)

```
cl. -fortunei ssp. fortunei (1/2)
    -                   -fortunei ssp. discolor (1/8)
Jerez-      -Lady Bessborough-
    -          -campylocarpum ssp. campylocarpum
    -Jalisco-               Elatum Group (1/8)
       -   -dichroanthum ssp. dichroanthum (1/8)
       -Dido-
          -decorum (1/8)
```

Fred Wynniatt g.  One of a family of 6.  Pale lemon yellow flowers.  (Edmund de Rothschild, reg. 1966)

```
g. & cl.
      -griersonianum (1/2)
Jeritsa-          -fortunei ssp. discolor (1/4)
      -Lady Bessborough-
                  -campylocarpum ssp. campylocarpum Elatum
                                  Group (1/4)
```

A tall shrub with good foliage of dull olive green.  Pale yellow
flowers with small crimson spot, in heavy trusses. (Rothschild,
1942) A. M. (RHS) 1951

```
cl.                    -griffithianum (1/4)
         -Jean Marie de Montague-
Jerome Kern-           -unknown (1/4)
      -sanguineum ssp. sanguineum var. haemaleum (1/2)
```

Tall, spreading plant; broad, glossy, light green leaves.  Rose
Bengal flowers, funnel-shaped, 3.5in(9cm) wide, in tight trusses
of 10-15. (Cecil S. Seabrook, 1965)

```
cl.                              -griffithianum (1/8)
                -George Hardy-
        -Mrs. Lindsay Smith-          -catawbiense (1/8)
Jersey Cream-        -Duchess of Edinburgh--unknown
        -                    -campylocarpum ssp. campylocarpum
        -unnamed hybrid-                            (1/4)
                    -unknown (1/2)
```

Zuiderzee g.  Compact, rounded trusses of funnel-shaped, cream-
colored flowers with crimson marks. (J. J. Crosfield, Embley
Park)  A. M. (RHS) 1939

```
cl.
            -wardii var. wardii (1/2)
Jervis Bay-          -fortunei ssp. discolor (1/4)
        -Lady Bessborough-
                    -campylocarpum ssp. campylocarpum
                            Elatum Group (1/4)
```

6ft(1.8m)        5°F(-15°C)        M        3/2

Hawk g.  Thought to be the best of the Hawks,  then came Crest.
Still a fine yellow, with a deep red spot from wardii. (Rothschild)    A. M. (RHS) 1951    See color illus. ARS J 36:4 (Fall
1982): cover.

```
g.    -dichroanthum ssp. dichroanthum (1/2)
      -                    -griffithianum (1/16)
Jester-          -Kewense-
      -      -Aurora-          -fortunei ssp. fortunei (5/16)
      -Naomi-      -thomsonii ssp. thomsonii (1/8)
            -fortunei ssp. fortunei
```

Plant low and spreading; flowers yellow with pink spots. (Rothschild, n.d.)

```
cl.
        -caucasicum (1/2)
Jewess-
        -unknown (1/2)
```

An old hybrid still in cultivation.  Buds of light violet
opening white with brown spots on upper lobe. (Emil Liebig,
1859; sold by Standish & Noble, before 1860)

```
g. & cl.
        -griersonianum (1/2)
Jibuti-          -arboreum ssp. arboreum (1/4)
    -Gill's Triumph-
                -griffithianum (1/4)
```

Rose-colored, campanulate flowers in conical trusses.  Tall, upright plant. (Rothschild)  A. M (RHS) 1949

```
cl.
        -lutescens (1/2)
Jill-
        -unknown (1/2)
```

1.5ft(.45m)        -15°F(-26°C)        M

Ball-shaped terminal flower clusters of 1-4 trusses, each 2- to
4-flowered, with a total of about 14 flowers.  Greenish yellow
buds opening primrose yellow and fading lighter, widely funnel-
shaped, 1.5in(3.8cm) wide.  Very free-flowering.  Plant broader
than tall; leaves held 1 year.  See also Lois Jean. (Seed from
Otto Prycl; raised by William Fetterhoff; reg. 1980)

```
cl.              -fortunei ssp. fortunei (1/4)
        -Ruby F. Bowman-          -griffithianum (1/8)
Jim Drewry-              -Lady Bligh-
        -elliottii (1/2)          -unknown (1/8)
```

5ft(1.5m)        10°F(-12°C)        M        4/3

Plant of slightly open but upright habit with mid-sized leaves.
Flowers chrysanthemum crimson with darker spotting, funnel-shaped, about 27 in large trusses. (John S. Druecker, 1971)

```
cl.
            -neriiflorum ssp. neriiflorum Euchaites Group
Jiminy Cricket-                              (1/2)
            -microgynum Gymnocarpum Group (1/2)
```

1ft(.3m)        0°F(-18°C)        VE

Dwarf plant with leaves 2in(5cm) by 1in(2.5cm), elliptic, dark
green.  Flowers are Turkey red, funnel-campanulate, 2.5in(6.4
cm) across; rounded trusses of 13. (Del W. James, 1962)

```
cl.          -wardii var. wardii (3/4)
    -Carolyn Grace-
Jimmy-          -unknown (1/4)
    -wardii var. wardii
```

6ft(1.8m)        0°F(-18°C)        M

Plant of vigorous growth habit with good foliage.  Ruffled
flowers, white with light flush of green in throat, flat, 4.5in
(10.8cm) across. (The Bovees, 1963)

```
cl.          -dichroanthum ssp. dichroanthum (1/4)
        -Fabia-
Jingle Bells-    -griersonianum (1/4)
        -              -campylocarpum ssp. campylocarpum
        -Ole Olson-              Elatum Group (1/4)
                -fortunei ssp. discolor (1/4)
```

3ft(.9m)        -5°F(-21°C)        M        4/4

Low-growing plant; dense foliage.  Orange flowers fading yellow,
with the red throat retained. (Halfdan Lem cross; James A. El-
liot, intro. & reg. 1974)    See color illus. VV p. 132.

```
g.    -dichroanthum ssp. dichroanthum (1/2)
      -                    -fortunei ssp. discolor (1/8)
Jiusepa-    -Lady Bessborough-
      -Day  -          -campylocarpum ssp. campylocarpum
      Dream-                      Elatum Group (1/8)
            -griersonianum (1/4)
```

Flowers a blend of biscuit and yellow, in loose trusses.  Plant
medium-sized, compact. (Rothschild)  A. M. (RHS) 1949

```
g.          -campanulatum ssp. campanulatum (1/4)
    -Boddaertianum-
Jo-          -arboreum ssp. cinnamomeum var. album (1/4)
    -smithii (1/2)
```

An introduction by E. J. P. Magor, 1920.

```
        -dichroanthum ssp. dichroanthum (1/2)
cl.    -Jasper-          -fortunei ssp. discolor (1/8)
    -      -Lady      -
Jo Ann -    Bessborough-campylocarpum ssp. campylocarpum
Newsome-                      Elatum Group (1/8)
    -      -dichroanthum ssp. dichroanthum
    -Dido-
        -decorum (1/4)
```

Trusses of 13-14 flowers, signal red with calyx of same color,
5-lobed. (H. L. Larson, 1983)

```
cl.          -caucasicum (1/4)
    -Nobleanum-
Joan Bye-          -arboreum ssp. arboreum (1/4)
    -          -campylocarpum ssp. campylocarpum (1/4)
    -Unique-
        -unknown (1/4)
```

Trusses hold 16 campanulate flowers, maize yellow flushed with
pink. (V. J. Boulter, 1963)

cl.
            -Marion--unknown
Joan Langdon-
            -Sir Joseph Whitworth--unknown (a ponticum hybrid?)

Flowers lilac, blotched light olive green, in trusses of 15. (G.
Langdon, 1976)

                                -griffithianum (1/4)
                        -Loderi-
cl.         -unnamed hybrid-    -fortunei ssp. fortunei (1/8)
            -               -unknown (5/8)
Joan Ramsden-                   -griffithianum
            -               -Snowdrop-
            -unnamed hybrid-        -unknown
                        -unknown

Pure white flowers, 3in(7.6cm) across, held in large trusses.
(Cross by Sir John Ramsden; reg. by Sir William Pennington-Ramsden, 1965)

                    -griersonianum (1/4)
cl.         -Matador-
            -               -strigillosum (1/4)
Joan Scobie-                -thomsonii ssp. thomsonii (1/8)
            -   -Shilsonii-
            -Gaul-          -barbatum (1/8)
                -elliottii (1/4)

Flowers held in trusses of up to 15 with 5 joined lobes, claret
rose. Leaves are narrowly elliptic 6in(15cm) long, with
undersurface showing traces of light brown, woolly indumentum.
(Gen. Harrison) A. M. (RHS) 1972

cl.
                -unknown (1/2)
Joan Thompson-
                -macabeanum (1/2)

Flowers open spinel red, fading lighter. (W. M. Spry, 1968)

g.
        -lacteum (1/2)
Joanita-
        -campylocarpum ssp. caloxanthum (1/2)

Typical crimson eye of lacteum appears in daffodil yellow, bell-
shaped flowers, from rich orange yellow buds. Plant low, rounded, with distinctive broad, dark green leaves. See also Costa
del Sol. (Rothschild, 1942)

g. & cl.
        -lacteum (1/2)
Jocelyne-
        -calophytum var. calophytum (1/2)

Trusses hold 22 flowers, white with a tinge of cream, and a red
blotch in throat. Plant tall; dull green foliage. (Rothschild,
1942) A. M. (RHS) 1954; F. C. C. (RHS) 1956    See color illus.
PB pl. 22.

g.
        -williamsianum (1/2)
Jock-
        -griersonianum (1/2)

3ft(.9m)        -5°F(-21°C)        EM        2/3

Dark rosy pink flowers, orange tint in throat, 3in(7.6cm) wide,
cover the plant in small loose trusses. Habit dense, spreading; dark green leaves. Does best in full sun. (Stirling Maxwell, 1939)

            -wardii var. wardii (1/4)
        -Hawk-           -fortunei ssp. discolor (5/16)
cl.     -Lady Bessborough-
        -                   -campylocarpum ssp. campylocarpum
Jock -                          Elatum Group (3/16)
Scott-      -fortunei ssp. discolor
        -unnamed-
        - hybrid-unknown (1/8)
        -unnamed-                   -fortunei ssp. discolor
          hybrid-     -Lady
        -       -Bessborough-campylocarpum ssp.
            -Lindberg-          campylocarpum Elatum Group
                -           -dichroanthum ssp. dichroanthum
                -Dido-                           (1/16)
                -decorum (1/16)

Flower buds are apricot, opening salmon pink with a light yellow flush. (John Waterer, Sons & Crisp, 1974)

cl.
        -racemosum (1/2)
Jodi-
        -moupinense (1/2)

2ft(.6m)        0°F(-18°C)        E        4/3

Leaves are small, 1.5in(3.8cm) by .75in(2cm). Small flowers
1.5in(3.8cm) across, pink with a little spotting, blooming at
branch tips in clusters of up to 6 buds, each 4-flowered. (Halfdan Lem cross; reg. by Mrs. Mae K. Granston, 1977)

cl.
        -Catalgla--catawbiense var. album Glass (1/2)
Joe Gable-
        -wardii var. wardii (1/2)

5ft(1.5m)        -10°F(-23°C)        M        3/3

Plant habit dense, compact and wider than tall. Flowers of pale
ivory, funnel-campanulate, to 3.5in(8.9cm) across; trusses of 13
-15. (Joseph B. Gable cross, 1954; intro. 1968; reg. 1972)

                    -griersonianum (1/4)
cl.     -Vulcan's Flame-    -griffithianum (3/8)
        -               -Mars-
Joe Kruson-             -unknown (3/8)
        -   -griffithianum
        -Mars-
            -unknown

3ft(.9m)        -5°F(-21°C)        M

Funnel-shaped flowers, cardinal red, held in trusses of 13.
Plant medium compact, semi-dwarf, as broad as tall, with
revolute, pointed leaves, dark green. (Henry Yates, reg. 1972)

cl.
        -lacteum (1/2)
John Barr-                      -caucasicum (1/8)
Stevenson-          -Dr. Stocker-
        -Logan Damaris-         -griffithianum (1/8)
            -campylocarpum ssp. campylocarpum (1/4)

Flowers of lemon yellow. (J. B. Stevenson, 1962)

cl.
        -johnstoneanum (1/2)
John Bull-
        -edgeworthii (bullatum) (1/2)

Pale pink flowers flushed cream. Flowered in 1957 from seed
sown 1953. (Noble)

```
                    -griffithianum (1/8)
cl.              -Mars-
             -Vulcan-    -unknown (1/8)
John C. White-    -griersonianum (1/4)
              -fortunei ssp. fortunei (Gable's form) (1/2)
```

5ft(1.5m)         -10°F(-23°C)         EM

Flowers openly funnel-shaped, 3.5in(9cm) across, 6-lobed, medium neyron rose edged lighter to light carrot red, in dome-shaped trusses of 13. Floriferous.  Plant habit wider than tall, well-branched;  leaves 6.25in(15.8cm) long, held 2 years.  (Cross by George Ring, III; reg. by Col. and Mrs. Raymond Goodrich, 1980)

```
                  -wardii var. wardii (1/4)
          -                      -griffithianum (1/32)
        -Idealist-          -Kewense-
cl.   -(selfed)-    -Aurora-      -fortunei ssp. fortunei
      -         -Naomi-     -                    (5/32)
John  -            -      -thomsonii ssp. thomsonii (1/16)
Caller-          -fortunei ssp. fortunei
      -    -Lady   -fortunei ssp. discolor (1/8)
      -      -Bessborough-
      -Jalisco-      -campylocarpum ssp. campylocarpum
       Orange-                      Elatum Group (1/8)
         -    -dichroanthum ssp. dichroanthum (1/8)
          -Dido-
             -decorum (1/8)
```

5ft(1.5m)         5°F(-15°C)          M

Chartreuse flowers on an open plant of strong growth habit. See also Sail Wing. (J. Waterer, Sons & Crisp, 1975)

```
                        -catawbiense (1/8)
g.            -Grand Arab-
      -unnamed hybrid-    -arboreum ssp. arboreum
John Coutts-    -griffithianum (1/4)      (1/8)
      -griersonianum (1/2)
```

4ft(1.2m)         0°F(-18°C)         L         4/3

Flowers glowing salmon pink, deeper in throat, very smooth texture; large open trusses.  Growth habit dense, a little sprawly; plant completely covered  by long, pointed foliage.  (RBG, Kew, 1948)

```
cl.   -Blue Peter--unknown (5/8)
      -           -arboreum ssp. arboreum (1/8)
John Dosser-    -Doncaster-
      -Corry -       -unknown
      -Koster-         -griffithianum (1/8)
        -George Hardy-
              -catawbiense (1/8)
```

Trusses hold 19 flowers, purple violet with paler center and black blotch on upper lobe. (Raised by K. Van de Ven; intro. by D. Dosser, 1974)

cl.

John Foster Dulles     See Years of Peace

```
cl.
          -wardii var. wardii (1/2)
John Harris-
          -macabeanum (1/2)
```

Loose trusses of 14-16 flowers, pale primrose yellow with staining of maroon deep in throat, broadly campanulate, 2.5in(6.4cm) wide.  Leaves elliptic-orbicular, medium green. (Cross by John Harris; intro. by Dr. D. F. Booth)  A. M. (RHS) 1983

```
cl.
          -arboreum ssp. arboreum (1/2)
John Holms-
          -barbatum (1/2)
```

Duke of Cornwall g.  Scarlet flowers. (Gibson of Rhu) A. M. (RHS) 1957

```
cl.
          -sperabile var. sperabile (1/2)
John Marchand-
          -moupinense (1/2)
```

An interesting cross between the lepidote (scaly) and  elepidote (non-scaly) species.  Flowers deep rose pink, open-campanulate, 2in(5cm) wide, in trusses of 3-4. Dwarf plant; broadly elliptic leaves 1.5in(3.8cm) long, without indumentum.   (John Marchand cross; raiser Collingwood Ingram)  A. M. (RHS) 1966

cl.

John R. Elcock--form of fortunei ssp. discolor Houlstonii Group

5ft(1.5m)         -5°F(-21°C)         M         3/3

Trusses hold 8-10 flowers, rose purple fading to light yellow in throat, funnel-campanulate, 3.2in(8cm) wide; leaves to 3.5in(8.9 cm) long, oblong-elliptic. (Collector unknown; raised by Crown Estate, Windsor)  A. M. (RHS) 1977

cl.

John Skrentny--form of arboreum ssp. arboreum

5ft(1.5m)         10°F(-12°C)         E         4/3

Globe-shaped truss holds 16 flowers, clear cherry red, bell-shaped, to 2in(5cm) wide.  Leaves 8in(20.3cm) long, dark green, with light fawn indumentum. (Collector John Skrentny; Ben Lancaster, reg. 1966)

```
cl.
          -arboreum ssp. arboreum (1/2)
John Tremayne-
          -griffithianum (1/2)
```

Beauty of Tremough g.  Cross made by Tremayne.

```
cl.
          -catawbiense (1/2)
John Walter-
          -arboreum ssp. arboreum (1/2)
```

5ft(1.5m)         -10°F(-23°C)         ML         2/3

Ruffled flowers of crimson red cover the compact plant.  Leaves slightly rough in texture, dull olive green. A vigorous grower; buds freely. (J. Waterer, before 1860)

```
---
          -catawbiense (1/2)
John Waterer-
          -unknown (1/2)
```

Purplish red flowers; still in cultivation. (J. Waterer, before 1860)

```
cl.
          -griffithianum (1/2)
John Willis Fleming-
          -unknown (1/2)
```

Synonyms J. W. Fleming, Glory.  Flowers of rose.  (W. H. Rogers)

cl.

John Wister     See Janet Blair

```
g. & cl.
          -johnstoneanum (double form) (1/2)
Johnnie Johnston-
          -tephropeplum (1/2)
```

Flowers of Tyrian rose, staining of darker rose. (Sir Edward Bolitho, 1941)

cl.                          -griffithianum (1/4)
          -Jean Marie de Montague-
Johnny Bender-                -unknown (1/4)
       -                  -dichroanthum ssp. scyphocalyx (1/4)
          -Indiana-
                    -kyawii (1/4)

4.5ft(1.35m)          -5°F(-21°C)          M          3/5

Flowers bright currant red, darker spotting dorsally, of good
substance, in round tight trusses. Glossy dark green leaves 6in
15.2cm) long, heavily textured. (Cross by C. S. Seabrook, 1960;
reg. by John Eichelser, 1980)    See color illus. HG p. 111.

cl.
          -yakushimanum ssp. yakushimanum (1/2)
Johnny Rose-
          -unnamed hybrid--unknown (1/2)

Light orchid pink flowers, flushed a deeper shade. (J. Waterer,
Sons & Crisp, 1975)

g.
          -dichroanthum ssp. dichroanthum (1/2)
Jordan-
          -griffithianum (1/2)

Medium-sized plant; pale orange flowers with a touch of pink
opening in early mid-season. (Rothschild, 1942)

---
               -minus var. minus Carolinianum Group (1/2)
Joseph Dunn*
               -racemosum (1/2)

Lovely pink flowers in late midseason. Bush 4ft(1.2m) tall and
3ft(.9m) wide in 10 years; leaves 1in(2.5cm) long, a good dark
green all year. (Dr. G. David Lewis)

cl.
               -ponticum (1/2)
Joseph Whitworth-
               -unknown (1/2)

Synonym Sir Joseph (J.) Whitworth? Flowers of dark purple lake,
with dark spotting. (J. Waterer, before 1867)

g.
          -wardii var. wardii (1/2)
Josephine-    -fortunei ssp. discolor (1/4)
          -Ayah-
               -facetum (eriogynum) (1/4)

Terracotta pink buds open to large, bell-shaped, cream-colored
flowers, lightly spotted red in throat and flushed pink. Plant
of medium size; long, narrow, glossy leaves. (Rothschild, 1942)

                         -caucasicum (3/16)
cl.            -Jacksonii-       -caucasicum
         -Goldsworth-      -Nobleanum-            (1/8)
Josephine- Yellow  -              -arboreum ssp. arboreum
V. Cary  -       -campylocarpum ssp. campylocarpum (1/4)
         -      -brachycarpum ssp. brachycarpum (1/4)
         -unnamed-                -catawbiense (1/16)
            hybrid-        -Atrosanguineum-
               -Mrs. P.  -              -unknown (1/8)
               den Ouden-         -arboreum ssp. arboreum
                         -Doncaster-
                              -unknown

2ft(.6m)          -15°F(-26°C)          ML

Light orchid pink flowers with heavy apricot spotting in dorsal
lobe, funnel-shaped, 2in(5cm) wide, 5 frilled lobes; ball-shaped
trusses of 12. Plant broader than tall, well-branched; elliptic
leaves of dark olive green, held 2 years. See also Strawberries
and Cream, William P. Cary, Cary's Yellow, Cary's Cream. (E. A.
Cary, 1980)

cl.
          -macabeanum (1/2)
Joy Bells-    -falconeri ssp. falconeri (1/4)
          -Fortune-
               -sinogrande (1/4)

A New Zealand hybrid. Cream-colored flowers suffused chartreuse, with small crimson blotch in throat. (Cross by Joseph
Joyce; reg. by Mrs. W. J. Hayes, 1979)

g.
          -wardii var. wardii (1/2)
Joyance-
          -dichroanthum ssp. scyphocalyx (1/2)

Cream flowers with pale orange spots, in small trusses. Medium-
sized plant. Midseason. (Rothschild, 1942)

cl.
     -macabeanum (1/2)
Joyce-    -falconeri ssp. falconeri (1/4)
     -Fortune-
          -sinogrande (1/4)

Trusses hold 20-22 flowers, 7-lobed, cream suffused chartreuse,
with small crimson blotch, about 3in(7.6cm) across. (Mrs I. Mc-
Kenzie, cross 1969; raiser J. Joyce; Mrs. W. Hayes, reg. 1984)

cl.         -aureum var. aureum (chrysanthum) (1/4)
     -unnamed-
Joyce - hybrid-maximum (1/4)       -neriiflorum ssp. nerii-
Harris*                  -unnamed-           florum (1/16)
     -         -Phyllis Ballard- hybrid-dichroanthum ssp. dichro-
     -unnamed-                           anthum (1/16)
       hybrid-         -fortunei ssp. discolor (1/8)
               -Clark's White--catawbiense var. album (1/4)

Delft rose buds opening light French rose with darker margins,
yellow throat, and specks of bright yellow green. Plant of medium size; hardy to -15°F(-26°C). (Weldon E. Delp)

cl.          -maximum (1/2)
       -Midsummer-
Joyce Lyn*        -unknown (1/4)
       -              -maximum
       -unnamed hybrid-
                    -wardii var. wardii (1/4)

A parent of Oh Joyce. Flowers of peach, pink and yellow. (Weldon E. Delp)

cl.
          -fortunei ssp. discolor (1/2)
Joyce Rickett-         -catawbiense (1/4)
          -Madame Carvalho-
                    -unknown (1/4)

White flowers with pink margins. (Hillier, 1945)

cl.  -fortunei ssp. fortunei (1/2)
     -              -fortunei ssp. discolor (1/8)
Joyful-    -Lady
     -    -Bessborough-campylocarpum ssp. campylocarpum
     -Jalisco-                Elatum Group (1/8)
       -    -dichroanthum ssp. dichroanthum (1/8)
          -Dido-
               -decorum (1/8)

Fred Wynniatt g. Flowers of deep cream, flushed carmine pink on
outer lobes with a central red line down each lobe. Plant tall,
compact. First shown at the Chelsea Flower Show, 1966. (Rothschild)

cl.
          -Blue Ensign--unknown (1/2)
Juan de Fuca-
          -ponticum (1/2)

6ft(1.8m)          10°F(-12°C)          L

Buds of bishop's violet and flowers of lilac with dark red spotting and blotch, 3in(7.6cm) across, in trusses of 12; large foliage. (H. L. Larson, cross 1964;  reg. by Northwest Ornamental
Horticultural Society, 1977)

```
g.                -griffithianum (1/4)
           -Isabella-
Juanita-          -auriculatum (1/4)
        -griersonianum (1/2)
```

Pink flowers, flushed reddish salmon at base of throat. (Collingwood Ingram, 1939)

```
g. & cl.           -griffithianum (1/4)
            -Loderi Venus-
Jubilee-           -fortunei ssp. fortunei (1/4)
 Queen -      -decorum (1/4)
         -Rose du-          -maximum (1/8)
      Barri -Standishii-         -unnamed-catawbiense
                    -Altaclarense- hybrid-      (1/32)
                    -         -ponticum (1/32)
                    -arboreum ssp. arboreum
                                      (1/16)
```

Lobe tips tinged rose on opening, then pure white flowers.  See
also Margaret Rose. (Loder; Lady Loder)   A. M. (RHS) 1935

```
                       -griffithianum (1/8)
                 -Dorothea-
cl.      -Rochelle-      -decorum (1/8)
         -           -catawbiense (1/8)
Judith Anne*    -Kettledrum-
         -              -unknown (1/8)
         -Catalgla--catawbiense var. album Glass (1/2)
```

Buds roseine purple open rhodamine purple and lighter, throat of
Tyrian purple, flecks of ruby red.  Medium-sized plant; hardy to
-15°F(-26°C). (Weldon E. Delp)

```
cl.
      -fortunei ssp. fortunei (1/2)
Judy-
      -campylocarpum ssp. campylocarpum (1/2)
```

Funnel-campanulate flowers, 3.5in(8.9cm) wide,  of primrose yellow with a small red blotch. (Del W. James, intro. 1956)

```
          -wardii var. wardii (1/2)
          -              -griffithianum (1/32)
cl.   -Idealist-       -Kewense-
      -          -Aurora-      -fortunei ssp. fortunei
Judy  -       -Naomi-    -thomsonii ssp. thomsonii (1/16)
Clarke-             -fortunei ssp. fortunei (5/32)
      -   -wardii var. wardii
      -Hawk-         -fortunei ssp. discolor (1/8)
           -Lady  -
        Bessborough-campylocarpum ssp. campylocarpum
                                 Elatum Group (1/8)
```

Loose trusses of 8-12 flowers, 5-lobed, saucer-shaped, to 4in
(10.2cm) across; flowers light primrose yellow, slightly deeper
in throat; calyx flushed red purple. Leaves elliptic-ovate, to
3.6in(9cm) long.  (John Clarke cross, 1965;  exhibited by Anne,
Countess of Rosse; Nymans Garden; reg. 1983)   A. M. (RHS) 1983

```
cl.
           -maximum (1/2)
Judy Spillane-
           -Dexter hybrid--unknown (1/2)
```

4ft(1.2m)        -5°F(-21°C)         L

Late flowers of light purplish pink, paler in center, with prominent blotch of strong yellow green, openly funnel-shaped, 2.5in
(6.4cm) across, fragrant; spherical trusses of 14.  Plant rounded, with dense foliage; flat olive green leaves held 2-4 years.
(Dr. John Wister, John Tyler Arboretum, reg. 1980)    See color
illus. ARS Q 32:3 (Summer 1978): p. 170.

```
               -griffithianum (1/8)
cl.        -Loderi-
      -Albatross-     -fortunei ssp. fortunei (1/8)
Julia -          -fortunei ssp. discolor (1/2)
Grothaus-          -fortunei ssp. discolor
      -Golden Belle-     -dichroanthum ssp. dichroanthum (1/8)
                -Fabia-
                      -griersonianum (1/8)
```

6ft(1.8m)         0°F(-18°C)          L

Leaves 6in(15.2cm) long.  Flowers fragrant, ruffled, 5in(12.7cm)
wide, 7-lobed, peach with white edging, small brown blotch, in
trusses of 12. (Mrs. L. Grothaus, 1975)  C. A. (ARS) 1975   See
color illus. ARS Q 31:1 (Winter 1977): p. 34.

```
g.
      -griersonianum (1/2)
Juliana-          -griffithianum (1/4)
      -Queen Wilhelmina-
                 -unknown (1/4)
```

Scarlet flowers.  See also Captain Blood. (Collingwood Ingram,
1939)

```
cl.
               -griffithianum (1/2)
Julie--Loderi (selfed)-
               -fortunei ssp. fortunei (1/2)
```

Large white flowers, suffused sulphur, 8 per truss.  A. M. (RHS)
1944

```
               -griffithianum (1/2)
cl.      -Marinus Koster-
      -              -unknown (1/4)
Julie Titcomb-           -maximum (1/8)
      -           -Halopeanum-
      -Snow Queen-          -griffithianum
      -           -griffithianum
             -Loderi-
               -fortunei ssp. fortunei (1/8)
```

6ft(1.8m)         0°F(-18°C)          M         3/3

Flowers 4.5in(11.5cm) wide,  carmine outside, blush pink within,
slight crimson spotting.  Large trusses hold about 16 flowers.
Leaves 7in(18cm) by 2.25in(5.7cm) wide. (H. L. Larson)  P. A.
(ARS) 1958

```
           -brookeanum var. gracile (1/4)
      -Taylori-
      -           -jasminiflorum
Juliet-    -Princess Alexandra-     -jasminiflorum (3/16)
      -           -Princess-
      -teysmanii (1/2)        Royal -javanicum (1/16)
```

Vireya hybrid.  Primrose yellow flowers. (Veitch, 1891)

```
cl.

Juliet      See Sham's Juliet
```

```
               -arboreum ssp. arboreum (1/8)
cl.        -Doncaster-
      -Thunderstorm-      -unknown (3/8)
Julischka-         -unknown
      -Koichiro Wada--yakushimanum ssp. yakushimanum (1/2)
```

Trusses of 16-20 flowers, 5-lobed, deep spirea red, shading to
rose Bengal, and to very light neyron rose in throat.  Leaves
lanceolate-obovate and hairy. (Cross by H. Hachmann; reg. by G.
Stuck, 1983)

```
cl.               -griffithianum (1/4)
            -Isabella-
July Fragrance-      -auriculatum (1/4)
            -diaprepes (1/2)
```

Frilled flowers, white with crimson stain at base (showing
pale rose on outside), funnel-shaped, 4.3in(11cm) wide, strongly
fragrant; loose trusses of 8-10. Leaves to 10.5in(26cm), bronze
flush when young. (Hillier & Sons, 1968)

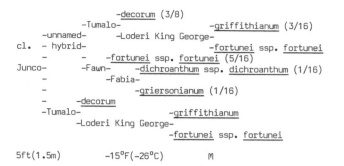

```
                        -decorum (3/8)
           -Tumalo-                    -griffithianum (3/16)
   -unnamed-        -Loderi King George-
cl. - hybrid-                    -fortunei ssp. fortunei
    -           -fortunei ssp. fortunei (5/16)
Junco-    -Fawn-   -dichroanthum ssp. dichroanthum (1/16)
    -         -Fabia-
    -                 -griersonianum (1/16)
    -    -decorum
    -Tumalo-                    -griffithianum
       -Loderi King George-
                        -fortunei ssp. fortunei
```

5ft(1.5m)          -15°F(-26°C)          M

Very fragrant white flowers held in trusses of 6-8. Plant well-
branched, broad; ivy green leaves held 2 years. (Raised by Del
W. James; reg. by Mrs. Del James for Hendricks Park, Eugene, OR,
1980)

```
                        -griffithianum (1/16)
                 -Loderi-
           -Albatross-    -fortunei ssp. fortunei
cl.  -Marie   -                           (1/16)
     -Antoinette-    -fortunei ssp. discolor (1/4)
Jungfrau-    -        -fortunei ssp. discolor
     -         -Ariel-
     -unknown    -Memoir--unknown (5/8)
```

Flowers creamy white in throat shading to pale Tyrian rose, 3.75
in(9.5cm) across; huge conical trusses hold up to 32. Plant of
tall, compact habit; dark green leaves. One of the very few Exbury creations without a complete record of parentage. (Rothschild) A. M. (RHS) 1966

```
            -campylocarpum ssp. campylocarpum
g.    -Penjerrick-         Elatum Group (1/8)
   -Amaura-    -griffithianum (1/8)
Juno-   -griersonianum (1/2)
    -         -neriiflorum ssp. neriiflorum (1/4)
   -F. C. Puddle-
            -griersonianum
```

Pale pink flowers. (Lord Aberconway, intro. 1941)

```
---                    -griffithianum (1/4)
     -Mrs. E. C. Stirling-
Jupiter-                -unknown (1/4)
     -unknown (1/2)
```

Flowers of soft lilac rose. (J. Waterer, Sons & Crisp)

```
cl.
        -campylocarpum ssp. campylocarpum (1/2)
Just Dreamin'-
        -wardii var. wardii (1/2)
```

Trusses of 16 flowers, medium primrose yellow with darker yellow
center. (G. Langdon, 1980)

```
---
            -japonicum var. japonicum (metternichii) (1/2)
Justizrath Stein*
            -unknown (1/2)
```

Flowers intense salmon pink, with dark markings. (Seidel,
intro. 1873)

```
g.
    -elliottii (1/2)
Jutland-                    -fortunei ssp. discolor (1/8)
     -          -Norman Shaw-      -catawbiense (1/16)
     -Bellerophon-         -B. de Bruin-
     -                          -unknown (1/16)
         -facetum (eriogunum) (1/4)
```

6ft(1.8m)          5°F(-15°C)          L          4/2

Intense red, waxy flowers flecked very dark red, widely campanulate; large, compact, dome-shaped trusses of 19. Handsome
foliage of dark rich green. (Rothschild, 1942) A.M. (RHS) 1942

```
cl.          -catawbiense (1/4)
         -Sefton-
K. D. Harris*   -unknown (1/2)
     -               -ponticum (1/4)
         -Purple Splendour-
                 -unknown
```

Buds of beetroot purple open to dark greyed purple, with specks
of cardinal red. Plant of medium size; hardy to -15°F(-26°C).
Weldon E. Delp)

cl.

K.S.W.     See KSW (as one word)

cl.

Kaiser Wilhelm     Parentage unknown

A parent of Homer. Carmine red flowers. Probably hybridized by
the Seidels, before 1890.

```
cl.
            -Marion--unknown (3/4)
     -Edith Boulter-    -campylocarpum ssp. campylocarum
     -          -Unique-               (1/8)
Kalimna-          -unknown
     -                    -griffithianum (1/8)
     -         -Queen Wilhelmina-
     -Unknown Warrior-         -unknown
            -Stanley Davies--unknown
```

Color is medium neyron rose to a paler rose, with light yellow
brown on upper lobe. (V. J. Boulter, 1972)     See color illus.
ARS Q 29:3 (Summer 1975): p. 170.

```
cl.        -Koichiro Wada--yakushimanum ssp. yakushimanum
     -Morgenrot-    -griffithianum (1/4)          (1/2)
     -      -Spitfire-
Kalinko*          -unknown (1/4)
     -      -griffithianum
     -unnamed-Mars-
       hybrid-   -unknown
          -Koichiro Wada--yakushimanum ssp. yakushimanum
```

Rose pink flowers. (Hans Hachmann; intro. 1983)

cl.

Kallistos*--form of nuttallii var. stellatum

A parent of Alf Bramley, q.v.

```
cl.
     -arboreum ssp. arboreum (1/2)
Kaponga-
     -Ivery's Scarlet--unknown (1/2)
```

A New Zealand hybrid with red flowers. (Bernard Holland, 1979)

```
                                -fortunei ssp. discolor
                    -Ladybird-                        (3/16)
            -Diva-          -Corona--unknown (9/16)
cl.    -Seattle Gold-   -griersonianum (1/8)
   -              -            -fortunei ssp. discolor
Karen -           -Lady
Triplett-         Bessborough-campylocarpum ssp.
   -                          campylocarpum Elatum Group
       -late flowering yellow--unknown              (1/8)
```

Trusses of 10-12 flowers, aureolin yellow, 4.5in(11.5m) wide, 6-lobed. (H. L. Larson, reg. 1982)

```
                                    -griffithianum (1/8)
cl.             -Queen Wilhelmina-
      -Britannia-               -unknown (3/8)
Karin-          -Stanley Davies--unknown
      -williamsianum (1/2)
```

3ft(.9m)          -15°F(-26°C)          EM          4/4

Shrub is rounded, compact. Large pink flowers, saucer-shaped, fringed; trusses of 8-9. See also Linda. (Experimental Station, Boskoop)  Gold Medal (Boskoop) 1966; A. M. (RHS) 1968

```
g. & cl.
        -griersonianum (1/2)
Karkov-        -arboreum ssp. arboreum (1/4)
      -Red Admiral-
               -thomsonii ssp. thomsonii (1/4)
```

5ft(1.5m)        15°F(-10°C)          EM          3/3

Large round trusses of 16 flowers, carmine rose lightly spotted, funnel-shaped, with frilled margins. Vigorous plant of medium size and habit. (Rothschild, 1943)  A. M. (RHS) 1947

```
cl.           -Corona--unknown (1/4)
      -Bow Bells-
Karl  -          -williamsianum (1/4)
Hoschna-         -dichroanthum ssp. dichroanthum (1/4)
      -Jasper   -         -fortunei ssp. discolor (1/8)
      (Exbury form)-Lady
               Bessborough-campylocarpum ssp. campylocar-
                          carpum Elatum Group (1/8)
```

5ft(1.5m)        -5°F(-21°C)          M

Funnel-shaped flowers, barium yellow, 2in(5cm) across, in loose trusses. Plant of shapely habit; small glossy leaves. (C. S. Seabrook, 1967)

```
---                          -griffithianum (1/4)
            -Mrs. E. C. Stirling-
Kate Greenaway-              -unknown (3/4)
            -unknown
```

Synonym Niobe. Flowers soft rose, veined red, fringed. (Waterer, Sons & Crisp)

```
cl.
      -catawbiense (1/2)
Kate Waterer-
      -unknown (1/2)
```

5ft(1.5m)        -10°F(-23°C)          ML          2/3

Flowers of rose pink with a golden center. Plant fairly upright and compact. (Offered by J. Waterer, before 1890)  See color illus. VV p. 103.

```
cl.              -wardii var. wardii (1/4)
            -Hawk-               -fortunei ssp. discolor
Katharine Fortescue-  -Lady Bessborough-       (1/8)
   -                            -campylocarpum ssp.
      -griffithianum (1/2)      campylocarpum Elatum
                                         Group (1/8)
```

Trusses of 12 flowers, light yellow paling to off-white with a slight blotch of red in the throat, openly funnel-shaped, about 4.5in(11.5cm) wide, 5-lobed. Leaves undulate, up to 5in(12.7cm) long. (L. S. Fortescue)  P. C. (RHS) 1975

```
cl.
        -fortunei ssp. discolor (1/2)
Katherine Dalton-
        -smirnowii (1/2)
```

5ft(1.5cm)        -15°F(-26°C)          EM          3/4

Very large, pale lavender pink flowers from brighter pink buds. Excellent plant habit; dense foliage with scant indumentum. (J. B. Gable, exhibited 1937; reg. 1958)

```
cl.

Kathmandu--form of traillianum var. dictyotum
```

4ft(1.2m)         0°F(-18°C)          M          4/3

Leaves have dense, cinnamon brown indumentum beneath. Flowers white with crimson blotch and upper lobe spotting, openly bell-shaped, 2.3in(6cm) wide; tight trusses of 19. (Edmund de Rothschild)  A. M. (RHS) 1965

```
cl.         -spinuliferum (1/2)
Kathryn Reboul-
      -racemosum (1/2)
```

3ft(.9m)          -5°F(-21°C)          EM

Leaves are small and the flowers are only 1in(2.5cm) across, pale yellow with salmon pink blushing, held in trusses of 14. (Cross by Donald Hardgrove; reg. by Adele Reboul, 1975)

```
cl.
      -catawbiense
Kathryna-                  and perhaps unknown hybrids
      -maximum
```

6ft(1.8m)        -15°F(-26°C)          ML

Pink flowers with a small yellow blotch, to 2.5in(6.4cm) across; 17-flowered trusses. Leaves 4.25in(10.8cm) long. (W. D. Smith, 1978)

```
cl.
      -Corona--unknown (5/8)
Kathy Doll-   -fortunei ssp. discolor (1/4)
      -Dondis-        -arboreum ssp. arboreum (1/8)
              -Doncaster-
                     -unknown
```

4ft(1.2m)         -5°F(-21°C)          L

Plant upright; smooth, dark green leaves, to 4.5in(11.5cm) long. Flowers are waxy, pink, and flattened. (Rudolph Henny, 1964)

```
            -catawbiense var. album (1/4)
cl. -unnamed-
    - hybrid-fortunei ssp. discolor (1/4)
Katja-                            -ponticum (1/16)
    -                    -Michael Waterer-
    -            -Prometheus-         -unknown (5/16)
    -Madame de Bruin-      -Monitor--unknown
    -                     -arboreum ssp. arboreum (1/8)
         -Doncaster-
                -unknown
```

4.5ft(1.35m)        -10°F(-23°C)          ML

Bright rose buds opening to vivid pink, shading paler in throat, widely funnel-campanulate, 2in(5cm) wide, of heavy substance; 16 held in ball-shaped trusses. Upright, rounded plant, as wide as tall; glossy leaves 6in(152cm) long, held 2 years. (A. Raustein, 1976)

'Hong Kong' by Leach
Photo by Greer

'Hoppy' by Waterer & Crisp
Photo by Greer

'Hudson Bay' by Leach
Photo by Leach

'Hudson Bay & Malta' (group) by Leach
Photo by Leach

'Hugh Koster' by M. Koster
Photo by Greer

'Humboldt' by Seidel
Photo by Schmalscheidt

'Hyperion' by Aberconway
Photo by Greer

'Ibex' by Rothschild
Photo by Greer

'Ightham' by Reuthe
Photo by Greer

'Ilam Alarm' by Stead
Photo by Greer

'Imperial' by Greer
Photo by Greer

'Indiana' by Rothschild
Photo by Greer

'Iola' by Rothschild
Photo by Greer

'Ivan D. Wood' by Mrs. R. J. Coker
Photo by Greer

'Ivory Coast' by Leach
Photo by Leach

'Ivory Tower' by Leach
Photo by Leach

'Jackwill' by Hobbie
Photo by Greer

'Janet' by Rothschild
Photo by Greer

'Janet Blair' by Leach
Photo by Greer

'Java' by Leach
Photo by Leach

'Jennie Dosser' by Lillie Dosser
Photo by Greer

'Jerome Kern' by Seabrook
Photo by Greer

'Jingle Bells' by Lem/J. A. Elliott
Photo by Greer

'Joanita' by Rothschild
Photo by Greer

```
                       -arboreum ssp. arboreum (1/16)
                 -Beauty of-
             -Norman- Tremough-griffithianum (5/16)
             - Gill -
cl.   -Anna-        -griffithianum
      -     -                      -griffithianum
Katrina-  -Jean Marie de Montague-
      -                             -unknown (3/8)
      -              -ponticum (1/4)
      -Purple Splendour-
                      -unknown
```

4.5ft(1.35m)          -10°F(-23°C)          ML

Flowers magenta rose with black blotch and spotting, openly funnel-shaped, 3in(7.6cm) across, 5 wavy lobes; ball-shaped trusses
of 10. Free-flowering. Plant upright, open; dark green, glossy
leaves held 3 years. (Elsie Watson, 1983)


```
cl.              -smirnowii (1/4)
     -unnamed hybrid-
Katydid*         -yakushimanum ssp. yakushimanum (1/4)
      -     -wardii var. wardii (1/4)
      -Crest-              -fortunei ssp. discolor (1/8)
            -Lady Bessborough-
                             -campylocarpum ssp. campylocarpum
                                         Elatum Group (1/8)
```

Plant of medium size with flowers sap green shaded from pale to
deep. Hardy to -15°(-26°C). (Weldon E. Delp)


```
cl.
     -catawbiense (1/2)
Kaulbach-
     -unknown (1/2)
```

White flowers with lilac margins, yellowish brown markings. (T.
J. R. Seidel, 1926)


```
                       -griffithianum (3/8)
cl.        -Queen Wilhelmina-
    -Britannia-            -unknown (3/8)
Kay-        -Stanley Davies--unknown
    -              -griffithianum
     -Loderi King George-
                    -fortunei ssp. fortunei (1/4)
```

5ft(1.5m)          0°F(-18°C)          M

Flowers are phlox pink with red spots, bell-shaped. (Marshall
Lyons, 1972)


```
             -dichroanthum ssp. dichroanthum (1/2)
cl.   -Jasper-         -fortunei ssp. discolor (1/8)
     -     -Lady   -
Kay   -     Bessborough-campylocarpum ssp. campylocarpum
Kirsten-                            Elatum Group (1/8)
     -              -arboreum ssp. arboreum (1/8)
     -        -Doncaster-
     -Belvedere-      -unknown (1/8)
            -dichroanthum ssp. dichroanthum
```

Flowers pale amber yellow, 5-lobed, in trusses of 11-12. Smooth
leaves, 5.25in(13.3cm) long. Exterior color in registration is
99C, blue in RHS Colour Chart. (H. Larson; reg. 1982)


```
cl.              -Moser's Maroon--unknown (1/4)
      -Romany Chai-
Kay Logan-         -griersonianum (1/4)
      -         -forrestii ssp. forrestii Repens Group (1/4)
       -Elizabeth-
                -griersonianum (1/4)
```

Plant large and upright; rather small leaves. Currant red tubular flowers, 2.5in(6.4cm) long by 1.5in(3.8cm) wide, crinkly,
in loose trusses. Late midseason. (C. S. Seabrook, 1969)


```
          -fortunei ssp. fortunei (1/4)
cl.   -Fawn-   -dichroanthum ssp. dichroanthum (1/8)
      -    -Fabia-
Kayla Rae-      -griersonianum (1/8)
      -     -wardii ssp. wardii (1/4)
       -Crest-         -fortunei ssp. discolor (1/8)
             -Lady Bessborough-
                              -campylocarpum ssp. campylo-
                                 carpum Elatum Group (1/8)
```

3ft(.9m)          0°F(-18°C)          M

Trusses 7in(17.8cm) high, of 14 flowers, 7-lobed, crimson with
cardinal red deep in throat. Plant wider than high, with stiff,
upright branches; spinach green leaves held 2 years. (Dr. Carl
Phetteplace cross, 1968; reg. by Audrey Holmeide, 1985)


```
cl.
        -campylocarpum ssp. campylocarpum (1/2)
Keay Slocock-
        -hardy hybrid--unknown (1/2)
```

Flowers pale yellow, flushed salmon or creamy white, with slight
blotch. (W. C. Slocock)


cl.

Keillour Castle* --form of ambiguum

"Hardiest yellow triflorum, the series with flowers mostly
2in(5cm) in diameter, not carried in trusses. All are extremely
free flowering in mid-spring." Catalog of Glendoick Gardens.


```
g.
     -keiskei (1/2)
Keiskarb-
     -arboreum ssp. arboreum (1/2)
```

White flowers shaded to rose, with crimson blotch and two short
lines of crimson spots. (Cross by E. J. P. Magor, 1918)


```
g.
     -keiskei (1/2)
Keiskrac-
     -racemosum (1/2)
```

A pink-flowered hybrid. (E. J. P. Magor, 1926)


```
cl.
     -Pygmalion--unknown
Kelley-     -haematodes ssp. haematodes
     -unnamed-      -fortunei ssp. fortunei
      hybrid-Wellfleet-
                  -decorum
```

6ft(1.8m)          0°F(-18°C)          M

Exact combination of above parentage is unknown. Flowers strong
purplish red with spotting, in spherical trusses of 10. Olive
green leaves held 2 years. (C. O. Dexter cross; named by Charles Herbert; reg. by Gertrude Wister, 1983)


cl.

Ken Janeck--form of yakushimanum ssp. yakushimanum
           (formerly called metternichii var. yakushimanum)

3ft(.9m)          -15°F(-26°C)          M          5/5

Flowers fuchsine pink fading white, upper lobe stippled fern
green, openly funnel-shaped, 2.5in(6.4cm) wide; compact trusses
of 13-17. Compact plant twice as wide as high; dark green foliage heavily floccose-tomentose beneath. (Kenneth Janeck, 1965)
A. E. (ARS) 1969   See color illus. HG p. 25.

```
                      -fortunei ssp. discolor (3/8)
                 -Lady
cl.     -Day Dream-Bessborough-campylocarpum ssp. campylocarpum
        -          -                           Elatum Group (1/8)
Kenhelen-          -griersonianum (3/8)
        -          -fortunei ssp. discolor
        -Margaret-       dichroanthum ssp. dichroanthum (1/8)
          Dunn  -Fabia-
                      -griersonianum
```

Flowers held in loose trusses of 8-10, 6-lobed, tubular funnel-
shaped, burnt orange in bud, opening to apricot with lemon
throat.  Very late season.  (Wilbur Graves cross; raised by Mr.
& Mrs. Kenneth Janeck, reg. 1965)

```
cl.
          -orbiculare ssp. orbiculare (1/2)
Kenlis-
          -meddianum var. meddianum (1/2)
```

Ten-flowered trusses of neyron rose.  (Headfort)   A. M. (RHS)
1948

```
g. & cl.        -haematodes ssp. haematodes (1/4)
          -Hiraethlyn-
Kenneth-        -griffithianum (1/4)
        -       -forrestii ssp. forrestii Repens Group (1/4)
        -Elizabeth-
                -griersonianum (1/4)
```

A plant with trusses of 8-10 flowers, colored geranium lake.
(Lord Aberconway)  A. M. (RHS) 1949

```
cl.
          -brachycarpum ssp. brachycarpum (1/2)
Kentucky Cardinal-
          -Essex Scarlet--unknown (1/2)
```

4ft(1.2m)       -15°F(-26°C)       ML       2/3

Synonym The Cardinal.  Small, very dark red flowers.  Attractive
dark green foliage; plant of open habit.  (Joseph B. Gable, 1958)

```
cl.
          -barbatum (1/2)
Kernick Gem-    -fortunei ssp. fortunei (1/4)
          -Luscombei-
                -thomsonii ssp. thomsonii (1/4)
```

Flowers rich pink, with dark spotting on lower petals.  (R. Gill
& Son)   A. M. (RHS) 1928

---

```
          -smirnowii (1/2)
Kesselringii-
          -ponticum (1/2)
```

A natural hybrid, 1910.

```
cl.
          -catawbiense (1/2)
Kettledrum-
          -unknown (1/2)
```

5ft(1.5m)       -20°F(-29°C)       ML

A very hardy hybrid, but habit and foliage are inferior.
Flowers are purplish crimson.  Often used as a parent for
hardiness.  (A. Waterer, 1877)

```
cl.
    -Koichiro Wada--yakushimanum ssp. yakushimanum (1/2)
Kevin-         -dichroanthum ssp. dichroanthum (1/8)
     -Fabia-
     -Jade-   -griersonianum (1/8)
        -Corona--unknown (1/4)
```

4.5ft(1.35m)       0°F(-18°C)       M

Leaves indumented and convex, 5in(12.7cm) long.  Flowers deep
pink shading to yellowish pink with margins and exterior darker,
up to 2.25in(5.7cm) across; trusses of about 14.  (Robert Bovee
cross; reg. by Sorensen & Watson, 1975)

---

Kew Pearl      Parentage unknown

A parent of Warburton.  Pink flowers, edged in rose.  (Origin
RBG, Kew Gardens?)

```
g.        -griffithianum (1/4)
       -Kewense-
Kewarb-   -fortunei ssp. fortunei (1/4)
       -arboreum ssp. arboreum (blood red form) (1/2)
```

Blood red flowers.  (E. J. P. Magor cross, 1918)

```
g.        -griffithianum (1/4)
       -Kewense-
Kewdec-   -fortunei ssp. fortunei (1/4)
       -decorum (1/2)
```

White flowers.  See also  Kewdec White Lady.  (E. J. P. Magor,
1913)

```
cl.            -griffithianum (1/4)
            -Kewense-
Kewdec White Lady-   -fortunei ssp. fortunei (1/4)
            -decorum (1/2)
```

Kewdec g.  Flowers are white with a tinge of green or pale
crimson within, held in trusses of about 9.  (E. J. P. Magor,
1913) A. M. (RHS) 1938

```
g.
    -griffithianum (1/2)
Kewense-
    -fortunei ssp. fortunei (1/2)
```

6ft(1.8m)       0°F(-18°C)       M

Pink in bud, blush white when open, and fragrant.  Not as large
as the more famous Loderi g. but has much vigor and may be a bit
more hardy.  Shouldn't the Loderis really be called the Kewense
g. since this was introduced 13 years earlier?  (RBG, Kew, 1888)

```
g.        -griffithianum (1/4)
       -Kewense-
Kewxen-   -fortunei ssp. fortunei (1/4)
       -         -adenogynum (1/4)
       -Xenosporum (detonsom)-
                -unknown (1/4)
```

Violet rose flowers with ochre spotting.  (E. J. P. Magor, 1927)

```
               -Catalgla--catawbiense var. album Glass
          -Gosh-                          (1/8)
          -Darn-        -decorum ? (3/32)
          -      -      -Caroline-
     -unnamed- -Mrs. H. R.-   -brachycarpum ssp.
     - hybrid-  Yates-         brachycarpum ? (3/32)
cl.  -      -      -unknown (3/16)
     -      -             -decorum ?
Keystone*  -      -Caroline-
     -  -Mrs. H. R. Yates-   -brachycarpum ssp.
     -                       brachycarpum ?
     -                  -unknown
     -wardii var. wardii (1/2)
```

Flowers of medium to pale primrose yellow.  Plant medium-sized.
(Weldon E. Delp)

g. & cl.
```
   -elliottii (1/2)
Kiev-        -thomsonii ssp. thomsonii (1/4)
  -Barclayi -              -arboreum ssp. arboreum (1/8)
    Robert Fox-Glory of Penjerrick-
                              -griffithianum (1/8)
```

6ft(1.8m)        5°F(-15°C)        M        3/2

White anthers contrast sharply in this darkest of all the Exbury
reds; flowers of waxy texture with sooty spotting in the 3 upper
lobes; trusses of about 12. A vigorous, tall, if rather lanky,
plant. (Rothschild; reg. 1943)  A. M. (RHS) 1950   See color
illus. PB pl. 59.

cl.

Kildonan--form of magnificum (K-W 9200)

5ft(1.5m)        10°F(-12°C)        VE        4/5

Leaves up to 18in(46cm) long and 8in(20cm) wide, covered below
with thin, pale indumentum. Fuchsine pink flowers, tubular-campanulate, 2.75in(7cm) long by 2.2in(5.5cm) wide; 30 in compact,
dome-shaped trusses. (Brodick Castle Gardens)   F. C. C. (RHS)
1966

g. & cl.
```
      -elliottii (1/2)
Kilimanjaro-        -Moser's Maroon--unknown (1/4)
       -Dusky Maid-
                  -fortunei ssp. discolor (1/4)
```

5ft(1.5m)        0°F(-18°C)        ML        5/2

Very large flowers of luminous currant red, spotted chocolate
inside, funnel-shaped, wavy-edged; spherical, compact trusses of
about 18. Attractive deep green foliage. (Rothschild, 1943)
F. C. C. (RHS) 1947   See color illus. PB pl. 33.

cl.
```
   -campylogynum (1/2)
Kim-
   -campylogynum Cremastum Group (1/2)
```

1ft(.3m)        -5°F(-21°C)        M        4/4

(Two forms of the same species do not make a hybrid.)  Delightful flowers like little yellow lanterns, from pink buds; tubular-campanulate, .75in(1.9cm) wide and long. Excellent compact
plant, wider than tall; jade green foliage. (Mr. & Mrs. J. F.
Caperci, 1966)  A. E. (ARS) 1973   See color illus. ARS Q 28:1
(Winter 1974): p. 35;  ARS J 37:3 (Summer 1983): cover.

cl.
```
   -williamsianum (1/2)
Kimberly-
   -fortunei ssp. fortunei (1/2)
```

3ft(.9m)        -10°F(-23°C)        EM        3/4

Of interest all winter with moss green foliage and bright purple bud scales. Foliage is hidden when the pastel pink flowers
bloom in profusion; flowers 3.5in(8.9cm) wide, fading white, in
open trusses. (Harold E. Greer)  P. A. (ARS) 1963

cl.
```
      -unknown (1/2)
Kimberton-    -wardii var. wardii (1/4)
      -Crest-          -fortunei ssp. discolor (1/8)
         -Lady -
         Bessborough-campylocarpum ssp. campylocarpum
                                        Elatum Group (1/8)
```

5ft(1.5m)        -5°F(-21°C)        ML

Fragrant flowers of rhodamine pink with chartreuse green dorsal
blotch, the reverse spirea red, 3.5in(8.9cm) wide, 6 wavy lobes;
trusses of 12. Plant, upright, rounded; glossy leaves retained
2 years. (Cross by Lewis Bagoly; reg. by Charles Herbert, 1979)

cl.
```
            -williamsianum (1/4)
      -Kimberly-
Kimbeth-      -fortunei ssp. fortunei (1/4)
     -          -forrestii ssp. forrestii Repens Group (1/4)
      -Elizabeth-
            -griersonianum (1/4)
```

3ft(.9m)        5°F(-21°C)        EM        4/4

Flowers deep neyron rose, funnel-shaped, 2.5in(6.4cm) wide, 5-
lobed; flat trusses of 3-5. Plant rounded, broader than tall;
medium green leaves held 2 years, new growth bronze. (Harold E.
Greer, 1979)

cl.

Kinabalu Mandarin--form of brookeanum

Synonym Mandarin. Vireya. A striking red flower with a bright
yellow throat, in trusses of 18. (Collected & raised by Mr. and
Mrs. E. F. Allen)  F. C. C. (RHS) 1970

cl.
```
   -yakushimanum ssp. yakushimanum (1/2)
King Bee-
   -tsariense (1/2)
```

2.5ft(.75m)        5°F(-15°C)        EM

Flowers 2.25in(5.7cm) across with 5 wavy lobes, opening pale
rose fading to white, with scarlet spotting and stripes of light
scarlet. Plant upright; leaves heavily indumented below. (Warren E. Berg, 1983)  C. A. (ARS) 1984

cl.

King Edward     See King Edward VII

g.
```
      -javanicum (1/2)
King Edward VII-
      -teysmanni (1/2)
```

Synonym King Edward. Vireya hybrid. Deep yellow flowers. Similar but superior to Exquisite, q.v.  A. M. (RHS) 1901

cl.
```
                        -arboreum ssp. zeylanicum
         -Ilam Alarm-                      (1/8)
   -Scarlet King-      -griffithianum (1/8)
King Elliott-      -griersonianum (1/4)
   -elliottii (1/2)
```

Flowers currant red, with darker edges; center of corolla lighter scarlet. (E. W. E. Butler, NZ, 1972)

cl.
```
   -heliolepis var. heliolepis (1/2)
King Fisher-
   -maddenii ssp. maddenii (polyandrum) (1/2)
```

Flowers of pale lilac, tubular-campanulate, to 3in(7.6cm) long,
in loose trusses. (Sir John Ramsden, Muncaster Castle, 1965)

cl.
```
   -griffithianum (1/2)
King George-
   -unknown (1/2)
```

Bright red, gloxinia-shaped flowers.  (Otto Schulz cross, 1892;
intro. by C. V. van Nes & Sons, 1896)

cl.
```
      -Nereid
King of Jordan-Tally Ho
      -fortunei ssp. discolor
```

3ft(.9m)        -10°F(-23°C)        ML

The exact combination of the above parentage is uncertain. Big
dome-shaped trusses 7in(17.8cm) high, of 10 flowers, chrome yellow shading to edges of dawn pink. Plant rounded, broader than
high, with leaves held 2 years. See also Kyoto Coral. (Cross
by Else Frye ?; reg. by Bernice I. Jordan, 1977)


cl.
                -fortunei ssp. discolor (1/2)
King of Shrubs-      -dichroanthum ssp. dichroanthum (1/4)
                -Fabia-
                     -griersonianum (1/4)

4ft(1.2m)          0°F(-18°C)          ML          4/2

Synonym Orange Azor. One of the best orange flowers, 4in(10cm)
wide, of good substance, in lax trusses of 9-10. Yellow stripes
from the throat flare and merge with orange at edges; inner mar-
gin banded rose. Plant wider than tall, open; dull green leaves.
(E. Ostbo, 1958)  P. A. (ARS) 1950   See color illus. VV p. 62.


                      -smirnowii (1/4)
cl.    -unnamed hybrid-                          -catawbiense
       -               -Parsons Grandiflorum-       (1/16)
King Tut-       -America-                  -unknown
       -               -dark red hybrid--unknown (7/16)
       -               -catawbiense (red form) (1/4)
       -unnamed hybrid-
                      -unknown

5ft(1.5m)          -20°F(-29°C)          M          3/3

Bright, deep pink flowers with yellowish brown blotch, 2.5in(6.4
cm) across; conical trusses. Leaves 5in(12.7cm) long. (Anthony
M. Shammarello, 1958)   See color illus. LW pl. 88;  VV p. xii.


g. & cl.
          -dichroanthum ssp. dichroanthum (1/2)
Kingcup-       -auriculatum (1/4)
      -Bustard-          -campylocarpum ssp. campylocarpum
             -Penjerrick-              Elatum Group (1/8)
                     -griffithianum (1/8)

4ft(1.2m)          -5°F(-21°C)          M

Waxen, tubular-shaped flowers of gleaming Indian yellow, held in
loose, flat-topped trusses. Plant compact, rather low. (Rothschild)  A. M. (RHS) 1943


cl.

Kingdom Come--form of eclecteum var. eclecteum (K-W 6869)

Trusses hold 5-7 flowers, white flushed very light sap green
with slight spotting of purple in upper throat. (Collector
Frank Kingdon-Ward; raiser Col. S. R. Clarke; reg. by R. N. S.
Clarke)  A. M. (RHS) 1978


cl.

Kingdon-Ward Pink* --form of edgeworthii

A fine hardy form of this species grown at Bodnant. A. M. (RHS)
1946


cl.
          -arboreum ssp. arboreum ? (1/2)
Kingianum*
          -unknown (1/2)

A parent of Janet Bee. Once considered a species, it is now
regarded as an arboreum hybrid. See The Rhododendron Handbook,
1980 (RHS) under "Synonyms".


g.
          -arboreum ssp. zeylanicum (1/2)
Kingking-          -griffithianum (1/4)
        -Mrs. Randall-
         Davidson  -campylocarpum ssp. campylocarpum
                         (Hooker's form) (1/4)

A hybrid by E. J. P. Magor, crossed 1915.

cl.
          -racemosum (F 19404) (1/2)
Kinglet-
        -Finch--rubiginosum Desquamatum Group (1/2)

Racemes of 3-4 blooms along stems, fuchsine pink, exterior phlox
pink; flowers tubular bell-shaped to 1.25in(3.2cm) across, .75in
(1.9cm) long. Plant 7ft(2.1m) tall, upright; dark green leaves
to 3in(7.6cm) long. (Rudolph Henny, 1964)

cl.
          -unknown (3/4)
King's Destiny-          -ponticum (1/4)
          -Purple Splendour-
                         -unknown

3ft(.9m)          -10°F(-23°C)          ML

Flowers of deep lavender with darker purple blotch and spotting,
openly funnel-shaped, 4.5in(11.5cm) across, 5-6 waxy lobes; 11-
flowered tall trusses. Plant habit upright, well-branched, almost as wide as tall; narrowly elliptic leaves 6in(15.2cm) long,
held 2 years. (Mrs. Halsey A. Frederick, Jr., 1978)

cl.
                    -ponticum (1/2)
          -Purple Splendour-
King's Favor-          -unknown (1/2)
          -          -mauve seedling--unknown
          -Arthur Bedford-
                    -ponticum

3ft(.9m)          -10°F(-23°C)          ML

Dark purple buds opening imperial purple, with a golden yellow
spotted blotch; flowers openly funnel-shaped, 3.25in(8.3cm) wide
with 5 frilled lobes, in ball-shaped trusses of 12. Plant habit
rounded, well-branched, as broad as tall; dark green leaves 8in
(20.3cm) long by 2in(5cm) across, held 2 years. (Mrs. Halsey A.
Frederick, Jr., 1978)

cl.
          -campylocarpum ssp. campylocarpum (1/2)
King's Ransom-
          -unknown (1/2)

A Canadian hybrid with light yellow flowers in trusses of 10-14.
Throat has a dark chocolate purple blotch. (Cross by Stanley
Irvine; reg. by Mr. & Mrs. H. H. McCuaig, 1974)

cl.
          -insigne (1/2)
King's Ride-
          -yakushimanum ssp. yakushimanum (1/2)

Trusses are 7in(17.8cm) tall and 6in(15.2cm) across, holding 17
flowers, white flushed phlox pink, speckled with brown. Plant
vigorous, compact; leaves 4.5in(11.5cm) long, dull dark green.
(Crown Estate Commissioners, 1972)   H. C. (RHS) 1975

cl.

Kingston     See next page

cl.
          -fortunei ssp. discolor (1/2)
Kingsway-
          -arboreum ssp. zeylanicum (1/2)

A tall, compact, late-flowering plant. Flowers white, blushed
pink, with deep pink throat; large shapely trusses. (Edmund de
Rothschild)  P. C. (RHS) 1960

```
                    -dichroanthum ssp. dichroanthum (1/8)
             -Dido-
       -Lem's-      -decorum (1/8)
       -Cameo-          -Beauty of-griffithianum (5/32)
cl.    -       -     -Tremough-
       -       -     -Norman-          -arboreum ssp. arboreum
Kingston- -Anna- Gill -griffithianum               (1/32)
       -       -         -griffithianum
       -           -Jean Marie-
       -           de Montague-unknown (9/16)
       -Polynesian Sunset--unknown
```

6ft(1.8m)          5°F(-15°C)          M

Plant upright, rounded, as wide as high; glossy, dark green foliage, held 3 years; new leaves terracotta brown.  Flowers carrot
red  fading to orange buff, edged in rhodonite red, 3.5in(8.9cm)
across, of good substance, 7 wavy lobes, in lax trusses of 8-10.
Floriferous.  (Audrey Holmeide & Alice Poot Smith, 1983)

```
g.                   -griffithianum (1/4)
       -Mrs. Kingsmill-
Kingthom-            -campylocarpum ssp. campylocarpum (1/4)
       -thomsonii ssp. thomsonii (1/2)
```

A hybrid by E. J. P. Magor.

```
cl.
       -smirnowii (1/2)
Kinloss-
       -unknown (1/2)
```

Flowers of a striking shade of cerise pink.  Midseason.  (Knap
Hill Nursery, reg. 1968)

```
cl.
```

Kirsty--form of adenogynum Adenophorum Group (R 59636)

White flowers, suffused medium spirea red with darker spotting
in upper throat, funnel-shaped, 5-lobed, with red calyx; trusses
of 12.  Dark green leaves 4.3in(11cm) long, with thick plastered
indumentum below.  (Collected by Rock; raiser Col. S. R. Clarke;
shown by R. N. S. Clarke)  A. M. (RHS) 1976

```
cl.          -Moser's Maroon--unknown (3/4)
       -Grenadier-
Kismet-           -elliottii (1/4)
       -Pygmalion--unknown
```

Red-flowered.  (Rudolph Henny, reg. 1958)

```
cl.
       -catawbiense (1/2)
Kissena-
       -unknown (1/2)
```

Purple flowers.  (Parsons)

```
cl.
          -cinnabarinum ssp. cinnabarinum (1/2)
Kit Corynton-
          -maddenii ssp. maddenii (1/2)
```

Flowers of light bronze yellow, darkening at base, with sepals
flushed jasper red.  (Gen. Harrison, 1972)   P. C. (RHS) 1971

```
g.
       -lutescens (1/2)
Kittiwake-
       -edgeworthii (1/2)
```

Erect, compact plant with pale yellow flowers, faintly fragrant.
(J. C. Williams, 1933)

```
cl.
          -griersonianum (1/2)
Kitty Cole-             -fortunei ssp. discolor (1/4)
       -unnamed white hybrid-
                        -unknown (1/4)
```

Rose-colored flowers.  (Raised by Lowinsky; intro. by Sir James
Horlick; reg. 1962)

```
                       -neriiflorum ssp. neriiflorum
cl.        -unnamed hybrid-                    (1/4)
       -              -strigillosum (1/4)
Klassy's Pride-       -griffithianum (1/8)
       -              -Loderi-
       -unnamed-      -fortunei ssp. fortunei (1/8)
       hybrid-thomsonii ssp. thomsonii (1/4)
```

5ft(1.5m)          -10°F(-23°C)          EM

Currant red flowers with black spotting, funnel-shaped, 2in(5cm)
wide, 5 wavy lobes; 14-flowered flat trusses.  Plant almost as
broad as tall; leaves 5.25in(13.3cm) long, held 3 years.  (Cross
by B. Nelson; reg. by Dr. C. G. Heller, 1979)

```
                           -griffithianum (1/8)
cl.        -Queen Wilhelmina-
       -Britannia-         -unknown (7/8)
Kluis Sensation-    -Stanley Davies--unknown
       -unknown
```

6ft(1.8m)          -5°F(-21°C)          ML          3/3

A dense, compact plant; deep green, heavy-textured leaves, edges
slightly concave.  Flowers waxy bright scarlet, funnel-shaped, 2
in(5cm) across, crinkled edges; compact spherical trusses of 17-
18.  Shrub vigorous, sun-tolerant, floriferous.  (A. Kluis, 1946)
See color illus. ARS Q 28:3 (Summer 1974): p. 170;  VV p. 60.

```
cl.
       -griffithianum (1/2)
Kluis Triumph-
       -unknown (1/2)
```

6ft(1.8m)          0°F(-18°C)          ML          2/3

Flowers terracotta red, lighter in throat and along midrib, with
black spotting on upper petal, campanulate, 2.5in(6.4cm) across;
15-18 in dome-shaped trusses.  Plant upright, vigorous, broader
than tall; dull dark green leaves 7.25in(18.5cm) long.  (Anthony
Kluis, reg. 1958)  A. M. (Wisley Trials) 1969

```
cl.
```

Knaphill*--form of campanulatum ssp. campanulatum

4ft(1.2m)          -5°F(-21°C)          EM          4/3

A large shrub; obovate leaves covered with thick, rusty brown
indumentum below.  Fine lavender blue, campanulate flowers; perhaps the closest to blue of any elepidote.  (Rothschild)  A. M.
(RHS) 1925

```
                       -dichroanthum ssp. dichroanthum (1/8)
              -Fabia-
cl.    -Old Copper-    -griersonianum (1/4)
       -         -         -griffithianum (3/16)
       -         -     -Mars-
Knight's-    -Vulcan-    -unknown (1/16)
 Beauty -         -griersonianum
       -         -Sir Charles Butler--fortunei ssp. fortunei
       -Van Nes -         -griffithianum               (1/4)
       -Sensation-Halopeanum-
                       -maximum (1/8)
```

4ft(1.2m)          10°F(-12°C)          ML

Flower buds strong purplish red opening to a strong purplish
pink, with spotting on upper lobes.  A compact shrub with light
green foliage.  (Cross by Walter Elliott; reg. by Edgar Knight,
1980)

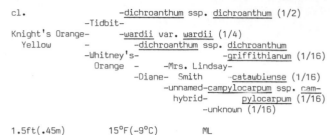

cl.                        -dichroanthum ssp. dichroanthum (1/2)
                   -Tidbit-
Knight's Orange-       -wardii var. wardii (1/4)
    Yellow     -         -dichroanthum ssp. dichroanthum
            -Whitney's-            -griffithianum (1/16)
             Orange   -  -Mrs. Lindsay-
                -Diane-  Smith    -catawbiense (1/16)
                    -unnamed-campylocarpum ssp. cam-
                       hybrid-        pylocarpum (1/16)
                           -unknown (1/16)

1.5ft(.45m)         15°F(-9°C)         ML

Flowers medium orange yellow  with strong yellowish pink edges
and stripes down center of lobe, openly funnel-shaped, about 3in
(7.6cm) across, 5 wavy petals; lax trusses of 10-11.  Free-flowering.  Plant twice as wide as tall; dull olive green leaves retained 3 years.  (Edgar L. Knight, reg. 1980)

cl.                   -Monsieur Thiers--unknown (1/4)
            -J. H. Van Nes-
Knight's Pink-        -griffithianum (1/2)
     -             -griffithianum
          -Loderi King George-
                    -fortunei ssp. fortunei (1/4)

5ft(1.5m)        15°F(-10°C)          E

Trusses hold 14 flowers, neyron rose with light orchid pink in
throat, and stripe down center of each lobe.  Plant broader than
tall with leaves dull dark green, held 2 years.  (Cross by Dr.
Ben Briggs; raised by Edgar L. Knight, reg. 1981)

cl.
            -Pygmalion--unknown (1/2)
Knowle Brilliant-
            -venator (1/2)

Scarlet flowers.  (Thacker, 1942)

cl.
            -wardii var. wardii (1/2)
Knowle Yellow-        -campylocarpum ssp. campylocarpum
       -Letty Edwards-            Elatum Group (1/4)
                -fortunei ssp. fortunei (1/4)

Clear yellow flowers.  (Thacker, 1940)


---


Koenig Albert     See Konig Albert

g.
            -falconeri ssp. falconeri (1/2)
Koenig Carola-
            -ponticum (1/2)

Lavender flowers with purple throat.  (Ludiecke, 1926)

g.                    -falconeri ssp. falconeri (1/4)
            -Koenig Carola-
Koenigdis-            -ponticum (1/4)
            -fortunei ssp. discolor (1/2)

Flowers of a light heliotrope, one lobe spotted olive green; 8-
lobed corolla.  (E. J. P. Magor, 1913)

cl.

Koichiro Wada--form of yakushimanum ssp. yakushimanum

1.5ft(.45m)         -15°F(-26°C)         EM        5/5

Deep rose in bud, opening apple blossom pink then white.  Heavily indumented foliage.  A parent of many hybrids.  (Selected by
Wada; intro. by RHS Gardens, Wisley; reg. 1969)  F. C. C. (RHS)
1947  See col. ill. ARS Q 35:2 (Spring 1981): p. 74; VV p. 80.

cl.                 -catawbiense (1/4)
            -Humboldt-
Kokardia-       -unknown (1/2)
       -               -fortunei ssp. fortunei (1/4)
            -Direcktor E. Hjelm-
                       -unknown

Trusses of 12-17 flowers, soft mauve pink, marked with ruby red,
blotched dark brown, 5-lobed.  Leaves hairy, elliptic-oval.  See
also Hachmann's Ornament.  (Cross by H. Hachmann; G. Stuck, reg.
1983)  See color illus. ARS J 39:2 (Spring 1985): p. 62.

g.           -griffithianum (1/4)
    -Isabella-
Kola-       -auriculatum (1/4)
    -       -griesronianum (1/4)
     -Tally Ho-
            -facetum (eriogynum) (1/4)

Pink flowers on well-shaped trusses.  A tall plant  of rather
loose habit.  Late-blooming.  (Rothschild, intro. 1947)


---

            -falconeri ssp. eximium (1/2)
Konig Albert-
            -unknown (1/2)

Synonym Koenig Albert.  Cream-colored flowers.
---
            -grande (1/2)
Konig Albert*
            -Album Hybridium--unknown (1/2)

Rose-colored flowers.  (Seidel catalog, 1880)


---

            -Moliere--unknown (3/4)
Koningen Louise-                      -griffithianum
       -                -George Hardy-      (1/8)
        -Mrs. Lindsay Smith-        -catawbiense (1/8)
                    -Duchess of Edinburgh--unknown

(M. Koster & Sons)

cl.
            -campylocarpum ssp. campylocarpum (1/2)
Koster's Cream-
            -unknown (1/2)

Flowers of creamy yellow with a green blotch.  (M. Koster & Sons)

cl.
    -Koichiro Wada--yakushimanum ssp. yakushimanum (1/2)
Kristin-        -Corona--unknown (1/4)
      -Bow Bells-
            -williamsianum (1/4)

4.5ft(1.35m)         0°F(-18°C)         M

Flowers pale pink  with faint red spotting, 2.5in(6.4cm) across;
trusses of 14.  Leaves 3in(7.6cm)long  by 2in(5cm) wide.  (Cross
by Robert Bovee; intro. by Sorenson & Watson, 1975)

cl.

KSW     See next page

                    -griffithianum (1/8)
cl.         -Queen Wilhelmina-
        -Britannia-       -unknown (3/8)
Kubla Khan-     -Stanley Davies--unknown
       -            -dichroanthum ssp. dichroanthum
        -Goldsworth Orange-               (1/4)
                -fortunei ssp. discolor (1/4)

5ft(1.5m)         -5°F(-21°C)         ML        5/3

Mandarin red flowers, fading to fire red; a very large blotch of
currant red extends to rays of spots of the same color.  Calyx
very large.  (L. Brandt; reg. 1962)  See color illus HG p. 80.

```
                -fortunei ssp. discolor (1/4)
  -King of Shrubs-      -dichroanthum ssp. dichroanthum (1/8)
  -           -Fabia-
cl. -                    -griersonianum (1/8)
  -                          -griffithianum (9/32)
KSW *                  -Beauty of-
  -          -Norman-Tremough -arboreum ssp. arboreum
  -              - Gill -                        (1/32)
  -        -Anna--        -griffithianum
  -      -     -          -griffithianum
  -Walloper-    -Jean Marie -
  -             - de Montague-unknown (3/16)
  -                    -griffithianum
             -Marinus Koster-
                        -unknown
```

5ft(1.5m)          -5°F(-21°C)          ML          3/3

Flowers are pink with an orange tone.  Foliage is lush and full.
(Halfdan Lem)


cl.
    -vernicosum (1/2)
Kulu-                    (natural hybrid: R 18139)
    -unknown (1/2)

6ft(1.8m)          -10°F(-23°C)          M

Fragrant Venetian pink flowers, flushed deeper to French rose in
the throat, openly funnel-shaped, to 4in(10.2cm) across, 7 wavy
lobes; dome-shaped trusses of 7-10.  Floriferous.  Showy red
bracts on new growth.  Plant rounded, as wide as tall; leaves of
dark olive green, held 2 years.  See also Mount Siga.  (Seed
collected by Dr. Rock, 1929; raiser Joseph Gable; intro. 1958;
reg. by Caroline Gable, 1979)


cl.
    -rupicola var. chryseum (1/2)
Kunming-
    -minus var. minus (white form) Carolinianum Group (1/2)

3ft(.9m)          -15°F(-26°C)          M

Flowers barium yellow, midribs lighter, margins aging to pale
brick red, dorsal spotting of light mimosa yellow; 9-10 floral
buds form terminal clusters of 34 flowers, each 1in(2.5cm) wide.
Floriferous.  Plant rounded, as broad as tall, well-branched;
closely spaced scales cover stem and underside of leaf.  (David
G. Leach, cross 1974; intro. 1978; reg. 1981)


cl.
            -phaeopeplum (1/2)
Kurt Herbert Adler-
            -lochiae (1/2)

3ft(.9m)          25°F(-4°C)          Winter

Vireya hybrid.  Plant rounded, as broad as tall, with new growth
orange brown (from dense scales).  Flat trusses of 7-8 flowers,
narrow tubular funnel-shaped, 1.5in(3.8cm) wide by 2.75in(7cm)
long, very fragrant, mandarin red with deeper throat.  (T. Lelliott cross; reg. by Strybing Arboretum, 1974)


cl.
            -fortunei ssp. discolor
Kyoto Coral -Nereid
            -Tally Ho
            -unknown

2.5ft(.75m)          -10°F(-23°C)          M

The exact combination of the above parentage is unknown.  Large
flowers with crepe-like lobes, geranium lake to Delft rose, held
in dome-shaped trusses of 12.  Plant upright, semi-dwarf, with
leaves bearing reddish brown indumentum.  See also King of
Jordan.  (Bernice I. Jordan, reg. 1977)


cl.

L. L. Liebig     Parentage unknown

Synonym Ludwig Leopold Liebig.  A parent of Burgemeester Aarts.
Flowers scarlet red, blackish red markings.  (Emil Liebig, before 1880; credited to Jean Bracke in the Register)


cl.

La Mar     Parentage unknown

A parent of the Exbury hybrid Fairylight.


```
                -griffithianum (1/4)
            -Mars-
cl.   -Vulcan-    -unknown (3/8)
La Verne-    -griersonianum (1/4)
      -              -griffithianum
      -         -unnamed hybrid-
      -Mrs. Furnival-         -unknown
      -                   -caucasicum (1/8)
        -unnamed hybrid-
                    -unknown
```

3.5ft(1m)          -10°F(-23°C)          EM

Buds dark rose, opening to flowers of rose Bengal with deep red
blotch, 5-lobed, 4in(10.2cm) wide; trusses of 9-11.  Plant habit
upright, as broad as tall; dark green leaves.  (J. Freeman Stephens, reg. 1975)


cl.

LaBar's White--white form of catawbiense

5ft(1.5m)          -20°F(-29°C)          ML          3/3

Much like other white forms of this species, but these white
flowers have a light yellow marking in the throat.  (LaBar's
Nursery, 1959)


cl.
                    -catawbiense (1/4)
        -Catawbiense Album-
Labrador-              -unknown (1/4)
        -Mist Maiden--yakushimanum ssp. yakushimanum (1/2)

4ft(1.2m)          -25°F(-32°C)          ML

Flowers 2in(5cm) across, white centers shading to edges of rhodamine purple, in trusses of 24.  Upright plant, 2/3 as wide as
high; leaves with light indumentum beneath, held 2 years.  (Rudy
Behring, cross, 1972; reg. 1985)


cl.

Lackamas Blue--form of augustinii ssp. augustinii

6ft(1.8m)          -5°F(-21°C)          EM          3/3

Lavender blue flowers, flat, to 3.25in(8.3cm) across; trusses of
3-4.  Plant upright; medium green leaves, 3in(7.6cm) long.  (B.
Lancaster, 1964) P. A. (ARS) 1963


cl.
                    -wardii var. wardii ? (1/2)
Lackamas Cream--form of Chlorops-
                    -vernicosum ? (1/2)

3.5ft(1m)          -5°F(-21°C)          M          3/3

Flowers primrose yellow to rich cream, rayed mahogany, 7-lobed,
10-12 per truss.  Easily propagated.  Leaves bluish green, 4in
(10.2cm) by 2in(5cm).  (Benjamin Lancaster)     P. A. (ARS) 1962

cl.
                    -Essex Scarlet--unknown (1/2)
Lackamas Firebrand-
                    -griersonianum (1/2)

3.5ft(1m)          -5°F(-21°C)          VL

Plant upright, bushy, holding leaves 5.5in(14cm) long with tan
indumentum beneath.  Globe-shaped trusses hold 18 flowers, bell-
shaped, currant red, to 2.5in(6.4cm) wide.  (Benjamin Lancaster,
1967)


                                        -griffithianum
cl.                      -Queen Wilhelmina-        (1/8)
                 -Earl of Athlone-              -unknown (3/8)
Lackamas Glory-              -Stanley Davies--unknown
              -thomsonii ssp. thomsonii (1/2)

3ft(.9m)           -5°F(-21°C)             M

Sturdy, slow-growing plant with dark green lustrous leaves, held
4 years.  Flowers cardinal red of thick substance, wax-like,
campanulate, to 3in(7.6cm) wide, with persistent calyx of thomsonii; rounded trusses of 10-12 long-lasting flowers.  See also
Lackamas Ruby.  (Benjamin Lancaster, 1963)


cl.                   -wardii var. wardii ? (1/4)
                 -Chlorops-
Lackamas Gold-        -vernicosum ? (1/4)
             -wardii var. wardii (1/2)

3ft(.9m)           -5°F(-21°C)          M      3/3

Compact, neat habit.  Leaves medium-sized, narrow, bluish green.
Flowers bell-shaped, up to 3in(7.6cm) across, primrose yellow;
trusses of 12-15.  (Benjamin Lancaster, 1964)  P. A. (ARS) 1962


cl.

Lackamas Ruby      Parentage as Lackamas Glory

3ft(.9m)           -5°F(-21°C)             E

Cardinal red, bell-shaped, waxy flowers, to 3in(7.6cm) wide; 15-
17 in rounded trusses.  Persistent leaves, to 6in(15.2cm) long.
See also Lackamas Glory.  (Benjamin Lancaster, 1963)


cl.                       -ponticum (1/4)
                 -Purple Splendour-
Lackamas Sovereign-       -unknown (1/4)
            -             -griersonianum (1/4)
                 -Tally Ho-
                          -facetum (eriogynum) (1/4)

4ft(1.2m)          -5°F(-21°C)             L

Open bell-shaped flowers, Tyrian purple, 3.5in(8.9cm) wide; upright round trusses of about 16.  Plant bushy, erect; leaves 6in
(15cm) long.  (Benjamin Lancaster, 1963)


cl.                  -wardii var. wardii ? (1/4)
                 -Chlorops-
Lackamas Spice-      -vernicosum ? (1/4)
              -diaprepes (1/2)

6ft(1.8m)          0°F(-18°C)          M      3/3

Flowers light yellow to creamy white, rayed mahogany, flat bell-
shaped, to 4in(10.2cm) wide; graceful trusses of 10-12; spicy
fragrance.  Plant sturdy, rounded; leaves 7in(18cm) long.  (Benjamin Lancaster, 1963)  P. A. (ARS) 1962


cl.
     -lacteum (1/2)
Lacs-
     -sinogrande (1/2)

Pale cream flowers with a dark crimson blotch.  (Cross by E. J.
P. Magor, 1936)


cl.
        -lacteum (1/2)
Lacsino-
        -sinogrande (1/2)

White flowers with a dark blotch.  See also Lacs.  (E. J. P. Magor cross, 1913)


g.                 -fortunei ssp. fortunei (1/4)
           -Luscombei-
Lactcombei-        -thomsonii ssp. thomsonii (1/4)
           -lacteum (1/2)

Yellow flowers, flushed pink.  (J. B. Stevenson cross, 1930; intro. 1951)


cl.                   -campylocarpum ssp. campylocarpum (1/4)
               -Moonstone-
Ladies' Choice-       -williamsianum (1/4)
    -         -wardii var. wardii (1/4)
         -Hawk-          -fortunei ssp. discolor (1/8)
                 -Lady Bessborough-
                                 -campylocarpum ssp.
                            campylocarpum Elatum Group
                                                    (1/8)
3ft(.9m)           10°F(-12°C)             M

Flowers primrose yellow with red spotting in throat, openly funnel-shaped 3in(7.6cm) across, 6 wavy lobes; lax trusses of 9-10.
Plant rounded, branching well, as broad as tall; glossy spinach
green leaves 3in(7.6cm) long, held 2 years.  (James A. Elliott,
reg. 1983)


cl.                          -maximum (1/4)
           -Lady Clementine Mitford-
Ladifor-                      -unknown (1/4)
        -fortunei ssp. fortunei (1/2)

Pink flowers.  (Joseph B. Gable, before 1958)


cl.            -unnamed-yakushimanum ssp. yakushimanum (1/4)
               - hybrid-
Lady Adam Gordon-     -dichroanthum ssp. dichroanthum (1/4)
               -unnamed-wardii var. wardii (1/4)
                 hybrid-
                       -decorum (1/4)

3ft(.9m)           0°F(-18°C)             M

Pale shell pink flowers in trusses of 12.  (A. F. George, Hydon
Nurseries, 1976)


cl.

Lady Alice Fitzwilliam      Parentage unknown

5ft(1.5m)          20°F(-7°C)          EM      3/3

Parentage probably like Fragrantissimum; similiar growth habit,
but bushier, more erect, open; leaves 3.5in(8.9cm) long, concave
margins, of dull rich green.  Rosy buds open to white flowers,
stained pink in throat, funnel-shaped, to 5in(12.7cm) wide, with
a nutmeg fragrance; lax trusses of 2-4.  Heat-tolerant.  (Origin
unknown)  F. C. C. (RHS) 1881

cl.
                -maximum (1/2)
Lady Annette de Trafford-
                -unknown (1/2)

Synonym Black-eyed Susan. Pink flowers with a dark eye. Blooms late. (A. Waterer, 1874)

cl.
         -dichroanthum ssp. dichroanthum (1/4)
      -Dido-
Lady April-    -decorum (1/4)
        -williamsianum (1/2)

1.5ft(.45m)        -10°F(-23°C)        EM

Very slow-growing plant, rounded, wider than tall; new foliage bronze. Flowers of heavy substance, white lightly flushed with spinel red and fading to translucent pink, 2.75in(7cm) wide; lax trusses of about 5. Free-flowering. (Arthur Childers, 1978)

cl.
        -catawbiense (1/2)
Lady Armstrong-
        -unknown (1/2)

6ft(1.8m)        -20°F(-29°C)      ML      4/3

Synonym Annedore? which is credited to Seidel by German sources. Very hardy old hybrid of good growth habit. Flowers of carmine rose with white throat. (A. Waterer, before 1870)

g. & cl.        -ciliatum (1/4)
      -Rosy Bell-
Lady Berry-      -glaucophyllum var. glaucophyllum (1/4)
   -        -cinnabarinum ssp. cinnabarinum (1/4)
      -Royal Flush-
          -maddenii ssp. maddenii (1/4)

5ft(1.5m)        5°F(-15°C)      ML      4/3

Growth habit open and upright; aromatic bluish green leaves, 3in (7.6cm) long. Tubular flowers 3in(7.6m) long, 2in(5cm) across, rosy opal inside and rich jasper red outside; pendant trusses of 8. (Rothschild, 1935) A. M. (RHS) 1937; F. C. C. (RHS) 1949

g. & cl.
        -fortunei ssp. discolor (1/2)
Lady Bessborough-
        -campylocarpum ssp. campylocarpum Elatum Group
                                (1/2)
6ft(1.8m)        -5°F(-21°C)      ML      3/3

Large loose trusses of pale yellow to ivory flowers, more yellow in the throat. Growth habit is open and upright. A parent of many fine hybrids. See also Montreal, Roberte, Ole Olson, Ottawa. (Rothschild) F. C. C. (RHS) 1933

cl.
        -griffithianum (1/2)
Lady Bligh-
        -unknown (1/2)

5ft(1.5m)        -5°F(-21°C)      ML      4/3

Buds open dark strawberry pink, fading to pastel pink, making a beautiful two-toned effect; flowers widely campanulate, to 3in (7.6cm) across. Shrub spreading, rounded, wider than tall; medium green leaves, 6in(15.2cm) long, held 2 years. (C. B. van Nes) A. M. (RHS) 1934 See color illus. ARS Q 29:3 (Summer 1975): cover; VV p. 42.

cl.                -fortunei ssp. fortunei (1/4)
      -Pilgrim-          -arboreum ssp.arboreum
Lady Bowes Lyon-   -Gill's Triumph-         (1/8)
    -                -griffithianum (1/8)
    -yakushimanum ssp. yakushimanum (1/2)

Large globe-shaped trusses of 20 flowers, white, tinged rose pink darkening to phlox pink. (Cross by Francis Hanger, RHS Garden, Wisley) A. M. (RHS) 1962

cl.          -cinnabarinum ssp. cinnabarinum Roylei Group
  -Lady Rosebery-                  (1/4)
Lady-        -Royal Flush-cinnabarinum ssp. cinnabarinum
Byng-        (pink form)              (1/8)
   -              -maddenii ssp. maddenii (1/8)
    -davidsonianum (1/2)

Rose pink flowers. (Thacker, 1942)

cl.

Lady C. Walsh     See Lady Clementine Walsh

cl.
        -auriculatum (1/2)
Lady Catherine-
        -Corona--unknown (1/2)

Flowers flushed rose, rusty speckling within; fragrant. Late. (Sir John Ramsden cross, 1922; intro. 1936) A. M. (RHS) 1936

g. & cl.
        -cinnabarinum ssp. cinnabarinum Roylei Group
Lady Chamberlain-     -cinnabarinum ssp.   (1/2)
      -Royal Flush-    cinnabarinum (1/4)
      (orange form)-maddenii ssp. maddenii (1/4)

5ft(1.5m)        10°F(-12°C)      ML      4/3

Plant upright, with slender willowy branches, bluish green new foliage. Flowers fleshy, bright orange to salmon pink, tubular to long trumpet shape, 3in(7.6cm) long by 1.5in(3.8cm) wide; 3-6 in drooping trusses. See also Apricot Lady Chamberlain, Bodnant Yellow, Chelsea, Exbury Lady Chamberlain, Gleam, Golden Queen, Ivy, Lady Chamberlain Seville, Oriflamme, Salmon Trout. (Rothschild) F. C. C. (RHS) 1931 See color illus. PB pl. 36.

cl.

Lady Chamberlain Seville     Parentage as above

Lady Chamberlain g. Bright orange flowers. (Rothschild, 1930)

cl.                -phaeopeplum (1/4)
      -unnamed hybrid-
Lady Clare-         -lochiae (1/4)
    -leucogigas (1/2)

Vireya hybrid. Trusses of 10-12 flowers, tubular-campanulate, of light empire rose. Height to 7ft(2.1m) (Cross by P. Sullivan; raiser R. M. Withers; M. Baldwin, reg. 1984)

cl.

Lady Clement Walsh     See Lady Clementine Walsh

cl.
                -maximum (1/2)
Lady Clementine Mitford-
                -unknown (1/2)

5ft(1.5m)        -5°F(-21°C)      ML      3/3

The description of the flowers of the registered plant differs from those found and pictured elsewhere: in the I.R.R. the color is white, margins flushed medium purplish rose, spotted magenta rose and bronze yellow; other sources describe the flower as soft peach pink, darker at edges, with slight yellow eye. All agree on the handsome foliage, glossy green covered with silvery hairs when young; and a plant vigorous, upright, broader than tall, free-flowering. (A. Waterer, 1870) See color illus. VV p. 53.

cl.

Lady Clementine Walsh    Parentage unknown

Synonyms  Lady Clement Walsh, Lady C. Walsh.  Flowers pale pink
or white, spotted green above; blush edged in pink. (J. Waterer)
A. M. (RHS) 1902


cl.
                -catawbiense (1/2)
Lady Clermont-
                -unknown (1/2)

Flowers rosy red or light red, with very dark blotch.  (A. Waterer)  F. C. C. (RHS) 1865


---

Lady Constance    Parentage unknown

Medium-sized flowers of deep rose.  A. M. (RHS) 1923


cl.
                -griffithianum (1/2)
Lady de Rothschild-
                -Sappho--unknown (1/2)

5ft(1.5m)        -5°F(-21°C)        M        4/2

A beautiful white flower flushed pink with a crimson blotch and
faint red spotting, held in large trusses.  Not to be confused
with the plant Mrs. Lionel de Rothschild, also a Waterer hybrid.
(A. Waterer)  A. M. (RHS) 1952


cl.
            -facetum  (1/2)
Lady Digby-
            -strigillosum (1/2)

Large shrub  with shape and leaves like strigillosum; leaves indumented, to 10in(25cm) long.  Flowers blood red, campanulate, 8
to a truss.  Midseason. (Lord Digby) A. M. (RHS) 1946


cl.
                    -maximum (1/2)
Lady Eleanor Cathcart-
                    -arboreum ssp. arboreum (1/2)

6ft(1.8m)        -15°F(-26°C)        ML        2/2

A hardy plant that is heat-resistant, somewhat compact, rounded.
Flowers clear pink with a purple blotch. (Waterer, before 1850)


cl.
                -arboreum ssp. arboreum (1/2)
Lady Falmouth-
                -unknown (1/2)

Flowers rose-colored, black blotch. (J. Waterer, before 1875)


cl.
                -catawbiense (1/2)
Lady Grey Egerton-
                -unknown (1/2)

5ft(1.5m)        -15°F(-26°)        ML        2/2

Flowers pale lilac, held in tight conical trusses. (A. Waterer,
before 1888)


cl.                                    -griffithianum (1/4)
                    -unnamed hybrid-
Lady Gwendoline Broderick-            -unknown (3/4)
                    -Bacchus--unknown

Flowers dark pink, heavily spotted with red; loose trusses. (C.
B. van Nes & Sons, before 1922)


cl.

Lady Harcourt    Parentage unknown

A parent of Gondolier.  Pink flowers, spotted with crimson. (J.
Waterer)


cl.
                -griffithianum (1/2)
Lady Hillingdon-
                -unknown (1/2)

Pale mauve flowers with yellow markings. (J. Waterer)


cl.
                -campylocarpum ssp. campylocarpum (1/2)
Lady Horlick-        -griffithianum (1/4)
                -Loderi-
                        -fortunei ssp. fortunei (1/4)

Trusses hold 7 flowers, white, lightly scented,  and 6in(15.2cm)
across. (Sir James Horlick, Isle of Gigha, reg. 1962)


g.
            -diaprepes (1/2)
Lady Jean-
            -facetum (eriogynum) (1/2)

A cross made by Lord Stair; intro. 1943.


cl.                -cinnabarinum ssp. cinnabarinum Roylei Group
            -Lady    -                            (1/4)
Lady    -Rosebery-            -cinnabarinum ssp. cinnabarinum
Lawrence-        -Royal Flush-                        (1/8)
            -        (pink form)-maddenii ssp. maddenii (1/8)
            -ambiguum (1/2)

Coppery rose flowers. (Thacker, 1942)


cl.                -catawbiense (1/4)
                -Cynthia-
Lady Longman-        -griffithianum  (1/4)
            -                        -maximum (1/4)
            -Lady Eleanor Cathcart-
                            -arboreum ssp. arboreum (1/4)

5ft(1.5m)        -5°F(-21°C)        ML

Handsome large trusses of large flowers, vivid rose on a pale
ground, with a chocolate eye.  Plant heat-tolerant; leaves distinctively veined. (Harry White, Sunningdale Nurseries)


cl.                        -griersonianum (1/4)
                -Tally Ho-
Lady Malcolm Stewart-        -facetum (eriogynum) (1/4)
            -                        -ponticum (1/4)
                -Purple Splendour-
                            -unknown (1/4)

Rich pink flowers. (Adams-Acton, 1942)

```
                                        -griffithianum (1/32)
                              -Kewense-
                     -Aurora-      -fortunei ssp. fortunei
cl.        -Naomi-        -thomsonii ssp. thomsonii (1/16)
         -Carita-     -fortunei ssp. fortunei (5/32)
Lady          -campylocarpum ssp. campylocarpum (1/4)
Malmesbury-
              -wardii var. wardii (1/2)
```

Flowers of light primrose yellow, darker in the throat, with a deeper yellowish green eye; trusses of 12-14. (Edmund de Rothschild, 1982)

cl.

Lady Mar    Parentage unknown

A parent of Fairy Light.

g. & cl.
```
         -griffithianum (1/2)
Lady Montagu-
         -thomsonii ssp. thomsonii (1/2)
```

6ft(1.8m)        10°F(-12°C)        EM        4/3

A tall shrub with very large, loose trusses; flowers large, dark rose pink, deeper outside, and of good substance. Ovate leaves. (Rothschild) A. M. (RHS) 1931

```
cl.                          -maximum (1/4)
         -Lady Eleanor Cathcart-
Lady of June-                -arboreum ssp. arboreum (1/4)
         -decorum (1/2)
```

6ft(1.8m)        -5°F(-21°C)        L

Flowers are strong purplish pink fading lighter, with distinct greenish yellow spots, held in trusses of 16. Plant grows upright, spreading broader than tall, with olive green foliage. (C. O. Dexter cross; named by Dr. John Wister; reg. by Gertrude Wister, 1980)  See color illus. LW pl. 17.

```
                  -griersonianum (1/4)
         -Mrs. Horace-     -griffithianum (13/32)
         - Fogg    -Loderi-
cl.      -           Venus-fortunei ssp. fortunei (1/8)
         -                      -griffithianum
Lady of-                -Beauty of-
Spain -          -Norman- Tremough-arboreum ssp. arboreum
         -       - Gill -                      (1/32)
         -    -Anna-     -griffithianum
         -    -     -     -griffithianum
         -Point  -  -Jean Marie -
         Defiance-   de Montague-unknown (3/16)
         -                -griffithianum
         -Marinus Koster-
                      -unknown
```

5ft(1.5m)        0°F(-18°C)        M

Funnel-shaped flowers, 6in(15cm) across, crimson fading to carmine rose, held in tall, conical trusses of 12-15. Plant is upright, broader than tall, with glossy, yellow green leaves, held 2 years. See also Canadian Beauty, Sierra Sunrise. (John G. Lofthouse, 1981)

```
cl.
         -campylocarpum ssp. campylocarpum (1/2)
Lady Primrose-
         -unknown (1/2)
```

4ft(1.2m)        0°F(-18°C)        EM        3/3

Flowers clear primrose yellow with red spots, held in a compact ball-shaped truss. Plant habit dense; medium-sized light green leaves. Plant grows slowly, and needs shade. (W. C. Slocock) A. M. (Exbury Trials) 1933

```
cl.     -souliei (1/2)
                     -decorum ? (1/8)
Lady Rae-    -Caroline-
         -Robert -     -brachyanthum ssp. brachyanthum ? (1/8)
         -Allison-fortunei ssp. discolor (1/4)
```

8ft(2.4m)        -5°F(-21°C)        M

Plant habit is upright candelabra structure; leaves glossy, dark green, 8in(20.3cm) long. Flowers saucer-shaped, color mother of pearl, fading white in an open truss. (Joseph Gable cross; reg. 1969 by T. Coleman Andrews)

```
         -yakushimanum ssp. yakushimanum (1/2)
cl.    -          -lacteum (1/8)
       -          -          -griffithianum
Lady   -     -Lionel's-      -Kewense-    (3/64)
Romsey-      -Triumph -  -Aurora-  -fortunei ssp.
       -     -          -          fortunei
       -     -    -Naomi-    -thomsonii ssp.
       -Elizabeth de-     -          thomsonii (3/32)
       -Rothschild -    -fortunei ssp. fortunei
       -                      -griffithianum
       -               -Kewense-
       -         -Aurora-      -fortunei ssp.
       -Exbury Naomi-  -      fortunei (15/64)
                       -thomsonii ssp. thomsonii
                -fortunei ssp. fortunei
```

Trusses hold 17-20 very pale greenish white flowers, with faint olive yellow spotting in upper throat. (Edmund de Rothschild) A. M. ((RHS) 1982

```
g. & cl.
         -cinnabarinum ssp. cinnabarinum Roylei Group (1/2)
Lady          -cinnabarinum ssp. cinnabarinum (1/4)
Rosebery-Royal Flush-
         (pink form)-maddenii ssp. maddenii (1/4)
```

5ft(1.5m)        5°F(-15°C)        M        4/3

Similar to Lady Chamberlain except the selected clones are more pink than orange due to the use of the pink form of Royal Flush as the male parent. Other named forms with only minor variations are Lady Rosebery Dalmeny, Lady Rosebery Etna, Lady Rosebery Pink Dawn, Lady Rosebery Pink Delight. Plant upright, of willowy habit, taller than wide; smooth medium green leaves 3in(7.6cm) long, scaly below. Flowers shell pink, shaded pink at base, tubular, fleshy, held in very lax trusses of 6-8. (Rothschild) A. M. (RHS) 1930; F. C. C. (RHS) 1932  See color illus. F p. 49.

cl.

Lady Rosebery Pink Dawn    Parentage as above

Lady Rosebery g. Another selected form, registered by Lord Digby, 1954.

cl.

Lady Rumbold    Parentage unknown

A parent of Grenada, an Exbury hybrid.

```
g.       -griersonianum (1/2)
         -                -griffithianum (1/8)
Lady Stair-     -Loderi-
         -Albatross-fortunei ssp. fortunei (1/8)
                  -fortunei ssp. discolor (1/4)
```

A hybrid by Lord Stair, 1939.

cl.

Lady Stuart of Wortley    Parentage uncertain. May be hybrid of griffithianum.

5ft(1.5m)        0°F(-18°C)        ML        3/2

Upright and open growth habit with large leaves. Drooping
trusses hold big, glowing pink flowers. (M. Koster & Sons, 1909)
A. M. (RHS) 1933

g. & cl.

       -fortunei ssp. discolor (1/2)
Ladybird-
       -Corona--unknown (1/2)

6ft(1.8m)        -5°F(-15°C)       VL        4/3

Extremely vigorous plant, forming a huge bush; glossy leaves of
rich green. Very large dome-shaped trusses of very large flowers, in soft shades of coral pink with a darker eye, speckled
yellow inside. (Rothschild) A. M. (RHS) 1933

cl.

      -johnstonianum (1/2)
Laerdal-
      -dalhousiae ssp. dalhousiae (1/2)

Pure white flowers. (Sir Edward Bolitho, 1947)

g.

        -minus var. minus Carolininanum Group (1/2)
Laetevirens-
          -ferrugineum (1/2)

3ft(1.2m)        -15°F(-26°C)        L         3/3

Synonym Wilsoni. Rosy pink flowers on a compact plant with narrow leaves. Sun-tolerant. (Origin unknown)

cl.

    -macabeanum (1/2)
Lagg-
    -magnificum (1/2)

Flowers very light sap green, with staining deep in throat of a
light magenta rose, shading much darker. (Brodick Castle Gardens, 1973)

cl.

            -griffithianum (1/2)
      -Lady Bligh-
Lake Labish-               -unknown (1/4)
    -                   -griffithianum
      -Loderi Vernus-
             -fortunei ssp. fortunei (1/4)

5ft(1.5m)        5°F(-15°C)        ML        4/2

Growth habit upright, open, wider than tall; leaves 5in(12.7cm)
long, medium green. Strawberry rose flowers, campanulate, 3.5in
(8.9cm) wide; tall trusses of about 17. (Rudolph Henny) P. A.
(ARS) 1955

cl.

Lakeside--form of trichostomum Ledoides Group

Rounded compact trusses 1.12in(2.8cm) wide and .9in(2.2cm) high,
of 11 flowers about .5in(1.2cm) across, white flushed light rhodamine purple, tubular-shaped, 5-lobed. Aromatic leaves with
brown scales, narrowly oblong, about .7in(1.8cm) long. (Crown
Estate Commissioners) A. M. (RHS) 1972

cl.

    -neriiflorum ssp. neriiflorum (1/2)
Lal Kapra-
      -sanguineum ssp. sanguineum var. sanguineum (1/2)

Red flowers. (Sir Edward Bolitho, 1962)

cl.

Lalique       Parentage uncertain--possibly a seedling of
                          Loderi or griffithianum

New Zealand hybrid. Trusses of 14-15 flowers, neyron rose shading lighter. (Mrs. A. G. Holmes, 1979)

cl.                      -cinnabarinum ssp. cinnabarinum
      -Lady Chamberlain-          Roylei Group (1/4)
Lambourn-              -         -cinnabarinum ssp. cinnabarinum
    -maddenii          -Royal-                   (1/8)
     ssp. maddenii     Flush-maddenii ssp. maddenii (5/8)

Lax trusses of 6-7 flowers, light Persian rose with upper lobe
stained amber, reverse darker rose; tubular corolla 2.5in(6.4cm)
long by 3in(7.6cm) across. Leaves 6in(15.2in) long. (Crown Estate, Windsor) P. C. (RHS) 1962

g.

    -campanulatum ssp. campanulatum (1/2)
Lamellen-
      -griffithianum (1/2)

Flowers pale mauve. (E. J. P. Magor, Lamellen, 1943)

cl.

      -dichroanthum ssp. dichroanthum (1/2)
Lamellen Dante-
        -facetum (eriogynum) (1/2)

Trusses of 8-9 flowers loosely held, open-campanulate, 2in(5cm)
across, scarlet with deep shading along margins. Throat spotted with burnt orange. (Cross by E. J. P. Magor; intro. by Maj.
E. W. M. Magor) P.C. (RHS) 1974

cl.                              -griffithianum (1/8)
            -Queen Wilhelmina-
      -Britannia-              -unknown (5/8)
Lamplighter-        -Stanley Davies--unknown
     -                    -fortunei ssp. fortunei (1/4)
      -Madame Fr. J. Chauvin-
            -unknown

5ft(1.5m)        -5°F(-21°C)        M         4/3

Large, sparkling light red flowers with a salmon glow; tall conical trusses of 10-12. Plant compact, rounded; medium green foliage, 6in(15.2cm) long. (M. Koster & Sons) F. C. C. (Boskoop)
1955    See color illus. VV p. 128.

g.

    -griersonianum (1/2)
Lancer-
    -hookeri (1/2)

Blood red flowers. (Lord Aberconway, 1950)

cl.                      -griffithianum (1/4)
      -Queen Wilhelmina-
Langley Park-            -unknown (3/4)
      -Stanley Davies--unknown

5ft(1.5m)        -5°F(-21°C)        M         4/3

Deep red flowers, 2.5in(6.4cm) wide of rather thin texture need-
ing some protection from sun; large dome-shaped trusses of 12-
15. Plant bushy, broader than tall; dark green leaves 6in(15.2
cm) long, folded along midrib, do not yellow in sunlight. (C.
P. van Nes & Sons, before 1922)

cl.

     -fortunei ssp. fortunei (1/2)
Langworth-
      -Sappho--unknown (1/2)

A tall, vigorous plant, wider than high; dark, dull green leaves
to 8in(20.3cm) long.  Large white flowers with throat streaked
greenish brown, funnel-shaped, 4in(10.2cm) wide; conical, rather
lax trusses of 16.  (Slocock Nurseries)    H. C. (Wisley Trials)
1960;   A. M. (RHS) 1962

```
                                      -thomsonii ssp. thom-
cl.                    -Bagshot Ruby-       sonii (1/8)
          -Princess Elizabeth-        -unknown (3/8)
Lanny Pride*          -unknown
      -          -smirnowii (1/4)
          -unnamed hybrid-
                         -yakushimanum ssp. yakushimanum (1/4)
```

Flowers bright guardsman red, on a medium-sized plant.  See also
Jennifer Harris.  (Weldon E. Delp)

```
cl.
          -haematodes ssp. haematodes (1/2)
Lanyon-
          -elliottii (1/2)
```

Marshall g.  This clone exhibited by Col. Bolitho, Cornwall, UK.

```
cl.
        -Marion--unknown (11/16)
Laraine-                              -griffithianum (1/16)
Langdon-               -Queen Wilhelmina-
      -         -Unknown-            -unknown
        -unnamed-Warrior-Stanley Davies--unknown
           hybrid-
                    -arboreum ssp. arboreum (1/4)
```

Trusses of 19 flowers, medium neyron rose.  (G. Langdon, 1982)

```
g.           -dichroanthum ssp. dichroanthum (1/4)
        -Astarte-         -campylocarpum ssp. campylocarpum
Largo-      -Penjerrick-                Elatum Group (1/8)
     -               -griffithianum (1/8)
        -neriiflorum ssp. neriiflorum Euchaites Group (1/2)
```

Red-flowered hybrid.  (Lord Aberconway, 1946)

```
cl.
        -taggianum ? (1/2)
Lartag-
        -unknown (1/2)
```

6ft(1.8m)        15°F(-10°C)          VE        4/4

Grown for years in the Pacific Northwest as R. taggianum but is
evidently a hybrid, from H. L. Larson.  Rose pink in bud, opening pure white, a small yellow blotch, funnel-shaped, 3in(7.6cm)
wide, fragrant; loose trusses of 4.  Very floriferous.  Spreading, drooping bush; leaves 5in(12.7cm) long.    (Dr. & Mrs. P. J.
Bowman, named & intro. 1966)

```
cl.          -dichroanthum ssp. dichroanthum (1/4)
        -Fabia-
Lascaux-       -griersonianum (1/4)
        -wardii var. wardii Litiense Group (1/2)
```

Buds a blend of red and orange; flowers are fleshy, bell-shaped,
barium yellow with crimson blotch.  Petaloid calyx of same color
as flower.  (Cross by Francis Hanger, 1947,  RHS Garden, Wisley)
A. M. (RHS) 1954

```
cl.          -griffithianum (1/4)
        -Mars-
Last Chance-  -unknown (1/4)
        -facetum (eriogynum) (1/2)
```

Flowers 3.5in(8.9cm) across, rosy claret shaded rose in center;
trusses of about 14.  Emerald green leaves 6in(15.2cm) long, and
curly.  (Rudolph Henny)  P. A. (ARS) 1957

```
cl.            -catawbiense (white form) (3/8)
          -Belle Heller-         -catawbiense
Last Hurrah-          -Madame Carvalho-
          -aureum var. aureum (1/2)   -unknown (1/8)
```

2ft(.6m)          -10°F(-23°C)        May or Sept.    4/5

Synonym Athens.  Flowers almost-white tinged faint yellow green,
some dorsal spotting of greyed chartreuse, flat saucer-shaped,
2.75in(7cm) across, 5-lobed; ball-shaped trusses of 13.  Plant
well-branched, 3 times wider than tall; dark olive green leaves,
of heavy texture, retained 3 years.  In many climates this clone
blooms almost fully in early fall, and there intended as an autumn flowering rhododendron.  (David G. Leach, intro. 1974; reg.
1981)

```
cl.
          -fortunei ssp. discolor (1/2)
Last Rose-     -griersonianum (1/4)
          -Tally Ho-
                    -facetum (eriogynum) (1/4)
```

Rose madder flowers held in well-rounded trusses of 7-9; upper
lobes freckled orange, staining throat with orange flush.  Plant
8ft(2.4m) tall; leaves to 8in(20.3cm) long.    (Mr. & Mrs. E. J.
Greig, 1963)

```
cl.

Latifolia--form of ferrugineum
```

2ft(.6m)          -15°F(-26°C)          M          3/4

The wider-leaf form of the Swiss alpine rose, colored rose pink.

```
g.
        -souliei (1/2)
Latona-
        -dichroanthum ssp. dichroanthum (1/2)
```

A hybrid with flowers cream to pink.  (Lord Aberconway, 1933)

```
g. & cl.
              -griersonianum (1/2)
Laura Aberconway-     -thomsonii ssp. thomsonii (1/4)
              -Barclayi-      -arboreum ssp. arboreum
                    -Glory of -              (1/8)
                    Penjerrick-griffithianum (1/8)
```

6ft(1.8m)        10°F(-12°C)          M          2/2

Plant of rather open habit; thick, leathery, dark green leaves.
Flowers geranium lake, funnel-shaped, 3.5in(8.9cm) across, mar-
gins frilled; loose trusses of about 9.  (Lord Aberconway, 1933)
A. M. (RHS) 1944

```
cl.                       -griffithianum (1/8)
          -Geoffrey Millais-
        -Countess -          -unknown (1/4)
Laura Marie-of Athlone-              -catawbiense
      -          -Catawbiense Grandiflorum-    (1/8)
        -ponticum (1/2)              -unknown
```

5ft(1.5m)          -5°F(-21°C)          M

Mauve flowers openly bell-shaped, 3in(7.6cm) wide, ruffled; full
rounded trusses.  Plant habit medium compact; leaves 4in(10cm)
long, medium green.  (Benjamin Lancaster, 1971)

```
cl.
        -Moser's Maroon--unknown (1/2)
Laurago-
        -yakushimanum ssp. yakushimanum (1/2)
```

4ft(1.2m)          5°F(-15°C)          M

Flowers of neyron rose with dark olive yellow dorsal spotting, openly funnel-shaped, 3in(7.6cm) across, 5 wavy lobes, fragrant; conical trusses of 16-18. Floriferous. Plant rounded, as broad as tall; dull green leaves with golden buff indumentum beneath, held 4-5 years. (Dr. David Goheen, 1983)

```
                        -caucasicum (1/4)
cl.         -Boule de Neige-                    -catawbiense (1/8)
        -                   -unnamed hybrid-
Laurel Pink-                              -unknown (1/8)
        -                    -neriiflorum ssp. neriiflorum (1/4)
        -F. C. Puddle-
                        -griersonianum (1/4)

2ft(.6m)         -20°F(-29°C)          EM         4/4
```

Flowers of neyron rose, to 2in(5cm) wide, 6-7 lobes, frilled, in rounded trusses of 10-12. Plant mound-shaped, wider than tall; small round leaves 1.75in(4.5cm) long. (Mrs. J. F. Knippenberg, 1967)

```
cl.
       -minus var. minus Carolinianum Group (3/4)
Laurie-        -minus var. minus Carolinianum Group
       -P. J. M.-
                    -dauricum var. sempervirens (1/4)

2ft(.6m)          -20°F(-29°C)          EM
```

Light pink to white flowers sometimes semi-double; consistently heavy blooming only a few days later than P. J. M. Corolla has golden dorsal spotting, light purple stripes on reverse; openly funnel-shaped, 1.75in(4.5cm) wide, 5 wavy lobes. Terminal clusters hold 28-35 flowers. Plant slow-growing, spreads as broad as tall; glossy dark green small foliage, aromatic when crushed, turns copper bronze in winter. (E. Mezitt, Weston Nurseries, reg. 1983)

```
cl.
       -griersonianum (1/2)
Lava Flow-
       -sanguineum ssp. didymum ? (1/2)
```

Low and compact, with pointed dark green leaves. Up to 10 flowers per truss, scarlet red, spotted crimson. Very late. (Sunningdale Nurseries, 1955)

```
cl.                      -decorum (1/4)
            -unnamed hybrid-
Lavender Charm-          -griffithianum (1/4)
        -                   -catawbiense (1/4)
        -Purpureum Elegans-
                        -unknown (1/4)

6ft(1.8m)       -5°F(-21°C)          ML
```

Large leaves, 7in(17.8cm) by 2.5in(6.4cm). Flowers pale pink, shadowed with lavender, a small dark blotch, the reverse much deeper; trusses of about 14. (Cross by Joseph Gable; reg. by C. Herbert, 1976)

```
cl.
       -fortunei ssp. fortunei (1/2)
Lavender Girl-          -catawbiense (1/4)
        -Lady Grey Egerton-
                        -unknown (1/4)

5ft(1.5m)        -5°F(-21°C)          M         3/4
```

Flowers of pale lavender with edges rosy mauve, fading at center with golden brown spotting on upper lobe; corolla fully expanded 2.5in(6.4cm) across, in dome-shaped trusses of 18. Fragrant and floriferous. Vigorous plant, wider than high; leaves slightly cupped, held 2-3 years. (W. C. Slocock)   A. M. (RHS) 1950; F. C. C. (Wisley Trials) 1967   See color illus. VV p. iv.

```
cl.
                -catawbiense (red form) (1/2)
Lavender Queen*          -caucasicum (1/4)
        -Boule de Neige-          -catawbiense (1/8)
                        -unnamed hybrid-
                                 -unknown (1/8)

5ft(1.5m)        -10°F(-23°C)          M         2/4
```

Plant of good growth habit, sturdy and bushy; glossy dark green foliage. Flowers of light bluish lavender, faint brown blotch, about 2in(5cm) wide, slightly ruffled, in rounded trusses. Easy to propagate. See also Prize. (A. M. Shammarello)

```
cl.
       -diaprepes (1/2)
Lavender Time-          -ponticum (1/4)
        -Purple Splendour-
                        -unknown (1/4)
```

A hybrid by Gen. Harrison, 1956.

```
cl.                      -russatum (1/4)
        -unnamed hybrid-
Lavendula*          -saluenense (1/4)
        -rubiginosum (1/2)
```

Small, compact lepidote with aromatic foliage and large lavender flowers. (Dietrich Hobbie)

```
cl.                 -aurigeranum (1/4)
        -unnamed hybrid-          -phaeopeplum (1/8)
Lazarus-          -Dr. Herman Sleumer-
        -koneri (1/2)          -zoelleri (1/8)

2.5ft(.75m)          32°F(0°C)          Dec.-Apr.
```

Vireya hybrid. Very fragrant flowers, azalea pink changing to Spanish orange in throat, held in lax trusses of 8-10. (Peter Sullivan raiser; reg. by William A. Moynier, 1983)

```
cl.
       -dichroanthum ssp. dichroanthum (1/2)
Leaburg-          -campylocarpum ssp. campylocarpum
        -Penjerrick-          Elatum Group (1/4)
                -griffithianum (1/4)

3ft(.9m)          5°F(-15°C)          EM         4/3
```

Leaves are small, 2in(5cm) by 1in(2.5cm). Flowers are a bright, waxy, blood red, up to 3in(7.6cm) across and are held in flat trusses. (Dr. Carl Phetteplace, reg. 1958) P. A. (ARS) 1956

```
cl.

Leachii--form of maximum
```

"...a curiosity with compact growth, undulating leaves and globular flower buds which is said to come more or less true from seed." P. Cox, p. 250, The Larger Species of Rhododendron. (1st ed., 1979)

```
cl.                      -griffithianum 1/2)
Leah Yates--Mars (selfed)-
                        -unknown (1/2)

4ft(1.2cm)          -15°F(-26°C)          ML
```

Flowers of medium magenta with a large white dorsal flare having 2 rays of greyed mustard spots, widely funnel-campanulate, about 3in(7.6cm) across. Plant upright, slightly broader than tall; dark grass green leaves, 7.5in(19cm) long, held 2 years. (Cross by Henry Yates, 1956; reg. by Mrs. Yates, 1976)

g.
  -dichroanthum ssp. apodectum (1/2)
Leda-
   -griersonianum (1/2)

A medium-sized shrub with vermilion flowers. Midseason.  (Lord Abcrconway, 1933)


```
                   -sanguineum ssp. didymum (1/8)
              -Carmen-
cl.  -unnamed-   -forrestii ssp. forrestii Repens Group (1/8)
    - hybrid-      -haematodes ssp. haematodes (1/8)
Leeann-    -Choremia-
    -              -arboreum ssp. arboreum (1/8)
    -              -griffithianum (1/4)
    -Gill's Crimson-
                      -unknown (1/4)
```

3ft(.9m)          10°F(-12°C)          E

Ruby red flowers with light blotch of maroon spots, widely funnel campanulate, 4.75in(12cm) across; lax trusses of 8.  Plant rounded, well-branched; dark green leaves held 3 years. (Cross by Carl G. Heller; raiser Shirley Lent; reg. 1979)


cl.
   -catawbiense (1/2)
Lee's Best Purple*
   -unknown (1/2)

6ft(1.8m)          -20°F(-29°C)          L          2/3

Handsome foliage--smooth, dark and glossy.  Flowers of deep purple on a hardy old hybrid.  (Lee)


cl.
   -catawbiense (1/2)
Lee's Dark Purple-
   -unknown (1/2)

6ft(1.8m)          -15°F(-26°C)          ML          2/3

An old reliable, still grown in the colder areas.  Bush compact, rounded, like catawbiense in habit; foliage dark, slightly wavy.  Flowers royal purple, brownish markings within, in dense rounded trusses.  (Lee, before 1851)


cl.
   -caucasicum (1/2)
Lee's Scarlet-
   -unknown (1/2)

4ft(1.2m)          -5°F(-21°C)          VE          3/3

Flowers rosy crimson, fading to bright pink.  The earliest of all rhododendrons, flowering before Christmas in parts of Great Britain, and in open weather onwards to March.  (Lee)


```
              -Corona--unknown (1/4)
cl.     -Bow Bells-
              -williamsianum (1/4)
Legal Johnny-              -dichroanthum ssp. scypho-
    -              -Socrianum-              calyx (1/8)
    -unnamed hybrid-       -griffithianum (1/8)
    -              -wardii var. wardii (1/8)
              -Rima-
                  -decorum (1/8)
```

3ft(.9m)          0°F(-18°C)          M

Flowers of crushed strawberry red with creamy yellow centers, edges of rhodonite red; trusses about 12-flowered.  See also Bow Street, Caroline de Zoete, Cheapside.  (A. F. George, Hydon Nurseries, 1976)


cl.
```
        -forrestii ssp. forrestii Repens Group (1/4)
    -Jaipur-
Leilie-    -meddianum var. meddianum (1/4)
    -      -haematodes ssp. haematodes (1/4)
    -May Day-
          -griersonianum (1/4)
```

Semi-dwarf plant, 2ft(.6m) by 3ft(.9m) in 16 years, compact with leaves 7in(17.8cm) long, heavy brown indumentum below.  Flowers orient red, 2.5in(6.4cm) wide, held in trusses of 9.  (Lester E. Brandt, cross 1949; reg. 1966)


cl.
```
        -decorum (1/2)
Lemon Bells-    -dichroanthum ssp. dichroanthum (1/4)
        -Fabia-
          -griersonianum (1/4)
```

Large flowers of pinkish yellow, bell-shaped, 5-6 per lax truss.  Dwarf plant 2.5ft(.75m) high; leaves up to 6in(15cm) long.  Midseason.  (Rudolph Henny, cross 1943; reg. 1958)


g.
   -rigidum (caeruleum) (1/2)
Lemon Bill-
   -cinnabarinum ssp. xanthocodon Concatenans Group (1/2)

Introduced by Lord Aberconway, 1943.  See also Peace.


```
              -wardii var. wardii (1/2)
       -Crest-              -fortunei ssp. discolor (1/8)
       -    -Lady Bessborough-
cl.    -                   -campylocarpum ssp. campylocarpum
Lemon  -                             Elatum Group (1/8)
Custard-                   -George-griffithianum
       -       -Mrs. Lindsay-Hardy -          (1/32)
       -         - Smith      -    -catawbiense(1/32)
       -    -Mrs. Betty-        -Duchess of Edinburgh--
       -     - Robertson-           unknown (1/8)
       -unnamed-        -unnamed-campylocarpum ssp. campylo-
        hybrid-         - hybrid-          carpum (1/16)
        -                   -unknown
        -wardii var. wardii
```

5ft(1.5m)          10°F(-6°C)          EM

Waxy flowers buff yellow with no markings, openly funnel-campanulate, 4in(10.2cm) across,  slightly fragrant; lax trusses of 8-12.  Free-flowering.  Plant rounded, well-branched, wider than tall; dark green leaves held 2 years. (Walter Elliott, 1981)


cl.
```
          -campylocarpum ssp. campylocarpum (1/4)
    -Moonstone-
Lemon Drop-    -williamsianum  (1/4)
    -unknown (1/2)
```

1ft(.3m)          -5°F(-21°C)          M

Compact plant with orbicular leaves.  Brilliant light yellow green flowers, 2.5in(6.4m) across.  (Bovees, 1962)


```
                      -dichroanthum ssp. dichroanthum
              -Goldsworth-                    (1/8)
       -Hotei-  Orange -fortunei ssp. discolor (1/8)
cl.    -    -              -souliei (1/8)
       -    -unnamed hybrid-
Lemon Float-              -wardii var. wardii (1/8)
       -              -yakushimanum ssp. yakushimanum (1/8)
       -      -White -
       -unnamed-Wedding-yakushimanum ssp. makinoi (1/8)
        hybrid-
          -lacteum (1/4)
```

Trusses of 15-20 flowers, chartreuse green.  Plant habit low and compact with dark foliage.  Sets flower buds when very young.  Easy to propagate.  (John G. Lofthouse, 1979)

cl.
            -minus var. minus Carolinianum Group (yellow) (1/2)
Lemon Girl-
            -lutescens (1/2)

3ft(.9m)           -10°F(-23°C)        EM

Terminal flower clusters  of 1-3 trusses, each 7-14  flowers, of
light mimosa yellow, dorsal spotting lemon yellow, 1.25in(3.2cm)
wide, 5-lobed.  Free-flowering.  Upright plant nearly as broad
as high, with stiff branches; dark olive green, narrowly elliptic leaves, 3.25in(8.3cm) long.  (Louis B. Mraw, intro. 1978;
reg. 1983)


cl.
                    -wardii var. wardii. (1/2)
Lemon Lodge--Prelude (selfed)-
                    -fortunei ssp. fortunei (1/2)

Primrose yellow flowers, with a few tiny spots deep in throat.
(Pukeiti Rhododendron Trust, NZ, reg. 1972)


cl.                 -laetum (1/4)
            -unnamed hybrid-
Lemon Minuet-               -gracilentum (3/4)
            -gracilentum

Vireya hybrid.  Trusses of 4-5 flowers, tubular campanulate, reflexed lobes, light canary yellow.  Height up to 16in(.4m). (G.
L. S. Snell, Australia, reg. 1984)


cl.
            -xanthostephanum (1/2)
Lemon Mist-
            -leucaspis (1/2)

2ft(.6m)           10°F(-12°C)        VE         3/3

A compact bush almost twice as wide as tall, with narrow, medium
green leaves.  Bright greenish yellow flowers cover the plant in
early spring; corolla is open funnel-shaped, 1.5in(3.8cm) wide.
(Robert W. Scott, 1968)  A. E. (ARS) 1969


cl.                 -caucasicum (1/8)
            -Dr. Stocker-
      -Damaris-           -griffithianum (1/8)
Lemonade*    -campylocarpum ssp. campylocarpum (1/4)
      -     -wardii var. wardii (1/4)
      -Crest-           -fortunei ssp. discolor (1/8)
            -Lady        -
            Bessborough-campylocarpum ssp. campylocarpum
                               Elatum Group (1/8)

6ft(1.8m)          -5°F(-21°C)         M          4/4

Attractive plant habit and foliage, glossy green rounded leaves.
Flowers bright yellow, beautifully textured.  (J. E. Eichelser)


cl.

Lem's Cameo    See upper right.


cl.
            -fortunei ssp. discolor (1/2)
      -Ole Olson-
Lem's Goal-           -campylocarpum ssp. campylocarpum
      -     -griersonianum (1/4)         Elatum Group (1/4)
      -Azor-
            -fortunei ssp. discolor

5ft(1.5m)          0°F(-18°C)          ML         3/3

Dull, light green, pointed foliage  on a plant of upright habit,
not too open. Flowers of soft creamy apricot, shading darker in
throat; trusses of 6-7.  (Halfdan Lem, 1958)  P. A. (ARS) 1952
Parentage above from American Rhododendron Hybrids, c1980.

              -dichroanthum ssp. dichroanthum (1/4)
      -Dido-
cl.   -     -decorum (1/4)
      -                       -griffithianum (5/16)
Lem's Cameo-      -Beauty of-
      -     -Norman- Tremough-arboreum ssp. arboreum (1/16)
      -           - Gill -
      -Anna-      -griffithianum
      -                 -griffithianum
      -Jean Marie -
            de Montague-unknown (1/8)

5ft(1.5m)          5°F(-15°C)          M          5/3

An exceptional hybrid, with glowing flowers of apricot cream and
pink, with a small scarlet dorsal blotch; corolla widely funnel-
campanulate, 3.5in(9cm) wide, 6-7 wavy lobes; large dome-shaped
trusses of about 20.  Floriferous.  Foliage deep shiny green in
summer, bright bronzy red when new; elliptic leaves 5in(12.7cm)
long.  Difficult to propagate.  (Halfdan Lem, intro. 1962; reg.
1975)    S. P. A. (ARS) 1971    See color illus.:  ARS Q 25:4
(Fall 1971): p. 230; ARS Q 29:1 (Winter 1975): cover;  HG p. 66;
K p. 148.


                    -griffithianum (9/16)
            -Beauty of-
      -Norman- Tremough-arboreum ssp. arboreum
cl.      - Gill -                     (1/16)
      -Anna-      -griffithianum
Lem's  -     -           -griffithianum
Monarch*   -Jean Marie -
      -     -de Montague-unknown (3/8)
      -                 -griffithianum
      -Marinus Koster-
                  -unknown

6ft(1.8m)          -5°F(-21°C)             4/4

Huge trusses of pink-edged cream flowers.  Attractive large foliage on a strong-stemmed plant. May be the same as Pink Walloper, q.v.  (Halfdan Lem)    C. A. (ARS) 1971    See color illus.
HG p. 130.


                    -fortunei ssp. discolor (1/4)
cl.   -King of Shrubs-    -dichroanthum ssp. dichroanthum
      -           -Fabia-                       (1/8)
Lem's Salmon*             -griersonianum (3/8)
      -                 -griffithianum (1/16)
      -     -unnamed-Loderi-
      -Flame- hybrid-    -fortunei ssp. fortunei (1/16)
      -     -Corona--unknown (1/8)
      -griersonianum


An unregistered hybrid by Halfdan Lem.


                              -griffithianum
                    -Queen Wilhelmina-    (5/16)
cl.         -Britannia-         -unknown (9/16)
      -Burgundy-        -Stanley Davies--unknown
Lem's  -     -           -ponticum (1/8)
Stormcloud-      -Purple Splendour-
      -     -griffithianum      -unknown
      -Mars-
            -unknown

5ft(1.5m)          -15°F(-26°C)        ML         4/3

Large flowers of glossy red, paling at center, with a light dorsal blotch, a flattened corolla.  (Cross by Halfdan Lem; reg. by
Britt M. Smith, 1980)


cl.

Len Beer--form of glaucophyllum var. glaucophyllum (B L & M 315)

3ft(.9m)           0°F(-18°C)          EM         3/2

Truss holds 4-8 white flowers.  (Collector L. Beer;  named by
Kathleen Dryden in memory of Len Beer of the Beer, Lancaster &
Morris Expedition, Nepal, 1971; reg. 1978)

'Jodi' by Lem/Granston
Photo by Greer

'John Coutts' by RGB, Kew Gardens
Photo by Greer

'Joseph Whitworth' by Waterer
Photo by Greer

'Juan de Fuca' by Larson
Photo by Greer

'Judy Spillane' by Wister
Photo by West

'Kalimna' by V. J. Boulter
Photo by Greer

'Kaponga' by B. Holland (NZ)
Photo by Greer

'Karin' by Exp. Sta., Boskoop
Photo by Greer

'Katherine Dalton' by Gable
Photo by Knierim

'Kevin' by Bovee
Photo by Greer

'Kilimanjaro' by Rothschild
Photo by Greer

'Kim' by Caperci
Photo by Greer

'Kimberly' by Greer
Photo by Greer

'Kimbeth' by Greer
Photo by Greer

'King's Milkmaid' by W. King (NZ)
Photo by Greer

'Kubla Khan' by Brandt
Photo by Greer

'Kunming' by Leach
Photo by Leach

'Kurt Herbert Adler' by Lilliott
Photo by Greer

'Ladies' Choice' by J. A. Elliott
Photo by Greer

'Lady Armstrong' by Waterer, 1870
Photo by Leach

'Lady Berry' by Rothschild
Photo by Greer

'Lady de Rothschild' by A. Waterer
Photo by Greer

'Lady Dorothy Ella' by unknown
Photo by Greer

'Lady Longman' by Harry White
Photo by Leach

'Lady of June' by Dexter/Wister
Photo by West

cl.
    -pubescens (1/2)
Lenape-
    -keiskei (1/2)

3ft(.9m)          -10°F(-23°C)        E         3/4

A well-branched, shapely plant; narrow, rich green, hairy leaves
2in(5cm) long; terminal clusters of small, light yellow flowers.
Floriferous and easy to propagate.  Similar to Chesapeake, q.v.
See also Hockessin.  (G. Guy Nearing, intro. 1950; reg. 1958)

cl.
        -edgeworthii (bullatum) (1/4)
    -White Wings-
Lencret-          -ciliicalyx (1/4)
    -ciliatum (1/2)

Cream-colored flowers, fading to white.  (Adams-Acton, 1942)

g. & cl.                   -griffithianum (1/8)
    -Queen Wilhelmina-
  -Britannia-          -unknown (3/8)
Leo-      -Stanley Davies--unknown
  -elliottii (1/2)

5ft(1.5m)        -5°F(-21°C)         ML        4/3

Heavy, flaming red flowers are set in tight trusses of up to 25.
Blooms later than most reds.  Plant of good habit with very dark
green foliage. (Rothschild) A. M. (RHS) 1948  See color illus.
PB pl. 28; VV p. 87.

cl.
        -Corona--unknown (1/4)
    -Bow Bells-
Leo Friedman-      -williamsianum (1/4)
    -strigillosum (1/2)

Small, very compact plant; small, very smooth leaves. Flowers
of rose madder, campanulate, 1.25in(3.2cm) wide. Early. (Cecil
S. Seabrook, reg. 1969)

cl.
   -Corona--unknown (5/8)
Leona-      -fortunei ssp. discolor (1/4)
  -Dondis-      -arboreum ssp. arboreum (1/8)
    -Doncaster-
      -unknown

5ft(1.5m)        0°F( 18°C)         ML        4/3

Flowers an intense rich pink, lightly spotted darker rose, very
open in form, 3in(7.6cm) wide; dome-shaped trusses.  Plant open-
growing; medium green leaves, 5in(12.7cm) long. (Rudolph Henny,
1958)

cl.
      -brachyanthum ssp. brachyanthum (1/2)
Leonard Messel-
    -unknown (1/2)

Flowers held in loose trusses of 4-5, primrose yellow, lightly
flecked greenish brown, campanulate, about 1in(2.5cm) across, 5-
lobed.  Elliptic leaves 2in(5cm) long, slightly fragrant, light
scaly indumentum beneath.  (Collector Kingdon-Ward; raiser Leonard Messel; exhibited by the Countess of Rosse, Nymans)  A. M.
(RHS) 1966

cl.          -griersonianum (3/4)
  -Tally Ho-
Leonardo-      -facetum (eriogynum) (1/4)
  -griersonianum

Vermilion flowers.  (Adams-Acton, 1937)

cl.                       -maximum (1/4)
      -Halopeanum-
Leonardslee Brilliant-      -griffithianum (1/4)
    -thomsonii ssp. thomsonii (1/2)

Gem g.  Flowers of deep red.  (Sir E. Loder)

cl.                       -thomsonii ssp. thomsonii (3/4)
    -Ascot Brillant-
Leonardslee Flame-      -unknown (1/4)
    -thomsonii ssp. thomsonii

Red Star g.  (Sir E. Loder)

g.                           -ciliatum (1/4)
    -unnamed hybrid-
Leonardslee Gem of the Woods-      -unknown (1/4)
    -virgatum ssp. virgatum (1/2)

White flowers.  (At Leonardslee, 1912)  A. M. (RHS) 1925

cl.                       -maximum (1/4)
    -Halopeanum-
Leonardslee Gertrude-      -griffithianum (1/2)
  -          -griffithianum
    -Loderi Pink-
    Diamond  -fortunei ssp. fortunei (1/4)

White Lady g.  White flowers with pink markings.  (Sir E. Loder)

        -maximum (1/4)
cl.      -Standishii-      -unnamed-cataubiense
  -          -          - hybrid-      (1/16)
Leonardslee Giles-      -Altaclarense-      -ponticum
  -          -          (1/16)
  -          -arboreum ssp. arboreum
    -griffithianum (1/2)      (1/8)

6ft(1.8m)        0°F(-18°C)         M         4/3

Cameo pink buds opening pale pink and fading white; campanulate
flowers in very large trusses.  Vigorous, sturdy, large-leaved
plant, which prefers partial shade. (Sir Edmund Loder)   A. M.
(RHS) 1948

cl.

Leonardslee Lemon     See Leonardslee Yellow

cl.
     -thomsonii ssp. thomsonii (1/2)
Leonardslee Peach-
    -campylocarpum ssp. campylocarpum (1/2)

Exminster g.  Flowers of orange pink.  (Sir E. Loder)

cl.

Leonardslee-campylocarpum ssp. campylocarpum (1/2)
 Primrose  -
    -arboreum ssp. cinnamomeum var. album (1/2)

Earl of Morley g.  Primrose yellow flowers dotted on upper half
with small maroon spots. (Sir Edmund Loder)   A. M. (RHS) 1933

cl.
    -lindleyi (1/2)
Leonardslee Yellow-
    -nuttallii (sinonuttallii) (1/2)

Synonym Leonardslee Lemon.  A plant for the  cool greenhouse.
Flowers soft Naples yellow in the throat, paling to creamy white
margins, tubular funnel-campanulate, 5 joined lobes 3.8in(9.5cm)
across, in trusses of 3-4.  Narrowly elliptic leaves, both surfaces scaly, glossy dark green, to 7in(17.8cm) long.  (Crossed
& raised by G. R. Loder; exhibited Sir Giles Loder) A. M. (RHS)
1964; F. C. C. (RHS) 1980   [N.B. The 1965 entry describes the
flower as  pale lemon yellow with a light uranium green throat.
"This is the same plant."]

cl.

Leonarran     Parentage unknown

A flowering plant for the cool greenhouse with loose trusses
holding 2-3 flowers tubular-shaped, 3in(7.5cm) long, white with
reverse flushed rose Bengal. (Origin Brodick Castle; raised and
shown by Sir Giles Loder)  A. M. (RHS) 1979

```
g. & cl.
        -auriculatum (1/2)
Leonore-
        -kyawii (1/2)
```

Very late blooming, very large plant. Tall loose trusses of 12
huge raspberry red flowers. Narrow, hairy, rich green leaves in
big rosettes. Not too hardy. (Rothschild, 1947)   A. M.
1948

```
---
        -arboreum ssp. cinnamomeum var. album (1/2)
Leopardi-
        -unknown (1/2)
```

White flowers with slight red tint and large red spots. (T.
Methven & Son, 1868)

```
cl.
        -Mira--unknown (1/2)
Leopold-
        -catawbiense (1/2)
```

[Mira by Stevenson too recent to be used by Seidel]   Flowers of
dark purple violet marked yellow brown. (T. J. R. Seidel, 1909)

```
cl.
            -lepidotum (1/2)
Lepidoboothii-
            -boothii (1/2)
```

Yellowish white flowers tinged pink and green. (E. J. P. Magor)
A. M. (RHS) 1919

```
g. & cl.
            -campylocarpum ssp. campylocarpum Elatum Group
Letty Edwards-                                        (1/2)
            -fortunei ssp. fortunei (1/2)
```

5ft(1.5m)        0°F(-18°C)         M         3/3

Rounded, rather compact bush; medium green leaves 5in(12.7cm)
long. Pale pink buds open primrose yellow to cream. (A. M. variety has clear yellow flowers with red throat) Rounded trusses
of 9-11. (Col. S. R. Clarke) A. M. (RHS) 1946; F. C. C. (Wisley Trials) 1948

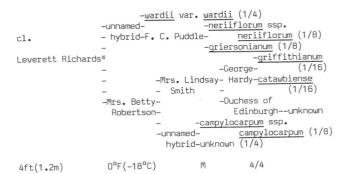

```
                  -wardii var. wardii (1/4)
            -unnamed-        -neriiflorum ssp.
cl.         - hybrid-F. C. Puddle-    neriiflorum (1/8)
            -                 -griersonianum (1/8)
                                       -griffithianum
Leverett Richards*          -George-           (1/16)
            -        -Mrs. Lindsay- Hardy-catawbiense
            -        - Smith    -             (1/16)
            -Mrs. Betty-        -Duchess of
              Robertson-         Edinburgh--unknown
                          -campylocarpum ssp.
                    -unnamed-   campylocarpum (1/8)
                    hybrid-unknown (1/4)
```

4ft(1.2m)        0°F(-18°C)         M         4/4

Flowers more yellow than sister Virginia Richards, but similar
in multiple color tones of pink and apricot. (William Whitney)

cl.

Lhasa--form of nuttallii (sinonuttallii) (L S & E 12117)

One for the cool greenhouse with flowers white, flushed yellow.
(Crown Estate Commissioners) P. C. (RHS) 1971

```
g.                      -fortunei ssp. fortunei (1/4)
            -Faggetter's Favourite-
Libelle-                -unknown (1/4)
        -williamsianum (1/2)
```

Bright rose-colored flowers. (Dietrich Hobbie, 1952)

```
g.                -sperabile var. sperabile (1/4)
            -Eupheno-
Liberty-          -griersonianum (1/4)
        -barbatum (1/2)
```

Deep red flowers. (Lord Aberconway, 1950)

```
cl.
        -lochiae (?) (1/2)
Liberty Bar-
        -aurigeranum (1/2)
```

Vireya hybrid. Trusses of 10-15 flowers, tubular funnel-shaped,
medium jasper red. Height of plant 5ft(1.5m). (Cross by D. B.
Stanton; J. Rouse, reg. 1984)

```
g.                -griersonianum (1/4)
            -Lancer-
Lifeguard-        -hookeri (1/4)
        -arboreum ssp. arboreum (1/2)
```

A hybrid with red trusses. (Lord Aberconway, intro. 1950)

```
cl.
        -diaprepes (1/2)
Lilac Time-            -ponticum (1/4)
        -Purple Splendour-
                      -unknown (1/4)
```

Sister to Lavender Time; both intro. by Gen. Harrison, 1956.

```
---
        -griffithianum (1/2)
Lilian-
        -Mangles hybrid--unknown (1/2)
```

Red flowers fading to blush. (W. C. Slocock)

```
cl.
        -racemosum (1/2)
Lilian Harvey-
        -Hatsugiri (Kurume azalea) (1/2)
```

Azaleodendron. Trusses of 18 flowers of very pale lavender,
lightly flushed with pinkish mauve at midribs and into throat, a
speckling of soft magenta; corolla 1.2in(3cm) wide, openly funnel-shaped, wavy margins. Floriferous. Plant upright, broader
than tall; small glossy leaves. (P. W. Hartijzer, 1983)  H. C.
(RHS) 1976; A. M. (Wisley Trials) 1983

```
g.
        -arboreum ssp. arboreum (red form) (1/2)
Lilianae-        -thomsonii ssp. thomsonii (1/4)
        -Shilsonii-
                -barbatum (1/4)
```

Blood red flowers shaded carmine. A. M. (RHS) 1914

cl.
             -<u>grande</u> (1/2)
Lillian Deans-
             -<u>protistum</u> var. <u>giganteum</u> (1/2)

Flowers held in a tight truss 10in(25.4cm) wide, rose pink buds opening to pale cream, darker in throat. When registered plant was 9ft(2.7m) high and 15ft(4.5m) wide, with leaves 12in(30.4cm) long, 4in(10.2cm) wide. (Raiser James Deans; R. G. Deans, reg. 1969)

cl.            -<u>griffithianum</u> (5/16)
    -Solent Queen-
    -         -<u>fortunei</u> ssp. <u>discolor</u> (1/4)
Lillian-         -<u>griffithianum</u>
Hodgson-        -Mars-
  -      -Vulcan-    -unknown (1/16)
  -           -<u>griersonianum</u> (1/4)
  -Old Copper-    -<u>dichroanthum</u> ssp. <u>dichroanthum</u> (1/8)
        -Fabia-
           -<u>griersonianum</u>

Flowers with shell pink edges and a center of light orange yellow; 10-flowered trusses. (R. C. Rhodes, B.C., Canada, 1979)

cl.
      -Koichiro Wada--<u>yakushimanum</u> ssp. <u>yakushimanum</u>
Lillian Peste-                    (1/2)
      -unnamed W. E. Whitney hybrid--unknown (1/2)

2ft(.6m)        0°F(-18°C)       M

Plant of dense habit, twice as wide as tall; dark almond green leaves, to 4.75in(12cm) long, with heavy ochre indumentum below. Buds dark rose pink, opening coral pink at edges and shading to carrot red in throat, reverse shaded pink, tubular-campanulate, of good substance, 2.5in(6.4cm) across, in spherical trusses of 25-30. Floriferous. (Fred Peste, 1981)

           -<u>auriculatum</u> (1/4)
--- -unnamed hybrid-    -<u>griffithianum</u> (1/4)
  -         -Alice-
Lily*           -unknown (1/4)
  -              -Corona--unknown
  -      -unnamed hybrid-
  -Mrs. Donald Graham-      -<u>griersonianum</u> (1/8)
        -      -<u>griffithianum</u>
        -Loderi-
           -<u>fortunei</u> ssp. <u>fortunei</u> (1/8)

Synonym White Lily. A white-flowered hybrid, by Endre Ostbo. P. A. (ARS) 1952 (Two Ostbo hybrids originally "Lilies", received a P. A. from the ARS: Lily [unregistered] as White Lily, 1952, and Lily #3 [registered as Edna McCarty] in 1959; the parentage of the latter is Alice x <u>auriculatum</u>)

cl.
      -<u>griersonianum</u> (1/2)
Lily Dache-
    -<u>fortunei</u> ssp. <u>discolor</u> (1/2)

Azor g. Salmon pink flowers. (J. B. Stevenson)

cl.
      -<u>fortunei</u> ssp. <u>discolor</u> (1/2)
Lily Maid-
    -unknown (1/2)

7ft(2.1m)       5°F(-15°C)      L

Tubular funnel-shaped flowers, fragrant, pale pink, 3.5in(8.9 cm) across, with 7 frilled lobes; tall trusses of 16. Plant of upright habit; leaves 8.75in(22cm) long, held 3 years. (B. Nelson cross; reg. by Florence Putney, 1975)

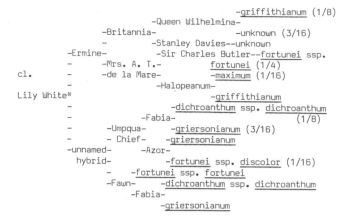

cl.
                      -<u>griffithianum</u> (1/8)
                 -Queen Wilhelmina-
            -Britannia-     -unknown (3/16)
         -        -Stanley Davies--unknown
        -Ermine-      -Sir Charles Butler--<u>fortunei</u> ssp.
      -      -Mrs. A. T.-   <u>fortunei</u> (1/4)
cl.    -      -de la Mare-   -<u>maximum</u> (1/16)
      -        -Halopeanum-
Lily White*    -           -<u>griffithianum</u>
      -      -<u>dichroanthum</u> ssp. <u>dichroanthum</u>
      -        -Fabia-           (1/8)
      -    -Umpqua-  -<u>griersonianum</u> (3/16)
      -    - Chief-  -<u>griersonianum</u>
    -unnamed-    -Azor-
      hybrid-      -<u>fortunei</u> ssp. <u>discolor</u> (1/16)
         -<u>fortunei</u> ssp. <u>fortunei</u>
        -Fawn-   -<u>dichroanthum</u> ssp. <u>dichroanthum</u>
           -Fabia-
             -<u>griersonianum</u>

A medium-sized plant with flowers opening rhodonite red, slowly fading to white. (Weldon E. Delp)

---
       -<u>arboreum</u> ssp. <u>arboreum</u> (1/2)
Limbatum-
    -unknown (1/2)

A parent of Quapp, a hybrid by Seidel. Flowers white or pale blush, edges crimson. (Standish & Noble, before 1870)

           -unnamed-<u>brachycarpum</u> ssp. <u>brachycarpum</u> (1/8)
         - hybrid-   -<u>wardii</u> var. <u>wardii</u> (1/16)
        -    -Crest-      -<u>fortunei</u> ssp. <u>discolor</u>
        -    -Lady -          (1/32)
        -        -Bessborough-<u>campylocarpum</u> ssp.
    -unnamed-       <u>campylocarpum</u> Elatum Gp. (1/32)
    - hybrid-        -Parsons   -<u>catawbiense</u>
    -      -       -Grandiflorum-   (1/32)
    -      -unnamed-America-   -unknown (9/64)
    -      - hybrid-    -dark red hybrid--unknown
    -      -      -<u>griffithianum</u> (3/64)
cl.  -    -Cindy-    -   -Mars-
    -      Lou-     -Blaze-  -unknown
Limeade*  -          -<u>catawbiense</u> (red form)(1/32)
    -         -<u>griffithianum</u>
    -      -Mars-
    -    -Red Brave-  -unknown
    -        -Parsons   -<u>catawbiense</u>
    -      -      -Grandiflorum-
    -      -America-    -unknown
    -         -dark red hybrid--unknown
    -<u>aureum</u> var. <u>aureum</u> (<u>chrysanthum</u>) (1/4)
  unnamed hybrid-
        -<u>maximum</u> (1/4)

Light chartreuse green buds opening to pale yellow. (Weldon E. Delp)

cl.
      -<u>catawbiense</u> var. <u>album</u> (1/2)
Limelight-
        -<u>fortunei</u> ssp. <u>fortunei</u> (1/4)
   -unnamed hybrid-
           -<u>wardi</u> var. <u>wardii</u> (1/4)

4ft(1.2m)     -25°F(-32°C)    ML

Pale yellow flowers with a strong yellowish green blotch, saucer-shaped and open like <u>wardii</u>, 3.5in(8.9cm) wide, 7-lobed, 12-15 in full ball-shaped trusses. Leaves 5in(12.7cm) long and 2.5in(6.4cm) across. Has bloomed after -32°F(-36°C). See also Ivory Tower. (David G. Leach, 1963)

                    -<u>griffithianum</u> (1/8)
g.         -Queen Wilhelmina-
     -Britannia-       -unknown (3/8)
Limerick-    -Stanley Davies--unknown
    -<u>dichroanthum</u> ssp. <u>dichroanthum</u> (1/2)

See clones Margela, Piccaninny. (Earl of Limerick, 1948)

```
                               -griffithianum (1/8)
cl.            -Queen Wilhelmina-
      -Britannia-                    -unknown (3/8)
Linda-          -Stanley Davies--unknown
        -williamsianum (1/2)
```

3ft(.9m)         -15°F(-26°C)         M         3/3

Rounded, compact plant.  Rose red flowers, open-campanulate, in
trusses of 7-8.  Broadly ovate leaves to 3in(7.6cm) long.  See
also Karin.  (Experimental Station, Boskoop)  A. M. (Boskoop)
1968

```
cl.                  -fortunei ssp. discolor (1/4)
        -Lady Bessborough-
Lindberg-            -campylocarpum ssp. campylocarpum
      -     -dichroanthum ssp.       Elatum Group (1/4)
        -Dido-  dichroanthum (1/4)
              -decorum (1/4)
```

4ft(1.2)         5°F(-15°C)         ML         4/3

Jalisco g.  Similar to others of this grex, but has yellow flowers.  (Rothschild, 1954)

```
cl.
        -lindleyi (1/2)
Lindbull-
        -edgeworthii (bullatum) (1/2)
```

Fragrant white flowers, yellow at base.  (E. J. P. Magor, 1926)

```
g.
        -lindleyi (1/2)
Lind-dal-
        -dalhousiae var. dalhousiae (1/2)
```

White flowers, tinged with yellow.  (Cross by Lord Aberconway,
1927)

```
                             -griffithianum (1/8)
                  -George Hardy-
cl.   -Mrs. Lindsay Smith-      -catawbiense (1/8)
    -               -Duchess of Edinburgh--unknown (1/4)
Lingos-               -fortunei ssp. discolor (1/8)
    -         -Lady Bessborough-
      -Jalisco-          -campylocarpum ssp. campylocarpum
    Goshawk-            Elatum Group (1/8)
           -dichroanthum ssp. dichroanthum (1/8)
              -Dido-
                -decorum (1/8)
```

Flowers are cream colored, shading to sulphur yellow in throat.
(Collingwood Ingram, 1972)

```
cl.      -griffithianum (1/4)
      -Loderi-
Linley-    -fortunei ssp. fortunei (1/4)
      -unknown (1/2)
```

Soft pink flowers blotched carmine at base.  (Messel)  A. M.
(RHS) 1927

```
cl.              -griffithianum (1/4)
              -Loderi-
Linley Sambourne-  -fortunei ssp. fortunei (1/4)
              -unknown (1/2)
```

Rich pink flowers blotched carmine at base.  (Messel)  A. M.
(RHS) 1928

```
                         -griffithianum (1/8)
g.                -Queen Wilhelmina-
              -Britannia-           -unknown (3/8)
Linswegeanum-       -Stanley Davies--unknown
        -forrestii ssp. forrestii Repens Group (1/2)
```

Flowers deep scarlet red.  (Dietrich Hobbie, 1946)

```
cl.         -wardii var. wardii (1/4)
        -Hawk-           -fortunei ssp. discolor (1/8)
Lionel  -  -Lady Bessborough-
Fortescue-           -campylocarpum ssp. campylocarpum
        -Ellestee--wardii var.        Elatum Group (1/8)
                 wardii (1/2)
```

Flowers are medium yellow shading lighter to edges, with dark
blotch of ruby red.  (L. S. Fortescue, 1977)

```
g. & cl.
        -lacteum (1/2)
Lionel's-              -griffithianum (1/16)
 Triumph-       -Kewense-
      -    -Aurora-     -fortunei ssp. fortunei (5/16)
      -Naomi-   -thomsonii ssp. thomsonii (1/8)
           -fortunei ssp. fortunei
```

6ft(1.8m)         -5°F(-21°C)         M         4/4

Rose pink in bud, opening to rich Cornish cream flowers, opaline
pink margins fading later; up to 18 large flowers in magnificent
trusses.  Considered by Exbury to be of highest quality.  (Rothschild)  A. M. (RHS) 1954; F. C. C. (RHS) 1974   See color illus.
ARS J 27:4 (Fall 1973): p. 238.

```
cl.        -dichroanthum ssp. dichroanthum (1/4)
        -Fabia-
Lipstick-    -griersonianum (1/4)
      -     -fortunei ssp. discolor (1/4)
        -Dondis-    -arboreum ssp. arboreum (1/8)
            -Doncaster-
                -unknown (1/8)
```

Deep pink flowers on a semi-dwarf plant.  (Rudolph Henny, 1958)

```
cl. -Catalgla--catawbiense var. album Glass (1/2)
    -              -decorum (1/8)
Lisa-      -unnamed hybrid-
    -     -          -griersonianum (1/8)
      -Madonna-            -catawbiense (1/16)
          -    -Parsons Grandiflorum-
        -America-          -unknown (3/16)
              -dark red hybrid--unknown
```

6ft(1.8m)         -15°F(-26°C)         ML         3/3

White flowers with a light sap green blotch, 4in(10.2cm) across,
in large trusses of about 18.  Elliptic leaves 8in(20.3cm) long,
4in(10.2cm) wide.  (Joseph B. Gable, 1964)  P. A. (ARS) 1962

```
                         -catawbiense (1/8)
cl.           -Parsons Grandiflorum-
      -Nova Zembla-          -unknown (3/8)
Lissabon-       -hardy red hybrid--unknown
        -williamsianum (1/2)
```

Flowers of deep rhodonite red with margins cardinal red.  (Joh.
Bruns, W. Germany, 1972)

```
cl.
        -campylogynum Cremastum Group (1/2)
Little Amy-
        -campylogynum (1/2)
```

1ft(.3m)         0°F(-18°C)         ML

(Two forms of the same species do  not make a hybrid.)  Flowers
orient pink, openly campanulate, .75in(1.9cm) across, 5-lobed,
held in trusses of 6 nodding flowers.  Floriferous.  Plant well-
branched, twice as wide as tall; leaves 1.3in(3.5cm) long, aromatic, held 1 year.  (J. Caperci, 1977)  See color illus. ARS
Q 31:2 (Spring 1977): p. 102.

cl.
       -augustinii ssp. augustinii (1/2)
Little Augie*
       -unknown (1/2)

3ft(.9m)        5°F(-15°C)         EM        3/3

Plant of compact habit, bearing bright blue flowers. (Barber, California ?)

---
       -malayanum (1/2)
Little Beauty-                    -brookeanum var. gracile (1/8)
  -        -Duchess of-
  -Monarch- Edinburgh-longiflorum (lobbii) (1/8)
    -              -jasminiflorum (3/16)
  -Princess-              -jasminiflorum
  Alexandra-Princess-
      Royal  -javanicum (1/16)

Vireya hybrid. Bright red flowers.  A. M. (RHS) 1896

cl.
      -neriiflorum ssp. neriiflorum (1/2)
Little Ben-
      -forrestii ssp. forrestii Repens Group (1/2)

2ft(.6m)        10°F(-12°C)        EM        3/3

A low-growing, dense plant with small, stiff, dark green leaves. Deep scarlet, waxy, bell-shaped flowers held in clusters.  (C. R. Scrase-Dickins) F. C. C. (RHS) 1937

cl.
      -forrestii ssp. forrestii Repens Group (1/2)
Little Bert-
      -neriiflorum ssp. neriiflorum Euchaites Group (1/2)

2ft(.6m)        10°F(-12°C)        EM        3/3

Elliptic ovate leaves, up to 3in(7.6cm) long, both ends rounded. Flowers bell-shaped with spreading lobes, 2in(5cm) wide, scarlet crimson, hanging in clusters of 4-5, in rather loose trusses. (C. R. Scrase-Dickins) A. M. (RHS) 1939    See color illus. VV p. 15.

cl.
      -williamsianum (1/2)
Little Bill-                    -griffithianum (3/8)
    -Coombe Royal?-
  -Lady Stuart-              -unknown (1/8)
  of Wortley -griffithianum

3ft(.9m)        5°F(-15°C)         EM        2/3

Plant dome-shaped, medium compact; narrow leaves with reddish petioles. Trumpet-shaped flowers of deep rose pink, held in lax trusses. (Robert Wallace, 1934)

cl.
      -haematodes ssp. haematodes (1/4)
  -Humming Bird-
Little Birdie-              -williamsianum (1/4)
  -forrestii ssp. forrestii Repens Group (1/2)

A dwarf with blood red flowers.  See also Little Bobbie, Little Janet. (Rudolph Henny, reg. 1958)

cl.
      -haematodes ssp. haematodes (1/4)
  -Humming Bird-
Little Bobbie-              -williamsianum (1/4)
  -forrestii ssp. forrestii Repens Group (1/2)

A dwarf by Rudolf Henny.  See also Little Birdie, Little Janet. (Intro. 1956)

cl.
      -dichroanthum ssp. dichroanthum (1/4)
  -Fabia-
Little Dragon-    -griersonianum (1/4)
  -venator (1/2)

3ft(.9m)        5°F(-15°C)         M        3/3

Attractive foliage, red-stemmed leaves, fiery red flowers. (Benjamin F. Lancaster, 1958)

cl.
      -forrestii ssp. forrestii Repens Group (1/2)
Little Ernie-    -haematodes ssp. haematodes (1/4)
  -May Day-
      -griersonianum (1/4)

An upright dwarf plant  with flowers of Turkey red, in trusses of 8-10.  Early. (Ernest V. Allen, 1962)

cl.
      -sanguineum ssp. didymum (1/4)
  -Carmen-
Little Gem-    -forrestii ssp. forrestii Repens Group (1/4)
  -elliottii  (K-W 7725) (1/2)

2ft(.6m)        0°F(-18°C)         M        4/4

Flowers of dark currant red, widely funnel-campanulate, 2in(5cm) wide, of heavy substance; lax trusses of 7-8.  Plant wider than tall, well-branched; glossy dark green leaves with light greyish brown indumentum, held 3 years. (W. Whitney cross; reg. by G. & A. Sather, 1976)  P.A. (ARS) 1962    See color illus. VV p. 84.

cl.
      -forrestii ssp. forrestii Repens Group (1/2)
Little Glendoe-
      -arboreum ssp. delavayi var. delavayi (1/2)

Cardinal red flowers, campanulate, 2.5in(6.4cm) wide, in trusses of 7-9. Rounded, compact plant 28in(.7m) in 14 years, broader than tall; leaves to 3in(7.6cm) long, without indumentum.  Very early.  Hardy to 15°F(-9°C).  (L. E. Jury, raiser; intro. by Ms E. G. Johnstone, NZ, reg. 1968)

cl.
      -impeditum (1/2)
Little Imp*
      -unknown (1/2)

3ft(.9m)        0°F(-18°C)         EM        3/3

More vigorous than the species impeditum. Plant is covered with blue purple flowers. Leaves have a silvery blue cast like some other forms of this species. (Barber)

cl.
      -haematodes ssp. haematodes (1/4)
  -Humming Bird-
Little Janet-              -williamsianum (1/4)
  -forrestii ssp. forrestii Repens Group (1/2)

A dwarf plant with red flowers.  See also Little Birdie, Little Bobbie. (Rudolph Henny, reg. 1958)

cl.
      -forrestii ssp. forrestii Repens Group (1/2)
Little Joe-    -haematodes ssp. haematodes (1/4)
  -May Day-
      -griersonianum (1/4)

1ft(.3m)        5°F(-15°C)         EM        3/3

Plant compact, wider than tall; elliptic dark green leaves 1.5in (3.8cm) long.  Flowers bright currant red, waxy, in lax trusses of 3-4.  Requires some shade; propagates easily. (Lester Brandt, intro. 1951)

cl.                              -ciliatum (1/8)
            -unnamed hybrid-
        -Lucy Lou-                    -leucaspis (3/8)
Little Lou-     -leucaspis
            -valentinianum (1/2)


1ft(.3m)          5°F(-15°C)            E


Very low and spreading, 8in(20.3cm) high and 12in(30cm) broad at
6 years; narrow leaves to 1.75in(4.5cm) long.  Yellow buds opening greenish yellow, tinged apricot; flowers 1.75in(4.5cm) wide.
(Maurice Sumner, reg. 1964)   P. A. (ARS) 1963


cl.
            -haematodes ssp. haematodes (1/2)
Little Minx-    -williamsianum (1/4)
            -Jock-
                -griersonianum (1/4)


Very slow-growing, flat dwarf: 2in(5cm) high and 6in(15cm) wide
in 10 years.  Leaves 2in(5cm) long, 1in(2.5cm) wide. Flowers of
Tyrian rose, funnel-shaped, 1.25in(3.2cm) across; loose trusses
of about 6.  (Vernon Wyatt, 1966)


cl.            -sanguineum ssp. didymum (1/4)
            -Carmen-
Little Nemo-    -forrestii ssp. forrestii Repens Group (1/4)
-                       -haematodes ssp. haematodes
    -Choremia (F.C.C. form)-             (1/4)
                    -arboreum ssp. arboreum (1/4)


Small dwarf only 8in(20cm) high by 15in(38cm) across at 9 years.
Flowers in trusses of 7, Turkey red, large calyx of same color.
See also Honore Hacanson, Fred Robbins.  (L. E. Brandt, 1965)


cl.
            -forrestii ssp. forrestii Repens Group (1/2)
Little Patty-       -haematodes ssp. haematodes (1/4)
            -Humming Bird-
                    -williamsianum (1/4)


Orient red flowers held in trusses of 4-5.  (Rudolph Henny, reg.
1962)   P. A. (ARS) 1958?


                            -griffithianum (1/16)
                -Queen Wilhelmina-
cl.         -Earl of-                -unknown (3/16)
    -unnamed-Athlone-Stanley Davies--unknown
Little- hybrid-     -dichroanthum ssp. dichroanthum (1/8)
Peep -      -Fabia-
-               -griersonianum (1/8)
    -forrestii ssp. forrestii Repens Group (1/2)


Blood red flowers on a prostrate dwarf.  (Rudolph  Henny, 1956)


cl.
            -decorum (1/2)
Little Pudding-     -dichroanthum ssp. dichroanthum (1/4)
            -Fabia-
                -griersonianum (1/4)


5ft(1.5m)          5°F(-15°C)            ML           3/3


Slow-growing; rather asymmetrical in habit; dark green, elliptic leaves with thin indumentum beneath.  Camellia rose flowers
shading to coral in throat, campanulate, 4in(10.2cm) across, 5-7
lobes; 10-13 in loose terminal umbels. (R. Henny, 1958)   P. A.
(ARS) 1953


                            -griffithianum
                -Queen Wilhelmina-       (1/16)
cl.         -Earl of Athlone-            -unknown (3/16)
    -unnamed-                -Stanley Davies--unknown
Little- hybrid-     -dichroanthum ssp. dichroanthum (1/8)
 Sheba-     -Fabia-
-               -griersonianum (1/8)
    -forrestii ssp. forrestii Repens Group (1/2)


1ft(.3m)          5°F(-15°C)            M           3/4


Low-growing, prostrate plant; dark green leaves, 2.25in(5.7cm)
long.  Blood red flowers, 2in(5cm) wide, 3 in a terminal umbel.
Wind-tolerant; easily propagated. (Rudolph Henny, 1958)   P. A.
(ARS) 1954


cl.                   -sanguineum ssp. didymum (1/4)
            -Arthur Osborn-
Little Trooper-        -griersonianum (1/2)
-           -dichroanthum ssp. dichroanthum (1/4)
            -Fabia-
                -griersonianum


2ft(.6m)          0°F(-18°C)            L


Dwarf plant, compact, broader then tall with leaves dull olive
green, 3in(7.6cm) long.  Flowers are ruffled, deep blood red,
held in trusses of 8.  (George L. Baker, 1967)


cl.
            -fortunei ssp. fortunei (1/2)
Little White Dove-
                -yakushimanum ssp. yakushimanum (1/2)


3ft(.9m)          5°F(-15°C)            M           4/4


Light rose pink buds, white flowers, openly campanulate, 2.25in
(5.7cm) across, of heavy substance; rounded trusses of about 12.
Plant rounded, dense; glossy dark green leaves held 3 years. (B.
Lancaster cross; intro. by James Elliott, 1974)


cl.
            -lochiae (1/2)
Littlest Angel-
            -pauciflorum (1/2)


Vireya hybrid.  Trusses of 4 flowers, tubular campanulate, Turkey red.  Plant 16in(.4m) tall; elliptic leaves about 1.6in(4cm)
long.  (G. L. S. Snell, reg. 1984)


cl.                   -griffithianum (1/2)
            -Loderi King-
Livonia Lindsley- George   -fortunei ssp. fortunei (1/4)
             -griffithianum
                -Mars-
                    -unknown (1/4)


4ft(1.2m)          -5°F(-21°C)            M


Flowers orchid pink, fading slightly, in large conical trusses
of 24.  Plant mound-shaped; leaves 5in(12.7cm) long by 2in(5cm)
wide.  (Raised by A. O. Wright, Jr. & Sr.; intro. by A. Lindsley, Sr.; reg. 1969)


cl.


Liz Ann*--form of sargentianum


1.25ft(.38m)          -5°F(-21°C)            M           4/3


Synonym Pink Maricee.  An excellent dwarf with small Daphne-like
flowers opening pastel pink, fading lighter.  See also Maricee.
(J. F. Caperci)


cl.                   -decorum (1/4)
            -unnamed hybrid-
Liza's Yellow-             -fortunei ssp. discolor (1/4)
            -unknown (1/2)


2ft(.6m)          5°F(-15°C)            L


Trusses of 12 flowers, buttercup yellow fading to light Indian
yellow at edges, and fire red in the throat.  Semi-dwarf plant,
branching well; bronze new growth.  (Cross by Grady Barefield;
reg. by Mary W. Barefield, 1980)

cl.
  -minus var. minus Carolinianum Group (white form) (1/2)
Llenroc*
  -Cornell Pink--mucronulatum (1/2)

Profusion of very light pink to white flowers with a yellow eye.
Early. Semi-evergreen with small leaves, green in summer, turning red orange in autumn. Broad, upright plant growing 7in(17.8
cm) in a year. (Edmund Mezitt, Weston Nurseries)


cl.

Loch Eck--form of vernicosum

5ft(1.5m)  -15°F(-26°C)  M  3/3

A fine form of this species with trusses of 12 flowers, 3.25in
(8.3cm) wide, saucer-shaped and pure white. (Younger Botanic
Garden, Argyll) A. M. (RHS) 1964


cl.
    -spinuliferum (1/2)
Lochinch Spinbur-
    -burmanicum (1/2)

Spinbur g. Loose-growing plant with long leaves, scaly beneath.
Tubular-shaped flowers of light yellow, stained pink, in loose
trusses. (Lord Stair) A. M. (RHS) 1957


cl.
  -lochae (1/2)
Lochmin-
  -jasminiflorum (1/2)

Vireya hybrid. Trusses of 7 flowers, light neyron rose, with a
darker tube. (Cross by T. Lelliot, reg. by R. Cutten, 1981)


cl.    -Blue Peter--unknown (5/8)
  -John Dosser-    -caucasicum (1/8)
      -Corry Koster-
Lockington Jane-      -unknown
 -  -campylocarpum ssp. campylocarpum (1/4)
 -Unique-
    -unknown

Trusses of 15 flowers, funnel-shaped, white with spotting of red
purple. Height about 3ft(.9m); obovate leaves. (D. Dosser, Australia, reg. 1984)


      -griffithianum (13/32)
  -Loderi King-
  - George -fortunei ssp. fortunei (5/32)
  -Morio-     -griffithianum
cl. - -    -Loderi-
 - - -Albatross-  -fortunei ssp. fortunei
Lockington- -Cup Day-  -fortunei ssp. discolor (1/16)
 Pride -  -  -elliottii (1/16)
  -  -Fusilier-
  -    -griersonianum (1/16)
  -    -griffithianum
  -Mrs. E. C. Stirling-
      -unknown (1/4)

Trusses hold 13 flowers, light purple with red center and red
spots on upper lobe. See also Marg Sawers. (D. Dosser, 1978)


g.  -dichroanthum ssp. dichronathum (1/4)
 -Fabia-
Lodabia- -griersonianum (1/4)
 -  -griffithianum (1/4)
 -Loderi-
  -fortunei ssp. fortunei (1/4)

Rose-colored flowers. (Lord Aberconway, 1946)


cl.    -griffithianum (1/4)
  -Loderi-
Lodauric Iceberg- -fortunei ssp. fortunei (1/4)
  -auriculatum (1/2)

6ft(1.8m)  0°F(-18°C)  L  4/3

Lodauric g. Late-flowering, pure white and fragrant. Abundant
foliage of light grass green. (W. C. Slocock, 1946)


g.
 -griffithianum (1/2)
Loderi-
 -fortunei ssp. fortunei (1/2)

6ft(1.8m)  0°F(-18°C)  M  5/3

A large tree-like shrub, of strong stature. Flowers very large,
lily-shaped, sweetly scented, in enormous trusses; colors range
from pastel pinks to pure white, depending on selection. Register of 1958 shows 32 named clones. Beautiful foliage. Vigorous
but needs some shelter. See also Kewense. (Sir Edmund Loder,
Leonardslee, 1901)


cl.

Loderi Cougar  See Cougar


cl.
    -griffithianum (1/2)
Loderi Game Chick-
    -fortunei ssp. fortunei (1/2)

6ft(1.8m)  0°F(-18°C)  M  5/3

Loderi g. Flowers are pale pink with a faint blotch. (Sir Edmund Loder, 1901)


cl.
  -griffithinum (1/2)
Loderi Julie-
  -fortunei ssp. fortunei (1/2)

Loderi g. The Register did not credit this to Loder: an omis-
sion, or was it created by another hybridizer? Flowers cream,
suffused with sulphur yellow; the closest to a 'yellow' Loderi.
A. M. (RHS) 1944


cl.
    -griffithianum (1/2)
Loderi King George-
    -fortunei ssp. fortunei (1/2)

6ft(1.8m)  0°F(-18°C)  M  5/3

Loderi g. Considered the best white form in this family. Pink
buds open to flowers flushed pale pink which later become white,
with a green basal flash; blossoms are 6in(15.2cm) wide, openly
funnel-campanulate, in 9- to 10-flowered trusses; leaves narrowly oblong, 10.5in(26cm) long. Delightfully fragrant. A manifi-
cent shrub that in time may become a tree-sized plant, with some
protection from wind, sun, and subzero cold. (Sir Edmund Loder,
1901) A. M. (RHS) 1968  See color illus. F p. 51; VV p.
26, 118.


cl.      -griffithianum (1/2)
 -Loderi Pink Diamond-
Loderi Olga-    -fortunei ssp. fortunei (1/2)
 -    -griffithianum
 -Loderi King George-
    -fortunei ssp. fortunei

6ft(1.8m)  -5°F(-21°C)  M

Trusses of 10 flowers, very fragrant, clear white with throat of
Dresden yellow. (Lester E. Brandt, 1968)

cl.
                    -griffithianum (1/2)
Loderi Pink Diamond-
                    -fortunei ssp. fortunei (1/2)

6ft(1.8m)          0°F(-18°C)          M          5/3

Loderi g.  Flowers pastel shell pink, and fragrant, otherwise
similiar to Loderi King George. (Sir E. Loder, 1901)  F. C. C.
(RHS) 1914


cl.
                    -griffithianum (1/2)
Loderi Pretty Polly-
                    -fortunei ssp. fortunei (1/2)

5ft(1.5m)          0°F(-18°C)          M          5/5

Loderi g.  Shown simply as "pink" in the Register of 1958.  Sir
Giles Loder reported it extinct; Greer's Guidebook to Available
Rhododendrons, 1982, lists it. (Sir Edmund Loder, 1901)


cl.                                    -griffithianum (1/2)
                    -Loderi King George-
Loderi Princess Marina-              -fortunei ssp. fortunei
-                     -griffithianum      (1/2)
                    -Loderi Sir Edmund-
                              -fortunei ssp. fortunei

6ft(1.8m)          0°F(-18°C)          M          5/3

Loderi g.  These pale pink flowers in trusses of 12  shade to
white with pink patches. (Sir Edmund Loder)  A. M. (RHS) 1948


cl.
                    -griffithianum (1/2)
Loderi Sir Edmund-
                    -fortunei ssp. fortunei (1/2)

Loderi g.  Flowers blush pink. (Sir Edmund Loder)  A. M. (RHS)
1930


cl.
                    -griffithianum (1/2)
Loderi Sir Joseph Hooker-
                    -fortunei ssp. fortunei (1/2)

Loderi g.  Flowers of deep shell pink, with conspicuous veining.
(Sir Edmund Loder cross, 1901)   A. M. (RHS) 1973


cl.
                    -griffithianum (1/2)
Loderi Superlative-
                    -fortunei ssp. fortunei (1/2)

6ft(1.8m)          0°F(-18°C)          M          5/3

Loderi g.  Giant  white flowers with the interior flushed a pale
yellow or pink. Fragrant. Other forms may exist. (Sir Edmund
Loder, 1901)


cl.
                    -griffithianum (1/2)
Loderi Venus-
                    -fortunei ssp. fortunei (1/2)

6ft(1.8m)          0°F(-18°C)          M          5/3

Loderi g.  The flowers are soft pink,  and extremely fragrant.
(Sir Edmund Loder, 1901)   See color illus. ARS Q 31:3 (Summer
1977): p. 170.


cl.
                    -griffithianum (1/2)
Loderi White Diamond-
                    -fortunei ssp. fortunei (1/2)

6ft(1.8m)          0°F(-18°C)          M          5/3

Loderi g.  Flowers of ivory white, with a blotch. (Sir Edmund
Loder, 1901)


cl.
          - ? arboreum ssp. cinnamomeum var. album (1/2)
Loder's White-
          -griffithianum (1/2)

A different parentage is given in Hillier's manual, as follows:

                              -catawbiense (1/8)
                    -Album Elegans-
          -unnamed hybrid-              -unknown (1/8)
Loder's White-              -griffithianum (1/2)
          -                     -griffithianum
                    -White Pearl-
                              -maximum (1/4)

5ft(1.5m)          0°F(-18°C)          M          4/3

Plant compact, broader than tall;  bright green leaves 6in(15cm)
long.  Trusses large, conical.  Buds a delicate pink, opening to
slightly fragrant white flowers with pink edges and a tinge of
yellow in the throat. Often a parent. (J. H. Mangles, before
1884) A. M. (RHS) 1911; A. G. M. (RHS) 1931   See color illus.
ARS J 39:1 (Winter 1985): p. 7;  F p. 52;  VV p. i.


cl.
          -catawbiense var. album (3/4)
Lodestar-              -catawbiense var. album
          -Belle Heller-              -catawbiense (1/8)
                    -Madame Carvalho-
                              -unknown (1/8)

5ft(1.5m)          -20°F(-29°C)          ML          3/2

Flowers pale lilac fading to white, with a strongly spotted dorsal blotch of dark greenish yellow, in full trusses of 15; corolla widely funnel-shaped, 3.2in(8cm) wide.  Plant wider than
tall  with superior foliage density; elliptic convex leaves 5in
(12.7cm) long.   See also  Swansdown.   (David G.  Leach, 1965)
C. A. (ARS) 1981; A. E. (ARS) 1982   See color illus. HG p. 66.


g.                    -facetum (1/4)
          -Jacquetta-
Lodetta-              -griersonianum (1/4)
-              -griffithianum (1/4)
          -Loderi-
                    -fortunei ssp. fortunei (1/4)

A hybrid with scarlet rose flowers. (Lord Aberconway, 1946)


cl.
          -rex ssp. arizelum (1/2)
Logan Belle-
          -hodgsonii (1/2)

Large, compact truss with flowers opening cream flushed pink,
fading to cream, with purple blotch. (Hambro)  P. C. (RHS) 1960


cl.                    -caucasicum (1/4)
          -Dr. Stocker-
Logan Damaris-              -griffithianum (1/4)
          -campylocarpum ssp. campylocarpum (1/2)

4ft(1.2m)          5°F(-15°C)          EM          4/3

Damaris g.  Flowers of clear lemon yellow, 3in(7.6cm) across, in
shapely, rather loose trusses of 9-11.  Plant upright, vigorous;
narrow, rich green leaves 5in(12.7cm) long.  2in(5cm) wide.  See
also Cornish Cracker, Townhill Damaris.  (J. B. Stevenson)  A.
M. 1948   See color illus. F p. 41.

'Ladybird' by Rothschild
Photo by Greer

'Lady of Spain' by Lofthouse
Photo by Greer

'Last Hurrah' by Leach
Photo by Leach

'Laura Aberconway' by Aberconway
Photo by Greer

'Laurago' by Goheen
Photo by Goheen

'Lee's Scarlet' by Lee
Photo by Greer

'Lem's Cameo' by Lem
Photo by Greer

'Lem's Goal' by Lem
Photo by Greer

'Lem's Stormcloud' by Lem/B. M. Smith
Photo by Greer

'Leonardslee Giles' by E. Loder
Photo by Greer

'Letty Edwards' by S. R. Clarke
Photo by Greer

'Leverett Richards' by Whitney
Photo by Greer

'Lila Pedigo' by Sather
Photo by Greer

'Lily' by Ostbo
Photo by Greer

'Limelight' by Leach
Photo by Leach

'Linda' by Exp. Sta., Boskoop
Photo by Greer

'Little Dragon' by Lancaster
Photo by Greer

'Little Gem' by Whitney/Sather
Photo by Greer

'Little White Dove' by Lancaster/J. A. Elliott
Photo by Greer

'Livonia Lindsley' by Wright/Lindsley
Photo by Greer

'Lodabia' by Aberconway
Photo by Greer

'Loderi Game Chick' by E. Loder
Photo by Greer

'Loderi Pink Diamond' by E. Loder
Photo by Greer

'Loderi White Diamond' by E. Loder
Photo by Greer

cl.
    -pemakoense (1/2)
Lois-
    racemosum (1/2)

1.5ft(.45m)          -5°F(-21°C)        E         3/4

Flowers funnel-shaped, .75in(1.9cm) broad by 1in(2.5cm) long, 5-lobed, of a soft purplish pink, in a profusion of terminal clusters of about 12. Floriferous. Well-branched plant, wider than tall; medium green leaves 1.5in(3.8cm) long, held 2 years. (H. E. Greer, intro. 1962; reg. 1980)

cl.
    -lutescens (1/2)
Lois Jean-
    -unknown (1/2)

2ft(.6m)           -15°F(-26°C)          M

Terminal clusters, ball-shaped, hold about 14 flowers of primrose yellow fading lighter, 1/4 of each lobe shell pink. Corolla widely funnel-shaped, 1.6in(4cm) wide, 6 wavy lobes. Plant rounded, well-branched, wider than tall; small scaly leaves held 1 year. See also Jill. (Seed from Otto Prycl; raised by W. M. Fetterhoff, reg. 1981)

cl.
    -arboreum ssp. arboreum (1/4)
  -Cornubia-
     -thomsonii ssp. thomsonii (1/8)
Loki-    -Shilsonii-
  -unknown (1/2)    -barbatum (1/8)

Blood red flowers. (Lady Loder) A. M. (RHS) 1933

cl.
    -griffithianum (1/4)
    -Loderi-
Lollipop-    -fortunei ssp. fortunei (1/4)
    -williamsianum (1/2)

2.5ft(.75m)         -5°F(-21°C)         E

Flowers rose pink, of heavy texture, up to 4in(10cm) across, in trusses of 7-9. Plant is broader than tall. (Raised by Endre Ostbo; reg. by Arthur O. Wright, 1964)

cl.
    -williamsianum (1/2)
Lollipop Lace-
    -unknown (1/2)

Trusses of 3 flowers, rich pink with paler, very frilly lobes, 3.75in(9.5cm) wide; a striped effect on exterior. Ovate leaves 3.25in(8.3cm) long. (L. Jury cross; reg. by F. M. Jury, 1982)

cl.
    -griffithianum (1/4)
  -Queen Wilhelmina-
London-    -unknown (3/4)
  -Stanley Davies--unknown

Flowers pink suffused bright crimson; exterior deep crimson. See also Britannia, Earl of Athlone, Langley Park, Trilby, Unknown Warrior. (C. B. van Nes & Sons, before 1922)

cl.
    -Grand Marquis--unknown
Longbeach Dream-
    -unknown

Clear white flowers in trusses of 21. (Mrs. G. E. Grigg, 1982)

cl.
    -burmanicum (1/2)
Longbow-
    -johnstoneanum (1/2)

Trusses of 8-12 flowers, light mimosa yellow; upper lobe has a flare of aureolin yellow. (R. Scorborio, 1982)

          -caucasicum (1/8)
      -Cunningham's White-
   -Vernus-    -ponticum (white form) (1/8)
cl.  -    -     -catawbiense (red form) (1/8)
   -   -red catawbiense-
Longwood-    hybrid   -unknown (1/8)
   -       -griffithianum (1/8)
   -   -Loderi King George-
  -Olympic Lady-    -fortunei ssp. fortunei
    -williamsianum (1/4)     (1/8)

4.5ft(1.35m)        0°F(-18°C)          M         3/3

Ruby red buds opening to lilac pink flowers spotted dark red in throat, roseine purple outside, openly campanulate, 2.5in(6.4cm) across, 6 wavy lobes; dome-shaped trusses of 4-7. Floriferous. Plant rounded, well-branched; medium green leaves held 2 years. (Cross by David G. Leach; raised & intro. by Longwood Gardens; reg. 1976)

      -wardii var. wardii (croceum) (1/4)
cl.  -Charley-  -dichroanthum ssp. dichroanthum (3/16)
  -    -Fabia-
Lonny-    -griersonianum (3/16)
  -    -fortunei ssp. fortunei (1/8)
  -   -Fawn-  -dichroanthum ssp. dichroanthum
  -unnamed-  -Fabia-
    hybrid-    -griersonianum
  -         -caucasicum (1/16)
  -   -Dr. Stocker-
   -Damaris-    -griffithianum (1/16)
    -campylocarpum ssp. campylocarpum (1/8)

5ft(1.5m)          -10°F(-23°C)         EM

Plant upright, as broad as tall with leaves 7in(17.8cm) long, held 2 years. Flowers funnel-shaped, chartreuse green in throat shading to neyron rose as edging of lobes, held in trusses 10in(25.4cm) across. (Del James raiser; reg. by Ray James for Hendricks Park, Eugene, Oregon, 1980)

cl.
    -griffithianum (1/2)
Lord Lambourne-
    -unknown (1/2)

Red flowers. (Otto Schulz cross, 1892; intro. by C. B. van Nes & Sons, 1896)

cl.

Lord Milner     Parentage unknown

A parent of an Exbury hybrid, Good Cheer.

cl.

Lord Palmerston     See Cynthia

cl.

Lord Roberts     Parentage unknown

5ft(1.5m)          -15°F(-26°C)         ML        3/3

A parent of several hybrids. Plant vigorous, rounded, upright, hardy; foliage glossy green and crinkled. Flowers deep red with a black blotch, 2in(5cm) across, in tight ball-shaped trusses. Easily propagated. (B. Mason, reg. 1958)   See color illus. VV title page.

cl.
    -lindleyi (1/2)
Lord Stair-
    -taggianum (1/2)

White flowers with a pale orange stain at the base of the two upper lobes. (Lord Stair, Lochinch) A. M. (RHS) 1952

cl.
```
          -griffithianum (1/2)
Lord Swaythling-
          -unknown (1/2)
```

Flowers pink, neyron rose inside with dark maroon spots.  (Otto Schulz cross, 1890; intro. by C. B. van Nes & Sons, 1896)


cl.
```
          -Duchess of Teck--unknown (1/2)
Lord Wolseley-
          -javanicum (1/2)
```

Vireya hybrid, with red orange flowers.  A parent of Boule D'Or. (Origin unknown)


cl.
```
          -forrestii ssp. forrestii Repens Group (1/2)
Lori Eichelser-        -Corona--unknown (1/4)
          -Bow Bells-
                    -williamsianum (1/4)
```

2ft(.6m)        -5°F(-21°C)        EM        4/4

One of the best dwarfs--dense growth habit, twice as wide as tall; leaves deep jade green, broadly elliptic.  Flowers campanulate, 2.5in(6.4cm) wide; trusses of 3-4 cover the little plant with cherry pink bells. (L. Brandt cross; K. Janeck, reg. 1967)


cl.
```
     -souliei (1/2)
Lorien-
     -wardii var. wardii (1/2)
```

Flowers are a very light sap green.  (J. J. Crosfield, 1973)


cl.

Lost Horizon--form of principis Vellereum Group (K-W 5656)

5ft(1.5m)        -5°F(-21°C)        E        3/4

Flowers pink fading to white, flushed pink with carmine spots on interior, to 1.6in(4cm) wide; compact, rounded trusses of 15-22. Narrowly elliptic leaves, dark green with fawn indumentum below. (Kingdon-Ward, collector; raised by Col. S. R. Clarke; reg. by R. N. S. Clarke)  A. M. (RHS) 1976


g.
```
          -dichroanthum ssp. scyphocalyx Herpesticum Group
     -Metis-                                        (1/4)
Lotis-    -griersonianum (1/4)
     -    -griffithianum (1/4)
     -Loderi-
          -fortunei ssp. fortunei (1/4)
```

Orange red flowers.  (Lord Aberconway, Bodnant, 1946)


cl.
```
          -Mrs. Tritton--unknown
Louis Pasteur-
          -Visount Powerscourt--unknown
```

5ft(1.5m)        -5°F(-21°C)        ML        3/1

Two-toned flowers, brilliant light crimson margins shading to white centers, deep pink exterior; large, dense rounded trusses. Plant has spreading growth habit. Difficult to propagate. (L. J. Endtz & Co., 1923)


cl.

Louvecienne--form of rigidum

6ft(1.8m)        -5°F(-21°C)        EM        4/3


Flowers of pale rose with a speckled eye of rose Bengal, widely funnel-shaped, 3in(7.6cm) wide, 5-lobed, in trusses of 5-7.  Elliptic leaves 2in(5cm) long and 1in(2.5cm) wide. (Collector not known; raised by L. de Rothschild; exhibited by Edmund de Rothschild)  A. M. (RHS) 1975


cl.
```
          -chrysodoron (1/2)
Lovelock-
          -unknown (1/2)
```

Trusses of 5 flowers, light chartreuse green, darkening to yellow at base.  (Dunedin Rhododendron Group, New Zealand, 1977)


cl.
```
          -neriiflorum ssp. neriiflorum Euchaites Group
Lovely William-                                    (1/2)
          -williamsianum (1/2)
```

Deep rose pink flowers.  (Sir James Horlick, 1940)


---

Lowinsky's White     Parentage unknown

A parent of several hybrids.  A large, beautiful plant with very dark green, sharply pointed leaves.  Pale pink buds opening to pure white, unspotted flowers.  Blooms late to very late.    (T. Lowinsky; Sunningdale Nurseries, intro. before 1955)


```
                              -griffithianum (1/8)
---          -Queen Wilhelmina-
          -Britannia-          -unknown (3/8)
Loyalty-          -Stanley Davies--unknown
     -arboreum ssp. zeylanicum (1/2)
```

Red flowers.  (Williams at Lanarth)


cl.
```
          -ciliatum (1/4)
     -Cilpinense-
Lucil-          -moupinense (1/4)
     -lutescens (1/2)
```

Fine Feathers g.  Intro. by Lord Aberconway, 1942.


cl.
```
          -dichroanthum ssp. dichroanthum (1/4)
     -Jasper-          -fortunei ssp. discolor (1/8)
Lucille -     -Lady
Williamson-     Bessborough-campylocarpum ssp. campylocarpum
     -          -souliei (1/4)     Elatum Group (1/8)
          -Virginia Scott-
                    -unknown (1/4)
```

Flower buds currant red, opening lighter red, inside buttercup yellow; corolla 7-lobed, 3.5in(8.9cm) across.  Trusses 11-flowered.  Leaves 4.5in(11.5cm) long.  (H. L. Larson, 1982)


cl.

Lucky Hit--seedling of davidsonianum

Trusses of 6-8, small, saucer-shaped flowers, lilac rose with a blotch of red purple dots. (D. M. van Gelderen, Boskoop, 1969) Gold Medal (Boskoop) 1966


cl.
```
     -griersonianum (1/2)          -griffithianum
     -          -George Hardy-          (3/16)
Lucky Strike-     -Pink Pearl-          -catawbiense (3/16)
     -     -          -arboreum ssp. ar-
     -Countess-          -Broughtonii-     boreum (1/16)
     of Derby-                    -unknown (1/16)
     -          -catawbiense
          -Cynthia-
                    -griffithianum
```

4ft(1.2m)          10°F(-12°C)        ML         3/3

Flowers of deep salmon pink, funnel-shaped, 3in(7.6cm) wide, in
firm conical trusses of about 9. Dull green leaves, 6in(15.2cm)
long. Does best in partial shade. See also Anna Rose Whitney.
(T. Van Veen, Sr., reg. 1958)   See color illus. VV p. 59.


cl.

Lucy Elizabeth--form of formosum var. formosum Iteophyllum Gp.

4ft(1.2m)          15°F(-10°C)        ML         4/3

A plant for the cool greenhouse. White flowers flushed chrome
yellow in upper throat, funnel-shaped, 5-lobed, held in trusses
of 2-3. Dark green leaves, scaly on both surfaces. (Collector
unknown; raised & intro. by Mrs. Elizabeth McKenzie; reg. 1979)
A. M. (RHS) 1979

cl.                       -ciliatum (1/4)
          -unnamed hybrid-
Lucy Lou-                 -leucaspis (3/4)
          -leucaspis

3ft(.9m)          5°F(-15°C)         E          3/4

Flowers of pure snowy white. Attractive foliage, rounded leaves
of soft green. Very floriferous.   (H. L. Larson, reg. 1958)


cl.

Lucy Neal     See Genoveva

cl.                       -fortunei ssp. fortunei (1/4)
          -Ruby F. Bowman-        -griffithianum (1/8)
Lucy's Good Pink-         -Lady Bligh-
          -                       -unknown (1/8)
          -griersonianum (1/2)

6ft(1.8m)          -15°F(-26°C)              L

Flowers of Tyrian rose, widely funnel-shaped, 4.75in(12cm) wide,
7-lobed; large lax trusses of 7. Plant upright, well-branched;
narrowly elliptic spinach green leaves, held 3 years. (John S.
Druecker, intro. 1966; E. R. German, reg. 1976)


cl.          -yakushimanum ssp. yakushimanum (1/2)
Luise Verey-        -Margaret--unknown (1/4)
          -Glamour-
                   -griersonianum (1/4)

3ft(.9m)          0°F(-18°C)              M

Margaret may be a griffithianum hybrid. Rather flat pink flowers with crimson on upper throat; trusses hold 17-18. See also
Diana Pearson. (A. F. George, Hydon Nurseries, 1978)


                   -cinnabarinum ssp. cinnabarinum
g.     -Lady Rosebery-        Roylei Group (1/4)
       -               -              -cinnabarinum ssp.
Luminous-          -Royal Flush-        cinnabarinum (1/8)
       -           (pink form) -maddenii ssp. maddenii (1/8)
       -cinnabarinum ssp. xanthocoden Concatenans Group (1/2)

A hybrid with orange red flowers. (Lord Aberconway, 1950)

cl.      -griffithianum (1/2)
Lunar-
Queen-     -wardii var. wardii (1/4)
     -Hawk-           -fortunei ssp. discolor (1/8)
          -Lady Bessborough-
                   -campylocarpum ssp. campylocarpum
                                   Elatum Group (1/8)

Flowers cream-colored, 4in(10.2cm) across and 2.5in(6.4cm) long,
5-lobed; trusses of 9. Early midseason. (Gen. Harrison, 1967)

cl.
          -yakushimanum ssp. yakushimanum (1/2)
Lurline-           -griffithianum (1/8)
       -           -Queen Wilhelmina-
       -Unknown Warrior-           -unknown (3/8)
                   -Stanley Davies--unknown

2.5ft(.75m)        -5°F(-21°C)              M

Buds of dark rose madder, opening lighter rose and fading paler,
campanulate, to 2.5in(6.4cm) wide, frilled petals; large rounded
trusses of about 20. Compact plant, as broad as high; leathery
dark green leaves with fawn indumentum. (P. M. Brydon, 1969)


g.        -fortunei ssp. fortunei (1/2)
Luscombei-
          -thomsonii ssp. thomsonii (1/2)

Deep rose pink flowers with a ray of crimson, trumpet-shaped,
very large, in loose trusses. Tall, broadly dome-shaped bush.
See also Pride of Leonardslee. (Shown by G. Luscombe, c.1880)


cl.

Luscombei Leonardslee     See Pride of Leonardslee


cl.          -fortunei ssp. fortunei (1/2)
Luscombe's Sanguineum-
                   -thomsonii ssp. thomsonii (1/2)

Luscombei g.  Scarlet crimson flowers. (Luscombe)


g.        -dichroanthum ssp. dichroanthum (1/4)
     -Dante-
Lustre-    -facetum (eriogynum) (1/4)
     -griersonianum (1/2)

Scarlet flowers. (Lord Aberconway, intro. 1941)


---          -javanicum (5/8)
Luteo-roseum-           -jasminiflorum (3/8)
          -Princess Alexandra-        -jasminiflorum
                   -Princess Royal-
                            -javanicum

Vireya hybrid. Flowers soft pinkish yellow to salmon. F. C. C.
(RHS) 1886


cl.
     -catawbiense var. album (1/2)
     -                       -caucausicum (3/32)
Luxor-           -Jacksonii-        -caucasicum
     -      -Goldsworth-     -Nobleanum-
     -      - Yellow -             -arboreum ssp.
     -Goldfort-           -              arboreum (1/32)
     -           -campylocarpum ssp. campylocarpum (1/8)
     -fortunei ssp. fortunei (1/4)

6ft(1.8m)          -25°F(-32°C)        M         4/3

Synonym Morocco. Buds of neyron rose open to flowers of empire
yellow, openly funnel-shaped, 2.75in(7cm) wide, 5- to 7-lobed,
of good substance; spherical trusses hold 14-15. Sturdy plant,
broader than tall; dark green leaves retained 3 years. See also
Congo. (David G. Leach, intro. 1973; reg. 1983)


          -souliei (1/8)
g.     -Soulbut-
     -Vanessa-        -Sir Charles Butler--fortunei ssp.
Lycia-    -griersonianum (1/4)        fortunei (1/8)
     -elliottii (1/2)

Scarlet red. (Lord Aberconway, 1942)

```
                              -griffithianum (1/16)
                    -George Hardy-
          -Pink Pearl-             -catawbiense (1/16)
cl. -Antoon  -          -          -arboreum ssp. arboreum
    -Van Welie-        -Broughtonii-                  (1/16)
Lydia-          -unknown              -unknown (5/16)
    -                         -fortunei ssp. discolor (1/8)
    -        -Lady Bessborough-
    -Day Dream-               -campylocarpum ssp. campylocar-
      -griersonianum (1/4)         pum Elatum Group (1/8)

6ft(1.8m)         0°F(-18°C)        ML        4/3
```

Flowers of China rose with upper lobe lightly marked in cardinal red, rotate funnel-shaped, to 4in(10.2cm) across, 5-lobed; tight trusses of 12-14.   Upright plant; elliptic leaves to 6in(15cm). (Harold E. Greer, 1964)   C. A. (ARS) 1972

```
cl.           -griffithianum (1/4)
        -Alice-
Lyminge-      -unknown (3/4)
        -unknown
```

Bright pink flowers, similar to Alice. Leaves elliptic, glossy dark green, sparsely hairy beneath.   (A. E. Hardy cross; G. A. Hardy, reg. 1984)

```
cl.
        -unnamed hybrid--unknown (1/2)
Lympne-
        -elliottii (1/2)
```

A red-flowered hybrid.   (Maj. A. E. Hardy, intro.)  A. M. (RHS) 1975

```
cl.        -Loder's White, q.v. (2 possible diagrams)
        -C.I.S.-   -dichroanthum ssp. dichroanthum
Lyta Way-     -Fabia-
        -            -griersonianum
        -macrophyllum

4ft(1.2m)         15°F(-10°C)        M
```

Flowers carmine rose with shading of carmine at edges, cardinal red spotting; corolla openly funnel-shaped, 2.5in(6.5cm) across, 5 wavy lobes; flattened trusses of 13 flowers.  Plant upright, about as wide as high; leaves held 2 years. (Cross by Carl Tuomala, 1969; reg. by Eugene R. German, 1984)

```
cl.
        -thomsonii ssp. thomsonii (1/2)
Mabel-
        -unknown (1/2)
```

An apricot-colored thomsonii.   (R. Gill & Son)

```
cl.
        -Blue Steel--impeditum (3/4)
Macaw-       -Blue Steel--impeditum
      -Night Sky-          -russatum (1/8)
              -Russautinii-
                         -augustinii ssp. augustinii (1/8)
```

Truss holds 7 flowers, 6-lobed, bluebird blue with a small white eye. Leaf lanceolate, sparsely scaly.  (J. P. C. Russell, 1983)

```
cl.
        -meddianum var. meddianum (1/2)
Machrie-
        -unknown (1/2)
```

Moderately compact trusses hold about 15 flowers,  5-lobed, currant red.  3.5in(8.9cm) across. Calyx funnel-shaped, almost the same color. (Brodick Castle)  A. M. (RHS) 1965

```
cl.
        -racemosum (1/2)
Macopin-
        -unknown (1/2)

3ft(.9m)         -20°F(-29°C)        M
```

Flowers pale lavender, in globular clusters. Glossy, dark green leaves, 2in(5cm) by 1in(2.5cm).  (G. Guy Nearing, 1963)

```
cl.

Madame A. Moser  or
Madame Albert Moser      Parentage unknown
```

A parent of Edwin O. Weber. Flowers pale mauve, shading white at base, striking gold blotch on upper petal, funnel-shaped, wavy margins; high conical trusses. Free-flowering. Plant tall and vigorous; glossy green leaves. (Moser? A. Waterer?)  A. M. (Wisley Trials) 1954

```
cl.
            -catawbiense (1/2)   (white form?)
Madame Carvalho-
            -unknown (1/2)
```

Flowers are white with greenish spots, in ball-shaped trusses. Dark green foliage; leaves 5in(12.7cm) long, 2.5in(6.4cm) wide. Very bud hardy. (J. Waterer, 1866) See color illus. VV p. 101.

```
cl.

Madame Colijn*      Parentage unknown
```

A parent of Betelgeuse.

```
                            -ponticum (1/8)
cl.            -Michael Waterer-
        -Prometheus-            -unknown (5/8)
Madame de Bruin-    -Monitor--unknown
        -            -arboreum ssp. arboreum (1/4)
        -Doncaster-
                -unknown

4ft(1.2m)         -10°F(-23°C)        ML        3/2
```

A sturdy, vigorous shrub, taller than wide; dark green, pointed leaves  with conspicuous pale green midrib, 5in(12.7cm) long and 2in(5cm) broad. Cerise red flowers; ball-shaped trusses. Needs some shade; roots easily. (M. Koster, 1904)    See color illus. VV p. 39.

```
cl.
            -fortunei ssp. fortunei (1/2)
Madame Fr. J. Chauvin-
            -unknown (1/2)

5ft(1.5m)         -10°F(-23°C)        ML        4/3
```

Upright, compact plant; light green leaves 7in(17.8cm) long by 1.75in(4.5cm) wide.  Flowers rich rosy pink,  paler in throat, with a small red blotch, and held in round trusses. (M. Koster, 1916)   A. M. (RHS) 1933   See color illus. VV p. 33.

```
g.                    -arboreum ssp. arboreum (1/4)
                -Doncaster-
Madame G. Verde Delisle-    -unknown (1/4)
                -griffithianum (1/2)
```

Bright pink flowers. (Lowinsky)   A. M. (RHS) 1919

```
cl.

Madame Guillemot    See Monsieur Guillemot
```

cl.
```
            -caucasicum (1/2)
Madame Jeanne Frets-
            -unknown (1/2)
```

A parent of Rouge de Mai. Early-blooming flowers of rose with a red blotch. (C. Frets & Son)

cl.

Madame Linden      Parentage unknown

A parent of several hybrids by Seidel. Flowers purplish violet. (J. Linden, 1873)

cl.
```
            -racemosum (1/2)
Madame Loth-
            -Kurume azalea (1/2)
```

Azaleodendron with 8-12 funnel-shaped flowers per truss, Persian rose. (Raiser W. H. Hardijzer; intro. by J. van Gelderen, 1965)

cl.
```
            -catawbiense (1/2)
Madame Masson-
            -ponticum (1/2)
```

5ft(1.5m)          -10°F(-23°C)          M          3/3

A large, well-shaped, dense shrub, with shiny dark green leaves. Flowers are white with a golden yellow basal blotch, rather flat with lobes separated, a star-like effect. (Bertin, 1849)    See color illus. F p. 52; VV p. 33.

cl.
```
            -campanulatum ssp. campanulatum (1/2)
Madame Nauen-
            -unknown (1/2)
```

A parent of Viola, by Seidel.

g.
```
            -oreodoxa var. oreodoxa (1/2)
Madame Omei-
            -forrestii ssp. forrestii Repens Group (1/2)
```

Flowers of deep rose. (Dietrich Hobbie, 1954)

cl.
```
            -caucasicum (1/2)
Madame Wagner-
            -unknown (1/2)
```

5ft(1.5m)          -25°F(-32°C)(?)          ML          3/3

Similar to Vincent van Gogh and Rainbow, but with tighter truss, and hardier. Flowers have rose margins and white centers. Very old hybrid. Has been confused with Princess Mary of Cambridge, which is not the same. (J. Macoy)

g.
```
            -maddenii ssp. maddenii (1/2)
Maddchart-
            -chartophyllum var. praecox ? (1/2)
```

Flowers open very pale lilac and fade white. (E. J. P. Magor, 1921)

cl.
```
            -Catalgla--catawbiense var. album Glass (1/4)
          -                      -catawbiense (1/8)
      -Tony's-      -Catawbiense Album-
      - Gift -Belle -              -unknown (1/8)
Madhatter*   -Heller-              -catawbiense
          -          -white catawbiense-
          -                 hybrid       -unknown
          -Cunningham's Sulphur--caucasicum (1/2)
```

Buds of pale purplish pink open to white flowers with purple stamens. (Weldon E. Delp)

cl.
```
      -fortunei ssp. fortunei ? (1/2)
Madison Hill-
      -unknown (1/2)
```

5ft(1.5m)          -5°F(-21°C)          M

Plant rounded, broader than tall; leaves 5.5in(14cm) long, yellowish green, held 2 years. Fragrant flowers of strong purplish pink, paler in throat, without markings, 4in(10.2cm) wide, 5-6 wavy lobes; spherical trusses of 15. (Cross by Charles O. Dexter; named by John Wister & reg. by Gertrude Wister for Swarthmore College, 1982)

g.
```
      -griffithianum (1/2)
Madonna-
      -Mrs. Messel--decorum (1/2)
```

White flowers; hardy. (Messel)   A. M. (RHS) 1931

cl.
```
            -decorum (1/4)
      -unnamed hybrid-
      -            -griersonianum (1/4)
Madonna-                    -catawbiense (1/8)
      -      -Parsons Grandiflorum-
      -America-              -unknown (3/8)
            -dark red hybrid--unknown
```

A parent of Lisa. White flowers with yellow throat. (Joseph B. Gable, 1947)

cl.
```
      -griffithianum (1/4)
      -Mars-                        -catawbiense (3/16)
      -      -unknown   -Parsons Grandiflorum-
Madras-      -America-              -unknown (9/16)
      -      -      -dark red hybrid--unknown
      -Fanfare-            -catawbiense
            -Kettledrum-
                  -unknown
```

4ft(1.2m)          -15°F(-26°C)          M

Synonym Bengal. Plant twice as broad as high; elliptic leaves 4.75in(12cm) long, held 2 years. Flowers cardinal red with dorsal blotch of maroon, widely funnel-shaped, 2.75in(7cm) wide, 5 wavy lobes; conical trusses of 19-21. Free-flowering. (David G. Leach, intro. 1973; reg. 1983)

cl.
```
      -Dexter #L-1--unknown (7/8)
Madrid-                      -catawbiense (1/8)
      -      -Parsons Grandiflorum-
      -America-              -unknown
            -dark red hybrid--unknown
```

5ft(1.5m)          -20°F(-29°C)          ML

Plant rounded, broader than tall; dark green leaves, 5in(12.7cm) long, held 2 years. Flowers roseine purple with a bold dorsal blotch of ruby red, widely funnel-shaped, 3.25in(8.3cm) across; spherical trusses of 12-14. (David G. Leach, intro. 1973; reg. 1983)    See color illus. HG p. 130, as Seville (synonym).

cl.
```
                  -thomsonii ssp. thomsonii (1/4)
      -Barclayi Robert Fox-            -arboreum ssp. arboreum
Maestro-                  -Glory of -            (1/8)
      -                  Penjerrick-griffithianum (1/8)
      -williamsianum (1/2)
```

Dark red flowers, suffused pink. (Gen. Harrison, reg. 1962) P. C. (RHS) 1969; A. M. (RHS) 1975

cl.
```
      -aberconwayi (or hybrid thereof) (1/2)
Magic Moments-
      -yakushimanum ssp. yakushimanum (1/2)
```

White flowers with a large, bright maroon blotch on upper lobe; trusses of 16-22. Floriferous. Glossy green foliage. (John G. Lofthouse, reg. 1979)

cl.                     -wardii var. wardii (1/4)
           -unnamed hybrid-
Magnagloss-            -fortunei ssp. fortunei (1/4)
           -unknown (1/2)


5ft(1.5m)          -15°F(-26°C)          L


Well-branched plant, as broad as tall; yellow green, glossy foliage, held 2 years. Flowers lilac orchid with conspicuous blotch
of greyed red, funnel-shaped, 2.5in(6.4cm) wide; ball-shaped
trusses of 12. (G. Guy Nearing, reg. 1973)


cl.

Magnificum* --form of cinnabarinum ssp. cinnabarinum Roylei Gp.

Two other plants by this name are registered; both parentage unknown, by J. Waterer, and Byls before 1839. In this one, flowers are exceptionally large, orange red. (Reuthe) A. M. (RHS)
1918


g.                            -ciliatum (1/4)
           -Countess of Haddington-
Magniflorum-                  -dalhousiae var. dalhousiae
           -edgeworthii (1/2)                          (1/4)


Exhibited by Parker, 1917.


cl.            -keiskei (1/4)
        -Chink-
Mah Jong-  -trichocladum (1/4)
        -valentinianum (1/2)


Light chartreuse green flowers. (Crown Estate, Windsor) P. C.
(RHS) 1971


                              -griffithianum (1/16)
                      -George-
           -Mrs. Lindsay- Hardy-catawbiense (1/16)
           -Harvest- Smith      -
cl.        - Moon -       -Duchess of Edinburgh-unknown (1/4)
           -                    -campylocarpum ssp. campylocarpum
Maharani-  -unnamed hybrid-                         (1/8)
           -                -unknown
           -         -campylocarpum ssp. campylocarpum
           -Letty Edwards-          Elatum Group (1/4)
                     -fortunei ssp. fortunei (1/4)


Truss of 16 flowers, 5-lobed, greyed white tinged orchid pink,
shading inwards to ruby red with blotch nearly the same. Leaves
elliptic-oval, glabrous. See also Simona. (Cross by H. Hachmann; reg. by G. Stuck, 1983)


cl.
           -brookeanum var. gracile (1/2)
Maiden's Blush-              -jasminiflorum (3/8)
           -Princess Alexandra-          -jasminiflorum
                     -Princess Royal-
                              -javanicum (1/8)


Vireya hybrid. Flowers cream and pink.   F. C. C. (RHS) 1876


cl.
           -campylocarpum ssp. campylocarpum (1/2)
Maiden's Blush-
           -griffithianum (1/2)


Mrs. Randall Davidson g. Flowers pale cream, pink margins. See
also Mrs. Kingsmill. (Loder)


g.
     -haematodes ssp. haematodes (1/2)
Major-
     -thomsonii ssp. thomsonii (1/2)


Brilliant scarlet flowers set in the large thomsonii calyx. The
plant is medium-sized and rather compact. (Rothschild, 1947)

                    -fortunei ssp. discolor (1/4)
              -Roberte-
              -          -campylocarpum ssp. campylocarpum
              -                         Elatum Group (1/8)
     -Coromandel-       -souliei (1/16)
cl.  -          -Soulbut-
     -                    -Sir Charles Butler--fortunei
Malabar-      -Vanessa-              ssp. fortunei (1/16)
     -                   -griersonianum (3/16)
     -Marmora-    -dichroanthum ssp. dichroanthum (5/16)
              -fortunei ssp. discolor
          -Margaret-    -dichroanthum ssp. dichroanthum
          Dunn  -Fabia-
                   -griersonianum


Mimosa yellow flowers, deepening in the throat. (James Russell,
Sunningdale Nurseries, 1972)


cl.            -arboreum ssp. arboreum (1/4)
        -Gill's Triumph-
Malahat-           -griffithianum (1/4)
        -strigillosum (1/2)


Trusses of 14 flowers, guardsman red, spotted deeper red, 2.75in
(7cm) across, 6-lobed. Leaves to 6in(15.2cm), with sparse coppery brown indumentum. (H. L. Larson, reg. 1983)


                              -thomsonii ssp. thomsonii
              -Ascot Brilliant-               (1/8)
              -              -unknown (3/16)
cl.  -J. G. Millais-    -George-griffithianum (1/16)
     -              - Hardy-
Malcolm-       -Pink Pearl-    -catawbiense (1/16)
  Allan -           -           -arboreum ssp. ar-
     -              -Broughtonii-   boreum (1/16)
     -                       -unknown
     -neriiflorum ssp. neriiflorum (1/2)


Deep red flowers. (Sir James Horlick, 1931)


cl.            -griffithianum (1/4)
        -Loderi King George-
Malemute-          -fortunei ssp. fortunei (1/4)
        -unnamed orange hybrid by Lem--unknown (1/2)


5ft(1.5m)          10°F(-12°C)          EM


Buds spirea red opening phlox pink with darker blotch and edges,
openly funnel-shaped, 5in(12.7cm) wide, 7 wavy lobes, fragrant;
lax 10-flowered trusses. Floriferous. Plant rounded, wider
than tall; ivy green leaves 6in(15cm) long, held 2 years. (Cross
by Halfdan Lem, 1966; reg. by J. Elliott, 1978)


cl.            -auriculatum (1/4)
        -unnamed hybrid-
Maletta-          -fortunei ssp. discolor (1/4)
     -            -ungernii (1/4)
        -unnamed hybrid-
                -unknown (1/4)


5ft(1.5m)          5°F(-15°C)          ML


Flowers cream in center with edges of old gold, openly funnel-
campanulate, to 3.25in(8.3cm) across, frilled; pyramidal trusses
of 13-15. Plant rather compact; leaves to 6in(15.2cm). (Henry
R. Yates, intro. 1970; reg. 1972)   See color illus. LW pl. 93.


cl.                     -racemosum (1/4)
              -Conemaugh-
Malta--Pioneer (selfed)-    -mucronulatum (1/4)
     (Gable)    -unknown (1/2)


4ft(1.2m)          -25°F(-32°C)          E          4/4


Plant rounded, broader than tall, twiggy; red-stemmed leaves 2.5
in(6.4cm) long, held 2 years . Flowers mauve pink flushed deeper amaranth rose, semi-double, widely funnel-shaped, 1.75in(4.5
cm) across; ball-shaped trusses of 2-6. Floriferous. (David G.
Leach, cross 1956; reg. 1981)

```
cl.
     -souliei (1/2)
Mamie-          -Corona--unknown (1/4)
     -Bow Bells-
                -williamsianum (1/4)
```

Compact plant to 3.5ft(1m) tall; elliptic leaves 3.5in(8.9cm) long. Flowers fuchsine pink, campanulate, 3in(7.6cm) wide, in upright trusses of 7-11. Midseason. (Rudolph Henny, 1963)

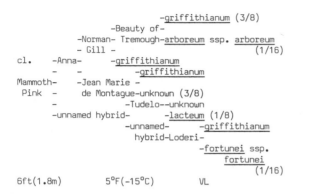

```
                           -griffithianum (3/8)
               -Beauty of-
          -Norman- Tremough-arboreum ssp. arboreum
          - Gill -                       (1/16)
cl.  -Anna-    -griffithianum
     -              -griffithianum
Mammoth-   -Jean Marie -
 Pink -      de Montague-unknown (3/8)
     -          -Tudelo--unknown
     -unnamed hybrid-        -lacteum (1/8)
                  -unnamed-     -griffithianum
                  hybrid-Loderi-
                             -fortunei ssp.
                               fortunei
                                    (1/16)
6ft(1.8m)      5°F(-15°C)         VL
```

Plant upright, well-branched with arching branches; dark grass green leaves 9in(22.9cm) long, held 3 years. Flower buds cardinal red, opening to light neyron rose with blotch of ruby red, 4.75in(12cm) wide, 6-7 wavy lobes, fragrant; truss 13-flowered. (Grady E. Barefield, cross 1958; reg. by Mary Barefield, 1981)

```
                    -griffithianum (1/8)
cl.        -Queen Wilhelmina-
      -Britannia-          -unknown (3/8)
Man of War-    -Stanley Davies--unknown
     -                -griffithianum (1/4)
      -Loderi King George-
                    -fortunei ssp. fortunei (1/4)
```

Vigorous plant with large leaves, 10in(25.4cm) by 4in(10cm). Rose red flowers with dark spots, held in conical trusses of 15. (Marshall Lyons, 1972)

```
              -griffithianum (1/8)
cl.      -Mars-
     -Vulcan-    -unknown (1/8)
Manda Sue-     -griersonianum (1/4)
     -      -campylocarpum ssp. campylocarpum (1/4)
     -Elspeth-
          -hardy hybrid--unknown (1/4)

3ft(.9m)      0°F(-18°C)        M      4/3
```

Plant of compact habit; bright green leaves, 4in(10.2cm) long. Flowers shell pink with red margins and yellowish center, in upright trusses of 12-14. (George L. Baker, reg. 1971)

```
g.
     -haematodes ssp. haematodes (1/2)
Mandalay-
     -venator (1/2)
```

Plant about 4ft(1.2m) tall, of compact habit. Showy deep red flowers in loose trusses. Early midseason. (Rothschild, 1947)

```
cl.       -campylocarpum ssp. campylocarpum
     -Letty Edwards-          Elatum Group (1/4)
Mandarin-          -fortunei ssp. fortunei (1/4)
     -griersonianum (1/2)
```

Flowers of orange salmon. (Thacker, 1946)

```
cl.
Mandarin    See Kinabalu Mandarin (E. F. Allen)
```

```
cl.
     -lacteum (1/2)
Mandate-
     -thomsonii ssp. thomsonii (1/2)
```

Clear pink flowers, 3.5in(8.9cm) wide, in trusses of 10. Early. (Gen. Harrison, reg. 1967)

```
                         -griffithianum (1/8)
                    -George-
          - Betty  -Hardy -catawbiense (1/8)
          - Wormald-
     -Scandinavia-      -unknown (7/32)
cl.     -            -George-    -griffithianum
     -         -Hugh -Hardy -catawbiense
Manderley-     Koster-       -arboreum ssp.
     -          -unnamed-Doncaster- arboreum (1/32)
     -          -hybrid -    -unknown
     -                     -unknown
     -    -dichroanthum ssp. dichroanthum (1/4)
     -Fabia-
          -griersonianum (1/4)

4ft(1.2m)      -5°F(-21°C)        M      3/4
```

Cardinal red, campanulate flowers with darker spotting. Reddish new foliage on a compact plant. (Slootjes, of Boskoop, 1969) A. M. (Boskoop) 1965; A. M. (Wisley Trials) 1983

```
cl.
     -griffithianum (1/2)
Manglesii-          -catawbiense (1/4)
     -unnamed white hybrid-
                    -unknown (1/4)
```

An early Veitch hybrid. White flowers, spotted pink. F. C. C. (RHS) 1885

```
cl.                    -catawbiense (1/4)
     -Catawbiense Boursault-
Mannie Weber-          -unknown (3/4)
     -Madame Albert Moser--unknown

4ft(1.2m)      -5°F(-21°C)        ML
```

Flowers of imperial purple, upper lobe spotted fern green, flat saucer-shaped, 3.5in(8.9cm) wide, 5 frilled lobes; trusses hold 14. Upright plant; very dark green leaves held 3 years. (Edwin O. Weber, 1974)

```
cl.
     -tephropeplum (1/2)
Manon-
     -ciliatum (1/2)
```

Pure white flowers. (Adams-Acton, 1942) Another Manon (?) "a cross of Sappho", shown in Fairweather, p. 53.

```
              -dichroanthum ssp. dichroanthum (5/16)
          -Dido-
cl.   -    -decorum (5/16)
     -                    -fortunei ssp. discolor
Manor Hill-        -Lady  -           (1/16)
     -        -Bessborough-campylocarpum ssp.
     -    -Jalisco-       campylocarpum Elatum Gp.
     -    -    -          -dichroanthum ssp. dichroanthum (1/16)
     -unnamed-   -dichroanthum ssp. dichroanthum
      hybrid-  -Dido-
     -          -decorum
     -yakushimanum ssp. yakushimanum (1/4)
```

Very dwarf, compact plant. Buttercup yellow flowers with paler edges, some staining of Chinese yellow, 1.75in(4.5cm) wide, in trusses of 8-9. Leaves dark green above and sparsely covered with brown indumentum below. (Crown Estate, Windsor)    A. M. (RHS) 1974

g.
       -falconeri ssp. falconeri (1/2)
Mansellii-
       -grande (1/2)

See also Muriel.  Shown by Downie, 1875.

g.
      -Pauline--unknown (1/2)
Marathon-
      -elliottii (1/2)

Plant of moderate height, spreading habit, and rather tender.
Bright red flowers in late midseason.  (Rothschild, 1947)

cl.
      -auriculatum (1/2)
Marbella-
      -griffithianum (1/2)

A hybrid registered by Collingwood Ingram, 1972.

cl.
      -caucasicum citrinum ? (1/2)
March Sun-      -campylocarpum ssp. campylocarpum (1/4)
      -Moonstone-
        -williamsianum (1/4)

1.5ft(.45m)      -5°F(-21°C)      E

Flowers lemon yellow with small crimson blotch in throat, campanulate, 2.25in(5.7cm) wide, 5-lobed; tight trusses of 12.
Very compact plant; olive green leaves with yellow midrib, 2.5in
(6.4cm) long.  (Rollin G. Wyrens, 1965)   P. A. (ARS) 1963

cl.
      -maximum (1/2)
Marchioness of Lansdowne-
      -unknown (1/2)

5ft(1.5m)      -15°F(-26°C)      L      3/2

Flowers of soft violet rose with a striking dark blotch, in neat
domed trusses; "a proud beauty".  Plant habit spreading, rather
open.  A long blooming period; heat-tolerant.  (A. Waterer, before 1915)    See color illus. VV p. 107.

cl.

Marchioness of Tweeddale      Parentage unknown

Large rosy crimson flowers; upper lobes paler with yellow spots.
A. M. (RHS) 1906

cl.
      -campylocarpum ssp. campylocarpum (3/4)
Marcia-      -campylocarpum ssp. campylocarpum
      -Gladys-
        -fortunei ssp. fortunei (1/4)

4ft(1.2m)      0°F(-18°C)      EM      4/3

Upright, slow-growing plant; medium green leaves 3.5in(8.9cm)
long.  Flowers rich primrose yellow with small red eye, campanulate; rounded compact trusses.  Needs some shade.  (Lord Swathling)    F. C. C. (RHS) 1944

cl.
      -Koichiro Wada--yakushimanum ssp. yakushimanum (1/2)
Mardi Gras-      -souliei (1/8)
      -   -Soulbut-
      -Vanessa-      -Sir Charles Butler--fortunei ssp.
        -griersonianum (1/4)      fortunei (1/8)

2.5ft(.75m)      -5°F(-21°C)      EM

Pale pink flowers shading irregularly into white, edges and reverse a strong purplish pink; corolla openly funnel-shaped, 2.5
in(6.4cm) wide, 5 frilled lobes.  Ball-shaped trusses of 11-12.
Plant rounded, wider than tall; dark green leaves with reddish
brown indumentum, held 3-4 years.  (Robert Bovee cross; reg. by
Bovees Nursery, 1976)

      -griffithianum (13/32)
      -Loderi King George-
      -      -fortunei ssp. fortunei (5/32)
      -Morio-      -griffithianum
cl.      -      -Loderi-
      -    -Albatross-  -fortunei ssp. fortúnei
Marg  -  -Cup Day-  -fortunei ssp. discolor (1/16)
Sawers-    -    -elliottii (1/16)
      -    -Fusilier-
      -      -griersonianum (1/16)
      -      -griffithianum
      -Mrs. E. C. Stirling-
        -unknown (1/4)

Light neyron rose flowers held in trusses of 18.  See also Lockington Pride.  (D. Dosser, Australia, reg. 1978)

---

Margaret--unknown

A parent of several hybrids.  Large trusses of widely funnel-
shaped flowers, rose-colored buds opening to white with a blackish crimson flare.  (Origin unknown--Lowinsky?)

cl.
      -hybrid unknown (1/2)
Margaret Alice-      -forrestii ssp. forrestii Repens Group
      -Elizabeth-      (1/4)
        -griersonianum (1/4)

3ft(.9m)      -10°F(-23°C)      EM

Plant wider than tall, branches decumbent; dark green leaves 3.5
in(8.9cm) long.  Flowers Tyrian rose with small red streaks in
throat, openly funnel-shaped, 3.25in(8.3cm) wide, 6 waxy lobes;
spherical trusses of about 8.  (J. Freeman Stephens, reg. 1973)

cl.
      -campylocarpum ssp campylocarpum (1/2)
Margaret Bean-      -griffithianum (1/8)
      -    -Loderi King-
      -Esmeralda- George  -fortunei ssp. fortunei (1/8)
        -neriiflorum ssp. neriiflorum (1/4)

Yellow flowers, fringed in pink.  (RBG, Kew, 1935)

cl.
      -yakushimanum ssp. yakushimanum (1/4)
      -Bambi-    -dichroanthum ssp. dichroanthum
Margaret Cutten-  -Fabia  -      (1/8)
      -    Tangerine-griersonianum (1/8)
      -arboreum ssp. arboreum (1/2)

Trusses hold 21 flowers of bright scarlet.  (R. Cutten, 1981)

g. & cl.
      -fortunei ssp. discolor (1/2)
Margaret Dunn-    -dichroanthum ssp. dichroanthum (1/4)
      -Fabia-
        -griersonianum (1/4)

Flowers apricot pink, flushed shell pink, openly funnel-shaped,
3.5in(8.9cm) wide, in loose trusses of 8-9.  Floriferous.  Narrow elliptic leaves, 6in(15.2cm) long.  See also Golden Belle,
Margaret Dunn Talisman. (Swaythling)   A. M. (RHS) 1946

'Lodestar' by Leach
Photo by Leach

'Lois' by Greer
Photo by Greer

'Lonny' by D. W. James
Photo by Greer

'Lori Eichelser' by Brandt
Photo by Greer

'Luscombei' by Luscombe, 1880
Photo by Greer

'Luxor' by Leach
Photo by Leach

'Lydia' by Greer
Photo by Greer

'Madame Wagner' (*caucasicum* ✕) by Macoy
Photo by Greer

'Madhatter' by Delp
Photo by Delp

'Madras' by Leach
Photo by Leach

'Madrid' by Leach
Photo by Leach

'Malta' by Leach
Photo by Leach

'Manda Sue' by G. Baker
Photo by Greer

'Manderlay' by Slootjes
Photo by Greer

'Mannie Weber' by Weber
Photo by Greer

'Manon' by Adams-Acton
Photo by Greer

'Mardi Gras' by Bovee
Photo by Greer

'Margaret Dunn Talisman' by R. Henny
Photo by Greer

'Marion' by Cheal
Photo by Greer

'Marinus Koster' by M. Koster
Photo by Greer

'Markeeta's Flame' by Mrs. H. Beck
Photo by Greer

'Markeeta's Prize' by Flora Markeeta Nurs.
Photo by Greer

'Marquis of Lothian' by Wm. Martin, 1880
Photo by Greer

'Martha Robbins' by Brandt
Photo by Greer

cl.

Margaret Dunn Talisman     Parentage as above

5ft(1.5m)          0°F(-18°C)          L          3/3

Synonym Talisman. Margaret Dunn g. Flowers funnel-shaped, narrow, apricot colored, in flattened trusses. Medium green leaves
to 6in(15.2cm) long  (Rudloph Henny, reg. 1958)

cl.
                    -auriculatum (1/2)
Margaret Falmouth-
                    -griffithianum (1/2)

Isabella g. White flowers with inner throat heavily marked in
cardinal red, 5-7 frilled lobes, 4.5in(11.5cm) across; trusses
of about 10. Leaves up to 9in(22.9cm) long. (Raised by Col. G.
H. Loder;  exhibited by the Hon. H. E. & Mrs. Boscawen)  A. M.
(RHS) 1968

g.
                    -wardii var. wardii (1/2)
Margaret Findlay-
                    -griersonianum (1/2)

Flowers are almost white with a red stain at base of throat.
(Sir James Horlick, raiser; shown by Crown Estate Commissioners)
P. C. (RHS) 1974

cl.
                -Pink Dexter--unknown (3/4)
Margaret Knight-                    -ponticum (1/4)
                -Purple Splendour-
                                    -unknown

3.5ft(1m)          0°F(-18°C)          M

Plant of compact habit, wider than tall; dark green leaves 5.5in
(14cm) long. Flowers light violet purple edged in medium violet,
with blotch of ruby red, 3.5in(8.9cm) wide; 14-flowered trusses.
(Clarence Loeb, 1971)

cl.
        -Marion--unknown (13/16)         -griffithianum (1/16)
Margaret-                         -George Hardy-
  Mack   -        -Pink Pearl-          -catawbiense (1/16)
         -Annie E.-        -          -arboreum ssp. arboreum
           Endtz -        -Broughtonii-                (1/16)
                 -unknown          -unknown

Flowers soft lilac pink fading paler to center, upper lobe scant
spotting of deep rose, widely campanulate with 5 frilled lobes.
Leaves rich green, margins slightly concave.  (V. J. Boulter,
1965)

cl.

Margaret Mead--form of veitchianum

4ft(1.2m)          20°F(-6°C)          M-L          4/3

A plant for the cool greenhouse. White flowers flushed slightly
with orange red in upper throat, openly funnel-shaped, 5-lobed,
to 3.5in(8.9cm) wide. Dull green leaves, scaly below.  (Origin
unknown; exhibited by Geoffrey Gorer) A. M. (RHS) 1978

cl.                    -griffithianum (1/4)
          -Loderi Venus-
Margaret-              -fortunei ssp. fortunei (1/4)
   Rose  -      -decorum (1/4)
         -Rose du-              -maximum (1/8)
           Barri -Standishii-          -unnamed-catawbiense
                     -Altaclarense- hybrid-          (1/32)
                          -          -ponticum (1/32)
                          -arboreum ssp. arboreum
                                                (1/16)

Jubilee Queen g.  (Loder)

cl.            -decorum (1/4)
          -unnamed hybrid-
Margaret-              -griersonianum (1/4)
Victoria-                        -catawbiesnes (1/8)
        -              -Parsons Grandiflorum-
        -America-                        -unknown (3/8)
                    -dark red hybrid--unknown

2.5ft(.75m)          -15°F(-26°C)          ML

Synonym Degram 15-59. Flowers moderate purplish pink with dark
orangeyellow spotting, in dome-shaped trusses of 8-10. An upright plant, as broad as high, leaves dark yellowish green, held
2 years. Plant wood is brittle and needs some winter protection
from snow and ice. (Joseph Gable; reg. by R. K. Peters, 1983)

cl.                        -griffithianum (1/8)
          -Queen Wilhelmina-
        -Britannia-              -unknown (3/8)
Margela-        -Stanley Davies--unknown
        -dichroanthum ssp. dichroanthum (1/2)

Limerick g. Flowers orange brown, shading to wide band of geranium lake at margins. (Earl of Limerick)  A. M. (RHS) 1948

cl.            -catawbiense (3/4)
          -Mrs. Milner-
Margot-            -unknown (1/4)
          -catawbiense

Flowers dark purplish red with brown markings.  (T. J. R. Seidel cross, 1910)

---
          -Mucronatum--unknown (1/2)
Margot-
          -micranthum (1/2)

Vivid magenta pink flowers. The I.I.R. of 1958 called this an
azaleodendron. (Collingwood Ingram, 1937)

cl.                    -griffithianum (1/8)
          -Mars-
        -Vulcan-    -unknown (1/8)
Margret Einarson-        -griersonianum (1/2)
        -    -griersonianum
        -Azor-
                -fortunei ssp. discolor (1/4)

2ft(.6m)          5°F(-15°C)          L

Shrub rounded, well-branched, slightly broader than high; boat-
shaped, light green leaves retained 2 years; new growth coppery
bronze. Flowers glowing geranium lake with deep rose stripes on
reverse, openly funnel-shaped, 2.5in(6.4cm) wide, 6-lobed; round
trusses of 12. Floriferous. (Mrs. Sigrid Laxdall, reg. 1977)

cl.
          -catawbiense (1/2)
Marguerite-
          -unknown (1/2)

Deep pink flowers. (Paul Bosley, Sr., before 1958)

cl.
      -late-flowering hybrid--unknown
Marham*
      -Mary Waterer--unknown

Pink flowers with a red eye and speckled blotch. (Knap Hill,
1967)

cl.
          -lacteum (1/2)
Marianne Hardy-
          -fortunei ssp. discolor (1/2)

Creamy white flowers with blotch of yellow orange in throat, held in trusses of 11-15. (Cross by L. de Rothschild; intro. by Maj. A. E. Hardy & G. A. Hardy)  F. C. C. (RHS) 1982

cl.

Maricee--form of sargentianum

2ft(.6m)          -5°F(-21°C)          M          4/3

Dwarf and twiggy with shiny leaves typical of this species, but faster growing and more floriferous. Flowers small, creamy white and forming delicate miniature trusses.  See also Liz Ann. (James F. Caperci, 1962)  A. M. (Wisley Trials) 1983   See color illus. ARS J 37:3 (Summer 1983): p. 127.

```
                               -griffithianum (1/8)
g.                    -Loderi-
              -Albatross-      -fortunei ssp. fortunei (1/8)
Marie Antoinette-     -fortunei ssp. discolor (1/2)
              -         -fortunei ssp. discolor
                  -Ariel-
                      -Memoir--unknown (1/4)
```

White flowers blushed pink and flecked yellow green, in enormous conical trusses.  Plant tall, upright, blooming late midseason. (Rothschild, 1945)

```
cl.             -catawbiense (1/4)
         -unnamed hybrid-
Marie Giasi-             -unknown (1/4)
         -maximum (1/2)
```

5ft(1.5m)         -25°F(-32°C)          M

Plant broader than tall; dark green leaves held 3 years.  Ruby red buds opening deep spirea red, spotted with beetroot purple, 3in(7.6cm) wide, 5 heavily ribbed lobes, in 17-flowered trusses. Floriferous.  (W. David Smith, intro. 1980; reg. 1983)

```
cl.
         -vernicosum (1/2)
Marie    -           -wardi var. wardii (1/8)
Tietjens- -Crest-          -fortunei ssp. discolor (1/16)
     -    -    -Lady-
     -Full-  Bessborough-campylocarpum ssp. campylo-
     Moon-                carpum Elatum Group (1/16)
     -                      -griffithianum (1/32)
     -     -Mrs.   -George Hardy-
     -Harvest-Lindsay-           -catawbiense (1/32)
      Moon - Smith -Duchess of Edinburgh--unknown(1/8)
         -       -campylocarpum ssp. campylo-
     -unnamed hybrid-            carpum (1/16)
                      -unknown
```

3ft(.9m)         -10°F(-23°C)          ML

Flowers soft spirea red without markings, openly funnel-shaped, 3.5in(8.9cm) across, 7 wavy lobes; spherical trusses of 6-7. Floriferous.  Plant upright, wider than high; glossy leaves 6.5in(16.5cm) long, held 2 years. (Cross by Lewis Bagoly, 1969; raiser Marie Tietjens; reg. 1976)

```
g.             -caucasicum (1/4)
         -Dr. Stocker-
Mariloo-           -griffithianum (1/4)
         -lacteum (1/2)
```

6ft(1.8m)         10°F(-12°C)          EM          4/3

Named for Mrs. Lionel de Rothschild.  Called the superior lacteum hybrid of Exbury.  Bell-shaped flowers, pale lemon yellow, flushed green and tubular campanulate on opening; the green tint fades and the tube expands later.  Trusses very large, perfectly shaped; leaves about 6in(15.2cm) long, with variations among the forms; young foliage cold-sensitive.  Considered "a superlative plant."  See also Mariloo Eugenie, Mariloo Gilbury. (Rothschild, 1941)   See color illus. PB pl. 29.

cl.

Mariloo Eugenie     Parentage as above

6ft(1.8m)         10°F(-12°C)          EM          4/3

Mariloo g.  Similiar to the above plant, but the flowers are pale cream or creamy white, with a few small crimson spots.  The magnificent trusses hold up to 17 flowers.  (Rothschild)  A. M. (RHS) 1950    See color illus. PB pl. 30.

cl.

Mariloo Gilbury     Parentage as above.

6ft(1.8m)         10°F(-12°C)          EM          4/3

Mariloo g.  Also similiar to the above.  These flowers are pale creamy pink with dark pink stripes on reverse of lobes. (Rothschild)  A. M. (RHS) 1943

```
cl.          -elliottii (1/4)
      -Fusilier-
Marilyn-     -griersonianum (1/2)
     -     -dichroanthum ssp. dichroanthum (1/4)
      -Fabia-
         -griersonianum
```

A low, dwarf plant with narrow leaves, 2.5in(6.4cm) long.  Large flowers red to scarlet, with orange tint.  (Rudolph Henny, 1965)

cl.

Marine--form of augustinii ssp. augustinii

6ft(1.8m)         10°F(-12°C)          EM          3/3

Growth habit typical of the species--plant rather slender until mature.  Flattened flowers of deep lavender blue, spotted purple on upper petal; trusses of 3.  Leaves medium green, 2in(5cm) by 1in(2.5cm).  (James Barto selection; raised & intro. by The Bovees; reg. 1962)  P. A. (ARS) 1960

```
            -Lady      -fortunei ssp. discolor (3/16)
            -Bessborough-
cl.  -Day Dream-        -campylocarpum ssp. campylocarpum
     -        -griersonianum (1/4)    Elatum Group (3/16)
Mariner-     -wardii var. wardii (1/8)
     -       -Hawk-
     -       - -Lady    -fortunei ssp. discolor
     -unnamed-  Bessborough-
      hybrid-        -campylocarpum ssp. campylocarpum
         -    -wightii (1/8)    Elatum Group
          -China-
              -fortunei ssp. fortunei (1/8)
```

Trusses of 15-20 flowers, unfading primrose yellow with a greenish yellow blotch, ochre in bud. (A. F. George, Hydon Nurseries, 1966)

```
            -dichroanthum ssp. dichroanthum (1/8)
         -Astarte-    -campylocarpum ssp. campylocar-
         -    -Penjerrick-    pum Elatum Group (1/16)
g.  -Solon-         -griffithianum (3/16)
     -   -    -griffithianum
Marinka-  -Sunrise-
     -   -    -griersonianum (3/8)
     -    -griersonianum
      -Rapture-
         -arboreum ssp. zeylanicum (1/4)
```

Pale rose flowers.  (Lord Aberconway, intro. 1950)

cl.
                -griffithianum (1/2)
Marinus Koster-

                -unknown (1/2)

6ft(1.8m)          -5°F(-21°C)          M          4/3

Rich crimson buds opening rose pink, heavy brown spotting on upper petals; flowers campanulate, to 5in(12.7cm) wide, in huge
trusses of 10-12. Floriferous. Shrub upright, vigorous; leaves
shiny dark green, 7in(17.8cm) long. Sun-tolerant, pest-resistant, easily propagated. (M. Koster & Sons, 1937) A. M. (RHS)
1937; F. C. C. (RHS) 1948


                            -griffithianum (1/8)
                  -George Hardy-
cl.   -Pink Pearl-          -catawbiense (3/8)
      -            -        -arboreum ssp. arboreum (1/8)
Marion-      -Broughtonii-
      -                     -unknown (3/8)
      -                  -catawbiense
      -Catawbiense Grandiflorum-
                            -unknown

Large pink flowers. (Felix & Dijkhuis)


cl.


Marion     Parentage unknown

Often a parent. Tyrian rose flowers, margins frilled and upper
petal spotted orange, reverse Tyrian rose. Reportedly resembles
the Marion from Felix & Dijkhuis. (Cheal & Son)    A. M. (RHS)
1955


                                -griffithianum (1/8)
cl.                   -George Hardy-
            -Mrs. L. A. Dunnett-          -catawbiense (1/8)
Marion Koster-                   -unknown (1/4)
            -griersonianum (1/2)

Red flowers; late. (M. Koster & Sons)


cl.            -griffithianum (1/4)
          -Alice-
Marion Street-      -unknown (1/4)
          -yakushimanum ssp. yakushimanum (1/2)

Compact plant, with indumentum like yakushimanum; deep pink buds
opening to lighter pink flowers, white in throat, green spotted,
2in(5cm) wide. (Cross by John Street; reg. by Frederick Street,
1965) A. M. (RHS) 1978


                        -griffithianum (7/16)
                  -Beauty of-
              -Norman- Tremough-arboreum ssp. arboreum (1/16)
              - Gill -
cl.    -Anna-     -griffithianum -griffithianum
       -      -Jean Marie de Montague-
Marj   -                  -unknown (1/4)
Anderson-              -griffithianum
       -          -Mars-
       -Captain Jack-   -unknown
                  -facetum (eriogynum) (1/4)

3ft(.9m)        5°F(-15°C)          M

Flowers 3.5in(8.9cm) wide, 5-6 lobes, fragrant, light orient
pink with edging of medium neyron rose; in 11-flowered trusses
8.5in(21.6cm) across. Leaves medium to large, glossy dark green
and held 2 years. (Cross by Grady Barefield; reg. by Mary Barefield, 1983


                -dichroanthum ssp. dichroanthum (1/8)
          -Fabia-
cl.    -Iviza-   -griersonianum (1/8)
                -auriculatum (1/8)
Marjorie-   -Bustard-       -campylocarpum ssp. campylo-
Baird -         -Penjerrick- carpum Elatum Group (1/16)
      -                      -griffithianum (1/16)
      -campylocarpum ssp. campylocarpum (1/2)

Deep yellow flowers with a maroon blotch, tubular-campanulate.
Plant 6ft(1.8m) in 10 years, with leaves 5.5in(14cm) long. (H.
L. Larson, 1965)


                            -griffithianum (1/16)
                  -George Hardy-
cl.      -Mrs. Lindsay-      -catawbiense (1/16)
      -Diane-  Smith  -Duchess of Edinburgh--unknown (1/4)
Mark -       -          -campylocarpum ssp.
Henny-    -unnamed hybrid-  campylocarpum (1/8)
      -                  -unknown
      -williamsianum (1/2)

A dwarf plant 1.5ft(.45m) tall; dark green leaves 3.5in(8.9cm)
long. Flowers cherry red to currant red, funnel-campanulate, in
trusses of 7-8. (Cross by Rudolph Henny; intro. & reg. by Leona
Henny, 1965)


cl.                     -griffithianum (1/4)
          -Jean Marie de Montague-
Mark Twain-             -unknown (1/4)
       -          -dichroanthum ssp. scyphocalyx (1/4)
          -Indiana-
             -kyawii (1/4)

A large open plant, but shapely, with very large, smooth leaves.
Flowers currant red, campanulate, 3in(7.6cm) across, in tight,
high trusses, 8in(20cm) in diameter. (C. S. Seabrook, 1967)


cl.


Markeeta--form of open pollinated adenopodum

5ft(1.5m)          -5°F(-21°C)          EM          3/4

Flowers of very light neyron rose fading almost white, with tan
spotting, widely funnel-campanulate, 3in(7.6cm) across, 5 wavy
lobes; conical trusses of 16-18. Plant upright, almost as broad
as tall; narrow concave leaves, 7in(17.8cm) long, with fawn indumentum, greyed white tomentum on new growth. Leaves held 3-4
years. (Dr. David Goheen, 1983)


                  -griffithianum (9/16)
cl.    -Loderi Venus-
       -          -fortunei ssp. fortunei (1/4)
Markeeta's-             -griffithianum
 Flame -        -Beauty of-
       -    -Norman- Tremough-arboreum ssp. arboreum (1/16)
       -Anna- Gill -
       -        -griffithianum
       -                    -griffithianum
       -Jean Marie de Montague-
                            -unknown (1/8)

5ft(1.5m)          -5°F(-21°C)          M          4/4

Much like Markeeta's Prize and from the same parents. Deep fir
green leaves, red leaf stems. Flowers neyron rose with dorsal
spotting of rose red, openly funnel-shaped, 5in(12.7cm) across,
5 wavy lobes; ball-shaped trusses of 10-12. Floriferous. Plant
upright; leaves held 2 years. See also Markeeta's Prize. (Cross
by Mrs. Howard Beck; raiser Flora Markeeta Nurseries; intro. &
reg. by Allen Korth, 1978)


cl.


Markeeta's Prize     Parentage as above

4ft(1.2m)          -5°F(-21°C)          M          5/4

Plant broader than tall; leathery dark green leaves 6in(15.2cm)
long.  Scarlet red flowers 5in(12.7cm) wide, 12 per truss.  See
also Markeeta's Flame.  (Flora Markeeta Nursery, 1967)   See
color illus. HG p. 110.


cl.
        -catawbiense (1/2)
Markgraf*
        -unknown (1/2)

Flowers of dark violet, with brownish yellow to greenish yellow
markings; frilled edges.  (T. J. R. Seidel, intro. 1910)


cl.
        -dichroanthum ssp. dichroanthum  (5/8)
Marmora-        -fortunei ssp. discolor (1/4)
        -Margaret-    -dichroanthum ssp. dichroanthum
         Dunn  -Fabia-
                      -griersonianum (1/8)

A parent of Malabar.  Flowers bright orange yellow flushed rose
at edges.  (Sunningdale Nurseries, 1955)


g.              -arboreum ssp. arboreum (1/4)
      -Red Admiral-
Maroze-          -thomsonii ssp. thomsonii (1/4)
      -meddianum var. meddianum (1/2)

Tubular-shaped, waxy red flowers.  (J. B. Stevenson cross, 1937;
intro. 1946)


cl.
                -thomsonii ssp. thomsonii (1/2)
Marquis of Lothian-
                -griffithianum (1/2)

Trusses hold 7-10 flowers colored crimson and carmine rose with
a darker flush.  (Crossed before 1880 by William Martin; reg. by
Dunedin Rhododendron Group, New Zealand, 1977)


cl.
      -griffithianum (1/2)
Mars-
      -unknown (1/2)

4ft(1.2m)           -10°F(-23°C)        ML        4/3

Flowers deep true red, unspotted, waxen, widely bell-shaped with
white stamens, held in neat rounded trusses.  Plant rather slow-
growing and compact; narrow, ribbed, dark green leaves.  Not as
hardy as Nova Zembla, but a fine red for many cold areas.  Needs
afternoon shade; roots easily.   (Waterer, before 1875)    A. M.
(RHS) 1928;  F. C. C. (RHS) 1935   See color illus. VV p. 76.


---
          -Betsy Trotwood--unknown (3/4)
Mars Novus*    -griffithianum (1/4)
          -Mars-
                -unknown

A parent of several hybrids.  Flowers fiery scarlet red, lighter
in center.  (Seidel, before 1880)


g.
        -haematodes ssp. haematodes (1/2)
Marshall-
        -elliottii (1/2)

Low, spreading plant; dark green leaves with slight fawn indumentum.  Waxy, bright scarlet, funnel-shaped flowers in medium-
sized trusses.  Midseason.  See also Lanyon.  (Rothschild, 1947)


cl.                 -campylocarpum ssp. campylocarpum
          -Ole Olson-                Elatum Group (1/4)
Marshall Lyons-      -fortunei ssp. discolor (1/4)
              -Loderi King George-  -griffithianum (1/4)
                             -fortunei ssp. fortunei (1/4)

Flowers phlox pink with deeper shades and a small red blotch; of
open shape.  Plant of open habit; pointed oval leaves 6.5in(16.5
cm) long.  (Marshall Lyons, 1972)


cl.                        -griffithianum (1/8)
                    -Loderi-
        -unnamed hybrid-   -fortunei ssp. fortunei (1/8)
Marshmallow-        -unknown (3/4)
        -unknown

Trusses of 10 flowers, orchid pink with a faint red purple flare
in throat.  Leaves 6in(15.2cm) long.  (Mrs. R. J. Coker, 1982)


cl.
        -azalea occidentale (1/2)
Martha Isaacson-                 -Corona--unknown (1/8)
        -          -unnamed hybrid-
        -Mrs. Donald-            -griersonianum (1/8)
         Graham   -   -griffithianum (1/8)
                  -Loderi-
                        -fortunei ssp. fortunei (1/8)

Parentage as in Van Veen.  A fragrant azaleodendron with unusual
foliage; sun-tolerant.  Leaves tinged deep maroon, a good contrast to white flowers striped in pink.   (E. Ostbo, reg. 1958)
P. A. (ARS) 1956


cl.        -griffithianum (1/4)
        -Mars-
Martha May-  -unknown (1/4)
        -        -griersonianum (1/4)
        -Tally Ho-
                -facetum (eriogynum) (1/4)

Ten flowers per upright truss, clear red.  Tall plant of medium
spread, leaves 6in(15.2cm) long, dark green.  (R. Henny, 1965)


cl.
        -forrestii ssp. forrestii Repens Group (1/2)
Martha Robbins-
        -sperabile var. sperabile (1/2)

1.5ft(.45m)          0°F(-18°C)        EM        4/4

A delightful plant.  Bright red flowers, 2.25in(5.7cm) wide, in
trusses of 4; floriferous.  Plant twice as wide as high; foliage
like repens but slightly larger, more vigorous; leaves glossy,
dark green, 2in(5cm) long.  (Lester Brandt, 1971)


cl.
        -burmanicum (1/2)
Martha Wright-    -edgeworthii (1/4)
        -Fragrantissimum-
                -formosum var. formosum (1/4)

Plant was 5ft(1.5m) high by 4ft(1.2m) wide when registered in
1967.  Flowers creamy white with yellow center, fragrant, in
flat trusses of 4.  (Mr. & Mrs. Maurice Sumner)  Award as the
best new hybrid in California, 1963.


cl.

Martin Hope Sutton     Parentage unknown

A parent of Nuneham Park.  Flowers of rich rose scarlet.   (A.
Waterer, before 1915)

cl.
```
      -racemosum (1/2)
Martine-
          -evergreen azalea--unknown (1/2)
```

2.5ft(.75m)          -5°F(-21°C)          EM          4/3

An azaleodendron similar to Ria Hardijzer. Clear fuchsine pink flowers, darker spotting on upper lobe, 1.25in(3.2cm) wide, funnel shaped; flowers in 5-10 lateral trusses of 2-4, forming a compact pseudo-terminal truss. Shiny foliage.    (W. Hardijzer, 1965)

cl.
```
            -Blue Peter--unknown (13/16)
Martin's Pride-
                        -ponticum (1/8)
          -Purple Splendour-
        -unnamed-              -unknown
         hybrid-            -griffithianum
              -            -Queen Wilhelmina-      (1/16)
              -Trilby-              -unknown
                    -Stanley Davies--unknown
```

6ft(1.8m)          0°F(-18°)          ML

Flowers of strong reddish purple, lightly spotted yellow brown, openly funnel-shaped, 3in(7.6cm) wide; rounded trusses of 12-16. Floriferous. Plant as broad as tall; deep green leaves retained 2-3 years. (Cross by Martin Wapler; reg. by Del Loucks, 1977)

cl.
```
                        -catawbiense (1/8)
            -Atrosanguineum-
        -Atrier-              -unknown (1/8)
Mary Belle-    -griersonianum (1/4)
        -          -decorum (1/4)
        -Dechaem-
                  -haematodes ssp. haematodes (1/4)
```

5ft(1.5m)          -15°F(-26°C)          M          4/4

Flowers opening light salmon pink from deeper buds, fading golden peach, with basal blotch of cardinal red. Flowers of much substance, 4in(10.2cm) wide, very ruffled edges; flattened 10-flowered trusses. Plant fairly compact.    (Jos. B. Gable, 1964) P. A. (ARS) 1962    See color illus. LW pl. 54; VV p. 106.

cl.
```
            -haematodes ssp. haematodes (1/2)
Mary Briggs-          -forrestii ssp. forrestii Repens Group
        -Elizabeth-                              (1/4)
                  -griersonianum (1/4)
```

8in(20cm)          0°F(-18°C)          M

Blood red flowers, funnel-campanulate, 2in(5cm) across, 8-10 in compact trusses. Plant twice as broad as high; dark green foliage, 3in(7.6cm) long. (Vernon Wyatt, 1967)

cl.
```
            -yakushimanum ssp. yakushimanum (1/2)
Mary Chantry-                    -griffithianum (1/8)
        -              -Queen Wilhelmina-
        -Britannia-                    -unknown (3/8)
                      -Stanley Davies--unknown
```

2.5ft(.75m)          0°F(-18°C)          ML

Buds of neyron rose open to fragrant flowers, bright neyron rose shading lighter in throat, held in trusses of 12. Plant semi-dwarf, as wide as tall, with leaves held 3 years, deep yellow green. See also Rosy Dream. (Lester Brandt cross, 1966; intro. by Griswold Nursery, 1973; reg. by Linda Malland, 1984)

cl.
```
          -fortunei ssp. fortunei (1/2)
Mary Cowan*
          -diaprepes (1/2)
```

Unregistered hybrid by Benjamin F. Lancaster. A parent of Vistoso, q.v.

cl.
```
            -fortunei ssp. fortunei (3/4)
Mary D. Black-              -fortunei ssp. fortunei
          -Madame Fr. J. Chauvin-
                        -unknown (1/4)
```

Pink flowers with a red blotch.  (H. L. Larson, 1956)

cl.
```
          -griffithianum (1/4)
        -Angelo-
Mary Drennen-    -fortunei ssp. discolor (1/4)
        -wardii var. wardii (1/2)
```

Seven-lobed flowers, light aureolin yellow, 4.9in(12.5cm) wide; trusses of 11-12. Leaves 8in(20.3cm) long. (H. L. Larson, 1983)

cl.
```
          -racemosum (1/2)
Mary Fleming-
          -keiskei (1/2)
```

2.5ft(.75m)          -25°F(-32°C)          E          4/3

Small, shapely plant reaching 4ft(1.2m) in 20 years; leaves narrow, pointed, 2in(5cm) long, bronze in winter. Flowers bisque yellow streaked and blotched with salmon, in abundant clusters. (G. Guy Nearing, 1972) A. E. (ARS) 1973

cl.
```
        -fortunei ssp. fortunei (cream form) (1/2)
Mary Garrison*
        -vernicosum (R 18139) (1/2)
```

4ft(1.2m)          -10°F(-23°C)          EM          4/4

Openly campanulate flowers, a blend of salmon yellow and brownish red, fading to creamy yellow. Difficult to propagate, but a good parent of yellows.  (Joseph B. Gable)    See color illus. LW pl. 55.

cl.
```
        -neriiflorum ssp. neriiflorum Euchaites Group (1/2)
Mary Greig-
        -souliei (1/2)
```

Flowers are rose madder, 2.5in(6.4cm) across. (Mrs. E. J. Greig, 1962)

cl.
```
                      -Corona--unknown (1/8)
            -unnamed-
            - hybrid-griersonianum (1/8)
        -Mrs. Donald-    -griffithianum (1/8)
        - Graham    -Loderi-
Mary Harmon-              -fortunei ssp. fortunei (1/8)
        -azalea occidentale (1/2)
```

An azaleodendron with funnel-shaped flowers 2in(5cm) wide, striped pink on outside, fragrant; 15-flowered trusses. (Endre Ostbo, reg. 1962)  A. E. (ARS) 1958

cl.
```
        -yakushimanum ssp. yakushimanum (1/2)
        -                    -griffithianum (1/16)
Mary Jane-          -Queen Wilhelmina-
        - -Britannia-                -unknown (3/16)
        -Leo-          -Stanley Davies--unknown
          -elliottii (1/4)
```

3ft(.9m)          0°F(-18°C)          M

Flowers of carmine rose, held in ball-shaped trusses of 12. Semi-dwarf plant has leaves with deep fawn indumentum, retained 2-3 years. See also Verna Carter, Vallerie Kay. (Mrs. Le Vern Freidman, 1974)

cl.
                 -dwarf white hybrid--unknown (3/4)
Mary Kittel*                               -catawbiense (1/8)
              -                 -Atrosanguineum-
              -Mrs. P. den Ouden-              -unknown
              -                     -arboreum ssp. arboreum
                   -Doncaster-                      (1/8)
                             -unknown

Low, compact, slow-growing plant, forming a mound wider than
high; large, bright green foliage all year.  Very floriferous.
Pink flowers, in medium-sized trusses. (Edmund V. Mezitt, cross
1963)

cl.                         -Sir Charles Butler--fortunei ssp.
          -Mrs. A. T. de la Mare-               fortunei (1/4)
Mary         -                     -maximum (1/8)
Lucille-              -Halopeanum-
       -             -griersonianum (1/4) -griffithianum (5/32)
       -Earl of        -hardy hybrid--unknown (3/16)
         Donoughmore-                     -griffithianum
              -unnamed-              -George-
             hybrid-Mrs. L. A.- Hardy-catawbiense (1/32)
                    Dunnell  -
                            -unknown

5ft(1.5m)          -10°F(-23°C)          EM

Flowers of heavy substance, slightly fragrant, rose opal, with
dark red blotch in trusses of 9-15.  Plant is leafy, as broad as
high, with leaves dark green, 6in(15cm) by 2in(5cm). (Mr. &
Mrs. Freeman Stephens, 1975)

cl.                      -griffithianum (1/4)
       -Loderi King George-
Mary Mayo-               -fortunei ssp. fortunei (1/4)
     -Ostbo Y 3--unknown (1/2)

Flowers a blend of pinks, throat suffused brilliant yellow, to
4.5in(11.5cm) wide, frilled; trusses of 11.  Plant 3ft(.9m) tall
with leaves to 6.5in((16.5cm) long. (The Bovees, 1962)    P. A.
(ARS) 1960

cl.
           .-griersonianum (1/2)
Mary McQuilken-               -campylocarpum ssp. campylocarpum
             -Souvenir of W.-                          (1/4)
                 C. Slocock  -unknown (1/4)

Small flowers, white with throat stained port wine red, held in
lax trusses.  Blooms late midseason. (Sir James Horlick, 1962)

cl.               -decorum (1/4)
         -unnamed hybrid-
Mary Oleri-          -fortunei ssp. discolor (1/4)
       -          -griersonianum (1/4)
         -Tally Ho-
               -facetum (eriogynum) (1/4)

4ft(1.2m)          -10°F(-23°C)          M

Flowers of good substance, yellowish white with pink shadings on
opening, throat a light greyed yellow green, 4in(10.2cm) wide, 7
lobes; conical trusses of 13.  Fragrant and floriferous.  Plant
upright, rounded, as wide as high, with leaves held 3 years.
(Alfred A. Raustein, 1983)

cl.
            -fortunei ssp. discolor (1/2)
Mary Roxburghe-          -arboreum ssp. zeylanicum (1/4)
             -St. Keverne-
                         -griffithianum (1/4)

6ft(1.8m)          -5°F(-21°C)          ML          2/1

Sir Frederick Moore g.  A large, spreading plant with large pink
flowers, darker in throat, held in large trusses. (Rothschild,
intro. 1954)     See color illus. PB pl. 26.

cl.
                 -campylocarpum ssp. campylocarpum (1/2)
Mary Swaythling-
                 -fortunei ssp. fortunei (1/2)

Gladys g.  Soft sulphur yellow flowers without markings, held in
large trusses. (Lord Swaythling)  A. M. (RHS) 1934

                      -fortunei ssp. discolor (1/8)
              -Lady    -
              -Bessborough-campylocarpum ssp. campylocarpum
cl.       -Jalisco-                Elatum Group (1/8)
          -          -dichroanthum ssp. dichroanthum (1/4)
Mary Tasker-     -Dido-
          -          -decorum (1/8)
          -     -fortunei ssp. fortunei (1/4)
          -Fawn-     -dichroanthum ssp. dichroanthum
              -Fabia-
                 -griersonianum (1/8)

Flowers are dawn pink with edges of Naples yellow and a throat
of currant red. (H. R. Tasker, New Zealand, 1979)

cl.
            -fortunei ssp. discolor
Mary Tranquillia-Nereid, q.v.
            -Tally Ho, q.v.
            -Autumn Gold, q.v.

3ft(.9m)          -10°F(-23°C)          L

The exact combination of the above parentage is unknown. Flowers edged in neyron rose, with throat shades of buttercup yellow
and fine dorsal spotting of coral red and light scarlet; flowers
of heavy substance, about 4in(10.2cm) across, in rounded trusses
of 10-11.  Floriferous.  Plant erect, slightly wider than high;
leaves with thin brown indumentum when young. (Mrs. Bernice I.
Jordan, cross 1963; intro. & reg. 1977)

cl.

Mary Waterer          Parentage unknown

A parent of Marham.  Bright rich pink flowers with paler center
and buff spots. (Knap Hill Nursery, 1955)

                   -catawbiense (1/4)
cl.       -Pink Twins-
          -          -haematodes ssp. haematodes (1/4)
Mary Yates-          -griffithianum (1/4)
          -          -Mars-
          -Leah Yates-     -unknown (1/4)
              -Mars (as above)

4ft(1.2m)          -10°F(-23°C)          M

Flowers orchid pink, shading lighter to center, throat lightly
flushed pale yellow with a few small gold spots, openly funnel-
shaped, 2.5in(6.4cm) across, 7-lobed; spherical trusses hold 10.
Floriferous.  Plant twice as wide as high, decumbent branches;
dark green leaves held 2 years. (Cross by Henry Yates; reg. by
Maletta Yates, 1977)

cl.
     -fortunei ssp. discolor (1/2)
Maryke-     -dichroanthum ssp. dichroanthum (1/4)
      -Fabia-
          -griersonianum (1/4)

5ft(1.5m)          -5°F(-21°C)          ML          4/3

Flowers a beautiful pastel blend of light pink and yellow with a
touch of yellow in the throat, 3in(7.6cm) across, 7-lobed; dome-
shaped trusses of 13-15.  Floriferous.  Plant erect, full; elliptic, dull green leaves, 5in(12.7cm) long.  Of his own hybrids
this was Mr. Van Veen's favorite. (Theodore Van Veen, Sr., intro. 1955; reg. 1974)     See color illus. VV p. 56.

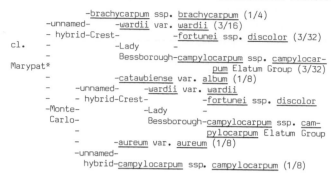

```
                    -brachycarpum ssp. brachycarpum (1/4)
       -unnamed-      -wardii var. wardii (3/16)
       - hybrid-Crest-          -fortunei ssp. discolor (3/32)
cl.    -                 -Lady
       -                 Bessborough-campylocarpum ssp. campylocar-
Marypat*                                  pum Elatum Group (3/32)
       -                 -catawbiense var. album (1/8)
       -      -unnamed-      -wardii var. wardii
       -      - hybrid-Crest-          -fortunei ssp. discolor
       -Monte-                -Lady
       Carlo-            Bessborough-campylocarpum ssp. cam-
       -                                pylocarpum Elatum Group
       -      -aureum var. aureum (1/8)
       -unnamed-
          hybrid-campylocarpum ssp. campylocarpum (1/8)
```

Flowers chartreuse green, with an eye of brilliant yellow green
specks.  (Weldon E. Delp)

```
cl.            -fortunei ssp. fortunei (1/4)
       -Ruby Bowman-          -griffithianum (1/8)
Marzo-            -Lady Bligh-
       -ririei (1/2)          -unknown (1/8)
```

6ft(1.8m)          5°F(-15°C)          VE

Magenta flower buds open to spirea red, fading to roseine purple
with blotch of violet purple deep in throat.  Trusses 9in(23cm)
across, dome-shaped, hold 10-14 flowers.  Upright, rounded plant
as broad as tall; dark green leaves, concave, 7in(17.8cm) long,
held 2-3 years.  (Dr. David Goheen, 1982)

```
cl.
       -griersonianum (1/2)
Master Dick-          -arboreum ssp. arboreum (1/8)
       -          -Doncaster-
       -The Don-          -unknown (1/8)
          -griffithianum (1/4)
```

Flowers of carmine scarlet with brownish speckling.  (Crosfield)
A. M. (RHS) 1936

---

Master, The       See The Master

```
g. & cl.
       -griersonianum (1/2)
Matador-
       -strigillosum (1/2)
```

Tubular-shaped flowers, dark orange red, held in trusses of 7-8.
See also Exbury Matador.  (Lord Aberconway)  A. M. (RHS) 1945;
F. C. C. (RHS) 1946

```
cl.                    -fastigiatum (1/8)
       -Wachtung--sport of Ramapo-
       -unnamed-          -minus var. minus (5/8)
Matilda* hybrid-unknown (1/4)          Carolinianum Group

       -minus var. minus Carolinianum Group
```

A truly hardy good blue.  Reaches 2.5ft(.75m) in 10 years; early
vertical growth (1-2 years) fills in later.  Foliage typical of
the small-leaf lepidote.  (Leon Yavorsky)

```
cl.                    -griffithianum (1/8)
       -Queen Wilhelmina-
       -Britannia-          -unknown (7/8)
Maud Corning-      -Stanley Davies--unknown
       -Skyglow--unknown
```

5ft(1.5m)          -15°F(-26°C)          L

Flowers of pale purplish pink, with edges of orchid pink and sap
green throat, of good substance, 7 overlapping lobes, fragrant;
lax trusses hold 8.  Plant rounded, branching well; narrowly elliptic leaves  held 2 years.  (C. O. Dexter seedling; raised at
Holden Arboretum; reg. by Peter Bristol, 1983)

```
cl.
       -williamsianum (1/2)
Maureen-
       -          -campylocarpum ssp. campylocarpum
       -          -Ole Olson-          Elatum Group (1/8)
       -Lem's Goal-          -fortunei ssp. discolor (1/4)
       -          -griersonianum (1/8)
          -Azor-
             -fortunei ssp. discolor
```

Flowers very light neyron rose, tinged yellow at the base.  (H.
Lem cross; reg. 1973)

```
cl.
       -edgeworthii (bullatum) (1/2)
Maurice Skipworth-
       -burmanicum (1/2)
```

White flowers with faintest green tint, and yellow speckling at
base, in 5-flowered trusses.  (Cross by M. R. Skipworth; reg. by
Dunedin Rhododendron Group, New Zealand, 1977)

```
cl.
       -Marion--unknown
Mauve Satchel-
       -Sir Joseph Whitworth--unknown
```

Trusses hold 20 flowers, rose purple with lighter center,
speckled olive green.  See also Joan Langdon.  (G. Langdon,
1976)

```
                              -griffithianum (9/32)
                    -Loderi-
             -Albatross-          -fortunei ssp. fortunei
       -Cup-                              (7/32)
       -Day-          -fortunei ssp. discolor (1/16)
       -          -elliottii (1/16)
    -Midnight-   -Fusilier-
cl. -          -griersonianum (1/16)
    -          -          -ponticum (1/8)
Max  -          -Purple Splendour-
Marshland-          -unknown (3/16)
       -          -griffithianum
          -Pink -          -fortunei ssp. fortunei
    -Coronation-Shell-H. M. Arderne-
       Day  -          -unknown
       -          -griffithianum
          -Loderi-
             -fortunei ssp. fortunei
```

Trusses of 19 flowers, light magenta rose with a blotch of cardinal red.  See also Sue Lissenden.  (K. Van de Ven, 1981)

```
cl.            -caucasicum (1/4)
       -Chevalier F. de Sauvage-
Max Sye-          -hardy hybrid--unknown (3/4)
       -unknown
```

Flowers dark red with black blotch; early.  (C. Frets & Son, 1935)

```
cl.
       -maximum (1/2)
Maxdis-
       -fortunei ssp. discolor (1/2)
```

White flowers.  (G. Nearing & J. Gable)

```
cl.
       -maximum (1/2)
Maxecat*
       -catawbiense (1/2)
```

Pink flowers, blooming late.  Hardy to -25°F(-32°C).  (J. Gable)

---
```
                  -maximum (1/2)
Maxhaem Salmon*
                  -haematodes ssp. haematodes (1/2)

?            -15°F(-26°C)          LM          3/3
```

Synonym Maxhaem #7.  Salmon flowers, medium-sized, held in well-
filled trusses.  Floriferous.  Fine, dark green foliage.  The
haematodes pollen, "true form", from E. J. P. Magor in England.
(Joseph B. Gable)   See color illus. LW pl. 47.

cl.

Maximum Roseum     See Ponticum Roseum

---

Maximum Wellesleyanum     See Wellesleyanum

cl.
```
          -strigillosum (1/2)
Maxine -          -forrestii ssp. forrestii Repens Group (1/4)
Childers-Elizabeth-
                  -griersonianum (1/4)

3ft(.9m)          -5°F(-21°C)          E          3/4
```

Flowers cardinal red, widely funnel-campanulate, 2.5in(6.4cm)
wide, 5-lobed, of heavy substance; trusses of 8-10.  Shrub well-
branched, rounded, wider than tall; olive green leaves, light
orange brown indumentum beneath; new growth reddish bronze.  The
entire plant reflects strigillosum.  See also Maxine Margaret.
(Mrs. A. Childers cross; raiser Dr. C. Phetteplace; reg. 1977)

cl.
```
                  -strigillosum (1/2)
Maxine Margaret-          -forrestii ssp. forrestii Repens Group
          -Elizabeth-                                  (1/4)
                  -griersonianum (1/4)

2ft(.6m)          -15°F(-26°C)          E
```

Flowers currant red with 5 black nectaries, widely funnel-campanulate, 2in(5cm) across, 5-lobed, of heavy substance; trusses
of 8-11.  Plant well-branched, wider than tall; dark green
leaves, 3.5in(8.9cm) long, held 2 years.  See also Maxine Childers.  (Reg. by Maxine & Arthur Childers, 1977)

cl.

Maxwell T. Masters     Parentage unknown

Flowers of rosy crimson.  Parent of Armistice Day.  (A. Waterer,
before 1915)

g. & cl.
```
          -haematodes ssp. haematodes (1/2)
May Day-
          -griersonianum (1/2)
```

Flowers of brilliant orange scarlet, funnel-shaped, held in lax
trusses.  Plant of vigorous habit and rapid growth, wider than
tall; dark green leaves, with tan indumentum beneath.  Several
forms exist.  See also Bodnant May Day, Exbury May Day.  (A.
M. Williams)   A. M. (RHS) 1932    See color illus. ARS J 37:2
(Spring 1983): p. 75; F p. 54.

cl.
```
          -caucasicum (1/2)
May Firth-
          -unknown (1/2)
```

Large, pale pink flowers, held in loose trusses.  Early.  (Knap
Hill Nurseries, 1962)

g. & cl.
```
          -haematodes ssp. haematodes (1/4)
          -May Day-
May Morn-          -griersonianum (1/4)
          -beanianum (pink form) (1/2)
```

Funnel-shaped flowers, 2in(5cm) across, azalea pink, flushed on
edges with porcelain rose, held in trusses of 8-10.  (Lord Aberconway) A. M. (RHS) 1946

cl.
```
                  -fortunei ssp. discolor (1/4)
          -Cornish Loderi-
May Pink-          -griffithianum (1/2)
          -          -griffithianum
          -Loderi Sir Edmund-
                  -fortunei ssp. fortunei (1/4)
```

Flowers of fuchsine pink, shaded light purple.  (Loder, 1946)
A. M. (RHS) 1946

g.
```
          -fortunei ssp. fortunei (1/2)
May Queen-
          -unknown (1/2)
```

Flowers large, waxy, rich pink.  (Sir E. Loder, 1926)

cl.
```
                  -campylogynum Cremastum Group (1/4)
          -Candi-
May Schwartz-          -racemosum (1/4)
          -tephropeplum (1/2)

2ft(.6m)          5°F(-15°C)          E
```

Flowers fuchsine pink, the reverse darker and streaked, 1.25in
(3.2cm) across, 5 wavy lobes, of heavy substance; small terminal
inflorescences of 4-6.  Free-flowering.  Plant broader than
tall; glossy, narrowly elliptic leaves 1.4in(3.4cm) long, scaly
beneath.  (Robert W. Scott, reg. 1977)

cl.
```
                  -Corona--unknown (1/4)
          -Bow Bells-
          -          -williamsianum (1/4)
May Song-          -fortunei ssp. discolor (1/8)
          -          -Lady Bessborough-
          -Day -                  -campylocarpum ssp. campylo-
          Dream-                       carpum Elatum Group (1/8)
                  -griersonianum (1/4)

3ft(.9m)          0°F(-18°C)          M
```

Flowers pale pink with reverse flushed deeper pink, openly campanulate, to 3.5in(8.9cm) wide, ruffled; trusses of 9-11.  Plant
fairly compact; leaves 2.5in(6.4cm) long.  (Cross by G. Grace;
intro. by The Bovees, 1963; reg. 1972)

cl.
```
                  -griffithianum (1/4)
          -Queen Wilhelmina-
May Templar-          -unknown (3/4)
          -Stanley Davies--unknown
```

These parents also produced Britannia, Unknown Warrior and other hybrids.  Bright scarlet flowers.  (C. B. van Nes & Sons)

cl.
```
          -catawbiense var. album (1/2)
May Time-
          -Koichiro Wada--yakushimanum ssp. yakushimanum (1/2)

3ft(.9m)          -20°F(-29°C)          M          3/4
```

Flowers white with faint ochre dorsal spots, purplish pink flush
on outer ribs, widely funnel-campanulate, 2.5in(6.4cm) across, 5
wavy lobes; dome-shaped trusses of 12-14.  Plant well-branched,
wider than tall; dark green convex leaves, 4.25in(10.8cm) long,
held 3 years.  See also Anna H. Hall, Great Lakes, Pink Frosting, Spring Frolic.  (David G. Leach, 1973)

cl.
    -sutchuenense (1/2)
Maya-
    -ririei (1/2)

Synonym Sutchrir. Flowers pale mauve within, spotted deep
purple and blotched intense purple at base. Outside of flower
is a deeper shade. (E. J. P. Magor, 1933) A. M. (RHS) 1940

cl.
    -Marion--unknown (11/16)
Maybe-    -griersonianum (1/4)
  -Radium-              -griffithianum (1/16)
       -Earl of-Queen Wilhelmina-
      -Athlone-       -unknown
         -Stanley Davies--unknown

Trusses of 14 flowers, medium neyron rose with darker center.
See also Crazy Joe. (G. Langdon, 1980)

cl.
      -impeditum (3/4)
   -Blue Tit-
Mayfair-    -augustinii ssp. augustinii (1/4)
   -impeditum

Dense plant with attractive foliage. Lavender blue flowers open
later than Sapphire's. (Knap Hill Nursery, 1969)

cl.
            -racemosum (1/2)
Mayflower--Conestoga (selfed)-
            -minus var. minus Carolinianum
                  Group (1/2)

F2 seedling of Conestoga; parentage reversed in Livingston/West.
Pink flowers, blooming early; hardy to -20°F(-29°C). (Joseph B.
Gable, reg. 1958)

                -campylocarpum ssp. campylocarpum
       -Penjerrick-        Elatum Group (1/16)
g.  -Amaura-     -griffithianum (1/16)
   -Eros-    -griersonianum (5/8)
Mayros-    -haematodes ssp. haematodes (1/4)
   -May Day-
       -griersonianum

A red-flowered hybrid. (Lord Aberconway, 1946)

cl.
              -catawbiense (1/8)
      -Parsons Grandiflorum-
    -America-     -unknown (3/8)
May's Delight*   -dark red hybrid--unknown
    -yakushimanum ssp. yakushimanum (1/2)

4ft(1.2m)      -15°F(-26°C)     M     3/4

Rounded, compact truss of flowers, red in bud, fading to long-
lasting pink. Plant grows broader than tall; leaves have thin
plastered indumentum beneath. (Weldon Delp, intro. 1965)

cl.

Meadow Pond--form of wardii var. wardii (L & S 15764)

4ft(1.2m)      -5°F(-21°C)     M     4/3

Large primrose yellow flowers, heavily blotched crimson purple.
Early midseason. (Exhibited by Crown Estate, Windsor) A. M.
(RHS) 1963

cl.
            -catawbiense (1/2)
   -Mrs. C. S. Sargent-
Meadowbrook-     -unknown (1/2)
  -       -catawbiense
   -Everestianum-
      -unknown

5ft(1.5m)      -15°F(-26°C)    ML    3/3

Fair plant habit; grows as broad as tall. Medium-large foliage.
Flowers vibrant pink with a white blotch, green spots, frilled;
conical trusses of 22. (Paul Vossberg, reg. 1958) A. E. (ARS)
1973

cl.
      -burmanicum (1/2)
Meadowgold-     -xanthostephanum (1/4)
    -Lemon Mist-
        -leucaspis (1/4)

Flowers sulphur yellow, spotted on upper lobe, tubular funnel-
shaped, 1.5in(3.8cm) wide; terminal inflorescence up to 3 buds,
each of 4-6 flowers. Plant rounded, wider than tall; dark green
glossy leaves, 1.5in(3.8cm) long, scales beneath, held 2 years.
(Robert W. Scott, 1976)

cl.
          -caucasicum (1/4)
  -Boule de Neige-    -catawbiense (1/8)
Meath-      -unnamed hybrid-
  -          -unknown (1/8)
   -Koichiro Wada--yakushimanum ssp. yakushimanum (1/2)

3ft(.9m)      -5°F(-21°C)     ML

Flowers light mauve pink, darker at edges with light olive yellow dorsal blotch, openly funnel-shaped, 3.75in(9.5cm) wide, 7
waxy lobes, very fragrant; ball-shaped trusses of 9. Free-flowering. Plant upright; leaves to 4.25in(10.8cm) long, retained 3
years. (Dr. W. A. Reese, 1979)

cl.
   -falconeri ssp. falconeri (1/2)
Mecca-
   -niveum (1/2)

Tightly packed trusses of 42 flowers, white with light flush of
cyclamen purple and small deep purple stain in throat, 2in(5cm)
wide. Leaves 9.25(23.6cm) long; pale, mat-like indumentum below. (Mrs. Douglas Gordon) A. M. (RHS) 1965

g.          -arboreum ssp. arboreum (1/4)
  -Red Admiral-
Medea-      -thomsonii ssp. thomsonii (1/4)
  -sutchuenense (1/2)

A hybrid with red flowers. (E. J. P. Magor, intro. 1931)

g.
   -dichroanthum ssp. scyphocalyx (1/2)
Medusa-
   -griersonianum (1/2)

3ft(.9m)      -5°F(-21°C)     M     3/3

Flowers Spanish orange shading on lobes to mandarin red, heavily
speckled in light brown. Plant dense, compact; greyish green
leaves with light woolly indumentum. (Lord Aberconway, 1936)
See color illus. VV p. 111.

cl.
      -campylocarpum ssp. campylocarpum (1/2)
Meg Merrilees-
     -A. W. Hardy Hybrid--unknown (1/2)

Flowers of apricot and yellow, with a red blotch; also, a creamy
white shading to yellow. (Slocock, 1924)

cl.
   -Marion--unknown (1/2)
Melba-          -Sir Charles Butler--fortunei ssp. for-
   -Van Nos Sensation-            tunei (1/4)
    -         -maximum (1/8)
      -Halopeanum-
          -griffithianum (1/8)

Amaranth rose fading to very light solferine purple in center.
(V. J. Boulter, 1972)

g.                          -thomsonii ssp. thomsonii (1/4)
        -Ascot Brilliant-
Melissa-              -unknown (1/4)
        -arboreum ssp. arboreum (1/2)

Red flower with darker red spotting.  (E. J. P. Magor, 1932)

cl.
        -myrtifolium (kotschyi) (1/2)
Mcliz-
        -unknown (1/2)

1ft(.3m)         0°F(-18°C)          ML

Venetian pink flowers, slightly spotted currant red, openly campanulate, almost 1in(2.5cm) across, in lax inflorescences of 1-4
trusses, each 5- to 8-flowered.  Plant rounded, wider than tall;
elliptic leaves 1.25in(3.2cm) densely scaly beneath.
(J. F. Caperci, reg. 1977)

g.                      -griffithianum (1/4)
        -Loderi Venus-
Melody-              -fortunei ssp. fortunei (1/4)
     -                      -thomsonii sp. thomsonii (1/4)
        -General Sir John du Cane-
                            -fortunei ssp. discolor (1/4)

A tall, shapely plant.  Creamy pink flowers in well-packed
trusses. (Rothschild, 1950)

g.
        -dichroanthum ssp. dichroanthum (1/2)
Melrose-        -fortunei ssp. discolor (1/4)
        -Avocet-
                -fortunei ssp. fortunei (1/4)

Pink flowers.  Plant not highly rated at Exbury.  (Rothschild,
intro. 1947)

cl.

Memoir    Parentage unknown

A parent of Ariel.  Flowers lilac white with green spots. (A.
Waterer, before 1915)

cl.
        -campylocarpum ssp. campylocarpum (1/2)
Memorial -      -wardii var. wardii (1/4)
Kate Bagg-     -                      -griffithianum (1/32)
        -Idealist-            -Kewense-
                    -Aurora-        -fortunei ssp. fortunei
            -Naomi-    -thomsonii ssp.         (5/32)
            -           thomsonii (1/16)
                    -fortunei ssp. fortunei

Medium-sized plant with trusses of 12-18 flowers, bell-shaped,
clear Dresden yellow fading to primrose yellow; free of markings.  Handsome foliage like Idealist's. (R. Strauss)   A. M.
(RHS) 1964

cl.              -griffithianum (1/4)
        -King George-
Memory*      -unknown (1/4)
        -haematodes ssp. haematodes (1/2)

Gipsy King g.  Rich, dark red flowers, openly campanulate, with
the puckered calyx of haematodes. (Rothschild)    A. M. (RHS)
1945

                    -dichroanthum ssp. dichroanthum (1/8)
g.       -Fabia-
        -Erebus-    -griersonianum (3/8)
Menippe-     -griersonianum
     -       -griffithianum (1/4)
        -Loderi-
                -fortunei ssp. fortunei (1/4)

A pink-flowered hybrid.  (Lord Aberconway, intro. 1941)

                    -dichroanthum ssp. dichroanthum (1/8)
g.      -Astarte-            -campylocarpum ssp. campylocarpum
     -Ouida-      -Penjerrick-            Elatum Group (1/16)
Mera-    -griersonianum    -griffithianum (1/16)
        -           -neriiflorum ssp. neriiflorum (1/4)
        -F. C. Puddle-
                -griersonianum (1/2)

Pale pink flowers. (Lord Aberconway, 1946)

cl.
        -campylogynum (white form) (1/2)
Merganser-
        -luteiflorum (1/2)

A dwarf plant under 10in(25cm) tall by 17in(44cm) wide, compact,
vigorous, very free-flowering; leaves 1.4in(3.5cm) long, dull
dark green.  Flowers primrose yellow, funnel-shaped, 1.25in(3.2
cm) wide, in trusses of 3.   (Peter A. Cox)   H. C. (Wisley Trials) 1981

g.              -Mrs. Tritton--unknown (1/2)
        -Louis Pasteur-
Merkur-      -Viscount Powerscourt--unknown
        -williamsianum (1/2)

Flowers delicate rose, deeper at edges; large loose umbels.  (D.
Hobbie, 1953)

cl.
                    -griersonianum (1/2)
Merle Lee--Azor (selfed)-
                    -fortunei ssp. discolor (1/2)

Plant of spreading habit; leaves 6in(15cm) long by 1.75in(4.5cm)
wide.  Funnel-shaped flowers, 4in(10cm) across, rhodamine pink.
(B. J. Esch)  P. A. (ARS) 1954

g.
        -arboreum ssp. arboreum (1/2)
Merlin-      -sperabile var. sperabile (1/4)
        -Eupheno-
                -griersonianum (1/4)

Red-flowered hybrid.  (Lord Aberconway, 1950)

---

Mermaid    Parentage unknown

Very tall, vigorous plant, broader than high; dull, dark green
foliage.  Compact trusses, 9in(23cm) across, holding 8 flowers;
corolla light neyron rose, darker at wavy margins, campanulate,
4in(10.2cm) wide. (W. C. Slocock, reg. 1958)  A. M. (RHS) 1962

g.
        -lacteum (1/2)
Merops-
        -Cunningham's Sulphur--form of caucasicum (1/2)

Sulphur yellow flowers. (Cross by E. J. P. Magor, shown by Collingwood Ingram, 1939)

                -Lady         -fortunei ssp. discolor (1/8)
                    -Bessborough-
            -Lem's-        -campylocarpum ssp. campylocarpum
            -Goal -   -griersonianum     Elatum Group (1/16)
cl.  -Jingle-    -Azor-
    - Bells-            -fortunei ssp. discolor
Merrie-     -      -dichroanthum ssp. dichroanthum (1/8)
 Bells-    -Fabia-
     -          -griersonianum (3/16)
    -lacteum (1/2)

4ft(1.2m)         5°F(-15°C)          M

Buds currant red, flowers scarlet red, tubular bell-shaped, 5-
lobed, in lax trusses of 10.  Plant habit rounded, as broad as
tall; leaves dull dark green. (Britt M. Smith, 1980)

cl.                                              -griffithianum (1/16)
      -Marion--unknown (5/8)  -George Hardy-
Merry-      -Mrs. L. A. Dunnett-              -catawbiense (1/16)
   Lass-Rodeo-                        -unknown
         -griersonianum (1/4)

Trusses of 16 flowers, light neyron rose spotted red with creamy
yellow center.  See also Min Langdon.  (G. Langdon, 1980)


                                    -catawbiense (1/4)
cl.              -Mrs. C. S. Sargent-
                                    -unknown (5/8)
Merveille de Boskoop-                        -griffithianum
                        -Queen Wilhelmina-            (1/8)
            -Britannia-                -unknown
                    -Stanley Davies--unknown

Compact trusses of 18 flowers, Persian rose with dark brown
spotting.  (Felix & Dijkhuis, Boskoop, 1966)


                        -ciliatum (1/8)
            -Rosy Bell-
cl. -Lady Berry-        -glaucophyllum var. glaucophyllum (1/8)
    -          -          -cinnabarinum ssp. cinnabarinum (1/4)
Meta-         -Royal-
    -      Flush-maddenii ssp. maddenii (1/4)
    - -cinnabarinum ssp. cinnabarinum Roylei Group (1/4)
    -Ivy-            -cinnabarinum ssp. cinnabarinum
        -Royal Flush-
                -maddenii ssp. maddenii

6.5ft(1.9m)        10°F(-12°C)          M

Funnel-shaped flowers in clusters of 4,  Indian pink shading
darker, 3in(7.6cm) wide.  Well-branched, upright plant;  leaves
about 3.5in(8.9cm) long, covered beneath with yellow to golden
brown scales; held 2 years.  Foliage has spicy fragrance.  (Dr.
Carl G. Heller, 1980)


                            -catawbiense (1/8)
---                  -unnamed-
        -Altaclarense- hybrid-ponticum (1/8)
Meteor-        -arboreum ssp. arboreum (1/4)
        -catawbiense (1/2)

Flowers rosy crimson.  (Standish & Noble, before 1850)


cl.

Meteor*--form of barbatum

5ft(1.5m)        5°F(-15°C)          F        4/4

A scarlet-flowered form of the species. Leaves are bristly or
"bearded" (hence barbatum).  Mature plants have smooth, reddish
plum-colored bark.  (Slocock Nurseries)


g.
    -dichroanthum ssp. scyphocalyx Herpesticum Group (1/2)
Metis-
    -griersonianum (1/2)

A hybrid with orange red flowers.  (Lord Aberconway, 1941)


cl.

Metternianus--form of japonicum var. japonicum (metternichii)

A pink form of this species selected by Koichiro Wada and much
used by Dietrich Hobbie for hybridizing.  (Reg. 1958)


cl.                                      -griffithianum (1/4)
            -Mrs. E. C. Stirling-
Mevrouw P. A.-                -unknown (9/16)  -ponticum
   Colijn   -                    -Michael Waterer-    (1/16)
            -          -Prometheus-              -unknown
        -Madame de-          -Monitor--unknown
          Bruin    -          -arboreum ssp. arboreum (1/8)
                -Doncaster-
                    -unknown

4ft(1.2m)        -10°F(-23°C)        ML        3/3

Plant of compact habit; dark green leaves, 4.5in(11.5cm) long.
Clear pink flowers, held in conical trusses.  (M. Koster & Sons,
reg. 1958)

cl.
    -lindleyi (1/2)
Mi Amor-
    -nuttallii (1/2)

6ft(1.8m)        15°F(-10°C)        M        5/3

Open habit, with leaves dark green above, grey green underneath.
Very large flowers, as much as 6in(15cm) wide, bell-shaped,
white with yellow throat in trusses of about 6.  (Maurice H.
Sumner, 1962)  C. A. (ARS) 1969; A. M. (RHS) 1975

cl.
    -macabeanum (1/2)
Michael Beekman-
    -unknown (1/2)

Red flowers turning to yellow with age, held in trusses of 20.
See also Francesca Beekman, Diane Beekman, Belinda Beekman.  (J.
Beekman, Australia, 1980)

cl.
    -catawbiense
Michael Rice-          and possibly unknown hybrids
    -maximum

6ft(1.8m)        -15°F(-26°C)        E-M

Buds deep purplish red open to flowers deep purplish pink with
darker spots and edges, widely funnel-shaped, 3in(7.6cm) across,
5 wavy lobes, of good substance; spherical trusses of 15.  Plant
upright, with stiff branches; glossy olive green leaves held 3
years.  (W. David Smith, cross 1970; reg. 1979)


cl.
    -ponticum (1/2)
Michael Waterer-
    -unknown (1/2)

6ft(1.8m)        -15°F(-26°C)        ML        3/3

Flowers of magenta red fading to rosy crimson, funnel-shaped, in
a profusion of well-packed spherical trusses.  Plant habit good,
compact; medium green leaves 4in(10.2cm) long.  Easy to propagate.  (J. Waterer, before 1894)   See color illus. VV p. 23.

cl.
    -burmanicum (1/2)
Michael's Pride-
    -dalhousiae var. dalhousiae (1/2)

A plant for the cool greenhouse with attractive bronze new foliage.  Lime green buds, large waxy flowers of creamy yellow, tubular, to 5in(12.7cm) long;  lily-shaped and fragrant.  (Charles
Michael, 1964)

g.
    -griersonianum (1/2)
Michele-
    -Afghan--unknown (1/2)

Plant of moderate size, rather loose growth habit.  Red flowers
in well-shaped trusses, early midseason.  (Rothschild, 1947)

```
                      -Pygmalion--unknown (1/4)
cl.        -unnamed hybrid-
           -                -haematodes ssp. haematodes (1/4)
Microsplash*       -yakushimanum ssp. yakushimanum (Exbury form)
           -unnamed-                                    (3/8)
           hybrid-               -yakushimanum ssp. yakushimanum
                 -Serendipity-
                          -aureum var. aureum (1/8)
```

Flowers of greenish white with a blotch of brilliant greenish
yellow. (Weldon E. Delp)

```
cl.
           -sinogrande (1/2)
Middlemarch-
           -lacteum (1/2)
```

A full, rounded truss of 18-20 flowers, 8-lobed, creamy white in
bud, opening to white with red dorsal blotch, light spotting in
upper throat.  Corolla campanulate, 3.2in(8cm) wide, 8-lobed.
Leaves medium green, paler below, to 8in(20.3cm) long. (Raised
by Col. S. R. Clarke;  intro. & reg. by R. N. S. Clarke)  A. M.
(RHS) 1980

cl.

Midget--form of leucaspis

A dwarf, compact form of this species, 6in(15cm) high and 12in
(30.4cm) wide in 8 years.  White flowers, 1.5in(3.8cm) across,
with 10 black stamens. (The Bovees, reg. 1972)

```
                          -griffithianum (1/16)
                 -Loderi-
           -Albatross-    -fortunei ssp. fortunei (1/16)
cl.     -Cup Day-    -fortunei ssp. discolor (1/8)
        -         -     -elliottii (1/8)
Midnight-    -Fusilier-
        -              -griersonianum (1/8)
        -              -ponticum (1/4)
        -Purple Splendour-
                        -unknown (1/4)
```

5ft(1.5m)         -5°F(-21°C)            M

Flowers of soft red purple with heavy, very dark spotting on upper lobe and center, 5 wavy lobes; rounded trusses of 16.  Dark
green, glossy leaves. (K. Van de Ven, 1978)    See color illus.
ARS Q 31:3 (Summer 1977): p. 171.

```
cl.
           -maximum (1/2)
Midsummer-
           -unknown (1/2)
```

Rose pink flowers. (J. Waterer, Son & Crisp)

```
cl.                 -griffithianum (1/4)
                 -Isabella-
Midsummer Snow-       -auriculatum (1/4)
                 -diaprepes (1/2)
```

6ft(1.8m)         5°F(-15°C)              L

White flowers with green tinge in throat, trumpet-shaped, 4.75in
(12cm) across, 7 slightly frilled lobes, very fragrant; large
loose trusses of 8-10.  Buds and young shoots a bright yellowish
green. (Hillier & Sons, reg. 1968)

cl.

Midwinter--seedling of dauricum

4ft(1.2m)        -20°F(-29°C)            VE          4/3

Bright rose purple flowers with 5 spreading lobes in terminal
clusters of 2-6.  Semi-deciduous--sometimes retaining its leaves
through winter. (Crown Estate, Windsor)  A. M. (RHS) 1963;
F. C. C. (RHS) 1969

```
g.
          -griffithianum (3/4)
Mikado-          -thomsonii ssp. thomsonii (1/4)
       -Cornish Cross-
                    -griffithianum
```

Pale rose flowers. (Lord Aberconway, intro. 1950)

```
                          -thomsonii ssp. thomsonii (1/16)
                 -Shilsonii-
g.             -Bella-       -barbatum (1/16)
       -Rosefinch-   -griffithianum (3/8)
Milady-        -griersonianum (1/4)
       -           -griffithianum
       -Loderi-
              -fortunei ssp. fortunei (1/4)
```

Flowers deep rose pink. (Rothschild cross, 1935; intro. 1950)

```
cl.
             -minus var. minus Carolinianum Group (1/2)
Mildred Amateis-
             -edgeworthii (1/2)
```

3ft(.9m)        -5°C(-21°C)              M

White flowers, flushed pink. (Edmond Amateis, reg. 1958)

```
cl.                -fortunei ssp. fortunei (3/8)
        -Faggetter's Favourite-
Mildred-               -unknown (3/8)
Fawcett-                   -Corona--unknown
       -             -unnamed hybrid-
       -Mrs. Donald-         -griersonianum (1/8)
        Graham   -      -griffithianum (1/8)
                   -Loderi-
                         -fortunei ssp. fortunei
```

6ft(1.8m)        5°F(-15°C)              M

Leaves 6.25in(16cm) by 2in(5cm).  Flowers blush pink, with an
orange pink blotch, about 3in(7.6cm) across, and held in large
trusses. (Carl P. Fawcett, 1962)   P. A. (ARS) 1960

```
g.                      -brookeanum var. gracile (1/4)
        -Duchess of Edinburgh-
Militaire-              -longiflorum (lobbii) (1/4)
        -javanicum (1/2)
```

Vireya hybrid.  Scarlet crimson flowers.    F. C. C. (RHS) 1855

---

Milkmaid       Parentage unknown

A parent of Norfolk Candy, a hybrid by James Russell.  Flowers
yellow edged lilac with slight violet and brown speckling and an
olive brown flare.  Plant habit upright and straggly.  Blooms
late. (Harry White, Sunningdale Nurseries, before 1955)

```
cl.
           -insigne (1/2)
Mill Reef-
           -griersonianum (1/2)
```

Elongated foliage on a medium-sized plant.  Flowers of reddish
purple with darker red dorsal spotting. (A. F. George, Hydon
Nurseries, reg. 1972)

```
cl.                -decorum (1/4)
          -Decatros-             -catawbiense (1/8)
Millard Kepner-        -Atrosanguineum-
           -                      -unknown (1/8)
              -yakushimanum ssp. yakushimanum (1/2)
```

4ft(1.2m)         -5°F(-21°C)            EM

Flowers dark pink, fading to lighter pink with lilac undertone, openly funnel-shaped, to 3.5in(8.9cm) wide, 7 wavy lobes, very fragrant; spherical trusses of 15. Plant erect, well-branched; narrow dark green leaves 5.5in(14cm) long, held 3 years. (Cross by F. W. Schumacher; raised and reg. by Dr. M. E. Byrkit, 1975)

cl.
```
                -racemosum (1/2)
Millicent Scott-              -xanthostephanum (1/4)
                -Saffron Queen-
                              -burmanicum (1/4)
```

2.5(.75m)          5°F(-15°C)          E

Flowers buff-colored, up to 1in(2.5cm) across, with light red markings, blooming along the branches and at tips, in buds of up to 3 flowers. Floriferous. Well-branched plant, twice as wide as high; small rounded dark green leaves, dense golden brown scales beneath, held 2 years. New growth bronze green. (Robert W. Scott, 1977)

cl.
```
               -macabeanum (1/2)
Milton Hollard-
               -unknown (1/2)
```

Trusses hold 22 flowers, phlox pink in bud opening primrose yellow, flushed pink with red spotting. See also Thoron Hollard. (Bernard Hollard, New Zealand, 1979)

cl.
```
      -catawbiense (1/2)
Mims-
      -Diadema--unknown (1/2)
```

Flowers light purplish pink with yellow green markings. (T. J. R. Seidel, 1910)

g.
```
           -campylocarpum ssp. campylocarpum (1/4)
        -Anita-
Mimulus-    -griersonianum (1/2)
        -        -griersonianum
        -Rapture-
                 -arboreum ssp. zeylanicum (1/4)
```

Red flowers. (Lord Aberconway, intro. 1950)

cl.
```
                                          -griffithianum
           -Marion--unknown (5/8)   -George Hardy-      (1/16)
Min Langdon-    -Mrs. L. A. Dunnett-          -catawbiense
           -Rodeo-                  -unknown          (1/16)
                 -griersonianum (1/4)
```

Trusses of 17 flowers, light neyron rose with creamy yellow center. See also Merry Lass. (G. Langdon, 1980)

cl.
```
                               -catawbiense (1/8)
                -Parsons Grandiflorum-
           -Nova Zembla-              -unknown (3/8)
Minas Maid-            -hardy red hybrid--unknown
          -yakushimanum ssp. yakushimanum (1/2)
```

3ft(.9m)          -15°F(-26°C)          L

Flowers phlox pink with dorsal flecks of ruby, openly funnel-shaped, 2.25in(5.7cm) wide, 5 wavy lobes; ball-shaped trusses of 15. Floriferous. Plant rounded, slightly wider than tall; dark glossy leaves with moderate amount of greyed brown indumentum beneath. (Cross by George S. Swain; Dr. D. L. Craig, reg. 1979) See color illus. ARS Q 33:3 (Summer 1979): p. 170.

cl.
```
         -minus var. minus Carolinianum Group (1/2)
Minaura*
         -dauricum (1/2)
```

3ft(.9m)          -20°F(-29°C)          E

Leaves smaller than P.J.M., dark mahogany in winter. Flowers lavender pink. See also Olga Mezitt. (Weston Nurseries)

g.
```
                            -fortunei ssp. discolor (1/4)
        -Sir Frederick Moore-       -arboreum ssp. zeylanicum
Minerva-               -St. Keverne-                    (1/8)
                                    -griffithianum (1/8)
        -elliottii (1/2)
```

A tall plant bearing pink flowers in large trusses. (Rothschild, 1947)

---
```
        -javanicum (5/8)
Minerva-               -jasminiflorum (3/8)
        -Princess Alexandra-       -jasminiflorum
                           -Princess Royal-
                                           -javanicum
```

Vireya hybrid. Flowers light rose, with yellow or orange spots. (before 1865) F. C. C. (RHS) 1885

cl.
```
                    -griffithianum (1/8)
      -Loderi-
    -Albatross-     -fortunei ssp. fortunei (1/8)
Ming-        -fortunei ssp. discolor (1/4)
      -wardii var. wardii (1/2)
```

Flowers are uranium green with reddish blotch on upper lobe, 4in(10cm) wide. When registered plant was 5ft(1.5m) tall with leaves 6in(15cm) long. (Rudolph Henny, 1962)

cl.
```
              -dichroanthum ssp. scyphocalyx (1/4)
       -Medusa-
Ming Toy-     -griersonianum (1/4)
        -     -wardii var. wardii (1/4)
        -Crest-           -fortunei ssp. discolor (1/8)
               -Lady Bessborough-
                               -campylocarpum ssp. campylocarpum
                                            Elatum Group (1/8)
```

3ft(.9m)          -10°F(-23°C)          EM

Trusses of 7-9 flowers, yellow, shaded orange and transluscent. Plant as broad as tall with leaves 4in(10cm) long, held 2 years. (Arthur & Maxine Childers, reg. 1977)

cl.
```
        -tephropeplum (1/2)
Mini Bell-
        -unknown (1/2)
```

Trusses of up to 35 flowers, brick red with spotting on upper petal. (Cross by Stanley Irvine; reg. by H. H. McCuaig, 1974)

cl.
```
        -Koichiro Wada--yakushimanum ssp. yakushimanum (1/2)
Mini Gold-
         -aureum var. aureum (chrysanthum) (1/2)
```

1ft(.3m)          -25°F(-32°C)          EM

Dwarf plant 3 times broader than high, with slightly indumented leaves held 2-3 years. Flowers mimosa yellow, openly funnel-shaped, 2.4in(6cm) across, in domed trusses of 8. Floriferous. See also Astroglow, Gold Spread, Summit Gold. (Basil C. Potter, 1981)

cl.
```
         -maximum (1/2)
Mini White-
          -aureum var. aureum (chrysanthum) (1/2)
```

dwarf          -25°F(-32°C)          ML

Plant compact; dark green, ovate leaves, 4in(10cm) long. Flowers pure white, funnel-campanulate, 1.75in(4.5cm) wide; trusses of 15. See also Serendipity. (Basil C. Potter, reg. 1972)

g.
```
          -Corona--unknown (1/2)
Minnehaha-
          -souliei (1/2)
```

Plant low and spreading.  Flowers pink and waxy, in small loose
trusses, early midseason.  Not large, but showy.    (Rothschild,
1947)

g.
```
          -forrestii ssp. forrestii Repens Group (1/2)
Minstrel-         -neriiflorum ssp. neriiflorum (1/4)
          -Neriihaem-
                  -haematodes ssp. haematodes (1/4)
```

Deep red flowers.  (Lord Aberconway, intro. 1950)

cl.
```
                 -decorum ? (1/4)
          -Caroline-
Mint Julep*      -brachycarpum ssp. brachycarpum ? (1/4)
          -        -Catalgla--catawbiense var. album Glass (1/4)
          -Calsap-
                  -Sappho--unknown (1/4)
```

Buds of light purplish pink open to very pale purple, speckled
with strong greenish yellow.  Hardy to -15°F(-26°C).    (Weldon
E.  Delp)

cl.
```
                 -dichroanthum ssp. dichroanthum (1/2)
Minterne Apricot-
                 -griersonianum (1/2)
```

Fabia g.  (Exhibited by Lord Digby, 1952)

cl.
```
                    -arboreum ssp. arboreum (1/4)
             -Doncaster-
Minterne Berryrose-         -unknown (1/4)
             -dichroanthum ssp. dichroanthum (1/2)
```

Berryrose g.  (Lord Digby, 1952)

cl.
```
                    -cinnabarinum ssp. cinnabarinum (1/2)
Minterne Cinnkeys-
                    -keysii (1/2)
```

Cinnkeys g.  Truss of up to 30 tubular-shaped flowers held in
dense clusters, orange red. (Lord Digby, 1931) A. M.  (RHS) 1951;
F. C. C. (RHS) 1952

cl.

Mira*     Parentage unknown

A parent of Leopold and other hybrids by T. J. R. Seidel; before
1910.

g.
```
    -beanianum (1/2)
Mira-
    -meddianum var. meddianum (1/2)
```

Flowers deep red or reddish purple.  (J. B. Stevenson, 1951)

cl.
```
          -dichroanthum ssp. dichroanthum (3/8)
      -Fabia-
Mirage-   -griersonianum (3/8)
      -         -fortunei ssp. discolor (1/4)
      -Margaret Dunn-   -dichroanthum ssp. dichroanthum
                  -Fabia-
                       -griersonianum
```

Flat flowers with yellow throat, shaded orange on petals.
(Rudolph Henny cross, 1947; reg. 1958)

cl.
```
                   -Corona--unknown (1/4)
          -Bow Bells-
Mira-Mi Linda-         -williamsianum (1/4)
          -     -wardii var. wardii (1/4)
          -Hawk-         -fortunei ssp. discolor (1/8)
          -Lady   -
          Bessborough-campylocarpum ssp. campylocarpum
                                    Elatum Group (1/8)
5ft(1.5m)          -5°F(-21°C)          M
```

Plant of open habit with elliptic leaves, rounded at base, 2.25
in(5.7cm) long. Upright trusses hold 9 flowers, apricot shading
to Chinese coral, about 2.5in(6.4cm) across.  (A. R. Heineman,
1966)

cl.

Mirelle Vernimb--form of hemitrichotum

2ft(.6m)          -5°F(-21°C)          EM

Pink flowers, .75in(1.9cm) wide, 5-lobed, bud axillary with 2-3
flowers per bud near the ends of stems; 24 or more form compact
trusses 2in(5cm) across.  Plant 1ft(.3m) tall, 1.5ft(.45m) wide
at 6 years; narrow leaves 1in(2.5cm) long, dark green above and
densely scaly below, alternate along entire stem. (Selected by
Bryan Vernimb, reg. 1969)

g.
```
                  -griffithianum (3/4)
Miss Adelaide Clow-         -maximum (1/4)
                  -Halopeanum-
                          -griffithianum
```

Flowers white, flushed pink, with a few chocolate spots.  (Lowinsky, 1919)  A. M. (RHS) 1919

cl.
```
                 -campylocarpum ssp. campylocarpum
        -Ole Olson-          Elatum Group (1/4)
Miss Jack-      -fortunei ssp. discolor (1/4)
        -            -lacteum (1/4)
        -unnamed hybrid-         -campylocarpum ssp. campylo-
                  -Mary    -               carpum (1/8)
                  Swaythling-
                          -fortunei ssp. fortunei (1/8)
```

Compact plant of medium size; leaves 4.5in(11.5cm) long.  Ivory
flowers spotted red on upper lobe, widely campanulate, 4in(10.2
cm) across; compact trusses of 10-12.  (Del W. James, 1964)

cl.
```
                  -griffithianum (1/2)
Miss Noreen Beamish-
                  -unknown (1/2)
```

Deep pink flowers.  (M. Koster & Sons, 1823)

cl.
```
                     -griffithianum (1/4)
         -Loderi King George-
Miss Olympia-        -fortunei ssp. fortunei (1/4)
         -williamsianum (1/2)
4ft(1.2m)          0°F(-18°C)          EM          4/4
```

Flowers to 4in(10.2cm) wide, blush pink with throat deeper pink.
Leaves 5in(12.7cm) long.  (Endre Ostbo cross; Roy Clark, raised
intro. & reg. 1962)  P. A. (ARS) 1960

g.
```
          -griersonianum (3/4)
      -Azma-
Miss Pink-  -fortunei ssp. fortunei (1/4)
      -griersonianum
```

Exhibited by Bolitho, 1943.

cl.
          -decorum (1/2)
Miss Prim-
          -irroratum ssp. irroratum (1/2)

4ft(1.2m)          -5°F(-21°C)          E          3/3

White flowers with brilliant yellow green blotch, to 4in(10.2cm)
aross; trusses of about 15. Leaves 4in(10.2cm) long. (R. Bovee,
1960; reg. 1962)


cl.

Miss Street     See Alice Street


cl.
          -williamsianum (1/2)
Mission Bells-
          -orbiculare ssp. orbiculare (1/2)

4ft(1.2m)          -5°F(-21°C)          M          3/4

Plant compact, sun-tolerant; glossy leaves, to 4in(10.2cm) long.
Pale pink, slightly fragrant flowers, campanulate, 2.5in(6.4cm)
wide; lax trusses of 6-8. Easy to propagate. (Benjamin Lancaster, reg. 1958)  See color illus. VV p. 69.


cl.
    -hodgsonii (1/2)
Mist-     -falconeri ssp. falconeri (1/4)
    -Muriel-
          -grande (1/4)

Evening g. Flowers of dark pinkish mauve. (Sir Giles Loder,
1954)


cl.

Mist Maiden--form of yakushimanum ssp. yakushimanum

3ft(.9m)          -20°F(-29°C)          M          4/5

Slightly faster growing than other forms of this species. Buds
very deep pink, opening to apple blossom pink, then white; flowers openly funnel-shaped, 2.5in(6.4cm) across, 5 wavy lobes, in
ball-shaped trusses of 14-17. Plant broad, well-branched; stems
and winter buds grey. Leaves typical of the species; 5.75in(14.
6cm) long, with heavy greyed orange indumentum, held 5 years.
(Seed from Francis Hanger, RHS; selected and raised by David G.
Leach; reg. 1983)  See color illus. VV p. 80.


cl.
     -fortunei ssp. discolor (1/2)
Mistake-                    -griffithianum (1/8)
     -          -George Hardy-
     -Pink Pearl-          -catawbiense (1/8)
     -          -arboreum ssp. arboreum (1/8)
          -Broughtonii-
                    -unknown (1/8)

Flowers pale pink, suffused with rose madder. (Brig. J. M. J.
Evans)  A. M. (RHS) 1956


cl.
                         -williamsianum (1/2)
Misty Moonlight*--Kimberly (selfed)-
                         -fortunei ssp. fortunei (1/2)

4ft(1.2m)          -10°F(-23°C)          EM          3/4

A vigorous variety which grows into a compact plant. Foliage is
glossy green with reddish purple petioles, deeply colored stems
and buds. Abundant fragrant flowers open pale orchid. (Harold
E. Greer, intro. 1982)


cl.
          -unknown (3/4)
Misty Morn-     -ponticum (1/4)
          -Purple Splendour-
                    -unknown

5ft(1.5m)          -5°F(-21°C)          M

Flowers purple violet with rays of olive yellow, openly funnel-
shaped, 3.5in(8.9cm) across, 5 frilled lobes; trusses of 12-14.
Plant rounded, wider than tall; elliptic leaves held 2 years,
new growth bluish green. (Mrs. Halsey A. Frederick, Jr., 1976)


cl.
     -wardii var. wardii (1/2)
Mitzi-     -dichroanthum ssp. dichroanthum (1/4)
     -Fabia-
          -griersonianum (1/4)

Crimson to orange flowers with darker throat and blotch. Dwarf
plant; slow growing. (Rudolph Henny cross, 1946; reg. 1958)


                              -griffithianum (1/16)
                    -Queen  -
          -Britannia-Wilhelmina-unknown (13/16)
cl.          -          -
     -Burgundy-     -Stanley Davies--unknown
Mobur-     -          -ponticum (1/8)
     -     -Purple Splendour-
     -Moser's Maroon--unknown  -unknown

5ft(1.5m)          10°F(-12°C)          ML

Flowers held in trusses of 15, dahlia purple, spotted maroon.
(Cross by Halfdan Lem; reg. by Dr. Paul Bowman, 1962)


cl.
          -williamsianum (1/2)
Mochicha Bell-
          -unknown (1/2)

Trusses of 2-3 flowers, dog-rose pink outside, silvery pink inside. (I. A. Hayes, cross 1969; reg. 1980)


cl.
               -haematodes ssp. haematodes (1/4)
          -Humming Bird-
Mochicha Moon-     -williamsianum (1/4)
          -unknown (1/2)

Trusses of 4-5 flowers, cream flushed pale pink. (I. A. Hayes,
cross 1969; reg. 1980)


cl.
     -impeditum (1/2)
Moerheim-
     -unknown (1/2)

1ft(.3m)          -15°F(-26°C)          EM          3/4

Buds appear on very young plants; flowers open a pretty shade of
violet, 1.25in(3.2cm) across. Small foliage turns maroon in the
winter. (J. D. Ruys, Moerheim Nursery, Netherlands; reg. 1966)


cl.
          -Essex Scarlet--unknown (1/2)
Moerheim Jubilee-
          -forrestii ssp. forrestii Repens Group (1/2)

Elizabeth Hobbie g. (Dietrich Hobbie; intro. by Moerheim Nursery, Netherlands; reg. 1969)

cl.
           -williamsianum (1/2)
Moerheim's Pink-     -catawbiense (1/4)
         -Genoveva-
              -unknown (1/4)

Plant compact, vigorous, broader than tall; height at 10 years
about 3.5ft(1m); leaves of dull, dark green, 2.25in(5.7cm) long.
Flowers very pale pink, deeper near edges, widely funnel-shaped,
to 3.5in(8.9cm) across; spherical trusses of 8. Free-flowering.
(Raised by Dietrich Hobbie; shown by Royal Moerheim Nurseries)
A. M. (RHS) 1972     See color illus. F p. 54.


cl.

Moerhermii*--form of impeditum

1ft(.3m)          -10°F(-23°C)          EM          4/4

Lavender flowers, larger than usual for this dwarf species.


g. & cl.
      -dichroanthum ssp. dichroanthum (1/2)
Mohamet-    -griersonianum (1/4)
     -Tally Ho-
        -facetum (eriogynum) (1/4)

Large, bell-shaped, bright scarlet flowers with frilled margins
and large scarlet petaloid calyces; loose trusses of 5-6. Plant
of medium size, rather open; light green leaves with fawn indumentum beneath.  Clones grown in Northwest U.S. are deep salmon
pink and orange red; these often bloom in autumn.  (Rothschild)
A. M. (RHS) 1945


            -fortunei ssp. discolor (3/8)
        -Lady     -
cl. -Day Dream-Bessborough-campylocarpum ssp. campylocarpum
   -          -               Elatum Group (1/8)
Mohur-     -griersonianum (3/8)
  -        -fortunei ssp. discolor
   -Margaret Dunn-   -dichroanthum ssp. dichroanthum
       -Fabia-               (1/8)
         -griersonianum

Dream Girl g. Yellow flowers, spotted green.  May be extinct.
See also Gold Mohur, Golden Pheasant, Shah Jehan.   (Lester E.
Brandt, reg. 1968)  P. A. (ARS) 1955


cl.

Moliere     Parentage unknown

Flowers of a dull rose red, spotted black on upper petal.  (M.
Koster & Son)   A. M. (RHS) 1953


cl.           -griffithianum (1/4)
        -Loderi-
Mollie Coker-    -fortunei ssp. fortunei (1/4)
    -unknown (1/2)

Flowers roseine pink with purplish red blotch. (Mrs. R. J. Coker, NZ, reg. 1979)


cl.        -forrestii ssp. forrestii Repens Group (1/4)
     -Elizabeth-
Molly Ann-    -griersonianum (1/4)
    -garden hybrid--unknown (1/2)

2ft(.6m)          -10°F(-23°C)          EM          4/4

A dense, compact shrub, as broad as tall; glossy, dark green,
rounded leaves held 2-3 years. Flowers rosy crimson, funnel-
shaped, 2in(5cm) across, 5 wavy lobes; lax trusses of 7.  Plant
buds heavily; flowers of good substance, long-lasting. (Mrs. Le
Vern Freimann, 1974)


cl.           -griffithianum (1/8)
      -Kewense-
    -Aurora-     -fortunei ssp. fortuunei (5/8)
Molly Buckley-   -thomsonii ssp. thomsonii (1/4)
    -fortunei ssp. fortunei

Naomi g.  Flowers bright orchid pink, speckled brownish orange.
(Edmund de Rothschild, 1982)


cl.
     -yakushimanum ssp. yakushimanum (1/2)
Molly Miller-     -dichroanthum ssp. dichroanthum
   -Fabia Tangerine-         (1/4)
       -griersonianum (1/4)

3ft(.9m)          5°F(-15°C)          M

Flowers yellow with a tinge of rose.  Foliage of average size,
with some indumentum beneath.  See also Shrimp Girl. (John Waterer, Sons & Crisp, reg. 1972)


cl.
   -Koichiro Wada--yakushimanum ssp. yakushimanum (1/2)
Molly-         -griffithianum (1/8)
Smith-     -unnamed hybrid-
    -Mrs. Furnival-     -unknown (1/4)
       -          -caucasicum (1/8)
       -unnamed hybrid-
          -unknown

4ft(1.2m)          0°F(-18°C)          M

White flowers with prominent blotch of marigold orange, openly
funnel-shaped, 2.5in(6.4cm) across, 5-lobed; conical trusses of
12. Floriferous.  Plant rounded, as wide as tall; heavy, dark
green leaves with grey brown indumentum beneath, held 3-4 years.
(Cecil S. Smith, reg. 1982)


cl.
    -catawbiense var. album (1/2)
Monaco-         -dichroanthum ssp. dichroanthum (1/4)
   -unnamed hybrid-      -griffithianum (1/8)
        -unnamed hybrid-
             -auriculatum (1/8)

6ft(1.8m)          -20°F(-29°C)          VL

Flower buds spirea red, opening Dresden yellow with reverse very
light spirea red, bold olive yellow blotch and spotting; corolla
openly funnel-shaped, 3.25in(8.3cm) wide, 5 wavy lobes; trusses
flat, 19-flowered.  Plant rounded, as broad as tall; narrow  elliptic leaves 5.25in(13.8cm) long, held 1 year. (David G. Leach,
cross 1952; reg. 1984)


---           -brookeanum var. gracile (1/4)
   -Duchess of Edinburgh-
Monarch-         -longiflorum (lobbii) (1/4)
   -          -jasminiflorum (3/8)
    -Princess Alexandra-    -jasmimiflorum
         -Princess Royal-
            -javanicum (1/8)

Vireya hybrid. Yellow orange flowers. (Davies)  F. C. C. (RHS)
1882


cl.
      -cinnabarinum ssp. cinnabarinum (1/2)
Monica Wellington-
      -unknown (1/2)

Trusses of 4-5 flowers, tubular funnel-shaped, 2in(5cm) across,
inside deep rose purple, the reverse rayed with bright magenta;
Elliptic leaves 2in(5cm) long, glossy green above, scaly below.
(Cross by Mrs. Roza Stevenson; raised & reg. by Hydon Nurseries)
A. M. (RHS) 1980

'Martine' by W. Hardijzer
Photo by Greer

'Mary Fleming' by Nearing
Photo by Greer

'Mary Mayo' by Bovee
Photo by Greer

'Mary Tasker' by Tasker (NZ)
Photo by Greer

'Marzo' by Goheen
Photo by Goheen

'Max Marshland' by K. Van de Ven
Photo by Greer

'May Time' by Leach
Photo by Leach

'Max Sye' by Frets, C.
Photo by Greer

'Meadowgold' by R. W. Scott
Photo by Greer

'Merganser' by Cox
Photo by Greer

'Michael's Pride' by C. Michael
Photo by Greer

'Midnight' by Van de Ven
Photo by Greer

'Mist Maiden' by Leach
(Not a hybrid; form of *yakushimanum*)
Photo by Leach

'Misty Moonlight' by Greer
Photo by Greer

'Mollie Coker' by Mrs. R. J. Coker
Photo by Greer

'Molly Smith' by C. Smith
Photo by C. Smith

'Monaco' by Leach
Photo by Leach

'Monsieur Guillemot' by Moser
Photo by Greer

'Montego' by Leach
Photo by Leach

'Mood Indigo' by Brandt
Photo by Greer

'Moon Mist' by Leach
Photo by Leach

'Mount Mazama' by Grace/Grothaus
Photo by Greer

'Mount Mitchell' by Leach
(Not a hybrid; form of *maximum*.)
Photo by Leach

'Mrs. Bernice Baker' by Larson
Photo by Greer

```
                                -griffithianum (1/8)
cl.             -Queen Wilhelmina-
        -Britannia-                     -unknown (5/8)
Monique-          -Stanley Davies--unknown
        -                     -ponticum (1/4)
        -Purple Splendour-
                          -unknown
```

3ft(.9m)          -5°F(-21°C)          M

Small plant with narrow leaves, medium-sized light purple flowers. (Milton R. Nelson, intro. 1970; reg. 1972)

```
                                -caucasicum (1/8)
cl.                   -Cunningham's White-
                -Rocket-                -ponticum (white) (1/8)
Monique Behring-        -catawbiense (1/4)
                -yakushimanum ssp. yakushimanum (1/2)
```

1ft(.3m)          -25°F(-32°C)          ML

Plant well-branched, as broad as high; leaves dark yellow green,
narrowly elliptic, light greyed yellow indumentum below. Flowers neyron rose with darker rose dorsal spotting, openly funnel-
shaped, 2.5in(6.5cm) wide, 5 wavy lobes; conical trusses hold 13
-15. (Rudy Behring, Ontario, 1983)

```
cl.      -griffithianum (1/4)
    -Mars-
Moniz*  -unknown (5/8)            -catawbiense (1/8)
    -            -Parsons Grandiflorum-
    -America-                    -unknown
                -dark red hybrid--unknown
```

A parent of Heatarama.

```
cl.
    -thomsonii ssp. thomsonii (1/2)
Monk-
    -arboreum ssp. delavayi var. delavayi (1/2)
```

Dark red flowers. (Sir James Horlick, 1934)

cl.

Monsieur Guillemot      Parentage unknown

6ft(1.8m)          -10°F(-23°C)          L          2/3

Synonym Madame Guillemot. Upright, fairly compact plant; glossy
dark green leaves. Dark rose flowers in compact trusses. Floriferous. (Moser et Fils)  See color illus. VV p. 68.

cl.

Monsieur Thiers      Parentage unknown

A parent of J. H. van Nes. Flowers light red. (J. Macoy)

```
cl.                         -griffithianum (1/4)
        -Mrs. E. C. Stirling-
Monstrous-                  -unknown (1/2)
        -                -smirnowii (1/4)
        -unnamed hybrid-
                      -unknown
```

Flowers rose pink  or lavender pink.   (C. Waterer, Son & Crisp)
A. M. (RHS) 1925

```
cl.
        -pubescens (1/2)
Montchanin-
        -keiskei (1/2)
```

3ft(.9m)          -25°F(-32°C)          EM          2/2

Shapely plant habit with small foliage, and a profusion of small
white flowers. See also Hockessin. (G. Guy Nearing, 1958)

```
                -catawbiense var. album (1/4)
cl.  -unnamed hybrid-        -wardii var. wardii (1/8)
    -            -Crest-        -fortunei ssp. discolor
Monte-                -Lady -                   (1/16)
Carlo-                Bessborough-campylocarpum ssp. campy-
    -                          locarpum Elatum Gp. (1/16)
    -            -aureum var. aureum (1/4)
    -unnamed-
    hybrid-campylocarpum ssp. campylocarpum (1/4)
```

2.5ft(.75m)          -10°F(-23°C)          M

Ball-shaped trusses of 16 flowers, pale empire yellow flushed
orchid pink, faint dorsal blotch of Naples yellow; corolla widely funnel-shaped, 2.25in(5.7cm) wide, with 5 wavy lobes. Free-
flowering. Plant rounded, nearly as wide as tall; dark yellow
green, glossy leaves, held 2 years. (David G. Leach cross, 1966;
reg. 1981)

```
cl.        -catawbiense (1/4)
    -Sefton-
Montego-    -unknown (1/2)
    -            -ponticum (1/4)
    -Purple Splendour-
                -unknown
```

5ft(1.5m)          -15°F(-26°C)          M

Openly funnel-shaped flowers, 3.25in(8.3cm) wide, violet purple
with green blotch, aging to Indian lake; ball-shaped trusses of
18. Plant twice as broad as tall; stems of current growth near-
pink. Elliptic leaves 5.75in(1.6cm) long, dark Winchester green
and held 2 years. (David G. Leach, cross 1957; reg. 1983)

```
cl.
    -fortunei ssp. discolor (1/2)
Montreal-
    -campylocarpum ssp. campylocarpum Elatum Group (1/2)
```

6ft(1.8m)          5°F(-15°C)          M          3/2

Lady Bessborough g. A tall plant  with pink buds opening deep
cream; flowers funnel-shaped with wavy edges. Blooms midseason.
(Rothschild, 1933)   See color illus. F p. 49.

*MOOD INDIGO (?)  SEE PHOTO*

```
cl.
        -fortunei ssp. fortunei (1/2)
Moon Mist-        -wardii var. wardii ? (1/4)
        -Lackamas Cream--Chlorops-
                          -vernicosum ? (1/4)
```

Plant 3ft(.9m) tall by 3.5ft(1m) wide at 10 years; leaves elliptic, 4in(10.2cm) long. Bell-shaped flowers of Dresden yellow,
deeper in throat with mahogany rays, 3.5in(8.9cm) across; graceful upright trusses of 10-12. Midseason. (B. Lancaster, 1965)

```
cl.  -catawbiense var. album (1/2)
    -        -neriiflorum ssp. neriiflorum (1/8)
Moon Mist*    -unnamed-
    -        - hybrid-dichroanthum ssp. dichroanthum (1/8)
    -unnamed-
      hybrid-fortunei ssp. discolor (1/4)
```

6ft(1.8m)          -15°F(-26°C)          ML          5/2

Very large, waxy, white flowers flushed pale pink, with a golden
blotch. (David G. Leach, intro. 1968)

```
cl.
        -Catalgla--catawbiense var. album Glass (1/2)
Moon Shot*
        -wardii var. wardii (1/2)
```

?          -10°F(-23°C)          EM          4/4

Creamy white flowers; dark green foliage. Name of hybrid may be
found as one word also. (Joseph B. Gable) See color illus. LW
pl. 48.

```
                       -griffithianum (9/16)
            -Kewense-
g.          -Aurora-       -fortunei ssp. fortunei (5/16)
      -Naomi-      -thomsonii ssp. thomsonii (1/8)
Moonbeam-     -fortunei ssp. fortunei
      -griffithianum
```

Flowers pale primrose yellow in shapely trusses on a tall plant.
Midseason. (Rothschild, 1947)

```
g.                     -griffithianum (1/4)
           -Loderi Venus-
Moonglow-              -fortunei ssp. fortunei (1/4)
       -              -fortunei ssp. discolor (1/4)
           -Lady Bessborough-
                      -campylocarpum ssp. campylocarpum
                                    Elatum Group (1/4)
```

Tall plant, blooming in midseason. Pink flowers in well-shaped
trusses. (Rothschild, 1946)

```
cl.
                  -macabeanum (1/2)
Moonlight Sonata-
                  -Pamela--unknown (1/2)
```

Trusses of 24 flowers, very light yellow, stained  phlox pink
with ruby red blotch. (Raised by New Zealand Rhododendron Assoc.
and reg. by Mrs. J. Kerr, 1979)

```
                  -griffithianum (1/4)
cl.       -Mars-
      -       -unknown (3/8)
Moonlight Tango-           -fortunei ssp. discolor (1/8)
      -       -Ladybird-
          -Diva-         -Corona--unknown
                  -griersonianum (1/4)
```

Tall, showy trusses of cardinal red flowers in late season. A
tall plant with leaves 5in(12.7cm) long. (Rudolph & Leona Henny, 1965)

```
cl.
      -augustinii ssp. chasmanthum (1/2)
Moonrise-             -cinnabarinum ssp. cinnabarinum Roylei Group
      -Lady                                       (1/4)
      Chamberlain-           -cinnabarinum ssp. cinnabarinum
                  -Royal Flush-                   (1/8)
                  (orange form)-maddenii ssp. maddenii (1/8)
```

Cream flowers, suffused pale lilac pink; 6-flowered truss. (Gen.
Harrison cross, 1949; intro. 1955)

```
cl.           -campylocarpum ssp. campylocarpum (1/4)
      -Souvenir of W.-
Moonrise- C. Slocock    -unknown (1/4)
      -             -griffithianum (1/4)
      -Loderi King George-
                  -fortunei ssp. fortunei (1/4)
```

Crossed, raised & reg. by the A. A. Wrights, before 1958.

```
                  -campylocarpum ssp. campylocarpum (1/8)
                  -unnamed-
g. & cl. -Adriaan- hybrid-unknown (1/4)
      - Koster-                  -griffithianum (1/16)
Moonshine-             -George Hardy-
      -       -Mrs. Lindsay-          -catawbiense (1/16)
      -        Smith   -Duchess of Edinburgh--unknown
      -wardii var. wardii Litiense Group (1/2)
```

Compact trusses of 16 flowers, shallow rotate, primrose yellow,
darker on upper lobe with throat stained a dark crimson blotch.
See 3 other selected clones below. (Cross by Francis Hanger,
RHS Gardens, Wisley) A. M. (RHS) 1952

cl.

Moonshine Bright      Parentage as above

Moonshine g.  Rich yellow flowers in trusses of 20-22.  Thin
glossy foliage. (RHS Garden, Wisley, 1952)

```
cl.           -wardii var. wardii (1/4)
      -Jervis Bay-          -fortunei ssp. discolor
Moonshine-     -Lady Bessborough-                (1/8)
  Crescent-              -campylocarpum ssp. campy-
      -wardii var. wardii          locarpum Elatum Gp. (1/8)
      Litiense Group (1/2)
```

4ft(1.2m)           0°F(-18°C)          M

(Not Moonshine g.) Plant upright; leaves dull green. Flowers
primrose yellow, in compact dome-shaped trusses of 14-16. (RHS
Garden, Wisley, 1962)  A. M. (Wisley Trials) 1960

cl.

Moonshine Glow      For parentage diagram see Moonshine

Moonshine g. Openly funnel-shaped flowers in trusses of 9, of a
yellow that is close to uranium green.  Dull green leaves, 4.25
in(10.8cm) long, 2.25in(5.7cm) wide. (RHS Garden, Wisley)  H.C.
(RHS) 1957

cl.

Moonshine Supreme      For parentage diagram see Moonshine

Moonshine g.  Flowers in trusses of 15, primrose yellow with
darker staining on upper lobes, broadly campanulate. (RHS Gardens, Wisley) A. M. (RHS) 1953   See color illus. F p. 55.

```
g.
      -campylocarpum ssp. campylocarpum (1/2)
Moonstone-
      -williamsianum (1/2)
```

3ft(1.5m)           -5°F(-21°C)          EM          3/4

Compact, rounded shrub, wider than tall; dense foliage of flat,
smooth, oval leaves, 2.5in(6.4cm) long.  Orange rose buds open
to pale creamy yellow flowers, lightly tinged pink, openly campanulate, 2.5in(6.4cm) across; lax trusses of 3-5. Floriferous.
(J. C. Williams, 1933)     See color illus. ARS Q 31:3 (Summer
1977): p. 171; F p. 55; VV p. 85.

---

```
      -wardii var. wardii
Moontide-
      -Loder's White--See Loder's White, 2 possible diagrams
```

4ft(1.2m)           0°F(-18°C)          EM          3/2

Growth habit rather open but rounded, upright. Leaves flat, oblong, medium green, to 4in(10.2cm) long. White flowers, funnel-
shaped, over 3in(7.6cm) wide, held in full, rounded trusses of
16. (Rudolph Henny, reg. 1958)   P. A. (ARS) 1955

```
                  -King of-fortunei ssp. discolor (3/8)
                  - Shrubs-       -dichroanthum ssp. dichro-
cl.   -Holy Moses-       -Fabia-        anthum (1/16)
      -                              -griersonianum (1/16)
Moonwax-          -Souvenir of Anthony Waterer--unknown (1/4)
      -                  -griffithianum (1/8)
      -          -Loderi-
      -Albatross-       -fortunei ssp. fortunei (1/8)
               -fortunei ssp. discolor
```

6ft(1.8m)           0°F(-18°C)          M

Flowers 4in(10.2cm) wide, 7-lobed, fragrant, Naples yellow center with light mauve edging, unmarked; dome-shaped trusses hold
12. Plant broader than tall, with stiff, upright branches; narrow elliptic leaves 6.5in(16.5cm) long, held 2 years. (Cross by
Halfdan Lem; L. L. Newcomb, reg. 1981)   C. A. (ARS) 1983   See
color illus. ARS J 38:4 (Fall 1984): p. 171.

cl.
     -konorii (1/2)
Moonwind-              -Pink Delight--unknown (1/4)
    -Nancy Adler Miller-
                  -jasminiflorum (1/4)

2ft(.6m)        32°F(0°C)        Nov-Feb

Vireya hybrid. Very fragrant flowers of chrome yellow, tubular funnel-shaped, 1.75in(1.9cm) wide and long, 5 wavy lobes; dome-shaped trusses of 12. Plant well-branched, rounded, as broad as tall; leaves 3.25in(8.3cm) long, golden brown scales beneath, held 3 years. See also Aravir, Shasta. (Cross by Peter Sullivan; raiser William Moynier; reg. 1980)

cl.
     -konorii (1/2)
Moonwood-              -Pink Delight (Veitch clone)--unknown
    -Nancy Adler Miller-               (1/4)
                  -jasminiflorum (1/4)

2ft(.6m)        32°F(0°C)        Nov.-Feb.

Vireya hybrid. Flowers very fragrant, tubular-shaped, yellowish white with chrome yellow throat; trusses of 12. Plant rounded, as wide as tall; leaves held 3 years. See also Shasta. (Cross by Peter Sullivan; William A. Moynier, reg. 1979)

                -griffithianum (3/8)
cl.  -Isabella-
   -           -auriculatum (1/4)
Morawen-             -griffithianum
   -        -Loderi-
   -Shepherd's-    -fortunei ssp. fortunei (1/4)
    Delight -      -fortunei ssp. fortunei
       -Luscombei-
              -thomsonii ssp. thomsonii (1/8)

Phlox pink flowers with upper 3 petals slightly marked with dark pink spots. (Heneage-Vivian)  A. M. (RHS) 1950

cl.
     -morii (1/2)
Morfar-
     -oreodoxa var. fargesii (1/2)

Flowers pinkish white with crimson spots. (Collingwood Ingram, 1949)

cl.
     -Koichiro Wada--yakushimanum ssp. yakushimanum (1/2)
Morgenrot-        -griffithianum (1/4)
(Morning -Spitfire-
  Red)        -unknown (1/4)

Dark red buds, opening to spinel red flowers, shading inside to light neyron rose, 5-lobed, in trusses of 16-18. Rounded, compact bush; leaves oblanceolate-obovate, hairy. Hardy to -8°F (-22°C). (H. Hachmann cross; reg. by C. Stuck, 1983) See color illus. ARS J 39:2 (Spring 1985): p. 63.

              -griffithianum (5/16)
cl. -Loderi King George-
   -           -fortunei ssp. fortunei (5/16)
Morio-            -griffithianum
   -         -Loderi-
   -  -Albatross-    -fortunei ssp. fortunei
  -Cup Day-     -fortunei ssp. discolor (1/8)
     -      -elliottii (1/8)
     -Fusilier-
        -griersonianum (1/8)

Color is a very light greenish white. (H. Van de Ven, 1972)

cl.
         -yakushimanum ssp. yakushimanum (1/2)
Morning Cloud-     -griersonianum (1/4)
      -Springbok-       -ponticum (1/8)
         -unnamed hybrid-
               -unknown (1/8)

3ft(.9m)       -5°F(-21°C)      M

Compact plant with white flowers flushed a very light lavender. See also Hydon Ball. (A. F. George, Hydon Nurseries, 1972) A. M. (RHS) 1971

cl.
         -dichroanthum ssp. dichroanthum (1/4)
      -Fabia-
Morning Frost-   -griersonianum (1/4)
     -unnamed hybrid--unknown (1/2)

White flowers, flushed red purple, with greenish yellow spotting on upper lobe. (John Waterer, Sons & Crisp, reg. 1972)

cl.
         -yakushimanum ssp. yakushimanum (1/2)
Morning Magic-     -griersonianum (1/4)
      -Springbok-       -ponticum (1/8)
         -unnamed hybrid-
               -unknown (1/8)

3ft(.9m)       -5°F(-21°C)      M

Vigorous, compact plant, wider than high; dense foliage with leaves 3.5in(8.9cm) long. White flowers, flushed neyron rose, pronounced spotting of orange buff on upper lobes, campanulate, to 2.5in(6.4cm) wide, wavy margins; globular trusses of 16. See also Hydon Ball, Hydon Globe, Hydon Glow, Hydon Hunter, Morning Cloud. (A. F. George, Hydon Nurseries, 1972) H. C. (RHS) 1977; A. M. (RHS) 1976

g.              -griffithianum (1/4)
        -Sunrise-
Morning Star-    -griersonianum (1/4)
      -williamsianum (1/2)

Pale pink flowers. (Lord Aberconway, 1942)

cl.

Morocco--form of coriaceum

5ft(1.5m)     5°F(-15°C)     EM     3/4

A species with long leaves, 9in(23cm) by 3.5in(9cm) broad, covered below with pale fawn indumentum. Trusses hold 16-20 white flowers with blotch of deep crimson. (Crown Estate, Windsor, 1962)

cl.

Morocco (D. Leach)     See Luxor

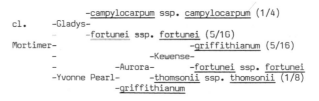

         -campylocarpum ssp. campylocarpum (1/4)
cl.  -Gladys-
   -      -fortunei ssp. fortunei (5/16)
Mortimer-            -griffithianum (5/16)
   -         -Kewense-
   -     -Aurora-    -fortunei ssp. fortunei
  -Yvonne Pearl-   -thomsonii ssp. thomsonii (1/8)
        -griffithianum

Large, heavy trusses of 12 flowers, white, suffused yellow with crimson blotch in throat; corolla campanulate, 5in(12.7cm) wide, exterior stained pink. Leaves 5.5in(14cm) long. (Crown Estate, Windsor) A. M. (RHS) 1964

cl.
     -elliottii (1/2)
Morvah-
     -wattii (1/2)

Spherical trusses of 20 flowers, Turkey red with spotting on upper lobes.  (Sir Edward Bolitho, reg. 1962)  A. M. (RHS) 1956;
F. C. C. (RHS) 1959

```
cl.
          -ambiguum (1/2)
Mosaique-          -cinnabarinum ssp. cinnabarinum (1/4)
          -Cinnkeys-
                   -keysii (1/4)
```

Small flowers of pale yellow, bright red at base, with 5 narrow
lobes; trusses of about 12.  Small, dense plant, blooming late
midseason.  (Rothschild)  A. M. (RHS) 1945

```
cl.

Moser's Maroon    Parentage unknown, but may be maximum hybrid

5ft(1.5m)         -10°F(-23°C)         ML          2/3
```

A parent of many hybrids.  Growth habit vigorous, rather sprawling; new foliage an attractive red.  Small tight trusses hold
very small flowers of dark wine red, spotted with black.  (Moser
& Fils)  A. M. (RHS) 1932   See color illus. F p. 56.

```
cl.
     -megeratum (1/2)
Moth-
     -boothii Mishmiense Group (1/2)
```

Silver Ray g.  Flowers are lemon yellow, heavily spotted brown.
(Lord Aberconway)  A. M. (RHS) 1955

```
cl.
          -hippophaeiodes var. hippophaeiodes (1/2)
Mother Greer*
          -unknown (1/2)

1.5ft(.45m)       -10°F(-23°C)         ML          3/4
```

Similiar in shape to the blue flowers of hippophaeiodes, but
larger and  the plant is much more compact.  Blooms later than
most "blue" rhododendrons.  Cross made by Harold Greer's mother.
(Harold E. Greer, intro. 1982)   See color illus. HG p. 143.

```
cl.
                                   -griffithianum (1/4)
Mother                  -George Hardy-
of Pearl--sport of Pink Pearl-     -catawbiense (1/4)
                  -              -arboreum ssp. arbor-
                  -Broughtonii-               eum (1/4)
                               -unknown (1/4)

6ft(1.8m)         -5°F(-21°C)          M          4/3
```

Rich pink buds opening delicate pink, fading to pure white with
very faint brownish spots on upper lobe; pink stamens, and a few
pink streaks on exterior.  Large flowers, vigorous plant, good
foliage. (J. Waterer, 1925)  A. M. (RHS) 1930   See color illus.
VV p. 104.

```
---
                   -griffithianum (aucklandii) (1/2)
Mottled Aucklandii-
                   -unknown (1/2)
```

Shell pink flowers.  (R. Gill & Son)

```
cl.
          -pemakoense (1/2)
Moukoense-
          -moupinense (1/2)
```

White flowers, flushed with pink.  (Lester E. Brandt, 1953)

```
cl.
                   -yakushimanum ssp. yakushimanum (1/4)
          -Bambi-          -dichroanthum ssp. dichroanthum
Moulin Rouge-     -Fabia                              (1/8)
          -         -Tangerine-griersonianum (1/8)
          -arboreum ssp. arboreum (1/2)
```

Turkey red flowers held in trusses of 16.  (R. Cutten, 1981)

```
cl.                            -Sir Charles Butler--fortunei
          -Van Nes Sensation-          ssp. fortunei (1/4)
Mount Clearview-               -        -maximum (1/8)
                              -Halopeanum-
          -Purple  -ponticum (1/4)   -griffithianum (1/8)
            Splendour-
                    -unknown (1/4)

5ft(1.5m)         0°F(-18°C)          M
```

Flowers of imperial purple with beetroot purple throat, dark
greyed purple dorsal spotting, openly funnel-shaped, 4in(10.2cm)
across, 5-lobed; dome-shaped trusses of 14.  Rounded plant with
stiff branches, as wide as tall; leaves held 3 years.  (Lloyd L.
Newcomb, 1981)

```
cl.

Mount Everest*--form of cinnabarinum ssp. cinnabarinum
```

A tall, free-flowering form of this species, with small rounded
leaves; tubular-shaped red flowers with yellow edges.  (Slocock)

```
cl.
          -campanulatum ssp. campanulatum (1/2)
Mount Everest-
          -griffithianum (1/2)
```

Narrow campanulate flowers of pure white with brown markings in
throat, in conical trusses.  Very free-flowering.  Plant large,
vigorous, dense.  (W. C. Slocock, 1930)  A. M. (RHS) 1953; F. C.
C. (Wisley Trials) 1958; A. G. M. (RHS) 1969

```
cl.          -fortunei ssp. fortunei (1/4)
          -Fawn-   -dichroanthum ssp. dichroanthum (1/8)
          -    -Fabia-
Mount Hood-     -griersonianum (1/8)
          -    -wardii var. wardii (1/4)
          -Crest-          -fortunei ssp. discolor (1/8)
               -Lady Bessborough-
                              -campylocarpum ssp. campylo-
                               carpum Elatum Group (1/8)

6ft(1.8m)         -5°F(-21°C)          M
```

Fragrant flowers, white with chartreuse shading in throat, 6-
lobed, openly campanulate, 3.5in(8.9cm) wide; conical trusses of
11-13.  Upright plant, branching moderately; grey green leaves,
5.5in(14cm) long, held 2 years.  (Carl H. Phetteplace, 1975)

```
cl.                            -griffithianum (3/8)
          -            -Queen Wilhelmina-
          -Britannia-          -unknown (3/8)
Mount Mazama-    -Stanley Davies--unknown
          -    -griffithianum
          -Loderi-
               -fortunei ssp. fortunei (1/4)

5ft(1.5m)         0°F(-18°C)          ML          4/4
```

Flowers fuchsia red with brighter red spotting; an unusual flower with dark red nectar pouches and stripes on reverse.  Corolla
openly funnel-shaped, 4.5in(11.5cm) across, 5 wavy lobes.  Plant
wider than tall, rounded; leaves held 2 years.  (George Grace,
intro. 1967; reg. by Molly & Louis Grothaus, 1980)

```
cl.

Mt. Mitchell--form of maximum

6ft(1.8m)         -25°F(-32°C)         VL          3/3
```

From the mountains of North Carolina,  an erratic plant--flowers
sometimes pink, sometimes red, or lighter.  An excess of red
pigment may also produce streaked stems, leaves. (Gable, 1958)


cl.


Mount Mitchell--form of <u>maximum</u>

5ft(1.5m)          -25°F(-32°C)          VL          3/3

Buds red, opening to flowers suffused strong pink with a yellow-
ish green dorsal blotch, exterior stained red; trusses of 16.
Leaves 6in(15.2cm) long. (Raised by David G. Leach, seed from
Warren Baldsiefen; intro. 1964)


cl.

            -<u>laetum</u> (1/2)
Mount Pire-
            -<u>javanicum</u> (1/2)

3ft(.9m)          32°F(0°C)          Aug.-Sept.

Vireya hybrid.  Flowers tubular funnel-shaped, nasturtium orange
tubes, Saturn red lobes, in trusses 5.5in(14cm) across.  Willowy
branches; leaves held 3 years. (Cross by D. Stanton; raiser Dr.
E. C. Brockenbrough; reg. by William A. Moynier, 1981)


cl.
            -<u>vernicosum</u> (R 18139) (1/2)
Mount Siga-                                 (natural hybrid)
            -unknown (1/2)

5ft(1.5m)          -5°F(-21°C)          EM

Flowers orient pink, slightly deeper in throat, with faint spots
and dorsal blotch of peach, openly funnel-shaped, 4in(10.2cm)
wide, 7 wavy lobes, slightly fragrant.  Plant rounded, upright;
leaves held 2 years. (Seed from Rock; raised & intro. by Joseph
Gable, 1958; reg. by Caroline Gable, 1979)


cl.


Mount Wilson--<u>yakushimanum</u> ssp. <u>yakushimanum</u> (selfed)

Flowers are pale pink, fading to white with yellow spotting.
(P. G. Valder, Australia, 1972)


cl.                   -<u>griffithianum</u> (1/4)
                -Dorothea-
Mountain Aura-        -<u>decorum</u> (1/4)
                -red hybrid--unknown (1/2)

3ft(.9m)          -15°F(-26°C)          ML          4/4

Flowers flax blue with a white center, 4in(10.2cm) across, in
rounded trusses 7in(17.8cm) across.  Leaves 6in(15.2cm) long and
half as wide.  See also Mountain Glow, Mountain Queen. (G. Guy
Nearing, reg. 1970)


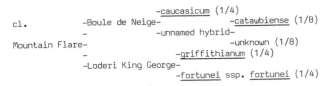

cl.       -Boule de Neige-                 -<u>catawbiense</u> (1/8)
          -                   -unnamed hybrid-
Mountain Flare-                            -unknown (1/8)
                              -<u>griffithianum</u> (1/4)
          -Loderi King George-
                              -<u>fortunei</u> ssp. <u>fortunei</u> (1/4)

Buds light jasper red opening pale empire rose, with greenish to
purplish spotting; flowers openly funnel-shaped, 4.5in(11.5cm)
wide, fragrant; spherical trusses of 15.  Plant as broad as high
with stiff, upright branches; leaves 7in(17.8cm) long, held 2
years. (G. Guy Nearing, intro. 1972; reg. 1973)


cl.                   -<u>griffithianum</u> (1/4)
                -Dorothea-
Mountain Glow-        -<u>decorum</u> (1/4)
                -red hybrid--unknown (1/2)

3ft(.9m)          -15°F(-26°C)          ML

Densely branching plant; oval leaves to 6in(15.2cm) long.  Flowers reddish purple, 4in(10.2cm) across, in trusses 6in(15.2cm)
wide.  See also Mountain Aura, Mountain Queen. (G. Guy Nearing,
reg. 1968)


cl.                   -<u>griffithianum</u> (1/4)
                -Dorothea-
Mountain Queen-       -<u>decorum</u> (1/4)
                -red hybrid--unknown (1/2)

4ft(1.2m)          -15°F(-26°C)          ML          4/4

Oval leaves, to 7in(17.8cm) long.  Slightly fragrant flowers up
to 4in(10.2cm) across, rose-colored, held in round trusses. (G.
Guy Nearing, 1968)


cl.
            -Stanley Davies--unknown (1/2)
Mountain Star-
            -<u>yakushimanum</u> ssp. <u>yakushimanum</u> (1/2)

The trusses hold 20-30 flowers, rose pink. (Frederick Street,
reg. 1962)    P. C. (RHS) 1959


cl.                         -Essex Scarlet--unknown (1/4)
                -Beau Brummel-
Mouton Rothschild-          -<u>facetum</u> (<u>eriogynum</u>) (1/4)
                -<u>elliottii</u> (1/2)

Red, waxy, bell-shaped flowers in conical trusses.  Floriferous,
late-blooming. (E. de Rothschild)   A. M. (RHS) 1958


g.
            -<u>cinnabarinum</u> ssp. <u>xanthocodon</u> Concatenans Group (1/2)
Mozari-
            -<u>ambiguum</u> (1/2)

Flowers in pastel shades of yellow, pink, etc. (J. B. Stevenson
cross, 1941; intro. 1951)


---
                      -<u>griffithianum</u> (1/2)
Mrs. A. C. Kenrick-
                      -unknown (1/2)

Flowers deep rose pink with darker spots. (M. Koster & Sons ?)
A. M. (RHS) 1925


cl.                   -<u>diaprepes</u> (1/4)
                -Polar Bear-
Mrs. A. J. Holden-    -<u>auriculatum</u> (1/4)
          -           -<u>fortunei</u> ssp. <u>discolor</u> (1/4)
          -Evening-   -<u>dichroanthum</u> ssp. <u>dichroanthum</u>
          Glow  -Fabia-                                (1/8)
                      -<u>griersonianum</u> (1/8)

5ft(1.5m)          -5°F(-21°C)          L

Long leaves, rather narrow, 7.5in(19cm) by 2in(5cm).  Flowers 7-
lobed, nicely scented, up to 5.25in(13.3cm) across, sunny yellow
and spotted light green with a pink edge, held in trusses of 10.
(A. John Holden, 1979)


cl.                   -<u>griffithianum</u> (1/2)
Mrs. A. M. Williams-
                      -unknown (1/2)

A large, dense plant with round trusses of bright scarlet and a
flare of black spots. (Otto Schulz cross; intro. by C. B. van
Nes, 1896) A. M. (RHS) 1926; A. M. (Wisley Trials) 1933; F.C.C.
(Wisley Trials) 1954

cl.             -arboreum ssp. arboreum (1/4)
          -Doncaster-
Mrs. A. R. Bede-      -unknown (3/4)
          -unknown

Very large truss of rich rose scarlet flowers with pale
interior. Much admired at Chelsea show, 1923. (M. Koster)


cl.              -griffithianum (1/4)
          -Loderi-
Mrs. A. R. McEwan-      -fortunei ssp. fortunei (1/4)
          -unknown (1/2)

6ft(1.8m)      -5°F(-21°C)      EM      3/3

Growth habit is compact and roundish. Foliage glossy green,
waxy and quite large, 7.5in(19cm) by 2.5in(6.4cm). Flowers
5in(12.7cm) across, fuchsine pink to Persian rose with a white
throat, held in dome-shaped trusses. (Cross by Herbert Ihrig;
intro. by University of Washington, 1958)   A. E. (ARS) 1956


cl.
             -Sir Charles Butler--fortunei ssp. fortunei
Mrs. A. T. de la Mare-      -maximum (1/4)      (1/2)
          -Halopeanum-
                -griffithianum (1/4)

5ft(1.5m)      -15°F(-26°C)      M      3/3

A plant that can take full exposure in many climates. Buds of
blush pink opening to delicately fragrant, large white flowers
with green spots; large, dome-shaped trusses hold 12-14. (C. B.
van Nes)   A. M. (RHS) 1958   See color illus. ARS Q 35:1 (Winter 1981): p. 11;  VV p. 66.


                -griffithianum (1/16)
             -Kewense-
cl.          -Aurora-      -fortunei ssp. fortunei (5/16)
       -Naomi  -      -thomsonii ssp. thomsonii (1/8)
Mrs. Alice-Nautilus-fortunei ssp. fortunei
  Blyskal  -
          -Catanea--catawbiense (1/2)

6ft(1.8m)      -15°F(-26°C)      M

Buds are roseine purple opening to very light purple flowers
with a creamy white throat, held in trusses of 12-17. (Walter
Blyskal raiser; reg. by Mrs. James Walton, 1983)


cl.
          -fortunei ssp. fortunei (1/2)
Mrs. Anthony Waterer-
          -unknown (1/2)

White flowers with yellow blotch on a tall growing shrub.
(Waterer before 1915)


---
          -fortunei ssp. fortunei (1/2)
Mrs. Arderne-
          -unknown (1/2)

Flowers waxy, cherry red.


g.

Mrs. Arthur Evans    See Mrs. Arthur Fawcus


g.
          -caucasicum (1/2)
Mrs. Arthur Fawcus-      -griffithianum (1/4)
          -Kewense-
                -fortunei ssp. fortunei (1/4)

Synonym Mrs. Arthur Evans. Flowers of pale yellow. (Knap Hill
Nursery, 1946)


---
             -campylocarpum ssp. campylocarpum (1/2)
Mrs. Ashley Slocock-
          -hardy hybrid--unknown (1/2)

A shrub of medium height with cream flowers suffused apricot.
(Slocock Nurseries, 1933)


cl.              -griffithianum (1/4)
          -Dawn's Delight-
Mrs. Bernice Baker-      -unknown (1/4)
          -fortunei ssp. fortunei (1/2)

6ft(1.8m)      -5°F(-21°C)      M      4/3

Plant habit rather open; attractive foliage. Rose pink flowers,
lighter at center, with pale midline markings down 5 wavy lobes;
large trusses. (Larson, reg. 1958)  See color illus. VV p. 79.


             -decorum (1/4)
cl.       -unnamed hybrid-
       -             -fortunei ssp. discolor (1/4)
Mrs. Betty Hager-            -Michael-ponticum (1/16)
       -             -Waterer-
       -      -Prometheus-      -unknown (5/16)
    -Madame de-      -Monitor--unknown
      Bruin  -      -arboreum ssp. arboreum
          -Doncaster-             (1/8)
                -unknown

4ft(1.2m)      0°F(-18°C)      M

Deep pink flowers with a red blotch, dorsal spotting, 3in(7.6cm)
across; spherical trusses of 15. Leaves 5.5in(14cm) long.  (A.
Raustein, 1974)


                -griffithianum (1/8)
cl.          -George Hardy-
       -Mrs. Lindsay-      -catawbiense (1/8)
Mrs. Betty- Smith   -Duchess of Edinburgh--unknown (1/2)
  Robertson-      -campylocarpum ssp. campylocarpum (1/4)
       -unnamed hybrid-
             -unknown

4ft(1.2m)      5°F(-15°C)      M      3/3

Medium-sized, creamy yellow flowers with a red blotch, in dome-
shaped trusses. Compact growth habit, with interesting rough-
textured foliage; leaves 4in(10.2cm) long.  (M. Koster & Sons,
cross 1920)  See color illus. VV p. 114.


cl.

Mrs. Butler    See Sir Charles Butler


---
          -souliei (1/2)
Mrs. Byron Scott-
          -unknown (1/2)

Flowers straw yellow. (Seed from England; raised & distributed by H. L. Larson; reg. 1958)


cl.

Mrs. C. S. Sargent    See Mrs. Charles S. Sargent


cl.              -griffithianum (1/4)
          -Princess Juliana-
Mrs. C. B. van Nes-      -unknown (3/4)
          -Florence Sarah Smith--unknown

5ft(1.5m)      0°F(-18°C)      M      3/3

Rose red buds opening almost red, fading to a pretty soft pink.
Plant has rather open habit; medium green, shiny leaves. (C. B.
van Nes, reg. 1958)  See color illus. VV p. 112.

```
                                      -maximum (1/8)
                         -Halopeanum-
g.            -Snow Queen-            -griffithianum (1/2)
              -              -        -griffithianum
Mrs. C. Whitner-            -Loderi-
              -                      -fortunei ssp. fortunei (3/8)
              -                      -griffithianum
              -Loderi Sir Edmund-
                                     -fortunei ssp. fortunei
```

A large plant for the light woodlands. Large white flowers suffused magenta with a darker basal stain, trumpet-shaped, in conical trusses of 15.  (Sir Giles Loder)  A. M. (RHS) 1963

```
cl.
                    -decorum (1/2)
Mrs. Carter Glass-
                    -Catalgla--catawbiense var. album Glass (1/2)
```

White flowers.  Parentage reversed in ARH.  (Joseph B. Gable)

```
cl.                         -griffithianum (1/4)
                  -Coombe Royal-
Mrs. Charles E. Pearson-        -unknown (1/2)
              -                         -catawbiense
              -Catawbiense Grandiflorum-        (1/4)
                                        -unknown
```

```
6ft(1.8m)          -10°F(-23°C)          M          3/4
```

Vigorous plant, with lush deep green foliage; tolerates heat and
sun.  Widely funnel-shaped flowers of pale pinkish mauve, fading
paler, with heavy chestnut brown spotting; very large, conical
trusses.  (M. Koster & Sons, 1909)  A. M. (RHS) 1933; F. C. C.
(RHS) 1955    See color illus. VV p. 34.

```
cl.
                    -catawbiense (1/2)
Mrs. Charles S. Sargent-
                    -unknown (1/2)
```

```
6ft(1.8m)          -25°F(-32°C)          ML          3/3
```

Synonym Mrs. C. S. Sargent.  Plant rounded, moderately compact.
Flowers dark carmine rose  with yellow spotting in throat, wavy
margins; compact, dome-shaped trusses.  One of the best of the
"ironclads".  Somewhat hard to root.  (A. Waterer, 1888)    See
color illus. VV p. 88.

```
cl.
```

Mrs. Davies Evans     Parentage unknown

A vigorous plant of compact habit; free-flowering. Imperial purple flowers with white blotch and yellow spots, funnel-shaped,
with frilled margins; compact globular trusses.  (A. Waterer before 1915)  A. M. (Wisley Trials) 1958

```
cl.                       -Corona--unknown (1/4)
                  -unnamed hybrid-
Mrs. Donald Graham-             -griersonianum (1/4)
              -         -griffithianum (1/4)
              -Loderi-
                       -fortunei ssp. fortunei (1/4)
```

```
5ft(1.5m)          5°F(-15°C)          L          3/2
```

Flowers of Tyrian rose with spinel red line on reverse of lobes;
trusses of 8-10.  Leaves 7in(17.8cm) long, 2in(5cm) wide; plant
upright, rather open. (Seed from Rose, England; raised by Endre
Ostbo)    P. A. (ARS) 1954

```
cl.
                  -griffithianum (1/2)
Mrs. E. C. Stirling-
                  -unknown (1/2)
```

```
6ft(1.8m)          -5°F(-21°C)          M          4/3
```

Ruffled flowers of blush pink shaded mauve, with long up-curved
stamens.  Floriferous and sun-tolerant.   Growth habit upright
when young, later spreading, open.   (J. Waterer)  A. M. (RHS)
1906)   See color illus. VV p. 11, p. 86.

```
                     -griersonianum (1/4)
cl.      -Jibuti-                -arboreum ssp. arboreum (1/8)
            -Gill's Triumph-
Mrs. Eddy-                      -griffithianum (1/8)
         -                      -griersonianum (1/4)
         -Gladys Rillstone-
                     -unknown (1/4)
```

An Exbury hybrid with trusses of 15 flowers, of light spirea red
heavily speckled light ruby red, an eye of darkest red purple.
(Edmund de Rothschild, cross 1952; reg. 1982)

```
---                           -griffithianum (1/4)
                  -unnamed hybrid-
Mrs. Edwin Hillier-           -unknown (3/4)
                  -Monsieur Thiers--unknown
```

Dark pink flowers.  (C. B. van Nes & Sons)

```
cl.
Mrs. Elizabeth Titcomb     See Elizabeth Titcomb
```

```
                     -griffithianum (5/32)
cl.       -Kewense-
      -Aurora-      -fortunei ssp. fortunei (9/32)
Mrs.  -     -thomsonii ssp. thomsonii (10/32)
Ena Agius-     -wardii var. wardii (8/32)
      -Idealist-                -griffithianum
      -                      -Kewense-
      -      -Aurora-      -fortunei ssp. fortunei
      -Naomi-     -thomsonii ssp. thomsonii
      -fortunei ssp. fortunei
```

Trusses hold 10 flowers, very light yellow green.  (Crossed
1952; reg. 1982, Edmund de Rothschild)

```
                              -griffithianum (1/8)
                  -unnamed-
cl.            -Mrs. Furnival- hybrid-unknown (1/4)
          -           -          -caucasicum
Mrs. Erna Heyderhoff-      -unnamed hybrid-      (1/8)
          -                      -unknown
          -Catalgla-catawbiense var. album Glass (1/2)
```

```
4ft(1.2m)          -10°F(-23°C)          ML
```

White flowers with prominent blotch of cardinal red, pale pink
margins; trusses of about 21.  Plant as broad as tall, with dark
green, glossy leaves, held 3 years.  (Cross by James P. Bevry;
raiser Henry Heyderhoff; reg. 1965)

```
cl.
              -fortunei ssp. fortunei (1/2)
Mrs. Frank S. Baker-           -griffithianum (1/4)
              -Dawn's Delight-
                              -unknown (1/4)
```

Pink flowers.  (Exhibited by Larson, 1942)

```
cl.
              -fortunei ssp. fortunei (1/2)
Mrs. Fred Paul-
              -unknown (1/2)
```

Light pink flowers.  (Paul)

```
cl.                      -griffithianum (1/4)
           -unnamed hybrid-
Mrs. Furnival-              -unknown (1/2)
           -                   -caucasicum (1/4)
           -unnamed hybrid-
                           -unknown
```

4ft(1.2m)        -10°F(-23°C)      ML       5/5

Considered one of the best. Often a parent of other hybrids.
Light rose pink flowers paler at center, with conspicuous sienna
blotch, crimson markings, widely funnel-shaped, held in dome-
shaped trusses. Handsome, compact plant; dark green leaves, 4in
(10.2cm) long. (A. Waterer, 1920)  A. M. (Exbury Trials) 1933;
F. C. C. (Wisley Trials) 1948; A. G. M. (RHS) 1969   See color
illus. F p. 57; HG p. 131; VV front cover, p. 8.

```
---
              -edgeworthii (3/4)
Mrs. G. Reuthe-          -edgeworthii
           -Fragrantissimum-
                           -formosum var. formosum (1/4)
```

G. Reuthe, 1935.

```
cl.                      -griffithianum (1/4)
           -Coombe Royal-
Mrs. G. W. Leak-           -unknown (1/2)
           -                   -caucasicum (1/4)
           -Chevalier Felix de Sauvage-
                                   -unknown
```

6ft(1.8m)        0°F(-18°C)       EM       4/3

Tall, vigorous plant with smooth olive green leaves. Flowers of
clear light pink with a striking brownish red blotch, crimson
markings; large, compact, conical trusses. Roots easily; grows
in sun or shade. (M. Koster & Sons)  F. C. C. (RHS) 1934   See
color illus. F p. 56; VV p. 26, 49.

```
cl.
           -macabeanum (1/2)
Mrs. George Huthnance-
           -unknown (1/2)
```

Trusses hold 16 flowers, primrose yellow with red flecking inside upper lobe. (G. Huthnance, 1981)

```
cl.
           -minus var. minus Carolinianum Group (1/2)
Mrs. George O. Clark-
               -moupinense (1/2)
```

Pink flowers. (G. O. Clark)

```
cl.
           -griffithianum (1/2)
Mrs. George Paul-
           -unknown (1/2)
```

Flowers blush pink changing to white. (A. Waterer)

```
cl.
              -decorum ? (1/4)
           -Caroline-
Mrs. H. R. Yates*    -brachycarpum ssp. brachycarpum ?
           -unknown (1/2)                   (1/4)
```

5ft(1.5m)        -15°F(-26°C)      M        3/4

Synonym Caroline Cream. Flowers open pale mauve, fading to
cream. Parent of Yates' Best, q.v. (Joseph B. Gable)

```
cl.

Mrs. H. Stocker    Parentage unknown
```

Flowers brilliant scarlet with maroon blotch. (M. Koster &
Sons, before 1923) A. M. (RHS) 1948

```
cl.
              -fortunei ssp. fortunei (1/2)
Mrs. H. T. Krug-
              -unknown (1/2)
```

6ft(1.8m)        -15°F(-26°C)          E

Buds deep pink, opening heather pink, edges shading to very pale
sap green; flowers of good substance, openly funnel-shaped, 4in
(10.2cm) across, 7-lobed, in compact domed trusses of about 12.
Plant upright, broader than tall; dark green leaves retained  3
years. (Harold Krug, 1980)

```
cl.                      -griffithianum (1/4)
           -Princess Juliana-
Mrs. Harold Terry-         -unknown (3/4)
           -Florence Smith--unknown
```

Soft rose pink flowers. (C. B. van Nes & Son)  A. M. (RHS) 1948

```
---
           -multicolor (1/2)
Mrs. Heal-           -Duchess of-brookeanum var. gracile
           -          - Edinburgh-                (1/8)
           -Princess Beatrice-      -longiflorum (lobbii) (1/8)
                       -Princess -jasminiflorum (3/16)
                        Alexandra-         -jasminiflorum
                              -Princess-
                              Royal  -javanicum (1/16)
```

Vireya hybrid. White flowers tinged with delicate shade of pink.
(Veitch)  F. C. C. (RHS) 1894

```
cl.                      -decorum (1/4)
           -unnamed hybrid-
Mrs. Helen Dunker-         -fortunei ssp. discolor (1/4)
           -       -wardii var. wardii (1/4)
           -Crest-          -fortunei ssp. discolor (1/8)
               -Lady     -
               Bessborough-campylocarpum ssp. campylo-
                           carpum Elatum Group (1/8)
```

2.5ft(.75m)        -5°F(-21°C)          M

Fragrant white flowers of good substance, openly funnel-shaped,
3.5in(8.9cm) across, 7 wavy lobes; large dome-shaped trusses of
9. Free-flowering. Plant upright, rounded, as broad as tall;
dark, glossy, yellowish green leaves, held 2 years. (Alfred A.
Raustein, 1983)

```
cl.
           -Mrs. J. J. Crosfield--unknown (3/4)
Mrs. Helen Koster-              -catawbiense (1/4)
               -Catawbiense Grandiflorum-
                           -unknown
```

4ft(1.2m)        -5°F(-21°C)       ML       3/3

Flowers light orchid with a purplish red blotch. Foliage deep
green. (M. Koster & Sons)

```
cl.                      -elliottii (1/4)
           -Fusilier-
Mrs. Helen Weyerhauser-     -griersonianum (1/2)
           -     -griersonianum
           -Jean-
               -decorum (1/4)
```

Pink-flowered hybrid. (H. L. Larson, 1952)

```
cl.
           -grande (1/2)
Mrs. Henry Agnew-
           -arboreum ssp. cinnamomeum var. album (1/2)
```

Flowers white, fringed pink. (Mangles, 1915)

cl.
          -<u>arboreum</u> ssp. <u>arboreum</u> ? (1/2)
Mrs. Henry Shilson-
          -<u>barbatum</u> ? (1/2)

Rounded loose trusses of 15-16 white flowers, suffused light
roseine purple. Narrow, elliptic leaves, to 4.5in(11cm) long,
dark green above, silvery grey below. (Raised by Samuel Smith;
shown by the Hon. H. E. Boscawen) A. M. (RHS) 1973

---
          -<u>griersonianum</u> (1/2)
Mrs. Horace Fogg-        -<u>griffithianum</u> (1/4)
        -Loderi Venus-
                  -<u>fortunei</u> ssp. <u>fortunei</u> (1/4)

5ft(1.5m)        0°F(-18°C)        ML      4/2

A pretty, compact plant that does well in open sun. Flowers
very large, silvery pinkish rose, stained deeper rose red in the
throat. Smooth, elliptic, olive green leaves. (Cross by Ridge-
way; reg. by H. L. Larson, 1958) P. A. (ARS) 1964    See color
illus. VV p. 98.

cl.          -unnamed hybrid--unknown (1/2)
 -                -<u>griffithianum</u> (1/16)
Mrs. Howard Phipps-      -Kewense-
 -       -Aurora-    -<u>fortunei</u> ssp. <u>fortunei</u>
 -Naomi-    -              (5/16)
 -       -<u>thomsonii</u> ssp. <u>thomsonii</u> (1/8)
        -<u>fortunei</u> ssp. <u>fortunei</u>

Flowers medium orchid pink, in compact trusses of 13, blooming
midseason. When registered plant was 5ft(1.5m) tall, 3ft(.9m)
wide, compact; matte leaves 7in(17.8cm) long. (Howard Phipps,
intro. 1958; reg. 1972)

cl.

Mrs. J. C. Williams    Parentage unknown

7ft(2.1m)        -15°F(-26°C)        L      4/3

Ruffled white flowers, small blotch of dark red spots, funnel-
shaped, to 2.5in(6.4cm) wide; compact ball-shaped trusses of 18-
19. Floriferous. Plant upright, dense, wider than high; leaves
medium dull green, slightly rough. (A. Waterer) A. M. (Wisley
Trials) 1960

cl.
        -<u>diaprepes</u> (1/2)
Mrs. J. Comber-
        -<u>decorum</u> (1/2)

White flowers tinged yellow at base. (Messel) A. M. (RHS) 1932

cl.

Mrs. J. G. Millais    Parentage unknown

6ft(1.8m)        -5°F(-21°C)        ML      5/3

Synonym Mrs. John Millais. White flowers, orchid tint on opening with very large golden blotch. Young plants heavily budded.
Mature plant has good stature. (A. Waterer, reg. 1958)   See
color illus. HG p. 65.

---
               -<u>catawbiense</u> (1/4)
        -Carl Mette-
Mrs. J. H. van Nes-    -unknown (1/2)
 -             -<u>griffithianum</u> 1/4)
     -Princess Juliana-
           -unknown

Pink flowers, spotted. (C. B. van Nes & Sons)

cl.

Mrs. J. J. Crosfield    Parentage unknown

A parent of Mrs. Helen Koster. Flowers pale rose with a crimson
blotch. (M. Koster & Sons)

cl.
        -<u>thomsomii</u> ssp. <u>thomsonii</u> (1/2)
Mrs. James Horlick-    -<u>caucasicum</u> (1/4)
      -Dr. Stocker-
             -<u>griffithianum</u> (1/4)

Flowers soft rose pink. (Sir James Horlick, cross 1931; flowered 1940)

cl.
        -<u>maximum</u> (1/2)
Mrs. John Clutton-
        -unknown (1/2)

White flowers with small yellow green blotch. (A. Waterer) F.
C. C. (RHS) 1865

cl.
        -<u>griersonianum</u> (1/2)
Mrs. John Crawford-
        -<u>fortunei</u> ssp. <u>discolor</u> (1/2)

Vigorous plant of spreading habit. Flowers of rose madder, 3.5
in(8.9cm) wide, crimson throat, dark spots on upper lobes; 14 to
a truss. Floriferous. (Sir James Horlick, reg. 1963)

cl.

Mrs. John Millais    See Mrs. J. G. Millais

g.
        -<u>griffithianum</u> (1/2)
Mrs. Kingsmill-
        -<u>campylocarpum</u> ssp. <u>campylocarpum</u> (1/2)

Flowers yellow fading to cream. See also Mrs. Randall Davidson,
Maiden's Blush. (Mangles) A. M. (RHS) 1911

cl.                  -<u>griffithianum</u> (1/4)
        -George Hardy-
Mrs. L. A. Dunnett-    -<u>catawbiense</u> (1/4)
      -unknown (1/2)

Very late blooming; flowers of rosy pink, with lighter centers.
(M. Koster & Sons cross, 1909)

cl.
        -<u>wardii</u> var. <u>wardii</u> (1/2)
Mrs. Lammot Copeland-    -<u>souliei</u> ? (1/4)
      -Virginia Scott-
             -unknown (1/4)

6ft(1.8m)        0°F(-18°C)        L      4/3

Flowers clear yellow, campanulate, 3.5in(8.9cm) across, held in
trusses of 15. Large foliage. (H. L. Larson, 1971)   See color
illus. HG p. 68.

g. & cl.                -<u>catawbiense</u> (1/4)
         -B. de Bruin-
Mrs. Leopold de Rothschild-    -unknown (1/4)
        -<u>griersonianum</u> (1/2)

5ft(1.5m)        -5°F(-21°C)        M

A compact plant of medium size, blooming profusely in midseason.
Flowers light red to scarlet with unusual near-orange ground,
spotted fawn, a widely expanded corolla. Stiffly pointed glossy
green leaves. (Rothschild) A. M. (RHS) 1933

cl.                        -griffithianum (1/4)
                    -George Hardy-
Mrs. Lindsay Smith-            -catawbiense (1/4)
                    -Duchess of Edinburgh--unknown (1/2)

6ft(1.8m)          0°F(-18°C)          ML          4/1

Parentage as shown in the Register; in the Rhododendron Yearbook
(1929) R. W. Wallace of Tunbridge Wells said one of its parents
is Loder's White. Large, flat white flowers, lightly spotted in
red on upper lobe; large upright trusses. Plant open, sometimes
pendant; light green leaves. (M. Koster & Sons, 1910)   A. M.
(RHS) 1933

cl.

Mrs. Lionel de Rothschild      Parentage unknown

5ft(1.5m)          0°F(-18°C)          M          4/2

Not shown as grex in the register, and the color is called pink,
but the clone sold in the Northwest USA has a white flower with
a blotch of crimson spots, rated above. Hillier's Manual calls
it white, edged apple blossom pink; Slocock's catalog says it's
white with red spots. And yours ?  Plant is compact, with upright branches and large, firm trusses of widely funnel-shaped
flowers. (A. Waterer)   A. M. (RHS) 1931

g.                               -catawbiense (1/8)
                    -unnamed hybrid-
         -Altaclarense-          -ponticum (1/8)
Mrs. Loudon-         -arboreum ssp. arboreum (1/4)
         -                 -maximum (1/4)
              -unnamed hybrid-
                         -unknown (1/4)

Flowers carmine or light bright rose, with all petals spotted.
(Standish & Noble, 1850)

cl.
              -campylocarpum ssp. campylocarpum (1/2)
Mrs. Mary Ashley-
              -unknown (1/2)

4ft(1.2m)          5°F(-15°C)          EM          3/3

Similar to Unique, with salmon pink flowers shaded cream, funnel-shaped, in domed trusses. Plant fairly compact; leaves medium green, oval, 3in(7.6cm) long. (W. C. Slocock, reg. 1958)

---
              -fortunei ssp. fortunei (1/2)
Mrs. Matthew Rose-
              -unknown (1/2)

Pink flowers. (Paul)

cl.

Mrs. Messel--form of decorum

White flowers. (Messel)   A. M. (RHS) 1923

cl.
         -catawbiense (1/2)
Mrs. Milner-
         -unknown (1/2)

A parent of several Seidel hybrids: Mims, Plusch, Quendel, etc.
Crimson flowers. (A. Waterer, before 1900)

cl.
              -yakushimanum ssp. yakushimanum (1/2)
Mrs. Muriel Carfrae-
              -unknown (1/2)

2.5ft(.75m)          -5°F(-21°C)          M

Deep pink buds, flowers phlox pink, widely funnel-campanulate,
2in(5cm) wide, 5 wavy lobes, of heavy substance; conical trusses
of 12-15. Floriferous. Plant rounded, broader than tall; elliptic leaves, held 3 years. (Cross by Warren Berg; raised by
J. F. Caperci; Ken Trainor, intro. & reg. 1975)

cl.

Mrs. P. D. Williams      Parentage unknown

5ft(1.5m)          -10°F(-23°C)          ML          3/2

Flowers ivory white with a large golden brown blotch, of medium
size; compact rounded trusses. Young plant may sprawl, sturdier later; very dark green, smooth, narrow leaves to 5.5in(14cm)
long. (A. Waterer)   A. M. (RHS) 1936

cl.                        -catawbiense (1/4)
              -Atrosanguineum-
Mrs. P. den Ouden-            -unknown (1/2)
         -              -arboreum ssp. arboreum (1/4)
              -Doncaster-
                    -unknown

5ft(1.5m)          -15°(-26°C)          M          (1/4)

Plant of compact habit. Flowers of deep crimson, in great profusion. (H. den Ouden & Son, intro. 1925)

cl.                              -griffithianum (1/8)
                    -Loderi-
              -Albatross-      -fortunei ssp. fortunei (1/8)
Mrs. Paul B. Smith-      -fortunei ssp. discolor (1/4)
              -Pygmalion--unknown (1/2)

Flowers of opalescent pink, with a red throat. (H. L. Larson,
reg. 1958)

cl.                    -griffithianum (1/4)
              -Loderi-
Mrs. Percy McLaren-      -fortunei ssp. fortunei (1/4)
              -pink seedling--unknown (1/2)

Flowers of pale shell pink, with frilled lobes. (Dr. R. W. Medlicott, New Zealand, reg. 1979)

cl.

Mrs. Philip Martineau      Parentage unknown

6ft(1.8m)          -5°F(-21°C)          L          4/2

Rose pink flowers fading lighter with a pale yellow eye; large
rounded trusses. Growth habit sprawling. (Knap Hill Nursery)
A. M. (RHS) 1933; F. C. C. (RHS) 1936

cl.
              -Catalgla--catawbiense var. album Glass (1/2)
Mrs. Powell Glass-
              -decorum (1/2)

5ft(1.5m)          -20°F(-29°C)          ML          3/2

Synonym Anne Glass. Tall, vigorous, compact plant. Pure white
flowers in large and rather loose trusses, framed by dark green
leaves. (Joseph B. Gable, exhibited 1949)

cl.

Mrs. R. G. Shaw      Parentage unknown

A parent of Bonfire. Flowers blush pink, with a dark eye. (A.
Waterer, before 1915)

cl.

Mrs. R. S. Holford        Parentage unknown

A parent of Flare and Hypatia, both Exbury hybrids. Rosy salmon flowers flushed strawberry, lightly spotted with crimson, widely funnel-shaped, 2.5in(6.4cm) across, frilly-edged, in large tight trusses. Plant upright, rounded, medium to tall, hardy to -15°F(-26°C); medium green leaves, 5in(12.7cm) long. Midseason to late. (A. Waterer, 1866)

g.

```
                    -griffithianum (1/2)
Mrs. Randall Davidson-
                    -campylocarpum ssp. campylocarpum
                                   (Hooker's form) (1/2)
```

Several forms: white, creamy yellow, pink. See also Maiden's Blush, Mrs. Kingsmill. (Mangles, before 1884)

---

```
                    -arboreum ssp. arboreum (1/2)
Mrs. Richard Gill-
                    -unknown (1/2)
```

Flowers bright salmon rose; bright red splashing in throat. (R. Gill & Son)

cl.

```
              -ponticum (1/2)
Mrs. S. J. Beale-
              -unknown (1/2)
```

Lilac-colored flowers. (M. Koster & Sons)

cl.

Mrs. T. H. Lowinsky*        Exact parentage unknown: a combination
                            of catawbiense, maximum and ponticum

5ft(1.5m)         -15°F(-26°C)        L        4/3

According to Bean (Trees and Shrubs... v. 3, p. 879) this is not the same hybrid as Mrs. Tom H. Lowinsky, which received an A. M. in 1919. Mauve buds open to white flowers tinged mauve, fading white, with a striking orange brown blotch, widely funnel-shaped and 3in(7.6cm) across; compact trusses of 14. Floriferous and vigorous. Dark green leaves about 4.5in(11.5cm) long. (Waterer, before 1917)

Mrs. Thistleton Dyer        See Mrs. W. T. Thistleton-Dyer

cl.

```
              -griffithianum (3/4)
Mrs. Tom H. Lowinsky-
                     -maximun (1/4)
              -Halopeanum-
                     -griffithianum
```

5ft(1.5m)         -15°F(-26°C)        EM        4/3

Sometimes confused with Mrs. T. H. Lowinsky. Rounded and rather compact plant; very dark green, glossy leaves 6in(15.2cm) long, somewhat convex. Flowers open blush, fading white, with reddish brown blotch; rounded trusses. Heat-tolerant. (Cross probably by A. Waterer; intro. by Lowinsky) A. M. (RHS) 1919 See color illus. HG p. 65; VV p. 81.

cl.

Mrs. Tritton        Parentage unknown

A parent of Louis Pasteur. Flowers crimson with light center. (J. Waterer)

cl.

```
              -campylocarpum (1/2)
Mrs. W. C. Slocock-
              -unknown (1/2)
```

A compact shrub with flowers apricot pink fading to buff. Midseason. (Slocock Nurseries) A. M. (RHS) 1929

cl.

```
              -fortunei ssp. fortunei (1/2)
Mrs. W. R. Coe-
              -unknonwn (1/2)
```

6ft(1.8m)        -5°F(-21°C)        ML        3/4

Flowers deep bright pink with crimson throat, over 4in(10.2cm) wide; very large, dome-shaped trusses. Glossy, dark green foliage. (C. O. Dexter, 1925-42; reg. 1958)

cl.

```
                            -griffithianum (1/4)
                -Princess Juliana-
Mrs. W. R. Dykes*                 -unknown (1/2)
                -                  -catawbiense (1/4)
                -Carl Mette-
                            -unknown
```

5ft(1.5m)        -10°F(-23°C)        ML        3/3

Flowers are strong rose red, with foliage of matte citrus green. (C. B. van Nes) [The 3 Prinses Juliana in I.R.R., are azaleas.]

---

```
                    -fortunei ssp. fortunei (1/2)
Mrs. W. T. Thistleton-Dyer-
                    -unknown (1/2)
```

Synonym Mrs. Thistleton Dyer. Flowers soft pink. (George Paul)

cl.

```
                    -maximum (1/4)      -catawbiense (1/16)
        -Standishii-          -unnamed-
Mrs. Walter-        -Altaclarense- hybrid-ponticum (1/16)
 Burns   -                    -
        -                    -arboreum ssp. arboreum (1/8)
        -griffithianum (1/2)
```

6ft(1.8m)        0°F(-18°C)        EM        3/3

Plant vigorous, upright; dark green leaves. Pale pink flowers, deeper margins, rose blotch. (T. Lowinsky) A. M. (RHS) 1931

cl.

```
                                    -griffithianum
                    -Queen Wilhelmina-           (1/8)
          -Earl of -              -unknown (3/8)
Mrs. Wayne W. Keyes- Athlone-Stanley Davies--unknown
                    -
                    -fortunei ssp. fortunei (1/2)
```

Deep red flowers. (H. L. Larson, 1958)

---

```
              -fortunei ssp. fortunei (1/2)
Mrs. Webley-
              -unknown (1/2)
```

Flowers of light rose, spotted. (Paul)

cl.

```
                    -griffithianum (1/2)
Mrs. William Watson-
                    -unknown (1/2)
```

A tall plant bearing white flowers with violet spots. (A. Waterer) A. M. (RHS) 1925

cl.

```
        -mucronulatum (1/2)
Mucram-
        -ambiguum (1/2)
```

Flowers lavender pink. (Joseph B. Gable, before 1958)

---

Mucronatum      Parentage unknown

A parent of Margot, an azealodendron.  Formerly mucronatum, a species; it is now considered a hybrid.

cl.
                -minus var. minus (1/2)
Mugby Junction-
                    -unknown (1/2)

Pink flowers spotted brown in throat, 5-lobed, funnel-shaped, in trusses of 5-8.  Leaves to 2.25in(5.7cm) long, densely covered with reddish brown scales beneath.  (Collector unknown; raised by R. N. S. Clarke)  A. M. (RHS) 1984

cl.
              -ciliatum (1/2)
Multiflorum-
                -virgatum ssp. virgatum (1/2)

Like the species virgatum, the blush pink flowers are held in twos all up the stems.  Olive green leaves, long and narrow.  (Waterer, Sons & Crisp, reg. 1958)

cl.
               -ponticum (1/2)
Multimaculatum-
                -brachycarpum ssp. brachycarpum (1/2)

5ft(1.5m)         -25°F(-32°C)         ML        2/3

A very hardy plant praised in the 19th century and "still worth growing"; white flowers with orange red spots.  (J. Waterer, before 1860)  See color illus. in F. Street's Hardy Rhododendrons pl. IVb.

cl.
      -maximum (1/2)
Mum-
      -unknown (1/2)

Flowers white with lemon yellow eye; compact bush.  (J. Waterer, 1897)

cl.

Mumtaz-i-Mahal    Extinct hybrid reg. by L. E. Brandt, 1966.

cl.
                -cinnabarinum ssp. cinnabarinum (1/2)
Muncaster Bells-
                  -cinnabarinum ssp. xanthocodon Concatenans Group
                                                              (1/2)

Flowers apricot, up to 3in(7.6cm) long, tubular bell-shaped, in loose trusses.  (Raised by Sir John Ramsden; intro. by Sir William Pennington-Ramsden; reg. 1965)

cl.                               -griffithianum (3/8)
                -unnamed hybrid-
                -              -unknown (1/4)
Muncaster Icicle-            -griffithianum
                -      -Loderi-
                -Lodauric-   -fortunei ssp. fortunei (1/8)
                         -auriculatum (1/4)

Fragrant white flowers with a red throat, funnel-shaped, up to 4in(10.2cm) across, in large loose trusses.  (Raised by Sir John Ramsden; intro. by Sir William Pennington-Ramsden; reg. 1965)

cl.
                -campanulatum ssp. campanulatum (1/2)
Muncaster Mist-
                  -floribundum (1/2)

A compact shrub with campanulate flowers in conical trusses, blue with darker markings.  (Raised by Sir John Ramsden; intro. by Sir William Pennington-Ramsden; reg. 1965)

cl.
              -orbiculare ssp. orbiculare (1/2)
Muncaster Ruby-
              -thomsonii ssp. thomsonii (1/2)

Medium-sized rose pink flowers, bell-shaped.  (Raised by Sir John Ramsden; intro. by Sir William Pennington-Ramsden, reg. 1965)

cl.
                -cinnabarinum ssp. cinnabarinum (1/2)
Muncaster Trumpet-
                  -maddenii ssp. maddenii (1/2)

Tubular bell-shaped flowers in loose trusses, white with apricot throat.  (Raised by Sir John Ramsden; intro. by Sir William Pennington-Ramsden, reg. 1965)

cl.         -caucasicum (1/4)
        -Nobleanum-
Mundai-     -arboreum ssp. arboreum (1/4)
            -campylocarpum ssp. campylocarpum (1/4)
        -Unique-
            -unknown (1/4)

Flower color magenta.  See also Joan Bye, Shirley Scott.  (V. J. Boulter, 1965)

g.              -neriiflorum ssp. neriiflorum (1/4)
     -F. C. Puddle-
Mureun-         -griersonianum (1/4)
     -barbatum (1/2)

Deep red flowers.  (Lord Aberconway, intro. 1946)

cl.
      -falconeri ssp. falconeri (1/2)
Muriel-
      -grande (1/2)

Mansellii g.  Creamy white flowers with a dark crimson blotch at the base.  (Loder)  A. M. (RHS) 1925

g.                -griersonianum (1/4)
        -Griercalyx-
Muriel Holman-    -megacalyx (1/4)
              -maddenii ssp. maddenii (1/2)

Tremayne; exhibited in 1941.

cl.          -griffithianum
        -Loderi-
Muriel Messel-  -fortunei ssp. fortunei
        -Loder's White, q.v. for 2 possible parentage
                           diagrams

Buds bright pink; flowers delicate pink to white.  (Messel)  A. M. (RHS) 1929

                        -griffithianum (5/16)
              -Beauty of-
        -Norman- Tremough-arboreum ssp. arboreum (1/16)
        - Gill -
cl.  -Anna-      -griffithianum
     -     -      -griffithianum
Muriel-  -Jean Marie -
Pearce-  de Montague-unknown (1/8)
     -       -elliottii (1/4)
       -Fusilier-
              -griersonianum (1/4)

6ft(1.8m)         10°F(-12°C)         EM

Plant upright, well-branched; dark green leaves, narrowly elliptic, 4in(10.2cm) long, held 2 years.  Flowers shaded rose madder with much red spotting on upper 3 lobes, openly funnel-shaped, 3.5in(8.9cm) wide; conical trusses of 14-15.  (Cross by Halfdan Lem; reg. by Owen Pearce, 1973)

cl.
     -Marion--unknown (3/4)
Murraba-       -ponticum (1/4)
    -Purple Splendour-
          -unknown

Flowers deep mallow purple, shading lighter. (V. J. Boulter, 1972)


cl.
      -fortunei ssp. fortunei (1/2)
Mutter Emma-
      -unknown (1/2)

Trusses hold 12 flowers, bright spirea red shading lighter, heavily spotted oxblood red. (B. Leendertz, W. Germany, 1976)


cl.
        -decorum (1/4)
    -Decsoul-
Muy Lindo-   -souliei (1/4)
  -      -griffithianum (1/4)
    -Isabella-
        -auriculatum (1/4)

Flushed pink buds opening to white flowers, fragrant, funnel-campanulate, 5.5in(14cm) across, 7-8 lobes; elongated trusses of 14-18. A large shrub; dark green leaves 11in(28cm) long by 3in (7.6cm) wide. (Collingwood Ingram, cross 1952; reg. 1963) A. M. (RHS) 1969


cl.
             -veitchianum (1/2)
My Lady--Forsterianum (selfed)-
             -edgeworthii (1/2)

2.5ft(.75m)      15°F(-10°C)      E    4/3

Compact plant, as wide as high; glossy dark green leaves, bullate, 3in(7.6cm) long. Flowers white, blushed pink with a pale yellow throat, very open-rotate, 2.5in(6.4cm) wide, held in open trusses. (Mr. & Mrs. Maurice Sumner, intro. 1967; reg. 1972)


cl.
      -carneum (1/2)
My Pretty One-
      -moupinense (1/2)

2ft(.6m)     5°F(-15°C)     VE    3/4

A rambling dwarf shrub with mahogany-colored bark, peeling at all seasons. Flowers apple blossom pink fading off-white, shallow trumpet-shaped, to 3.5in(89cm) wide, in trusses of 3. (Del W. James cross; raised by Arthur A. Childers; reg. 1965)


cl.
    -Nigrescens--unknown
Myrte-
    -Mira--unknown

Mira by Stevenson was not available at this date. Purple violet flowers with reddish brown markings. (T. J. R. Seidel cross, 1910)


---
      -minus var. minus (1/2)
Myrtifolium-
    -hirsutum (1/2)

3ft(.9m)     -15°F(-26°C)     L    3/5

Broadly dome-shaped shrub; tolerant of heat and sun. Small foliiage deep bronze red in winter, rich dark green in summer. Purplish rose flowers, spotted crimson, in terminal racemes. Late. (Origin unknown; shown 1917) See color illus. HG p. 79.


cl.
      -campylogynum Myrtilloides Group (1/2)
Myrtle Bells-
      -glaucophyllum var. glaucophyllum (1/2)

Small flowers in clusters of 5-8, bell-shaped, 5-lobed, deep violet purple, red purple spotting in upper throat. (H. E. Hawden cross; shown by R. Strauss) P. C. (RHS) 1969


              -arboreum ssp. nilagiricum (1/8)
          -Noyo Chief-
cl. -Rubicon-     -unknown (5/16)
    -        -elliottii (3/8)
Myrtle-  -Kilimanjaro-   -Moser's Maroon--unknown
Manson-        -Dusky Maid-
  -           -fortunei ssp. discolor
  -     -elliottii
  -Kilimanjaro-   -Moser's Maroon--unknown
        -Dusky Maid-
            -fortunei ssp. discolor (3/16)

Trusses of 25 flowers 2.75in(7cm) wide, deep blood red with dark spotting. Leaves narrowly elliptic, glabrous. (R. C. Gordon, 1984)


cl.
        -thomsonii ssp. thomsonii (1/4)
    -Barclayi-     -arboreum ssp. arboreum (1/8)
Mystic-    -Glory of -
     -     Penjerrick-griffithianum (1/8)
    -williamsianum (1/2)

Clear pink flowers. (Gen. Harrison, cross 1950; intro. 1957)


cl.
      -fortunei ssp. discolor (1/2)
Namu-        -griersonianum (1/4)
  -Mrs. Horace Fogg-     -griffithianum (1/8)
          -Loderi Venus-
            -fortunei ssp. fortunei (1/8)

Truss of 13-14 flowers, 6-7-lobed, color roseine purple. (H. L. Larson, reg. 1983)


cl.
      -aurigeranum (1/2)
Nan Cutten-     -phaeopeplum (1/4)
  -Dr. Herman Sleumer-
        -zoelleri (1/4)

Vireya hybrid. Trusses of 9 flowers, buttercup yellow. (Cross by Tom Lelliot; raised by Ron Cutton; reg. 1982)


g.
      -fortunei ssp. fortunei (1/2)
Nanceglos-
      -elliottii (1/2)

A hybrid by Bolitho. (1945)


cl.
      -Pink Delight--unknown (1/2)
Nancy Adler Miller*
      -jasminiflorum (1/2)

An unregistered vireya hybrid at Strybing Arboretum. A parent of Aravir, Clipsie, Moonwind, Shasta.


cl.
         -smirnowii (1/4)
     -unnamed hybrid-
Nancy Behring-     -fortunei ssp. fortunei (1/4)
    -Mist Maiden--yakushimanum ssp. yakushimanum (1/2)

3.5ft(1.05m)     -20°F(-29°C)     M

Fragrant flowers, very light neyron rose with pea green spotting
in throat, openly funnel-shaped, 2.3in(6cm) wide, 5 frilled
lobes; conical trusses of 14.  Upright plant, well-branched and
broad; dark yellow green leaves held 3 years. (R. Behring, 1983)

```
                    -dichroanthum ssp. dichroanthum (1/4)
          -Goldsworth-
      -Hotei- Orange   -fortunei ssp. discolor (1/8)
cl. -      -            -souliei (1/8)
               -unnamed hybrid-
Nancy-              -wardii var. wardii (1/8)
Evans-        -dichroanthum ssp. dichroanthum
   -     -Dido-
   -     -      -decorum (1/8)
   -Lem's-        -Beauty of-griffithianum (5/32)
    Cameo-   -Norman- Tremough-
      -      - Gill -         -arboreum ssp. arboreum (1/32)
        -Anna-    -griffithianum
           -              -griffithianum
            -Jean Marie de Montague-
                          -unknown (1/16)
```

3ft(.9m)          10°F(-12°C)          M

Orange red buds  open to amber yellow flowers, hose-in-hose, of
heavy substance, openly funnel-shaped, 2.5in(6.4cm) wide, 5 or 6
wavy lobes; spherical trusses of 19.  Free-flowering.  Plant as
broad as tall, rounded;  medium green leaves held 2+ years.  New
foliage brownish.  (Dr. E. C. Brockenbrough, reg. 1983)

cl.
```
          -cinnabarinum ssp. xanthocodon Concatenans Group (1/2)
Nancy   -
Fortescue-Lady Alice Fitzwilliam--unknown (1/2)
```

Trusses hold 11-13 flowers, rich chrome yellow.  (Cross by L. S.
Fortescue; reg. by Keith Wiley  for the Fortescue Garden Trust,
1980)

---
```
          -pemakoense (1/2)
Nancy Read-    -racemosum (1/4)
      -Racil-
            -ciliatum (1/4)
```

Light pink flowers.  (H. L. Larson, 1953)

cl.

Nanette     Parentage unknown

Flowers blush pink, dark blotch.   (W. C. Slocock)   A. M. (RHS)
1933

cl.
```
            -forrestii ssp. forrestii Repens Group
         -Elizabeth-                        (1/4)
Nanie Garrett-     -griersonianum (1/4)
         -arboreum ssp. arboreum (1/2)
```

Claret rose flowers in trusses of 18.   (D. Dosser, 1980)

cl.
```
   -smirnowii (1/2)
Nansen-
   -unknown (1/2)
```

Pale lilac rose flowers.  (M. Koster & Sons)

cl.

Nanum*--form of hanceanum Nanum Group

1ft(.3m)     5°F(-15°C)     EM     4/4

Numerous dwarf plants bear the label "var. nanum",  but this one
has the small leaves and bright yellow flowers.   Compact, dome-
shaped plant.

```
                    -griffithianum (1/8)
g. & cl.    -Kewense-
     -Aurora-      -fortunei ssp. fortunei (5/8)
Naomi-     -thomsonii ssp. thomsonii (1/4)
     -fortunei ssp. fortunei
```

5ft(1.5)        -5°F(-21°C)          M          4/3

A large, sturdy, well-filled shrub, producing an abundance of
sweetly scented flowers of tender pink with yellow undertones,
with a ray of faint brown, widely open, 7-lobed, in big, rounded
trusses.  Handsome foliage of rich green, rounded at both ends.
A parent of many other fine hybrids.  This grex, named after his
youngest daughter, is the finest  that L. de Rothschild raised,
and was his favorite.  All the Naomi clones are described sep-
arately, including Paris and Exbury Naomi, q.v.   (Rothschild,
1926)  A. M. (RHS) 1933

cl.

Naomi Astarte     Parentage as above

5ft(1.5m)        -10°F(-23°C)          M          4/4

Naomi g. Shrub of fairly compact habit; leaves 7in(17.9cm) long
and rounded at both ends.  Flowers light pink, shading to a soft
yellow toward center and deeper yellow in throat.   (Rothschild)
See color illus. ARS Q 27:3 (Summer 1973): p. 171.

cl.

Naomi Carissima     Parentage as above

6ft(1.8m)        -10°F(-23°C)          EM          2/3

Naomi g.  Flowers pale pink, flushed creamy white.  (Rothschild)
See color illus. PB pl. 64.

cl.

Naomi Early Dawn     Parentage as above

6ft(1.8m)        -10°F(-23°C)          EM          3/3

Naomi g.  This form has pale pink flowers.  (Rothschild)

cl.

Naomi Glow     Parentage as above

6ft(1.8m)        -10°F(-23°C)          EM          4/4

Naomi g.  Large flowers glowing deep pink, darker in throat, in
dome-shaped trusses. The only Naomi of one clear, single color.
(Rothschild)   See color illus. F p. 58;  PB pl. 64;  VV p. 63.

cl.

Naomi Hope     Parentage as above

6ft(1.8m)        -10°F(-23°C)          EM          4/3

Naomi g.  Flowers pink, tinged with mauve.  (Rothschild)   See
color illus. PB pl. 64.

cl.

Naomi Nautilus     Parentage as above

6ft(1.8m)        -5°F(-21°C)          M          4/4

Naomi g. The flowers are rose, flushed pale orange with a soft
greenish yellow throat.  Foliage and plant habit make this one
of the  best of the Naomi group. (Rothschild) A. M. (RHS) 1938

cl.

Naomi Nereid     Parentage as above

6ft(1.8m)        -10°F(-23°C)          EM          4/4

Naomi g.  Very fine foliage like that of Naomi Nautilus.  Flowers of lavender and yellow (The Register) or, crimson and lavender (Rothschild Rhododendrons).  (Rothschild)     See color
illus. PB pl. 13.

cl.

Naomi Pink Beauty      Parentage as above

6ft(1.8m)          -10°F(-23°C)          EM          4/4

Naomi g.  Flowers satiny pink.  (Rothschild)  See color illus.
PB pl. 64.

cl.

Naomi Pixie     Parentage as above

6ft(1.8m)          -10°F(-23°C)          EM          4/4

Naomi g.  Plant habit rather compact.  Large flowers of bright
pink with a rich crimson stain in throat, held in dome-shaped
trusses.  Long-lasting flowers.  Not in cultivation at Exbury.
(Rothschild)  ARS Q 27:3 (Summer 1973): p. 170.

cl.

Naomi Stella Maris      Parentage as above

6ft(1.8m)          -10°F(-23°C)          EM          4/4

Naomi g.  Flowers slightly larger, trusses fuller, and leaves
longer than in the other clones.  Flowers buff, shaded lilac
pink.  (Rothschild)  F. C. C. (RHS) 1939


cl.   -Margaret Dunn-     -fortunei ssp. discolor (1/4)
      -              -dichroanthum ssp. dichroanthum (1/8)
Naranja-            -Fabia-
      -             -griersonianum (3/8)
      -Tally Ho-    -griersonianum
                    -facetum (eriogynum) (1/4)

Flowers are vermilion to jasper red.  (Lester E. Brandt, 1963)

cl.
      -aurigeranum (1/2)
Narnia-
      -zoelleri (1/2)

5ft(1.5m)          32°F(0°C)          all year

Vireya hybrid.  Flowers of heavy substance, tubular funnel-
shaped, 3in(7.6cm) across, saffron yellow with red tips, in
trusses 9in(23cm) across.  Plant upright with willowy branches;
leaves 5in(12.7cm) by 2.5in(6.4cm).  (Cross by Tom Lelliott;
reg. by William A. Moynier, 1978)    See color illus. ARS Q 35:1
(Winter 1981): p. 7.

cl.                   -augustinii ssp. augustinii (1/4)
          -Blue Diamond-           -intricatum (1/8)
Nasca Blossom-        -Intrifast-
          -unknown (1/2)           -fastigiatum (1/8)

3ft(.9m)          -5°F(-21°C)          EM

Trusses hold 5-6 flowers, pale rose-tinged purple with the upper lobe spotted orange brown.  (I. A. Hayes, reg. 1982)


cl.

Nassau Red*--form of maximum

Flowers bright rose pink.

cl.
      -griersonianum (1/2)
Nasturtium-       -griffithianum (1/4)
         -Loderi-
                 -fortunei ssp. fortunei (1/4)

East Knoyle g.  Nasturtium red flowers.  (Sir James Horlick)


cl.             -haematodes ssp. haematodes (1/4)
      -Hummingbird-
Nathaniel-       -williamsianum (1/4)
      -          -forrestii ssp. forrestii Repens Group (1/4)
      -Elizabeth-
                 -griersonianum (1/4)

A hybrid from British Columbia.  Cherry red flowers in trusses
of six.  (Robert C. Rhodes, reg. 1979)


g.
      -calophytum var. calophytum (1/2)
Nausicaa-         -arboreum ssp. arboreum (1/4)
      -Gill's Triumph-
                     -griffithianum (1/4)

A hybrid introduced by E. J. P. Magor.


cl.
      -decorum (1/2)
Neapolitan-      -dichroanthum ssp. dichroanthum (1/4)
      -Fabia-
              -griersonianum (1/4)

Flowers salmon pink outside with orange pink base.  (Sunningdale
Nurseries, 1955)


cl.
      -campanulatum ssp. campanulatum (1/2)
Neat-0-
      -Koichiro Wada--yakushimanum ssp. yakushimanum (1/2)

3ft(.9m)          5°F(-15°C)          EM

Buds of neyron rose opening lighter, shading to very pale rose
in throat, spotted neyron rose; flowers campanulate, 1.25in(3.2
cm) wide; of good substance, 5 wavy lobes; spherical trusses of
20-25. Plant rounded, as broad as tall; leaves held 3-4 years,
with heavy greyed orange indumentum below; greyed white tomentum
on new growth.  (Dr. David Goheen, reg. 1982)


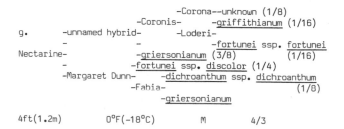

g.         -unnamed hybrid-      -Corona--unknown (1/8)
      -            -Coronis-     -griffithianum (1/16)
      -                          -Loderi-
Nectarine-                                -fortunei ssp. fortunei
      -                          -griersonianum (3/8)    (1/16)
      -                          -fortunei ssp. discolor (1/4)
      -Margaret Dunn-      -dichroanthum ssp. dichroanthum
                     -Fabia-                          (1/8)
                            -griersonianum

4ft(1.2m)          0°F(-18°C)          M          4/3

An unusual orange-colored flower of lovely tones.  Leaves are
long and narrow, dull olive green.  (Lester E. Brandt, reg. 1958)


g.
   -dichroanthum ssp. dichroanthum (1/2)
Neda-
   -Cunningham's Sulphur--caucasicum (1/2)

Orange flowers, flushed pink.  (Lord Aberconway, 1933)

g.
    -<u>griersonianum</u> (1/2)
Nehru-      -<u>barbatum</u> (1/4)
   -Huntsman-
          -<u>campylocarpum</u> ssp. <u>campylocarpum</u> Elatum Group
                                (1/4)

Showy trusses of rich scarlet flowers  on a tall, upright plant.
Blooms early to midseason.  (Rothschild, 1946)

cl.                   -<u>decorum</u> (1/4)
      -unnamed hybrid-
Nelle S. -         -<u>fortunei</u> ssp. <u>discolor</u> (1/4)
Barefield-     -<u>lacteum</u> (1/4)
     -unnamed-
       hybrid-     -<u>griffithianum</u> (1/8)
       -Loderi-
          -<u>fortunei</u> ssp. <u>fortunei</u> (1/8)

6ft(1.8m)       5°F(-15°C)       L

Spinel red buds opening light neyron rose, mimosa yellow throat
with greyed orange spotting, reverse lighter rose; flowers openly funnel-shaped, 5in(12.7cm) across, 7-lobed; large ball-shaped
trusses of 10-12.  Plant habit rounded, erect; narrow medium
green leaves, 8.5in(21.6cm) long, held 2 years. (Grady E. Barefield, intro. 1973; reg. 1983)

cl.

Neon Light    Parentage unknown

A parent of Callight.

cl.
     -<u>catawbiense</u> var. <u>album</u> (1/2)
Nepal*        -<u>wightii</u> (1/4)
    -unnamed hybrid-
              -<u>fortunei</u> ssp. <u>fortunei</u> (1/4)

6ft(1.8m)     -25°F(-32°C)    M    4/4

Rose pink buds opening to white flowers. (David G. Leach, 1972)

cl.

Nepal--form of <u>cinnabarinum</u> ssp. <u>cinnabarinum</u> (L S & H 21283)

5ft(1.5m)     5°F(-15°C)    EM-VL    4/4

Trusses hold 4-8 tubular flowers, 1.25in(3.2cm) across by 2in(5
cm) long, of soft aureolin yellow, deepening near base to medium
rhodonite red. Oblanceolate leaves, scaly beneath.  (Seed from
Ludlow, Sherriff & Hicks; raised by J. B. Stevenson; intro. &
reg. by Hydon Nurseries) A. M. (RHS) 1977

cl.

Nepal*--form of <u>niveum</u>

5ft(1.5m)     5°F(-15°C)    M    3/3

A large shrub; young shoots covered with light tan indumentum.
Tight trusses of rich purple, campanulate flowers. (Origin unknown)  F. C. C. (RHS) 1979

                  -<u>souliei</u> (1/16)
            -Soulbut-
cl.     -Vanessa-     -Sir Charles Butler--<u>fortunei</u>
    -Eudora-    -<u>griersonianum</u>  ssp. <u>fortunei</u> (1/16)
Neptune-   -<u>facetum</u> (1/4)   (1/8)
  -      -<u>griffithianum</u> (1/4)
    -Loderi-
       -<u>fortunei</u> ssp. <u>fortunei</u> (1/4)

Eulo g.  Flowers are pale pink. (Lord Aberconway, 1941)

g.
     -<u>neriiflorum</u> ssp. <u>neriiflorum</u> (1/2)
Nereid-
      -<u>dichroanthum</u> ssp. <u>dichroanthum</u> (1/2)

2ft(.6m)       0°F(-18°C)    ML    3/3

Compact plant with small, dark green, oblong leaves. Flowers of
peachy salmon pink with faint orange spotting, waxy, bell-shaped
and 1in(2.5cm) wide; lax trusses of 7-8.  (E. F. Wilding, 1934)

g.
     -<u>dichroanthum</u> ssp. <u>apodectum</u> (1/2)
Neriiapo-
      -<u>neriiflorum</u> ssp. <u>neriiflorum</u> (1/2)

Flowers orange red. (E. J. P. Magor, 1929)

g.
     -<u>arboreum</u> ssp. <u>arboreum</u> (1/2)
Neriiarb-
      -<u>neriiflorum</u> ssp. <u>neriiflorum</u> (1/2)

Flowers of clear red  with slight spotting of a darker red. (E.
J. P. Magor, 1928)

g.
     -<u>neriiflorum</u> ssp. <u>neriiflorum</u> (1/2)
Neriihaem-
      -<u>haematodes</u> ssp. <u>haematodes</u> (1/2)

Waxy flowers, blood red and unspotted.  (E. J. P. Magor, 1927)

g.             -<u>arboreum</u> ssp. <u>arboreum</u> (1/4)
     -Choremia-
Nerissa-      -<u>haematodes</u> ssp. <u>haematodes</u> (1/2)
  -      -<u>neriiflorum</u> ssp. <u>neriiflorum</u> (1/4)
    -Neriihaem-
       -<u>haematodes</u> ssp. <u>haematodes</u>

A Bodnant hybrid with blood red flowers. (Lord Aberconway, 1945)

cl.
    -<u>barbatum</u> (1/2)
Nestor-
    -<u>thomsonii</u> ssp. <u>thomsonii</u> (1/2)

Cardinal red flowers with a few darker markings in upper throat,
campanulate, 2.5in(5.7cm) across, 5-lobed; about 13 per truss.
Elliptic leaves dark green above, paler below. (Sir Edmund Loder cross; shown by the Hon. H. E. Boscawen, High Beeches) A. M.
(RHS) 1969

cl.
    -<u>fortunei</u> ssp. <u>fortunei</u> (1/2)
Nestucca-
    -<u>yakushimanum</u> ssp. <u>yakushimanum</u> (1/2)

2ft(.6m)     -10°F(-23°C)    M

White flowers, brown traces in throat, bowl-shaped, 5in(12.7cm)
wide, in dome-shaped trusses of 12-15.  Plant compact, rigid,
wider than high; dark green leaves  to 5in(12.7cm) long. (Seed
from F. Hanger; grown & intro. by Cecil C. Smith, 1960)    P. A.
(ARS) 1950

                        -<u>griffithianum</u> (1/8)
cl.               -George Hardy-
      -Mrs. L. A. Dunnett-     -<u>catawbiense</u> (1/8)
Netty Koster-        -unknown (1/2)
  -         -<u>griersonianum</u> (1/4)
    -unnamed hybrid-
        -unknown

Openly funnel-shaped flowers, crimson with darker spots on upper
petal. (M. Koster & Sons) A. M. (RHS) 1945

'Mrs. Donald Graham' by Rose/Ostbo
Photo by Greer

'Mrs. Helen Koster' by M. Koster
Photo by Greer

'Mrs. Horace Fogg' by Ridgeway/Larson
Photo by Greer

'Mrs. Lammot Copeland' by Larson
Photo by Greer

'Mrs. Lindsay Smith' by M. Koster
Photo by Greer

'Mrs. Mary Ashley' by Slocock
Photo by Greer

'Mrs. P. D. Williams' by A. Waterer
Photo by Greer

'Mrs. Philip Martineau' by Knap Hill
Photo by Greer

'Mrs. W. C. Slocock' by Slocock
Photo by Greer

'My Lady' by M. Sumner
Photo by Greer

'My Pet' by Whitney/Sather
Photo by Greer

'Myrtifolium' by unknown
Photo by Greer

'Nancy Evans' by Brockenbrough
Photo by Greer

'Naomi Nautilus' by Rothschild
Photo by Greer

'Naomi Pixie' by Rothschild
Photo by Greer

'Nectarine' by Brandt
Photo by Greer

'Nepal' by Leach
Photo by Leach

'Nereid' by Wilding
Photo by Greer

'Nestucca' by Hanger/C. Smith
Photo by Greer

'New Moon' by Slocock
Photo by Greer

'Newcomb's Sweetheart' by Newcomb
Photo by Greer

'Nicholas' by Rothschild
Photo by Greer

'Nightwatch' by Van de Ven
Photo by Greer

'Nile' (Syn.'Sahara') by Leach
Photo by Greer

```
             -wardii var. wardii (1/4)
         -                      -griffithianum (3/32)
     -Idealist-        -Kewense-
cl. -         -   -Aurora-    -fortunei ssp. fortunei
     -         -Naomi-   -thomsonii ssp. thomsonii (3/16)
New  -         -fortunei ssp. fortunei (15/32)
Comet-                   -griffithianum
     -              -Kewense-
     -   -Aurora-    -fortunei ssp. fortunei
     -Exbury-   -thomsonii ssp. thomsonii
       Naomi -fortunei ssp. fortunei
```

Emerald Isle g. Flowers of mimosa yellow, flushed pale pink, openly funnel-shaped; large, heavy, spherical trusses. Blooms early to midseason. (RHS Garden, Wisley) A. M. (RHS) 1957

cl.

New Guinea Gold--form of aurigeranum

A vireya species with trusses bearing 8-10 flowers, rich maize yellow. (RBG, Kew) A. M. (RHS) 1981

```
cl.
        -yakushimanum ssp. yakushimanum (1/2)
New Hope-   -elliottii (1/4)
      -Kiev-      -thomsonii ssp. thomsonii (1/8)
         -Barclayi -       -arboreum ssp. arboreum (1/16)
           Robert Fox-Glory of  -
                   Penjerrick-griffithianum (1/16)
```

3ft(.9m)        -5°F(-21°C)           EM

Flowers dark pink fading lighter at center, darker pink spotting on upper lobe and the reverse dark pink, openly funnel-shaped, 3.25in(8.3cm) across; spherical trusses of 10. Plant rounded, as wide as high; yellow green leaves, 5in(12.7cm) long, felted fawn indumentum beneath. (Cross by Bovee & Mayo Rare Seed Co.; intro. & reg. by Charles Herbert, 1976)

```
cl.                    -campylocarpum ssp. campylocarpum
         -Mrs. W. C. Slocock-                    (1/4)
New Moon-                -unknown (1/4)
         -fortunei ssp. fortunei (1/2)
```

Cream buds tinged mauve pink; white flowers tinged primrose yellow on throat and upper petal. (W. C. Slocock) A. M. (Wisley Trials) 1953

```
cl.              -griffithianum (1/2)
       -Lady Bligh-
New Romance-       -unknown (1/4)
       -        -griffithianum
       -Loderi Venus-
               -fortunei ssp. fortunei (1/4)
```

Buds of rose pink, opening to white flowers, 2.5in(6.4cm) wide. Leaves to 5in(12.7cm) long. Late season. (Rudolph Henny cross; reg. by Leona Henny, 1966)

```
                           -griffithianum (9/32)
                 -Beauty of-
             -Norman- Tremough-arboreum ssp. arboreum
              - Gill -                       (1/32)
          -Anna-    -griffithianum
cl.       -    -        -griffithianum
    -Pink    -  -Jean Marie -
Newcomb's -Walloper-  de Montague-unknown (3/16)
Sweetheart-     -        -griffithianum
        -      -Marinus Koster-
        -decorum (1/2)         -unknown
```

4ft(1.2m)        0°F(-18°C)           M

Flowers of light pinkish mauve darkening with age to orchid pink and edged in solferino purple, blotched in roseine purple; corolla openly funnel-shaped, 4in(10.2cm) wide, 6-lobed, fragrant. Ball-shaped trusses of 11. Plant wider than tall; leaves held 2 years. (Lloyd L. Newcomb, 1981) C. A. (ARS) 1983

```
cl.
     -catawbiense (1/2)
Newport-
     -unknown (1/2)
```

5ft(1.5m)        -10°F(-23°C)           ML

Flowers of an unusual plum purple, held in very tight trusses. Plant habit rather open; fine foliage. "May have been grown by Warren from seed from North Cemetery"--O. Pride. [Butler, Pa.?] (Warren Stokes, reg. 1958)

cl.

Nez Perce Chief      See The Chief

```
cl.
     -ponticum (1/2)
Nicholas-
     -unknown (1/2)
```

5ft(1.5m)        -10°F(-23°C)           ML           4/2

One of the few Exbury hybrids whose parentage is uncertain, but the polished dark green, pointed leaves suggest ponticum as a parent. Flowers of petunia purple, paler towards center, throat and center of upper segment white, spotted green; corolla openly funnel-shaped, 3.3in(8cm) wide, 5-lobed. Trusses of about 19, closely packed, on a broad, tall shrub. (Edmund de Rothschild, 1966) A. M. (RHS) 1965  See color illus. PB pl. 56.

```
cl.
      -wardii var. wardii (1/2)
Nigel Marshall-
        -unknown (1/2)
```

Light yellow flowers. (The National Trust, Mount Stewart, reg. 1983)

cl.

Night Editor--form of russatum

4ft(1.2m)        5°F(-15°C)           E

Synonym Betty's Purple Russatum. Upright, twiggy plant, with stiff branches; leaves narrowly elliptic, 2in(5cm) long, strongly revolute, scaly beneath, and held 1-2 years. Spectrum violet flowers, edged and spotted violet, openly funnel-shaped, 1.5in (3.8cm) wide, 5-lobed; small spherical trusses of 6-7. (Betty Sheedy, reg. 1981)

```
cl.
     -Blue Steel--impeditum (1/2)
Night Sky-      -russatum (1/4)
     -Russautinii-
             -augustinii ssp. augustinii (1/4)
```

Trusses of 5 flowers, deep violet blue, paler in throat. Lanceolate leaves, sparsely scaly. (J. P. C. Russell, 1983)

*NIGHTWATCH    ?   SEE PHOTO*

cl.

Nigrescens      Parentage unknown

A parent of Myrte, a hybrid by Seidel. Flowers dark plum color. (A. Waterer, before 1871)

cl.

Nikko      See next page

```
cl.
     -catawbiense var. album (1/2)
Nile-
     -wardii var. wardii (L S & T 5679) (1/2)
```

6ft(1.8m)        -20°F(-29°C)           EM

Primrose yellow flowers with dorsal blotch of currant red, of good substance, 3in(7.6cm) across, 5-lobed, widely funnel-shaped; spherical trusses of 13-14. Plant rounded, as broad as tall; dark yellow green, glossy leaves retained 2 years. (David G. Leach, intro. 1973; reg. 1983)

```
                 -cataubiense var. album (1/2)
       -unnamed-       -wardii var. wardii (1/4)
cl. - hybrid-Crest-           -fortunei ssp. discolor (1/16)
 -             -Lady      -
Nikko-         Bessborough-campylocarpum ssp. campylocarpum
 -                                       Clatum Group (1/16)
 -                   -cataubiense var. album
 -         -Ivory Tower-      -wardii var. wardii
 -unnamed-            -unnamed-
    hybrid-             hybrid-fortunei ssp. fortunei (1/16)
     -           -cataubiense var. album
       -unnamed-        -wardii var. wardii
          hybrid-unnamed-
                   hybrid-decorum (1/16)
```

5ft(1.5m)          -15°(-26°C)          M

Brick red buds open to primrose yellow flowers, darker in throat, widely funnel-campanulate, 3.2in(8cm) wide, 5-6 lobed; dome-shaped trusses of 14. Plant habit rounded, slightly wider than tall; dark yellow green, narrow leaves, 4.75in(12cm) long, held 1-2 years. (David G. Leach, reg. 1983)

```
                    -maximum (1/8)
               -Halopeanum-
cl.   -Snow Queen-         -griffithianum (1/2)
 -          -       -griffithianum
Nimbus-        -Loderi-
 -                   -fortunei ssp. fortunei (1/8)
 -                   -fortunei ssp. discolor (1/4)
       -Cornish Loderi-
                   -griffithianum
```

Plant broader than tall, of vigorous, upright habit; leaves medium dull green, 7.5in(19cm) long. Cream buds flushed dawn pink, opening white blushed pale pink and fading white, funnel-shaped, 4in(10.2cm) across, slightly wavy margins; compact, spherical trusses, 8-flowered. Very floriferous.    (Knap Hill Nurseries) A. M. (RHS) 1965;  F. C. C. (Wisley Trials) 1967

```
g.
     -Polka Dot--irroratum ssp. irroratum (1/2)
Nimrod-
     -calophytum var. calophytum (1/2)
```

Long broad leaves of calophytum and tubular bell-shaped flowers of irroratum. Delicate pink blossoms, heavily spotted brown, in many large, well-filled trusses. Early, at Exbury. (Edmund de Rothschild, 1963)

```
cl.
     -fortunei ssp. fortunei (1/2)
Nina-
     -unknown (1/2)
```

5ft(1.5m)          -20°F(-29°C)          M

Fragrant flowers, spirea red with heavy ruby red blotch circling the throat; trusses 10-flowered. Plant twice as broad as tall; leaves 5in(12.7cm) by 2.5in(6.4cm). (Origin uncertain; raised, & intro. by Edward T. Wytovich; reg. 1979)

```
g.
     -fortunei ssp. discolor (1/2)
Ninette-
     -yakushimanum ssp. makinoi (1/2)
```

Clear, pale pink flowers. (J. B. Stevenson, 1936)

---

Niobe     See Kate Greenaway

---

```
cl.
     -decorum (1/2)
Nissequogue-
       -Pygmalion--unknown (1/2)
```

3ft(.9m)          -5°F(-21°C)          M

Buds dark spirea red, flowers lighter red with dorsal spotting of grey brown in throat, widely funnel-campanulate, 3in(7.6cm) wide, of good substance; dome-shaped trusses of 10. Free-flowering. Plant upright, well-branched, as broad as tall; glossy dark green leaves held 3 years. (A. A. Raustein, reg. 1983)

```
      -Boule de Neige Improved -caucasicum (1/4)
cl.  - Boule de Neige (selfed)-        -cataubiense (1/8)
 -                        -hardy hybrid-
No Way*                        -unknown (1/8)
 -       -flavidum var. flavidum (1/4)
   -unnamed-      -cinnabarinum ssp. cinnabarinum Roylei
     hybrid-Lady  -                    Group (1/8)
       Rosebery-    -cinnabarinum ssp. cinnabarinum
         -Royal-                      (1/16)
         Flush-maddenii ssp. maddenii (1/16)
         (pink form)
```

Pale yellow flowers, with a red blotch.  Hardy to at least -15°F (-26°C).  (Weldon E. Delp)

---

```
     -javanicum (1/2)
Nobilius-
     -teysmanni (1/2)
```

Vireya hybrid.  Parentage as given in I.R.R.  Large flowers of deep golden yellow.  A. M. (RHS) 1896

```
g.
     -caucasicum (1/2)
Nobleanum-
     -arboreum ssp. arboreum (1/2)
```

Early and showy.  Brilliant rosy scarlet buds, opening rich rose flushed white inside, with a few crimson spots, widely funnel-shaped; compact trusses.  Slow-growing shrub; leaves dark green, with plastered buff indumentum beneath.  One of the earliest man-made hybrids.  Several forms exist; see entries below this. (A. Waterer, Knap Hill, about 1832-1835) A. G. M. (RHS) 1926

```
cl.
        -caucasicum (1/2)
Nobleanum Album-
        -arboreum ssp. arboreum (1/2)
```

Nobleanum g.  White flowers.  (Knap Hill)

```
                           -ponticum (1/8)
---              -unnamed hybrid-
         -Altaclarense-        -cataubiense (1/8)
Nobleanum Bicolor-     -arboreum ssp. arboreum (1/4)
         -cataubiense (1/2)
```

Flowers deep rose with white throat.  (Standish & Noble, 1850)

```
cl.
        -caucasicum (1/2)
Nobleanum Coccineum-
        -arboreum ssp. arboreum (1/2)
```

5ft(1.5m)          0°F(-18°C)          VE          2/2

Nobleanum g.  A large, conical shrub with bell-shaped, deep rose flowers, with some basal spotting of dark crimson.  Another form has scarlet flowers of medium size, in small rounded trusses, on a rather sprawling bush.  (A. Waterer, 1835)

cl.
                    -caucasicum stramineum ? (1/2)
Nobleanum Lamellen-
                    -arboreum ssp. arboreum (1/2)

Deep red flowers on a hybrid by E. J. P. Magor, intro. 1932.


cl.
                        -caucasicum (1/2)
Nobleanum Silberaad's Early-
                    -arboreum ssp. arboreum (1/2)

Nobleanum g.  Flowers pale rose and pure white; very early.
(Knap Hill)


cl.
                    -caucasicum (1/2)
Nobleanum Venustum-
                    -arboreum ssp. arboreum (1/2)

5ft(1.5m)          0°F(-18°C)          VE          2/2

Nobleanum g.   Densely leafy bush, broad and rounded. Compact
trusses of funnel-shaped flowers of shining rose pink, of medium
size. In England it flowers in late winter, even in a mild Dec-
eember. Raised by William Smith near Kingston, Surrey, in 1829.
A. G. M. (RHS) 1969; A. M. (RHS) 1973

                    -griersonianum (1/4)
g.      -Laura    -          -thomsonii ssp. thomsonii (1/8)
        -Aberconway-Barclayi-          -arboreum ssp. arboreum
Nocturne-                    -Glory of  -                 (1/16)
        -                    Penjerrick-griffithianum (5/16)
        -          -griffithianum
        -Loderi-
                    -fortunei ssp. fortunei (1/4)

A red-flowered hybrid.  (Lord Aberconway, 1950)

cl.                          -catawbiense (1/4)
                -red catawbiense hybrid-
Nodding Bells*               -unknown (1/4)
        -                    -forrestii ssp. forrestii
        -unnamed hybrid-     Repens Group (1/4)
                             -griersonianum (1/4)

2ft(.6m)          -15°F(-26°C)          EM

A spreading bush with outer branches arching downward.  Cherry
red flowers, openly campanulate.  (Edmund Amateis)


cl.                -caucasicum (1/4)
                -Dr. Stocker-
Noele Boulter-     -griffithianum (1/4)
        -          -arboreum ssp. arboreum (1/4)
        -Cornubia-          -thomsonii ssp. thomsonii (1/8)
                -Shilsonii-
                    -barbatum (1/8)

An upright shrub with 10-14 flowers per truss, neyron rose in
bud, stained ruby red in the throat.  (Flower color not given)
(V. J. Boulter, 1963)


cl.
        -Marion (Cheal)--unknown (5/8)
Noila-          -diaprepes (1/4)
        -Lilac Time-          -ponticum (1/8)
                -Purple Splendour-
                    -unknown

Fuchsine pink flowers, heavily spotted with yellow on the upper
lobes.  (V. J. Boulter, 1965)


cl.
        -cinnabarinum ssp. cinnabarinum Roylei Group (1/2)
Nonesuch-
        -maddenii ssp. maddenii (1/2)

Flowers of pale apricot, fading cream.  (Adams-Acton, 1942)


cl.
        -Marion--unknown (1/2)
Noorook-
        -neriiflorum ssp. neriiflorum (1/2)

Flowers Delft rose, overall light spotting.   (Boulter, 1972)


---                          -maximum (1/4)
                -unnamed hybrid-
Norbitonense Aureum-          -ponticum (1/4)
                -azalea molle (1/2)

Azaleodendron. Synonym Smithii Aureum. Flowers soft tawny yellow with darker markings, in a rhododendron-like truss, blooming
late. Plant habit neat, compact.  (W. Smith of Norbiton, 1830)


cl.                          -maximum (1/4)
                -unnamed hybrid-
Norbitonense Broughtonianum-  -ponticum (1/4)
                -azalea molle (1/2)

Azaleodendron. Synonym Broughtonii Aureum. Flowers soft yellow
with orange yellow spots, about 2.5in(6.4cm) wide; small round
trusses.  Foliage rough-textured, rather sparse. (Origin not
known; raised by W. Smith of Norbiton, c.1830)   F. C. C. (RHS)
1935     See color illus. VV p. 129.


g.
        -Essex Scarlet--unknown (1/2)
Norderney-
        -williamsianum (1/2)

Flowers deep rose.  (Dietrich Hobbie, 1947)


cl.
        -Milkmaid--unknown (1/2)
Norfolk Candy-     -dichroanthum ssp. dichroanthum (5/16)
        -Marmora-          -fortunei ssp. discolor (1/8)
                -Margaret-          -dichroanthum ssp.
                Dunn  -Fabia-          dichroanthum
                             -griersonianum (1/16)

Very light Chinese yellow flowers, with an orange brown  blotch.
(J. Russell, reg. 1972)


                             -fortunei ssp. discolor (3/8)
                -Roberte-
                -          -campylocarpum ssp. campylo-
        -Coromandel-        carpum Elatum Group (3/16)
        -          -          -Sir Charles Butler--
        -          -          -Soulbut- fortunei ssp.
        -          -Vanessa-  -          fortunei (1/32)
        -Malabar-          -          -souliei (1/32)
        -          -          -griersonianum (5/32)
cl.     -          -dichroanthum ssp. dichroanthum (7/32)
        -          -Marmora-          -fortunei ssp. discolor
Norfolk-          -Margaret-          -dichroanthum ssp.
Johnny-          Dunn  -Fabia-          dichroanthum
        -                    -griersonianum
        -          -fortunei ssp. discolor
        -          -Belle-
        -Spanish-  -campylocarpum ssp. campylocarpum Elatum
        Galleon-          -fortunei ssp. discolor          Group
                -Margaret-          -dichroanthum ssp dichroanthum
                Dunn  -Fabia-
                    -griersonianum

Trusses of 9-10 flowers 2in(5cm) across, 6-lobed, bright canary
yellow,  dorsal spotting of rich crimson, deeper in throat.   At
27 years height was 5ft(1.5m).  (J. P. C. Russell, reg. 1983)


cl.      -griffithianum (1/4)
        -Mars-
Norlen-  -unknown (1/4)
        -yakushimanum ssp. yakushimanum (1/2)

5ft(1.5m)          -20°F(-29°C)          M

Flowers are azalea pink with peppermint stripes down center of
lobes on opening, fading white. Corolla openly funnel-shaped, 2
in(5cm) wide, with 5 frilled lobes; ball-shaped trusses of 20.
Plant well-branched, as broad as tall; convex, glossy leaves
with some greyed orange indumentum, held 3 years. (Crossed by
by Maurice Hall, 1968; reg. 1980)

cl.
                -unnamed hybrid--unknown (1/2)
Norma Hodge-    -dichroanthum ssp. dichroanthum (1/4)
            -Fabia-
                    -griersonianum (1/4)

A hybrid from British Columbia. Flowers of empire rose, with a
basal blotch of greyed red/Turkey red, radiating in a spotted
flare of Delft rose on 4 upper petals; trusses of 8. (Cross by
Eric Langton; reg. by Robert C. Rhodes, 1979)

cl.                                 -catawbiense (1/8)
                -Parsons Grandiflorum-
Norman -Nova Zembla-               -unknown (3/8)
Behring-                 -red hybrid--unknown
        -Pink Parasol--yakushimanum ssp. yakushimanum (1/2)

1ft(.3m)        -25°F(-32°C)         L

Flowers light mallow purple, with darker edges and yellow green
spotting, openly funnel-shaped, 2.2in(5.5cm) across, 5-lobed, in
dome-shaped trusses of 18. Plant well-branched, broader than
tall; leaves with greyed orange indumentum, held 2 years. (Rudy
Behring cross, 1974; reg. 1985)

cl.
                -calophytum var. calophytum (1/2)
Norman Colville-
                -rex ssp. arizelum (1/2)

White flowers with blotch of dark greyed violet, widely funnel-
campanulate, to 2in(5cm) wide and long, 5-lobed; rounded trusses
of 16-18. Leaves narrowly elliptic, dark green above, brown
felted indumentum below. (N. R. Colville cross; intro. by Mrs.
Norman Colville)  A. M. (RHS) 1979

cl.                      -griffithianum (3/4)
            -Beauty of Tremough-
Norman Gill-             -arboreum ssp. arboreum (1/4)
            -griffithianum

6ft(1.8m)        5°F(-15°C)         ML        5/3

Very large flowers, white with a red basal blotch, held in tall
trusses. Plant habit vigorous; leaves 5in(12.7cm) long. Does
best in partial shade. (R. Gill & Son) A. M. (RHS) 1922; A. M.
(Exbury Trial) 1936

g. & cl.
            -fortunei ssp. discolor (1/2)
Norman Shaw-             -catawbiense (1/4)
            -B. de Bruin-
                        -unknown (1/4)

A tall but compact hybrid with phlox pink flowers held in shapely trusses. Blooms in late season. (Rothschild)  A. M. (RHS)
1926

cl.
        -Newburyport Beauty--unknown
Normandy-                           (Dexter hybrids)
        -Newburyport Belle--unknown

3.5ft(1.05m)        -20°F(-29°C)         ML

Flowers strong, bright rose pink, flushed deeper on edges, with
dorsal spotting of tangerine orange, openly funnel-shaped, 2.8
in(7.4cm) across, with 5-6 wavy lobes. Plant rounded, broader
than tall; elliptic dark green leaves, held 2 years. (David G.
Leach, cross 1968; reg. 1983)

cl.
            -minus var. minus Carolinianum Group
Northern Lights*             (white form) (1/2)
            -Arctic Pearl--dauricum (white form) (1/2)

Plant 4ft(1.2m) high and 3.5ft(1m) broad, at 10 years. Flowers
white, with a well-defined lavender blush on margins, making a
bicolor effect. Very early; blooms a week after P.J.M. (Dr. G.
David Lewis, intro. c.1983)

cl.
            -Laetevirens (Wilsonii) (1/2)
Northern Rose*
            -Cornell Pink--mucronulatum (pink form) (1/2)

3ft(.9m)        -20°F(-29°C)         VE

Slightly darker pink than Cornell Pink. A slow-growing dwarf
with long pointed leaves. (Dr. Robert Ticknor cross; intro. by
The Bovees Nursery)

cl.
            -venator (1/2)
Northern Rover-       -arboreum ssp. arboreum (1/4)
            -Doncaster-
                    -unknown (1/4)

Blood red flowers. (Thacker, 1942)

cl.
            -fortunei ssp. discolor (1/2)
Northern Star-               -griffithianum (1/8)
        -        -Loderi-
        -Lodauric-       -fortunei ssp. fortunei (1/8)
            Iceberg-auriculatum (1/4)

5ft(1.5m)        -5°F(-21°C)         L

A large, compact plant. White flowers with a greenish throat,
cinnamon-colored anthers, widely funnel-shaped, 7-lobed, and
fragrant. See also Southern Cross, Gipsy Moth, Starcross. (A.
F. George, Hydon Nurseries, 1968)

cl.                      -catawbiense (1/4)
            -Parsons Grandiflorum-
Nova Zembla-             -unknown (3/4)
            -hardy red hybrid--unknown

5ft(1.5m)        -25°F(-32°C)         M        4/3

Very showy red flowers in well-filled trusses. A vigorous grower with good foliage. Probably the best hardy red that is
widely available. Heat-tolerant. (M. Koster & Sons, 1902) See
color illus. VV p. 3.

cl.             -arboreum ssp. nilagiricum ? (1/4)
        -Noyo Chief-
Noyo Brave-         -unknown (1/4)
        -Koichiro Wada--yakushimanum ssp. yakushimanum (1/2)

2.5ft(.75m)        0°F(-18°C)         M        4/4

Leaves 5in(12.7cm) by 2in(5cm). Flowers 2.5in(6.4cm) wide, of
mandarin red fading with age, a very small glowing red blotch.
Ball-shaped trusses hold 22. (Cecil Smith, 1978)   See color
illus. K p. 88.

cl.     -arboreum ssp. nilagiricum ? (1/2)
        -
Noyo Chief-unknown (1/2)

5ft(1.5m)        10°F(-12°C)         M        3/4

Formerly called a form of arboreum kingianum. Outstanding foliage of parsley green, very glossy, deeply ribbed, 7in(17.8cm)
long, plastered fawn tomentum beneath. Flowers of a clear rose
red, broadly campanulate, 2.5in(6.4cm) wide, in compact trusses
of 16. (Originated at Reuthe Nursery, England; raised by Lester
E. Brandt; named & reg. by Dr. & Mrs. P. J. Bowman, 1966) C. A.
(ARS) 1966    See color illus. ARS Q 26:3 (Summer 1972): cover;
VV p. 29.

cl.
        -Koichiro Wada--yakushimanum ssp. yakushimanum (1/2)
Noyo Maiden-        -arboreum ssp. nilagiricum ? (1/4)
        -Noyo Chief-
                -unknown (1/4)

3ft(.9m)        5°F(-15°C)        EM

Plant rounded, broader than tall; leaves heavily indumented, and
held 3 years. Clear pink buds open to white flowers without
markings, openly campanulate, 2.5in(6.4cm) across, 5-lobed, of
good substance; rounded trusses of 15-19. Floriferous. (Cecil
C. Smith, reg. 1983)


cl.
      -catawbiense var. album (1/2)
Nuance-                -neriiflorum ssp. neriiflorum (1/8)
      -        -unnamed-
      -unnamed- hybrid-dichroanthum ssp. dichroanthum (1/8)
         hybrid-
              -fortunei ssp. discolor (1/4)

5ft(1.5m)        -15°F(-26°C)        ML        4/4

Flowers pale orange yellow, margins broadly suffused strong purplish pink, strong greenish yellow spotting; corolla openly funnel shaped, 2.75in(7cm) wide, 5-lobed, of heavy substance; in
dome-shaped trusses of 14. Plant rounded, slightly wider than
tall; elliptic medium green leaves held 2 years. See also Bali.
(David G. Leach, intro. 1968; reg. 1973)    See color illus. LW
pl. 99.


              -griffithianum (1/8)
cl.      -Kewense-
    -Aurora-        -fortunei ssp. fortunei (1/8)
Nubar-      -thomsonii ssp. thomsonii (1/4)
    -campylocarpum ssp. campylocarpum (1/2)

Buds colored light rhodonite red opening to very light yellow
green, deepening to light chartreuse green with center speckles
of deepest claret rose. (Edmund de Rothschild, 1981)


cl.      -williamsianum (1/4)
    -Cowslip-
Nugget-        -wardii var. wardii (1/4)
    -chamaethomsonii var. chamaethauma (1/2)

Trusses of 8 flowers, empire yellow, flushed peach. (Lester E.
Brandt, 1962)


---

Numa--Of the 5th generation of 5 species: javanicum, jasmini-
florum, brookeanum, indicum, and multicolor.

Flowers orange red with throat a deeper shade. A. M. (RHS) 1895


cl.                -griffithianum (1/4)
      -unnamed hybrid-
Nuneham Park-            -unknown (3/4)
      -Martin Hope Sutton--unknown

Red flowers. (C. B. van Nes & Sons, before 1922)


cl.
      -nuttallii (1/2)
Nutcracker-
      -maddenii ssp. maddenii (polyandrum) (pink form) (1/2)

Trusses of 7 flowers, tubular bell-shaped, up to 4.4in(11cm)
across, 5-lobed, creamy white with deep yellow throat, strongly
scented. Leaves elliptic, dark green, densely scaly beneath.
(G. A. Hardy) A. M. (RHS) 1984


cl.
      -sanguineum ssp. didymum (1/2)
Nutmeg-
      -griersonianum (1/2)

Hardier and sturdier grower than Arthur Osborne, q.v. Also more
numerous but smaller flowers, ruby-colored and trumpet-shaped.
(Haworth-Booth, Farall Nurseries, reg. 1969)


cl.

Nutmegacalyx*--form of megacalyx

4ft(1.2m)        15°F(-10°C)        EM-L        4/3

Named by Kingdon-Ward for the fragrance, usually likened to nutmeg. Very large flowers, pure white flushed pink. Too tender
to grow outdoors even in London. (Shown by Adm. Heneage-Vivian
of Clyne Park, Swansea) A. M. (RHS) 1937


g.
    -forrestii ssp. forrestii Repens Group (1/2)
Nymph-        -dichroanthum ssp. dichroanthum (1/8)
    -Astarte-        -campylocarpum ssp. campylocarpum
    -Largo-      -Penjerrick-        Elatum Group (1/16)
    -        -griffithianum (1/16)
    -neriiflorum ssp. neriiflorum Euchaites Group (1/4)

1ft(.3m)        5°F(-15°C)        E        3/3

A dwarf of spreading habit, with dark green leaves and deep red
flowers. (Lord Aberconway, Bodnant, 1940)


                        -catawbiense (1/8)
cl.        -Parsons Grandiflorum-
    -America-                -unknown (3/8)
Oakton-      -dark red hybrid--unknown
    -yakushimanum ssp. yakushimanum (1/2)

5ft(1.5m)        -10°F(-23°C)        M

Buds of medium magenta opening phlox pink, lighter in lobe centers, with striking pink pistil; flowers openly funnel-campanulate, 2.5in(6.4cm) across, waxy, 5-lobed, in dome-shaped trusses
of 13. Plant habit upright, taller than wide; very dark green,
dull leaves with slight tan indumentum, held 2+ years. White
tomentum on new growth. (Cross by F. W. Schumacher; reg. by G.
Ring III, 1983)


        -fortunei ssp. fortunei (1/4)
cl. -Fawn-        -dichroanthum ssp. dichroanthum (3/8)
    -      -Fabia-
Oasis-        -griersonianum (1/8)
    -      -dichroanthum ssp. dichroanthum
    -Dido-
        -decorum (1/4)

Flowers of barium yellow with overlay of coral pink, widely
funnel-shaped, 4in(10cm) wide, held in trusses of 9. (Crossed
by Del James; reg. by Mrs. Ray James, 1972)


        -thomsonii ssp. thomsonii (1/2)
g.  -Barclayi-                -arboreum ssp. arboreum (1/8)
    -      -Glory of Penjerrick-
Oberon-                -griffithianum (1/8)
    -        -thomsonii ssp. thomsonii
    -Thomwilliams-
        -williamsianum (1/4)

A hybrid with flowers of deep rose. (Lord Aberconway, 1950)

cl.

Obovatum*--form of <u>lepidotum</u>

1.5ft(.45m)          -5°F(-21°C)          EM          3/3

This form of the species has large leaves and greenish yellow
flowers. (Slocock Nurseries)

```
                          -intricatum (1/8)
               -Intrifast-
cl.      -Blue  -           -fastigiatum (1/8)
         -Diamond-augustinii ssp. augustinii (3/8)
Oceanlake-              -impeditum (3/8)
    -            -Blue Tit-
         -Sapphire-      -augustinii ssp. augustinii
               -impeditum
```

2.5ft(.75m)          -5°F(-21°C)          EM          4/4

Deep violet blue flowers, flattened, 1in(2.5cm) wide, in trusses
of 8. Dense foliage, 1in(2.5cm) long. Sun-tolerant. Blooms
later than other blues but flowers last longer. (Arthur Wright,
Sr. & Jr., 1966)

cl.

Ocelot--form of <u>parmulatum</u> (K-W 5875)

3ft(.9m)          5°F(-15°C)          EM          3/4

Loose trusses hold 3-6 flowers, very light sap green, each lobe
with a deeper colored central band, upper throat heavily spotted
reddish purple; corolla tubular-campanulate, 2.2in(5.5cm) across
and 2in(5cm) long. (Seed from Kingdon-Ward; raised by E. J. P.
Magor; intro. & reg. by Gen. Harrison) A. M. (RHS) 1977

```
                          -griffithianum (1/8)
g. & cl.     -Loderi King George-
      -Akbar-                  -fortunei ssp. fortunei (1/8)
Octavia-     -fortunei ssp. discolor (1/4)
      -griersonianum (1/2)
```

Plant medium-sized with good growth habit. Red flowers in well-
packed trusses. Blooms late midseason. (Rothschild, 1947)

```
g.
      -dichroanthum ssp. dichroanthum (1/2)
Octopus-
      -kyawii (1/2)
```

Low and widely spreading plant. Orange flowers in loose trusses
in late season. (Rothschild, 1950)

```
cl.           -wightii (1/4)
         -China-
Odd Ball-    -fortunei ssp. fortunei (1/4)
    -           -wardii var. wardii (1/4)
         -Goldbug-    -dichroanthum ssp. dichroanthum (1/8)
               -Fabia-
                     -griersonianum (1/8)
```

5ft(1.5m)          10°F(-12°C)          E

Cardinal red buds open to brick red flowers edged in jasper red,
with the rest of the corolla spotted bright guardsman red, tubular campanulate, 3in(7.6cm) wide and 2in(5cm) long, with 7 wavy
lobes. Colors change slowly until corolla is biscuit-colored,
the spots and nectaries terracotta. Lax 12-flowered trusses.
Plant well-branched, as broad as tall; dark green leaves held 2
years. See also Oh-Too. (Walter Elliott, reg. 1981)

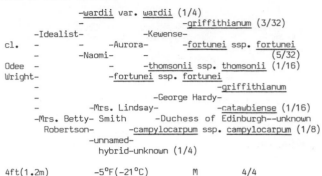

```
                     -wardii var. wardii (1/4)
             -                 -griffithianum (3/32)
         -Idealist-        -Kewense-
cl.  -           -Aurora-      -fortunei ssp. fortunei
     -           -Naomi-     -           (5/32)
Odee -                 -thomsonii ssp. thomsonii (1/16)
Wright-      -fortunei ssp. fortunei
     -                 -griffithianum
     -           -George Hardy-
     -      -Mrs. Lindsay-      -catawbiense (1/16)
     -Mrs. Betty- Smith     -Duchess of Edinburgh--unknown
       Robertson-      -campylocarpum ssp. campylocarpum (1/8)
             -unnamed-
             hybrid-unknown (1/4)
```

4ft(1.2m)          -5°F(-21°C)          M          4/4

Compact plant; waxy green leaves 3.5in(8.9cm) long. Peach buds
opening chartreuse with carmine red spotting in throat; flowers
4.5in(11.5in) wide, slightly ruffled, in trusses of 12-15. (A.
O. Wright, Sr. & Jr., reg. 1965) P. A. (ARS) 1966 See color
illus. K p. 117; VV p. 75.

---

```
      -ponticum (1/2)
Odoratum-
      -azalea periclymenoides (nudiflorum) (1/2)
```

4ft(1.2m)          -15°F(-26°C)          ML          2/3

An azaleodendron with fragrant orchid lilac flowers. (Thompson's
Nursery, before 1875)

```
                          -campylocarpum ssp. campylocarpum
               -Penjerrick-          Elatum Group (1/16)
g.     -Amaura-      -griffithianum (1/16)
     -Eros-      -griersonianum (5/8)
Oedipus-     -griersonianum
     -           -facetum (eriogynum) (1/4)
     -Jacquetta-
             -griersonianum
```

A scarlet-flowered hybrid. (Lord Aberconway, 1941)

```
                     -smirnowii (1/4)
cl.     -unnamed hybrid-
     -           -yakushimanum ssp. yakushimanum (1/4)
Oh Joyce*           -maximum (1/4)
             -Midsummer-
     -Joyce Lyn-      -unknown (1/8)
     -           -maximum
             -unnamed-
             hybrid-wardii var. wardii (1/8)
```

Medium-sized bush, hardy to -15°F(-26°C); wide, pale peach flowers, pink-tinged ruffled edges, a few gold dorsal spots. (Delp)

```
cl.
      -yakushimanum ssp. yakushimanum (1/2)
Ohio Pink-
      -unknown (1/2)
```

3ft(.9m)          -20°F(-29°C)          ML

Dome-shaped trusses with 12-13 flowers, deep purplish pink with
a paler blotch spotted olive brown, of good substance, widely
funnel-campanulate, 2.5in(6.4cm) across, slightly fragrant.
Plant wider than high, well-branched; glossy olive green leaves
with scant golden brown indumentum, new growth light brown tomentose. See also Ohio Pink Blush, Ohio Pink Glo, Ohio White.
(Raised & intro. by John Ford; reg. by Secrest Arboretum, 1984)

```
cl.
      -yakushimanum ssp. yakushimanum (1/2)
Ohio Pink Blush-
      -unknown (1/2)
```

2.5ft(.75m)          -15°F(-26°C)          ML

Flowers. very pale purplish pink with light purple shading at edges, sparse strong yellow spotting in throat, widely funnel-shaped, 2.5in(6.4cm) wide, frilled, slightly fragrant; trusses of 7-13, dome-shaped. Floriferous. Plant rounded, broader than tall; leaves with slight tan indumentum, gray tomentose on new growth. (Raised & intro. by John Ford; reg. by Secrest Arboretum, 1984)

cl.
       -yakushimanum ssp. yakushimanum (1/2)
Ohio Pink Glo-
       -unknown (1/2)

4ft(1.2m)        -15°F(-26°C)        ML

Flowers strong purplish pink with reverse strong purplish red, widely funnel-campanulate, 2in(5cm) across, with 5 wavy lobes; trusses of 15-18. Floriferous, slightly fragrant. Plant upright, taller than wide; leaves held 2 years, light brown tomentum on new foliage. (Raised & intro. by John Ford; reg. by Secrest Arboretum, 1984)

cl.
       -yakushimanum ssp. yakushimanum (1/2)
Ohio White-
       -unknown (1/2)

3ft(.9m)         -15°F(-26°C)        ML

White flowers with small, vivid yellow blotch in throat, widely funnel-campanulate, 2in(5cm) across, frilled, 5-lobed; 8-10 in dome-shaped trusses. Plant habit rounded, wider than tall; new foliage has light brown tomentum. (Raised & intro. by John Ford; reg. by Secrest Arboretum, 1984)

cl.        -wightii (1/4)
  -China-
Oh-Too-    -fortunei ssp. fortunei (1/4)
  -        -wardii var. wardii (1/4)
  -Goldbug-    -dichroanthum ssp. dichroanthum (1/8)
    -Fabia-
      -griersonianum (1/8)

4ft(1.2m)        -15°F(-26°C)        M         4/3

Interesting multicolored flowers: Red purple buds open to rhodonite red, maturing to buttercup yellow with a salmon pink throat and heavy spotting of currant red on upper corolla; mature flowers edged in rhodonite red. Corolla 2.5in(6.4cm) wide, of heavy substance; rather lax trusses of 14. Plant rounded, as wide as high; glossy, moss green leaves held 2 years. (Cross by W. Elliott; raiser W. Whitney; reg. by G. & A. Sather, 1975) See color illus. ARS Q 30:4 (Fall 1976): p. 239.

cl.

Oki Island* --form of japonicum var. japonicum (metternichii)

"One of the best wild forms of this species in Japan. A fine foliaged plant with rose-colored flowers and a neat habit." From catalog of Glendoick Gardens.

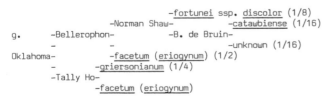

g.      -Bellerophon-
             -Norman Shaw-
                   -fortunei ssp. discolor (1/8)
                     -catawbiense (1/16)
Oklahoma-     -B. de Bruin-
 -        -facetum (eriogynum) (1/2)
            -unknown (1/16)
       -griersonianum (1/4)
  -Tally Ho-
    -facetum (eriogynum)

Late-blooming, and not very hardy. Currant red flowers, lightly spotted darker red on upper lobe, with black anthers and stigma; large trusses of 20-22. (Cross of Lionel de Rothschild; intro. by Major A. E. Hardy)  A. M. (RHS) 1975

cl.                -griffithianum (1/8)
      -Mars-
   -Vulcan-    -unknown (1/8)
Old Copper-    -griersonianum (1/2)
 -        -dichroanthum ssp. dichroanthum (1/4)
   -Fabia-
      -griersonianum

5ft(1.5m)        -5°F(-21°F)        L        4/3

Blossoms of a unique copper color. Large campanulate flowers in loose trusses; long, dark green leaves on an upright plant. An attractive hybrid that is heat-tolerant. (Theodore Van Veen, Sr., reg. 1958)  See color illus. VV p. 15.

cl.
     -catawbiense (1/2)
Old Port-
     -unknown (1/2)

5ft(1.5m)        -15°F(-26°C)        M        3/3

Two hybrids are grown under the name of Old Port. The above has flowers of dark wine red or plum, with blackish crimson markings and glossy leaves like catawbiense. The plant of medium size and dense habit. The other may be a ponticum hybrid, of deeper wine purple, without blotching, and glossy smooth foliage. Both plants are vigorous and sturdy. Also Butler Port in App. D. (A. Waterer, 1865)

cl.        -griersonianum (1/4)
   -Azor-
Old Spice-    -fortunei ssp. fortunei (1/4)
  -decorum (1/2)

4ft(1.2m)        -5°F(-21°C)        ML        3/3

Vigorous, upright plant; leaves to 7in(17.8cm) long by 2in(5cm) wide. Fragrant flowers, phlox pink shading to apricot, openly campanulate, 3.5in(9cm) wide, and held in trusses of 12. (Benjamin Lancaster, 1963)

cl.
     -catawbiense (1/2)
Oldewig-
     -unknown (1/2)

Ruby red flowers with lighter center and reddish brown markings on light background. (T. J. R. Seidel, intro. 1912)

cl.
     -campylocarpum ssp. campylocarpum Elatum Group (1/2)
Ole Olson-
     -fortunei ssp. discolor (1/2)

5ft(1.5m)        0°F(-18°C)        ML        3/3

Synonym Old Olson. Lady Bessborough g. Plant habit open, upright; medium green leaves, 4in(10.2cm) long. Flowers of pale yellow, 3.5in(8.9cm) wide, held in loose trusses. (Joseph Gable; intro. by H. Lem)

cl.
     -griersonianum (1/2)
Oleanda-    -arboreum ssp. arboreum (1/4)
   -Doncaster-
      -unknown (1/4)

Compact trusses hold 8-10 flowers, each 3in(7.6cm) across, colored claret rose, blooming in late season. (Haworth-Booth, Farall Nurseries, reg. 1968)

cl.                -griffithianum (1/8)
      -George Hardy-
 -Mrs. Lindsay Smith-    -catawbiense (1/8)
Olga-        -Duchess of Edinburgh--unknown (1/4)
 -        -dichroanthum ssp. dichroanthum (1/4)
  -Dido-
    -decorum (1/4)

Truss of 12 flowers, openly funnel-shaped, very light Chinese yellow, flushed azalea pink towards margins, flecked orange red on upper segments. (Slocock Nurseries, 1982) H. C. (Wisley Trials) 1982

cl.
```
            -minus var. minus Carolinianum Group (1/2)
Olga Mezitt-
            -minus var. minus (1/2)
```

3ft(.9m)          -15°F(-26°C)          EM

Not really a hybrid, as two forms of the same species were crossed. Clear phlox pink flowers, openly funnel-shaped, 1.5in (3.8cm) across, 5-lobed; small dome-shaped trusses of about 12 cover the bush. Plant upright, well-branched; shiny leaves, 2in(5cm) long, bright green in summer, mahogany in winter. (Edmund V. Mezitt, Weston Nurseries, reg. 1983)

cl.
```
                -griffithianum (1/4)
            -Mars-
Olin O. Dobbs-  -unknown (1/2)
            -            -ponticum (1/4)
            -Purple Splendour-
                        -unknown
```

4ft(1.2m)          -15°F(-26°C)          M          5/3

Waxy flowers of deep reddish purple with brown spotting, openly funnel-shaped, 2.5in(6.4cm) across, of heavy substance, 5-lobed; dome-shaped trusses of 12-15. Floriferous. Plant upright; 5in (12.7cm) leaves, held 3 years. (Cross of O. O. Dobbs; intro. by H. E. Greer; reg. 1979)    See color illus. HG p. 142.

cl.
```
        -moupinense (1/2)
Olive-
        -dauricum (1/2)
```

4ft(1.2m)          -15°F(-26°C)          VE          3/3

Flowers are orchid pink with darker spots scattered over the base of the upper lobe; blossoms held singly or in twos. (Stirling Maxwell) A. M. (RHS) 1942

cl.
```
            -oreotrephes (1/2)
Olive Judson-
        -Lady      -cinnabarinum ssp. cinnabarinum Roylei
        Chamberlain-                        Group (1/4)
        -          -cinnabarinum ssp. cinnabarinum
        -Royal-                              (1/8)
            Flush-maddenii ssp. maddenii (1/8)
```

Flowers rosy carmine with darker markings in the throat. (G. A. Judson, reg. 1973)

cl.
```
        -griffithianum (1/4)
    -Loderi-
Oliver-    -fortunei ssp. fortunei (1/4)
    -unknown (1/2)
```

Deep pink flowers. (Messell) A. M. (RHS) 1933

cl.
```
                    -catawbiense (1/4)
        -Charles Dickens-
Oliver Twist-           -unknown (1/2)
        -yakushimanum ssp. yakushimanum (1/2)
```

Semi-dwarf plant as wide as tall; leathery leaves matted with tan indumentum. Up to 20 flowers in globe-shaped trusses, clear fuchsine pink, ruffled, 2.5in(6.4cm) across. Midseason. At 5 years plant was 16in(40.6cm) tall and wide. (Benjamin Lancaster, reg. 1969)

---

```
                    -fortunei ssp. discolor (1/8)
g.          -Lady Bessborough-
    -Day Dream-          -campylocarpum ssp. campylo-
Olympia-        -                carpum Elatum Group (1/8)
    -           -griersonianum (1/4)
    -elliottii (1/2)
```

Shrub of average size and good growth habit; deep green foliage. Flowers brilliant red, fragrant, in large, well-shaped trusses, in midseason. A tender child of elliottii. (Rothschild, 1947)

cl.
```
            -dichroanthum ssp. dichroanthum (1/4)
        -Fabia-
Olympic Blondie-    -griersonianum (1/4)
        -unknown (1/2)
```

An additional 9 clones were named from this cross; listed below. (Roy W. Clark)

| | | |
|---|---|---|
| Olympic Brave | Olympic Chinook | Olympic Knight |
| Olympic Brownie | Olympic Hose-In-Hose | Olympc Maid |
| Olympic Chimes | Olympic Hunter | Olympic Miss |

The following are presumed extinct (1980):

| | | |
|---|---|---|
| Olympic Blondie | Olympic Brownie | Olympic Maid |
| Olympic Brave | Olympic Hunter | Olympic Miss |

No information currently available on these hybrids:

| | | |
|---|---|---|
| Olympic Chimes | Olympic Chinook | Olympic Hose-in-Hose |

cl.
```
            -dichroanthum ssp. dichroanthum (1/4)
        -Fabia-
Olympic Knight-    -griersonianum (1/4)
        -unnamed hybrid--unknown (1/2)
```

5ft(1.5m)          5°F(-15°C)          M          3/3

Dark red flowers with prominent blackish red blotch. Attractive foliage. (Roy W. Clark, 1958) P. A. (ARS) 1960

g.
```
                    -griffithianum (1/4)
        -Loderi King George-
Olympic Lady-       -fortunei ssp. fortunei (1/4)
        -williamsianum (1/2)
```

3ft(.9m)          -5°F(-21°C)          EM          3/4

Cup-shaped flowers, opening pale pink and fading white, 4in(10.2 cm) wide, in lax trusses of 4-5. Plant compact, broader than tall, and covered with flowers; smooth, flat, dark green leaves. (Endre Ostbo cross; R. W. Clark, reg. 1958)    A. M. (RHS) 1977

cl.
```
            -griersonianum (1/2)
Olympic Quinault-
        -unknown (1/2)
```

Cross by Layritz Nursery; introduced by Roy W. Clark.

cl.
```
                    -Moser's Maroon--unknown (1/4)
        -Romany Chai-
Olympic Sweetheart-    -griersonianum (1/4)
        -fortunei ssp. discolor (1/2)
```

Hybrid by Roy W. Clark.

cl.

Olympic Vicki Reine    See Vicki Reine

g.
```
     -catacosmum (1/2)
Omar-
     -beanianum (1/2)
```

Deep red flowers. (J. B. Stevenson cross, 1939; intro. 1951)

```
---
      -catawbiense ? (1/2)
Omega-
      -unknown rose-colored hybrid (1/2)
```

Light ruby red with red brown or yellow green markings, on a light background. (T. J. R. Seidel, 1912)

cl.

Omo--form of hyperythrum

3ft(.9m)          -15°F(-26°C)          EM          3/4

Loose, rounded trusses of 8-10 white flowers, 5-lobed, funnel-campanulate. Glossy, dark green leaves to 5.25in(13.3cm) long. (Shown by Collingwood Ingram) A. M. (RHS) 1976

```
                            -dichroanthum ssp. dichroanthum (1/8)
                  -Dido-
                  -       -decorum (1/8)        -griffithianum
          -Lem's-              -Beauty of-            (7/32)
cl.       -Cameo-   -Norman Gill- Tremough-arboreum ssp.
          -         -           -              arboreum (1/32)
One Thousand-  -Anna-      -griffithianum
  Butterflies-                    -griffithianum
          -          -Jean Marie de Montague-
          -                            -unknown (1/2)
          -              -Jan Dekens--unknown
          -Pink Petticoats-              -griffithianum
          -                   -Queen
          -Britannia- Wilhelmina-unknown
                      -
                      -Stanley Davies--unknown
```

5ft(1.5m)          0°F(-18°C)          ML

Flowers neyron rose, shading lighter within, cardinal red blotch and flare, butterfly-shaped. Plant upright, well-branched, almost as wide as tall; leaves dark yellowish green, held 3 years. (John G. Lofthouse cross, 1975; reg. 1981) See color illus. ARS J 36:2 (Spring 1983): p. 118.

```
                -fortunei ssp. discolor (1/4)
       -Golden-      -dichroanthum ssp. dichroanthum (1/4)
cl.    - Belle-Fabia-
       -               -griersonianum (1/8)
Ooh Gina-       -dichroanthum ssp. dichroanthum
       -    -Dido-
       -         -decorum (1/8)
       Lom's                   -griffithianum (5/32)
       Cameo-          -Beauty of-
          -    -Norman- Tremough-arboreum ssp. arboreum
          -    - Gill -                         (1/32)
          -Anna-      -griffithianum
          -              -griffithianum
             -Jean Marie-
             de Montague-unknown (1/16)
```

3ft(.9m)          5°F(-15°C)          ML

Flowers orient red shading to camellia rose with blotch of cardinal red, openly funnel-shaped, 3.5in(8.9cm) wide, 7-lobed; 15-flowered trusses. Slightly fragrant, floriferous. Plant rounded, broader than tall; spinach green leaves retained 3-4 years. (Raiser C. R. Burlingame; reg. 1983)

cl.
```
       -yakushimanum ssp. yakushimanum (1/2)
Ooh-La-La-        -Jan Dekens--unknown      -griffithianum
       -Pink     -         -Queen Wilhelmina-      (1/16)
       Petticoats-Britannia-            -unknown (7/16)
                      -Stanley Davies--unknown
```

1.5ft(.45m)          0°F(-18°C)          ML

Synonym Hot Pants. Flowers of 6 frilled lobes open neyron rose fading white, dorsal yellow flare, openly funnel-shaped, 3in(7.6 cm) wide; spherical trusses of 22-26. Plant as broad as tall, with decumbent branches. Dark yellow green leaves held 3 years; new growth with pale grey green tomentose. (John G. Lofthouse, cross, 1966; reg. 1981)

cl.

Opal Fawcett          Parentage unknown

5ft(1.5m)          0°F(-18°C)

Leaves are quite large. Flowers very pale pink fading to white, up to 3.5in(9cm) across, 5-lobed, slightly fragrant; trusses of 16. (Cross by Endre Ostbo; reg. by C. P. Fawcett, 1958) A. E. (ARS) 1958

cl.
```
       -davidsonianum (Exbury pink form) (1/2)
Open Dawn-              -ciliatum (1/4)
       -Countess of Haddington-
                          -dalhousiae var. dalhousiae (1/4)
```

Buds of rose opal flushed cream, opening to camellia rose flowers, fading white with pink flush, and dorsal flare of greenish yellow spots; fragrant. Oblong leaves with brown scales. (R. C. Gordon, reg. 1984)

cl.
```
      -yunnanense (1/2)
Openwood-
      -unknown (1/2)
```

Flowers of mauve lavender, speckled red. (Knap Hill, 1962)

cl.
```
    -javanicum (7/8)
Ophelia-       -javanicum
       -Princess Alexandra-       -jasminiflorum (1/8)
                    -Princess Royal-
                          -javanicum
```

Vireya hybrid. Flowers creamy buff, margins soft pinkish mauve. (Before 1855) A. M. (RHS) 1889

```
                            -fortunei ssp. discolor (3/8)
cl.          -Lady Bessborough-
        -Day Dream-        -campylocarpum ssp. campylocar-
Ophir-       -griersonianum        pum Elatum Group (1/8)
        -         -fortunei ssp. discolor
        -Margaret Dunn-   -dichroanthum ssp. dichroanthum (1/8)
                -Fabia-
                    -griersonianum (3/8)
```

Dream Girl g. Clone selected by Lester E. Brandt.

cl.
```
    -thomsonii ssp. thomsonii (1/2)
Oporto-
    -sanguineum ssp. sanguineum var. haemaleum (1/2)
```

Loose trusses hold 5-7 flowers, color between cardinal red and oxblood red. Corolla campanulate, 2in(5cm) across, 5-lobed, and waxy in appearance. (Collingwood Ingram) A. M. (RHS) 1967

cl.

Orange Bells          See Orange Marmalade

cl.

Orange Bill* --form of <u>cinnabarinum</u> ssp. <u>xanthocodon</u> Concatenans
                                              Group (K-W 5874)
A compact plant with beautiful apricot orange flowers and fine
foliage.  Named by Frank Kingdon-Ward who first saw it in the
Himalayas.

cl.


Orange Cross    See Fred Hamilton


                                    -<u>ciliatum</u> (1/16)
                         -Rosy Bell-
                -Lady -          -<u>glaucophyllum</u> var. <u>glaucophyllum</u>
                -Berry-                                        (1/16)
cl.  -unnamed-        -        -<u>cinnabarinum</u> ssp. <u>cinnabarinum</u>
     - hybrid-   -Royal-                                  (3/16)
Orange -        -   Flush-<u>maddenii</u> ssp. <u>maddenii</u> (3/16)
Delight-    -<u>cinnabarinum</u> ssp. <u>xanthocodon</u> (1/4)
     -       -<u>cinnabarinum</u> ssp. <u>cinnabarinum</u> Roylei Gp. (1/4)
     -Lady  -       -<u>cinnabarinum</u> ssp. <u>cinnabarinum</u>
       Rosebery-Royal Flush-
                         -<u>maddenii</u> ssp. <u>maddenii</u>


7ft(2.1m)        10°F(-12°C)        M


Funnel-shaped flowers in clusters of 6, nasturtium orange; corolla 2in(5cm) wide and long, 5-lobed.  Plant well-branched, upright, half as broad as tall; glossy, knobbed leaves, scaly beneath, held 3 years.  Foliage has spicy fragrance. (Dr. Carl G.
Heller, reg. 1980)


                              -<u>neriiflorum</u> ssp. <u>neriiflorum</u>
                    -Nereid-                         (1/16)
            -unnamed-      -<u>dichroanthum</u> ssp.
            - hybrid-            <u>dichroanthum</u> (1/16)
cl.    -Blondie-     -<u>fortunei</u> ssp. <u>discolor</u> (1/8)
       -      -        -<u>maximum</u> (1/8)
Orange Honey*    -Russell Harmon-
       -                    -<u>catawbiense</u> (1/8)
       -Catalgla--<u>catawbiense</u> var. <u>album</u> Glass (1/2)


5ft(1.5m)        -20°F(-29°C)        L


Flowers of salmon pink, like those of Mary Belle, in 18-flowered
trusses.  Plant habit rather open; attractive foliage. (Orlando
S. Pride, 1977)


cl.
           -<u>yakushimanum</u> ssp. <u>yakushimanum</u> (1/2)
Orange Marmalade-          -<u>wardii</u> var. <u>wardii</u> (1/4)
            -Mrs. Lammot-             -<u>souliei</u> ? (1/8)
            Copeland  -Virginia Scott-
                                   -unknown (1/8)


3.5ft(1m)        0°F(-18°C)        M


Synonym Orange Bells.  Turkey red buds open to rich amber yellow
flowers flushed salmon pink, in lax trusses of 10-12; corolla
campanulate, 2.5in(6.4cm) across, of good substance, 5-lobed.
Free-flowering.  Plant rounded, as broad as tall, well-branched;
elliptic leaves, 4.5in(11.5cm) long, held 2 years. (Cross by H.
L. Larson; raised by J. A. Davis; reg. 1983)


g.
         -<u>fortunei</u> ssp. <u>discolor</u> Houlstonii Group (1/2)
Orbhoulst-
         -<u>orbiculare</u> ssp. <u>orbiculare</u> (1/2)


Pink flowers. A. M. (RHS) 1932


---
         -<u>arboreum</u> ssp. <u>arboreum</u> (1/2)
Orbicarb-
         -<u>orbiculare</u> ssp. <u>orbiculare</u> (1/2)


Synonym Rotundarb.  A plant with brilliant red flowers, darker
at base. (E. J. P. Magor, 1936)


cl.
         -<u>griersonianum</u> (1/2)
Orchard-     -<u>arboreum</u> ssp. <u>zeylanicum</u> (1/4)
       -Ilam Alarm-        -<u>griffithianum</u> (1/8)
            -unnamed hybrid-
                          -unknown (1/8)


Flowers scarlet. (W. T. Stead, 1950)


cl.
         -azalea <u>occidentale</u> (1/2)
Oregon Queen-
         -<u>macrophyllum</u> (1/2)


5ft(1.5m)        -10°F(-23°C)        ML        2/3


An azaleodendron with light pink flowers, originating in those
areas of Oregon where the two species are common. A natural hybrid.  (Reg. 1958)


g.
         -Gladys Rillstone--unknown
Oregonia-
         -<u>griersonianum</u> (1/2)


Plant of medium size and good growth habit.  Flowers of deep,
rich pink, in shapely trusses.  Blooms late midseason. (Rothschild, 1947)


g.
         -<u>oreotrephes</u> (1/2)
Oreoaug-
         -<u>augustinii</u> ssp. <u>augustinii</u> (1/2)


Flowers of light lilac with greenish brown spots. (E. J. P. Magor, 1932)


g.
         -<u>oreotrephes</u> (1/2)
Oreocinn-
         -<u>cinnabarinum</u> ssp. <u>cinnabarinum</u> (1/2)


Violet rose flowers with two lines of brownish pink inside; another form has flowers of soft apricot.  Beautiful new foliage
of an unusual blue green, the color retained all summer. (E. J.
P. Magor, 1926)    See color illus. ARS J 39:3 (Summer 1985):
p. 150.


cl.
         -<u>oreotrephes</u> ? (1/2)
Oreoroyle*
         -<u>cinnabarinum</u> ssp. <u>cinnabarinum</u> Roylei Group ? (1/2)


Parentage shown is only what is suggested by the name.  A parent
of two Bodnant hybrids: Florescent, Rover.


g.
              -<u>thomsonii</u> ssp. <u>thomsonii</u> (1/4)
       -Shilsonii-
Orestes-      -<u>barbatum</u> (1/4)
       -<u>griersonianum</u> (1/2)


A tall plant with large, dark scarlet flowers. (Lord Aberconway, intro. 1941)


cl.            -<u>campylocarpum</u> ssp. <u>campylocarpum</u> (1/4)
         -A. Gilbert-
Organdie-        -<u>fortunei</u> ssp. <u>discolor</u> (1/4)
         -<u>dichroanthum</u> ssp. <u>scyphocalyx</u> Herpesticum Group (1/2)


Icarus g.  Flowers 5-lobed, lemon yellow, with pink margins and
pink stain at base. Plant compact, and a little less than medium-sized. (Rothschild)  A. M. (RHS) 1947

```
                          -neriiflorum ssp. neriiflorum (1/4)
g.    -F. C. Puddle-
      -              -griersonianum (1/4)
Oriana-              -griffithianum (1/8)
      -    -Loderi-
      -Coreta-       -fortunei ssp. fortunei (1/8)
              -arboretum ssp. zeylanicum (1/4)
```

Red flowers.  (Lord Aberconway, 1946)

```
cl.
              -Duchess of Cornwall--unknown (1/2)
Orient Express-
              -calophytum var. calophytum (1/2)
```

Trusses are rounded and firm, holding up to 20 flowers, white
flushed ruby red with a blotch of deeper ruby red in the throat.
Corolla widely funnel-canpanulate, 2.5in(6.4cm) across, 5-lobed.
Long, dark green leaves, with a light coat of brown indumentum
beneath.  (R. N. S. Clarke, reg. 1979)  A. M. (RHS) 1979

```
cl.
              -aurigeranum (1/2)
Oriental Orange-
              -unknown (1/2)
```

Vireya hybrid.  Trusses of 15 flowers of tangerine orange.  (R.
Lelliott cross; reg. by R. Cutten, 1982)

```
cl.
         -cinnabarinum ssp. cinnabarinum Roylei Group (1/2)
Oriflamme-             -cinnabarinum ssp. cinnabarinum (1/4)
         -Royal Flush -
          (orange form)-maddenii ssp. maddenii (1/4)
```

5ft(1.5m)           10°F(-12°C)          ML          4/3

Lady Chamberlain g.  Fine orange red flowers.  (Rothschild,
1930)

```
cl.
```

Original--form of caucasicum var. album

Pinkish white flowers with variegated foliage.  (Standish &
Noble, 1850)

```
cl. -griersonianum (1/2)
    -                            -griffithianum (1/8)
Orion-        -Mrs. E. C. Stirling-
    -         -                   -unknown (9/32)
    -Madame P. A.-                        -ponticum
        Colijn   -              -Michael Waterer-  (1/32)
              -        -Prometheus-       -unknown
              -Madame  -        -Monitor--unknown
               de Bruin-        -arboreum ssp. arboreum
                    -Doncaster-              (1/16)
                         -unknown
```

Introduced by Reuthe Nurseries.

```
---
      -fortunei ssp. discolor (1/2)
Orion-            -fortunei ssp. fortunei (1/4)
    -H. M. Arderne-
                 -unknown (1/4)
```

Flowers rose pink with red eye.  (Waterer, Sons & Crisp, 1920)

```
                        -neriiflorum ssp. neriiflorum (1/8)
g.        -F. C. Puddle-
    -Phoebus-           -griersonianum (1/8)
Orion-      -haematodes ssp. haematodes (1/4)
    -       -griffithianum (1/4)
    -Loderi-
         -fortunei ssp. fortunei (1/4)
```

Rose pink flowers with a dark eye.  (Lord Aberconway, 1950)

```
---
         -azalea viscosum (1/2)
Ornatum-
         -ponticum (1/2)
```

Azaleodendron.  Dark scarlet flowers; another form, sulphur yellow with orange spots.  (Gowan, 1832)

```
                        -neriiflorum ssp. neriiflorum (1/8)
g.        -Nereid-
    -Euryalus-      -dichroanthum ssp. dichroanthum (1/8)
Ortega-     -griersonianum (1/2)
    -       -griffithianum (1/4)
    -Sunrise-
         -griersonianum
```

Pale rose flowers.  (Lord Aberconway, 1941)

```
cl.       -dichroanthum ssp. dichroanthum (1/4)
        -Fabia-
Ostbo's Copper-  -griersonianum (1/4)
        -                -campylocarpum ssp.
        -Mrs. W. C. Slocock-      campylocarpum (1/4)
                 -unknown (1/4)
```

Copper red flowers.  (Endre Ostbo, reg. 1958)

```
                                -ponticum (1/16)
                        -Michael Waterer-
                -Prometheus-       -unknown (5/16)
cl.     -Madame de-      -Monitor--unknown
      - Bruin   -        -arboreum ssp. arboreum (1/8)
Ostfriesland-   -Doncaster-
                        -unknown
      -forrestii ssp. forrestii Repens Group (1/2)
```

Scarlet red flowers.  (Dietrich Hobbie, 1949)

```
         -sanguineum ssp. didymum (1/4)
g.   -Carmen-
     -     -forrestii ssp. forrestii Repens Group (1/4)
Othello-              -griffithianum (1/16)
     -           -unnamed hybrid-
     -    -Armistice Day-       -unknown (3/16)
     -unnamed-      -Maxwell T. Masters--unknown
        hybrid-griersonianum (1/4)
```

Hybrid introduced in 1954 by Lester E. Brandt.

```
cl.          -sanguineum ssp. didymum (1/4)
      -Arthur Osborn-
Othello-      -griersonianum (1/4)
      -facetum (eriogynum) (1/2)
```

The Moor g.  Flowers cardinal red.  See also Tremeer.  (Lord
Aberconway, intro. 1941)

```
cl.
      -fortunei ssp. discolor (1/2)
Ottawa-
      -campylocarpum ssp. campylocarpum Elatum Group (1/2)
```

6ft(1.8m)        -5°F(-21°C)          ML          3/3

Lady Bessborough g.  Similiar to others of this family.  Flowers
orange pink, fading white.  (Rothschild, 1933)

```
cl.          -campylocarpum ssp. campylocarpum (1/4)
        -Butterfly-       -catawbiense (1/8)
Otto Homdahl-   -Mrs. Milner-
        -ponticum (1/2)   -unknown (1/8)
```

4ft(1.2m)        -5°F(-21°C)          ML

Flowers openly funnel-shaped, 3in(7.6cm) across, of deep creamy
yellow with crimson flecks on upper lobes; rounded trusses of 15
-18.  Plant habit compact; leaves 4in(10.2cm) long, with dark
maroon petioles.  (I. Owen Ostbo, 1969)

cl.
     -augustinii ssp. augustinii (1/2)
Oudijk's Favorite-
     -unknown (1/2)

Flowers of campanula violet, 1.75in(4.5cm) across.  (Fa. le Feber & Co., Booskoop)  Flora Nova, Silver Medal (Boskoop) 1958


cl.
     -Essex Scarlet--unknown (1/2)
Oudijk's Sensation-
     -williamsianum (1/2)

Flowers of bright Tyrian rose, darker pink margins and spotting
on upper lobes, widely funnel-campanulate, to 3in(7.6cm) long;
loose trusses of 5-7.  Small, dense, rounded shrub; leaves cordate, sharply pointed.  Midseason.  (Cross by Dietrich Hobbie;
intro. by Fa. Le Feber & Co., Booskoop; reg. 1965)  Gold Medal,
Flora Nova (Boskoop) 1958 & 1961; Gold Medal, Floriade (Rotterdam) 1960


g. & cl.  ·  -dichroanthum ssp. dichroanthum (1/4)
  -Astarte-   -campylocarpum ssp. campylocarpum
Ouida-  -Penjerrick-    Elatum Group (1/8)
      -griffithianum (1/8)
  -griersonianum (1/2)

Flowers flushed pink rose, darker at base.  (Cross by Lord Aberconway, 1930)  A. M. (RHS) 1936


g. & cl.
   -calophytum var. calophytum (1/2)
Our Kate-
   -macabeanum (1/2)

First flowered 15 years after sowing seed.  Beautiful flowers,
openly campanulate, very light pink flushed deeper pink on margins, with sparkling ruby throat; large, impressive lax trusses
of 20.  Shrub is a still-growing tree of 15ft(4.5m) and should
become much taller; leaves 11in(28cm) long by 4in(10cm) wide,
loosely indumented beneath.  Named for Lionel de Rothschild's
eldest daughter.  (Rothschild)  A. M. (RHS) 1963
[Descriptions from Rothschild Rhododendrons.]


cl.      -xanthostephanum (1/4)
   -Saffron Queen-
Owen Pearce-   -burmanicum (3/4)
    -burmanicum

4ft(1.2m)   20°F(-7°C)   EM

Buds sulphur yellow, opening lighter yellow; flowers rotate, 2in
(5cm) in diameter, held in open trusses.  Plant habit compact;
leaves 3in(7.6cm) long. (Mr. & Mrs. Maurice Sumner, intro. 1968;
reg. 1972)


cl.    -catawbiense (1/4)
 -Catawbiense Album-
Oz-    -unknown (1/4)
-  -dichroanthum ssp. dichroanthum (1/4)
 -Fabia-
   -griersonianum (1/4)

3ft(.9m)   5°F(-15°C)   M

Flowers carmine rose at edges shading through lemon yellow to
Naples yellow in throat and lobe centers, spotted orange on dorsal lobe; corolla openly funnel-shaped, 2.5in(6.4cm) across, 5-
lobed; dome-shaped trusses of 11-15. Floriferous. Plant as
broad as tall, well-branched; glossy dark leaves held 3-4 years.
(Cross by Halfdan Lem; raised by Grady Barefield; reg. by Mary
W. Barefield, 1980)


g.
   -minus var. minus Carolinianum Group (1/2)
P. J. M.*
   -dauricum (1/2)

4ft(1.2m)   -25°F(-32°C)   E   4/4

Cold-hardy and heat-tolerant.  Small rounded leaves are green in
summer and mahogany-colored in winter.  Flowers early, a bright
lavender pink.  Several selected forms available. (P. J.
Mezitt, cross 1939; intro. 1959)  See color illus. HG p. 100;
VV p. 115.


cl.
   -minus var. minus Carolinianum Group (1/2)
P. J. Mezitt*
   -dauricum (1/2)

A clone selected from the P.J.M. grex.  Flowers of medium rosy
purple, slight throat markings deeper color, openly campanulate,
1.75in(4.5cm) wide, 5-lobed; small trusses of 4-9.  Leaves elliptic, to 2.5in(6.6cm) long, lightly scaly beneath, aromatic.
(Raised by  Weston Nurseries, USA; exhibited by  Crown Estate,
Windsor)  P. C. (RHS) 1967;  A. M. (RHS) 1972


cl.     -griffithianum (1/4)
   -Loderi Venus-
Pacific Glow-   -fortunei ssp. fortunei (1/4)
   -strigillosum (1/2)

5ft(1.5m)   10°F(-12°C)   E   4/4

Flowers of clear unmarked pink with darker throat, openly funnel-shaped, 3.25in(8.3cm) wide, 6-lobed; ball-shaped trusses of
15-17.  Floriferous.  Plant compact, rounded, as broad as tall;
narrowly elliptic leaves to 5.5in(14cm) long, dull green, with
moderate brown indumentum.  (H. L. Larson cross; reg. by Evelyn
Jack, 1979)  See color illus. ARS Q 34:1 (Winter 1980): p. 35.
(Plate; painting by Mary Comber Miles)


cl.    -wardii var. wardii (1/4)
  -Mrs. Lammot-  -souliei (?) (1/8)
Pacific Gold- Copeland  -Virginia Scott-
-     -unknown (5/8)
   -yellow hybrid--unknown

3.5ft(1m)   0°F(-18°C)   M

Bright orange red buds open to funnel-shaped flowers, aureolin
yellow with pale yellow margins, 3in(7.6cm) across, with 5 wavy
lobes; dome-shaped trusses of 8-12.  Plant nearly as broad as
tall; elliptic, glossy leaves held 2 years; new growth mahogany
green. (Cross by H. L. Larson; raised by J. A. Davis; reg. 1984)


      -fortunei ssp. discolor (1/4)
cl. -King of Shrubs- -dichroanthum ssp. dichroanthum
-    -Fabia-    (5/16)
Pacific -    -griersonianum (3/16)
Princess-   -fortunei ssp. fortunei (1/8)
-  -Fawn- -dichroanthum ssp. dichroanthum
-unnamed-  -Fabia-
 hybrid-  -griersonianum
-   -dichroanthum ssp. dichroanthum
  -Dido-
   -decorum (1/8)

Orient pink flowers with a center of orange buff.  Leaves to
4.75in(12cm) long. (Hybridizer in the USA, unknown; reg. by Mrs.
R. J. Coker, New Zealand, 1982)


cl.     -griffithianum (1/4)
  -Loderi King George-
Pacific Rim-   -fortunei ssp. fortunei (1/4)
  -macabeanum (1/2)

Orchid pink buds opening to very light salmon flowers, 8-lobed,
with conspicuous brown anthers.  Leaves 6.8in(17.5cm) long.  (H.
L. Larson, reg. 1982)

cl.
       -wardii var. wardii (1/2)
Page Boy-               -campylocarpum ssp. campylocarpum
     -Mrs. W. C. Slocock-                (1/4)
                 -unknown (1/4)

A round and compact shrub with trusses holding up to 13 flowers
colored cream, touched with pink. (Gen. Harrison, reg. 1965)

g.            -calophytum var. calophytum (1/4)
     -Calrose-
Pageant-         -griersonianum (1/4)
  -         -griffithianum (1/4)
     -Loderi-
         -fortunei ssp. fortunei (1/4)

A hybrid with pale pink flowers. (Lord Aberconway, intro. 1950)

cl.

Painted Snipe*--form of orbiculatum

Vireya. A plant for the cool greenhouse. White flowers with
long, slender, tubular corolla; fragrant. Firm trusses hold up
to 24 flowers. Leaves 3.25in(8.3cm) long, 2in(5cm) wide, with a
dark red midrib. (Collected and raised by Mr. & Mrs. E. F. Allen) A. M. (RHS) 1970

                     -catawbiense (1/4)
             -Mrs. C. S. Sargent-
cl.     -Meadowbrook-        -unknown (1/4)
  -      -         -catawbiense
Painted Star-        -Everestianum-
  -               -unknown
  -     -campylocarpum ssp. campylocarpum (1/4)
    -Anita-
      -griersonianum (1/4)

5ft(1.5m)       -5°F(-21°C)      ML

Cream-colored flowers with a reddish throat. (Donald Hardgrove,
reg. 1958)

cl.
       -catawbiense (1/2)
Pale Perfection-
       -unknown (1/2)

5ft(1.5m)       -15°F(-26°C)      ML      2/4

Narrow, medium-sized leaves on a plant twice as broad as tall.
Flowers rose-colored with distinctive dark blotch, to 2.5in(6.4
cm) across; trusses of about 13. (Richard Wyman cross; R. Ticknor, intro. & reg. 1967)

cl.                -griffithianum (1/4)
    -Loderi King George-
Pall Mall-           -fortunei ssp. fortunei (1/4)
  -       -dichroanthum ssp. scyphocalyx (1/4)
    -Socrianum-
        -griffithianum (1/4)

Trusses of 10-12 flowers, deep pink with darker edges of rose
red. (Cross by Mrs. Roza Stevenson; intro. by Hydon Nurseries,
1976)

g.        -williamsianum (1/4)
    -Pallida-
Palla-      -griffithianum (1/4)
    -griersonianum (1/2)

Flowers of neyron rose. (Lord Aberconway, intro. 1942)

cl.
      -davidsonianum (1/2)
Pallescens-
      -racemosum (1/2)

A natural hybrid. Small shrub; scaly leaves; terminal inflorescence of pale pink, margins white or carmine. A.M. (RHS) 1933

g.
      -williamsianum (1/2)
Pallida-
      -griffithianum (1/2)

A hybrid with white flowers flushed pink. (Lord Aberconway,
1933)

cl.

Palma--form of parmulatum (K-W 5875)

3ft(.9m)       5°F(-15°C)      EM      3/3

Trusses of 3-7 flowers, tubular-campanulate, 5-lobed, very light
green white with deeper central bands and heavy spotting.
Leaves oblong-oval, glabrous. (Collector Kingdon-Ward; raiser
Lord Aberconway) A. M. (RHS) 1983

cl.           -fortunei ssp. fortunei (1/4)
    -Ruby F. Bowman-      -griffithianum (1/8)
Paloma-        -Lady Bligh-
  -            -unknown (1/8)
  -floccigerum (1/2)

6ft(1.8m)       0°F(-18°C)      M

Dome-shaped trusses, 6in(15cm) across, of 14-16 flowers, Venetian pink with margins dawn pink, dorsal spotting guardsman red.
Flowers openly campanulate, 2.5in(6.4cm) across, 5-lobed, of
good substance. Plant over half as wide as tall; leaves 5.5in
(14cm) long, held 3 years. (Dr. David W. Goheen cross, 1966;
reg. 1985; intro. by Dr. Lansing Bulgin, 1983)

---

Pamela     Parentage unknown

A parent of Moonlight Sonata. Flowers deep blush. (Standish &
Noble, before 1860)

cl.
      -griffithianum (1/2)
Pamela Fielding-
      -red hybrid--unknown (1/2)

Flowers scarlet to white blush with exterior carmine pink;
margins pale cerise. (C. B. van Nes & Sons, before 1922)

             -thomsonii ssp. thomsonii (1/4)
cl.    -unnamed hybrid-
  -        -unknown (1/2)
Pamela Love-       -griffithianum (1/8)
  -       -Loderi-
  -unnamed hybrid-   -fortunei ssp. fortunei (1/8)
         -unknown

6ft(1.8m)       0°F(-18°C)      E

Flat-topped trusses 6in(15cm) across with funnel-shaped flowers,
colored rose Bengal. (Melvin V. Love, 1965)

                 -catawbiense (3/16)
          -Parsons
       -America-Grandiflorum-unknown (3/8)
     -       -
cl.  -Fanfare-     -dark red hybrid--unknown
  -       -      -catawbiense
Panama-    -Kettledrum-
  -         -unknown
  -    -Catalgla--catawbiense var. album Glass (1/4)
  -unnamed-        -dichroanthum ssp. dichroanthum
    hybrid-     -unnamed-            (1/16)
      -unnamed- hybrid-griersonianum (1/16)
      hybrid-   -fortunei ssp. discolor (1/16)
      -unnamed-
      hybrid-Corona--unknown

5ft(1.5m)        -15°F(-26°C)        ML        3/2

Cardinal red flowers with 5 wavy lobes, spotted darker, held in
dome-shaped trusses of 10-13. Plant twice as broad as tall with
leaves retained 2 years, dark yellow green. (David G. Leach
cross, 1962; intro. 1976; reg. 1983)

```
          -dichroanthum ssp. dichroanthum (1/4)
     -Fabia-
Pandora-     -griersonianum (1/4)
     -facetum (eriogynum) (1/2)
```

Flowers are light rose, veined in white with chocolate spotting.
(Lord Aberconway, 1941)

```
g. & cl.
     -G. A. Sims--unknown (1/2)
Panoply-
     -facetum (eriogynum) (1/2)
```

Flowers rose claret with darker spots on upper petals.  See also
Red Chief. (Col. S. R. Clarke) A. M. (RHS) 1942

cl.

Papillon--form of augustinii ssp. rubrum (bergii) (F 25914)

Flowers strongly flushed mallow purple, paler towards rim, upper
corolla spotted medium purplish red, widely funnel-shaped, 2in
(5cm) across, 5-lobed; clusters of 5-7.  Leaves 2.5in(6.5cm)
long, both surfaces lightly scaly.  (Collector George Forrest;
raiser Col. S. R. Clarke; shown by R. N. S. Clarke) A. M. (RHS)
1978

```
          -Goldsworth-dichroanthum ssp. dichroanthum
          - Orange -                          (1/8)
cl.   -Hotei-      -fortunei ssp. discolor (1/8)
      -            -souliei (1/8)
Paprika Spiced*  -unnamed hybrid-
      -            -wardii var. wardii (1/8)
      -Tropicana (Brandt)--unknown ? (1/2)
```

3ft(.9m)        0°F(-18°C)        M

Tropicana may be  Britannia X Goldsworth Orange. The flowers are
a delicious mixture of orange, peach, gold, and yellow, liberal-
ly peppered with paprika red. The calyx also is speckled. (Dr.
E. C. Brockenbrough)

```
                    -griffithianum (1/8)
cl.      -Queen Wilhelmina-
     -Britannia-          -unknown (3/8)
Paracutin-   -Stanley Davies--unknown
     -       -griersonianum (1/4)
     -Tally Ho-
          -facetum (eriogynum) (1/4)
```

4ft(1.2m)        -5°F(-21°C)        ML

Bright cardinal red flowers, 3.5in(8.9cm) wide; tall trusses of
about 23. Named "Best new hybrid, 1963, Seattle, WA." Named
after the recently created Mexican volcano. (Lester Brandt, reg.
1968)

```
               -arboreum ssp. arboreum (5/32)
          -Cardinal-      -thomsonii ssp. thomsonii
g.   -Red Ensign-   -Barclayi-            (1/16)
     -       -griersonianum -    -arboreum ssp.
Paragon-          (1/4)   -Glory of -    arboreum
     -                Penjerrick-
     -    -griffithianum          -griffithianum
     -Loderi-                          (9/32)
          -fortunei ssp.
               fortunei (1/4)
```

Rose-colored flowers. (Lord Aberconway, intro. 1950)

```
          -griffithianum (1/8)
cl.
     -Aurora-     -fortunei ssp. fortunei (5/8)
Paris-   -thomsonii ssp. thomsonii (1/4)
     -fortunei ssp. fortunei
```

6ft(1.8m)        -5°F(-21°C)        M        3/3

Naomi g. Named later than other Naomi clones. Of good habit,
nearly as broad as tall, and it grows to at least 20ft(6m). The
dark green foliage covers the plant from the ground up. Flowers
are light purple rose, 3in(7.6cm) across, in loose trusses of 10
or 12. As a garden plant, Mr. Barber says it is "...perhaps
the finest in this distinguished company". (Rothschild)  P. C.
(RHS) 1968

```
cl.
     -valentinianum (1/2)
Parisienne-
     -burmanicum (1/2)
```

3ft(.9m)        15°F(-10°C)        EM        2/2

Plant of dense habit, with a maximum height of 4ft(1.2m).  Many
small loose trusses of slightly fragrant yellow flowers, blooming in April at Exbury. (Rothschild, 1947)

cl.

Parker's Pink        Parentage unknown

5ft(1.5m)        -25°F(-32°C)        ML        3/2

Synonym Parker #1 PP. Flowers of bright deep pink heavily spotted in dark red, fading to white in throat, 3.5in(8.9cm) across,
fragrant; 12-flowered trusses. Plant as broad as tall, branching well; leaves held 2 years. (C. O. Dexter, John Parker; Paul
Vossberg, intro. 1959) A. E. (ARS) 1973    See color illus. ARS
Q 25:3 (Summer 1971): cover; LW pl. 3.

cl.

Parkside--form of charitopes ssp. charitopes

1ft(.3m)        -5°F(-21°C)        M        3/3

Flowers orchid pink, suffused and heavily spotted with darker
red purples, campanulate, 1.5in(3.8cm) across, 5-lobed, in clusters of 3-5.  Leaves 2.25in(5.7cm) long, dark green above, scaly
beneath.  (Collector unknown; raised by Crown Estate, Windsor)
A. M. (RHS) 1979

```
cl.
     -catawbiense (1/2)
Parsons Gloriosum-
     -unknown (1/2)
```

5ft(1.5m)        -25°F(-32°C)        ML        2/2

Old and hardy. Plant upright, compact; deep green leaves about
6in(15.2cm) long. Flowers orchid lavender, shaded pink, held in
compact conical trusses. (Intro. about 1850 by Samuel Parsons,
Long Island, NY; raiser A. Waterer)    See color illus. VV p. 74.

```
cl.
     -catawbiense (1/2)
Parsons Grandiflorum-
     -unknown (1/2)
```

5ft(1.5m)        -25°F(-32°C)        ML        2/3

Habit and foliage rated fair; the purplish rose flowers are more
attractive than the above indicates, and long-lasting. Very old
and hardy. (Intro. by Samuel Parsons, before 1875; raised by A.
Waterer)

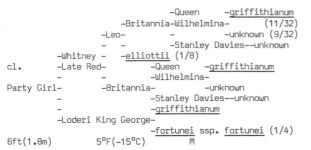

```
                        -Queen    -griffithianum
            -Britannia-Wilhelmina-      (11/32)
        -Leo-            -          -unknown (9/32)
        -    -          -Stanley Davies--unknown
    -Whitney -    -elliottii (1/8)
cl.     -Late Red-        -Queen    -griffithianum
    -         -        -Wilhelmina-
Party Girl-    -Britannia-        -unknown
    -                 -Stanley Davies--unknown
    -                 -griffithianum
    -Loderi King George-
                 -fortunei ssp. fortunei (1/4)
6ft(1.8m)        5°F(-15°C)           M
```

Flowers light spirea red with neyron rose margins, greyed orange
in throat, fragrant, 4in(10.2cm) wide, 6-lobed; 10-12 per truss.
Floriferous. Plant rounded, broader than tall; glossy, narrow
leaves 6.5in(16.5cm) long, held 2 years. (Fred Peste, reg. 1983)

```
                -dichroanthum ssp. dichroanthum (1/8)
            -Dido-
            -    -decorum (1/8)    -arboreum ssp.
    -Lem's Cameo-        -Beauty of- arboreum (1/32)
cl.    -          -Norman- Tremough-
    -          -    - Gill -    -griffithianum (9/32)
Party  -        -Anna-    -griffithianum
Package-        -        -griffithianum
    -        -Jean Marie -
    -         de Montague-unknown (3/16)
    -        -fortunei ssp. fortunei (1/4)
    -Ruby F. Bowman-        -griffithianum
                -Lady Bligh-
                      -unknown
5ft(1.5m)        0°F(-18°C)           M
```

Crimson buds opening carmine rose, fading lighter, with star-
shaped stain deep in throat, widely funnel-shaped, 4in(10.2cm)
across, with 7 wavy and frilled lobes; conical trusses of about
15. Plant upright, nearly as broad as high,; leaves held 1.5
years; new growth bronze with dark red petioles. (Cross 1976 by
John G. Lofthouse; reg. 1984)

```
                    -griffithianum (1/8)
            -unnamed hybrid-
cl.    -Mrs. Furnival-    -unknown (1/4)
    -        -        -caucasicum (1/8)
Party Pink-        -unnamed hybrid-
    -                -unknown
    -catawbiense var. album (1/2)
5ft(1.5m)        -20°F(-29°C)      M        5/5
```

Flowers 3in(7.6cm) wide, openly funnel-shaped, of purplish pink
shading lighter in the center, with conspicuous dorsal spotting
of strong yellow. Ball-shaped truss 6in(15cm) across, 18-flow-
ered. Plant broad, branching well, with leaves widely elliptic,
held 3 years. See also Cyprus, Persia. (David G. Leach, 1973)
C. A. (ARS) 1981; A. E. (ARS) 1982; S. P. A. (ARS) 1983   See
color illus. ARS J 38:2 (Spring 1984): cover.

cl.

Pastel    See Vanessa Pastel

```
                    -decorum (1/8)
            -unnamed-
    -unnamed- hybrid-fortunei ssp. discolor (1/8)
cl.    - hybrid-    -fortunei ssp. fortunei (1/8)
    -    -unnamed-    -wardii var. wardii (1/16)
Pastor Dunker-    hybrid-unnamed-
    -        - hybrid-dichroanthum ssp.
    -                dichroanthum (1/16)
    -Jean Marie -griffithianum (1/4)
     de Montague-unnamed-
            -unknown (1/4)
3ft(.9m)        0°F(-18°C)           M
```

Flowers of heavy substance, 3in(7.6cm) across, fragrant, pink,
throat faint yellow, up to 15 per truss. Plant semi-dwarf,
broader than high with leaves 5in(12.7cm) by 2in(5cm), retained
3 years. (Alfred A. Raustein, 1979)

cl.

Patricia--form of campylogynum Charopaeum Group

1ft(.3m)        -10°F(-23°C)        EM        3/3

Small, rounded, dark green leaves, paler beneath. Flowers magnolia purple, to 1.25in(3.2cm) across. (J. F. Caperci, 1962)

```
cl.            -thomsonii ssp. thomsonii (1/4)
    -General Sir -
Patricia-John du Cane-fortunei ssp. discolor (1/4)
Harewood-        -griffithianum (1/4)
    -Loderi King George-
                -fortunei ssp. fortunei (1/4)
```

Heather pink flowers suffused darker externally, held in trusses
of 10-11. (Cross by Geoffrey Hall; Lord Harewood, intro. 1979)

```
cl.
    -macabeanum (1/2)
Patricia Lee-        -griffithianum (1/4)
    -Loderi King George-
                -fortunei ssp. discolor (1/4)
2ft(.6m)        -5°F(-21°C)           EM
```

Plant twice as broad as high; moss green leaves to 4.75in(12cm)
long, held 1 year. Fragrant flowers, yellow with green tinge,
widely funnel-campanulate, 3.5in(8.9cm) across, 7-lobed, in
trusses of 10. (Carl H. Phetteplace cross; raised & reg. by R.
C. Sparks)

```
cl.
    -yakushimanum ssp. yakushimanum (1/2)
Patricia's Day-
    -dwarf yellow-flowered plant from Windsor--unknown
                            (1/2)
3ft(.9cm)        0°F(-18°C)           M
```

Flowers of soft lavender, deeper veining of pinkish lavender.
(John Waterer, Sons & Crisp, 1975)

```
cl.
    -Yaku Fairy--keiskei (1/2)
Patty Bee-
    -fletcherianum (1/2)
1.5ft(.45m)        15°F(-10°C)        EM        5/4
```

Flowers clear yellow, with 5 wavy lobes, openly funnel-shaped,
to 2in(5cm) across, in lax trusses of 6. Plant wider than tall,
well-branched; small dark green leaves held 2 years. Sun- and
heat-tolerant. (Warren E. Berg cross, 1970; reg. 1977)   C. A.
(ARS) 1983; A. E. (ARS) 1984   See color illus. ARS J 38:2
(Spring 1984): p. 103.

```
cl.        -haematodes ssp. haematodes (3/4)
        -May Day-
Paul Detlefsen-    -griersonianum (1/4)
        -haematodes var. haematodes
```

Compact plant, 2ft(.6m) high in 10 years. Leaves dark, glossy,
indumented. Flowers orient red, 2.5in(6.4cm) across, in loose
trusses. Midseason.  See also Harry von Tilzer, Steven Foster.
(C. S. Seabrook, 1967)

```
cl.        -Margaret--unknown (1/4)
        -Glamour-
Paul Lincke-    -griersonianum (1/4)
    -strigillosum (1/2)
1.5ft(.45m)        5°F(-15°C)        E        4/3
```

Plant height also given as 4ft(1.2m). Leaves like <u>strigillosum</u> but wider. Campanulate flowers, currant red, 2.5in(6.4cm) across, in tight, upstanding trusses. (C. S. Seabrook, reg. 1965)  See color illus. HG p. 110.

cl.
               -<u>maximum</u> (1/2)
Paul Vossberg-
               -<u>thomsonii</u> ssp. <u>thomsonii</u> (1/2)

4ft(1.2m)          -10°F(-23°C)          L

Openly funnel-shaped flowers, cardinal red with dark garnet blotch, 2.5in(6.4cm) wide with 5 wavy lobes; spherical trusses of 13-15. Floriferous. Plant rounded, as broad as tall, well-branched; spinach green leaves 5.5in(14cm) long, held 2 years. (Mrs. J. F. Knippenberg, intro. 1975; reg. 1984)

```
                                      -fortunei ssp. discolor (1/8)
g.                        -Norman Shaw-           -catawbiense (1/16)
       -Bellerophon-            -B. de Bruin-
Paulette-              -                       -unknown (1/16)
       -              -facetum (eriogynum) (1/4)
       -kyawii (1/2)
```

Shrub of average size and good growth habit; deep green foliage. Bright crimson flowers in well-filled trusses, blooming late. A rather tender plant. (Rothschild, 1950)


g.


Pauline      Parentage unknown

Funnel-shaped crimson flowers with a striking dark eye. Plant of medium height. (Raised by T. Lowinsky; intro. by Lionel de Rothschild)  A. M. (RHS) 1933; A. M. (Wisley Trials) 1957

cl.
            -<u>fortunei</u> ssp. <u>fortunei</u> (1/2)
Pauline Bralit-
            -<u>catawbiense</u> (low white form) (1/2)

3ft(.9m)          -10°F(23°C)          M

Rhodonite red buds  opening to fragrant white flowers, shading light pink at 2 dorsal lobe edges, with 3 greyed orange rays in throat. Flowers of good substance, openly funnel-shaped, 2.5in (6.4cm) wide, in ball-shaped trusses of 9-12. Floriferous. Upright, well-branched; leaves of heavy texture, held 2 years. (Edmund Mezitt, cross 1958; intro. 1965; reg. 1963)

```
                              -fortunei ssp.
                       -Koster's-   fortunei (3/16)
            -unnamed- Choice -unknown (1/32)
            - hybrid-      -griffithianum
        -Palmer-      -Loderi-
      -unnamed-      -          -fortunei ssp. fortunei
cl.   - hybrid-      -wardii var. wardii (croceum) (1/8)
      -                        -griffithianum (5/32)
Pawhuska*    -Loderi King George-
      -                    -fortunei ssp. fortunei
      -              -catawbiense (1/4)
      -Madame Masson-
                    -ponticum (1/4)
```

5ft(1.5m)          -5°F(-21°C)          ML

Beautiful white trusses on a plant with large glossy foliage. See also Sugar and Spice. (Tom & Emma Bowhan)

cl.
    -<u>rigidum</u> (<u>caeruleum</u>) (white form) (1/2)
Peace-
    -<u>cinnabarinum</u> ssp. <u>xanthocodon</u> Concatenans Group (1/2)

Lemon Bill g.  Flowers broadly funnel-shaped, white, faintly flushed  pale rose, in trusses of 6. (Lord Aberconway)  A. M. (RHS) 1946

cl.
         -Scintillation--unknown (1/2)
Peach Brandy-
           -<u>haematodes</u> ssp. <u>haematodes</u> (1/2)

4ft(1.2m)          -5°F(-21°C)          M

Flowers strong pink with slight dark red spotting, openly funnel-shaped, 3in(7.6cm) across, with 5-6 wavy lobes; ball-shaped trusses of 13. Floriferous. Plant rounded, as broad as tall; elliptic leaves held 2 years. Plant size shown above was "reduced by excessive taking of cuttings". (Cross at Scott Horticultural Foundation; described by Gertrude Wister; reg. by Tyler Arboretum, 1982)

cl.
        -<u>neriiflorum</u> ssp. <u>neriiflorum</u> (1/4)
    -Nereid-
Peach Lady-    -<u>dichroanthum</u> ssp. <u>dichroanthum</u> (1/4)
    -<u>fortunei</u> ssp. <u>discolor</u> (1/2)

4ft(1.2m)          -5°F(-21°C)          ML          3/3

Upright, rounded plant; leaves medium green. Flowers peach-colored, edged camellia rose, with yellow eye, to 4in(10.2cm) wide; rounded trusses of 12. Heat- and sun-tolerant. (Seed from Rose of England; raised & reg. by Benjamin F. Lancaster, 1958)

cl.
    -Apricot No. 3--unknown (1/2)
Peach Loderi-          -<u>griffithianum</u> (1/4)
    -Loderi King George-
            -<u>fortunei</u> ssp. <u>fortunei</u> (1/4)

Flat-topped trusses of 11 trumpet-shaped flowers, apricot pink spotted maroon in throat. Plant grew to 2.5ft(.75m) in 7 years with leaves 6in(15.2cm) long. (Rollin G. Wyrens, 1965)

cl.
    -<u>catawbiense</u> var. <u>album</u> (1/2)
Peach Parfait-          -<u>dichroanthum</u> ssp. <u>dichroanthum</u>
    -          -unnamed-                    (1/8)
    -unnamed- hybrid-<u>griffithianum</u> (1/8)
        hybrid-
          -<u>auriculatum</u> (1/4)

5ft(1.5m)          -15°F(-26°C)          L          4/2

Roseine purple buds open to flowers of lighter color with primrose yellow stripes, Chinese yellow dorsal spotting, openly funnel-campanulate, 2.5in(6.4cm) across; trusses of 13-15. Free-flowering. Plant broader than tall; dark green leaves retained 2 years. See also Duet. (David G. Leach, cross 1952; reg. 1983) See color illus. <u>American Horticulturist</u> 52:4, 1973, p. 18.

cl.


Peachblow--possibly a form of <u>hirtipes</u> (K-W 5659)

3ft(.9m)?          0°F(-18°C)          E          3/3

Plant of compact habit; light green leaves 3in(7.6cm) long, yellow midrib. Apricot flowers, fading primrose yellow. Best in partial shade. (Raised by J. Barto; intro. by the Wrights)

```
                              -neriiflorum ssp. neriiflorum
cl.                 -Nereid-                      (1/8)
       -unnamed hybrid-      -dichroanthum ssp. dichroanthum
Peaches  -              -fortunei ssp. discolor      (1/8)
and Cream*       -maximum (1/4)          (1/4)
       -Russell-
         Harmon-catawbiense (1/4)
```

6ft(1.8m)          -10°F(-23°F)          ML

A combination of pale pink and yellow flowers on a plant of good habit. (David G. Leach cross; intro. by Orlando S. Pride, 1970)

'Normandy' by Leach
Photo by Leach

'Noyo Brave' by C. Smith
Photo by Greer

'Noyo Maiden' by C. Smith
Photo by C. Smith

'Nuance' by Leach
Photo by Leach

'Oceanlake' by Wright
Photo by Greer

'Olga Mezitt' by Mezitt
Photo by Greer

'Olive' by Maxwell
Photo by Greer

'Oreocinn' by Magor
Photo by Greer

'Ostfriesland' by Hobbie
Photo by Greer

'Oudijk's Favorite' by Fa. le Feber (Boskoop)
Photo by Greer

'P.J.M.' by Mezitt
Photo by Greer

'Pacific Queen' by Coker
Photo by Greer

'Pania' by NZ Rhod. Assoc.
Photo by Greer

'Paprika Spiced' by Brockenbrough
Photo by Greer

'Parsons Gloriosum' by Parsons, 1850
Photo by Greer

'Parsons Grandiflorum' by Parsons
Photo by Leach

'Party Pink' by Leach
Photo by Leach

'Peace' by Aberconway
Photo by Greer

'Peach Parfait' by Leach
Photo by Leach

'Peach Brandy' by Wister
Photo by West

'Peeping Tom' by Wright
Photo by Greer

'Peking' by Leach
Photo by Leach

'Pelopidas' by Waterer
Photo by Greer

'Penjerrick' by Samuel Smith
Photo by Greer

```
cl.                    -catawbiense (3/8)
          -Mrs. C. S. Sargent-
Pearce's-                    -unknown (1/2)
American-                              -catawbiense
Beauty -              -Atrosanguineum-
          -Dr. H. C. Dresselhuys-     -unknown
                              -arboreum ssp. arboreum
                    -Doncaster-                 (1/8)
                              -unknown
```

6ft(1.8m)        0°F(-18°C)        VL

(The above parentage is uncertain.)  Flowers of good substance,
3in(7.6cm) wide, 5-lobed, medium beetroot purple with olive
yellow spotting; ball-shaped trusses of 18.  Plant as broad as
tall, well-branched; dark olive leaves, 7in(17.8cm) long, held 2
-3 years.  (Intro. by Reginald A. Pearce, late 1930s; reintro.
& reg., Elsie Watson, 1983)

```
                    -dichroanthum ssp. dichroanthum (1/8)
cl.           -Dido-
          -Ice Cream-    -decorum (1/8)
Pearl Diver-        -fortunei ssp. discolor (1/4)
          -Moser's Maroon--unknown (1/2)
```

Compact trusses of 18-20 flowers, pink  with a yellow eye.  (A.
F. George, Hydon Nurseries, 1966)

```
cl.
          -Loder's White (2 possible parentage diagrams--
Pedlinge-              see Loder's White)
          -decorum (Wilson 1782)
```

Leaves 7.5in(19cm) by 3in(7.6cm), narrowly elliptic.  White
flowers with a greenish flush, 7-lobed, held in full trusses of
10-12.  (Maj. A. E. Hardy)  A. M. (RHS) 1967

```
                    -sanguineum ssp. didymum (1/8)
               -Carmen-
cl.      -unnamed-    -forrestii ssp. forrestii Elatum Group
         - hybrid-                                (1/8)
Peekaboo*     -          -campylocarpum ssp. campylocarpum
              -Moonstone-                              (1/8)
         -          -williamsianum (1/8)
         -elliottii (1/2)
```

2ft(.6m)        -5°F(-21°C)        EM        3/4

An attractive compact plant with lovely rounded leaves of spring
green.  Glowing blood red, waxy, bell-shaped flowers, pendant.
(William E. Whitney)  See color illus. VV p. 6.

```
cl.
          -wardii var. wardii (1/2)
Peeping Tom-                    -griffithianum (1/8)
         -          -unnamed hybrid-
         -Mrs. Furnival-         -unknown (1/4)
         -          -caucasicum (1/8)
              -unnamed hybrid-
                    -unknown
```

3ft(.9m)        0°F(-18°C)        M        4/3

White flowers with a deep plum purple eye, shallow-campanulate,
to 2.5in(6.4cm) across; compact trusses of 10-12.  Plant rounded
and rather compact; ovate leaves 3in(7.6cm) long, 1.25in(3.2cm)
wide.  (Arthur Wright, Sr. & Jr., reg. 1966)

```
cl.


Peggy     Parentage unknown
```

Pink flowers.  (Waterer, Sons & Crisp)  A. M. (RHS) 1940

```
                                        -griffithianum (1/16)
cl.                      -George Hardy-
          -unknown (13/16) -Pink -     -catawbiense (1/16)
Peggy  -              -Pearl-     -arboreum ssp. arbor-
Bannier-Antoon van Welie-    -Broughtonii-         eum (1/16)
                    -unknown     -unknown
```

Large pyramidal trusses hold 20-22 fringed flowers, colored
Tyrian rose.  (Adr. van Nes)  A. M. (Boskoop) 1960

```
          -unnamed-catawbiense var. album (1/2)
cl.  - hybrid-   -wardii var. wardii (1/4)
     -     -Hawk-     -fortunei ssp.. discolor
Peking-     -Lady -
     -          Bessborough-campylocarpum ssp. campy-
     -                         locarpum Elatum Gp. (1/8)
     -unnamed-LaBar's White--catawbiense var. album
          hybrid-   -wardii var. wardii
          -Crest-          -fortunei ssp. discolor
               -Lady
               Bessborough-campylocarpum ssp. campylo-
                              carpum Elatum Group
```

5ft(1.5m)        -15°F(-26°C)        M        4/3

Buds strong yellowish pink, opening brilliant greenish yellow,
with dorsal blotch and sparse spotting of dark red.  Flowers of
heavy substance, openly funnel-shaped, 6-lobed; globular trusses
of about 15.  Upright, well-branched plant; dark green, slightly
concave leaves retained 2 years.  (David G. Leach, reg. 1973)

```
cl.
     -Glory of Keston--unknown (1/2)
Pele-     -haemodes ssp. haemodes (1/4)
     -Mandalay-
          -venator (1/4)
```

Plant spreading, nearly as wide as tall; barbatum type bark and
and foliage.  Flowers Turkey red with dark nectaries, in trusses
of 23-25.  (Lester E. Brandt, 1965)

```
cl.
     -catawbiense (1/2)
Pelopidas-
     -unknown (1/2)
```

Light crimson flowers.  (J. Waterer)

```
cl.
          -pemakoense (1/2)
Pematit Cambridge-     -impeditum (1/4)
               -Blue Tit-
                    -augustinii ssp. augustinii (1/4)
```

Plant 27in(.68m) tall by 38in(.95m) wide, compact, vigorous;
deep green leaves 1in(2.5cm) long.  Flowers of light mineral
violet, slightly darker margins and throat, funnel-shaped, 1.6in
(4cm) across; small globular trusses of 3-4.  Very floriferous.
(G. Reuthe, Fox Hill Nurseries)  H. C. (Wisley Trials) 1981

```
cl.
     -griersonianum (1/2)          -griffithianum (1/8)
Penalverne-          -Queen Wilhelmina-
     -Earl of Athlone-          -unknown (3/8)
               -Stanley Davies--unknown
```

Radium g.  (Bolitho)

```
cl.
     -jasminiflorum (1/2)
Pendance-
     -christianae (1/2)
```

Vireya hybrid.  Trusses of 9-12 white flowers flushed very pale
pink, throat white.  Leaves lanceolate.  (D. B. Stanton cross;
J. Clyde Smith, reg. 1983)

cl.                    -lochiae (1/4)
          -unnamed hybrid-
Pendragon-                  -laetum (1/4)
          -macgregoriae (1/2)

Vireya hybrid.  Trusses of 4-9 flowers, tubular funnel-shaped,
maize yellow, edged bright Saturn red.  (D. B. Stanton cross;
J. Clyde Smith, reg. 1983)

g.
          -griersonianum (1/2)
Penelope-          -facetum (1/4)
          -Dragon Fly-
                  -auriculatum (1/4)

Red flowers with dark spots.  (Waterer, Sons & Crisp, 1935)

cl.
          -thomsonii ssp. thomsonii (1/2)
Pengaer-
          -griffithianum (1/2)

Cornish Cross g.  Large, handsome flowers of crimson red.  (Lle-
welyn, 1911)  A. M. (RHS) 1911

cl.
          -concinnum var. pseudoyanthinum (1/2)
Penheale Blue-
          -russatum (1/2)

Semi-dwarf plant, wider than tall, vigorous, compact; glossy,
small dark leaves.  Very compact trusses of 20 flowers, wisteria
blue flushed spinel red, 1in(2.5cm) across.  Floriferous.  (Lt.-
Col. N. R. Colville)  A. M. (RHS) 1974; F. C. C. (RHS) 1981

g. & cl.
          -campylocarpum ssp. campylocarpum Elatum Group (1/2)
Penjerrick-
          -griffithianum (1/2)

6ft(1.8m)          15°F(-10°C)          E          4/3

Considered one of the most beautiful of all hybrids. Perfect
bell-shaped flowers may be white, creamy yellow or pink, some
fragrant; several forms are distributed.  A large shrub, fine
foliage, smooth coppery bark.  Plant habit upright; leaves 5in
(12.7cm) long. (Samuel Smith, Head Gardener, Penjerrick)  A. M.
(RHS) 1923

g.
          -orbiculare ssp. orbiculare (1/2)
Penllyn-
          -griffithianum (1/2)

Flowers of shell pink.  (Lord Aberconway, intro. 1933)

cl.

Penny     See Indian Penny

cl.          -griffithianum (1/4)
          -Mars-
Pennywise-   -unknown (1/4)
          -Koichiro Wada--yakushimanum ssp. yakushimanum (1/2)

4ft(1.2m)          -5°F(-21°C)          ML

Cardinal red buds opening to carmine rose flowers with 5 frilled
lobes, 2in(5cm) across; ball-shaped trusses of 17. Plant taller
than wide, upright; leaves held 3 years,  new growth light tan
tomentose.  (Dr. W. A. Reese, 1977)

```
  *  *  *  *  *  *  *  *  *  *  *  *  *  *  *  *  *  *
  * Hybridizers, registrants, raisers, etc., with their *
  *          locations are listed in App. B             *
  *  *  *  *  *  *  *  *  *  *  *  *  *  *  *  *  *  *
```

cl.
          -lochiae (1/2)
Penrose-
          -aurigeranum (1/2)

Vireya hybrid.  Trusses of 8-10 flowers, scarlet red, tubular
funnel-shaped, 2.5in(6.4cm).  Leaves oblanceolate to elliptic.
(D. B. Stanton, cross; J. Clyde Smith, reg. 1983)

cl.
          -campylocarpum ssp. campylocarpum (1/2)
Penrose Atkinson-
          -unknown (1/2)

Almond pink flowers shaded orange.  (Slocock)

cl.
          -williamsianum (1/2)
Pensive-
          -irroratum ssp. irroratum (1/2)

A rounded shrub with cordate leaves.  Flowers pale pink, lightly
spotted, 2.75in(7cm) across, 11 per truss. (Gen. Harrison, reg.
1965)

cl.
          -Moser's Maroon--unknown (1/2)
Peppermint Stick-
          -yakushimanum ssp. yakushimanum (1/2)

2.5ft(.75m)          0°F(-18°C)          ML

Plant rounded, wider than tall; narrow leaves 6in(15cm) long,
held 4 years.  White flowers edged rose, the center appearing
star-shaped, sparse brownish spotting on upper lobe; 21 flowers,
2in(5cm) wide, per conical truss.  (Cross by B. Lancaster; reg.
by Thomas McGuire, 1973)  C. A. (ARS) 1974

                   -dichroanthum ssp. dichroanthum (1/8)
cl.       -Fabia-
          -Goldbug-          -griersonianum (1/8)
Pepperpot-     -wardii var. wardii (1/4)
          -unnamed orange hybrid of Lem--unknown (1/2)

3ft(.9m)          15°F(-10°C)          EM

Flowers straw yellow heavily spotted Jasper red on dorsal lobes,
lighter elsewhere, campanulate, 1.5in(3.8cm) wide, 5-lobed, in
lax trusses of 8-10.  Plant rounded, as wide as tall; leaves
spinach green, retained 3 years, with new growth reddish brown.
(James A. Elliott, reg. 1978)

                   -neriiflorum ssp. neriiflorum (1/8)
g.        -F. C. Puddle-
          -Ethel-          -griersonianum (1/8)
Pera-     -forrestii ssp. forrestii Repens Group (1/4)
          -campylocarpum ssp. campylocarpum Elatum Group
          -Penjerrick-                              (1/4)
                   -griffithianum (1/4)

A hybrid with rose-colored flowers.  (Lord Aberconway, 1946)

g.
          -pemakoense (1/2)
Pera-
          -racemosum (1/2)

1ft(.3m)          -5°F(-21°C)          E          3/3

Vigorous plant habit, compact and twiggy with small oblong
leaves.  Prolific blooms, 1.5in(3.8cm) wide, lilac pink; up to 6
held in flat-topped trusses.  (Halfdan Lem, 1958)

cl.
          -yakushimanum ssp. yakushimanum (1/2)
Percy Wiseman-                   -dichroanthum ssp. dichroanthum
          -Fabia          -                              (1/4)
          Tangerine (selfed)-griersonianum (1/4)

3ft(.9m)          5°F(-15°C)          M

Vigorous, compact plant, wider than high; dark green leaves 3in (7.6cm) long. Flowers funnel-shaped, 2in(5cm) across, of cream deepening at base, lightly blushed pale pink, dull orange spots on upper segment; spherical trusses of 13-15. Floriferous. (J. Waterer, Sons & Crisp, reg. 1971)   H. C. (RHS) 1977

```
cl.                 -fortunei ssp. discolor (1/4)
        -Lady Bessborough-
Perdita-            -campylocarpum ssp. campylocarpum Elatum
        -souliei (1/2)                               Group (1/4)
```

5ft(1.5m)          -5°F(-21°C)          M          3/4

Halcyone g. A shapely plant; foliage and flowers like souliei. Buds of rose; cup-shaped, pale pink flowers fading milky white, a few red spots; lax, flat-topped trusses of about 8. (Rothschild, 1940)   A. M. (RHS) 1948   See color illus. PB pl. 52.

```
g. & cl.
          -souliei (1/2)
Peregrine-     -wardii var. wardii (1/4)
        -Hawk-     -fortunei ssp. discolor (1/8)
            -Lady      -
            Bessborough-campylocarpum ssp. campylocarpum
                                    Elatum Group (1/8)
```

From Crown Lands, 1951.

```
cl.
        -fortunei ssp. fortunei (1/2)
Perfume*
        -unknown (1/2)
```

May be sister seedling of Heavenly Scent. Flowers medium pink, faint red spotting on upper lobe; delightful spicy fragrance. Rounded, well-branched plant; leaves held 2 years. Best in partial shade. (W. E. Whitney; intro. by Anne & Ellie Sather)

```
cl.
          -mucronulatum (1/2)
Perla Rosa-
          -ciliatum (1/2)
```

Trusses hold 1-5 buds, each with 2-3 flowers of soft lilac pink. (Research Station for Arboriculture, 1982)

```
cl.      -yakushimanum ssp. yakushimanum (1/4)
    -Bambi-          -dichroanthum ssp. dichroanthum (1/8)
Perri -    -Fabia    -
Cutten-     Tangerine-griersonianum (1/8)
    -arboreum ssp. arboreum (1/2)
```

Flowers are bright claret rose with an enlarged pink calyx. (R. L. Cutten, 1976)

```
cl.

Perry Wood--seedling of faberi ssp. prattii (W 3958)
```

5ft(1.5m)          0°F(-18°C)          M          3/4

Leaves narrowly elliptic, 8in(20.3cm) long, with light brown plastered indumentum beneath. White flowers, strong flush of purplish red in throat, 2in(5cm) wide, 5-lobed, 16-18 per truss. (Collector Wilson; shown by Maj. A. E. Hardy)   A. M. (RHS) 1967

```
g.
          -forrestii ssp. forrestii Repens Group (1/2)
Persephone-
          -floccigerum ssp. floccigerum (1/2)
```

Flowers of orange red. (Lord Aberconway, 1941)

```
                       -souliei (1/8)
g.        -Soulbut-
        -Vanessa-       -Sir Charles Butler--fortunei ssp.
Perseus-    -griersonianum (1/2)      fortunei (1/8)
        -        -facetum (eriogynum) (1/4)
        -Jacquetta-
                 -griersonianum
```

Rosy salmon pink flowers. (Lord Aberconway, 1941)

```
              -cinnabarinum ssp. cinnabarinum
cl.    -Lady              Roylei Group (3/4)
       -Chamberlain-       -cinnabarinum ssp. cinna-
Perseverance-    -Royal Flush -          barinum (1/8)
        -        (orange form)-
        -                -maddenii ssp. maddenii
        -cinnabarinum ssp. cinnabarinum       (1/8)
                       Roylei Group
```

Flowers apricot, fading to coral. (Adams-Acton, 1942)

```
                           -griffithianum (1/8)
                -unnamed hybrid-
cl.   -Mrs. Furnival-      -unknown (1/4)
      -        -        -caucasicum (1/8)
Persia-        -unnamed hybrid-
      -catawbiense var. album (1/2) -unknown
```

5ft(1.5m)          -20°F(-29°C)          ML          4/4

Purple buds open to lighter flowers of good substance with bold spotting of light olive green, held in trusses 6.75in(17cm) across. Plant conical, spreading as broad as tall. See also Cyprus, Party Pink. (David G. Leach cross, 1953; intro. 1976; reg. 1982)   See color illus. ARS J 36:1 (Winter 1982): cover.

```
cl.

Persil*      Parentage unknown
```

Vigorous, upright, compact habit with leaves 3in(7.6cm) long, light glossy green. Truss globular-shaped, holding 20 flowers, funnel-shaped, buttercup yellow with white blotch on upper lobe. (Origin Knap Hill; raised by by W. C. Slocock)   H. C. (RHS) 1966

```
cl.
        -dichroanthum ssp. dichroanthum (1/2)
Persimmon-
        -elliottii (1/2)
```

4ft(1.2m)          10°F(-12°C)          M          3/3

Golden Horn g. Similiar to Golden Horn but without double calyx and more compact. Flowers waxy orange red with dark speckling. (Rothschild, 1945)   See color illus. PB pl. 11.

```
cl.

Peter Barber      See Jalisco Jubilant
```

Name (Peter Barber) removed from Register; invalid registration.

```
                           -catawbiense (1/8)
cl.          -Parsons Grandiflorum-
        -Nova Zembla-              -unknown (3/8)
Peter Behring-      -red hybrid--unknown
        -Mist Maiden--yakushimanum ssp. yakushimanum (1/2)
```

3ft(.9m)          -25°F(-32°C)          L

Conical trusses of 14 flowers, light Tyrian purple with dorsal spotting of olive yellow, rose red ribs down lobe centers on reverse; corolla 2.5in(6.4cm) wide, with 5 wavy lobes. Plant upright, almost as broad as tall; leaves with patchy golden brown indumentum, held 2 years. See also Norman Behring. (Rudy Behring, cross 1972; reg. 1985)

cl.
       -strigillosum (1/2)
Peter Faulk-
       -unknown (1/2)


2.5ft(.75m)          10°F(-12°C)          VE

Ball-shaped trusses of 16 flowers, bright cherry red with scattered darker spots, funnel-shaped, 2.5in(6.4cm) wide, 5-7-lobed.
Floriferous.  Plant broader than tall; dark green leaves with
patchy tawny indumentum beneath.  (Cross by Peter Faulk; raiser
Arthur Hanson; Allen P. Johnson, reg. 1981)

cl.                       -griffithianum (1/4)
      -George Hardy-
      -               -catawbiense (1/4)
Peter Koster-                    -arboreum ssp. arboreum
      -               -Doncaster-              (1/8)
      -unnamed hybrid-       -unknown (3/8)
             -unknown

5ft(1.5m)          -5°F(-21°C)          ML          2/3

Flowers bright magenta red with lighter margins, trumpet-shaped,
held in solid trusses. Sturdy, bushy plant; flat leaves. (M.
Koster & Sons, cross 1909)


cl.                    -griffithianum (1/4)
      -Loderi-
   -Albatross-       -fortunei ssp. fortunei (1/4)
Petia-       -fortunei ssp. discolor (1/4)
   -       -griersonianum (1/4)
   -Sarita Loder-       -griffithianum
      -Loderi-
          -fortunei ssp. fortunei

This cross of 2 well-known hybrids produced a truss of 12 flowers, white flushed amaranth rose, corolla 4in(10cm) across. (By
Francis Hanger, RHS Garden, Wisley) A. M. (RHS) 1962


cl.
      -christianae (1/2)
Petra-
      -jasminiflorum (1/2)

Vireya hybrid.  Trusses of 5-11 flowers, palest rose pink shading to white at base. (Cross by T. Lelliott; reg. by E. F. Allen, 1979)


cl.
      -pemakoense (1/2)
Phalarope-
      -davidsonianum (1/2)

Vigorous, upright, compact habit; glossy leaves 1in(2.5cm) long;
4-6 flowers, openly funnel-shaped, translucent white, slightly
flushed mauve.  Very floriferous. (Raiser Peter A. Cox) P. C.
(RHS) 1968;  A. M. (Wisley Trials) 1983


             -souliei (1/16)
          -Soulbut-
     -Vanessa-       -Sir Charles Butler--fortunei
g.       -Eudora-       -       ssp. fortunei (1/16)
   -       -       -griersonianum (1/2)
Phantasy-       -facetum (eriogynum) (1/4)
   -              -dichroanthum ssp. dichroanthum (1/8)
   -       -Fabia-
   -Erebus-       -griersonianum
        -griersonianum

A hybrid with red flowers. (Lord Aberconway, 1942)

cl.
      -griffithianum (1/2)
Phantom-
      -hybrid unknown (1/2)

Shell pink buds opening to white flowers, with a wine red blotch
at the base. (Mrs. R. J. Coker, 1979)


cl.
Phantom Rock--form of sanguineum ssp. sanguineum var. haemaleum
                                  (R 11049)

Trusses hold 4-6 flowers, 5-lobed, tubular-campanulate, ruby
red.  Leaves are dark, glossy above, grey-felted beneath. (Seed
collected by Joseph Rock; raised by Col. S. R. Clarke; shown by
R. N. S. Clarke) F. C. C. (RHS) 1981

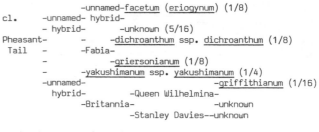

           -unnamed-facetum (eriogynum) (1/8)
cl.    -unnamed- hybrid-
    - hybrid-       -unknown (5/16)
Pheasant-       -dichroanthum ssp. dichroanthum (1/8)
  Tail  -    -Fabia-
    -       -griersonianum (1/8)
    -       -yakushimanum ssp. yakushimanum (1/4)
    -unnamed-                 -griffithianum (1/16)
      hybrid-       -Queen Wilhelmina-
      -Britannia-              -unknown
          -Stanley Davies--unknown

3ft(.9m)          0°F(-18°C)          M

Flowers rose pink, with a lighter center. (John Waterer, Sons &
Crisp, reg. 1975)


          -dichroanthum ssp. dichroanthum (1/8)
     -Astarte-       -campylocarpum ssp.
     -       -Penjerrick-campylocarpum Elatum Gp. (1/16)
  -Phidias-       -       -griffithianum (1/16)
g. -       -       -neriiflorum ssp. neriiflorum (1/8)
Phico-       -F. C. Puddle-
  -       -griersonianum (1/8)
  -       -williamsianum (1/4)
  -Cowslip-
     -wardii var. wardii (1/4)

Flowers are cream-colored flushed pink. (Lord Aberconway, 1941)


g. &       -dichroanthum ssp. dichroanthum (1/4)
cl.    -Astarte-       -campylocarpum ssp. campylocarpum
   -       -Penjerrick-              Elatum Group (1/8)
Phidias-       -griffithianum (1/8)
   -       -neriiflorum ssp. neriiflorum (1/4)
  -F. C.Puddle-
     -griersonianum (1/4)

Flowers are clear rich red. (Lord Aberconway) A. M. (RHS) 1938


cl.                       -griffithianum (1/4)
     -Mrs. E. C. Stirling-
Philip Waterer-       -unknown (1/2)
    -       -maximum (1/4)
    -unnamed hybrid-
        -unknown

Conical trusses of funnel-shaped flowers, soft rose with darker
veins; wavy edges. (J. Waterer, Sons & Crisp) A. M. (RHS) 1924


cl.                       -campylocarpum ssp. campylocarpum
     -Lady Primrose-              (1/4)
Phillippa Howells-       -unknown (1/4)
     -griersonianum (1/2)

Crimson buds open to pink and yellow flowers. (A. Howells, 1968)


cl.       -dichroanthum ssp. dichroanthum (1/4)
     -Fabia-
Philomene-       -griersonianum (1/4)
    -       -auriculatum (1/4)
    -Bustard-       -campylocarpum ssp. campylocarpum
     -Penjerrick-              Elatum Group (1/8)
        -griffithianum (1/8)

5ft(1.5m)          5°F(-15°C)          ML          3/3

Iviza g.  Plant of medium size, with glossy spoon-shaped leaves. Flowers straw-colored flushed pink, with darker throat. (Rothschild, 1941)

```
                           -dichroanthum ssp. dichroanthum (1/8)
            -Astarte-          -campylocarpum ssp.
            -         -Penjerrick- campylocarpum Elatum Group
g.   -Phidias-              -griffithianum          (1/16)
     -                    -neriiflorum ssp. neriiflorum (1/8)
Phisun-    -F. C. Puddle-
     -                -griersonianum (3/8)
     -     -griffithianum (5/16)
     -Sunrise-
              -griersonianum
```

Rose-colored flowers. (Lord Aberconway, intro. 1946)

```
g.                -neriiflorum ssp. neriiflorum (1/4)
       -F. C. Puddle-
Phoebus-          -griersonianum (1/4)
       -haematodes ssp. haematodes (1/2)
```

Bell-shaped flowers, bright blood red on a spreading shrub of medium size.  Early.  See also Dragon.  (Lord Aberconway, intro. 1941)

```
g.                -griffithianum (1/4)
       -Dawn's Delight-
Phoenix-          -unknown (1/4)
       -     -griersonianum (1/4)
       -Tally Ho-
              -facetum (eriogynum) (1/4)
```

Plant of average size and rather open habit.  Fine trusses of waxy red flowers, in midseason.  (Rothschild, 1950)

```
g.      -campylocarpum ssp. campylocarpum (1/2)
Phryne-      -griffithianum (1/4)
       -Loderi-
              -fortunei ssp. fortunei (1/4)
```

Cream-colored flowers.  (Lord Aberconway, 1933)

```
cl.
       -Powell Glass--catawbiense var. album (1/2)
Phyllis-
       -yakushimanum ssp. yakushimanum (1/2)
```

2ft(.6m)          -15°F(-26°C)          ML

Flowers white with strong reddish purple edging and shading, openly funnel-shaped, 2in(5cm) across, in ball-shaped trusses of 13-16.  Plant rounded, well-branched; glossy olive green leaves held 3 years.  (R. G. Shanklin, reg. 1983)

```
cl.                -neriiflorum ssp. neriiflorum (1/4)
              -Nereid-
Phyllis Ballard-     -dichroanthum ssp. dichroanthum (1/4)
              -fortunei ssp. discolor (1/2)
```

Grass green leaves with yellow midveins and petioles.  Bronzy orange red to coral red flowers in late season.  See also Coral. (Endre Ostbo, reg. 1958) P. A. (ARS) 1956

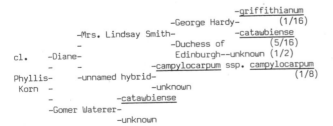

```
                              -griffithianum
                  -George Hardy-     (1/16)
       -Mrs. Lindsay Smith-       -catawbiense
       -                -Duchess of     (5/16)
cl.  -Diane-          Edinburgh--unknown (1/2)
     -              -campylocarpum ssp. campylocarpum
Phyllis-   -unnamed hybrid-            (1/8)
Korn -              -unknown
     -         -catawbiense
     -Gomer Waterer-
              -unknown
```

5ft(1.5m)          -15°F(-26°C)          M

Synonym Karl Korn.  White flowers, a large currant red blotch which fades primrose yellow, of heavy substance, 4in(10.2cm) wide, 5-lobed; trusses of 12.  Plant upright, branching well; glossy leaves held 2 years.  (Cross by Robert Korn; reg. by W. J. Shranger, 1981)

```
cl.
       -spinuliferum (1/2)
Pia Lehmann-
       -veitchianum Cubittii Group (1/2)
```

4ft(1.2m)          15°F(-10°C)          EM

Flowers greenish white at base of tube, pale yellow shading through white to rose pink edges, upper throat spotted pink and yellow.  Trusses of 2-4.  (A. F. George, Hydon Nurseries, 1978)

```
                              -griffithianum (1/8)
cl.          -Queen Wilhelmina-
       -Britannia-          -unknown (3/8)
Piccaninny-       -Stanley Davies--unknown
       -dichroanthum ssp. dichroanthum (1/2)
```

Limerick g.  Orange flowers tinged cherry at margins.  (Earl of Limerick) A. M. (RHS) 1956

```
cl.
       -Catalgla--catawbiense var. album Glass (1/2)
Pickering-       -fortunei ssp. fortunei (1/4)
       -unnamed hybrid-
              -campylocarpum ssp. campylocarpum (1/4)
```

6ft(1.8m)          -5°F(-21°C)          ML

Trusses of 8-10 flowers, light rose with deeper edges, 3 upper lobes pale yellow, with darker spotting, throat yellow; corolla widely campanulate, to 3in(7.6cm) wide.  Free-flowering.  Plant upright, slightly wider than tall; glossy leaves.  (Charles Herbert, reg. 1977)

---

Picotee Roseum (Picotee Rosea)     Parentage unknown

Bright rose flowers with intense black blotch.  (Veitch) F.C.C. (RHS) 1863

```
cl.
       -campanulatum ssp. campanulatum (1/2)
Pictum-
       -maximum ? (1/2)
```

Flowers white to pale lilac with dark spots.  (J. Waterer, 1839)

```
                              -catawbiense (1/8)
                  -unnamed-
---      -Altaclarense- hybrid-ponticum (1/8)
              -arboreum ssp. arboreum (1/4)
Picturatum-       -     -maximum (1/4)
       -unnamed hybrid-
              -unknown (1/4)
```

Flowers bright rose, spotted crimson or blush white with chocolate blotch?  (A. Waterer; dist. by Standish & Noble, 1850)

```
*  *  *  *  *  *  *  *  *  *  *  *  *  *  *  *  *  *  *  *
* Hybrids of unknown parentage are described briefly in App. D *
*  *  *  *  *  *  *  *  *  *  *  *  *  *  *  *  *  *  *  *
```

```
                            -wardii var. wardii (1/8)
                 -unnamed-         -neriiflorum ssp.
cl.      -Virginia- hybrid-F. C. Puddle-   neriiflorum (1/16)
         -Richards-                    -griersonianum (1/16)
         -          -                             -griffithianum
         -          -                    -George-          (1/32)
Pieces of-          -          -Mrs. Lindsay- Hardy-catawbiense
 Eight   -          -          - Smith     -              (1/32)
         -          -Mrs. Betty-           -Duchess of
         -          Robertson -             Edinburgh--unknown
     -unknown (5/8)          -      -campylocarpum ssp.
                            -unnamed- campylocarpum (1/16)
                            hybrid-
                                   -unknown

5ft(1.5m)          0°F(-18°C)          EM
```

Flowers 7-lobed, 4.5in(11.5cm) wide, fragrant, of pale yellow,
spotted orange; conical trusses of 10. Plant rounded, as broad
as tall; glossy medium green leaves held 2 years. (Roy Kersey,
1977)


cl.

Pied Piper     See Arnold Piper


```
g.                          -fortunei ssp. fortunei (1/4)
             -Madame Fr. J. Chauvin-
Pierre Laplace-                     -unknown (1/4)
             -williamsianum (1/2)
```

Flowers delicate rose.  (Dietrich Hobbie, 1952)


```
cl.
             -caucasicum (1/2)
Pierre Moser-
             -unknown (1/2)
```

Flowers light pink.  Early season.  (Moser & Fils, 1914)   See
color illus. F p. 59.


```
cl.       -keiskei (1/2)
Pikeland-
          -campylogynum (Tower Court form) (1/2)

1ft(.3m)          -5°F(-21°C)          EM
```

Plant rounded, as wide as tall; leaves 2.25in(5.7cm) long, scaly
below, held 3 years. Flowers pale rose with darker edges and
spotting, 1.25in(3.2cm) across, 5-lobed; inflorescence of about
6 trusses, each 6- to 7-flowered.  Free-flowering. (Charles Herbert, reg. 1976)


```
g. & cl.
            -fortunei ssp. fortunei (1/2)
Pilgrim-             -arboreum ssp. arboreum (1/4)
       -Gill's Triumph-
                    -griffithianum (1/4)

5ft(1.5m)          -5°F(-21°C)          M          4/4
```

Sturdy open plant, size increases rapidly; long, narrow, dark
green leaves. Large rounded trusses hold huge trumpet-shaped
flowers of rich clear pink, sparsely marked darker. (G. H. Johnstone)   A. M. (RHS) 1926   See color illus. F p. 52; VV p. xii.


```
                          -campylocarpum ssp. campylocarpum
                -Penjerrick-           Elatum Group (1/16)
g.       -Amaura-        -griffithianum (1/16)
        -Eros-   -griersonianum (3/8)
Pimpernel-  -griersonianum
        -arboreum ssp. arboreum (1/2)
```

Flowers are deep red.  (Lord Aberconway, intro. 1950)


```
cl.
         -campylocarpum ssp. campylocarpum (1/2)
Pinafore-      -griffithianum (1/4)
        -Loderi-
               -fortunei ssp. fortunei (1/4)
```

Barbara g.  Similiar to Barbara, but with white flowers blushed
pink toward the edges.  (Rothschild, 1948)


```
cl.
         -laetum (1/2)
Pindi Peach-
         -phaeopeplum (1/2)
```

Vireya hybrid.  Trusses of 4 flowers, saffron yellow. (T. Lelliot cross; R. L. Cutten, reg. 1981)


```
cl.
         -laetum (1/2)
Pindi Pearl-
         -phaeopeplum (1/2)
```

Vireya hybrid.  Trusses of 6 flowers, yellow orange flushed with
poppy red.  (T. Lelliot cross; R. L. Cutten, reg. 1981)


cl.

Pineapple     Parentage unknown

Parent of an Exbury hybrid, Geisha.


```
                              -dichroanthum ssp. dichroanthum
cl.       -Goldworth Orange-                   (1/8)
       -Hotei-               -fortunei ssp. discolor (1/8)
Pineapple-     -            -souliei (1/8)
 Delight -     -unnamed hybrid-
                            -wardii var. wardii (1/8)
          -Tropicana (Brandt)--unknown (1/2)

2.5ft(.75m)          0°F(-18°C)          E
```

Flowers amber yellow shading lighter at edges, throat lemon yellow, scant greyed yellow spotting; corolla widely funnel-shaped,
4in(10.2cm) wide, of good substance, 7-lobed; spherical trusses
of 12-14. Plant rounded, as broad as tall; dark green, bullate
leaves held 3 years. (Dr. E. C. Brockenbrough, cross 1971; Greg
Kesterson intro.; reg. 1983)


cl.

Pink Baby* --form of pumilum (K-W 6961)

Has both single and double flowers of a delicate shell pink or
pinkish mauve.  A rock garden dwarf plant growing to 1ft(.3m).
Name first used by Frank Kingdon-Ward.   (Shown by Lord Swaythling, Townhill Park)  A. M. (RHS) 1935


```
cl.            -griffithianum (1/4)
          -unnamed hybrid-
Pink Beauty-        -unknown (1/4)
          -         -catawbiense (1/4)
          -John Walter-
                    -arboreum ssp. arboreum (1/4)
```

Flowers pink to pale rose pink.  (C. B. van Nes, before 1922)


```
cl.
         -williamsianum (1/2)              -griffithianum
Pink Bountiful-              -Queen    -          (1/16)
         -                   -Wilhelmina-unknown (3/16)
         -             -Britannia-
         -Linswegeanum-      -Stanley Davies--unknown
                      -forrestii ssp. forrestii Repens Gp.
                                                (1/4)

3ft(.9m)          0°F(-18°C)          M
```

Bell-shaped flowers, wine red, lighter towards the center. Plant vigorous, compact; leaves medium dull green. (Cross by D. Hobbie; reg. by Slocock, 1969)    A. M. (RHS) 1975

cl.                    -maximum (1/4)
               -Halopeanum-
Pink Bride-            -griffithianum (3/4)
               -griffithianum

A tall plant with bell-shaped flowers, blushed pale. (Loder) A. M. (RHS) 1931

cl.
               -Poot--unknown (1/2)
Pink Brightness-
               -williamsianum (1/2)

Trusses hold 7-9 flowers, 2.5in(6.4cm) long, pale pink outside, phlox pink inside with darker edges. (Raised by D. Hobbie; intro. by Le Feber, and reg. 1965)

cl.                    -caucasicum (1/4)
          -Boule de Neige-            -catawbiense (5/8)
Pink Cameo-            -unnamed hybrid-
          -                         -unknown (1/8)
          -catawbiense (red form)

5ft(1.5m)        -20°F(-29°C)        ML        3/3

Lovely flesh pink flowers with a deeper pink blotch. Extremely hardy hybrid with good-looking foliage. (Anthony M. Shammarello, 1955)

cl.
          -yakushimanum ssp. yakushimanum (1/2)
Pink Cherub-          -arboreum ssp. arboreum (1/4)
               -Doncaster-
                         -unknown (1/4)

Plant compact, vigorous, 2.25ft(.68m) high by 4.5ft(1.4m) wide; leaves dull dark green. Flowers white, flushed light fuchsia purple on edges, widely funnel-shaped, 2in(5cm) across; spherical trusses of 15-20. Floriferous. (J. Waterer, Sons & Crisp) A. M. (Wisley Trials) 1968

cl.

Pink Chiffon--form of fortunei ssp. fortunei (R 30)

6ft(1.8m)        -5°F(-21°C)        M        3/3

Flowers Tyrian rose on edges, lighter pink inside with reddish orange markings, fragrant, funnel-campanulate, to 5in(12.7cm) wide; tall trusses of 9-12. (Seed from Rock; selected by Ruth M. Hansen; reg. 1968)

          -fortunei ssp. fortunei (1/4)
cl. -Fawn-     -dichroanthum ssp. dichroanthum (1/8)
  -     -Fabia-
Pink-          -griersonianum (1/4)
Crepe-               -campylocarpum ssp. campylocarpum
  -     -Ole Olson-               Elatum Group (1/8)
  -Lem's-          -fortunei ssp. discolor (1/4)
     Goal-          -griersonianum
          -Azor-
               -fortunei ssp. discolor

6ft(1.8m)        0°F(-18°C)        ML

Plant habit is open, upright and vigorous with leaves 7in(17.8cm) long, medium dark green. Trusses hold 7-9 flowers, 7-lobed, light purplish pink with edges and outside strong purplish red. See also Pink Divinity. (Arthur & Maxine Childers, 1972)

cl.
               -oreodoxa var. oreodoxa (1/2)
Pink Crest-
               -thomsonii ssp. thomsonii (1/2)

Trusses hold 6-8 flowers, colored rose Bengal; narrowly elliptic leaves 4.25in(10.8cm) long. (A. F. George cross; P. J. Urlwin-Smith, reg. 1982)

cl.

Pink Delight*        Parentage unknown

A parent of Nancy Adler Miller, a vireya. Hybridized by Veitch in the nineteenth century.    See color illus. ARS Q 27:1 (Winter 1973): p. 34.

g.
               -arboreum ssp. arboreum (1/2)
Pink Delight-
               -unknown (1/2)

Clear deep pink, or white edged deep rose pink. (R. Gill & Son) A. M. (RHS) 1926

cl.               -fortunei ssp. fortunei (1/4)
       -Fawn-     -dichroanthum ssp. dichroanthum (1/8)
Pink   -     -Fabia-
Divinity-          -griersonianum (1/4)
       -               -campylocarpum ssp. campylocarpum
       -     -Ole Olson-               Elatum Group (1/8)
       -Lem's-          -fortunei ssp. discolor (1/4)
          Goal-     -griersonianum
               -Azor-
                    -fortunei ssp. discolor

5ft(1.5m)        0°F(-18°C)        M

A plant with open, loose habit, leaves 7in(17.8cm) long. Trusses hold 9 flowers colored pale pink with a deep yellowish pink throat. See also Pink Crepe. (Arthur & Maxine Childers, 1972)

cl.
          -fortunei ssp. discolor (1/2)
Pink Domino-
          -hardy hybrid--unknown (1/2)

Carmine pink flowers spotted yellow. (Waterer, Sons & Crisp) A. M. (RHS) 1925

---
               -calostrotum ssp. calostrotum (1/2)
Pink Drift-
               -polycladum Scintillans Group (1/2)

1.5ft(.45m)        -10°F(-23°C)        EM        4/4

Neat, compact habit; small, strongly aromatic foliage of cinnamon bronze. Flowers rich lavender rose in tidy clusters. (H. White, Sunningdale Nursery, intro. before 1955)

cl.               -caucasicum (1/4)
          -Boule de Neige-          -catawbiense (5/8)
Pink Flair-          -unnamed-
          -               hybrid-unknown (1/8)
          -catawbiense (red form)

5ft(1.5m)        -20°F(-29°C)        ML        3/3

Bushy, compact plant; leathery dark green leaves, 3.5in(8.9cm) long. Flowers pastel pink with conspicuous red blotch. (A. M. Shammarello, reg. 1972)

* * * * * * * * * * * * * * * * * * * *
* Species used in hybrids of this volume are listed in App. A. *
* * * * * * * * * * * * * * * * * * * *

cl.
       -catawbiense var. album (1/2)
Pink Flourish-          -decorum (1/8)
      -          -unnamed-
      -unnamed- hybrid-griffithianum (1/8)
       hybrid-          -catawbiense (1/8)
           -unnamed-
           hybrid-unknown (1/8)

6ft(1.8m)          -25°F(-32°C)          M          3/3

Leaves 5in(12.7cm) long. Flowers pale pink, edged brighter pink
with small yellow brown blotch, 3.5in(8.9cm) wide, 5-lobed; pyramidal trusses of 15. Bloomed after -32°F(-36°C). (D. G. Leach,
reg. 1962)   See color illus. ARS Q 25:2 (Spring 1971): p. 100.

cl.
      -racemosum (1/2)
Pink Fluff-
      -davidsonianum (1/2)

3ft(.9m)          5°F(-15°C)          EM          4/3

Very similiar to the selected form of davidsonianum, Ruth Lyons,
but more compact in growth habit. Flowers open light pink then
turn to a very clear deep pink,  openly funnel-shaped, 3.5in(8.9
cm) wide; terminal clusters of 2-6 trusses, each 2- or 3-flower-
ed. Floriferous. Difficult to propagate. (H. E. Greer, 1979)

                -griffithianum (1/8)
cl.          -Mars-
      -unnamed hybrid-          -unknown (1/8)
Pink Fondant*          -Catalgla--catawbiense var. album
      -wardii var. wardii (1/2)          (1/4)

5ft(1.5m)          -15°F(-26°C)          ML

Fine pink flowers of excellent substance, and very good foliage.
(Joseph B. Gable cross; intro. by Orlando S. Pride, 1970)

g.                                        -griffithianum (1/4)
      -Miss Noreen Beamish-
Pink Fragrance-          -unknown (1/4)
      -fortunei ssp. discolor (1/2)

Flowers shell pink. (Thacker, 1940)

cl.
      -macabeanum (1/2)
Pink Frills-
      -unknown (1/2)

Rose Bengal buds opening to cream, with frilled rose Bengal margins, a prominent blotch; trusses of 20. See also Eyestopper.
(George Huthnance, New Zealand, reg. 1981)

cl.

Pink     -catawbiense var. album (1/2)
Frosting-
      -Koichiro Wada--yakushimanum ssp. yakushimanum (1/2)

4ft(1.2m)          -25°F(-32°C)          M          4/4

Bright pink buds opening light pink, darker on reverse; flowers
rotate-campanulate, to 2in(5cm) across, 14 per truss. Compact
plant, broader than tall; obovate leaves 4in(10.2cm) long, with
thin indumentum. See also Anna H. Hall, Great Lakes, May Time,
Spring Frolic. (David G. Leach, reg. 1964)

cl.
      -yakushimanum ssp. yakushimanum (1/2)
Pink Ghost-
      -Pauline--unknown (1/2)

White flowers flushed very pale pink, with greyed orange spot-
ting on lower segments, 2in(5cm) wide; spherical trusses of 15.
Plant upright, vigorous, slightly wider  than tall--4ft(1.2m).
See also Renoir, Telstar. (RHS Garden, Wisley) H. C. (RHS) 1972

cl.
      -yunnanense (1/2)
Pink Gin-          -cinnabarinum ssp. cinnabarinum Roylei Group
      -Lady          -cinnabarinum ssp.          (1/4)
      Rosebery-Royal Flush-          cinnabarinum (1/8)
      (pink form)-
             -maddenii ssp. maddenii (1/8)

Trusses hold 4-5 flowers, 1.5in(3.8cm) by 1.5in(3.8cm), 5-lobed,
light solferino purple shading to peach in the center. (E. G.
Millais Nursery) P. C. (RHS) 1981

cl.
      -catawbiense (Gable's red) (1/2)
Pink Globe-          -griersonianum (1/4)
      -unnamed hybrid-
            -fortunei ssp. fortunei (1/4)

6ft(1.8m)          -15°F(-26°C)          ML

Rosy pink flowers with a small cardinal red blotch, 3.5in(8.9cm)
wide; ball-shaped trusses hold 12. Plant upright, as  broad as
tall; leaves 6in(15cm) long, retained 3 years. (G. Guy Nearing,
1973)

cl.
      -irroratum ssp. irroratum (1/2)
Pink Glory-          -griffithianum (1/4)
      -Loderi-
      -fortunei ssp. fortunei (1/4)

White Glory g.  A tall plant with trumpet-shaped flowers, blush
pink and stained rose madder. (Lady Loder) A. M. (Wisley Trials) 1940

cl.
      -Antoon van Welie, q.v.
Pink Goliath-Professor J. H. Zaayer, q.v.
      -Annie E. Endtz, q.v.

Three hybrids of Pink Pearl have been crossed, but the exact
combination is unknown. Firm trusses of 15 flowers, each flower
4in(10.2cm) across, pure deep pink, a lighter blotch, greenish
yellow stripes. Plant habit spreading; hardy in normal winters;
good for forcing. (P. van Nes) Gold Medal (Boskoop) 1958;  Cer-
tificate First Class (Boskoop) 1959; Gold Medal (Rotterdam) 1960

cl.

Pink Ice     See Pink Sherbet

cl.
      -catawbiense (red form) (1/2)
Pink Icing*          -catawbiense (white form) (1/4)
      -Catalode-          -griffithianum (1/8)
           -Loderi King George-
                -fortunei ssp. fortunei
                    (1/8)
5ft(1.5m)          -15°F(-26°C)          ML

Good foliage; ruffled bright pink flowers, fading to off-yellow.
(Orlando S. Pride, intro. 1950)    See color illus.  ARS Q 34:3
(Summer 1980): p. 142.

cl.          -neriiflorum ssp. neriiflorum (1/4)
      -Nereid-
Pink Mango*          -dichroanthum ssp. dichroanthum (1/4)
      -fortunei ssp. discolor (1/2)

5ft(1.5m)          -10°F(-23°C)          ML

Good plant habit with flowers a stunning coral pink. Difficult
to propagate. (David G. Leach cross; raised by Orlando S. Pride,
intro. 1970)  See color illus. ARS Q 34:3 (Summer 1980): p. 143.

cl.

Pink Maricee     See Liz Ann

cl.
                    -griersonianum (1/2)
Pink Mermaid--Azor (selfed)-
                    -fortunei ssp. discolor. (1/2)

6ft(1.8m)          5°F(-15°F)          L

Trusses of 9 flowers, funnel-shaped, 4.5in(11.5cm) wide, colored
dawn pink edged with China rose, deeper in tube. Leaves are 6in
(15.2cm) by 1.75in(4.5cm). (B. J. Esch) P. A. (ARS) 1954


cl.


Pink Panther--form of mucronulatum (selfed)

6ft(1.8m)          -10°F(-23°C)          VE

Dark red buds open to deep pink flower with 5 wavy lobes, widely funnel-shaped. Leaves typical of the species, 2.25in(5.7cm)
long. (Selfed by E. K. Egan; raised by Dr. G. D. Lewis; reg.
1977) See color illus. ARS Q 31:3 (Summer 1977): p. 170; ARS J
37:2 (Spring 1983): p. 70.


cl.


Pink Parasol--form of yakushimanum ssp. yakushimanum

3.5ft(1m)          -20°F(-29°C)          EM          4/5

Plant over twice as broad as tall; dense foliage of flat leaves
4.25in(10.8cm) long, with heavy tan indumentum, held 4 years.
Buds of vibrant pink open pale mauve pink, aging clear pink to
white; flowers rotate-campanulate, 3in(7.6cm) wide; trusses of
15. See also Mist Maiden. (David G. Leach, reg. 1968)


cl.


Pink Parfait     Parentage unknown

A pink-flowered azaleodendron. (Joseph Senko, reg. 1962) P. A.
(ARS) 1961


cl.                    -griffithianum (1/4)
               -George Hardy-
Pink Pearl-            -catawbiense (1/4)
         -            -arboreum ssp. arboreum (1/4)
               -Broughtonii-
                       -unknown (1/4)

6ft(1.8m)          -5°F(-21°C)          M          3/3

Often called the standard by which all pinks should be judged.
Deep pink in bud, opening soft pink, fading to blush and paler
at edges, with a ray of reddish brown spots; in large conical
trusses. Plant strong-growing, rather bare at base; leaves 5in
(12.7cm) long. A parent of many hybrids. (J. Waterer) A. M.
(RHS) 1897; F. C. C. (RHS) 1900: A. G. M. (RHS) 1952  See color
illus. F p. 59; VV p. 80.


cl.
          -callimorphum ssp. callimorphum (1/2)
Pink Pebble-
          -williamsianum (1/2)

A hybrid with clear pink flowers. (Gen. Harrison, intro. 1954)
A. M. (RHS) 1975


cl.                         -brookeanum var. gracile
              -Duchess of Edinburgh-          (1/4)
         -                    -longiflorum (lobbii) (1/4)
Pink Perfection-    -jasminiflorum (3/8)
              -Princess -          -jasminiflorum
              -Alexandra-Princess Royal-
                              -javanicum (1/8)

Vireya hybrid. Pale pink flowers, tinged lilac pink. (L. Van
Houtte, 1958) See color illus. ARS J 36:1 (Winter 1982): p. 11.


cl.
          -Jan Dekens--unknown (7/8)
Pink Petticoats-          -griffithianum (1/8)
         -          -Queen Wilhelmina-
         -Britannia-          -unknown
                    -Stanley Davies--unknown

5ft(1.5m)          0°F(-18°C)          M          4/4

Flowers China rose at margins, lighter at center, openly funnel-
shaped, to 2.5in(6.4cm) across, frilled picotee edges; tall compact trusses of about 32. Upright, dense plant, slightly wider
than tall; leaves 8in(20.3cm) long. (John G. Lofthouse, 1966)
C. A. (ARS) 1971  See color illus. VV p. 96.


                    -griersonianum (1/8)
cl.        -Sarita Loder-    -griffithianum (1/16)
     -unnamed-          -Loderi-
Pink   - hybrid-          -fortunei ssp. fortunei (1/16)
Prelude-     -calophytum var. calophytum (1/4)

         -macabeanum (1/2)

6ft(1.8m)          10°F(-12°C)          VE

Pink buds opening to flowers of very pale purple, faintly tinged
pink when aging, throat circled by greyed purple blotch; corolla
2.25in(5.7cm) wide, of good substance, 6-lobed. Plant as broad
as tall; glossy leaves 10in(25cm) long with light indumentum,
tinted orange in new growth. (Elsie Watson, reg. 1985)


cl.
          -Catalgla--catawbiense var. album Glass (1/2)
Pink Punch-
          -fortunei ssp. fortunei F2 (1/2)

3ft(.9m)          -10°F(-23°C)          M

Flowers pink fading to pale pink edges, with a flare speckled
gold, funnel-campanulate, 2.75in(7cm) wide, 5-lobed; pyramidal
trusses of 15. Plant habit compact. (Henry R. Yates, reg. 1971)


cl.


Pink Queen     Parentage unknown

A Knap Hill hybrid, reg. 1972.  A. M. (RHS) 1972


cl.
          -fortunei ssp. fortunei (1/2)
Pink Rosette-          -griffithianum (1/4)
          -Mrs. E. C. Stirling-
                    -unknown (1/4)

Flowers pinkish white, suffused fuchsine pink. (Evans) A. M.
(RHS) 1956


cl.
          -griffithianum (1/2)
Pink Shell-          -fortunei ssp. fortunei (1/4)
          -H. M. Arderne-
                    -unknown (1/4)

Pale pink flowers. (Lowinsky) A. M. (RHS) 1923


cl.
          -yakushimanum ssp. yakushimanum (Exbury form) (1/2)
Pink Sherbet-
          -unknown (1/2)

3ft(.9m)          0°F(-18°C)          M

Synonym Pink Ice. Flowers are spirea red, paler in throat, fading to near-white, in ball-shaped trusses of 8-12. Plant rounded, as broad as high; elliptic leaves held 3 years, heavy silver
indumentum beneath. (H. L. Larson cross; J. Davis, reg. 1983)

cl.                    -ciliatum (1/2)
        -Cilpinense-
Pink Silk-             -moupinense (1/2)
        -Cilpinense (as above, but selfed and reselfed)

Clusters of 3 flowers, very light neyron rose.  (A. Teese, 1981)


cl.
              -racemosum (1/2)
Pink Snowflakes-
              -moupinense (1/2)

1.5ft(.45m)          0°F(-18°C)          VE          4/4

Buds red, opening to white flowers flushed soft pink, with darker spotting, openly funnel-shaped, in terminal clusters.    Plant
broader than high; very free-flowering.  Small glossy leaves are
bronze red when new.  (Robert W. Scott, 1969)  See color illus.
ARS Q 29:1 (Winter 1975): p. 35.


cl.
           -catawbiense (1/2)
Pink Twins*
           -haematodes ssp. haematodes (1/2)

4ft(1.2m)          -15°F(-26°C)          ML          4/3

Unusual flowers of light shrimp pink, hose-in-hose, fleshy, in
trusses of 15.  Plant compact, slow-growing, broader than tall;
ovate elliptic leaves, emerald green with yellow petioles. (Joseph B. Gable)    See color illus.  ARS Q 29:1 (Winter 1975):
p. 35; LW pl. 56; VV p. 13.


                                  -griffithianum (9/16)
                 -Beauty of Tremough-
             -Norman-                    -arboreum ssp. arboreum
cl.   -Anna- Gill -griffithianum                         (1/16)
      -    -          -griffithianum
Pink  -   -Jean Marie
Walloper*  de Montague-unknown (3/8)
      -              -griffithianum
     -Marinus Koster-
                     -unknown

6ft(1.8m)          -5°F(-21°C)          M          4/4

May be the same as Lem's Monarch.  Very large trusses of satin
pink.  Large leaves, deep green with reddish leaf stems.  New
growth stems often larger than a man's finger, but the strength
is needed to hold up the large, heavy trusses. See also Red
Walloper, Point Defiance.  (Cross by Halfdan Lem; selected by
Dr. Frank Mossman)


cl.
          -yakushimanum ssp. yakushimanum (1/2)
Pink Wonder-
          -unknown (1/2)

2.5ft(.75m)          0°F(-18°C)          L

Strong reddish purple flowers,  slight dorsal yellow green spotting, 1.6in(4cm) wide; dome-shaped trusses of 12-13.  Rounded
plant, broader than tall; leathery leaves held 2 years, new foliage light brown tomentose. See also Ohio Pink, Ohio Pink Blush,
Ohio Pink Glo, Ohio White, Secrest Pink. (John E. Ford; Secrest
Arboretum, reg. 1984)


cl.
          -aberconwayi (1/2)
Pinkerton-
          -unknown (1/2)

Pink flowers.  (Knap Hill Nursery, 1975)


                                     -catawbiense (1/4)
                       -Mrs. C. S.-
                -Meadowbrook- Sargent  -unknown (1/4)
cl.      -unnamed-         -          -catawbiense
         - hybrid-        -Everestianum-
Pinkie Price-            -              -unknown
         -          -fortunei ssp. fortunei (1/2)
         -sibling of the above

Flowers deep purplish pink shading to almost white with 2 green
blotches, 3.75in((9.5cm) wide, in compact trusses of about 10.
Leaves elliptic, 4.75in(12cm) long.  (Howard P. Phipps, 1973)


cl.             -catawbiense (1/2)
      -unnamed hybrid-
Pinnacle-         -unknown (1/2)
      -             -catawbiense
      -unnamed hybrid-
                  -unknown

6ft(1.8m)          -20°F(-29°C)          ML          4/3

Vibrant pink flowers  with a delicate citron yellow blotch, held
in conical trusses. A shapely plant with good foliage.  Bloomed
after -28°F(-34°C) 1984 & 1985. (Anthony M. Shammarello, intro.
1955; reg. 1958)    See color illus.  ARS Q 27:2 (Spring 1973):
cover; LW pl. 91; VV p. 85.


g.
      -arboreum ssp. arboreum (1/2)
Pioneer-         -sanguineum ssp. didymum (1/4)
      -Arthur Osborn-
                  -griersonianum (1/4)

A hybrid with red flowers.  (Lord Aberconway, intro. 1942)


cl.             -racemosum (1/4)
      -Conemaugh-
Pioneer*         -mucronulatum (1/4)
      -unknown (1/2)

5ft(1.5m)          -20°F(-29°C)          VE          4/3

Parentage as in Leach.  Very early flowering: a heavy blooming
of mauve pink  flowers 1in(2.5cm) wide. Plant upright, multiple stems; small leaves, semi-deciduous. Propagates easily. (J.
B. Gable, 1952)  See color illus. LW pl. 38; VV p. 114.


cl.             -caucasicum (1/4)
      -Dr. Stocker-
Pipaluk-         -griffithianum (1/4)
      -williamsianum (1/2)

Very floriferous.  Flowers opening slightly pink, turning white
when fully open, each 3in(7.6cm) across, held in trusses of 8.
(Gen. Harrison, 1968)


cl.
      -lowndesii (1/2)
Pipit-                         a natural hybrid
      -lepidotum (1/2)

1ft(.3m)          0°F(-18°C)          EM          3/3

Small, flat-faced flowers of light creamy pink; very small foliage. Difficult to grow. (Seed from Stainton, Sykes & Williams
expedition, Nepal, 1954; Peter A. Cox, raiser; reg. 1971)


cl.
      -valentinianum (1/2)
Piquante-
      -leucaspis (1/2)

Trusses of 3.  Flowers 2in(5cm) across, pale yellow, darker in
the center.  (Gen. Harrison, reg. 1962)

```
g.                -catawbiense (1/4)
        -B. de Bruin-
Pirate-         -unknown (1/4)
      -meddianum var. meddianum (1/2)
```

A tall plant.  Well-formed trusses of deep red flowers, in midseason.  (Rothschild, 1940)

```
cl.
        -Koichiro Wada--yakushimanum ssp. yakushimanum (1/2)
Pirouette-      -Jan Dekens--unknown (7/16)
      -Pink  -                -griffithianum
      Petticoats-      -Queen Wilhelmina-        (1/16)
             -Britannia-        -unknown
                        -Stanley Davies--unknown
```

Large, ball-shaped trusses of up to 28 frilled flowers, flecked
pink, fading to white.  Compact plant with dark green foliage.
Best in open shade.  (John G. Lofthouse, reg. 1973) Best of the
Vancouver Show award, 1981

```
g.
        -forrestii ssp. forrestii Repens Group (1/2)
Pixie-
        -dichroanthum ssp. apodectum (1/2)
```

A hybrid with orange red flowers.  (Lord Aberconway, intro. 1941)

```
cl.
        -augustinii ssp. augustinii (3/4)
Plaineye-      -impeditum (1/4)
      -Saint Tudy-
              -augustinii ssp. augustinii
```

Lavender blue flowers, unmarked, held in trusses of 4-8.  (Gen.
Harrison, 1956)

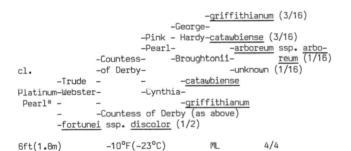

```
                            -griffithianum (3/16)
                        -George-
                    -Pink - Hardy-catawbiense (3/16)
                    -Pearl-        -arboreum ssp. arbo-
               -Countess-   -Broughtonii-    reum (1/16)
cl.            -of Derby-        -unknown (1/16)
      -Trude -        -    -catawbiense
Platinum-Webster-   -Cynthia-
Pearl* -        -        -griffithianum
      -        -Countess of Derby (as above)
      -fortunei ssp. discolor (1/2)
```

6ft(1.8m)        -10°F(-23°C)        ML        4/4

Strong-growing plant forming many very large buds.  Pearl pink
flowers with a dark rose blotch at base; corolla rather open and
ruffled.  Long, leek green leaves; sturdy stems.  Best in light
shade.  (H. E. Greer, intro. 1982)    See color illus. HG p. 77.

```
g.
        -dichroanthum ssp. dichroanthum (1/2)
Pleiades-
        -dichroanthum ssp. scyphocalyx (1/2)
```

Flowers orange with red.  (Reuthe, 1941)

```
cl.                -ponticum (1/2)
        -Purple Splendour-
Plum Beautiful*        -unknown (1/2)
      -        -mauve seedling--unknown
      -A. Bedford-
              -ponticum
```

5ft(1.5m)        -5°F(-21°C)        ML        4/3

Flowers of a delightful plum purple with a deep blackish purple
flare.  Medium large, glossy green leaves cover the plant.  (H.
E. Greer, intro. 1982)

```
cl.
Plum Warner* --form of campylogynum Myrtilloides Group
```

1ft(.3m)        -10°F(-23°C)        M        3/4

Named by Kingdon-Ward, this form of campylogynum has small plum-
colored flowers, bell-shaped, like "sculptured wax."  Considered
a rock garden plant. A pink form received an A. M. (RHS) 1925; a
rose magenta form the F. C. C. (RHS) 1945, both from Exbury.

```
cl.                -catawbiense (3/4)
        -Mrs. Milner-
Plusch-        -unknown (1/4)
        -catawbiense
```

Ruby red flowers  with faint reddish brown markings on a lighter
background.  (T. J. R. Seidel, intro. 1913)

```
cl.
Poet's Lawn--form of hodgsonii
```

5ft(1.5m)        10°F(-12°C)        EM        3/4

Makes a large shrub or small tree with 25 flowers held in compact, rounded trusses of white shaded rhodomine purple.  Leaves
12in(30.4cm) long with thin brown tomentum.  (Crown Estate,
Windsor)  A. M. (RHS) 1964

```
                            -griffithianum (9/16)
                    -Beauty of Tremough-
cl.        -Norman-        -arboreum ssp. arboreum
      -Anna- Gill -griffithianum        (1/16)
Point  -    -        -griffithianum
Defiance-   -Jean Marie -
      -    de Montague-unknown (3/8)
      -        -griffithianum
      -Marinus Koster-
              -unknown
```

6ft(1.8m)        0°F(-18°C)        M        4/4

Walloper g.  Large flowers of heavy substance, 4.5in(11.5cm)
wide, white, edged red on all lobes, without blotch; tall, compact trusses of 17.  Tall, vigorous plant; leaves 7.25in(18.5cm)
long, dark matte green above, lighter beneath.  See also Lem's
Monarch, Pink Walloper, Red Walloper.  (Halfdan Lem cross; reg.
by the Tacoma Park Dept., 1970)

```
g. & cl.
        -diaprepes (1/2)
Polar Bear-
        -auriculatum (1/2)
```

6ft(1.8m)        -5°F(-21°C)        L        3/3

Large flowers, trumpet-shaped,  white with a light green throat,
a lily-like fragrance,  and held in large, rather loose trusses.
Plant vigorous, eventually wider than high; under woodland con-
ditions may reach 30ft(9m).  Handsome, heavily veined leaves, to
7in(17.8cm) long.  Grows well in full sun but prefers some
shade.  (J. B. Stevenson, intro. 1926)   F. C. C. (RHS) 1946

```
cl.                -sanguineum ssp. didymum (1/4)
        -Red Cap-
Polar Cap-      -facetum (eriogynum) (1/4)
      -        -diaprepes (1/4)
      -Polar Bear-
              -auriculatum (1/4)
```

Trusses hold 7 flowers loosely, 5-lobed, funnel-campanulate,
spinel red with some faint darker markings.  Leaves dark green
sparsely covered with brown woolly indumentum.  (P. J. Urlwin-
Smith)  A. M. (RHS) 1974

cl.

Polar Crest      Parentage as above

Trusses hold about 9 flowers, the inner corolla spirea red  bar-
red crimson, and rim flushed crimson. (P. J. Urlwin-Smith, 1979)

cl.

Polar Dawn     Parentage as above

Flower color is carmine.  (P. J. Urlwin-Smith, reg. 1974)

cl.

Polar Glow      Parentage as above

Trusses hold 7 funnel-campanulate flowers, color very light neyron rose, heavily stained deep neyron rose.  Leaves narrowly elliptic, dark green above, underneath sparsely covered with brown
woolly indumentum.  See also Polar Cap, Polar Crest, Polar Dawn.
(P. J. Urlwin-Smith)  P. C. (RHS) 1974

cl.
          -smirnowii (1/2)
Polar Star-         -arboreum ssp. arboreum (1/4)
          -Doncaster-
                    -unknown (1/4)

Red flowers.  (M. Koster & Sons cross, 1902)


---                    -diaprepes (1/4)
          -Polar Bear-
Polar Sun-             -auriculatum (1/4)
          -facetum (1/2)

Pink flowers.  (Lord Digby cross, 1939; intro. 1955)


cl.
        -Koichiro Wada--yakushimanum ssp. yakushimanum (1/2)
Polaris*     -catawbiense (1/4)
        -Omega-
                -unknown (1/4)

Buds carmine red, opening to carmine pink ruffled flowers, with
the throat paler.  Unusually floriferous even as young plant; as
it matures each branch forms 3-5 buds.  Plant compact, rounded,
hardy to -17°F(-27°C).  (H. Hachmann, cross 1963; intro. 1978)
Gold Medal (Boskoop) 1980

cl.

Polka Dot--form of irroratum ssp. irroratum

5ft(1.5m)        5°F(-15°C)         EM          4/3

White flowers suffused pink and heavily spotted deep purple.
(Rothschild)  A. M. (RHS) 1957


                        -dichroanthum ssp. scyphocalyx (1/8)
              -Socrianum-
cl.  -unnamed-        -griffithianum (1/8)
  - hybrid-          -wardii var. wardii (3/8)
Polly -      -Rima-
Clarke-          -decorum (1/8)
    -    -wardii var. wardii
    -Crest-            -fortunei ssp. discolor (1/8)
        -Lady-
            Bessborough-campylocarpum ssp campylocarpum Elatum
                                                    Group (1/8)
5ft(1.5m)        5°F(-15°C)           M

Flowers of light, soft yellow, from deeper yellow buds.  Trusses
of 10 flowers.  (A. F. George; reg. by Hydon Nurseries, 1978)


cl.

Polly Peachum      See Emily Mangles


cl.
        -maddenii ssp. maddenii (polyandrum) (1/2)
Polyroy*
        -cinnabarinum ssp. cinnabarinum Roylei Group (1/2)

Flowers of apricot and red.  Late season.  (Sir John Ramsden)


cl.
          -grande (1/2)
Pomo Princess-
          -arboreum ssp. arboreum (1/2)

6ft(1.8m)          15°F(-10°C)          VE

Parentage uncertain, but probably as shown above. Trusses of 13
flowers, carmine red fading lighter to white in throat.  Corolla widely funnel-campanulate, 2in(5cm) across, of 10 wavy lobes.
Plant upright, broad, with leaves having brown indumentum,  held
2 years.  (Seed from RHS, Wisley; raisers John S. Druecker & E.
R. German, reg. 1978)


cl.
          -ponticum (1/2)
Ponticum Roseum-
          -maximum (1/2)

Synonym Maximum Roseum.  A tall, very hardy plant with flowers
pinkish lilac, long narrow leaves. (Origin  unknown; propagated
& distributed by O. S. Pride and others in Pennsylvania)


---

Poot     Parentage unknown

A parent of Pink Brightness.  Red flowers.  (Dietrich Hobbie)


cl.              -forrestii ssp. forrestii Repens Group
          -Elizabeth-                         (1/4)
Popacatapetl-        -griersonianum (1/4)
          -Compactum Multiflorum--unknown (1/2)

Flowers of Delft rose, with upper segments spotted chrysanthemum
crimson, funnel-shaped, 5in(13cm) wide; 5-7 in open dome-shaped
trusses.  Plant very free-flowering, vigorous, 2.25ft(.68m) tall
and broader than high; medium dull green leaves.  (G. Reuthe)
H. C. (RHS) 1981


cl.
        -macgregoriae (1/2)
Popcorn-
        -loranthiflorum (1/2)

Vireya hybrid.  Trusses of 10-14 flowers, tubular funnel-shaped,
chrome yellow and white.  Height about 3ft(.9m).  (J. Rouse
cross; reg. by Australian Rhododendron Society, 1984)


                    -griffithianum (1/8)
cl.      -Loderi-
    -Lodauric-     -fortunei ssp. fortunei (1/8)
Popeye*   -auriculatum (1/4)
    -              -ponticum (1/4)
    -Purple Splendour-
                -unknown (1/4)

Flowers open dark mauve fading to white with a large eye of deep
purple.  Vigorous plant with leaves deep cedar green, of heavy
texture.  (Dr. Frank Mossman cross; intro. by Harold E. Greer)

```
cl.                     -maximum (1/4)
          -Russell Harmon-
Poppinjay-              -catawbiense (1/4)
               -dichroanthum ssp. dichroanthum (1/4)
          -Jasper-          -fortunei ssp. discolor (1/8)
               -Lady   -
          Bessborough-campylocarpum ssp. campylocarpum
                                         Elatum Group (1/8)
```

5ft(1.5m)          -15°F(-26°C)          VL          4/3

Buds are strong red opening to flowers of strong orange, openly
campanulate, 2.75in(7cm) across, of unusually heavy substance;
ball-shaped trusses of 17.  Plant 50% broader than tall, rather
open; elliptic, undulant leaves of dull dark green.  (David G.
Leach, reg. 1972)

```
cl.
          -mucronulatum (1/2)
Portent-
          -ciliatum (1/2)
```

White flowers.  (David G. Leach, 1955)

```
g.                -brookeanum var. gracile (1/4)
          -Taylori-       -jasminiflorum (3/16)
Portia-   -Princess -           -jasminiflorum
          Alexandra-Princess Royal-
          -teysmannii (1/2)          -javanicum (1/16)
```

Vireya hybrid.  Flowers primrose yellow.  (Veitch, 1891)

```
g. & cl.
          -neriiflorum ssp. neriiflorum Euchaites Group (1/2)
Portia-
          -strigillosum (1/2)
```

The flowers are dark crimson scarlet, faintly spotted within.
(Lord Aberconway)  A. M. (RHS) 1935;  F. C. C. (RHS) 1947

```
cl.
          -augustinii ssp. chasmanthum (1/2)
Portrait-      -cinnabarinum ssp. cinnabarinum
     -Lady   -          Roylei Group (1/4)
     Chamberlain-      -cinnabarinum ssp. cinnabarinum
          -Royal Flush -                   (1/8)
          (orange form)-maddenii ssp. maddenii (1/8)
```

Trusses of 6 flowers, each 2in(5cm) across, of pale oyster pink.
(Gen. Harrison, reg. 1962)

```
cl.
          -cinnabarinum ssp. cinnabarinum Roylei Group (1/2)
Postling-
          -cinnabarinum ssp. cinnabarinum Blandfordiiflorum Group
                                                        (1/2)
```

5ft(1.5m)          5°F(-15°C)          EM-VL          4/4

The crossing of two forms of the same species does not a  hybrid
make.  (Maj. A. E. Hardy)  A. M. (RHS) 1975

```
cl.
          -yakushimanum ssp. yakushimanum (1/2)
Posy-          -arboreum ssp. nilagiricum (1/8)
          -Noyo Chief-
     -College Pink-          -unknown (3/8)
               -unknown
```

Trusses of 22 flowers, rose pink fading to rose madder then to
neyron rose.  Red spotting on  upper lobe.  (R. C. Gordon, 1982)

```
cl.          -haematodes ssp. haematodes (1/4)
     -Thor-          -dichroanthum ssp. dichroanthum (1/8)
Potlatch*  -Felis-
     -          -facetum (eriogynum) (1/8)
          -unknown (1/2)
```

3ft(.9m)          5°F(-15°C)          M          3/4

A small shrub with dense thick foliage; leaves indumented.  Very
large trusses of bright scarlet flowers.  (Roy W. Clark)

```
cl.                     -griffithianum (1/8)
          -Queen Wilhelmina-
     -Earl of Athlone-     -unknown (3/8)
Pow Wow-          -Stanley Davies--unknown
     -     -dichroanthum ssp. dichroanthum (1/4)
          -Fabia-
          -griersonianum (1/4)
```

Bright waxy red flowers with a large convex calyx. (Rudolph Henney cross, 1944; reg. 1958)

```
cl.
          -yakushimanum ssp. yakushimanum (1/2)
Powder Mill Run-     -griffithianum (1/4)
          -Mars-
               -unknown (1/4)
```

3ft(.9m)          -5°F(-21°C)          ML

Leaves are large and narrow.  Flowers deep pink, shading to
white in the center and held in ball-shaped trusses of up to 24.
(Cross by Lewis Bagoly; reg. by Charles Herbert, 1976)

cl.

Powell Glass--Catalgla--white variant of catawbiense

5ft(1.5m)          -15°F(-26°C)          M          3/3

A fifth generation from seed, presumed to have stabilized so
that seed from it would continue to produce pure white flowers.
Selected from seed of Catalgla (Gable).  Attractive foliage on a
vigorous, hardy plant.  (Edmond Amateis, 1962)

```
                    -catawbiense var. album (3/16)
          -Lodestar-     -catawbiense var. album
          -     -Belle -     -catawbiense (1/16)
          -     -Heller-Madame  -
     -R. O. Delp-          -Carvalho-unknown (1/4)
     -     -               -catawbiense
     -     -     -Atrosanguineum-
cl. -     -     -Atrier-          -unknown
     -     -Mary -     -griersonianum (1/16)
Power-     Belle-     -decorum (1/16)
 Base*     -Dechaem-
     -          -haematodes ssp. haematodes (1/16)
     -          -thomsonii ssp. thomsonii
     -     -Bagshot Ruby-               (1/16)
     -     -Princess -     -unknown
     -Lanny-Elizabeth-unknown
      Pride-          -smirnowii (1/8)
          -unnamed hybrid-
               -yakushimanum ssp. yakushimanum (1/8)
```

Strong purplish red buds open to flowers of purplish pink with
specks of deep purplish red.  (Weldon E. Delp)

```
cl.
          -ciliatum (1/2)
Praecox-
          -dauricum (1/2)
```

4ft(1.2)          -5°F(-21°C)          VE          3/3

Purple crimson buds opening to rosy lilac flowers in clusters of
2-3, each 3-4 flowered, widely funnel-shaped, 2in(5cm) across.
Glossy dark leaves paler beneath, 2in(5cm) long. (Isaac Davies,
1861; exhibited by The Hon. H. E. Boscawen)   Commemded 1861;
A. G. M. (RHS) 1926;  F. C. C. (RHS) 1978    See color illus. F
p. 60;  VV p. 120.

```
                         -valentinianum (1/8)
cl.               -Eldorado-
          -unnamed hybrid-     -johnstoneanum (1/8)
Prairie Gold-          -leucaspis (1/2)
          -          -xanthostephanum (1/4)
          -Lemon Mist-
               -leucaspis
```

2ft(.6m)          15°F(-10°C)          E

Yellow flowers, about 1.5in(3.8cm) wide, terminal inflorescence of 1-3 trusses, each 3-4 flowered. Free-flowering. Plant twice as wide as tall, with decumbent branches; gray green leaves with minute white hairs, scaly beneath, 1.25in(3.2cm) long, held 1 year. (Robert W. Scott, 1976)

```
                    -fortunei ssp. discolor (3/8)
cl. -Lady        -
    -Bessborough-campylocarpum ssp. campylocarpum Elatum Group
Prawn-                                              (1/4)
    -                      -dichroanthum ssp. dichroanthum
    -          -Goldsworth-                              (1/8)
    -Tortoiseshell- Orange   -fortunei ssp. discolor
      Wonder       -
                    -griersonianum (1/4)
```

A late-blooming hybrid with flowers colored salmon shrimp. (W. C. Slocock, Goldsworth Nursery, reg. 1967)

```
g. & cl.
        -wardii var. wardii (1/2)
Prelude-
        -fortunei ssp. fortunei (1/2)
```

5ft(1.5m)          -5°F(-21°C)          M          4/3

The last cross made by Lionel de Rothschild, exhibited 9 years after his death. Shrimp pink or coral buds open to creamy white flowers shading to primrose, openly cup-shaped; flattened globular trusses hold 10. Plant is compact with foliage like wardii, leathery, glossy deep green, and has the vigor of fortunei. (Rothschild, cross 1942) A. M. (RHS) 1951

```
cl.
        -maximum (1/2)          -griffithianum (1/8)
President Kennedy-          -George Hardy-
        -Pink Pearl-          -catawbiense (1/8)
        -             -arboreum ssp. arboreum
        -Broughtonii-                      (1/8)
                    -unknown (1/8)
```

6ft(1.8m)          -15°F(-26°C)          ML

Trusses of 20 flowers, pure white with blotch of pale greenish yellow. (Richard A. Fennichia, 1965)

```
cl.
        -catawbiense (1/2)
President Lincoln-
        -unknown (1/2)
```

6ft(1.8m)          -25°F(-32°C)          ML          2/3

Synonyms Herbert Parsons, Betsy Parsons, Bertha Parsons. Not to be confused with Abraham Lincoln. Very hardy and adaptable. Flowers lavender pink with bronze blotch, held in tight dome-shaped trusses. (Samuel Parsons, before 1871)

```
                    -fortunei ssp. discolor
            -Lady    -              (1/16)
            -Bessborough-campylocarpum ssp. campy-
        -Jalisco-               locarpum Elatum Gp. (1/16)
    -unnamed-      -dichroanthum ssp. dichroanthum
cl.    - hybrid-  -Dido-                      (1/16)
    -        -          -decorum (1/16)
President-    -yakushimanum ssp. yakushimanum (1/4)
Point    -              -wardii var. wardii ? (1/16)
    -          -Chlorops-
    -    -Lackamas-    -vernicosum ? (1/16)
    -unnamed- Spice -diaprepes (1/8)
      hybrid-
            -lacteum (1/4)
```

3ft(.9m)          5°F(-15°C)          M

Spinel red buds open to flowers of very light chrome yellow, 3in (7.6cm) wide, 7 wavy lobes, in large dome-shaped trusses of 14. Plant is low, upright, broader than high; leaves of dull dark green, held 3 years. (Cross by Cecil C. Smith; raised by Alice Poot Smith; reg. 1983)

```
cl.                    -mauve seedling--unknown (1/2)
            -A. Bedford-
Pride of Kings-          -ponticum (1/2)
            -              -ponticum
            -Purple Splendour-
                            -unknown
```

4ft(1.2m)          -10°F(-23°C)          M

Flowers of light purple, heavy dark brown blotch, throat satiny royal purple, openly funnel-shaped, 3.25in(8.9cm) wide; spherical trusses of 9. Upright plant, nearly as broad as tall; dark green leaves held 2 years. See also King's Favor. (Mrs. Halsey A. Frederick, Jr., 1978)

```
cl.
            -fortunei ssp. fortunei (1/2)
Pride of Leonardslee-
            -thomsonii ssp. thomsonii (1/2)
```

6ft(1.8m)          0°F(-18°C)          EM          3/4

Luscombei g. Plant upright, rounded, sturdy; shiny leaves 4.5 in(10.8cm) long. Rather slow to set buds. Flowers deep pink, shaded mauve, in loose trusses. Luscombei cross first made by Luscombe, exhibited 1880. (Sir Edmund Loder)

```
            -wardii var. wardii (1/8)
            -              -griffithianum 1/64)
        -Idealist-      -Kewense-
    -unnamed-    -Aurora-      -fortunei ssp. for-
cl. - hybrid-    -Naomi-      tunei (5/64)
    -        -      -thomsonii ssp. thomsonii
    -        -      -fortunei ssp. fortunei (1/32)
Prima-    -unknown (1/4)
Donna-          -fortunei ssp. discolor (1/8)
    -      -Lady
    -Jalisco -Bessborough-campylocarpum ssp. campylocarpum
     (selfed)-                    Elatum Group (1/8)
    -      -dichroanthum ssp. dichroanthum (1/8)
        -Dido-
            -decorum (1/8)
```

4ft(1.2m)          5°F(-15°C)          M

Plant of compact habit, with good foliage. Flowers chartreuse green, a brown blotch in the throat. (J. Waterer, Sons & Crisp, 1971)

```
---                    -brookeanum var. gracile (1/4)
        -Maiden's Blush-      -jasminiflorum (3/16)
Primrose-          -Princess-      -jasminiflorum
    -          Alexandra-Princess-
    -teysmannii (1/2)      Royal -javanicum (1/16)
```

Vireya hybrid. A cool greenhouse plant. F. C. C. (RHS) 1888

```
cl.
            -caucasicum (1/2)
Prince Camille de Rohan-
            -unknown (1/2)
```

5ft(1.5m)          -15°F(-26°C)          EM          3/3

Very ruffled flowers, light pink with deep reddish pink blotch. Foliage slightly twisted, with lighter green areas on leaves. (Cross by Waelbrouck, 1855; intro. by J. Verschaffelt, 1865)

```
cl.            -forrestii ssp. forrestii Repens Group (1/4)
        -Elizabeth-
Prince of-      -griersonianum (1/4)
 McKenzie-              -griffithianum (1/8)
    -      -Queen Wilhelmina-
    -Earl of-              -unknown (3/8)
     Athlone-
            -Stanley Davies--unknown
```

2.5ft(.75m)          -5°F(-21°C)          EM

Scarlet red flowers of good substance, 2in(5cm) across, 5 wavy
lobes; 4- to 6-flowered trusses, 6in(15cm) wide. Plant rounded
with decumbent branches; leathery leaves retained 2 years. (Carl
H. Phetteplace cross; reg. 1983)

```
g.
                -jasminiflorum (3/4)
Princess Alexandra-            -jasminiflorum
                  -Princess Royal-
                              -javanicum (1/4)
```

Vireya hybrid. Tubular, waxy, white flowers; fragrant. Blooms
all year in Los Angeles area. (Veitch)  F. C. C. 1865

```
cl.
             -ciliatum (1/2)
Princess Alice-
             -edgeworthii (1/2)
```

Leaves rich dark green. Flowers long, narrow, lax; white when
fully open, with slight yellow eye, and a distinctive pink line
down the back of each lobe in bud. Sweetly fragrant. (Veitch)
F. C. C. 1862

```
cl.
             -hanceanum Nanum Group (1/2)
Princess Anne-
             -keiskei (1/2)
```

2ft(.6m)         -10°F(-23°C)         EM          4/4

Synonym Golden Fleece. Flowers of a beautiful yellow with faint
greenish tint, campanulate, 2.75in(7cm) wide, held in trusses of
8. Very floriferous. Compact plant habit; foliage can change to
various shades of bronze depending on climate. (W. S. Reuthe,
raiser; intro. by G. Reuthe, ltd.)  A. M. (RHS) 1978; F. C. C.
(Wisley Trials) 1983    See color illus. F p. 60.

```
g.
                           -brookeanum var. gracile
                 -Duchess of Edinburgh-      (1/4)
Princess Beatrice-         -longiflorum (lobbii) (1/4)
                -         -jasminiflorum (3/8)
                  -Princess Alexandra-    -jasminiflorum
                                   -Princess-
                                   Royal  -javanicum (1/8)
```

Vireya hybrid. Flowers of cream and pink. F. C. C. (RHS) 1884

```
---
                 -javanicum (5/8)
Princess Christian-      -brookeanum var. gracile (1/4)
                  -Princess -          -jasminiflorum (1/8)
                  Frederica-Princess Royal-
                                   -javanicum
```

Vireya hybrid. Flowers yellow to yellow orange, or, white with
large orange blotch, black spotting. (?) (J. Waterer) F. C. C.
(RHS) 1883

```
cl.                       -thomsonii ssp. thomsonii (1/4)
                  -Bagshot Ruby-
Princess Elizabeth-       -unknown (3/4)
                  -unknown
```

6ft(1.8m)        -15°F(-26°C)         ML          3/3

Erect plant; its fairly sparse branches rather asymmetrical, for
an unusual form; dark leaves on thick reddish shoots, somewhat
pendant, bullate. Deep crimson flowers, paler at base, lightly
spotted, in tall conical trusses. (Waterer, Sons & Crisp, 1928)
A. M. (RHS) 1933

```
g.
                  -brookeanum var. gracile (1/2)
Princess Frederica-
                  -           -jasminiflorum (1/4)
                   -Princess Royal-
                              -javanicum (1/4)
```

Vireya hybrid. Yellow flowers. (Veitch, 1891)

```
g.
             -ciliatum (1/2)
Princess Helena-
             -edgeworthii (1/2)
```

Fragrant white flowers, tinged rose. (J. A. Henry, 1862)

```
---
             -jasminiflorum (1/2)
Princess Helena*
             -unknown (1/2)
```

Vireya hybrid. Pink flowers.  F. C. C. (RHS) 1865

```
cl.
              -griffithianum (1/2)
Princess Juliana
              -unknown (1/2)
```

6ft(1.8m)        5°F(-15°C)          EM          3/3

Buds vivid rose opening to soft rose pink, frilly flowers, fading white, in tight trusses. Vigorous, spreading plant; glossy
dark leaves. (Otto Schultz, cross 1890) A. M. (RHS) 1910   See
color illus. VV p. 26.

```
g.
              -dalhousiae var. dalhousiae (1/2)
Princess Leopold*
              -formosum var. formosum (1/2)
```

Flowers white, suffused rose. (J. A. Henry, 1862)

```
g.
              -brookeanum var. gracile (1/2)
Princess Leopold*
              -longiflorum (lobbii) (1/2)
```

Vireya hybrid. Reddish orange flowers. (Davies)  F. C. C.
(RHS) 1876

```
cl.

Princess Mary of Cambridge     Parentage unknown
```

5ft(1.5m)        -25°F(-32°C)         ML          3/3

Has been confused with Madame Wagner, but not the same.  White
with rose edges. Similar to Vincent Van Gogh and Rainbow only
with tighter truss and more hardy. (J. Waterer, before 1871)

```
---
              -campylocarpum ssp. campylocarpum (1/2)
Princess of Orange-        -caucasicum (1/4)
                  -Prince Camille de Rohan-
                              -unknown (1/4)
```

Shrub of neat, rounded habit. Funnel-shaped, waxy pink flowers,
with crimson markings and nectaries, crinkled margins, in loose
trusses. Early. (Veitch)

```
cl.
              -jasminiflorum (1/2)
Princess Royal-
              -javanicum (1/2)
```

Vireya hybrid. Parent of Duke of Teck, and others. Large pale
pink flowers, shaded yellow inside. (Henderson & Son, 1863)

```
---

Princess William of Wurtemberg     Parentage unknown
```

Flowers white, spotted crimson.  F. C. C. (RHS) 1894

cl.

Prinses Marijke      Parentage unknown

Deep pink flowers, a lighter phlox pink inside; outside of petals rose madder. (Felix & Dijkhuis, 1948) A. M. (Boskoop) 1948

```
cl.                    -caucasicum (1/4)
          -Boule de Neige-              -catawbiense (1/8)
Prize-               -unnamed hybrid-
     -                         -unknown (1/8)
          -catawbiense (red form) (1/2)
```

4ft(1.2m)         -20°F(-29°C)         ML         3/2

Flowers of clear shrimp pink  with a yellowish brown blotch, 2in
5cm) across, held in spherical trusses.  Leaves 4in(10.2cm) long
by 1.75cm(4.5cm) wide. (Anthony M. Shammarello, 1955)

```
cl.                       -catawbiense (1/4)
                 -Everestianum-
Professor Amateis-            -unknown (1/4)
     -                -Sir Charles Butler--fortunei ssp.
          -Van Nes  -                fortunei (1/4)
          Sensation-           -maximum (1/8)
                        -Halopeanum-
                              -griffithianum (1/8)
```

3ft(.9m)          -15°F(-26°C)         ML

Openly campanulate flowers, 3.5in(8.9cm) across, dawn pink, held
in medium-sized trusses.  Plant rounded, as broad as tall; ovate
leaves 5in(12.7cm) long.    (Edmond Amateis cross; intro. by Warren Baldsiefen; reg. 1972)

```
cl.                       -arboreum ssp. arboreum (1/4)
                 -Doncaster-
Professor F. Bettex-         -unknown (1/2)
     -                       -catawbiense (1/4)
          -Atrosanguineum-
                         -unknown
```

6ft(1.8m)        5°F(-15°C)         EM         3/2

Good-looking foliage and brilliant red flowers. (H. den Ouden &
Son, intro. 1925; reg. 1958)

```
                             -griffithianum (1/8)
                  -George Hardy-
cl.         -Pink Pearl-         -catawbiense (1/8)
     -             -        -arboreum ssp. arboreum
Professor Hugo-    -Broughtonii-               (3/8)
de Vries  -              -unknown (3/8)
     -          -arboreum ssp.  arboreum
          -Doncaster-
                    -unknown
```

5ft(1.5m)        -5°F(-21°C)         ML         3/3

Very much like Pink Pearl, but darker.  Buds rich rose opening
to lilac rose flowers with reddish brown ray, funnel-shaped, in
large, conical trusses.  Good foliage. (L. J. Endtz & Co., reg.
1958; raised & exhibited by Major A. E. Hardy) A. M. (RHS) 1975

```
                             -griffithianum (1/8)
                  -George Hardy-
cl.         -Pink Pearl-         -catawbiense (3/8)
     -             -        -arboreum ssp. arboreum (1/8)
Professor J.-      -Broughtonii-
H. Zaayer  -              -unknown (3/8)
     -                   -catawbiense
          -red catawbiense hybrid-
                         -unknown
```

5ft(1.5m)        -5°F(-21°C)         ML         3/2

Light red flowers in dome-shaped trusses. Plant vigorous, very
sprawly; large, twisted, light green leaves. (L. J. Endtz, 1958)

---

```
              -fortunei ssp. fortunei (1/2)
Profusion-
              -unknown (1/2)
```

Flowers rosy pink. (Paul) A. M. (RHS) 1896

```
cl.                    -ponticum (1/4)
          -Michael Waterer-
Prometheus-              -unknown (3/4)
          -Monitor--unknown
```

Scarlet crimson flowers. (C. Noble)

```
cl.
              -strigillosum (1/2)
Promise of Spring-
              -arboreum ssp. arboreum (1/2)
```

Very early blooming season.  Flowers of fluorescent rose, on a
plant of medium size. Good foliage. (John G. Lofthouse, 1973)

```
cl.
          -saluenense ssp. chameunum Prostratum Group (1/2)
Prostigiatum-
          -fastigiatum (1/2)
```

Deep purple flowers. (E. J. P. Magor) A. M. (RHS) 1924

```
g.
          -saluenense ssp. chameunum Prostratum Group (1/2)
Prostsal-
          -saluenense var. saluenense (1/2)
```

Flowers are pink violet, slightly spotted red. (E. J. P. Magor,
1926)

```
                      -Marion--unknown (17/32)
          -Joan Langdon-
cl.        -         -Sir Joseph Whitworth--unknown
     -          -griffithianum (5/64)
Proud Cis-       -Pink  -George-griffithianum
     -          -Shell-Mrs. L. A.- Hardy-
     -    -Rosabel-  - Dunnett -    -catawbiense
-unnamed-      -              -unknown      (1/64)
  hybrid-    -griersonianum (1/8)
          -arboreum ssp. arboreum (1/4)
```

Trusses of 17 flowers, bright neyron rose. (G. Langdon, 1982)

```
g.
     -Sir Charles Butler--fortunei ssp. fortunei (1/2)
Psyche-
     -williamsianum (1/2)
```

Pink flowers. (Dietrich Hobbie, 1950)

```
cl.
     -leucaspis (1/2)
Ptarmigan-
     -orthocladum var. microleucum (1/2)
```

1ft(.3m)         -5°F(-21°C)         EM         4/3

Pure white flowers, broadly funnel-shaped, 1in(2.5cm) wide, in
terminal clusters of several trusses, each 2 or 3-flowered. Compact, spreading plant; leaves to 1in(2.5cm) long, densely scaly
beneath. (E. H. M. & P. A. Cox) F. C. C. (RHS) 1965

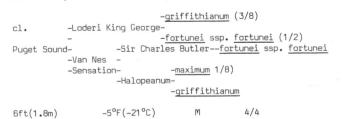

```
                        -griffithianum (3/8)
cl.       -Loderi King George-
     -              -fortunei ssp. fortunei (1/2)
Puget Sound-      -Sir Charles Butler--fortunei ssp. fortunei
     -Van Nes  -
     -Sensation-         -maximum 1/8)
               -Halopeanum-
                        -griffithianum
```

6ft(1.8m)        -5°F(-21°C)         M         4/4

'Persia' by Leach
Photo by Leach

'Peter Faulk' by Faulk/Hanson/Johnson
Photo by Greer

'Peter Koster' by M. Koster
Photo by Greer

'Phyllis Korn' by Korn/Shranger
Photo by Greer

'Pink Cameo' by Shammarello
Photo by Greer

'Pink Flourish' by Leach
Photo by Leach

'Pink Frosting' by Leach
Photo by Leach

'Pink Parasol' by Leach
Photo by Leach

'Pinnacle' by Shammarello
Photo by Leach

'Pipit' by Cox
Photo by Greer

'Platinum Pearl' by Greer
Photo by Greer

'Pleasant Dream' by Whitney/Sather
Photo by Greer

'Point Defiance' by Lem
Photo by Greer

'Polar Bear' by J. B. Stevenson
Photo by Greer

'Polynesian Sunset' by Whitney/Sather
Photo by Greer

'Poppinjay' by Leach
Photo by Leach

'Potlatch' by R. W. Clark
Photo by Greer

'Prairie Gold' by R. W. Scott
Photo by Greer

'President Lincoln' by Parsons
Photo by Greer

'Princess Alexandra' by Veitch, F.C.C. 1865
Photo by Greer

'Princess Elizabeth' by Waterer
Photo by Greer

'Princess Mary of Cambridge' (parents unknown) by Waterer
Photo by Greer

'Professor Hugo de Vries' by Endtz
Photo by Greer

'Puget Sound' by R. W. Clark
Photo by Greer

Very large, fragrant, ruffled flowers, pink with slight tinge of lilac, in tall, shapely trusses. Vigorous plant habit; glossy dark leaves on rosy petioles. (Roy W. Clark,  reg. 1958)

cl.

Pukeiti--form of <u>protistum</u> var. <u>giganteum</u>

5ft(1.5m)          15°F(-10°C)          E-EM          3/4

Trusses hold up to 30 flowers, solferino purple fading to fuchsia purple. (Collected by Kingdon-Ward, North Burma, 1953; selected by G. F. Smith, reg. 1979)

```
g.
            -catawbiense (5/8)
Pulchellum-                        -catawbiense
        -               -unnamed hybrid-
        -Altaclarense-                 -ponticum (1/8)
                     -arboreum ssp. arboreum (1/4)
```

Rosy pink flowers with white throat. (Standish & Noble, 1850)

```
---
            -arboreum ssp. arboreum (1/2)
Pulcherrimum-
            -caucasicum (1/2)
```

White flowers. (J. Waterer, Sons & Crisp; exhibited by Waterer, 1835)

```
                     -catawbiense var. album (7/16)
          -Lodestar-      -catawbiense var. album
          -      -Belle -      -catawbiense (1/16)
          -      -Heller-Madame -
     -R. O.-            -Carvalho-unknown (1/16)
     - Delp-                  -catawbiense
     -      -            -Atrosanguineum-
cl.  -      -   -Atrier-            -unknown
     -      -Mary -      -griersonianum (1/16)
Pure -      Belle-      -decorum (1/16)
Elegance*      -Dechaem-
     -            -haematodes ssp. haematodes (1/16)
     -                  -neriiflorum ssp. neriiflorum
     -            -Nereid-                  (1/16)
     -Phyllis-      -dichroanthum ssp. dichroanthum
-unnamed-Ballard-fortunei ssp. discolor (1/8)      (1/16)
     hybrid-
            -Clark's White--catawbiense var. album
```

Flowers medium to deep purplish pink, with a light yellow blotch containing specks of vivid red. (Weldon E. Delp)

```
cl.            -brookeanum var. gracile (1/4)
          -Taylori-                  -jasminiflorum (3/16)
     -            -Princess Alexandra-            -jasminiflorum
Purity-                        -Princess Royal-
     -                              -javanicum (1/16)
     -teysmannii (1/2)
```

Vireya hybrid. Pale yellow or primrose yellow flowers. (Veitch, 1891?)  A. M. (RHS) 1894; A. M. (RHS) 1899

```
cl.
          -edgeworthii (1/2)
Purity-
          -formosum var. formosum (1/2)
```

Flowers white with yellow eye. (A. Waterer, before 1871)  F. C. C. (RHS) 1888

```
                     -griffithianum (5/16)
            -Beauty of-
          -Norman- Tremough-arboreum ssp. arboreum (1/16)
          - Gill -
cl.  -Anna-      -griffithianum
     - -            -griffithianum
Purple-      -Jean Marie de Montague-
 Anna -                  -unknown (1/8)
     -            -campylocarpum ssp. campylocarpum Elatum Group
     -Ole Olson-                              (1/4)
            -fortunei ssp. discolor (1/4)
```

4ft(1.2m)          5°F(-15°C)          L

Flowers funnel-shaped, 3.5in(8.9cm) across, with 5 wavy lobes, lilac purple with edging of violet purple; dome-shaped trusses of 11-12. Plant upright, rounded; elliptic, medium green leaves 5.5in(14cm) long, held 3 years. (Grady Barefield cross, 1961; Mary Barefield, intro. 1976; reg. 1983)

```
cl.
          -ponticum (1/2)
Purple Elegance*
          -unknown (1/2)
```

5ft(1.5m)          -15°F(-26°C)          ML          3/3

A hardy, rapid-growing, dense hybrid that will stand sun. Showy purple red flowers.

cl.

Purple Emperor--<u>uniflorum</u> var. <u>imperator</u> (K-W 6884)

Named by Kingdon-Ward. A tender, creeping plant for the rock garden. Aromatic foliage; small flowers of bright purple. (Exhibited by Lord Swaythling)  A. M. (RHS) 1934

```
cl.
          -Moser's Maroon--unknown
Purple Emperor-
          -unknown
```

6ft(1.8m)          -5°F(-21°C)          ML          2/2

A tall plant of open habit. Leaves 5.5in(14cm) by 1.5in(2.5cm). Flowers dark purple with black dots on upper petal. Propagates easily. (Knap Hill Nursery)  A. M. (RHS) 1953

```
cl.
          -augustinii ssp. augustinii (3/4)
Purple Eye-            -impeditum (1/4)
          -Saint Tudy-
                     -augustinii ssp. augustinii
```

Mauve blue flowers with a purple spot. (Gen. Harrison, 1955)

cl.

Purple Fake--form of <u>concinnum</u> var. <u>pseudoyanthinum</u>

5ft(1.5m)          -5°F(-21°C)          M          4/3

Synonym Fake. Flowers orchid purple with crimson spots, tubular campanulate, 2in(5cm) across. Leaves 3in(7.6cm) long by 2in(5cm) wide. See also Chief Paulina. (Rudolph Henny, 1962)

```
cl.
          -fastigiatum (1/2)
Purple Gem-
          -minus var. minus Carolinianum Group (1/2)
```

2ft(.6m)          -20°F(-29°C)          EM          3/4

Parentage may be the reverse of the above. Plant rounded; medium green scaly leaves .75in(1.9cm) long; new foliage a beautiful blue. Small flowers of light purple, or brilliant purple in another form, in many small trusses. Similiar to Ramapo but not as dwarf. See also Ramapo. (Original cross by Joseph B. Gable; raised by G. Nearing, then D. Hardgrove, in the 1950s)       See color illus. LW pl. 64.

cl.
            -unnamed hybrid--unknown (1/2)
Purple Heart-                 -ponticum (1/2)
            -Purple Splendour-
                        -unknown

5ft(1.5m)          -5°F(-21°C)              M

Flowers of violet, with a yellow green blotch and markings. (J.
Waterer, Sons & Crisp, 1972)


                            -griffithianum (1/8)
cl.             -Queen Wilhelmina-
        -Britannia-              -unknown (5/8)
Purple Lace-        -Stanley Davies--unknown
        -                  -ponticum (1/4)
            -Purple Splendour-
                        -unknown

5ft(1.5m)          -5°F(-21°C)              ML          4/4

Parentage as in Van Veen. A rival for Purple Splendour? Flowers fringed, deep purple red. Glossy green foliage. (From Boskoop, via England; reg. 1969)  See color illus. VV p. 119.


cl.                     -ponticum (1/4)
            -Purple Splendour-
Purple Opal-            -unknown (3/4)
            -unknown

Trusses of 12-14 flowers, bell-shaped, rich purple violet.
Height about 3ft(.9m); elliptic leaves to 4in(10.2cm) long. (V.
J. Boulter cross; reg.by F. Boulter, 1984)


cl.

Purple Pillow--form of russatum

Low, compact form of this species. Leaves are small, glossy.
Flowers are amethyst violet, 1.25in(3.2cm) long. Many forms of
this species exist. (Fa. J. Streng, Jr., Boskoop, 1965)


cl.
            -ponticum (1/2)
Purple Splendour-
            -unknown (1/2)

5ft(1.5m)          -5°F(-21°C)              ML          4/3

King  of the royal purples. Large, ruffled flowers--one of the
darkest purples, and the black blotch makes it seem even deeper;
dome-shaped to spherical trusses of many flowers. Bush compact,
sturdy; dark green, convex leaves with a depressed midrib. Easy
to propagate; grows in sun or shade. (A. Waterer,  before 1900)
A. M. (RHS) 1931; A. G. M. (RHS) 1969    See color illus. F p.
60, 69;  VV p. 40.


cl.

Purpurellum--cinnabarinum ssp. cinnabarinum (L S & T 6349)

Unusual plum purple, tubular-shaped flowers on a slender shrub.
Attractive foliage.  See also Nepal.


cl.
            -catawbiense (1/2)
Purpureum Elegans-
            -unknown (1/2)

5ft(1.5m)          -25°F(-32°C)              ML          2/3

One of the old ironclads and still considered a good plant, with
blue purple flowers and nice foliage. (H. Waterer, before 1850)
See color illus. VV p. 6.


cl.
            -catawbiense (1/2)
Purpureum Grandiflorum-
            -unknown (1/2)

5ft(1.5m)          -20°F(-29°C)          ML          2/3

Plant of good growth habit; large, convex leaves.  Flowers medium purple with golden orange blotch. (H. Waterer, before 1850)


cl.

Pygmalion         Parentage unknown

5ft(1.5)          -5°F(-21°C)          ML          3/2

Crimson scarlet flowers,  spotted black. Foliage of dark olive
green. (John Waterer, Sons & Crisp) A. M. (RHS) 1933


cl.            -campylocarpum ssp. campylocarpum (1/4)
        -Moonstone-
Pygmy*        -williamsianum (1/4)
        -        -sanguineum ssp. sanguineum var. didymoides (1/4)
        -Carmen-
            -forrestii ssp. forrestii Repens Group (1/4)

1ft(.3m)          0°F(-18°C)          EM          4/3

A  widely spreading dwarf plant, entirely covered with dark red,
bell-shaped flowers held in loose trusses. (Skonieczny)


g.                -arboreum ssp. arboreum (1/4)
        -Doncaster-
Pyramus-        -unknown (1/4)
        -        -griffithianum (1/4)
        -Loderi-
            -fortunei ssp. fortunei (1/4)

A hybrid intro. in 1933 by Lord Aberconway.


cl.
    -haematodes ssp. haematodes (1/2)
Pyrex-
    -facetum (1/2)

Scarlet flowers. (Reuthe, 1945)


g.                -thomsonii ssp. thomsonii (1/2)
        -Exminister-
Pyrope-        -campylocarpum ssp. campylocarpum (1/4)
        -            -thomsonii ssp. thomsonii
        -Cornish Cross-
            -griffithianum (1/4)

A red-flowered hybrid. (Lord Aberconway, 1932)


g.                -souliei (1/4)
        -Sulphur Yellow-
Quadroon-        -campylocarpum ssp. campylocarpum (1/4)
        -griffithianum (1/2)

A tall plant with good growth habit. Pale cream-colored flowers
held in rather loose trusses. (Rothschild, 1950)

cl.
            -catawbiense (1/2)
Quadroona -
(Quadrona)-unknown (1/2)


Flowers purplish red with a bronze blotch. (J. Waterer, before
1874)

g.
       -hyperythrum (1/2)
Quaker Girl-             -griffithianum (1/8)
         -Loderi-
    -Avalanche-     -fortunei ssp. fortunei (1/8)
         -calophytum var. calophytum (1/4)

A tall hybrid and a small species produced a medium-sized plant of rather open growth habit, bearing large white flowers spotted with green, and held in good trusses. (Rothschild, 1950)

cl.
          -ciliicalyx (1/4)
     -Else Frye-
Quala-A-Wa-Loo-    -unknown (1/4)
      -johnstoneanum (1/2)

2.5ft(.75m)       15°F(-10°C)       E

Plant upright, twice as broad as tall; leaves 2.75in(7cm) long, glossy, with densely hairy margins, scaly beneath. Flowers empire yellow with medium red blotch, openly funnel-shaped, 3.25in (8.3cm) wide; small trusses of 4-5. (Dr. Richard Anderson, cross 1967; raised & intro. by H. J. Braafladt, reg. 1979)

cl.

Quarry Wood--form of trichostomum Ledoides Group

2.5ft(.75m)       -5°F(-21°C)      M       4/3

White flowers flushed light rhodamine purple. (Mrs. Martyn Simmons) A. M. (RHS) 1971

cl.

Quartz--form of rex (R 03800 or R 18234)

6ft(1.8m)       5°F(-15°C)      EM     3/4

Leaves 12in(30.4cm) long by 4in(10cm) wide, covered beneath with brown tomentum. Flowers 3in(7.6cm) across, tubular bell-shaped, white with pinkish blue tinge and crimson blotch. (Crown Estate, Windsor, reg. 1962) A. M. (RHS) 1955

cl.
     -wardii var. wardii (3/4)
Quasar-      -wardii var. wardii
        -             -griffithianum (1/32)
    -Idealist-     -Kewense-
           -Aurora-    -fortunei ssp. fortunei
        -Naomi-     -thomsonii ssp. thomsonii (1/16)
           -fortunei ssp. fortunei (5/32)

4ft(1.2m)       10°F(-12°C)      EM

Deep shell pink buds opening to canary yellow flowers with ruby red blotch in throat, 5-lobed, wavy, 2.75in(7cm) wide; flattened trusses of 8. Plant nearly as broad as tall; elliptic yellow green leaves held 3 years. (H. A. Short, cross 1970; reg. 1979)

g.
     -leucaspis (1/2)
Quaver-
     -sulfureum (1/2)

3ft(.9m)       0°F(-18°C)      E     3/4

Creamy primrose yellow flowers, widely funnel-campanulate, about 1in(2.5cm) wide, held loosely in 4- to 6-flowered trusses. Very floriferous. Foliage dense, woolly, deep green. (Rothschild, intro. 1950) A. M. (RHS) 1968 See color illus. F p. 62.

---
       -niveum (1/2)
Queen Alexandra-
      -grande (argenteum) (1/2)

Cream-colored flowers, tinged mauve.

cl.
            -brachycarpum ssp. brachycarpum (1/4)
     -unnamed hybrid-
Queen Anne's-      -catawbiense (1/4)
     -      -fortunei ssp. fortunei (1/4)
     -white hybrid-
      -unknown (1/4)

5ft(1.5m)     0°F(-18°C)     EM     4/4

Plant rounded, as wide as high; glossy olive green leaves held 2 years. Flowers open pale violet, fading quickly to clear white, no markings; corolla of heavy substance, 2.5in(6.4cm) across; no stamens--fused and petaloid, .75in(1.9cm) long. Many spherical trusses of 13 gardenia-like flowers. (Dr. H. T. Skinner cross, 1959; reg. 1979) See col. ill. ARS Q 33:4 (Fall 1979): p. 238.

cl.
     -dichroanthum ssp. dichroanthum (1/2)
Queen Bee-
     -unknown (1/2)

Compact plant 2ft(.6m) tall and 2.5ft(.75m) wide when registered. Tubular bell-shaped flowers, 3in(7.6cm) long, burnt orange red, held in lax trusses of 6-7, terminal umbels. (Dr. William Corbin, reg. 1962)

cl.
         -wardii var. wardii (1/2)
       -         -griffithianum (1/32)
     -Idealist-     -Kewense-
     -    -Aurora-   -fortunei ssp.
Queen    -          fortunei (5/32)
       -    -Naomi-    -thomsonii ssp. thomsonii
Elizabeth II-      -              (1/16)
         -fortunei ssp. fortunei
     -   -wardii var. wardii
     -Crest-     -fortunei ssp. discolor (1/8)
       -Lady
       Bessborough-campylocarpum ssp. campylocarpum
                  Elatum Group (1/8)

Flowers of very pale chartreuse green, openly funnel-shaped, 4.5 in(11.5cm) across, 7-lobed, in trusses of 10-12. Leaves 5.5in(14 cm) long by 2.5in(6.4cm) across, no indumentum. (Crown Estate, Windsor) A. M. (RHS) 1967

cl.
     -Corona--unknown (1/2)
Queen Mab-
     -souliei (1/2)

Compact trusses of 12 flowers, 2.5in(6.4cm), Tyrian rose. (Lester Brandt, reg. 1962)

cl.
     -Marion--unknown (3/4)
Queen Mary-      -catawbiense (1/4)
     -Mrs. C. S. Sargent-
          -unknown

5ft(1.5m)     -15°F(-26°C)    M    3/4

Lovely rose pink flowers. Plant habit sturdy; glossy, leatherlike leaves. (Felix & Dijkhouis, 1950) F. C. C. (Boskoop) 1948 See color illus. VV p. 38.

cl.

Queen Mother     See The Queen Mother

cl.
          -griffithianum (9/16)
    -Loderi Venus-
    -      -fortunei ssp. fortunei (1/4)
Queen   -          -griffithianum
Nefertiti-        -Beauty of Tremough-
    -    -Norman Gill-    -arboreum ssp.
    -    -          arboreum (1/16)
    -Anna-       -griffithianum
               -griffithianum
     -Jean Marie de Montague-
          -unknown (1/8)

5ft(1.5m)          15°F(-10°C)          M

Ball-shaped trusses 8.5in(21.6cm) across, of 12 flowers, light
neyron rose, darker edging and reverse, spotted Indian lake.
Plant almost as wide as tall; juniper green leaves held 2 years.
(Mrs. H. Beck cross, 1960; raiser Flora Markeeta Nursery;  Allan
C. Korth, reg. 1977)

g.
                -Pink Bell--unknown (1/2)
Queen o' the May-
                -griersonianum (1/2)

A parent of Chaste.  White-flowered.  (C. Smith;  exhibited by
Hamilton, before 1930)

g. & cl.
            -meddianum var. meddianum (1/2)
Queen of Hearts-
            -Moser's Maroon--unknown (1/2)

6ft(1.8m)          -5°F(-21°C)          ML          4/3

Deep red buds opening to glowing, dark crimson flowers with
white stamens, red calyx, and peppered with small black spots on
upper lobes; domed trusses of 16.  Strongly veined dark leaves
to 4.5in(11.5cm) long, thin pale indumentum beneath.  A shrub of
striking appearance; one of Lionel de Rothschild's last crosses.
(Rothschild)  A. M. (RHS) 1949     See color illus. PB pl. 34.

              -wardii var. wardii (1/2)
              -                    -griffithianum (1/32)
cl.   -Idealist-         -Kewense-
      -       -        -Aurora-     -fortunei ssp. fortunei
Queen of-     -Naomi-   -                     (5/32)
McKenzie-     -        -thomsonii ssp. thomsonii (1/16)
      -        -fortunei ssp. fortunei
      -   -wardii var. wardii
      -Crest-        -fortunei ssp. discolor (1/8)
          -Lady    -
          Bessborough-campylocarpum ssp. campylocarpum
                           Elatum Group (1/8)

3ft(.9m)          -5°F(-21°C)          M

Flowers of canary yellow with sulphur yellow throat and garnet
brown blotch, the reverse marigold orange; corolla of good substance, saucer-shaped, to 5.5in(14cm) wide; flat trusses of 9-
11.  Floriferous.  Plant upright, rounded; parsley green leaves
held 2 years.  (Cross by C. Phetteplace; reg. by C. & E. Phetteplace, 1982)

---
                -javanicum (5/8)
Queen of the Roses-           -jasminiflorum (3/8)
          -Princess Alexandra-      -jasminiflorum
                        -Princess-
                        Royal  -javanicum

Vireya hybrid. Flowers pinkish yellow to salmon. (Veitch, 1891)

---
                -javanicum (5/8)
Queen of the Yellows-
          -            -brookeanum var. gracile (1/4)
          -Princess -      -jasminiflorum (1/8)
          Frederica-Princess-
                        Royal  -javanicum

Vireya hybrid. Flowers yellow to yellow orange. F. C. C. (RHS)
1866

cl.

Queen  -fortunei ssp. fortunei (1/2)
Souriya-            -campylocarpum ssp. campylocarpum (1/4)
      -unnamed hybrid-
                  -unknown (1/4)

Sweetly fragrant flowers, 7-lobed,  pale ochre edged with lilac.
Plant medium  to tall; blooms midseason.   (W. C. Slocock, 1937)
A. M. (Wisley Trials) 1957

                              -catawbiense (1/8)
---          -unnamed hybrid-
          -Altaclarense-          -ponticum (1/8)
Queen Victoria-          -arboreum ssp. arboreum (1/4)
          -unknown (1/2)

Flowers deep claret purple.  (Standish & Noble, 1850)  F. C. C.
(RHS) 1882

---
              -brookeanum var. gracile (1/2)
Queen Victoria-
              -longiflorum (lobbii) (1/2)

Vireya hybrid.  Flowers yellow orange, or, primrose yellow to a
pale salmon.  (Veitch, exhibited 1891)

cl.
          -griffithianum (1/2)
Queen Wilhelmina-
          -unknown (1/2)

6ft(1.8m)          10°F(-12°C)          EM          4/3

Carmine buds  opening to very large, deep rosy  scarlet flowers,
fading to rose pink, widely campanulate, in lax trusses.  Plant
stiffly branched. (Schulz, 1890; intro. by C. B. van Nes & Sons,
1896)

cl.
          -souliei (1/2)
Queen's Wood-
          -aberconwayi (1/2)

White saucer-shaped flowers  suffused a light red purple, with a
darker blotch; up to 8 in loose trusses.  (Cross by T. H. Findlay for Crown Estate Commissioners)  A. M. (RHS) 1972

cl.
      -catawbiense (3/4)
Quendel-         -catawbiense
      -Mrs. Milner-
                -unknown (1/4)

Purplish red flowers, with very dark brown markings.  (T. J. R.
Seidel, c.1914)

cl.
      -catawbiense (1/2)
Querele-
      -unknown (1/2)                         —

Flowers of dark violet, with a few markings.  (T. J. R. Seidel
cross, 1906; intro. 1914)

g. & cl.
      -Red Night--unknown (1/2)
Querida-
      -elliottii (1/2)

Shining deep scarlet flowers, the reverse darker, widely campanulate, in tight trusses of about 16.  A tall shrub; large, deep
green leaves. (Edmund de Rothschild, 1950)  A. M. (RHS) 1952

```
                -brachycarpum ssp. brachycarpum (1/4)
    -unnamed-      -wardii var. wardii (1/8)
cl. - hybrid-Crest-        -fortunei ssp. discolor (1/16)
            -Lady
Quest*          -Bessborough-campylocarpum ssp. campylocarpum
-                                    Elatum Group (1/16)
-               -maximum (1/8)
-        -Opal-
-unnamed-       -catawbiense (1/8)
    hybrid-                           -caucasicum (3/32)
-               -Jacksonii-          -caucasicum
    -Goldsworth-          -Nobleanum-
        Yellow  -                 -arboreum ssp. arbo-
                                           reum (1/32)
                -campylocarpum ssp. campylocarpum (1/8)
```

Pale yellow flowers, speckled with red. Plant of medium size. (Weldon E. Delp)

```
                        -griffithianum (1/8)
cl.         -Queen Wilhemina-
    -Britannia-              -unknown (3/8)
Quinella-       -Stanley Davies--unknown
-               -haematodes ssp. haematodes (1/4)
    -May Day-
            -griersonianum (1/4)
```

3ft(.9m)        0°F(-18°C)       M       3/3

Flowers geranium red, and appear double with the calyx as large as the flower. Leaves dull green, rather narrow. (Rudolph Henney, reg. 1958)

```
g. & cl.
    -catawbiense (1/2)
Quinte*
    -unknown (1/2)
```

Flowers of lilac pink lighter in throat, with greenish markings. (Seidel, cross 1906; intro. 1914)

```
g.
    -ciliatum (1/2)
Quiver-
    -lutescens (1/2)
```

A tender dwarf plant of small leaves, with cream-colored flowers in small clusters. Floriferous. (Rothschild; reg. 1950)

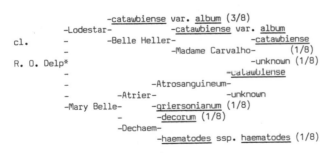

```
                -catawbiense var. album (3/8)
        -Lodestar-           -catawbiense var. album
cl. -       -Belle Heller-          -catawbiense
R. O. Delp*     -Madame Carvalho-       (1/8)
-                       -unknown (1/8)
-                       -catawbiense
-                   -Atrosanguineum-
-           -Atrier-               -unknown
    -Mary Belle-    -griersonianum (1/8)
-               -decorum (1/8)
        -Dechaem-
                -haematodes ssp. haematodes (1/8)
```

Flowers of peach, pink, and salmon. Plant of medium size; hardy to -15°F(-26°C). (Weldon E. Delp)

```
cl.
    -chrysodoron (1/2)
R. W. Rye-
    -johnstoneanum (1/2)
```

Primrose yellow flowers, gradually deepening in throat, pendulous, in trusses of 4. (Cross by Lord Stair, 1938; intro. 1951) A. M. (RHS) 1951

```
                -catawbiense (1/8)
cl.     -Parsons Grandiflorum-
    -America-              -unknown (3/8)
Rachter*    -dark red hybrid--unknown
    -forrestii ssp. forrestii Repens Group (1/2)
```

Scarlet red flowers. (Dietrich Hobbie, 1947)

```
cl.
    -racemosum (1/2)
Racil-
    -ciliatum (1/2)
```

3ft(.9m)        -5°F(-21°C)       E       3/2

Plant habit open when young, compact later; dark green leaves, 1.75in(4.5cm) long. Clusters of 3-4 small, shell pink flowers cover the bush. Easily rooted. (Shown by N. S. Holland, 1937)

```
                -souliei (1/8)
g.      -Soulbut-
    -Vanessa-    -Sir Charles Butler--fortunei ssp.
Radiance-   -griersonianum (3/4)        fortunei (1/8)
    -griersonianum
```

Deep rose pink flowers. (Lord Aberconway, intro. 1936)

```
g. & cl.        -dichroanthum ssp. dichroanthum (1/4)
        -Fabia-
Radiant Morn-   -griersonianum (1/4)
-               -griffithianum (1/4)
        -Sunrise-
            -griersonianum (1/4)
```

Porcelain rose, suffused geranium pink. (Lord Aberconway) A. M. (RHS) 1951

```
g.
    -griersonianum (1/2)
Radium-             -griffithianum (1/8)
        -Queen Wilhelmina-
    -Earl of Athlone-       -unknown (3/8)
        -Stanley Davies--unkonwn
```

5ft(1.5m)       5°F(-15°C)       ML      3/3

Rounded, loose trusses of geranium scarlet flowers. Plant habit rather open; glossy leaves 5in(12.7cm) long. Easily propagated. See also Penalverne. (J. J. Crosfield, 1936)

```
cl.
    -calostrotum ssp. keleticum Radicans Group (1/2)
Radmosum-
    -racemosum (1/2)
```

Rosy mauve flowers. (Thacker, exhibited 1946)

cl.

Rae Berry--form of trichostomum

2.5ft(.75m)     -5°F(-21°C)       M       3/3

Deciduous shrub with straggly habit; leaves fringed with hair. Flowers about 1in(2.5cm) long, strong pink. (Raiser Mrs. A. C. U. Berry; intro. by The Bovees; reg. 1972)

```
                    -catawbiense (1/8)
            -unnamed hybrid-
---  -Altaclarense-     -ponticum (1/8)
-           -
Raeanum-    -arboreum ssp. arboreum (1/4)
-               -maximum (1/4)
-unnamed hybrid-
        -unknown (1/4)
```

Rosy crimson flowers with black spots. (Standish & Noble, exhibited 1850)

cl.
      -racemosum (1/2)
Raeburn-
        -tephropeplum (1/2)

Flowers of light yellowish copper.  (Collingwood Ingram, 1955)

cl.

Rae's Delight--form of japonicum var. pentamerum (degronianum)

2ft(.6m)            -10°F(-23°C)            EM            4/3

Semi-dwarf plant, more than twice as broad as high, compact,
rounded, with new growth having grey brown indumentum.  Flowers
are roseine purple, held in trusses of 12.  (Seed from Japan;
raiser Mrs. A. C. U. Berry; reg. by Portland Chapter, ARS, 1980)

---
        -griffithianum (1/2)
Ragged Robin-
         -unknown (1/2)

Crimson cerise flowers, fading lighter red.  (J. Waterer, Sons &
Crisp)

cl.
      -A. W. Hardy Hybrid--unknown (1/2)
Rainbow-
        -griffithianum (1/2)

6ft(1.8m)          0°F(-18°C)              M            4/3

Upright plant, as wide as tall; glossy foliage of slightly rough
texture.  Beautiful flowers edged in deep pink, with white centers.  (W. C. Slocock, 1928)   See color illus. VV p. 94.

cl.

Raja*--form of brookeanum var. gracile

Vireya with flowers of 5 joined lobes, tubular funnel-shaped,
2in(5cm) long, 3in(7.6cm) across, held in trusses of 5-11,
colored yellow and fragrant.  (Collected as small seedling,
raised and shown by Mr. & Mrs. E. F. Allen)  F. C. C. (RHS) 1972

cl.
      -neriiflorum ssp. neriiflorum  Euchaites Group (1/2)
Rajah-
      -thomsonii ssp. thomsonii (1/2)

Flowers 5-lobed, 2.75in(7cm) across, of dark bright red  without
markings, and held in trusses of 12.  (Gen. Harrison, reg. 1962)

```
                 -Pink -catawbiense (1/8)
              -Twins-
cl.      -Sunset-     -haematodes ssp. haematodes (1/8)
         - Yates-                -griffithianum (1/8)
Ralph Pape-    -Leah Yates--Mars (selfed)-
         -                        -unknown (1/8)
         -  -Catalgla--catawbiense var. album Glass (1/4)
         -Pink -
         Punch-fortunei ssp. fortunei (1/4)
```

3ft(.9m)           -15°F(-26°C)            ML

Ball-shaped trusses of 13-17 flowers, deep red with  spirea red
throat and faint olive spotting, 3in(7.6cm) across, with 5 wavy
lobes. Floriferous. Plant almost twice as broad as high, with
arching branches; very dark green elliptic leaves, 4.5in(11.5cm)
long, retained 2 years. (Henry Yates cross, 1968; reg. by  Mrs.
Maletta Yates, 1984)

cl.
```
           -cerasinum (red form) (1/2)
Ralph Purnell-         -arboreum ssp. arboreum (1/4)
           -Doncaster-
                   -unknown (1/4)
```

Flowers dark red, 5-lobed, in compact trusses of 20.  (Intro. by
Thacker, 1942; reg. 1962)

cl.
      -fastigiatum (1/2)
Ramapo-
      -minus var. minus Carolinianum Group (1/2)

2ft(.6m)          -25°F(-32°C)            EM            3/4

A nearly spherical plant that grows in sun or shade, more compact in full sun.  Leaves almost circular, 1in(2.5cm) long; new
growth dusty grey blue, in winter a deep metallic color.  Flowers 1in(2.5cm) across, of lively pinkish violet, in great abundance.  (G. Guy Nearing, exhibited 1940)    See color illus. VV
p. 52.

```
                         -neriiflorum ssp. neriiflorum (1/8)
               -F. C. Puddle-
g.     -Ethel-          -griersonianum (1/8)
       -     -forrestii ssp. forrestii Repens Group (1/4)
Ramillies-          -thomsonii ssp. thomsonii (1/4)
       -    -Shilsonii-
       -Redwing-      -barbatum (1/8)
       -          -thomsonii ssp. thomsonii
          -Barclayi-       -arboreum ssp. arboreum
               -Glory of -           (1/16)
               Penjerrick-griffithianum (1/16)
```

Blood red flowers.  (Lord Aberconway, 1941)

cl.
```
        -barbatum (1/4)
   -Huntsman-
Ramona-    -campylocarpum ssp. campylocarpum Elatum Group
   -neriiflorum ssp. neriiflorum (1/2)          (1/4)
```

6ft(1.8m)          -15°F(-26°C)            ML

Flowers 2in(5cm) long, campanulate, and deep red, held in flat
trusses.  (Raised by Sir John Ramsden; reg. by Sir William Pennington-Ramsden, Muncaster Castle, 1965)

cl.

Ramsey Gold*--form of keiskei

2ft(.6m)          -10°F(-23°C)            EM            4/4

A yellow-flowered selection by G. Guy Nearing, after raising
several generations of seedlings.

cl.
```
           -pubescens (1/4)
     -Chesapeake-
Ramsey Tinsel-    -keiskei (1/4)
     -unknown (1/2)
```

1ft(.3m)          -10°F(-23°C)            EM

A dwarf with leaves 1in(2.5cm) long.  Flowers ocher yellow, 1in
(2.5cm) wide, held in clusters of 3-4; corolla of 8-10 very narrow, separate petals.  See also Elam, Elsmere.  (G. Guy Nearing,
intro. 1963; reg. 1972)

cl.

Random Harvest--fortunei ssp. discolor Houlstonii Group (W 648A)

5ft(1.5m)          -5°F(-21°C)            M            2/3

Trusses hold 10-12 flowers, white tinged pink with some lettuce
green in upper throat, funnel-campanulate, 7- or 8-lobed. Foliage dark green above, paler beneath. (Collector E. H. Wilson;
raised & reg. by R. N. Stephenson Clarke)   A. M. (RHS) 1981

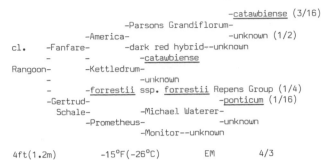

```
                                    -catawbiense (3/16)
                    -Parsons Grandiflorum-
            -America-                 -unknown (1/2)
cl.  -Fanfare-      -dark red hybrid--unknown
     -        -           -catawbiense
Rangoon-     -Kettledrum-
     -                    -unknown
     -       -forrestii ssp. forrestii Repens Group (1/4)
     -Gertrud-                        -ponticum (1/16)
       Schale-            -Michael Waterer-
          -Prometheus-                  -unknown
                         -Monitor--unknown
```

4ft(1.2m)        -15°F(-26°C)         EM        4/3

Plant almost twice as wide as tall at maturity; elliptic, dark
green leaves to 3.75in(9.5cm) long, retained 2 years. Flowers
medium to dark red, widely funnel-shaped, 2in(5cm) across; dome-
shaped tusses of 7-8. Dark red winter buds and leaf petioles.
(David G. Leach, 1973)


cl.
            -griffithianum (1/2)
Raoul Millais-
            -unknown (1/2)

Very large trusses of salmon pink flowers with a pale center and
faintly spotted upper lobe. Fades lighter without changing hue.
(M. Koster, before 1922)   A. M. (RHS) 1935


cl.
        -catawbiense (1/2)
Raphael-
        -unknown (1/2)

Flowers dark ruby red, with dark brown markings.   A registered
hybrid of the same name and unknown parentage by  A. Waterer is
dated 1868. (T. J. R. Seidel, cross 1907; intro. 1915)


g.
        -griersonianum (1/2)
Rapture-
        -arboreum ssp. zeylanicum (1/2)

Flowers porcelain rose. (Lord Aberconway, 1950)


                        -arboreum ssp. arboreum (1/8)
cl.      -Gill's Triumph-
     -Apache-           -griffithianum (1/8)
Rascal-    -thompsonii ssp. thompsonii (1/4)
        -neriiflorum ssp. neriiflorum Euchaites Group (1/2)

Deep red flowers. (Gen. Harrison cross, 1947; intro. 1957)


cl.
            -yakushimanum ssp. yakushimanum (1/2)
Raspberry Ripple
            -Pauline--unknown (1/2)

White flowers lightly flushed pink towards wide rose red edges,
upper lobe flecked dark purplish red. Trusses of 16. (RHS Garden, Wisley; reg. 1982)


cl.              -griffithianum (1/4)
                -Mars-
Raspberry Sherbet-    -unknown (1/4)
                 -Catanea--catawbiense (1/2)

6ft(1.8m)        -15°F(-26°C)         ML

Flowers spirea red with greyed white blotch, widely funnel-cam-
panulate, 2.75in(.7cm) wide, 5-lobed; conical trusses of 17-18.
Upright plant habit; medium grass green leaves retained 2 years.
(Walter J. Blyskal, reg. 1980)


cl.                -griffithianum (1/4)
        -Leah Yates--Mars (selfed)-
Ravels-                -unknown (1/4)
     -          -catawbiense (1/4)
        -Pink Twins-
                -haematodes ssp. haematodes (1/4)

Flowers neyron rose with darker veins, a large lime green dorsal
blotch, openly funnel-shaped, 3.5in(8.9cm) wide; conical trusses
of 12. Free-flowering. Shrub upright, broader than tall; dense
medium green foliage, held 2 years. See also Shaazam. (Henry
Yates; intro. by Mrs. Maletta Yates; reg. 1977)


cl.
     -fortunei ssp. fortunei (21/32)
Ray-    -wardii var. wardii (1/4)
    -        -           -griffithianum (1/32)
    -Idealist-      -Kewense-
    -        -Aurora-   -fortunei ssp. fortunei
    -Naomi-  -thomsonii ssp. thomsonii (1/16)
            -fortunei ssp. fortunei

5ft(1.5m)        0°F(-18°C)         M        3/3

Pale yellow flowers, 4in(10.2cm) wide, in lax spherical trusses
of 10-12. Roots easily. (D. James, reg. 1958) P. A. (RHS) 1956


cl.      -campylocarpum ssp. campylocarpum (1/2)
        -Unique-
Raydel-    -unknown (1/2)
     -                 -campylocarpum ssp. campylocarpum
        -Mrs. W. C. Slocock-
                    -unknown

6ft(1.8m)        -10°F(-23°C)         E

Mimosa yellow flowers, throat tinged chartreuse, edges of neyron
rose, openly funnel-shaped, 2.5in(6.4cm) wide; spherical trusses
of 12-14. Floriferous. Plant rounded, as broad as tall; elliptic leaves held 2 years; new growth light green, red-tipped. (D.
W. James, cross 1950; reg. by Amy A. Hitchcock, 1980)


cl.
        -spinuliferum (1/2)
Razorbill-
        -unknown (1/2)

3ft(.9m)        0°F(-18°C)         EM        3/3

Flowers light rose pink, with variable darker pink overtones,
small, tubular-shaped; conical compact trusses of 12. Very floriferous. Vigorous plant, broader than tall; dark green leaves
1.75in(4.5cm) long, hair-fringed, reverse scaly. (Peter A. Cox)
P. C. (RHS) 1976; A. M. (RHS) 1981; F. C. C. (RHS) 1983


cl.  -Mrs. J. G. Millais--unknown (1/2)
     -                -fortunei ssp. discolor (1/16)
Razzle-      -Lady      -
Dazzle*      -Bessborough-campylocarpum ssp. campylo-
     -      -Jalisco-          carpum Elatum Group (1/16)
     -            -dichroanthum ssp. dichroanthum
     -Cheyenne-  -Dido-                    (1/16)
           -          -decorum (1/16)
           -      -griffithianum (1/8)
           -Loderi-
                -fortunei ssp. fortunei (1/8)

6ft(1.8m)        -5°F(-21°C)         ML        4/3

Large trusses of flowers with a deep pinkish maroon flare on the
upper petal. Petal edges open with a deeper pink picotee which
contrasts with lighter pink center. Medium-sized leaves, fern
green, cover the entire plant. See also Great Scott, White Gold.
(Harold E. Greer, intro. 1982) See color illus. HG p. 112.

cl.
    -<u>catawbiense</u> (1/2)
Rebe*
    -<u>smirnowii</u> (1/2)

Ruby red flowers, pale green markings.   (T. J. R. Seidel, cross
1907; intro. 1915)


cl.
    -Anton van Welie, q.v.
Record-Professor J. H. Zaayer, q.v.
    -Annie E. Endtz, q.v.

The exact combination of the above parentage is uncertain.
Parents were all Pink Pearl hybrids. Large pyramidal trusses
carry up to 14 flowers, deep pink with slightly darker veining,
aging to light pink. (P. van Nes, reg. 1962)   A. M. 1958 (after trial at Boskoop)


g.
          -<u>arboreum</u> ssp. <u>arboreum</u> (1/2)
Red Admiral-
          -<u>thomsonii</u> ssp. <u>thomsonii</u> (1/2)

Early blooming, tall, tree-like plant with campanulate flowers
of glowing red. (J. C. Williams, Caerhays Castle, before 1958)


g.
        -<u>grande</u> (1/2)
Red Argenteum-
          -<u>arboreum</u> ssp. <u>arboreum</u> (1/2)

A parent of Diogenes. (Exhibited, 1926)


cl.
     -Essex Scarlet--unknown (1/2)
Red Bells-
     -<u>williamsianum</u> (1/2)

Bell-shaped purplish red flowers in trusses of 6. Compact plant
habit; ovate leaves dark green, paler below. See also Oudjik's
Sensation. (Dietrich Hobbie cross; Le Feber & Co., Boskoop, intro. 1966; reg. 1969)


g.         -<u>neriiflorum</u> ssp. <u>neriiflorum</u> (1/4)
    -F. C. Puddle-
Red Brocade-       -<u>griersonianum</u> (1/4)
      -<u>arboreum</u> ssp. <u>arboreum</u> (1/2)

Deep red. (Lord Aberconway, 1943)


g.
     -<u>sanguineum</u> ssp. <u>didymum</u> (1/2)
Red Cap-
     -<u>facetum</u> (<u>eriogynum</u>) (1/2)

Flowers in late season, blood red or deep plum red. See also
Borde Hill Red Cap, Exbury Red Cap, Townhill Redcap. (J. B. Stevenson, 1935)


                  -<u>catawbiense</u> (1/8)
cl.         -Parsons Grandiflorum-
       -America-            -unknown (3/8)
Red Carpet-     -dark red hybrid--unknown
       -<u>forrestii</u> ssp. <u>forrestii</u> Repens Group (1/2)

Flowers guardsman red, tinged black around edges, changing to
post office red towards throat, veined darker, campanulate, 2.25
in(5.7cm) across; lax trusses hold 4. Floriferous. Plant 2.7ft
(.8m) tall and 5.4ft(1.6m) broad, at time of award. (Dietrich
G. Hobbie cross, 1945; Walter Schmalscheidt, reg. 1983   A. M.
(Wisley Trials) 1983


cl.
       -G. A. Sims--unknown (1/2)
Red Chief-
       -<u>facetum</u> (<u>eriogynum</u>) (1/2)

Panoply g. Red flowers. (Sir Giles Loder, 1948)


cl.            -<u>griersonianum</u> (1/4)
       -Tally Ho-
Red Cloud-      -<u>facetum</u> (<u>eriogynum</u>) (1/4)
      -Corona--unknown (1/2)

5ft(1.5m)        5°F(-15°C)        ML       3/3

Open growth habit; medium green leaves, 6.5in(16.5cm) long, pale
beneath. Flowers funnel-campanulate, 3.5in(9cm) across, colored
claret rose to scarlet inside; dome-shaped trusses of 18. Easy
to propagate. (Rudolph Henny)   P. A. (ARS) 1953


cl.

Red Collar--form of <u>edgeworthii</u> (K-W 20840)

4ft(1.2m)       15°F(-10°C)       EM       4/4

Trusses have 3-5 flowers, white suffused pink on the three upper
lobes, deeper on reverse as a diffused central band. Also some
light yellow orange spotting in throat. Leaves dark green,
glossy. (Collector Kingdon-Ward; raised & reg. by Sir Giles
Loder) F. C. C. (RHS) 1981


                   -<u>griffithianum</u> (1/8)
cl.         -Queen Wilhelmina-
       -Britannia-       -unknown (3/8)
Red Crest-     -Stanley Davies--unknown
       -<u>arboreum</u> ssp. <u>delavayi</u> var. <u>delavayi</u> (1/2)

Flowers cardinal red, slightly spotted on upper lobe. See also
Fire Prince, Fire Walk. (K. Van de Ven, 1973)


cl.
     -Moser's Maroon--unknown (1/2)
Red Cross*     -<u>griersonianum</u> (1/4)
     -Tally Ho-
          -<u>facetum</u> (<u>eriogynum</u>) (1/4)

A vigorous plant, with large flowers of dark scarlet red. (Slocock Nurseries)


cl.
       -<u>griersonianum</u> (1/2)
Red Dragon-
       -<u>thomsonii</u> ssp. <u>thomsonii</u> (1/2)

Hecla g. A parent of Rowena. (Lord Aberconway, 1943)


             -<u>arboreum</u> ssp. <u>arboreum</u> (5/16)
g.     -Cardinal-   -<u>thomsonii</u> ssp. <u>thomsonii</u> (1/8)
-         -Barclayi-     -<u>arboreum</u> ssp. <u>arboreum</u>
Red Ensign-         -Glory of  -
-               Penjerrick-<u>griffithianum</u> (1/16)
     -<u>griersonianum</u> (1/2)

Flowers rose red. (Lord Aberconway, 1942)


cl.         -<u>ponticum</u> (1/2)
    -Anah Kruschke-
Red Eye*       -unknown (1/2)
-            -<u>ponticum</u>
    -Purple Splendour-
          -unknown

5ft(1.5m)       -10°F(-23°C)      ML      4/4

A dependable, sun-tolerant variety, that buds young, propagates
easily. Flowers purplish red with green gold eye slowly turning
red. Plant rounded; dense glossy foliage. (H. E. Greer, intro.
1982)   See color illus. HG p. 144.

cl.
      -<u>catawbiense</u> (1/2)
Red Frilled-
        -unknown (1/2)

3.5ft(1.05m)          -15°F(-26°C)          M

Synonym Low Red Frilled. Flowers funnel-shaped, 5-lobed, very light red purple, reverse Tyrian rose, in conical trusses of 12-15. Plant 2/3 as wide as tall; leaves glossy dark green, elliptic, held 2 years. (E. Mezitt, Weston Nurseries, reg. 1983)

cl.
          -<u>maximum</u> (1/4)
      -Halopeanum-
Red Glow-      -<u>griffithianum</u> (1/4)
     -<u>thomsonii</u> ssp. <u>thomsonii</u> (1/2)

Gem g. Bright red flowers. (Loder, 1949)

              -<u>thomsonii</u> ssp. <u>thomsonii</u> (1/2)
g.      -Chanticleer-
   -           -<u>facetum</u> (eriogynum) (1/4)
Red Hackle-      -<u>thomsonii</u> ssp.<u>thomsonii</u>
   -      -Shilsonii-
   -      -      -<u>barbatum</u> (1/8)
   -Redwing-      -<u>thomsonii</u> ssp. <u>thomsonii</u>
      -Barclayi-      -<u>arboreum</u> ssp. <u>arboreum</u>
         -Glory of -      (1/16)
        Penjerrick-<u>griffithianum</u> (1/16)

Deep red flowers. (Lord Aberconway, 1946)

cl.
           -<u>catawbiense</u> (1/4)
     -Atrosanguineum-
Red Head*      -unknown (1/4)
     -<u>griersonianum</u> (1/2)

5ft(1.5m)        -5°F(-21°C)        M        3/3

Red flowers that prefer afternoon shade. (Joseph B. Gable)

cl.
             -<u>griffithianum</u> (1/8)
cl.      -Queen Wilhelmina-
   -Britannia-      -unknown (3/8)
Red Hot-      -Stanley Davies--unknown
   -      -<u>forrestii</u> ssp. <u>forrestii</u> Repens Group (1/4)
   -Elizabeth-
      -<u>griersonianum</u> (1/4)

Flowers of orient red, shaded mandarin red in throat. (Rudolph Henny, reg. 1962)

cl.

Red Imp--<u>haematodes</u> ssp. <u>haematodes</u>

A dwarf, compact plant with small leaves. Flowers medium red outside, lighter inside. Midseason. (The Bovees, reg. 1962)

cl.
          -<u>neriiflorum</u> ssp. <u>neriiflorum</u>
cl.      -F. C. Puddle-      (1/8)
   -Phoebus-      -<u>griersonianum</u> (1/8)
Red Lacquer-      -<u>haematodes</u> ssp. <u>haematodes</u> (1/4)
   -<u>forrestii</u> ssp. <u>forrestii</u> Repens Group (1/2)

A dwarf prostrate plant; elliptic leaves 2in(5cm) long. Campanulate flowers, post office red, 1.7in(4.5cm) long, held in clusters of 5. (Collingwood Ingram)  A. M. (RHS) 1967

cl.
        -<u>griersonianum</u> (1/4)
   -Tally Ho-
Red Lion-      -<u>facetum</u> (eriogynum) (1/4)
   -<u>catawbiense</u> (Gable's red) (1/2)

5ft(1.5m)        -10°F(-23°C)        L

Upright, open plant; elliptic leaves with thin greyed indumentum, held 2 years. Guardsman red flowers, faintly spotted, 2in (5cm) across; lax trusses of 10. (G. Guy Nearing, reg. 1973)

          -<u>griffithianum</u> (3/8)
cl.      -Loderi King George-
   -      -<u>fortunei</u> ssp. <u>fortunei</u> (1/4)
Red Loderi*      -<u>griffithianum</u>
   -      -Queen Wilhelmina-
   -Earl of Athlone-      -unknown (3/8)
      -Stanley Davies--unknown

Parent of Halfdan Lem. (Halfdan Lem)

cl.
      -<u>elliottii</u> (1/2)
Red Majesty-
     -<u>strigillosum</u> (1/2)

5ft(1.5m)        10°F(-12°C)        E        4/3

Plant broader than tall; leaves to 8in(20.3cm) long. Flowers of currant red, heavily spotted, campanulate, to 3in(7.6cm) wide, in rounded trusses of 10. (H. L. Larson, reg. 1965)

cl.
        -<u>elliottii</u> (1/4)
   -Fusilier-
Red Mill-      -<u>griersonianum</u> (1/4)
   -Ruddy--unknown (1/2)

Tall, compact plant; leaves 6in(15.2cm) long. Currant red flowers, in trusses of 16. (Marshall Lyons, reg. 1972)

cl.

Red Night      Parentage unknown

A parent of an Exbury hybrid, Querida.

            -Beauty of-<u>griffithianum</u> (5/16)
      -Norman Gill- Tremough-
cl.      -Anna-      -<u>arboreum</u> ssp. <u>arboreum</u>
   -      -      -<u>griffithianum</u>      (1/16)
Red Olympia-      -Jean Marie de Montague--unknown (1/8)
   -      -<u>elliottii</u> (1/4)
   -Fusilier-
      -<u>griersonianum</u> (1/4)

5ft(1.5m)        5°F(-15°C)        M        5/3

Cardinal red buds opening to lighter red flowers, spotted black, the reverse medium red, openly cup-shaped, 4.25in(10.8cm) wide; domed trusses of 13-16. Plant upright, half as broad as tall; glossy dark leaves 7.5in(18.5cm) long, held 3 years. See also Muriel Pearce. (Halfdan Lem cross; reg. by L. J. Pierce, 1975)

          -<u>dichroanthum</u> ssp. <u>dichroanthum</u> (1/16)
      -Fabia-
    -unnamed-      -<u>griersonianum</u> (1/16)
cl.  -unnamed- hybrid-<u>haematodes</u> ssp. <u>haematodes</u> (1/8)
 - hybrid-      -<u>griffithianum</u> (5/16)
Red  -      -Queen Wilhelmina-
Paint-      -Earl of-      -unknown (7/16)
   -      Athlone-Stanley Davies--unknown
   -      -<u>griffithianum</u>
   -Jean Marie de Montague-
      -unknown

4ft(1.2)        0°F(-18°C)        E

Plant upright, well-branched; narrow, dark green, glossy leaves, 5in(12.7cm) long, held 4 years. Flowers clear bright red, a few tiny spots, openly funnel-shaped, 3in(7.6cm) across, 5-lobed, in spherical trusses of 10-14. (Cross by Dr. David Goheen; reg. by Charles McNew, 1974)

          -<u>williamsianum</u> (1/4)
cl.      -Wilbar-      -<u>thomsonii</u> ssp. <u>thomsonii</u> (1/8)
   -      -Barclayi-      -<u>arboreum</u> ssp. <u>arboreum</u>
Red Pantaloons-      -Glory of -      (1/16)
   -      Penjerrick-<u>griffithianum</u> (1/16)
   -<u>chamaethomsonii</u> var. <u>chamaethauma</u> (1/2)

1.5ft(.45m)        15°F(-10°C)        E

Hose-in-hose flowers of heavy substance, currant red, held in
lax trusses of 4. Dwarf plant, rounded, as wide as tall with
decumbent branches. See also Rose Pantaloons. (Dr. Carl G.
Heller, reg. 1977)

```
                              -Jan Dekens--unknown (13/16)
              -Pink Petticoats-
cl.     -              -           -Queen  -griffithianum
        -              -Britannia-Wilhelmina-      (3/16)
Red Petticoats*        -              -unknown
        -              -Stanley Davies--unknown
        -                    -griffithianum
        -      -Queen Wilhelmina-
        -Britannia-              -unknown
              -Stanley Davies--unknown
```

Excellent tight trusses of pure red, frilly flowers. Foliage of
average quality and quantity. (John G. Lofthouse)

```
           -forrestii ssp. forrestii Repens Group (1/2)
Red Poll-     -meddianum var. meddianum (1/4)
        -Rocket-
              -strigillosum (1/4)
```

Bright ruby red flowers in trusses of 8. Parentage confirmed by
Hydon Nurseries. (A. F. George, Hydon Nurseries, reg. 1974)

```
cl.          -dichroanthum ssp. dichroanthum (1/4)
        -Golden Horn-
Red Puff-      -elliottii (1/4)
        -Catanea--catawbiense var. album (1/2)
```

2ft(.6m)        -15°F(-26°C)        ML

Red flowers 2in(5cm) across, speckled darker red, in 16-flowered
ball-shaped trusses 5in(12.7cm) wide. Growth habit dwarf, decumbent; leaves borne in drooping rosettes, and held 2 years. (G.
Guy Nearing, 1973)

```
g.             -griffithianum (3/8)
        -Gill's Crimson-
Red Queen-      -unknown (1/4)
        -      -thomsonii ssp. thomsonii (1/4)
        -Barclayi-      -arboreum ssp. arboreum (1/8)
              -Glory of  -
              Penjerrick-griffithianum
```

Rounded trusses of 10 cardinal red flowers, openly funnel-campanulate, 2.2in(5.5cm) wide, 5-lobed. Leaves to 6in(15cm) long,
of dull dark green, paler underneath. (Cross by The Second Lord
Aberconway, Bodnant)    A. M. (RHS) 1980

cl.

Red River        For parentage see Red Sea, next column

5ft(1.5m)        -20°F(-29°C)        L

Flowers bright red, shading pink toward an almost white center,
with yellow dorsal flare; full pyramidal trusses. Large leaves.
(David G. Leach, reg. 1984)

```
cl.          -Moser's Maroon--unknown (1/4)
        -Bibiana-
Red Rock-      -arboreum ssp. arboreum (1/4)
        -elliottii (1/2)
```

Gibraltor g. Flowers cardinal red, brighter in throat, upper
lobes lightly spotted black, widely funnel-campanulate, to 3.5in
(8.9cm) wide, 5-lobed; full rounded trusses of 24. Narrowly elliptic dark green leaves, to 8in(20.3cm) long. (Edmund de
Rothschild)    A. M. (RHS) 1970

```
                      -thomsonii ssp.
              -Ascot Brilliant-  thomsonii (5/8)
              -              -unknown (3/16)
        -J. G. Millais-        -griffithianum (1/16)
g.      -         -George Hardy-
        -         -Pink -    -catawbiense (1/16)
Red Rover-      Pearl-      -arboreum ssp.
        -         -Broughtonii-  arboreum (1/16)
        -                    -unknown
        -thomsonii ssp. thomsonii
```

Small flowers of deep crimson red. (Whitaker, 1939)

```
cl.      -thomsonii ssp. thomsonii (1/4)
     -Barclayi-      -arboreum ssp. arboreum (1/8)
Red Rum-    -Glory of -
     -      Penjerrick-griffithianum (1/8)
     -forrestii ssp. forrestii Repens Group (1/2)
```

Loose trusses of 4-5 flowers, scarlet red, funnel campanulate,
2in(5cm) wide, 5-lobed. Leaves 3.3in(8.5cm) long, medium green,
oblong. (Crown Estate, Windsor)    A. M. (RHS) 1974

```
cl.
        -Mount Mitchell (Leach)--maximum (1/2)
        -         -griffithianum (1/8)
Red Sea-    -Mars-
        -      -unknown (9/32)
        -unnamed-          -Parsons    -catawbiense (3/32)
        hybrid-    -America- Grandiflorum-
        -         -         -unknown
        -Fanfare-    -dark red hybrid--unknown
              -         -catawbiense
              -Kettledrum-
                    -unknown
```

5ft(1.5m)        -25°F(-32°C)        L

Flowers light red purple shading through rose red to near white
in throat, widely funnel-shaped, 2.75in(7cm) across, 5- to 6-
lobed; upright trusses of 20. Floriferous. Plant broader than
tall; dull green leaves 4.5in(11.5cm) long, held 2 years. (David
G. Leach, reg. 1984)

```
cl.          -griersonianum (3/4)
        -Azor-
Red Snapper*    -fortunei ssp. discolor (1/4)
        -griersonianum
```

Flowers pale geranium with darker edges.    Plant of loose habit.
(Rudolph Henny cross, 1946; intro. 1956)

```
                        -griffithianum (1/8)
              -George Hardy-
---     -Mrs. Lindsay Smith-      -catawbiense (1/8)
        -            -Duchess of Edinburgh--unknown
Red Star-        -Moliere--unknown (3/4)
```

Bright scarlet flowers in loose trusses. (M. Koster & Sons)

```
g.             -thomsonii ssp. thomsonii (3/4)
        -Ascot Brilliant-
Red Star-      -unknown (1/4)
        -thomsonii ssp. thomsonii
```

Velvety crimson flowers, spotted black on upper petal. (Loder)

```
                    -griffithianum (1/8)
              -Queen  -
cl.     -Earl of -Wilhelmina-unknown (3/8)
        - Athlone-
Red Tape-      -Stanley Davies--unknown
        -      -dichroanthum ssp. dichroanthum (1/4)
        -Fabia-
              -griersonianum (1/4)
```

Clear bright red flowers. (Rudolph Henny, cross 1944)

```
cl.            -griffithianum (1/4)
          -Mars-
Red Torch-     -unknown (3/4)
          -unknown
```

Trussesof 16-23 flowers, cardinal red  with small black blotch.
(Research Station for Woody Nursery Crops, Boskoop, 1979)


```
cl.              -elliottii (1/4)
          -Fusilier-
Red Velvet-     -griersonianum (1/4)
          -williamsianum (1/2)
```

2.5ft(.75m)         0°F(-18°C)          M          3/3

Plant broader than tall; leaves to 4.5in(11.5cm) long.  Flowers
coral red, campanulate. An attractive plant, like an improved
Jock, q.v.   (H. L. Larson, 1964)


```
                          -griffithianum (9/16)
               -Beauty of-
            -Norman- Tremough-arboreum ssp. arboreum (1/16)
cl.         - Gill -
      -Anna-       -griffithianum
Red    -    -              -griffithianum
Walloper* -Jean Marie de Montague-
      -      -griffithianum      -unknown (3/8)
       -Marinus-
        Koster-unknown
```

6ft(1.8m)         -5°F(-21°C)          M          4/4

Deep rose red in bud, opens slowly through shades of pink to a
beautiful pastel pink.  Matures to a strong plant of large stature and handsome foliage.   See also Lem's Monarch and Point Defiance.  (Halfden Lem)


```
cl.        -griffithianum (1/4)
       -Mars-
Redberth-    -unknown (1/4)
        -Koichiro Wada--yakushimanum ssp. yakushimanum (1/2)
```

3ft(.9m)         -5°F(-21°C)          ML

Plant upright, well-branched; leaves very variable in size and
shape, to 6in(15cm) long.  Flowers of strong red, openly funnel-shaped, 3in(7.6cm) wide; conical trusses of 20. (Dr. W. A.
Reese, 1976)


cl.


Redpoll     See Red Poll


```
            -dichroanthum ssp. dichroanthum (1/4)
g.    -Dante-
   -       -facetum (eriogynum) (1/2)
Redskin            -souliei (1/16)
   -           -Soulbut-
   -    -Vanessa-      -Sir Charles Butler--fortunei ssp.
   -Eudora-    -griersonianum (1/8)    fortunei (1/16)
       -facetum (eriogynum)
```

Scarlet flowers. (Lord Aberconway, 1950)


```
                      -fortunei ssp. discolor (1/4)
cl.        -Lady Bessborough-
   -                  -campylocarpum ssp. campylocarpum
Redskin Chief-                    Elatum Group (1/4)
   -       -haematodes ssp. haematodes (1/4)
       -May Day-
            -griersonianum (1/4)
```

Red flowers. (Sunningdale Nurseries, 1955)


```
cl.
       -haematodes ssp. haematodes (3/4)
Redwax-     -haematodes ssp. haematodes
    -May Day-
             -griersonianum (1/4)
```

2.5ft(.75m)         5°F(-15°C)          M          3/4

Flowers orient red and unspotted, funnel-campanulate, 2.25in(5.7
cm) wide, in loose trusses of 3-5.  Plant dense, spreading; dark
green leaves, 3in(7.6cm) long with heavy tan indumentum beneath.
Sometimes listed as Red Wax. (R. Henny, 1956)  P. A. (ARS) 1958


```
cl.           -thomsonii ssp. thomsonii (1/2)
     -Shilsonii-
Redwing-       -barbatum (1/4)
    -        -thomsonii ssp. thomsonii
    -Barclayi-      -arboreum ssp. arboreum (1/8)
             -Glory of  -
           Penjerrick-griffithianum (1/8)
```

Flowers pale rose, with a darker stripe down the center of each
lobe. (Lord Aberconway)  F. C. C. (RHS) 1937


```
cl.
   -minus var. minus Carolinianum Group (1/2)
Regal*
   -dauricum (1/2)
```

4ft(1.2m)         -25°F(-32°C)          E          4/4

One of the selected forms of P.J.M. propagated and distributed
by Weston Nurseries.   Grows 4-7 inches per year.  Blooms after
Victor and before Elite; strong lavender pink flowers.


cl.


Reginald Childs--form of uvarifolium

4ft(1.2m)         5°F(-15°C)          E          4/4

White flowers suffused light magenta with large blotch of bright
cardinal red in trusses of 20.  Leaves glossy dark green with
ash grey plastered indumentum below. (Collector unknown; raised
by RBG of Kew, from Wakehurst Pl.)  A. M. (RHS) 1976


```
cl.
       -dichroanthum ssp. scyphocalyx (1/2)
Reginald Farrer-
        -fortunei ssp. discolor (1/2)
```

Apricot orange flowers. (Thacker, 1942)


```
cl.
       -dendricola Taronense Group (1/2)
Reine Long-      -ciliicalyx (1/4)
       -Else Frye-
             -unknown (1/4)
```

5ft(1.5m)         20°F(-7°C)          EM

Very fragrant, large, open white flowers, streaked  rose  with a
large yellow blotch, held in loose trusses of 4-5. (Dr. Paul J.
Bowman cross; intro. by E. H. Long, 1964)


```
cl.           -haematodes ssp. haematodes (1/4)
       -May Day-
Relaxation-  -griersonianum (1/4)
       -Koichiro Wada--yakushimanum ssp. yakushimanum (1/2)
```

3ft(.9m)         -5°F(-21°C)          M

Flowers cardinal red with dark red basal nectaries, campanulate,
2in(5cm) across, 5-lobed; lax trusses of 12-15.  Rounded plant,
as wide as tall; elliptic leaves held 2-3 years, with heavy indumentum, biscuit-colored. (Cecil C. Smith, reg. 1983)

```
g.              -dichroanthum ssp. dichroanthum (1/4)
          -Dante-
Rembrandt-    -facetum (eriogynum) (1/4)
          -          -griffithianum (1/4)
          -Sunrise-
                    -griersonianum (1/4)
```

Scarlet flowers.  (Lord Aberconway, 1950)


```
g.
      -valentinianum (1/2)
Remo-
      -lutescens (1/2)
```

2.5ft(.75m)          10°F(-12°C)          EM          3/2

Compact habit with narrow leaves, as lutescens.  Flowers in
loose clusters as valentinianum; buds vivid lemon yellow, flowers bright yellow bells.  For mild climates. (J. B. Stevenson,
Tower Court, 1943)  See color illus. F p. 62.


```
g.
      -forrestii ssp. forrestii Repens Group (1/2)
Remus-
      -beanianum (1/2)
```

Carmine flowers. (Lord Aberconway, intro. 1946)


```
                        -griffithianum (3/8)
              -George Hardy-
          -Peter -              -catawbiense (1/8)
---       -Koster-          -arboreum ssp. arboreum
          -    -unnamed-Doncaster-                 (1/16)
Renaissance-  hybrid-       -unknown (3/16)
          -        -unknown
          -                   -griffithianum
          -Loderi King George-
                        -fortunei ssp. fortunei (1/4)
```

Flowers light red. (J. Barto cross; Mrs. H. Fogg named & intro.)


```
cl.          -catawbiense (1/2)
      -Annedore-
Renata-       -unknown (1/2)
      -           -catawbiense
      -Mrs. Milner-
                    -unknown
```

Glowing dark purplish red flowers,  with very faint brown markings. (T. J. R. Seidel cross, 1907; intro. 1915)


```
g.
              -arboreum ssp. arboreum (1/2)
Rendall's Scarlet-
              -ponticum (1/2)
```

Bright red flowers. (Lord Aberconway, 1946)


```
cl.
                    -Catalgla--catawbiense var. album
Renee         -                Glass (1/2)
Shirey*--Caltwins (selfed)-    -catawbiense (1/4)
                    -Pink -
                    -Twins-haematodes ssp. haematodes
                                        (1/4)
```
Light purplish pink buds opening to flowers of strong purplish
pink with lighter shading and specks of vivid yellow green. (W.
E. Delp)


```
cl.
      -elliottii (1/2)
Renhaven-                    -dichroanthum ssp. dichroanthum (1/8)
      -           -Fabia-
      -Umpqua Chief-    -griersonianum (1/4)
      -         -griersonianum
            -Azor-
                  -fortunei ssp. discolor (1/8)
```

---

4.5ft(1.35m)          -5°F(-21°C)          EM          4/3

Compact plant; very dark, lanceolate leaves to 6in(15.2cm) long.
Flowers glowing dark red lightly speckled, funnel-campanulate, 3
in(7.6cm) wide; lax trusses of 6-8. (Del James, reg. 1958)
P. A. (ARS) 1955


```
cl.
      -Pauline--unknown (1/2)
Renoir-
      -yakushimanum ssp. yakushimanum (1/2)
```

3ft(.9m)          0°F(-18°C)          M          3/3

Flowers of neyron rose with a white throat and crimson spots on
upper lobe, deeply campanulate; rounded trusses of 11. Floriferous. See also Pink Ghost, Telestar. (Francis Hanger cross,
RHS Garden, Wisley)  A. M. (RHS) 1961


```
g.
      -forrestii ssp. forrestii Repens Group (1/2)
Reparm-
      -parmulatum (1/2)
```

Pale rose flowers. (Lord Aberconway, intro. 1946)


g.

Replet     See Riplet


```
g. & cl.
      -lacteum (1/2)
Repose-
      -fortunei ssp. discolor (1/2)
```

Much like lacteum in size and  growth habit. Large, campanulate
flowers of creamy white flushed green, giving the effect of pale
lemon; large trusses of 18. See also Welcome Stranger. (Intro.
by Edmund de Rothschild, 1950; exhibited by Slocock) A. M. (RHS)
1956    See color illus. PB pl. 27.


```
g. & cl.
      -facetum (eriogynum) (1/2)
Retreat-
      -Lowinsky hybrid--unknown (1/2)
```

Tall plant of open habit. Red flowers shading to pink,  in full
trusses. (Rothschild, 1936)


```
cl.
              -Sappho--unknown
Return to Paradise-
              -unknown
```

5ft(1.5m)          -10°F(-23°C)          M

Fragrant white flowers with golden tan spotting, in dome-shaped
trusses of 16. Upright plant, over half as wide as high; leaves
glossy yellowish green, retained 2 years. (Mrs. Halsey A. Frederick, Jr., reg. 1978)


cl.

Reuthe's Purple--form of lepidotum

1ft(.3m)          0°F(-18°C)          EM          4/4

Elliptic foliage .8in(2cm) long, densely scaly beneath, aromatic; flowers bright purple violet, rotate-campanulate, 1.5in(3.8
cm) across, in small clusters. A Himalayan species. (Collector
unknown; raised by G. Reuthe)  A. M. (RHS) 1967


```
cl.
      -nuttalli (1/2)
Reuthe's Reward-
      -lindleyi (1/2)
```

Large flowers of creamy white with apricot blotch. Foliage like
lindleyi. (G. Reuthe, Fox Hill Nurseries, 1971)

g. & cl.
```
        -forrestii ssp. forrestii Repens Group (1/2)
Reve Rose-        -Corona--unknown (1/4)
        -Bow Bells-
                -williamsianum (1/4)
```

2.5ft(.75m)        -5°F(-21°C)        EM        3/4

Beautiful little plant with bright green leaves, 1.5in(3.8cm)
long.  Small rose-colored flowers, in loose trusses of about 5.
(Lester E. Brandt, reg. 1958)

cl.
```
        -griersonianum (1/2)
Revell's Red-
        -unknown (1/2)
```

Deep red flowers.  (Knap Hill Nursery, cross 1935; reg. 1963)

cl.
```
            -griffithianum (1/2)
Rev. R. W. Carew Hunt-
            -unknown (1/2)
```

Red flowers.  (Otto Schulz cross, 1890; intro. by C. B. van Nes
Sons, 1896)

g.
```
        -auriculatum (1/2)
Reverie-        -arboreum ssp. zeylanicum (1/4)
        -St. Keverne-
                -griffithianum (1/4)
```

A large, rather compact plant with large flowers, white blushed
pink, in late season.  (Rothschild; intro. 1950)

cl.
```
        -neriiflorum ssp. neriiflorum
Review Order-        Euchaites Group (1/2)
        -        -haematodes ssp. haematodes (1/4)
        -May Day-
                -griersonianum (1/4)
```

Lax trusses of bell-shaped flowers, blood red with faint brown
spotting.  Dense shrub, leaves with heavy tomentose beneath.
(Lord Stair) A. M. (RHS) 1954

cl.
```
        -cinnabarinum ssp. cinnabarinum (Roylei Group) (3/4)
Revlon-        -cinnabarinum ssp. cinnabarinum Roylei Group
        -Lady        -        -cinnabarinum ssp. cinnabarinum
        Chamberlain-Royal Flush -                (1/8)
                (orange form)-maddenii ssp. maddenii (1/8)
```

Perseverance g.  Tall, upright plant; glossy foliage densely
scaly beneath.  Flowers are gleaming, slender, waxen trumpets of
bright carmine red, in clusters of about 7.  Midseason.  (Rothschild)  A. M. (RHS) 1957

```
                -griersonianum (1/4)
g.        -Dorinthia-        -haematodes ssp. haematodes (1/8)
        -        -Hiraethlyn-
Rhapsody-                -griffithianum (3/8)
        -        -griffithianum
        -Loderi-
                -fortunei ssp. fortunei (1/4)
```

Rose flowers.  (Lord Aberconway, 1950)

cl.

Rhodoland's Silver Mist--form of yakushimanum ssp. yakushimanum

1.5ft(.45m)        -20°F(-29°C)        M

Plant twice as wide as high; very shiny leaves, 4in(10.2cm) long
with heavy tan indumentum beneath; held 5 years.  Flowers white
with greenish spots, tubular campanulate, 2in(5cm) wide, in flat
trusses of 11-13. (Seed collected in Yakushima, Japan, by Arthur
and Maxine Childers; reg. 1978)

g.
```
        -aperantum (1/2)
Rhythm-
        -forrestii ssp. forrestii Repens Group (1/2)
```

Flowers deep red to pale rose.  (Lord Aberconway, 1946)

cl.
```
        -racemosum (1/2)
Ria Hardijzer-
        -Hinodegiri--Kurume azalea (1/2)
```

2.5ft(.75m)        -5°F(-21°C)        EM        4/4

Evergreen azaleodendron.  Vibrant pink flowers, widely funnel-
shaped, to 1.2in(3.2cm) across, in small trusses of 19.  Floriferous.  Compact, vigorous plant, as broad as tall;  leaves .75in
(1.9cm) long, dark green tinged red.  (Hardijzer, 1965)    H. C.
(RHS) 1969;  A. M. (RHS) 1974

---
```
        -fortunei ssp. fortunei (1/2)
Richard Gill-
        -thomsonii ssp. thomsonii (1/2)
```

Flowers of deep rose.  (R. Gill & Son)  A. M. (RHS) 1920

cl.
```
        -auriculatum (1/2)
Richard Gregory-
        -kyawii (1/2)
```

Tubular bell-shaped flowers in loose trusses of 10, pink inside
with crimson blotch, slightly scented.  Leaves 6in(15.2cm) long
by 2.5in(6.4cm) wide, with slight floccose indumentum.  (R. M.
Gregory cross; reg. by A. M. Williams, 1969)

cl.
```
        -Catalgla--catawbiense var. album Glass
Richie*--Calsap (selfed)-                (1/2)
        -Sappho--unknown (1/2)
```

Vivid purple buds open to flowers of very pale purple, edged in
strong purple with a blotch of deep purplish red.  Plant of medium size, hardy to -25°F(-32°C).  (Weldon E. Delp)

g.
```
            -dichroanthum ssp. dichroanthum (1/4)
        -Golden Horn-
Rickshaw-        -elliottii (1/4)
        -        -fortunei ssp. discolor (1/4)
        -Lady Bessborough-
                -campylocarpum ssp. campylocarpum
                        Elatum Group (1/4)
```

Bright flowers of beige and orange, deeper in throat;  small
double calyx.  Well-filled, medium-sized trusses on a compact,
medium-sized bush.  Late season.  (Rothschild)  P. C. (RHS) 1958

cl.
```
        -Metternianus--japonicum var. japonicum (metternichii)
Rijneveld-        -griersonianum (1/4)        (1/2)
        -unnamed hybrid-
                -unknown (1/4)
```

Very hardy plant. Trusses of 6-10 flowers, coral pink, in early
midseason.  (Raiser Dietrich Hobbie; intro. by V. van Nes, reg.
1969) Gold Medal (Boskoop) 1966

cl.
```
        -racemosum (1/2)
Rik*
        -keiskei (1/2)
```

Mature plant 2.5ft(.75m) high;  very hardy.  Blooms just after
P.J.M.  White flowers, edged in pink fading white.  Fine winter
color with deep red stems and foliage.  (Leon Yavorsky)

g.
   -wardii var. wardii (1/2)
Rima-
    -decorum (1/2)

A parent of Bow Street. Creamy yellow flowers.  (J. B. Stevenson, 1941)

cl.                    -catawbiense (1/4)
    -Mrs. Milner-
Rinaldo-              -unknown (1/4)
    -smirnowii (1/2)

A parent of Gorlitz, a German hybrid by Joh. Bruns.  Flowers of
ruby red, markings of light green. (T. J. R. Seidel, cross 1907;
intro. 1915)

cl.
    -Darigold--unknown (1/2)
Ring of-       -wardii var. wardii (1/4)
 Fire* -       -              -griffithianum (1/32)
    -Idealist-        -Kewense-
    -       -Aurora-       -fortunei ssp. fortunei
      -Naomi-      -                 (5/32)
    -       -thomsonii ssp. thomsonii (1/16)
      -fortunei ssp. fortunei

4ft(1.2m)        0°F(-18°C)         ML

A most striking flower of yellow with a narrow band of orange
red around the margins.  Plant vigorous, compact; dense foliage.
(Willard & Margaret Thompson cross; Harold Greer, intro. 1984)

cl.
  -Newburyport Beauty--unknown
Rio-                    (Dexter hybrids)
  -Newburyport Belle--unknown

3ft(.9m)         -20°F(-29°C)        M

Salmon pink flowers with yellow throat, 3.25in(8.3cm) wide, in
globular trusses of 16. Plant much wider than tall; glossy yellowish green leaves, retained 2 years.  See also Normandy. (D.
G. Leach, cross 1968; reg. 1983)

        -dichroanthum ssp. dichroanthum (1/4)
cl. -Goldsworth Orange-
  -                 -fortunei ssp. discolor (1/4)
Ripe-                -griffithianum (1/16)
Corn-           -Kewense-
  -       -Aurora-       -fortunei ssp. fortunei (5/16)
  -Exbury-      -thomsonii ssp. thomsonii (1/8)
    Naomi-fortunei ssp. fortunei

The color is maize, slightly tinged pink.  (W. C. Slocock, 1967)

g.
  -forrestii ssp. forrestii Repens Group (1/2)
Riplet-           -campylocarpum ssp. campylocarpum Elatum Gp.
  -Letty Edwards-                               (1/4)
      -fortunei ssp. fortunei (1/4)

2ft(.6m)         0°F(-18°C)         EM

Originally meant to be Replet (repens X Letty).  Low, spreading,
heavy-stemmed plant; dark leaves to 2.5in(6.4cm) long.  Flowers
crimson fading to salmon, 4in(10.2cm) wide; trusses of 5-7.  Another form has much yellow. (H. Lem)  P. A. (ARS) 1961  See
color illus. ARS Q 31:1 (Winter 1977): p. 35;  K p. 92.

---

Riviera Beauty--form of arboreum ssp. arboreum

Carmine pink flowers.  (R. Gill & Son, before 1958)

cl.
    -minus var. minus Carolinianum Group (pink form) (1/2)
Robby*    -mucronulatum (1/4)
    -Mucram-
      -ambiguum (1/4)

A compact plant with white flowers, larger than one expects from
the parentage. (Weldon E. Delp, intro. 1965)

cl.                -decorum ? (1/4)
    -Caroline-
Robert Allison-       -brachycarpum ssp. brachycarpum ? (1/4)
    -fortunei ssp. discolor (1/2)

5ft(1.5m)        -10°F(-23°C)         ML        4/3

Pink flowers with golden throat, to 3.5in(8.9cm) wide, fragrant;
flat-topped trusses. Floriferous.  Large, waxy green foliage on
a vigorous upright plant.  (Joseph B. Gable, reg. 1958)

cl.
    -arboreum ssp. zeylanicum (1/2)
Robert Balch-
    -elliottii (K-W 19083) (1/2)

Trusses hold 10-16 flowers, currant red.  (R. W. Balch cross;
reg. by Dunedin Rhododendron Group, 1977)

cl.
    -catawbiense var. compactum (1/2)
Robert Forsythe-    -griffithianum (1/4)
    -Mars-
      -unknown (1/4)

4ft(1.2m)        -15°F(-26°C)         ML

A dense, broad plant spreading as wide as high, with leaves 5in
(12.7cm) long, deep green.  Flowers deep rose with brown blotch.
(Samuel Baldanza cross, 1958; reg. 1972)

g.
    -griffithianum (1/2)
Robert Fortune-       -fortunei ssp. fortunei (1/4)
    -H. M. Arderne-
      -unknown (1/4)

Flowers pink with crimson markings at base.  A. M. (RHS) 1922

cl.
    -Everchoice--unknown
Robert Huber-
    -unknown

4ft(1.2m)        -5°F(-21°C)         M

Probably a cross of 2 Dexter hybrids.  Dark pink buds open to
orchid pink flowers with darker pink edges, spotting of Spanish
orange; spherical trusses of 14.  Plant upright, almost as broad
as tall; glossy yellowish green leaves.  (S. Everitt cross; reg.
by Charles Herbert, 1979)

cl.
    -lacteum (1/2)
Robert Keir-       -fortunei ssp. fortunei (1/4)
    -Luscombei-
      -thomsonii ssp. thomsonii (1/4)

Flowers pale yellow, flushed pale pink in throat and dark pink
outside. (J. B. Stevenson, 1951)  A. M. (RHS) 1957

cl.                -haematodes ssp. haematodes (1/4)
    -May Day-
Robert Louis-    -griersonianum (1/4)
 Stevenson -      -dichroanthum ssp. dichroanthum (1/4)
  -       -                 -griffithianum (1/32)
    -Jester-        -Kewense-
  -       -Aurora-       -fortunei ssp. fortunei
  -Naomi-      -thomsonii ssp. thomsonii (1/16)
    -fortunei ssp. fortunei (5/32)

2ft(.6m)         0°F(-18°C)         L        4/4

Plant habit open; leaves small, dark green, very shiny, grooved
at veins.  Small blood red flowers, funnel-campanulate, in loose
trusses.  Blooms later than most reds.  (Cecil S. Seabrook, 1967)


cl.
                -calostrotum ssp. keleticum (1/2)
Robert Seleger*
                -unknown (1/2)


1ft(.3m)          -5°F(-21°C)          EM          3/3


Low habit with tiny leaves, like keleticum.  Plant covered with
rose purple flowers.  (Dietrich Hobbie cross; offered by Greer)


cl.                      -griffithianum (3/8)
                -Marinus Koster-
Robert Verne-            -unknown (1/4)
            -        -fortunei ssp. fortunei (1/4)
            -Pilgrim-             -arboreum ssp. arboreum (1/8)
                -Gill's Triumph-
                             -griffithianum


5ft(1.5m)          10°F(-12°C)          L


Flowers of heavy substance, Tyrian rose with prominent maroon
spotting.  Upright plant, half as wide as tall.  More compact if
grown in sun.  See also Sofus Eckrem.  (Sigrid Laxdall cross,
1963; reg. 1979)


cl.
                -christianae (1/2)
Robert Withers-
                -aequabile (1/2)


Vireya hybrid.  Trusses of 6-8 flowers of Saturn red.  (Cross by
Craig Faragher; intro. and reg. by Graham Snell, 1982)


cl.
        -fortunei ssp. discolor (1/2)
Roberte-
        -campylocarpum ssp. campylocarpum Elatum Group (1/2)


6ft(1.8m)          -5°F(-21°C)          ML          3/3


Lady Bessborough g.  The plant was given the Christian name of
Lady Bessborough.  Flowers are rose pink, speckled red.  (Rothschild)   F. C. C. (RHS) 1936


g.
                -calophytum var. calophytum (1/2)
Robin Hood-
                -sutchenense (1/2)


A parent of Carol Jean.  (J. C. Williams, 1933)


cl.        -Catalua--catawbiense var. album Glass (1/2)
        -                        -campylocarpum ssp. campylo-
Robin Leach-            -unnamed-                carpum (1/16)
        -          -Adriaan- hybrid-unknown (1/8)
        -        - Koster-                -griffithianum
    -unnamed-        -            -George-           (1/32)
        hybrid-    -Mrs. Lindsay- Hardy-catawbiense
        -            Smith     -           (1/32)
        -                    -Duchess of
        -williamsianum (1/4)  Edinburgh--unknown


5ft(1.5m)          -20°F(-29°C)          EM          4/4

Much broader than tall with dense foliage.  Ovate leaves medium
green, turning convexly, sharply downward.  Flowers 3in(7.6cm)
wide, openly bell-shaped, white with faint red spotting; globular full trusses of 7.  See also Applause.  (D. G. Leach, reg.
1972)   See color illus. ARS J 36:1 (Winter 1982): p. 2.


cl.
                -lochiae (1/2)
Rob's Favourite-
                -macgregoriae (1/2)


Vireya hybrid.  Flowers tubular funnel-shaped, 5-lobed, colored vermilion, with a red calyx; trusses of 10-12.  Plant height
about 5ft(1.5m).  (O. S. Blumhardt, reg. 1984)


g.
        -barbatum (1/2)
Roc-
        -fulgens (1/2)


Red flowers.  (J. B. Stevenson cross, 1929; intro. 1950)


cl.            -griffithianum (1/4)
        -Dorothea-
Rochelle-       -decorum (1/4)
        -            -catawbiense (1/4)
        -Kettledrum-
                -unknown (1/4)


4.5ft(1.35m)          -10°F(-23°C)          ML          4/4

Slightly fragrant flowers, 4in(10.2cm) across, rose-colored with
a strawberry red blotch, velvet-textured; 7-flowered trusses of
unusual beauty.  (G. Guy Nearing, 1970)   See color illus. HG p.
112; LW pl. 60.


cl.

Rock Rose--form of racemosum (R 59578)


Axillary flowers form a many-flowered raceme, color roseine purple fading to white in throat, some greenish red and scarlet
spotting.  Leaves aromatic, 1in(2.5cm) across.  (J. B. Stevenson)  A. M. (RHS) 1970


cl.
        -meddianum var. meddianum (1/2)
Rocket-
        -strigillosum (1/2)


Bell-shaped flowers, glowing blood red, in flat-topped trusses.
Early season.  (J. B. Stevenson, Tower Court)  A. M. (RHS) 1954


cl.            -caucasicum (1/4)
        -Cunningham's White-
Rocket-            -ponticum (white form) (1/4)
        -catawbiense (red form) (1/2)


5ft(1.5m)          -15°F(-26°C)          EM          3/4

Medium-sized foliage, thick, heavily veined, glossy.  Frilled
flowers 2.5in(6.4cm) wide, vibrant coral pink, blotched scarlet,
in conical trusses.  Heat- and sun-tolerant.  (Anthony M. Shammarello, 1955)   See color illus. VV p. 67.


cl.                -caucasicum (1/4)
        -Dr. Stocker-
Rocketfire-            -griffithianum (1/4)
    -      -wardii var. wardii (1/4)
        -Hawk-            -fortunei ssp. discolor (1/8)
            -Lady
            Bessborough-campylocarpum ssp. campylocarpum
                                Elatum Group(1/8)


Flowers sulphur yellow, in midseason.  Dwarfish plant; leaves to
5.5in(14cm) long, dark green.  (Rudolph & Leona Henny, 1965)

```
                         -wardii var. wardii (1/4)
             -Crest-          -fortunei ssp. discolor (1/16)
             -     -Lady-
             -     Bessborough-campylocarpum ssp.campylocar-
             -                            pum Elatum Gp. (1/16)
         -unnamed-             -wardii var. wardii
         - hybrid-                        -griffithianum
         -     -     -Ideal-          -Kewense-          (1/16)
cl.      -     -     - -ist-    -Aurora-          -fortunei ssp..
         -     -     -    -Naomi-     -thomsonii ssp...(1/32)
         -     -Odee-          -fortunei ssp. fortunei
         -     Wright-                   -griffithanum
         -     -         -Mrs.    George-
Rockhill-      -         - Lindsay- Hardy-catawbieniense
Ivory -        -         - Smith -           (1/64)
Ruffles-       -Mrs. Betty-     -Duchess of Edinburgh
         -     - Robertson-     --unknown (1/16)
         -     -         -campylocarpum ssp.
         -     -   unnamed-campylocarpum (1/32)
         -     -        hybrid-
         -     -                -unknown
         -                -fortunei ssp. fortunei (15/64)
         -     -Fawn-    -dichroanthum ssp. dichroanthum
         -     -    -Fabia-                      (1/16)
         -     -         -griersonianum (1/8)
         -Skipper-          -griersonianum
         -     -Sarita-         -griffithianum
         -     - Loder-Loderi-
         -Penny-          -fortunei ssp. fortunei
         -              -wardii var. wardii
         -       -                -griffith...
         -    -Idealist-          -Kewense-
         -           -Aurora-    -fortunei...
         -      -Naomi-    -thomsonii ssp.
         -           -       thomsonii
         -           -fortunei ssp. fortunei
```

5ft(1.5m)          10°F(-12°C)          E

Flowers 5in(12.7cm) across with 7 wavy lobes, throat primrose
yellow shading outward to yellowish white with Turkey red
spotting, in trusses 9in(23cm) across, holding 15. Broad plant,
branching well, 3/4 as wide as tall; ivy green leaves retained 3
years. (James C. Brotherton, reg. 1983)

cl.

Rockhill Parkay     Parentage as above

5ft(1.5m)          10°F(-12°C)          E

Flowers 4.5in(11.5cm) across, 6-lobed, sulphur yellow in throat
shading to creamy white with Turkey red eye, held in ball-shaped
trusses of 15-17. Upright plant, 3/4 as wide as tall with
parsley green leaves held 2 years. (J. C. Brotherton, reg. 1983)

cl.

Rockhill Sunday Sunrise     Parentage as above

5ft(1.5m)          10°F(-12°C)          E

Flowers 5in(12.7cm) across, 5-lobed, primrose yellow with red
blotch in throat. Upright plant with stiff branches, 3/5 as
wide as high with leaves ivy green, retained 3 years. (James C.
Brotherton, reg. 1983)

```
cl.          -caucasicum (1/4)
       -Boule de Neige-
Rococo-              -catawbiense (1/8)
       -         -unnamed hybrid-
       -                      -unknown (1/8)
       -fortunei ssp. fortunei (1/2)
```

5ft(1.5m)          -25°F(-32°C)          EM          3/4

May be exinct. Plant as broad as tall, of superior foliage density with leaves 4.5in(11.5cm) by 1.5in (3.8cm). Flowers fluted
and ruffled, up to 2.25in(5.7cm) across, lilac pink with deeper
edging and yellow green rays, held in trusses of 12. (Cross by
Joseph Gable; reg. by David G. Leach, 1965)

```
                              -griffithianum (1/8)
cl.           -George Hardy-
       -Mrs. L. A. Dunnett-    -catawbiense (1/8)
Rodeo-            -unknown (1/4)
       -griersonianum (1/2)
```

A parent of Merry Lass. Red flowers with an orange flush. (M.
Koster & Sons)

```
                              -catawbiense (1/8)
cl.         -Parsons Grandiflorum-
       -Nova Zembla-          -unknown (5/8)
Roland-       -hardy red hybrid--unknown
       -         -ponticum (1/4)
       -Purple Splendour-
                 -unknown
```

Flowers in trusses of 18-21, moderate purplish red with dark red
spotting. Rather glossy dark green leaves. (H. Hachmann cross,
1967; reg. by G. Stuck, 1984)

cl.

Roland Cooper--form of campanulatum ssp. campanulatum

Leaves 6in(15.2cm) long, covered below with brown indumentum.
Flowers fragrant, white, spotted with crimson and held in
trusses of about 11. (Collector R. E. Cooper; raised at RBG,
Edinburgh) A. M. (RHS) 1964

```
cl.                  -griffithianum (1/4)
       -Loderi King George-
Roma Sun-            -fortunei ssp. fortunei (1/4)
       -Ostbo Y3--unknown (1/2)
```

6ft(1.8m) [at 24 yrs.]          0°F(-18°C)          M

Upright plant, as broad as tall; medium green leaves rather bullate, 8.25in(21cm) long. Frilled fragrant flowers 4.5in(11.5cm)
wide, yellowish pink becoming light yellow green, edges of deep
pink which fades, reverse fades from deep pink to light yellow;
full flattened trusses of 10. Floriferous. (R. M. Bovee cross;
Sorensen & Watson, intro. & reg. 1975)

```
cl.               -arboreum ssp. arboreum (1/4)
       -Glory of Penjerrick-
Romala-            -griffithianum (1/4)
       -thomsonii ssp. thomsonii (1/2)
```

Barclayi g. Crimson flowers. (Barclay Fox)

```
cl.
       -dichroanthum ssp. dichroanthum (1/2)
Roman Pottery-
       -griersonianum (1/2)
```

Fabia g. Pale orange flowers with coppery lobes. (J. J. Crosfield, 1934)

```
g. & cl.
       -Moser's Maroon--unknown (1/2)
Romany Chai-
(Gypsy     -griersonianum (1/2)
 Children)
```

5ft(1.5m)          0°F(-18°C)          ML

Upright, rather open plant, with large dark green leaves. Flowers of rich terracotta speckled with brown, held in substantial
trusses. Heat-tolerant. Red forms exist, as Empire Day, q.v.
(Rothschild)  A. M. (RHS) 1932

'Pure Elegance' by Delp
Photo by Delp

'Pygmalion' by Waterer
Photo by Greer

'Pygmy' (Moonstone × Carmen) by Skonieczny
Photo by Greer

'Queen Anne's' (double-flowered) by Skinner
Photo by Greer

'Queen o' the May' by C. Smith, before 1930
Photo by Greer

'Queen of Hearts' by Rothschild
Photo by Greer

'Queen of McKenzie' by Phetteplace
Photo by Greer

'Racil' by N. S. Holland
Photo by Greer

'Radiant Morn' by Aberconway
Photo by Greer

'Ramapo' by Nearing
Photo by Greer

'Rangoon' by Leach
Photo by Leach

'Razorbill' by Cox
Photo by Greer

'Red Cloud' by R. Henny
Photo by Greer

'Red Eye' by Swenson/Greer
Photo by Greer

'Red Head' by Gable
Photo by Greer

'Red Majesty' by Larson
Photo by Greer

'Red Olympia' by Lem/Pierce
Photo by Greer

'Red River' by Leach
Photo by Leach

'Red Sea' by Leach
Photo by Leach

'Red Velvet' by Larson
Photo by Greer

'Red Walloper' by Lem
Photo by Greer

'Redwax' by R. Henny
Photo by Greer

'Relaxation' by C. Smith
Photo by C. Smith

'Renee Shirey' by Delp
Photo by Delp

g. & cl.
        -Moser's Maroon--unknown (1/2)
Romany Chal -
(Gypsy Girl)-facetum (eriogynum) (1/2)

6ft(1.8m)        5°F(-15°C)        L        4/3

Upright, vigorous, tall plant; dark green, recurved leaves 6.5in
(16.5cm) long, with slight indumentum. Bell-shaped flowers of a
rich, true, glowing red, faintly spotted brown; very large round
trusses. Needs shade. (Rothschild) A. M. (RHS) 1932; F. C. C.
(RHS) 1937   See color illus. VV p. 58.


cl.
      -griersonianum (1/2)
Romany Rye-        -souliei (1/4)
     -Souldis-
            -fortunei ssp. discolor (1/4)

Trusses of deep blue, holding 9 flowers, each 4.5in(11.5cm) wide
and 3in(7.6cm) long. (Gen. Harrison, reg. 1962)


cl.
    -kyawii (1/2)
Romarez-
     -griersonianum (1/2)

Late-blooming, dark red flowers. (J. B. Stevenson cross, 1932;
intro. 1953) A. M. (RHS) 1953


                       -arboreum ssp. arboreum
             -Glory of Penjerrick-      (1/16)
cl.    -Romala-          -griffithianum (1/16)
  -Coer-     -thomsonii ssp. thomsonii (1/8)
Romari-    -arboreum ssp. delavayi var. delavayi (1/4)
   -meddianum var. meddianum (1/2)

Red flowers. (J. B. Stevenson cross, 1939)


cl.
      -oreodoxa var. oreodoxa (1/2)
Romberpark-      -arboreum ssp. arboreum (1/4)
     -Doncaster-
          -unknown (1/4)

Ronsdorfer Fruhbluhende g. A hybrid of good habit with flowers
of broad funnel-form, white outside, pinkish toward base, 8-
lobed. Blooms early. See also Westfalenpark. (Crossed by A.
Arends; intro. by G. D. Bohlje Nurseries, W. Ger.; reg. 1969)


                  -catawbiense (3/8)
cl. -red catawbiense hybrid-
   -            -unknown (5/8)
Romeo-             -catawbiense
   -   -Parsons Grandiflorum-
   -America-        -unknown
     -dark red hybrid--unknown

6ft(1.8m)        -20°F(-29°C)        ML        3/3

Tall, vigorous, shapely plant; leathery dark leaves, 4in(10.2cm)
long. Flowers blood red with darker blotch, funnel-shaped, 2.5
in(5cm) across; globular trusses. (A. M. Shammarello, reg. 1972)


cl.      -wardii var. wardii (1/4)
  -Inamorata-
Romy-    -fortunei ssp. discolor (1/4)
  -Mrs. J. G. Millais--unknown (1/2)

Compact plant, spreading broader than tall, very free-flowering
with 13 per truss, primrose yellow. (E. G. Millais)   H. C.
(RHS) 1974


cl.
     -Marion--unknown (1/2)
Ron Langdon-      -griersonianum (1/4)
      -unnamed hybrid-
            -fortunei ssp. fortunei (1/4)

Flowers in trusses of 13, medium neyron rose slightly spotted.
See also Glen Bruce, Spring Fiesta. (G. Langdon, 1975)


g.        -arboreum ssp. arboreum (1/4)
  -Cornubia-     -thomsonii ssp. thomsonii (1/8)
Rona-    -Shilsonii-
  -       -barbatum
  -barbatum (5/8)

Deep red flowers. (Lord Aberconway, 1946)


cl.       -arboreum ssp. arboreum (1/4)
    -Cornubia-     -thomsonii ssp. thomsonii (3/8)
Rona -    -Shilsonii-
Martin-       -barbatum (3/8)
      -thomsonii ssp. thomsonii
   -Shilsonii-
      -barbatum

Fine trusses of dark red flowers, medium-sized, in March. (Sir
James Horlick, Isle of Gigha, reg. 1962)


cl.
   -hodgsonii (1/2)
Ronald-
   -sinogrande (1/2)

Heavy trusses of 30 flowers, white stained pink, with some variable darker shading on the reverse. (The Gibsons, Glenarn, reg.
1958)   A. M. (RHS) 1958


cl.
    -Blue Peter--unknown (3/4)
Ronkonkoma-     -ponticum (1/4)
   -Purple Splendour-
        -unknown

3ft(.9m)        -10°F(-23°C)        ML

Plant rounded, almost as wide as high; glossy elliptic leaves to
7in(17.8cm) long, held 3 years. Dark violet purple buds opening
to amethyst violet flowers, darker edges and reverse, 3in(7.6cm)
across, frilly, 5-lobed; spherical trusses of 22. (A. Raustein,
1977)


g.
             -oreodoxa var. oreodoxa (1/2)
Ronsdorfer Fruhbluhende-    -arboreum ssp. arboreum (1/4)
          -Doncaster-
            -unknown (1/4)

Flowers rose to purple. See also Westfalenpark. (G. Arends,
reg. 1958)


g. & cl.      -griffithianum (1/2)
     -Loderi-
Rosa Bonheur-   -fortunei ssp. fortunei (1/4)
    -      -thomsonii ssp. thomsonii (1/4)
     -Cornish Cross-
         -griffithianum

Deep pink flowers. (Loder)


cl.
    -yakushimanum ssp. makinoi (1/2)
Rosa Perle-        -Queen     -griffithianum
          -Britannia-Wilhelmina-    (1/16)
     -Kluis Triumph-     -unknown (7/16)
        -unknown  -Stanley Davies--unknown

Flowers deep pink. (Joh. Bruns, 1972)

```
cl.                 -griffithianum (1/4)
        -Pink Shell-            -fortunei ssp. fortunei (1/8)
Rosabel-        -H. M. Arderne-
-                          -unknown (1/8)

        -griersonianum (1/2)
```

Large, loose trusses of trumpet-shaped flowers, rose red in bud,
opening pale pink with darker veining, pale salmon pink with a
dark eye when fully open. (Knap Hill, 1865)  A. M. (RHS) 1936

```
g. & cl.
         -oreodoxa var. fargesii (1/2)
Rosalind-
         -thomsonii ssp. thomsonii (1/2)
```

Flowers of rose pink. (Lord Aberconway) A. M. (RHS) 1938

```
cl.
              -neriiflorum ssp. neriiflorum (phoenicodum)
Rosalind of Arden-                                  (1/2)
              -williamsianum (1/2)
```

Flowers deep rose red. (Thacker, 1946)

```
          -arboreum ssp. arboreum (1/4)
g.   -Cornubia-            -thomsonii ssp. thomsonii (1/8)
-            -Shilsonii-
Rosamond-           -barbatum (1/8)
-       -griffithianum (1/4)
    -Loderi-
          -fortunei ssp. fortunei (1/4)
```

Clear pink flowers. (Loder, 1934)

```
cl.                    -griffithianum (1/4)
          -George Hardy-
Rosamund Millais-         -catawbiense (1/4)
-          -arboreum sp. arboreum (1/4)
     -Doncaster-
          -unknown (1/4)
```

Flowers cerise, blotched purple. (M. Koster & Sons, before 1922)
A. M. (RHS) 1933

```
---
         -caucasicum (1/2)
Rosamundii-
         -unknown (1/2)

4ft(1.2m)      -5°F(-21°C)         VE
```

Often compared to Christmas Cheer, this one flowers later, with
pinker flowers, smoother leaves. Blooming in March, the compact
plant seems to be covered with rosy snowballs. (Standish & Noble, mid-1800s)  See color illus. VV p. 130.

```
cl.  -Laura   -griersonianum (1/4)
     -Aberconway-         -thomsonii ssp. thomsonii (1/8)
Rosanita-      -Barclayi-         -arboreum ssp. arboreum
-               -Glory of -               (1/16)
-               Penjerrick-griffithianum (1/16)
     -yakushimanum ssp. yakushimanum (1/2)
```

Trusses of 12 flowers, 5-lobed, light neyron rose with a deeper
stripe down center of each lobe. Narrow leaves 5.5in(14cm) long
have woolly indumentum beneath. (J. F. J. McQuire, reg. 1983)

```
---
    -javanicum (5/8)
Rose-        -jasminiflorum (3/8)
  -Princess -        -jasminiflorum
    Alexandra-Princess-
            Royal  -javanicum
```

Vireya hybrid. Flowers pinkish yellow to salmon. (Veitch, 1891)

```
cl.
Rose Bay--synonym maximum (not a selected form)

---
         -thomsonii ssp. thomsonii (1/2)
Rose Beauty-
         -unknown (1/2)
```

Flowers rosy pink with a few markings on upper lobe. (R. Gill)

```
g.
    -decorum (1/2)
Rose du-        -maximum (1/4)          -catawbiense
 Barri -Standishii-          -unnamed hybrid-     (1/16)
          -Altaclarense-          -ponticum (1/16)
              -arboreum ssp. arboreum (1/8)
```

Rose-colored flowers. (Loder)

```
cl.
    -racemosum (1/2)
Rose Elf-
    -pemakoense (1/2)
```

White flowers blushed pink, campanulate. Dwarf plant, compact,
spreading; small, shiny, dark green leaves. Floriferous; sun-
tolerant. (Lancaster) P.A. (ARS) 1954   See col. ill. VV p. 70.

```
cl.
    -fortunei ssp. fortunei (1/2)
Rose Haines-
    -unknown (1/2)

5ft(1.5m)        0°F(-18°C)         M
```

Flowers of heavy substance, 5-lobed, frilled, cyclamen purple
with sap green throat; in trusses of 9-11. Upright plant, wider
than tall; dark green leaves held 2-3 years. (Cross by C. English, Jr.; intro. by C. R. Burlingame 1967; reg. 1977)

```
cl.
    -Koichiro Wada--yakushimanum ssp. yakushimanum (1/2)
Rose Imp-
    -unknown (1/2)
```

Trusses of 12 flowers, light purplish pink with red flare in upper throat. Light tan indumentum on new foliage. Bush of 1.7ft
(.5m) at 10 years. (Pukeiti Rhododendron Trust; reg. by G. F.
Smith, 1984)

```
cl.                 -griersonianum (1/4)
          -Tally Ho-
Rose Lancaster-     -facetum (eriogynum) (1/4)
          -yakushimanum ssp. yakushimanum (1/2)

4ft(1.2m)        5°F(-15°C)         ML
```

Crimson buds, neyron rose flowers of heavy substance, 3in(7.6cm)
across; ball-shaped trusses hold 18-30. Plant rounded; leaves
dark green with heavy tan indumentum, held 3 years. (B. Lancaster cross, 1962; Dr. David Goheen named, reg. 1978)

```
---
         -cinnabarinum ssp. cinnabarinum (1/2)
Rose Mangles-
         -maddenii ssp. maddenii (1/2)
```

A parent of Cock of the Walk. Flowers of rose pink. (Mangles)

```
---
         -griffithianum (1/2)
Rose Newcomb-
         -unknown (1/2)
```

Flowers white with pink edges. (Mangles)

```
cl.                -griersonianum (1/4)
            -Tally Ho-
Rose of China-          -facetum (eriogynum) (1/4)
        -fortunei ssp. discolor (1/2)
```

4ft(1.2m)          0°F(-18°C)          L          3/3

Plant upright, rounded; leaves 7in(17.8cm) long. Flowers clear
China rose, campanulate, 4in(10.2cm) wide, in trusses of 10-12.
Sun-tolerant. (Cross by Lem; intro. by B. Lancaster; reg. 1958)

```
cl.          -fortunei ssp. discolor (1/2)
Rose Pageant-
            -Corona--unknown (1/2)
```

6ft(1.8m)          -5°F(-21°C)          VL

Full, upright trusses of 12 flowers, 5-lobed, deep neyron rose
with white throat. Leaves 9in(23cm) long. (Rudolph & Leona
Henny, reg. 1965)

```
cl.            -williamsianum (1/4)
        -Wilbar-          -thomsonii ssp. thomsonii (1/8)
Rose    -      -Barclayi-          -arboreum ssp. arboreum
Pantaloons-          -Glory of  -            (1/16)
        -          -Penjerrick-griffithianum (1/16)
        -chamaethomsonii var. chamaethauma (1/2)
```

1.5ft(.45m)          15°F(-10°C)          E

Hose-in-hose flowers of heavy substance, tubular funnel-shaped,
spinel red, in lax trusses of 5. Plant as broad as high; leaves
2in(5cm) long, glossy yellow green, held 3 years. See also Red
Pantaloons. (Dr. Carl H. Heller, reg. 1977)

---

Rose Perfection          Parentage unknown

A parent of Sidona. Deep red buds open clear pink. (R. Gill)

---

```
            -javanicum (5/8)
Rose Perfection-          -jasminiflorum (3/8)
            -Princess Alexandra-          -jasminiflorum
                    -Princess Royal-
                            -javanicum
```

Vireya hybrid. Flowers of rose pink.  F. C. C. (RHS) 1886

```
cl.          -dichroanthum ssp. dichroanthum (1/4)
        -Dido-
Rose Point-    -decorum (1/4)
    -williamsianum (1/2)
```

3ft(.9m)          0°F(-18°C)          EM          3/4

Plant rounded, dense, wider than tall; dull yellowish green
leaves with red petioles, held 2 years. Spirea red buds opening
lighter red with some shading, to 3in(7.6cm) wide, 7-lobed, held
in trusses of 7. (Cross by H. Lem; raiser L. Pierce; reg. by J.
A. Elliott, 1980)

```
cl.
Rose Red     Parentage unknown
```

A parent of Jack Lyons.

```
cl.            -ciliicalyx (1/4)
        -Else Frye-
Rose Scott-          -unknown (1/4)
    -          -johnstoneanum (1/4)
    -unnamed hybrid-
                -veitchianum Cubittii Group (1/4)
```

Plant 2.75ft(.8m) at 7 years, well-branched; glossy dark green
leaves, bullate, 3.5in(8.9cm) long, held 2 years. White flowers
blotched and marked in shades of pink, 4.25in(10.8cm) wide, fragrant; trusses of 3-7. Floriferous.    (Robert W. Scott, 1977)

```
cl.
        -griersonianum (1/2)
Rose Splendour-          -ponticum (1/4)
            -Purple Splendour-
                    -unknown (1/4)
```

4ft(1.2m)          0°F(-18°C)          ML          4/3

Compact, upright, rounded plant; dark green leaves with slight
indumentum. Flowers magenta rose with geranium lake eye, bell-
shaped, to 4in(10.2cm) across; rounded trusses of 12. Sun-tolerant. (B. F. Lancaster, 1958)

```
                    -griffithianum (3/8)
cl.          -Queen Wilhelmina-
        -Britannia-          -unknown (3/8)
Roseann-      -Stanley Davies--unknown
    -              -griffithianum
        -Loderi Venus-
                -fortunei ssp. fortunei (1/4)
```

6ft(1.8m)          5°F(-15°C)          M          4/4

Synonym Tatoosh. Flowers 4.5in(11.5cm) across, rose pink spotted crimson on upper lobe, a "frosted" appearance; about 17 per
truss. Leaves 8.5in(21.6cm) long, 2.5in(6.4cm) wide, with red
petioles. (Rudolph Henny, reg. 1958)    P. A. (ARS) 1956

```
                -thomsonii ssp. thomsonii (1/8)
g. & cl.      -Shilsonii-
        -Bella-          -barbatum (1/8)
Rosefinch-      -griffithianum (1/4)
    -griersonianum (1/2)
```

Deep rose red with bluish tinge. (Lord Aberconway) A. M. 1938

```
cl.          -orbiculare ssp. orbiculare (1/2)
Rosemary Chipp-    -griffithianum (1/4)
        -Loderi-
            -fortunei ssp. fortunei (1/4)
```

Pale rose pink flowers. (RBG, Kew)  A. M. (RHS) 1928

```
cl.          -facetum (eriogynum) (3/4)
Rosenkavalier-          -griersonianum (1/4)
        -Tally Ho-
                -facetum (eriogynum)
```

Trusses of 15 flowers, scarlet with red spotting on upper lobes,
4in(10.2cm) across. Leaves have fawn indumentum below. Best in
partial shade. (RHS Garden, Wisley)   A. M. (RHS) 1959

```
cl.
```

Rosenoble--form of arboreum ssp. cinnamomeum var. roseum

Loose, rounded trusses of 16-18 flowers, 5-lobed, tubular-
campanulate, light roseine purple with darker veining. Leaves
up to 8in(20.3cm) long, grey plastered indumentum underneath.
(Intro. by Gill about 1910; shown by The Hon. H. E. Boscawen)
A. M. (RHS) 1973

```
cl.          -catawbiense (1/2)
Roseum Elegans-
        -unknown (1/2)
```

6ft(1.8m)          -25°F(-32°C)          ML          2/4

A vigorous, fast-growing plant with good foliage. Flowers rose lilac, rather small, held in dome-shaped trusses. Very popular variety with some Eastern U.S. nurserymen and landscapers. See also English Roseum. (A. Waterer, before 1851) See color ill. VV p. 105.

cl.

Roseum Pink    See English Roseum

cl.
                -catawbiense (1/2)
Roseum Superbum-
                -unknown (1/2)

6ft(1.8m)       -20°F(-29°C)        ML          2/3

Similar to Roseum Elegans, but pinker flowers of purplish rose, on a full plant. (A. Waterer, before 1865) See color illus. VV p. 81.

cl.

Rosevallon--form of neriiflorum ssp. neriiflorum

Bright red flowers on a plant with habit and leaf shape of neriiflorum, but with a striking purplish red underleaf. Selfed seeds have reproduced this character, although registered as a hybrid of neriiflorum by Caerhays Castle, Cornwall. (Shown by Crown Estate, Windsor) A. M. (RHS) 1975

cl.

Rosey Ball--form of rubiginosuum

6ft(1.8m)       -5°F(-21°C)         EM

Flowers 5-lobed, 1.1in(3cm) wide, strong purplish pink with red spots; ball-shaped inflorescence made of several trusses of 6-8 flowers each. Plant upright, rather leggy; glossy olive green leaves, scaly beneath, held 2 years. (Seed from 1948 Rock expedition; raiser Dr. Carl H. Phetteplace; reg. 1974)

cl.
          -decorum (1/2)
Rosina-                    -wardii var. wardii (1/4)
Lukach-      -Prelude-
     -unnamed-         -fortunei ssp. fortunei (1/8)
      hybrid-   -wardii var. wardii
         -Crest-          -fortunei ssp. discolor (1/16)
             -Lady
             Bessborough-campylocarpum ssp. campylo-
                              carpum Elatum Group (1/16)

3ft(.9m)        0°F(-18°C)          M

Funnel-shaped flowers 2in(5cm) across, 5-lobed, fragrant, primrose yellow in trusses of 8. Upright plant, nearly as broad as high; glossy yellow green leaves 5.5in(14cm) long, held 2 years. (Cross by Alfred A. Raustein, 1972; reg. 1983)

cl.                    -catawbiense (1/2)
       -Purpureum Elegans-
Roslyn-               -unknown (1/2)
      -           -catawbiense
       -Everestianum-
                    -unknown

4ft(1.2m)       -10°F(-23°C)        ML          4/3

Ruffled flowers of purplish violet with paler center, 2.25in(6.4 cm) wide, 5-lobed; spherical trusses of 12. Compact plant; dark green leaves held 2 years. (P. Vossburg, 1972) A. E. (ARS) 1973

cl.

Ross Bigler    Parentage includes catawbiense, maximum
               and perhaps unknown hybrids.

6ft(1.8m)       -15°F(-26°C)        M

Flowers deep purplish pink with red purple blotch and edging, to 2.5in(6.4cm) across, of good substance; spherical trusses of 19. Plant rounded, as wide as tall; leaves 4.5in(11.5cm) long, dull olive green, held 2 years. (W. David Smith, reg. 1979)

cl.             -elliottii (1/4)
          -Fusilier-
Ross Maud-     -griersonianum (1/4)
          -    -campylocarpum ssp. campylocarpum (1/4)
          -Unique-
               -unknown (1/4)

Bright neyron rose flowers. (Alfred Bramley, Australia, 1975)

cl.
          -ciliatum (1/2)
Rosy Bell-
          -glaucophyllum var. glaucophyllum (1/2)

2ft(.6m)        -5°F(-21°C)         VE          3/3

Plant of spreading habit, rather open; pale green leaves, scaly on top, 2.5in(6.4cm)long. Rose pink, bell-shaped flowers 1.5in (3.8cm) across, in small trusses of 4-5. Flaky dark brown bark. (Isaac Davies) A. M. (RHS) 1894

cl.
          -macgregoriae (orange form) (1/2)
Rosy Chimes-
          -gracilentum (1/2)

Vireya hybrid. Trusses of 4-6 flowers, pendulous, 1in(2.5cm) across, of bright salmon pink. Shiny, dark green leaves, scaly. (O. S. Blumhardt, reg. 1984)

cl.
          -yakushimanum ssp. yakushimanum (1/2)
Rosy Dream-                    -griffithianum (1/8)
          -         -Queen Wilhelmina-
          -Britannia-             -unknown (3/8)
                    -Stanley Davies--unknown

2ft(.6m)        0°F(-18°C)          M

Cardinal red buds open to funnel-shaped flowers of spinel red, paler in throat, 2.5in(6.4cm) across, in domed trusses of 12-16. Plant almost twice as wide as tall; dull green leaves with silvery indumentum aging to cinnamon. See also Mary Chantry. (H. L. Larson, cross 1969; intro. by J. A. Davis, 1980; reg. 1984)

cl.
          -griersonianum (1/2)
Rosy Fido-     -griffithianum (1/4)
          -Loderi-
               -fortunei ssp. fortunei (1/4)

Rose-colored flowers. (Sir James Horlick, 1930)

cl.
          -souliei (1/2)
Rosy Morn-     -griffithianum (1/4)
          -Loderi-
               -fortunei ssp. fortunei (1/4)

Rich pink buds opening to pale salmon pink flowers with a darker eye, saucer-shaped, in rounded trusses. Attractive foliage like souliei. Late season. (Credited to Rothschild in the Register; raised by Harry White, Sunningdale Nurseries) A. M. (RHS) 1931 See color illus. PB pl. 55.

cl.                   -caucasicum (1/4)
         -Dr. Stocker-
Rosy Queen-          -griffithianum (1/4)
         -thomsonii ssp. thomsonii (1/2)

Asteroid g.  Selected by Sunningdale Nurseries; exhibited 1934.


cl.
         -sanguineum ssp. didymum (1/2)
Rothatte*
         -williamsianum (1/2)

3ft(.9m)          -5°F(-21°C)          M          3/3

Deep rose flowers give the plant the appearance of a "red hat".
Attractive plant habit and foliage. (Cross by Dietrich Hobbie;
offered by H. E. Greer)


                                   -griffithianum (1/16)
                         -George Hardy-
               -Mrs. Lindsay-          -catawbiense (1/16)
cl.       -Diane- Smith  -Duchess of Edinburgh--unknown
-          -          -          -campylocarpum ssp. campylocarpum
Rothenburg-   -unnamed hybrid-                              (1/8)
                              -unknown (1/4)
         -williamsianum (1/2)

Flowers of light primrose yellow, funnel-shaped. Large leaves,
apple green and very glossy all year. (J. Bruns, 1972)    See
color illus. F p. 63.


g.
         -orbiculare ssp. orbiculare (1/2)
Rotundarb-
         -arboreum ssp. arboreum (1/2)

Synonym Orbicarb. Brilliant carmine, darker at base. (E. J. P.
Magor, 1919)


g. & cl.
         -Lowinsky hybrid--unknown (1/2)
Rouge-
         -elliottii (1/2)

Very large flowers, very bright crimson, and very late blooming,
on a tall but tender offspring of elliottii.(Rothschild) A. M.
(RHS) 1950


                                   -arboreum ssp. arboreum (1/8)
                         -Doncaster-
cl.       -Professor -          -unknown (1/2)
          - F. Bettex-          -catawbiense (1/8)
Rouge de Mai-          -Atrosanguineum-
-          -                         -unknown
-          -                    -caucasicum (1/4)
         -Madame Jeanne Frets-
                              -unknown

Trusses of 20 flowers, China rose with dark brown blotch, 3.5in
(8.9cm) across. (D. Hendriksen, reg. 1965) A. M. (Boskoop) 1961


cl.

Rouge et Noir--form of sperabile var. weihsiense (K-W 7124)

Loose trusses hold 11 flowers, tubular-campanulate, deep crimson
with darker spotting in throat. Leaves with thin, fawn, woolly
indumentum. Plant 13ft(3.9m) at registration. (Grown by S. R.
Clarke; reg. by R. N. S. Clarke)  A. M. (RHS) 1985


cl.

Round Wood--form of lanigerum (silvaticum) (K-W 6258)

4ft(1.2m)          5°F(-15°C)          EM          4/3


Small tree with oblong lanceolate leaves, dark green above with
whitish indumentum below. Rounded, compact trusses of cherry
red flowers. (Crown Estate, Windsor) A. M. (RHS) 1951


g.
     -Oreoroyle--unknown (1/2)
Rover-          -cinnabarinum ssp. cinnabarinum (1/4)
     -Royal Flush-
               -maddenii ssp. maddenii (1/4)

Flowers rose pink. (Lord Aberconway, 1946)


                              -neriiflorum ssp. neriiflorum (1/8)
cl.       -Neriihaem-
     -Hebe-          -haematodes ssp. haematodes (1/8)
Rowena-   -williamsianum (1/4)
-          -griersonianum (1/4)
     -Red Dragon-
               -thomsonii ssp. thomsonii(1/4)

Loose trusses of funnel-shaped flowers, salmon red, to 3in(7.6
cm) across. Late midseason. (Sir John Ramsden cross; intro. by
Sir William Pennington-Ramsden, reg. 1965)


cl.
         -brachycarpum ssp. brachycarpum (1/2)
Rowland-          -catawbiense (1/8)
P. Cary-          -Atrosanguineum-
     -Van der Hoop-               -unknown
-               -arboreum ssp. arboreum (1/8)
               -Doncaster-
                         -unknown (1/4)

6ft(1.8m)          -20°F(-29°C)          L

Buds dark pink, flowers blush pink with light tan spotting, in
spherical trusses of 19-20. Plant rounded, nearly as broad as
tall; dark green leaves with thin brown indumentum; new growth
silvery green. (Edward A. Cary, 1974)


cl.
         -burmanicum (1/2)
Roy Hudson-
         -nuttallii (1/2)

5ft(1.5m)          ?          E

Upright, well-branched plant; leaves olive green, very scaly on
upper surface, held 2-3 years. Fragrant flowers, white with
saffron yellow throat, 3.5in(8.9cm) wide; terminal inflorescence
of many buds, each 6-8 flowered. Floriferous. (Howard W. Kerrigan), 1970)  See color illus. ARS J 39:3 (Summer 1985): p. 122.


                         -griersonianum (1/4)
cl.       -Azor-
         -fortunei ssp. discolor (1/2)
Royal Anne-          -neriiflorum ssp. neriiflorum (1/8)
-          -Nereid-
     -unnamed-          -dichroanthum ssp. dichroanthum (1/8)
         hybrid-fortunei ssp. discolor

Loose trusses of 8-10 flowers, buttercup yellow suffused vermilion, spotted dull sap green. (E. J. Greig, Royston, 1962)


cl.       -cinnabarinum ssp. cinnabarinum Roylei Group
     -Lady  -          -cinnabarinum ssp.
Royal -Rosebery-Royal Flush-   cinnabarinum (1/8)     (3/4)
Beauty-          -maddenii ssp. maddenii (1/8)
     -cinnabarinum ssp. cinnabarinum Roylei Group

Flowers of various shades of red. (Lord Digby, cross 1936; intro. 1953)


cl.
         -elliottii (1/2)
Royal Blood-
         -Rubens--unknown (1/2)

Exceptional trusses of 36 flowers, cardinal red with dark blood
red spotting. (Francis Hanger, RHS Garden, Wisley) A. M. 1954

```
                    -cinnabarinum ssp. cinnabarinum Roylei Group
cl.      -Lady -                                    (1/4)
        -Rosebery-          -cinnabarinum ssp. cinnabarinum
Royal Blush-      -Royal Flush-                        (1/8)
         -                    -maddenii ssp. maddenii (1/8)
         -cinnabarinum ssp. cinnabarinum Blandfloriiflorum
                                        Group (1/2)
```

Flowers of crimson and gold.  (Adams-Acton, 1935)

```
              -Essex Scarlet--unknown (1/4)
        -unnamed-
Royal   - hybrid-facetum (eriogynum) (3/8)
Coachman-            -griersonianum (1/8)
            -Tally Ho-
        -unnamed-        -facetum (eriogynum)
          hybrid-        -neriiflorum ssp. neriiflorum
        -Sunshine-  Euchaites Group (1/8)
            -beanianum (1/8)
```

5ft(1.5m)          5°F(-15°C)          ML

Brilliant red flowers with darker spotting over all petals.  (J. Waterer, Sons & Crisp)

```
                    -facetum (eriogynum) (1/8)
            -unnamed-     -dichroanthum ssp. dichroanthum
        -unnamed- hybrid-Fabia-                (1/4)+3/16
cl.    - hybrid-          -griersonianum (1/4)+3/16
        -        -        -haematodes ssp. haematodes (1/8)
Royal   -    -May Day-
Dragoon-                -griersonianum
        -Fabia   -dichroanthum ssp. dichroanthum
        -unnamed-Tangerine-
          hybrid-        -griersonianum
        -     -griffithianum (1/8)
            -Mars-
              -unknown (1/8)
```

5ft(1.5m)          5°F(-15°C)          M

Dark blood red flowers.  (J. Waterer, Sons & Crisp, intro. 1975)

```
g.
            -cinnabarinum ssp. cinnabarinum (1/2)
Royal Flush-
            -maddenii ssp. maddenii (1/2)
```

6ft(1.8m)          5°F(-15°C)          M          4/3

Two forms--pink and yellow--both, flowers of heavy substance, tubular-campanulate, on a tall, upright plant, with dark green leaves and clusters of long waxy trumpets.  The pink form has wide, rather flattened flowers of soft pink; the yellow opens as champagne tinged pink, turning to cream flushed pale yellow. (J. C. Williams, Caerhays Castle, before 1930)  See color illus. ARS J 39:3 (Summer 1985): p. 151; HG p. 65.

```
cl.             -Moser's Maroon--unknown (1/4)
        -Romany Chai-
Royal Mail-       -griersonianum (1/2)
        -         -griersonianum
        -Tally Ho-
              -facetum (eriogynum) (1/4)
```

Bushy plant with silvery young shoots and scarlet flowers in late season.  (W. C. Slocock, 1940)

```
cl.           -Kaiser Wilhelm--unknown (1/2)
        -Homer-
Royal Pink-    -Agnes--unknown
        -williamsianum (1/2)
```

3ft(.9m)          -15°F(-26°C)          M          4/4

Rounded, vigorous, compact plant with good foliage; broadly elliptic leaves.  Flowers about 3in(7.6cm), openly funnel-shaped, pale fuchsine pink inside, the reverse darker; tight spherical trusses. (Dietrich Hobbie cross; intro. by Le Feber Nursery in 1965) Silver Medal (Boskoop) 1958;  Gold Medal (Rotterdam) 1960 See color illus. HG p. 130.

```
cl.
        -Moser's Maroon--unknown
Royal Star*
        -unknown
```

Full, well-rounded trusses of fine deep purple flowers, with a very dark blotch.  Plant of medium size; foliage recurved. (Donald Hardgrove)

```
cl.         -elliottii (1/2)
        -Jutland-        -Norman-fortunei ssp. discolor (1/16)
        -       -Bellerophon- Shaw -    -catawbiense
Royal   -                -B. de Bruin-        (1/32)
Windsor-                -unknown (9/32)
        -        -facetum (eriogynum) (1/8)
        -        -elliottii
        -Royal Blood-
            -Rubens--unknown
```

Firm rounded trusses of 22-24 flowers, 1.6in(4cm) wide, rich red purple, darker spotting in throat.  Narrow dark green leaves, to 5.25in(13.3cm) long.  Hardy to 0°F(-18°C). (Crown Estate, Windsor) A. M. (RHS) 1975

cl.

Royale--form of orbiculare ssp. orbiculare

Trusses of 8 flowers, 7-lobed, Persian rose.  Leaves orbicular, glabrous.  Tall plant. (Lady Adam Gordon, reg. 1983)

```
g.          -yunnanense (1/4)
        -Yunncinn-
Royalty-    -cinnabarinum ssp. cinnabarinum (1/2)
        -        -cinnabarinum ssp. cinnabarinum
        -Royal Flush-
              -maddenii ssp. maddenii (1/4)
```

Loose, pendant trusses hold 4-6 flowers, signal red at base fading to neyron rose with blotch of orange red in upper throat. Leaves elliptic, 2.25in(5.7cm) long, densely covered with brown pitted indumentum. (Lord Aberconway, 1944) A. M. (RHS) 1968

```
g.      -maddenii ssp. maddenii (1/2)
Roylmadd-
        -roylei var. magnificum ? (1/2)
```

Flowers are carmine lake. (E. J. P. Magor, 1920)

```
cl.
            -auriculatum (1/2)
Royston Festival-
            -kyawii (1/2)
```

Trusses of 15 flowers, light neyron rose with Delft rose center. (Cross by E. J. & Mary Greig; reg. by Alleyne R. Cook, 1981)

```
cl.
            -auriculatum (1/2)
Royston Reverie-    -dichroanthum ssp. dichroanthum (1/4)
            -Fabia-
              -griersonianum (1/4)
```

Trusses hold 8 flowers, light yellow with edges of light brick red.  (E. J. & Mary Greig cross; reg. by Alleyne R. Cook, 1981)

```
cl.
        -forrestii ssp. forrestii Repens Group (1/2)
Royston Red*
        -thomsonii ssp. thomsonii (1/2)
```

2.5ft(.75m)          0°F(-18°C)          E          4/4

Very attractive bright blood red flowers show color earlier than most "reds". (Harry White cross; named by Alleyne Cook)

```
                         -fortunei ssp. discolor (1/4)
cl.        -Last Rose-
           -          -          -griersonianum (1/8)
Royston Rose-          -Tally Ho-
           -                     -facetum (eriogynum) (1/8)
           -auriculatum (1/2)
```

Flowers of China rose, in trusses of 15. (Cross by E. J. & Mary Greig; reg. by Alleyne R. Cook, 1981)

```
cl.                -auriculatum (1/2)
                            -fortunei ssp. discolor (1/4)
Royston Summertime-
                   -Last Rose-          -griersonianum (1/8)
                            -Tally Ho-
                                     -facetum (eriogynum) (1/8)
```

Trusses of 10 white flowers with edges very light neyron rose, a flare of greyed red. (Cross by E. J. & Mary Greig, reg. by Alleyne R. Cook, 1981)

```
cl.


Roza Harrison     See Roza Stevenson
```

```
cl.                          -griffithianum (1/4)
                -Loderi Sir Edmund-
Roza Stevenson-              -fortunei ssp. fortunei (1/4)
                -wardii var. wardii (1/2)
```

Brunhilde g. Flowers saucer-shaped with 7 joined petals, 4.5in (11.5cm) across, held in full trusses of 10-12, deep lemon yellow, darker in bud. (Mr. & Mrs. J. B. Stevenson; reg. 1970) F. C. C. (RHS) 1968 (Shown as Mrs. Roza Harrison by RBG, Kew)

```
g.         -wardii var. wardii (1/2)
Rozamarie-          -campylocarpum ssp. campylocarpum
          -Penjerrick-           Elatum Group (1/4)
                    -griffithianum (1/4)
```

Large loose trusses of bell-shaped flowers, clear yellow, blooming in midseason. Compact, medium-sized plant with attractive foliage. (J. B. Stevenson, 1951)

```
                               -arboreum ssp. arboreum (1/8)
cl.        -Glory of Penjerrick-
           -Romala-                -griffithianum (1/8)
Rozie-          -thomsonii ssp. thomsonii (1/4)
           -fortunei ssp. discolor (1/2)
```

Flowers of deep rose pink. (J. B. Stevenson cross, about 1919)

```
cl.


Rubaiyat--form of arboreum ssp. arboreum

6ft(1.8m)          10°F(-12°C)          E          3/4
```

Trusses are 5in(12.7cm) wide, holding 21 bright red flowers which attracted attention to this tender, tree-like species over 150 years ago. Leaves 7in(17.8cm) long by 2in(5cm) wide, glossy, plastered with silvery indumentum below. (Edmund de Rothschild) A. M. (RHS) 1968

```
cl.


Rubens     Parentage unknown

3ft(.9m)          -5°F(-21°F)          ML          2/2
```

A parent of Royal Blood. Flowers deep rich red with white stamens, dark spotting on upper lobe; tight, dome-shaped trusses. Plant broader than tall; flat leathery leaves 5in(12.7cm) long, held several years. (A. Waterer, before 1865)

```
cl.


Rubescens     Parentage unknown
```

A parent of Sir Richard Carton. Red flowers.

```
cl.            -arboreum ssp. nilagiricum (kingianum) (1/4)
        -Noyo Chief-
Rubicon-       -unknown (3/8)
        -              -elliottii (1/4)
        -Kilimanjaro-       -Moser's Maroon--unknown
                      -Dusky Maid-
                               -fortunei ssp. discolor (1/8)
```

Trusses hold 17-18 flowers of cardinal red, spotted black inside upper lobes. (R. C. Gordon, New Zealand, reg. 1979)

```
g.     -sanguineum ssp. didymum (1/2)
Rubina-       -griersonianum (1/4)
       -Tally Ho-
              -facetum (eriogynum) (1/4)
```

A low and spreading plant; leaves with fawn indumentum beneath. Dark red, trumpet-shaped flowers, blooming in late summer. (J. J. Crosfield, 1938)

```
cl.


Rubrum Davidian*--form of augustinii ssp. augustinii
```

A form of this species with red flowers; blooms earlier in the season than the typical species. (See Bean, Trees and Shrubs… v. 3, p. 603)

```
cl.            -fortunei ssp. fortunei (1/2)
Ruby F. Bowman-       -griffithianum (1/4)
              -Lady Bligh-
                       -unknown (1/4)

5ft(1.5m)          -5°F(-21°C)          M          4/4
```

Long-lasting, large trusses of 15, rose pink with flower centers shaded deeper pink, throat ruby red. Leaves flat, 7in(17.8cm) long. Sun-tolerant. (Dr. Paul J. Bowman, 1958) P. A. (ARS) 1951 See color illus. VV p. 59.

```
g.         -forrestii ssp. forrestii Repens Group (1/4)
        -Charm-          -thomsonii ssp. thomsonii (1/8)
Ruby Gem-  -Shilsonii-
        -              -barbatum (1/8)
        -haematodes ssp. haematodes (1/2)
```

Red flowers. (Lord Aberconway, 1943)

```
                         -sanguineum ssp. didymum (1/8)
                    -Carmen-
cl.     -unnamed-       -forrestii ssp. forrestii Repens Group
        - hybrid-       -forrestii ssp. forrestii Repens
Ruby Hart-  -Elizabeth-              Group (1/4)
        -              -griersonianum (1/8)
        -elliottii (1/2)

3ft(.9m)          0°F(-18°C)          EM          4/4
```

Dark red flowers 1.5in(3.8cm) wide, in lax trusses of 7. Free-flowering. Plant well-branched, as broad as tall; glossy, dark green leaves with grey brown indumentum, held 3 years. (William E. Whitney, cross c.1956; reg. by the Sathers, 1976) See color illus. VV p. 48.

```
g.     -griersonianum (1/2)
Ruddigore-
       -arboreum ssp. delavayi var. delavayi (1/2)
```

Scarlet flowers. (Lord Aberconway, 1948)

---

Ruddy     Parentage unknown

A parent of Rcd Mill.  Flowers deep red.  (Origin unknown)


cl.
              -Little Red Riding Hood--unknown
Rudolf Friml-
              -unknown

Plant 6ft(1.8m) at 10 years; light green, medium-sized leaves.
Flowers solferino purple, funnel-shaped, 3.5in(8.9cm) wide, in
tight trusses.  Blooms late midseason.  (C. S. Seabrook, 1964)


cl.                  -dichroanthum ssp. dichroanthum (1/4)
                -Fabia-
Rudolph's Orange-    -griersonianum (1/4)
                -        -orbiculare ssp. orbiculare (1/4)
                -Temple Belle-
                         -williamsianum (1/4)

4ft(1.2m)         0°F(-18°C)         ML         3/3

Plant sturdy, wider than tall; leaves 2.75in(7cm) long, held one
year.  Flowers open-campanulate, 2in(5cm) across,  light orange
shaded pink inside and out; lax trusses of 6.  (R. Henny cross;
intro. & reg. by James F. Caperci, 1977)


cl.              -griersonianum (1/4)
              -Azor-
Rudy's Candy-    -fortunei ssp. discolor (1/4)
              -Corona--unknown (1/2)

Plant to 3.5ft(1m) tall; leaves to 5.5in(14cm) long.  Flowers 3
in(7.6cm) wide, two-toned, deep pink outside, 6 or 7 lobes; tall
trusses of 13-17, midseason.  (Rudolph & Leona Henny, reg. 1970)


cl.

Rudy's Fairy Tale
                     See Fairy Tale (Henny)
Rudy's Fairytale


                                      -griffithianum
                       -Queen Wilhelmina-      (1/8)
cl.              -Britannia-              -unknown  (3/8)
Rudy-Leona--Quinella-      -Stanley Davies--unknown
        (selfed)-      -haematodes ssp. haematodes
                -May Day-                   (1/4)
                      -griersonianum (1/4)

Broad plant, up to 3ft(.9m) high with leaves 6in(15cm) by
2.5in(6.4cm).  Trusses hold 6-10 flowers, dark currant red,
semi-double to double.  (Rudolph & Leona Henny, 1969)


cl.
         -catawbiense (1/2)
Ruffles*
         -unknown (1/2)

5ft(1.5m)         -15°F(-26°C)         ML         3/3

Ruffled flowers of light orchid, blooming later than most; apple
green leaves held by straw-colored petioles.  (Origin unknown)


cl.

Ruffles and Frills     See upper right.


g.
    -forrestii ssp. forrestii Repens Group (1/2)
Rufus-
       -sanguineum ssp. sanguineum var. sanguineum (1/2)

Red flowers.  (Lord Aberconway, 1941)


                -yakushimanum ssp. yakushimanum (1/4)
         -Hydon Glow-      -griersonianum (1/8)
cl.      -         -Springbok-      -ponticum (1/16)
         -              -unnamed-
Ruffles  -              hybrid-unknown (1/8)
-nd Frills-      -dichroanthum ssp. dichroanthum (1/8)
         -   -Dido-
         -      -decorum (1/8)
         -Lem's-                         -griffithianum
         Cameo-        -Beauty of Tremough-        (5/32)
         -     -Norman-              -arboreum ssp.
         -Anna- Gill -griffithianum       arboreum (1/32)
                       -griffithianum
              -Jean Marie -
              de Montague-unknown

6ft(1.8m)         -5°F(-21°C)              M

Flowers openly funnel-shaped, frilled, 6-lobed, fragrant, throat
Naples yellow,  shading through orient pink at lobe margins,  to
jasper red in dorsal throat, with spotting; the reverse suffused
azalea pink.  Trusses dome-shaped with 12 flowers.  New foliage
bronze; leaves of heavy texture, held 3 years.  (Wildfong, 1984)


g.
         -russatum (1/2)
Russautinii-
         -augustinii ssp. augustinii (1/2)

5ft(1.5m)         -10°F(-23°C)         EM         4/3

Lepidote hybrid with small foliage as russatum.  Flowers in
trusses of 2-5, near to augustinii in size, slate blue with
darker eye.  Flowers profusely at early age.  (Sir John Ramsden,
1937)


cl.
         -maximum (1/2)
Russell Harmon*
         -catawbiense (1/2)

6ft(1.8m)         -25°F(-32°C)         L         4/3

Excellent, large, high truss of 25 flowers, magenta pink.  Open
plant habit.  Very hardy and parent of other very hardy hybrids.
(A natural hybrid (?) intro. by LaBar's Rhododendron Nursery)


---

         -catawbiense (1/2)
Russellianum-
         -arboreum ssp. arboreum (1/2)

A large shrub or small tree, with rounded trusses of crimson red
flowers.  Blooms early to midseason.  (Russell, 1831)


cl.
              -arboreum ssp. arboreum (1/2)
Russellianum Tigrinum-
              -unknown (1/2)

Red flowers, spotted.  (T. Methven, 1868-69)


cl.
              -arboreum ssp. arboreum (1/2)
Russellianum Superbum-
              -unknown (1/2)

Red flowers.  (T. Methven, 1868-69)


g.
         -russatum (1/2)       -intricatum (1/8)
Rustic Maid-           -Intrifast-
       -Blue Diamond-       -fastigiatum (1/8)
                   -augustinii ssp. augustinii (1/4)

2.5ft(.75m)         -5°F(-21°C)         EM         4/3

Flowers of deep, vibrant lilac blue.  (Collingwood Ingram, 1945)
P. C. (RHS) 1972

```
                         -neriiflorum ssp. neriiflorum (1/8)
g.        -E. C. Puddle-
   -Ethel-          -griersonianum (1/8)
Ruth-      -forrestii ssp. forrestii Repens Group (1/4)
    -           -thomsonii ssp. thomsonii (1/4)
    -Barclayi-           -arboreum ssp. arboreum (1/8)
               -Glory of Penjerrick-
                         -griffithianum (1/8)
```

Blood red flowers.  (Lord Aberconway, 1946)

```
cl.                              -maximum (1/4)
          -Marchioness of Lansdowne-
Ruth A. Weber-            -unknown (1/2)
              -         -catawbiense (1/4)
              -Old Port-
                    -unknown
```

4ft(1.2m)        -5°F(-21°C)          ML

Violet-colored flowers with light purple spotting, 3.75in(9.5cm)
across, 5-lobed; spherical trusses of 17.  Floriferous.  Upright
plant; dark green leaves held 3 years.   (Edwin O. Weber, 1974)

```
cl.
          -yakushimanum ssp. yakushimanum (1/2)
Ruth Davis-
          -japonicum var. japonicum (metternichii) (1/2)
```

2.5ft(.75m)        -10°F(-23°C)          M

Cherry red buds open to white flowers, reverse pink until fully
open, held in ball-shaped trusses of 17.  Semi-dwarf plant,
broader than tall with distinctive winter foliage, glossy dark
green. (Joseph Gable cross; reg. by Ross B. Davis, Jr., 1979)

```
          -wardii var. wardii (1/4)
cl.  -Idealist-                 -griffithianum (1/32)
   -      -              -Kewense-
Ruth  -       -Aurora-      -fortunei ssp. fortunei
Hansen-       -Naomi-      -thomsonii ssp. thomsonii (1/16)
   -               -fortunei ssp. fortunei (13/32)
   -    -wightii (1/4)
      -China-
          -fortunei ssp. fortunei
```

4ft(1.2m)        0°F(-18°C)          ML        4/3

Moderately compact plant, 30in(.76m) at 7 years; medium green
leaves 6in(15.2cm) long.  Flowers primrose yellow, paler at margins, to 4.5in(11.5cm) wide, 6-lobed, in conical trusses of 13.
(Dr. Carl W. Phetteplace, reg. 1969)

cl.

Ruth Lyons--form of davidsonianum

5ft(1.5m)        0°F(-18°C)          M        4/3

An upright plant with leaves that fold upwards from the midrib.
Flowers bright pink, up to 1.25in(3.2cm) across, in flat-topped
trusses of 6-7.  (James Barto raiser; reg. by M. G. Lyons, 1962)
P. A. (ARS) 1961

```
                         -griffithianum (1/8)
cl.          -Queen Wilhelmina-
      -Britannia-            -unknown (3/8)
Ruth Wick-      -Stanley Davies--unknown
      -yakushimanum ssp. yakushimanum (1/2)
```

1.5ft(.45m)        0°F(-18°C)          M

Flowers neyron rose at margins, lighter in center and throat, no
markings, funnel-shaped, 1.75in(4.5cm) wide, 5-lobed; spherical
trusses of 10-14.  Plant rounded, 1/3 broader than tall; elliptic, heavily textured leaves held 2 years. (Seed from the ARS
Exchange, cross by Mrs. V. Manenica; raised by Mr. & Mrs. Wick,
reg. 1983)

```
g.                -thomsonii ssp. thomsonii (1/4)
        -Cornish Cross-
Ruthelma-            -griffithianum (1/2)
        -           -griffithianum
        -Loderi Pink-
       Diamond  -fortunei ssp. fortunei (1/4)
```

Flowers of reddish pink.  See also Coralia, H. Whitner.  (Sir E.
Loder)

```
cl.                -dichroanthum ssp. dichroanthum (1/4)
      -Goldsworth Orange-
Sabine-            -fortunei ssp. discolor (1/4)
      -      -haematodes ssp. haematodes (1/4)
      -May Day-
            -griersonianum (1/4)
```

Flowers rose Bengal with throat spotted  Venetian pink. (Joh.
Bruns, 1972)

```
g.
        -Rose Perfection--unknown (1/2)
Sadonia*
        -fortunei ssp. fortunei (1/2)
```

Rose pink flowers.  (Lord Aberconway cross, 1926; intro. 1933)

```
cl.       -arboreum ssp. arboreum (3/4)
   -Cornubia-          -thomsonii ssp. thomsonii (1/8)
Sady-      -Shilsonii-
   -           -barbatum (1/8)
   -arboreum ssp. arboreum
```

Flowers similar to Cornubia, cardinal red, blooming from end of
May to September in Victoria, Australia.  (Alfred Bramley, 1964)

```
cl.                -xanthostephanum (1/4)
          -Saffron Queen-
Saffron Prince-      -burmanicum (3/4)
          -burmanicum
```

Trusses of 8 flowers, brilliant yellow green in bud, opening to
brilliant greenish yellow, openly funnel-campanulate, 2in(5cm)
across; floriferous.  Plant rounded, 3ft(.9m) at 5 years; leaves
2.5in(6.4cm) long, held 3 years.  (M. & F. Sumner, reg. 1985)

```
cl.
          -xanthostephanum (1/2)
Saffron Queen-
          -burmanicum (1/2)
```

4.5ft(1.35m)        20°(-7°C)          EM        4/3

Growth habit upright, open; leaves glossy green, narrow, to 3in
(7.6cm) long.  Flowers sulphur yellow with dark spotting, 2.5in
(6.4cm) wide; lax trusses of 8-9.  (Charles Williams)    A. M.
(RHS) 1948    See color illus. HG 68; VV p. 123.

cl.

Sahara    See Nile

These two names have been used by Leach for the same clone.  The
valid, registered name is Nile.

cl.

Sail Wing    See next page.

```
cl.
          -impeditum (1/2)
Saint Breward-
          -augustinii ssp. augustinii (1/2)
```

Small, rounded, dense, very hardy bush, wider than high; floriferous. Flowers openly campanulate, 2.3in(6cm) across, sea lavender violet, darker at edges; tight spherical trusses of about
26. Cutting from original Lamellen plant grew to 4.75ft(1.4m) in
15 years. (Cross by E. J. P. Magor; intro. by Gen. Harrison)
F. C. C. (RHS) 1962

```
                  -wardii var. wardii (1/4)
                  -                    -griffithianum (1/32)
cl. -Idealist-          -Kewense-
    -(selfed)-   -Aurora-    -fortunei ssp. fortunei
Sail-      -Naomi-    -thomsonii ssp. thomsonii (1/16)
Wing-        -fortunei ssp. fortunei (5/32)
    -            -fortunei ssp. discolor (1/8)
    -     -Lady
    -Jalisco-Bessborough-campylocarpum ssp. campylocarpum
      Orange-                    Elatum Group (1/8)
          -     -dichroanthum ssp. dichroanthum (1/8)
          -Dido-
              -decorum (1/8)
```

White flowers with blush pink tinge and a yellow throat. (John
Waterer, Sons & Crisp, reg. 1975)

```
g.                 -griffithianum (1/4)
     -unnamed hybrid-
St. George-            -unknown (1/2)
     -            -fortunei ssp. fortunei (1/4)
    -H. M. Arderne-
              -unknown
```

Light crimson buds, opening paler, distinctly veined in pale
crimson. (Waterer, Sons & Crisp, 1932)  A. M. (RHS) 1946

```
cl.
          -arboreum ssp. zeylanicum (1/2)
Saint Keverne-
          -griffithianum (1/2)
```

Bright red flowers with a few dark brown dots. (P. D. Williams)
A. M. (RHS) 1922

```
cl.
        -augustinii ssp. augustinii (3/4)
Saint Kew-     -impeditum (1/4)
       -Saint Breward-
                 -augustinii ssp. augustinii
```

Trusses of 11-13 clustered flowers, violet blue with traces of
darker shading at margins, 1.75in(4.5cm) wide. Narrow pointed
leaves to 2in(5cm) long. (Gen. Harrison)   P. C. (RHS) 1970

```
cl.          -souliei (1/4)
       -Souldis-
Saint Mabyn-     -fortunei ssp. discolor (1/4)
       -unknown (1/2)
```

Trusses of 10 flowers 4in(10.2cm) across, saucer-shaped, white
tinged yellow inside and palest pink outside, top petal spotted
red. (Gen. Harrison cross, 1921; reg. 1967)

```
cl.              -brachycarpum ssp. brachycarpum (1/4)
       -unnamed hybrid-
Saint Mary's-    -catawbiense (1/4)
       -         -fortunei ssp. fortunei (1/4)
       -unnamed hybrid-
                 -unknown (1/4)
```

6ft(1.8m)        -10°F(-23°C)        M

Lavender buds; fragrant flowers, pale violet fading pure white;
ball-shaped trusses of 9-13. Upright, tall plant, as broad as
high; leaves 5in(12.7cm) long, glossy olive green, best grown in
sun. (Crossed early 1930s by Dr. Henry T. Skinner; reg. 1979)

```
cl.              -augustinii ssp. augustinii (1/4)
       -Saint Tudy-
Saint Merryn-    -impeditum (3/4)
       -impeditum
```

2.5ft(.75m)        -5°F(-21°C)        EM        3/3

A small compact hybrid with aster violet flowers, funnel-shaped,
about 1in(2.5cm) wide; clusters of 2-4. Plant twice as broad as
tall. (Gen. Harrison)  A. M. (RHS) 1970;  F. C. C. (RHS) 1973;
A. M. (Wisley Trials) 1983

```
cl.
     -russatum (1/2)
Saint Minver-      -impeditum (1/4)
     -Saint Breward-
             -augustinii ssp. augustinii (1/4)
```

Violet blue flowers; a low, compact plant. (Gen. Harrison, reg.
1973)

```
cl.
     -impeditum (1/2)
Saint Tudy-
     -augustinii ssp. augustinii (1/2)
```

3ft(.9m)        -5°F(-21°C)        EM        3/3

A vigorous, upright plant; floriferous. Lobelia blue, campanu-
late flowers, 2in(5cm) wide, in dome-shaped trusses of 14. (E.
J. P. Magor, raiser; intro. by Gen. Harrison)  A. M. (RHS) 1960;
F. C. C. (RHS) 1973

```
cl.
     -lochiae (1/2)
Saint Valentine-
     -gracilentum (1/2)
```

Vireya hybrid. Trusses of 3-5 flowers, tubular-campanulate, of
bright post office red. Elliptic leaves. (T. Lelliot cross;
raiser P. Sullivan; Australian Rhododendron Society, reg. 1984)

```
                    -intricatum (1/8)
cl.            -Intrifast-
       -Blue Diamond-     -fastigiatum (1/8)
Saint Wenn-     -augustinii ssp. augustinii (1/4)
       -polycladum Scintillans Group (1/2)
```

The flowers are mauve. (Gen. Harrison, reg. 1973)

```
cl.            -Dexter hybrid--unknown
       -Helen Everitt-
Sally Fuller-      -Dexter hybrid--unknown
       -Honeydew (Dexter)--unknown
```

5ft(1.5m)        -20°F(-29°C)        M

Flowers of very heavy substance, 7-lobed, 4in(10.2cm) wide, very
fragrant, white with a slight rose flush, and a golden glow in
throat, held in trusses of 7. Plant broader than tall; leaves
medium green, held 2 years. See also Fran Labera. (Henry &
Selma Fuller 1961; reg. 1978)

```
cl.           -williamsianum (1/4)
       -Jock-
Salmon Jubilee-   -griersonianum (1/4)
       -unknown (1/2)
```

Soft scarlet buds open to salmon-colored flowers in trusses of
5-9. (Lady Marion Philipps, Picton Castle, S. Wales, 1977)

```
cl.
       -cinnabarinum ssp. cinnabarinum Roylei Group (1/2)
Salmon Trout-      -cinnabarinum ssp. cinnabarinum (1/4)
       -Royal Flush -
       (orange form)-maddenii ssp. maddenii (1/4)
```

5ft(1.5m)        10°F(-12°C)        ML        4/3

Lady Chamberlain g. Described in the Register as "the color of
freshly cooked salmon trout." (Rothschild, 1930)

g.                          -griffithianum (3/4)
      -Queen Wilhelmina-
Salome-                     -unknown (1/4)
      -griffithianum

Flowers of rose red.  (Lord Aberconway, 1950)

cl.
          -griersonianum (1/2)
Saltwood-                        -maximum (1/4)
          -Lady Clementine Mitford-
                                 -unknown (1/4)

Openly funnel-shaped flowers, 4.3in(11cm) wide, 7-lobed, of deep
neyron rose, reverse carmine rose; loose trusses of 8-9. Leaves
6in(15cm) long, dark green above, pale below. (Maj. A. E. Hardy)
A. M. (RHS) 1965

                    -sanguineum ssp. didymum (1/4)
g.  -Arthur Osborn-
    -               -griersonianum (1/4)
Salus-              -griffithianum (1/8)
    -     -Loderi-
    -Coreta-        -fortunei ssp. fortunei (1/8)
            -arboreum ssp. zeylanicum (1/4)

Flowers of deep red.  (Lord Aberconway, 1941)

g.
          -griffithianum (1/2)
Salutation-
          -lacteum (1/2)

Tall, attractive bush, rather tender, bearing large, well-filled
trusses of cream-colored flowers, midseason.  (Rothschild; 1953)

cl.

Samisen--yakushimanum ssp. yakushimanum (Exbury form selfed)

3ft(.9m)          -15°F(-26°C)          ML

Flowers of good substance, light neyron rose edges shading paler
rose at base, widely campanulate, 1.75in(4.5cm) wide; trusses of
12-18.  Plant well-branched, 50% wider than tall; leaves with
plastered rust indumentum, held 3 years. (Seed from Ben Nelson;
grown by Mary Louisa B. Hill; reg. 1984)

cl.          -griffithianum (1/4)
          -Mars-
Sammetglut-  -unknown (5/8)            -catawbiense (1/8)
(Velvet  -         -Parsons Grandiflorum-
  Glow) -Nova Zembla-              -unknown
                    -dark red hybrid--unknown

Trusses of 16-21 flowers, 5-lobed, bright cardinal red without
markings, conspicuous anthers of very pale peach. Floriferous.
Plant upright, hardy to -17°F(-27°C); elliptic, hairy leaves.
Sold by some U.S.A. nurseries as Crimson Classic. See also Hach-
mann's Feuerschein, Blinklicht.  (Cross by H. Hachmann; reg. by
G. Stuck, 1983)

cl.
      -azalea occidentale (1/2)
Samurai-      -elliottii (1/4)
      -Fusilier-
              -griersonianum (1/4)

Azaleodendron.  Trusses hold 28-30 flowers, neyron rose with a
red throat.  Bloomed 1963 from seeds planted 1950. (Lester E.
Brandt, reg. 1969)

cl.                   -macgregoriae (1/4)
      -unnamed hybrid-
San Gabriel-           -aurigeranum (1/4)
      -               -phaeopeplum (1/4)
      -Dr. Herman Sleumer-
                        -zoelleri (1/4)

2ft(.6m) [13 yrs.]          32°F(0°C)          Mar.-Dec.

Vireya hybrid. Trusses of 8 tubular funnel-shaped flowers, pale
yellow flecked with pink, 2.25in(5.7cm) wide, fragrant, 5-lobed.
Shrub wider than high; flat yellow green leaves, scaly beneath,
held 3 years.  (T. Lelliott cross; W. A. Moynier, reg. 1985)

cl.
      -souliei (1/2)
Sandling-       -fortunei ssp. discolor (1/4)
      -Lady -
      Bessborough-campylocarpum ssp. campylocarpum Elatum
                                               Group (1/4)
5ft(1.5m)          -5°F(-21°C)          M          3/3

Halcyone g.  Narrow leaves 7in(17.8cm) long, dark above, paler
beneath.  Frilled flowers, 3.5in(8.9cm) wide, of rhodamine pink,
throat flushed amber, darker spots; compact trusses of about 12.
(Major A. E. Hardy, Sandling Park)  A. M. (RHS) 1965

                         -catawbiense (9/64)
                    -Omega-
cl.  -unnamed hybrid-    -unknown (6/32)
                    -wardii var. wardii (5/8)
Sandra-    -wardii var. wardii
      -    -                         -griffithianum
      -    -          -George-            (1/64)
      -unnamed-    -Mrs. Lindsay- Hardy-catawbiense
        hybrid-       - Smith
          -     -Diane-        -Duchess of
          -     -      -            Edinburgh--unknown
          -Alice-   -    -campylocarpum ssp.
          Street-   -unnamed-      campylocarpum (1/32)
            -       hybrid-unknown
          -wardii var. wardii

Reverse cross of Goldkrone.  Trusses of 9-11 flowers, 5-lobed,
medium neyron rose shading lighter and tinted light amber yellow. Elliptic, hairy leaves. (Cross by H. Hachmann; reg. by G.
Stuck, 1983)

g.
      -sanguineum ssp. sanguineum var. sanguineum (1/2)
Sangreal-
      -griersonianum (1/2)

Introduced by the Marquess of Headfort, 1937.

cl.
          -griersonianum (1/2)
Santa Claus-
          -unknown (1/2)

Flowers a deep bright red.  (Gen. Harrison, 1965)

cl.

Sappho     Parentage unknown

6ft(1.8m)          -5°F(-21°C)          M          3/2

Often a parent.  Mauve buds open to pure white flowers with a
conspicuous blotch of violet overlaid with blackish purple. Medium-sized flowers, widely funnel-shaped, held in large conical
trusses.  Plant habit open, rather leggy; leaves olive green,
narrow.  Mentioned in the earliest surviving nursery catalogue,
for 1847.  (A. Waterer)  A. G. M. (RHS) 1969; A. M. (RHS) 1974
See color illus. F p. 64;  VV p. 25, 106.

cl.
          -Sappho--unknown (1/2)
Sappho's Choice-
          -Catanea--catawbiense (1/2)

3ft(.9m)          5°F(-15°)          M

Flowers white with beetroot purple flare of spotting in dorsal
lobe, held in dome-shaped trusses of 17. Flowers similar to
those of Sappho, but plant habit is superior and spreads well,
as broad as tall. (Cross by Emil V. Bohnel; raised by Lonnie M.
Player; reg. 1984)

cl.
```
        -Zella--unknown (1/2)
Sarah Hardy-                    -griffithianum (1/8)
     -              -Isabella-
     -Muy Lindo-        -auriculatum (1/8)
     -                -decorum (1/8)
             -unnamed hybrid-
                     -souliei (1/8)
```

White flowers, lightly flushed pink in trusses of 11-12. (Cross by Collingwood Ingram; intro. & reg. by G. A. Hardy, 1982)

```
                        -griffithianum (1/8)
                -Queen   -
cl.        -Britannia-Wilhelmina-unknown (5/8)
     -
Sarah Jane-         -Stanley Davies--unknown
     -                -yakushimanum ssp. yakushimanum (1/4)
     -unnamed hybrid-
             -unnamed hybrid--unknown
```

Flowers light pink deepening at edges to rose red. (John Waterer, Sons & Crisp, cross 1964; reg. 1975)

g.
```
        -fortunei ssp. discolor (1/2)
Sardis-        -catawbiense (1/4)
     -C. S. Sargent-
             -unknown (1/4)
```

A parent of Firedrake. (RBG, Kew)

```
                -griersonianum (3/8)
cl.        -Sarita Loder-     -griffithianum (1/8)
     -                -Loderi-
Sarita Coker-            -fortunei ssp. fortunei (1/8)
     -                -fortunei ssp. discolor (1/4)
     -King of Shrubs-     -dichroanthum ssp. dichroanthum
             -Fabia-                        (1/8)
                     -griersonianum
```

Trusses of 10 flowers, 7-lobed, deep neyron rose in bud, opening lighter with brighter red throat. (Cross by Mrs. R. J. Coker; intro. by A. G. Holmes; reg. by Mrs. R. J. Coker, 1983)

g. & cl.
```
        -griersonianum (1/2)
Sarita Loder-   -griffithianum (1/4)
        -Loderi-
        -fortunei ssp. fortunei (1/4)
```

4ft(1.2m)        5°F(-15°C)        ML        4/3

Dark crimson buds, opening to deep rose or salmon flowers, in loose trusses.  Plant habit upright, open; long pointed leaves. Needs some protection from sun, frost. (Col. G. H. Loder, High Beeches) A. M. (RHS) 1934

g.
```
        -sargentianum (1/2)
Sarled-
        -trichostomum Ledoides Group (1/2)
```

Slow-growing, compact, dwarf lepidote. Rounded trusses of small tubular flowers, creamy white, pink in bud, daphne-like.  Hardy in Ontario.  (Collingwood Ingram) A. M. (RHS) 1974  See color illus. Rhod. Soc. of Can. Bull. 7:2 (1978): p.15.

cl.
```
        -Blue Peter--unknown (1/2)
Sasha-        -fortunei ssp. fortunei (1/4)
     -unnamed hybrid-        -wardii var. wardii (1/8)
                     -unnamed-
                     hybrid-dichroanthum ssp. dichroanthum
                                             (1/8)
```
3ft(.9m)        -5°F(-21°C)        M

Leaves 5in(12.9cm) by 2.5in(6.4cm), retained 3 years.  Flowers white shaded to medium reddish purple and held in trusses of 14. (Alfred A. Raustein, 1980)  See color illus. ARS J 36:3 (Summer 1982): p. 102.

cl.
```
        -caucasicum (1/4)
     -Boule de Neige-        -catawbiense (1/8)
Satin-        -hardy unnamed hybrid-
     -                -unknown (1/8)
     -catawbiense (red form) (1/2)
```

Flowers shrimp pink.  (A. M. Shammarello, before 1958)

```
            -fortunei ssp. fortunei (37/64)
cl.   -Ray-      -wardii var. wardii (1/8)
     -   -           -griffithianum (5/64)
     -  -Idealist-        -Kewense-
     -        -  -Aurora-        -fortunei ssp. fortunei
Satin -        -Naomi-    -thomsonii ssp. thomsonii
Bouquet-        -fortunei ssp. fortunei        (1/32)
     -
     -                -griffithianum
     -                -George-
     -   -Mrs. Lindsay- Hardy-catawbiense (1/16)
     -Dot-  Smith
     -        -        -Duchess of Edinburgh--unknown (1/8)
     -fortunei ssp. fortunei
```

4ft(1.2m)        0°F(-18°C)        ML

Light green leaves 6in(15cm) long. Flowers off-white with yellow green centers, held in ball-shaped trusses of 10-12.  (Arthur & Maxine Childers, reg. 1972)

cl.
```
        -Koichiro Wada--yakushimanum ssp. yakushimanum (1/2)
Satin Cloud-
        -unknown (1/2)
```

Plant 3ft(.9m) high at 10 years.  Trusses of 12-13 flowers, of light purplish pink spotted red in throat.  Thick tan indumentum beneath young leaves.  (Pukeiti Rhododendron Trust, G. F. Smith, reg. 1984)

```
                        -souliei (1/16)
                -Soulbut-
        -Vanessa-        -Sir Charles Butler--fortunei ssp.
g.   -Radiance-        -griersonianum        fortunei (3/16)
     -        -griersonianum (3/8)
Satire-        -griffithianum (1/8)
     -   -Loderi-
     -Coreta-   -fortunei ssp. fortunei
             -arboreum ssp. arboreum (1/4)
```

Bright red flowers.  (Lord Aberconway, 1950)

g.
```
        -haematodes ssp. haematodes (1/4)
     -Choremia-
Satyr-        -arboreum ssp. arboreum (1/4)
     -strigillosum (1/2)
```

Red flowers.  (Lord Aberconway, 1941)

cl.
```
        -calophytum var. calophytum (1/2)
Sausalito-        -griffithianum (1/4)
     -Loderi Venus-
             -fortunei ssp. fortunei (1/4)
```

6ft(1.8m)        -5°F(-21°C)        E        4/3

Light pink flowers with 1in(2.5cm) red blotch on upper lobe, reverse strong pink, to 4in(10.2cm) wide; trusses of 11-14. Dark green leaves 6in(15.2cm) long.  (John Henny, 1968)  P. A. (ARS) 1967

```
g.        -griersonianum (3/4)
     -Azor-
Saxa-     -fortunei ssp. discolor (1/4)
     -griersonianum
```

Flowers of pale salmon rose. (Lord Aberconway, 1946)

```
                               -griffithianum (1/4)
                   -George Hardy-
cl.     -Betty Wormald-          -catawbiense (1/4)
     -               -red hybrid--unknown (3/8)
Scandinavia-                     -griffithianum
     -               -George Hardy-
     -Hugh Koster-              -catawbiense
               -               -arboreum ssp. arboreum (1/8)
                    -Doncaster-
                               -unknown

3ft(.9m)        0°F(-18°C)        ML        3/2
```

Plant broader than tall; leaves 7in(18cm) by 2in(5cm) of medium
green. Funnel-shaped dark crimson flowers with black blotch, in
large dome-shaped trusses. (M. Koster & Sons) A. M. (RHS) 1950

```
cl.
               -laetum (1/2)
Scarlet Beauty-                           -brookeanum var.
     -               -Duchess of Edinburgh-  gracile (1/8)
          -Triumphans-                    -longiflorum (1/8)
                    -javanicum (1/4)
```

Vireya hybrid. Trusses of 13 flowers; corolla orange buff, lobes
bright poppy red. (Don Stanton cross; reg. by G. Langdon, 1982)

```
cl.            -griffithianum (3/8)
          -Mars-
Scarlet Blast-    -unknown (3/8)
     -               -griffithianum
     -unnamed-Mars-
       hybrid-    -unknown
                    -catawbiense var. rubrum (1/4)

5ft(1.5m)        -20°F(-29°C)        M        4/2
```

Plant rather open, broader than tall; leaves dark green, convex,
elliptic. Flowers rotate-campanulate, 2.6in(6.6cm) wide, dark
red shading lighter, blotched light pinkish yellow; firm pyramidal trusses of 17. See also Inca Chief. (David G. Leach, reg.
1972)

```
---                           -brookeanum var. gracile (1/4)
               -Duchess of Edinburgh-
Scarlet Crown-                -longiflorum (lobbii) (1/4)
          -javanicum (1/2)
```

Vireya hybrid. Flowers of scarlet crimson. F. C. C. (RHS) 1883

```
cl.                    -catawbiense (1/2)
          -unnamed red hybrid-
Scarlet Glow-               -unknown (1/2)
     -                    -catawbiense
     -unnamed red hybrid
                         -unknown

5ft(1.5m)        -20°F(-29°C)        ML        3/2
```

Brick red, funnel-shaped flowers, 2.5in(6.4cm) across, held in
conical trusses of 15. Plant compact; dark green leaves to 3.5
in(9cm) long. (A. M. Shammarello, cross 1962; reg. 1972 & 1976)

```
g.                    -arboreum ssp. zeylanicum (1/4)
          -Ilam Alarm-          -griffithianum (1/8)
Scarlet King-     -unnamed hybrid-
     -griersonianum (1/2)    -unknown (1/8)

5ft(1.5m)        5°F(-15°C) ?        ML
```

Rich scarlet flowers on a nice-looking shrub. Does well in hot
climates. See also Orchard. (W. T. Stead, 1950)

```
cl.               -fortunei ssp. fortunei (1/4)
          -Luscombei-
Scarlet Lady-          -thomsonii ssp. thomsonii (1/4)
          -haematodes ssp. haematodes (1/2)
```

A parent of Glasgow Glow. (Stirling Maxwell, 1936)

```
cl.
          -neriiflorum ssp. neriiflorum (1/2)
Scarlet Nymph-
          -strigillosum (1/2)

3ft(.9m)        5°F(-15°C)        EM
```

Rounded plant, broader than tall; leaves 4in(10.2cm) long, with
loose brownish indumentum. Campanulate blood red flowers, 1.5in
(3.8cm) wide, in rounded trusses. (H. L. Larson, reg. 1965)

```
cl.          -forrestii ssp. forrestii Repens Group (3/4)
          -Elizabeth-
Scarlet-          -griersonianum (1/4)
Tanager-
          -forrestii ssp. forrestii Repens Group
```

A small shrub bearing intense scarlet flowers. (Collingwood Ingram; reg. 1970)

```
cl.
          -Essex Scarlet--unknown (1/2)
Scarlet Wonder-
          -forrestii ssp. forrestii Repens Group (1/2)

2ft(.6m)        -10°F(-23°C)        M        4/4
```

Plant compact, vigorous, over twice as wide as high; dense foliage of medium dark green leaves, heavy-textured, with a slight
sheen. Campanulate flowers, wavy-edged, of bright cardinal red
with faint brown markings, 5-7 in loose trusses. (Dietrich
Hobbie cross; intro. by Le Feber & Co.; reg. 1965)   Silver Gilt
Medal (Rotterdam) 1960;   Gold Medal (Boskoop) 1961   See color
illus. ARS Q 34:2(Spring 1980): p. 70;  F p. 25;  HG p. 144;  VV
p. 50.

```
g.          -thomsonii ssp. thomsonii (1/2)
Scarlett-               -griffithianum (1/8)
 O'Hara -     -Queen     -
     -Langley-Wilhelmina-unknown (3/8)
       Park  -
               -Stanley Davies--unknown
```

Loose trusses of waxy red flowers on a handsome bush. (Sir James
Horlick cross, 1932; exhibited 1942)

```
cl.          -yakushimanum ssp. yakushimanum ( 1/2)
Schamenek's Glow-
          -smirnowii (1/2)

2ft(.6m)        -10°F(-23°C)        ML        4/4
```

Well-branched plant, broader than tall;  dark glossy leaves with
heavy fawn indumentum, held 3 years; new growth light green, tomentose. Flowers 3.5in(8.9cm) wide, pale pink to white, chartreuse blotch; spherical trusses of 12. Floriferous. (Cross by
John Schamenek; intro. & reg. by Marie Tietjens, 1976)

```
cl.          -catawbiense (1/2)
Scharnhorst-
          -unknown (1/2)
```

A parent of Grafin Kirchbach. Flowers shade from dark ruby red
to purplish red, with dark brown markings; frilled. (T. J. R.
Seidel, cross c.1908; intro. 1916)

cl.              -arboreum ssp. arboreum (1/4)
        -Doncaster-
Schiller-     -unknown (1/4)
      -                -griffithianum (1/4)
         -George Hardy-
                    -catawbiense (1/4)

Flowers light rose, fringed.  (M. Koster & Sons cross, 1909)

cl.
           -Mrs. J. G. Millais--unknown (5/8)
Schneehukett-          -caucasicum (1/8)
(Snow-       -          -Viola-
  bouquet) -Bismarck-      -unknown
                     -catawbiense (1/4)

Trusses hold 18-26 flowers, 5-6-lobed, clear white blotched ruby
red, about 3.2in(8cm) wide.  Elliptic leaves. (Cross by H. Hachmann; reg. by G. Stuck, 1983)

cl.
             -catawbiense (1/4)
Schneekrone-Humboldt-
 (Snow-      -        -unknown (1/4)
  crown)  -Koichiro Wada--yakushimanum ssp. yakushimanum (1/2)

Trusses hold 12-17 flowers, 5-lobed, opening light neyron rose
fading to clear white with soft pink tinge, ruby red spotting.
Elliptic, hairy leaves.  (Cross by H. Hachmann; G. Stuck, reg.
1983)

cl.
          -Mrs. J. G. Millais--unknown (1/2)
Schneewolke-
(Snowcloud)-Koichiro Wada--yakushimanum ssp. yakushimanum (1/2)

Pollen parent may be Sammetglut x Koichiro Wada--yakushimanum
ssp. yakushimanum.  White flowers, 5-lobed,  held in trusses of
12-17.  Hairy leaves, revolute margins.  Hardy to -13°F(-24°C).
(H. Hachmann cross, 1968; intro. by G. Stuck, 1982)

cl.
        -griffithianum (1/2)
Schubert-
        -unknown (1/2)

Flowers pale orchid, fringed.  (M. Koster & Sons)

cl.
          -catawbiense ? (1/2)
Schuylkill-                        (A natural hybrid?)
           -decorum ? (1/2)

2.5ft(.75m)        -5°F(-21°C)          EM

Leaves 5.75in(14.6cm) long.  Pink flowers fading to creamy white
with some darker pink spotting, pink stripe down center of each
lobe; rounded trusses of 18.  (Charles Herbert, reg. 1977)

cl.

Scintillation     Parentage unknown

5ft(1.5m)          -10°F(-23°C)          M        4/4

Often a parent. Distinctive leaves, deep shiny green, waxy texture, on a shrub broader than tall.  Strong stems support large
trusses of 15 flowers of pastel pink, flared golden bronze markings in throat.  One of the most popular  of the Dexter hybrids.
(Cross by C. O. Dexter,  1925-42; raised by  Paul Vossberg; reg.
1973)   A. E. (ARS) 1973     See color illus. ARS Q 29:3 (Summer
1975) p. 171; HG p. 78;  K p. 40; LW pl. 8,9; VV p. 32, cover.

cl.
        -catawbiense (1/2)
Scipio-
        -unknown (1/2)

Purplish red flowers.  (A. Waterer, before 1871)

cl.          -griffithianum (1/4)
        -Loderei-
Sea Mist-     -fortunei ssp. fortunei (1/4)
        -sutchuenense (1/2)

Seagull g.  Dark pink flowers.  (Loder)

                          -griffithianum (1/8)
                  -George Hardy-
cl.    -Mrs. Lindsay-      -catawbiense (1/8)
       - Smith    -Duchess of Edinburgh--unknown (1/4)
Sea Spray-              -fortunei ssp. discolor (1/8)
              -Lady
        -Jalisco-Bessborough-campylocarpum ssp. campylocarpum
           Goshawk-                    Elatum Group (1/8)
        -      -dichroanthum ssp. dichroanthum (1/8)
          -Dido-
             -decorum (1/8)

White flowers with a distinctive reddish brown flare.  (Collingwood Ingram, reg. 1973)

g. & cl.      -griffithianum (1/4)
         -Loderi-
Seagull-     -fortunei ssp. fortunei (1/4)
        -sutchuenense (1/2)

Loose trusses of 12-15 pure white flowers speckled crimson.
Leaves dull green with light brown indumentum, 8.5in(21.6cm)
long, 3in(7.6cm) wide. See also Sea Mist, Seamew. (Lady Loder,
1938)  A. M. (RHS) 1938; F. C. C. (RHS) 1976

cl.
          -dichroanthum ssp. dichroanthum (1/2)
Sealing Wax-      -haematodes ssp. haematodes (1/4)
          -May Day-
                -griersonianum (1/4)

Vermilion flowers.  (Sunningdale Nurseries, 1955)

cl.          -griffithianum (1/4)
        -Loderi-
Seamew-     -fortunei ssp. fortunei (1/4)
        -sutchuenense (1/2)

Seagull g.   Flowers pure white, or pale pink.   (Loder)   A. M.
(RHS) 1940

cl.

Searchlight     Parentage unknown

A parent of Autumn Beauty, by F. Lovegrove, Australia.

cl.
        -campylocarpum ssp. campylocarpum (1/2)
Sea-Shell-
        -unknown (1/2)

Flowers creamy yellow with a chocolate blotch.  (W. C. Slocock)

cl.
        -Moser's Maroon--unknown (1/2)
Sea-Tac-
        -williamsianum (1/2)

2.5ft(.75m)        0°F(-18°C)          E

Leaves 2.5in(6.4cm) long by 1.5in(3.8cm) wide.  Buds dark greyed
red; dark cardinal red flowers, 2.5in(6.4cm) wide, in trusses of
7.  (H. L. Larson, reg. 1977)

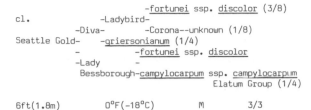

```
cl.                        -fortunei ssp. discolor (3/8)
              -Ladybird-
         -Diva-         -Corona--unknown (1/8)
Seattle Gold-  -griersonianum (1/4)
         -           -fortunei ssp. discolor
         -Lady
           Bessborough-campylocarpum ssp. campylocarpum
                                        Elatum Group (1/4)
```

6ft(1.8m)        0°F(-18°C)        M        3/3

Compact plant habit; long, slender leaves. Flowers warm light
yellow with brown markings, held in well-filled trusses. (Cross
by Halfdan Lem; intro. by Don McClure; reg. 1958)

```
cl.          -lacteum (1/4)
        -unnamed-              -griffithianum (1/16)
Seattle- hybrid-      -Loderi-
  Queen -     -Lodauric-    -fortunei ssp. fortunei (1/16)
        -              -auriculatum (1/8)
        -campylocarpum ssp. campylocarpum (1/2)
```

Ivory flowers. (Halfdan Lem cross; intro. by Donald McClure,
before 1958)

```
cl.
                -leucaspis (1/2)
Seattle Springtime-
                -mucronulatum (1/2)
```

3ft(.9m)        5°F(-15°C)        E        3/2

Growth habit open; small leaves, medium olive green. Flowers of
white, flushed amaranth rose, 1.25in(3.2cm) across.    (Brian O.
Mulligan, intro. 1954)

cl.

Second Attempt--form of callimorphum ssp. callimorphum

3ft(.9m)        -5°F(-21°C)        M        3/4

Trusses of 4-5 flowers, white with large blotch of greyed
purple, reverse flushed fuchsia purple. Leaves broadly elliptic
1.75in(4.5cm) long, dark glossy green. (Crown Estate) A. M. 1980

```
cl.                    -wardii var. wardii (Barto form) (3/4)
        -unnamed hybrid-        -A. W. Hardy Hybrid--unknown
Second   -        -Devonshire-                    (1/8)
Honeymoon*       Cream  -campylocarpum ssp. campylo-
        -                          carpum (1/8)
         -wardii var. wardii
```

4ft(1.2m)        5°F(-15°C)        M        4/4

Flowers yellow with a red blotch. Compact plant with heavy dark
green foliage. See also Honeymoon. (William E. Whitney)

```
cl.
        -yakushimanum ssp. yakushimanum (1/2)
Secrest Pink-
        -unknown (1/2)
```

2.5ft(.75m)        -15°F(-26°C)        M

Plant nearly twice as wide as tall; dull olive green leaves held
2 years; new growth has light brown tomentose. Buds strong purplish red opening to flowers light purplish pink with small dark
blotch, reverse darker pink, 2.5in(5cm) wide; trusses of 11-17.
(John E. Ford, Secrest Arboretum; reg. 1984)

```
cl.
            -griersonianum (1/2)
Secretary of State-        -haematodes ssp. haematodes (1/4)
            -Grosclaude-
                    -facetum (eriogynum) (1/4)
```

Large trusses of blood red flowers. (Cross by The Rt. Hon. Michael Noble; reg. by Sir George Campbell, 1965)

```
cl.
        -griersonianum (1/2)
Seemly-            -fortunei ssp. discolor (1/4)
        -rose-colored hybrid-
                    -unknown (1/4)
```

Large rose-colored flowers, in late season. (Sir James Horlick,
reg. 1962)

```
cl.                            -griffithianum (1/16)
                        -George-
                    -Pink - Hardy-catawbiense (1/16)
cl.             -Pearl-        -arboreum ssp. arbo-
        -Professor - -Broughtonii-      reum (3/16)
Seestadt  -Hugo de Vries-        -unknown (3/16)
Bremerhaven -        -        -arboreum ssp. arboreum
(Seaport...)-        -Donaster-
                    -        -unknown
        -insigne (1/2)
```

Trusses of 8-10 flowers, pink with a darker blotch. Plant 5ft
(1.5m) high at registration; dark green elliptic leaves. (Cross
by W. Bruns; reg. by Joh. Bruns, 1985)

```
cl.
        -catawbiense (1/2)
Sefton-
        -unknown (1/2)
```

4ft(1.2m)        -20°F(-29°C)        ML        3/2

Low, spreading, straggly growth habit. Dark maroon flowers and
trusses are larger than those on most catawbiense hybrids. (A.
Waterer, 1881) See color illus. ARS Q 26:4 (Fall 1972): cover.

```
cl.
        -cinnabarinum ssp. cinnabarinum Blandfordiiflorum Group
Selig-                                    (1/2)
        -calophytum var. calophytum (1/2)
```

Flowers pink, fading to salmon pink and orange on tube.   (Lady
Loder, 1937) F. C. C. (RHS) 1937

```
cl.
        -Lackamas Blue--augustinii ssp. augustinii (3/4)
Senora-                    -intricatum (1/8)
Meldon-        -Intrifast-
        -Blue Diamond-        -fastigiatum (1/8)
                    -augustinii ssp. augustinii
```

5ft(1.5m)        5°F(-15°C)        E

Wisteria blue flowers with a few green spots, to 5in(12.7cm)
wide, held in terminal clusters; floriferous and fragrant. Upright rounded plant, well-branched; small yellowish green leaves
held 2-3 years. (Dr. David Goheen intro. 1975; reg. 1982)

```
cl.                        -griffithianum (1/4)
        -Loderi King George
Senorita-                -fortunei ssp. fortunei (1/4)
        -Ostbo Y3--unknown (1/2)
```

5ft(1.5m)        5°F(-15°C)        ML

Leaves 6in(15.2cm) long, 2in(5cm) wide. Ruffled flowers, lilac
pink, yellowish pink throat, upper lobe spotted brown, 4in(10.2
cm) across; trusses of 14. (The Bovees, reg. 1962)

```
g.              -sperabile var. sperabile (1/4)
        -Eupheno-
Sentinel-        -griersonianum (1/4)
        -        -thomsonii ssp. thomsonii (1/4)
        -Barclayi-        -arboreum ssp. arboreum (1/8)
              -Glory of -
                Penjerrick-griffithianum (1/8)
```

Blood red flowers. (Lord Aberconway, 1950)

cl.
            -leucaspis  (1/2)
September Snow-
            -edgeworthii (1/2)

White flowers in trusses of 4-6.  (B. W. Campbell, reg. 1981)

```
                   -Lady      -campylocarpum ssp. campylocarpum
                   -Bessborough-              Elatum Group (1/8)
        -Jalisco-             -fortunei ssp. disolor (1/8)
cl.   -    -      -dichroanthum ssp. dichroanthum (1/8)
-        -Dido-
Serena-           -decorum (1/8)
-                   -souliei (1/8)
-        -Soulbut-
-Vanessa-          -Sir Charles Butler-fortunei ssp. fortunei
-griersonianum (1/4)                                      (1/8)
```

Flowers of bright carmine pink, cherry at base of corolla, reverse is pale pink suffused cherry. (Francis Hanger cross; RHS
Garden, Wisley)  A. M. (RHS) 1956


cl.

Serenade*--form of davidsonianum

6ft(1.8m)          0°F(-18°C)          EM          4/3

A tall shrub with dark green lanceolate leaves; flowers a very
clear pink.  Hardier than the F. C. C. form of this species.


```
cl.                      -maximum (1/4)
             -Russell Harmon-
Serenata-                -catawbiense (1/4)
-          -dichroanthum ssp. dichroanthum (1/4)
-unnamed-           -fortunei ssp. discolor (1/8)
   hybrid-unnamed-
            hybrid-campylocarpum ssp. campylocarpum (1/8)
```

5ft(1.5m)          -15°F(-26°)          ML          4/3

Leaves grow obliquely at a 45° angle on this plant of superior
foliage density and spreading habit. Flowers of heavy substance
to 2.5in(6.4cm) across, light orange yellow with a strong orange
blotch and a petaloid calyx spotted orange; large trusses of 17.
See also Tahiti. (David G. Leach, reg. 1965)


cl.
       -yakushimanum ssp. yakushimanum (1/2)
Serendipity-
       -aureum var. aureum (chrysanthum) (1/2)

1ft(.3m)          -25°F(-32°C)          M

Plant compact and spreading; dark green leaves 2.75in(7cm) long.
Flowers pale primrose yellow, widely capanulate, 2.3in(6.4cm)
wide; trusses of 9-10. (Basil C. Potter, reg. 1972)


cl.
       -campylocarpum ssp. campylocarpum
Serin-
       -Loder's White, q.v. for 2 possible parentage diagrams

Albino g.  Grown by Collingwood Ingram.


cl.

Sesame--form of pachytrichum

6ft(1.8m)          -5°F(-21°C)          E          2/2

Grows over 10ft(3m) high.  White flowers, some tinged shades of
solferino purple, capanulate, in lax trusses of 8. (Lord Aberconway, Bodnant)  A. M. (RHS) 1963


---
       -edgeworthii (1/2)
Sesterianum-
       -unknown (1/2)

Flowers creamy white.  The pollen parent may have been formosum
var. formosum, or Gibsonii--unknown. (Rinz, before 1862)

g. & cl.
   -spinuliferum (1/2)
Seta-
     -moupinense (1/2)

5ft(1.5m)          5°F(-15°C)          E          4/3

Upright plant; small-leaved. Narrowly campanulate flowers, pink
with much paler base, the reverse darker pink, in umbels. (Lord
Aberconway, intro. 1933)  A. M. (RHS) 1933; F. C. C. (RHS) 1960
See color illus.  ARS Q 35:3 (Summer 1981): p. 142.

cl.

Seven Devils*--form of macrophyllum

5ft(1.5m)          -5°F(-21°C)          M-L          2/3

Deep wine red flowers, widely bell-shaped, in domed trusses of
about 20. Lanceolate leaves. Crosses with elepidotes; a parent
of Sparkling Burgundy. (Collected by William Magness, near Sev-
en Devils Rd., Bandon, Ore.; intro. by Dr. Frank Mossman)

```
cl.                          -griffithianum (1/4)
          -Loderi Sir Joseph Hooker-
Seven Stars-              -fortunei ssp. fortunei
-                                           (1/4)
          -yakushimanum ssp. yakushimanum (1/2)
```

3ft(.9m)          0°F(-18°C)          M

Flowers white with dawn pink flush, campanulate, 2in(5cm) wide;
dome-shaped trusses hold 15.  Very free-flowering.  Plant vigorous, upright; matte green leaves. (Crown Estate, Windsor, 1966)
H. C. (RHS) 1965;  A. M. (RHS) 1967;  F. C. C. (RHS) 1974    See
color illus. F p. 65.

cl.

Seventh Heaven--form of sutchuenense (W 1232)

5ft(1.5m)          -10°F(-23°C)          VE          4/3

Trusses of 14-16 flowers, white in throat suffused lilac, many
small ruby spots.  Leaves 9.25in(23cm) long, dark green above,
paler beneath. (Collector probably E. H. Wilson; raiser  R. N.
Stephenson Clarke)  A. M. (RHS) 1978

```
cl.               -catawbiense (1/4)
      -Pink Twins-
Shaazam-          -haematodes ssp. haematodes (1/4)
-                          -griffithianum (1/4)
      -Leah Yates--Mars (selfed)-
                          -unknown (1/4)
```

4.5ft(1.35m)          -10°F(-23°C)          ML

Flowers neyron rose with yellow blotch, shading lighter to
edges, 7-lobed, 3.5in(9cm) across, in ball-shaped trusses of 10.
See also Mary Yates, Sunset Yates, Ravels, Yates' Red  (Henry
Yates cross; Mrs. Yates, reg. 1977)  See color illus. LW pl. 92.

```
cl.
         -elliottii (1/2)
Shadow Secretary-     -griersonianum (1/4)
              -Sarita-          -griffithianum (1/8)
                 Loder-Loderi-
                          -fortunei ssp. fortunei (1/8)
```

Signal red flowers in large racemose umbels of 17-18. Leaves
glabrous, 10in(25.4cm) long, 2.5in(6.4cm) wide. (Cross by Rt.
Hon. M. A. C. Noble; reg. by Sir George Campbell, 1966)

'Revlon' by Rothschild
Photo by Greer

'Ring of Fire' by Thompson
Photo by Greer

'Rio' by Leach
Photo by Leach

'Robin Leach' by Leach
Photo by Leach

'Rochelle' by Nearing
Photo by Greer

'Rockhill Parkay' by Brotherton
Photo by Brotherton

'Rockhill Sunday Sunrise' by Brotherton
Photo by Brotherton

'Rosabel' by Waterer
Photo by Greer

'Rosalind' by Aberconway
Photo by Greer

'Rose Lancaster' by Lancaster/Goheen
Photo by Greer

'Rose Mangles' by Mangles
Photo by Greer

'Roseann' by R. Henny
Photo by Greer

'Royal Flush' (Pink form) by J. C. Williams
Photo by Greer

'Roslyn' by Vossberg
Photo by Greer

'Royal Pink' by Hobbie
Photo by Greer

'Royal Purple' by H. White
Photo by Greer

'Russautinii' by Ramsden
Photo by Greer

'Saint Merryn' by Harrison
Photo by Greer

'Sarita Loder' by G. Loder
Photo by Greer

'Sardis' by RBG, Kew
Photo by Greer

'Scarlet Blast' by Leach
Photo by Leach

'Scarlet King Pines' by Edgar Stead (NZ)
Photo by Greer

'Serenata' by Leach
Photo by Leach

'Serendipity' by Potter
Photo by Greer

                        -fortunei ssp. discolor (3/8)
cl.      -Lady Bessborough-
      -Day -              -campylocarpum ssp. campylocarpum
Shah -Dream-                              Elatum Group (1/8)
Jehan-    -griersonianum (3/8)
      -          -fortunei ssp. discolor
      -Margaret-    -dichroanthum ssp. dichroanthum (1/8)
        Dunn -Fabia-
                 -griersonianum

Dream Girl g. Plant of compact habit with rounded leaves.
Flowers of warm chrome yellow with burnt orange spotting. See
also Gold Mohur, Golden Pheasant, Mohur.  (L. E. Brandt, 1966)

---                        -griffithianum (1/4)
           -Mrs. E. C. Stirling-
Shakespeare-              -unknown (3/4)
           -Essex Scarlet--unknown

Bright rose pink flowers.  (M. Koster & Sons, 1916)

                      -griffithianum (3/8)
             -Beauty of-
         -Norman- Tremough-arboreum ssp. arboreum (1/8)
          - Gill -
        -Anna-     -griffithianum
cl.     -       -     -griffithianum
        -    -Jean Marie -
Shalom-     de Montague-unknown (7/16)
       -                   -griffithianum
               -George Hardy-
       -       -Pink -      -catawbiense (1/16)
       -Antoon van-Pearl-    -arboreum ssp. arboreum
         Welie   - -Broughtonii-
                 -unknown      -unknown

5ft(1.5m)        -5°F(-21°C)        ML

Leaves 4.75in(12cm) long. Flowers white, shaded rose, with darker rose flare, 3.5in(8.9cm) across. Flattened trusses hold 16.
(Halfdan Lem cross; reg. by N. E. Hess, 1974)  C. A. (ARS) 1973

cl.
      -keiskei (1/2)
Shamrock-
      -hanceanum Nanum Group (1/2)

1.5ft(.45m)       5°F(-15°C)         E

Tubular funnel-shaped flowers, 1.25in(3cm) across by 1in(2.5cm)
long, 5-lobed, of an unusual chartreuse color with slight yellow
spotting; small trusses of 8-9. Plant well-branched, over twice
as wide as high. (Dr. Robert L. Ticknor, cross 1971; reg. 1978)
See color illus. ARS Q 32:4 (Fall 1978); cover; K p. 128.

                -unnamed-catawbiense (7/16)
          -Pinnacle- hybrid-
cl.       -     -     -unknown (5/16)
          -     -unnamed-catawbiense
Sham's Candy-   hybrid-
          -         -unknown
          -              -caucasicum (1/8)
          -      -Boule de Neige-  -catawbiense
          -Pink Cameo-    -unnamed-
          -              hybrid-unknown
                 -catawbiense (red form) (1/8)

5ft(1.5m)       -20°F(-29°C)       ML        4/3

Synonym Candy. Upright plant about as broad as tall; glossy elliptic leaves held 2 years. Flowers strong purplish pink with
a yellow green blotch, 2.75in(7cm) wide; conical trusses of 12.
(Anthony M. Shammarello, 1975)

cl.                 -caucasicum (1/4)
          -Boule de Neige-   -catawbiense (1/8)
Sham's Juliet-      -unnamed-
          -         hybrid-unknown (1/8)
          -catawbiense (red form) (1/2)

4.5ft(1.35m)       -20°F(-29°C)        ML        3/3

Synonym Juliet. Compact plant; olive green leaves 3.5in(8.9cm)
long. Flowers apple blossom pink, with brown blotch, 2.5in(6.4
cm) wide, in conical trusses. (A. M. Shammarello, reg. 1972)

cl.

Sham's Pink      Parentage as above

4ft(1.2m)        -20°F(-29°C)        ML        3/2

Medium-sized, pea green foliage. Flowers 2.5in(6.4cm) wide, of
rose pink, darker at edges with a light red blotch, in spherical
trusses. (Anthony M. Shammarello, reg. 1972)

                -catawbiense (3/8)
cl.      -Kettledrum-
          -unknown (5/8)              -catawbiense
Sham's Ruby-
           -Parsons Grandiflorum-
          -America-              -unknown
             -dark red hybrid--unknown

3ft(.9m)        -20°F(-29°C)        ML        2/2

Synonym Ruby. Flowers strong purplish red with a slight blotch
of deeper tone, openly funnel-shaped, 2.25in(5.7cm) wide, in 16-
flowered trusses. Plant upright, almost as wide as tall; glossy
olive green leaves. (A. M. Shammarello, reg. 1958 as Ruby; reg.
1976 as Sham's Ruby)  See color illus. LW pl. 90.

                    -unnamed-griffithianum (1/16)
                     - hybrid-
           -Mrs. Furnival-   -unknown (5/8)
cl.    -unnamed-         -unnamed-caucasicum (1/16)
       - hybrid-         hybrid-
Shanghai-      -              -unknown
       -    -catawbiense var. album (1/4)
       -unnamed mauve seedling, gold-blotched--unknown

6ft(1.8m)        -10°F(-23°C)        ML        4/2

Plant wider than tall, branching moderately; medium green elliptic leaves, held 2 years. Flowers widely funnel-shaped, 3.75in
(9.5cm) wide, white flushed pink on perimeter of lobes, blotch
of strong orange yellow; corolla lobes recurved; trusses spherical, about 18-flowered, 7.25in(18.5cm) across. (David G. Leach,
reg. 1973)

g. & cl.
                  -thomsonii ssp. thomsonii (1/4)
          -General Sir -
Shangri La-John du Cane-fortunei ssp. discolor (1/4)
          -
          -griffithianum (1/2)

Big white flowers flushed pink, held in large compact trusses on
an erect, vigorous shrub. Late midseason. (Rothschild)   P. C.
(RHS) 1965

cl.
     -soulei (1/2)
Sharon-              -griffithianum (1/4)
      -Loderi King George-
                  -fortunei ssp. fortunei (1/4)

Flowers white with a crimson blotch. (Del James)   P. A. (ARS)
1955

cl.

Sharpie     See next page.

cl.
     -konorii (1/2)
Shasta-              -Pink Delight (Veitch) unknown (1/4)
      -Nancy Adler Miller-
                  -jasminiflorum (1/4)

2.5ft(.75m)         32°F(0°C)         Dec.-Feb.

Vireya hybrid.  Flowers very fragrant, white, narrowly tubular, 2.25in(5.7cm) wide, 2.5in(6.4cm) long; rounded trusses of 6-10. Floriferous.  Plant wider than high; glossy, concave leaves held 4 years.  (P. Sullivan cross, Strybing Arboretum; reg. by W. A. Moynier, 1980)

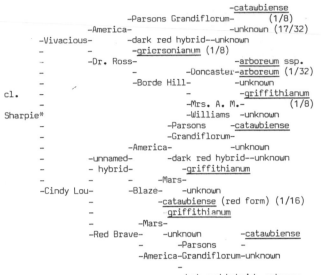

```
                                          -catawbiense
                       -Parsons Grandiflorum-    (1/8)
                   -America-                    -unknown (17/32)
         -Vivacious-        -dark red hybrid--unknown
          -          -      -griersonianum (1/8)
          -          -Dr. Ross-            -arboreum ssp.
          -          -        -Doncaster-arboreum (1/32)
          -          -Borde Hill-      -unknown
cl.       -          -      -          -griffithianum
          -                 -Mrs. A. M.-        (1/8)
Sharpie*                    -Williams  -unknown
          -                 -Parsons   -catawbiense
          -                 -Grandiflorum-
          -          -America-          -unknown
          -   -unnamed-  -dark red hybrid--unknown
          -    - hybrid-    -griffithianum
          -            -Mars-
       -Cindy Lou-  -Blaze-  -unknown
          -         -      -catawbiense (red form) (1/16)
          -         -       -griffithianum
          -         -Mars-
       -Red Brave-  -unknown       -catawbiense
          -         -Parsons
          -America-Grandiflorum-unknown
                    -
                    -dark red hybrid--unknown
```

Vivid red buds open to flowers of strong cardinal red, veined on reverse in vivid red, stamens and pistil also vivid red.  Plant medium-sized, hardy to -15°F(-26°C).  (Weldon E. Delp)

cl.
        -griffithianum (1/2)
Sheffield Park-
        -unknown (1/2)

Pink flowers with scarlet crimson edges and crimson red exterior. (C. B. van Nes & Sons)

g. & cl.
          -elliottii (1/2)
Sheila Moore-
          -decorum (1/2)

Rose pink flowers with darker spotting on upper lobes; 12-flowered trusses.  See also Cerisette. (Lord Digby, 1948)    A. M. (RHS) 1948

cl.
          -fortunei sp. discolor (1/2)
Sheila Osborn-      -griffithianum (1/4)
          -Strategist-        -catawbiense (1/8)
                   -John Waterer-
                            -unknown (1/8)

Rose pink flowers. (RBG, Kew, 1928)  A. M. (RHS) 1928

cl.
          -elliottii (1/2)
Shellaby-
          -fortunei ssp. fortunei (1/2)

Nanceglos g. (RHS Garden, Wisley, 1954)

g.
                 -griffithianum (1/4)
             -Loderi-
Shepherd's Delight-  -fortunei ssp. fortunei (1/2)
          -      -fortunei ssp. fortunei
          -Luscombei-
                 -thomsonii ssp. thomsonii (1/4)

(Intro. by Heneage-Vivian, Clyne Castle, Wales, 1927)

cl.              -Moser's Maroon--unknown (3/4)
          -Grenadier-
Shepway-        -elliottii (1/4)
          -G. A. Sims--unknown

Firm, rounded trusses of 15-17 flowers, dark red, shading lighter to throat with darker spotting, 3in(7.6cm) wide.  Leaves dark green, with traces of brownish woolly indumentum.  (Maj. A. E. Hardy)  A. M. (RHS) 1976

                 -fortunei ssp. fortunei (1/4)
          -Fawn-  -dichroanthum ssp. dichroanthum (1/4)
cl.       -   -Fabia-
          -        -griersonianum (1/8)
Sherill-            -fortunei ssp. discolor (1/8)
          -   -Lady -
          -   -Bessborough-campylocarpum ssp. campylocarpum
          -Jalisco-          Elatum Group (1/8)
          -   -dichroanthum ssp. dichroanthum
          -Dido-
              -decorum (1/8)

6ft(1.8m)         -10°F(-23°C)         M

Very fragrant flowers, 6in(15.2cm) wide, barium yellow with a large blotch of currant red in throat, 7-lobed.  Upright plant, almost as wide as tall; ivy green leaves 9in(23cm) long, held 3 years.  (Del W. James cross, 1959; reg. by Mrs. James for Hendricks Park, Eugene, OR, 1980)

---
                 -arboreum ssp. cinnamomeum var. album (1/2)
Sherwoodeanum-
                 -catawbiense (1/2)

Lilac rose flowers, heavily spotted.  (J. Waterer, before 1865)

g. & cl.
          -barbatum (1/2)
Shilsonii-
          -thomsonii ssp. thomsonii (1/2)

Early-blooming, campanulate, waxy blood red flowers with faint dark brown markings.  Large shrub or small tree with reddish plum bark like barbatum, rather small, rounded leaves like thomsonii. (Richard Gill, 1900)  A. M. (RHS) 1900

                 -dichroanthum ssp. dichroanthum
              -Fabia-
cl.     -C. I. S.-    -griersonianum
          -    -
Shirley -    -Loder's White, q.v. for 2 possible
Creelman-              parentage diagrams
          -    -unnamed hybrid--unknown
          -Apricot-          -fortunei ssp. discolor
          Nectar -   -Lady  -
              -    -Bessborough-campylocarpum ssp. campy-
              -Jalisco-          locarpum Elatum Group
                   -    -dichroanthum ssp. dichroanthum
                   -Dido-
                       -decorum

4ft(1.2m)         5°F(-15°C)         M

Turkey red buds open to flowers of spinel red, 7-lobed, 3.25in (8.9cm) wide; spherical trusses of 7-14.  Floriferous.  Plant upright, as wide as tall; dark leaves with scant brown indumentum beneath.  (C. C. Barrow cross; raised & intro. by Alice P. Smith; reg. 1983)

cl.
          -strigillosum (1/2)
Shirley Rose Lent-
          -praevernum (1/2)

4.5ft(1.35m)         10°F(-12°C)         VE

Plant rounded, well-branched; narrow, dark olive green leaves held 3 years; new growth reddish. Deep pink flowers with a few darker spots, tubular campanulate, 2.25in(5.7cm) wide; spherical trusses of 13. (B. Nelson cross; Dr. C. G. Heller, reg. 1977)

```
cl.                  -caucasicum (1/4)
            -Nobleanum-
Shirley Scott-          -arboreum ssp. arboreum (1/4)
            -          -campylocarpum ssp. campylocarpum (1/4)
            -Unique-
                  -unknown (1/4)
```

Saturn red in bud, opening to barium yellow flowers, in trusses of about 16. (V. J. Boulter, 1963)

```
cl.              -dichroanthum ssp. dichroanthum (3/8)
        -Fabia-
        -          -griersonianum (3/8)
Shirley-Jean-          -dichroanthum ssp. dichroanthum
        -    -Fabia-
        -    -          -griersonianum
        -Otis-              -griffithianum (1/16)
        Hyde-        -Loderi-
            -Albatross-          -fortunei ssp. fortunei (1/16)
                    -fortunei ssp. discolor (1/8)
```

Trusses of 6-7 flowers, 5-lobed, bright azalea pink heavily spotted in yellow ochre; calyx also bright azalea pink. Leaves about 4.8in(12.5cm) long. (H. L. Larson, reg. 1982)

```
cl.
        -hippophaeoides var. hippophaeoides Fimbriatum Group
Shooting-                                        (1/2)
 Star* -racemosum (1/2)
```

3ft(.9m)        -5°F(-21°C)        E        3/3

Long spires of lavender pink cover this fast-growing dwarf; must be trimmed back after blooming to keep compactness. Leaves 3-4 times longer than wide. Does well in open sun; young plants are heavily budded. (H. E. Greer, intro. 1982)

```
cl.

Shortoff--dwarf form of maximum!
```

3ft(.9m)        -20°F(-29°C)        VL

Trusses 3in(7.6cm) wide by 4in(10.2cm) high, holding 18-20 white flowers, 5-lobed. Plant 2/3 as wide as tall, 40% of the size of typical species [original plant about 8ft(2.4m) high at age 44] and flowers later; leaves held 4 years. (Collected on Shortoff Mt., NC, & raised by Henry Wright; intro. by C. Towe; reg. 1985)

```
g.        -campylocarpum ssp. campylocarpum (1/2)
Shot Silk-
        -dichroanthum ssp. dichroanthum (1/2)
```

A shrub of moderate height, with flowers of yellowish pink and orange. Blooms midseason. (Sir John Ramsden, 1933)

```
cl.
        -yakushimanum ssp. yakushimanum (Exbury form) (1/2)
Show Boat-        -decorum (1/4)
        -Tumalo-                  -griffithianum (1/8)
            -Loderi King George-
                      -fortunei ssp. fortunei (1/8)
```

4ft(1.2m)        -5°F(-21°C)        M        4/4

Plant compact, rounded, well-branched; olive green leaves with scant buff indumentum, held 2-3 years. Pink buds, white flowers with small yellow green blotch, 7 wavy lobes, 2.75in(7cm) wide. (Dr. Carl Phetteplace, reg. 1975)

```
cl.
        -yakushimanum ssp. yakushimanum (1/2)
Shrimp Girl-        -dichroanthum ssp. dichroanthum (1/4)
        -Fabia Tangerine-
                  -griersonianum (1/4)
```

3ft(.9m)        0°F(-18°C)        M        2/3

Plant habit compact, foliage good quality. Flowers of soft ney-ron rose. Slow-growing. See also Molly Miller. (John Waterer, Sons & Crisp, 1971)

```
cl. -yakushimanum ssp. yakushimanum (1/2)
    -        -Lady        -fortunei ssp. discolor (3/16)
Si Si*        -Bessborough-
    -    -Day        -campylocarpum ssp. campylocarpum
    -Gold -Dream-griersonianum        Elatum Group (1/16)
    Mohur-        -fortunei ssp. discolor
        -Margaret-        -dichroanthum ssp. dichroanthum (1/16)
        Dunn    -Fabia-
                  -griersonianum (3/16)
```

5ft(1.5m)        -15°F(-26°C)        ML        4/3

Strong red buds open to strong and light pinks; throat pale yellow with red blotch. Plant vigorous, sturdy. (Weldon E. Delp)

```
cl.
        -catawbiense var. album (1/2)
Sia-
        -yakushimanum ssp. yakushimanum (1/2)
```

Trusses of 19-20 flowers, broadly funnel-campanulate, 5-lobed; buds roseine purple opening to phlox pink, aging to white. Narrowly elliptic to elliptic leaves with orange brown indumentum; new shoots with tan tomentum. Height at 30 years, 4ft(1.2m). (David G. Leach cross, 1952; reg. 1983)

```
* * * * * * * * * * * * * * * * * * * * * * *
* Hybrids of unknown parentage are described briefly in App. D *
* * * * * * * * * * * * * * * * * * * * * * *
```

```
cl.
        -catawbiense var. album (1/2)
Siam-
        -yakushimanum ssp. yakushimanum (1/2)
```

4ft(1.2m)        -20°F(-29°C)        ML        4/4

Well-branched plant, broader than high; glossy, elliptic, convex leaves 4in(10cm) long, orange brown indumentum beneath. Roseine purple buds open light phlox pink aging to white; flowers widely funnel-campanulate, 2in(5cm) wide, in rounded trusses of 19-20. (David G. Leach, cross 1952; intro. 1975; reg. 1982)

```
cl.

Sidlaw--form of vernicosum var. euanthum ?
```

5ft(1.5m)        -15°F(-26°C)        M        2/3

Leaves 4in(10cm) long, dark green above, paler below. Trusses hold 9-12 flowers, white flushed light purple with darker veining, and a large blotch of Indian lake. (E. H. M. & P. A. Cox, Glendoick Gardens) P. C. (RHS) 1969

```
g.
        -Rose Perfection (R. Gill) unknown (1/2)
Sidonia-
        -fortunei ssp. fortunei (1/2)
```

Rose pink. (Lord Aberconway, 1933)

```
* * * * * * * * * * * * * * * * * * * * * * *
* Species used in hybrids of this volume are listed in App. A. *
* * * * * * * * * * * * * * * * * * * * * * *
```

```
                -griersonianum (1/4)
    -Mrs. Horace-           -griffithianum (13/32)
    - Fogg     -Loderi Venus-
cl. -                       -fortunei ssp. fortunei (1/8)
    -                           -griffithianum
Sierra-                   -Beauty of-
Beauty-              -Norman- Tremough-arboreum ssp. arboreum
    -               - Gill -                        (1/32)
    -        -Anna-      -griffithianum
    -        -    -          -griffithianum
    -Point   -    -Jean Marie -
    Defiance-      de Montague-unknown (3/16)
            -                -griffithianum
            -Marinus Koster-
                        -unknown
```

5ft(1.5m)          0°F(-18°C)          ML

Red buds open to flowers of very light pink, edges and 2 dorsal rays darker, 5 wavy reflexed lobes; dome-shaped trusses of about 18. Plant rounded, as broad as high; yellowish green leaves to 9in(23cm) long, held 3 years. See also Lady of Spain, Canadian Beauty. (John G. Lofthouse, reg. 1983)

```
                -wardii var. wardii (1/4)
cl.         -Crest-          -fortunei ssp. discolor (1/8)
    -          -Lady     -
Sierra del Oro-    Bessborough-campylocarpum ssp. campylo-
    -                               carpum Elatum Group (1/8)
    -lacteum (1/2)
```

3.5ft(1.05m)       0°F(-18°C)          VE

Trusses lasting 4 weeks, 7in(17.8cm) across, of 15 light yellow flowers with 5 wavy, reflexed lobes, 3.5in(8.9cm) wide. Plant broader than tall, with arching branches; leaves held 1.5 years. (John G. Lofthouse, intro. 1975; reg. 1982)

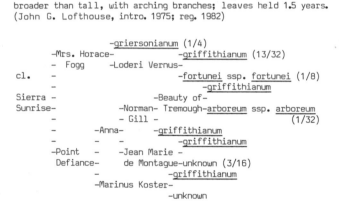

```
                -griersonianum (1/4)
    -Mrs. Horace-           -griffithianum (13/32)
    - Fogg     -Loderi Vernus-
cl. -                       -fortunei ssp. fortunei (1/8)
    -                           -griffithianum
Sierra -                  -Beauty of-
Sunrise-              -Norman- Tremough-arboreum ssp. arboreum
    -               - Gill -                        (1/32)
    -        -Anna-      -griffithianum
    -        -    -          -griffithianum
    -Point   -    -Jean Marie -
    Defiance-      de Montague-unknown (3/16)
            -                -griffithianum
            -Marinus Koster-
                        -unknown
```

6ft(1.8m)          0°F(-18°C)          ML          5/4

A large, handsome plant, with dark green, pointed leaves curving upward from the midrib. Large, ruffled flowers, pale pink to white, narrowly edged in rose pink, with a small reddish blotch; conical trusses one foot tall. (John G. Lofthouse, reg. 1975)

cl.

Sierra Sunset      See next column upper right

```
cl.         -wardii var. wardii (1/4)
    -Crest-                  -fortunei ssp. discolor (1/8)
Sierra  -   -Lady Bessborough-
Treasure-               -campylocarpum ssp. campylocarpum
    -lacteum (1/2)                   Elatum Group (1/8)
```

5ft(1.5m)          0°F(-18°C)          E

Fragrant flowers to 4.5in(11.5cm) across, yellow with 3 maroon blotches in throat, 5 wavy, reflexed lobes; about 13 in conical trusses 8in(20cm) wide. Plant as broad as high; dark green concave leaves, held 1.5 years. (Cross by John G. Lofthouse, 1967; raised by Dr. R. C. Rhodes; reg. 1985)

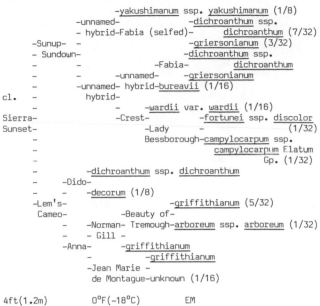

```
                -yakushimanum ssp. yakushimanum (1/8)
    -unnamed-           -dichroanthum ssp.
    - hybrid-Fabia (selfed)-      dichroanthum (7/32)
-Sunup- -                 -griersonianum (3/32)
- Sundown-               -dichroanthum ssp.
    -               -Fabia-         dichroanthum
    -        -unnamed-     -griersonianum
    -unnamed- hybrid-bureavii (1/16)
cl. -    hybrid-
    -          -wardii var. wardii (1/16)
Sierra-         -Crest-        -fortunei ssp. discolor
Sunset-          -Lady     -                   (1/32)
    -           Bessborough-campylocarpum ssp.
    -                       campylocarpum Elatum
    -                               Gp. (1/32)
    -       -dichroanthum ssp. dichroanthum
    - -Dido-
    - -    -decorum (1/8)
    -Lem's-            -griffithianum (5/32)
    Cameo-       -Beauty of-
    -       -Norman- Tremough-arboreum ssp. arboreum (1/32)
    - - Gill -
    -Anna-     -griffithianum
    -              -griffithianum
    -Jean Marie -
    de Montague-unknown (1/16)
```

4ft(1.2m)          0°F(-18°C)          EM

Strong salmon buds open to carmine flowers, edged shell pink. Plant compact, nearly as broad as high with deep green leaves held 2.5 years. (John G. Lofthouse, reg. 1984)

cl.

Sigismund Rucker      Parentage unknown

Late-blooming flowers of unusual rich cherry magenta with a black flare, in medium-sized trusses. (A. Waterer) F. C. C. (RHS) 1872

```
cl.         -catawbiense (1/4)
    -Atrosanguineum-
Signal-         -unknown (1/4)
 Horn -             -caucasicum (3/16)
    -       -Jacksonii-       -caucasicum
    -Goldsworth-     -Nobleanum-            (1/16)
    Yellow  -           -arboreum ssp. arboreum
        -campylocarpum ssp. campylocarpum (1/4)
```

5ft(1.5m)          -15°F(-26°C)          ML

Leaves 7in(17.8m) long, 2in(5cm) wide. Flowers rosy pink with darker edges, and a distinctive spotted red blotch, to 2in(5cm) across; lax trusses of about 8. (G. Guy Nearing, reg. 1973)

```
cl.         -griffithianum (3/8)
    -Marinus Koster-
Sigrid-         -unknown (1/4)
    -     -fortunei ssp. fortunei (1/4)
    -Pilgrim-        -arboreum ssp. arboreum (1/8)
        -Gill's Triumph-
                -griffithianum
```

4.5ft(1.35m)       5°F(-15°C)          L

Plant upright, moderately branched; dark green leaves retained 2 years. Pink flowers fading paler, with darker edges and reverse, in trusses of 12; floriferous. Flowers of heavy substance and very weather-resistant. (Sigrid Laxdall, reg. 1977)

cl.

```
Silberwolke-Koichiro Wada--yakushimanum ssp. yakushimanum (1/2)
  (Silver-  -           -catawbiense (1/4)
   cloud)  -Album Novum-
                    -unknown (1/4)
```

Flowers of very light purple, spotted yellow green, the reverse shaded darker; trusses of 12-16. Hairy ovate leaves, to 4in(10 cm) long. (Cross by H. Hackmann, 1963; reg. by G. Stuck, 1983)

cl.
                    -japonicum var. pentamerum (degronianum) (1/2)
Silent Surprise-
                    -aureum var. aureum (1/2)

Dwarf        -25°F(-32°C)          EM

A plant only 9in(23cm) high in 11 years, but 25in(64cm) broad!
Buds pink, opening to primrose yellow flowers without markings,
2.25in(5.7cm) wide, in dome-shaped trusses of 7-9. Floriferous.
Leaves to 3in(7.6cm), held 2-3 years. (Basil C. Potter, 1983)


                                -catawbiense (1/8)
cl.           -Charles Dickens-
     -Dr. V. H. Rutgers-              -unknown (3/8)
Silja-               -Lord Roberts--unknown
     -viscidifolium (1/2)

Cardinal red flowers held in trusses of 6-8. (Dietrich G.
Hobbie cross; intro. by B. Leendertz; reg. 1976)


cl.
     -leucaspis (1/2)
Silkcap-          -ciliatum (1/4)
     -Cilpinense-
                  -moupinense (1/4)

Pure white flowers. (Crown Lands, 1951)


cl.
     -caucasicum (1/2)
Silver Bells-
     -williamsianum (1/2)

2ft(.6m)          -5°F(-21°C)          E

Flowers pure silvery white, bell-shaped, ruffled, 2.5in(6.4cm)
wide, in trusses of 6-7. Leaves glossy dark green, oval, 2.5in.
long; plant about twice as broad as high. (B. Lancaster, 1966)


cl.
                    -yakushimanum ssp. yakushimanum
          -unnamed hybrid-                    (1/4)
Silver Doctor-          -unknown (3/4)
          -yellow hybrid--unknown

White flowers tinged light red. (J. Waterer, Sons & Crisp, reg.
1975)


          -campylocarpum ssp. campylocarpum (1/4)
cl.    -Mrs. W. C.-
     - Slocock  -unknown (5/16)
Silver -
Jubilee-          -Pink -griffithianum (1/4)
     -               -Shell-          -fortunei ssp. fortunei
     -Coronation-     -H. M. Ardenne-               (3/16)
       Day    -          -unknown
              -     -griffithianum
              -Loderi-
                     -fortunei ssp. fortunei

Flowers chartreuse green fading to pale greenish white, crimson
markings in upper throat, in trusses of 14. (A. F. George cross
in 1967; Hydon Nurseries, intro. and reg. 1978)


g.
          -megeratum (1/2)
Silver Ray-
          -boothii Mishmiense Group (1/2)

Orange flowers. See also Moth. (Lord Aberconway, 1950)


                    -dichroanthum ssp. dichroanthum (1/8)
          -Fabia-
     -unnamed-     -griersonianum (1/4)
cl.   - hybrid-          -fortunei ssp. discolor (3/16)
     -          -unnamed-
Silver -     hybrid-facetum (eriogynum) (1/8)
Sixpence-               -fortunei ssp. discolor
     -          -unnamed-
     -unnamed- hybrid-unknown (1/16)
     -unnamed- hybrid-
       hybrid-     -griersonianum
          -     -wardii var. wardii (1/8)
          -unnamed-
          hybrid-yakushimanum ssp. yakushimanum (1/8)

White flowers flushed empire rose, spotted yellow. (John
Waterer, Sons & Crisp, reg. 1975)


                    -yakushimanum ssp. yakushimanum (1/8)
          -unnamed-
          - hybrid-     -dichroanthum ssp.
          -     -Fabia-          dichroanthum (3/32)
     -Sunup- -     (selfed)-griersonianum (3/32)
     -Sundown-          -dichroanthum ssp.
cl.   -     -     -Fabia-     dichroanthum
     -     -     -unnamed-     -griersonianum
Silver -     -unnamed- hybrid-bureavii (1/16)
Trumpets-     hybrid-     -wardii var. wardii (1/16)
     -          -Crest-          -fortunei ssp. discolor
     -               -Lady    -               (1/32)
     -               Bessborough-campylocarpum ssp.
     -                         campy. Elatum Gp.
     -yellow Whitney hybrid--unknown (1/2)          (1/32)

3ft(.9m)          0°F(-18°C)          M

Buds plum red open to trumpet-shaped flowers, plum color fading
to white with blotched throat. Leafy calyx, spotted as flower.
Trusses 9in(23cm) across hold 7-10 flowers. Plant semi-dwarf
with arching branches and dark green leaves, nearly as wide as
tall. (John G. Lofthouse cross, 1979; reg. 1984)


          -yakushimanum ssp. yakushimanum (1/2)
     -unnamed-          -griffithianum (3/16)
cl.   - hybrid-     -Queen  -
     -     -Britannia-Wilhelmina-unknown (3/16)
Silver-     -
 Twist-          -Stanley Davies--unknown
     -          -griffithianum
     -     -Loderi-
     -unnamed-     -fortunei ssp. fortunei (1/8)
       hybrid-
          -yakushimanum ssp. yakushimanum

Creamy white flowers. (John Waterer, Sons & Crisp, reg. 1975)


cl.
          -edgeworthii (1/2)
Silver Wedding-
          -veitchianum Cubittii Group (1/2)

Loose clusters of 3 flowers, 5-lobed, 5in(12.7cm) across, white
flushed with shades of cardinal red, upper throat suffused and
spotted greyed yellow. Leaves 4.5in(11.5cm) long, glossy, scaly
beneath. (G. A. Hardy cross; shown by Maj. A. E. Hardy)  P. C.
(RHS) 1978


cl.
          -myrtifolium (1/2)
Silverburn-
          -telmateium (1/2)

Lepidote hybrid. Flowers orchid purple, held in compact trusses
of 3-4. (J. N. J. Hartley, intro. 1951)  A. M. (RHS) 1958

cl.

Silvia--form of laninerum (silvaticum)

4ft(1.2m)          5°F(-,5°C)          E          4/4

Crimson to pink  flowers  suffused white, in large trusses of 30,
blooming very early. Lance-shaped leaves with grey brown indumentum beneath. (Hambro)  A. M. (RHS) 1954


cl.
                -lacteum (1/2)
Simbu Sunset-
                -zoelleri (1/2)

Vireya hybrid. Trusses of 4-6 flowers, funnel-shaped, 5-lobed,
burnt orange shading to maize yellow in center of lobes and
throat; outside of tube brilliant orange.  Young stems deep red.
(T. Lelliott cross; grown by E. B. Perrott; G. F. Smith, reg.
1984)


cl.
        -fortunei ssp. fortunei (1/2)
        -                 -fortunei ssp. discolor (1/8)
Simita-     -Lady
        -     -Bessborough-campylocarpum ssp. campylocarpum
        -Jalisco-                 Elatum Group (1/8)
        -     -dichroanthum ssp. dichroanthum (1/8)
        -Dido-
                -decorum (1/8)

Fred Wynniatt g. Tall, compact habit.  Flowers  openly campanulate, of maize yellow with a brown eye, slight brown spotting;
petaloid calyx to 1in(2.5cm) long. Flat-topped trusses hold 10
flowers, late midseason.  (Edmund de Rothschild, reg. 1967)


                        -griffithianum (1/16)
                -George-
            -Mrs. Lindsay- Hardy-catawbiense (1/16)
        -Harvest- Smith       -
cl.  - Moon  -         -Duchess of Edinburgh--unknown (1/4)
        -     -         -campylocarpum ssp. campylocarpum (1/8)
Simona-     -unnamed-
        -       hybrid-unknown
        -             -campylocarpum ssp. campylocarpum
        -Letty Edwards-          Elatum Group (1/4)
                -fortunei ssp. fortunei (1/4)

Trusses of 12 flowers, 5-lobed, very light neyron rose, edged
phlox pink, blotched ruby red.  See also Maharani. (H. Hachmann
cross, 1964; reg. by G. Stuck, 1983)

                        -dichroanthum ssp. dichroanthum (1/8)
                -Astarte-
                -         -campylocarpum ssp.
                -       -Penjerrick- campylocarpum Elatum Gp.
            -Solon-         -                (1/16)
g.      -     -                 -griffithianum (7/16)
        -     -         -griffithianum
Simplicity-  -Sunrise-
        -             -griersonianum (1/8)
        -     -griffithianum
        -Loderi-
                -fortunei ssp. fortunei (1/4)

Brilliant red flowers. (Lord Aberconway, 1950)


g.
            -G. A. Sims--unknown (3/4)
Simsodour-                 -ponticum (1/4)
            -Purple Splendour-
                        -unknown

Deep, rich  purple flowers,  with a black  blotch.  (Col. G. H.
Loder, 1938)


cl.
        -concinnum (1/2)
Sinbad-         -cinnabarinum ssp. cinnabarinum Roylei Group
        -Lady  -                                (1/4)
        Chamberlain-Royal-cinnabarinum ssp. cinnabarinum (1/8)
                Flush-
                    -maddenii ssp. maddenii (1/8)

(Crown Lands; exhibited 1952)


cl.
        -griffithianum (1/2)
Sincerity-
        -unknown (1/2)

Plant habit and foliage similar to griffithianum. Flowers 5- to
7-lobed, deep pink fading to near-white, 5in(12.7cm) wide; conical trusses hold 9. (Gen. Harrison, reg. 1965)


                    -Parsons   -catawbiense (3/16)
                -America-Grandiflorum-
                -       -                -unknown (1/2)
cl.     -Fanfare-   -dark red hybrid--unknown
                -           -catawbiense
Singapore-  -Kettledrum-
        -           -unknown
        -     -forrestii ssp. forrestii Repens Group (1/4)
        -Gertrud-                 -ponticum 1/16)
        Schale-         -Michael Waterer-
            -Prometheus-           -unknown
                -Monitor--unknown

1ft(.3m)          -15°F(-26°C)          M

Plant well-branched, almost twice as broad as high; medium green
elliptic leaves held 2 years. Flowers widely campanulate, 2.25
in(5.7cm) across, medium to strong red; dome-shaped trusses hold
about 13.  See also Flamenco, Rangoon,  Small Wonder. (David G.
Leach, reg. 1973)


cl.
            -dichroanthum ssp. scyphocalyx (1/2)
Sir Arthur -   -dichroanthum ssp. dichroanthum (1/4)
Conan Doyle-Jasper-
            -Lady     -fortunei ssp. discolor (1/8)
            Bessborough-
                    -campylocarpum ssp. campylocarpum
                            Elatum Group (1/8)

A small but not dwarf plant; bluish green rounded leaves.  Campanulate flowers, Spanish orange with pink edges, unfading, 2.5
in(5cm) wide in flat trusses of 8-10.   (C. S. Seabrook, 1967)


cl.

Sir Charles Butler--form of fortunei ssp. fortunei

Synonym Mrs. Butler.  Pale mauve to delicate pearl rose.  Often
a parent of hybrids. (G. Paul)


cl.

Sir Charles Lemon--form of arboreum ssp. cinnamomeum var. album

5ft(1.5m)          5°F(-15°C)          EM          4/5

Considered a natural hybrid of arboreum and campanulatum by
plant explorer A. D. Shilling.  Very handsome foliage with
bright cinnamon brown indumentum; the  new leaves unfold white.
Flowers are ivory white, faintly spotted in the throat, in large
rounded trusses. (Lord Aberconway, from seed of Sir J. Hooker,
intro. 1937)  See color illus. VV p. 52.


g. & cl.
            -fortunei ssp. fortunei (1/2)
Sir Frederick Moore-
        -             -arboreum ssp. zeylanicum (1/4)
            -Saint Keverne-
                    -griffithianum (1/4)

6ft(1.8m)          -5°F(-21°C)          ML

A large rhododendron with large trusses of large flowers, clear pink and fragrant. Named for the Director of the Glasnevin Botanic Gardens, Dublin. (Rothschild, 1935)  A. M. (RHS) 1937; F. C. C. (RHS) 1972

---
```
                    -javanicum (1/2)
Sir George Holford-
                    -unknown (1/2)
```

Vireya hybrid. Orange yellow flowers, shaded red on margins. A. M. (RHS) 1930    See color illus. ARS Q 30:2 (Spring 1976): p. 102.

```
cl.                      -fortunei ssp. fortunei (1/4)
                 -Luscombei-
Sir George Sansom-       -thomsonii ssp. thomsonii (1/4)
                 -lacteum (1/2)
```

Lactcombei g. Tight, round trusses of 15 flowers, pale yellow, prominently tinged pink, campanulate, 3in(7.6cm) wide. (Geoffrey Gorer) A. M. (RHS) 1965

```
cl.
                 -griffithianum (1/2)
Sir Isaac Balfour-
                 -unknown (1/2)
```

Reddish pink flowers. (Otto Schulz, raised c.1890; intro. by C. B. van Nes)

```
cl.
           -fortunei ssp. fortunei (1/2)
Sir James-
           -unknown (1/2)
```

5ft(1.5m)          -10°F(- 23°C)          M          3/3

Light pink flowers; floriferous. Medium green foliage. Probably a Dexter hybrid. (Joseph B. Gable)

```
cl.
                   -falconeri ssp. falconeri (1/2)
Sir Jamie Darling-
                   -sinogrande (1/2)
```

Trusses of 31 flowers, 8-lobed, cream with calyx of pale buff. Leaves obovate, 15.6in(39cm) by 6.8in(75cm), with pale buff in-indumentum. (J. S. Basford cross, 1961; Brodick Castle, Isle of Arran, Scotland, reg. 1983)

---
```
                  -ponticum (1/2)
Sir John Broughton-
                  -unknown (1/2)
```

Light carmine red flowers. (Before 1867)

```
g. & cl.
                 -Corona--unknown (1/2)
Sir John Ramsden-
                 -thomsonii ssp. thomsonii (1/2)
```

Above parentage as in Register, 1958. Hillier shows pollen parent either campylocarpum or wardii. Lax trusses hold 10 large flowers, carmine with pale pink margins. A large plant with a vigorous loose habit. (Waterer, Sons & Crisp) A. M. (RHS) 1926; A. M. (Wisley Trials) 1948; F. C. C. (Wisley Trials) 1955

---

Sir Joseph Whitworth          Parentage unknown

A parent of Joan Langdon. (Waterer)

```
---                          -griffithianum (1/4)
                -unnamed hybrid-
Sir Richard Carton-          -unknown (3/4)
                -Rubescens--unknown
```

Flowers of crimson scarlet with cream-colored stamens and style. (C. B. van Nes & Sons)

```
g. & cl.      -haematodes ssp. haematodes (1/4)
        -Choremia-
Siren-        -arboreum ssp. arboreum (1/4)
        -griersonianum (1/2)
```

Bell-shaped, waxy, brilliant red flowers, with darker spotting, in rounded trusses. Dark, glossy foliage on a small shrub that blooms midseason. (Lord Aberconway, 1942) A. M. (RHS) 1942

```
cl.
        -maddenii ssp. crassum (1/2)
Sirius-
        -cinnabarinum ssp. cinnabarinum Roylei Group (1/2)
```

Cross by the Messers G. Reuthe; shown by Major A. E. Hardy. P. C. (RHS) 1973

cl.

Sirunke Orange--form of macgregoriae

Selected form of a vireya species. Orange flowers, held in 24-flowered umbel 4.5in(11.5cm) across. Leaves to 5in(12.7cm) long and 1.75in(4.5cm) wide, both surfaces with small black scales. (Collected & shown by Michael Black)  P. C. (RHS) 1969

```
                  -fortunei ssp. fortunei (25/64)
         -Fawn-       -dichroanthum ssp. dichroanthum (1/8)
cl.   -     -Fabia-
      -           -griersonianum (1/4)
Skipper-                   -griersonianum
      -     -Sarita Loder-       -griffithianum (5/64)
      -     -           -Loderi-
      -Indian-               -fortunei ssp. fortunei
      Penny-    -wardii var. wardii (1/8)
      -     -                 -griffithianum
         -Idealist-      -Kewense-
         -     -Aurora-       -fortunei ssp. for-
         -                          tunei
         -Naomi-    -thomsonii ssp. thomsonii
         -                          (1/32)
         -fortunei ssp. fortunei
```

5ft(1.5m)          0°F(-18°C)          EM          5/3

Compact plant; leaves 6in(15.2cm) long. Flowers empire yellow, paler at center, widely campanulate, to 5in(12.7in) wide; compact rounded trusses of 14. (Del James; C. Thompson, reg. 1972)

```
cl.
        -fortunei ssp. fortunei (1/2)
Skokomish-                       -catawbiense (1/8)
        -              -Atrosanguineum-
        -Dr. H. C. Dresselhuys-      -unknown (1/4)
        -                    -arboreum ssp. arboreum
                     -Doncaster-             (1/8)
                     -unknown
```

Parker, before 1958.

cl.

Skyglow     Parentage unknown

A parent of Maud Corning. Poor leaf color and growth habit. Fragrant flowers, 6-lobed, up to 3in(7.6cm) across, peach-colored, edged pink with a pale greenish yellow blotch, in flattened trusses of 12. (Seed from Veitch, of England; C. Dexter cross; reg. by Heritage Plantation, 1978)  See color illus. LW pl. 18.

cl.

Sleea--form of martinianum

Pink-flowered.  (From G. Reuthe; raiser  Mrs. Kenneth, Argyll)
P. C. (RHS) 1964

```
cl.
        -yakushimanum ssp. yakushimanum (1/2)
Sleepy-                 -arboreum ssp. arboreum (1/4)
     -Doncaster (selfed)-
                       -unknown (1/4)
```

3ft(.9m)          -10°F(-23°C)          M

A plant of compact habit.  Flowers are lavender pink with brown
blotch on upper lobe.  See also Bashful, Pink Cherub, Sneezy.
(J. Waterer, Sons & Crisp, 1971)

```
cl.
                      -catawbiense (1/2)
Slippery Rock*--(Newport F4)-
                      -unknown (1/2)
```

Vivid purplish red buds open to flowers of medium pink, edged in
vivid purplish red, with white upper lobe; brilliant greenish
yellow speckling.   Plant of medium size; hardy to -25°F(-32°C).
(Weldon E. Delp)

```
cl.
        -pemakoense (1/2)
Small Gem*
        -leucaspis (1/2)
```

1.5ft(.45m)          -5°F(-21°C)          E          3/3

Both parents are evident, leucaspis in the flower, pemakoense in
the plant.  Buds heavily and when very young.  Opening buds are
pink, turning white with a slight blush of soft pink. (Frisbee)

```
                                    -catawbiense (3/16)
                     -Parsons Grandiflorum-
                 -America-               -unknown (1/2)
      -Fanfare-      -dark red hybrid--unknown
cl.   -     -          -catawbiense
      -     -Kettledrum-
Small -          -unknown
Wonder-                          -ponticum (1/16)
      -              -Michael Waterer-
      -          -Prometheus-         -unknown
      -Gertrud Schale-      -Monitor--unknown
                     -forrestii ssp. forrestii Repens Gp. (1/4)
```

3ft(.9m)          -20°F(-29°C)          EM          4/4

Flowers dark red with lighter small centers, openly campanulate,
2in(5cm) across, held in globe-shaped trusses of 7.  Leaves narrowly elliptic.  Plant very densely foliaged, broader than tall.
See also Rangoon.  (David G. Leach, 1973)   See color illus. ARS
J 36:1 (Winter 1982): p. 6;  HG p. 110.

```
cl.
        -wardii var. wardii (1/4)
    -Hawk-              -fortunei ssp. discolor (1/8)
Smew-   -Lady Bessborough-
    -              -campylocarpum ssp. campylocarpum
    -soulei (1/2)                    Elatum Group (1/8)
```

Pink buds changing to white flowers suffused with pink; trusses
of 6-8 flowers.  (Crown Estate, Windsor)  P. C. (RHS) 1961

```
g.
        -smirnowii (1/2)
Smirnauck-
        -griffithianum (aucklandii) (1/2)
```

Pure white flowers.  (E. J. P. Magor, 1924)

cl.

```
            -arboreum ssp. arboreum (1/2)
Smithii Album-
            -unknown (1/2)
```

White flowers, spotted.  See also Boddaertianum.


cl.

Smithii Aureum     See Norbitonense Aureum

cl.

Smoke--form of augustinii ssp. augustinii

6ft(1.8m)          -5°F(-21°C)          EM          4/3

Beautiful clear blue form, from the Trewithen Garden.


```
                                         -griffithianum
                             -Queen Wilhelmina-    (1/16)
cl.          -Britannia-              -unknown (13/16)
       -Burgundy-       -Stanley Davies--unknown
Smokey #9*    -            -ponticum (1/8)
       -          -Purple Splendour-
       -Moser's Maroon--unknown  -unknown
```

6ft(1.8m)          -5°F(-21°C)          ML          3/3

Fir  green leaves, heavily veined.  Leafy flower buds open to
deep purple flowers, in full trusses.  (Halfdan Lem)

```
cl.               -ponticum (3/8)
     -Purple Splendour-
Smyrna-               -unknown (3/8)
     -              -campylocarpum ssp. campylocarpum (1/4)
     -Constant Nymph-          -ponticum
                     -Purple Splendour-
                             - unknown
```

Large lavender flowers, without markings.  (Knap Hill Nursery,
reg. 1968)

```
cl.
        -yakushimanum ssp. yakushimanum (1/2)
Sneezy-         -arboreum ssp. arboreum (1/4)
       -Doncaster-
               -unknown (1/4)
```

3ft(.9m)          -10°F(-23°C)          M          3/3

Compact plant habit.  Red buds open to soft pink flowers, darker
at edges, with deep red blotch on upper lobe.  (J. Waterer, Sons
& Crisp, reg. 1971)

```
cl.
        -pemakoense (1/2)
Snipe-
        -davidsonianum (1/2)
```

2ft(.6m)          5°F(-15°C)          EM          3/3

White flowers suffused and spotted in light orchid pinks.  Plant
upright, twiggy; glossy light green leaves, both surfaces scaly.
(P. A. Cox, Glendoick Gardens, reg. 1978)  A. M. (RHS) 1975

```
cl.
        -Loder's White, q.v. for 2 possible
Snow Bells-              parentage diagrams
        -williamsianum
```

1.5ft(.45m)          0°F(-18°C)          E

Plant broader than tall, dense; oval leaves 2in(5cm) long. Flowers pure white, openly campanulate, 2.5in(6.4cm) wide; trusses
of 5-7.  (Benjamin F. Lancaster, 1968)

cl.
```
            -arboreum ssp. arboreum (1/2)
Snow Bunting-
            -sutchuenense (1/2)
```

White flowers. (J. C. Williams)


cl.
```
      -souliei (1/2)
Snow Cap-                    -griffithianum (1/8)
     -           -Loderi White-
  -unnamed-  Diamond   -fortunei ssp. fortunei (1/8)
    hybrid-
            -williamsianum (1/4)
```

4ft(1.2m)        -5°F(-21°C)        M        3/3

Shell pink buds opening to pure white flowers without markings,
3.25in(8.3cm) wide, 7-lobed, in trusses of 7; floriferous. Plant
rounded, as broad as tall; leaves held 3 years. (William Whitney; reg. by G. & A. Sather, 1976)


cl.      -fortunei ssp. fortunei (1/4)
```
      -Fawn-       -dichroanthum ssp. dichroanthum (1/8)
Snow -  -Fabia-
Crest-      -griersonianum (1/8)
   -     -wardii var. wardii (1/4)
    -Crest-
        -Lady      -fortunei ssp. discolor (1/8)
        Bessborough-
                -campylocarpum ssp. campylocarpum
                              Elatum Group (1/8)
```

Buds of China rose open to white flowers lightly tinged French
rose, 5-7 lobed, held in trusses 8in(20cm) wide, of 9-11 flowers. Upright plant, about as wide as tall; yellow green leaves
held 2 years. (Arthur & Maxine Childers, cross 1959; reg. 1978)


cl.
```
      -moupinense (1/2)
Snow Fairy-
      -mucronulatum ? (1/2)
```

Mrs. Roza M. Stevenson.    P. C. (RHS) 1957


```
                        -fortunei ssp. discolor (3/8)
                -Lady Bessborough-
cl. -Day Dream-         -campylocarpum ssp. campylo-
   -          -                    carpum Elatum Group (1/8)
Snow -         -griersonianum (1/4)
Goose-          -dichroanthum ssp. dichroanthum (1/8)
   -        -Dido-
  -Ice Cream-    -decorum (1/8)
            -fortunei ssp. discolor
```

Funnel-shaped flowers, white in bud, opening white with a large
yellow blotch, in trusses of 10. See also Anne George. (A. F.
George, Hydon Nurseries, reg. 1966)


cl.
```
      -leucaspis (1/2)
Snow Lady-
      -ciliatum ? (1/2)
```

2.5ft(.75m)        0°F(-18°C)        M        4/4

Compact spreading shrub; medium green leaves to 3in(7.6cm) long,
setose. Fragrant flowers to 2in(5cm) wide, white with dark anthers, in lax trusses. Buds heavily even in shade. Cuttings
from plant in W. L. Tucker collection, who obtained it from Lionel de Rothschild as leucaspis. (B. Lancaster, before 1958)
P. A. (ARS) 1955     See color illus. HG p. 99; VV p. 95.


cl.
```
      -dalhousiae var. dalhousiae (1/2)
Snow Mantle-       -edgeworthii (1/4)
      -Fragrantissimum-
                  -formosum var. formosum (1/4)
```

Trusses hold 3 flowers, white with a gold throat fading out. (R.
C. Gordon, New Zealand, reg. 1982)


g. & cl.             -maximum (1/4)
```
      -Halopeanum-
Snow Queen-       -griffithianum (1/2)
      -           -griffithianum
       -Loderi-
            -fortunei ssp. fortunei (1/4)
```

5ft(1.5m)        5°F(-15°C)        M        4/2

Deep pink buds opening to large, funnel-shaped pure white flowers with tiny red basal blotch; large dome-shaped trusses.  One
of the best whites. Vigorous plant of open habit; leaves to 8in
(20.3in) long, won't withstand sun. (Sir Edmund Loder)   A. M.
(RHS) 1934;  A. M. (Wisley Trials) 1946


cl.
```
      -fortunei ssp. discolor (3/4)
Snow Shimmer-       -maximum (1/4)
      -unnamed hybrid-
                  -fortunei ssp. discolor
```

6ft(1.8m)        -5°F(-21°C)        L

Flowers 4in(10cm) wide, 7-lobed, very fragrant, white with pale
yellow glow at center; spherical trusses of 10. Plant broader
than tall; olive green leaves, new growth greenish brown. (John
& Gertrude Wister; Scott Horticultural Foundation of Swarthmore
College, 1980) See col. illus. ARS Q 34:3 (Summer 1980): cover.


cl.
```
            -leucaspis (1/4)
      -Snow Lady-
Snow Sprite-       -ciliatum ? (1/4)
      -moupinense (1/2)
```

1ft(.3m)        5°F(-15°C)        E

Plant as wide as high, more compact than its parents. Leaves of
glossy green, 1.5in(3.8cm) long. Pure white flowers, 5-lobed,
with chocolate brown anthers, openly campanulate; trusses of 3-
5. (B. F. Lancaster & R. Whalley, reg. 1969)


g.
```
      -griffithianum (1/2)
Snow White-
      -fortunei ssp. fortunei (1/2)
```

Kewense g. or Loderi g. Pure white flowers, very open. (Lowinsky) A. M. (RHS) 1923


cl.
```
      -sutchuenense (1/2)
Snowball-       -griffithanum (1/4)
      -Loderi Pink Diamond-
                  -fortunei ssp. fortunei (1/4)
```

White flowers. ("Not from Leonardslee"--Sir Giles Loder)


---
```
      -griffithianum (1/2)
Snowdrop-
      -unknown (1/2)
```

Pure white flowers. A parent of Joan Ramsden.

cl.
          -yakushimanum ssp. yakushimanum (1/2)
Snowstorm-       -Corona--unknown (5/16)
     -Cary Ann-           -griffithianum (1/16)
     -        -Mars-
     -Vulcan-     -unknown
              -griersonianum (1/8)

Pure white flowers  with rose spotting, a contrast to dark green
foliage. Plant vigorous and compact with horizontal new growth.
(John G. Lofthouse, reg. 1973)


cl.
            -ririei (1/2)
Snowy River-
            -niveum (1/2)

Rounded trusses of 13-15 flowers, lilac with deeper veining and
flushing, 1.75in(4.5cm) wide.  Leaves 6in(15.2cm) long, with
thinly plastered fawn indumentum below.  (Cross by the 2nd Lord
Aberconway; intro. by Lord Aberconway)   A. M. (RHS) 1974


cl.           -sanguineum ssp. sanguineum var. sanguineum
     -Sangreal-                                   (1/4)
Snugbug-        -griersonianum (1/4)
     -tsariense (1/2)

4ft(1.2m)        -5°F(-21°C)         M

Openly funnel-shaped  flowers, 3in(7.6cm) long, 5-lobed, opening
cherry red fading to carmine  rose, held  in trusses 5in(12.7cm)
across.  Upright plant with willowy branches; leaves held 3
years.  New growth has silvery tomentose.  (Milton Wildfong,
reg. 1983)


g.
     -dichroanthum ssp. scyphocalyx (1/2)
Socrianum-
     -griffithianum (1/2)

A parent of Bow Street, Cheapside, Pall Mall.  Apricot-colored
flowers.  (J. B. Stevenson, 1951)


cl.
          -yakushimanum ssp. yakushimanum (1/2)
Soft Shadows-
          -argyrophyllum ssp. argyrophyllum (1/2)

Trusses of 12-15 flowers, 5-lobed, carmine rose fading to pure
white.  Leaves 3.5in(8.9cm) long, with fawn, suede-like indumentum beneath. (F. M. Jury, New Zealand, reg. 1982)


cl.            -griffithianum (3/8)
     -Marinus Koster-
Sofus -           -unknown (1/4)
Eckrem-    -fortunei ssp. fortunei (1/4)
     -Pilgrim-    -arboreum ssp. arboreum (1/8)
              -Gill's -
              Triumph-griffithianum

5ft(1.5m)          10°F(-12°C)          L

Phlox pink buds open to rose pink flowers with deeper pink rims,
4.5in(11.5cm) wide; conical trusses of 10-11.  Plant moderately
branched; glossy dark leaves, 8in(20.3cm) long, held 2 years.
See also Robert Verne.  (Sigrid Laxdall, 1979)


cl.
     -campylocarpum ssp. caloxanthum (1/2)
Solarium-              -caucasicum (1/8)
     -        -Dr. Stocker-
     -Damaris-         -griffithianum (1/8)
          -campylocarpum ssp. campylocarpum (1/4)

Flowers clear yellow, spotted crimson.  (Sunningdale Nurseries,
1955)


             -facetum (eriogynum) (1/8)
          -unnamed-
cl.   -unnamed- hybrid-unknown (5/16)
     - hybrid-    -dichroanthum ssp. dichroanthum (1/8)
Soldier-    -Fabia-
Palmer-         -griersonianum (1/8)
     -       -yakushimanum ssp. yakushimanum (1/4)
     -unnamed-        -Queen  -griffithianum (1/16)
     hybrid-        -Wilhelmina-
          -Britannia-          -unknown
              -Stanley Davies--unknown

3ft(.9m)         0°F(-18°C)          M

A compact plant  with flowers of brick red, spotted red.  (John
Waterer, Sons & Crisp, reg. 1975)


cl.
          -griffithianum (1/2)
Solent Queen-
          -fortunei ssp. discolor (1/2)

6ft(1.8m)          5°F(-15°C)          ML          3/3

Angelo g.  Flowers blush pink with a green ray, fading to almost
pure white, 5.5in(14cm) wide, fragrant.  Similiar to the other
clones of the group.  (Rothschild, 1933)   A. M. (RHS) 1939


cl.

Solent Snow       Parentage as above

Angelo g.  White flowers, green markings.  (Rothschild, 1933)

cl.

Solent Swan       Parentage as above

Angelo g.  Pure white flowers, no markings.  (Rothschild, 1955)


                      -griffithianum (1/8)
cl.           -Queen Wilhelmina-
     -Britannia-          -unknown (3/8)
Solitude-        -Stanley Davies--unknown
     -         -elliottii (1/4)
     -Fusilier-
              -griersonianum (1/4)

Glowing red flowers.  (Rudolph Henny, before 1958)


          -dichroanthum ssp. dichroanthum (1/4)
g.  -Astarte-        -campylocarpum ssp. campylocarpum
     -      -Penjerrick-          Elatum Group (1/8)
Solon-        -griffithianum (3/8)
     -      -griffithianum
     -Sunrise-
          -griersonianum (1/4)

Pale rose flowers.  (Lord Aberconway, 1941)


cl.
     -dichroanthum ssp. dichroanthum (1/2)
Solveig-
     -griersonianum (1/2)

Fabia g.  (Origin unknown.  Incorrectly credited in the Register
to the RHS Garden, Wisley)


cl.            -ponticum (1/4)
     -Purple Splendour-
Sonata-            -unknown (1/4)
     -dichroanthum ssp. dichroanthum (1/2)

3ft(.9m)          0°F(-18°C)          L

Orange flowers, edged with burgundy.  Sun-tolerant plant; very
dense foliage.  (G. Reuthe, 1949)  H. C. (Wisley Trials) 1959

cl.
       -russatum (1/2)
Songbird-       -impeditum (1/4)
     -Blue Tit-
            -augustinii ssp. augustinii (1/4)

3ft(.9m)     -5°F(-21°C)     EM     3/4

Clusters of violet blue bell-shaped flowers. Plant habit strong and tidy; small glossy leaves. (Sir James Horlick, 1954)  A. M. (RHS) 1957   See color illus. F p. 64.

cl.
       -russatum (1/2)
Songster-       -impeditum (1/4)
     -Blue Tit-
           -augustinii ssp. augustinii (1/4)

A small, compact shrub covered with small blue flowers. Blooms early. (Sir James Horlick, 1957)

---
      -wardii var. wardii (1/2)
Sonia*          -arboreum ssp. cinnamomeum var. album
   -     -unnamed-                (1/8)
-Constance- hybrid-griffithianum (1/8)
   -
       -auriculatum (1/4)

Ivory white flowers.  (Rothschild?)

cl.
       -caucasicum (1/2)
Sophia Gray-
     -unknown (1/2)

A parent of Buxum, Gipsy Maid. Fimbriated pink flowers, spotted with burnt umber. (M. Koster & Sons, before 1922)

g.
     -souliei (1/2)
Soularb-
   -arboreum ssp. arboreum (1/2)

A deep cherry red flower.  (E. J. P. Magor, 1926)

g.
     -souliei (1/2)
Soulbut-
  -Sir Charles Butler--fortunei ssp. fortunei (1/2)

White tinged pink, broad crimson spotting. (E. J. P. Magor, 1926)

g.
     -souliei (1/2)
Souldis-
  -fortunei ssp. discolor (1/2)

Large flowers of blush pink fading to off-white, with a small crimson basal blotch, saucer-shaped, in loose trusses on a tall plant. Blooms midseason to late season.  See also Endre Ostbo, Exbury Souldis.  (E. J. P. Magor, 1927)  P. A. (ARS) 1954

g.
     -souliei (1/2)
Soulkew-     -griffithianum (1/4)
  -Kewense-
       -fortunei ssp. fortunei (1/4)

Flowers are blush white, pink outside. (E. J. P. Magor, 1926)

g.
     -souliei (1/2)
Soulking-
   -arboreum ssp. zeylanicum (1/2)

Crimson pink flowers. (E. J. P. Magor, 1926)

cl.  -unnamed hybrid-
          -fortunei ssp. discolor (1/4)
    -      -decorum (1/4)
Southern-            -griffithianum (1/32)
 Belle -          -Kewense-
    -     -Aurora-   -fortunei ssp. fortunei
    -   -Naomi-  -thomsonii ssp. thomsonii (1/16)
  -Carita-
     -     -fortunei ssp. fortunei (5/32)
      -campylocarpum ssp. campylocarpum (1/4)

5ft(1.5m)     5°F(-15°C)     ML

Jasper red buds open to fragrant flowers, primrose yellow fading to light Naples yellow, held in dome-shaped trusses 6.5in(16.5cm) across, of 11 flowers. Upright plant, 3/4 as broad as tall; leaves 5in(12.7cm) long, held 2 years. (Cross by Grady E. Barefield, 1959; reg. by Mary Barefield, 1981)

cl.
      -lindleyi (1/2)
Southern Cloud-
     -nuttallii (1/2)

Trusses hold 3-11 flowers, neyron rose outside, salmon inside with Indian yellow base. Elliptic leaves, 7.5in(19cm) long. (A. Bramley cross, before 1965; reg. by J. F. Wilson, 1983)

cl.
      -fortunei ssp. discolor (1/2)
Southern Cross-     -griffithianum (1/8)
   -     -Loderi-
   -Lodauric-   -fortunei ssp. fortunei (1/8)
    Iceberg-
       -auriculatum (1/4)

Plant vigorous, upright, compact, free-flowering. Large funnel-campanulate flowers, white with bronze throat; spherical trusses of 10-12. (A. F. George, Hydon Nurseries, 1955)  H. C. (Wisley Trials) 1969

cl.
      -arboreum ssp. arboreum (1/4)
Souvenir de -Doncaster-
D. A. Koster-    -unknown (3/4)
     -Charles Waterer--unknown

A profusion of dark scarlet flowers with dark spotting.  (D. A. Koster, raiser; intro. by Wezelenberg & Son)  A. M. (RHS) 1922

                -griffithianum (1/8)
       -George Hardy-
cl.    -Pink -    -catawbiense (3/8)
      -Pearl-    -arboreum ssp. arboreum (3/8)
Souvenir de -  -Broughtonii-
Dr. S. Endtz-      -unknown (1/8)
          -catawbiense
     -John Walter-
        -arboreum ssp. arboreum

Buds of rose, opening to rich pink flowers marked with a crimson ray, widely funnel-shaped, 3.25in(8.3cm) wide; domed trusses of 15-17. Plant vigorous, broader than tall; dark green leaves 6.5 in(16.5cm) long.  (L. J. Endtz of Boskoop) A. M. 1924; A. G. M. 1969; F. C. C. (Wisley Trials) 1970   See color illus. F p. 64.

---
      -Crown Princess-brookeanum var. gracile (1/4)
Souvenir de - of Germany -      -jasminiflorum (1/8)
J. H. Mangles-       -Princess Royal-
     -javanicum       -javanicum (5/8)

Vireya hybrid. Chrome orange flowers. F. C. C. (RHS) 1888 See color illus. ARS Q 32:3 (Summer 1978): p. 171.

cl.
                      -griffithianum (1/2)
Souvenir de Madame-
 J. H. van Nes    -unknown (1/2)

Scarlet flowers. (Otto Schulz, raiser, 1890; intro. by C. B. van
Nes & Sons, 1896)

cl.

Souvenir of Anthony Waterer      Parentage unknown

A parent of several hybrids.  Flowers dark rosy red or salmon,
yellow blotch.  (A. Waterer)

cl.
                   -campylocarpum ssp. campylocarpum (1/2)
Souvenir of W. C. Slocock-
                   -unknown (1/2)

4ft(1.2m)         -5°F(-21°C)         M         3/3

Similar to Unique.  Plant compact and attractive.  Buds of deep
apricot pink opening pale primrose yellow, lightly flushed apricot, and pink on reverse; rounded trusses.  Medium green leaves
have a characteristic slight twist.  (Slocock)  A. M. (RHS) 1935
See color illus. F p. 65;  VV p. 14.

                 -fortunei ssp. discolor (1/2)
cl.          -Belle-
             -         -campylocarpum ssp. campylocarpum
Spanish Galleon-                Elatum Group (1/4)
             -             -fortunei ssp. discolor
             -Margaret Dunn-    -dichroanthum ssp.
                    -Fabia-        dichroanthum (1/8)
                          -griersonianum (1/8)

Flowers butter yellow.  (Sunningdale Nurseries, 1955)

cl.
            -elliottii (1/2)
Spanish Glory-    -dichroanthum ssp. dichroanthum (1/4)
            -Fabia-
                -griersonianum (1/4)

4ft(1.2m)         10°F(-12°C)         M

Plant habit open; dark green leaves, 6in(15.2cm) long.  Flowers
campanulate, red, in trusses of 7.  (Mr. & Mrs. M. Sumner, 1971)

cl.             -fortunei ssp. fortunei (1/4)
            -Pilgrim-    -arboreum ssp.
            -       -Gill's Triumph-      arboreum (1/8)
Sparkle Plenty-           -griffithianum (1/8)
            -        -sanguineum ssp. didymum (1/4)
                -Red Cap-
                    -facetum (eriogynum) (1/4)

Flowers light yellow to white with vermilion spots, held in tall
trusses.  Medium upright growth habit.  (R. & L. Henny, 1964)

                    -facetum (eriogynum) (1/8)
                -unnamed-
                - hybrid-unknown (5/16)
cl.      -unnamed-    -dichroanthum ssp. dichroanthum (1/8)
         - hybrid-Fabia-
Sparkler-          -griersonianum (1/8)
         -      -yakushimanum ssp. yakushimanum (1/4)
         -unnamed-                    -griffithianum (1/16)
             hybrid-      -Queen Wilhelmina-
                 -Britannia-          -unknown
                    -Stanley Davies--unknown

3ft(.9m)          0°F(-18°C)          ML

Plant of compact habit.  Large flowers of rich bright red, nonfading and floriferous.  See also Borderer, Pheasant Tail, Sold-
ier Palmer.  (J. Waterer, Sons & Crisp, reg. 1971)

cl.                    -ponticum (1/4)
            -Purple Splendour-
Sparkling Burgundy-      -unknown (1/4)
            -Seven Devils--macrophyllum (1/2)

5ft(1.5m)          10°F(-12°C)          M

Flowers openly funnel-shaped, 2in(10cm) wide,  soft lilac purple
throat with touch of white, some aster violet spotting; trusses
dome-shaped, 17-flowered.  Plant habit open, upright; ivy green
leaves retained 3 years. (Dr. Frank Mossman cross, 1973; J. Brotherton, reg. 1983)

cl.
            -fortunei ssp. discolor (1/2)
Sparkling Jewel-      -fortunei ssp. fortunei (1/4)
            -unnamed hybrid-
                -unknown (1/4)

6ft(1.8m)          -5°F(-21°C)          ML

Very fragrant flowers, white with pale yellowish green spotting,
openly funnel-shaped, 4in(10.2cm) wide, 7-lobed; ball-shaped
trusses of 11.  Plant upright, branching moderately; leaves held
2 years.  See also Sunlit Snow.  (John & Gertrude Wister for A.
Scott Horticultural Foundation, Swarthmore College; reg. 1980)

cl.

Spatter Paint--form of irroratum ssp. irroratum

5ft(1.5m)          5°F(-15°C)          VE

Venetian pink flowers with heavy spotting over most of corolla.
Campanulate flowers 2in(5cm) wide, 5-lobed; flattened trusses of
13.  Floriferous.  Upright, well-branched plant; handsome foli-
age, leaves held 4 years; new growth has red bracts.  (Seed from
Rock, 1948 Expedition; raised by J. Henny, C. C. Smith; reg. by
Meldon Kraxberger, 1979)   See color illus HG p. 27.

cl.
         -fortunei ssp. discolor (1/2)
Spectra-
         -unknown (1/2)

Flowers flushed pink, fading to white.  Late season.  (Endre
Ostbo; intro. by C. P. Fawcett; reg. 1962)

cl.                -maximum (1/4)
         -Russell Harmon-          (natural hybrid)
Spellbinder-          -catawbiense (1/4)
         -          -calophytum var. calophytum (1/4)
         -Robin Hood-
                -sutchuenense (1/4)

6ft(1.8m)          -15°F(-26°C)          E          4/5

Flowers of heavy substance, light purplish pink, dorsal spotting
of deep purple red, widely campanulate, 3.5in(8.9cm) wide; globular trusses of 16.  Floriferous.  Plant well-branched, rounded;
densely furnished with glossy leaves of unusual size--9.5in(24
cm) long.  (David G. Leach, 1975)

cl.
         -macabeanum (1/2)
Spiced Honey-    -campylocarpum ssp. campylocarpum (1/4)
         -Unique-
                -unknown (1/4)

Trusses of 16-18 flowers, 7-lobed, apricot  in bud and flushed
pink on tube, opening barium yellow  with deep maroon blotch.
Elliptic leaves with woolly grey indumentum and yellow petioles,
6in(15cm) long.  (G. F. Smith, New Zealand, reg. 1984)

cl.
      -<u>scabrifolium</u> var. <u>spiciferum</u> (1/2)
Spicil-      -<u>ciliatum</u> (1/4)
    -Cilpinense-
        -<u>moupinense</u> (1/4)

Flowers light roseine purple fading to very light mallow purple. (Hybridizer unknown; intro. by H. G. Rutland; reg. by Dunedin Rhododendron Group, 1980)


g.
      -<u>spinuliferum</u> (1/2)
Spinbur-
    -<u>burmanicum</u> (1/2)

See Lochinch Spinbur. (Lord Stair, 1950)


cl.
      -<u>spinuliferum</u> (1/2)
Spinulosum-
    -<u>racemosum</u> (1/2)

4ft(1.2m)     5°F(-15°C)     M    4/2

Light pink buds and dark pink flowers for an attractive two-tone effect. Narrowly campanulate flowers, with protruding anthers, grow along stems and in pompom-like small trusses. Plant puts on the long growth of <u>racemosum</u>. See also Exbury Spinulosum. (RBG, Kew) A. M. (RHS) 1944


cl.
      -<u>griffithianum</u> (1/2)
Spitfire-
    -unknown (1/2)

6ft(1.8m)     0°F(-18°C)     ML

Tall, compact shrub with fine growth habit, good foliage. Flowers deep chrysanthemum crimson with dark brown blotch. Considered one of the best late reds. (A. Kluis, 1946)


cl.
      -<u>triflorum</u> (1/2)
Spring Dance-
    -unknown (1/2)

Similar to the species <u>triflorum</u>. (Raiser James E. Barto; intro. & reg. by The Bovees, 1962)


cl.
            -<u>catawbiense</u> (1/2)
    -unnamed hybrid-
Spring Dawn*      -unknown (1/2)
   -        -<u>catawbiense</u>
   -Mrs. Charles S. Sargent-
           -unknown

5ft(1.5m)     -20°F(-29°C)     EM    3/3

Flowers strong rosy pink with a warm golden yellow blotch. Very hardy plant, vigorous and attractive. (A. M. Shammarello) See color illus. LW pl. 83; VV p. 73.


cl.
              -<u>minus</u> var. <u>minus</u> Carolin-
   -Laetevirens (Wilsoni)-   ianum Group (1/4)
Spring Delight-      -<u>ferrugineum</u> (1/4)
    -unknown (1/2)

2ft(.6m)     -20°F(-29°C)     EM

Synonym A.M.S. Plant well-branched, broader than tall; olive green leaves held 2 years. Flowers of pale purplish pink fading to white, with .5in light olive brown spot in dorsal throat, 1.5 in(3.8cm) across, 5 wavy lobes. The last hybrid registered by the late Tony Shammarello. (A. M. Shammarello, intro. 1979; reg. 1981) Color illus. in Flower & Garden, Dec.-Jan. issue, 1981.


cl.
      -Marion--unknown (1/2)
Spring Fiesta-      -<u>griersonianum</u> (1/4)
    -unnamed hybrid-
           -<u>fortunei</u> ssp. <u>fortunei</u> (1/4)

Trusses of 12 flowers, light neyron rose. (G. Langdon, 1976)


cl.
      -unknown (1/2)
Spring Fling*
    -<u>russatum</u> (1/2)

2.5ft(.75m)     -10°F(-23°C)     E    4/3

An abundance of dark electric blue flowers. Growth habit compact; small deep green leaves, reddish bronze in winter. (Intro. 1982 by Harold E. Greer)


cl.
        -<u>catawbiense</u> var. <u>album</u> (1/2)
Spring Frolic-
    -Koichiro Wada--<u>yakushimanum</u> ssp. <u>yakushimanum</u>
                        (1/2)
5ft(1.5m)     -25°F(-32°C)     EM    4/5

Clear pink buds open to white flowers, reflecting the <u>yakushimanum</u> parentage; unusually floriferous. Plant broader than tall; very dense foliage of elliptic leaves, held 3 years; new growth farinose, brown, several weeks. See also Pink Frosting. (D. G. Leach, 1972) See color illus. ARS Q 25:2 (Spring 1972): p. 100.


cl.
                 -<u>caucasicum</u> (1/4)
    -Cunningham's White-
Spring Glory-        -<u>ponticum</u> (white form) (1/4)
   -        -<u>catawbiense</u> (1/4)
   -red hybrid-
        -unknown (1/4)

5ft(1.5m)     -15°F(-26°C)     EM    4/2

Plant compact, vigorous, grows well in sun or shade. Flowers of clear pink, deeper at edges, large crimson blotch. (Anthony M. Shammarello, 1955) See color illus. HG p. 112; VV p. 84.


cl.
      -Marion--unknown (3/4)
Spring Glow-        -<u>catawbiense</u> (1/4)
   -Mrs. C. S. Sargent-
        -unknown

Compact trusses of 16 flowers, each 3.5in(9cm) wide, magenta with brown spotting. (Fa. Felix & Dijkhuis) Flora Nova, Silver Medal (Boskoop) 1958


cl.
      -Essex Scarlet--unknown (1/2)
Spring Magic-
    -<u>forrestii</u> ssp. <u>forrestii</u> Repens Group (1/2)

Vigorous and free-flowering, compact and spreading habit with dark green leaves. Flowers in trusses of 5-7, deep currant red with slight spotting on upper lobe. (Dietrich Hobbie, raiser; intro. by Primrose Hill Nursery) A. M. (Wisley Trials) 1969


cl.
        -<u>catawbiense</u> (1/4)
   -red hybrid-
Spring Parade-    -unknown (1/4)
   -        -<u>caucasicum</u> (1/4)
   -Cunningham's White-
        -<u>ponticum</u> (white form) (1/4)

4.5ft(1.35m)     -20°F(-29°C)     M    3/2

Clear scarlet red flowers, funnel-shaped, 2.5in(6.4cm) wide, in globular trusses. Upright plant, as broad as high; dark green, recurved leaves 4in(10.2cm) long. (A. M. Shammarello, reg. 1962) See color illus. HG p. 142.

cl.
                -<u>aureum</u> var. <u>aureum</u> (<u>chrysanthum</u>) (1/2)
Spring Snow-
                    -<u>japonicum</u> var. <u>pentamerum</u> (<u>metternichii</u> var. <u>penta-
                                                          merum</u>) (1/2)

5ft(.75m)          -5°F(-21°C)          E

Plant rounded, about twice as wide as tall; leaves 2.5in(6.4cm)
long, with thin tan indumentum, held 3 years.  Light pink buds
opening white, unmarked; flowers 2in(5cm) wide, in trusses of 7-
10.  (Crossed at University of Washington Arboretum; Brian Mulligan, reg. 1978)

cl.                        -<u>racemosum</u> (1/4)
                -unnamed hybrid-
Spring Song-               -<u>keiskei</u> (3/4)
                -<u>keiskei</u>

2.5ft(.75m)          -10°F(-23°C)          EM

Light yellow flowers, aging to salmon apricot.  (D. Hardgrove,
exhibited 1951)

cl.

Spring Sonnet--form of <u>vernicosum</u> (R 59625)

5ft(1.5m)          -15°F(-26°C)          M          3/3

Loose trusses of 6-11 white flowers with sparse ruby spotting in
throat, the reverse tinged rose, 2.75in(7cm) wide, 7-lobed.  Upper surface of leaves waxy (vernicosum = varnished).  (J. Rock,
collector; raised & intro. by Lord Aberconway)  A. M. (RHS) 1976

cl.
                -<u>williamsianum</u> (1/2)
Spring Sun*
                -unknown (1/2)

3ft(.9m)          0°F(-18°C)          ML          3/3

Compact plant habit with rounded leaves as <u>williamsianum</u>.
Flowers light yellow.  (D. Hobbie cross; offered by H. E. Greer)

cl.
                -<u>griersonianum</u> (1/2)
Springbok-                -<u>ponticum</u> (1/4)
                -unnamed hybrid-
                          -unknown (1/4)

4ft(1.2m)          0°F(-18°C)          ML

An open-growing plant with large trusses of about 18 flowers,
vivid carmine pink, spotted crimson.  (A. F. George, 1964)

                              -<u>griersonianum</u> (3/8)
                          -Azor-
cl.        -Umpqua Chief-    -<u>fortunei</u> ssp. <u>discolor</u> (1/8)
           -            -    -<u>dichroanthum</u> ssp. <u>dichroanthum</u>
Springfield-        -Fabia-                              (1/4)
           -                -<u>griersonianum</u>
           -    -<u>fortunei</u> ssp. <u>fortunei</u> (1/4)
           -Fawn-    -<u>dichroanthum</u> ssp. <u>dichroanthum</u>
                -Fabia-
                    -<u>griersonianum</u>

5ft(1.5m)          5°F(-15°C)          L          4/3

Flowers orange pink with slight crimson dorsal spotting and ring
at base of throat, 4in(10.2cm) wide; flat trusses of about 11.
Plant upright; fir green leaves 4in(10.2cm) long.  (Cross by Del
James; Harold E. Greer, reg. 1979)    See color illus. ARS Q 33:4
(Fall 1979): p. 239; K p. 116.

cl.
                -<u>praevernum</u> (1/2)
Springtime-
                -<u>arboreum</u> ssp. <u>cinnamomeum</u> var. <u>album</u> (1/2)

White flowers shading to pale greenish yellow in throat.  (Gill,
1945)  A. M. (RHS) 1945

cl.                            -<u>wardii</u> var. <u>wardii</u> (1/4)
            -Mrs. Lammot Copeland-        -<u>souliei</u> ? (1/8)
Spun Gold-                    -Virginia Scott-
            -                              -unknown (5/8)
            -late flowering deep yellow--unknown

3ft(.9m)          0°F(-18°C)          ML

Amber yellow flowers with small red dorsal spots, openly campanulate, 3in(7.6cm) across; lax trusses of 12-14.  Upright plant,
almost as broad as tall, well-branched; matte green leaves held
2 years.  (H. L. Larson cross; Woodrow Robertson, intro. 1980;
reg. 1983)

cl.
        -<u>fortunei</u> ssp. <u>fortunei</u> (1/2)
Stacia-                -<u>catawbiense</u> (1/4)
        -Everestianum-
                    -unknown (1/4)

5ft(1.5m)          -15°F(-26°C)          ML

Plant wider than tall; spinach green leaves, 5.8in(15cm) long,
held 2 years.  Fragrant flowers to 3.25in(8.3cm) across, cobalt
violet, fading in throat, faint blotch of uranium green; spherical trusses of 12.  (J. Druecker cross; reg. by E. German, 1976)

cl.                        -<u>campylocarpum</u> ssp. <u>campylocarpum</u>
                -Letty Edwards-        Elatum Group (1/4)
Stadt Westerstede-        -<u>fortunei</u> ssp. <u>fortunei</u> (1/4)
  (The City...)    -            -<u>wardii</u> var. <u>wardii</u> (1/4)
                -unnamed hybrid-
                        -unknown (1/4)

Trusses of 10-13 flowers, 5-lobed, light yellow with a divided
red basal blotch.  (Cross by J. Bohlje; Walter Schmalscheidt,
reg. 1984)

---
            -<u>griffithianum</u> (1/2)
Standishi-
            -unknown (1/2)

Flowers white with red spots.  (Veitch)

---
            -<u>maximum</u> (1/2)            -<u>catawbiense</u> (1/8)
Standishii-            -unnamed hybrid-
            -Altaclarense-            -<u>ponticum</u> (1/8)
                    -<u>arboreum</u> ssp. <u>arboreum</u> (1/4)

Violet crimson flowers, black spots.  (Standish & Noble, 1850)

                    -<u>elliottii</u> (1/4)
cl.        -Fusilier-
            -            -<u>griersonianum</u> (1/4)
Stanford-                -<u>fortunei</u> ssp. <u>discolor</u> (1/8)
            -        -Lady        -
            -Jalisco-Bessborough-<u>campylocarpum</u> ssp. <u>campylocarpum</u>
                Goshawk-                Elatum Group (1/8)
                -            -<u>dichroanthum</u> ssp. <u>dichroanthum</u> (1/8)
                    -Dido-
                        -<u>decorum</u> (1/8)

Crossed and intro. by Maj. A. E. Hardy.    A. M. (RHS) 1975

cl.
        -<u>falconeri</u> ssp. <u>falconeri</u> (1/2)
Stanley-
        -<u>sinogrande</u> (1/2)

Trusses hold 27-28 flowers, very light sap green fading greenish
white, with basal blotch of magenta on upper lobes.  (Edmund de
Rothschild, reg. 1982)

cl.

Stanley Davies     Parentage unknown

4ft(1.2m)          -10°F(-23°C)          ML          4/4

A parent of Britannia and many other hybrids.  Fiery red flowers with black markings, held in compact trusses of 16. Broadly
dome-shaped, densely leafy plant, slow-growing; midrib markedly
sunken.   (Davies of Ormskirk, England, 1890)


cl.
                -glaucophyllum var. glaucophyllum (1/2)
Stanley Perry-
                -trichostomum Radinum Group (1/2)

Frilled flowers, 5-lobed, .75in(2cm) across, old rose color,
held in trusses of 10.  (Thacker of Warwickshire, reg. 1962)


cl.
                -yakushimanum ssp. yakushimanum (1/2)
Stanley Rivlin-          -elliottii (1/4)
                -Royal Blood-
                        -Rubens--unknown (1/4)

3ft(.9m)          0°F(-18°C)          M

Buds cardinal red opening paler red, the outside heavily flushed
spirea red, the interior spotted crimson.   (A. F. George, Hydon
Nurseries, 1972)


cl.
                -grandiflora--form of christianae (1/2)
Stanton's Glory-
                -aurigeranum (1/2)

Vireya hybrid. Trusses of 8-10 flowers of marigold orange,  reverse light tangerine orange. (Don Stanton cross; Graham Snell,
reg. 1982)


cl.
        -fortunei ssp. fortunei (1/2)
Stanway-     -Lady     -fortunei ssp. discolor (1/8)
        -       -Bessborough-
        -Jalisco-         -campylocarpum ssp. campylocarpum
        -                               Elatum Group (1/8)
        -        -dichroanthum ssp. dichroanthum (1/8)
                -Dido-
                        -decorum (1/8)

Fred Wynniatt g.  Possibly the best of the group.  Flowers of
rich dark yellow, the throat deeper yellow, also the stripe down
each lobe.   Named for the birthplace of the late Fred Wynniatt.
(Rothschild)   A. M. (RHS) 1971


---                           -brookeanum var. gracile (1/4)
                -Crown Princess-           -jasminiflorum (1/8)
Star of India- of Germany  -Princess Royal-
             -                            -javanicum
              -javanicum (5/8)

Vireya hybrid.  Flowers yellow to orange.  (Veitch, 1891)


cl.
        -jasminiflorum (1/2)
Star Posy-
        -unknown (1/2)

Vireya hybrid.  Trusses of 10-15 flowers, salverform, light but
vivid pink.  Glossy leaves; small, dark, scattered scales when
young.   (T. Lelliott cross; raiser E. B. Perrott; O. Blumhardt,
reg. 1984)


            -yakushimanum ssp. yakushimanum (1/2)
        -unnamed-              -griffithianum (3/16)
cl. - hybrid-        -Queen
     -         -Britannia-Wilhelmina-unknown (3/16)
Star -          -
Shine-              -Stanley Davies--unknown
     -          -griffithianum
     -      -Loderi-
     -unnamed-      -fortunei ssp. fortunei (1/8)
        hybrid-yakushimanum ssp. yakushimanum

Plant of compact habit; flowers of amaranth rose. (John Waterer, Sons & Crisp, reg. 1971)  A. M. (RHS) 1977


cl.
        -davidsonianum (Exbury form) (1/2)
Star Trek-
        -Ruth Lyons--davidsonianum (1/2)

3ft(.9m)          -15°F(-26°C)          E

Crossing 2 forms of the same species creates another form, not
a hybrid.   Terminal spherical inflorescence, 4in(10.2cm) wide,
of 3-5 trusses, each with 3-4 purple flowers.  Plant as broad as
tall; glossy, dark green leaves, 2.25in(5.7cm) long.  (Arthur A.
Childers, reg. 1977)


cl.
        -Moser's Maroon--unknown (3/4)
Starburst*              -ponticum (1/4)
        -Purple Splendour-
                         -unknown

6ft(1.8m)          5°F(-15°C)          L          4/3

Flower has gold blotch surrounded by white on a background of
deep reddish purple.  Plant upright, open; deep green leaves on
red petioles.  (H. E. Greer, intro. 1982)


cl.     -fortunei ssp. discolor (1/2)
        -              -griffithianum (1/8)
Starcross-       -Loderi-
        -Lodauric-       -fortunei ssp. fortunei (1/8)
             Iceberg-auriculatum (1/4)

Buds soft rose, flowers blush pink with slight bronze markings
in throat,  fragrant, in large compact trusses.   See also Gipsy
Moth, Northern Star, Southern Cross.  (A. F. George, Hydon Nurseries, reg. 1968)


cl.
        -griffithianum (1/2)
Starfish-
        -unknown (1/2)

Flowers bright pink with dark spotting, starfish-shaped.  Plant
of medium height.  (Waterer, Sons & Crisp, before 1922)


                -fortunei ssp. discolor
cl.     -Ladybird-
        -Diva-      -Corona--unknown (1/8)
Starlet-  -griersonianum (1/4)
        -williamsianum (1/2)

2.5ft(.75m)          5°F(-15°C)          EM          4/4

Elliptic-ovate leaves 3in(7.6cm) long.  Flowers funnel-rotate,
4in(10.2cm) wide,  rose madder to crimson; loose trusses of 5-6.
(H. Lem cross; C. P. Fawcett, reg. 1964)     P. A. (ARS) 1963


cl.
        -minus var. minus Carolinianum Group (1/2)
Starlight-
        -leucaspis (1/2)

White flowers.  (Donald Hardgrove, 1953)

cl.                        -catawbiense (red form) (1/4)
            -unnamed hybrid-
Starry Eyed-              -fortunei ssp. discolor (1/4)
        -        -griersonianum (1/4)
            -Azor-
                    -fortunei ssp. discolor (1/4)

Pink flowers with a red blotch.  (Hardgrove, 1955)

cl.


Starry Night      See Gletschernacht


cl.
            -catawbiense (1/2)
Stella-
        -unknown (1/2)

Synonym Stella Waterer.  Flowers lilac rose, spotted dark chocolate.  (A. Waterer)  F. C. C. (RHS) 1865


cl.
            -forrestii ssp. forrestii Repens Group (1/2)
Stephanie-
        -unknown (1/2)

2ft(.6m)        0°F(-18°C)          M          3/3

Compact, small-leaved plant.  Flowers light red.  (W. Whitney)


cl.
                -williamsianum (1/2)
Stephanie Wilson-
                -unknown (1/2)

Trusses hold 9 flowers, pale seafoam green fading to very light
neyron rose. (Edmund de Rothschild, intro. 1979; reg. 1982)


                        -griffithianum (1/8)
cl.             -Queen   -
        -Britannia-Wilhelmina-unknown (3/8)
Stephen-               -
 Clarke-       -Stanley Davies--unknown
     -         -fortunei ssp. discolor (1/4)
    -Autumn-    -dichroanthum ssp. dichroanthum (1/8)
     Gold -Fabia-
                -griersonianum (1/8)

5.5ft(1.65m)        10°F(-12°C)          M

Carmine buds open to coral pink flowers with edges neyron rose,
fading empire yellow; trusses hold about 11 flowers. Plant upright, wider than tall; leaves held 3 years. (Mrs. J. H. Clarke;
intro. by Clarke Nurseries; S. Clarke, reg. 1981)


cl.             -haematodes ssp. haematodes (3/4)
        -May Day-
Steven Foster-   -griersonianum (1/4)
        -haematodes ssp. haematodes

Rounded plant, 2ft(.6m) high in 10 years; leaves dark green and
glossy, some indumentum. Blood red, campanulate flowers to 2.75
in(7cm) wide, in loose trusses. Early. See also Harry von Tilzer, Paul Detlefsen. (C. S. Seabrook, reg. 1967)

cl.
                Parentage includes catawbiense, maximum, and
Stewart Manville    perhaps unknown hybrids

7ft(2.1m)        -15°F(-26°C)          ML

Upright, spreading plant; leaves medium olive green, held 2
years. Flowers medium purplish red spotted in deep greenish
yellow, 3in(7.6cm) wide, 5-lobed, floriferous; spherical trusses
of 14. (W. David Smith, reg. 1980)

cl.
            -unknown garden hybrid (1/2)
Stockholm*
            -williamsianum (1/2)

Flowers of delicate pink.  (Joh. Bruns, about 1942)


cl.
            -Catalgla--catawbiense var. album Glass (1/2)
Stockholm-
        -decorum (1/2)

5ft(1.5m)        -20°F(-29°C)          E          3/3

Flowers widely funnel-campanulate, 2.1in(5.4cm) wide, white with
2 dorsal rays of strong greenish yellow; spherical trusses of
14. Floriferous. Plant wider than high; glossy, dark green elliptic leaves, held 2 years. (David G. Leach; reg. 1974)


cl.

Stonehurst--form of lanigerum (silvaticum)

Flowers a light shade of cherry red, campanulate, 2.5in(6.4cm)
wide; tight globular trusses of 35. Plant typical of species;
thick indumentum beneath leaves. (R. Strauss) A. M. (RHS) 1961

cl.
            -griersonianum (1/2)
Stoplight-       -arboreum ssp. arboreum (1/4)
        -Cornubia-       -thomsonii ssp. thomsonii (1/8)
            -Shilsonii-
                    -barbatum (1/8)

6ft(1.8m)        10°F(-12°C)          EM          4/2

Oblong leaves 5.5in(15.5cm) long. Flowers geranium red, faintly
blotched, funnel-campanulate, 3.5in(8.9cm) wide, 5-lobed; about
13 per truss. (Rudolph Henny, reg. 1958)  P. A. (ARS) 1951


cl.
            -griffithianum (1/2)
Strategist-       -catawbiense (1/4)
        -John Waterer-
                -unknown (1/4)

Broad, bright green leaves. Well-shaped blush pink flowers,
deep rose at edges, pale yellow spots in throat, in tall conical
trusses. A large, late-blooming plant. (J. Waterer)


                        -caucasicum (3/16)
cl.             -Jacksonii-       -caucasicum
            -Goldsworth-       -Nobleanum-
            - Yellow -                -arboreum ssp.
Strawberries-        -                    arboreum (1/8)
and Cream  -         -campylocarpum ssp. campylocarpum (1/4)
            -        -brachycarpum ssp. brachycarpum (1/4)
            -unnamed-                -catawbiense
             hybrid-       -Atrosanguineum-       (1/16)
            -Mrs. P. Den-               -unknown (1/8)
             Ouden       -arboreum ssp.
                    -Doncaster-   arboreum
                        -unknown

3ft(.9m)        -15°F(-26°C)          ML

Flowers of neyron rose fading to empire yellow, 1.75in(4.5cm)
wide, 5-lobed; spherical trusses of 12. Plant rounded, nearly
as broad as tall; yellowish green leaves held 2 years. See also
Cary's Cream, Cary's Yellow, Josephine V. Cary, William P. Cary.
(Edward A. Cary, reg. 1980)


cl.
            -aberconwayi (1/2)
Streatley-
        -yakushimanum ssp. yakushimanum (1/2)

'Seta' by Aberconway
Photo by Greer

'Sham's Ruby' by Shammarello
Photo by Leach

'Shanghai' by Leach
Photo by Leach

'Shooting Star' by Greer
Photo by Greer

'Show Boat' by Phetteplace
Photo by Greer

'Shrimp Girl' by J. Waterer & Crisp
Photo by Greer

'Siam' by Leach
Photo by Leach

'Sierra Sunrise' by Lofthouse
Photo by Greer

'Sir Frederick Moore' by Rothschild
Photo by Greer

'Skipper' by James/Thompson
Photo by Greer

'Small Wonder' by Leach
Photo by Leach

'Smokey #9' by Lem
Photo by Greer

'Snipe' by Cox
Photo by Greer

'Snow Cap' by Whitney/Sather
Photo by Greer

'Snow Queen' by E. Loder
Photo by Greer

'Snow Shimmer' by Wister
Photo by West

'Solent Swan' by Rothschild
Photo by Greer

'Solitude' by R. Henny
Photo by Greer

'Sonata' by Reuthe
Photo by Greer

'Souvenir de J. H. Mangles' by unknown, F.C.C. 1888
Photo by Greer

'Spanish Glory' by Sumner
Photo by Greer

'Sparkling Jewel' by Wister
Photo by West

'Spellbinder' by Leach
Photo by Leach

'Spinulosum' by RBG, Kew Gardens
Photo by Greer

White flowers shaded with rosy pink, some light red spotting and
red margins, 3.5in(8.9cm) wide; lax trusses of 9.  Magenta buds
contrast nicely with the open flowers.   (Crown Estate, Windsor)
A. M. (RHS) 1964


```
cl.
         -edgeworthii (1/2)
Suave-
         -bullatum (sic) (1/2)
```

An old hybrid still in cultivation.  Flowers campanulate, pure
white; reverse pink shading to white.  (Emil Liebig, 1883)


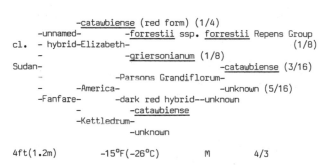

```
                     -griffithianum (1/8)
cl.         -Loderi-
      -Albatross-    -fortunei ssp. fortunei (1/8)
Success-      -fortunei ssp. discolor (1/4)
      -         -elliottii (1/4)
        -Fusilier-
                    -griersonianum (1/4)
```

Flowers are colored rose Bengal.  See also Cup Day.  (Karel Van
de Ven, 1965)


```
                -catawbiense (red form) (1/4)
        -unnamed-          -forrestii ssp. forrestii Repens Group
cl.  - hybrid-Elizabeth-                            (1/8)
Sudan-    -              -griersonianum (1/8)
     -                          -catawbiense (3/16)
     -          -Parsons Grandiflorum-
     -   -America-          -unknown (5/16)
     -Fanfare-     -dark red hybrid--unknown
     -             -catawbiense
        -Kettledrum-
                   -unknown
```

4ft(1.2m)        -15°F(-26°C)         M         4/3

Flowers of cherry red blotched with very light mimosa yellow, 5-
lobed, openly funnel-shaped, 2.25in(5.7cm) wide; domed trusses
of 14.  Plant wider than tall, with decumbent branches; leaves
dark yellowish green, 4in(10.2cm) long, held 2 years.  (David G.
Leach, intro. 1976; reg. 1982)


```
cl.
                    -griffithianum (1/2)
Sue--Loderi King George (selfed)-
                    -fortunei ssp. fortunei (1/2)
```

6ft(1.8m)       0°F(-18°C)          M         4/4

A beautiful shrub with all the Loderi characteristics of plant
habit, foliage and flowers.  Very tall, narrow trusses of deep
pink, fragrant flowers, 5in(12.7cm) wide. (Del James, reg. 1958)


```
cl.         -yakushimanum ssp. yakushimanum (1/4)
       -Bambi-              -dichroanthum ssp. dichroanthum
Sue Cutten-   -Fabia Tangerine-                  (1/8)
       -                   -griersonianum (1/8)
        -arboreum ssp. arboreum (1/2)
```

Light scarlet flowers.  (R. Cutten, reg. 1981)


```
cl.
         -fortunei ssp. fortunei (1/2)
Sue Gordon-
          -unknown (1/2)
```

Trusses of 11 flowers, fuchsia purple.  (Lady Adam Gordon, reg.
1981)


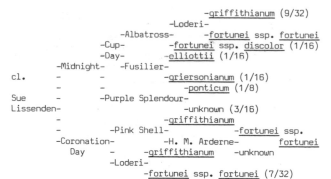

```
                          -griffithianum (9/32)
                 -Loderi-
          -Albatross-    -fortunei ssp. fortunei
       -Cup-        -fortunei ssp. discolor (1/16)
       -Day-        -elliottii (1/16)
cl.  -Midnight-  -Fusilier-
     -         -         -griersonianum (1/16)
     -         -              -ponticum (1/8)
Sue    -    -Purple Splendour-
Lissenden-            -unknown (3/16)
       -             -griffithianum
       -      -Pink Shell-         -fortunei ssp.
      -Coronation-       -H. M. Arderne-      fortunei
       Day   -     -griffithianum  -unknown
             -Loderi-
                -fortunei ssp. fortunei (7/32)
```

Scarlet flowers, in trusses of 19.  See also Max Marshland.  (K.
Van de Ven, 1981)


```
cl.
        -haematodes ssp. haematodes (1/2)
Suede-
       -bureavii (1/2)
```

Red flowers fading to deep rose, with brown speckles in throat.
(G. A. Judson, reg. 1973)


```
                                -fortunei ssp.
                      -Koster's-   fortunei (3/16)
               -unnamed- Choice -unknown (1/32)
cl.            - hybrid-    -griffithianum (5/32)
          -Palmer-    -Loderi-
     -unnamed-    -        -fortunei ssp. fortunei
     - hybrid-   -wardii var. wardii (croceum) (1/8)
Sugar and-       -griffithianum
 Spice*  -    -Loderi King George-
         -            -fortunei ssp. fortunei
         -       -catawbiense (1/4)
         -Madame Masson-
                 -ponticum (1/4)
```

5ft(1.5m)        -5°F(-21°C)           ML

Flowers creamy white, with a golden brown blotch on upper lobe.
Leaves deep waxy green.  See also Pawhuska.  (Tom & Emma Bowhan)


```
cl.
         -decorum (1/2)
Sugar Daddy-
          -unknown (1/2)
```

5ft(1.5m)        0°F(-18°C)           EM

Plant habit upright, stiff.  Flowers of amaranth rose, fading
white, with a green blotch in throat; compact, rounded trusses.
(John Waterer, Sons & Crisp, reg. 1971)


```
cl.

Sugar Pink      See next page
```
See next page


```
cl.         -campylocarpum ssp. campylocarpum (1/4)
        -Moonstone-
Sugar Plum-  -williamsianum (1/4)
        -       -wardii var. wardii (1/4)
        -Caroline Grace-
                 -unknown (1/4)
```

3ft(.9m)         0°F(-18°C)           EM        3/3

Deep pink flowers, widely funnel-shaped, to 3in(7.6cm) wide, in
lax trusses of 5-8.  Dense, small-leaved plant.  (Arthur Wright,
Sr. & Jr., 1964)

```
                                    -griffithianum (3/16)
                        -George Hardy-
            -Pink -                 -catawbiense (3/16)
        -Countess-Pearl-            -arboreum ssp. arboreum
        -of Derby-    -Broughtonii-                   (1/16)
    -Trude -         -              -unknown (3/16)
cl. -Webster-            -catawbiense
    -           -       -Cynthia-
Sugar-          -       -griffithianum
 Pink-      -Countess of Derby (as above)
    -           -fortunei ssp. fortunei (1/8)
    -       -Fawn-     -dichroanthum ssp. dichroanthum (1/16)
  -unnamed-     -Fabia-
    hybrid-            -griersonianum (3/16)
         -               -Pink Bell--unknown
           -Queen o' the May-
                        -griersonianum
```

6ft(1.8m)        -5°F(-21°C)          M        4/4

Flowers cotton candy pink with light brown spotting, 5in(12.7cm)
across, about 12 per truss, which may be 1ft(.3m) tall.  Smooth,
deep green leaves, 7.5in(19cm) long.  (H. E. Greer, 1979)    See
color illus. ARS Q 28:1 (Winter 1974): p. 34; HG p. 79; K p. 73.

cl.
        -sulfureum (1/2)
Sulfmeg-
        -megeratum (1/2)

Lepidote hybrid.  Flowers pale sulphur yellow,  in clusters of
2-3. (E. J. P. Magor cross, 1924; intro. 1940) A. M. (RHS) 1940

g.
        -souliei (1/2)
Sulphur Yellow-
            -campylocarpum ssp. campylocarpum (1/2)

J. C. Williams cross.  Exhibited by Lord Swaythling, 1939.

```
                            -catawbiense (1/8)
cl.         -Parsons Grandiflorum-
        -America-           -unknown (9/16)
Sumatra-    -dark red hybrid--unknown
        -               -forrestii ssp. forrestii Repens Group
        -Gertrud Schale-                        (1/4)
        -               -Michael-ponticum (1/16)
            -Prometheus-Waterer-
                    -           -unknown
                    -Monitor--unknown
```

2ft(.6m)        -15°F(-26°C)         M        4/5

Dome-shaped trusses of 8-10 cardinal red flowers, 2.6in(6.5cm)
across, openly funnel-shaped, 5 wavy lobes.  Plant over twice as
broad as high, with very dense foliage; leaves yellow green, 3in
(7.6cm) long.  (David G. Leach, intro. 1973; reg. 1982) See col-
or illus. ARS J 36:1 (Winter 1982): p. 3; LW pl. 98.

cl.
        -maximum (1/2)
Summer Rose-         -Moser's Maroon--unknown (1/4)
        -Romany Chai-
                -griersonianum (1/4)

4ft(1.2m)        -10°F(-23°C)          L

Flowers deep pink, with dark reddish purple spotting  on dorsal
lobes, widely funnel-shaped, 2.5in(6.4cm) wide, 5-lobed; spherical trusses of 8-12.  Upright plant, wider than tall; narrowly
elliptic leaves, thin tan indumentum; new growth silver grey tomentose, new stems red.  (Dr. R. Ticknor, cross; raised & intro.
by E. Mezitt, Weston Nurseries, 1973; reg. 1980)

cl.                      -auriculatum (1/4)
        -unnamed hybrid-
Summer Sequel-          -fortunei ssp. discolor (1/4)
        -Catalgla--catawbiense var. album Glass (1/2)

Trusses of 16-17 flowers, pale purplish pink, lobe edges and reverse darker pink, dorsal blotch and spotting of strong greenish
yellow; corolla 3.3in(8.5cm) across.  Very late season.  Height
at 22 years, 7ft(2.1m). (David G. Leach, reg. 1984)

cl.
        -maximum (1/2)
Summer Snow-            -ungernii (1/4)
        -unnamed hybrid (selfed)-
                    -auriculatum (1/4)

6ft(1.8m)        -15°F(-26°C)          VL        4/2

White flowers  with small rayed dorsal blotch of strong greenish
yellow, funnel-campanulate, 4in(10.2cm) across; dome-shaped,
rather loose trusses of about 11.  Flowers not obscured by new
growth.  Vigorous plant, as wide as tall; leaves variable, to
9.5in(24cm) long.  (David G. Leach, reg. 1970)

```
        -maximum (1/4)
    -Summer-             -ungernii (1/8)
cl. - Snow -unnamed hybrid-
    -               -auriculatum (1/8)
Summer              -griffithianum (1/16)
Splendor-           -unnamed-
    -       -Mrs. Furnival- hybrid-unknown (1/8)
    -       -       -       -caucasicum
    -Party Pink-        -unnamed hybrid-    (1/16)
    -                   -unknown
        -catawbiense var. album (1/4)
```

Heather pink buds opening to orchid pink flowers, paler throat,
with bold dorsal blotch of orange buff, 3.25in(8.3cm) wide; 13
per truss.  Shrub 6.2ft(1.9m) in 16 years; elliptic leaves 5.75
in(14.6cm) long.  Very late season.  (David G. Leach, reg. 1984)

cl.
        -maximum (1/2)
Summer Summit-          -auriculatum (1/4)
        -unnamed hybrid-
                -fortunei ssp. discolor (1/4)

Phlox pink buds, opening to white flowers flushed orchid pink at
base, and exterior, aging white, with dark olive yellow dorsal
spotting, 5-lobed, 2.5in(6.4cm) across.  Shrub 11.3ft(3.4m) in
26 years; elliptic leaves 6in(15.2cm) long.  Very late season.
(David G. Leach, reg. 1985)

g.
        -wardii var. wardii (1/2)
Summer's Dawn-
        -garden hybrid--unknown (1/2)

Yellow flowers.  (Collingwood Ingram; exhibited 1936)

cl.

Summertime--form of maximum

6ft(1.8m)        -25°F(-32°C)          L

White flowers suffused roseine purple with upper throat yellow
green, held in trusses of 22-24.  Tall-growing plant with leaves
5.5in(14cm) long, dark green.  (Collector unknown; exhibited by
the Crown Estate, Windsor) A. M. (RHS) 1974

cl.
        -yakushimanum ssp. yakushimanum (Exbury form) (1/2)
Summit Gold-
        -aureum var. aureum (chrysanthum) (1/2)

1ft(.3m)        -25°F(-32°C)          EM

Dwarf plant, 3 times broader than high, branching well. Flowers
primrose yellow, 2.25in(5.7cm) wide in dome-shaped trusses of 9.
See also Astroglow, Mini Gold, Gold Spread. (B. C. Potter, 1981)

```
                  -campylocarpum ssp. caloxanthum (1/4)
cl.      -Solarium-                    -caucasicum (1/16)
-           -          -Dr. Stocker-
Sun Chariot-    -Damaris-              -griffithianum (1/16)
                      -campylocarpum ssp. campylocarpum
          -wardii var. wardii (1/2)                   (1/8)
```

Yellow flowers. (Raised by L. de Rothschild; intro. by Waterer,
Sons & Crisp; exhibited by Sunningdale Nurseries, 1955)

```
cl.                -Nereid
         -King of Jordan-Tally Ho
Sun Flame-             -fortunei ssp. discolor
         -Whitney hybrid--unknown
```

2.5ft(.75m)         -5°F(-21°C)          M

(Exact combination of the parentage of King of Jordan is uncertain.) Trusses of 9-13 fragrant flowers, 7-lobed, throat mimosa
yellow shading to orient pink at edges. Semi-dwarf plant, much
broader than tall with leaves 5.5in(14cm) long, yellow green,
retained 3 years. (Mrs. Bernice I. Jordan, 1977)

```
cl.       -Parker's Pink--unknown (1/2)
Sun Gleam-                    -decorum (1/8)
             -unnamed hybrid-
          -Clara  -            -fortunei ssp. discolor (1/8)
          Raustein-            -wardii var. wardii (1/16)
          -       -unnamed-
          -unnamed- hybrid-dichroanthum ssp.
             hybrid-          dichroanthum (1/16)
                 -fortunei ssp. fortunei (1/8)
```

2ft(.6m)          -5°F(-21°C)          M

Cardinal red buds open to white flowers, light red edges with
exterior rose Bengal, 2in(5cm) wide, 5 wavy-frilled lobes; globular trusses of 12. Plant rounded, as broad as tall; leaves 4.5
in(11.5cm) long, held 3 years. (Alfred A. Raustein, reg. 1983)

```
cl.
     -wardii var. wardii ? (1/2)
Sundance*
     -unknown (1/2)
```

4ft(1.2m)         -5°F(-21°C)          ML         4/3

Plant of compact habit; leaves of deep bluish green. Buttery
yellow flowers. (Raiser, William Whitney)

```
cl.
       -augustinii ssp. augustinii (Tower Court form) (1/2)
Sundari-
       -trichanthum (Tower Court form) (1/2)
```

5ft(1.5m)         -5°F(-21°C)          EM

Plant upright, well-branched; narrow elliptic leaves 3.35in(8.3
cm) long, scaly beneath, held 2-3 years. Frilled flowers, widely funnel-shaped, zygomorphic, 2.5in(6.4cm) wide, of violet with
pale center, light yellow green spots in throat; trusses of 4-5.
Floriferous. (Howard A. Short, reg. 1977)

```
              -griersonianum (1/2)
g.   -Dorinthia-          -haematodes ssp. haematodes (1/8)
-           -Hiraethlyn-
Sundor-                   -griffithianum (3/8)
-        -griffithianum
         -Sunrise-
              -griersonianum
```

Flowers scarlet salmon. (Lord Aberconway, 1946)

```
g.                     -griffithianum (3/4)
      -Loderi Pink Diamond-
Sunkist-                  -fortunei ssp. fortunei (1/4)
      -griffithianum
```

White flowers, suffused pink. (Loder)

```
cl.
      -fortunei ssp. discolor (1/2)
Sunlit Snow-         -fortunei ssp. discolor (1/4)
      -unnamed hybrid-
                 -unknown (1/4)
```

6ft(1.8m)         0°F(-18°C)          L

Very fragrant, funnel-shaped flowers, 4in(10.2cm) wide, of light
purplish pink to blush, to bright yellow green in throat; about
9 in flattened trusses. Plant broader than tall; leaves yellowish green, held 2 years. (Scott Horticultural Foundation,
Swarthmore College, Dr. & Mrs. J. C. Wister; reg. 1981)

```
cl.
Sunningdale*--form of hippophaeoides var. hippophaeoides
```

3ft(.9m)          -25°F(-32°C)          E          4/3

Synonyn Haba Shan form. Deep lavender blue flowers .75in(2cm)
wide, 11 per dense truss, opening slowly over a long period.
Small, grey green, aromatic leaves. (Collector, Forrest; raised
by Sunningdale Nurseries) A. G. M. (RHS) 1925

```
cl.               -dichroanthum ssp. dichroanthum (3/8)
          -Golden Horn-
Sunningdale-        -elliottii (1/4)
  Apricot  -            -fortunei ssp. discolor (1/4)
          -Margaret Dunn-   -dichroanthum ssp. dichroanthum
               -Fabia-
                   -griersonianum (1/8)
```

Apricot-colored flowers. (Sunningdale Nurseries; exhibited by
Hamilton Smith, 1955)

```
cl.
Sunningdale Blue*--form of impeditum
```

1ft(.3m)          -15°F(-26°C)          EM          4/4

Plant very dwarf, with flowers of pale violet blue. Young foliage blue green. (Sunningdale Nurseries)

```
cl.
    -christianae (1/2)
Sunny-
    -macgregoriae (1/2)
```

Vireya hybrid. Flowers tubular funnel-shaped, vivid yellow and
strong yellowish pink, in trusses of 8-15. Shrub to 7ft(2.1m)
at registration. (Origin unknown; reg. by G. L. S. Snell, 1985)

```
g. & cl.
      -griffithianum (1/2)
Sunrise-
      -griersonianum (1/2)
```

6ft(1.8m)         5°F(-15°C)          ML          4/2

Large flowers opening deep pink, fading with age to white except
margins and base. Upright plant as broad as tall; medium green,
dull leaves 6.5in(16.5cm) long. More than one form distributed.
(Lord Aberconway, intro. 1933) F. C. C. (RHS) 1942

```
cl.
Sunrise-Sunset     See Sunup-Sundown
```

```
cl.             -griffithianum (1/2)
      -Queen Wilhelmina-
Sunset-            -unknown (1/4)
-        -griffithianum
      -Loderi-
           -fortunei ssp. fortunei (1/4)
```

Flowers of pale yellow shaded pink, or white flushed pale pink.
(G. Loder; exhibited 1931) A. M. (RHS) 1931

```
                         -griffithianum (1/8)
cl.            -Queen Wilhelmina-
       -Earl of -                      -unknown (3/8)
Sunset- Athlone-Stanley Davies--unknown
     -         -dichroanthum ssp. dichroanthum (1/4)
       -Fabia-
                  -griersonianum (1/4)
```

Flowers red, with a white stripe on each lobe of calyx.  Decumbent branching  on a very straggly plant.  (Rudolph Henny cross, 1944)

```
cl.
            -griersonianum (1/2)
Sunset Lake-             -arboreum ssp. arboreum (1/4)
         -unnamed hybrid-
                         -decorum (1/4)
```

Trusses of 8-12 flowers, light crimson to crimson lake with a deep red throat.  See also Sunset Queen.  (Oswald Blumhardt, New Zealand, 1974)

```
cl.
            -griersonianum (1/2)
Sunset Queen-           -arboreum ssp. arboreum (1/4)
         -unnamed hybrid-
                         -decorum (1/4)
```

Large, loose trusses of rich coral rose flowers.  See also  Sunset Lake.  (Oswald Blumhardt, 1974)

```
cl.              -catawbiense (1/4)
        -Pink Twins-
Sunset Yates-          -haematodes ssp. haematodes (1/4)
       -                         -griffithianum (1/4)
          -Leah Yates--Mars (selfed)-
                                     -unknown (1/4)
```

3.5ft(1m)        -10°F(-23°C)        ML

Flowers of heavy substance, 3.5in(8.3cm) wide, neyron rose shading paler, large medium red blotch, 5-lobed; spherical trusses of 15.  Floriferous.  Plant broader than high, arching branches; dark grass green leaves held 2 years.  (Henry Yates; intro. & reg. by Mrs. Henry Yates, 1977)

```
cl.
          -neriiflorum ssp. neriiflorum Euchaites Group (1/2)
Sunshine-
          -beanianum (1/2)
```

Bright dark red flowers.  (Gen. Harrison, 1955)

```
                    -campylocarpum ssp. campylocarpum
              -Ole Olson-              Elatum Group (1/4)
cl.    -Alice  -      -fortunei ssp. discolor (1/4)
       -Franklin-         -griffithianum (1/8)
Sunspray-      -Loderi King George-
       -                         -fortunei ssp. fortunei
       -  -wardii var. wardii (1/4)            (1/8)
       -Crest-            -fortunei ssp. discolor
              -Lady Bessborough-
                         -campylocarpum ssp. campylocarpum
                                          Elatum Group
```

6ft(1.8m)        -5°F(-21°C)        M

Flowers funnel-campanulate, 5in(12.7cm) wide with 6 wavy lobes, pale primrose yellow edges, deeper at center and throat; trusses flat, of 9-10.  Plant upright, moderately branched, taller than broad; dark green leaves held 2 years. (Crossed & raised by W. J. Swenson; intro. by H. E. Greer; reg. 1979)  See color illus. HG p. 98.

```
cl.
Sunte Rock--form of praestans (coryphaeum)  (R 59480)
```

5ft(1.5m)        -5°F(-21°C)        EM        3/4

Flowers of 6-8 lobes, 2.5in(6.4cm) across, of transparent white streaked rose red, upper throat blotched ruby red. Oblanceolate leaves to 12in(30.5cm) long, plastered silvery brown indumentum beneath.  (Collected by Rock; raiser Geoffrey Gorer; reg. 1970) P. C. (RHS) 1969

```
cl.
                    -sutchuenense (1/2)
Sunte Rose--form of Geraldii-
                    -praevernum (1/2)
```

Flowers light rhodamine purple with reddish purple blotch and spotting in upper throat.  (Geoffrey Gorer) A. M. (RHS) 1971

```
                -yakushimanum ssp. yakushimanum (1/4)
cl.    -unnamed-    -dichroanthum ssp. dichroanthum (3/16)
       - hybrid-Fabia-
Sunup- -                -griersonianum (3/16)
  Sundown-               -dichroanthum ssp. dichroanthum
       -           -Fabia-
       -    -unnamed-    -griersonianum
     -unnamed- hybrid-bureavii (1/8)
       hybrid-   -wardii var. wardii (1/8)
           -Crest-        -fortunei ssp. discolor (1/16)
               -Lady -
            Bessborough-campylocarpum ssp. campylo-
                          carpum Elatum Group (1/16)
```

2.5ft(.75m)        0°F(-18°C)        E

Buds blood red open to flowers empire rose fading to shell pink in trusses 6in(15cm) across, of 8-10 flowers. Semi-dwarf plant, compact, spreading much broader than tall with glossy yellow green leaves, retained 3 years. (John G. Lofthouse cross, 1975; reg. 1982)

```
                           -griffithianum
                 -Queen Wilhelmina-      (1/16)
g.           -Britannia-          -unknown (3/16)
    -Linswegeanum-      -Stanley Davies--unknown
Suomi-      -forrestii ssp. forrestii Repens Group (1/4)
    -Metternianus--japonicum var. japonicum (metternichii) (1/2)
```

Red flowers.  (Dietrich Hobbie, 1953)

```
cl.          -ciliicalyx (1/4)
       -Else Frye-
Super Jay-    -unknown (1/4)
       - johnstoneanum (1/2)
```

2.5ft(.75m)        15°F(-10°C)        E

Plant broader than tall; medium grass green leaves 2.75in(7cm) long, scaly beneath. Flowers fragrant, 3.5in(8.9cm) wide, white with orange dorsal blotch; small trusses of 2-5. (Cross by Richard Anderson; reg. by H. J. Braafladt, 1979)

```
cl.
          -yakushimanum ssp. yakushimanum (1/2)
Super Star-    -arboreum ssp. arboreum (1/4)
       -Doncaster-
              -unknown (1/4)
```

Pink-flowered hybrid, from West Germany.  (Joh. Bruns, 1972)

```
                    -dichroanthum ssp. dichroanthum (1/8)
            -Goldsworth-
cl.    -Hotei- Orange  -fortunei ssp. discolor (1/8)
       -     -           -souliei (1/8)
Supergold-  -unnamed hybrid-
       -                 -wardii var. wardii (1/8)
       -     -lacteum (1/4)
       -Joanita-
              -campylocarpum ssp. caloxanthum (1/4)
```

5ft(1.5m)          0°F(-18°C)          EM

Spherical trusses of 15 flowers, 5-6-lobed, medium orange yellow
with small red orange blotch, 2.5in(6.4cm) across; floriferous.
Plant upright, with stiff branches, about as wide as tall; deep
green, glossy, convex leaves, 3.5in(8.9cm) long. (John G. Lofthouse cross, 1974; reg. 1983)

g.
        -falconeri ssp. falconeri (1/2)
Surprise-
        -thomsonii ssp. thomsonii (1/2)

Mauve flowers with a black spot.  See also Faltho, J. E. Harris.
(J. Waterer, before 1867; exhibited by Loder, 1937)

cl.          -unnamed-facetum (eriogynum)
             - hybrid-       -dichroanthum ssp. dichroanthum (1/8)
Surrey Heath-      -Fabia-
             -              -griersonianum (1/8)
             -       -yakushimanum ssp. yakushimanum (1/4)
             -unnamed-      -Queen     -griffithianum (1/16)
             hybrid-       -Wilhelmina-
                   -Britannia-          -unknown (3/16)
                         -Stanley Davies--unknown

3ft(.9m)          0°F(-18°C)          M

Flowers of rose pink with lighter edging, on a conpact plant.
See also Venetian Chimes.  (J. Waterer, Sons & Crisp, reg. 1975)

cl.
        -campanulatum ssp. campanulatum (1/2)
Susan-
        -fortunei ssp. fortunei (1/2)

6ft(1.8m)          -5°F(-21°C)          M          4/4

Large, bushy plant with excellent foliage--glossy, dark green,
flat leaves 5.5in(14cm) long, with a little brown indumentum beneath.  Rosy lavender buds opening cool lilac blue, paler near
center and spotted maroon on upper lobe, in large loose trusses.
(Raiser J. C. Williams, Caerhays Castle; intro. W. C. Slocock)
A. M (RHS) 1930;   A. M. (Wisley Trials) 1948;   F. C. C. (RHS)
1954    See color illus. F. p. 66; VV p. 49.

cl.
        -maximum (1/2)
Susan Kay-
        -wardii var. wardii (1/2)

1.5ft(.45m)          -10°F(-23°C)          L

Plant compact, twice as wide as tall; leaves glossy, yellowish
green, held 2 years.  Flowers primrose yellow with light yellow
green dorsal blotch and spotting, 1.75(4.5cm) across; rather lax
dome-shaped trusses of 15. (William M. Fetterhoff,  reg. 1981)

cl.

Susan   -Loder's White, q.v. for 2 possible parentage diagrams.
Lonsdale-
        -Kew Pearl--unknown

White flowers tinged light spirea red heavily spotted on all
lobes with Indian lake.  See also Warburton.  (H. D. Rose, 1973)

cl.
        -haematodes ssp. haematodes (1/2)
Sussex Bonfire-       -thomsonii ssp. thomsonii (1/4)
        -Cornish Cross-
                -griffithianum (1/4)

Deep blood red flowers with darker calyx.  Matures at 5ft(1.5m).
(Loder of Leonardslee) A. M. (RHS) 1934

g.
        -barbatum (1/2)
Sutchbarb-
        -sutchuenense (1/2)

An introduction by E. J. P. Magor, 1936.

cl.          -catawbiense (1/4)
        -Camich-          -ponticum (1/8)
        -       -Michael Waterer-
Suzy Bell-          -unknown (1/8)
        -       -forrestii ssp. forrestii Repens Group (1/4)
        -Elizabeth-
                -griersonianum (1/4)

1.5ft(.45m)          -5°F(-21°C)          M

Synonym Tinker Bell.  Flowers openly campanulate, 2in(5cm) wide,
of Tyrian purple without markings; lax trusses of 7-9.  Floriferous.  Plant broader than tall, arching branches; leaves to 3.5
in(8.9cm), held 2 years; new growth bronze, tan tomentum. (J. B.
Gable & H. Yates; reg. by Mrs. Maletta Yates, 1980)

cl.          -souliei (1/2)
Swallowfield-
        -yakushimanum ssp. yakushimanum (1/2)

Flowers white with light reddish purple blotch and spotting on
upper throat, openly campanulate, 3in(7.6cm) wide; trusses of 6-
7. Floriferous.  Plant vigorous, upright, compact, broader than
tall; leaves medium green, 3in(7.6cm) long. (Crown Estate, Windsor)  H. C. (Wisley Trials) 1981

cl.          -ponticum (1/4)
        -Purple Splendour-
Swamp Beauty-          -unknown (1/4)
        -          -griffithianum (1/4)
        -Loderi Superlative-
                -fortunei ssp. fortunei (1/4)

5ft(1.5m)          10°F(-12°C)          ML

Flowers openly funnel-shaped, 4in(10.2cm) wide, 5-6 wavy lobes,
roseine purple with maroon spotting, shading to white in throat,
and large maroon blotch.  Trusses 7in(17.8cm) by 7in., ball-
shaped, with 18 flowers.  Plant upright, rounded, nearly as wide
as tall; leaves held 2 years.  (James A. Elliott, reg. 1983)

cl.          -auriculatum (1/2)
Swan Lake-       -griffithianum (1/4)
        -Godesberg-
                -unknown (1/4)

Blanc-mange g.  Plant tall, and equally as wide;  large matte
green leaves.  White flowers, similar to the clone Blanc-mange.
(Rothschild; intro. 1955)

cl.          -catawbiense var. album (3/4)
        -Belle Heller-          -catawbiense (1/8)
Swansdown-       -Madame Carvalho-
        -          -unknown (1/8)
        -catawbiense var. album

5ft(1.5m)          -25°F(-32°C)          ML          4/4

Plant compact, densely foliaged; elliptic leaves 5in(12.7in)
long, retained 2-3 years.  Flowers white, sometimes very pale
pink, with bold dorsal blotch of strong yellow, 3in(7.6cm) wide,
open funnel-shaped; pyramidal trusses of 20.  See also Lodestar.
(David G. Leach, 1966) C. A. (ARS) 1982; A. E. (ARS) 1983  See
color illus. K. p. 144.

cl.

Sweet Bay--form of trichostomum Radinum Group

Trusses of 25 flowers, deeply lobed, color Tyrian rose suffused
white, giving a pleasing pink effect.  Leaves 1.5in(3.8cm) long,
underside densely scaly. (Crown Estate, Windsor, reg. 1963)
A. M. (RHS) 1960

cl.
           -inconspicuum (1/2)
Sweet Mac-
           -macgregoriae (1/2)

Natural vireya hybrid from seed collected in the wild. Trusses
hold 7-9 flowers, azalea pink. (Raiser L. Searle, New Guinea;
intro. by Graham Snell, reg. 1982)

cl.
                -ponticum (1/2)
Sweet Simplicity-
                -unknown (1/2)

5ft(1.5m)          -5°F(-21°C)        ML         2/3

Bushy plant, almost round; leaves 5in(12.7cm) long, glossy dark
green, slightly waved. Deep pink buds opening to ruffled flowers
flushed white, edged in pink, 2.5in(6.4cm) wide; rounded, well-
filled trusses. (J. Waterer, Sons & Crisp, 1922)

cl.
           -Jan Dekens--unknown
Sweet Sixteen-
           -unknown

5ft(1.5m)          -5°F(-21°C)        M          4/3

Flowers of orchid pink with a deeper pink picotee edging, 4.75in
(12cm) wide; about 12 in spherical trusses. Plant covered with
large, curled leaves. Difficult to propagate. (W. Whitney
cross; intro. & reg. by Sather, 1976)    See color illus. ARS Q
30:4 (Fall 1976): p. 238.

cl.                -williamsianum (1/4)
           -Cowslip-
Sweetie Pie-       -wardii var. wardii (1/4)
           -forrestii ssp. forrestii Repens Group (1/2)

Loose trusses of 6 flowers, pale cream flushed pink. Compact,
dwarf plant with leaves 1in(2.5cm) long, rounded, dark green.
Early season. (Del W. James, reg. 1962)

cl.
     -yakushimanum ssp. yakushimanum (1/2)
Swen-      -griffithianum (1/4)
      -Mars-
           -unknown (1/4)

4ft(1.2m)          -10°F(-23°C)        M

Flowers of heavy substance, Tyrian purple with lighter center, a
white flare, openly funnel-shaped, 2.75in(7cm) across; rounded
trusses of 15. Plant as broad as high; leaves with moderate red
brown indumentum beneath, held 4 years. (Willard Swenson cross,
1970; intro. by Arthur & Maxine Childers; reg. 1978)

cl.

Swinhoe--form of floribundum

6ft(1.8m)          10°F(-12°C)        M          3/3

Leaves crinkled above, and have woolly indumentum beneath. Lax
trusses hold about 8 large purple flowers with a crimson blotch
in the throat, frilled margins. (Edmund de Rothschild)  A. M.
(RHS) 1963

---
     -javanicum (1/2)
Sybil-
     -unknown (1/2)

Vireya hybrid. Clear rose pink with centers fading to white.
Rich green foliage with red stems. (Rothschild, 1938)

cl.
           -edgeworthii (1/2)
Sylvania-
           -formosum var. inequale (1/2)

Trusses of 3-6 flowers, funnel-shaped, up to 4.8in(12cm) across,
white with a small yellow orange blotch; strongly fragrant.
Leaves glossy green above, scaly below. (Raiser J. S. Basford;
G. A. Hardy, reg. 1984)  F. C. C. (RHS) 1984

cl.                      -Monsieur Thiers--unknown (1/2)
               -J. H. van Nes-
Sylvia V. Knight-        -griffithianum (1/4)
           -             -Corona--unknown
               -Bow Bells-
                         -williamsianum (1/4)

5ft(1.5m)          10°F(-12°C)        EM

Plant upright, well-branched; leaves 4.5in(11.5cm) long, elliptic, dark yellowish green, held 1-2 years. Flowers openly campanulate, 3.75in(9.5cm) wide, spinel red with white flares in
each lobe, small red dorsal blotch; trusses of 10. (Bruce Briggs
cross; reg. by Edgar L. Knight, 1979)

g.
                  -sanguineum ssp. didymum (1/4)
       -Arthur Osborn-
Symphony-          -griersonianum (1/2)
       -          -neriiflorum ssp. neriiflorum (1/4)
       -F. C. Puddle-
                  -griersonianum

Red flowers. (Lord Aberconway, 1941)

g.
       -arboreum ssp. cinnamomeum var. album (1/2)
Syonense-
       -catawbiense (1/2)

Flowers delicate pink, spotted crimson. Knightian Medal to Ive-
son, 1849.

cl.
           -ciliatum (1/2)
Tacoma Maiden-
           -ciliicalyx (1/2)

3ft(.9m)          10°F(-12°C)        E

Pure white, fragrant flowers in April. Plant of open habit. (L.
Brandt, raiser; intro. by Carl P. Fawcett; reg. 1961)

cl.
     -griersonianum (1/2)
Taffy-
     -unknown (1/2)

A rounded, close-growing, floriferous shrub. Flowers of glowing
geranium lake, in trusses of 11. (Seed from L. de Rothschild,
1929; raised by Sir James Horlick; reg. 1964)

cl.                 -maximum (1/4)
     -unnamed hybrid-
Tahiti-             -catawbiense (1/4)
     -             -dichroanthum ssp. dichroanthum (1/4)
     -unnamed-     -fortunei ssp. discolor (1/8)
       hybrid -unnamed-
              hybrid -campylocarpum ssp. campylocarpum (1/8)

4.5ft(1.35m)        -15°F(-26°C)        ML         4/3

Flowers of orange pink lobes with yellow center and rust-colored
blotch, to 3in(7.6cm) across, with conspicuous calyx; trusses of
10. Plant wider than tall; leaves 4.5in(11.5cm) long. See also
Serenata. (David G. Leach, reg. 1962)

cl.

Taiping--seedling of the Triflorum series (?)

Probably a hybrid.  Trusses hold 9-11 flowers, Persian rose,
lighter inside with yellow markings on upper lobes.  Tall,
upright plant with leaves 2in(5cm) long, dark green. (James E.
Barto selection; raiser Ruth M. Hansen; reg. 1968)


cl.          -griersonianum (1/4)
     -Tally Ho-
Taku*        -facetum (eriogynum) (1/4)
     -yakushimanum ssp. yakushimanum (1/2)

4ft(1.2m)         0°F(-18°C)          M          3/4

Flowers of medium pink, fading lighter.  Leaves thickly covered
with indumentum. (G. Clark)


cl.
          -moupinense (1/2)
Talavera-
          -sulfureum (1/2)

3ft(.9m)          5°F(-15°C)          EM         4/3

Golden yellow flowers. (F. J. Williams)  F. C. C. (RHS) 1963


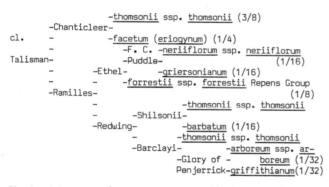

```
                     -thomsonii ssp. thomsonii (3/8)
           -Chanticleer-
cl.        -            -facetum (eriogynum) (1/4)
           -            -F. C. -neriiflorum ssp. neriiflorum
Talisman-           -Puddle-                          (1/16)
    -        -Ethel-    -griersonianum (1/16)
    -        -     -forrestii ssp. forrestii Repens Group
    -Ramilles-                                   (1/8)
    -        -           -thomsonii ssp. thomsonii
    -        -Shilsonii-
    -Redwing-        -barbatum (1/16)
    -        -        -thomsonii ssp. thomsonii
           -Barclayi-    -arboreum ssp. ar-
                    -Glory of -    boreum (1/32)
                    Penjerrick-griffithianum(1/32)
```

Blood red flowers. (Lord Aberconway, 1950)


cl.

Talisman     See Margaret Dunn Talisman


g. & cl.
          -griersonianum (1/2)
Tally Ho-
          -facetum (eriogynum)

6ft(1.8m)        10°F(-12°C)          L          4/3

Upright, rather dense, rounded plant; bullate leaves of medium
green 6in(15.2cm) long, with twisted margins.  Brilliant scarlet
flowers in lax trusses.  Easily propagated. (J. J. Crosfield)
F. C. C. (RHS) 1933


```
                     -fortunei ssp. discolor (1/4)
               -Lady
cl.   -Jalisco-Bessborough-campylocarpum ssp. campylocarpum
      -Eclipse-                       Elatum Group (1/4)
Tan   -        -     -dichroanthum ssp. dichroanthum (1/4)
Crossing-      -Dido-
               -        -decorum (1/4)
      -Jalisco Goshawk--(as above)
```

Trusses hold 10-12 flowers, 7-lobed, very light canary yellow,
with orange filaments, brown stamens.  Dark green, oblanceolate
leaves, 8.25in(21cm) long. (Crown Estate, Windsor, intro. 1950)
P. C. (RHS) 1965; A. M. (RHS) 1976


cl.
          -decorum (1/2)
Tanana-
          -yakushimanum ssp. yakushimanum (Exbury form selfed) (1/2)

2.5ft(.75m)       -10°F(-23°C)          M

Trusses of 12-15 flowers, white with yellowish green spotting,
slightly fragrant, 3in(7.6cm) across, 7-lobed.  Plant rounded,
broader than tall; leaves without indumentum, 3in(7.6cm) long,
held 4 years. (Arthur & Maxine Childers cross, 1963; reg. 1979)


cl.          -griersonianum(1/2)
     -Rapture-
Tara-        -arboreum ssp. zeylanicum (1/4)
     -       -griersonianum
     -Tally Ho-
          -facetum (eriogynum) (1/4)

A shapely free-flowering bush; leaves 5.5in(14cm) long.  Flowers
funnel-campanulate, 2.25in(5.7cm) long by 1.5in(3.8cm) wide, of
soft Turkey red, lightly spotted; compact trusses of 17. (Gen.
Harrison, reg. 1962)  A. M. (RHS) 1964


g.
          -falconeri ssp. falconeri (1/2)
Taranto-
          -falconeri ssp. eximium (1/2)

Two forms of the same species crossed by Admiral  A. W. Heneage-
Vivian, Clyne Castle, Wales.  Introduced in 1940.


g.
          -catacosmum (1/2)
Tasco-
          -griersonianum (1/2)

Plant about 5ft(1.5m) tall and broad, floriferous.  Leaves have
cinnamon brown indumentum below.  Lax trusses hold about 6 widely campanulate flowers  of deep scarlet, with extended calyces.
(Edmund de Rothschild, 1966)   P. C. (RHS) 1967     See color
illus. PB pl. 9.


cl.                           -javanicum (1/8)
               -Sir George Holford-
     -unnamed hybrid-         -unknown (1/8)
Tashbaan-        -leucogigas (1/4)
     -aurigeranum (1/2)

3ft(.9m)         32°F(0°C)          Feb.-Mar.

Vireya hybrid.  Flowers are tubular funnel-shaped, 3in(7.6cm)
across, 5-lobed, not fragrant, orange buff, trusses 6.5in(16cm)
across.  Plant upright, broad as tall, with leaves held 3 years.
(Cross by Peter Sullivan; reg. by William A. Moynier, 1981)


cl.

Tatoosh     See Roseann


cl.          -griffithianum (1/4)
     -Jean Marie de Montague-
Taurus-      -unknown (1/4)
     -strigillosum (1/2)

6ft(1.8m)        -5°F(-21°C)          EM         4/4

A magnificent shrub, vigorous, full in shape.  Leaves pointed,
deep green, held 3 years.  Deep red buds in winter open to campanulate flowers 3.5in(8.9cm) across, orient red petals and Turkey to cherry red throat,  black speckling on upper petal.   See
also Grace Seabrook. (Dr. Frank Mossman, 1972)  C. A. (ARS) 1972
See color illus. K p. 133.

```
---
            -brookeanum var. gracile (1/2)
Taylori-                    -jasminiflorum (3/8)
        -Princess Alexandra-        -jasminiflorum
                        -Princess Royal-
                                -javanicum (1/8)
```

Vireya hybrid.  Pink flowers.  (Byls, before 1874; exhibited by
Veitch, 1891)  F. C. C. (RHS) 1935

```
cl.
    -brachyanthum ssp. hypolepidotum (1/2)
Teal-
    -fletcherianum (1/2)
```

3ft(.9m)          5°F(-15°C)          M          4/4

Widely campanulate flowers in trusses of 5-8, showy, clear light
yellow.  Leaves narrowly elliptic, 2.25in(5.7cm) long.  Reddish
peeling bark. (Peter A. Cox)   A. M. (RHS) 1977

```
            -Loder's White, q.v. for 2 possible diagrams
cl.    -C. I. S.-        -dichroanthum ssp. dichroanthum
       -        -Fabia-
Ted Drake-            -griersonianum
       -      -wardii var. wardii
       -Crest-        -fortunei ssp. discolor
            -Lady
            Bessborough-campylocarpum ssp. campylocarpum
                                    Elatum Group
```

5ft(1.5m)          10°F(-12°C)          ML

Plant upright, well-branched; glossy, dark yellow green leaves
5in(12.7cm) long, held 2 years. Flowers of barium yellow, brick
red rays in throat, reverse tinged scarlet.  Lobe tips cut off,
making a circular hoop-skirt effect; spherical trusses of 10-13.
(Dr. C. H. Phetteplace cross; raiser E. F. Drake; Diane A. Lux,
reg. 1981)

```
cl.
    -griersonianum (1/2)
Ted Greig-            -campylocarpum ssp. campylocarpum (1/4)
        -unnamed hybrid-
                    -fortunei ssp. discolor (1/4)
```

A slender shrub of 5ft(1.5m) at 10 years; trusses of 9-10 flowers colored geranium lake, lighter on outside of corolla and
edges. (Royston Nursery, before 1962)

```
cl.

Ted Waterer     Parentage unknown
```

Flowers of blush lilac, in compact trusses. (J. Waterer, before
1922)  A. M. (RHS) 1925

```
                -griffithianum (9/32)
        -Jean Marie-
cl.    -de Montague-unknown (1/4)
       -                    -griffithianum
Tell Taylor-            -Kewense-
       -        -Aurora-        -fortunei ssp. fortunei
       -    -Naomi-    -                    (5/32)
       -Carita-    -    -thomsonii ssp. thomsonii (1/16)
            -    -fortunei ssp. fortunei
            -campylocarpum ssp. campylocarpum (1/4)
```

A very large, open plant; very large, glossy leaves.  Crimson
flowers, funnel-shaped, 3.5in(8.9cm) wide; rather tight trusses.
Midseason. (C. S. Seabrook, 1965)

```
cl.
    -Pauline--unknown (1/2)
Telstar-
    -yakushimanum ssp. yakushimanum (1/2)
```

Flowers of rose madder  paling almost to white inside, the upper
throat oxblood red, funnel-shaped, 2.25in(5.7cm) long by 2.75in
(7cm) wide.  Glossy dark green leaves 5in(12.7cm) wide. See also Pink Ghost, Renoir.  (F. Hanger cross, RHS Garden, Wisley,
1951)  H.  C. (Wisley Trials) 1962;  A. M. (Wisley Trials) 1966

```
cl.        -griffithianum (1/4)
      -Mars-
Tempest-   -unknown (1/4)
       -        -dichroanthum ssp. dichroanthum (1/4)
       -Fabia-
            -griersonianum (1/4)
```

5ft(1.5m)          -5°F(-21°C)          L          4/3

Upright growth habit on an attractive plant.  Flowers of fine
bright red, to 3in(7.6cm) across, in loose trusses of 8-10. (A.
A. Wright, Sr. & Jr., 1963)

```
g.
        -orbiculare ssp. orbiculare (1/2)
Temple Belle-
        -williamsianum (1/2)
```

3ft(.9m)          -5°F(-21°C)          EM          3/4

Plant neat, rounded, compact; cordate-oval matte green leaves on
pinkish petioles.  Flowers campanulate, rich rose pink, in small
loose trusses.  Pest-resistant.  (RBG, Kew, 1916)  See color
illus. F p. 34, 67.

```
            -Lady    -fortunei ssp. discolor (1/8)
cl.    -Day -Bessborough-
       -Dream-        -campylocarpum ssp. campylocarpum
Temptation*            Elatum Group (1/8)
       -    -griersonianum (1/4)
       -unknown (1/2)
```

4ft(1.2m)          0°F(-18°C)          ML          3/3

Flowers of creamy peach, with a darker center.  Red  bud scales
give color all winter. (J. E. Eichelser)

```
cl.        -dichroanthum ssp. dichroanthum (1/4)
       -Fabia-
Tensing-   -griersonianum (1/2)
       -        -Moser's Maroon--unknown (1/4)
       -Romany Chai-
            -griersonianum
```

Narrowly campanulate flowers of camellia rose shading into tinge
of orange in the throat; large trusses.  Midseason. (Cross by
Francis Hanger, RHS Garden, Wisley)  A. M. (RHS) 1953

```
cl.            -phaeopeplum (1/4)
      -Dr. Herman Sleumer-
Terebinthia-        -zoelleri (1/4)
       -Pink Delight (Veitch)--unknown (1/2)
```

4ft(1.2m)          32°F(0°C)          Sept.-Nov.

Vireya hybrid.  Flowers tubular funnel-shaped, fragrant, carmine
red with neyron rose throat; conical trusses of 6-10, 4.5in(11.5
cm) across.  Floriferous.  Plant upright, with stiff branches;
dull green leaves 3in(7.6cm) long, held 2-3 years.  (Cross by T.
Lelliot; raiser Peter Sullivan; William A. Moynier, reg. 1981)

```
cl.
    -minus var. minus Carolinianum Group (1/2)
Terry Herbert-
    -augustinii ssp. augustinii (1/2)
```

3ft(.9m)          -5°F(-21°C)          EM

Plant rounded, as wide as tall; small medium green leaves, scaly
beneath.  Flowers widely funnel-shaped, 2in(5cm) wide, clear orchid purple, no markings; terminal inflorescence, 7-12 flowers
with as many as 4-5 buds each.  (Charles Herbert, 1977)

```
g. & cl.    -ciliatum (1/4)
      -Praecox-
Tessa-    -dauricum (1/4)
      -moupinense (1/2)
```

4ft(1.2m)          -5°F(-21°C)          E          3/3

Lilac pink flowers with a ray of crimson spots, 2in(5cm) wide, in loose flattened umbels. Vigorous plant with reddish bronze bark, small emerald green leaves with convex edges. Does well in sun or shade. (J. B. Stevenson) A. M. (RHS) 1935; A.G.M. (RHS) 1969

cl.

Tessa Bianca       Parentage as above (Has been shown reversed)

4ft(1.2m)          -5°F(-21°C)          E          4/4

Tessa g. White flowers flushed pale pink, with throat tinged a light yellow. Plant habit compact. An improved Tessa? (Lester Brandt, 1965)

cl.

Tessa Roza       Parentage as above

4ft(1.2m)          0°F(-18°C)          E          4/3

Tessa g. Slightly deeper pink than Tessa, spotted deep carmine. (J. B. Stevenson) A. M. (RHS) 1953

cl.

```
                -campylocarpum ssp. campylocarpum (1/2)
Thacker's Sulphur-
                -Cunningham's Sulphur--caucasicum (1/2)
```

Compact trusses of 15 flowers, bell-shaped, 5-7 lobes, sulphur yellow with blotch of red. (Thacker, reg. 1962)

```
                -neriiflorum ssp. neriiflorum (1/8)
g.          -Nereid-
    -Euryalus-      -dichroanthum sp. dichroanthum (1/8)
Thais-      -griersonianum (1/4)
    -        -griffithianum (1/4)
        -Loderi-
                -fortunei ssp. fortunei (1/4)
```

Pale pink flowers. (Lord Aberconway, 1941)

---

```
        -javanicum (3/4)
Thalia-          -jasminiflorum (1/4)
    -Princess Royal-
                -javanicum
```

Vireya hybrid. Flowers ivory white to creamy pink. (Veitch, 1891)

cl.

```
    -ciliatum (1/2)
Thalia-
    -ciliicalyx (1/2)
```

(Lester E. Brandt, 1954)

```
g.              -thomsonii ssp. thomsonii (1/4)
    -Ascot Brilliant-
Thalia-              -unknown (1/4)
    -griffithianum (1/2)
```

A flower of mauve rose with a large crimson blotch and spots. (E. J. P. Magor, reg. 1958)

cl.

The Bride--form of caucasicum var. album (selfed)

6ft(1.8m)          -15°F(-26°C)          ML          2/3

An old selected form with white flowers blotched yellow; a good growth habit. (Standish & Noble, before 1850) F. C. C. (RHS) 1871

cl.

The Cardinal       See Kentucky Cardinal

cl.

```
                -Moser's Maroon--unknown (1/4)
        -Romany Chal-
The Chief-      -facetum (1/4)
        -elliottii (1/2)
```

6ft(1.8m)          0°F(-18°C)          ML

Synonym Nez Perce Chief. A tall plant of open habit; elliptic leaves 8.5in(21cm) long. Dome-shaped trusses of 16 oxblood red flowers, campanulate-recurved, to 3.75in(9.5cm) across. (Arthur & Maxine Childers, reg. 1972)

cl.

```
        -glaucophyllum var. glaucophyllum (1/2)
The Clown-
        -unknown (1/2)
```

Trusses of 3-9 flowers, white flushed brownish red with dark spotting in upper throat. (Origin unknown, acquired from G. Reuthe; intro. by R. N. S. Clarke) A. M. (RHS) 1982

```
g. & cl.        -arboreum ssp. arboreum (1/4)
        -Doncaster-
The Don-        -unknown (1/4)
    -griffithianum (1/2)
```

Considered a fine plant with flowers of strawberry red fading to pink. Height medium to tall. (Lowinsky) A. M. (RHS) 1920

```
cl.        -falconeri ssp. falconeri (1/4)
        -Muriel-
The Dowager-      -grande (1/4)
    -arboreum ssp. arboreum (1/2)
```

Tight round trusses of 24-26 flowers, widely funnel-campanulate, margins very pale orchid pink, corolla suffused with tints of purplish pink, lightly spotted, 2in(5cm) across. Narrow leaves 9.5in(24.3cm) long, silver grey indumentum below. (Raised by the Dowager Lady Loder; shown by Miss E. Godman) A. M. (RHS) 1968

```
cl.        -griffithianum ((1/4)
        -Loderi-
The Dream-      -fortunei ssp. fortunei (1/4)
    -unknown (1/2)
```

Trusses hold up to 10 flowers, rhodamine pink shading lighter. (Mrs. R. J. Coker cross, 1957; reg. 1979)

---

```
        -catawbiense (red form) (hybrid?)
The General-
        -catawbiense (red form) (hybrid?)
```

5ft(1.5m)          -20°F(-29°C)          ML          3/4

Crimson flowers with a dark red blotch; upright trusses. Leaves dark green. (Anthony M. Shammarello, 1955)

---

```
                -grffithianum (1/2)
The Hon. Jean Marie de Montague-
                -unknown (1/2)
```

5ft(1.5m)          -5°F(-21°C)          M          4/4

Often called "Jean Marie" or "Jean Marie de Montague". The red by which all newly introduced reds are judged. Large, bright scarlet flowers in dome-shaped trusses of 10-14; budding on very young plants. Bush compact, spreading; dense, thick, sun-tolerant foliage, deep emerald green. (C. B. van Nes & Sons, before 1958) See color illus. HG p. 111; VV p. 21.

```
---                      -catawbiense (red form) (1/4)
          -unnamed hybrid-
The Master-              -unknown (1/4)
        -                      -caucasicum  (1/4)
          -Cunningham's White-
                               -ponticum (white form) (1/4)
```

Rose pink flowers.  (A. M. Shammarello, exhibited 1955)

```
cl.            -wightii (1/4)
          -China-
The Master-     -fortunei ssp. fortunei (1/2)
        -              -campylocarpum ssp. campylocarpum
          -Letty Edwards-            Elatum Group (1/4)
                         -fortunei ssp. fortunei
```

Large pink flowers with dark red basal blotch; very large globular trusses of 14. Vigorous, upright plant, broader than tall;
dull, dark green leaves, 8.5in(21.6cm) long. (Slocock Nurseries,
1948)  H. C. (Wisley Trials) 1964; A. M. (Wisley Trials) 1966

```
g.                 -sanguineum ssp. didymum (1/4)
          -Arthur Osborn-
The Moor-              -griersonianum (1/4)
          -facetum (eriogynum) (1/2)
```

Red flowers. See also Tremeer.  (Lord Aberconway, 1941)

```
cl.            -souliei (1/4)
          -Halcyone-
The Queen-       -Lady      -fortunei ssp. discolor (1/8)
  Mother  -      Bessborough-
          -                   -campylocarpum ssp. campylocarpum
          -aberconwayi (1/2)              Elatum Group (1/8)
```

Trusses hold 9-10 flowers, magenta rose paling to light rhodamine purple. Leaves 3.25in(8.3cm) long, dark green. (Crown Estate, Windsor)  A. M. (RHS) 1968

```
---           -catawbiense (white form) (1/2)
The Virgin-            -oreodoxa var. fargesii (1/4)
          -unnamed hybrid-
                          -unknown (1/4)
```

White flowers.  (Abbott, before 1958)

```
cl.            -campylocarpum ssp. campylocarpum
          -Penjerrick-            Elatum Group (3/8)
Theale-         -griffithianum (1/4)
        -      -wardii var. wardii (1/4)
          -Crest-          -fortunei ssp. discolor (1/8)
          -Lady
          -Bessborough-campylocarpum ssp. campylocarpum
                                  Elatum Group
```

Open trusses of 10 flowers, primrose yellow with slight flush of
cardinal red in throat, widely campanulate, 2.25in(5.7cm) wide.
Crown Estate, Windsor, reg. 1967)  A. M. (RHS) 1966

```
cl.
          -griersonianum (1/2)
Thelma-                     -griffithianum (1/8)
       -             -unnamed hybrid-
          -Armistice Day-            -unknown (3/8)
                         -Maxwell T. Master--unknown
```

Darlene g.  Flowers funnel-shaped, 4in(10.2cm) wide, 5-lobed,
near geranium lake in color, in well-filled trusses. Leaves 6-7
in(17.8cm) long.  (Halfdan Lem, reg. 1962)  P. A. (ARS) 1959

```
g.                 -Moser's Maroon--unknown (1/4)
          -Romany Chal-
Theresa-              -facetum (eriogynum) (1/4)
          -griersonianum (1/2)
```

See Festival.  (Stavordale; exhibited 1950)

```
---
          -javanicum (5/8)
Thetis-              -brookeanum var. gracile (1/4)
        -Princess Frederica-        -jasminiflorum (1/8)
                         -Princess Royal-
                                  -javanicum
```

Vireya hybrid.  Flowers yellow to yellow orange.  F. C. C. (RHS)
1887

```
cl.
          -moupinense (1/2)
Thicket-
          -seinghkuense (1/2)
```

Trusses of 3-4 flowers, chartreuse green with throat spotted
Spanish orange.  (Crown Estate, Windsor)  P. C. (RHS) 1960

cl.

Thimble--seedling of campylogynum Cremastum Group

```
1ft(.3m)          -10°F(-23°C)          EM          3/4
```

Leaves 1in(2.5cm) long by .5in(1.3cm) wide, lightly covered with
brown scaly indumentum. Bell-shaped flowers held in clusters of
1-3, 5-lobed, salmon pink. (Collector unknown; exhibited by Collingwood Ingram; reg. 1967)  A. M. (RHS) 1966

```
g.
          -thomsonii ssp. thomsonii (1/2)
Thomaden-
          -adenogynum (1/2)
```

An introduction by E. J. P. Magor, 1927.

```
g.
          -sanguineum ssp. sanguineum var. haemaleum (1/2)
Thomaleum-
          -thomsonii ssp. thomsonii (1/2)
```

Dark crimson flowers.  (Intro. by Collingwood Ingram; reg. 1958)

```
                         -zoelleri (5/8)
cl.        -unnamed hybrid-
          -unnamed-          -lochiae (1/8)
Thomas- hybrid-
Becket-     -zoelleri
        -             -aurigeranum (1/4)
          -unnamed hybrid-
                    -zoelleri
```

```
5ft(1.5m)          32°F(0°C)          Sept.-Nov.
```

Vireya hybrid.  Flowers tubular funnel-shaped, 5-lobed, not
fragrant, orange red, in lax trusses 6.5in(16.5cm) across.
Plant is upright, leaves held 2 years, glossy, yellow green.
(Cross by Peter Sullivan; reg. by William A. Moynier, 1981)

```
cl.
          -forrestii ssp. forrestii Repens Group (1/2)
Thomas Church-     -campylocarpum ssp. campylocarpum (1/4)
          -Moonstone-
                    -williamsianum (1/4)
```

Dwarf plant 3ft(.9m) by 3ft. with leaves 2in(5cm) long. Trusses
hold 5 flowers, light orpiment (golden) orange when open, darker in bud.  (Lester E. Brandt, 1968)

```
g.
          -thomsonii ssp. thomsonii (1/2)
Thomdeton-              -adenogynum (1/4)
          -Xenosporum (detonsum)-
                         -unknown (1/4)
```

An introduction by E. J. P. Magor, 1941.

g.
```
              -thomsonii ssp. thomsonii (1/2)
Thomking-                    -griffithianum (1/4)
          -Mrs. Randall Davidson-
                             -campylocarpum ssp. campylocarpum
                                        (Hooker's form) (1/4)
```

An introduction of E. J. P. Magor, reg. 1958.


g.
```
              -thomsonii ssp. thomsonii (1/2)
Thomwilliams-
              -williamsianum (1/2)
```

2ft(.6m)        -5°F(-21°C)        EM        3/4

Deepest shade of rose pink. Foliage like that of williamsianum.
See also Bodnant Thomwilliams. (E. J. P. Magor, intro. 1927)


g. & cl.
```
      -haematodes ssp. haematodes (1/2)
Thor-      -dichroanthum ssp. dichroanthum (1/4)
      -Felis-
            -facetum (eriogynum) (1/4)
```

3ft(.9m)        5°F(-15°C)        M        4/4

Parentage may be reversed. Flowers of geranium lake, with large
calyces. Plant compact; foliage indumented. (L. Brandt, 1963)


cl.
```
            -macabeanum (1/2)
Thoron Hollard-
            -unknown (1/2)
```

Phlox pink in bud, opening creamy white, outside of lobes with
pink lines and striking red blotch in throat. See also Milton
Hollard. (Bernard Hollard, New Zealand, reg. 1979)


cl.
```
         -oreodoxa var. fargesii (1/2)
Throstle-
         -morii (1/2)
```

Hybrid by Collingwood Ingram.  P. C. (RHS) 1970


```
                     -fortunei ssp. discolor (1/4)
cl.   -Lady Bessborough-
                     -campylocarpum ssp. campylocarpum
Throstle*                          Elatum Group (1/4)
-                    -souliei (1/8)
-         -Soulbut-
-Vanessa-            -fortunei ssp. fortunei (1/8)
         -griersonianum (1/4)
```

A hybrid by Mrs. M. Rabbets, Ringwood, Hants.


cl.
```
            -haematodes ssp. haematodes (1/2)
         -May Day-
Thumbelina-         -griersonianum (1/2)
         -May Day (as above)
```

2.5ft(.75m)        5°F(-15°C)        M

Leaves 2in(5cm) by 1in(2.5cm) on a loose, open-growing plant.
Flowers small, bell-shaped, deep rose, with flounced calyx; 8-9
per truss. See also Tinkerbell. (Tressa McMurry, 1965)


cl.
```
              -arboreum ssp. arboreum (1/4)
         -Doncaster-
Thunderstorm-         -unknown (3/4)
         -unknown
```

Blood red flowers with white stamens, in tall trusses. Medium-
sized plant. (W. C. Slocock, cross 1930; A. M. (Wisley Trials)
1955


cl.
```
         -decorum ? (1/4)
      -Golden Jubilee-
Tiara-               -unknown (1/4)
     -               -griffithianum (1/4)
     -Loderi King George-
                        -fortunei ssp. fortunei (1/4)
```

4ft(1.2m)        5°F(-15°C)        E

Fragrant, white flowers with chartreuse green throat, held in
compact, erect trusses. Light green leaves up to 9in(23cm) long
by 4in(10cm) wide. (Melvin V. Love; reg. 1965)


cl.
```
            -caucasicum (1/8)
      -Viola-
   -Bismarck-     -unknown (1/8)
Tibet-      -catawbiense (1/4)
   -williamsianum (1/2)
```

Plant medium low, compact; ovate leaves 3in(7.6cm) long. Flower pinkish in bud, white in full bloom. (D. Hobbie; intro. by
Gebr. Boer, reg. 1969)  Gold Medal (Boskoop) 1966


cl.
```
                -griffithianum (1/4)
      -Jean Marie de Montague-
Tick-Tock-               -unknown (1/4)
         -williamsianum (1/2)
```

Rose pink flowers, funnel-campanulate, in rounded trusses of 10.
Low, compact plant; leaves 2.75in(7cm) long. (H. E. Greer, 1963)


cl.
```
      -dichroanthum ssp. dichroanthum (1/2)
Tidbit-
      -wardii var. wardii (1/2)
```

3ft(.9m)        5°F(-15°C)        M        3/4

Vigorous plant, a little wider than high, rather straggly; dark
green, glossy, pointed leaves 3.25in(8.3cm) long. Flowers first
flushed with cherry red, turning straw yellow and deepening with
age, campanulate, 2in(5cm) wide, of good substance; trusses of 6
to 7; floriferous. (Rudolph Henny, reg. 1958) P. A. (ARS) 1957
H. C. (RHS) 1977 [Shown by E. Cox]  See color illus. VV p. 95.


g.
```
      -yunnanense (chartophyllum) (1/2)
Tiepolo-            -ciliatum (1/8)
-         -Praecox-
-Clodion-          -dauricum (1/8)
         -cinnabarinum ssp. cinnabarinum Roylei Group (1/4)
```

Magenta flowers. (Adams-Acton, 1934)


cl.
```
                     -racemosum (1/8)
               -Conemaugh-
   -Anna    -Pioneer-      -mucronulatum (1/8)
Tiffany-Baldsiefen-     -unknown (1/4)
   -          -Pioneer (as above)
   -keiskei (1/2)
```

2.5ft(.75m)        -15°F(-26°C)        M        3/3

An attractive dwarf. Star-shaped, reflexed flowers primarily
pink with mixture of apricot and yellow in throat. Good-looking
foliage. (Warren Baldsiefen, reg. 1972)  See color illus. HG
p. 130.


cl.
```
         -taylorii (1/2)
Tiffany Rose-
         -leptanthum (1/2)
```

Vireya hybrid. Trusses of 4-6 flowers, tubular bell-shaped,
neyron rose. Plant height 14in(.35m); elliptic leaves 2.3in(6cm)
long. (Peter Sullivan cross; raiser R. M. Withers; reg. by G.
L. S. Snell, 1984)

cl.
```
      -dichroanthum ssp. dichroanthum (1/2)
Tiger-              -fortunei ssp. fortunei (1/8)
    -        -Luscombei-
    -Cremorne-          -thomsonii ssp. thomsonii (1/8)
              -campylocarpum ssp. campylocarpum (1/4)
```

Shrub vigorous, compact, as broad as high; leaves medium green,
3in(7.6cm) long.  Yellow flowers tinged orange red, heavy spotting of greyed purple on upper segment, 2.75in(7cm) wide; loose,
flattened trusses of 4-6. (Reuthe, reg. 1971)   H. C. (Wisley
Trials) 1970

```
                                -catawbiense (1/8)
                    -Parsons Grandiflorum-
              -America-                  -unknown (1/2)
cl.      -unnamed-      -dark red hybrid--unknown
       - hybrid-yakushimanum ssp. yakushimanum (1/4)
Tim Craig*      -griffithianum (1/8)
        -Mars-
    -Moniz-  -unknown              -catawbiense
    -        -Parsons Grandiflorum-
        -America-              -unknown
                  -dark red hybrid--unknown
```

Deep red buds open to strong red flowers with touches of red and
white in the throat; white stamens. (Weldon E. Delp)

```
          -griffithianum (5/32)
        -Mars-
cl. -Vulcan-  -unknown (7/32)
    -    -griersonianum (1/2)
Tim -      -griersonium
Flint-    -Azor-
    -unnamed-  -fortunei ssp. discolor (1/8)
      hybrid-              -Queen    -griffithianum
                    -Britannia-Wilhelmina-
          -C. P. Raffill-    -      -unknown
                  -      -Stanley Davies--unknown
                  -griersonianum
```

5ft(1.5m)        5°F(-15°C)        ML

Fragrant flowers, color between Turkey red and currant red, 2.6
in(6.6cm) wide, fragrant, 5 or 6-lobed; spherical trusses of 11.
Rounded plant, nearly as broad as tall; leaves with light golden
brown indumentum, held 3 years. (Grady Barefield cross; reg. by
Mary M. Barefield, 1981)

cl.

Tinker Bell (Gable/Yates)    See Suzy Bell

cl.
```
        -catawbiense (5/8)
Tinker Hill-                      -decorum (1/8)
        -          -unnamed hybrid-
        -Lavender Charm-          -griffithianum (1/8)
              -              -catawbiense
              -Purpureum Elegans-
                    -unknown (1/8)
```

5ft(1.5m)        -5°F(-21°C)        EM

Plant upright, well-branched; leaves glossy, dark yellow green,
4.25in(10.8cm) long.  Magenta buds opening to neyron rose flowers with ruby red blotch and dorsal spotting, 4in(10.2cm) wide,
fragrant, 7-lobed, in spherical trusses of 13. (Charles Herbert,
1977)

cl.
```
                  -haematodes ssp. haematodes (1/2)
Tinkerbell--May Day (selfed)-
                  -griersonianum (1/2)
```

Low-growing plant, compact but spreading, 2ft(.6m) by 3ft(.9m)
in 8 years. Flowers rose pink, openly bell-shaped, in trusses
of 8-9. Blooms in midseason. (Tressa McMurry, 1965)

```
      ---
          -maximum (1/2)
Tintoretto-
          -unknown (1/2)
```

Flowers light pink with reddish blotch. (C. Frets & Son)

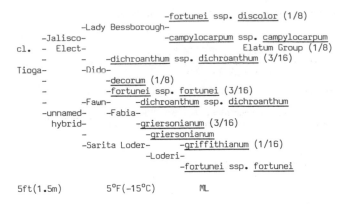

```
                  -fortunei ssp. discolor (1/8)
          -Lady Bessborough-
      -Jalisco-          -campylocarpum ssp. campylocarpum
cl. - Elect-                    Elatum Group (1/8)
    -    -    -dichroanthum ssp. dichroanthum (3/16)
Tioga-    -Dido-
    -        -decorum (1/8)
    -        -fortunei ssp. fortunei (3/16)
    -    -Fawn-    -dichroanthum ssp. dichroanthum
    -unnamed-    -Fabia-
      hybrid-        -griersonianum (3/16)
                  -griersonianum
            -Sarita Loder-    -griffithianum (1/16)
                  -Loderi-
                    -fortunei ssp. fortunei
```

5ft(1.5m)        5°F(-15°C)        ML

Plant well-branched, willowy, wider than tall; leaves 4in(10cm)
long, dark green, held 3 years. Flowers primrose yellow lightly
spotted reddish brown, to 4in(10.2cm) across, 7-lobed, in lax
trusses of 8; floriferous. (Cross by Del James; reg. by W. V.
Joslin, 1974)

cl.
```
        -griffithianum (3/4)
Tip-the-Wink-    -griffithianum
      -Kewense-
            -fortunei ssp. fortunei (1/4)
```

6ft(1.8m)        0°F(-18°C)        M        5/3

No additional information located about this hybrid by Col. G.
H. Loder, High Beeches. It is closely related to the Loderi hybrids by Sir Edmund Loder of Leonardslee.   F. C. C. (RHS) 1936

cl.
```
        -fortunei ssp. discolor (1/4)
    -Beckyann-
Tish-    -campylocarpum ssp. campylocarpum (1/4)
    -        -fortunei ssp. fortunei (Gable's cream form)
    -unnamed hybrid-              (1/4)
            -vernicosum (R 18139) (1/4)
```

4ft(1.2m)        -5°F(-21°C)        ML

Plant upright, as broad as tall; glossy leaves 5.5in(14cm) long.
Flowers 3.25in(8.3cm) wide, Naples yellow deeper in throat, unmarked, fragrant, 6-lobed; spherical trusses of 10. (Henry Yates
cross; reg. by Mrs. Yates, 1976)

cl.

Titan--form of brookeanum

A selected form of this vireya. Flowers orange fading to light
azalea pink. (E. F. Allen, reg. 1972)

```
        -facetum (eriogynum) (1/4)
cl. -unnamed-        -dichroanthum ssp. dichroanthum (1/4)
    - hybrid-Fabia  -
Titian-        Tangerine-griersonianum (1/4)
Beauty-
        -yakushimanum ssp. yakushimanum (1/4)
    -unnamed-        -dichroanthum ssp. dichroanthum
      hybrid-Fabia  -
        Tangerine-griersonianum
```

3ft(.9m)        0°F(-18°C)        M        3/3

Plant has neat, erect, compact habit; small foliage. Flowers of
Turkey red. (John Waterer, Sons & Crisp, reg. 1971)

cl.
```
                    -auriculatum (1/2)
Titness Aladdin-
                    -griersonianum (1/2)
```

Rose-colored trusses in late season.  (Sir James Horlick, 1962)

g.
```
                    -griersonianum (1/2)
Titness Beauty-     -griffithianum (1/4)
            -Loderi-
                    -fortunei ssp. fortunei (1/4)
```

(Sir James Horlick, 1937)

```
cl.
                                        -griffithianum
                        -George Hardy-          (7/16)
                    -Pink -         -catawbiense (3/16)
                    -Pearl-         -arboreum ssp. arbor-
        -Countess of-    -Broughtonii-       eum (1/16)
cl.     - Derby   -              -unknown (1/16)
        -                   -catawbiense
Titness Belle-      -Cynthia-
        -                   -griffithianum
        -       -griffithianum
        -Loderi-
                    -fortunei ssp. fortunei (1/4)
```

Large flowers of deep rose, fading pink, in large trusses.  (Sir James Horlick cross, 1930; intro. 1942)

cl.
```
                    -griffithianum (1/4)
            -Loderi-
Titness Delight-    -fortunei ssp. fortunei (1/4)
            -unknown (1/2)
```

Bell-shaped flowers  in trusses of 11, light pink fading white, and  slightly scented.  (Sir James Horlick, 1962)

cl.
```
                -barbatum (1/2)
Titness Park-
                -calophytum var. calophytum (1/2)
```

Trusses of 18 flowers, varying shades of phlox pink with chocolate blotch in throat, and some chocolate spotting.  (Sir James Horlick cross; raised  & exhibited  by the Crown Estate)  A. M. (RHS) 1954;  F. C. C. (RHS) 1974

cl.
```
                    -thomsonii ssp. thomsonii (1/4)
            -Shilsonii-
Titness Perfection-     -barbatum (1/4)
            -praevernum (1/2)
```

An early flowering hybrid, rose pink with a deep throat.  (Sir James Horlick, 1962)

cl.
```
                    -thomsonii ssp. thomsonii (1/4)
            -Shilsonii-
Titness Pink-       -barbatum (1/4)
            -sutchuenense (1/2)
```

Early-flowering, deep rosy pink.  (Sir James Horlick, 1962)

cl.
```
                    -griffithianum (1/4)
            -Loderi-
Titness Saucy-      -fortunei ssp. fortunei (1/4)
            -unknown (1/2)
```

Trusses of 12 flowers, each 3in(7.6cm) across, slightly scented, apple blossom pink fading to white.  Midseason.  (Sir James Horlick, 1962)

cl.
```
                    -dichroanthum ssp. dichroanthum (1/4)
            -Fabia-
Titness Scarlet-    -griersonianum (1/4)
            -neriiflorum ssp. neriiflorum (1/2)
```

Loose trusses of dark reddish orange scarlet flowers.  Midseason.  (Sir James Horlick, reg. 1962)

cl.
```
                -basilicum (1/2)
Titness Triumph-
                -montroseanum (mollyanum) (1/2)
```

Pink flowers.  (Sir James Horlick cross, 1939; intro. 1956)

cl.
```
                -wightii (1/2)
Tittenhurst-
                -barbatum (1/2)
```

Primrose yellow flowers.  (White)  A. M. (RHS) 1933

cl.
```
                    -Corona--unknown (1/2)
Tittenhurst Belle-
                -griffithianum (1/2)
```

Flowers rose pink.  (Lowinsky)  A. M. (RHS) 1925

cl.   -Patricia--campylogynum Charopeum Group (1/2)
```
      -
To Bee-Yaku Fairy--keiskei (1/2)
```

1ft(.3m)          -10°F(-23°C)          EM

Venetian pink flowers with claret rose spotting, 5 wavy-frilled lobes, 1.5in(3.8cm) wide, in clusters of 3-5; floriferous.  Bush twice as broad as high, rounded; small, elliptic, convex leaves, held 2 years.  (Warren E. Berg cross, 1972; reg. 1983)

cl.
```
                -elliottii (1/4)
        -Fusilier-
Tod B.  -       -griersonianum (1/4)
Galloway-   -dichroanthum ssp. dichroanthum (1/4)
        -Jasper-         -fortunei ssp. discolor (1/8)
            -Lady     -
        Bessborough-campylocarpum ssp. campylocarpum
                            Elatum Group (1/8)
```

Medium-sized plant, open, rather flat and spreading; light green leaves.  Flowers of varying light pink and yellow shades, showy double calyx; flattened trusses of 8-12.  (C. S. Seabrook, 1965)

cl.
```
        -Pygmalion--unknown (1/2)
Todmorden-          -haematodes ssp. haematodes (1/4)
        -unnamed hybrid-         -fortunei ssp. fortunei (1/8)
                        -Wellfleet-
                    (Dexter#8)-decorum (1/8)
```

5ft(1.5m)          -5°F(-21°C)          M          3/3

The exact combination of the above is uncertain.  Deep purplish pink margins  shading lighter in lobe centers and throat, fading almost white for a bicolor effect; flowers of good substance, 5-7 wavy-frilled lobes, 3.5in(8.9cm) wide; spherical trusses of 8-15.  Plant upright, rounded, as broad as high; leaves held 2 years. (C. O. Dexter cross, before 1943; reg. by the A. H. Scott Horticultural Foundation, Swarthmore College, 1983)    See color illus. ARS Q 26:2 (Spring 1972); p. 93; LW pl. 10.

cl.

Tofino     See next page

cl.
```
        -griffithianum (1/4)
    -Mars-
Tokatee-    -unknown (1/4)
    -williamsianum (1/2)
```

2.5ft(.75m)          0°F(-18°C)          EM

Plant rounded, twice as wide as high; small leaves held 4 years.  Deep pink flowers, very pale lobe centers and throat, lightly spotted rose, 2.75in(7cm) across; lax trusses of 7.  (B. F. Lancaster cross; reg. by Lewis C. Grothaus, 1975)  C. A. (ARS) 1972

```
                -dichroanthum ssp. dichroanthum (7/32)
            -Dido-
        -     -decorum (3/16)
    -Lem's-                      -griffithianum (5/32)
    -Cameo-           -Beauty of-
    -     -Norman- Tremough-arboreum ssp. arboreum (1/32)
    -     -  - Gill -
    -     -Anna-   -griffithianum
cl. -     -        -griffithianum
    -           -Jean Marie -
Tofino-         de Montague-unknown (1/16)
    -                   -fortunei ssp. discolor (5/32)
    -               -Lady
    -             -Bessborough-campylocarpum ssp. campylo-
    -     -Jalisco-          carpum Elatum Group (3/32)
    -     -     -dichroanthum ssp. dichroanthum
    -     -  -Dido-
    -unnamed-      -decorum
      hybrid-      -wardii var. wardii (1/16)
    -       -Crest-          -fortunei ssp. discolor
    -     -     -Lady
    -unnamed-   Bessborough-campylocarpum ssp.
      hybrid-          campylocarpum Elatum Gp.
    -              -fortunei ssp. discolor
          -King of Shrubs-   -dichroanthum ssp.
                    -Fabia-        dichroanthum
                    -griersonianum (1/32)
```

6ft(1.8m)        0°F(-18°C)        ML

Red buds open to light Dresden yellow flowers, neyron rose markings, 4in(10.2cm) wide; dome-shaped trusses of 15-16. Plant upright, about as broad as tall; glossy convex leaves, dark yellow
green, heavy-textured, held 2 years. (John Lofthouse, reg. 1983)

```
                -facetum (1/4)
cl. -unnamed-    -dichroanthum ssp. dichroanthum (1/8)
    - hybrid-Fabia-
Tolkien-    -griersonianum (1/8)
    -     -yakushimanum ssp. yakushimanum (1/4)
    -unnamed-                      -griffithianum (1/16)
      hybrid-     -Queen Wilhelmina-
    -Britannia-          -unknown (3/16)
          -Stanley Davies--unknown
```

Plant has growth habit of yakushimanum. Cardinal red  flowers.
(Cross by P. Wiseman, 1958; reg. by D. E. Mayers, 1985)

```
                -griersonianum (1/4)
cl. -Sarita Loder-  -griffithianum (1/8)
    -         -Loderi-
Tolo-             -fortunei ssp. fortunei (1/4)
    -     -lacteum (1/4)
    -unnamed-           -campylocarpum ssp. campylocarpum
      hybrid-Mary Swathling-          (1/8)
                    -fortunei ssp. fortunei
```

A hybrid by Del W. James, before 1958.

```
                    -souliei ? (1/8)
               -Virginia-
    -unnamed- Scott  -unknown (1/8)
cl.     - hybrid-    -Ole -campylocarpum ssp. campylocar-
    -       -Alice -Olson-       pum Elatum Group (1/16)
Tom     -     Franklin-  -fortunei ssp. discolor (1/16)
Ethrington-     -          -griffithianum (1/16)
    -         -Loderi  -
    -           King George-fortunei ssp. fortunei
    -yakushimanum ssp. yakushimanum (1/2)          (1/16)
```

5ft(1.5m)        10°F(-12°C)        E

Upright plant, moderately branched; elliptic leaves 5in(12.7cm)
long, held 3 years. Rosy buds opening to fragrant primrose yellow flowers, unmarked, 3.75in(9.5m) wide, 8-lobed; dome-shaped
trusses of 12. (H. L. Larson, reg. 1979)

cl.
    -racemosum (1/2)
Tom Koenig-
    -keiskei (1/2)

3ft(.9m)        -10°F(-23°C)        EM        3/3

Narrow leaves, 2in(5cm) long. Pale pink flowers in dense clusters like racemosum, but larger. (G. Guy Nearing; intro. by T.
W. Koenig, 1970)

cl.

Tom Spring Smythe--form of dalhousiae var. dalhousiae
                            (T.S.S. 32)

Collected by T. Spring-Smyth, although the name is incorrectly
spelled in the official registration.  Intro. by Maj. A. E.
Hardy as a cool greenhouse plant. A. M. (RHS) 1974

```
cl.        -griffithianum (1/4)
        -Mars-
Tom Thumb*   -unknown (1/4)
        -       -catawbiense (red form) (1/4)
        -Cathaem #1-
              -haematodes ssp. haematodes (1/4)
```

2.5ft(.75m)        -15°F(-26°C)        M        4/4

Very slow-growing plant with distinctive leaf--indented at mid-
section.  Flowers of good, clear red, a few small pale markings
at base, widely funnel-shaped, 5-lobed (joined most of length);
rounded trusses of 10-12.  (Joseph B. Gable)  See color illus.
LW pl. 52.

```
cl.
    -sanguineum ssp. didymum (1/2)
Tom Thumb-
    -thomsonii ssp. thomsonii (1/2)
```

At 25 years shrub was 3ft(.9m) high.  Umbels of 4 flowers borne
low on plant, beneath the leaves.  Flowers tubular-campanulate,
deep crimson, 1.5in(3.8cm) long; obovate leaves to 2.75in(7cm).
(Cross by E. J. P. Magor; reg. by Maj. E. W. M. Magor, 1962)

```
---                      -arboreum ssp. arboreum (1/4)
                    -Doncaster-
Tom Willis Flemming-      -unknown (3/4)
                    -unknown
```

Deep red flowers.  (W. C. Slocock)

```
g.           -haematodes ssp. haematodes (1/4)
        -Choremia-
Tomahawk-    -arboreum ssp. arboreum (1/4)
        -hookeri (1/2)
```

Blood red.  (Lord Aberconway, 1950)

```
                              -dichroanthum ssp. dichro-
cl.           -Goldsworth Orange-        anthum (1/8)
        -Tortoiseshell-          -fortunei ssp. fortunei
Tombola- Wonder   -griersonianum (1/2)          (1/8)
        -              -Moser's Maroon--unknown (1/4)
        -Romany Chai-
                    -griersonianum
```

The trusses hold up to 16 flowers, bright reddish pink with a
blotch of orange brown. (Research  Station for Woody  Nursery
Crops, Boskoop, 1979)

```
cl.          -dichroanthum ssp. dichroanthum (1/4)
    -unnamed hybrid-
Tomeka-          -griersonianum (1/4)
    -decorum (1/2)
```

5ft(1.5m)        10°F(-12°C)        M

Flowers vermilion with an orange glow,  red veins, dorsal spots,
and nectaries.  Lax 5- to 9-flowered trusses.  Leaves long, narrow. (Del James cross, 1964; Hadley Osborn, reg. 1979)

```
g.                           -griffithianum (1/4)
                         -Mars-
Tommie*--Tom Thumb (selfed)-    -unknown (1/4)
                     -            -catawbiense (red form)
                         -Cathaem #1-              (1/4)
                              -haematodes ssp. haema-
                                              todes (1/4)


semi-dwarf        -10°F(-23°C)          M          3-4/3
```

Flowers red, pink, or pinkish salmon.  Grex includes: Bonnie,
Little Bonnie, Mary Yates, and the Tommies #1-#12. (J. B. Gable)

```
cl.                          -griffithianum (1/4)
                         -Mars-
Tommie #2*--Tom Thumb (selfed)-    -unknown (1/4)
                     -            -catawbiense(red form)
                         -Cathaem-              (1/4)
                              -haematodes ssp. haema-
                                              todes (1/4)


semi-dwarf        -10°F(-23°C)          M          4/3
```

Tommie g.  The flowers are bicolor; in the LW photograph they're
soft salmon red, with centers of lobes cream, upper lobe spotted
a dark color, 5 divided lobes; rather small flattened trusses of
6-7. The ARS picture shows white flowers with wide pink margins,
pale spotting, 5 joined lobes, very rounded trusses of about 12.
Who has whom?  (Joseph B. Gable)  See color illus. ARS J 36:1
(Winter 1982): p. 10; LW pl. 49.

```
cl.             -neriiflorum ssp. neriiflorum (1/4)
            -Daphne-             -arboreum ssp. arboreum (1/8)
Tomyris-       -Red Admiral-
       -                    -thomsonii ssp. thomsonii (5/8)
            -thomsonii ssp. thomsonii
```

Claret red, bell-shaped flowers of good substance, held in loose
trusses of 5-7. (E. J. P. Magor cross; named and reg. by Maj. E.
W. M. Magor, 1963)

```
---
        -decorum (1/2)
Tony-
        -unknown (1/2)
```

White flowers. (Messel) A. M. (RHS) 1930

```
cl.             -caucasicum (1/4)
            -Boule de Neige-             -catawbiense (5/8)
Tony-                   -unnamed hybrid-
       -catawbiense (red form)          -unknown (1/8)


4ft(1.2m)          -20°F(-29°C)          ML          4/3
```

A handsome low-growing hybrid with crinkly foliage and bright
cherry red flowers.  See also Lavender Queen, Pink Cameo, Pink
Flair,  Sham's Juliet,  Sham's Pink.  (A. M. Shammarello, 1955)
See color illus. ARS Q 25:2 (Spring 1971): p. 100; VV p. 110.

cl.

Tony Shilling--form of arboreum ssp. cinnamomeum var. roseum

Trusses of 25 flowers, medium rose pink, spotted and streaked
crimson. (Collector unknown; intro. by RBG, Kew from  Wakehurst
Place)  F. C. C. (RHS) 1974

```
                         -dichroanthum ssp. dichroanthum
                -Goldsworth-                      (1/8)
cl.         - Orange  -fortunei ssp. discolor (1/8)
        -Hotei ? -        -souliei (1/8)
Top Banana-        -unnamed hybrid-
       -                     -wardii var. wardii (1/8)
            -unknown (1/2)


3.5ft(1m)          5°F(-15°C)          EM
```

Brick red buds open to empire yellow flowers, unmarked, openly
funnel-shaped, 2.5in(6.4cm) wide, 6-lobed; spherical trusses of
17.  Plant upright, taller than wide; glossy dark leaves held 2
years. (William Whitney; intro. by Anne Sather; reg. 1985)  See
color illus. ARS J 38:3 (Summer 1984): p. 115.

```
cl.                    -griffithianum (1/4)
            -Loderi King George-
Topaz-                     -fortunei ssp. fortunei (1/2)
      -                      -fortunei ssp. fortunei
            -Faggetter's Favourite-
                           -unknown (1/4)
```

Trusses of 7-10 flowers in early season, phlox pink outside,
white inside.  Upright, tall plant with leaves 7in(17.8cm) by
2.5in(6.4cm).  (Rudolph & Leona Henny, 1963)

```
cl.                              -griffithianum (1/4)
                          -George-
Topsvoort--sport of Pink Pearl- Hardy-catawbiense (1/4)
   Pearl     -              -arboreum ssp.
                  -Broughtonii- arboreum (1/4)
                          -
                          -unknown (1/4)
```

Similar to Pink Pearl, q.v.,  but this has pale pink flowers and
red frilled edges. (Topsvoort Nursery, Holland, before 1958)

```
                              -griffithianum (1/4)
cl.         -Queen Wilhelmina-
      -Britannia-              -unknown (3/8)
Torch-         -Stanley Davies--unknown
      -              -griersonianum (1/4)
            -Sarita Loder-     -griffithianum
                          -Loderi-
                              -fortunei ssp. fortunei (1/8)
```

Flowers rosy orange with darker throat. (Slocock)

TORLONIANUM  (PHOTO P340)

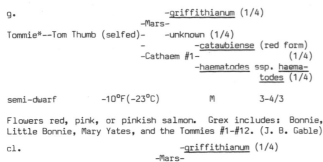

```
                                  -griffithianum
                          -Queen  -     (1/32)
                  -Britannia-Wilhelmina-unknown (3/32)
                  -              -
           -Wilgen's-         -Stanley Davies--unknown
           - Ruby   -              -catawbiense (1/16)
        -unnamed-         -John Walter-
cl.     - hybrid-                  -arboreum ssp. arbor. (1/16)
                         -haematodes ssp. haematodes (1/4)
Tornado-        -May Day-
       -              -griersonianum (1/4)
       -                  -haematodes ssp. haematodes
       -              -May Day-
        -Billy Budd-         -griersonianum
                   -elliottii (1/4)
```

Flowers are currant red, 9-14 per truss.  (Research Station for
Woody Nursery Crops, Boskoop, 1979)

```
                              -sanguineum ssp. didymum (1/8)
g.              -Arthur Osborn-
        -Toreador-              -griersonianum (5/8)
Torsun-         -griersonianum
       -              -griffithianum (1/4)
        -Sunrise-
                -griersonianum
```

Rose red flowers. (Lord Aberconway, 1946)

```
g.                         -dichroanthum ssp. dichroanthum
            -Goldsworth Orange-                      (1/4)
Tortoiseshell-              -fortunei ssp. discolor (1/4)
           -griersonianum (1/2)
```

See also Champagne. (Slocock exhibited, 1946)

```
cl.             -dichroanthum ssp. dichroanthum (1/4)
            -Goldsworth-
Tortoiseshell- Orange  -fortunei ssp. discolor (1/4)
   Orange     -
            -griersonianum (1/2)
```

Tortoiseshell g. Large flowers of deep orange; blooms midseason
to late season. (Slocock Nurseries, 1945)

cl.

Tortoiseshell Salome     Parentage as above

Tortoiseshell g.  Large flowers of biscuit and pink.   Has sometimes been called Biscuit?  (Slocock, 1946)

cl.                                -dichroanthum ssp. dichroanthum
          -Goldsworth Orange-                              (1/4)
Tortoiseshell-                     -fortunei ssp. discolor (1/4)
  Wonder    -griersonianum (1/2)

4ft(1.2m)          -5°F(-21°C)          ML          4/3

Tortoiseshell g.  Flowers a blend of orange salmon  with salmon
pink.   (W. C. Slocock)   A. M. (RHS) 1947

cl.                     -Metternianus--japonicum var. japon-
     -First Step (selfed)-          icum (metternichii) (1/8)
Tosca*                   -arboreum ssp. arboreum (1/4)
  -                      -griersonianum (1/4)
     -Vulcan's Flame-    -griffithianum (1/4)
                   -Mars-
                        -unknown (1/8)

A heat-resistant hybrid.  Ruffled flowers warm rosy pink at margins, shading lighter toward the center; a long, white, recurved
pistil.  Attractive medium green foliage.  (Koichiro Wada)   See
color illus. ARS J 37:4 (Fall 1983): p. 194.

cl.                           -griffithianum (1/16)
                   -Kewense-
          -Aurora-          -fortunei ssp. fortunei (5/16)
     -Exbury Naomi-      -thomsonii ssp. thomsonii (1/8)
Tosca-                  -fortunei ssp. fortunei
     -wardii var. wardii Litiense Group (1/2)

Rounded trusses of 10 flowers, shallowly campanulate, 3.75in(9.5
cm) wide, opening primrose yellow from buds of pale pink.  (Cross
by Francis Hanger, RHS Garden, Wisley, 1962)  A. M. (RHS) 1959

cl.
        -ferrugineum (1/2)
Tottenham-
        -unknown (1/2)

2ft(.6m)          0°F(-18°C)          ML          3/3

Pale pink flowers.  Suggested for the novice by Cox.  (Moerheim)

cl.
          -agastum (1/2)
Touch of Class-      -griersonianum (1/4)
          -Tally Ho-
                -facetum (eriogynum) (1/4)

Flower buds are deep jasper red, opening a lighter shade of red.
(Mrs. R. J. Coker, reg. 1980)

cl.
        -griffithianum (1/2)
Touchstone-
        -unknown (1/2)

Large loose trusses of very large, bell-shaped flowers, mottled
deep pink. (J. J. Crossfield, Embley Park)  A. M. (RHS) 1937;
F. C. C. (RHS) 1939   See color illus. ARS J 36:4 (Fall 1982):
p. 139.

cl.
        -minus var. minus (white form) Carolinianum Group (1/2)
Tow Head-
        -ludlowii (1/2)

1ft(.3m)          -15°F(-26°C)          EM          4/4

Plant twice as wide as tall; leaves 1.75in(4.5cm) long, glossy,
scaly on both surfaces.  Flowers brilliant greenish yellow with
orange yellow spotting on dorsal lobe, 1.5in(3.2cm) wide, 5 per
bud; blooms before leaves appear.  (David G. Leach, reg. 1968)

---
          -catawbiense (5/8)          -catawbiense
Towardii-            -unnamed hybrid-
          -Altaclarense-          -ponticum (1/8)
               -arboreum ssp. arboreum (1/4)

Flowers light rose or rosy lilac.  See also Blandyanum.  (Standish & Noble, 1850)

cl.                                -griffithianum
                    -Queen Wilhelmina-     (1/16)
          -Britannia-          -unknown (3/16)
     -C. P. Raffill-      -Stanley Davies--unknown
     -               -griersonianum (3/8)
Towhee-              -sanguineum ssp. didymum (1/8)
     -          -Red Cap-
     -unnamed-     -facetum (eriogynum) (1/4)
     hybrid-      -griersonianum
          -Tally Ho-
               -facetum (eriogynum)

4.5ft(1.35m)          5°F(-15°C)          ML          3/3

Leaves 4.5in(11.5cm) long by 1.5in(3.8cm) wide.  Bright scarlet
red, waxy flowers, 3.5in(8.9cm) across, in open-topped trusses
of 10.  (Del W. James, reg. 1968)   P. A. (ARS) 1956

cl.                          -griffithianum (1/4)
          -Loderi King George-
Townhill Albatross-          -fortunei ssp. fortunei
          -fortunei ssp. discolor (1/2)          (1/4)

Albatross g.  Flowers fuchsine pink, almost white center.  (Lord
Swathling)  A. M. (RHS) 1945

cl.                          -fortunei ssp. fortunei (1/4)
               -Luscombei-
Townhill Cremorne-          -thomsonii ssp. thomsonii (1/4)
               -campylocarpum ssp. campylocarpum (1/2)

Cremorne g.  A nicely shaped, medium-sized plant with trusses of
about 12 flowers, Chinese yellow with rosy coral margins.  (Lord
Swaythling, Townhill Park)   A. M. (RHS) 1947

cl.                          -caucasicum (1/4)
               -Dr. Stocker-
Townhill Damaris-          -griffithianum (1/4)
               -campylocarpum ssp. campylocarpum (1/2)

Damaris g.  (Exhibited by Lord Swaythling, 1948)

cl.
               -sanguineum ssp. didymum (1/2)
Townhill Redcap-
               -facetum (eriogynum) (1/2)

Redcap g.  Flowers oxblood red, turning scarlet with back-light-
ing.  (Lord Swathling)  A. M. (RHS) 1945

cl.
          -Blue Peter--unknown (1/2)
Traci Suzanne-          -griffithianum (1/4)
          -Loderi King George-
                    -fortunei ssp. fortunei (1/4)

6ft(1.8m)          5°F(-15°C)          L

Plant rounded  with willowy branches; narrow leaves 6in(15.2cm)
long, held 3 years.  Buds light cyclamen purple, opening pale orchid spotted green on upper lobe; flowers to 3.5in(8.9cm) wide,
fragrant, in spherical trusses of 12.  Floriferous.  (E. L. Kaiser cross; reg. by Mrs. Mae Granston, 1977)

'Spring Frolic' by Leach
Photo by Leach

'Springfield' by James/Greer
Photo by Greer

'Starburst' by Greer
Photo by Greer

'Starcross' by A. F. George
Photo by Greer

'Stephanie' by Whitney
Photo by Greer

'Stephen Clarke' by Mrs. J. H. Clarke
Photo by Greer

'Stockholm' by Leach
Photo by Leach

'Sudan' by Leach
Photo by Leach

'Sue' by James
Photo by Greer

'Sugar and Spice' by Bowhan
Photo by Greer

'Sumatra' by Leach
Photo by Greer

'Summer Snow' by Leach
Photo by Leach

'Sunlit Snow' by Wister
Photo by West

'Sunrise' by Aberconway
Photo by Greer

'Swamp Beauty' by J. A. Elliott
Photo by Greer

'Sweet Simplicity' by J. Waterer
Photo by Greer

'Tahiti' by Leach
Photo by Leach

'Talavera' by F. J. Williams
Photo by Greer

'Tally Ho' by Crosfield
Photo by Greer

'Ted's Orchid Sunset' by T. Fawcett
Photo by Greer

'Tessa' by J. B. Stevenson
Photo by Greer

'Tessa Bianca' by Brandt
Photo by Greer

'The General' by Shammarello
Photo by Greer

'The Queen Mother' by Crown Estate, Windsor
Photo by Greer

cl.
      -yakushimanum ssp. yakushimanum (1/2)
Tracigo-
      -sperabile var. sperabile (1/2)

3ft(.9cm)          5°F(-15°C)          M          3/3

Plant upright, rounded, broader than tall; dark green leaves, greyed orange indumentum beneath, held 4 years. Flowers of good substance, 2in(5cm) wide, in many dome-shaped trusses of 12-15, produce a delightful apple blossom effect with bright red buds, flower exteriors pink, white interiors. (Dr. David Goheen, 1983)

                    -campylocarpum ssp. campylo-
           -Penjerrick- carpum Elatum Group (1/16)
g.    -Amaura-        -griffithianum (1/16)
  -Eros-      -griersonianum (3/8)
Trafalgar-  -griersonianum
-        -fortunei ssp. fortunei (1/4)
  -Luscombei-
      -thomsonii ssp. thomsonii (1/4)

Rose pink flowers. (Lord Aberconway, 1950)

                    -griffithianum (1/8)
cl.         -unnamed hybrid-
  -Mrs. Furnival-       -unknown (3/4)
Trail Blazer-      -    -caucasicum (1/8)
-        -unnamed hybrid-
  -Sappho--unknown     -unknown

5ft(1.5m)          -10°F(-23°C)          M          4/4

Blossoms like Mrs. Furnival, pink with striking blotch, but here the blotch is ruby red; flowers openly funnel-shaped, 3in(7.6cm) wide, in globular trusses of 19. Compact, vigorous plant, twice as wide as tall; glossy dark leaves held 3 years. (A. A. Wright, Sr. & Jr., reg. 1979)

cl.
    -unknown (1/2)
Travis L.-        -griffithianum (1/8)
-      -Loderi-
  -Albatross-    -fortunei ssp. fortunei (1/8)
        -fortunei ssp. discolor (1/4)

4ft(1.2m)          -5°(-21°C)          EM

Upright plant habit with arching branches; long, narrow leaves, smooth and dark green. Flowers to 4.5in(11.5cm) wide, of ivory white with pale greenish white blotches, 7 wavy lobes; globular trusses of 8-12. (Clifford Cannon, reg. 1981)

g.
    -forrestii ssp. forrestii Repens Group (1/2)
Treasure-
    -williamsianum (1/2)

2ft(.6)          -5°F(-21°C)          EM          3/4

A mounded shrub with gnarled branches; neat, rounded dark green leaves, bronze when young. Nodding campanulate flowers of deep rose, 2in(5cm) across. (J. J. Crosfield, 1937)

---
    -arboreum ssp. arboreum (1/2)
Trebah Gem-
    -griffithianum (1/2)

Soft pink flowers. (R. Gill & Son)

cl.
    -griffithianum (1/2)
Treetops-
    -arboreum ssp. arboreum (1/2)

Beauty of Tremough g. Loose trusses of about 10 flowers, white flushed reddish purple shades, widely funnel-campanulate, 3.25in (8.3cm) wide, 5-lobed. Dull, medium green leaves, 5.5in(14cm) long. (Lord Aberconway, reg. 1981)  A. M. (RHS) 1981

cl.
    -arboreum ssp. arboreum (1/2)
Trelawny-
    -barbatum (1/2)

Flowers deep carmine with paler throat, darker spots. (R. Gill) A. M. (RHS) 1936

cl.        -sanguineum ssp. didymum (1/4)
  -Arthur Osborn-
Tremeer-      -griersonianum (1/4)
  -facetum (eriogynum) (1/2)

The Moor g. (Gen Harrison, 1935)

cl.
      -azalea occidentiale (1/2)
Tressa McMurry-
    -ponticum (1/2)

3ft(.9m)          0°F(-18°C)          ML

Azaleodendron. Plant upright, as wide as tall; dark green, narrow leaves 3in(7.6cm) long. Widely funnel-campanulate flowers, 1.5in(3.8cm) wide and long, of fuchsine pink with sienna blotch, flat trusses of 10-18. Floriferous. (Mrs. T. McMurry, 1977)

cl.
    -calophytum var. calophytum (1/2)
Tretawn-
    -arboreum ssp. cinnamomeum var. album (1/2)

Trusses of 14 flowers, white shading to phlox pink in throat with blotch of Indian lake in upper throat. (Cross by E. J. P. Magor; reg. by Maj. E. W. M. Magor, 1976)  A. M. (RHS) 1976

cl.        -cinnabarinum ssp. cinnabarinum
  -Full House-      Blandfordiiflorum Group (1/4)
Trewithen-    -maddenii ssp. maddenii (1/4)
 Orange -
    -cinnabarinum ssp. xanthocodon Concatenans Group (1/2)

Loose, tubular-shaped, deep orange brown flowers in pendant trusses, early to mid-season. An erect plant with sea green foliage. (Raised at Trewithen) F. C. C. (RHS) 1950

cl.

Triangle--form of cinnabarinum ssp. tamaense (K-W 21003)

Flowers axillary, in loose clusters of 24-26, flushed with purple shades, white deep in throat. Leaves dull dark green, 2in (5cm) by 1.25in(3cm), lightly scaly. (Collector F. Kingdon-Ward; raised by Maj. A. E. Hardy)  P. C. (RHS) 1978

cl.  -fortunei ssp. fortunei (1/2)
-        -fortunei ssp. discolor (1/8)
Trianon-  -Lady Bessborough-
-    -    -campylocarpum ssp. campylocar-
  -Jalisco-      pum Elatum Group (1/8)
      -dichroanthum ssp. dichroanthum (1/8)
    -Dido-
      -decorum (1/8)

Fred Wynniatt g. Tall, compact plant; flowers rose pink with a golden yellow throat. (Edmund de Rothschild, 1966)

g.
    -triflorum (1/2)
Triaur-
    -xanthostephanum (1/2)

A yellow-flowered hybrid by E. J. P. Magor, 1920.

                  -griffithianum (1/8)
cl.      -Queen Wilhelmina-
  -Britannia-      -unknown (3/8)
Trident-    -Stanley Davies--unknown
  -facetum (eriogynum) (1/2)

Firetail g. Dark red flowers. (Shown by Loder, 1948)

```
cl.                      -griffithianum (1/4)
         -Queen Wilhelmina-
Trilby-                  -unknown (3/4)
         -Stanley Davies--unknown


5ft(1.5m)        -15°F(-26°C)       ML        3/4
```

Same parents as Britannia.  Flowers deep crimson with dark markings; greyish green leaves contrast with bright red stems.  Sun-
tolerant.  (C. B. van Nes & Sons)  See color illus. VV p. 103.


```
                  -unnamed-dichroanthum ssp. scyphocalyx (1/8)
cl.      -Calcutta- hybrid-
         -         -     -kyawii (1/8)
Trinidad-        -Catalgla--cataubiense var. album Glass (1/4)
         -                -maximum (1/8)
         -     -Russell Harmon-
         -Tahiti-           -cataubiense (1/8)
         -        -dichroanthum ssp. dichroanthum (1/8)
              -unnamed-    -fortunei ssp. discolor (1/16)
              hybrid-unnamed-
                         hybrid-campylocarpum ssp. campylocarpum
                                                         (1/16)
4ft(1.2m)        -20°F(-29°C)        ML
```

Spherical trusses of 14 flowers, sharply defined bright cherry
red corolla perimeter, ivory center, sparse greyed yellow dorsal
spotting, 2.75in(7cm) across.  Free-flowering.  Plant wider than
tall; dark green leaves held 3 years.  (D. G. Leach, reg. 1982)


```
cl.
       -Powell Glass (selfed)--cataubiense (white form) (1/2)
Trinity-
       -yakushimanum ssp. yakushimanum (1/2)

2.5ft(7.5m)       -25°F(-32°C)        ML
```

Plant about as wide as tall; leaves held 3 years; new growth has
thin orange tan indumentum.  Flowers white with faint green dorsal spotting, shaded pink on edges and reverse on opening, 2.75
in(7cm) across, slightly fragrant; globular trusses of 14.  (Orlando S. Pride, reg. 1979)


```
g.
             -dichroanthum ssp. dichroanthum (1/2)
Tristan Thacker-
             -G. A. Sims--unknown (1/2)
```

Flowers salmon orange.  (Thacker, 1944)


```
---
        -javanicum (1/2)
Triton-          -multicolor (1/4)
       -unnamed hybrid-
                 -unknown (1/4)
```

Vireya hybrid.  Flowers salmon rose with pale  yellow  throat.
A. M. (RHS) 1900


```
---                         -brookeanum var. gracile (1/4)
           -Duchess of Edinburgh-
Triumphans-              -longiflorum (lobbii) (1/4)
           -javanicum (1/2)
```

Vireya hybrid.  Scarlet crimson flowers.  (Byls, before 1871)
F. C. C. (RHS) 1883   See color illus. ARS Q 33:2 (Spring 1979):
p. 103.


```
g.
     -thomsonii ssp. thomsonii (3/4)
Troas-          -thomsonii ssp. thomsonii
     -Cornish Cross-
                 -griffithianum (1/4)
```

Early-blooming crimson flowers in larger truss than thomsonii,
but not as large as that of Cornish Cross.  Foliage like thomsonii, slightly larger.  (Sir John Ramsden, 1942)


```
g.                    -neriiflorum ssp. neriiflorum (1/4)
         -F. C. Puddle-
Trojan-            -griersonianum (1/4)
       -         -thomsonii ssp. thomsonii (1/4)
         -Shilsonii-
                   -barbatum (1/4)
```

Dark red flowers.  (Lord Aberconway, 1950)


```
cl.
         -javanicum (1/2)
Tropic Fanfare-
         -lochiae (1/2)
```

Vireya hybrid.  Trusses of 8-10 flowers, azalea pink.  (Crossed
and raised by Don Stanton; intro. by Graham Snell, reg. 1982)


```
cl.
       -laetum (1/2)
Tropic Glow-
       -zoelleri (1/2)
```

Vireya hybrid.  Trusses of 4-6 flowers, funnel-shaped, 5-lobed,
golden yellow shaded orange red.  Height about 8ft(2.4m); elliptic leaves, to 4in(10cm) long.  (T. Lelliot cross; raiser E. B.
Perrot; reg. by O. Blumhardt, 1984)


```
cl.
          -aurigeranum (1/2)
Tropic Summer-
          -macgregoriae (1/2)
```

Vireya hybrid.  Trusses of 10-12 flowers, light saffron yellow
with some lobes of Saturn red.  (Cross by Donald Stanton; intro.
& reg. by G. Snell, 1982)


```
cl.                -jasminiflorum (3/8)
          -Princess -            -jasminiflorum
Tropic Tango-Alexandra-Princess Royal-
          -                    -javanicum (1/8)
          -laetum (1/2)
```

Vireya hybrid.  Trusses of 5-8 flowers, tubular funnel-shaped,
strong orange yellow and reddish orange.  Shrub 5ft(1.5m) when
registered. (Cross by J. Rouse, 1978; G. L. S. Snell, reg. 1985)


```
cl.
      -Lady Alice Fitzwilliam--unknown (1/2)
Tropicana-
        -maddenii ssp. crassum (1/2)
```

White waxy flowers, strongly fragrant, larger and opening wider
than the species maddenii ssp. crassum.  Quite hardy.  (Michael
Haworth-Booth, reg. 1963)


```
                              -sanguineum ssp. didymum (1/8)
g.          -Arthur Osborn-
       -Toreador-          -griersonianum (3/8)
Troubadour-    -griersonianum
       -arboreum ssp. arboreum (1/2)
```

Blood red flowers.  (Lord Aberconway, 1950)


```
                            -griffithianum (3/8)
                  -George Hardy-
            -Pink -          -cataubiense (3/8)
cl.   -Countess-Pearl-       -arboreum ssp. arboreum (1/8)
      -of Derby-   -Broughtonii-
Trude-      -              -unknown (1/8)
Webster-    -         -cataubiense
       -       -Cynthia-
       -              -griffithianum
       -Countess of Derby--(as above)

5ft(1.5m)       -10°F(-23°C)       M        5/4
```

The first plant to be given the Superior Plant Award by the ARS. Extra-large, upright trusses of 14; flowers a clear shade of pink, upper lobe spotted, 5in(12.7cm) wide, 5-lobed. Leaves of medium green, rather glossy, to 7in(17.8cm) long, with a slight twist. (Harold E. Greer, 1961)   S. P. A. (ARS) 1971   See color illus. ARS Q 26:2 (Spring 1972): p. 92; ARS Q 31:2 (Spring 1977): cover;  HG p. 131;  K p. 156.

```
                          -arboreum ssp. arboreum (1/8)
cl.        -Doncaster-
      -Belvedere-       -unknown (1/8)
Trula-      -dichroanthum ssp. dichroanthum (1/2)
    -       -dichroanthum ssp. dichroanthum
    -Jasper-             -fortunei ssp. discolor (1/8)
         -Lady Bessborough-
                         -campylocarpum ssp. campylocarpum
                                         Elatum Group (1/8)

5ft(1.5m)          10°F(-12°C)
```

Plant well-branched, as wide as high; dark green leaves, narrowly elliptic, 4.5in(11.5cm) long, held 3 years.  Flowers medium yellow orange with green spotting in dorsal sector, edging and exterior jasper red, 3in(7.6cm) across, with 5-6 wavy lobes; in ball-shaped trusses of 10. Floriferous.  (H. L. Larson, 1979)

```
cl.               -yakushimanum ssp. yakushimanum (1/4)
         -White Wedding-
Truly Fair-          -yakushimanum ssp. makinoi (1/4)
      -War Paint--elliottii (1/2)

2.5ft(.75)        0°F(-18°C)          M
```

Flowers open pink and fade white, heavy brownish spotting and flare, 1.75in(4.5cm) wide, frilled, 5-lobed; globular trusses of 18-21.  Floriferous.  Plant vigorous, much broader than tall; elliptic, pointed leaves have heavy felt-like indumentum. (John G. Lofthouse cross, 1969; reg. 1981)

```
cl.               -campylocarpum ssp. campylocarpum (1/4)
       -Mrs. W. C. Slocock-
Truth-      -         -unknown (1/4)
     -          -decorum (1/4)
     -unnamed hybrid-
                    -griffithianum (1/4)
```

Trusses hold up to 12 flowers, each 4.5in(11.5cm) across by 2.5in(6.4cm) long, pale pink buds opening white. (Gen. Harrison, 1962)

```
cl.            -edgeworthii (bullatum) (1/4)
    -unnamed hybrid-
Tui-          -unknown (3/4)
    -unknown
```

Trusses of 3-5 flowers, sulphur yellow in bud opening white with red flush and a yellow flare. (Pukeiti Rhododendron Trust, New Zealand; intro. and reg. by G. F. Smith, 1979)

```
cl.
    -decorum (1/2)
Tumalo-          -griffithianum (1/4)
     -Loderi King George-
                    -fortunei ssp. fortunei (1/4)

5ft(1.5m)      0°F(-18°C)       M      4/3
```

Plant compact, spreading; light green leaves 6.5in(16.5cm) long. Fragrant, ruffled white flowers, green at base, suffused pale chartreuse when opening,  5-6in(12.7-15.2cm) across, in large rounded trusses; blooms so heavily that partial disbudding may be needed. (Del W. James, reg. 1958)  P. A. (ARS) 1955

```
                                -griffithianum
              -George Hardy-
cl. -Mrs. Lindsay Smith-      -catawbiense
     -              -Duchess of Edinburgh--unknown
Tutu-      -dichroanthum ssp. dichroanthum
    -    -Fabia-
    -C.I.S.-    -griersonianum

              -Loder's White, q.v. for 2 possible parentage diagrams

3ft(.9m)        5°F(-15°C)         ML
```

Compact plant, as broad as high; elliptic leaves about 7in(18cm) long, light green.  Flowers 4in(10.2cm) across, pale buff shading to delicate empire yellow; rather loose, spherical trusses of 8-9. (Edwin K. Parker, reg. 1971)

```
              -fortunei ssp. fortunei (1/4)
     -unnamed -     -griffithianum (1/8)
cl.   - hybrid -Alice-
     -(selfed)   -unknown (1/8)
Twilight-              -fortunei ssp. discolor
 Pink* -       -Lady     -           (1/8)
     -       -Bessborough-campylocarpum ssp. campy-
     -    -Jalisco-          locarpum Elatum Gp.(1/8)
     -    -        -dichroanthum ssp. dichroanthum
     -Comstock-   -Dido-                 (3/16)
     -        -          -decorum (1/16)
     -        -dichroanthum ssp. dichroanthum
     -Jasper-       -fortunei ssp. discolor
         -Lady    -
         Bessborough-campylocarpum ssp. campy-
                       locarpum Elatum Group

4ft(1.2m)       -5°F(-21°C)        M       4/4
```

Plant of compact, rounded habit, with dense apple green foliage. The flower has a large calyx, giving the appearance of a double. Color is a warm pink that seems to glow at twilight. (Harold E. Greer, intro. 1982)   See color illus. HG p. 67.

```
cl.
    -racemosum (1/2)
Twinkles-
    -scabrifolium var. spiciferum (1/2)

4ft(1.2m)       0°F(-18°C)        EM       3/2
```

Growth habit strong and vigorous; decumbent tendency responds well to pruning.  Leaves dark green above, grey beneath, 1.25in (3.2cm) long.  Flowers small, light pink, appearing along the stems. Floriferous. (A. A. Wright, Sr. & Jr., 1958)

```
              -catawbiense (1/4)
cl.      -Pink Twins-
     -          -haematodes ssp. haematodes (1/4)
Twins Candy-            -griffithianum (1/4)
     -          -Loderi Venus-
     -Cotton Candy-       -fortunei ssp. fortunei
          -Marinus-griffithianum       (1/8)
            Koster-
                 -unknown (1/8)

5ft(1.5m)       -5°F(-21°C)        EM
```

Plant upright, rounded; glossy leaves 5in(12.7cm) long. Flowers 4in(10.2cm) wide, 7-lobed, vibrant pink shading lighter at center, upper 2 lobes spotted cardinal red; ball-shaped trusses of about 18. (Charles Herbert, 1978)

```
cl.
    -minus var. minus Carolinianum Group (1/2)
Twister*
    -dauricum (1/2)
```

Blooms with P.J.M.; colors of flowers and winter foliage closely resemble P.J.M.  Small, pointed leaves have an unusual twist. Plant well-branched, of good growth habit, and very hardy. (Dr. G. David Lewis)

```
cl.                        -griffithianum (1/2)
            -Loderi King George-
Two Kings-                 -fortunei ssp. fortunei (1/4)
        -                  -griffithianum
            -King George-
                           -unknown (1/4)
```

Trusses hold 12-13 flowers, white, flushed with shades of
rhodamine pink.  (Cross by Col. G. H. Loder; reg. by The Hon. H.
E. Boscawen, 1982)

```
            -griersonianum (1/4)
cl. -Esquire-
    -              -unknown (1/4)
Tyee-        -wardii var. wardii (1/4)
    -        -                    -griffithianum (1/32)
    -Idealist-         -Kewense-
    -        -Aurora-         -fortunei ssp. fortunei (5/32)
             -Naomi-    -thomsonii ssp. thomsonii (1/16)
                   -fortunei ssp. fortunei
```

5ft(1.5m)          5°F(-15°C)            M

Primrose yellow flowers with uranium green throat, 4in(10.2cm)
across, in lax trusses of 11.  Leaves 4in(10.2cm) by 1.5in(4cm).
(Del W. James, 1960)  A. E. (ARS) 1960

```
cl.
          -nuttallii (1/2)
Tyermannii-
          -formosum var. formosum (1/2)
```

Large, lax trusses of large flowers, lily-shaped, tinged outside
with green and brown, aging to pure white, with much yellow in
throat; sweetly fragrant.  Dark glossy green leaves; main trunk
of this handsome plant has rich brownish bark.  (Tyermann)
F. C. C. (RHS) 1925

```
            -griersonianum (1/4)
g.   -Karkov-         -arboreum ssp. arboreum (1/8)
     -     -Red Admiral-
Tzigane-              -thomsonii ssp. thomsonii (1/8)
     -                   -griffithianum (1/8)
     -          -King George-
     -Gipsy King-         -unknown (1/8)
          -hematodes ssp. haematodes (1/4)
```

Crimson flowers with a much enlarged crimson calyx, for a hose-
in-hose effect, held in shapely trusses of about 11.  Leaves are
sharply pointed, to 8in(20.3cm) long, with thin fawn indumentum.
Plant of good habit, rather spreading; blooms early midseason.
(Rothschild; intro. 1955)

```
cl.                    -dichroanthum ssp. dichroanthum (1/4)
         -Goldsworth Orange-
Ulrike-                -fortunei ssp. discolor (1/4)
         -       -haematodes ssp. haematodes (1/4)
         -May Day-
               -griersonianum (1/4)
```

Flowers bright claret rose, with darker edges.   (Joh. Bruns, W.
Germany, reg. 1972)

```
cl.              -dichroanthum ssp. dichroanthum (1/4)
         -Fabia-
Umpqua Chief-    -griersonianum (1/2)
         -       -griersonianum
         -Azor-
               -fortunei ssp. discolor (1/4)
```

Vermilion flowers, with large calyces.  Plant of medium compactness.  (Del W. James, before 1958)

```
g.
     -ungerii (1/2)
Ungerio-
     -facetum (eriogynum) (1/2)
```

An introduction by E. J. P. Magor, 1933.

```
cl.
      -campylocarpum ssp. campylocarpum (1/2)
Unique-
      -unknown (1/2)
```

4ft(1.2m)         -5°F(-21°C)          FM          3/4

Leaves clover green, glossy, oblong,  on a dense, rounded plant.
Bright pink buds open to buttery cream flowers on rose red pedicels; compact, dome-shaped trusses.  Young flowers often flushed
pink.  Does well in warm climates, with some shade.  (W. C. Slocock)   A. M. (RHS) 1934; F. C. C. (RHS) 1935   See color illus.
F p. 67; V V p. 61, 62.

```
cl.                    -griffithianum (1/4)
            -Queen Wilhelmina-
Unknown Warrior-       -unknown (3/4)
            -Stanley Davies--uknown
```

Plant vigorous, upright, rather open; pointed dark green leaves.
Crimson buds open to deep rose red flowers with a few dark brown
markings, widely campanulate, to 3in(7.6cm) wide; dome-shaped
trusses of 12.  Sun-tolerant; early blooming.  See also Britannia, Earl of Athlone, Langley Park, London, C. B. van Nes.  (C.
B. van Nes, before 1922)  See color illus. VV p. 53.

```
                              -griffithianum
cl.                -Queen Wilhelmina-       (1/8)
            -Earl of Athlone-       -unknown (3/8)
Ursula Siems-         -Stanley Davies--unknown
          -forrestii ssp. forrestii Repens Group (1/2)
```

Flowers intense scarlet carmine, translucent, of good substance,
in umbels of 6-10.  (Dietrich Hobbie, 1951)

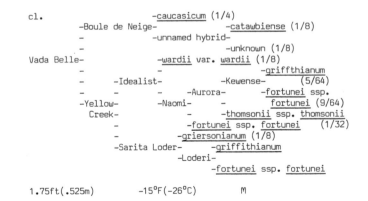

```
cl.              -caucasicum (1/4)
       -Boule de Neige-         -catawbiense (1/8)
       -            -unnamed hybrid-
       -                   -unknown (1/8)
Vada Belle-       -wardii var. wardii (1/8)
       -     -            -griffthianum
       -  -Idealist-    -Kewense-     (5/64)
       -  -    -   -Aurora-  -fortunei ssp.
       -Yellow-  -Naomi-   -    fortunei (9/64)
        Creek-        -thomsonii ssp. thomsonii
       -        -fortunei ssp. fortunei  (1/32)
       -           -griersonianum (1/8)
       -Sarita Loder-    -griffithianum
              -Loderi-
                   -fortunei ssp. fortunei
```

1.75ft(.525m)        -15°F(-26°C)          M

Flowers light primrose yellow with yellow green blotch and some
spotting, widely funnel-shaped, about 2in(5cm) wide; ball-shaped
trusses of 9-12.  Plant broader than tall; glossy leaves retained 2 years.  (William Fetterhoff cross, 1960; reg. 1981)

```
g. & cl.
      -valentinianum (1/2)
Valaspis-
      -leucaspis (1/2)
```

A dwarf of spreading habit with loose trusses of widely campanulate flowers of pale yellow (another form bright yellow).  Very
early season.  (Lord Aberconway)   A. M. (RHS) 1935

g.            -souliei (1/4)
    -Latona-
Valda-          -dichroanthum ssp. dichroanthum (1/4)
     -haematodes ssp. haematodes (1/2)

Flowers rosy red. (Lord Aberconway, 1941)


cl.

Valewood Pink      Parentage unknown

Flowers shell pink with darker margins. (Mangles) A. M. (RHS)
1934


g.            -dichroanthum ssp. dichroanthum (1/4)
     -Astarte-          -campylocarpum ssp. campylocarpum
Valiant-    -Penjerrick-          Elatum Group (1/4)
                    -griffithianum (1/4)
     -haematodes ssp. chaetomallum (1/2)

Red flowers. (Lord Aberconway, 1946)


cl.
          -yakushimanum ssp. yakushimanum (1/2)
     -                         -griffithianum (1/16)
Vallerie Kay-          -Queen Wilhelmina-
     -    -Britannia-          -unknown (3/16)
     -Leo-          -unknown
          -elliottii (1/4)

1ft(.3m)        0°F(-18°C)          M

Bright carmine pink flowers of heavy substance, fading lighter,
3in(7.6cm) across, lobes frilled, wavy; ball-shaped trusses hold
20. Plant upright, almost as broad as tall. Elliptic leaves of
medium green with fawn indumentum, held 2-3 years; new growth is
mouse grey. See also Mary Jane and Verna Carter. (Mrs. LeVerne
Freimann, intro. 1971; reg. 1974)


cl.
          -Essex Scarlet--unknown (1/2)
Valley Creek-
          -fortunei ssp. fortunei (1/2)

6ft(1.8m)        -5°F(-21°C)          ML

Upright plant with stiff branches; narrowly elliptic leaves 6.75
in(17cm) long, held 3 years. Flowers deep rosy pink with dark
blotch, to 3.5in(8.9cm) wide, 6-7-lobed, slightly fragrant; flat
trusses of 10.    (Cross by J. Gable; intro. by C. Herbert, 1976)


cl.               -catawbiense (1/4)
     -Atrosanguineum-
Valley Forge-          -unknown (1/4)
     -               -fortunei ssp. fortunei (1/4)
     -unnamed hybrid-
               -williamsianum (1/4)

6ft(1.8m)        -5°F(-21°C)          ML

Flowers medium spinel red, spotted cardinal red, heavily on dorsal lobe, 3.75in(9.5cm) wide, 7-lobed; trusses of 15-18. Plant
upright, rounded, well-branched; glossy leaves 6in(15.2cm) long,
retained 2 years. (Charles Herbert, reg. 1976)


cl.
          -azalea occidentale (1/2)
Valley Sunrise*          -ponticum (1/4)
          -Purple Splendour-
               -unknown (1/4)

Azaleodendron. Flowers have a sensational orange blotch glowing
against an orchid background. (Dr. R. L. Ticknor cross; intro.
by H. E. Greer, 1983) See color illus. HG p. 100.


g. & cl.
          -moupinense (1/2)
Valpinense-
          -valentinianum (1/2)

Flower buds are deep primrose yellow, opening a lighter shade.
(Lord Aberconway)    A. M. (RHS) 1943


cl.               -griffithianum (1/8)
          -Queen Wilhelmina-
     -Britannia-          -unknown (3/8)
Vampire-     -Stanley Davies--unknown
     -     -dichroanthum ssp. dichroanthum (1/4)
     -Fabia-
          -griersonianum (1/4)

Red flowers. (Wright, 1951) P. A. (ARS) 1951


cl.
     -griersonianum (1/2)          -griffithianum (3/16)
               -George Hardy-
Van-          -Pink Pearl-          -catawbiense (3/16)
     -          -          -arboreum ssp. arboreum
     -Countess of-          -Broughtonii-          (1/16)
     Derby     -          -unknown (1/16)
          -     -catawbiense
          -Cynthia-
               -griffithianum

6ft(1.8m)        -5°F(-21°C)          ML          4/4

Flowers of heavy substance, deep pink, heavily spotted dark purplish red on dorsal lobe, 4.5in(11.5cm) wide; rounded trusses of
13-15. Plant as broad as tall; dark leaves held 2 years. (Van
Veen, Sr., P. Griebnow; reg. by H. E. Greer, 1979)


cl.               -catawbiense (1/4)
     -Charles Dickens-
Van den Broeke-          -unknown (3/4)
     -Lord Roberts--unknown

Crimson flowers. (H. den Ouden & Sons cross, 1912; intro. 1925)


cl.               -catawbiense (1/4)
     -Atrosanguineum-
Van der Hoop-          -unknown (1/2)
     -          -arboreum ssp. arboreum (1/4)
     -Doncaster-
          -unknown

A parent of Rowland P. Cary. Flowers dark crimson rose. (H.
den Ouden & Sons, 1925)


g.               -thomsonii ssp. thomsonii (1/4)
     -Chanticleer-
Van Huysum-          -facetum (eriogynum) (1/4)
     -               -ponticum (1/4)
     Purple Splendour-
               -unknown (1/4)

Deep red flowers. (Adams-Acton, 1935)


---               -griffithianum (1/4)
          -Queen Wilhelmina-
Van Nes Glory-          -unknown (3/4)
          -Stanley Davies--unknown

Red flowers. Same parentage as Britannia, Unknown Warrior, etc.
(C. B. van Nes & Sons)


cl.
          -Sir Charles Butler--fortunei ssp. fortunei
Van Nes Sensation-          -maximum (1/4)          (1/2)
          -Halopeanum-
               -griffithianum (1/4)

5ft(1.5m)        -5°F(-21°C)          M          4/4

Very large dome-shaped trusses of wide, funnel-shaped flowers, a light orchid pink with lighter centers, 7-lobed and quite fragrant. Sturdy, compact plant with attractive foliage. See also Admiral Piet Hein. (C. B. van Nes) See color illus. VV p. 5.

cl.
        -griersonianum (1/2)
Van Veen-
        -Pygmalion--unknown (1/2)

6ft(1.8m)          -5°F(-21°C)          ML          3/2

Flowers of rich, clear, dark red, 3in(7.6cm) across, in rounded trusses of 7. Plant habit rather open; long dark green leaves. (T. Van Veen, Sr., intro. 1956) See color illus. VV p. vi.

cl.                            -catawbiense (1/4)
                -Charles Dickens-
Van Weerden Poelman-           -unknown (3/4)
                -Lord Roberts--unknown

Crimson flowers. (H. den Ouden & Sons cross, 1912; intro. 1925)

g. & cl.
            -souliei (1/4)
        -Soulbut-
Vanessa-       -SirCharles Butler--fortunei ssp. fortunei (1/4)
        -griersonianum (1/2)

5ft(1.5m)          5°F(-15°C)          ML

Flowers of soft pink, spotted carmine at the base, in handsome trusses. Spreading, shapely, slow-growing bush; foliage glossy dark green. (Lord Aberconway cross, 1924) F. C. C. (RHS) 1929

cl.                  -souliei (1/4)
                -Soulbut-
Vanessa Pastel-       -Sir Charles Butler--fortunei ssp. for-
                -griersonianum                  tunei (1/4)

5ft(1.5m)          5°F(-15°C)          ML          3/3

Buds open brick red, changing to apricot to deep cream with darker bronze yellow in throat, exterior of orange buff suffused rose pink, pale peach. A much pinker form also exists. Flowers to 4.5in(10.8cm), 5-lobed, about 8 per truss. Plant rather upright; pointed moss green leaves 5in(12.7cm) long. (Lord Aberconway cross, 1930) A. M. (RHS) 1946; F. C. C. (Wisley Trials) 1971 See color illus. ARS J 37:2 (Spring 1983): p. 67.

g.
        -venator (1/2)
Vanguard-
        -griersonianum (1/2)

3ft(.9m)          15°F(-10°C)          EM          3/2

Plant bushy, broader than tall; dark green leaves, 5.5in(14cm) long. Bright scarlet red flowers in lax trusses. Blooms at an early age; floriferous. (Lord Headfort, Co. Meath, Ire., 1940)

                -souliei (1/8)
g.          -Soulbut-
        -Vanessa-       -Sir Charles Butler--fortunei ssp.
Vanity-       -griersonianum (1/4)          fortunei (1/8)
        -       -dichroanthum ssp. dichroanthum (1/4)
        -Dante-
                -facetum (eriogynum) (1/4)

Flowers of scarlet. (Lord Aberconway, 1946)

cl.
        -griffithianum (1/2)
Vanity-
        -hardy hybrid--unknown (1/2)

Flowers open pale pink, fading to pearly white, narrow edging of rose madder, funnel-shaped, 4in(10.2cm) across; trusses hold 11. Vigorous, upright plant; glossy leaves to 7in(18cm) long. Very free-flowering. (Slocock Nursery) H. C. (RHS) 1962

g.                      -souliei (1/8)
                -Soulbut-
            -Vanessa-       -SirCharlesButler--fortunei ssp.
Vanity Fair-       -griersonianum (1/4)          fortunei (1/8)
            -facetum (eriogynum) (1/2)

Flowers of deep red. (Lord Aberconway, 1941)

cl.
        -thomsonii ssp. thomsonii (1/2)
Vantom-
        -Van Nes hybrid ? --unknown (1/2)

Pink flowers. (Exhibited 1938)

g.                      -souliei (1/8)
                -Soulbut-
            -Vanessa-       -Sir Charles Butler--fortunei ssp.
Vanven-       -griersonianum (1/4)          fortunei (1/8)
        -venator (1/2)

Deep red flowers. (Lord Aberconway, 1946)

cl.

Variegatum*--form of ponticum

4ft(1.2m)          -10°F(-23°C)          L          2/3

Lavender flowers with variegated foliage. A vigorous plant from Ireland, distributed by Rhododendron Species Foundation.

cl.

Variegatum*--form of ferrugineum

2ft(.6m)          -15°F(-26°C)          M          3/4

Leaves have a thin border of creamy white; new growth has rust-colored scales. Flowers are rose, or rose scarlet, in clusters of 6-12. Plant is a slow-growing dwarf. (Bean, v. III, p. 660)

g.                  -sanguineum ssp. didymum (1/4)
        -Carmen-
Varna-       -forrestii ssp. forrestii Repens Group (1/4)
        -williamsianum (1/2)

Yellow flowers, flushed rose. (Lord Aberconway, 1946)

g.
        -aperantum (1/2)
Vasco-
        -stewartianum (1/2)

Red flowers. (Lord Aberconway, 1946)

cl.
        -catawbiense (1/2)
Vauban-
        -unknown (1/2)

Flowers mauve with a bronze yellow blotch. (A. Waterer)

cl.
        -impeditum (1/2)
Veesprite-
        -racemosum (1/2)

1.5ft(.45m)          -10°F(-23°C)          EM

Very small leaves, .75in(2cm) by .5in(1.3cm); flowers also very small, .75in(2cm) across, of Persian rose, in terminal clusters of 3-5 flowers each. (R. Forster, Horticultural Research Institute of Ontario, Can.)

g.         -dichroanthum ssp. dichroanthum (1/4)
     -Fabia-
Vega-      -griersonianum (1/4)
     -haematodes ssp. haematodes (1/2)

Scarlet flowers.  (Lord Aberconway, 1941)

cl.           -wardii var. wardii (3/4)
       -Prelude-
Vellum-       -fortunei ssp. fortunei (1/4)
       -wardii var. wardii

Flowers of very light chartreuse, held in trusses of 6.  (Edmund
de Rothschild, reg. 1980)

                                   -griffithianum (1/8)
                       -unnamed hybrid-
cl.        -Pink Beauty-           -unknown (1/8)
        -           -           -catawbiense (1/8)
Velma Rozetta-        -John Walter-
        -                      -arboreum ssp. arboreum
        -           -griersonianum (3/8)        (1/8)
        -unnamed hybrid-   -griersonianum
                      -Azor-
                           -fortunei ssp. discolor (1/8)

Hybrid by William Whitney, before 1958.

cl.
        -venator (1/2)
Venapens-
        -forrestii ssp. forrestii Repens Group (1/2)

Dwarf plant with deep red flowers in midseason.  (Sir John Ramsden, 1940)

g.
        -venator (1/2)
Venco-              -griffithianum (1/8)
     -     -Loderi-
     -Coreta-       -fortunei ssp. fortunei (1/8)
            -arboreum ssp. zeylanicum (1/4)

Crimson scarlet flowers.  (Lord Aberconway, 1946)

cl.            -facetum (eriogynum) (1/4)
        -unnamed-    -dichroanthum ssp. dichroanthum (1/8)
Venetian- hybrid-Fabia-
  Chimes -         -griersonianum (1/8)
        -          -yakushimanum ssp. yakushimanum (1/4)
        -unnamed-       -Queen    -griffithianum (1/16)
          hybrid-Britannia-Wilhelmina-
                     -          -unknown (3/16)
                     -Stanley Davies--unknown

2ft(.6m)         0°F(-18°C)         M         4/3

Plant  vigorous, compact, wider than high;  leaves to 4in(10.2cm)
long.  Flowers campanulate, 4.5in(11.5cm) across, carmine rose
with dark spotting;  ball-shaped trusses of 11.  See also Surrey
Heath.  (John Waterer, Sons & Crisp, 1971)

                          -unnamed-griffithianum (3/16)
                          - hybrid-
                   -Mrs. Furnival-    -unknown (1/4)
                   -          -          -caucasicum (1/16)
cl.   -Party Pink-           -unnamed hybrid-
      -            -                      -unknown
Venice-          -catawbiense var. album (1/4)
      -            -yakushimanum ssp. yakushimanum (1/4)
      -unnamed hybrid-   -griffithianum
                    -Mars-
                        -unknown

4ft(1.2m)        -20°F(-29°C)        ML

Flowers delicate pale pink, flushed darker, faint dorsal spotting of yellow, openly funnel-shaped, 3.25in(8.3cm) wide;  spherical trusses of 22.  Plant rounded, as wide as tall;  leaves held
2-3 years.  (David G. Leach, cross 1965;  reg. 1984)

                   -campylocarpum ssp. campylocarpum
g.        -Penjerrick-          Elatum Group (1/8)
     -Amaura-       -griffithianum (1/8)
Venus-     -griersonianum (1/4)
      -facetum (eriogynum) (1/2)

Flowers bright red or salmon red.  (Lord Aberconway, 1936)

                     -wardii var. wardii (1/8)
          -unnamed-       -neriiflorum ssp. nerii-
          - hybrid-F. C. Puddle-        florum (1/16)
          -                 -griersonianum (3/32)
       -Virginia-              -griffithianum
       -Richards-          -George-
cl.    -        -Mrs. Lindsay- Hardy-catawbiense
       -        -    - Smith -           (1/32)
Vera   -    -Mrs. Betty-      -Duchess of Edinburgh
Elliott-      Robertson-        --unknown
       -                  -campylocarpum ssp. campylo-
       -          -unnamed-           carpum (1/16)
       -          hybrid-unknown (1/8)
       -fortunei ssp. fortunei (1/2)

6ft(1.8m)         0°F(-18°C)         ML

Rose-colored flowers, lightly spotted orange red, 7-lobed, 4.5in
(11.5cm) across,  held in trusses of up to 10.  Leaves 6.5in(16.5
cm) by 2.5in(6.4cm).  (Walter Elliott cross, 1966;  reg. 1977)

                     -griffithianum (1/8)
cl.        -Loderi-
       -Albatross-    -fortunei ssp. fortunei (1/8)
Vera Hawkins-    -fortunei ssp. discolor (1/4)
       -      -dichroanthum ssp. dichroanthum (1/4)
       -Fabia-
            -griersonianum (1/4)

6ft(1.8m)         5°F(-15°C)         ML

Flowers porcelain rose veined with rose, dark orange dots with
an overlay of pale yellow on upper petal, to deep red in throat.
Flowers 4in(10.2cm) wide, campanulate, in trusses of 12.  Plant
broader than tall;  leaves 6in(15.2cm) long.  (H. L. Larson, 1964)

cl.
        -yakushimanum ssp. yakushimanum (1/2)
        -                 -griffithianum (1/16)
Verna Carter-        -Queen Wilhelmina-
       -    -Britannia-        -unknown (3/16)
       -Leo-      -unknown
          -elliottii (1/4)

3ft(.9m)         0°F(-18°C)         M

Pale pink buds open to white flowers 2in(5cm) wide, 5 wavy and
frilled lobes,  slightly fragrant,  in ball-shaped trusses of 12.
Plant rounded, almost as wide as tall;  medium green leaves with
fawn indumentum, mouse-colored new growth.  See also Mary Jane,
Vallerie Kay.  (Mrs. LeVerne Freimann, reg. 1974)

cl.             -fortunei ssp. discolor (1/4)
       -Lady  -
Verna  -Bessborough-campylocarpum ssp. campylocarpum
Phetteplace-            Elatum Group (1/4)
       -yakushimanum ssp. yakushimanum (Exbury Form) (1/2)

6ft(1.8m)         -5°F(-21°C)         ML         4/3

Plant rounded, as broad as tall;  olive green leaves, 5in(12.7cm)
long.  Flowers 3.5in(8.9cm) wide, ivory shaded pink with vivid
red dorsal blotch, the reverse pink diffusing through corolla;
globular trusses of 13-15.  (Dr. Carl H. Phetteplace, 1975)

cl.             -caucasicum (1/4)
      -Cunningham's White-
Vernus-         -ponticum (white form) (1/4)
      -         -catawbiense (1/4)
      -unnamed red hybrid-
              -unknown (1/4)

5ft(1.5m)          -25°F(-32°C)          VE          3/3

One of the earliest of the very hardy hybrids.  An abundance of
light pink flowers with darker centers, about 2in(5cm) across;
leaves 5in(12.7cm) by 1.75in(4.5cm).  Good plant habit.  (A. M.
Shammarello cross; raised & intro. by David G. Leach; reg. 1962)
See color illus. VV p. 46.

cl.
                    -campylocarpum ssp. campylocarpum (1/2)
Veronica Milner*          -neriiflorum ssp. neriiflorum (1/4)
                    -Little Ben-
                              -forrestii ssp. forrestii Repens
                                                  Group (1/4)

Flowers of rose madder.  (E. J. Greig, Royston Nursery, 1962)

                                        -catawbiense
                    -Parsons Grandiflorum-          (1/16)
cl.          -Nova Zembla-          -unknown (1/16)
        -unnamed-          -red hybrid--unknown (1/8)
Veronika- hybrid-Catalgla--catawbiense var. album Glass (1/4)
Peiffer -          -fortunei ssp. discolor (1/4)
        -Lady Bessborough-
                    -campylocarpum ssp. campylocarpum
                                        Elatum Group (1/4)

1ft(.3m)          -25°F(-32°C)          ML

Plant as broad as high; dark green leaves to 4in(10.2cm) long.
Dome-shaped trusses of 15 flowers, light cyclamen purple, without markings, 5 wavy-frilled lobes; floriferous.  (Rudy Behring
cross, 1974; reg. 1985)

g.
          -campylocarpum ssp. campylocarpum (1/2)
Verrocchio-
          -neriiflorum ssp. neriiflorum (1/2)

Amber-colored flowers.  (Adams-Acton, 1936)

cl.
          -pseudochrysanthum (1/2)
Veryan Bay*
          -williiamsianum (1/2)

A dense, rounded shrub, with almost-circular sea green leaves
and many large, clear pink flowers.  Blooms early midseason.

cl.
          -williamsianum (1/2)
Vesper Bells-
          -albertsenianum (1/2)

1ft(.3m)          E

Rose Bengal buds opening to lighter rose bell-shaped flowers, 3
in(7.6cm) wide, in trusses of 6-9.  Plant almost twice as broad
as tall, thickly branched; oval leaves 2.5in(6.4cm) long.  (Benjamin F. Lancaster, 1967)

                    -ponticum (1/8)
g.          -Michael Waterer-
    -Prometheus-          -unknown (3/8)
Vesta-          -Monitor--unknown
    -williamsianum (1/2)

Flowers of deep rose.  (Dietrich Hobbie, 1949)

cl.
          -catawbiense (white form) (1/2)
Vestale-
          -unknown (1/2)

White flowers.  (David G. Leach, 1954)

g.
          -griersonianum (1/2)
Vestris-
          -orbiculare ssp. orbiculare (1/2)

Deep rose flowers.  (Lord Aberconway, 1946)

cl.
          -catawbiense (1/2)
Vesuvius-
          -arboreum ssp. arboreum (1/2)

Synonym Grand Arab.  Flowers bright orange scarlet, shaded violet
on upper segment.  (J. Waterer, 1867)

cl.
          -griersonianum (3/4)
Vesuvius-          -Moser's Maroon--unknown (1/4)
    -Romany Chai-
          -griersonianum

Flowers brilliant vermilion red, unspotted; well-shaped trusses
of 12-15.  Small, compact plant with foliage like griersonianum.
Very late, and hardier than most late reds.  (Sunningdale)

                    -soulei (1/16)
          -Soulbut-
    -Vanessa-          -Sir Charles Butler--fortunei (1/16)
g.  -Etna-          -griersonianum (1/4)
    -          -dichroanthum ssp. dichroanthum (1/8)
Veta-    -Fabia-
    -          -griersonianum
    -venator (1/2)

Rose flowers.  (Lord Aberconway, 1946)

cl.
          -diaprepes (1/2)
Vibrant-
          -wardii var. wardii (1/2)

The reverse cross of Vigil.  Trusses of 14 flowers, 4in(10.2cm)
wide, saucer-shaped, 6-7 frilled lobes, greenish white and without marks.  Young foliage bronze.  (Gen. Harrison, reg. 1962)

cl.

Vicki Reine          Parentage unknown

5ft(1.5m)          -5°F(-21°C)          ML          4/3

Synonym Olympic Vickie Reine.  Flowers bicolor, deep rose red on
margins, fading white in throat, rose exterior, to 4.5in(11.5cm)
across; trusses of about 12.  Plant as wide as high; dark green,
deeply veined leaves 7.5in(19cm) long.  (Roy Clark, 1963)  C. A.
(ARS) 1971    See color illus HG p. 67.

                              -griffithianum (1/4)
                    -Pink Shell-          -fortunei ssp.
cl.          -          -H. M. Arderne-          fortunei
    -Coronation Day-          -unknown
Victor -          -          -griffithianum
Boulter-          -Loderi-
    -          -fortunei ssp. fortunei (3/16)
    -unknown (9/16)

Trusses of 12-14 flowers, funnel-shaped, magenta rose with ruby
red blotch, the reverse darker magenta.  Plant 4ft(1.2m) at registration; elliptic leaves 3.5in(8.9cm) long.  (V. J. Boulter
cross; F. Boulter, reg. 1983)

cl.          -griersonianum (1/4)
          -Mrs. Horace Fogg-          -griffithianum (1/8)
Victor Herbert-          -Loderi-
          -          Venus-fortunei ssp. fortunei
          -strigillosum (1/2)          (1/8)

Crimson campanulate flowers, 3in(7.6cm) across; flat trusses of
12-15.  Plant at 9 years 3.5ft(1m) tall, 4ft(1.2m) wide; foliage
like strigillosum, but broader.  Early.  (C. S. Seabrook, 1964)

cl.                             -campylocarpum ssp. campylo-
                      -Moonstone-                    carpum (3/4)
Victoria de Rothschild-          -williamsianum (1/4)
                      -campylocarpum ssp. campyocarpum

Trusses of 5-7 flowers, light yellow with greenish tinge. (Edmund de Rothschild, reg. 1980)

---
              -dalhousiae var. dalhousiae (1/2)
Victorianum-
              -nuttallii (1/2)

4ft(1.2m)          20°F(-7°C)          EM          5/2

Plant erect and may be leggy. Beautiful flowers, creamy yellow
in bud, opening pure white, in lax trusses of 5.  Reverse cross
of Edinense. (Pince, before 1871)

cl.
       -wardii var. wardii (1/2)
Vienna-                    -griffithianum (1/16)
       -          -Kewense-
       -     -Aurora- -fortunei ssp. fortunei (5/16)
       -Naomi-     -thomsonii ssp. thomsonii (1/8)
              -fortunei ssp. fortunei

5ft(1.5m)          5°F(-15°C)          M          4/3

Idealist g.  Pale yellow flowers on a plant like Idealist.  Exbury cross; selection from seedlings at Knap Hill. (Reg. 1964)
P. C. (RHS) 1962

                         -dichroanthum ssp. dichroanthum (1/8)
                   -Dido-
       -Lem's -     -decorum (1/8)          -griffithianum
cl.    - Cameo-          -Beauty of Tremough-
       -     -     -Norman-          -arboreum ssp. ar-
Viennese-  -Anna- Gill -griffithianum (1/4)  boreum (1/32)
  Waltz -     -          -griffithianum
       -          -Jean Marie  -
       -          de Montague-unknown (15/32)
       -Pink     -Jan Dekens-unknown          -griffithianum
        Petticoats-          -Queen Wilhelmina-
                   -Britannia-          -unknown
                        -Stanley Davies--unknown

5ft(1.5m)          0°F(-18°C)          M

Buds rhodonite red opening to fragrant flowers of light red with
orange brown spotting, 6-7 lobes; trusses 9in(23cm) wide holding
up to 34 flowers, densely packed.  Plant 3/4 as broad as tall;
dark green leaves held 2 years. (John C. Lofthouse cross, 1976;
reg. 1984)

                                   -catawbiense (1/8)
                   -Parsons Grandiflorum-
              -America-          -unknown (11/16)
       -unnamed-     -dark red hybrid--unknown
cl.    - hybrid-          -griffithianum (3/16)
       -          -Queen Wilhelmina-
Viet Vet*     -Britannia-          -unknown
       -          -Stanley Davies--unknown
       -          -catawbiense
       -          -Parsons Grandiflorum-
       -unnamed-America-          -unknown
         hybrid-     -dark red hybrid--unknown
              -     -griffithianum
                -Mars-
                   -unknown

Deep red buds open to flowers of strong red  with white throats.
Plant of medium size; hardy to -15°F(-26°C). (Weldon E. Delp)

cl.
       -wardii var. wardii (1/2)
Vigil-
       -diaprepes (1/2)

Reverse cross of Vibrant.  Truss has 15 flowers, each 3.5in(9cm)
across, saucer-shaped, 7-lobed, colored cream with basal spots.
(Gen. Harrison, 1962)

                   -dichroanthum ssp. dichroanthum (1/8)
             -Astarte-          -campylocarpum ssp.
             -     -Penjerrick- campylocarpum Elatum Gp. (1/16)
g.   -Solon-          -griffithianum (3/16)
     -     -griffithianum
Viking-  -Sunrise-
     -          -griersonianum (1/8)
     -          -thomsonii ssp. thomsonii (1/4)
     -Chanticleer-
              -facetum (eriogynum) (1/4)

Flowers rosy red. (Lord Aberconway, 1950)

cl.

Vin Rose--form of cinnabarinum ssp. cinnabarinum Roylei Group

5ft(1.5m)          5°F(-15°C)          M          3/3

Pendulous, open-tubular, waxy flowers of plum crimson. Slender,
upright plant. (Crown Estate, Windsor)  A. M. (RHS) 1953

cl.

Vincent van Gogh      Parentage unknown

A parent of Dorothy Peste Anderson.  Bright cerise red flowers,
striped white. (M. Koster & Sons, 1939)

cl.
       -keiskei (1/2)
Vinestar-
       -racemosum (1/2)

Flowers pale canary yellow, flecked with orange brown in throat.
(Horticultural Research Institute  of Ontario, Can., cross 1961;
intro. & reg, 1977)

cl.   -yakushimanum ssp. yakushimanum (1/2)
      -                    -fortunei ssp. discolor
Vintage-          -Lady    -          (1/16)
  Rose -     -Jalisco-Bessborough-campylocarpum ssp. campylo-
       -     -Eclipse-          carpum Elatum Gp. (1/16)
       -unnamed-     -dichroanthum ssp. dichroanthum
         hybrid-     -Dido-          (1/16)
       -          -decorum (1/16)
       -          -elliotii (1/8)
       -Fusilier-
              -griersonianum (1/8)

2ft(.6m)          0°F(-18°C)          LM

Plant 50% wider than high, vigorous, compact, floriferous; dull
dark green leaves 4.1in(10.5cm) long. Flowers 2.3in(6cm) wide,
white lightly tinged rose, deeper in throat, slight red spotting
on upper segment. (John Waterer, Sons & Crisp, reg. 1975) H. C.
(RHS) 1979

cl.          -campanulatum ssp. campanulatum (1/4)
     -Madame Nauen-
Viola-          -unknown (1/4)
     -          -catawbiense (1/4)
     -unnamed hybrid-
              -ponticum (1/4)

Parentage as in Walter Schmalscheidt; listed in the Register  as
a caucasicum hybrid. A parent of Albert, Bismarck, Tibet.
Flowers porcelain white. (T. J. R. Seidel cross, 1872)

cl.
       -ponticum (1/2)
Violet Gose-     -sutchuenense (1/4)
         -Geraldii-
              -praevernum (1/4)

Flowers of mauve to pansy violet, in trusses of 17, early midseason.  Compact plant, 4ft(1.2m) in 10 years; leathery, dark
green leaves. (Melvin V. Love, reg. 1965)

cl.
                    -caucasicum (1/2)
Violet Parsons-
                         -thomsonii ssp. thomsonii (1/2)

Profuse trusses of salmon pink or peach pink flowers.   Plant of
moderate size.  (Offered by Slocock)

---                                -brookeanum var. gracile (1/4)
          -Duchess of Edinburgh-
Virgil-                    -longiflorum (lobbii) (1/4)
     -                     -jasminiflorum (3/8)
     -Princess Alexandra-          -jasminiflorum
                         -Princess Royal-
                              -javanicum (1/8)

Vireya hybrid.  Primrose yellow flowers.   A. M. (RHS) 1889

cl.

Virgin*--form of catawbiense

Pure white flowers in small- to medium-sized trusses.  Plant upright, broad, slow-growing; foliage medium.   (Weston Nurseries)

---

Virgin, The      See The Virgin

cl.
                    -Koichiro Wada--yakushimanum ssp. yakushimanum
Virginia Anderson-      -Corona--unknown (1/4)          (1/2)
               -Bow Bells-
                    -williamsianum (1/4)

3ft(.9m)          0°F(-18°C)          ML

Plant rounded, wider than tall; dark leaves with slight indumentum, held 2-3 years.  Flowers 3in(7.6cm) across, pale purplish
pink shading to white, reverse striped purplish red, 5-lobed; 12
per truss. Floriferous. (Cross by the Bovees; reg. by Sorensen
& Watson, 1975)

cl.
                 -neriiflorum ssp. neriiflorum (1/2)
Virginia Carlyon-
                 -thomsonii ssp. thomsonii (1/2)

Dense, hemispherical trusses of about 12 flowers, 5-lobed, of
strong red, without markings. (Grown by E. T. R. Carlyon; reg.
by Mrs. G. Carlyon, 1985)

cl.
                         -racemosum (1/4)
            -unnamed hybrid-
Virginia Delito*         -mucronulatum (pink form) (1/4)
            -minus var. minus Carolinianum Group (1/2)

Sister seedling of Alice Swift, q.v. Differs chiefly in bearing
larger flowers of deep pink, and growing taller, to 3.5ft(1m) in
10 years. (Leon Yavorsky)

                         -maximum (1/8)
                 -Maxecat-
          -unnamed-       -catawbiense (1/8)
cl.       - hybrid-   -dichroanthum ssp. dichroanthum (1/8)
          -    -Jasper-          -fortunei ssp. discolor
Virginia-             -unnamed hybrid-          (1/16)
Leach -                         -campylocarpum ssp. campy-
     -                                      locarpum (1/16)
     -               -catawbiense var. album (1/4)
          -unnamed hybrid-
                         -griersonianum (1/4)

4ft(1.2m)          -15°F(-26°C)          ML          4/3

Flowers of vibrant greenish yellow edged in strong purplish pink
with dorsal lobe deeper yellow, a faint blotch of orange brown
spots.  Flowers openly funnel-shaped, 3in(7.6cm) wide, partially
hose-in-hose; pyramidal trusses hold 18. Shrub wider than tall;
dark green leaves 4.5in(11.5cm) long. (David G. Leach, 1972)

g. & cl.       -wardii var. wardii (1/4)
      -unnamed-      -neriiflorum ssp. neriiflorum (1/8)
      - hybrid-F. C. Puddle-
Virginia-          -griersonianum (1/8)
Richards-          -George-griffithianum (1/16)
       -Mrs. Lindsay- Hardy-
      -Mrs. Betty- Smith   -catawbiense (1/16)
        Robertson-      -Duchess of Edinburgh--unknown
                    -               (1/4)
          -unnamed-campylocarpum ssp. campylocarpum
               hybrid-          (1/8)
                    -unknown

4ft(1.2m)          -5°F(-21°C)          M          4/4

Flowers open pale yellow with a pink blush and crimson blotch,
and fade Chinese yellow; the garden effect is peach or apricot.
Flowers 4.5in(11.5cm) wide, in trusses of about 12.  Plant compact, with dark glossy foliage.  Other forms exist. (William
Whitney, intro. 1965; reg. by the Sathers, 1976)   See color illus. ARS J 31:3 (Summer 1984): p. 115; HG p. 68;  VV p. 2.

cl.
          -souliei ? (1/2)
Virginia Scott-
          -unknown (1/2)

Yellow flowers.  (H. L. Larson)

cl.                    -ciliatum (1/4)
      -Countess of Haddington-
Virginia-          -dalhousiae var. dalhousiae (1/4)
  Stewart-
      -nuttallii (1/2)

5ft(1.5m)          25°F(-3°C)          EM

Plant upright, wider than tall; glossy, dark, elliptic leaves,
scaly beneath, 4in(10.2cm) long. Very fragrant white flowers,
with a quickly fading yellow stain in throat, about 3.5in(8.9cm)
wide, in flat trusses of 7. (Cross by H. W. Kerrigan; intro. by
Coulter Stewart, 1975)

cl.
        Antoon van Welie, q.v.
Virgo   Professor J. H. Zaayer, q.v.
        Annie E. Endtz, q.v.

Exact  combination of the parentage is unknown, but each  parent
is a Pink Pearl hybrid.  Pyramidal trusses hold 17-18 flowers,
fringed, 3.5in(8.9cm) wide, slightly pink when first open, then
fading white, with large reddish brown spotted blotch.  See also
Pink Goliath, Record. (P. van Nes, 1962)  A. M. (Boskoop) 1959

cl.
     -oreotrephes(1/2)
Virtue-
    -Lady     -cinnabarinum ssp. cinnabarinum Roylei Group
     Chamberlain-                    (1/4)
          -          -cinnabarinum ssp. cinnabarinum
          -Royal Flush -               (1/8)
          (orange form)-maddenii ssp. maddenii (1/8)

Mauve flowers, 2.25in(5.7cm) across, held in trusses of about 6.
(Gen. Harrison, reg. 1962)

cl.
     -racemosum (1/2)
Viscount-
 Linley -ciliatum (1/2)

Similar to Racil in color and shape of truss. (Alfred Bramley,
1963)

cl.

Viscount Powerscourt      Parentage unknown

A parent of Louis Pasteur. Purplish red flowers with dark spotting.  (J. Waterer)  A. M. (RHS) 1906

```
cl.    -unnamed-fortunei ssp. fortunei (7/16)
       - hybrid-
Vistoso-      -diaprepes (3/8)
       -           -fortunei ssp. fortunei
       -    -Mary Cowan-
    -unnamed-          -diaprepes
      hybrid-              -griffithianum (1/16)
       -           -Loderi-
      -Avalanche-      -fortunei ssp. fortunei
                   -calophytum var. calophytum (1/8)


6ft(1.8m)        0°F(-18°C)        M
```

Trusses 10in(25cm) across, of 12-14 fragrant flowers, empire
rose with darker rose spotting, openly funnel-campanulate, 5.5in
(14cm) wide, 7 wavy lobes; plant upright, half as broad as high;
dark green leaves held 3 years. (Cross by Dr. David Goheen,
1962; reg. 1985; intro. by Dr. Lansing Bulgin)

```
                             -catawbiense (1/8)
cl.         -Parsons Grandiflorum-
       -America-              -unknown (1/2)
Vivacious-    -dark red hybrid--unknown
       -     -griersonianum (1/4)  -arboreum ssp. arboreum
      -Dr. Ross-        -Doncaster-              (1/16)
           -Borde Hill-       -unknown
                       -         -griffithianum (1/16)
                     -Mrs. A. M.-
                       Williams -unknown

3.5ft(1.m)      -5°F(-21°C)       ML
```

Flowers of heavy substance 2.5in(6.4cm) wide, shaded medium cardinal red; 10 or more in spherical trusses. Rounded plant wider
than tall; medium green elliptic leaves held 2 years. (Cross by
Ray Forster; Horticultural Research Institute, Ont., reg. 1976)

```
                      -catawbiense (1/8)
---          -unnamed hybrid-
    -Altaclarense-              -ponticum (1/8)
Vivid-           -arboreum ssp. arboreum (1/4)
    -           -maximum (1/4)
    -unnamed hybrid-
                  -unknown (1/4)
```

Flowers rose, or bright purplish rose. (Standish & Noble, 1850)

```
cl.
       -Koichiro Wada--yakushimanum ssp. yakushimanum (1/2)
Vivid-O-
       -arboreum ssp. arboreum (red form) (1/2)

4ft(1.2m)      5°F(-15°C)       VE
```

Synonym Estacada. Cardinal red buds open to crimson to carmine
rose flowers, openly campanulate, 2.5in(6.4cm) wide; dome-shaped
trusses of 18-20; floriferous. Upright plant, broader than tall;
elliptic leaves with greyed white indumentum, held 3 years; new
growth white tomentose. (Dr. D. Goheen cross, 1967; reg. 1983)

```
cl.
       -laetum (1/2)
Vladimir-                    -brookeanum var. gracile
Bukovsky-        -Crown Princess-              (1/8)
     -Souvenir de  - of Germany  -Princess-jasminiflorum
       J. H. Mangles-           Royal -            (1/16)
       -                           -javanicum (5/16)
              -javanicum

3ft(.9m)       32°F(0°C)        Dec.-Jan.
```

Vireya hybrid. Tubular flowers 2.25in(5.7cm) across, 1.75in(4.5
cm) long, of nasturtium red with Chinese yellow in throat and on
reverse, 5-lobed; dome-shaped trusses of 7-12. Plant rounded,
almost twice as broad as tall; glossy concave leaves, scaly,
held 3 years. (T. Lelliott cross; raiser P. Sullivan; reg. by
W. A. Moynier, 1984)

```
                     -campylocarpum ssp. campylocarpum
              -Penjerrick-          Elatum Group (1/16)
g.       -Amaura-        -griffithianum (5/16)
      -Eros-     -griersonianum (3/8)
Voltaire-   -griersonianum
      -         -griffithianum
      -Loderi-
            -fortunei ssp. fortunei (1/4)

Flowers red or pink. (Lord Aberconway, 1950)
```

```
                      -griffithianum (1/8)
cl.         -George Hardy-
      -Mrs. Lindsay Smith-       -catawbiense (1/8)
Vondel-          -Duchess of Edinburgh--unknown (1/2)
      -          -campylocarpum ssp. campylocarpum (1/4)
      -unnamed hybrid-
              -unknown

Pale yellow flowers. (M. Koster & Sons)
```

```
                       -griffithianum (1/8)
cl.         -Queen Wilhelmina-
      -Britannia-              -unknown (3/8)
Voodoo-      -Stanley Davies--unknown
      -          -haematodes ssp. haematodes (1/4)
      -May Day-
            -griersonianum (1/4)
```

Flowers funnel-campanulate, 2.5in(6.4cm) wide, rose red shading
to cardinal red in throat; trusses of 10-12. Semi-dwarf plant;
leaves 4.5in(11.5cm) long, recurved margins, tan indumentum beneath. (Rudolph Henny, reg. 1958)   P. A. (ARS) 1952

```
cl.      -griffithianum (1/4)
      -Mars-
Vulcan-   -unknown (1/4)
      -griersonianum (1/2)

5ft(1.5m)       -10°F(-23°C)       ML       4/4
```

Registered as a clone, but several forms are known. Bright fire
red flowers and dark green pointed leaves on a rounded plant.
See also Vulcan's Flame. (Waterer, Sons & Crisp, 1938)   A. M.
(Wisley Trials) 1957   See color illus. VV p. 12.

```
cl.              -griersonianum (1/4)
      -Vulcan's Flame-  -griffithianum (1/8)
Vulcan's Bells-       -Mars-
      -               -unknown (1/8)
      -williamsianum (1/2)

2ft(.6m)       0°F(-18°C)        EM        3/4
```

Plant a sturdy mound, wider than tall, densely foliaged; leaves
oval, dark green, 2.5in(6.4cm) long. Flowers of clear rose red,
openly campanulate, 2.5in(6.4cm) wide; trusses of 6-8. (Benjamin F. Lancaster, 1967)

```
cl.
          -griersonianum (1/2)
Vulcan's Flame-  -griffithianum (1/4)
      -Mars-
            -unknown (1/4)

5ft(1.5m)       -15°F(-26°C)       ML       4/4
```

Reverse cross of Vulcan. Similar to Vulcan but may be a bit
more hardy. Bright red flowers, to 3in(7.6cm) wide; trusses of
12-15. Vigorous, rather compact plant; leaves deep yew green,
red petioles. Appropriately, it does well in hot climates. (B.
F. Lancaster, reg. 1958)   See color illus. VV p. 100.

g.
           -haematodes ssp. haematodes (1/2)
W.F.H.-          -griersonianum (1/4)
        -Tally Ho-
                 -facetum (eriogynum) (1/4)

Small plant with flowers funnel-shaped, brilliant scarlet.
Named after W. F. Hamilton, head gardener at Pylewell Park, Lymington. (Whitaker, 1941)

---
              -griffithianum (1/2)
W. H. Forster-
              -unknown (1/2)

Bright red flowers.

cl.
           -griffithianum (1/4)
      -Loderi-
W. Leith-     -fortunei ssp. fortunei (1/4)
      -decorum (1/2)

White Ensign g.  Flowers ivory, tinged faint greenish yellow,
lobes recurved.  Sturdy, erect plant with large foliage, upright
trusses.  Named after the Head Gardener, Clyne Castle, Swansea.
(Heneage-Vivian)  A. M. (RHS) 1935

cl.
              -fastigiatum (1/2)
Wachtung*--sport of Ramapo-
                    -minus var. minus Carolinianum Group
                                                    (1/2)
A parent of Matilda, q.v.  (Leon Yavorsky)

cl.

Wakehurst--form of rubiginosum

6ft(1.8m)         0°F(-18°C)         EM         3/3

Campanulate flowers 2in(5cm) across, mallow purple with crimson
spots on upper lobe, held in trusses of 25. (Sir Henry Price)
A. M. (RHS) 1960

                        -griffithianum (9/16)
            -Beauty of-
           - Tremough-arboreum ssp. arboreum
        -Norman Gill-                    (1/16)
g.   -Anna-        -griffithianum
     -             -griffithianum
Walloper*   -Jean Marie de Montague-
     -                    -unknown (3/8)
     -         -griffithianum
     -Marinus Koster-
                  -unknown

6ft(1.8m)         -5°F(-21°C)         M         4/3

Flowers deep pink, in huge trusses. Large, vigorous plant,
heavily foliaged.  See also Pink Walloper, Point Defiance, Red
Walloper.  (Halfdan Lem)

cl.
              -griersonianum (1/2)
Walluski Chief-
              -macrophyllum (1/2)

Flowers of soft rose red with darker specks, 3in(7.6cm) across,
campanulate; trusses of 15. Upright, rounded plant, 6ft(1.8m)
at 10 years; leaves 7in(17.8cm) long.  (G. L. Baker, reg. 1965)

cl.
              -minus var. minus (white form) Carolinianum Group
Wally (Vallya)*                                        (1/2)
              -Cornell Pink--mucronulatum (1/2)

A mass of soft pink flowers, very early. Wide, upright plant,
growing 7in(17.8cm) a year; semi-evergreen.  Small, dark green,
pointed leaves turn orange yellow in autumn.  Suited for sunny,
exposed locations.  (Edmund V. Mezitt, Weston Nurseries)

cl.
           -Lu Shan--fortunei ssp. fortunei (1/2)
Wally Zeglat-
           -unknown (1/2)

Large plant, with lacy flowers  deep rose purple fading lighter,
brown speckled blotch.  Blooms midseason. (Raiser, W. Zeglat; J.
G. Lofthouse, reg. 1973)    Best of Show, Vancouver, 1970

                  -wardii ssp. wardii (1/2)
            -Crest-          -fortunei ssp. discolor (3/16)
                   -Lady   -
            -          Bessborough-campylocarpum ssp.
        -Lemon -              campylocarpum Elatum Group
        -Custard-                           -griffith.
        -       -                 -George-   (1/64)
        -       -         -Mrs. Lindsay- Hardy-
cl.     -       -          - Smith   -     -cat.(1/64)
        -       -Mrs. Betty-          -Duchess of
Walt    -       -Robertson -           Edinburgh--unk.
Elliott-    -unnamed-        -campylocarpum ssp.
        -       hybrid-    -unnamed- campylocarpum (1/32)
        -         -        hybrid-
        -         -             -unknown (1/16)
        -         -wardii var. wardii
        -   -wardii var. wardii
        -Crest-          -fortunei ssp. discolor
            -Lady Bessborough-
                        -campylocarpum ssp. campylocarpum
                                        Elatum Group (3/16)
6ft(1.8m)         0°F(-18°C)         M

Fragrant flowers, funnel-shaped, 6-lobed, Naples yellow edged in
light Venetian pink, 5in(12.7cm) wide; lax trusses of 7-10. Upright, open plant, broader than tall; yellow green leaves 7in(18
cm) long.  (Walter Elliott cross, 1969; Fred Peste, reg. 1982)

g.           -catawbiense (1/4)
      -John Walter-
Waltdis-      -arboreum ssp. arboreum (1/4)
      -fortunei ssp. discolor (1/2)

Large flowers of deep clear rose, slightly fragrant; large tight
umbels.  (Dietrich Hobbie, 1952)

cl.

Walter Curtis     Parentage includes catawbiense, maximum
                    and perhaps unknown hybrids

7ft(2.1m)         -15°F(-26°C)         M

Upright, spreading plant; olive green leaves, slightly hairy, a
thin brown indumentum beneath.  Flowers medium purplish pink
with brilliant yellow green spotting, 3in(7.6cm) wide, 5-lobed;
globular trusses of 18. Floriferous.  (W. David Smith, 1979)

cl.

Walter Maynard*--form of ciliicalyx

4ft(1.2m)         15°F(-10°C)         M         4/3

Trusses hold 3-4 white flowers, a stripe of soft red purple from
base to rim of each lobe fading to edges,  upper throat flushed
yellow green, 3.3in(8.5cm) wide.  Glossy, dark green leaves, 3.5
in(8.9cm) long, scaly underneath. (Exhibited by Geoffrey Gorer,
Sunte House)  A. M. (RHS) 1975

cl.          -minus var. minus Carolinianum Group (3/4)
      -Wilsoni-
Waltham*      -ferrugineum (1/4)
      -minus var. minus Carolinianum Group

Plant low, compact, growing 2 to 4in(5-10cm) per year; foliage
small, dark green.  Clear pink flowers, many small trusses. (Dr.
R. L. Ticknor cross; intro. by E. Mezitt, Weston Nurseries)

```
cl.          -dichroanthum ssp. dichroanthum (1/2)
       -Fabia-
Wantage-      -griersonianum (1/4)
       -           -dichroanthum ssp. dichroanthum
       -Dido-
              -decorum (1/4)
```

Two feet (.6m) high and 6ft(1.8m) wide in 16 years; narrowly elliptic leaves 4.5in(11.5cm) long. Flowers widely funnel-campanulate, 2in(5cm) across, rowan berry red, held loosely in trusses
of 6-8.  (Crown Estate, Windsor)  P. C. (RHS) 1970

```
cl.          -griffithianum (1/4)
       -Mars-
War Dance-   -unknown (3/4)
       -Pygmalion--unknown
```

4ft(1.2m)          -10°F(-23°C)          M

Trusses of 17-21 flowers, bright currant red with black dorsal
blotch and  spotting, 2.75in(7cm) wide, with 5 wavy lobes. Plant
much broader than high; dark green leaves 5in(12.7cm) long.  (M.
E. Hall cross, 1970; intro. by Edward J. Brown; reg. 1979)

```
                   -griersonianum (1/8)
             -Tally Ho-
      -unnamed-      -facetum (eriogynum) (1/8)
cl. - hybrid-                    -griffithianum (3/16)
     -       -           -Queen Wilhelmina-
War -     -Britannia-            -unknown (9/16)
Lord-     -
     -              -Stanley Davies--unknown
     -                       -griffithianum
     -        -Queen Wilhelmina-
     -Britannia-            -unknown
            -Stanley Davies--unknown
```

Handsome trusses of deep red flowers, in May.  (Knap Hill, 1967)

cl.

War Paint--form of elliottii

Crimson scarlet flowers.  (Selected by Del James)  P. A. (ARS)
1956

```
cl.
       -Loder's White, q.v.--2 possible parentage diagrams
Warburton-
       -Kew Pearl--unknown
```

White flowers tinged light spirea red, lightly spotted on upper
lobes with Indian lake.  See also Susan  Lonsdale. (H. D. Rose,
Australia, reg. 1971)

```
cl.          -griffithianum (3/8)
       -Tip-the-Wink-      -griffithianum
Warden Wink*      -Kewense-
       -               -fortunei ssp. fortunei (1/8)
       -wardii ssp. wardii (1/2)
```

Shown by Mr. Mansfield, Gardener to Col. G. Loder, High Beeches,
1939.

```
                   -fortunei ssp. discolor (1/4)
            -Lady     -
            -Bessborough-campylocarpum ssp. campylocarpum
cl.   -Jalisco-               Elatum Group (1/4)
   -      -      -dichroanthum ssp. dichroanthum (1/8)
Warfield-    -Dido-
   -           -decorum (1/8)
   -      -wardii var. wardii (1/4)
   -Crest-      -fortunei ssp. discolor
       -Lady     -
       Bessborough-campylocarpum ssp. campylocarpum
                            Elatum Group
```

Flowers very light primrose yellow, deeper in center and throat,
marked in currant red, widely funnel-campanulate, to 3.75in(9.5
cm) across; 10-12 in rounded trusses.  Leaves 5.5in(14cm) long.
(Crown Estate, Windsor) A. M. (RHS) 1970

```
cl.          -Moser's Maroon--unknown (1/2)
       -Romany Chal-
Warlock-      -facetum (eriogynum) (1/4)
       -           -ponticum (1/4)
       -Purple Splendour-
              -unknown
```

6ft(1.8m)          0°F(-18°C)          L

Plant broader than tall, with decumbent branches; leaves dark
green, 5.5in(14cm) long, held 3 years; new growth maroon.  Flowers of 5 waxy lobes, 3.5in(8.9cm) wide, dark reddish purple with
black dorsal blotch and spotting; flat trusses hold 14.  (Raiser
C. W. Bledsoe; D. L. Bledsoe, reg. 1976)

```
cl.               -dichroanthum ssp. dichroanthum (1/4)
       -unnamed hybrid-
Warm Glow-        -unknown (3/4)
       -Vida--unknown
```

3ft(.9m)          0°F(-18°C)          ML          4/4

Campanulate flowers, 2.5in(6.4cm) wide, pale orange darker in
throat with slight dark red spotting, reddish orange outside, in
lax trusses of 10-12.  Plant as wide as tall; narrowly elliptic
leaves 4.25in(10.8cm) long.  (Harold E. Greer, reg. 1979)

```
cl.   -dichroanthum ssp. dichroanthum (3/8)
       -Dido-
Warm -   -decorum (1/4)
Spring-   -fortunei ssp. fortunei (1/4)
       -Fawn-   -dichroanthum ssp. dichroanthum
            -Fabia-
               -griersonianum (1/8)
```

Plant of medium size and compactness; leaves 8in(20.3cm) long.
Rounded trusses of 8-10 flowers, Indian yellow flushed pink, 4in
(10.2cm) long.  (Del W. James, 1961)

← WAR PAINT (108)

```
                   -griersonianum (5/16)
            -unnamed-              -haematodes ssp. haema-
            - hybrid-    -Choremia-           todes (9/32)
cl.   -unnamed-       -Siren-    -arboreum ssp. arboreum
    - hybrid-         -griersonianum           (1/32)
Water -      -    -dichroanthum ssp. dichroanthum (1/8)
Cricket-     -Fabia-
     -           -griersonianum
     -        -haematodes ssp. haematodes
       -Grosclaude-
              -facetum (eriogynum) (1/4)
```

4ft(1.2m)          0°F(-18°C)          M

White flowers, flushed rose where the petals meet.  (John Waterer, Sons & Crisp, reg. 1975)

```
cl.              -ferrugineum (1/2)
Waterer's Hybridium-
              -unknown (1/2)
```

Rose pink flowers.  (Waterer)

```
cl.     -cinnabarinum ssp. cinnabarinum (1/2)
Waterfall-
       -maddenii ssp. crassum (Cooper's form) (1/2)
```

Inner corolla of flower is very light orchid pink,  darkening in
the throat to bright rhodamine pink.  (Lord Aberconway, Bodnant)
A. M. (RHS) 1971

```
cl.
       -laetum (1/2)
Wattle Bird-
       -aurigeranum (1/2)
```

Vireya hybrid.  Trusses of 7-9 flowers, tubular campanulate, to
2.25in(5.7cm) wide, deep Indian yellow.  Ovate leaves, 4.75in(12
cm) long.  Height 6.4ft(2m)  (J. Rouse, reg. 1984)

cl.

Waxen Bell--form of <u>campanulatum</u> ssp. <u>campanulatum</u>

Leaves 3in(7.6cm) long, orange brown indumentum below. Flowers
in loose trusses of 13, phlox purple, spotted with dark purple.
(RBG, Edinburgh)   A. M. (RHS) 1965


cl.         -<u>griffithianum</u> (1/4)
      -Loderi-
Waxeye-     -<u>fortunei</u> ssp. <u>fortunei</u> (1/4)
      -unknown (1/2)

Trusses of 12-13 flowers, white with a very light green flare.
(Mr. & Mrs. A. G. Holmes, New Zealand, reg. 1979)


                      -<u>dichroanthum</u> ssp. <u>dichroanthum</u> (1/8)
            -Astarte-         -<u>campylocarpum</u> ssp.
            -          -Penjerrick- <u>campylocarpum</u> Elatum Group
      -Phidias-        -                        (1/16)
g.    -        -              -<u>griffithianum</u> (1/16)
      -        -              -<u>neriiflorum</u> ssp. <u>neriiflorum</u> (1/8)
Wayfarer-   -F. C. Puddle-
      -.            -<u>griersonianum</u> (1/8)
      -          -<u>thomsonii</u> ssp. <u>thomsonii</u> (1/4)
      -Chanticleer-
                  -<u>facetum</u> (<u>eriogynum</u>) (1/4)

Red flowers. (Lord Aberconway, intro. 1950)


cl.
      -<u>calophytum</u> var. <u>calophytum</u> (1/2)
Wayford-           -<u>arboreum</u> ssp. <u>arboreum</u> (1/4)
      -Gill's Triumph-
                  -<u>griffithianum</u> (1/4)

A plant of magnificent foliage--elliptic leaves 13in(33cm) long,
4.5in(11.5cm) wide, with sparse grey tomentum beneath.  Trusses
large, heavy, of about 24 openly campanulate flowers, white with
dark crimson outside, 3in(7.6cm) across. (E. J. P. Magor; shown
by Crown Estate, Windsor)  A. M. (RHS) 1957; F. C. C. (RHS) 1976


cl.                              -<u>maximum</u> (1/4)
            -Lady Clementine Mitford-
Weber's Pride-                   -unknown (1/2)
      -                   -<u>catawbiense</u> (1/4)
            -Kate Waterer-
                        -unknown

5ft(1.5m)          -5°F(-21°C)          M

Flowers of rhodamine purple with orange blotches, saucer-shaped,
3.25in(8.3cm) across, in trusses of 17. Upright, well-branched
plant; bright green leaves held 3 years. (Edwin O. Weber, 1974)


g.
      -Sir Charles Butler--<u>fortunei</u> ssp. <u>fortunei</u> (1/2)
Wega-
      -<u>williamsianum</u> (1/2)

Pink flowers. (Dietrich Hobbie)


cl.
            -<u>lacteum</u> (1/2)
Welcome Stranger-
            -<u>fortunei</u> ssp. <u>discolor</u> (1/2)

Repose g.  Flowers of primrose yellow, mimosa yellow in throat,
greenish red markings on upper lobes, campanulate, 7-lobed, 2.75
in(7cm) across; large, flat-topped trusses. Elliptic leaves to
7in(17.8cm) long. (L. de Rothschild; shown by Maj. A. E. Hardy)
A. M. (RHS) 1977


                        -<u>campylocarpum</u> ssp. <u>campylocarpum</u>
g. &        -Penjerrick-        Elatum Group (1/16)
cl.   -Amaura-      -<u>griffithianum</u> (1/16)
      -Eros-     -<u>griersonianum</u> (3/8)
Welkin-   -<u>griersonianum</u>
      -<u>haematodes</u> ssp. <u>hematodes</u> (1/2)

Trusses of about 4 flowers, varying in color from geranium lake
to Delft rose. (Lord Aberconway)   A. M. (RHS) 1946; F. C. C.
(RHS) 1951


---
            -<u>maximum</u> (1/2)
Wellesleyanum-
            -<u>catawbiense</u> (1/2)

Synonym Maximum Wellesleyanum. Flowers white, tinged light rose.
(A. Waterer, 1880)


cl.
      -<u>fortunei</u> ssp. <u>fortunei</u> (1/2)
Wellfleet*
      -<u>decorum</u> (1/2)

Dexter hybrid; parent of Acclaim, Aronimink, Todmorden, and
others.


cl.
      -<u>decorum</u> (1/2)
Welshpool-
      -Koichiro Wada--<u>yakushimanum</u> ssp. <u>yakushimanum</u> (1/2)

3ft(.9m)          -5°F(-21°C)          M

Plant upright, half as broad as tall; leaves 3.5in(8.9cm) long,
slight tan indumentum, held 3 years. Fragrant flowers 2in(5cm)
wide, white with faint yellow dorsal blotch and spots, 6-lobed;
globular trusses of 10. Floriferous. (Dr. Whildin Reese, 1977)


cl.         -<u>thomsonii</u> ssp. <u>thomsonii</u> (1/4)
      -Cornish Cross-
Wendy-      -<u>griffithianum</u> (1/4)
      -<u>williamsianum</u> (1/2)

4ft(1.2m)          5°F(-15°C)          EM          3/3

A hybrid with rounded leaves and cherry-colored trusses that
make a showy display in early spring. (Lester E. Brandt, 1962)


cl.         -<u>forrestii</u> ssp. <u>forrestii</u> Repens Group
      -Elizabeth-                        (1/4)
Wendy Lonsdale-      -<u>griersonianum</u> (1/4)
            -<u>neriiflorum</u> ssp. <u>neriiflorum</u> Euchaites Group
                              (1/2)

Flowers of cherry red. (H. D. Rose, Australia, 1973)


cl.
      -<u>arboreum</u> ssp. <u>arboreum</u> ? (1/2)
Werei-
      -<u>barbatum</u> ? (1/2)

Flowers rose pink. (Samuel Smith, Penjerrick)  A. M. (RHS) 1921.


cl.         -<u>haematodes</u> ssp. <u>haematodes</u> (1/4)
      -Humming Bird-
Werrington-      -<u>williamsianum</u> (1/4)
      -<u>forrestii</u> ssp. <u>forrestii</u> Repens Group (1/2)

(Williams, 1953)


                        -<u>catawbiense</u> (1/8)
g.          -Charles Dickens-
      -Dr. V. H. Rutgers-      -unknown (3/8)
Weser-            -Lord Roberts--unknown
      -<u>williamsianum</u> (1/2)

Rose-colored flowers. (Dietrich Hobbie, 1951)

cl.
    -maximum (1/2)
Westdale-
    -fortunei ssp. discolor (1/2)

6ft(1.8m)      -5°F(-21°C)    VL

Synonym Frontier. Deep purplish pink buds open to flowers pale to light purplish pink, slight twin flare of reddish brown, reverse of each lobe has stripe of pink, 2.25in(5.7cm) across; 20 per ball-shaped truss. Plant rounded, as wide as high; leaves 7.25(18cm) long. (John C. Wister cross; reg. by Gertrude Wister for Scott Horticultural Foundation, Swarthmore College, 1980)

cl.
    -oreodoxa var. oreodoxa (1/2)
Westfalenpark-    -arboreum ssp. arboreum (1/4)
    -Doncaster-
      -unknown (1/4)

Ronsdorfer Fruhbluhende g. Trusses hold up to 15 flowers, broad funnel-form, 5-7 lobes, dark pink with brown markings in throat. Blooms early. Grown in Botanic Gardens of Dortmund and Westfalen-Park. (Cross by G. Arends; intro. by G. D. Bohlje Nurseries, W. Germany, reg. 1969)

cl.

Westhaven--form of aberconwayi

4ft(1.2m)      -5°F(-21°C)    M

Plant of open spreading habit, as broad as tall with dark green leaves 3in(7.6cm) long, held 4 years. White flowers with a few maroon spots, held in trusses of 15. (Thomas J. McGuire, 1972)

cl.
    -javanicum (1/2)
Weston Glow-
    -lochiae (1/2)

Vireya hybrid. Trusses of 7-12 flowers, 5-lobed, nasturtium red. Height about 3ft(.9m). (P. Sutton, reg. 1983)

cl.
    -minus var. minus Carolinianum Group
    -P.J.M.-        (1/4)
Weston's Pink Diamond-    -dauricum var. sempervirens (1/4)
    -Cornell Pink--mucronulatum (1/2)

5ft(1.5m)      -10°F(-23°C)    E

Rose Bengal buds open to flat, saucer-shaped flowers, light fuchsia purple, held in ball-shaped trusses of 8-12. Upright plant, well branched, 2/3 as broad as high with leaves glossy yellow green, retained 1 year. (Edmund V. Mezitt cross, 1964; intro. by Weston Nurseries, 1977; reg. 1983)

cl.
    -fortunei ssp. discolor (1/2)
Westward Ho-
    -hardy hybrid--unknown (1/2)

Deep pink flowers with crimson in throat. (Slocock Nurseries, 1932)

cl.
    -fortunei ssp. discolor (1/4)
    -Sir Frederick-    -arboreum ssp. zeylanicum
Weybridge- Moore    -St. Keverne-    (1/8)
-    -griffithianum (1/8)
    -yakushimanum ssp. yakushimanum (1/2)

Plant of low, spreading habit, vigorous, compact; leaves 4.5in (11.5cm) long, glossy dark green. Flowers of pale pink, almost white margins, the reverse near Tyrian rose, 2.3in(6cm) across; globular trusses of 12. (Cross by Francis Hanger, RHS Garden, Wisley) H. C. (Wisley Trials) 1964

g.

WFH    See W.F.H. (Filed at start of W's)

cl.
    -Westbury--unknown (3/4)    -catawbiense (1/4)
Wheatley-    -Mrs. C. S. Sargent-
    -Meadowbrook-    -unknown
    -    -catawbiense
    -Everestianum-
    -unknown

6ft(1.8m)      -15°F(-26°C)    M    3/4

Synonym H. Phipps #2. Fragrant rose pink flowers with yellow green rays in throat, 3.25in(8.3cm) wide, frilled; ball-shaped trusses hold 16. Plant as broad as high, well-branched; leaves 7in(17.8cm) long, held 2 years. (Cross by Howard Phipps; Paul Vossberg, intro. 1958; reg. 1972) See color illus. ARS Q 26:2 (Spring 1972): cover; VV p. 109.

cl.
    -souliei (1/2)
Whimsey-    -Corona--unknown (1/4)
    -Bowbells-
    -williamsianum (1/4)

Dwarf plant with flowers camellia rose in throat shading to Egyptian buff. (Rudolph Henny, intro. 1961)

cl.
    -williamsianum (1/2)
Whisperingrose*    -forrestii ssp. forrestii Repens Group
    -Elizabeth-    (1/4)
    -griersonianum (1/4)

3ft(.9m)      0°F(-18°C)    EM    4/4

Compact plant; small, rounded leaves with red petioles. Flowers campanulate, of carnation rose. Floriferous. Buds young. (H. E. Greer, intro. 1982)

cl.
    -campylocarpum ssp. campylocarpum
    -Albino-
    -    -Loder's White, q.v. for 2 possible parentage
White Beauty-    diagrams
    -    -griffithianum
    -Loder's Pink Diamond-
    -fortunei ssp. fortunei

Pure white flowers, 12 to a truss. A. M. (RHS) 1945

cl.
    -unnamed-smirnowii (1/8)    -catawbiense
    - hybrid-    -Parsons Grandiflorum-    (9/32)
    -King Tut-    -America-    -unknown
White-    -    -dark red hybrid--unknown (3/32)
Bird-    -catawbiense
    -Koichiro Wada--yakushimanum ssp. yakushimanum (1/2)

2.5ft(.75m)      -25°F(-32°C)    ML

Synonym Snow Bird. Plant rounded, almost as wide as tall; dark green leaves with much orange brown indumentum on young leaves, held 3 years; new growth tomentose. Pink buds to white flowers with faint green spots, 2.25in(5.7cm) wide; ball-shaped trusses of 14. Floriferous. (Orlando S. Pride, reg. 1979)

cl.
    -campanulatum ssp. campanulatum (1/2)
White Campanula*
    -unknown (1/2)

White flowers of good substance. Broad, dark green leaves.

cl.
    -fortunei ssp. fortunei (1/2)
White Cloud-
    -hardy hybrid--unknown (1/2)

Register shows this as a griffithianum hybrid; Slocock's catalog
shows as above.  Large flowers of pure white, held in large flat
trusses.  (Slocock Nurseries, 1934)

```
                      -wardii var. wardii (1/4)
cl.        -Crest-                    -fortunei ssp. discolor (1/8)
       -       -Lady Bessborough-
White Crest-                      -campylocarpum ssp. campylo-
       -                                  carpum Elatum Group (1/8)
       -hyperythrum (1/2)
```

Pink in bud, opening white.  (Raised at Exbury Gardens; grown by
Brig. C. E. Lucas Phillips; reg. 1970)

```
cl.
           -scopulorum (1/2)
White Doves-
           -formosum var. inaequale (1/2)
```

White flowers with a yellow green mark in the throat; trusses of
4 flowers.  (F. M. Jury, reg. 1982)

```
g.
           -decorum (1/2)
White Ensign-        -griffithianum (1/4)
           -Loderi-
                    -fortunei ssp. fortunei (1/4)
```

(Exhibited by Heneage-Vivian, Wales, 1937; reg. 1958)

```
g. & cl.
           -irroratum ssp. irroratum (1/2)
White Glory-      -griffithianum (1/4)
           -Loderi-
                   -fortunei ssp. fortunei (1/4)
```

Large funnel-shaped flowers in loose trusses of 15-18; cream-
colored buds opening white.  Early season.  See also Pink Glory.
(Dowager Lady Loder cross)  A. M. (RHS) 1937

```
cl. -Mrs. J. G. Millais--unknown (1/2)
       -                 -Lady       -fortunei ssp. discolor (1/16)
White-              -Bessborough-campylocarpum ssp. campylo-
   Gold-     -Jalisco-                carpum Elatum Group (1/16)
       -       -       -     -dichroanthum ssp. dichroacnthum
    -Cheyenne-      -Dido-                               (1/16)
       -       -            -decorum (1/16)
       -       -     -griffithianum (1/8)
       -       -Loderi-
                     -fortunei ssp. fortunei (1/8)
```

5ft(1.5m)        -5°F(-21°C)        ML        4/3

Flowers widely funnel-shaped, 3.5in(8.9cm) across, pure white
with a large dorsal flare of brilliant yellow; dome-shaped
trusses of 12.  Plant rounded, as broad as tall; medium green
leaves.  See also Great Scott, Razzle Dazzle.  (H. Greer, 1979)

```
cl.
           -yakushimanum ssp. yakushimanum (1/2)
White Gull-
           -unknown (1/2)
```

3ft(.9m)         -5°F(-21°C)        ML

Plant rounded, wider than tall; glossy, convex leaves with tan
indumentum, held 3 years; new growth tomentose.  Buds of light
pink open to frilled white flowers with faint touch of pink at
edges, chartreuse green dorsal spotting, 2.75in(7cm); spherical
trusses of 20.  (Charles Herbert, reg. 1979)

cl.

White Lace--form of racemosum

4ft(1.2m)        -5°F(-21°C)        EM        4/3

Pure white with no pink tinge.  (P. A. Cox)  A. M. (RHS) 1974

```
g.              -maximum (1/4)
       -Halopeanum-
White Lady-           -griffithianum (1/2)
       -                        -griffithianum
       -Loderi Pink Diamond-
                             -fortunei ssp. fortunei (1/4)
```

Pure white flowers.  See Leonardslee Gertrude.  (Loder)

```
cl.
       -calophytum var. calophytum (1/2)
White -                -caucasicum (3/16)
Mustang-       -Jacksonii-      -caucasicum
    -Goldsworth-       -Nobleanum-              (1/16)
      Yellow  -                -arboreum ssp. arboreum
              -campylocarpum ssp. campylocarpum (1/4)
```

Synonym Debutante.  Plant 7ft(2.1m) tall when registered; leaves
10in(25.4cm) long.  Flowers to 3.5in(9cm) wide, white blotched
dawn pink.  Very early.  (Rudolph & Leona Henny, reg. 1964)

cl.

White Nes Loderi     See Emily Allison

```
cl.                            -griffithianum (1/4)
             -Loderi King George-
White Olympic Lady-         -fortunei ssp. fortunei
             -williamsianum (1/2)              (1/4)
```

4ft(1.2m)        0°F(-18°C)        E        4/4

Olympic Lady g.  Low, spreading plant, broader than tall; leaves
smooth, ovate, medium dark green, 3in(7.6cm) long.  White flowers faintly blushed pink, 4.5in(11.5cm) wide, in lax trusses of
about 14.  (Endre Ostbo cross; intro. by Roy N. Clark; reg. 1958)
P. A. (ARS) 1960; A. M. (RHS) 1977

cl.

White Pearl     See Halopeanum

```
cl.
          -williamsianum (3/4)      -griffithianum (1/8)
White Pippin-        -Loderi King-
          -White Olympic- George   -fortunei ssp. fortunei
            Lady       -williamsianum              (1/8)
```

Trusses of 7 flowers, 5-lobed, very light green white, 2.5in(6.4
cm) across.  Leaves rounded-elliptic, 2in(5cm) long, 1.2in(3cm)
wide.  (H. L. Larson, reg. 1983)

cl.

White Plains--form of alutaceum var. iodes

3ft(.9m)         -10°F(-23°C)        E-M        3/4

Trusses of 17-18 flowers, 5-lobed, funnel-campanulate, white
with reverse shading to light sap green.  Leaves dull green, underneath heavily felted with rust-colored indumentum.  (George
Forrest collector; raised by Col. S. R. Clarke; exhibited by R.
N. S. Clarke)  A. M. (RHS) 1978

```
cl.
        -fortunei ssp. discolor (1/2)
White Queen-
        -campylocarpum ssp. campylocarpum (1/2)
```

5ft(1.5in)       -5°F(-21°C)        ML

Leaves 6.5in(16.5cm) by 3in(7.6cm).  Flowers 4in(10.2cm) across,
white spotted deep chocolate on upper petal, 7-lobed, in trusses
of 10.  (Charles Herbert, 1967)

'Thomwilliams' by Magor
Photo by Greer

'Thor' by Brandt
Photo by Greer

'Tim Craig' by Delp
Photo by Delp

'Top Hat' by Whitney/Briggs
Photo by Greer

'Topsvoort Pearl' by Topsvoort Nurs.
Photo by Greer

'Torlonianum' by unknown
Photo by Greer

'Tortoiseshell Orange' by Slocock
Photo by Greer

'Tottenham' by Moerheim
Photo by Greer

'Tow Head' by Leach
Photo by Leach

'Tracigo' by Goheen
Photo by Goheen

'Trail Blazer' by Wright
Photo by Greer

'Treasure' by Crossfield
Photo by Greer

'Trewithen Orange' by Trewithen
Photo by Greer

'Trinidad' by Leach
Photo by Leach

'Triumphans' by Byles, 1871
Photo by Greer

'Trude Webster' by Greer
Photo by Greer

'Trula' by Larson
Photo by Greer

'Tumalo' by James
Photo by Greer

'Tyermannii' by Tyermann
Photo by Greer

'Umpqua Chief' by James
Photo by Greer

'Valaspis' by Aberconway
Photo by Greer

'Valley Sunrise' by Ticknor
Photo by Greer

'Van' by Van Veen/Griebnow
Photo by Greer

'Venice' by Leach
Photo by Leach

cl.                     -caucasicum (1/4)
          -Dr. Stocker-
White Robe-           -griffithianum (1/4)
          -           -dichroanthum ssp. dichroanthum (1/4)
          -Fabia-
                      -griersonianum (1/4)

3ft(.9m)          5°F(-15°C)          M

Plant broader than tall; leaves to 7in(17.8cm) long. Pale yellow buds opening to white flowers 3in(7.6cm) across, openly campanulate; upright trusses. (Rudolph & Leona Henny, reg. 1965)

cl.
          -Corry Koster--unknown (1/2)
White Samite-     -griffithianum (1/4)
          -Loderi-
                -fortunei ssp. fortunei (1/4)

White flowers. A. M. (RHS) 1932

cl.
          -decorum (1/2)              -griffithianum (1/8)
White Swan-           -George Hardy-
     -Pink Pearl-              -catawbiense (1/8)
          -              -arboreum ssp. arboreum (1/8)
                -Broughtonii-
                      -unknown (1/8)

6ft(1.8m)          -5°F(-21°C)          M          4/3

Extra large, perfectly formed trusses of satin white flowers, a
green basal blotch. Plant upright, rather compact; dusty green
concave leaves, 6in(15cm) long. (John Waterer, Sons & Crisp)
A. M. (RHS) 1937; F.C.C. (RHS) 1957

cl.

White Velvet--form of yakushimanum ssp. yakushimanum

2ft(.6m)          -5°F(-21°C)          M

Pink buds opening to light pink fading white, tan specks on upper lobe; flowers slightly fragrant, 2in(5cm) wide; trusses hold
16. Plant broader than tall; dark green leaves with tan indumentum, retained 3 years. (Seed from Japan by Edgar Greer; raiser Cyril H. Ward; intro. and reg. 1974)

cl.
          -Loder's White, q.v. for 2 possible parentage diagrams
White Way-                 -griffithianum
     -                -George Hardy-
     -Mrs. Lindsay Smith-          -catawbiense
                -Duchess of Edinburgh--unknown

White flowers with a faint yellow blotch. (Lord Digby, 1952)
A. M. (RHS) 1952

cl.
          -yakushimanum ssp. yakushimanum (Exbury form) (1/2)
White Wedding-
          -yakushimanum ssp. makinoi (1/2)

2ft(.6m)          -5°F(-21°C)          M

Plant 10in(25cm) high, 14in(35.5cm) broad at 2.5 years, covered
with trusses; small leaves heavily indumented. Frilled flowers,
opening pink and fading white, upper lobe speckled, 2.5in(6.4cm)
wide, 5-lobed; 15 per compact truss, 3 trusses to a terminal inflorescence. (John G. Lofthouse, reg. 1971)

cl.
          -edgeworthii (bullatum) (1/2)
White Wings-
          -ciliicalyx (1/2)

Very large flowers in neat trusses of 4-5, pure white with
yellow flare, sweetly fragrant. Plant of compact habit with
large leaves, as edgeworthii. (Scrase-Dickins) A. M. (RHS) 1939

cl.

Whitebait--form of sargentianum (white form)

1ft(.3m)          -5°F(-21°C)          EM          3/4

A very slow-growing dwarf with shiny leaves and small creamy
yellow flowers. (E. H. M. & P. A. Cox) A. M. (RHS) 1966

cl.

White's Cunningham's Sulphur          See Cunningham's Sulphur

cl.

White's Favourite--form of dichroanthum ssp. dichroanthum
                                        (F 6761)

Trusses of 4-5 flowers, 5-lobed, orange red with edges deeper,
1.25in(3.2cm) wide. Lanceolate leaves 3.5in(8.9cm) long; young
leaves with silky grey indumentum. Plant about 5ft(1.5m) tall.
(Raiser Harry White, Sunningdale Nurseries; reg. by J. P. C.
Russell, 1983)

cl.               -Pauline--unknown (1/2)
          -Grenadine-
Whitmoor-          -griersonianum (1/4)
          -               -ellottii (1/4)
          -Royal Blood-
                -Rubens--unknown

Flowers of medium cardinal red, spotted in upper throat. (Crown
Estate, Windsor) A. M. (RHS) 1971

                              -grffithianum (3/16)
                    -Queen Wilhelmina-
cl.       -Britannia-          -unknown (9/16)
          -Leo-          -Stanley Davies--unknown
Whitney - -elliottii (1/4)
Late Red*               -griffithanum
          -               -Queen Wilhelmina-
          -Britannia-          -unknown
                -Stanley Davies--unknown

A parent of Party Girl, q.v. (William Whitney)

cl.
          -yakushimanum ssp. yakushimanum (1/2)
Wickham's Fancy-
          -unnamed hybrid--unknown (1/2)

White flowers, slightly tinted with pink. (John Waterer, Sons
& Crisp, intro. 1975)

               -dichroanthum ssp. dichroanthum
               -Fabia-
cl.   -C. I. S.-     -griersonianum
      -          -Loder's White, q.v. for 2 possible
Wickiup*               parentage diagrams
      -               -griffithianum
      -unnamed hybrid-
               -unknown

5ft(1.5m)          5°F(-15'C)          M          4/3

A hybrid with multi-colored flowers--orange, apricot, and lemon
yellow, with touch of cherry red in the throat. Attractive foliage. (James A. Elliott)

                              -ponticum (1/16)
                    -Michael-
               -Prometheus-Waterer-unknown (5/16)
g.    -Madame de Bruin-          -Monitor-unknown
      -               -arboreum ssp. arboreum (1/8)
Wiekhoff-          -Doncaster-
               -unknown
      -williamsianum (1/2)

Deep rose flowers. (Dietrich Hobbie, 1947)

cl.
        -minus var. minus Carolinianum Group (1/2)
Wigeon-
        -Gigha--calostrotum ssp. calostrotum (1/2)

Pink flowers with deeper spots.  (P. A. Cox)  A. M. (RHS) 1982

g.
        -williamsianum (1/2)
Wilbar-            -thomsonii ssp. thomsonii (1/4)
     -Barclayi-                -arboreum ssp. arboreum (1/8)
              -Glory of Penjerrick-
                               -griffithianum (1/8)

4ft(1.2m)           5°F(-15°C)          E          3/3

Dark pink flowers.  Plant low and compact, spreading wider than
tall; dark green, ovate leaves.  (Lord Aberconway, 1946)

                          -thomsonii ssp. thomsonii
cl.                      -Bagshot-                (1/8)
       -Princess Elizabeth- Ruby  -unknown (3/8)
Wild Card*                   -unknown
         -maximum (white form) (1/2)

Bright red buds open to pale pink flowers with margins of strong
purplish 'red.  Plant medium-sized; hardy to -15°F(-26°C).  (Weldon E. Delp)

                  -dichroanthum ssp. dichroanthum (1/8)
          -Astarte-          -campylocarpum ssp.
          -        -Penjerrick- campylocarpum Elatum Group
g.    -Phidias-      -                    (1/16)
      -       -               -griffithianum (1/16)
Wilfred-      -          -neriiflorum ssp. neriiflorum (1/8)
      -     -F. C. Puddle-
      -                    -griersonianum (1/8)
      -williamsianum (1/2)

The flower is pink.  (Lord Aberconway, 1946)

                            -griffithianum (1/8)
cl.                -Queen Wilhelmina-
           -Britannia-              -unknown (3/8)
Wilgen's Ruby-       -Stanley Davies--unknown
           -              -catawbiense (1/4)
           -John Walter-
                    -arboreum ssp. arboreum (1/4)

5ft(1.5m)          -15°F(-26°C)          ML          4/3

Flowers very deep red with a brown blotch.  Attractive foliage;
hardy, compact plant.  (A. C. van Wilgen)    F. C. C. (Boskoop)
1951   See color ills. VV p. 125.

cl.
          -Wilgen's Ruby--unknown (1/2)
Wilgen's Surprise*
          -williamsianum (1/2)

Trusses 4.5in(11.5cm) across, of 7-8 flowers, a light China rose
with  deeper rose throat.  Vigorous plant, much wider than tall;
dull light green leaves,  3in(7.6cm) long.  (Van Wilgen's Nurseries)  H. C. (RHS) 1974

                            -griffithianum (1/8)
cl.               -Queen Wilhelmina-
           -Britannia-              -unknown (3/8)
Willbrit-        -Stanley Davies--unknown
         -williamsianum (1/2)

Flowers held in trusses of 6-8, fleshy, dark pink, inside China
rose, paler toward edges.  (Raised by Dietrich Hobbie; intro. by
Le Feber, Boskoop; reg. 1965)  Gold Medal (Rotterdam) 1960

cl.
          -catawbiense (1/2)
William Austin-
          -unknown (1/2)

Dark purple crimson flowers, spotted.  (J. Waterer, before 1915)

cl.                           -catawbiense (1/4)
               -English Roseum-
William Fetterhoff*            -unknown (1/4)
              -yakushimanum ssp. yakushimanum (1/2)

Compact plant with flowers of nonfading clear light pink.  Hardy
to -15°F(-26°C).  (Cross by William Fetterhoff; raiser Orlando S.
Pride; intro. 1975)

cl.
          -griffithianum (1/2)
William Fortescue-
          -campylocarpuym ssp. campylocarpum Elatum
                                       Group (1/2)

Penjerrick g.  Buds suffused light rose red,  opening to flowers
light greenish white throughout; trusses of 10-11.  (L. S. Fortescue cross, 1958; reg. by Fortescue Garden Trust, 1981)

cl.            -campylocarpum ssp. campylocarpum (1/4)
      -Gladys-
William-      -fortunei ssp. fortunei (3/8)
  King -         -griffithianum (1/8)
      -       -Loderi-
      -Ilam Cream-     -fortunei ssp. fortunei
          -unknown (1/4)

Flowers primrose yellow.  Shrub 7ft(2.1m) tall, at registration.
(Cross by Mrs. L. M. King, 1958; reg. by S. M. King, 1985)

cl.                           -catawbiense (1/4)
               -Atrosanguineum-
William Montgomery-            -unknown (1/2)
          -griersonianum (1/2)

Atrier g.  Clear scarlet flowers; hardy to -5°F(-21°C).  (Joseph
B. Gable, 1946)

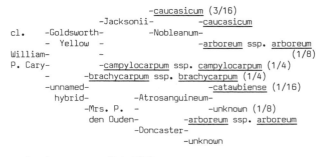

cl.   -Roman Pottery-          -dichroanthum ssp. dichroanthum (3/8)
      -           -griersonianum (1/4)
William-                 -dichroanthum ssp. dichroanthum
  Moss -          -Goldsworth-
      -      - Orange  -fortunei ssp. discolor (1/8)
      -Golden Fleece-           -griffithianum (5/32)
      -              -Kewense-
      -       -Aurora-     -fortunei ssp. for-
      -Yvonne-    -           tunei (1/32)
      -     -thomsonii ssp. thomsonii
      -griffithianum          (1/16)

Flowers azalea pink to shrimp pink in trusses of 7-11.  (William
Moss, reg. 1974)

                          -caucasicum (3/16)
              -Jacksonii-      -caucasicum
cl.   -Goldsworth-    -Nobleanum-
      - Yellow -              -arboreum ssp. arboreum
William-      -                     (1/8)
P. Cary-     -campylocarpum ssp. campylocarpum (1/4)
        -brachycarpum ssp. brachycarpum (1/4)
      -unnamed-                -catawbiense (1/16)
       hybrid-      -Atrosanguineum-
            -Mrs. P. -       -unknown (1/8)
           den Ouden-   -arboreum ssp. arboreum
                  -Doncaster-
                       -unknown

2ft(.6m)           -15°F(-26°C)          M

Pink buds open to flowers of empire yellow with much spotting of
yellow green, held in trusses of 16.  Plant rounded, as broad as
tall;  leaves 3.5in(9cm) long, dull yellow green,  held 2 years.
See also  Cary's Cream, Cary's Yellow, Josephine V. Cary, Strawberries and Cream.  (Edward A. Cary, 1980)

```
                            -arboreum ssp. arboreum (1/8)
cl.             -Doncaster-
        -Belvedere-          -unknown (1/8)
William-        -dichroanthum ssp. dichroanthum (1/2)
Pattman-    -dichroanthum ssp. dichroanthum
        -Fabia-
                -griersonianum (1/4)
```

Salmon orange flowers, 5-lobed, held in trusses of 10. (Thatcher, 1962)

```
cl.
                    -dichroanthum ssp. dichroanthum (1/2)
William W. Drysdale-        -griersonianum (1/8)
                    -unnamed-Azor-
                        hybrid-    -fortunei ssp. discolor (1/8)
                            -unknown (1/4)
```

Orange-colored flowers. (Thacker, 1944)

```
cl.                -thomsonii ssp. thomsonii (1/4)
            -Ascot Brilliant-
William Watson-        -unknown (1/4)
            -griffithianum (1/2)
```

Flowers reddish outside, shell pink inside. (Gill) A. M. (RHS) 1925

```
cl.
                -nuttallii (1/2)
William Wright Smith-
                -veitchianum (1/2)
```

These bell-shaped flowers are 5.5in(14cm) across, white with a distinct orange tinge, and frilled lobes. (RBG, Edinburgh; reg. 1962)   F. C. C. (RHS) 1960

```
---
            -arboreum ssp. arboreum (1/2)
Williamsonii-
            -azalea molle (1/2)
```

Azaleodendron. Flowers almost white, tinted pale lilac. (B. S. Williams & Son) A. M. (RHS) 1890

```
                        -thomsonii ssp.
                -Cornish Cross-    thomsonii (1/16)
            -unnamed-        -griffithianum (3/16)
cl.     -Harold- hybrid-wardii var. wardii (1/8)
        - Heal -        -griffithianum
Willy-Nilly-    -Loderi King-
        -        George  -fortunei ssp. fortunei (1/8)
        -williamsianum (1/2)
```

Creamy white flowers. (Collingwood Ingram, reg. 1972)

```
g.
        -ciliatum (1/2)
Wilsoni-
        -glaucophyllum ssp. glaucophyllum (1/2)
```

Two hybrids are known by this name; the other Wilsoni is registered as Laetevirens, q.v.  Pale rose flowers. (Origin unknown)

```
cl.
        -souliei (1/2)
Win Paul-        -fortunei ssp. discolor (1/8)
    -    -Ladybird-
        -Diva-    -Corona--unknown (1/8)
            -griersoniaum (1/4)
```

```
4.5ft(1.35m)        0°F(-18°C)        EM
```

Plant as broad as tall; leaves 5in(12.7cm) long. Flowers campanulate, 3.5in(8.9cm) wide, medium lilac rose with paler center and brick red spots; small round trusses. (H. L. Larson, 1972)

```
g.            -minus var. minus Carolinianum Group (1/4)
        -Conestoga-
Windbeam-        -racemosum (1/4)
        -unknown (1/2)
```

```
4ft(1.2m)        -25°F( -32°C)        EM        4/3
```

More than one form. Very hardy plant with apricot pink or white flowers 1in(2.5cm) wide. Foliage small, round, rather aromatic on a plant that's half as wide as tall, a little straggly. (G. Guy Nearing, 1958)   C. A. (ARS) 1972; A. E. (ARS) 1973   See color illus. ARS Q 27:4 (Fall 1973): p. 239; K p. 101; LW pl. 61, 62; VV p. 40.

```
cl.
            -minus var. minus (white form) Carolinianum Group
Windle Brook-        -ciliatum (1/4)            (1/2)
        -Cilpinense-
                -moupinense (1/4)
```

White flowers with pale green spots. (Crown Estate Commissioners) P. C. (RHS) 1971

```
            -dichroanthum ssp. dichroanthum (5/16)
cl.  -Dido-
    -    -decorum (5/16)        -fortunei ssp. discolor
Windles-        -Lady    -            (1/16)
    -            -Bessborough-campylocarpum ssp. campylo-
    -    -Jalisco-            carpum Elatum Group (1/16)
    -unnamed-    -dichroanthum ssp. dichroanthum
        hybrid-    -Dido-
    -            -decorum
    -yakushimanum ssp. yakushimanum (1/4)
```

Yellow orange flowers, in trusses of 6-8. (Crown Estate, Windsor, 1974)

```
                    -griffithianum (1/8)
cl.            -Queen Wilhelmina-
        -Britannia-        -unknown (5/8)
Windlesham Scarlet-    -Stanley Davies--unknown
    -        -arboreum ssp. arboreum (1/4)
        -Doncaster-
            -unknown
```

Compact, dome-shaped trusses 7in(17.8cm) deep, of 12-16 flowers, cardinal red with black spots. Vigorous, upright plant with leaves 6.5in(16.5cm) long, dull dark green. (W. Fromow Nurseries) A. M. (RHS) 1968

```
---        -ponticum ? (1/2)
Windsor Lad-
        -unknown (1/2)
```

```
4.5ft(1.35m)        -10°F(-23°C)        ML        4/3
```

Bluish purple flowers with a bold golden eye. Narrow foliage of medium length, on a well-shaped plant. (Knap Hill Nurseries)

```
cl.
        -formosum var. formosum (1/2)
Winifred Merson-
        -burmanicum (1/2)
```

Uranium green in bud, opening to white flowers with chrome yellow flush. (T. Lelliott, Australia, 1967)

```
cl.

Winifred Murray--form of cephalanthum ssp. cephalanthum
```

```
2ft(.6m)        -5°F(-21°C)        M        4/3
```

Loose trusses of 8 flowers, tubular-shaped, .25in(.64cm) across, neyron rose fading to white. Leaves up to .5in(1.3cm) long, densely covered with flaky brown scales. (Mrs. K. Dryden) A. M. (RHS) 1979

```
cl.    -griffithianum (1/4)
    -Loderi-
Wink-    -fortunei ssp. fortunei (1/4)
        -campylocarpum ssp. campylocarpum (1/4)
    -Mrs. Mary Ashley-
            -unknown (1/4)
```

Pink flowers. (Rudolph Henny)   P. A. (RHS) 1960

```
                -elliottii (1/4)
cl.       -Fusilier-
      -              -griersonianum (1/4)
Winkfield-      -Lady    -fortunei ssp. discolor (1/8)
      -        -Bessborough-
      -Jalisco-          -campylocarpum ssp. campylocarpum
        Elect -                      Elatum Group (1/8)
            -    -dichroanthum ssp. dichroanthum (1/8)
              -Dido-
                  -decorum (1/8)
```

Pinkish yellow flowers, upper lobes stained mandarin red on the
reverse with throat tinged and spotted crimson. (Crown Estate,
Windsor)  A. M. (RHS) 1958

```
cl.
      -Marion--unknown     -griffithianum (1/4)
Winning-       -Pink Shell-     -fortunei ssp.
 Post  -Coronation-    -H.M.Arderne-   fortunei (3/16)
       Day    -            -unknown (9/16)
             -      -griffithianum
              -Loderi-
                  -fortunei ssp. fortunei
```

Truss of 12 flowers, bright rose Bengal. (V. J. Boulter, 1978)

```
g. & cl.       -haematodes ssp. haematodes (1/4)
      -Humming Bird-
Winsome-          -williamsianum (1/4)
      -griersonianum (1/2)
```

3ft(.9m)        0°F(-18°C)        EM        3/4

Buds remain reddish all winter and open into rosy cerise flowers
in great abundance. Small, compact plant; small pointed leaves.
(Lord Aberconway, 1939)  A. M. (RHS) 1950

```
cl.
      -Marion--unknown (1/2)
Winter-         -Sir Charles Butler--fortunei ssp. fortunei
Beauty-Van Nes -       -maximum (1/8)          (1/4)
        Sensation-Halopeanum-
                    -griffithianum (1/8)
```

Rose purple flowers, conspicuous yellowish brown spotting on the
upper lobe. (V. J. Boulter, 1972)

cl.

Winter Brightness--form of mucronulatum

5ft(1.5m)        -15°F(-26°C)        VE        3/3

Rich purplish red flowers 1.5in(3.8cm) across on a deciduous
plant. Leaves 3in(7.6cm) long by 1.5in(3.8cm) wide. (Crown
Estate, Windsor)  F. C. C. (RHS) 1957

cl.

Winter Favourite      Parentage unknown

A parent of Denise.

```
cl.
      -mucronulatum (1/2)
Winterset-
      -unknown (1/2)
```

6ft(1.8m)        -15°F(-26°C)        VE

Open pollinated; collector, L. B. Grothaus. Foliage typical of
the species; semi-deciduous. Flowers pink, with a yellow green
flare, 2.75in(7cm) wide, in small trusses of 4-5. Floriferous.
(Dr. E. Brockenbrough, raiser; reg. 1976)  C. A. (ARS) 1975

```
cl.
      -yakushimanum ssp. yakushimanum (1/2)
Wishmoor-     -griersonianum (1/4)
      -Springbok-
          -ponticum (1/4)
```

Tight, rounded trusses hold 12-14 flowers, 7-lobed, bell-shaped,
orange red in bud opening to very light primrose yellow. Leaves
4in(10cm) long, lightly covered with indumentum below. (Crown
Estate, Windsor) A. M. (RHS) 1972

```
cl.
      -williamsianum (1/2)
Wisp*
      -Spatter Paint--irroratum ssp. irroratum (1/2)
```

4ft(1.2m)        0°F(-18°C)        E        4/4

Fast-growing and compact plant with bronze-colored new growth;
leaves mature to deep fir green. Flowers candy cane pink with
darker spotting over entire flower. (Intro. by Harold E. Greer)

```
                  -arboreum ssp. arboreum (1/8)
            -Doncaster-
cl.             -unknown (1/4)
   -unnamed-      -neriiflorum ssp. neriiflorum (1/8)
   - hybrid-Nereid-
Witch -          -dichroanthum ssp. dichroanthum (1/8)
Doctor-      -griffithianum (1/8)
   -      -Mars-
   -Vulcan-   -unknown
      -griersonianum (1/4)
```

4ft(1.2m)        0°F(-18°C)        ML        3/3

Heavy spotting over the entire flower. Cardinal red with
exceptionally large calyx. Foliage is deep fir green on this
attractive plant. (Cross by Lem; intro. by James Elliott, 1974)

```
cl.       -griffithianum (1/4)
      -Mars-
Witchery*   -unknown (1/4)
      -facetum (eriogynum (1/2)
```

5ft(1.5m)        -5°F(-21°C)        ML        4/3

Glowing fiery red flowers. New growth covered with silvery tomentum. (Rudolph Henny; not reg. due to name conflict)

```
cl.          -catawbiense (1/4)
      -Catawbiense Album-
Wizard-          -unknown (1/4)
    -    -dichroanthum ssp. dichroanthum (1/4)
      -Fabia-
          -griersonianum (1/4)
```

4ft(1.2m)        -10°F(-23°C)        ML        4/3

Flowers apricot and buff with edges of old rose, yellow streaks
flaring from throat, large calyx of apricot; trusses hold 14-20.
Leaves 5.5in(14cm) long, sharply pointed. (Halfdan Lem, 1962)
A. E. (ARS) 1959

```
cl.          -griffithianum (1/4)
      -Alice-
Wonderland-   -unknown (1/4)
      -auriculatum (1/2)
```

Funnel-shaped flowers held in compact trusses of 12, each
3.5in(9cm) across, creamy pink in bud, opening white with a
yellowish tinge. (Slocock)  A. M. (Wisley Trials) 1958

```
cl.
      -brachyanthum ssp. hypolepidotum (1/2)
Woodchat-
      -ludlowii (1/2)
```

A yellow-flowered hybrid intro. by P. A. Cox, 1978. (Reg. 1982)

```
cl.          -forrerstii ssp. forrestii Repens Group (1/4)
      -Elizabeth-
Woodcock-      -griersonianum (1/4)
      -hyperythrum (1/2)
```

Raiser: The Director, RHS Garden, Wisley.  P. C. (RHS) 1971

```
                   -yakushimanum ssp. yakushimanum (1/4)
        -unnamed-                        -griffithianum (1/16)
cl.     - hybrid-           -Queen
        -      -Britannia-Wilhelmina-unknown (3/16)
Woodpecker-
        -                     -Stanley Davies--unknown
        -        -griersonianum (1/4)
        -unnamed-      -Lady     -fortunei ssp. discolor
          hybrid-     -Bessborough-               (1/16)
              -Jalisco-         -campylocarpum ssp. campy-
               Elect -          locarpum Elatum Gp. (1/16)
                      -    -dichroanthum ssp. dichroanthum
                      -Dido-                         (1/16)
                          -decorum (1/16)
```

Plant of stiff habit with flowers primrose yellow fading lighter towards edges. Brown blotch in throat. (John Waterer, Sons & Crisp, reg. 1971)

```
cl.            -souliei (1/4)
          -Halcyone-          -fortunei ssp. discolor (1/8)
Woodside-     -Lady     -
        -    Bessborough-campylocarpum ssp. campylocarpum
        -aberconwayi (1/2)                Elatum Group (1/8)
```

Pink and white flowers. (Crown Estate, Windsor) P. C. (RHS) 1967

```
cl.
     -ludlowii (1/2)
Wren-
     -Yaku Fairy--keiskei (1/2)

dwarf        0°F(-18°C)        EM        4/4
```

Flowers 5-lobed, clear yellow. Leaves deep shiny green, scaly below. Habit prostrate, but mound-forming. (Cross by Peter A. Cox, 1971; reg. 1983)

```
cl.          -minus var. minus Carolinianum Group (1/4)
        -Conestoga-
Wyanokie-        -racemosum (1/4)
        -unknown (1/2)

3ft(.9m)        -25°F(-32°C)        M        3/4
```

An abundance of small white flowers on a small-leaved bush. See also Windbeam. (G. Guy Nearing, before 1958) See color illus. LW pl. 63.

```
g.
     -cinnabarinum ssp. xanthocodon (1/2)
Xanroy-
     -cinnabarinum ssp. cinnabarinum Roylei Group (1/2)
```

Orange red flowers. (Lord Aberconway, 1946)

```
g.                    -adenogynum (1/4)
     -Xenosporum (detonsum)-
Xenarb-                  -unknown (1/4)
     -arboreum ssp. arboreum (1/2)
```

Flowers are deep cherry color. (E. J. P. Magor, 1926)

```
g.
     -Helene Schiffner--unknown (1/2)
Xenia-
     -Sir Charles Butler--fortunei ssp. fortunei (1/2)
```

White flowers, crimson lines at base. (Lowinsky)   A. M. (RHS) 1919

```
cl.
        -fortunei ssp. fortunei (1/2)
Xenophile-        -griersonianum (1/4)
          -Tally Ho-
                  -facetum (eriogynum) (1/4)
```

Flowers are widely funnel-campanulate, magenta with staining of darker shade and heavy spotting of deep crimson, held in trusses of 12. (R. Strauss) A. M. (RHS) 1964

```
cl.
                -adenogynum (1/2)
Xenosporum (detonsum)-
                -unknown (1/2)
```

Xenosporum was an old name for detonsum, formerly called a species but now considered a hybrid of adenogynum. The International Registrar has entered the old name as a properly registered hybrid name.

```
cl.
     -yakushimanum ssp. yakushimanum (1/2)
Yakday-     -haematodes ssp. haematodes (1/4)
     -May Day-
            -griersonianum (1/4)
```

Leaves with tan indumentum; flowers deep cherry red, fading to pale pink. (Crossed by New Zealand Rhododendron Assoc.; reg. by Mrs. R. Pinney, 1982)

```
cl.

Yaku Angel--form of yakushimanum ssp. yakushimanum

2.5ft(.75m)        -15°F(-26°C)        M
```

Buds light purplish pink opening to white flowers with light brown spotting, held in trusses of 13-17. Dwarf plant spreading twice as wide as tall; deep green leaves with tan indumentum, retained 3 years; new growth white-tomentose. (Selection of Ernest Allen; intro. by H. E. Greer; reg. 1976) C. A. (ARS) 1975

```
cl.                          -wardii var. wardii ? (1/4)
        -Lackamas Cream--Chlorops-
Yaku Cream-               -vernicosum ? (1/4)
        -Koichiro Wada--yakushimanum ssp. yakushimanum (1/2)

2ft(.6m)        -10°F(-23°C)        M
```

Plant a bushy mound, wider than tall; dark green leaves 4in(10.2 cm) long. Mimosa yellow flowers turning primrose yellow, to 2in (5cm) across, campanulate, 6-lobed; trusses of 12. (Benjamin F. Lancaster, reg. 1968)

```
                        -smirnowii (1/8)
              -unnamed-     -Parsons      -catawbiense (1/32)
cl.           - hybrid-     -Grandiflorum-
        -King Tut-   -America-           -unknown (3/32)
Yaku    -           -       -dark red hybrid--unknown
Duchess-    -catawbiense (red form) (1/4)
        -Koichiro Wada--yakushimanum ssp. yakushimanum (1/2)

3ft(.9m)        -15°F(-26°C)        M        4/4
```

Red buds opening to flowers of deep pink with a lighter blotch, unspotted, aging lighter pink, 2.25in(5.7cm) wide, reverse deep purplish pink; spherical trusses of about 15. Medium-sized, indumented leaves. This cross produced a Royal Family of 6: see also p. 346. (A. M. Shammarello cross, 1961; reg. 1977)

```
cl.

Yaku Duke        Parentage as above

3ft(.9m)        -15°F(-26°C)        M        4/4
```

Plant upright, wider than high, arching branches; narrow olive green leaves with dark orange yellow indumentum; new growth silvery tomentose. Flowers open deep purplish pink with soft pink throat, aging paler, tubular funnel-shaped, of heavy substance, 5 wavy lobes; 14 in spherical trusses. (A. M. Shammarello, 1977)

cl.

<u>Yaku Fairy</u>--form of <u>keiskei</u>

1ft(.3m)          -10°F(-23°C)          EM          4/4

Differs only in height and habit from the type, being very slow-
growing and prostrate. Widely funnel-shaped, clear yellow flowers, 1.75in(4.5cm) across; trusses or 3-5. Red-tinged new
growth. (Raiser B. N. Starling, reg. 1971) A. M. (RHS) 1970

cl.
          -<u>smirnowii</u> (1/2)
<u>Yaku Frills</u>-
          -Koichiro Wada--<u>yakushimanum</u> ssp. <u>yakushimanum</u> (1/2)

1.5ft(.45m)          -10°F(-23°C)          M

Plant mound-shaped, wider than high; dark leaves, heavy tan indumentum. Pink buds open to frilly white flowers, 2.5in(6.4cm)
across; globular trusses of 12. See Crete. (Lancaster, 1969)

                         -<u>smirnowii</u> (1/8)
              -unnamed-          -Parsons     -<u>catawbiense</u>
cl.           - hybrid-          -Grandiflorum-      (1/32)
         -King Tut-      -America-          -unknown (3/32)
<u>Yaku King</u>-    -              -dark red hybrid--unknown
         -          -<u>catawbiense</u> (red form) (1/4)
         -Koichiro Wada--<u>yakushimanum</u> ssp. <u>yakushimanum</u> (1/2)

3ft(1.2m)          -15°F(-26°C)          M          4/4

Strong red buds opening to flowers of deep pink, blotched paler
pink, and fading lighter; reverse deep purplish pink, 2.25in(5.7
cm) across; ball-shaped trusses of about 18. Medium-sized, indumented leaves. (Anthony M. Shammarello, cross 1961; reg. 1977)

cl.
          -Moser's Maroon--unknown (1/2)
<u>Yaku Picotee</u>-
          -Koichiro Wada--<u>yakushimanum</u> ssp. <u>yakushimanum</u> (1/2)

2.5ft(.75m)          -10°F(-23°C)          M

Plant as wide as high; elliptic, recurved, dark leaves with matted indumentum beneath. Flowers rose Bengal, white at center,
rotate-campanulate, 2.75in(7cm) wide; ball-shaped trusses of 15.
(Benjamin F. Lancaster, reg. 1969)

                         -<u>smirnowii</u> (1/8)
              -unnamed-          -Parsons     -<u>catawbiense</u> (1/32)
cl.           - hybrid-America-Grandiflorum-
         -King Tut-      -          -unknown (3/32)
<u>Yaku</u> -      -              -dark red hybrid--unknown
<u>Prince</u>-      -<u>catawbiense</u> (red form) (1/4)
         -Koichiro Wada--<u>yakushimanum</u> ssp. <u>yakushimanum</u> (1/2)

3ft(.9m)          -15°F(-26°C)          M          4/4

Plant upright, rounded with arching branches; narrow olive green
leaves with dark orange yellow indumentum. Flowers 2.5in((6.4cm)
wide, opening strong purplish pink with paler blotch, dark red
orange spotting, the reverse strong purplish pink; ball-shaped
trusses of 14. Floriferous. (Anthony M. Shammarello reg. 1979)
See color illus. ARS Q 27:1 (Winter 1973): p. 35; LW pl. 89.

cl.

<u>Yaku Princess</u>     Parentage as above

3ft(.9m)          -15°F(-26°C)          M          4/4

Leaves 3.75in(9.5cm) long, indumented. Flowers of apple blossom
pink with a blush pink blotch and greenish spots, fading paler;
spherical trusses up to 15. (Anthony M. Shammarello, reg. 1977)

cl.

<u>Yaku Queen</u>     Parentage as above

3ft(.9m)          -15°F(-26°C)          M          4/4

Leaves medium-sized and indumented. Flowers pale pink with a
dim yellow blotch, fading white; held in trusses of 16. (A. M.
Shammarello, 1977) C. A. (ARS) 1981

cl.                    -<u>griersonianum</u> (1/4)
         -Rose Splendour-          -<u>ponticum</u> (1/8)
<u>Yaku Splendour</u>-          -Purple Splendour-
         -              -unknown (1/8)
         -Koichiro Wada--<u>yakushimanum</u> ssp. <u>yakushimanum</u>
                                        (1/2)

1.5ft(.45m)          -10°F(-23°C)          M

Globe-shaped dwarf plant as wide as high, with indumented foliage. Trusses of 18 flowers, phlox pink, paler inside, ruffled,
5-lobed. (Benjamin F. Lancaster, 1968)

                         -<u>griffithianum</u> (1/8)
              -Mars-
cl.           -Vulcan's Flame-     -unknown (1/8)
<u>Yaku Sunrise</u>-          -<u>griersonianum</u> (1/4)
         -Koichiro Wada--<u>yakushimanum</u> ssp. <u>yakushimanum</u> (1/2)

2ft(.6m)          -10°F(-23°C)          M

Openly campanulate flowers of rose madder, edges and the reverse
darker, 2.75in(7cm) wide, 5-lobed; compact trusses of 10. Plant
50% broader than high; deep green, slightly recurved leaves to 3
in(7.6cm) long. (Benjamin F. Lancaster, reg. 1967)

cl.          -<u>griffithianum</u> (1/4)
         -Mars-
<u>Yaku Warrior</u>-     -unknown (1/4)
         -<u>yakushimanum</u> ssp. <u>yakushimanum</u> (1/2)

1.5ft(.45m)          -10°F(-23°C)          M

Trusses of 18 flowers, Tyrian rose with lighter center, campanulate, 2in(5cm) wide; plant broader than high; dark, persistent, recurved leaves with thin indumentum. (B. Lancaster, 1968)

              -<u>cinnabarinum</u> ssp. <u>cinnabarinum</u> Roylei
         -                              Group (1/4)
g. -Lady Chamberlain-          -<u>cinnabarinum</u> ssp. <u>cinnabar</u>-
         -          -Royal Flush-          -<u>inum</u> (3/8)
Yama-          (orange form)-<u>maddenii</u> ssp. <u>maddenii</u> (1/8)
         -          -<u>yunnanense</u> (1/4)
         -Yunncinn-
              -<u>cinnabarinum</u> ssp. <u>cinnabarinum</u>

Yellowish plum-colored flowers. (Lord Aberconway, 1946)

cl.          -<u>campylocarpum</u> ssp. <u>campylocarpum</u> (1/4)
    -Letty Edwards-
Yamina-          -<u>fortunei</u> ssp. <u>fortunei</u> (3/4)
    -<u>fortunei</u> ssp. <u>fortunei</u>

Flowers uranium green, feathered red on upper lobe. (A. Teese,
Australia, 1964)

cl.
    -<u>yakushimanum</u> ssp. <u>yakushimanum</u> (1/2)
Yashmak-
    -<u>campylocarpum</u> ssp. <u>campylocarpum</u> (1/2)

Flowers light roseine purple with blotching and spotting of deep
currant red. (Collingwood Ingram, 1972)

```
cl.
            -catawbiense (red form) (1/2)
Yates' Albino-    -griffithianum (1/4)
        -Mars-
                -unknown (1/4)


4ft(1.2m)          -10°F(-23°C)          ML
```

Plant rounded, as broad as tall, arching branches; leaves apple green, 5.25in(13.3cm) long, held 2 years. Lax trusses of about 11 white flowers, a spotted blotch of medium apple green and red (few), 2.75in(7cm) wide. (Henry Yates; reg. by Mrs. Yates, 1976)

```
                    -decorum ? (1/8)
            -Caroline-
cl.    -Mrs. H. R.-         -brachycarpum ssp. brachycarpum ?
      - Yates   -unknown (1/4)                        (1/8)
Yates' Best-
        -Koichiro Wada--yakushimanum ssp. yakushimanum (1/2)


3ft(.9m)          -15°F(-26°C)          ML
```

Buds of rose opening to small pink flowers fading to white, of heavy substance, in ball-shaped trusses of 12. Rounded plant as wide as high, arching branches; medium-sized leaves with brownish indumentum, held 2 years. See also Yates' Second Best. (H. R. Yates, cross 1962; reg. by Mrs. H. R. Yates, 1977)

```
cl.                        -catawbiense (1/4)
        -Mrs. Charles S. Sargent-
Yates' Hazel-                  -unknown (1/4)
        -vernicosum (1/2)


5ft(1.5m)          -20°F(-29°C)          ML
```

Flowers of heavy substance, 3in(7.6cm) across, 6-lobed, of rhodamine pink with greyed yellow blotch; spherical trusses hold 12. Plant upright, as wide as tall; leaves 5.5in(14cm) long, elliptic, held 2 years. (Henry Yates; Mrs. Yates intro.; reg. 1977)

```
cl.                  -catawbiense (1/4)
        -Catawbiense Grandiflorum-
Yates' Purple-              -unknown (1/2)
      -             -ponticum (1/4)
        -Purple Splendour-
                    -unknown


3ft(.9m)          -5°F(-21°C)          ML
```

Imperial purple flowers, fading lighter, with plum purple blotch and spotting, to 2.5in(6.4cm) across; dome-shaped trusses hold 10-13. Plant upright, broad; elliptic leaves retained 2 years. (Henry Yates cross, 1965; reg. by Mrs. Yates, 1979)

```
cl.                        -griffithianum (1/4)
        -Leah Yates--Mars (selfed)-
Yates' Red-                    -unknown (1/4)
      -             -catawbiense (1/4)
        -Pink Twins-
                    -haematodes ssp. haematodes (1/4)


4ft(1.2m)          -5°F(-21°C)          ML
```

Sparkling cardinal red flowers shading to magenta in throat with greyed purple blotch and spotting, openly funnel-shaped, 3in(7.6 cm) wide, 5-lobed; conical trusses of 13-15. Plant broader than tall, well-branched; narrowly elliptic leaves 5in(12.7cm) long, held 2 years, petioles reddish in winter. (Henry Yates cross; Mrs. H. Yates intro. 1965; reg. 1976)

```
                    -decorum ? (1/8)
cl.            -Caroline-
      -Mrs. H. R. Yates-   -brachyanthum ? (1/8)
Yates' Second-            -unknown (1/4)
   Best    -Koichiro Wada--yakushimanum ssp. yakushimanum
                                          (1/2)
3ft(.9m)          -15°F(-26°C)          ML
```

Plant rounded, slightly wider than tall; glossy leaves 4.75in(12 cm) long, greyed orange indumentum below; new growth with orange brown tomentum. Buds rose, opening to pink flowers aging white, yellow dorsal spotting, 2.5in(6.4cm) across, of heavy substance, 5 wavy-frilled lobes; spherical trusses of 11. (Henry Yates; intro. by Mrs. Yates; reg. 1977)

```
            -Catalgla--catawbiense var. album Glass
cl.    -unnamed-       -wardii var. wardii        (1/4)
      - hybrid-unnamed-             (3/8)
Yates' Treena-      hybrid-unknown (1/8)
      -             -wardii var. wardii
      -unnamed-
        hybrid-yakushimanum ssp. yakushimanum (1/4)


3ft(.9m)          -10°F(-23°C)          ML
```

Buds rose shaded yellow open to canary yellow flowers with dorsal rays of burgundy spots, 3in(7.6cm) wide, 5 wavy to frilled lobes; domed trusses of 12-14. Upright plant, wider than tall, well-branched; cyprus green leaves to 4in(10.2cm) long, held 2 years. (Henry Yates cross; intro. by Mrs. Yates 1980; reg. 1984)

```
            -catawbiense (1/8)
        -Pink -
cl.    -Sunset-Twins-haematodes ssp. haematodes (1/8)
      - Yates-                      -griffithianum
Yates' Velvet-    -Leah Yates--Mars (selfed)-    (1/8)
  Charm  -                          -unknown (1/8)
        -Pink -Catalgla--catawbiense var. album Glass (1/4)
        -Punch-
            -fortunei ssp. fortunei (1/4)


2.5ft(.75m)          -10°F(-23°C)          ML
```

Plant rounded, broader than tall; dark yellowish green leaves to 4.75in(12cm) long, twisted, slight gold brown indumentum. Rose red flowers paler in center with slight greyed crimson markings, widely funnel-campanulate, 2.75in(7cm) wide; globular trusses of 22. (Henry Yates cross, 1968; intro. by Mrs. Yates; reg. 1982)

```
cl.
                        -catawbiense (1/2)
Years of Peace--Mrs. C. S. Sargent (selfed)-
                        -unknown (1/2)


5ft(1.5m)          -15°F(-26°C)          M
```

Synonym John Foster Dulles. Hardy, sun-tolerant flowers of pink with spotting on upper petals, 3in(7.6cm) across, in dome-shaped trusses of 6-10. Plant vigorous, upright with glossy dark green leaves. (E. J. Mezitt, Weston Nurseries, 1981)

```
cl.
        -Cunningham's Sulphur--form of caucasicum ? (1/2)
Yellow Bells-       -campylocarpum ssp. campylocarpum (1/4)
        -Moonstone-
                -williamsianum (1/4)


2ft(.6m)          -5°F(-21°C)          E
```

Glossy, wax-like leaves, 2.5in(6.4cm) long on a shrub as wide as tall. Bell-shaped, ruffled flowers of chrome yellow, 1.5in(3.8 cm) across, in trusses of 10-12. (Don Newkirk cross; intro. by Benjamin F. Lancaster; reg. 1967)

```
cl.


Yellow Bunting--form of fletcherianum


2.5ft(.75m)          5°F(-15°C)          EM          4/4
```

Form is typical of the species with primrose yellow flowers 1.75 in(4.5cm) wide, foliage of warm russet in winter. Compact shrub with oblong-lanceolate leaves, 2in(5cm) long. Formerly known as "Rock's valentinianum", its correct status was determined by Mr. H. H. Davidian. (Raised from a cutting from plant at RBG, Edinburgh, by E. H. M. & P. A. Cox) A. M. (RHS) 1964

```
                    -wardii var. wardii (1/4)
                    -
                                    -griffithianum (5/32)
cl.  -Idealist-           -Kewense-
  -        -       -Aurora-    -fortunei ssp. fortunei
Yellow-    -Naomi-     -                    (9/32)
 Creek-    -          -thomsonii ssp. thomsonii (1/16)
  -        -fortunei ssp. fortunei
  -        -griersonianum (1/4)
  -Sarita Loder-    -griffithianum
                  -Loderi-
                          -fortunei ssp. fortunei

6ft(1.8m)         5°F(-15°C)       EM        4/2
```

Plant of rather open habit; narrow, smooth, light green leaves
6in(15.2cm) long.  Flowers flat, flaring to 4in(10.2cm) across,
of primrose yellow in lax dome-shaped trusses of about 10.  Prefers some shade. (Del W. James, 1962)  P. A. (ARS) 1962

```
cl.
          -campylocarpum ssp. campylocarpum (1/2)
Yellow Dream-
          -wardii var. wardii (1/2)
```

Trusses of 14 flowers, light Dresden yellow with center deeper
yellow, spotted red. (G. Langdon, 1982)

```
cl.

Yellow Garland--form of xanthostephanum (F 21707)

4ft(1.2m)         15°F(-10°C)       EM        3/3
```

Plant may grow to 9ft(2.7m); leaves 1.5in(3.8cm) long, silvery
beneath.  Flowers aureolin yellow, 1in(2.5cm) wide, campanulate;
trusses hold 4-8.  Color may vary between plants. (Collected by
G. Forrest; Crown Estate, Windsor, reg. 1962)  A. M. (RHS) 1961

```
g.
          -sulfureum (1/2)
Yellow Hammer-
          -flavidum var. flavidum (1/2)

4ft(1.2m)          5°F(-15°C)       EM        4/3
```

Very deep yellow flowers .5in(1.27cm) wide, tubular campanulate;
in 3-flowered trusses.  Very floriferous; may bloom again in autumn.  Plant upright, open; small, light green scaly leaves. One
of the few yellows that does well in sunny locations. (At Caerhays before 1931; J. C. Williams, reg. 1958)

```
---                            -Duchess of Teck--unknown (1/4)
              -Lord Wolseley-
Yellow Perfection-     -javanicum (1/4)
              -teysmannii (1/2)
```

Vireya hybrid.   F. C. C. (RHS) 1888

```
                              -dichroanthum ssp. dichroanthum (1/8)
              -Goldsworth-
cl.     -Hotei- Orange   -fortunei ssp. discolor (1/8)
  -        -      -            -souliei (1/8)
Yellow  -     -unnamed hybrid-
Petticoats-                    -wardii ssp. wardii (3/8)

  -        -Pink     -Jan Dekens--unknown (7/32)
  -unnamed-Petticoats-               -griffithianum
     hybrid-      -           -Queen        (1/32)
           -         -Britannia-Wilhelmina-unknown
           -wardii var.       -
                   wardii      -Stanley Davies--unknown

4ft(1.2m)         0°F(-18°C)       M
```

Similar to Hotei, but larger trusses, more floriferous.  Habit
rounded, compact, as broad as high; dark green leaves, 4in(10cm)
long.  Deep yellow flowers, lighter at edges, unmarked,
3in(7.6cm) wide; trusses of 15. (John G. Lofthouse, reg. 1983)

```
cl.                              -wardii var. wardii (1/4)
          -Mrs. Lamont Copeland-      -souliei (1/8)
Yellow Pippin-              -Virginia-
          -                    Scott -unknown (1/8)
          -yakushimanum ssp. yakushimanum (1/2)
```

Flowers open carmine red, fading light aureolin yellow, center
sulphur yellow, spotted chartreuse green; trusses of 12.  Leaves
5.5in(14cm) long. (H. L. Larson, reg. 1983)

```
                              -wardii var. wardii (1/8)
                              -
                                      -griffithianum
              -Idealist-        -Kewense-   (9/128)
  -unnamed-     -       -Aurora-    -fortunei ssp.
  - hybrid-     -       -                fortunei
  -        -unnamed-    -Naomi-    -thomsonii ssp.
  -        - hybrid-    -             thomsonii (1/64)
  -        -       -    -fortunei ssp. fortunei
  -        -       -wardii var. wardii
cl.  -        -Hawk-    -fortunei ssp. discolor
  -        -Lady Bess-
Yellow-    -        borough-campylocarpum ssp. campy-
 Sally-    -                 locarpum Elatum Gp. (5/32)
  -        -     -campylocarpum ssp. campylocarpum (1/8)
  -        -Phyrne-    -griffithianum
  -        -Loderi-
  -        -     -fortunei ssp. fortunei (13/128)
  -        -     -fortunei ssp. discolor (5/32)
  -        -Lady Bessborough-
  -        -             -campylocarpum ssp. campylocarpum
  -Jalisco-                         Elatum Group
  -     -dichroanthum ssp. dichroanthum (1/8)
  Dido-
     -decorum (1/8)
```

Flowers  very light yellow, tinged with pale neyron rose.  (John
Waterer, Sons & Crisp, reg. 1975)

```
cl.
     -aberconwayi (1/2)
Yellow-     -Koichiro Wada--yakushimanum ssp. yakushimanum
Saucer-unnamed-                            (1/4)
       hybrid-     -dichroanthum ssp. dichroanthum (1/8)
         -Fabia-
          -griersonianum (1/8)

4ft(1.2m)         0°F(-18°C)       M
```

Newly opened flowers mottled flame orange, turning aureolin yellow with lemon yellow throat, small red dorsal spot; corolla of
good substance, rather flat, 3in(7.6cm) wide; in trusses of 8.
Floriferous.  Plant habit upright, about as broad as tall; glossy leaves held 2-3 years. (Cecil C. Smith, 1982)

```
cl.
     -keiskei (1/2)
Yellow Spring-
     -racemosum (1/2)

1ft(.3m)         -5°F(-21°C)       EM
```

Leaves dark yellowish green, scaly beneath, 2in(5cm) by 1in(2.5
cm), held 1 year.  Flowers 1in(2.5cm) across, 5-lobed, strong
neyron rose margins shading to yellow white in throat, yellowish
green spotting; terminal inflorescence of 8-10, each 3-5 flowered trusses, ball-shaped. (Charles Herbert, reg. 1976)

```
cl.
     -dichroanthum ssp. scyphocalyx (1/2)
Yellow-     -aureum var. aureum (chrysanthum) (1/4)
 Wolf -unnamed-          -sanguineum ssp. didymum (1/16)
      hybrid-   -Rubina-    -griersonianum (3/32)
       -unnamed-        -Tally Ho-
        hybrid-            -facetum (eriogynum)
         -     -dichroanthum ssp. dichroanthum (1/16)
          -Fabia-                    (1/16)
           -griersonianum
```

Semi-dwarf plant 2ft(.6m) tall; dark green leaves to 3in(7.6cm) long, paler below.  Fire red buds opening orange buff, stained peach on edges and reverse, 2in(5cm) wide; flat trusses of 5-7. (J. A. Witt & University of Washington Arboretum, reg. 1977)

```
g. & cl.
         -forrestii ssp. forrestii Repens Group (1/2)
Yeoman-        -haematodes ssp. haematodes (1/4)
     -Choremia-
               -arboreum ssp. arboreum (1/4)
```

1.5ft(.45m)         0°F(-18°C)           EM         3/3

A small sturdy plant with small deep green leaves, and bright red, waxy, bell-shaped flowers. (Lord Aberconway)  A. M. (RHS) 1947

```
                       -souliei (1/16)
               -Soulbut-
g.         -Vanessa-   -Sir Charles Butler--fortunei
      -Eudora-                        ssp. fortunei (1/16)
Yolande-        -griersonianum (1/8)
      -       -facetum (eriogynum) (1/4)
       -dichroanthum ssp. scyphocalyx (1/2)
```

A hybrid with rose-colored flowers. (Lord Aberconway, 1941)

```
cl.
      -Marion--unknown (1/2)
You Beaut-        -Sir Charles Butler--fortunei ssp.
      -Van Nes Sensation-          fortunei (1/4)
      -                -maximum (1/8)
                -Halopeanum-
                       -griffithianum (1/8)
```

Orchid pink flowers.  See also Melba and Winter Beauty.  (V. J. Boulter, 1972)

```
cl.              -cinnabarinum ssp. cinnabarinum
      -Lady Chamberlain-      Roylei Group (1/4)
Young  -            -cinnabarinum ssp.
Chamberlain-  -Royal Flush- cinnabarinum (1/8)
      -yunnanense (1/2)   -
                   -maddenii ssp. maddenii
                               (1/8)
```
Trusses of 7-9 flowers, 3in(7.6cm) wide, 5-lobed, white suffused pink with slight brown spotting on reverse; truss pendant, like Lady Chamberlain.  Elliptic leaves, 3.3in(8.5cm) long.  (D. B. Fox, reg. 1983)

```
cl.
         -cinnabarinum ssp. cinnabarinum (1/2)
Youthful Sin-
         -yunnanense (1/2)
```

These flowers are rhodamine purple.  (Lord Aberconway, Bodnant) A. M. (RHS) 1960

```
g.
      -cinnabarinum ssp. cinnabarinum (3/4)
Yucaba-        -yunnanense (1/4)
      -Yunncinn-
               -cinnabarinum ssp. cinnabarinum
```

Plum-colored flowers. (Lord Aberconway, 1946)

cl.

Yucondus--form of erythrocalyx (F 21741); now called a hybrid, perhaps of wardii var. wardii X selense

4ft(1.2m)         -5°F(-21°C)          E-M        2/3

Small trusses of flowers  deep neyron rose in bud, fading when open to near white,  variably stained and suffused by light rose pinks, widely funnel-campanulate, 2in(5cm) wide.  Elliptic, dark green leaves tend to conceal the flowers. (Collector George Forrest; raised by Col. S. R. Clarke; exhibited by R. N. S. Clarke) A. M. (RHS) 1979

```
cl.
      -minus var. minus Carolinianum Group (white form) (1/2)
Yukon-
      -dauricum var. album (1/2)
```

5ft(1.5m)         -20°F(-29°C)          E          4/3

White flowers, edges flushed lilac pink, aging white, 1.5in(3.8 cm) across, 5-lobed; dome-shaped terminal clusters of several 3-flowered trusses.  Plant rounded, well-branched, half as wide as tall; height if started from cuttings would be to 50% taller at 10 years.  Glossy yellowish green leaves 1.75in(4.3cm) long, tan scales below, held 2 years.  (D. Leach, intro. 1978; reg. 1983)

cl.

Yum Yum--form of tsariense (L & S 2858)

2ft(.6m)         0°F(-18°C)            M          3/5

Plant compact; leaves 1.5in(3.8cm) long, with persistent cinnamon brown indumentum beneath.  Carmine  buds open to white flowers, blushed phlox pink, 1.75in(3.8cm) wide, in trusses of 3-4. (Collectors Ludlow & Sherriff; Mrs. R. M. Harrison, reg. 1965) A. M. (RHS) 1964

```
g.
      -yunnanense (1/2)
Yunncinn-
      -cinnabarinum ssp. cinnabarinum (1/2)
```

Flowers violet rose, darker on outside; interior reddish brown (markings?) J. C. Williams also raised this cross earlier.  (E. J. P. Magor, intro. 1924)

```
              -griffithianum (5/8)
g.        -Kewense-
     -Aurora-     -fortunei ssp. fortunei (1/8)
Yvonne-     -thomsonii ssp. thomsonii (1/4)
     -griffithianum
```

6ft(1.8m)         5°F(-15°C)           EM         3/3

A tall, compact hybrid with flat trusses of very large flowers, widely funnel-shaped, white blushed pink, having a translucent quality.  One of the early crosses, LR 112; rediscovered after World War II.  (Rothschild)

```
                         -griffithianum (9/32)
                    -Loderi-
              -Albatross-   -fortunei ssp. fortunei
          -Cup Day-     -fortunei ssp. discolor (1/16)
      -       -        -elliottii (1/16)
     -Midnight-    -Fusilier-
cl.  -     -        -griersonianum (1/16)
     -     -        -ponticum (1/8)
Yvonne-    -Purple Splendour-
Davies-            -unknown (3/16)
     -        -griffithianum
     -    -Pink Shell-    -fortunei ssp. fortunei
     -    -       -H. M. Arderne-       (7/32)
     -Coronation-            -unknown
      Day    -    -griffithianum
             -Loderi-
                -fortunei ssp. fortunei
```

Trusses of 17 white flowers striped light neyron rose on reverse with a dark red center.  (Mrs. M. Davies, reg. 1980)

```
              -griffithianum (5/8)
cl.       -Kewense-
     -Aurora-     -fortunei ssp. fortunei (1/8)
Yvonne Dawn-    -thomsonii ssp. thomsonii (1/4)
     -griffithianum
```

6ft(1.8m)         5°F(-15°C)            M          3/3

Yvonne g.  The latest to flower, and the finest of this grex. Very large flowers of heavy substance, flesh pink fading almost white with age.  (Rothschild, n.d.)

cl.

Yvonne Opaline     Parentage as above

6ft(1.8m)          5°F(-15°C)          M          3/3

Yvonne g.  Very large, openly funnel-shaped flowers.  Buds of
deep pink open to flowers pale rose inside, deeper rose reverse.
(Rothschild)  A. M. (RHS) 1931    See color illus. ARS J 36:4
(Fall 1982): p. 135.

cl.

Yvonne Pearl     Parentage as above

6ft(1.8m)          5°F(-15°C)          M          3/3

Yvonne g.  Similiar to others of this group.  Flowers of good
substance, a pale pearly pink.  (Rothschild, 1925)

cl.

Yvonne Pride     Parentage as above

6ft(1.8m)          5°F(-15°C)          M          3/3

Yvonne g.  Flowers are very large funnels, 5in(12.7cm) across,
pale pink fading almost white.  Tall but compact plant.  (Rothschild)  A. M. (RHS) 1948

```
g.             -concinnum (1/4)
        -Daphne-
Zampa-         -augustinii ssp. augustinii (1/4)
      -              -neriiflorum ssp. neriiflorum (1/4)
        -F. C. Puddle-
                       -griersonianum (1/4)
```

A red-flowered hybrid.  (Lord Aberconway, 1941)

cl.
```
        -dichroanthum ssp. dichroanthum (1/2)
Zanna-
        -catawbiense var. compactum ? (1/2)
```

Orange pink flowers.  (Haworth-Booth, reg. 1962)

```
g.
        -catacosmum (1/2)
Zarie-
        -meddianum var. meddianum (1/2)
```

Red flowers.  (J. B. Stevenson cross, 1939; intro. 1951)

```
g. & cl.          -souliei (1/4)
        -Rosy Morn-        -griffithianum (1/8)
Zelia-            -Loderi-
Plumecocq-               -fortunei ssp. fortunei (1/8)
      -       -wardii var.wardii (1/4)
        -Crest-            -fortunei ssp. discolor (1/8)
        (FCC -Lady Bessborough-
         form)           -campylocarpum ssp. campylo-
                          carpum Elatum Group (1/8)
```

Large, open, saucer-like flowers of yellow with a pink blush, on
tall, handsome trusses.  Foliage and plant habit like Crest, the
flowers resemble souliei.  Named after Mme. Plumecocq, spon-
sor and organizer of the Valenciennes international flower show;
plant shown there in 1962.  (E. de Rothschild, reg. 1968)   See
color illus. ARS J (Fall 1982): p. 139.

cl.

Zella     Parentage unknown

A parent of Sarah Hardy.  Large, Loderi-type flowers, pale shell
pink, fading later; well-shaped, compact trusses.  (Collingwood
Ingram)  P. C. (RHS) 1964

```
                              -griffithianum (1/4)
                     -George Hardy-
cl.     -Mrs. Lindsay Smith-      -catawbiense (1/8)
        -                 -Duchess of Edinburgh--unknown (1/4)
Zellindo-            -griffithianum
        -            -Isabella-
        -Muy Lindo-      -auriculatum (1/8)
                    -              -decorum (1/8)
                    -unnamed hybrid-
                              -souliei (1/8)
```

A hybrid by Collingwood Ingram, reg. 1972.

```
                              -campylocarpum ssp. campylocarpum
                     -Penjerrick-           Elatum Group (1/16)
g.        -Amaura-     -griffithianum (1/16)
      -Eros-      -griersonianum (5/8)
Zenobia-     -griersonianum
      -              -sanguineum ssp. didymum (1/4)
        -Arthur Osborn-
                -griersonianum
```

A hybrid with deep red flowers.  (Lord Aberconway, 1941)

```
g.
      -souliei (1/2)
Zora-
      -wardii var. puralbum (1/2)
```

White flowers.  (J. B. Stevenson cross, 1934; intro. 1951)

```
                              -griffithianum (1/8)
g.                   -George Hardy-
        -Mrs. Lindsay Smith-      -catawbiense (1/8)
 Zuiderzee-          -Duchess of Edinburgh--unknown (1/2)
(Zuyderzee)          -campylocarpum ssp. campylocarpum (1/4)
        -unnamed hybrid-
                -unknown
```

3ft(.9m)          5°F(-15°C)          M          3/2

Synonym Zuyder Zee.  A compact shrub, broader than high; leaves
very light green, easily burned.  Pale yellow, campanulate flowers, spotted red at base.  Requires some shade.  See also Green-
finch, Jersey Cream.  (M. Koster)  A. M. (RHS) 1936

cl.

Zuyder Zee     See Zuiderzee

```
                              -arboreum ssp. arboreum (1/8)
cl.        -Glory of -
    -Barclayi -Penjerrick-griffithianum (1/8)
    -Robert Fox-
Zyxya-           -thomsonii ssp. thomsonii (1/4)
      -          -forrestii ssp. forrestii Repens Group (1/4)
        -Elizabeth-
                -griersonianum (1/4)
```

3ft(.9m)          5°F(-15°C)          E

Narrowly ovate leaves 4in(10.2cm) long, no indumentum.  Flowers
widely funnel-campanulate, 3in(7.6cm) across, clear blood red,
in trusses of 5-7.  (Gen. Harrison, reg. 1966)  A. M. (RHS) 1956

| REVISED NAME | FORMER NAME | REVISED NAME | FORMER NAME |
|---|---|---|---|
| aberconwayi | aberconwayi | chrysodoron | chrysodoron |
| adenogynum | adenogynum | ciliatum | ciliatum |
| adenogynum Adenophorum Group | adenophorum | ciliicalyx | ciliicalyx |
| adenopodum | adenopodum | cinnabarinum ssp. cinnabarinum | cinnabarinum var. cinnabarinum |
| agastum | agastum | Blandfordiiflorum Group | var. blandfordiiflorum |
| albertsenianum | albertsenianum | Roylei Group | var. roylei |
| alutaceum var. iodes | iodes | cinnabarinum ssp. tamaense | tamaense |
| alutaceum var. russotinctum | russotinctum | cinnabarinum ssp. xanthocodon | xanthocodon |
| ambiguum | ambiguum | Concatenans Group | concatenans |
| annae Laxiflorum Group | laxiflorum | concinnum | concinnum |
| anthopogon ssp. anthopogon | anthopogon var. anthopogon | coriaceum | coriaceum |
| ssp. hypenanthum | hypenanthum | coryanum | coryanum |
| aperantum | aperantum | crinigerum var. crinigerum | crinigerum var. crinigerum |
| araiophyllum | araiophyllum | cuffeanum | cuffeanum |
| arboreum ssp. arboreum | arboreum ssp. arboreum | dalhousiae var. dalhousiae | dalhousiae |
| arboreum ssp. cinnamomeum | arboreum ssp. arboreum | dauricum | dauricum var. dauricum |
| var. album | var. album | davidsonianum | davidsonianum |
| var. cinnamomeum | var. cinnamomeum | decorum | decorum |
| var. roseum | var. roseum | dendricola Taronense Group | taronense |
| arboreum ssp. delavayi | | diaprepes | diaprepes |
| var. delavayi | delavayi | dichroanthum ssp. dichroanthum | dichroanthum ssp. dichroanthum |
| arboreum ssp. nilagiricum | arboreum ssp. nilagiricum | ssp. apodectum | ssp. apodectum |
| arboreum ssp. zeylanicum | zeylanicum | dichroanthum ssp. scyphocalyx | dichroanthum ssp. scyphocalyx |
| argyrophyllum | argyrophyllum | Herpesticum Group | ssp. herpesticum |
| ssp. argyrophyllum | var. argyrophyllum | diphrocalyx | diphrocalyx |
| argyrophyllum ssp. hypoglaucum | hypoglaucum | eclecteum var. eclecteum | eclecteum var. eclecteum |
| argyrophyllum ssp. nankingense | argyrophyllum var. nankingense | edgeworthii | edgeworthii (bullatum) |
| augustinii ssp. augustinii | augustinii var. augustinii | elliottii | elliottii |
| augustinii ssp. chasmanthum | augustinii var. chasmanthum | eudoxum var. eudoxum | eudoxum ssp. eudoxum |
| augustinii ssp. rubrum | bergii | faberi ssp. prattii | prattii |
| aureum var. aureum | aureum | facetum (eriogynum) | facetum (eriogynum) |
| auriculatum | auriculatum | falconeri ssp. eximium | eximium |
| auritum | auritum | falconeri ssp. falconeri | falconeri |
| barbatum | barbatum | fastigiatum | fastigiatum |
| basilicum | basilicum | ferrugineum | ferrugineum |
| beanianum | beanianum | flavidum var. flavidum | flavidum var. flavidum |
| beesianum | beesianum | fletcheranum (fletcherianum) | fletcheranum (fletcherianum) |
| boothii Mishmiense Group | mishmiense | floccigerum ssp. floccigerum | floccigerum var. floccigerum |
| brachyanthum ssp. brachyanthum | brachyanthum var. brachyanthum | floribundum | floribundum |
| brachyanthum ssp. hypolepidotum | brachyanthum var. hypolepidotum | formosum var. formosum | formosum |
| brachycarpum ssp. brachycarpum | brachycarpum | Iteophyllum Group | iteophyllum |
| bureavii | bureavii | formosum var. inaequale | inaequale |
| burmanicum | burmanicum | forrestii ssp. forrestii | forrestii var. forrestii |
| callimorphum ssp. callimorphum | callimorphum | Repens Group | var. repens |
| calophytum var. calophytum | calophytum | forrestii ssp. papilliatum | var. tumescens |
| calostrotum ssp. calostrotum | calostrotum | fortunei ssp. discolor | discolor |
| calostrotum ssp. keleticum | | Houlstonii Group | houlstonii |
| Radicans Group | radicans | fortunei ssp. fortunei | fortunei |
| campanulatum ssp. campanulatum | campanulatum var. campanulatum | fulgens | fulgens |
| campylocarpum ssp. caloxanthum | caloxanthum | fulvum | fulvum |
| campylocarpum ssp. campy- | campylocarpum var. campy- | glaucophyllum | glaucophyllum |
| locarpum | locarpum | var. glaucophyllum | var. glaucophyllum |
| Elatum Group | var. elatum | glischrum ssp. rude | rude |
| campylogynum | campylogynum var. campylogynum | grande | grande |
| campylogynum Charopaeum Group | var. charopaeum | griersonianum | griersonianum |
| campylogynum Cremastum Group | var. cremastum | griffithianum | griffithianum |
| campylogynum Myrtilloides Group | var. myrtilloides | haematodes ssp. chaetomallum | chaetomallum var. chaetomallum |
| carneum | carneum | haematodes ssp. haematodes | haematodes |
| catacosmum | catacosmum | hanceanum | hanceanum var. hanceanum |
| catawbiense | catawbiense | Nanum Group | var. nanum |
| caucasicum | caucasicum | heliolepis var. heliolepis | heliolepis |
| cephalanthum ssp. cephalanthum | cephalanthum var. cephalanthum | hemitrichotum | hemitrichotum |
| cerasinum | cerasinum | hemsleyanum | hemsleyanum |
| chamaethomsonii var. chamae- | chamaethomsonii var. chamae- | hippophaeoides var. | |
| thomsonii | thomsonii | hippophaeoides | hippophaeoides |
| var. chamaethauma | var. chamaethauma | Fimbriatum Group | fimbriatum |
| charitopes ssp. charitopes | charitopes | hirsutum | hirsutum |
| charitopes ssp. tsangpoense | tsangpoense var. tsangpoense | hirtipes | hirtipes |
| Curvistylum Group | var. curvistylum | hodgsonii | hodgsonii |

| REVISED NAME | FORMER NAME | REVISED NAME | FORMER NAME |
|---|---|---|---|
| hookeri | hookeri | polylepis | polylepis |
| hyperythrum | hyperythrum | ponticum | ponticum |
| impeditum | impeditum | praestans | praestans |
| insigne | insigne | praevernum | praevernum |
| intricatum | intricatum | principis Vellereum Group | vellereum |
| irroratum ssp. irroratum | irroratum | protistum var. giganteum | giganteum |
| japonicum var. japonicum | metternichii | pruniflorum | tsangpoense var. pruniflorum |
| japonicum var. pentamerum | degronianum | pseudochrysanthum | pseudochrysanthum |
| johnstoneanum | johnstoneanum | pubescens | pubescens |
| keiskei | keiskei | pumilum | pumilum |
| keysii | keysii var. keysii | racemosum | racemosum |
| kyawii | kyawii | recurvoides | recurvoides |
| lacteum | lacteum | rex ssp. arizelum | arizelum |
| lanigerum | lanigerum | rex ssp. fictolacteum | fictolacteum |
| lepidotum | lepidotum | rigidum | rigidum |
| leucaspis | leucaspis | ririei | ririei |
| lindleyi | lindleyi | roxieanum var. roxieanum | roxieanum var. roxieanum |
| loundesii | loundesii | rubiginosum | rubiginosum |
| ludlowii | ludlowii | rubiginosum Desquamatum Group | desquamatum |
| luteiflorum | luteiflorum | rufum | rufum |
| lutescens | lutescens | rupicola var. chryseum | chryseum |
| macabeanum | macabeanum | russatum | russatum |
| macrophyllum | macrophyllum | saluenense ssp. chameunum | |
| maculiferum ssp. anhweiense | anhweiense |   Prostratum Group | prostratum |
| maddenii ssp. crassum | crassum | saluenense ssp. saluenense | saluenense |
| maddenii ssp. maddenii | maddenii | sanguineun ssp. sanguineum | |
| maddenii ssp. maddenii | polyandrum |   var. sanguineum | sanguineum ssp. sanguineum |
| magnificum | magnificum | sanguineum ssp. sanguineum | |
| mallotum | mallotum |   var. didymoides |   ssp. consanguineum |
| martinianum | martinianum |   var. didymoides |   ssp. didymoides |
| maximum | maximum |   ssp. didymum |   ssp. didymum |
| meddianum var. meddianum | meddianum var. meddianum |   var. haemaleum |   ssp. haemaleum |
|   var. atrokermesinum |   var. atrokermesinum | sargentianum | sargentianum |
| megacalyx | megacalyx | scabrifolium var. spiciferum | spiciferum |
| megeratum | megeratum | scopulorum | scopulorum |
| mekongense var. mekongense | mekongense | searsiae | searsiae |
|   Viridescens Group | viridescens | seinghkuense | seinghkuense |
| mekongense var. melinanthum | melinanthum | selense ssp. selense | selense var. selense |
| micranthum | micranthum | sidereum | sidereum |
| microgynum Gymnocarpum Group | gymnocarpum | sinogrande | sinogrande var. sinogrande |
| minus var. chapmanii | chapmanii | smirnowii | smirnowii |
| minus var. minus | minus | smithii | smithii |
|   Carolinianum Group | carolinianum var. carolinianum | souliei | souliei |
| mollicomum | mollicomum var. mollicomum | sperabile var. sperabile | sperabile var. sperabile |
| montroseanum | montroseanum | spinuliferum | spinuliferum |
| morii | morii | stewartianum | stewartianum var. stewartianum |
| moupinense | moupinense | strigillosum | strigillosum |
| mucronulatum | mucronulatum var. mucronulatum | sulfureum | sulfureum |
| myrtifolium | myrtifolium | sutchuenense | sutchuenense var. sutchuenense |
| neriiflorum ssp. neriiflorum | neriiflorum ssp. neriiflorum | taggianum | taggianum |
|   Euchaites Group |   ssp. euchaites | telmateium | telmateium |
| nivale ssp. nivale | | temenium var. gilvum | |
| niveum | niveum |   Chrysanthum Group | temenium var. chrysanthum |
| nuttallii | nuttallii | tephropeplum | tephropeplum |
| orbiculare ssp. orbiculare | orbiculare | thayeranum | thayeranum |
| oreodoxa var. fargesii | erubescens | thomsonii ssp. thomsonii | thomsonii var. thomsonii |
| oreodoxa var. fargesii | fargesii | traillianum var. dictyotum | dictyotum |
| oreodoxa var. oreodoxa | oreodoxa | trichanthum | trichanthum |
| oreotrephes | oreotrephes | trichocladum | trichocladum |
| orthocladum var. microleucum | microleucum | trichostomum | trichostomum var. trichostomum |
| pachytrichum | pachytrichum |   Ledoides Group |   var. ledoides |
| parmulatum | parmulatum |   Radinum Group |   var. radinum |
| pemakoense | pemokoense | triflorum | triflorum var. triflorum |
| phaeochrysum var. phaeochrysum | dryophyllum | tsariense | tsariense |
| phaeochrysum var. phaeochrysum | phaeochrysum | ungernii | ungernii |
| pocophorum var. pocophorum | pocophorum | uniflorum var. imperator | imperator |
| polycladum | compactum | uvariifolium | uvariifolium var. uvariifolium |
| polycladum | polycladum | valentinianum | valentinianum |
|   Scintillans Group | scintillans | | |

| REVISED NAME | FORMER NAME |
|---|---|
| veitchianum | veitchianum |
|  Cubittii Group | cubittii |
| venator | venator |
| vernicosum | vernicosum |
| vesiculiferum | vesiculiferum |
| virgatum ssp. oleifolium | oleifolium |
| virgatum ssp. virgatum | virgatum |
| viscidifolium | viscidifolium |
| wardii var. puralbum | puralbum |
| wardii var. wardii | wardii |
|  Litiense Group | litiense |
| wasonii | wasonii var. rhododactylum |
| wattii | wattii |
| wightii | wightii |
| williamsianum | williamsianum |
| xanthostephanum | xanthostephanum |
| yakushimanum ssp. makinoi | makinoi |
| yakushimanum ssp. yakushimanum | yakushimanum |
| yungningense | yungningense |
| yunnanense | yunnanense |
|  Hormophorum Group | hormophorum |
| zaleucum | zaleucum |

## BIBLIOGRAPHY

Bean, W. J. _Trees_ _and_ _Shrubs_ _Hardy_ _in_ _the_ _British_ _Isles_, V. III, 8th. ed.,  M. Bean & John Murray (Publishers), London, 1976.

Chamberlain, D. F. _A Revision of Rhododendron, Notes from the Royal Botanic Garden_, Edinburgh, Vol. 39, No. 2, 1982.

Clarke, J. Harold, ed. _Rhododendron_ _Information_, American Rhododendron Society, Portland, OR, 1980.

Cowles, Eveleth C. _The_ _Dexter_ _Estate:_ _Its_ _Gardens_ _and_ _Gardeners_, Massachusetts Chapter, American Rhododendron Society, 1980.

Cox, Peter A. _The_ _Larger_ _Species_ _of_ _Rhododendron_, B. T. Batsford, London, 1979.

Cox, Peter A. _The_ _Smaller_ _Rhododendrons_, Timber Press, Portland, OR, 1985.

Cullen, J. A. _A Revision of Rhododendron, Notes from the Royal Botanic Garden_, Edinburgh, Vol. 39, No. 1, 1980.

Davidian, H. H. _The_ _Rhododendron_ _Species,_ _V._ _1,_ _Lepidotes_, Timber Press, Portland, OR, 1982.

F       Fairweather, Christopher. _Rhododendrons_ _and_ _Azaleas_ _for_ _Your_ _Garden_, Floraprint, Calverton, Nottingham, England, 1980.

I.R.R.  Fletcher, H. R., comp. _The_ _International_ _Rhododendron_ _Register_, Royal Horticultural Society, Vincent Square, London, 1958.

HG      Greer, Harold E. _Greer's_ _Guidebook_ _to_ _Available_ _Rhododendron_ _Species_ _and_ _Hybrids_,  Offshoot Publications, Eugene, OR, 1980.

_Hillier's_ _Manual_ _of_ _Trees_ _and_ _Shrubs_, 5th ed., Hillier Nurseries, Ampfield House, Ampfield, Romsey, Hampshire, England, 1981.

Huse, Robert D., & Kenneth L. Kelly. _A_ _Contribution_ _Toward_ _Standardization_ _of_ _Color_ _Names_ _in_ _Horticulture_, American Rhododendron Society, Portland, OR, 1985

Kelly, Kenneth L., & Deane B. Judd. _Color:_  _Universal_ _Language_ _and_ _Dictionary_ _of_ _Names_, National Bureau of Standards, Washington, D. C., 1976.

K       Kraxberger, Meldon, ed. _American_ _Rhododendron_ _Hybrids_, American Rhododendron Society, Portland, OR, 1980.

Leach, David G. _Rhododendrons_ _of_ _the_ _World_ _and_ _How_ _to_ _Grow_ _Them_, Charles Scribner's Sons, New York, 1961.

LW      Livingston, Philip A., & Franklin H. West, eds. _Hybrids_ _and_ _Hybridizers:_ _Rhododendrons_ _and_ _Azaleas_ _for_ _Eastern_ _North_ _America_, Harrowood Books, Newton Square PA, 1978.

Nelson, Pat, comp., with Marlene Buffington, Nadine Henry. _Selected_ _Rhododendron_ _Glossary_, 3-D Publications, Graham, WA, 1982.

PB      Phillips, E. C. Lucas & Peter N. Barber. _The_ _Rothschild_ _Rhododendrons:_  _A_ _Record_ _of_ _the_ _Gardens_ _at_ _Exbury_, rev. ed., Cassell, London, 1970.

Rhodes, Janet. _A_ _Guide_ _to_ _Selecting_ _and_ _Enjoying_ _Rhododendrons_ _in_ _America_, The Bramble Bush, Oak Harbor, WA, 1982.

Royal Horticultural Society. _Royal_ _Horticultural_ _Society_ _Color_ _Chart_, The Society, Vincent Square, London, 1966.

RHS     Royal Horticultural Society. _Rhododendron_ _Species_ _in_ _Cultivation;_ _The_ _Rhododendron_ _Handbook_, The Society, Vincent Square, London, 1980.

Royal Horticultural Society, Rhododendron Group. _The_ _Rhododendron_ _Handbook,_ _Part_ _Two,_ _Rhododendron_ _Hybrids_, The Society, Vincent Square, London, 1969.

WS      Schmalscheidt, Walter. _Rhododendron-Zuchtung_ _in_ _Deutschland_, W. Schmalscheidt, Oldenberg, West Germany, 1980.

Sleumer, Herman. _An_ _Account_ _of_ _Rhododendrons_ _in_ _Malesia_, "Partly a reprint from _Flora_ _Malesiana_, ser. I, vol. 6, part 4", Jan. 1966.

Street, Frederick. _Hardy_ _Rhododendrons_, Collins, London, 1954.

VV      Van Veen, Ted. _Rhododendrons_ _in_ _America_, rev. ed., Binford & Mort, Portland, OR, 1980.


PERIODICALS AND YEARBOOKS:

ARS     American Rhododendron Society. _Quarterly_ _Bulletin_ 1958-1981.  Portland, OR.
                        _Journal_, 1982-84

Royal Horticultural Society. _Rhododendron_ _Year_ _Book_ 1958-1971.  Vincent Square, London.

                        _Rhododendrons_, 1972/73.
                        _Rhododendrons_ _with_ _Magnolias_ _and_ _Camellias_, 1974-84

Catalogs and price lists of rhododendron nurseries.

Abbot, Frank L., Saxton's River, VT
Aberconway, The Rt. Hon. Lord, Bodnant, Tal-y-Cafn, Colwyn Bay,
  Clwyd, North Wales, UK
Acland, Cuthbert (d.) Stagshaw, Ambleside, Cumbria, UK
Adams-Acton, G. Murray (d.) Beach House, Cooden Beach, Sussex, UK
Allen, Mr. & Mrs. E. F., Felcourt, Copcock, Ipswich, Suffolk, UK
Allen, Ernest V., Eugene, OR
Amateis, Edmond (d.) Brewster, NY
Anderson, Edw. 2903 Huntington Pl., Longview, WA 98632
Anderson, Dr. Richard, Eureka, CA
Anderson, Mrs. Stephen E., Bellevue, WA
Anne, Countess of Rosse, Nymans, Handcross, W. Sussex, UK
Arends, G. (d.) Wuppertal, W. Ger.
Arsen, F., 13 Vermont St., Lindenhurst, Long Island, NY 11757
Arthur H. Scott Horticultural Foundation,
  Swarthmore College, Swarthmore, PA 19081
Austin, Ray, Washougal, WA
Australian Rhododendron Society, Box 21, Olinda 3788
  Victoria, Australia

Bacher, John (1883-1961) Portland, OR
Bagoly, Lewis, 700 Terphanny Lane, Strafford, PA 19087
Bailey, J. & J. E., 8536 NE 26th, Bellevue, WA 98004
Bain, Mrs. Irene, Glenview, Rd. 1, Napier, NZ
Baker, George L. (d.) Astoria, OR
Balch, R. W., Dunedin, NZ
Baldanza, Samuel (d.) Benton Harbor, MI
Baldsiefen, Warren (d. 1974) Bellvale, NY
Baldwin, M., Mountain Highway, Boronia, Victoria, Australia
Banks, W. L. & R. A., Hergest Croft, Kington, Herefordshire, UK
Barber, Peter N., Otterwood Gate, Exbury Rd.,
  Beaulieu, Brockenhurst, Hants., UK
Barefield, Grady E. (d.) and Mary W. (d.) Seattle, WA
Barto, James E. (1881-1940) Eugene, OR
Basford, J. S., Brodick Castle Gardens, Isle of Arran, Scotland
Becales, Joseph, 36 Ivy Lane, Glen Mills, PA 19342
Beck, Mrs. Howard, Flora Markeeta Nursery, Edmunds, WA
Beekman, J., Sassafras, Victoria, Australia
Behring, Rudy, 74 Meredith Dr., St. Catharines, ON., Can.
Benmore--See Younger Botanic Garden
Berg, Warren E., Rt 1, Wren Ct., Port Ludlow, WA 98365
Berry, Mrs. A. C. U. (d.) Portland, OR
Bertin (Originator of R. Madame Masson, 1849)
Beury, James, Margate, NJ
Biltmore Gardens, Ashville, NC
Blaauw & Co., Boskoop, Holland
Black, Michael, Green Bank, Grasmere, Cumbria, UK
Bledsoe, D. (d.) Snohomish, WA
Blumhardt, Oswald, Rd. 9, Koromiko Nurseries, Whangarei, NZ
Blyskal, W. J., Box 134, Cherry Valley, NY 13320
Bodnant---See Aberconway
Boer, Gebr., Boskoop, Holland
Bohlje, G. D., Westerstede, Ger.
Bolitho, Lt. Col. Sir Edward H. W. (d.) Trengwainton,
  Penzance, Cornwall, UK
Bond, John D., Crown Estate, Windsor Great Park, Windsor, UK
Boot & Co., Boskoop, Holland
Borde Hill---See Clarke, R. N. Stephenson
Boscawen, The Hon. H. E., High Beeches, Handcross, Sussex, UK
Bosley, Paul, Sr., 9579 Mentor Ave., Mentor, OH 44060
Boulter, F., Everest Crescent, Olinda, Victoria, Australia
Boulter, V. J. & Sons, Olinda, Victoria, Australia
Bovee, Robert M. (d.) Portland, OR
Bowers, Clement Gray (d.) Binghamton, NY
Bowman, Dr. & Mrs. Paul J., Box 495, Fort Bragg, CA 95437
Braadfladt, H. J., 447 Excelsior Rd., Eureka, CA 95501
Bramley, Alfred, Perrins Creek Rd., Kallista,
  Victoria, Australia
Brandt, Lester E. (d.) Tacoma, WA
Brechtbill's Nursery, Eugene, OR

Briggs, Dr. Ben T., 825 Grant St., Shelton, WA 98584
Briggs, Bruce, 4407 Henderson Blvd., Olympia, WA 98501
Brockenbrough, Dr. E. C., 3630 Hunts Point Rd., Bellevue, WA
Brodick Castle Gardens, Isle of Arran, Scot., UK
Brotherton, James C., 20620 NE. Freedom Rd., Battleground, WA
Brown, Edward J., County St., Rt. 5, Lakeville, MA 02346
Browne of Upway, Victoria, Australia
Bruns, Joh., Bad Zwischenahn, W. Ger.
Brydon, P. H., 3025 Oakcrest Dr. NW., Salem, OR 97304
Brykit, Dr. Max E., 28 W. Potomac St., Williamsport, MD 21795
Buckland Monachorum--See Fortescue Garden Trust
Burlingame, C. R., Rt. 3, Box 1068, Hoquiam, WA 98335
Burns, Sidney, Ridge Rd. Muttontown, Syosset, Long Island, NY
Butler, Mr. & Mrs. M. K., 17427 Clover Rd., Bothell, WA 98011
Byls, J., Ghent, Belgium (Raised arboreum hybrids, 1860-75)

Caerhays Castle---See Williams, John Charles
Campbell, B. W., 209 Waireka Rd., Ravensbourne, Dunedin, NZ
Campbell, Sir George (d.) Crarae Lodge,
  Inveraray, Argyll, Scot., UK
Cannon, Clifford, 2705 Cain Rd., Olympia, WA 98501
Caperci, James F., 11436 SE. 208th St., Space #93, Kent, WA 98031
Carlyon, Miss G., Tregrehan, Par, Cornwall, UK
Cary, Edward A., 246 Boston Pike, Shrewsbury, MA 01545
Cavendish, Hugh, Holker Hall, Grange-Over-Sands, Cumbria, UK
Childers, Arthur A., Rhodoland Nursery, Vida, OR 97488
Childers, Maxine, 46451 McKenzie Highway, Vida, OR 97488
Christie, Sylvester (d.) Blackhills, Scotland
Clark, Roy W., 2101 Olympia Ave., Olympia, WA 98506
Clarke, R. N. Stephenson, Borde Hill, Haywards Heath, Sussex, UK
Clarke, Col. S. R. (d.) Borde Hill, Haywards Heath, Sussex, UK
Clarke, Dr. Harold J., 9750 Edwards Dr., Sun City, AZ 85351
Clarke, Steve, Clarke Nurseries, Long Beach, WA 98631
Clyne, Mrs. J. S., 38A Seddon St., Highfield, Timarau, NZ
Coe, William R. (d.) Oyster Bay, NY
Coker, Mrs. R. J., 129 Ilam Rd., Christchurch, NZ
Collinson, Peter (Intro. maximum to Britain, 1736)
Colville, Col. N. R. (d.) Penheale Manor, Launceston,
  Cornwall, UK
Colville, Mrs. N. R., Penheale Manor, Launceston, Cornwall, UK
Consolini, Anthony (d.) Sandwich, MA (Gardener to C. O. Dexter)
Cook, Alleyne R., 2117 Larson, North Vancouver, BC, Can.
Cooper, Roland Edgar (1890-1962) Collector
Coplen, M. G. (d.) Rockville, MD
Corbin, Dr. William, Portland, OR
Core, William, Silvercreek, WA
Corsock--See Ingall, Peter
Cottage Gardens Inc., 4992 Middle Ridge Rd., Perry, OH 44081
Cowles, John C., 745 Washington St., Wellesley, MA 02181
Cox, Peter A., Glendoick Gardens, Perth, Scot., UK
Craig, Dr. D. L., Research Station, Kentville, NS, Can.
Crarae Lodge--See Campbell, Sir George
Cribbs, Dalmas, Butler, PA
Crosfield, J. J. (d.) Embley Park, Romsey, Hampshire, UK
Crown Estate Commissioners, The Great Park, Windsor,
  Berkshire, UK
Crystal Springs Garden Committee, Portland Chapter,
  American Rhododendron Society, Portland, OR
Cutten, R. L., Ferny Creek, Victoria, Australia

Davies, Isaac (d.) Ormskirk, Lancashire, UK
Davis, Ross B., Jr., 404 Conestoga Rd., Wayne, PA 19087
Davis, Mrs. T., Sassafras, Victoria, Australia
Deans, James, Homebush, Christchurch, NZ
de Belder, R., Kalmthout, Belgium
de Longchamp, Roger, 540 Mt. Elam Rd., Fitchburg, MA 01420
Delp, Weldon E., Box 434, Harrisville, PA 16038
Del's Lane County Nursery, Eugene, OR 97401
den Ouden, H., Boskoop, Holland
Dexter, Charles O. (1862-1943) Sandwich, MA

Digby, Captain The Lord, Minterne, Dorchester, Dorset, UK
Dobbs, Olin O., Eugene, OR
Dosser, D., Droumana, Victoria, Australia
Dosser, Lillie, Centralia, WA
Dougall, Capt. Maitland (d.) Woodham, Woking, Surrey, UK
Drewry, James H. (d.) Fort Bragg, CA
Druecker, John S., Box 511, Fort Bragg, CA 95437
Dryden, Mrs. K., 30 Sheering Lower Rd.,
  Sawbridgeworth, Herts., UK
Dunedin Rhododendron Group, %57 Pilkington, Dunedin, NZ

Edinburgh--- See Royal Botanic Garden, Edinburgh, Scot., UK
Efinger, William, Brewster, NY
Egan, Ernest K., 40 Pease Rd., Woodbridge,
  New Haven, CT 06519
Eichelser, John (d.) Olympia, WA
Elliott, James A., Elliott Nursery, Rt. 4, Box 544, Astoria, OR
Elliott, Walter, 700 James Dr., Shelton, WA 98584
Embley Park--See Crosfield, J. J.
Endtz, L. J. & Co., Boskoop, Holland (Later named Blaauw & Co.)
English, Carl, Jr., Seattle, WA
Esch, Bernard J., Portland, OR
Evans, Bert (Head Gardener at Penjerrick, 1935-1970)
Everitt, Samuel A. (d.) Halesite, Huntington, Long Island, NY
Exbury Gardens---See Rothschild
Experimental Station, Boskoop, Holland
Farrer, Reginald (1880-1920)  Plant Explorer
Fawcett, Carl P., 8616 19th St. W., Tacoma, WA 98466
Felcourt--See Allen, E. F.
Felix and Dijkhuis, Boskoop, Holland
Fennichia, Richard A., 712 Bay Rd., West Webster, NY 14580
Fetterhoff, William F., 5279 Richland Rd., Gibsonia, PA 15044
Findlay, Hope (d.) Windsor Great Park
Flora Markeeta Nursery, Edmunds, WA
Ford, John E. (d. 1984) Secrest Arboretum, Wooster, OH
Forster, R. Ray, Horticultural Research Institute of Ontario,
  Vineland Station, ON, Can.
Fortescue, L. S. (d. 1981) Garden House, Buckland Monachorum,
  Yelverton, Devon, UK
Fortescue Garden Trust, Garden House, Buckland Monachorum,
  Yelverton, Devon, UK
Fortune, Robert (1812-80) Plant Explorer
Fox, Barclay (1873-1930) Son of Robert Fox,
  Penjerrick, Cornwall, UK
Fox, D. B., Bullwood Nurs., 54 Woodlands Rd., Hockley, Essex, UK
Foxhill Nurseries--See Reuthe, E. W.
Fraser, George (d.) Vancouver, BC, Can.
Frederick, Mrs. Halsey A., Jr., 530 Fisher's Rd., Bryn Mawr, PA
Freimann, Mrs. LeVern, 1907 38th St., Bellingham, WA 88225
Frets, C., Boskoop, Holland
Fromow, W., Windlesham, Surrey, UK
Frye, Else (d.) Seattle, WA
Fuller, Henry & Selma, The Shores, 1700 3rd Ave. W., Brandon, FL

Gable, Caroline, Stewartstown, PA, 17363
Gable, Joseph B. (1886-1972) Stewartstown, PA
Gatke, R. M. (d.) Salem, OR
Genista Gardens, Olinda, Victoria, Australia
George, A. F., Hydon Nurs., Hydon Heath, Godalming, Surrey, UK
German, Eugene R., Box 454, Fort Bragg, CA 95437
Gibson, A. C. (d.) and J. F. A. (d.) Glenarn,
  Rhu, Dumbartonshire, Scot., UK
Gigha, Isle of---See Horlick, Sir James N.
Gill, Richard & Sons (before 1900) Penryn, Cornwall, UK
Gillies, George, Supt., Marshall Field Estate, Oyster Bay, NY
Gillis, Jeanie, Horticulturist, Heritage Plantation,
  Sandwich, MA 02563
Girard, Peter, Sr. (d.) Geneva, OH
Glenarn---See Gibson, A. C.
Glendoick Gardens---See Cox, Peter
Glenkinglas, The Rt. Hon. The Lord, Strone House,

Cairndow, Argyll, Scot. UK
Glennie, W. E., Glenrose, Blicks Rd., R. D. 2,  Blenheim, NZ
Godsall, Robert, Head Gardener, Muncaster Castle,
  Ravenglass, Cumbria, UK
Goheen, Dr. David, Box 826, Camas, WA 98607
Golden, Albert, 117 Parker Ave., San Francisco, CA 94118
Goodrich, Col. & Mrs. R. H., 10015 Saddle Rd., Vienna, VA 22180
Gordon, Lady Adam, Hethersett, Littleworth Cross,
  Seale, Surrey, UK  GU 10 1JL
Gordon, R. C., Rongoiti Rd., Taihape, NZ
Gorer, Geoffrey, Sunte House, Haywards Heath, Sussex, UK
Gowen, J. R. (d.) Gardener to the Earl of Carnarvon,
  Highclere Castle, Newbury, Berkshire, UK
Grace, George (1897-1974) Portland, OR
Grange, The---See Ingram, Capt. Collingwood
Granston, Mrs. Mae K., 14346 Bear Creek Rd. NE.,
  Woodinville, WA 98072
Grant, C. A., Kapunatiki, Orton Rd., Temuka, NZ
Greer, Edgar (1895-1972) Eugene, OR
Greer, Harold E., Greer Gardens,
  1280 Goodpasture Island Rd., Eugene, OR 97401
Gregory, R. M. (d.) Werrington Pk., Launceston, Cornwall, UK
Greig, Mrs. E. J., Royston Nursery, Royston, BC, Can.
Griebnow, Paul, 805 Fairview Ave. SE., Salem, OR 97302
Griswold, Mrs. W. O., Kirkland, WA
Grothaus, Mr. & Mrs. Lewis C., 12373 SW.,
  Boones Ferry Rd., Lake Oswego, OR 97034
Guitteau, Mr. & Mrs. Robt. G., 1955 Graham St., Eugene, OR
Guttormsen, W. L., 1233 SE., First, Canby, OR 97013

Haag, Russell & Velma, Rt. 1, Box 89, Brevard, NC 28712
Hachmann, Hans, Brunnenstrasse 68,
  2202 Barmstedtin Holstein, W. Germany
Hall, Geoffrey, Harewood House Gardens, Leeds, Yorkshire, UK
Hall, Maurice, 135 Norlen Park, Bridgewater, MA 02324
Hancock, M. L. (d.) Mississauga, ON, Can.
Hanger, Francis (d.) Former Curator, RHS Gardens, Wisley, Surrey
Hansen, Ruth, 3514 N. Russett St., Portland, OR 97217
Hardgrove, Donald L. (d.) Merrick, Long Island, NY
Hardijzer, W. H., The Nurseries, Boskoop, Holland
Hardy, Maj. A. E., Sandling Park, Hythe, Kent, UK
Hardy, G. A., Hillhurst Farm, Hythe, Kent, UK
Harewood, Lord, Harewood House, Leeds, Yorkshire, UK
Harewood House Gardens---See Hall, Geoffrey
Harrison, Maj. Gen. E. G. W. W., Swallowfield Pk.,
  Swallowfield, Reading, Berkshire, UK
Harrison, Mrs. Roza (d.) Tremeer, St. Tudy, Cornwall, UK
Haworth-Booth, Michael, Farall Nurseries,
  Roundhurst Haslemere, Surrey, UK
Hayes, I. A., 2 Wellman Croft, Selly Oak, Birmingham, UK
Hayes, Mrs. Wesley, Centrewood Waimate, Box 29, S. Canterbury NZ
Headfort, Marquess of (d.) Headfort, Kells Co., Meath, Eire
Heineman, A. R., Milton, WA
Heller, Dr. Carl G. (d.) Poulsbo, WA 98370
Hendriksen, D. J., Boskoop, Holland
Heneage-Vivian, Adm. A. W. (d.) Clyne Castle, Swansea, Wales, UK
Henny, John, 8529 67th Ave. NE., Brooks, OR 97305
Henny, Mrs. Leona, 8991 75th Ave. NE., Rt. 6, Brooks, OR 97305
Henny, Rudolph (1909-1963) Brooks, OR
Herbert, Charles (d.) Phoenixville, PA
Heritage Plantation, Sandwich, MA (Formerly C. O. Dexter estate)
Hess, Nathaniel E., Sloanes Court, Sands Point, NY 11050
Heyderhoff, Henry, 348 Glen Wild Ave., Bloomingdale, NJ 07403
High Beeches--See Boscawen, The Hon. H. E.
Hillier & Sons, Winchester, Hants., UK
Hindla, Louis A., 986 Church St., Bohemia, NY 11716
Hinerman, Dr. D. L., 6800 Scio Church Rd., Ann Arbor, MI 48103
Hitchcock, Amy M., 1040 Ferry St., Eugene, OR 97401
Hobbie, Dietrich, Linswege Uber Westerstede, Oldenburg, W. Ger.
Hodgson, Mrs. Lillian, Vancouver, BC, Can.
Holden Arboretum, Sperry Rd., Mentor, OH 44060

'Vernus' by Shammarello/Leach
Photo by Leach

'Vida' by Steinmetz/Childers
Photo by Greer

'Vincent van Gogh' by M. Koster
Photo by Greer

'Virginia Leach' by Leach
Photo by Leach

'Virginia Richards' by Whitney
Photo by Greer

'Virgo' by P. van Nes
Photo by Greer

'Vivacious' by Forster
Photo by Greer

'Voodoo' by R. Henny
Photo by Greer

'Vulcan' by Waterer
Photo by Greer

'Vulcan's Bells' by Lancaster
Photo by Greer

'Waltham' by Ticknor/Mezitt
Photo by Greer

'Warm Glow' by Greer
Photo by Greer

'Warm Spring' by James
Photo by Greer

'Weber's Pride' by Weber
Photo by Greer

'Wheatley' by Phipps/Vossberg
Photo by Greer

'Whipped Cream' by Coker
Photo by Greer

'Whisperingrose' by Greer
Photo by Greer

'White Gold' by Greer
Photo by Greer

'White Swan' by Waterer
**Photo by Scott**

'Whitney's Late Orange' by Whitney/Sather
Photo by Greer

'Whitney's Orange' by Whitney/Sather
Photo by Greer

'Wickiup' by J. A. Elliott
Photo by Greer

'Wigeon' by Cox
Photo by Greer

'Wild Card' by Delp
Photo by Delp

Holden, Mrs. John, Rt. 4, Box 188, Shelton, WA 98584
Holden, John & Paul, Harstine Island Nursery, Shelton, WA
Hollard, Bernard, Kaponga, Taranki, NZ
Holman, Nigel T., Chyverton, Zelah, Truro, Cornwall, UK
Holmes, Mrs. A. G., Rakaia, NZ
Holmes, A. Graham, Holmeslee, Rakaia, Canterbury NZ
Hooftman, Hugo T., Boskoop, Holland
Horlick, Sir James N. (d.) Isle of Gigha, Argyll, Scot., UK
Horticultural Research Institute of Ontario,
 Vineland Station, ON, Can.
Housel, Mrs. D. Florine, Seattle, WA
Howard, Heman A., Box 197, South Wellfleet, MA 02663
Howells, A., Dickens Lane, Olinda, Victoria, Australia
Hughes, J. Hollis, Rt. 3, Box 264, Warrior, AL 35180
Huthnance, G., Carrington Rd., New Plymouth, NZ
Hydon Nurs., Clock Barn Lane, Hydon Heath, Godalming, Surrey, UK
Ihrig, Herbert (d.) Seattle, WA
Ingall, Peter (1899-1984) Corsock House, Castle Douglas, Scot.
Ingram, Capt. Collingwood (d. 1981)
 The Grange, Benenden, Cranbrook, Kent, UK
Institute of Ornamental Plant Growing,
 Caritasstraat 21, B-9230-Melle, Belgium
Irvine, Stanley (d.) Vancouver, BC, Can.

Jack, Evelyn, Univ. of British Columbia, Vancouver, BC, Canada
James, Delbert W. (1894-1963) Eugene, OR
James, Mrs. Ray, 3478 Hawthorne Ave., Eugene, OR 97402
Janeck, Kenneth, Tacoma, WA
Jarvis, Bernard R., 26924 Meridian E., Graham, WA 98338
Johnson, Allen P., 9370 SE. Cornell Rd., Port Orchard, WA 98366
Johnstone, Miss E. G., Little Glendoe, 43 Tolearne Ave.,
 Maori Hill, Dunedin, NZ
Johnstone, Maj. G. H. (d.) Trewithen, Grampound Rd., Cornwall, UK
Jordan, Bernice I., 1009 7th Ave. N., Tumwater, WA 98502
Joslin, W. V., Rt. 4, Box 338, Coos Bay, OR 97420
Jury, F. M., Tikorangi, Waitara, NZ
Jury, L. E., New Plymouth, NZ

Kehr, Dr. A. E., 240 Tranquility Pl., Hendersonville, NC
Kennedy, John F. Park, New Ross, Eire
Kenneth, Mrs. K. L., Tighnabrauaich, Ardrishaig, Argyll, Scot.
Kerr, Mrs. J., 17 Kitchener Square, Timaru, NZ
Kerr, Peter (Estate) Washougal, WA
Kerrigan, Howard W., 24249 2nd St., Hayward, CA 94541
Kersey, Roy J., Devon, PA
Kew, Royal Botanic Gardens, Richmond, Surrey, UK
Kilboggett--See Maskell, Sidney
King, S. M., 30 Devon Rd., Wanganui, NZ
Kingdon-Ward, Frank (1885-1958) Explorer and Writer
Kluis, Anthony, Boskoop, Holland (Later removed to USA)
Klupenger, Joseph, Rt. 2, Box 118, Aurora, OR 97002
Knap Hill Nurseries, Woking, Surrey, UK
Knight, Edgar L. (d.) Shelton, WA
Knight, Frank P. (d.) (1903-1985)
Knippenberg, Mrs. J. F., Laurelwood Gardens,
 736 Pines Lake Dr. W., Pines Lake, Wayne, NJ 07470
Koenig, Thomas W., 22 Rona St., Interlaken, NJ 07712
Kordus, Theodore, Jamesburg, NJ
Korn, Robert, 3700 NE. 9th Court, Renton, WA 98055
Korth, Allan C., 1465 38th Ave., Santa Cruz, CA 95060
Koster, D. A., Boskoop, Holland
Koster, M. & Sons, Boskoop, Holland
Koster, P. M., Bridgeton, NJ (Formerly of Boskoop)
Kraxberger, Mrs. M., 8450 Oleson Rd., Portland, OR 97223
Krug, H., 225 NW. 97th Ave., Portland, OR 97229
Kruschke, Franz (d.) Clackamas, OR

LaBar's Rhododendron Nursery, Stroudsburg, PA
Lamellen---See Magor, Maj. E. Walter M.
Lancaster, Benjamin F. (1892-1970) Camas, WA
Langdon, G., The Basin, 1405 Mountain Hyw., Victoria, Australia

Larson, Hjalmer L. (1897-1983) Tacoma, WA
Laxdall, Mrs. Dan (Sigrid) 3023 W. Alderwood, Bellingham, WA
Lawton, Lloyd H., 3941 Main Rd., Tiverton, RI 02878
Le Feber & Co., Boskoop, Holland
Leach, Dr. David G., 1894 Hubbard Rd., North Madison, OH 44054
Leendertz, B., Baumschule 'Heilmannshof' Krefeld-Traar, W. Ger.
Leith, W. (d.) Head Gardener, Glyne Castle, Swansea, UK
Lelliott, T., 15 Owen St., Boronia, Victoria, Australia
Lem, Halfdan (1886-1969)  Seattle, WA
Leonardslee--See Loder, Sir Giles
Lewis, Dr. G. David, 52 Glenwood Rd., Rt. 1, Colt's Neck, NJ
Lingholm--See Rochdale, Viscount
Linington, E. J., 11 Tamath Cr., Vancouver, BC, Can.
Lochinch--See Stair, Lord
Lock, Surgeon Capt. J. A. N., Lower Coombe Royal,
 Kingsbridge, Devon, UK
Lockhart, R. D., 25 Rubislaw Den N., Aberdeen, Scot., UK
Loder, Sir Edmund (d.) Leonardslee
Loder, Col. G. H. (d.) High Beeches
Loder, Gerald (d.) Wakehurst Place
Loder, Sir Giles, Bt., Ockenden House, Cuckfield, Sussex, UK
Loeb, Clarence, Puyallup, WA
Lofthouse, John G., 6649 Osler Ave., Vancouver, BC, Can.
Long, E. H., Oakland, CA
Longwood Gardens, Kennett Square, PA 19348
Louck's, Del, Del's Lane County Nursery, Eugene, OR
Love, Melvin V., Bellevue, WA
Lowinsky, T. (d.) Tittenhurst, Sunninghill, Berkshire, UK
Luenenschloss, Carl, 11 Brook Terrace, Fair Haven, NJ 07701
Lux, Mrs. Vincent A., Jr. (Diane) 1680 Monroe St.,
 Port Townsend, WA 98368
Lyons, Marshall W. (d.) Eugene, OR
Lyons, Ruth (d.) Renton, WA

McClure, Donald K., 4032 NE 95th, Seattle WA 98115
McCuaig, Mr. & Mrs. H. H., 5759 Newton Wynd, Vancouver, BC, Can.
MacEwen, Brig. Gen. D. (d.) Corsock House (See Ingall)
McGuinness, P. J., 14 Court St., West Haven, Conn 06516
McGuire, Thomas J., 9210 SW Westhaven Dr., Portland, OR 97225
Mackenzie, Mrs. Elizabeth, Hill Cottage,
 Fressingfield, By Diss, Norfolk, UK
McLaren, Henry (1879-1953) Bodnant (Later Lord Aberconway)
McLaren, Ruth, Agate Pass Garden, Box 288, Suquamish, WA
McLaughlin, C. A., Dunedin, NZ
McMurry, Tressa, Bellingham, WA
McNew, Charles C., 2710 Mt. Brynion Rd., Kelso, WA 98626
McQuire, J. F. J., Deer Dell, Botany Hill, Farnham, Surrey, UK
Magor, Maj. E. Walter M., Lamellen, St. Tudy, Cornwall, UK
Magor, Edward J. P. (1874-1941) Lamellen, St. Tudy, Cornwall, UK
Maloney, Francis, 3636 Corliss Ave. N., Seattle, WA 98103
Manenica, Mrs. Victor, 310 Hemlock Park Rd., Aberdeen, WA 98520
Mangles, Miss Clara (1846-1931) Littleworth
Mangles, James Henry (1832-1884) Valewood
Martin, Alexander, C., Laurel, MD
Maskell, Sidney, Kilbogget, Eire
Mauritsen, Richard, Kent, WA
Maxwell, Sir John Stirling, Pollock House, Glasgow, Scot., UK
Mayers, Dan E., Loth-Lorien, Wadhurst, Sussex, UK
Medlicott, Dr. R. W., Ashburn Hall, Dunedin, NZ
Messel, Lt. Col. L. C. R. (d.) Nymans, Handcross, Sussex, UK
Methven, T. (Before 1871) Edinburgh, Scot., UK
Mezitt, Edmund V., Weston Nurs., E. Main St., Hopkinton, MA
Mezitt, P. J. M. (d.) Hopkinton, MA
Michael, Charles (d.) Head Gardener to The Rt. Hon. Chas. W.
 Williams, Caerhays Castle, Gorran, Cornwall, UK
Michaud, Lawrence J., Issaquah, WA
Millais, E. G., Crosswater Farm, Churt, Farnham, Surrey, UK
Miller, Robert F., Star Route, Box 82, New Hope, PA 18938
Morris Arboretum, University of Pennsylvania,
 9414 Meadowbrook Ave., Philadelphia, PA 19118
Moser et Fils, Versailles, Fr.

Moss, William, Bryn Derwen Caerwys, Mold, Clwyd, Wales, UK
Mossman, Dr. Frank D., 1200 W. 39th St., Vancouver, WA 98660
Mount Stewart, Newtownards, County Down, UK
Moynier, William A., 2701 Malcolm Ave., Los Angeles, CA 90064
Mraw, L. B., 48 Shelbourne Dr., Trenton, NJ 08638
Mulligan, Brian O., 11632 106th Ave. NE., Kirkland, WA 98033
Muncaster Castle---See Pennington-Ramsden, Sir William
Murcott, Richard, Linden Lane, East Norwich, NY 11732

Nearing, G. Guy (1890-    ) Mahwah, NJ
Nelson, Benjamin (d.) Suquamish, WA
Nelson, Milton R., 5409 NW. Lincoln Ave., Vancouver, WA 98663
Newcomb, Lloyd L., 18432 Snohomish Ave., Snohomish, WA 98290
Newkirk, Donald, Camas, WA
Noble, Charles (d. before 1870)
Noble, Rt. Hon.--See Glenkinglas, The Rt. Hon.
Nymans Garden--See Anne, Countess of Rosse

Osborn, Hadley, Strybing Arboretum Society
 of Golden Gate Park, San Francisco, CA 94122
O'Shannassy, J., McGowans Rd., Donvale, Victoria 3111, Australia
Ostbo, Endre (1884-1958) Seattle, WA
Ostbo, I. Owen, Bellevue, WA

Pandrea, Joseph, 4516 S. 8th St., Tacoma, WA 98405
Parker, Edwin K., Rt. 2, Box 35, Astoria, OR 97103
Parker, John, Huntington, Long Island, NY
Parsons, Samuel B. (1819-1906) Flushing, Long Island, NY
Penheale Manor----See Colville, Mrs. Norman
Penjerrick---See Fox, Barclay
Pennington-Ramsden, Sir William, Muncaster Castle
 Ravenglass, Cumberland, UK
Perry, Mr. & Mrs. Matahiwi, Masterton, NZ
Peste, Fred, (d.) Shelton, WA
Peters, R. K., Box 142, Bendersville, PA 17306
Phetteplace, Dr. Carl H. (d.) Leaburg, OR
Philipps, Lady Marion, Picton Castle, Dyfed, South Wales, UK
Phipps, Howard, Post Rd., Westbury, Long Island, NY 11590
Pierce, Lawrence J., 900 University St., Apt. 10B, Seattle, WA
Pike, R. B., Lubec, ME
Pollard, Bill, Berkeley Horticultural Nursery, Berkeley, CA
Pot, Julian (d.) Chesterfield, OH
Potter, Basil C., 167 Lampman Ave., Port Ewen, NY 12466
Price, Sir Henry (d.) Wakehurst, Sussex, UK
Pride, Orlando S. (1905-1983) Butler, PA
Prycl, Otto, Rt. 1, New Stanton, PA 15672
Puddle, F. C., Former Curator, Bodnant, Tal-y-Cafn,
 Colwyn Bay, Clwyd, North Wales, UK
Puddle, F. C., Jr., Curator, Bodnant, Tal-y-Cafn,
 Colwny Bay, Clwyd, North Wales, UK
Pukeiti Rhododendron Trust, New Plymouth, NZ
Putney, Florence, Box 288, Suquamish, WA 98392

Rabbetts, Mrs. M., Beacon Hill Copse,
 Rockford, Ringwood, Hants., UK
Ramsden, Sir John (d.) Bulstrode Park & Muncaster Castle
Ranier Mountain Alpine Gardens, Seattle, WA
Raper, A., The Patch, Victoria, Australia
Raustein, A. A., 230 Union Ave., Holbrook, NY 11741
Rawinsky, G. B., Primrose Hill Nursery, Surrey, UK
Reboul, Adele, Box 373, St. James, Long Island, NY 11780
Reese, Dr. W. A., Redbarn Nursery, Box 215, Pennsburg, PA 18073
Reid, G. Albert, Rt. 1, Box 243, Blackman Rd.,
 Bargaintown, Linwood, NJ 08221
Reiley, H. E., 10210 Pine Tree Rd., Woodsborough, MD 21798
Research Station for Woody Nursery Crops, Boskoop, Holland
Reuthe, G., Foxhill Nurseries, Jackass Lane, Keston, Kent, UK
Rhodes, Dr. Robert C., Maple Ridge, BC, Can.
Ring, George W. III, 11400 Valley Rd., Fairfax, VA 22033
Rochdale, Viscount, Lingholm, Keswick, Cumbria, UK
Rock, Joseph Francis (1884-1962) Plant Explorer

Ross, Mrs. Henry, Ross Estate, Brewster, NY
Rothschild, Edmund de, Exbury Gardens, Southhampton, Hamps., UK
Rothschild, Lionel de (1882-1942) Founder of Exbury Gardens
Rouse, J., 8 Stonehaven Ct., Toorak, Victoria, Australia
Rowarth, R. L., 22 Bayview Ave., Upway, Victoria, Australia
Royal Botanic Garden, Edinburgh, Scot., UK
Royal Botanic Garden, Kew, Richmond, Surrey, UK
Royal Horticultural Society's Gardens, Wisley,
 Ripley, Woking, Surrey, UK
Royce, Mrs. Doris, Basket Neck Nursery, Box E, Remsenburg, NY
Royston Nursery, Royston, BC, Can.
Russell, J. P. C., The Dairies, Castle Howard, Yorkshire, UK
Rutherford, Mrs. Jean, Kelso, WA
Ruys, J. D., Moerheim Nursery, Dedemsvaart, Holland

Salley, Dr. Homer E., Box 315, Grand Rapids, OH 43522
Sandling Park---See Hardy, Maj. A. E.
Sather, Mrs. Anne, Box F., Brinnon, WA 98360
Saunders, Palmer F., 3030 Gilham Rd., Eugene, OR 97401
Savage, Glen, Portland, OR
Schamanek, John (d.) Philadelphia, PA
Schilling, A. D., Wakehurst Place Gardens, Ardingly,
 Haywards Heath, Sussex, UK
Schlaikjer, Mrs. Hugh C., Box 193, Halesite, Long Island, NY
Schmalscheidt, Walter, Ziegelhof-Strasse 85,
 2900 Oldenberg, W. Germany
Schroeder, Dr. H. Roland (d.) Evansville, Indiana
Schrope, Ned W., Rt. 2, Hegins, PA 17938
Schultz, Otto, Head Gardener, Royal Porcelain Factory,
 Berlin, Ger. (1890)
Schumacher, F. W., Spring Hill Rd., East Sandwich, MA 02537
Scott, Arthur H., Horticultural Foundation, Swarthmore, PA
Scott, Robert W., 9 Beverly Ct., Berkeley, CA 94707
Scrase-Dickins, C. R. (d.) Coolhurst, Horsham, Sussex, UK
Seabrook, Mrs. C. S., 345 Eldorado, Fircrest, WA 98466
Seabrook, Cecil S. (d.) Tacoma, WA
Seidel, T. J. (d.) Dresden, Ger.
Senko, Joseph F., Box 506, Cornelius, OR 97113
Shammarello, A. M. (1903-1982) South Euclid, OH
Shanklin, Robert G., Azalea Trace, Apt. 526,
 10100 Hillview Rd., Pensacola, FL 32504
Sheedy, Betty, 4335 SW. Twombly, Portland, OR 97201
Sherriff, George (1898-1982) Plant Explorer
Shilling, A. D., Wakehurst Place Gardens, Ardingly, Sussex, UK
Short, Howard A., A-11 350 Grow Ave., Bainbridge Island, WA
Shrauger, W. J., 140 Karr, Hoquiam, WA 98550
Sifferman, Karl, Seattle, WA
Sinclair, J. E., 617 NW. 175th St., Seattle, WA
Skinner, Dr. Henry T. (1907-1984) Hendersonville, NC
Slocock, M. O., Slocock Nursery, Woking, Surrey, UK
Slocock, Walter C., Goldsworth Nursery, Woking, Surrey, UK
Slonecker, Howard J., 15200 SE. Woodland Way, Milwaukie, OR
Slootjes, G. H., Boskoop, Holland
Smith, A. W., Horticultural Res. Institute of Ontario, Can.
Smith, A. P., 21505 President Point Rd., NE, Kingston, WA 98346
Smith, Britt M., 25809 124th Ave. SE., Kent, WA 98031
Smith, Cecil C., 5065 Ray Bell Rd. NE., St. Paul, OR 97137
Smith, Dr. Eric Ernest, Epsom, Surrey, UK
Smith, Graham F., Rd. 4, Taranaki, New Plymouth, NZ
Smith, Samuel, Head Gardener at Penjerrick, 1889-1935
Smith, W. David, Box 3250, Rd. 3, Spring Grove, PA 17362
Smith, William, Norbiton Common, Kingston, Surrey, UK
Snell, G. L. S., 970 Mountain Hwy., Boronia, Victoria, Australia
Snow's Ride Nurseries, Windlesham, Surrey, UK
Sorenson & Watson, Propriators of The Bovees, Portland, OR
Spady, Dr. H. A., 9460 Sunnyview Rd. NE., Salem, OR 97301
Sparks, Robert C., 1107 W. Harvard Ave., Shelton, WA 98584
Spring-Smyth, T. LeM, 14 Derwent Rd., New Milton, Hants., UK
Stair, Lord, Lochinch, Stranraer, Wigtownshire, Scot., UK
Standish & Noble, Nurs. owners before 1857, Windlesham, Surrey
Stanton, Donald B., Wollongong, N.S.W., Australia

Stanton, Ernest N., 21803 W. River Rd., Grosse Isle, MI 48138
Starling, B. N., Little Marles Cottage, Severs Green,
 Epping Upland, Essex, UK
Stead, L. Roland, Waterford, Amberley, North Canterbury, NZ
Steinmetz, Joseph, Springfield, OR
Stephens, J. Freeman, 896 Marine Dr., Bellingham, WA
Stevenson, John Barr (d.) Tower Court
Stevenson, Mrs. Roza---See Harrison, Mrs. Roza
Stewart, Coulter, 155 Broadmoor Ct., San Anselmo, CA
Stokes, Warren E. (d.) Butler, PA
Strauss, R., Stonehurst, Ardingly, Sussex, UK
Street, Frederick J., Heathermead Nurs., Woking, Surrey, UK
Streng, J., Boskoop, Holland
Strutt, E. A., Galloway House, Wigtown, Scot., UK
Strybing Arboretum, San Francisco, CA
Stuck, G., Hasselkamp 10, D-2300 Kronshagen/Keil, W. Ger.
Sullivan, Peter, 221 Rickland Ave., San Francisco, CA
Sumner, Maurice H., 350 Edgehill Way, San Francisco, CA
Sunningdale Nurseries, Windlesham, Surrey, UK
Sunrise Rhododendron Gardens, Bainbridge Island, WA 98110
Sunte House---See Gorer, Geoffrey
Sutton, P., 6 Mark Pl., Queanbeyan, New South Wales 2620, Aus.
Swain, George S., Agriculture Research Sta., NS, Can.
Swaythling, The Lord (d.) Bridley Manor, Worplesdon, Surrey, UK
Swenson, Willard J., 1005 Irvington Dr., Eugene, OR

Tacoma Metropolitan Pk., Point Defiance Rhod. Garden, Tacoma, WA
Tasker, H. R., Ashburton, NZ
Taylor, G., Burbank Seed Farms, Wyong 2259, N.S.W., Australia
Teese, A. J., Moore's Road, Monbulk, Victoria, Australia
Temple, Mrs. M., Koromako, The Downs, Rd. 21, Geraldine, NZ
Thacker, T. C., Knowle Nurseries, Warwick, Birmingham, UK
Thompson, Dr. Charles D., Eugene, OR
Ticknor, Dr. Robert L., 844 N. Holly, Canby, OR 97013
Tietjens, W. D., 52 Reeves Ave., Guilford, CT 06437
Tietjens, Marie, Blue Bell, PA
Tighnabrauaich, Ardrishaig, Argyll, Scot., UK
Tower Court---See Stevenson, John Barr
Trainor, Kenneth, 10502 32nd Ave., Seattle, WA
Tremeer--See Harrison, Maj. Gen. E. G. W. W.
Trewithen, St. Austell, Cornwall, UK
Tuomala, Carl, 20451 Lyta Way, Fort Bragg, CA 95437
Tyler, John J., Arboretum, Painter Rd., Lima, PA

University of Washington Arboretum, Seattle, WA
Urlwin-Smith, P. J., The Glade, Ascot, Berkshire, UK

Van de Ven, K., Monbulk Rd., Olinda, Victoria, Australia
van den Akker, C. A., Boskoop, Holland
van Gelderen, D. M., Boskoop, Holland
van Houtte, Louis (1810-1876) Belgium
van Nes, Adr. Boskoop, Holland
van Nes, C. B., Boskoop, Holland
van Nes, P., Boskoop, Holland
van Nes, Vuyk, Boskoop, Holland
Van Veen, Allen, 750 NW. 107th Ave., Portland, OR 97229
Van Veen, Theodore, Jr., 4201 SE. Franklin, Portland OR, 97206
Van Veen, Theodore, Sr. (1881-1961) Portland, OR
van Wilgen Nurseries, Boskoop, Holland
Veitch, Sir Harry James (1840-1924) London, England
Vernimb, Bryan, Rt. 2, Box 250, Howell, NJ 07731
Verschaffelt, Ambrose (1825-1886) Ghent, Belgium
Vossberg, Paul (d.) Westbury, Long Island, NY

Wada, Koichiro (1907-1981) Hakoneya Nursery, Yokohama, Japan
Wakehurst Place, Ardingly, Sussex, UK (Ext. of Kew Gardens)
Wallace, Robert (d. 1934) Tunbridge Wells, Kent, UK
Wapler, Martin, Eugene, OR
Ward--See Kingdon-Ward, Frank
Ward, Cyril H., 3111 NE. 49th St., Vancouver, WA 98663
Warren, Mrs. P. J., Dunedin, NZ

Waterer, Anthony (1853-97) Knap Hill, Woking, Surrey, UK
Waterer, Anthony, Jr. (d. 1924) Knap Hill, Woking, Surrey, UK
Waterer, F. Gomer (d.) Bagshot, Surrey, UK
Waterer, Hosea (d.) Owner of Knap Hill, 1842-53
Waterer, John, Sons & Crisp, Bagshot, Surrey, UK
Watson, Elsie M., 11530 Holmes Pt. Dr. NE., Kirkland, WA 98033
Weber, Edwin O., 14964 18th St. SW., Seattle, WA 98166
Wells, James S., 474 Nut Swamp Rd., Red Bank, NJ 07701
Weston Nurseries, E. Main St., Hopkinton, MA 01748
Wheeldon, Thomas J. Richmond, VA
Whitaker, W. I., Pylewell Park, Lymington, Hampshire, UK
White, G. Harry, Manager, Sunningdale Nurseries (1897-1936)
 Windlesham, Surrey, UK
Whitney, William E. (1894-1973) Brinnon, WA
Wick, Robert H., 8611 Fernwood NE., Olympia, WA 98506
Wildfong, Milton, Mission, BC, Can.
Wilding, Eustace Henry, Wexham Place,
 Stoke Poges, Buckinghamshire, UK
Wiley, K., The Garden House, Monachorum,
 Buckland Monachorum, Yelverton, Devon, UK
Willard, Rathbun, North Scituate, RI
Williams, A. M., Werrington Park, Launceston, Cornwall, UK
Williams, Charles, Caerhays Castle, Gorran, Cornwall, UK
Williams, John Charles (d. 1939) Caerhays Castle
Williams, F. Julian, Caerhays Castle, Gorran, Cornwall, UK
Williams, Michael, Lanarth, St. Keverne, Cornwall, UK
Williams, Percival D. (d.) Lanarth, St. Keverne, Cornwall, UK
Wilson, Dr. Ernest Henry (d.) Plant Explorer
Windsor Great Park--See Crown Estate
Wisley--See Royal Horticultural Society's Garden
Wister, John C. (d.) Swarthmore, PA
Wister, Mrs. John C. (Gertrude) 735 Harvard Ave., Swarthmore, PA
Witt, Joseph A. (d. 1984) Seattle, WA
Woodlands Nursery Garden, Carroll Ave., Ferndown, Dorset, UK
Wright, Arthur A., Jr., 1285 SE. Township Rd., Canby, OR 97013
Wright, Arthur A., Sr. (d.) Milwaukie, OR
Wyatt, Vernon, E. 3230, Hwy. 106, Union, WA 98592
Wyman, Richard M., Framingham, MA
Wyrens, Dr. Rollin G., 17600 Marine Dr., Stanwood, WA
Wytovich, E. T., Buck Laurel Acres, 497 1st St., Port Carbon, PA

Yates, Henry (d.) Frostburg, MD
Yates, Mrs. Henry (Maletta) Rt. 2, Box 268, Frostburg, MD 21532
Yavorsky, Leon, Bennett Rd., Freehold, N. J. 07728
Yelton, Dr. E. H., 330 W. Court St., Rutherfordton, NC 28139
Yoder, Elsie, Portland, OR
Young, Howard, Chadd's Ford, PA 19317
Younger Botanic Garden, Benmore by Dunoon, Argyll, Scot., UK
 (Extension of RBG, Edinburgh)

Zimmerman, Robert G., Rt. 1, Box 368, Chimacum, WA 98325

360

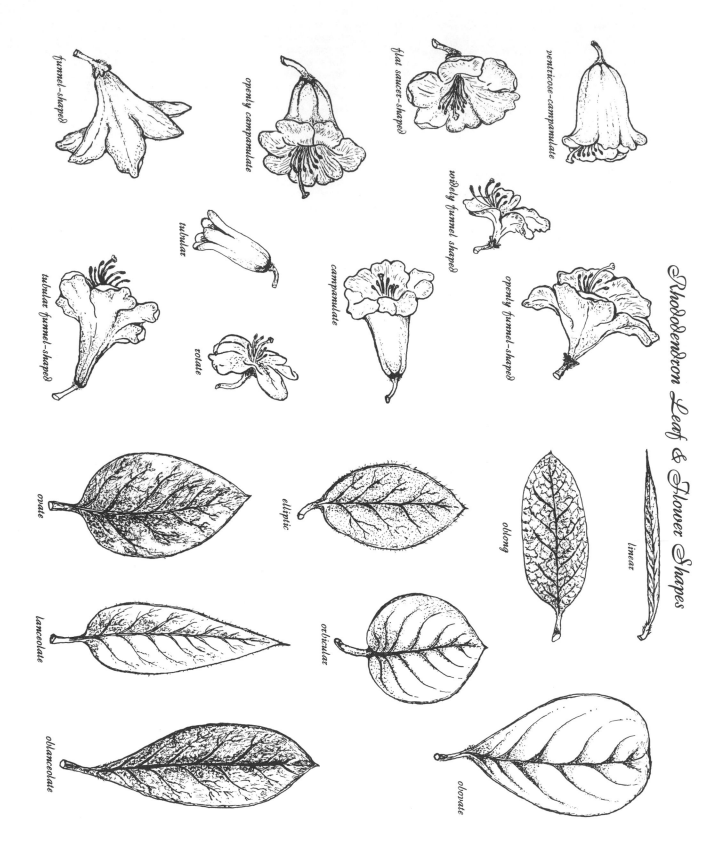

*funnel-shaped*

*openly campanulate*

*flat saucer-shaped*

*ventricose-campanulate*

*tubular*

*widely funnel shaped*

*tubular funnel-shaped*

*rotate*

*campanulate*

*openly funnel-shaped*

*Rhododendron Leaf & Flower Shapes*

*ovate*

*elliptic*

*oblong*

*linear*

*lanceolate*

*orbicular*

*oblanceolate*

*obovate*

By permission of Greer Gardens

| | |
|---|---|
| Acubifolium | ponticum |
| Aestivale | cinnabarinum ssp. cinnabarinum |
| Albiflorum Laciniatum | hirsutum |
| Albion Ridge | macrophyllum |
| Anna Strelow | annae Laxiflorum Group |
| Annapurna | anthopogon ssp. hypenanthum |
| Aola | valentinianum |
| Arctic Pearl | dauricum var. album |
| Ariel | javanicum |
| Ascreavie | maddenii ssp. maddenii |
| Ashcombe | veitchianum Cubittii Group |
| Ashleyi | maximum |
| Attar | decorum |
| Baby Mouse | campylogynum var. campylogynum |
| Bagshot Sands | lutescens |
| Bako | retivenum (brookeanum) |
| Barto Blue | augustinii ssp. augustinii |
| Barto Rose | oreodoxa var. fargesii |
| Beer Sheba | cerasinum |
| Benmore | montroseanum (mollyanum) |
| Bennan | meddianum var. atrokermesinum |
| Bergie | ciliatum |
| Berry's Bright Yellow | lutescens |
| Beryl Taylor | campylogynum var. campylogynum |
| Beta | lanigerum (silvaticum) |
| Betty Graham | anthopogon ssp. anthopogon |
| Bill Browning | mallotum |
| Blackhills | lacteum |
| Blackwater | spinuliferum |
| Blood Red | arboreum ssp. arboreum |
| Blue Cloud | augustinii ssp. chasmanthum |
| Blue Light | brachyanthum ssp. hypolepidotum |
| Blue Steel | impeditum |
| Bodnant Pink | arboreum ssp. cinnamomeum var. album |
| Bodnant Red | campylogynum Cremastum Group |
| Brodick | rex ssp. arizelum |
| Budget Farthing | oreodoxa var. fargesii |
| Butcher Wood | tephropeplum |
| Canton Consul | hanceanum Nanum Group |
| Caprice | augustinii ssp. augustinii |
| Carmelita | chamaethomsonii var. chamaethauma |
| Carse | irroratum ssp. irroratum |
| Catalgla | catawbiense var. album Glass |
| Catanea | catawbiense var. album |
| Cecil Nice | pocophorum var. pocophorum |
| Ceramic | wardii var. wardii |
| Chapel Wood | lanigerum (silvaticum) |
| Charisma | ciliicalyx |
| Cheiranthifolium | ponticum |
| Chelsea Chimes | coryanum |
| Cherry Brandy | cerasinum |
| Cherry Tip | rex ssp. fictolacteum |
| Chief Paulina | concinnum var. pseudoyanthinum |
| China Sea | longiflorum (lobbii) |
| Chinese Silver | argyrophyllum ssp. nankingense |
| Church Lane | yakushimanum ssp. yakushimanum |
| Coals of Fire | cerasinum |
| Coccinea | ferrugineum |
| Collingwood Ingram | tricostomum Ledoides Group |
| Colville | minus var. minus Carolinianum Group |
| Conroy (not a hybrid) | cinnabarinum Roylei crossed with cinnabarinum Concatenans |
| Convexum | yakushimanum ssp. yakushimanum |
| Copper | cinnabarinum ssp. xanthocodon Concatenans Group |
| Cornell Pink | mucronulatum |
| Corsock | phaeochrysum var. phaeochrysum |
| Cowtye | charitopes ssp. tsangpoense |
| Cox Apricot | racemosum |

| | |
|---|---|
| Crown Equerry | niveum |
| Cruachan | temenium var. gilvum Chrysanthum Gp. |
| Cunningham's Sulphur | caucasicum ? |
| Curtisii | multicolor |
| Cyril Berkeley | forrestii ssp. papillatum |
| Dalkeith | uniflorum var. uniflorum |
| Dalriada | japonicum var. pentamerum (degronianum) |
| Dame Edith Sitwell | lindleyi |
| Darjeeling | cinnabarinum ssp. cinnabarinum |
| Decimus | zoelleri |
| Deer Del | glaucophyllum var. glaucophyllum |
| Del James | taggianum |
| Demi-John | johnstoneanum |
| Doker-La | primuliflorum var. cephalanthoides ? |
| Doshong La | mekongense var. viridescens |
| Double Diamond | johnstoneanum |
| Down Under | lochiae |
| Easter Island | alutaceum var. russotinctum |
| Eleanor Black | konorii |
| Electra | augustinii ssp. augustinii |
| Elizabeth Bennett | formosum var. inaequale |
| Elizabeth David | burmanicum |
| Ellestee | wardii var. wardii |
| Elsa Reid | souliei |
| Elsie Louisa | macgregoriae |
| Epoch | minus var. minus Carolinianum Group |
| Eugene Cline | minus var. chapmanii |
| Exbury | praestans (coryphaeum) |
| Exbury Pink | souliei |
| Exbury's | davidsonianum |
| Exbury's | forrestii ssp. forrestii Repens Group |
| Exbury's | lutescens |
| Exbury's | recurvoides |
| Exbury's | saluenense ssp. saluenense |
| Exbury's | yakushimanum ssp. yakushimanum |
| Fair Sky | augustinii ssp. augustinii |
| Fake (Purple Fake) | concinnum var. pseudoyanthinum |
| Far Horizon | principis Vellereum Group |
| Fernhill | arboreum ssp. cinnamomeum var. roseum |
| Finch | rubiginosum Desquamatum Group |
| Fine Bristles | pubescens |
| Fischer | parryae |
| Fleurie | smithii |
| Folius Purpureus | ponticum |
| Folk's Wood | annae Laxiflorum Group |
| Foxy | fortunei ssp. fortunei |
| Frank Kingdon-Ward | glischrum ssp. rude |
| Frank Ludlow | dalhousiae var. dalhousiae |
| Gargantua | diaprepes |
| Gary Herbert | vernicosum |
| Geordie Sherriff | lindleyi |
| George Taylor | araiophyllum |
| Gerald Loder | japonicum var. pentamerum (degronianum) |
| Giant Rose | beesianum |
| Gigha | calostrotum ssp. calostrotum |
| Glen Cloy | luteiflorum |
| Glen Rosa | sidereum |
| Glendoick | racemosum |
| Goat Fell | arboreum ssp. arboreum |
| Golden Gate | zoelleri |
| Greenmantle | phaeochrysum var. phaeochrysum |
| Haba Shan | hippophaeoides var. hippophaeoides |
| Harkness | maximum var. leachii |
| Harp Wood | hodgsonii |
| Hayato | ellipticum (leiopodum) |
| Heane Wood | argyrophyllum ssp. hypoglaucum |
| Henry Shilson | arboreum ssp. arboreum |
| Herbert Mitchell | cerasinum |
| High Flyer | vesiculiferum |

| | | | |
|---|---|---|---|
| His Lordship | aberconwayi | Nassau Red | maximum |
| Ho Emma | japonicum var. japonicum (metternichii) | Nepal | cinnabarinum ssp. cinnabarinum |
| Hokkaido | dauricum | Nepal | niveum |
| Holker Hall | arboreum ssp. arboreum | Night Editor | russatum |
| Honey Wood | trichanthum | Nutmegacalyx | megacalyx |
| Hummel | rufum | Obovatum | lepidotum |
| Island Gem | oreotrephes | Ocelot | parmulatum |
| Island Sunset | zoelleri | Oki Island | japonicum var. japonicum (metternichii) |
| Ita | hirtipes | Omo | hyperythrum |
| Jack Hext | spinuliferum | Openwood | yunnanense (open pollinated) |
| Jade | campylogynum | Orange Bill | cinnabarinum ssp. xanthocodon |
| James Comber | sanguineum ssp. sanguineum var. haemaleum | | Concatenans Group |
| Jane Banks | ambiguum | Original | caucasicum var. album |
| Jaune | brachyanthum ssp. brachyanthum | Painted Snipe | orbiculatum |
| John R. Elcock | fortunei ssp. discolor Houlstonii Group | Pantagruel | diaprepes |
| John Skrentny | arboreum ssp. arboreum | Papillon | augustinii ssp. rubrum (bergii) |
| Kallistos | nuttallii var. stellatum ? | Parkside | charitopes ssp. charitopes |
| Kathmandu | traillianum var. dictyotum | Parvum | yakushimanum ssp. yakushimanum |
| Keillour Castle | ambiguum | Patricia | campylogynum Charopaeum Group |
| Ken Janeck | yakushimanum ssp. yakushimanum | Peachblow | hirtipes ? |
| Kildonan | magnificum | Perry Wood | faberi ssp. prattii |
| Kinabalu Mandarin (Mandarin) | retivenium (brookeanum) | Phantom Rock | sanguineum ssp. sanguineum var. haemaleum |
| Kingdom Come | eclecteum var. eclecteum | Pink Baby | pumilum |
| Kingdon-Ward Pink | edgeworthii | Pink Chiffon | fortunei ssp. fortunei |
| Kirsty | adenogynum Adenophorum Group | Pink Panther | mucronulatum |
| Knaphill | campanulatum ssp. campanulatum | Pink Parasol | yakushimanum ssp. yakushimanum |
| Koichiro Wada | yakushimanum ssp. yakushimanum | Planum | yakushimanum ssp. yakushimanum |
| Kulu | vernicosum | Plum Warner | campylogynum |
| Kyushu | japonicum var. japonicum (metternichii) | Poet's Lawn | hodgsonii |
| LaBar's White | catawbiense var. album | Polka Dot | irroratum ssp. irroratum |
| Lackamas Blue | augustinii ssp. augustinii | Powell Glass | catawbiense var. album |
| Lackamas Cream | chlorops (Now considered a hybrid.) | Pukeiti | protistum var. giganteum |
| Lakeside | trichostomum Ledoides Group | Purple Emperor | uniflorum var. imperator |
| Lancifolium | ponticum | Purple Fake (Fake) | concinneum var. pseudoyanthinum |
| Latifolia | ferrugineum | Purple Lustre | concinneum var. pseudoyanthinum |
| Leachii | maximum | Purple Pillow | russatum |
| Len Beer | glaucophyllum var. glaucophyllum | Purpurellum | cinnabarinum ssp. cinnabarinum |
| Lhasa | nuttallii var. stellatum ? | Quarry Wood | trichostomum Ledoides Group |
| Loch Eck | vernicosum | Quartz | rex ssp. rex |
| Lois Jean | lutescens (open pollinated) | Rae Berry | trichostomum |
| Lost Horizon | principis Vellereum Group | Rae's Delight | japonicum var. pentamerum (degronianum) |
| Louvecienne | rigidum | Raja | retivenum (brookeanum var. gracile) |
| Lucky Hit | davidsonianum | Ramsay Gold | keiskei |
| Lucy Elizabeth | formosum var. formosum Iteophyllum Gp. | Random Harvest | fortunei ssp. discolor Houlstonii Group |
| Magnificum | cinnabarinum ssp. cinnabarinum Roylei Group | Red Collar | edgeworthii |
| | | Red Imp | haematodes ssp. haematodes |
| Mahogany Red | mucronulatum | Reginald Childs | uvarifolium |
| Margaret Mead | veitchianum | Reuthe's Purple | lepidotum |
| Margum Castle | arboreum ssp. delavayi var. delavayi | Rhododactylum | wasonii |
| Maricee | sargentianum | Rhodoland's Silver Mist | yakushimanum ssp. yakushimanum |
| Marine | augustinii ssp. augustinii | Riviera Beauty | arboreum ssp. arboreum |
| Markeeta | adenopodum (open pollinated) | Rock Rose | racemosum |
| Meadow Pond | wardii var. wardii | Roland Cooper | campanulatum ssp. campanulatum |
| Meteor | barbatum | Rosenoble | arboreum ssp. cinnamomeum var. roseum |
| Metternianus | japonicum var. japonicum (metternichii) | Roseum | mucronulatum |
| Microleucum | orthocladum var. microleucum | Roseum | rex ssp. rex |
| Midget | leucaspis | Rosevallon | neriiflorum ssp. neriiflorum |
| Midwinter | dauricum Sempervirens Group | Rosey Ball | rubiginosum |
| Mirelle Vernimb | hemitrichotum | Round Wood | lanigerum (silvaticum) |
| Mist Maiden | yakushimanum ssp. yakushimanum | Royale | orbiculare ssp. orbiculare |
| Moerhermii | impeditum | Rubaiyat | arboreum ssp. arboreum |
| Montanum | yakushimanum ssp. yakushimanum | Rubeotinctum | johnstoneanum |
| Morocco | coriaceum | Rubrum Davidian | augustinii ssp. augustinii |
| Mount Everest | cinnabarinum ssp. cinnabarinum | Ruth Lyons | davidsonianum |
| Mount Mitchell | maximum | Salmon | campylogynum |
| Mount Wilson | yakushimanum ssp. yakushimanum | Scarlet Pimpernel | forrestii ssp. papillatum (var. tumescens) |
| Mrs. Butler | See Sir Charles Butler | Scarlet Runner | forrestii ssp. papillatum (var. tumescens) |
| Mrs. Messel | decorum | | |
| Nanum | hanceanum Nanum Group | Second Attempt | callimorphum ssp. callimorphum |

| | |
|---|---|
| Serenade | davidsonianum |
| Serenade | yunnanense |
| Sesame | pachytrichum |
| Seventh Heaven | sutchuenense |
| Shortoff | dwarf maximum ! |
| Sidlaw | vernicosum |
| Silvia | lanigerum (silvaticum) |
| Sir Charles Butler | fortunei ssp. fortunei |
| Sirunke Orange | macgregoriae |
| Sleea | martinianum |
| Smoke | augustinii ssp. augustinii |
| South Lodge | arboreum ssp. zeylanicum |
| Spatter Paint | irroratum ssp. irroratum |
| Spring Sonnet | vernicosum |
| Stonehurst | lanigerum (silvaticum) |
| Summer Skies | augustinii ssp. augustinii |
| Summertime | maximum |
| Sunningdale | hippophaeoides var. hippophaeoides |
| Sunningdale Blue | impeditum |
| Sunte Rock | praestans (coryphaeum) |
| Superbum | japonicum var. japonicum (metternichii) |
| Sweet Bay | trichostomum Radinum Group |
| Swinhoe | floribundum |
| The Bride | caucasicum var. album |
| Thimble | campylogynum Cremastum Group |
| Titan | retivenum (brookeanum) |
| Tom Spring Smythe | dalhousiae var. dalhousiae |
| Tony Schilling | arboreum ssp. cinnamomeum var. roseum |
| Tower Court | augustinii ssp. augustinii |
| Triangle | cinnabarinum ssp. tamaense |
| Unicolor | keysii |
| Variegatum | ferrugineum |
| Variegatum | ponticum |
| Vin Rose | cinnabarinum ssp. cinnabarinum |
| | Roylei Group |
| Virgin | catawbiense var. album |
| Wakehurst | rubiginosum Desquamatum Group |
| Walley's | augustinii ssp. augustinii |
| Walter Maynard | ciliicalyx |
| War Paint | elliottii |
| Ward's * | rex ssp. rex |
| Waxen Bell | campanulatum ssp. campanulatum |
| Westhaven | aberconwayi |
| White Lace | racemosum |
| White Plains | alutaceum var. iodes |
| White Velvet | yakushimanum ssp. yakushimanum |
| Whitebait | sargentianum |
| White's Cunningham's | |
| Sulphur | caucasicum ? |
| White's Favourite | dichroanthum ssp. dichroanthum |
| Windsor Park | souliei |
| Winifred Murray | cephalanthum ssp. cephalanthum |
| Winter Brightness | mucronulatum |
| Winter Sunset | mucronulatum |
| Yak-Ity-Yak | yakushimanum ssp. yakushimanum |
| Yaku Angel | yakushimanum ssp. yakushimanum |
| Yaku Fairy | keiskei |
| Yangtze Bend | uvariifolium |
| Yellow Bunting | fletcheranum |
| Yellow Fellow | mekongense var. mekongense |
| Yellow Garland | xanthostephanum |
| Yucondus | erythrocalyx (or hybrid of same) |
| Yum-Yum | tsariense |

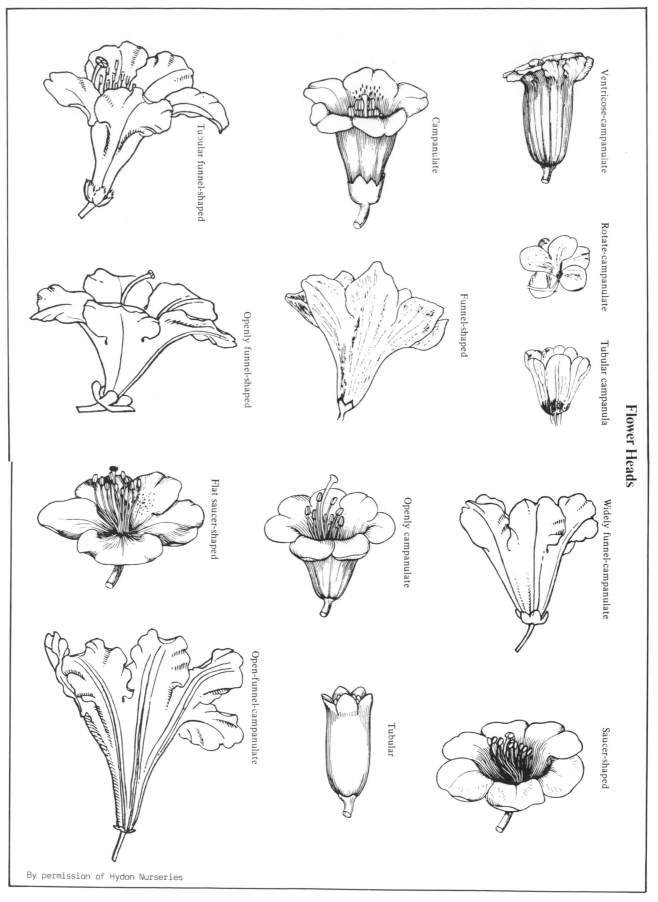

**Flower Heads**

Tubular funnel-shaped

Campanulate

Ventricose-campanulate

Rotate-campanulate

Openly funnel-shaped

Funnel-shaped

Tubular campanula

Flat saucer-shaped

Openly campanulate

Widely funnel-campanulate

Open-funnel-campanulate

Tubular

Saucer-shaped

cl. A. B. Mitford      Crimson. (A. Waterer, before 1915)
cl. A. C. Kendrick     Lilac rose, red purple blotch. (Koster)
--- A. G. Soames       Crimson red. (van Nes, before 1922)
cl. A. W. Coates       (Price, 1948)
cl. Abraham Dixon      Mauve, yellow eye. (Waterer, before 1915)
cl. Accomplishment     Red. (C. O. Dexter)
--- Aclandianum        Pink, scarlet spots. (A. Waterer, 1865)
--- Acutilibrum        (A. Veschaffelt, before 1857)
--- Adalbert           (T. P. Seidel, 1897-1908)
--- Adele              Deep plum color. (Standish/Noble, 1860)
cl. Adelphia           Red. (C. O. Dexter)
cl. Admirable          Light rose; late. (R. de Belder, 1969)
cl. Adonis             Rose, spotted. (Noble, before 1850)
--- Adrian             (A. Waterer, before 1875)
--- Afghan Chief       (Davies, before 1890)
g.  Agamemnon          Rose crimson. (A. Waterer, 1875)
cl. Agatha             Red. (C. O. Dexter/Tyler Arboretum)
g.  Agricola           Crimson lake. (T. Methven)
cl. Alandale           (Nearing & Gable)
--- Alane              (Before 1871)
g.  Alaric             Rosy purple. (Methven, before 1871)
cl. Alarich            Dark carmine, red marks. (Seidel)
g.  Alarm              Deep crimson/white center. (Waterer,1867)
--- Alatum             Bright rosy lake. (Noble, before 1850)
cl. Albertus
      Grandiflora      Pink, dark spots. (T. Methven, 1868)
cl. Albion             Rosy red. (T. Methven, 1868)
cl. Albrecht Durer     Bright red, fimbriated. (M. Koster)
--- Album Flavum       Blush white, orange yellow spots.
--- Album Hybridum     White flowers. (Before 1870)
--- Album Speciosum    (A. Verschaffelt, before 1854)
cl. Album
      Triumphans       White. (J. Waterer, before 1871)
--- Alburtus           White to pink, spotted. (Noble, 1850)
cl. Alexandrina        Pure white. (Standish/Noble, 1860)
cl. Alexina            Rosy lilac, black spots. (Noble, 1850)
cl. Alice Heye         Lilac rose, black blotch. (C. Frets)
cl. Alice Poore        Similar to Scintillation. (C. O. Dexter)
cl. All Beautiful      Pink, fragrant. (Kersey/Frederick, 1976)
cl. Alma               Lilac center with crimson edges.
--- Alonzo             Purple lake. (Standish/Noble, 1860)
--- Alstroemerioides    Rose, spotted. (Standish/Noble, 1860)
cl. Amazon             Pink. (T. Methven, before 1887)
--- Amethystina        Light pink, yellow blotch. (Methven)
--- Amphipyros         Deep red, black spots. (Standish, 1860)
cl. Amy Ann            Rose opal, late. (The Hennys, 1963)
--- Anadyomene         Blush white, faint spots. (Standish,1860)
cl. Angel Falls        White, yellow throat. (Kersy/Frederick)
cl. Angelina           White, yellow eye. (Standish/Noble, 1860)
--- Angeola            (Before 1857)
cl. Anica Bricogne     Pale mauve.
cl. Ann Willis
      Fleming          Deep pink, dark eye.
--- Anne Vervaef       (Before 1870)
cl. Anne's Delight     Yellow, red throat. (Whitney/Sather 1985)
cl. Annette            White, tinted. (Standish/Noble, 1860)
--- Annie Bricogne     (Bertin, before 1867)
--- Annie Dixwell      (A. Waterer, before 1875)
--- Anstie             (C. E. Heath, 1934)
cl. Antagonist         Deep purple red. (Standish/Noble, 1860)
cl. Antonia            Crimson, dark blotch. (Standish/Noble)
--- Apollo             Scarlet. (T. Methven)
--- Apology            Rose, dark blotch. (A. Waterer)
cl. Apple Blossom      White, flushed pink. (Dexter/Everitt)
cl. April Dream        Pink, large red blotch. (Whitney, 1976)
--- Archiduc Etienne   (A. Verschaffelt, 1866)
cl. Archimedes         Rose. (H. Waterer, before 1851)
cl. Arctic Tern        Pure white. (Cox, 1933)
cl. Ariadne            Pure white. (Standish/Noble, 1860)
cl. Ariel              Pink, yellow spots. (Standish/Noble,1860)
cl. Aristocrat         Pink, darker spots. (Koster, 1965)

cl. Arlequin           Red. (Dexter/Vossberg)
--- Arthur Helps       Dark crimson. (T. Methven)
cl. Arthurianum        (T. Methven, 1868)
cl. Artie Moon         Lavender, maroon blotch. (C. Loeb, 1966)
--- Asa Gray           (van Houtt, before 1874)
--- Athene             White, yellow blotch. (T. Methven)
cl. Atlas              Rose, dark spots. (Noble, before 1850)
cl. Atro-rubrum        Rose crimson. (J. Waterer, 1851)
cl. Attila             Claret purple. (T. Methven, 1867)
cl. Auchati-
      Barbaudum        (Offered by Sunningdale Nurs., 1955)
cl. Aucubaefolium      (T. Methven, 1868)
cl. Audilope           Pink
cl. Auguste Lemaire    Pinkish red. (Moser)
--- Augustum           (Byls, before 1875)
--- Augustus           Purplish crimson. (Standish/Noble, 1860)
--- Aunt Martha        (Roy W. Clark, Olympia, WA)
--- Aureum Punctatum   Primrose, spotted orange.
--- Aureum Superbum    Yellow with deep orange spots.
--- Aurora             Rose, yellow eye. (Standish/Noble, 1860)
g.  Austin Layard      Rosy crimson. (J. Waterer, 1873)
cl. B. W. Elliott      Clear rose with dark spots. (J. Waterer)
cl. Bahram             Pink, fading white. (Knap Hill)
cl. Bai Waterer        Scarlet rim, light center. (J. Waterer)
cl. Barber of Seville  Pale rose, black spots. (Standish/Noble)
--- Barclayanum        Reddish rose. (H. Waterer, before 1851)
--- Barnet Glory       Deep rosy red.
cl. Barnstable         Purplish pink; fragrant. (Dexter/Howard)
cl. Baron Adolphe
      de Rothschild    (Croux, 1900-1915)
cl. Baron Alphonse
      Malbet           (Moser)
cl. Baron Chandon      (Moser)
cl. Baron de Bruin     (A. Waterer, before 1915)
cl. Baron de Verdiere  (Croux, 1900-1913)
--- Baron Schroder     Plum, yellow green flare. (Waterer, 1915)
--- Baroness Bolsover  Rosy lilac, spotted.
g.  Baroness Lionel
      de Rothschild    Light red, darker edges. (Waterer, 1880)
--- Baroness von
      Panwitz          Rose with deeper margins.
--- Baronne Gabriel
      de St. Genois    (van Houtte, before 1875)
cl. Baronne Raissa
      Standersfkiold   (Croux, 1900-1913)
cl. Barto Alpine       Fuchsia purple. (Barto/Greer, 1964)
cl. Barto Ivory        Rich ivory. (Barto/Steinmetz)
cl. Barto Lavender     Orchid pink; fragrant. (Barto/Greer 1962)
cl. Bass River         Purplish pink. (Dexter/Brown/Wister)
--- Bassano            Crimson, heavily spotted. (M. Young)
cl. Beautiful Dreamer  Yellow, orange edges. (W. Whitney)
--- Beauty             Crimson, spotted.
--- Beauty of Kent     Purple, pale edges. (Stanley, 1860)
cl. Beethoven          Lilac red, purple blotch. (M. Koster)
cl. Bella              Lilac pink, orange marks. (Seidel, 1908)
cl. Bellona            Rosy pink. (Waterer/Crisp)
cl. Ben Moseley        Pink/deeper edges. (Dexter/Moseley)
cl. Beranger           White, red/brown spots. (Standish/Noble)
cl. Bernard
      Lauterback       Carmine red.
cl. Bertram Woodhouse
      Currie           Crimson, light center. (J. Waterer)
--- Betsy Trotwood     Rose red. (Standish/Noble, 1860)
cl. Betty Arrington    Rose pink, red flare. (Dexter/Arrington)
cl. Beverly Harvey     Persian rose. (Coen/Fawcett, 1966)
--- Bianchi            Salmon pink. (Maurice Young)
cl. Bicolor            Lilac, reddish purple. (Methven, 1851)
cl. Big Savage         (Yates, 1972)
cl. Bijou de Gand      (A. Verschaffelt, 1861)
cl. Bilderdijk         Carmine red.

cl. Black Beauty      Dark velvet crimson. (Slocock, 1930)
cl. Black Sport       Purple red. (Nelson/Bruce Briggs, 1980)
cl. Blackeyed Susan   Purplish lilac, black spots. (T. Methven)
--- Blanc Superb      White. (T. Methven, 1868)
--- Blatteum          Crimson claret. (J. Waterer, 1871)
cl. Blind Date        Pink and fragrant. (Whitney/Sather, 1975)
cl. Blue Beard        (Davies before 1890)
--- Blue Bell         (C. B. van Nes, 1932)
cl. Blue Frost        Purple/orange spots. (Whitney/Sather)
cl. Blue Gown         Deep lavender blue; early May.
cl. Bonzo             (Knap Hill, 1933)
cl. Boule de Feu      "Ball of Fire" (Davies, before 1890)
--- Bouquet de Flore  Purplish rose, black spots. (Standish/N.)
--- Bouquet Parfait   Pink, dark spots. (T. Methven, 1868)
cl. Brabantia         Deep blood red. (Standish/Noble, 1860)
--- Brayanum          Rosy scarlet. (H. Waterer, 1851)
--- Brebnerii         Rose lake. (Standish/Noble, 1860)
cl. Brennus           Crimson lake.
cl. Bridesmaid        White/variegated foliage. (Standish/N.)
--- British Queen     (Davies before 1871)
cl. Briton Ferry      Pink.
--- Brocole           (A. W. Heneage-Vivian, 1933)
cl. Bronze            White with pink edges. (Teese, 1981)
cl. Bronze Wing       White, flushed pink. (A. Teese, 1981)
--- Brookiana         (Before 1892)
cl. Brown Eyes        Pink, brown blotch. (Dexter, Bosley, Sr.)
--- Brutus            Purplish lake. (Standish/Noble, 1860)
g.  Bulstrode         Rosy crimson. (van Nes before 1922)
cl. Burgundy Cherry   Carmine/purple blotch. (Dexter)
cl. Busybody          White with greenish center. (A. Waterer)
--- Butlerianum       White, tinged pink. (J. Waterer)
cl. Butler Port       Dark purple. (W. Stokes/O. Pride)

cl. Cabaret           Spinel red, edged carmine. (Whitney)
--- Cabrera           Clear rosy crimson. (T. Methven, 1868)
cl. Cameronian        Azaleodendron, soft yellow pink.
cl. Canada            Bright bluish rose. (Greig/Caperci, 1977)
cl. Cannizara         Pale lilac rose. (M. Koster)
cl. Canon Furse       Rose, dark spots. (J. Waterer, 1922)
--- Cantatrice        White, lemon spots. (Standish/N. 1860)
cl. Caperci Special   Rosy lilac, red spots. (Caperci, 1977)
--- Caprice           Orchid pink, yellowish eye, frilled edge.
cl. Captain Beaumont  Pink, white stripes in center.
cl. Captain Webb      Deep lake. (M. Young)
cl. Cardiff           White with red blotch. (Gable/Reese 1980)
--- Cardinalis        Purplish crimson. (Standish/Noble, 1860)
--- Cariola           Pale lilac pink. (T. J. R. Seidel)
--- Carneum
     Versicolor       Yellow pink edging, finely spotted.
cl. Carolina          (Crown Lands, Windsor, 1950)
--- Caroline          Pure white. (Standish/Noble, 1860)
cl. Cary's Red        Same color as America. (Cary, 1974)
--- Cassandra         Purplish lake. (Standish/Noble, 1860)
--- Cassiope          (Rothschild, 1935)
cl. Cato              Rose, finely spotted.
--- Cavalier          Rosy lilac, black spots. (Noble, 1850)
--- Celestrial        Light pink. (Standish/Noble, 1860)
cl. Celia             Delicate rose madder.
--- Cerito            Deep purplish rose. (Noble, 1850)
cl. Cervantes         Pink. (M. Koster)
cl. Champagne         Pale gold. (C. O. Dexter, 1925-42)
--- Chancellor        Pure white. (Standish/Noble, 1860)
cl. Chancellor
     (Waterer's)      Light purple/spotted. (Standish/N., 1860)
cl. Charles Thorold   Purple, green yellow center. (A. Waterer)
--- Charles Truffant  Crimson and white. (Bertin, 1851)
cl. Charlestown       Pink, fragrant. (Dexter/Herbert, 1976)
cl. Chatham           Purplish pink. Fragrant. (Dexter/Wister)
--- Cherry Ripe       Cerise pink. (Knap Hill, 1955)
cl. Chiffon           Pink, deeper throat. (Whitney/Sather)
--- Childe Harold     Rose, white throat. (Standish/N., 1870)

--- Chinese Pink      (Shown by P. D. Williams, 1935)
cl. Chintz            Soft pink, ruby spots. (A. Waterer, 1931)
--- Chrissie          (Shown by T. McMeeken, 1895)
cl. Ciliocale         Small, white. (Seed from Abbe Delavay)
--- Cinnamomium
     Cunninghami      White, much spotted. (Standish/N., 1860)
cl. Cis Grey          (Shown by Milner, 1853)
cl. Clearbrook        Purplish pink. (Dexter/Gillies/Wister)
--- Clementine
     Lemaire          Pink yellow blotch. (Moser)
--- Cleopatra         Rose purple. (Standish/Noble, 1860)
cl. Climax            Crimson. (T. Methven, 1868)
cl. Colonel Coen      Fine deep purple. (E. Ostbo, 1958)
g.  Colonel Lynde     Pale purple. (Methven)
cl. Comet             Crimson. (Methven)
--- Compactum         Scarlet. (Standish/Noble before 1860)
--- Compeer           Purplish rose. (Noble, 1850)
cl. Compton's Scarlet Intense scarlet. (J. Waterer)
--- Comte Cavour      Rich plum color. (Standish/Noble, 1870)
cl. Comte de Gomer    White, edged crimson. (J. Waterer)
cl. Comtesse de
     Morello          Rose. (Standish)
cl. Concessum         Bright rose. (J. Byls, 1867)
--- Conchiflorum
     Striatum         Pale rose, splashes and spots. (Standish)
--- Congestum Aureum  Yellow.
--- Congestum Roseum  Light rose, spotted. (J. Waterer, 1875)
--- Conical Kate      Rosy crimson, yellow blotch. (Waterer)
--- Conspicuum        Crimson with black spots. (Noble, 1850)
cl. Constance Carson  Pale pink with yellow mark.
cl. Constance Terry   Deep pink and fimbriated.
--- Constellation     Deep rose. (Noble, 1850)
cl. Coral Star        White, edged pink.
--- Corine            White with yellow spots. (Standish/Noble)
cl. Corrector         Rosy crimson, dark spots.
cl. Correggio         Dark scarlet. (Standish/Noble, 1860)
cl. Cotterill         Deep coral, fimbriated and blotched.
cl. Countess          Delicate shell pink.
--- Countess de
     Morello          Soft rose, wavy edges. (Standish, 1861)
cl. Countess
     Fitzwilliam      Carmine rose, dark spots. (Fisher/Sibray)
cl. Countess of
     Beauchamp        Blush with crimson blotch.
cl. Countess of
     Cadogan          Pale pink. (J. Waterer)
cl. Countess of
     Clancarty        Light rosy crimson. (J. Waterer)
cl. Countess of
     Donoughmore      Marigold bright pink. (Fischer)
cl. Countess of
     Headfort         Lilac rose, spotted. (J. Waterer)
cl. Countess of
     Ilchester        Pale center, scarlet edge. (J. Waterer)
cl. Countess of
     Normanton        Pale mauve, fading white. (J. Waterer)
cl. Countess of
     Tankerville      Delicate rose. (J. Waterer before 1922)
cl. Countess of
     Wharnclyffe      Crimson. (Fisher/Sibray)
--- Countess of
     Wilton           Rosy crimson. (Rollison before 1874)
cl. Creamy Chiffon    Double creamy yellow. (Whitney/Briggs)
g.  Cresca            (Lord Aberconway, 1934)
cl. Crimson Glory     Many crimson flowers. (Cottage Gardens)
cl. Crosspatch        Pink, red throat. (Nelson/McLaren, 1975)
cl. Crown of Gold     Pink, darker stripes. (Frederick, 1978)
cl. Cruentum          Crimson lake. (A. Waterer before 1875)
cl. Cunningham's
     Blush            Lilac pink. (Cunningham)
--- Cupreum           Coppery orange, suffused pink.

| | | |
|---|---|---|
| --- | Currieanum | Dark purple or rosy lilac. (A. Waterer) |
| --- | Cyaneum | Purplish lilac. (M. Young) |
| --- | Daphne | White, yellow spots. (Standish/Noble) |
| --- | David Copperfield | Rose with peach center. (Noble, 1850) |
| cl. | Dean Hall | Pale neyron rose. (Maurice Hall, 1984) |
| cl. | Decorator | Scarlet with dark spots. (J. Waterer) |
| cl. | Defiance | Light rose, darker spots. (Methven, 1868) |
| cl. | Delicate Splendor | Purplish pink. (Dexter/Wister, 1981) |
| --- | Delicatum | White, maroon spots. (Standish/N., 1860) |
| --- | Delicatum Aureum | Light pink with orange spots. |
| cl. | Delta | Phlox pink. (Boot, Boskoop, 1965) |
| cl. | Dexter's Appleblossom | White, pink edging. (Dexter/Cowles, 1978) |
| cl. | Dexter's Apricot | Pink, yellow blotch. (As above) |
| cl. | Dexter's Brandy Green | Pink, green spotting. (As above) |
| cl. | Dexter's Brick Red | Pink, red blotch. (As above) |
| cl. | Dexter's Cream | Cream, shaded pink. (As above) |
| cl. | Dexter's Crown Pink | Pink, green blotch. (As above) |
| cl. | Dexter's Favorite | Pale pink. (Dexter/Amateis, reg. 1958) |
| cl. | Dexter's Giant Red | Pink, dark red throat. (Dexter/Cowles) |
| cl. | Dexter's Glow | Pink, dark red ring. (As above) |
| cl. | Dexter's Horizon | White, deep pink edges. (As above) |
| cl. | Dexter's Orange | Pink, brownish orange blotch. (As above) |
| cl. | Dexter's Peppermint | Pink, green blotch; fragrant. (As above) |
| cl. | Dexter's Pink Glory | Pink, spots of green and red. (As above) |
| cl. | Dexter's Spice | White, green spots; fragrant. (As above) |
| cl. | Dexter's Springtime | A bicolor of cream and pink. (As above) |
| cl. | Dexter's Vanilla | Cream white, pink edges. (As above) |
| cl. | Dexter's Victoria | Deep pink, brown blotch. (As above) |
| --- | Diadem | White, brown blotch. (T. Methven, 1892) |
| cl. | Dictator | Dark crimson/darker spots. (Waterer 1875) |
| cl. | Display | Rose, green blotch. (Kersey/Frederick) |
| cl. | Dr. Ferguson | Reddish purple. (T. Methven) |
| --- | Dr. Hogg | Dark red. (Before 1875) |
| cl. | Dr. Hooker | Rosy purple/white throat. (Standish/N) |
| cl. | Dr. Mill | (Royal Porcelain Factory, Berlin, 1890) |
| --- | Dr. Muller | Deep rose. (Standish, 1862) |
| cl. | Dr. S. Endtz | Rosy crimson. |
| cl. | Dorothy Fortescue | Dark cherry red. (J. Waterer, 1922) |
| --- | Dot | Azaleaodendron; rose crimson. |
| cl. | Double Date | Double-flowered red. (Whitney/Sather) |
| cl. | Double Dare | Yellow/pink margins. (Lem) |
| cl. | Dress Parade | Pink and fragrant. (Kersey/Frederick) |
| cl. | Duchess of Bedford | Deep rose, light marks. (J. Waterer) |
| cl. | Duchess of Buccleuch | Crimson. (T. Methven) |
| --- | Duchess of Cambridge | Cerise. (J. Waterer) |
| cl. | Duchess of Kent | Yellow orange/frilled. (Rothschild, 1982) |
| cl. | Duchess of Northumberland | White. (J. Waterer ?) |
| --- | Duchess of Sutherland | White or pale rose. (J. Waterer, 1868) |
| cl. | Duke of Malakoff | Blush, red spot. (Standish/Noble, 1860) |
| cl. | Duke of Norfolk | Rose. (Before 1871) |
| cl. | Duke of Portland | Bright scarlet. (J. Waterer before 1875) |
| cl. | Duleep Singh | Chocolate crimson. (J. Waterer) |
| cl. | Earl of Haddington | Clear rose. (J. Waterer) |
| cl. | Earl of Selkirk | Crimson. (T. Methven) |

| | | |
|---|---|---|
| cl. | Earl of Shannon | Deep crimson. (J. Waterer before 1867) |
| cl. | Ed de Concourt | Pink. (Moser) |
| cl. | Edgemont | Purple pink; fragrant. (Gillies/Wister) |
| cl. | Edwin Beinecke | Yellowish apricot. (Dexter/Young/Efinger) |
| cl. | Eileen | Pink, yellow blotch. (Waterer/Crisp) |
| cl. | El Camino | Dark pink. (Lem ?--Whitney/Sather, 1976) |
| --- | Elfrida | Pale rose, large blotch. (Standish/Noble) |
| cl. | Elizabeth Blackford | (E. Ostbo) |
| cl. | Elizabeth Lund | Yellow, maroon eye. (Mrs. T. Davis) |
| cl. | Elizabeth Red Foliage | Changeable red leaves. (Ostbo) |
| cl. | Elizabeth Sidamon-Eristoff | Orchid pink, red blotch. ((Phipps, 1973) |
| --- | Else Seidel | Pale pinkish white. (Seidel before 1906) |
| cl. | Elsie Waterer | White, red blotch. (Waterer/Crisp) |
| --- | Elvira | White, yellow blotch. (Standish/Noble) |
| --- | Emelie | Rosy crimson. |
| --- | Emilie | Rosy crimson. (Standish/Noble, 1860) |
| --- | Eminent | Rosy lilac. (Noble, 1850) |
| --- | Empress | Purplish crimson. (Noble, 1850) |
| cl. | Enchanted Evening | Peach, yellow throat. (Whitney/Sather) |
| cl. | Erasmus | Dark scarlet red. (M. Koster) |
| --- | Erectum | Rose crimson. (Standish before 1860) |
| cl. | Ereda Beekman | Light sap green. (J. Beekman, 1975) |
| --- | Erestium | Fine purple. (T. Methven, 1868) |
| cl. | Esmerelda | Pink, deeper edges. (M. Koster) |
| cl. | Esoteric | Deep pink. (G. O. Clark) |
| cl. | Estelka Gerster | Lilac rose, primrose spot. |
| cl. | Ethel Dupar | Cream yellow, pink blush. (Whitney/Dupar) |
| --- | Ethel Hall | White, scarlet edges. (M. Koster) |
| --- | Ethel Stocker | Deep pink, crimson buds. |
| --- | Euclid | Rosy red. (Noble, 1850) |
| --- | Eugenie | White, fine spots. (Standish/Noble, 1860) |
| cl. | Evening Star | White, dark eye. (Standish/Noble, 1860) |
| --- | Ewingii | White, green eye. (T. Methven, 1868) |
| cl. | Exalted Ruler | Violet, maroon blotch. (Lyons, 1972) |
| --- | Exquisite | White, buff spots. (J. Waterer, 1875) |
| cl. | Fabulous | Neyron rose, light center. (Whitney) |
| --- | Fair Helen | White, yellow spot. (A. Waterer, 1915) |
| --- | Fair Rosamond | Rosy pink. (Noble before 1850) |
| --- | Fairy Bell | (J. J. Crosfield, 1935) |
| --- | Fairy Queen | White/tinged pink. (Standish/N., 1860) |
| cl. | Farewell Party | White, green spots. (Kersey/Frederick) |
| cl. | Fascination | Dwarf plant, pink. |
| cl. | Faust | Puce, dark eye. (Standish/Noble, 1870) |
| cl. | Fee | Dark lilac, white filaments. (Seidel) |
| cl. | Ferelith | Mauve. (Knap Hill, 1962) |
| cl. | Festive Feast | Light purplish pink. (Dexter/Wister, 1980) |
| cl. | Fichte | Dark pink, yellow marks. (Seidel) |
| cl. | Figaro II | Intense red. (T. J. R. Seidel) |
| cl. | Fire | Rose red/dark spots. (Koster of Boskoop) |
| cl. | Firlefanz | Soft pink, few markings. (T.J.R.Seidel) |
| cl. | Florence | Pink. |
| cl. | Florence Nightingale | Rose, light center. (A. Waterer) |
| --- | Floretta | Cerise, white throat. (Standish/Noble) |
| cl. | Floriade | Vibrant red. (Adr. van Nes, 1962) |
| cl. | Florian | White, yellow spots. (Standish/Noble) |
| cl. | Fort Nisqually | Showy vibrant red. (R. Clark) |
| cl. | Fragrans Affinity | Orchid blue. |
| cl. | Francis B. Hayes | White, chocolate spots. (J. Waterer 1922) |
| cl. | Francis Dickson | Bright red or scarlet. (A. Waterer, 1865) |
| cl. | Frau Rosalie Seidel | White. (Seidel before 1916) |
| cl. | French Creek | Pink, green spots. (Herbert, 1970) |
| --- | Fumosum | Dark purple, darker spots. (Methven 1868) |
| cl. | Furvum | Very dark puce. (T. Methven, 1868) |
| cl. | G. B. Simpson | Bluish purple. (A. Waterer, 1915) |

| | | |
|---|---|---|
| --- | Galbannum | Rosy crimson, white dashes. (M. Young) |
| --- | Galceador | Lilac, gold eye. (Knap Hill) |
| cl. | Garibaldi | Bright light red/frilled. (Waterer, 1915) |
| cl. | Gay Hostess | Rose pink, green spot. (Kersey/Frederick) |
| --- | Gazelle | White, brown spots. (Standish/N., 1860) |
| cl. | Gemmatum | Scarlet, dark blotch. (Before 1874) |
| cl. | Gemmiferum | Azaleodendron. Magenta rose. (Methven) |
| cl. | Gemmosum | Scarlet rose, black spots. (T. Methven) |
| cl. | Genseric | Dark claret, spotted. (Standish/Noble) |
| cl. | Geoffrey Henslow | Bright crimson. (M. Koster) |
| cl. | George Bennington | Flower color is pink. (Paul) |
| cl. | George Paul | Crimson with spots. (A. Waterer) |
| cl. | George Watling | Primrose yellow. (Greig/Cook) |
| cl. | Georgia May | Rose pink, white center. (Core, 1964) |
| --- | Geraldine | Deep puce. (Standish/Noble, 1860) |
| cl. | Geranioides | Rosy crimson/dark spots. (Standish/Noble) |
| cl. | Gertrud | Bright carmine red, brown marks. (Seidel) |
| cl. | Giorgione | Rosy scarlet. |
| cl. | Glenda Farrell | Purplish red. (Dexter/Wister, 1983) |
| --- | Globosum | Rosy lilac. (Standish/Noble, 1860) |
| --- | Gloire de Boskoop | Dark crimson. |
| cl. | Gloriosum | Blush white. (J. Waterer, 1850) |
| cl. | Glow | (Crosfield, 1935) |
| cl. | Golden Fantasy | Cream, yellow center. (Kersey/Frederick) |
| cl. | Goldsworth Purple | Purple. (W. C. Slocock, 1935) |
| cl. | Goliath | Crimson. |
| cl. | Governor's Mansion | Imperial purple, orange blotch. (Dexter) |
| --- | Grace Darling | Lilac with dark spot. (Davies, 1890) |
| cl. | Graham Holmes | Red purple. (A. H. Holmes, 1982) |
| cl. | Great Eastern | Purple pink, yellow rays. (Dexter/Wister) |
| --- | Gulnare | Bright rose. (Standish/Noble, 1870) |
| cl. | Guttatum | White, green/red spots. (Standish/Noble) |
| cl. | Gwyneth Masters | Cardinal red, spotted. (Masters, 1982) |
| --- | Halcyon | Rosy lilac. (Noble before 1850) |
| cl. | Halesite | Strong purplish red. (Dexter/Wister) |
| cl. | Halesite Maiden | Light red. (Dexter/Schlaikjer, 1974) |
| --- | Hamachidon | Dwarf. Bright purple red. |
| cl. | Hamilcar | Amaranth red. |
| --- | Harry Bertram | Deep puce, spotted. (Standish/Noble) |
| cl. | Harry Goldney | Rose pink. |
| cl. | Hartley Luttrell | Rosy red. (Standish/Noble, 1870) |
| cl. | Haydn | Pink. |
| --- | Hector | Bright crimson. (J. Waterer, 1874) |
| --- | Helena | Rosy red. (Standish/Noble, 1870) |
| cl. | Helene Huber | Pink; fragrant. (Dexter/C. Herbert) |
| g. | Hendersonii | Purple, light center. (Henderson, 1865) |
| cl. | Henry Bohn | Crimson. (J. Waterer before 1875) |
| --- | Henry Drummond | Purplish crimson. (T. Methven) |
| cl. | Henry's Red | Deep dark red. (E. J. Mezitt) |
| --- | Herkules | Pale lilac to white. (Seidel, 1894) |
| cl. | Hermia | Rose madder, scented. (Rogers) |
| --- | Hermine Seidel | Purplish pink, dark red marks. (Seidel) |
| --- | Hespirus | Shaded pink. (Standish/Noble, 1860) |
| --- | Hester | White, brown spots. (Standish/Noble) |
| --- | Highland Mary | Blush, orange spots. (Standish/Noble) |
| cl. | Highlight | Truss of 27 red flowers. (Glennie, 1980) |
| cl. | Hilaria | Deep crimson. (Magor, 1963) |
| cl. | Hildebrand | Crimson. (Standish/Noble, 1870) |
| --- | Hildegard | White, dark purple marks. (Seidel) |
| cl. | Hillcrest | Strong red. (Esch/Briggs) |
| --- | His Majesty | Rose pink, dots deeper shade. (R. Gill) |
| cl. | Hogarth | Rosy Scarlet. (A. Waterer, 1871) |
| cl. | Holmeslee Bright Light | Red purple. (Graham Holmes, 1979) |
| cl. | Holmslee Greeneyes | Red purple, fading white. (Holmes, 1982) |
| --- | Holtei | Red, white throat. (Seidel) |
| --- | Holten | Carmine pink. (Seidel) |

| | | |
|---|---|---|
| cl. | Hon. John Boscawen | Cherry red. (J. Waterer, 1922) |
| cl. | Hon. Mrs. Mercer Henderson | Rose, black spots. (J. Waterer) |
| cl. | Hosea Waterer | Rich cerise pink. (Knap Hill, 1955) |
| --- | Hudibras | Deep pink. (Standish/Noble, 1860) |
| cl. | Hugh Wormald | Cerise, white stripe. (M. Koster, 1922) |
| --- | Hugo Richter | Pink, white stripes. (Seidel, 1894) |
| --- | Humboldti | Rose, white throat. (Standish/N., 1860) |
| cl. | Hunting Hill | Strong pink. (Dexter/Ross/Wister, 1980) |
| cl. | Hyperion | White, deep purple blotch. (A. Waterer) |
| cl. | Hyronimus Bosch | Light mauve. (M. Koster) |
| cl. | Ice Floe | Lavender, fading white. (Larson/Wildfong) |
| --- | Ida | White, red spots. (Standish/Noble, 1860) |
| cl. | Ignescens | Bright red/black spots. (Waterer, 1850) |
| cl. | Ilam Cerise | Crimson, white throat. (E. Stead, 1984) |
| cl. | Imbros | Rose Bengal. (E. de Rothschild, 1962) |
| cl. | Impact | Deep roseine purple. (D. Warren, 1972) |
| cl. | Independence Day | Red, dark spot. (A. Waterer, 1915) |
| --- | Ingomer | Deep rose. (Standish/Noble, 1870) |
| g. | Ingramii | White, lemon blotch. (Standish, 1860) |
| --- | Irene | Pink. (Gebr. Guldemond, Boskoop) |
| --- | Iris Lawrence | (Shown by Hugh Wormald, 1941) |
| --- | Isaac Davies | Scarlet. (Davies, 1890) |
| cl. | Isabel Meeres | Rosy crimson. (J. Waterer) |
| --- | Ivanhoe | Deep claret. (Standish/Noble, 1860) |
| cl. | Ivery's Scarlet | Brilliant red. |
| cl. | J. Fiala | Blush white, spotted. (From Belgium) |
| cl. | Jack Izod | Pale pink. (M. Koster) |
| --- | Jackmannii | Crimson, spotted. (T. Methven, 1869) |
| --- | Jacob Friederich Seidel | Deep purple. (Seidel, 1867) |
| cl. | James Bateman | Rosy scarlet. (A. Waterer, 1865) |
| --- | James Brigham | Reddish purple. |
| cl. | James Macintosh | Rosy scarlet. (A. Waterer. 1870) |
| --- | James Mangles | (Shown by R. E. Horsfall, 1934) |
| cl. | James Marshall Brooks | Scarlet, bronze spot. (A. Waterer, 1870) |
| cl. | James Mason | Light center/red edge. (J. Waterer , 1915) |
| cl. | James Nasmyth | Lilac, maroon blotch. (A. Waterer, 1870) |
| cl. | Jean Glennie | Loose truss, salmon. (Glennie 1980) |
| cl. | Jean Stearn | White, purple spots. (J. Waterer) |
| cl. | Jeanette Clarke | Red, truss of 12. (V. J. Boulter, 1984) |
| cl. | Jeanne d'Arc | White. (M. Koster) |
| cl. | Jenkinsii | Azaleodendron. Lemon, blushed pink. |
| --- | Jennie Deans | Deep blush, spotted. (Standish/N., 1860) |
| cl. | Jenny Lind | Neyron rose/carrot red flare. (Seabrook) |
| --- | Jessica | Lavender/dark eye. (Standish/Noble, 1860) |
| --- | Joan of Arc | Deep rosy pink. (Standish/Noble, 1860) |
| --- | John Gair | Rose, paler center. (Standish, 1860) |
| cl. | John Galsworthy | Maroon purple, yellow blotch. |
| cl. | John Henry Agnew | Pink, chocolate spots. (J. Waterer, 1922) |
| --- | Johnsonianum | Brilliant crimson. (Before 1851) |
| cl. | Joseph Martin | Light pink. (A. Waterer) |
| cl. | Joseph Martyr | Light pink, yellow blotch. (A. Waterer) |
| cl. | Josephine Everitt | Rosy pink. (Dexter) |
| cl. | Joy Ride | Pink/orange blotch. (Whitney/S., 1976) |
| --- | Jubar | Rose crimson, spots. (Standish/N., 1860) |
| --- | Jubilee | Recommended for Exbury trial. (Waterer) |
| --- | Julius Ruppell | Rose, carmine marks. (J. Waterer) |
| cl. | Kant | Yellow, pink in bud. (M. Koster) |
| cl. | Katherine Slater | Large truss, dusty pink. (Dexter/Mezitt) |
| cl. | Kathleen Fielding | Flower red, purple blotch. (M. Koster) |
| cl. | Kathleen Jane | Scarlet/black spotting. (Stephens, 1975) |
| cl. | Kathleen Wallace | Pale pink. (C. B. van Nes) |
| cl. | Kaulbach | White, edges lavender. (T. J. R. Seidel) |
| cl. | King of the Purples | Dark purple, spotted. (A. Waterer) |

| | | |
|---|---|---|
| --- | Lady Alice Peel | Dark rose, dark spots. (Standish/Noble) |
| --- | Lady Bowring | Shaded rose pink, spotted. |
| --- | Lady Bridport | Pure white. (Standish/Noble, 1860) |
| --- | Lady Cathcart | Bright clear rose, crimson spots. |
| cl. | Lady Decies | Light mauve/yellow eye. (J. Waterer 1922) |
| cl. | Lady Dorothy Neville | Dark purple, black spots. (Waterer, 1874) |
| cl. | Lady Easthope | Clear rose/dark spots. (J. Waterer, 1874) |
| cl. | Lady Emily Peel | Bright rose/chocolate spots. (J. Methven) |
| --- | Lady Ethel Edgar | Cerise red. (M. Koster) |
| --- | Lady Ethel Hall | Pink margins, pale interior. (M. Koster) |
| cl. | Lady Francis Crossley | Salmon or rosy pink. (A. Waterer, 1867) |
| cl. | Lady Galway | White, tinged mauve. |
| --- | Lady Gertrude Foljambe | Pink |
| cl. | Lady Godiva | White, yellow spots. (A. Waterer, 1875) |
| cl. | Lady Grenville | White, purple edges. (J. Waterer, 1875) |
| cl. | Lady Harcourt | Pink crimson spots. (J. Waterer) |
| cl. | Lady Howe | Rose. (J. Waterer) |
| cl. | Lady Knollys | Rose. (J. Waterer) |
| --- | Lady Mary Hood | Pure white. (Standish/Noble, 1860) |
| --- | Lady Mary Parker | Rose, edged vivid pink. |
| cl. | Lady Mosley | Pink, lighter center. (J. Waterer) |
| cl. | Lady of the Lake | Blush, well spotted. (T. Methven, 1968) |
| cl. | Lady Olive Guinness | White, dark spots. (J. Waterer) |
| --- | Lady Palmerston | Blush, orange yellow eye. (C. Noble) |
| --- | Lady Strangford | Blush, maroon spots. (C. Noble) |
| cl. | Lady Tankerville | Pink, light center. (J. Waterer, 1915) |
| cl. | Lady Truro | Rosy crimson. (J. Waterer) |
| --- | Lady Welch | Rose lilac. |
| cl. | Lady Winifred Herbert | Rose crimson, fringed. (Young) |
| --- | Laura | Crimson. (T. Methven, 1868) |
| cl. | Lawton's Chinese Red | Red flowers. (Lawton/Hoogendoorn Nurs.) |
| cl. | Lem's Fluorescent Pink | Pink/white. (Lem/Mrs. James Sinclair) |
| --- | Leo XIII | White, magenta edges. (Seidel, 1894) |
| --- | Leonora | Rosy pink, green spots. (Standish/Noble) |
| --- | Lessing | Rosy crimson, darker marking. |
| cl. | Liesbeth | Solferino purple. (H. T. Hooftman, 1965) |
| cl. | Little Red Riding Hood | Solferino purple. (Seabrook, 1964) |
| cl. | Littleworth Puffin | Mallow purple. (Mangles/Lady Adam Gordon) |
| --- | Littleworth Cream | (Shown by R. E. Horsfall, 1935) |
| cl. | Littleworth Flush | Rose Bengal. (Mangles/Lady Adam Gordon) |
| cl. | Locarno | Bengal rose/dark spots. (Koster, Boskoop) |
| cl. | Loeb's Moonlight | Rosy orchid. Semi-dwarf. (C. Loeb, 1966) |
| --- | Lord Brougham | Crimson. (T. Methven, 1868) |
| --- | Lord Byron | Rosy purple/black spots. (Standish/Noble) |
| cl. | Lord Eversley | Dark crimson, black spots. (J. Waterer) |
| cl. | Lord Fairhaven | Shrimp pink, cream blotch. (Knap Hill) |
| cl. | Lord Granville | Scarlet, tinged cerise. (C. Noble, 1870) |
| cl. | Lord John Russell | Pale rose/dark blotch. (A. Waterer, 1860) |
| cl. | Lord Selborne | Crimson. (J. Waterer) |
| --- | Lorenzo | White, yellow eye. (Standish/Noble, 1860) |
| cl. | Louis Chauvin | Rose pink, purple blotch. (Moser) |
| --- | Louis Phillippe | Crimson scarlet, black spot. (1867) |
| cl. | Love Story | Peach, fading yellow. (Whitney/Sather) |
| cl. | Lucidum | Rosy lilac. (H. Waterer, 1857) |
| --- | Lucien Linden | Crimson. (Linden) |
| --- | Lucifer | Bright scarlet. (Noble, 1855) |
| cl. | Luciferum | White. (J. Waterer before 1875) |
| --- | Lucy | Blush. (Standish/Noble, 1860) |
| --- | Lucy Neal | Claret, fine spots. (Standish/N., 1860) |
| cl. | Lynn Boulter | Deep to light red. (F. Boulter, 1984) |

| | | |
|---|---|---|
| --- | McHattianum | (Shown by T. Leslie, Edinburgh, 1906) |
| cl. | Maculatum Grandiflorum | Rich plum, spotted. (Standish/N., 1850) |
| cl. | Maculatum Nigrum | Deep plum/black eye. (Standish/N., 1860) |
| --- | Maculatum Nigrum Superbum | Large form of above. (Standish/N., 1870) |
| cl. | Maculatum Purpureum | Purple and spotted. (T. Methven, 1868) |
| cl. | Maculatum Roseum | Rose. (Before 1875) |
| cl. | Madame Auguste Pellerin | Cherry red, yellow eye. (Moser) |
| --- | Madame Bosio | Blush, fine spots. (Standish/N. 1870) |
| cl. | Madame Cochet | Lilac. (Bertin before 1888) |
| --- | Madame Edward Denny | Crimson, black blotch. (Knap Hill) |
| cl. | Madame Hugo T. Hooftman | Solferino purple. (H. T. Hooftman, 1966) |
| cl. | Madame Ida Rubenstein | Bright pink. (Moser) |
| cl. | Madame Jean Dupuy | Very pale lilac. (Moser) |
| cl. | Madame Jeanne Bois | Dark pink. |
| cl. | Madame Jules Porges | Pale orchid, gold blotch. (Moser, 1900) |
| cl. | Madame Nejedly | Bengal rose, dark spots. (Koster, Boskoop) |
| --- | Madame Penco | White, fine spots. (Standish/Noble, 1870) |
| cl. | Madame Pompidou | Salmon pink; truss of 28. (Kordus, 1972) |
| --- | Madame Sontag | White, pink edges. (Noble, 1855) |
| cl. | Madame Stoeffel | Light rose; double. (Whitney/Rutherford) |
| cl. | Madame van de Weyer | Rosy crimson. (J. Waterer, 1875) |
| q. | Madeline | (Lord Aberconway, Bodnant, 1935) |
| --- | Mademoiselle Victoire Balfc | White, spotted brown. (Standish/Noble) |
| cl. | Maggie Heyward | White, pink edge, yellow eye. (Waterer) |
| cl. | Maggie Stoffel | Rose, brown marks. (Whitney/Rutherford) |
| --- | Magnet | Rosy red. (Noble, 1855) |
| cl. | Magnificum | Crimson; early blooming. (J. Waterer) |
| --- | Magnificum | Purple. (Byls before 1839) |
| --- | Mahmoud | Lavender pink, large yellow eye. |
| --- | Maid of Athens | Rose, brown spot. (Standish/Noble, 1870) |
| --- | Majestic | Large crimson flower. (T. Methven, 1868) |
| --- | Majesticum | Deep rose. (Standish/Noble, 1860) |
| --- | Major Edwards | Pale pink. (Methven) |
| --- | Major Joicey | Dark crimson. (C. Noble) |
| cl. | Maletta | Cream center, gold edge. (Yates) |
| cl. | Mamie Doud Eisenhower | Truss of 30, pure pink. (Kordus, 1972) |
| cl. | Mammouth | Rosy purple. (Rogers before 1867) |
| cl. | Marcel Moser | Bright red. (Moser before 1914) |
| cl. | Marchioness of Downshire | White, spotted brown. (Street) |
| cl. | Marchioness of Londonderry | Pink. (W. C. Slocock ?) |
| --- | Marquerite | Blush, yellow spots. (Methven) |
| --- | Marian | French white/spotted brown. (Veitch 1862) |
| --- | Marie Stuart | Blush, maroon spot. (A. Waterer, 1875) |
| --- | Marmion | Purple crimson/shaded white. (Standish/N) |
| cl. | Marquis of Waterford | Bright pink, light center. (J. Waterer) |
| cl. | Martha Phipps | Pale yellow/suffused pink. (Phipps, 1973) |
| --- | Mary Blane | Claret. (Standish/Noble, 1860) |
| --- | Mary Forte | Dark purple. |
| cl. | Mary Frances Hawkins | (Gable, 1958) |
| --- | Mary Power | (Shown by R. E. Horsfall, 1936) |
| --- | Matchless | Rosy crimson, dark blotch. (T. Methven) |
| --- | Matilda | Blush, brown eye. (Standish/Noble, 1860) |
| cl. | Mauve Queen | Lilac, dark blotch. (M. Koster) |
| --- | Maximum Rubrum | Rose. (Standish/Noble, 1860) |

| | | |
|---|---|---|
| --- | Maximum Triumphans | Rose. |
| cl. | May Jeffery | Cerise pink, spotted. (Knap Hill, 1968) |
| cl. | May Moonlight | Yellowish pink; fragrant. (Dexter/Wister) |
| --- | Medora | Rose, much spotted. (Noble, 1855) |
| cl. | Melanthuma | Purplish crimson. (T. Methven, 1868) |
| cl. | Melton | Purple, dark center. (A. Waterer, 1915) |
| cl. | Memoir | Lilac white. (A. Waterer, 1915) |
| --- | Memoire de Dominique Vervaene | Pink, red blotch. (Vervaene, Belgium) |
| --- | Menziesii | Rosy pink/yellow blotch. (Standish/Noble) |
| --- | Meridian | Dark crimson, black blotch. (T. Methven) |
| --- | Meritorium | Light rose, yellow spots. (T. Methven) |
| cl. | Merry Mae White | White, purplish rays. (Hardgrove/Royce) |
| cl. | Message of Peace | White, tinged mauve. (Waterer/Crisp) |
| --- | Messenger | Purplish crimson, spotted. (Noble, 1855) |
| --- | Metallicum | Dull rosy crimson. (Veitch) |
| --- | Metaphor | Rose. (A. Waterer) |
| cl. | Meteor | Fiery red. (R. Gill) |
| --- | Methven's Scarlet | Deep red. (Methven) |
| cl. | Mexico | Red. (T. J. R. Seidel) |
| --- | Milkmaid | Yellow. (Harry White, Sunningdale, 1955) |
| --- | Minnie | Bluish white/chocolate blotch. (Standish) |
| --- | Minnie Waterer | White, crimson edges. |
| cl. | Mirabella | Intense red. (T.J.R.Seidel, intro. 1910) |
| --- | Mirabile | Rosy crimson. |
| --- | Mirandum | Light rose, spotted. (H. Waterer, 1851) |
| cl. | Miss Betty Stewart | Vermillion red. |
| --- | Miss Butler | Purplish rose, spotted. (Veitch, 1863) |
| cl. | Miss Edith Boscawen | White, black blotch. (A. Waterer) |
| --- | Miss Glyn | Pale rose. (Noble, 1855) |
| cl. | Miss Hilda de Trafford | Pink, yellow blotch. (J. Waterer) |
| cl. | Miss Jekyll | Blush, chocolate blotch. (A. Waterer) |
| cl. | Miss Mary Ames | Crimson. (A. Waterer before 1915) |
| --- | Miss Mercy Grogan | White and spotted. (Standish before 1875) |
| --- | Miss Pierce | White with a fine eye. (Standish/Noble) |
| --- | Miss Watson | Blush, spotted pink. (R. Gill) |
| --- | Mnemosyne | Rose. |
| --- | Modesty | Pale flush, fading. (A. Waterer) |
| cl. | Monsieur Alfred Chauchard | Magenta, dark blotch. (Moser) |
| cl. | Monsieur Felix Guyon | Rose, yellow blotch. (Moser) |
| cl. | Monsieur Guillemot | Dark rose; compact truss. (Moser) |
| cl. | Monsieur Tisserand | Red. (Moser) |
| cl. | Moon of Israel | Violet rose, spotted. (Rogers) |
| g. | Moonlight | Blush, dark spots. (T. McMeekin, 1895) |
| cl. | Moonlight Bay | Pale pink; fragrant. (Dexter/Wister) |
| --- | Morning Star | Violet, tinged white. (Standish/Noble) |
| cl. | Morning Sunshine | Primrose yellow. (Whitney/Sather, 1977) |
| --- | Moscow | Scarlet. (Collingwood Ingram) |
| cl. | Moses | Violet; pink brown marks. (Seidel, 1922) |
| --- | Mount Blanc | Pure white. (Standish/Noble, 1860) |
| cl. | Mountain Dew | Yellow green/brown blotch. (Waterer 1972) |
| --- | Mozart | Crimson, fine spots. (Standish/Noble) |
| --- | Mrs. Alistair McIntosh | Rose mauve, fading white. (Knap Hill) |
| cl. | Mrs. Arthur Hunnewell | Pink, primrose center. (A. Waterer, 1915) |
| cl. | Mrs. Arthur Walter | White with rose edge. (J. Waterer) |
| cl. | Mrs. Byrne | Cerise, heavily spotted. (Koster, 1923) |
| cl. | Mrs. Charles Thorold | Pink, yellow center. (A. Waterer, 1915) |
| --- | Mrs. Edward Denny | Crimson, black blotch. (Knap Hill) |

| | | |
|---|---|---|
| cl. | Mrs. F. Hankey | Salmon with spots. (A. Waterer) |
| cl. | Mrs. F. J. Kirchner | Cream/slightly spotted. (A. Waterer, 1915) |
| cl. | Mrs. Fitzgerald | Bright rosy scarlet. (J. Waterer, 1867) |
| cl. | Mrs. G. H. W. Heneage | Reddish purple/white center. (A. Waterer) |
| --- | Mrs. Hamilton | Heliotrope. (R. Gill, 1922) |
| cl. | Mrs. Harry Ingersoll | Rose lilac/green center. (A. Waterer 1877) |
| --- | Mrs. Hemans | Pinkish white/yellow spots. (Standish/N.) |
| cl. | Mrs. John Kelk | Late flowering, clear rose. (J. Waterer) |
| cl. | Mrs. John Penn | Light red, lighter center. (J. Waterer) |
| cl. | Mrs. John Waterer | Rosy crimson. (J. Waterer, 1865) |
| cl. | Mrs. Kenneth Wilson | Deep rose, frilled edges. (M. Koster) |
| cl. | Mrs. L. M. Hayes Palmer | Red, heavily blotched. (M. Koster) |
| --- | Mrs. Mangles | Violet white/spotted black. (Standish/N.) |
| cl. | Mrs. Mendel | Pink rayed white/yellow center. (Waterer) |
| cl. | Mrs. Peter Koster | Pale pink/reverse rose pink. (M. Koster) |
| --- | Mrs. Reuthe | Bright red, fading. |
| cl. | Mrs. Russell Sturgess | White, spotted chocolate. (J. Waterer) |
| cl. | Mrs. S. Simpson | White, finely spotted. (A. Waterer) |
| --- | Mrs. Samuel Wallrock | White, spotted red. (M. Koster) |
| cl. | Mrs. Shuttleworth | Scarlet, finely spotted. (A. Waterer) |
| --- | Mrs. Siddons | Rose, white throat. (Standish/Noble) |
| --- | Mrs. Standish | White, brown spots. (Standish) |
| cl. | Mrs. T. Wezelenburg | Dark scarlet. (K. Wezelenburg) |
| cl. | Mrs. Thomas Brassey | White, rosy purple marks. (J. Waterer) |
| cl. | Mrs. Thomas Wain | Pale rose/brown spots. (A. Waterer, 1875) |
| cl. | Mrs. Tom Agnew | White, lemon blotch. (J. Waterer, 1866) |
| --- | Mrs. W. H. Gaze | Wine red. (C. B. van Ness, 1922) |
| cl. | Mrs. W. R. Dykes | Strong rose red. |
| cl. | Mrs. William Agnew | Pale rose, purple blotch. (Waterer, 1875) |
| cl. | Mrs. William Bovill | Rose scarlet. (Waterer/Godfrey, 1874) |
| cl. | Mrs. Williams | Light rose. (J. Waterer, 1875) |
| cl. | Mundulum | Blush, pink spots. (T. Methven, 1868) |
| cl. | Murillo | (M. Koster) |
| cl. | My Pet | Pale yellow, red blotch. (Whitney/Sather) |
| cl. | N. N. Sherwood | Pink with gold center. |
| cl. | Nancy Sue | Magenta rose/gold spots. (Stephens, 1975) |
| cl. | Napoleon Baumann | Light rose/crimson spots. (Baumann, 1867) |
| --- | Narcissus | Rosy red, dark spots. (T. Methven, 1868) |
| cl. | Nathan Hale | Rose, darker blotch. (Dexter/Schlaikjer) |
| cl. | Neilsonii | Bright cherry red. (T. Methven, 1868) |
| --- | Nell Gwynne | Cerise pink, brown eye. (Waterer/Crisp) |
| --- | Nellie Moser | Lilac, pink, yellow and white. (Moser) |
| cl. | Nelly de Bruin | Pink, dark red blotch. (M. Koster) |
| --- | Nereus | Purple, spotted black. (T. Methven, 1868) |
| cl. | Newburyport Belle | Light purplish pink. (Dexter/Wister) |
| --- | Newton | Purplish pink. |
| cl. | Nina | Pink, dark red blotch. (Wytovich, 1979) |
| cl. | Ninon L'Enclos | Salmon pink. (From Belgium) |
| cl. | Nivaticum | White, yellow spots. (J. Waterer, 1873) |
| --- | Norah | Blush, pale cerise edges. (M. Koster) |
| cl. | Norma | White shaded pink. (Standish/Noble, 1860) |
| cl. | Norseman | Carmine/scarlet throat. (Lem/Elliott 1974) |
| cl. | Notabile | Bright rose. (J. Waterer, 1875) |
| cl. | Novelty | Pink, edged white. (A. Waterer) |
| cl. | Nuthatch | Azalea pink; bell-shaped. (R. Henny 1965) |
| cl. | Oculatum | Pink, dark spots. (J. Waterer) |
| cl. | Oculissimum | Rose, darker blotches. (J. Waterer, 1871) |

cl.  Omar Pasha            Purple lake. (J. Waterer, 1867)
cl.  Onsloweanum           Waxy blush/yellow eye. (J. Waterer, 1851)
---  Optimum               Rosy purple. (T. Methven, 1868)
---  Orangeade             Large, fringed, pale orange.
---  Ornamentum            Rosy scarlet. (T. Methven, 1868)
---  Orodes                Rosy scarlet. (T. Methven, 1868)
cl.  Ortrud                Pale lilac. (T. J. R. Seidel, 1912)
---  Osbornei              (Royal Botanic Garden, Kew, 1933)
---  Othello               Royal purple crimson. (Noble, 1850)
cl.  Owen Thomas           Crimson, spotted. (A. Waterer, 1915)


cl.  Pacific Queen         Jasper red. (Mrs. R. J. Coker, 1980)
cl.  Pamela-Louise         Light pink/reverse, deeper. (Hydon, 1974)
---  Pantherinum           Rose, dark spots. (T. Methven, 1868)
---  Papilionaceum         White, orange spots. (J. Waterer, 1871)
cl.  Pardolatum            Rosy lilac, spotted. (T. Methven, 1868)
cl.  Pause                 Intense carmine red. (Seidel, 1913)
---  Paxtonii              Rose, chocolate spots. (Standish/Noble)
cl.  Peachblow             Apricot. (Barto/Wright, Sr. & Jr., 1958)
cl.  Pearson's
     Elizabeth             Light blush, claret spots. (Methven 1868)
g.   Penelope              Rosy carmine. (R. Gill)
cl.  Penelope Vesuvius     Bright red. (R. Gill)
cl.  Penrice               Carrot red/empire rose. (Stanton)
---  Pentilly Scarlet      Red.
cl.  Perle Brillante       Light red, small dots. (Koster, Boskoop)
cl.  Perspicuum            White. (J. Waterer, 1851)
cl.  Peter Seidel          Lilac, white within. (Seidel, 1920)
---  Petraea               (Shown by Lady Martineau, 1934)
---  Philotis              Crimson. (T. Methven, 1868)
cl.  Picotee               White, picotee-edged. (A. Waterer)
---  Picturatum
     Superbum              Blush white, spotted. (Veitch, 1862)
cl.  Pieces of Gold        Cream/brown spots. (Mrs. Frederick, 1978)
cl.  Pieter de Hoogh       Red, magenta tinge.
cl.  Pince's Victoria      Purplish crimson. (T. Methven, 1868)
cl.  Pindar                Light purple violet. (Seidel, 1913)
cl.  Pink Cloud            Pink/yellow blotch. (Whitney/Sather, 1976)
cl.  Pinwheel              Yellow, gold center. (The Hennys, 1964)
---  Pizarro               Satiny rose. (Standish/Noble, 1870)
cl.  Pleasant Dream        Pink/orange blotch. (Whitney/Sather 1975)
---  Pluto                 Deep rose red. (Noble, 1850)
cl.  Polly Peachum         Rosy pink, red blotches.
cl.  Polynesian Sunset     Orange/spotting. (Whitney/Sather, 1976)
cl.  Ponticum Roseum       Pinkish lilac. Syn. Maximum Roseum.
---  Poussin               Deep rose. (Standish/Noble, 1850)
---  President
     Roosevelt             Red fading to white. Variegated foliage.
cl.  President van
     Den Hecke             Crimson/fine spots. (Standish/N., 1870)
---  Pretty Polly          Pink.
---  Pride of
     Cambridge             White, brownish blotch.
cl.  Pride of
     Glenrose              Red; large truss. (W. E. Glennie, 1980)
---  Pride of Kernick      Pink, scarlet marks. (R. Gill)
---  Pride of Tremough     (Montiford Longfield, Penryn, 1935)
---  Primulinum
     Elegans               Primrose yellow, pearl spots.
---  Prince Arthur         Plum color. (Standish/Noble, 1860)
---  Prince Consort        Bright rose, spotted. (Todman, 1861)
cl.  Prince Eugene         Rosy blush, edged pink. (Rinz, 1875)
---  Prince of Wales       Rose pink, light center. (Waterer, 1875)
cl.  Princeps              Pinkish lilac. (T. Methven, 1868)
cl.  Princess Amelia       Dark scarlet. (T. Methven, 1868)
---  Princess Ena          Mauve. (Before 1922)
cl.  Princess Frederick
     William               Blush, pink spots. (T. Methven, 1868)
---  Princess Hortense     Lilac rose, brown yellow eye. (Standish)
cl.  Princess Louise       Lilac white, green spot.

cl.  Princess Margaret
     of Connaught          Rose pink, dark blotch. (J. Waterer)
cl.  Prodigal's Return     Clear deep yellow. (Nelson)
cl.  Prof. Dr. Drude       Pale lilac, yellow marks. (Seidel, 1894)
cl.  Purple Lace           Much like Purple Splendour. From Boskoop.
cl.  Purple Prince         Purple/black blotch. (Sunningdale, 1956)

---  Queen Pomare          Chocolate, white throat. (Standish/Noble)
cl.  Quiet Quality         Purple pink/white center. (Dexter/Wister)
---  Quito                 Dark crimson red. (T. J. R. Seidel, 1914)

cl.  R. S. Field           Red or scarlet. (A. Waterer before 1871)
---  Rachel                Yellow/white spots. (Standish/N., 1870)
cl.  Rain of Gold          Pale yellow; fragrant. (Frederick, 1977)
cl.  Rainier Pink          Yellow, reverse pink. (Jordan, 1977)
cl.  Ralph Sanders         Purplish crimson. (A. Waterer, 1915)
cl.  Raphael               Dark lilac red. (T. J. R. Seidel)
cl.  Raphael               Rosy crimson/chocolate marks. (J. Waterer)
cl.  Ravigo                Orchid pink, yellow eye.
---  Rebecca               Blush, yellow brown spots. (Standish/N.)
cl.  Red Eagle             Deep scarlet. (Rogers)
---  Red Goblin            Bright orange scarlet.
cl.  Red Head              Orient red; leaves felted. (R. Henny 1965)
cl.  Redder Yet            Red, brown spots. (Leach/Pride)
cl.  Reedianum             Rose pink. (H. Waterer, 1851)
---  Regalia               Crimson. (T. Methven)
---  Regificum             Red or rosy scarlet. (J. Waterer, 1851)
---  Reginum               Deep blush. (Standish/Noble, 1870)
---  Regulator             Purple, red spots. (Noble, 1850)
---  Reine Amelie          Dark crimson. (Byls before 1858)
---  Remarkable            Lilac, dark eye. (Standish/Noble, 1860)
cl.  Rembrandt             Purple. (T. Methven, 1868)
cl.  Renidescens           Rosy crimson. (T. Methven, 1868)
---  Resplendent           Light pink, red spots. (T. Methven, 1868)
cl.  Return to
     Paradise              White, tan spots. (Mrs. Frederick, 1978)
cl.  Rhodoleucops          Bright rosy crimson. (T. Methven, 1868)
cl.  Richard Strauss       Cherry red.
---  Rienzi                Crimson/fine spots. (Standish/N., 1870)
cl.  Rifleman              Crimson. (Shown by Street)
cl.  RM 11                 Semidwarf red. (Whitney)
---  Robert Burns          Crimson, dark spots. (Standish/Noble)
---  Robert Croux          Dark red. (Croux, 1899)
cl.  Robustum              Lilac, dark spots. (T. Methven, 1868)
cl.  Robustum
     Cataubiense           Light purple, spotted. (T. Methven, 1868)
cl.  Robustum Hybridum     White, spotted. (T. Methven, 1868)
---  Rokeby                Rose, green spots. (Standish/Noble, 1860)
---  Romulus               Rosy purple. (Noble, 1850)
---  Ronda                 (Collingwood Ingram, 1944)
---  Rosetta               Bright rose/yellow eye. (Standish/N. 1870)
---  Roseum Argentinum     Rose. (Standish/Noble, 1870)
cl.  Rothsay               White/reverse blush. (R.W.Campbell 1977)
cl.  Rowallan Surprise     Spirea red. (D. Deans, 1982)
cl.  Royal Purple          Deep purple, yellow eye. (Harry White)
---  Royyun                (Shown by Mrs. Sebag-Montefiore, 1935)
cl.  Rudiger               Cerise red. (T. J. R. Seidel, 1900?)
---  Rudolph Seidel        Pink, light throat. (Seidel)
---  Ruth                  Bluish white, spotted. (Standish/Noble)

---  Sabrina               Rosy lilac. (Standish/Noble, 1870)
cl.  Sagamore Bayside      Light pink, yellow stripe.(Dexter/Wister)
cl.  Sagamore Bridge       Strong purplish pink. (Dexter/Wister)
---  St. Anthony           (P. D. Williams, 1935)
---  St. Probus            Intense crimson.
cl.  St. Simon             Rosy carmine, spotted. (A. Waterer)
---  Salmon Giant          Clear salmon pink. (R. Gill)
---  Salmon Perfection     Deep salmon. (R. Gill)
cl.  Salvini               Magenta rose. (Young before 1883)
cl.  Samuel Morley         Rosy crimson, light center. (J. Waterer)
---  Sandlefordianum       Rose, well marked. (J. Waterer, 1860)

--- Satanella            Rosy red.  (Noble, 1867)
--- Satellite            Crimson.  (T. Methven, 1868)
--- Seagull              Pale lavender, almost blue.  (R. Gill)
--- Selehurst Freak      (Shown by Dame Alice Godman, 1941)
cl. Shalimar            Lavender pink. (Vossberg/Schlaikjer 1974)
cl. Sheila Ann          Crimson, spotted black. (Caperci, 1970)
--- Shilsoneri           (Sir John Ramsden, 1935)
--- Shipwash Light       (Shown by Harrison, 1955)
cl. Sidney Herbert      Carmine, dark spots.  (J. Waterer, 1871)
--- Signet               Rosy purple.  (Noble before 1875)
--- Signor Lablanche     Rose purple, yellow spots. (Standish/N.)
cl. Silvis              Purple, yellow center. (A. Waterer, 1915)
--- Sinbad               Purplish lake/light blotch. (Standish/N.)
cl. Sir Arthur
    Guiness             Rose.  (J. Waterer)
cl. Sir Colin
    Campbell            Purplish rose, dark spots.  (J. Waterer)
--- Sir Harry Veitch     Magenta with a dark eye.
--- Sir Henry
    Havelock            Light red.  (Davies before 1890)
cl. Sir Henry
    Mildmay             Rosy crimson.  (J. Waterer, 1875)
cl. Sir Humphrey
    de Trafford         Rose yellow center.  (J. Waterer)
cl. Sir Isaac Newton    Crimson claret.  (A. Waterer, 1867)
cl. Sir James           (Dexter/Gable, 1958)
cl. Sir James Clark     Crimson/shaded purple. (J. Waterer, 1871)
--- Sir Jasper
    Goldthorpe          Mauve.  (Standish/Noble, 1870)
--- Sir John
    Franklin            Deep claret.  (Standish/Noble, 1860)
--- Sir John Moore       Crimson.  (1871)
--- Sir Laurence Peel    Rosy crimson.
--- Sir Richard
    Wallace             Rosy, yellow center.
cl. Sir Robert Peel     Crimson, dark spots.  (J. Waterer)
cl. Sir Thomas
    Sebright            Purple, bronze blotch. (A. Waterer, 1865)
--- Sir Walter Scott     Lilac, shaded white. (Standish/Noble)
cl. Sir William
    Armstrong           Dark crimson.  (J. Waterer, 1871)
--- Sir William
    Maxwell             Pink, beautifully spotted. (T. Methven)
cl. Skerryvore
    Monarch             Deep purplish pink.  (Dexter/Beinecke)
cl. Skyglow             Peach, edged pink. (Heritage Plant. 1978)
cl. Smithii             Fine bright scarlet.  (T. Methven)
--- Smithii Elegans      Crimson.  (T. Methven, 1868)
--- Snow Leopard         (Collingwood Ingram)
g.  Snowden             (Lord Aberconway, exhibited, 1935)
--- Snowball             White.  (Standish/Noble, 1860)
g.  Snowdon             (Lord Aberconway, 1935)
cl. Socrates            Rose pink, orange blotch.  (1867)
--- Sonia                White.
--- Sopia                Lavender blush, yellow eye. (Standish/N.)
--- Souvenir d'un Ami    Semi-double, violet rose.  (T. Methven)
--- Spectabilis          Salmon red.  (Veitch, 1862)
cl. Speculator          Deep rose, white throat. (Standish/Noble)
cl. Speculatum          Light pink/pale center. (Methven, 1868)
cl. Sphinx              Deep pink/red blotch. (Luenenschloss 1965)
cl. Stamfordianum       Claret, black blotch. (Methven, 1868)
--- Standard of
    Flanders            Blush rose/dark eye. (Standish/N., 1870)
--- Stanwellianum        Snowy white.  (T. Methven, 1868)
--- Star of Ascot        White, black spots.  (Standish, 1865)
cl. Star of Spring      Pink fading white. (Hardgrove/Royce, 1979)
--- Staving              Red.
--- Stephanotum          Crimson.  (T. Methven, 1868)
cl. Stephen Davies      Deep crimson.  (Davies)
--- Strawberry           (J. J. Crosfield, 1935)
--- Sultan               Rosy lake.  (Standish/Noble, 1870)

cl. Sultana             White or pale blush.  (J. Waterer, 1865)
cl. Summer Scandal      Pink, darker edges.  (Kersey/Frederick)
--- Sun Glow             (J. J. Crosfield, 1935)
cl. Sunny Day           Yellow/red rays.  (Whitney/Sather, 1985)
--- Superlative          Blush white.
cl. Sweet Sixteen       Orchid pink/darker edges. (Whitney/S. 1976)
cl. Sweet Sue           Pink, red specks.  (Waterer/Crisp, 1972)
cl. Sylph               Pink.  (A. Waterer before 1915)

cl. T. H. Kekewich      Crimson.  (J. Waterer)
cl  Tamerlane           Maroon.  (J. Waterer before 1865)
cl. Tan                 Biscuit, flushed peach.  (Dexter/du Pont)
--- Ted Marchand         (Shown by Collingwood Ingram)
cl. Terrific            Rose pink.  (Leach/Pride)
--- The Autocrat         Purple, dark eye.  (Standish/Noble)
--- The Bouncer          Crimson scarlet, dark eye.  (T. Methven)
cl. The Bridge          Spinel red; truss of 15.  (Holmes, 1979)
cl. The Cardinal        Syn. Kentucky Cardinal.  (Gable)
--- The Countess         White.  (Veitch)
--- The Czarina          Delicate lavender blush. (Standish/Noble)
--- The Gem              Blush, tipped rose.  (Standish/Noble)
--- The Giraffe          White, fine eye.  (Standish/Noble)
--- The Grenadier        Rose, spotted red.  (Standish /Noble)
--- The Hon. J.
    Boscawen            Crimson.
--- The King             Crimson.
--- The Lynx             Blush, good eye.  (Standish/Noble)
cl. The Maroon          Purplish red, chocolate blotch. (Waterer)
--- The Monitor          Vermilion.  (Standish/Noble, 1870)
--- The Painted Lady     Rose, blotched chocolate.  (T. Methven)
--- The Princess         Pure white.  (Standish/Noble, 1860)
--- The Queen            Blush, fading white.  (C. Noble, 1871)
--- The Saracen          Crimson lake.  (Standish/Noble, 1871)
--- The Saxon            Rich Carmine.  (Standish, 1861)
--- The Star             White, yellow eye.  (Standish/Noble)
--- The Vestal           Pure white.  (Standish/Noble, 1860)
--- The Warrior          Light crimson.  (J. Waterer, 1867)
--- Thomas Methven       Crimson, orange spots.  (T. Methven)
cl. Three Star          (Gable, 1958)
--- Thunderbolt          (On trial at Wisley from Lady Martineau)
--- Thunderhead          (Clark)
--- Timon                Rosy purple.  (Standish/Noble)
g.  Timoshenko          Scarlet.  (Collingwood Ingram, 1952)
cl. Tinicum             Pale pink; fragrant. (Dexter/Everitt)
cl. Tipoo Sahib         Dark chocolate.  (T. Methven, 1868)
--- Titania              Deep red, white center. (Standish/Noble)
cl. Titian              Rosy crimson. (H. Waterer, 1851)
cl. Toastmaster         Pale purplish pink.  (Dexter/Everitt)
cl. Todmorden           Pink and white.  (Dexter/Wister)
cl. Tom Everett         Deep purplish pink. (Dexter/Wells, 1985)
cl. Torlonianum         Purplish pink, yellow flare.
cl. Touch of Gold       Pale peach, gold blotch. (Mrs. Frederick)
--- Tregrehanii          (Shown by Capt. Carlyon, 1935)
cl. Tuesday's Child      Purplish pink. (Childers/Kraxberger)
--- Tulyar               Red, black blotch.  (A. Waterer)
cl. Turk                Bright light carmine red.  (Seidel, 1917)
--- Tyrian Queen         Rosy carmine.  (Rogers)

--- Una                 Blush, fine eye.  (Standish/Noble)
cl. Unique              Flesh, changes to buff.  (Wallace)
cl. Ursula Seidel       Pink, upper petal yellow/white. (Seidel)
--- Ute                 Carmine red.  (T. J. R. Seidel, 1918)

--- Valasquez           Cherry rose.  (T. Methven)
--- Valentina           Blush, yellow eye.  (Standish/Noble)
cl. Van Dyck            Rosy crimson.  (H. Waterer, 1851)
--- Vanguard            Bright crimson.  (Standish/Noble)
--- Varium              Deep pink.  (Knap Hill, 1955)
--- Vasari              Crimson, intensely spotted. (A. Waterer)
--- Vesper              Rosy pink, white throat.  (Noble, 1850)
--- Vesta               White.  (Standish/Noble, 1860)

```
---   Vesuvius          Blood red.  (Standish/Noble, 1870)
cl.   Vicomte
        Forceville       Dark crimson.  (J. Waterer)
cl.   Victor Frederick  Red/dark red blotch. (Lem/Sinclair, 1974)
---   Victor Hugo       Purplish crimson. (Standish ?  Waterer ?)
---   Victoria          Deep claret.  (Waterer before 1851)
cl.   Vida              Deep yellow.  (Steinmetz/Childers, 1965)
cl.   Viking Lady       Light mimosa yellow.  (Loeb, 1971)
cl.   Village Maid      Pale center, pink edges. (J. Waterer)
---   Villager          Rose.  (Noble before 1850)
cl.   Violet Parsons    Salmon pink.
---   Viscountess
        Elveden          Pink.  (W. C. Slocock, 1920)
cl.   Vivia Ward        Solferino purple, deep flare.  (Holmes)
cl.   Vivian Grey       Rose pink.  (J. Waterer)
---   Voltaire          Deep red.  (Standish/Noble, 1870)

---   W. A. Dunstan     Creamy white.  (M. Koster)
cl.   W. Downing        Dark puce.  (A. Waterer)
cl.   W. E. Gladstone   Carmine red. (J. Waterer)
cl.   W. H. Punchard    Plum color, yellow center.
---   W. Hozier         Pale pink.
---   W. Milton         Dark crimson.  (Young)
cl.   Walter Hunnewell  White, edged red purple. (Weston Nurs.)
cl.   Wareham           Purplish pink; fragrant. (Dexter/Wells)
cl.   Washington        (B. F. Lancaster, 1958)
---   Waterer           Salmon pink, orange tint.  Late.
cl.   Waterer's
        Victoria         Deep claret.  (T. Methven, 1868)
cl.   Wayne Pink        Spirea red.  (Dexter/Knippenberg, 1966)
cl.   Wesley Hayes      Pure white.  (Mrs. Wesley Hayes, 1972)
cl.   Westbury          Fragrant; purple pink to white.  (Dexter)
cl.   Weston            Rose, gold spot.  (Dexter/Mezitt, 1979)
cl.   Whipped Cream     White.  (Mrs. R. J. Coker, 1980)
cl.   White Sails       Pure white.  (Mrs. R. J. Coker, 1981)
cl.   Whitney Purple    Red purple, non-fading.  (Whitney/Sather)
cl.   Whitney's Dwarf
        Red              Buds red; blooms red.  (Whitney)
cl.   Whitney's Late
        Orange           Deep orange.  (William Whitney)
cl.   Whittenton        Purplish pink.  (Dexter/Wells, 1980)
cl.   Wianno            Purplish pink.  (Dexter/Wells/Wister)
cl.   Wilhelm Weise     China white.  (T. J. Seidel)
cl.   Willamette        (James, 1960)
cl.   Willard           Purplish red.  (Dexter/Wells/Wister)
cl.   William Cowper    Bright scarlet.  (J. Waterer, 1922)
---   William Downing   Puce, intense blotch.  (Standish/Noble)
---   William Godfrey   Blush white, yellow spot.
cl.   William R. Coe    Pink, gold yellow blotch. (Dexter/Coe)
cl.   Williams          Pink, large blotch.
---   Windsori          (Reuthe, 1851)
---   Winifred          (Shown by Loder, 1934)
cl.   Winifred Drake    Orange pink.  (Drake/ Zimmerman/Lux)
cl.   Winifred Kenna    Orchid, purple blotch.  (Loeb, 1965)
---   Winifred White    Pale pink, crimson edge.  (W. C. Slocock)
cl.   Winneconnet       Pale purple pink.  (Dexter/Wells, 1980)
cl.   Winning Ways      Purplish red.  (Dexter/Wister)
---   Winston Churchill Flower is scarlet.  (Kluis)
cl.   Wissahickon       Bright rose.  Sun-tolerant. (Dexter)
---   Withe Gem         White.
---   Withe Pearl       Blush pink, fading white.

---   Yanchart          (Collingwood Ingram, 1935)
cl.   Yo-Yo             Red, purple spots. (Whitney/Sather, 1976)

---   Zampa             Deep lake, violet tinge. (Standish/Noble)
---   Zanoni            Crimson lake, spotted.  (Standish/Noble)
---   Zealander         Salmon pink.  (Shown by Street)
---   Zenlenka          Shell pink, tipped darker shade.
cl.   Zephyr            Bright rosy red.
```

**Leaf Shapes**

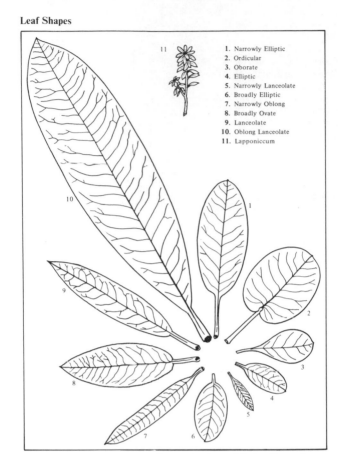

1. Narrowly Elliptic
2. Ordicular
3. Oborate
4. Elliptic
5. Narrowly Lanceolate
6. Broadly Elliptic
7. Narrowly Oblong
8. Broadly Ovate
9. Lanceolate
10. Oblong Lanceolate
11. Lapponiccum

By permission of Hydon Nurseries

RHODODENDRON HYBRIDS INTRODUCED BY:

Crystalaire Rhododendron Foundation
Weldon E. Delp, Hybridizer

Color designation below refers to the Royal Horticultural Society Colour Chart,
1966.


AA-17     (America x Ann Rutledge) Greyed/Purple #185-A and Red #53-A in bud
          opens to Red/Purple #57-A edged with Greyed/Purple #185-B, Greyed/
          Purple #185-A blotch, Size-Tall, HR-H1.

AA-36     (America x Ann Rutledge) Semi-petaloid, red/purple, Size-Tall, HR-H1.

Al Smith    (Madras x Cindy Lou) #1-85, Red #46-A in bud, opens to Red #45-A,
          Size-Med., HR-H2.

All East    [Gosh Darn x (Mrs. Yates x wardii)] #1-82, Yellow #4-C,B,A,
          Size-Med., HR-H2.

Andy Porter    [(Harrisville x Pink Twins) #2 x (smirnowii x catawbiense)
          #1] (hose & hose) #2-82, Red/Purple #59-C in bud, opens to Red/
          Purple #68-A,D, white upper lobe, Yellow/Green #154-B specks
          Size-Med., HR-H2.

Astrobrite    [Idealist x (Goldfort x Odee Wright)] #1-85, Yellow/Green #154-
          C,D, Yellow/Green #154-A in throat.

Atomic Blast    [(Caractacus x discolor) x yakusimanum] #15, Red/Purple #74-
          A in bud, opens to Red/Purple #74-A,B,C,D, Red/Purple #61-B
          specks.

Beesting    [Serendipity x (Phyllis Ballard x catawbiense Clark's White)] #1-
          85, Red #51-B in bud, opens to Orange #27-D, edged with Red #51-B,
          Yellow #8-C specks.

Betty White    [aureum (chrysanthum) x maximum) #1 x (Phyllis Ballard x
          catawbiense Clark's White) #1] #4-84, Yellow/Orange #19-B,C,
          edged with Red #51-B, Green/Yellow #1-B specks.

Big Top    (Catalgla x #5-Leach) Purple #75-A,B, white upper lobe with gold
          specks.

Billy Boy    [(America x Britannia) #2 x (America x Mars-Danik)] Greyed/Purple
          #185-A in bud, opens to Red/Purple #63-A,B,C, Green/Yellow #1-A
          on upper lobe, Yellow/Green #153-A specks.

Blast    (Russel Harmon x Evening Glow) #1-84, Red/Purple #65-D, edged with
          Red/Purple #68-D, yellow throat, copper specks, Size-Medium.

Bob AS    [(yakusimanum x Cindy Lou) #3 x (yakusimanum x Cindy Lou) #2] #1-82,
          Red/Purple #57-A in bud, opens to Red #55-A,B,C, bronze specks,
          Indumented, Size-Med., HR-H2.

Bob Danik    [carolinianum (pink form) Delp x P.J.M.] #8, red/purple, Size-
          Med., HR-H1.

Bon Bon    (Scott David x AA-36) petaloid, Red/Purple #73-A, Red/Purple #73-D
          stripes, Size-Med., HR-H2.

Born Free    [[(maximum x yakusimanum) x Adele's Yellow] #1-82 x (Bronze
          Wings x Adele's Yellow) #1] #1-85, white, Yellow/Green #154-A
          specks on upper three lobes.

Breathless    [(Mary Garrison x Mary Belle) #3-T.R. x (Si Si x Goody Goody) #6
          Delp] #2-85, Red #54-B in bud, opens to Yellow #6-B, Red #53-B
          in throat.

Brite Mite    (Faisa x Self) #1-81, Purple #78-B in bud, opens to Purple
          #78-D.

Bruce     [Carolyn Dana Lewis x (AA-17 x Cindy Lou) #1-78] #1-84, Red #53-A in
          bud, opens to Red #53-B,C, Red #53-A specks, Size-Med., HR-H2.

Calypso    [(America x yakusimanum) #1 x Frank Moniz = America x Mars] #1-85,
          Red #53-A in bud, opens to Red #53-C, white throat and
          stamens, Red #53-A specks.

Candy Man    [aureum (chrysanthum) x (Phyllis Ballard x catawbiense Clark's
          White)] #2-85, Red #51-B in bud, opens to Yellow #4-D, Yellow/
          Green #154-B upper lobe, Yellow/Green #154-A specks, Size-Small,
          HR-H2.

Carolyn Dana Lewis    [(Boule de Rose x Self) #1 x Lanny Pride] #1-80,
          frilled, Red #53-A,B,C, Size-Med., HR-H2.

Catamac    (Catalgla (green eye) x macabeanum) #1-83, Red #54-C in bud,
          opens to Yellow/White #158-D, Red/Purple #70-A in throat and
          specks, Size-Tall.

Cecil    [(Lackamus Spice x lacteum) x [aberconwayii x (yakusimanum x Fabia)]]
          #1-84 (Delp), Yellow #12-C,B.

Chapparell    (Nova Zembla x Dot's Cherry Jubilee), Red/Purple #68-B,C, maroon
          throat.

Chatterbee #1    (Ben Moseley x Besse Howells), Red/Purple #72-D, Red/Purple
          #75-C, maroon blotch.

Chatterbee #2    (Ben Moseley x Besse Howells), Red/Purple #73-A, maroon
          blotch.

Chatterbee #3    (Ben Moseley x Besse Howells), Red/Purple #67-A in bud, opens
          to Red/Purple #67-B,C,D, Red/Purple #61-A specks.

Chatterbox    [(smirnowii x yakusimanum) x Red Brave] #1-84, Red/Purple #68-A.

Cherry Crisp    [(smirnowii x yakusimanum) x Little Gem] #1, Red #53-B in bud
          opens to Red #53-C and Red #54-D, Indumented.

Cherry Float    [(brachycarpum x catawbiense) x AA-36]-petaloid, Red/Purple
          #66-A, Red/Purple #66-D stripes, Size-Med., HR-H2.

Christina Delp    (Catalgla (green eye) x Mrs. W.C.Slocock) #1-65, Red/
          Purple #70-B,C in bud, opens to Red/Purple #70-B,C,D,
          Yellow/Orange #20-A specks, Size-Med., HR-H2.

Christina's Sister    (Catalgla (green eye) x Mrs. W.C.Slocock) #2-65, Purple
          #77-C,D, edged with Purple #77-B, Yellow/Orange #20-A
          specks, Size-Med., HR-H2.

Cisa    [(yakusimanum x Gold Mohur) #2 x Serendipity] #10, shaded yellows,
          Indumented, Size-Dwarf, HR-H2.

Classy Lassy    [(Catalgla x Cadis) #1 x Self], Violet #84-A,B.

Color Coded    (Rochelle x Go Getter) #6, Red/Purple #68-B and Yellow/Orange
          #14-D in bud, opens to white, edged with Red/Purple #68-B,C,
          Yellow #11-B on upper lobe and in throat.

Colorvision    (Pink Cameo x Self) #1-85, Red/Purple #63-A in bud, opens to
          Red/Purple #63-B,C, Red/Purple #63-D in throat, Yellow/Green
          #153-B specks.

Cool Capers    [brachycarpum x (Calcutta x Tahiti)] Red #49-B,C,D, Yellow/
          Green #154-A blotch.

Cozy Posey    [brachycarpum x (chlorops x aureum (chrysanthum))] #1-85,
          Red #39-B and Yellow #11-B in bud, opens to Yellow #11-B,C,D,
          Yellow #2-A specks, Size-Medium.

'William Austin' by Waterer, 1915
Photo by Greer

'Winsome' by Aberconway
Photo by Greer

'Windsor Lad' by Knap Hill
Photo by Greer

'Wisp' by Sheedy/Skei/Greer
Photo by Greer

'Witch Doctor' by Lem/J. A. Elliott
Photo by Greer

'Witchery' by R. Henny
Photo by Greer

'Wizard' by Lem
Photo by Greer

'Xenophile' by Strauss
Photo by Greer

'Yates' Second Best' by Yates
Photo by Greer

'Yellow Saucer' by Cecil Smith
Photo C. Smith

'Youthful Sin' by Aberconway
Photo by Greer

'Yukon' by Leach
Photo by Leach

Crispy    [(smirnowii x yakusimanum) x Self] #1-84, Red/Purple #72-D,
          Purple #75-C, maroon specks, Indumented.

Crossbow    (Accomac x Besse Howells) #2-75, Red/Purple #66-C.

Crossfire    (Accomac x Besse Howells) #1-75, Red #54-A, Red/Purple #61-A
          specks.

Crystal Glo    (Midsummer x maximum (Delp's red form)) 1984, Red/Purple #63-D,
          edged with Red/Purple #63-B, green/yellow blotch, Size-Med.,
          HR-H1.

Crystalaire    (Midsummer x maximum (Delp's red form)) 1984, white with yellow
          blotch, Size-Med., HR-H1.

Cumulus    (Weldy x Serendipity) #1-82, white, Indumented, Size-Low, HR-H2.

Dainty Ruffles    (Tony's Gift x Lady Bligh) #3-81, Red/Purple #68-B,C,D, edged
          with Red/Purple #68-A, maroon specks, frilled, Size-Tall.

Dandy Andy    [(aureum (chrysanthum) x maximum) #1 x (Phyllis Ballard x cataw-
          biense Clark's White)] #1-85, Red #45-C in bud, opens to Orange/
          White #159-C, edged with Red #45-C, very slowly fades to white.

Dated Yellow    [brachycarpum x (chlorops x aureum (chrysanthum))] #1-75,
          Yellow #3-D and Orange/Red #35-D in bud, opens to Yellow #4-D.

David Harris    [(Sefton x Purple Splendour) #7 x Self] #1-83, Red/Purple
          #71-A, Greyed/Purple #183-A blotch, white stamens, Size-Med.,
          HR-H2.

D. D.    [(yakusimanum x Gold Mohur) #2 x Serendipity] #12, shaded yellows,
          Indumented, Size-Med., HR-H2.

Dream Whip    [Weldy x (aureum (chrysanthum) x maximum) #1] #1-85, Yellow
          #4-D, Yellow #4-B blotch, Yellow/Green #154-C specks.

Dusty Wiz    [R.O.Delp x Golden Star] #1-85, Greyed Red #182-D in bud, opens
          to Yellow/Orange #18-C,D, Red #47-C specks.

Dutch Treat    (Hello Dolly x Goldsworth Yellow) #1-85, Red #52-A in bud,
          opens to Red #51-B,C, Yellow/Green #154-B specks, flowers
          slowly fade to Yellow #8-D.

Edna Marie Porter    (Mary Belle x Dexter's Apple Blossom) fragrant, shaded
          pinks and yellow, Size-Tall, HR-H2.

Electra    (Bronze Wings x Donna Hardgrove) #2-79, Red/Purple #63-A,C, Red
          #46-A blotch.

Enchanter    (Annie Hardgrove x Anna Delp) #1-83, Red #45-A, Size-Tall, HR-H2.

Eye Appeal    (Lodestar x AA-17) #2 x (Lodestar x Dot's Cherry Jubilee), Red/
          Purple #62-A,B,C,D, large maroon blotch.

Faisa    (carolinianum (pink form) x scintillans) #1-76, Purple #78-B in bud,
          opens to Purple #78-C,D, Size-Med., HR-H1.

Falling BB    [Serendipity x [(Goldfort x chlorops) x (Hawk x Idealist]]
          #1-85, Yellow #4-B,C,D.

Fantasia #1    (Cindy Lou x Caltwins #5) #1-84, hose & hose, Red/Purple #63-
          C,D.

Fantasia #2    (Cindy Lou x Caltwins #5) #2-84, Red/Purple #63-C,D.

Fashion Tint    [Hotei x (Crest x macabeanum) #3-83, Yellow/Green #154-B,C,
          edged with Yellow/Green #154-D.

Felix    [(Sefton x Purple Splendour) #7 x (Blue Peter x Catalgla)] #1-82,
         purple, dark purple, Size-Med., HR-H2.

Feminique    (smirnowii x yakusimanum) #1 x [Catalgla x (discolor x Fabia)
             x (Mary Belle x Catalgla)] #2-84, Red/Purple #68-B,C,D, frilled.

Fireball    (Anne Hardgrove x Anna Delp) #3-84, Red #53-B, Size-Tall, HR-H2.

Firebolt    (Boule de Rose x Self) #1-85, Red #53-C in bud, opens to Red
            #53-B,C, Red #53-A specks, Red #56-D in throat.

Firefox    (Fireman Jeff x Anna Delp) #1-84, Red #46-B.

Fireking    (maximum x Gertrude Shale) #1-85, Red #46-A in bud, opens to
            Red #45-A, white on upper lobe and in throat, red stamens and
            pistil, white on tips.

Flashdance    (maximum (ivory white form) x Evening Glow) #1-84, Red #49-A,B,
              yellow throat, Size-Tall.

Four D's    (Hello Dolly x Caltwins #5) #1-84, Red/Purple #62-C,D, yellow
            throat.

Free Reign    [[R.O.Delp x [(brachycarpum x Crest) #4 x (Mars x America)]
              #1-82 Delp] #1-85, Yellow/White #158-C, edged with Red/Purple
              #70-C, Yellow/Green #154-C specks upper three lobes.

Gary Davis    (Viet Vet x Lanny Pride) Red #53-B,C, Size-Med., HR-H2.

Gen Schmidt    (America x yakusimanum) #7, Red/Purple #63-C,D, Size-Med.,
               HR-H2.

Golly Gee    (Donna Hardgrove x maximum) Red/Purple #66-D, edged with Red/
             Purple #66-A.

Goody Goody    (smirnowii x catawbiense) #1, Red/Purple #64-A,B, in bud,
               opens to Red/Purple #64-B,C, Yellow/Green #154-A blotch and
               specks, Size-Med., HR-H2.

Gosh Darn    (Catalgla x Mrs. H.R.Yates) F2-#1, pale yellow, Size-Med., HR-H2.

Groovy #1    (Tony's Gift x Mrs. Tom Lowinsky) #1-81, Red #56-D, gold and
             maroon blotch, Size-Tall.

Groovy #2    (Tony's Gift x Mrs. Tom Lowinsky) #2-81, Red #56-D, gold and
             maroon blotch, Size-Tall.

Gumdrop    (Cindy Lou x Slippery Rock) #1-84, Red/Purple #60-A,B in bud, opens
           to Red/Purple #66-A,B,C, Yellow/Green #154-A blotch, Size-Med.

Gutsy Gal    [(Harrisville x Pink Twins) x Goldsworth Yellow] #1-85, Red #48-A
             in bud, opens to Yellow #11-D, edged with Red #55-B, Red #52-A
             specks.

Had    (Newport F4) #209, blue/lavender, Size-Tall, HR-H1.

Handsome RIS    [(Mrs. J.G.Millais x La Bar's White) #1 x (Mrs. J.G.Millais x
                La Bar's White) #10] #1-85, Red/Purple #68-C.D and Red/Purple
                #65-D in bud, opens to white, Yellow/Green #151-C specks.

Heritage    [[(brachycarpum x catawbiense) x AA-36] x AA-36] Red/Purple
            #66-A,B,C,D, Size-Tall.

Hi Drama    [(brachycarpum x Crest) #4 x Bronze Wings] #1-85, Red/Purple
            #65-C,D in bud opens to white, Greyed/Red #182-A specks, tinge of
            Yellow #8-D in throat.

Hotline Beauty    [(Kettledrum x Jock) #1 x Mrs. C.S.Sargent] x (griersonianum
                  x Romany Chal) #8, Red/Purple #61-B in bud, opens to white,
                  edged with Red/Purple #67-A,B, Size-Tall.

Hotlips     (venator x Jenny Harris) #1-85, hose & hose, Red #46-A,B, in
            bud, opens to Red #43-A and Red #44-A.

Huggs      [(Tetraploid carolinianum x Watchung) #1 x Self] #1-85, Purple #78-B,
            and Purple #76-B,C,D.

Ididit #1     [(Caroline x #638 azalea) x #609 azalea] #2-84, Purple #76-A,B,
            azaleodendron, Size-Tall.

Ididit #2     [(Caroline x #638 azalea) x #609 azalea] #1-84, Red/Purple #72-B
            in bud opens to Red/Purple #77-B,C,D, Yellow/Green #148-B specks,
            deciduous azaleodendron, Size-Tall.

Indy Pink     (Serendipity x Lanny Pride) #1-83, Red #52-A in bud, opens to
            Red #50-D, Red #52-A specks, Indumented.

Inner Spirit     [R.O.Delp x (brachycarpum x Crest) #4 x Cindy Lou) #1-82]
            #1-85, Red/Purple #66-B,C in bud, opens to Red/Purple #65-D,
            Red/Purple #63-A freckles on all petals.

Incredible Hues     (Tony's Gift x open pollinated) #1-85, Red/Purple #72-D and
            Purple #75-C in bud, opens to Purple #75-A,B,C,D, Orange
            #26-B and Green/Yellow #1-B specks.

Ivory Palace     [Catalgla x (Morocco x R.O. Delp) #1-76, Red/Purple #62-D in
            bud, opens to Yellow/White #158-D, Purple #72-B stamens,
            Yellow/Green #1-B specks.

Jack Looye     (Si Si x Serendipity) #2-84, Orange/Red #32-D in bud, opens to
            Yellow/Orange #18-C,D, Size-Low, HR-H2.

Jack Milsom     (yakusimanum x Cindy Lou) #20, Red #52-A in bud, opens to Red
            #52-D, Size-Med., HR-H2.

Jacqueline Looye     (Sefton x Purple Splendour) #12, Red/Purple #71-A in bud,
            opens to Red/Purple #71-B and Red/Purple #72-B, edged
            with Red/Purple #72-A, Red/Purple #63-A on upper lobe,
            Greyed/Purple #187-A blotch, white stamens and pistil,
            Size-Med., HR-H2.

Jenny Harris     (Cindy Lou x Caltwins #5) #5-81, full hose & hose, Red/Purple
            #66-A in bud, opens to white, edged with Red/Purple #66-A,
            frilled, Size-Med., HR-H2.

Jennypoo     [Sweet Lulu x [(brachycarpum x Crest) #1 x Cindy Lou] #1-82]
            #1-85, Red #52-A in bud, opens to Red #52-D, edged with Red
            #52-A, maroon specks.

Jet Set     [(Weldy x Serendipity) #2 x maximum (white form collected in wilds
            at Jackson Center, Pa.)] #1-85, Yellow #2-C, Yellow/Green #154-B
            specks.

Joe     (Lanny Pride x Viet Vet) #1-83, Red/Purple #57-B,C, Size-Med., HR-H2.

Joylyn     (Sefton x Purple Splendour) #11, Red/Purple #71-A,B,C, Green/Yellow
            #1-B blotch, Yellow/Green #152-C specks, frilled, Size-Med., HR-H2

Jubilicious     (Lodestar x AA-17) #2 x (Lodestar x Dot's Cherry Jubilee) #2-81,
            Purple #75-D, edged with Red/Purple #72-C, Red/Purple #71-A
            blotch.

Judann     [brachycarpum x (Calcutta x Tahiti)] #1-84, Red #55-A,B,C,D, Yellow/
            Green #154-A blotch, Size-Tall.

Lanny Pride     [Princess Elizabeth x (smirnowii x yakusimanum)] #3, Red #45-
            A, Size-Med.

Leatherly     (yakusimanum x Cindy Lou) (with heat treatment; possibly tetra-
            ploid).

Lectrostatic      [Goody Goody x Anna Delp] #2-85, Red/Purple #67-B,D, Red/
                  Purple #64-B specks.

Lemon Drop        (maximum x vernicosum #18139) #1-85, Yellow #4-D, Yellow #4-B
                  blotch.

Lily White        [Ermine x (Umpqua Chief x Fawn)] #1-76, Red #51-A,B,C, grad-
                  ually fades to white, Size-Med.

Limeade       [[(brachycarpum x Crest) #1 x Cindy Lou] #1-82 Delp x (aureum
                  (chrysanthum) x maximum) #1 Delp] #1-85, Yellow/Green #154-C in
                  bud, opens to Yellow #2-C.

Livewire      [[(maximum rosa x C.S.Sargent) x (Pygmalion x catawbiense rubrum)
                  #1 x Goody Goody] #1-85, Red/Purple #63-A in bud, opens to Red/
                  Purple #63-D, heavily edged with Red/Purple #63-B, Yellow/Green
                  #153-A specks.

Looky Looky       (Sefton x Purple Splendour) #11, Red/Purple #74-A and Red/Purple
                  #72-A in bud, opens to Red/Purple #74-A, Red/Purple #73-D on
                  upper lobe, Green/Yellow #1-A specks, frilled.

Lovely Lady       (Lodestar x Catalgla) #1-82, white, Yellow/Green #151-A and
                  Orange/Red #35-A specks, HR-H2.

Lustrous Gem      [Gosh Darn x (Mrs. H.R.Yates x wardii "best yellow")] #1-85,
                  Yellow/White #158-C,D, edged with Red/Purple #73-B,C, touch
                  of Red #52-B in throat.

Lyall     (Crober's Apricot x Mary Belle) #1-85, Red #48-C,D, in bud, opens to
                  Red #37-B,C, Yellow #10-C in throat with Red #53-B specks.

Magic Plus        [[(Pygmalion x haematodes) x yakusimanum (Exbury form)] #1 x
                  Serendipity] #6-84, Red #52-A in bud, opens to Yellow/White
                  #158-D, edged with Red #52-D.

Magic Touch       [Bronze Wings x (maximum x wardii)] white edged with Red/
                  Purple #70-B, bronze/green in throat.

Making Waves      (Catalgla (green eye) x Evening Glow) #1-84, Red #50-B in bud,
                  opens to Red #50-C, Orange #28-D in throat.

Marge Danik      [carolinianum (pink form) Delp x P.J.M.] #3, red/purple, Size-
                  Med., HR-H1.

Marty     (Princess Elizabeth x Lanny Pride) #1-84, Red/Purple #63-C, Size-Med.,
                  HR-H2.

Mary Bean     [(Caractacus x discolor) x yakusimanum] #2, white, red throat,
                  pubescent, Size-Med., HR-H2.

Marypat     [(brachycarpum x Crest) #3 x Monte Carlo] #1-84, Yellow #2-C,
                  Yellow/Green #154-C blotch, bronze specks.

Masterblend     [(maximum x vernicosum #18139) #1 Delp x (aureum (chrysanthum)
                  x maximum) #1 Delp] #1-85, Red/Purple #60-B in bud, opens to
                  Red/Purple #62-B,C,D, Yellow #4-D in throat, Red/Purple
                  #61-A specks.

Matthew Weldon     (AA-17 x Goldsworth Yellow) #1-80, shaded pinks, Size-Med.,
                  HR-H2.

Maxsupreme      [(hyperythrum x maximum) #2-79] #1-85, Red/Purple #67-A in bud,
                  opens to Red/Purple #67-C,D, Yellow #10-C blotch, Yellow #10-A
                  specks.

May's Delight     (America x yakusimanum) #8, rose/pink, Size-Med., HR-H2.

Megan     [Tony's Gift x (Moser's Maroon x yakusimanum)] #2, Red/Purple #65-D,
                  edged with Red/Purple #70-B, Yellow/Green #154-B specks.

Microflare   [Tony's Gift x (<u>maximum</u> x <u>vernicosum</u> #18139)] #1-85, Red/Purple
             #73-D, edged with Purple #78-B, Yellow/Green #154-A specks,
             Purple #78-B stamens.

Microsplash  [[Pygmalion x <u>haematodes</u>) x <u>yakusimanum</u> (Exbury form)] #1 x
             Serendipity] #9-84, Green/White #157-D, Green/Yellow #1-D
             blotch.

Mikey   (Blue Peter x Ice Cube) #1-78, lavender/purple, Size-Med., HR-H1.

Mini Brite   (<u>racemosum</u> (apricot form) x <u>keiskei</u>) #2-85, Yellow #11-B
             and Yellow #10-D and Red #52-D, Red #52-B specks, large open
             flowers, Size-Low, HR-H2.

Mini Mode   [(<u>carolinianum</u> (pink form) x Pioneer) #20 x (Tetraploid <u>carolin-
            ianum</u> x <u>fastigiatum</u>)] #2-85, Purple #78-B,D, frilled open
            flowers, Size-Med., HR-H2.

Mini Pearl   [(Tetraploid <u>carolinianum</u> x Watchung) #1 x Self] #1-85, Pur-
             ple #76-A, Size-Med., HR-H2.

Mint Julep   [Caroline x (Catalgla x Sappho) #2] #1-83, Red/Purple #73-C in
             bud, opens to Red/Purple #73-D, Yellow/Green #153-B specks,
             flowers fade to white, Yellow/Green #154-A specks, Size-Med.,
             HR-H2.

Mitch   [(<u>yakusimanum</u> x Cindy Lou) x (<u>yakusimanum</u> x Henriette Sargent)]
        #5-83, shaded pinks, Indumented, Size-Med., HR-H2.

Moonburst   [[(<u>brachycarpum</u> x Crest) #1 x Cindy Lou] #1-82 x (<u>aureum</u> (<u>chry-
            santhum</u>) x <u>maximum</u>) #1] #1-85, Red #52-A in bud, opens to white,
            Yellow/Green #154-C specks, 24 flowers.

Moonglow   (Henry Yates x Catalgla) Red/Purple #68-C,D in bud, opens to Red/
           Purple #62-D, Yellow #6-A blotch, Yellow/Green #154-A specks,
           flowers fade to white.

Neala   [<u>aureum</u> (<u>chrysanthum</u>) x (Phyllis Ballard x <u>catawbiense</u> Clark's White)]
        #3-85 Yellow/Orange #18-B and Orange/Red #35-A,B, in bud, opens to
        Yellow #10-B,C, Yellow/Green #154-A specks, Size-Low, HR-H2.

Nebula   (Newport F5) Red/Purple #67-B, white upper lobe, Green #154-A specks,
         Size-Tall, HR-H1.

Nick   (Anna Delp x Little Gem) #1-83, Red #53-B, Size-Med., HR-H2.

Nobue   [(<u>smirnowii</u> x <u>catawbiense</u>) x (<u>yakusimanum</u> x Gold Mohur) #2] #13,
        shaded rose, Size-Med., HR-H2.

Noname   (Caroline x AA-17) #1-84, Red/Purple #68-C,D, maroon blotch, pleats
         on some petals.

Nubelle   (Newport F5) Red/Purple #71-C, white on upper three lobes, green
          specks, Size-Tall, HR-H1.

Oh Joyce   [(<u>smirnowii</u> x <u>yakusimanum</u>) Delp x Joyce Lyn] #2-84, red/purple
           Size-Med., HR-H2

Ollie   (<u>catawbiense</u> (dwarf red form) x <u>brachycarpum</u> #1) Red/Purple #68-
        A in bud, opens to Red/Purple #68-A,B,C, white upper lobe, Green/
        Yellow #1-C specks, HR-H2.

Ovation #1   (Tony's Gift x Lady Bligh) #1-84, Red/Purple #62-D, Red #53-A
             specks, Size-Tall.

Ovation #2   (Tony's Gift x Lady Bligh) #2-84, Red/Purple #62-D, Red/Purple
             #66-D specks, Size-Tall.

Pacesetter   [[R.O.Delp x [<u>vernicosum</u> #18139 x (Nereid x <u>discolor</u>)] Yellow
             bud-Haag's] #1-85, Yellow #3-D, Yellow #3-B,C, in throat.

Pana       [(smirnowii x yakusimanum) x Red Brave] #1-81, Red #53-B,C,D and Red
           #54-D, Size-Med., HR-H2.

Paper Moon     [(Charles Dickens x Atrosanguineum) #7 x yakusimanum] #1 x
               (fortunei x yakusimanum)] #11-74, white, red pistil, Size-Tall.

Peacemaker     (Double Dip x Serendipity) #1 x (yakusimanum x Gold Mohur) #2,
               Yellow/Orange #20-D and Red #48-C, Yellow/Green #154-B specks.

Peach Tree     (Gosh Darn x Si Si) #1-84, Orange #29-A in bud, opens to Orange
               #29-C,D, Size-Med.

Peggy Kirkpatrick     (Caroline x Peekaboo) #1, Color-unknown, Size-Med.,
                      HR-H3.

People Stopper     [[smirnowii x yakusimanum) x Mary Belle] #1 x Yaku Sunrise]
                   #1-85, Red #52-A in bud, opens to Red #49-D, edged with Red
                   #52-A.

Perfecta     (Henry R.Yates x Hotei) #1-84, white, copper/green specks, Size-
             Tall.

Pipe Dream     [(ponticum hybrid x Roslyn) #6-72 x Self] #1-84, Violet #84-A,
               B,C, pale green specks.

Plum Beauty     (Vivacious x Old Port) #1-85, Red/Purple #59-A in bud, opens to
                Red/Purple #61-A,B,C, Red #59-A specks.

Portage Road     (Cindy Lou x Self) #1-85, Red/Purple #59-A,B in bud, opens to
                 Red/Purple #59-B,C, Yellow/Green #152-D specks.

Posey Power     [Anna Delp x Goody Goody] #1-85, Red/Purple #63-A,B, Red/Purple
                #64-B specks.

Poverty Row     [Tony's Gift x (Moser's Maroon x yakusimanum)] #3-84, Red/
                Purple #69-D and Red/Purple #74-A in bud, opens to Red/Purple
                #69-D, edged with Red/Purple #74-A, Size-Tall.

Power Vibes     [(Susy x Anne Hardgrove) #1 x (#2 Red x Cindy Lou) #1] #1-85,
                Red/Purple #74-C,D, edged with Red/Purple #74-B, Red/Purple
                #59-A,B specks.

Priceless     (smirnowii x catawbiense) #2, Red/Purple #73-A,B, bronze and green
              blotch, HR-H2.

Protocol     (Tony's Gift x Mrs. Tom Lowinsky) #3-84, Red #56-D, gold/bronze
             blotch, Size-Tall.

Proud Performer     [Old Port x (America x yakusimanum) #19] #1-85, Red/Purple
                    #67-A in bud, opens to Red/Purple #67-B,C,D, Yellow/Green
                    #153-A specks.

Purplogic     [(Pinnacle x Goldsworth Orange) x Self] #1-85, Purple #78-A in
              bud, opens to Purple #78-C,D, bronze specks.

Racy     (Pink Cameo x Self) #1-84, Red/Purple #67-C, white upper lobe, Yellow/
         Green #154-A specks.

Racy Pink     (Princess Elizabeth x Crest) #1-85, Red/Purple #63-C,D, gold
              specks.

Rare Breed     (Cindy Lou x Self) #3-85, Red/Purple #60-A in bud, opens to
               Red/Purple #60-B,C,D, Yellow/Green #146-B specks.

RAS     [brachycarpum x (Calcutta x Tahiti)] #2-84, Red/Purple #62-B,C,D,
        yellow throat, Yellow/Green #154-A specks, Size-Tall.

Rave Revues     (Ostbo's Yellow x (Mrs.H.R.Yates x wardii)] #1-85, Greyed/Purple
                #185-C in bud, opens to white, Yellow/Green #154-A in throat,
                Yellow/Green #154-A specks.

Razmataz     (Midsummer x Ruddy Red Max) Delp, Red/Purple #66-B, Red/Purple
             #66-D in throat, yellow blotch.

Red Alert    (Bronze Wings x Donna Hardgrove) #3-85, frilled, Red #45-A, Red
             #53-A in throat.

Red Streak   (AA-17 x Cindy Lou) #1-78, Red #53-A and Red #46-A in bud, opens
             to Red #53-B and Red #46-B, white upper lobe, white tips on
             stamens, HR-H2.

Redwing    (Cindy Lou x Self) #2-85, Red/Purple #59-A in bud, opens to Red
             #53-A and Red #46-A,B, white stamens.

Renee Shirey    [[(Caltwins #2 x Self)] #1-85, Red/Purple #66-A and Red/
             Purple #68-D in bud, opens to Red/Purple #68-A,B,C,D, Yellow/
             Green #154-A specks, HR-H2.

Retroactive    [(Weldy x Serendipity) #2 x (Phyllis Ballard x catawbiense
             Clark's White)] #1-85, Red #50-B in bud, opens to Red #49-D,
             edged with Red #50-B, Green/Yellow #4-C in throat, truss
             slowly fades to white.

Rhapsody    [(smirnowii x yakusimanum) x (Calcutta x Tahiti)] #1-81, Red/
             Purple #62-A,B,C,D, Size-Med.

Rhodo Magic    [(Susy x Anne Hardgrove) x (#2 Red x Cindy Lou)] #2-85, Red/
             Purple #63-A and Red/Purple #67-A, edged with Red/Purple #72-
             B, Red/Purple #64-B specks.

Ritchie    (Catalgla x Sappho #3 x Self) #2-77, Purple/Violet #81-A
             in bud, opens to Purple #76-D, edged with Purple/Violet #80-A,B,
             Red/Purple #71-A blotch, Size-Med., HR-H1.

Robert Weldon    (Weldy x Serendipity) #7-83, yellow, Indumented, Size-Low,
             HR-H2.

R.O.Delp    (Lodestar x Mary Belle) #1-81, shaded pinks and peaches, Size-
             Med., HR-H2.

Sassy Lassie    [(carolinianum (pink form) x Pioneer) x (Tetraploid carolin-
             ianum x fastigiatum)] #1-85, Purple #78-A, Size-Med., HR-H2.

Saver    (Sefton x Purple Splendour) #1-84, Purple/Violet #82-A and Purple/
             Violet #81-A, red throat, dark maroon blotch, Size-Med.

Scott David    [(brachycarpum x catawbiense) x AA-36], petaloid, Red/Purple
             #61-B in bud, opens to Red/Purple #73-A,D, Red/Purple #57-A
             in throat and specks, Size-Tall, HR-H2.

Scott Harris    (Blue Peter x Ice Cube) #1-82, white, purple blotch, Size-Med.,
             HR-H2.

Scotty D    [Rocket x (Choice yakusimanum (Delp selection) x yakusimanum
             (Exbury form)] #1-81, Green/Yellow #1-C and Red #38-D in bud,
             opens to Yellow #4-D, Yellow/Green #154-A specks, Size-Med.

Scotty Lad    (Scott David x AA-36) #1-85, Red/Purple #60-A and Red/Purple
             #57-A in bud, opens to Red/Purple #63-B,C,D, Red/Purple #57-A
             in throat and specks, HR-H2.

Sensory Blast    (Janet Blair x Autumn Gold) #1-85, Red #52-B,C,D and Yellow/
             Orange #15-D in bud, opens to Yellow/Orange #15-D, edged with
             Red #49-D and Red #52-D, Orange/Red #35-A specks.

Serious Pink    (Cindy Lou x Caltwins #5) #15-81, Red/Purple #68-A,B,C, white
             upper lobe, gold specks, Size-Med.

Sharp Impression    [(carolinianum (pink form) x calostrotum (rose form) x
             Arctic Pearl] #2-76, Purple #78-C,D and Purple #76-B,C,
             Yellow/Orange #22-A specks, Size-Med., HR-H2.

Sharpie     (Vivacious x Cindy Lou) #1-83, Red #53-B,C and Red #46-C in bud,
            opens to Red #53-D veined on outer side of petals with Red #46-C,
            Red #46-B stamens and pistil, Size-Med., HR-H2.

Sheer Delight     (Tony's Gift x Cunningham's Sulphur) #1-85, Purple #78-A in
            bud, opens to Purple #78-B,C,D, white upper lobe, Yellow/
            Green #154-A specks.

Shockwaves     (Carolyn Dana Lewis x Lanny Pride) #1-85, Red #53-B in bud, opens
            to Red #53-B,C,D, Red/Purple #59-A specks, white stamens and
            pistil, white in throat, Yellow/Green #154-A at base of pistil.

Showtime    [(Pygmalion x haematodes) x yakusimanum (Exbury form)] #1-73,
            Red/Purple #57-A in bud, opens to Red/Purple #65-D, edged with
            Red/Purple #65-A, Size-Med., HR-H2

Si Barnes     [(Mrs. J.G.Millais x LaBar's White) #1 x (Mrs. J.G.Millais x La
            Bar's White) #11] #1-84, white, yellow/gold blotch, Size-Med.

Sidekick     (Boule de Rose x Self) #1, Red #53-A in bud, opens to Red #53-
            B,C,D, Red #53-A specks.

Silvery Moon     (Si Si x Serendipity) #4-84, peach fading to yellow, Size-
            Low, HR-H2.

Si Si     (yakusimanum x Gold Mohur) #2, Red #50-A in bud, opens to Red #50-
            C,D, Yellow #12-D in throat, Red #50-A blotch, Size-Med., HR-H2.

Sizzler     [(America x yakusimanum) #7 x Self] #1-84, Red #53-A,B, in bud,
            opens to Red/Purple #57-B,C,D, Red/Purple #57-A specks, Size-Med.

Skeeter Hill     [(brachycarpum x Crest) #4 x Bronze Wings] #1-82, shaded
            yellows, Size-Med., HR-H2.

Snapper     [Madras x (maximum x griersonianum)] #1-85, Red #53-B in bud,
            opens to Red/Purple #57-A,B,C, Red/Purple #61-B specks.

Snowflake     (Faisa x Self) #2-83, white.

Sonically Sound     [(Sefton x Purple Splendour) #7 x (Blue Peter x Powell
            Glass)] #1-85, Red/Purple #71-A in bud, opens to Red/Purple
            #65-D, heavily edged with Red/Purple #71-A,B, Yellow/Green
            #154-B specks.

Sonic Waves     (Opal x Goldsworth Yellow) #1, Red/Purple #70-C in bud, opens to
            Red/Purple #73-D, Yellow/Green #153-B specks.

Spectrum     (Naomi Exbury x Gargantua) #1-74 (Cecil Smith cross) Purple #77-
            B,C,D, Green/Yellow #1-A specks.

Spring Tonic     [(Phyllis Ballard x catawbiense Clark's White) x Pale Yellow
            Dexter] #1-84,  Red #53-B in bud, opens to Yellow #11-D,
            edged with Red #51-C, Yellow/Green #154-A in throat, Yellow/
            Green #154-A specks.

Stardust     (Tony's Gift x Cunningham's Sulphur) #2-85, Red/Purple #70-B
            in bud, opens to white, edged with Red/Purple #70-B, Yellow/
            Green #154-A specks.

Station Square     [brachycarpum x (Great Lakes x Sham's Juliet)] Red/Purple
            #62-B, copper specks.

Stokes' Bronze Wings     (maximum x catawbiense) pink, bronze blotch,
            Size-Tall, HR-H1

Stunning Array     [(yakusimanum x Mars) #1 x (Choice yakusimanum (Delp
            selection) x yakusimanum (Exbury form) #1] #1-85, Red/Purple
            #63-A in bud, opens to Red/Purple #63-D, edged with Red/
            Purple #63-B, Red/Purple #63-A specks.

<u>Subtle Beauty #1</u>    (Tony's Gift x Eldorado) #1-84, Purple #75-B,C,D, yellow/
          green blotch, red stamens and pistil, Size-Tall.

<u>Subtle Beauty #2</u>    (Tony's Gift x Eldorado) #2-84, Purple #75-B,C,D, yellow/
          green blotch, red stamens and pistil, Size-Tall.

<u>Suds</u>   (Weldy x Serendipity) #1-84, white, Size-Med., HR-H2.

<u>Sugar Baby</u>   [(<u>brachycarpum</u> x Crest) #4 x (Mars x America-Danik)] #1-84,
          Red/Purple #68-B,C.

<u>Sultry Peach</u>   [<u>aureum</u> (<u>chrysanthum</u>) x (Phyllis Ballard x <u>catawbiense</u> Clark's
          White)] #1-85, Red #48-C in bud, opens to Red #49-A,B,C,D,
          Size-Low, HR-H2.

<u>Sundex</u>   [Hello Dolly x (Catalgla x Lady Bessborough)] #1-84, Red #52-A in
          bud, opens to Red #52-D edged with Red #52-B, Yellow #4-B in throat,
          Yellow/Green #154-A specks, Size-Med.

<u>Sunny Crest</u>   (<u>brachycarpum</u> ssp. <u>tigerstedtii</u> x Crest) #3-75, Yellow #3-C in
          bud, opens to Yellow #3-C, edged with Yellow #3-D, Greyed/Orange
          #167-D specks.

<u>Sunnyview</u>   (Candidissimum x Slippery Rock) #1-84, Purple #75-A,B,C,D, pale
          gold blotch.

<u>Sunset Glo</u>   (Gosh Darn x Si Si) #1-85, Yellow #5-D in bud, opens to Yellow
          #11-D, Red #48-A in throat.

<u>Sunsheen</u>   [[(Sweet Lulu x [<u>vernicosum</u> #18139 x (Nereid x <u>discolor</u>)]] #1-85,
          Yellow #26-B,C in bud, opens to Yellow #10-B,C,D, tint of Orange
          #31-C.

<u>Super Q.T.</u>   (Bronze Wings x Donna Hardgrove) #1-84, Red/Purple #68-A,B,
          maroon blotch.

<u>Superstar</u>   [(Weldy x Serendipity) #2 x (<u>aureum</u> (<u>chrysanthum</u>) x <u>maximum</u>) #1]
          #1-85, white, Yellow #3-C blotch, Yellow/Green #154-A specks,
          Size-Med., HR-H2.

<u>Super Tuf</u>   [[(<u>maximum</u> x <u>yakusimanum</u>) #1 Delp x Adele's Yellow] #1-82 x
          (Bronze Wings x Adele's Yellow) #1-Delp] #3-85, Yellow #10-D in
          bud, opens to Yellow #8-D, Yellow/Green #154-A specks.

<u>Sweet Lulu</u>   [Gosh Darn x (Phyllis Ballard x <u>catawbiense</u> Clark's White) #1]
          #1-81, pale salmon fading to pale yellow, Size-Med.

<u>Sweet Medley</u>   (Earlene x Adele's Yellow) #1-85, Red/Purple #57-B in bud,
          opens to Red/Purple #62-D, edged with Red/Purple #57-B, pale
          yellow blotch.

<u>Swinger</u>   [Evening Glow x [<u>vernicosum</u> x (Nereid x <u>discolor</u>)]] #1-A Haag's,
          Yellow #4-A,B,C,D, Size-Med.

<u>Tangy</u>   [<u>maximum</u> x (Brookville x Mary Garrison)] #1-76, Red #56-D, edged with
          Red #56-A, Red #39-B blotch.

<u>Tender Tint</u>   [(<u>smirnowii</u> x <u>yakusimanum</u>) x Britannia) #2-76, Red/Purple #62-
          D, veined with Red/Purple #62-A, Size-Med.

<u>The Wild One</u>   (<u>yakusimanum</u> x Little Gem) #1-85, Red #53-B in bud, opens to
          Red #53-C,D.

<u>T-Time</u>   [Si Si x (<u>smirnowii</u> x <u>catawbiense</u>) #1] #1-84, Red/Purple #67-C,D,
          Size-Med., HR-H2.

<u>The Classic</u>   (Harvest Moon x Ice Cube) #2-80, white, yellow blotch on three
          lobes, Size-Med., HR-H2.

<u>The Finalist</u>   (Janet Blair x Autumn Gold) #1-72, Red #55-D, edged with Red
          #55-B, yellow in throat, salmon specks, Size-Med.

Tiger Print     [(Sefton x Purple Splendour) #7 x (Catalgla x Sappho)] #1-85,
                Red/Purple #72-A in bud, opens to Purple #76-D, edged with
                Purple #78-B, Yellow/Green #151-A specks on upper three lobes,
                Greyed/Purple #185-D specks on all petals.

Tigress    [Midsummer x (maximum x wardii)] white, edged with Red/Purple
           #68-A, salmon in throat.

Tin Lizzie    (Disca x Dorothy Amateis) Red/Purple #68-B,C, copper/gold in
              throat.

Tiny Bombshell #1    [(carolinianum (pink form) x davidsonianum) x Self] #1-85
                     Purple #78-B in bud, opens to Purple #78-C.

Tiny Bombshell #2    [(carolinianum (pink form) x davidsonianum) x Self] #2-85,
                     Red/Purple #72-B in bud, opens to Red/Purple #72-C.

Tiny Tina    [Bronze Wings x (maximum x wardii)] white, edged with Red/
             Purple #73-A, Green/Yellow #1-B specks, HR-H2.

Tiny Tyke    (racemosum (apricot form) x keiskei) #1-85, Yellow/White #158-
             B,C, and Red/Purple #62-A,B, Yellow/Green #145-B specks, Size-
             Low, HR-H2.

Tip Dip    [Midsummer x (maximum x wardii)] Red/Purple #68-D, edged with
           Red/Purple #68-A, yellow/salmon in throat.

Tols    (yakusimanum x catawbiense) Tolstead, Red/Purple #66-A in bud, opens
        to Red/Purple #65-D, edged with Red/Purple #66-B.

Tom Ring, M.D.    (Fireman Jeff x Anna Delp) #1-83, Red #46-A in bud, opens
                  to Red #45-A, Size-Tall.

Tony's Gift    (Catalgla x Belle Heller) = (Shammarello Hybrid S.E.100), white,
               Yellow/Green #144-B blotch, Size-Tall, HR-H1.

Too Lips    (Hello Dolly x Caltwins #5) #1-85, Red #50-A in bud, opens to Red
            #50-B,C, Yellow #4-B in throat, Yellow/Green #154-A specks.

Topper    [R.O.Delp x [(brachycarpum x Crest) #4 x (Mars x America)] #1-82]
          #1-85, Red/Purple #59-A and Red/Purple #60-A in bud, opens to Red/
          Purple #63-A,B,C, Red/Purple #64-A specks.

Topsy Turvy    (Scott David x AA-36) Red/Purple #61-C, Red/Purple #61-B
               freckles on upper lobe, Red/Purple #61-A in throat, Size-Med.,
               HR-H2.

Touch of Class    (Goody Goody x Si Si) #1-85, Red/Purple #64-B in bud, opens
                  to Red/Purple #64-C, slowly fades to white.

Toughy    [Harrisville x (Calcutta x Tahiti)] #1-84, Red/Purple #69-A,B,C,D,
          Yellow/Green #154-B specks, Size-Med., HR-H2.

Tower of Power    (Dr. Dresselhuys x Self) #1-47, Red/Purple #74-A in bud,
                  opens to Red/Purple #74-B, white upper lobe, Yellow/Green
                  #154-B specks, Size-Tall, HR-H1 (First cross of Weldon
                  Delp-made in 1947).

Tri-State    (Pink Cameo x Self) #2, Red/Purple #61-C, maroon specks,
             frilled, Size-Med., HR-H2.

Tuf Stuff    [(America x yakusimanum) #1 x Jean Marie de Montague] #1-85, Red/
             Purple #66-A, Red/Purple #61-B specks.

Twilighter    (Whitney's Orange x Catalgla) #1-85, Red/Purple #57-A in bud,
              opens to Red #56-D, edged with Red/Purple #57-B, Yellow #11-D
              on upper lobe.

Ultra-Violet    [Bronze Wings x (maximum x wardii)] Violet #82-A,B,C,D,
                HR-H2.

Vanity Fair    (Midsummer x Old Copper) white, edged with Red #54-B, bronze/
               green throat.

Velma #1    [[(smirnowii x yakusimanum) x Mary Belle] #1 x Polynesian
            Sunset] #1-83, shaded pinks and peaches, Size-Tall, HR-H3.

Velma #2    [[(smirnowii x yakusimanum) x Mary Belle] #1 x Polynesian Sunset]
            #1-83, Red/Purple #61-D, Red/Purple #58-D, yellow throat, copper
            specks, Size-Tall, HR-H3.

Vibes    [(smirnowii x yakusimanum) x Self] #1-84, Red/Purple #71-B,C, in
         bud, opens to Red/Purple #73-A,B,C, copper specks, Indumented,
         Size-Med.

Vibrant Vision    (Accomac x Besse Howells) #1-84, Red/Purple #58-B,C,D, Large
                  maroon blotch.

Viet Vet    [(America x Britannia) x (America x Mars)] #1-82, Red #53-A in bud,
            opens to Red #53-B,C,D, white throat, Red #53-A specks, Size-Med.,
            HR-H2.

Voletta    [(maximum x yakusimanum) x (maximum x wardii)] #1-84, Red/Purple
           #73-B,C,D, green throat.

Voyager    (Scott David x AA-36) semi-petaloid, Red/Purple #68-B, maroon
           throat.

WED    (Harrisville x Goldsworth Yellow) #1-84, Red #56-D, edged with Red #56-
       A, yellow blotch, salmon specks, Size-Tall.

Weldy    [(Pygmalion x haematodes) x yakusimanum (Exbury form)] #1-83,
         frilled, pink fading to ivory, Size-Med., HR-H2

Wendy Lyn    [Princess Elizabeth x (smirnowii x yakusimanum)] #17, Red #46-B,
             large white star in center, Size-Med.

Whimsical    [Si Si x (smirnowii x catawbiense)] #1-84, Red/Purple #73-D,
             edged with Red/Purple #73-B, Yellow/Green #154-A in throat,
             maroon specks.

Whirlwind    (Newport F4) #154 Red/Purple #78-A,B, and Red/Purple #66-A in bud,
             opens to Red/Purple #74-B,C,D, white upper lobe, Red/Purple #61-A
             specks, Size-Med., HR-H2.

Whirlybird    (Arthur Bedford x Ice Cube) #1-85, Purple #78-B in bud, opens to
              Purple #76-C,D, edged with Purple #76-A, Yellow/Green #153-B and
              Red/Purple #70-A specks.

Whispering Hope    [Harrisville x (Calcutta x Tahiti)] #1-84, Red/Purple #73-A
                   in bud, opens to Red/Purple #73-C,D, Yellow/Green #154-A
                   specks.

Whizette    (Princess Elizabeth x maximum (white form collected in wilds at
            Jackson Center, Pa.)) #1-84, Red #52-B,C,D, (Calyx) Size-Med.,
            HR-H2.

Wild Side    (yakusimanum x Cindy Lou) #1-85, Red/Purple #57-A in bud, opens
             to Red #55-A,B,C, white in throat, HR-H2.

William Bean    [(Caractacus x discolor) x yakusimanum] #1, white, red
                throat, pubescent, Size-Med., HR-H2.

Windjammer    [(maximum x yakusimanum) x Adele's Yellow] #1-82, x (Bronze
              Wings x Adele's Yellow) #1] #2-85, Yellow #2-C in bud, opens to
              Yellow #2-D, Yellow/Green #154-A specks.

Winning Ways #1    (Lodestar x Mary Belle) #1-84, Red/Purple #62-C,D, yellow
                   throat and pistil.

Winning Ways #2    (Lodestar x Mary Belle) #2-84, Red/Purple #62-C,D, yellow
                   throat and orange pistil.

Yellow Date    [(aureum (chrysanthum) x maximum) #1 x (Phyllis Ballard x
                   catawbiense Clark's White) #1] #3-84, yellow, Size-Med., HR-H2.

Yellow Ribbon   [Gosh Darn x (Mrs. H.R.Yates x wardii "best yellow")] #2-85,
                    Red #37-B and Yellow #4-A in bud, opens to Yellow #4-C,D.

Yellstar    [(brachycarpum ssp.tigerstedtii x Donna Hardgrove) #1-82 x (aureum
                (chrysanthum) x maximum) #1] #1-85, Yellow #4-D, Yellow/Green
                #154-B specks.

Ziggie    [(smirnowii x yakusimanum) x Britannia] #2-76, Red #55-D, veined
              with Red #55-C.

Zippy    (racemosum (apricot form) x keiskei) #2-85, Yellow #11-B, Yellow
             #10-D and a little Red #52-D, Red #52-B specks, large open flowers,
             Size-Low, HR-H2.

APPENDIX F
COLOR NAMES AND NUMBERS FROM TABLE OF CROSS REFERENCES, RHS COLOUR CHART, 1966

| | | | |
|---|---|---|---|
| absinthe | 145-A | cobalt blue | 101-A |
| aconite yellow | 14-C | copper brown | 175-A |
| almond green | 136-B | coral | 32-D |
| almond shell | 165-B | coral pink | 38-D |
| amaranth pink | 68-A | coral tint | 39-C |
| amaranth red | 63-A | corn husk | 24-D |
| amaranth rose | 65-A | cornflower blue | 95-A,B |
| amber yellow | 18-A,B,C,D | crimson | 52-A |
| amethyst | 79-C | crocus | 82-C |
| amethyst violet | 81-C 84-C | currant red | 46-A 47-A |
| amulet green | 126-A | cyclamen purple | 74-A,B |
| apricot | 24-C,D 168-D | Cyprus green | 140-B,C,D |
| apricot buff | 170-D | | |
| arras green | 139-A | dawn pink | 49-A,B |
| aster mauve | 78-C | Delft rose | 46-D 47-C |
| aster violet | 87-C | Dresden yellow | 5-A,B,C,D |
| attic rose | 178-A | Duesbury green | 131-D |
| aureolin | 12-A,B,C,D | Dutch vermilion | 40-A |
| azalea pink | 38-A,B 39-D 41-C | | |
| | | eggshell green | 133-C |
| baby pink | 50-D | Egyptian buff | 19-D |
| barium yellow | 10-A,B,C,D | emerald green | 134-A |
| battleship grey | 202-B,C | empire rose | 48-C,D 50-D |
| beech brown | 165-A | empire yellow | 11-A,D |
| beetroot purple | 71-A | enamel blue | 115-D |
| begonia | 39-B | ethyl blue | 112-D |
| beryl green | 131-C | Etruscan orange | 173-B |
| biscuit | 164-D | Etruscan rose | 175-D |
| bishops violet | 81-A,B | | |
| blood red | 45-D | faience blue | 111-C,D |
| blossom pink | 58-D | fire red | 33-B |
| bluebird blue | 94-A | flame | 33-B |
| bottle green | 135-A | flamingo | 37-A |
| brick dust | 178-D | forest green | 136-A |
| brick red | 35-A,B,C,D | French blue | 100-A,B,C,D |
| bronze yellow | 163-A | French rose | 49-D 50-C |
| bunting yellow | 9-A | fuchia | 74-A |
| burnt orange | 31-A,B | fuchia purple | 67-A,B |
| buttercup yellow | 15-A,B 16-C | | |
| butterfly blue | 106-A,B,C,D | gentian blue | 94-D |
| | | Georgian green | 139-B |
| cadmium orange | 23-A,B,C,D | geranium lake | 47-D |
| campanula violet | 82-C | geranium pink | 43-C |
| canary yellow | 9-A,B,C,D | golden brown | 164-A |
| capisicum red | 33-A | golden buff | 164-B |
| Capri blue | 120-A,B,C,D | grass green | 143-B |
| cardinal red | 53-A,B,C | green beetle | 135-B |
| carmine | 52-B | grenadine red | 33-C 34-D |
| carmine rose | 52-C,D | grotto green | 126-B |
| carnation green | 189-D | guardsman red | 45-A |
| carrot | 28-C | hazelnut | 166-C |
| carrot red | 29-A,B,C | heather pink | 60-D |
| cerulein blue | 107-A | henna | 176-A |
| champagne | 164-D | honeysuckle | 159-C |
| chartreuse green | 1-C,D 154-B | hyacinth blue | 91-A |
| chartreuse yellow | 2-D 154-D | imperial purple | 78-A,B |
| cherry | 45-C | Indian blue | 118-A,B,C |
| chessylite blue | 103-B | Indian Lake | 58-A 59-B |
| China rose | 58-B,D | Indian orange | 32-A |
| Chinese coral | 32-D | Indian pink | 179-C |
| Chinese jade | 189-B | Indian yellow | 17-A,B,C,D 19-C |
| Chinese yellow | 16-A 20-B,D | iris green | 147-C |
| chrome yellow | 14-D 15-C,D 158-A | | |
| chrysanthemum crimson | 185-B | jade | 124-A |
| chrysocolla green | 128-D 129-A,B | jade green | 124-A,B,C,D |
| cinnamon | 165-C | jasper red | 39-A,B,C |
| citron green | 151-B,C,D | jet black | 202-A |
| citrus yellow | 4-C | jewel green | 127-B |
| claret rose | 50-A,B,C | juniper | 133-A |

| | | | |
|---|---|---|---|
| langite | 121-D 123-B | porcelain green | 130-B |
| lemon | 13-A | post office red | 45-B |
| lemon tint | 4-D | primrose yellow | 4-A,B,C,D |
| lemon yellow | 13-A,B 14-C | prince's blue | 105-A,B |
| lettuce green | 144-A | | |
| lilac | 76-C | quartz green | 133-B |
| lilac purple | 70-A,B | | |
| lily green | 139-D | rhodamine pink | 62-A |
| limpid green | 135-D | rhodamine purple | 68-D |
| lobelia blue | 91-C | rhodonite red | 51-A,B,C,D |
| | | rose Bengal | 57-B,C 61-D |
| magenta | 66-A | rose pink | 54-D |
| magenta rose | 64-B,C,D 186-D | rose purple | 75-B,D |
| magnolia purple | 70-C,D | rose red | 58-B |
| maize yellow | 21-B,D | roseine purple | 68-A,B,C |
| majolica green | 127-A | rowan berry | 42-A |
| majolica yellow | 168-D | ruby | 59-A |
| mallow purple | 72-D 73-D | ruby red | 59-A 61-A 64-A |
| malmaison rose | 49-A | | |
| mandarin red | 40-B,C | saffron yellow | 19-B 21-A,B |
| marigold orange | 28-B,C | salmon | 27-A,B |
| Mars orange | 31-C,D | salmon pink | 34-D |
| mazarine blue | 101-C | sap green | 150-A,B,C,D |
| methyl violet | 85-C | satinwood | 164-C |
| Mexican pink | 174-D | Saturn red | 30-B,C |
| Mexican tan | 173-C | scarlet | 43-B,C,D |
| midnight | 103-A | scarlet red | 46-B |
| mimosa yellow | 8-A,B,C,D | Scheele's green | 143-A,B,C 144-B |
| mineral violet | 84-D | sea blue | 102-C,D |
| mistletoe | 136-C | shell pink | 37-C,D |
| | | shrimp red | 33-C,D |
| Naples yellow | 11-B,C | signal red | 43-A |
| nasturtium | 30-A | sky green | 144-D |
| nasturtium orange | 25-B,C,D | sky grey | 122-D |
| nasturtium red | 30-A 32-B | solferino purple | 65-B |
| Neptune green | 127-D | Spanish orange | 26-A,B,C,D |
| neyron rose | 55-A,B,C 56-A 58-C | spectrum blue | 105-C,D |
| nickle green | 130-A,B,C,D | spectrum orange | 28-A |
| | | spectrum violet | 82-B |
| olive yellow | 153-C | spinel red | 54-A,B,C,D |
| opaline green | 133-D | spirea red | 63-A,B,C,D |
| orange | 30-D | squirrel brown | 166-B |
| orange buff | 20-C 22-B,C,D | straw | 14-C |
| orchid pink | 62-C | straw yellow | 13-C |
| orient pink | 36-A,B,C,D | sulphur | 6-A |
| oriental blue | 110-A | sulphur yellow | 6-A,B,C,D |
| orpiment orange | 25-A | | |
| oxblood red | 183-B,C | tangerine | 25-B |
| | | tangerine orange | 24-A,B,C |
| | | terra cotta | 173-A |
| pansy | 83-A | tropic turquoise | 126-D |
| Paris green | 134-B,C,D | Turkey red | 46-C |
| pastel turquoise | 124-D | Tuscan yellow | 162-C |
| pastel yellow | 12-D | Tyrian purple | 57-A 61-C |
| pea green | 149-A,B,C,D | Tyrolite green | 127-C |
| peach | 29-C,D | union jack blue | 99-A |
| peach beige | 173-D | union jack red | 46-B |
| peach tint | 179-D | Venetian pink | 49-B,C |
| peridot | 147-B | verdigris | 129-C,D |
| Persian turquoise | 126-C | vermilion | 41-A,B |
| persimmon orange | 28-A | Veronese green | 141-D |
| petunia purple | 78-A | violet | 83-B |
| phlox pink | 62-B | violet purple | 77-A |
| phlox purple | 75-C | viridian green | 128-A,B,C |
| pistachio green | 139-C | yellow ochre | 22-A |
| plum | 79-A | water green | 139-D |
| plum purple | 79-A | willow green | 196-A |
| poppy | 33-A | Winchester green | 131-A |
| poppy red | 38-C 40-D | wisteria blue | 92-A,B,C,D |
| porcelain blue | 113-A,B | woodpecker green | 147-A |

| | | |
|---|---|---|
| Apricot Sherbet | Dido x Comstock | (Greer) |
| Bambino | (Britannia x yakushimanum) X Lem's Cameo | (Brockenbrough) |
| Blue Tit | (impeditum x augustinii) | (J. C. Williams) |
| Brenden King | Elisabeth Hobbie x Earl of Athlone | (S. M. King, NZ) |
| Brittenhill | The Hon. Jean Marie de Montague x Red Loderi | (Britt Smith) |
| Citation | Mrs. Charles S. Sargent x Swansdown | (Pride) |
| College Pink | Noyo Chief x unknown | (NZ Rhod. Assoc.) |
| Goldilocks | xanthostephanum x unknown | (Kerrigan) |
| Griergrande | griersonianum x grande | (unknown) |
| Gumdrop | Fabia x C. P. Raffill | (D. W. James) |
| Haemwill | haematodes x williamsianum | (unknown) |
| Homestead | griersonianum x Ilam Alarm | (Edgar Stead) |
| Ivory Coast | keiskei x Arctic Pearl--dauricum (white) | (Leach) |
| Jim Dandy | Mrs. Horace Fogg x hybrid unknown | (James Elliott) |
| King's Milkmaid | J. G. Millais x Gladys | (W. King) |
| Lady Dorothy Ella | nuttallii x lindleyi | (unknown) |
| Lila Pedigo | Crest x Odee Wright | (Sather) |
| Lily | Midnight x Coronation Day | (Van de Ven) |
| Mood Indigo | augustinii x unknown | (Brandt) |
| Nightwatch | Purple Splendour x Cup Day | (Karel Van de Ven) |
| Pania | dichroanthum x unknown | (NZ Rhod. Assoc.) |
| Polycinn | maddenii (polyandrum) x cinnabarinum | (unknown) |
| Scarlet King Pines | griersonianum x Ilam Alarm | (Edgar Stead) |
| Sperhook | sperabile x hookeri | (Edgar Stead) |
| Suave | edgeworthii x unknown | (unknown) |
| Tod's Orchid Sunset | Purple Splendour x Mrs. Donald Graham | (Ted Fawcett) |
| Vandec | Van Nes Sensation x decorum | (unknown) |
| Whitney's Orange | dichroanthum ? x Diane ? (reg. as unknown) | (Whitney/Sather) |